Chapter 4 Systems of Equations and Inequalities

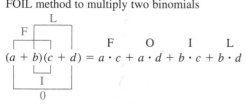

Consistent system — One Solution

Inconsistent system — No Solution

Dependent system — An Infinite Number of Solutions

Cramer's Rule: Given a system of equations of the form

$$a_1x + b_1y = c_1$$
$$a_2x + b_2y = c_2$$

then $x = \dfrac{\begin{vmatrix} c_1 & b_1 \\ c_2 & b_2 \end{vmatrix}}{\begin{vmatrix} a_1 & b_1 \\ a_2 & b_2 \end{vmatrix}}$ and $y = \dfrac{\begin{vmatrix} a_1 & c_1 \\ a_2 & c_2 \end{vmatrix}}{\begin{vmatrix} a_1 & b_1 \\ a_2 & b_2 \end{vmatrix}}$

A system of linear equations may be solved: (a) graphically: (b) by the substitution method, (c) by the addition or elimination method, (d) by determinants, or (e) by matrices.

$$\begin{vmatrix} a_1 & b_1 \\ a_2 & b_2 \end{vmatrix} = a_1 b_2 - a_2 b_1$$

Chapter 5 Polynomials

$a^m \cdot a^n = a^{m+n}$ $(a^m)^n = a^{m \cdot n}$

$a^m/a^n = a^{m-n}, a \neq 0$ $(ab)^m = a^m b^m$

$a^{-m} = \dfrac{1}{a^m}, a \neq 0$ $\left(\dfrac{a}{b}\right)^m = \dfrac{a^m}{b^m}, b \neq 0$

$a^0 = 1, a \neq 0$ $\left(\dfrac{a}{b}\right)^{-m} = \left(\dfrac{b}{a}\right)^m, a \neq 0, b \neq 0$

FOIL method to multiply two binomials

$$(a + b)(c + d) = a \cdot c + a \cdot d + b \cdot c + b \cdot d$$

F O I L

Square of a binomial:
$(a + b)^2 = a^2 + 2ab + b^2.$
$(a - b)^2 = a^2 - 2ab + b^2$

Product of the sum and difference of the same two terms (also called the difference of two squares):
$(a + b)(a - b) = a^2 - b^2$

Polynomial function:
$f(x) = a_n x^n + a_{n-1}x^{n-1} + a_{n-2}x^{n-2} + \cdots + a_1 x + a_0$

Quadratic function: $f(x) = ax^2 + bx + c, a \neq 0$

Chapter 6 Factoring

Difference of two squares: $a^2 - b^2 = (a + b)(a - b)$
Note: the sum of two squares $a^2 + b^2$ cannot be factored over the set of real numbers

Perfect square trinomials:
$a^2 + 2ab + b^2 = (a + b)^2, a^2 - 2ab + b^2 = (a - b)^2$

Sum of two cubes:
$a^3 + b^3 = (a + b)(a^2 - ab + b^2)$

Difference of two cubes:
$a^3 - b^3 = (a - b)(a^2 + ab + b^2)$

TO FACTOR A POLYNOMIAL

1. Determine if all the terms in the polynomial have a greatest common factor other than 1. If so factor out the GCF.
2. If the polynomial has two terms determine if it is a difference of two squares or a sum or difference of two cubes. If so factor using the appropriate formula.

3. If the polynomial has 3 terms determine if it is a perfect square trionomial. If so factor accordingly. If it is not, then factor the trinomial using the methods discussed in Section 6.2.
4. If the polynomial has more than 3 terms, then try factoring by grouping.
5. As a final step examine your factored polynomial to see if any factors listed have a common factor and can be factored further. If you find a common factor, factor it out at this point.

Standard form of a quadratic equation:
$ax^2 + bx + c = 0, a \neq 0$

Zero-factor property: If $a \cdot b = 0$, then either $a = 0$ or $b = 0$, or both a and $b = 0$.

Intermediate Algebra for College Students

THIS ANNOTATED INSTRUCTOR'S EDITION
IS EXACTLY LIKE YOUR STUDENT'S TEXT,
BUT IT CONTAINS ALL EXERCISE ANSWERS
DISPLAYED ON THE SAME PAGE.

Intermediate Algebra for College Students

Fourth Edition

Allen R. Angel
Monroe Community College

Prentice Hall

*Upper Saddle River
New Jersey 07458*

Sponsoring Editor:	Melissa S. Acuña
Editor-in-Chief:	Jerome Grant
Development Editor:	Ellen Credille
Director of Production and Manufacturing:	David W. Riccardi
Project Manager:	Robert C. Walters
Marketing Manager:	Jolene Howard
Copy Editor:	Barbara Zeiders
Interior Designer:	Geri Davis
Cover Designer:	Geri Davis
Creative Director:	Paula Maylahn
Art Director:	Amy Rosen
Manufacturing Manager:	Alan Fischer
Photo Researcher:	Rona Tuccillo
Photo Editor:	Lorinda Morris-Nantz
Supplements Editor:	Audra J. Walsh
Editorial Assistant:	April Thrower

©1996, 1992, 1988, 1985 by Prentice-Hall, Inc.

Simon & Schuster/A Viacom Company

Upper Saddle River, New Jersey 07458

Printed in the United States of America

10 9 8 7 6 5 4 3 2 1

ISBN 0-13-221244-7

Prentice-Hall International (UK) Limited, *London*
Prentice-Hall of Australia Pty, Limited, *Sydney*
Prentice-Hall Canada, Inc., *Toronto*
Prentice-Hall Hispanoamericana, S.A., *Mexico*
Prentice-Hall of India Private Limited, *New Delhi*
Prentice-Hall of Japan, Inc., *Tokyo*
Simon & Schuster Asia Pte. Ltd., *Singapore*
Editora Prentice-Hall do Brasil, Ltda., *Rio de Janeiro*

To my mother, Sylvia Angel-Baumgarten and to the memory of my father, Isaac Angel

Contents

Preface

This book was written for college students who have successfully completed a first course in elementary algebra. My primary goal was to write a book that students can read, understand, and enjoy. To achieve this goal I have used short sentences, clear explanations, and many detailed worked-out examples. I have tried to make the book relevant to college students by using practical applications of algebra throughout the text.

Features of the Text

FOUR-COLOR FORMAT: Color is used pedagogically in the following ways:

Important definitions and procedures are color screened.

Color screening or color type is used to make other important items stand out.

Artwork is enhanced and clarified with use of multiple colors.

The four-color format allows for easy identification of important features by students.

The four-color format makes the text more appealing and interesting to students.

READABILITY: One of the most important features of the text is its readability. The book is very readable, even for those with weak reading skills. Short, clear sentences are used and more easily recognized, and easy-to-understand language is used whenever possible.

ACCURACY: Accuracy in a mathematics text is essential. To insure accuracy in this book, mathematicians from around the country have read the galleys carefully for typographical errors and have checked all the answers.

CONNECTIONS: Many of our students do not thoroughly grasp new concepts the first time they are presented. In this text we encourage students to make connections. That is, we introduce a concept, then later in the text briefly reintroduce it and build upon it. Often an important concept is used in many sections of the text. Students are reminded where the material was seen before, or where it will be used again. This also serves to emphasize the importance of the concept. Important concepts are also reinforced throughout the text in the Cumulative Review Exercises and Cumulative Review Test.

PREVIEW AND PERSPECTIVE: This feature at the beginning of each chapter explains to the students why they are studying the material and where this material will be used again in other chapters of the book. This material helps students see the connections between various topics in the book, and the connection to real world situations.

VIDEOTAPE AND SOFTWARE ICONS: At the beginning of each section a videotape and software icon is displayed. These icons tell the student where material in this section can be found on the videotapes, saving your students time when they want to review this material on the videotapes or the tutorial software.

KEYED SECTION OBJECTIVES: Each section opens with a list of skills that the student should learn in that section. The objectives are then keyed to the appropriate portions of the sections with symbols such as **1**.

PRACTICAL APPLICATIONS: Practical applications of algebra are stressed throughout the text. Students need to learn how to translate application problems into algebraic symbols. The problem-solving approach used throughout this text gives students ample practice in setting up and solving application problems. The use of practical applications motivates students.

DETAILED WORKED-OUT EXAMPLES: A wealth of examples have been worked out in a step-by-step, detailed manner. Important steps are highlighted in color, and no steps are omitted until after the student has seen a sufficient number of similar examples.

STUDY SKILLS SECTION: Many students taking this course have poor study skills in mathematics. Section 1.1, the first section of this text, discusses the study skills needed to be successful in mathematics. This section should be very beneficial for your students, and should help them to achieve success in mathematics.

HELPFUL HINTS: The helpful hint boxes offer useful suggestions for problem solving and other varied topics. They are set off in a special manner so that students will be sure to read them.

COMMON STUDENT ERRORS: Errors that students often make are illustrated. The reasons why certain procedures are wrong are explained, and the correct procedure for working the problem is illustrated. These common student error boxes will help prevent your students from making those errors we see so often.

CALCULATOR CORNERS: The Calculator Corners, placed at appropriate intervals in the text, are written to reinforce the algebraic topics presented in the section and to give the student pertinent information on using the calculator to solve algebraic problems.

GRAPHING CALCULATOR CORNERS: Graphing Calculator Corners are placed at appropriate locations throughout the text. They reinforce the algebraic topics taught and sometimes offer alternate methods of working problems. This book is designed to give the instructor the option of using or not using a graphing calculator in their course. Each Graphing Calculator Corner contains graphing calculator exercises, whose odd answers appear in the answer section of the book. The illustrations shown in the Graphing Calculator Corners are from a Texas Instrument 82 calculator. The Graphing Calculator Corners generally do not give specific keystroke instructions so any graphing calculator may be used. The Graphing Calculator Corners are written assuming that the student has no prior graphing calculator experience.

EXERCISE SETS: Each exercise set is graded in difficulty. The early problems help develop the student's confidence, and then students are eased gradually into the more difficult problems. A sufficient number and variety of examples are given in each section for the student to successfully complete even the more difficult exercises. The number of exercises in each section is more than ample for student assignments and practice.

WRITING EXERCISES: Many exercise sets include exercises that require students to write out the answers in words. These exercises improve students' understanding and comprehension of the material. Many of these exercises involve problem solving and help develop better reasoning and critical thinking skills. Writing exercises are indicated by the symbol ✎.

CUMULATIVE REVIEW EXERCISES: All exercise sets (after the first two) contain questions from previous sections in the chapter and from previous chapters. These cumulative review exercises will reinforce topics that were previously covered and help students retain the earlier material, while they are learning the new material. For the students' benefit the Cumulative Review Exercises are keyed to the section where the material is covered.

GROUP ACTIVITY/CHALLENGE PROBLEMS: These exercises, which are part of every exercise set after Section 1.1, provide a variety of problems. Many were written to stimulate student thinking and to lead to interesting group discussions. Others provide additional applications of algebra or present material from future sections of the book so that students can see and learn the material on their own before it is covered in class. Others are more challenging than those in the regular exercise set. These problems may be assigned to your students individually or may be assigned as group exercises.

CHAPTER SUMMARY: At the end of each chapter is a chapter summary which includes a glossary and important chapter facts. The terms in the glossary are keyed to the page where they are first introduced.

REVIEW EXERCISES: At the end of each chapter are review exercises that cover all types of exercises presented in the chapter. The review exercises are keyed to the sections where the material was first introduced.

PRACTICE TESTS: The comprehensive end-of-chapter practice test will enable the students to see how well they are prepared for the actual class test. The Test Item File includes several forms of each chapter test that are similar to the student's practice test. Multiple choice tests are also included in the Test Item File.

CUMULATIVE REVIEW TEST: These tests, which appear at the end of each even-numbered chapter, test the students' knowledge of material from the beginning of the book to the end of that chapter. Students can use these tests for review, as well as for preparation for the final exam. These exams, like the cumulative review exercises, will serve to reinforce topics taught earlier.

ANSWERS: The *odd answers* are provided for the exercise sets and the Graphing Calculator Exercises. *All answers* are provided for the Cumulative Review Exercises, the Review Exercises, Practice Tests, and the Cumulative Practice Test. *Selected answers* are provided for the Group Activity/Challenge Problem exercises.

National Standards   Recommendations of the *Curriculum and Evaluation Standards for School Mathematics,* prepared by the National Council of Teachers of Mathematics, (NCTM) and *Crossroads in Mathematics: Standards for Introductory College Mathematics Before Calculus,* prepared by the American Mathematical Association of Two Year Colleges (AMATYC) were incorporated into this section.

Prerequisite   The prerequisite for this course is a working knowledge of elementary algebra. Although some elementary algebra topics are briefly reviewed in the text, students should have a basic understanding of elementary algebra before taking this course.

Modes of Instruction   The format and readability of this book lends itself to many different modes of instruction. The constant reinforcement of concepts will result in greater understanding and retention of the material by your students.

The features of the text and the large variety of supplements available make this text suitable for many types of instructional modes including:

- lecture
- modified lecture
- learning laboratory
- self-paced instruction
- cooperative or group study

Changes in the Fourth Edition   When I wrote the fourth edition I considered the many letters and reviews I got from students and faculty alike. I would like to thank all of you who made suggestions for improving the fourth edition. I would also like to thank the many instructors and students who wrote to inform me of how much they enjoyed and appreciated the text.

Some of the changes made in the fourth edition of the text include:

- *Preview and Perspective* has been added to the beginning of each chapter to give students connections between the material they are studying and the material they have or will study, and to associate the material with real life applications.
- The *Graphing Calculator Corners* that have been added throughout the text are used to reinforce algebraic concepts and to provide alternate methods to work problems. The text is designed so that instructors have the option of using, or not using, a graphing calculator with this book.
- There is much less overlap between Elementary Algebra material and Intermediate Algebra material in this edition of the series. Students get to study Intermediate Algebra material earlier in the semester than previously.
- More challenging problems have been added to many exercise sets.
- More varied exercises have been added to many exercise sets.
- Real life applications, including environmental and financial issues, have been added throughout the text.
- Additional detailed worked out examples have been added where needed.
- New and additional *Helpful Hint* and *Common Student Error* boxes have been added.
- More exercises that require a written response have been added.
- *Group Activity/Challenge Problems* exercises have been added to each exercise set. Many of these exercises are appropriate for group work and many will foster creative discussions among the members of the group. These exercises may be assigned to groups or to individual students in the class.
- More illustrations have been added throughout the book to help the visual learner.
- Chapter 1, *Basic Concepts,* has been reduced from seven sections to four sections.
- Graphing inequalities in one variable is now introduced earlier in the text.

- Equations of the form $|x| = |y|$ have been added to the section on solving absolute value equations.
- The sections previously titled *Slope of a Line* and *Slope-intercept and Point-slope Forms of a Linear Equation* have been combined into one section.
- A new section titled *Linear and Non-linear Functions* has been added to Chapter 3. This section will introduce students to basic non-linear functions (square root and absolute value functions) early in the course.
- Many real life graphs are used and interpreted throughout the text.
- A section titled *Solving Systems of Equations using Matrices* has been added to the text.
- The sections previously titled *Addition and Subtraction of Polynomials* and *Multiplication of Polynomials* have been combined into one section. More difficult problems have been added to exercise sets in the *Polynomials* chapter.
- The section previously titled *Using Factoring to Solve for a Variable in a Formula* has been removed. This material is now presented earlier in the book. More difficult problems have been added to the exercise set in the factoring chapter.
- The section titled *Applications of Radicals* has been removed and the applications have been added to other sections in the chapter.
- The section previously titled *The Square Root Function* has been removed. Material from this section now appears earlier in the book.
- The chapter on *Quadratic Functions and the Algebra of Functions* has been reorganized.
- Material on *writing equations quadratic in form* has been added to the text.
- The word "optional" has been removed from all sections.

Supplements to the Fourth Edition

For this edition of the book the author has personally coordinated the development of the *Student's Solution Manual* and the *Instructor's Solution Manual.* Experienced mathematics professors who have prior experience in writing supplements, and whose works have been of superior quality, have been carefully selected for authoring the supplements.

For Instructors

Annotated Instructor's Edition: Includes answers to every exercise on the same page.

Instructor's Solutions Manual: Contains solutions to all exercises not included in the Student's Solution manual.

Test Pro II: Allows users to generate tests by chapter or section number, choosing from thousands of test questions and hundreds of algorithms which generate different numbers for the same item. Editing and graphing capability are included.

Test Item File: Contains five tests per chapter.

For Students

MathPro Software: Carefully keyed section-by-section to the text, this software provides students with interactive feedback and help in solving exercises.

Each section contains Warm-up Exercises which allow the student unlimited practice and Practice Problems which are graded and recorded.

Videotapes: Closely tied to the book, these instructional tapes feature a lecture format with worked-out examples and exercises from each section of the book.

Student's Solutions Manual: Includes detailed step-by-step solutions to all exercises whose answers appear in the answer appendix. This includes answers to *odd-numbered* exercises and Graphing Calculator Exercises, and answers to *all* Cumulative Review Exercises, Review Exercises, Practice Test, and Cumulative Review Test. *Selected answers* for Group Activity/Challenge Problem exercises are also included.

Student's Study Guide: Includes additional worked-out examples, additional drill problems, and Practice Tests, and their answers. Important concepts are emphasized.

Acknowledgments

Writing a textbook is a long and time-consuming project. Many people deserve thanks for encouraging and assisting me with this project. Most importantly I would like to thank my wife Kathy, and sons, Robert and Steven. Without their constant encouragement and understanding, this project would not have become a reality. In addition, Robert did proofreading and answer checking.

I would like to thank Richard Semmler of Northern Virginia Community College, Marilyn Semrau of Monroe Community College, and Phyllis Barnidge and Lynda Steele of Laurel Technical Services for their conscientiousness and the attention to details they provided in checking galleys, pages, artwork and answers.

I would also like to thank my students and colleagues of Monroe Community College for their suggestions for improving the book. I would like to thank students and faculty from around the country for using the third edition and offering valuable suggestions for the fourth edition. I was especially pleased in receiving so many letters from students informing me how much they enjoyed using the book. Thank you for your kind words.

I would like to thank my editor at Prentice Hall, Melissa Acuña, my production editors Jack Casteel, Bob Walters, and Ellen Credille for their many valuable suggestions and conscientiousness with this project.

I would like to thank the following reviewers and proofreaders for their thoughtful comments and suggestions.

Larry Blevins, *Tyler Junior College*
Irene Doo, *Austin Community College*
D. Michael Hamm, *Brookhaven College*
Janie Jacks, *NEO A&M College*
Paula Jones, *Southwest Missouri State University*
Frances Leach, *Delaware Technical and Community College*
Wilene Leach, *Arkansas State University-Beebe*
Irwin Metviner, *State University of New York-Old Westbury*
Walter J. Orange, *University of Pittsburgh at Greensburg*
Smruti Patel, *Hutchinson Community College*
Dianne Phelps, *Sullivan County Community College*
Richard Semmler, *Northern Virginia Community College*
Marilyn Semrau, *Monroe Community College*
Katalin Szucs, *East Carolina University*

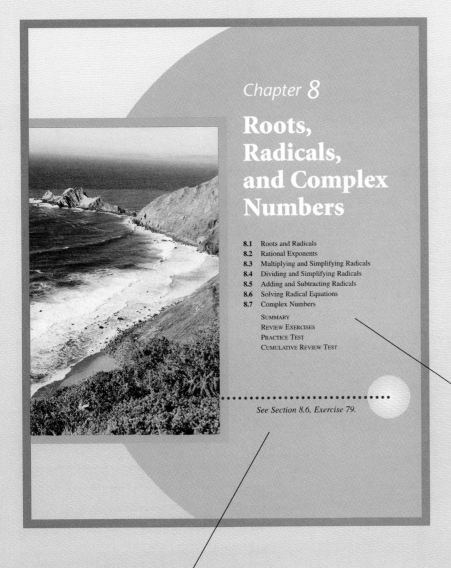

Chapter *8*

Roots, Radicals, and Complex Numbers

See Section 8.6, Exercise 79.

Intermediate Algebra for College Students 4/e is a significant revision of the Third Edition. The new design uses color pedagogically and helps increase the readability for the students and the instructors. The exercise sets have been enhanced and new applications have been added. The next few pages will highlight some of the new features of this edition of the text, as well as showing some important features that have made this text successful.

Each chapter opening page contains an outline of the contents of the chapter.

The photo application is taken from an exercise in the chapter, showing how the skills and concepts in the chapter relate to students' lives.

Preview and Perspective

We graphed equations containing square roots in Section 3.5. In this chapter we discuss roots and radicals further. We explain how to add, subtract, multiply, and divide radical expressions. We also introduce imaginary numbers and complex numbers.

Section 8.1 gives some basic concepts and definitions of roots. In Section 8.2 we change expressions from radical form to exponential form, and vice versa. We also work problems using rational exponents. The rules of exponents discussed in Sections 5.1 and 5.2 still apply to rational exponents and we use those rules again here.

In Section 8.4 we discuss rationalizing the denominator, which removes radicals from a denominator. Make sure that you understand the three requirements for a radical expression to be simplified, as discussed in Section 8.4. We learn to add and subtract radicals in Section 8.5. We also discuss rationalizing denominators further in this section.

In Section 8.6 we discuss how to solve equations that contain radical expressions. We will use these procedures again in Chapters 9 and 10. Section 8.6 also illustrates some applications of radical equations.

Imaginary numbers and complex numbers are introduced in Section 8.7. These numbers play a very important role in higher mathematics courses. We will be using imaginary and complex numbers throughout Chapter 9.

Every chapter begins with a Preview and Perspective, *giving students an overview of the chapter. This feature helps students see the connections between various concepts in the text and how these concepts relate to the real world.*

Each section begins with a list of the objectives for the section. The objectives are keyed to the appropriate sections in the text with icons.

8.1   Roots and Radicals

Tape
12

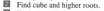
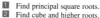
1 Find principal square roots.
2 Find cube and higher roots.
3 Evaluate radical expressions using absolute value.

In this chapter we expand on the concept of radicals introduced in Chapter 1, 3, and 4. So far we have discussed only square and cube roots. In this chapter we will also discuss higher roots.

In the expression $\sqrt{x}$, the $\sqrt{}$ is c[...]
sion within the radical sign is called [...]

The entire expression, including th[...] **cal expression.** Another part of the [...] the "root" of the expression. Square [...] roots is generally not written.

$$\sqrt{x}$$

Principal Square Root

1 Every positive number has two [...] root and a negative square root. [...] square root is written $\sqrt{x}$, and [...]

Great care has been taken to include practical applications of algebra. The problem-solving approach used in the text gives students needed practice in solving applied problems, and the real-life applications help motivate students.

EXAMPLE 14   Multiply $(5,600,000)(0.0002)$ by first converting the numbers to scientific notation. Write the answer without exponents.

Solution:
$$(5,600,000)(0.0002) = (5.6 \times 10^6)(2.0 \times 10^{-4})$$
$$= (5.6)(2.0) \times (10^6)(10^{-4})$$
$$= 11.2 \times 10^{6 + (-4)}$$
$$= 11.2 \times 10^2$$
$$= 1120$$

EXAMPLE 15   Divide $\dfrac{0.0000144}{0.003}$ by first converting the numbers to scientific notation. Write the answer without exponents.

Solution:
$$\frac{0.0000144}{0.003} = \frac{1.44 \times 10^{-5}}{3 \times 10^{-3}}$$
$$= \frac{1.44}{3} \times \frac{10^{-5}}{10^{-3}}$$
$$= 0.48 \times 10^{-5 - (-3)}$$
$$= 0.48 \times 10^{-5 + 3}$$
$$= 0.48 \times 10^{-2}$$
$$= 0.0048$$

EXAMPLE 16   On September 30, 1992, the U.S. public debt was approximately $4,065,000,000,000 (4 trillion 65 billion dollars). The U.S. population on that date was approximately 257,000,000.

(a) Find the average U.S. debt for every person in the United States (the per capita debt).

(b) On September 30, 1982, the U.S. debt was approximately $1,142,000,000,000. How much larger was the debt in 1992 than 1982?

(c) How many times greater was the debt in 1992 than 1982?

Solution:

(a) To find the per capita debt, divide the public debt by the population.

$$\frac{4,065,000,000,000}{257,000,000} = \frac{4.065 \times 10^{12}}{2.57 \times 10^8} \approx 1.58 \times 10^4 \approx 15,800$$

Thus, the per capita debt was about $15,800.

(b) We need to find the difference in the debt between 1992 and 1982.

$$4,065,000,000,000 - 1,142,000,000,000 = 4.065 \times 10^{12} - 1.142 \times 10^{12}$$
$$= (4.065 - 1.142) \times 10^{12}$$
$$= 2.923 \times 10^{12}$$
$$= 2,923,000,000,000$$

In addition to Calculator Corners, Intermediate Algebra includes Graphing Calculator Corners that *reinforce algebraic topics,* yet have been written assuming that the student has no graphing calculator experience.

Graphing Calculator Corner

One of the main features that set graphing calculators apart from other calculators is their ability to display one or more graphs on their screen. We will discuss this feature throughout the book.

When graphing on a grapher you must be sure that the x and y axes are set properly on your calculator. When you set the maximum and minimum values that you wish displayed on the x axis you are setting the "*domain,*" and when you set the maximum and minimum values that you wish displayed on the y axis you are setting the "range." You also need to provide the scale for the x and y axes. We will discuss the words *domain* and *range* further in Section 3.4 and throughout the book.

When you enter the domain and range you are defining what is to appear in the calculator's viewing **window** or display. On some graphing calculators the words *window variables* are used in place of the words domain and range. Below we show the graph of $y = x^3 - 2x + 3$ as displayed on the same calculator with two different axes instructions.

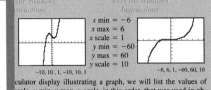

(or Window) structions		Axes (or Window) Instructions
	x min $= -6$	
	x max $= 6$	
	x scale $= 1$	
	y min $= -60$	
	y max $= 60$	
	y scale $= 10$	
$-10, 10, 1, -10, 10, 1$		$-6, 6, 1, -60, 60, 10$

culator display illustrating a graph, we will list the values of

---

Group Activity/ Challenge Problems

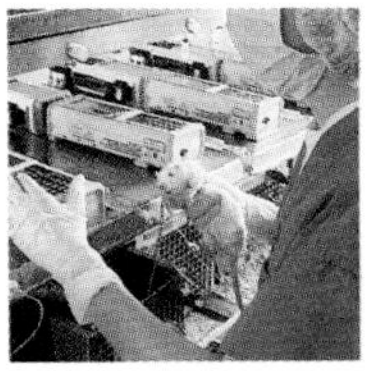

1. In an article published in the *Journal of Comparative Physiology and Psychology,* J. S. Brown discusses how we often approach a situation with mixed emotions. For example, when a person is asked to give a speech, he may be a little apprehensive about his ability to do a good job. At the same time, he would like the recognition that goes along with making the speech. J. S. Brown performed an experiment on trained rats. He placed their food in a metal box. He used that same box to administer small electrical shocks to the mice. Therefore, the rats "wished" to go into the box to receive food, yet did not "wish" to go into the box for fear of receiving a small shock. Using the appropriate apparatus, Brown arrived at the following relationships:

$$\text{pull (in grams) toward food} = -\frac{1}{5}d + 70 \qquad 30 < d < 172.5$$

$$\text{pull (in grams) away from shock} = -\frac{4}{3}d + 230 \qquad 30 < d < 172.5$$

where d is the distance in centimeters from the box (and food).

(a) Using the substitution method, find the distance at which the pull toward the food equals the pull awa...

(b) If the rat is placed ...

2. A *nonlinear system of e...* tion which is not linear ... 10.) Consider the follow... following questions.

Group Activity/Challenge Problems, *found in every exercise set, encourage critical thinking. Many exercises can be used as group learning or discussion problems. These exercises show extended applications of the mathematics in the section, and frequently will present material from later sections in the text.*

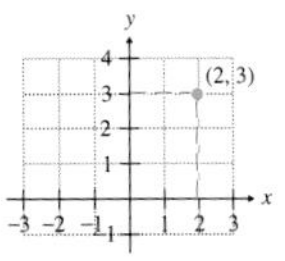

Figure 3.5 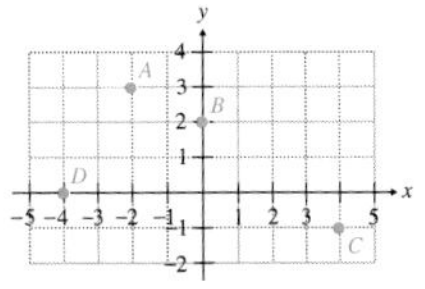 **Figure 3.6**

is plotted in Fig. 3.5. The phrase "the point corresponding to the ordered pair $(2, 3)$" is often abbreviated "the point $(2, 3)$." For example, if we write "the point $(-1, 5)$," it means the point corresponding to the ordered pair $(-1, 5)$. The ordered pairs, A at $(-2, 3)$, B at $(0, 2)$, C at $(4, -1)$, and D at $(-4, 0)$ are plotted in Fig. 3.6.

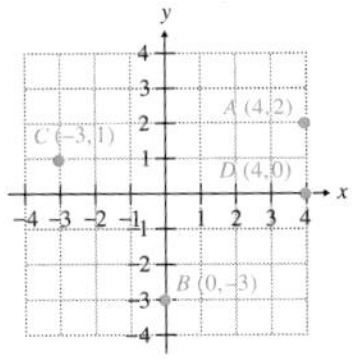

Example 3 Plot each of the following points on the same set of axes.

(a) $A(4, 2)$ (b) $B(0, -3)$ (c) $C(-3, 1)$ (d) $D(4, 0)$

Solution: See Fig. 3.7. Notice that when the x coordinate is 0, as in part (b), the point is on the y axis. When the y coordinate is 0, as in part (d), the point is on the x axis.

Figure 3.7

Distance between Two Points

Now we will see how to find the distance between any two points in a plane. After this, we will show how to find the midpoint of a given line segment. You need these two concepts to understand conic sections (Chapter 10).

To find the distance, d, between two points, we use the distance formula.

Distance Formula

The distance, d, between any two points (x_1, y_1) and (x_2, y_2) can be found by the distance formula

$$d = \sqrt{(x_2 - x_1)^2 + (y_2 - y_1)^2}$$

The distance between any two points will always be a positive number. Can you explain why? When finding the distance, it makes no difference which point we designate as point 1 (x_1, y_1) or point 2 (x_2, y_2). Note that the square of any real number will always be greater than or equal to zero. For example, $(5 - 2)^2 = (2 - 5)^2 = 9$.

Example 4 Determine the distance between the points $(-1, 5)$ and $(-4, 1)$.

Solution: First plot the points (Fig. 3.8). Call $(-1, 5)$ point 2 and $(-4, 1)$ point 1. Thus, (x_2, y_2) represents $(-1, 5)$ and (x_1, y_1) represents $(-4, 1)$. Now use the distance formula to find the distance, d.

$$\begin{aligned}
d &= \sqrt{(x_2 - x_1)^2 + (y_2 - y_1)^2} \\
&= \sqrt{[-1 - (-4)]^2 + (5 - 1)^2} \\
&= \sqrt{(-1 + 4)^2 + 4^2} \\
&= \sqrt{3^2 + 4^2} \\
&= \sqrt{9 + 16} = \sqrt{25} = 5
\end{aligned}$$

Thus, the distance between the points $(-1, 5)$ and $(-4, 1)$ is 5 units.

Figure 3.8

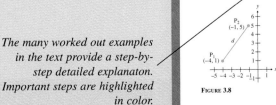

The many worked out examples in the text provide a step-by-step detailed explanaton. Important steps are highlighted in color.

End of the chapter Summary includes a glossary of key terms from the chapter and a list of the important rules, facts and procedures. This feature helps students review the key concepts and skills from the chapter.

Summary

GLOSSARY

Binomial (291): A two-term polynomial.

Degree of a polynomial (291): The same as the highest-degree term in the polynomial.

Descending order of the variable (291): Polynomial written so that the exponents on the variable decrease as terms go from left to right.

Monomial (291): A one-term polynomial.

Polynomial (290): A finite sum of terms in which all variables have whole-number exponents and no variable appear in a denominator.

Polynomial function (311): A function of the form $f(x) = a_n x^n + a_{n-1} x^{n-1} + a_{n-2} x^{n-2} + \dots + a_1 x + a_0$

Quadratic function (312): A function of the form $f(x) = ax^2 + bx + c, a \neq 0$.

Scientific notation (277): A form of writing large and small numbers as the product of a number greater than or equal to 1 and less than 10 and a power of 10.

Synthetic division (305): A shortened process of dividing a polynomial by a binomial of the form $x - a$.

Trinomial (291): A three-term polynomial.

IMPORTANT FACTS

Rules for exponents

1. $a^m \cdot a^n = a^{m+n}$ product rule   4. $a^0 = 1, a \neq 0$ zero exponent rule

2. $\dfrac{a^m}{a^n} = a^{m-n}, a \neq 0$ quotient rule   5. $(a^m)^n = a^{mn}$

$(ab)^m = a^m b^m$ power rules

3. $a^{-m} = \dfrac{1}{a^m}, a \neq 0$ negative exponent rule $\left(\dfrac{a}{b}\right)^m = \dfrac{a^m}{b^m}, b \neq 0$

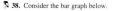

38. Consider the bar graph below.

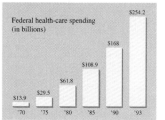

Federal health-care spending (in billions)

$254.2

$168

$108.9

$61.8

$29.5

$13.9

'70 '75 '80 '85 '90 '93

(a) 15.6 billion, 86.2 billion

(a) How much had health care spending increased from 1970 to 1975? From 1990 to 1993?

(b) Redraw this graph using the Cartesian coordinate system with dollars along the vertical axis and years along the horizontal axis. Your axes should be properly labeled.

(c) Does your graph give a different impression about federal health care spending? Explain your answer.

(d) Write a paragraph describing the trend in federal spending on health care from 1970 to 1993.

39. The information for the line graph that follows was provided by the U.S. Immigration and Naturalization Service.

Immigrants allowed into the U.S.

More than 60 million immigrants have been admitted to the U.S. since 1820. The past 10 years:

2,000,000
1,600,000
1,200,000
800,000
400,000
0

880,014

'84 '93

(c) Estim to the Expl

(d) Writ num State

40. The follo an averag about $50 tuition, ar public ins

— Private college Freedom Day: May 2
— Public college

1974 Tuition Freedom Day: March 14

1974 Tuition Freedom Day: Feb. 4

90
70
50
30
10
0

Number of working days

1974 '79 '84 '89 '94

Year

1994 Tuition Freedom Day: Feb. 14

(a) 73 days, 30 days

(a) In 1987, estimate the number of days that had to be worked to reach "Tuition Freedom Day" for a

---

The exercise sets are carefully developed. They are graded in difficulty to help students gain confidence and succeed with the more difficult exercises. The extensive number of examples in the text help prepare students to understand and work all the exercises.

Solution:

(a) $2^{-1} - 3 \cdot 2^{-2} = \dfrac{1}{2} - 3 \cdot \dfrac{1}{2^2} = \dfrac{1}{2} - 3 \cdot \dfrac{1}{4}$

$= \dfrac{1}{2} - \dfrac{3}{4} = \dfrac{2}{4} - \dfrac{3}{4} = -\dfrac{1}{4}$

(b) $-4^{-1} - 2 \cdot 3^{-1} - (4^2 \cdot 3)^0 = -1(4^{-1}) - 2 \cdot 3^{-1} - 1$

$= -1 \cdot \dfrac{1}{4} - 2 \cdot \dfrac{1}{3} - 1$

$= -\dfrac{1}{4} - \dfrac{2}{3} - 1$

$= -\dfrac{3}{12} - \dfrac{8}{12} - \dfrac{12}{12} = -\dfrac{23}{12}$

EXAMPLE 8 Simplify $\dfrac{(x^{-3})(2x^6)}{x^{-5}}$.

Solution: $\dfrac{(x^{-3})(2x^6)}{x^{-5}} = \dfrac{2x^{-3+6}}{x^{-5}} = \dfrac{2x^3}{x^{-5}} = 2x^8$

The following Helpful Hint is very important. Read it carefully.

Helpful Hint

Notice in Example 8 that we simplifed $\dfrac{2x^3}{x^{-5}}$ to $2x^8$. This result can be obtained by rewriting x^{-5} in the denominator as x^5 in the numerator

$$\dfrac{2x^3}{x^{-5}} = 2x^3 \cdot x^5 = 2x^8$$

When simplifying problems of this type, we can move a factor with a negative exponent from the numerator to the denominator, or from the denominator to the numerator, *by changing the sign of the exponent.*

If the same variable appears in both the numerator and denominator of an expression, we generally move the variable with the *lesser* exponent.

Examples

$\dfrac{x^{-5}}{x^{-3}} = \dfrac{1}{x^{-3} \cdot x^5} = \dfrac{1}{x^2}$ $\dfrac{z^{-2}}{z^{-6}} = z^{-2} \cdot z^6 = z^4$

$\dfrac{y^6}{y^{-3}} = y^6 \cdot y^3 = y^9$ $\dfrac{w^{-5}}{w^4} = \dfrac{1}{w^4 \cdot w^5} = \dfrac{1}{w^9}$

Now we will simplify an expression containing several variable factors.

$$\dfrac{x^{-6} y^{-3} z^{-4}}{x^{-4} y^{-7} z^2} = \dfrac{y^{-3} y^7}{x^{-4} x^6 z^2 z^4} = \dfrac{y^4}{x^2 z^6}$$

Helpful Hints boxes offer suggestions from problem solving and provide students with extra help on many topics.

To the Student

Algebra is a course that cannot be learned by observation. To learn algebra you must become an active participant. You must read the text, pay attention in class, and, most importantly, you must work the exercises. The more exercises you work, the better.

The text was written with you in mind. Short, clear sentences are used, and many examples are given to illustrate specific points. The text stresses useful applications of algebra. Hopefully, as you progress through the course, you will come to realize that algebra is not just another math course that you are required to take, but a course that offers a wealth of useful information and applications.

This text makes full use of color. The different colors are used to highlight important information. Important procedures, definitions, and formulas are placed within colored boxes.

The boxes marked **Helpful Hints** should be studied carefully, for they stress important information. The boxes marked **Common Student Errors** should also be studied carefully. These boxes point out errors that students commonly make, and provide the correct procedures for doing these problems.

Ask your professor early in the course to explain the policy on when the calculator may be used. Pay particular attention to the **Calculator Corners.** You should also read the **Graphing Calculator Corners** even if you are not using a graphing calculator in class. You may find the information presented here helps you better understand the algebraic concepts.

Other questions you should ask your professor early in the course include: What supplements are available for use? Where can help be obtained when the professor is not available? Supplements that may be available include: Student's Study Guide, Student's Solutions Manual, tutorial software, and video tapes, including a tape on the study skills needed for success in mathematics.

You may wish to form a study group with other students in your class. Many students find that working in small groups provides an excellent way to learn the material. By discussing and explaining the concepts and exercises to one another you reinforce your own understanding. Once guidelines and procedures are determined by your group, make sure to follow them.

One of the first things you should do is to read Section 1.1, Study Skills Needed for Success in Mathematics. Read this section slowly and carefully, and pay particular attention to the advice and information given. Occasionally, refer back to this section. This could be the most important section of the book. Carefully read the material on doing your homework and on attending class.

At the end of all exercise sets (after the first two) are **cumulative review exercises.** You should work these problems on a regular basis, even if they are not assigned. These problems are from earlier sections and chapters of the text, and they will refresh your memory and reinforce those topics. If you have a problem when working these exercises, read the appropriate section of the text or study your notes that correspond to that material. The section of the text where the Cumulative Review Exercises were introduced is indicated in brackets, [ ], to the left of the exercise. After reviewing the material, if you still have a problem, make an appointment to see your professor. Working the Cumulative Review Exercises throughout the semester will also help prepare you to take your final exam.

At the end of each exercise set are **Group Activity/Challenge Problem** exercises. These exercises often provide questions that create interesting discussions, present additional applications of algebra, help prepare you for future material or present more challenging questions. You may wish to try some of these questions even if they are not assigned.

At the end of each chapter are a **Summary,** a set of **Review Exercises,** and a **Practice Test.** Before each examination you should review these sections carefully and take the practice test. If you do well on the practice test, you should do well on the class test. The questions in the review exercises are marked to indicate the section in which that material was first introduced. If you have a problem with a review exercise question, reread the section indicated. You may also wish to take the **Cumulative Review Test** that appears at the end of every even-numbered chapter.

In the back of the text there is an **answer section** which contains the answers to the *odd-numbered* exercises and Graphing Calculator Exercises, *all* cumulative review exercises, review exercises, practice tests, and cumulative review tests. *Selected answers* are given for the Group Activity/Challenge Problem exercises. The answers should be used only to check your work.

I have tried to make this text as clear and error free as possible. No text is perfect, however. If you find an error in the text, or an example or section that you believe can be improved, I would greatly appreciate hearing from you. If you enjoy the text, I would also appreciate hearing from you.

Allen R. Angel

Intermediate Algebra for College Students

Chapter *1*

Basic Concepts

See Section 1.3, Example 12

Preview and Perspective

In this chapter we review many of the concepts that were presented in elementary algebra. We will be using these concepts throughout the course, so you must understand them. In Section 1.1, we discuss study skills that will help you in this and other mathematics courses. Study and apply this material. In Section 1.2 we discuss sets. Sets and their properties are unifying concepts of mathematics and will be used continuously in this and later mathematics courses you may take. We will apply the concepts of the union and intersection of sets when we discuss inequalities in Sections 2.5 and 2.6. The properties of real numbers discussed in Section 1.3 are used throughout the book, often without naming them. Pay particular attention to the definition of absolute value. Absolute value is an important concept that will be used in many sections of this book and in other mathematics courses. The rules in Section 1.4, "Order of Operations," are used in evaluating expressions and formulas. Knowledge of order of operations is also essential in checking solutions to equations. You must understand the order of operations to be successful in this course.

In this book we have Calculator Corners. Ask your instructor which calculator he or she recommends or requires for the course, and purchase it as soon as possible. Study your calculator manual and become proficient with the use of your calculator.

1.1 Study Skills for Success in Mathematics and Using a Calculator

Tape 1

1. Having a positive attitude.
2. Preparing for and attending class.
3. Preparing for and taking examinations.
4. Finding help.
5. Using a calculator.

You need to acquire certain study skills that will help you to complete this course successfully. These study skills will also help you succeed in any other mathematics courses you may take.

It is important for you to realize that this course is the foundation for more advanced mathematics courses. If you have a thorough understanding of algebra, you will find it easier to be successful in later mathematics courses.

Having a Positive Attitude

1. You may be thinking to yourself, "I hate math" or "I wish I did not have to take this class." You may have heard the term *math anxiety* and feel that you fall in this category. The first thing you need to do to be successful in this course is to change your attitude to a more positive one. You must be willing to give this course and yourself a fair chance.

Based on past experiences in mathematics, you may feel this is difficult. However, mathematics is something you need to work at. Many of you taking this course are more mature now than when you took previous mathematics courses. Your maturity and your desire to learn are extremely important and can make a tremendous difference in your ability to succeed in mathematics. I believe you can be successful in this course, but you also need to believe it.

Preparing for Class

PREVIEWING THE MATERIAL

Before class, you should spend a few minutes previewing any new material in the textbook. You do not have to understand everything you read yet. Just get a feeling for the definitions and concepts that will be discussed. This quick preview will help you to understand what your instructor is explaining during class. After the material is explained in class, read the corresponding sections of the text slowly and carefully, word by word, as explained below.

READING THE TEXT

A mathematics text is not a novel. Mathematics textbooks should be read slowly and carefully. If you do not understand what you are reading, reread the material. When you come across a new concept or definition, you may wish to underline it so that it stands out. This way, when looking for it later, it will be easier to find. When you come across a worked-out example, read and follow the example very carefully. Do not just skim it. Try working out the example yourself on another sheet of paper. Make notes of anything you do not understand to ask your instructor.

DOING HOMEWORK

Two very important commitments that you must make to be successful in this course are attending class and doing your homework regularly. Your assignments must be worked conscientiously and completely. Mathematics cannot be learned by observation. You need to practice what you have heard in class. By doing homework you truly learn the material.

Don't forget to check the answers to your homework assignments. Answers to the odd exercises are in the back of this book. In addition, the answers to all the cumulative review exercises, end-of-chapter review exercises, practice tests, and cumulative review tests are in the back of the book. Answers to selected group activity exercises are given.

If you have difficulty with some of the exercises, mark them and do not hesitate to ask questions about them in class. You should not feel comfortable until you understand all the concepts needed to work every assigned problem successfully.

When you do your homework, make sure that you write it neatly and carefully. Pay particular attention to copying signs and exponents correctly. Do your homework in a step-by-step manner. This way you can refer back to it later and still understand what was written. You can study your homework problems to review for tests.

ATTENDING AND PARTICIPATING IN CLASS

You should attend every class possible. Most instructors will agree that there is an inverse relationship between absences and grades. That is, the more absences you have, the lower your grade will be. Every time you miss a class, you miss important information. If you must miss a class, contact your instructor ahead of time and get the reading assignment and homework.

While in class, pay attention to what your instructor is saying. If you do not understand something, ask your instructor to repeat or explain the material. If you have read the upcoming material before class and have questions that have not been answered, ask your instructor. If you do not ask questions, your instructor will not know that you have a problem in understanding the material.

In class, take careful notes. Write numbers and letters clearly so that you can read them later. It is not necessary to write down every word your instructor says. Copy down the major points and the examples that do not appear in the text. You should not be taking notes so frantically that you lose track of what your instructor is saying. It is a mistake to believe that you can copy down material in class without understanding it and then figure it out later.

STUDYING

Study in the proper atmosphere. Study in an area where you are not constantly disturbed so that your attention can be devoted to what you are reading. The area where you study should be well ventilated and well lit. You should have sufficient desk space to spread out all your materials. Your chair should be comfortable. There should be no loud music to distract you from studying.

When studying, you should not only understand how to work a problem, you should also know why you follow the specific steps you do to work the problem. If you do not have an understanding of why you follow the specific process, you will not be able to solve similar problems.

TIME MANAGEMENT

It is recommended that students spend at least 2 hours studying and doing homework for every hour of class time. Some students require more time than others. Finding the necessary time to study is not always easy. The following are some suggestions that you may find helpful.

1. Plan ahead. Determine when you will have time to study and do your homework. Do not schedule other activities for these time periods. Try to space these periods evenly over the week.
2. Be organized so that you will not have to waste time looking for your books, pen, calculator, or notes.
3. Use a calculator to perform tedious calculations.
4. When you stop studying, clearly mark where you stopped in the text.
5. Try not to take on added responsibilities. You must set your priorities. If your education is a top priority, as it should be, you may have to cut the time spent on other activities.
6. If time is a problem, do not overburden yourself with too many courses. Consider taking fewer credits. If you do not have sufficient time to study, your understanding and your grade in all of your courses may suffer.

Examinations   3

STUDYING FOR AN EXAM

If you do some studying each day, you should not need to cram the night before an exam. If you wait until the last minute, you will not have time to seek the help you may need. To review for an exam:

1. Read your class notes.
2. Review your homework assignments.
3. Study the formulas, definitions, and procedures given in the text.
4. Read the Common Student Error boxes and Helpful Hint boxes carefully.
5. Read the summary at the end of each chapter.
6. Work the review exercises at the end of each chapter. If you have difficulties, restudy those sections. If you still have trouble, seek help.
7. Work the practice chapter test.

TAKING AN EXAM

Make sure that you get a good night's sleep the day before the test. If you studied properly, you should not have to stay up late the night before preparing for a test. Arrive at the exam site early so that you have a few minutes to relax before the exam. If you need to rush to get to the exam, you will start out nervous and anxious. After you receive the exam, do the following:

1. Carefully write down any formulas or ideas that you need to remember.
2. Look over the entire exam quickly to get an idea of its length and to make sure that no pages are missing. You will need to pace yourself to make sure that you complete the entire exam. Be prepared to spend more time on problems worth more points.
3. Read the test directions carefully.
4. Read each problem carefully. Answer each question completely and make sure that you have answered the specific question asked.
5. Work the questions you understand best first; then come back and work those problems you are not sure of. Do not spend too much time on any one problem.
6. Attempt each problem. You may be able to earn at least partial credit.
7. Work carefully and write clearly so that your instructor can read your work. Also, it is easy to make mistakes when your writing is unclear.
8. Check your work and your answers if you have time.
9. Do not be concerned if others finish the test before you. Do not be disturbed if you are the last to finish. Use all your extra time to check your work.

Finding Help   4

USING SUPPLEMENTS

This text comes with numerous supplements. The student's supplements are listed in the preface on page xv. Find out from your instructor early in the semester which of these supplements are available and which supplements might be beneficial for you to use. Reading supplements should not replace reading the text, but should enhance your understanding of the material. If you miss a class, you may want to review the videotape on the topic you missed before attending the next class.

SEEKING HELP

One thing I stress with my own students is to *get help as soon as you need it!* Do not wait! In mathematics, one day's material is often based on the previous day's material. So, if you don't understand the material today, you may not be able to understand the material tomorrow.

Where should you seek help? There are often a number of places to obtain help on campus. You should try to make a friend in the class with whom you can study. Often you can help one another. You may wish to form a study group with other students in your class. Discussing the concepts and homework with your peers will reinforce your own understanding of the material.

You should not hesitate to visit your instructor when you are having problems with the material. Be sure you read the assigned material and attempt the homework before meeting with your instructor. Come prepared with specific questions to ask.

Often other sources of help are available. Many colleges have a mathematics laboratory or a mathematics learning center where tutors are available to help students. Ask your instructor early in the semester if any tutors are available, and find out where the tutors are located. Then use these tutors as needed.

Calculators 5 Many instructors require their students to purchase and to use a calculator in class. You should find out as soon as possible which calculator, if any, your instructor expects you to use. If you plan on taking additional mathematics courses, you should determine which calculator will be required in those courses and consider purchasing that calculator for use in this course if its use is permitted by your instructor. Many instructors require a scientific calculator and some may require a graphing calculator (also called a *grapher*).

In this book we provide information about calculators in the Calculator Corners. Some Calculator Corners are for scientific calculators, others are for graphing calculators. A special supplement is available for those classes that base this course on the graphing calculator.

Exercise Set 1.1   **10.** at least two hours   **12.** slowly and carefully

Do you know all of the following information? If not, ask your instructor as soon as possible.

1. What is your instructor's name?
2. What are your instructor's office hours?
3. Where is your instructor's office located?
4. How can you best reach your instructor?
5. Where can you obtain help if your instructor is not available?
6. What supplements are available to assist you in learning?
7. Does your instructor recommend or require a specific calculator? If so, which one?
8. When can you use a calculator? Can it be used in class, on homework, on tests?
9. Do you know the name and phone number of a friend in class?
10. For each hour of class time, how many hours outside class are recommended for homework and studying?
11. List what you should do to be properly prepared for class. Do all the homework and preview the new material.
12. Explain how a mathematics textbook should be read.
13. Write a summary of the steps you should follow when taking an exam.
14. Having a positive attitude is very important for success in this course. Are you beginning this course with a positive attitude? It is important that you do!

15. You need to make a commitment to spend the time necessary to learn the material, to do the homework, and to attend class regularly. Explain why you believe this commitment is necessary to be successful in this course.

16. What are your reasons for taking this course?

17. What are your goals for this course?

18. Have you given any thought to studying with a friend or a group of friends? Can you see any advantages in doing so? Can you see any disadvantages in doing so?

1.2   Sets and Other Basic Concepts

Tape 1

1 Identify sets.
2 Identify subsets.
3 Identify and use inequalities.
4 Use set builder notation.
5 Find the union and intersection of sets.
6 Identify important sets of numbers.

We begin by introducing two important definitions. Letters called **variables** are used to represent numbers. The letters x, y, and z are often used for variables. However, other letters may be used. When presenting properties or rules, the letters a, b, and c, are often used as variables. The term **algebraic expression,** or simply **expression,** will be used often in the text. An expression is any combination of numbers, variables, exponents, mathematical symbols, and mathematical operations.

Sets   1 Sets are used in many areas of mathematics, so an understanding of sets and set notation is important. A **set** is a collection of objects. The objects in a set are called **elements** of the sets. Sets are indicated by means of braces, { }, and are often named with capital letters. When the elements of a set are listed within the braces, as illustrated below, the set is said to be in **roster form.**

$$A = \{a, b, c\}$$
$$B = \{\text{yellow, green, blue, red}\}$$
$$C = \{1, 2, 3, 4, 5\}$$

Set A has three elements, set B has four elements, and set C has five elements. The symbol $\in$ is used to indicate that an item is an element of a set. Since 2 is an element of set C, we may write $2 \in C$; this is read "2 is an element of set C." Note that 6 is not an element of set C. We may therefore write $6 \notin C$, which is read "6 is not an element of set C."

A set may be finite or infinite. Sets A, B, and C each have a finite number of elements and are therefore *finite sets.* In some sets it is impossible to list all the elements. These are *infinite sets.* The following set, called the set of **natural numbers** or **counting numbers,** is an example of an infinite set.

$$N = \{1, 2, 3, 4, 5, \ldots\}$$

The three dots after the last comma indicate that the set continues in the same manner indefinitely.

Another important infinite set is the integers. The set of **integers** follows.

$$I = \{\ldots, -4, -3, -2, -1, 0, 1, 2, 3, 4, \ldots\}$$

Notice that the set of integers includes both positive and negative integers and the number 0.

If we write

$$D = \{1, 2, 3, 4, 5, \ldots, 280\}$$

we mean that the set continues in the same manner until the number 280. Set D is the set of the first 280 natural numbers. D is therefore a finite set.

A special set that contains no elements is called the **null set,** or **empty set,** written { } or $\varnothing$. For example, the set of students in your class over the age of 150 is the null or empty set.

Subsets

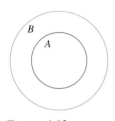

FIGURE 1.12

[2] Set A is a **subset** of set B, written $A \subseteq B$, if every element of set A is also an element of set B. Figure 1.1 illustrates two sets A and B, where set A is a subset of set B. *Note:* Every element that is in set A must also be in set B.

EXAMPLE 1   If the first set is a subset of the second set, insert $\subseteq$ in the shaded area between the two sets. If the first set is not a subset of the second set, insert the symbol $\not\subseteq$ (read "is not a subset of") between the two sets.

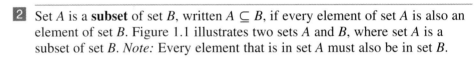

(a) $\{1, 2, 3\}$ ▪ $\{1, 2, 3, 4\}$ **(b)** $\{a, b, c, d\}$ ▪ $\{b, c, d, e, f, g\}$

Solution:

(a) $\{1, 2, 3\} \subseteq \{1, 2, 3, 4\}$

(b) $\{a, b, c, d\} \not\subseteq \{b, c, d, e, f, g\}$ Notice that *a* is in the first set but not the second set.

EXAMPLE 2   Determine if the set of natural numbers, $N = \{1, 2, 3, 4, 5, \ldots\}$ is a subset of the set of **whole numbers,** $W = \{0, 1, 2, 3, 4, 5, \ldots\}$.

Solution: Since every element in the set of natural numbers is also an element in the set of whole numbers, we may write

$$N \subseteq W$$

The natural numbers are a subset of the set of whole numbers.

COMMON STUDENT ERROR   Students commonly make the following errors when working with sets.

Correct	*Incorrect*
1. { } or $\varnothing$ is the empty set.	Students incorrectly write {0} to represent the empty set.
2. $3 \in \{1, 2, 3\}$	~~$3 \subseteq \{1, 2, 3\}$~~
	Since there are no braces around the 3, it is not a set. 3 is an element of the set, not a subset.
3. $\{3\} \subseteq \{1, 2, 3\}$	~~$\{3\} \in \{1, 2, 3\}$~~
	Since there are braces around the 3, {3} is a set. Thus {3} is a subset, not an element, of the set $\{1, 2, 3\}$.

Note that {0} is *not the empty set* since the set contains the element 0.

Inequalities

3 Before we introduce a second method of writing a set, called *set builder notation,* we will introduce the inequality symbols. The symbols used to indicate an inequality are $>$, $\geq$, $<$, $\leq$, and $\neq$.

> **Inequality Symbols**
>
> $>$ is read "is greater than."
>
> $\geq$ is read "is greater than or equal to."
>
> $<$ is read "is less than."
>
> $\leq$ is read "is less than or equal to."
>
> $\neq$ is read "is not equal to."

FIGURE 1.2

Inequalities can be explained using the real number line (Fig. 1.2). The number a is greater than the number b, $a > b$, when a is to the right of b on the number line (Fig. 1.3). We can also state that the number b is less than a, $b < a$, when b is to the left of a on the number line. The inequality $a \neq b$ means either $a < b$ or $a > b$.

Lesser   Greater

b a

FIGURE 1.3

EXAMPLE 3   Insert either $>$ or $<$ in the shaded area between the numbers to make the statement true.

(a) 7   3   **(b)** -6 -4 **(c)** 0 -2

Solution: Draw a number line illustrating the location of all the given values (Fig. 1.4).

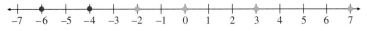

FIGURE 1.4

(a) $7 > 3$ Note that 7 is to the right of 3 on the number line.
(b) $-6 < -4$ Note that -6 is to the left of -4 on the number line.
(c) $0 > -2$ Note that 0 is to the right of -2 on the number line.

Remember that the symbol used in an inequality, if it is true, always points to the smaller of the two numbers.

We use the notation $x > 2$, read "x is greater than 2," to represent *all* real numbers greater than 2. We use the notation $x \leq -3$, read "x is less than or equal to -3," to represent all real numbers that are less than or equal to -3. The notation $-4 \leq x < 3$ means all real numbers that are greater than or equal to -4 and also less than 3. We discuss this notation in more detail later in the book. In the inequalities $x > 2$ and $x \leq -3$, the 2 and -3 are called **endpoints.** In the inequality $-4 \leq x < 3$, the -4 and 3 are the endpoints. The solution to inequalities that use either $<$ or $>$ do not include the endpoints, but the solutions to inequalities that use either $\leq$ or $\geq$ do include the endpoints. When inequalities are illustrated on the number line, a solid circle is used to show that the endpoint is included in the answer, and an open circle is used to show that the endpoints are not included. Below are some illustrations of how certain inequalities are indicated on the number line.

Inequality	*Inequality Indicated on the Number Line*
$x > 2$	![open circle at 2, line to the right] 2
$x \leq -3$	![line to the left, solid circle at −3] −3
$-4 \leq x < 3$	![solid circle at −4, open circle at 3] −4 3

Some students misunderstand the word *between.* The word *between* indicates that the endpoints are not included in the answer. For example, the set of natural numbers between 2 and 6 is {3, 4, 5}. If we wish to include the endpoints we can use the word *inclusive.* For example, the set of natural numbers between 2 and 6 inclusive is {2, 3, 4, 5, 6}.

Set Builder Notation

4 Now that we have introduced the inequality symbols we will discuss another method of indicating a set, called **set builder notation.** An example of set builder notation is

$$E = \{x \mid x \text{ is a natural number greater than 6}\}$$

This is read "Set E is the set of all x such that x is a natural number greater than 6." In roster form this set is written

$$E = \{7, 8, 9, 10, 11, \ldots\}$$

The general form of set builder notation is

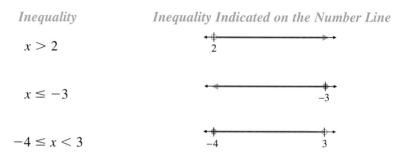

Two condensed ways of writing set $E = \{x \mid x \text{ is a natural number greater than 6}\}$ in set builder notation follow.

$$E = \{x \mid x > 6 \text{ and } x \in N\} \quad \text{or} \quad E = \{x \mid x \geq 7 \text{ and } x \in N\}$$

The set $A = \{x \mid -3 < x \leq 4 \text{ and } x \in I\}$ is the set of integers greater than -3 and less than or equal to 4. The set written in roster form is $\{-2, -1, 0, 1, 2, 3, 4\}$. Notice that the endpoint -3 is not included in the set but the endpoint 4 is included.

How do the sets $B = \{x \mid x > 2 \text{ and } x \in N\}$ and $C = \{x \mid x > 2\}$ differ? Can you write each set in roster form? Can you illustrate both sets on the number line?

Set B contains only the natural numbers greater than 2, that is, $\{3, 4, 5, 6, \ldots\}$. Set C contains not only the natural numbers greater than 2 but also fractions and decimal numbers greater than 2. If you attempted to write set C in roster form, where would you begin? What is the smallest number greater than 2? Is it 2.1 or 2.01 or 2.001? Since there is no smallest number greater than 2, this set cannot be written in roster form. Below we illustrate these two sets on the number line. We have also illustrated two other sets.

Set	*Set Indicated on the Number Line*
$\{x \mid x > 2 \text{ and } x \in N\}$	
$\{x \mid x > 2\}$	
$\{x \mid -1 \le x < 4 \text{ and } x \in I\}$	
$\{x \mid -1 \le x < 4\}$	

Another method of indicating inequalities, called **interval notation,** will be discussed in Section 2.5.

Union and Intersection of Sets

5 Just as *operations* such as addition and multiplication are performed on numbers, operations can be performed on sets. Now that we have discussed set builder notation we can discuss and define the set operations *union* and *intersection*.

The **union** of set A and set B, written $A \cup B$, is the set of elements that belong to either set A *or* set B. The union is formed by combining, or joining together, the elements in set A with those in set B.

Examples of Union of Sets

$A = \{1, 2, 3, 4, 5\}, \quad B = \{3, 4, 5, 6, 7\}, \quad A \cup B = \{1, 2, 3, 4, 5, 6, 7\}$
$A = \{a, b, c, d, e\}, \quad B = \{x, y, z\}, \quad\quad A \cup B = \{a, b, c, d, e, x, y, z\}$

In set builder notation we can express $A \cup B$ as

$$A \cup B = \{x \mid x \in A \quad or \quad x \in B\}$$

The **intersection** of set A and set B, written $A \cap B$, is the set of all elements that are common to both set A *and* set B.

Examples of Intersection of Sets

$A = \{1, 2, 3, 4, 5\}, \quad B = \{3, 4, 5, 6, 7\}, \quad A \cap B = \{3, 4, 5\}$
$A = \{a, b, c, d, e\}, \quad B = \{x, y, z\}, \quad\quad A \cap B = \{ \quad \}$

Note that in the last example sets A and B have no elements in common. Therefore, their intersection is the empty set. In set builder notation we can express $A \cap B$ as

$$A \cap B = \{x \mid x \in A \quad and \quad x \in B\}$$

Sets of Numbers   6   At this point we have all the necessary information to discuss important sets of real numbers.

Important Sets of Real Numbers

Real numbers	$\{x \mid x$ is a point on the number line$\}$
Natural or counting numbers	$\{1, 2, 3, 4, 5, \ldots\}$
Whole numbers	$\{0, 1, 2, 3, 4, 5, \ldots\}$
Integers	$\{\ldots, -3, -2, -1, 0, 1, 2, 3, \ldots\}$
Rational numbers	$\left\{\dfrac{p}{q} \middle\mid p \text{ and } q \text{ are integers}, q \neq 0\right\}$
Irrational numbers	$\{x \mid x$ is a real number that is not rational$\}$

Let us briefly look at the rational, irrational, and real numbers. A rational number is any number that can be represented as a quotient of two integers, with the denominator not zero.

Examples of Rational Numbers

$$\frac{3}{5}, \quad \frac{-2}{3}, \quad 0, \quad 1.63, \quad 7, \quad -12, \quad \sqrt{4}$$

Notice that 0, or any other integer, is also a rational number since it can be written as a fraction with a denominator of 1. For example, $0 = \frac{0}{1}$ and $7 = \frac{7}{1}$.

The number 1.63 can be written $\frac{163}{100}$ and is thus a quotient of two integers. Since $\sqrt{4} = 2$ and 2 is an integer, $\sqrt{4}$ is a rational number. *Every rational number when written as a decimal number will be either a repeating or a terminating decimal number.*

Examples of Repeating Decimals	*Examples of Terminating Decimals*
$\dfrac{2}{3} = 0.6666\ldots$ (the 6 repeats)	$\dfrac{1}{2} = 0.5$
$\dfrac{1}{7} = 0.142857142857\ldots$ (the block 142857 repeats)	$\dfrac{7}{4} = 1.75$

To show that a digit or group of digits repeat, we can place a bar above the digit or group of digits that repeat. For example, we may write

$$\frac{2}{3} = 0.\overline{6} \quad \text{and} \quad \frac{1}{7} = 0.\overline{142857}$$

Although $\sqrt{4}$ is a rational number, the square roots of most integers are not. Most square roots will be neither terminating nor repeating decimals when expressed as a decimal number, and are irrational numbers. Some irrational numbers are $\sqrt{2}, \sqrt{3}, \sqrt{5}$, and $\sqrt{6}$. Another irrational number is pi, π. When we give a decimal value for an irrational number, we are giving only an *approximation* of the value of the irrational number. The symbol $\approx$ means "is approximately equal to."

$$\pi \approx 3.14, \qquad \sqrt{2} \approx 1.41$$

The **real numbers** are formed by taking the *union* of the rational numbers with the irrational numbers. Therefore, any real number must be either a rational number or an irrational number. The symbol $\mathbb{R}$ is often used to represent the set of real numbers. Figure 1.5 illustrates various real numbers on the number line.

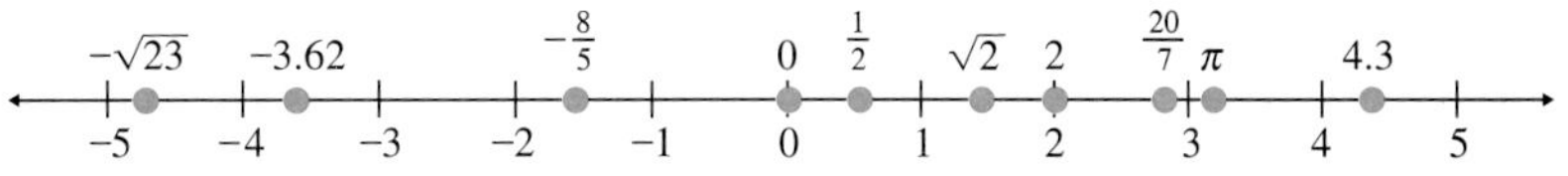

FIGURE 1.5

Figure 1.6 illustrates the relationship between the various subsets of the real numbers. Earlier we stated that the natural numbers were a subset of the set of whole numbers. In Figure 1.6(a), you see that the set of natural numbers is a subset of the set of whole numbers, of the set of integers, and of the set of rational numbers. Therefore, every natural number must also be a whole number, an integer, and a rational number. Using the same reasoning, we can see that the set of whole numbers is a subset of the set of integers and of the set of rational numbers, and that the set of integers is a subset of the set of rational numbers.

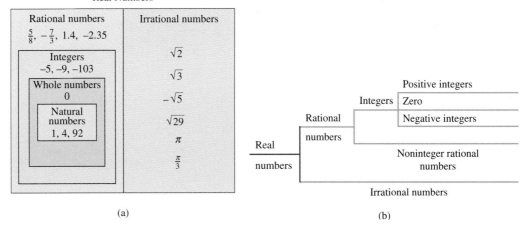

(a)   (b)

FIGURE 1.6

Using Figure 1.6(b) we see that the positive integers, 0, and the negative integers form the integers, that the integers and noninteger rational numbers form the rational numbers, and so on.

EXAMPLE 4   Consider the following set:

$$\left\{ -4,\ 0,\ \frac{3}{5},\ 2.7,\ \sqrt{3},\ -\sqrt{5},\ \frac{19}{6},\ 18,\ 4.62,\ -23,\ \pi \right\}$$

List the elements of the set that are:

(a) Natural numbers   **(b)** Whole numbers   **(c)** Integers
(d) Rational numbers   **(e)** Irrational numbers **(f)** Real numbers

Solution:

(a) 18

(b) 0, 18

(c) −4, 0, 18, −23

(d) −4, 0, $\frac{3}{5}$, 2.7, $\frac{19}{6}$, 18, 4.62, −23

(e) $\sqrt{3}$, $-\sqrt{5}$, π

(f) −4, 0, $\frac{3}{5}$, 2.7, $\sqrt{3}$, $-\sqrt{5}$, $\frac{19}{6}$, 18, 4.62, −23, π

Not all numbers are real numbers. Some numbers that we discuss later in the text that are not real numbers include complex numbers and imaginary numbers.

Exercise Set 1.2

Insert either ∈ or ∉ in the shaded area to make a true statement.

1. 6 ▓ {1, 3, 6, 9,} ∈

2. 4 ▓ {1, 2, 3} ∉

3. 0 ▓ {−1, 1, 3, 5} ∉

4. 9 ▓ {1, 2, 3, 4, . . . , 80} ∈

5. −3 ▓ {1, 2, 3, . . .} ∉

6. {3} ▓ {1, 2, 3, 4} ∉

7. {5} ▓ {1, 2, 3, 4, 5, 6} ∉

8. 0 ▓ {−3, −2, −1, 0, 1, 2, 3} ∈

Insert either ⊆ or ⊄ in the shaded area to make a true statement (see Example 1).

9. {2} ▓ {1, 2, 3, 4} ⊆

10. {0} ▓ {0, 1} ⊆

11. {5} ▓ {−1, −2, −3} ⊄

12. {8} ▓ {3, 4, 5, 6, . . .} ⊆

13. {72} ▓ {6, 7, 8, . . . , 70} ⊄

14. {−1} ▓ {−1, −2, −3} ⊆

15. {−1} ▓ {−4, −3, −2} ⊄

16. 3 ▓ {1, 2, 3} ⊄

17. 1.5 ▓ {1.3, 1.4, 1.5, 1.6, . . .} ⊄

18. $\left\{\frac{7}{3}\right\}$ ▓ $\left\{\frac{1}{3}, \frac{2}{3}, \frac{3}{3}, \frac{4}{3}, \ldots\right\}$ ⊆

19. $\left\{\frac{1}{10}\right\}$ ▓ $\left\{1, \frac{1}{2}, \frac{1}{3}, \frac{1}{4}, \ldots\right\}$ ⊆

20. {1.6} ▓ {0.1, 0.2, 0.3, 0.4, . . .} ⊆

List each set in roster form.

21. $A = \{x \mid 4 < x < 6 \text{ and } x \in N\}$ $A = \{5\}$

22. $B = \{x \mid x \text{ is an even integer between 2 and 8}\}$ $B = \{4, 6\}$

23. $C = \{x \mid x \text{ is an even integer greater than or equal to 6 and less than 10}\}$ $C = \{6, 8\}$

24. $D = \{x \mid x > 5 \text{ and } x \in N\}$ $D = \{6, 7, 8, 9, \ldots\}$

25. $E = \{x \mid x < 5 \text{ and } x \in W\}$ $E = \{0, 1, 2, 3, 4\}$

26. $F = \left\{x \mid -\frac{6}{5} \le x < \frac{15}{4} \text{ and } x \in N\right\}$ $F = \{1, 2, 3\}$

27. $H = \{x \mid x \text{ is a whole number multiple of 5}\}$

28. $I = \{x \mid x \text{ is an integer greater than } −5\}$

29. $J = \{x \mid x \text{ is an integer between } −6 \text{ and } −4.3\}$ $J = \{−5\}$

30. $K = \{x \mid x \text{ is a whole number between 3 and 4}\}$ $K = \{ \}$

27. $H = \{0, 5, 10, 15, 20, \ldots\}$ **28.** $I = \{−4, −3, −2, −1, \ldots\}$

Let N = the set of natural numbers, W = the set of whole numbers, I = the set of integers, Q = the set of rational numbers, H = the set of irrational numbers, and ℝ = the set of real numbers. Insert either ⊆ or ⊄ in the shaded area to make a true statement.

31. N ▓ W ⊆

32. W ▓ Q ⊆

33. I ▓ Q ⊆

34. W ▓ N ⊄

35. Q ▓ H ⊄

36. Q ▓ $\mathbb{R}$ ⊆

37. H ▓ $\mathbb{R}$ ⊆

38. Q ▓ I ⊄

39. Q ▓ N ⊄

40. $\mathbb{R}$ ▓ Q ⊄

41. H ▓ Q ⊄

42. I ▓ N ⊄

Answer true or false.

43. Some rational numbers are integers.   true

44. Every natural number is a whole number.   true

45. Every whole number is a natural number.   false

46. Every integer is a rational number.   true

47. Every rational number is an integer.   false

48. The union of the set of rational numbers with the set of irrational numbers forms the set of real numbers.   true

49. The intersection of the set of rational numbers and the set of irrational numbers is the empty set.   true

50. The set of natural numbers is a finite set.   false

51. The set of integers between 1 and 2 is the null set. true

52. The set of rational numbers between 1 and 2 is an infinite set.   true

53. Consider the set $\{-6, 4, \frac{1}{2}, \frac{5}{9}, 0, \sqrt{7}, \sqrt{5}, -1.23, \frac{99}{100}\}$. List the elements that are:

(a) Natural numbers   4

(b) Whole numbers   4, 0

(c) Integers $-6, 4, 0$

(d) Rational numbers $-6, 4, \frac{1}{2}, \frac{5}{9}, 0, -1.23, \frac{99}{100}$

(e) Irrational numbers $\sqrt{7}, \sqrt{5}$

(f) Real numbers $-6, 4, \frac{1}{2}, \frac{5}{9}, 0, \sqrt{7}, \sqrt{5}, -1.23, \frac{99}{100}$

54. Consider the set $\{2, 4, -5.33, \frac{9}{2}, \sqrt{7}, \sqrt{2}, -100, -7, 4.7\}$. List the elements that are:

(a) Whole numbers   2, 4

(b) Natural numbers   2, 4

(c) Rational numbers $2, 4, -5.33, \frac{9}{2}, -100, -7, 4.7$

(d) Integers $2, 4, -100, -7$

(e) Real numbers $2, 4, -5.33, \frac{9}{2}, \sqrt{7}, \sqrt{2}, -100, -7, 4.7$

(f) Irrational numbers $\sqrt{7}, \sqrt{2}$

Find $A \cup B$ and $A \cap B$ for each set A and B.

55. $A = \{1, 2, 3\}, B = \{2, 3, 4\}$
$A \cup B = \{1, 2, 3, 4\}, A \cap B = \{2, 3\}$

56. $A = \{2, 4, 6, 8\}, B = \{1, 3, 5, 7\}$
$A \cup B = \{1, 2, 3, 4, 5, 6, 7, 8\}, A \cap B = \{\ \}$

57. $A = \{-2, -4, -5\}, B = \{-1, -2, -4, -6\}$
$A \cup B = \{-1, -2, -4, -5, -6\}, A \cap B = \{-2, -4\}$

58. $A = \{-1, 0, 1\}, B = \{0, 2, 4, 6\}$
$A \cup B = \{-1, 0, 1, 2, 4, 6\}, A \cap B = \{0\}$

59. $A = \{\ \}, B = \{0, 1, 2, 3\}$
$A \cup B = \{0, 1, 2, 3\}, A \cap B = \{\ \}$

60. $A = \{2, 4, 6\}, B = \{2, 4, 6, 8, \ldots\}$ $A \cup B = \{2, 4, 6, 8, \ldots\}, A \cap B = \{2, 4, 6\}$

61. $A = \{0, 2, 4, 6, 8\}, B = \{1, 3, 5, 7\}$
$A \cup B = \{0, 1, 2, 3, 4, 5, 6, 7, 8\}, A \cap B = \{\ \}$

62. $A = \{1, 3, 5\}, B = \{1, 3, 5, 7, \ldots\}$
$A \cup B = \{1, 3, 5, 7, \ldots\}, A \cap B = \{1, 3, 5\}$

63. $A = \{0.1, 0.2, 0.3\}, B = \{0.2, 0.3, 0.4, 0.5, \ldots\}$
$A \cup B = \{0.1, 0.2, 0.3, 0.4, \ldots\}, A \cap B = \{0.2, 0.3\}$

64. $A = \{1, \frac{1}{2}, \frac{1}{4}, \frac{1}{6}, \ldots\}, B = \{\frac{1}{4}, \frac{1}{6}, \frac{1}{8}\}$
$A \cup B = \{1, \frac{1}{2}, \frac{1}{4}, \frac{1}{6}, \ldots\}, A \cap B = \{\frac{1}{4}, \frac{1}{6}, \frac{1}{8}\}$

Describe each set.

65. $A = \{1, 2, 3, 4, \ldots\}$ the set of natural numbers.

66. $B = \{5, 7, 9, 11, \ldots\}$ the set of odd natural numbers greater than or equal to 5

67. $C = \{8, 10, 12, \ldots, 30\}$

68. $A = \{a, b, c, d, \ldots, z\}$ the set of letters in the English alphabet

69. $B = \{\ldots, -5, -3, -1, 1, 3, 5, \ldots\}$ the set of odd integers

70. $C = \{$Alabama, Alaska, Arizona, $\ldots$, Wyoming$\}$

67. the set of even natural numbers greater than or equal to 8 and less than or equal to 30

70. the set of states in the United States

In Exercises 71 and 72, (a) write out how you would read each set; (b) write the set in roster form.

71. $A = \{x \mid x < 8 \text{ and } x \in N\}$

(a) Set A is the set of all x such that x is a natural number less than 8. (b) $A = \{1, 2, 3, 4, 5, 6, 7\}$

72. $B = \{x \mid x$ is one of the last five capital letters in the English alphabet$\}$

(a) Set B is the set of all x such that x is one of the last five capital letters in the English alphabet. (b) $B = \{V, W, X, Y, Z\}$

Insert either $<$ or $>$ in the shaded area to make the statement true (see Example 3).

73. $4 \ \ 2$ $>$ **74.** $6 \ \ 8$ $<$ **75.** $-2 \ \ 4$ $<$ **76.** $0 \ \ 4$ $<$

77. $-1 \ \ 1$ $<$ **78.** $-\frac{1}{2} \ \ -1$ $>$ **79.** $-3 \ \ -3.5$ $>$ **80.** $1 \ \ -3$ $>$

81. $-4 \ \ -2$ $<$ **82.** $-6 \ \ -1$ $<$ **83.** $-2 \ \ -5$ $>$ **84.** $-1 \ \ -4$ $>$

85. $-4.6 \ \ -4.7$ $>$ **86.** $-3 \ \ -3.5$ $>$ **87.** $1.1 \ \ 1.9$ $<$ **88.** $-1.1 \ \ -1.9$ $>$

89. $-952 \ \ -955$ $>$ **90.** $-780 \ \ -655$ $<$ **91.** $-\frac{7}{8} \ \ -\frac{8}{9}$ $>$ **92.** $-\frac{4}{7} \ \ -\frac{5}{9}$ $<$

Illustrate each set on the number line.

93. $\{x \mid x < 3\}$

94. $\{x \mid x \geq -1\}$

95. $\{x \mid x \geq 4\}$

96. $\{x \mid x \leq -2\}$

97. $\left\{x \mid -4 < x \leq \dfrac{3}{7}\right\}$

98. $\{x \mid -1.67 \leq x < 5.02\}$

99. $\{x \mid x > 2 \text{ and } x \in N\}$

100. $\{x \mid -1.75 \leq x \leq 2.4 \text{ and } x \in I\}$

101. $\left\{x \mid x < \dfrac{40}{9} \text{ and } x \in N\right\}$

102. $\left\{x \mid \dfrac{1}{2} < x \leq \dfrac{5}{9} \text{ and } x \in N\right\}$

Express in set builder notation the set of numbers indicated on the number line.

103. $\{x \mid x > -2\}$

104. $\{x \mid x \leq 5\}$

105. $\{x \mid x \geq -3 \text{ and } x \in I\}$

106. $\{x \mid x \leq -2 \text{ and } x \in I\}$

107. $\{x \mid -2 < x \leq 4.6\}$

108. $\{x \mid -4 \leq x < 7.6\}$

109. $\left\{x \mid x \leq -\dfrac{20}{3}\right\}$

110. $\{x \mid x \in I\}$

111. $\{x \mid -1 \leq x \leq 3 \text{ and } x \in I\}$

112. $\left\{x \mid -\dfrac{12}{5} < x \leq \dfrac{3}{10}\right\}$

113. (a) What does a bar above a digit or group of digits indicate? Write each fraction as a decimal number using a bar above the repeating digit or digits.

(b) $\dfrac{2}{3}$; $0.\overline{6}$ **(c)** $\dfrac{5}{9}$; $0.\overline{5}$ **(d)** $\dfrac{7}{6}$. $1.1\overline{6}$

114. Use the fact that $\dfrac{1}{3} = 0.\overline{3}$ and $\dfrac{2}{3} = 0.\overline{6}$ to write the fraction that is equivalent to each decimal number.

(a) $0.\overline{3}$ $\dfrac{1}{3}$ **(b)** $0.\overline{6}$ $\dfrac{2}{3}$ **(c)** $2.\overline{3}$ $2\dfrac{1}{3}$ **(d)** $2.\overline{6}$ $2\dfrac{2}{3}$

115. What is a variable? a letter used to represent a number

116. What is an expression?

117. What is the null or empty set?   a set that contains no element

118. Explain why every integer is also a rational number.

119. Describe the counting numbers, whole numbers, integers, rational numbers, irrational numbers and real numbers. Explain the relationships among the sets of numbers.

118. it can be written with a denominator of 1.

116. any combination of numbers, variables, exponents, mathematical symbols, and operations

113. (a) the digit or group of digits repeat

3. (a) the second set includes fractions and decimal numbers. **(b)** {3, 4, 5} **(c)** no

4. same number (both have cardinality, aleph-null—instructors may wish to assign this as a research project—can lead to an interesting group discussion)

Group Activity/ Challenge Problems

1. (a) Write the decimal numbers equivalent to $\dfrac{1}{9}, \dfrac{2}{9}, \dfrac{3}{9}$. $0.\overline{1}, 0.\overline{2}, 0.\overline{3}$

(b) Write the fractions equivalent to $0.\overline{4}, 0.\overline{5}, 0.\overline{6}$. $\dfrac{4}{9}, \dfrac{5}{9}, \dfrac{6}{9}$ or $\dfrac{2}{3}$

(c) What is $0.\overline{9}$ equal to? Explain how you determined your answer. 1

2. (a) Explain the difference between the following sets of numbers, $\{x \mid x > 1 \text{ and } x \in N\}$ and $\{x \mid x > 1\}$. $\{x \mid x > 1\}$ includes fractions and decimal numbers

(b) Write the first set given in roster form. {2, 3, 4, . . .}

(c) Can you write the second set in roster form? Explain your answer. no

3. Repeat Exercise 2 for the sets $\{x \mid 2 < x < 6 \text{ and } x \in N\}$ and $\{x \mid 2 < x < 6\}$.

4. In this section we discussed infinite sets. Consider the infinite set of natural numbers $\{1, 2, 3, 4, \ldots\}$ and the infinite set of whole numbers $\{0, 1, 2, 3, 4, \ldots\}$. Do you believe the set of whole numbers is greater than the set of natural numbers because the whole numbers contain the extra number 0, or do you believe the sets are the same size because they are both infinite sets? Explain your answer.

1.3   Properties of and Operations with Real Numbers

Tape 1

1 Evaluate absolute values.
2 Add real numbers.
3 Subtract real numbers.
4 Multiply real numbers.
5 Divide real numbers.
6 Use the properties of real numbers.

To succeed in algebra you must understand how to add, subtract, multiply, and divide real numbers. Before we can explain addition and subtraction of real numbers we need to discuss absolute value.

Two numbers that are the same distance from 0 on the number line but in opposite directions are called **additive inverses,** or **opposites,** of each other. For example, 3 is the additive inverse of -3, and -3 is the additive inverse of 3. The number 0 is its own additive inverse. The sum of a number and its additive inverse is 0. What are the additive inverses of -56.3 and $\dfrac{76}{5}$? Their additive inverses are

56.3 and $-\dfrac{76}{5}$, respectively. Notice that the additive inverse of a positive number is a negative number and the additive inverse of a negative number is a positive number.

Additive Inverse

For any real number a, its additive inverse is $-a$.

Consider the number -5. Its additive inverse is $-(-5)$. Since we know this number must be positive, this implies that $-(-5) = 5$. This is an example of the double negative property.

Double Negative Property

For any real number a, $-(-a) = a$.

By the double negative property $-(-7.4) = 7.4$ and $-\left(-\dfrac{12}{5}\right) = \dfrac{12}{5}$.

Absolute Value   1   The **absolute value** of a number is its distance from the number zero on the number line. The symbol $|\ |$ is used to indicate absolute value.

FIGURE 1.7

Consider the numbers 3 and −3 (see Fig. 1.7). Both numbers are 3 units from zero on the number line. Thus

$$|3| = 3 \quad \text{and} \quad |-3| = 3$$

EXAMPLE 1   Evaluate.

(a) $|4|$ **(b)** $|-1.6|$ **(c)** $|0|$

Solution:

(a) $|4| = 4$, since 4 is 4 units from zero on the number line.
(b) $|-1.6| = 1.6$, since −1.6 is 1.6 units from zero on the number line.
(c) $|0| = 0$.

The absolute value of any nonzero number will always be a positive number, and the absolute value of zero is zero.

To find the absolute value of a real number without using the number line, use the following definition.

> **Absolute Value**
>
> If a represents any real number, then
> $$|a| = \begin{cases} a & \text{if } a \geq 0 \\ -a & \text{if } a < 0 \end{cases}$$

The definition of absolute value indicates that the absolute value of any nonnegative number is the number itself, and the absolute value of any negative number is the additive inverse (or opposite) of the number. The absolute value of a number can be found by using the definition, as illustrated in the following examples.

$$|8| = 8 \qquad \text{Since 8 is greater than or equal to 0, its absolute value is 8.}$$
$$|0| = 0 \qquad \text{Since 0 is greater than or equal to 0, its absolute value is 0.}$$
$$|-12| = -(-12) = 12 \quad \text{Since } -12 \text{ is less than 0, its absolute value is } -(-12)$$
$$\text{or 12.}$$

By the definition, the absolute value of any real nonzero number must be a positive number.

EXAMPLE 2   Evaluate using the definition of absolute value.

(a) $|5|$ **(b)** $-|5|$ **(c)** $|-6.43|$ **(d)** $-|-6.43|$

Solution:

(a) $|5| = 5$, since $5 > 0$.
(b) We are finding the opposite of the absolute value of 5. Since the absolute value of 5 is positive, its opposite must be negative.

$$-|5| = -(5) = -5$$

(c) $|-6.43| = -(-6.43) = 6.43$, since $-6.43 < 0$

(d) We are finding the opposite of the absolute value of -6.43. Since the absolute value of -6.43 is positive, its opposite must be negative.

$$-|-6.43| = -(6.43) = -6.43$$

EXAMPLE 3   Insert $>$, $<$, or $=$ in the shaded area between the two values to make the statement true.

(a) $|-3|$ ▨ $|3|$ **(b)** $-|4|$ ▨ $|-3|$ **(c)** $-|-6|$ ▨ $|-5|$

Solution:

(a) $|-3| = |3|$, since both $|-3|$ and $|3|$ equal 3.
(b) $-|4| < |-3|$, since $-|4| = -4$ and $|-3| = 3$.
(c) $-|-6| < |-5|$, since $-|-6| = -6$ and $|-5| = 5$.

Addition of Real Numbers

2 We first discuss how to add two numbers with the same sign, either both positive or both negative, and then we will discuss how to add two numbers with different signs, one positive and the other negative.

> **To Add Two Numbers with the Same Sign (Both Positive or Both Negative)**
> Add their absolute values and place the common sign before the sum.

The sum of two positive numbers will be a positive number, and the sum of two negative numbers will be a negative number.

EXAMPLE 4   Add $-2 + (-5)$.

Solution: Since both numbers being added are negative, the sum will be negative. To find the sum, add the absolute values of these numbers and then place a negative sign before the value.

$$|-2| = 2$$
$$|-5| = 5$$

Now add the absolute values.

$$|-2| + |-5| = 2 + 5 = 7$$

Since both numbers are negative, the sum must be negative. Thus,

$$-2 + (-5) = -7$$

> **To Add Two Numbers with Different Signs (One Positive and the Other Negative)**
> Take the difference of the absolute values. The answer is positive if the positive number has the larger absolute value. The answer is negative if the negative number has the larger absolute value.

The sum of a positive number and a negative number may be either positive or negative or zero. The sign of the answer will be the same as the sign of the number with the larger absolute value.

EXAMPLE 5   Add $5 + (-9)$.

Solution: Since the numbers being added are of opposite signs, we find the difference of their absolute values. First take each absolute value.

$$|5| = 5$$
$$|-9| = 9$$

Now find the difference, $9 - 5 = 4$. The number -9 has a larger absolute value than the number 5, so their sum is negative.

$$5 + (-9) = -4$$

EXAMPLE 6   Evaluate.

(a) $-23 + 45$ **(b)** $-6.4 + (-8.5)$ **(c)** $\dfrac{5}{8} + \left(-\dfrac{4}{5}\right)$

Solution:

(a) $-23 + 45 = 22$

(b) $-6.4 + (-8.5) = -14.9$

(c) $\dfrac{5}{8} + \left(-\dfrac{4}{5}\right) = \dfrac{25}{40} + \left(-\dfrac{32}{40}\right) = \dfrac{25 + (-32)}{40} = \dfrac{-7}{40} = -\dfrac{7}{40}$

You must have a thorough understanding of addition of real numbers and be able to add real numbers that do not contain decimals or fractions quickly and mentally.

EXAMPLE 7   The Palau Trench in the Pacific Ocean lies 26,424 feet below sea level. The deepest trench, the Mariana Trench, is 9416 feet deeper than the Palau Trench. Find the depth of the Mariana Trench.

Solution: Consider distance below sea level to be negative. Therefore, the total depth is

$$-26,424 + (-9416) = -35,840 \text{ feet}$$

or 35,840 feet below sea level.

Subtraction of Real Numbers

③ Every subtraction problem can be expressed as an addition problem using the following rule.

> ### Subtraction of Real Numbers
> $$a - b = a + (-b)$$

This rule says that **to subtract b from a, add the opposite (or additive inverse) of b to a.**

SECTION 1.3 • PROPERTIES OF AND OPERATIONS WITH REAL NUMBERS **21**

The problem $5 - 7$ means $5 - (+7)$. To subtract $5 - 7$, add the opposite of $+ 7$, which is -7, to 5.

$$5 - 7 = 5 + (-7)$$

subtract positive add negative
7 7

Since $5 + (-7) = -2$, then $5 - 7 = -2$.

EXAMPLE 8 Evaluate.

(a) $6 - 10$ **(b)** $-8 - 4$

Solution:

(a) $6 - 10 = 6 + (-10) = -4$ **(b)** $-8 - 4 = -8 + (-4) = -12$

EXAMPLE 9 Evaluate $8 - (-10)$.

Solution: In this problem we are subtracting a negative number. The procedure to subtract remains the same.

$$8 - (-10) = 8 + 10 = 18$$

subtract negative add positive
10 10

Thus, $8 - (-10) = 18$.

By studying Example 9 and similar problems, we can see that for any real numbers a and b,

$$a - (-b) = a + b$$

We can use this principle to evaluate problems such as $8 - (-10)$ and other problems where we *subtract a negative quantity.*

EXAMPLE 10 Evaluate $-4 - (-12)$.

Solution: $-4 - (-12) = -4 + 12 = 8$

EXAMPLE 11 **(a)** Subtract 35 from -42.

(b) Subtract $-\dfrac{3}{5}$ from $-\dfrac{5}{9}$.

Solution:

(a) $-42 - 35 = -77$

(b) $-\dfrac{5}{9} - \left(-\dfrac{3}{5}\right) = -\dfrac{5}{9} + \dfrac{3}{5} = -\dfrac{25}{45} + \dfrac{27}{45} = \dfrac{2}{45}$

EXAMPLE 12 The highest point in the United States, Mt. McKinley in Alaska, is 20,320 feet above sea level. The lowest point in the United States, the Verdigris River in Kansas, is 680 feet below sea level. Find the difference in height between the two locations.

Solution: $20,320 - (-680) = 20,320 + 680 = 21,000$ feet

Addition and subtraction are often combined in the same problem, as in the following examples. Unless parentheses are present, if the expression involves only addition and subtraction we add and subtract from left to right. When parentheses are used, we add and subtract within the parentheses first; then we add and subtract from left to right.

EXAMPLE 13 Evaluate $-2 - (4 - 8) - 3$.

Solution:
$$-2 - (4 - 8) - 3 = -2 - (-4) - 3$$
$$= -2 + 4 - 3$$
$$= 2 - 3 = -1$$

EXAMPLE 14 Evaluate $6 - (-2) + (7 - 13) - 9$.

Solution:
$$6 - (-2) + (7 - 13) - 9 = 6 + 2 + (-6) - 9$$
$$= 8 + (-6) - 9$$
$$= 2 - 9 = -7$$

EXAMPLE 15 Evaluate $2 - |-3| + 4 - (6 - |-7|)$.

Solution: Begin by replacing the numbers in absolute value signs with their numerical equivalents; then evaluate.
$$2 - |-3| + 4 - (6 - |-7|) = 2 - 3 + 4 - (6 - 7)$$
$$= 2 - 3 + 4 - (-1)$$
$$= 2 - 3 + 4 + 1$$
$$= -1 + 4 + 1$$
$$= 3 + 1 = 4$$

Multiplication of Real Numbers

4 In a multiplication problem, the numbers or expressions that are multiplied are called **factors**. If $a \cdot b = c$, then a and b are factors of c. For example, since $2 \cdot 3 = 6$, both 2 and 3 are factors of 6.

The following rules are used in determining the sign of the product when two numbers are multiplied.

Multiplication of Real Numbers
1. The product of two numbers with **like** signs is a **positive** number.
2. The product of two numbers with **unlike** signs is a **negative** number.

EXAMPLE 16   Evaluate.

(a) $(4)(-3)$ **(b)** $(-16)\left(-\dfrac{1}{2}\right).$

Solution:

(a) $(4)(-3) = -12$ The numbers have unlike signs.

(b) $(-16)\left(-\dfrac{1}{2}\right) = 8$ The numbers have like signs, both negative.

EXAMPLE 17   Evaluate $\left(\dfrac{-3}{5}\right)\left|\dfrac{-6}{7}\right|.$

Solution: We know that $\left|\dfrac{-6}{7}\right| = \dfrac{6}{7}.$ Therefore,

$$\left(\dfrac{-3}{5}\right)\left|\dfrac{-6}{7}\right| = \left(\dfrac{-3}{5}\right)\left(\dfrac{6}{7}\right) = \dfrac{-3 \cdot 6}{5 \cdot 7} = -\dfrac{18}{35}$$

EXAMPLE 18   Evaluate $4(-2)(-3)(1).$

Solution: $4(-2)(-3)(1) = (-8)(-3)(1) = 24(1) = 24$

When multiplying more than two numbers, the product will be *negative* when there is an *odd* number of negative numbers. The product will be *positive* when there is an *even* number of negative numbers.

The multiplicative property of zero indicates that the product of 0 and any number is 0.

Multiplicative Property of Zero
For any number a, $a \cdot 0 = 0 \cdot a = 0.$

By the multiplicative property of zero, $5(0) = 0$ and $(-7.3)(0) = 0.$

EXAMPLE 19   Evaluate $9(5)(-2.63)(0)(4).$

Solution: If one or more of the factors is 0, the product is 0. Thus, $9(5)(-2.63)(0)(4) = 0.$ Can you explain why the product of any number of factors will be zero if any factor is 0?

Division of Real Numbers   ⑤ The rules for the division of real numbers are very similar to those for multiplication of real numbers.

Division of Real Numbers
1. The quotient of two numbers with **like** signs is a **positive** number.
2. The quotient of two numbers with **unlike** signs is a **negative** number.

EXAMPLE 20   Evaluate.

(a) $-24 \div 6$ **(b)** $-6.4 \div (-0.4)$

Solution:

(a) $\dfrac{-24}{6} = -4$ The numbers have unlike signs.

(b) $\dfrac{-6.4}{-0.4} = 16$ The numbers have like signs.

EXAMPLE 21   Evaluate $\dfrac{-3}{8} \div \left| \dfrac{-2}{5} \right|$.

Solution: Since $\left| \dfrac{-2}{5} \right|$ is equal to $\dfrac{2}{5}$, we write

$$\frac{-3}{8} \div \left| \frac{-2}{5} \right| = \frac{-3}{8} \div \frac{2}{5}$$

Now invert the divisor and proceed as in multiplication.

$$\frac{-3}{8} \div \frac{2}{5} = \frac{-3}{8} \cdot \frac{5}{2} = \frac{-3 \cdot 5}{8 \cdot 2} = -\frac{15}{16}$$

When the denominator of a fraction is a negative number, we usually rewrite the fraction with a positive denominator. To do this, we use the following fact.

$$\frac{a}{-b} = \frac{-a}{b} = -\frac{a}{b}$$

Thus, when we have a quotient of $\dfrac{1}{-2}$, we rewrite it as either $\dfrac{-1}{2}$ or $-\dfrac{1}{2}$.

Properties of Real Numbers

6   We have already discussed the double negative property and the multiplicative property of zero. Table 1.1 lists other basic properties for the operations of addition and multiplication on the real numbers.

For real numbers a, b, and c:	Addition	Multiplication
Commutative property	$a + b = b + a$	$ab = ba$
Associative property	$(a + b) + c = a + (b + c)$	$(ab)c = a(bc)$
Identity property	$a + 0 = 0 + a = a$ $\left(\begin{array}{c} 0 \text{ is called the } \textbf{additive} \\ \textbf{identity element} \end{array} \right)$	$a \cdot 1 = 1 \cdot a = a$ $\left(\begin{array}{c} 1 \text{ is called the } \textbf{multiplicative} \\ \textbf{identity element} \end{array} \right)$
Inverse property	$a + (-a) = (-a) + a = 0$ $\left(\begin{array}{c} -a \text{ is called the } \textbf{additive} \\ \textbf{inverse} \text{ or } \textbf{opposite} \text{ of } a \end{array} \right)$	$a \cdot \dfrac{1}{a} = \dfrac{1}{a} \cdot a = 1$ $\left(\begin{array}{c} 1/a \text{ is called the } \textbf{multiplicative} \\ \textbf{inverse} \text{ or } \textbf{reciprocal} \text{ of } a, a \neq 0 \end{array} \right)$
Distributive property (of multiplication over addition)	$a(b + c) = ab + ac$	

The distributive property applies when there are more than two numbers within the parentheses.

$$a(b + c + d + \cdots + n) = ab + ac + ad + \cdots + an$$

This expanded form of the distributive property is called the **extended distributive property.**

Note that the commutative property involves a change in *order,* and the associative property involves a change in *grouping.*

EXAMPLE 22   Name each property illustrated.

(a) $6 \cdot x = x \cdot 6$ **(b)** $(x + 2) + 3y = x + (2 + 3y)$

(c) $2x + 3y = 3y + 2x$ **(d)** $3x(y + 2) = 3x(y) + 3x(2)$

Solution:

(a) Commutative property of multiplication, change of order $6 \cdot x = x \cdot 6$

(b) Associative property of addition, change of grouping
$(x + 2) + 3y = x + (2 + 3y)$

(c) Commutative property of addition, change of order
$2x + 3y = 3y + 2x$

(d) Distributive property.

In Example 22(d) the expression $3x(y) + 3x(2)$ can be simplified to $3xy + 6x$ using the properties of the real numbers. Can you explain why?

EXAMPLE 23   Name each property illustrated.

(a) $4 \cdot 1 = 4$ **(b)** $x + 0 = x$

(c) $4 + (-4) = 0$ **(d)** $1(x + y) = x + y$

Solution:

(a) Identity property of multiplication

(b) Identity property of addition

(c) Inverse property of addition

(d) Identity property of multiplication

EXAMPLE 24   Write the additive inverse (or opposite) and multiplicative inverse (or reciprocal) of each of the following.

(a) -3 **(b)** $\frac{2}{3}$

Solution:

(a) The additive inverse is 3. The multiplicative inverse is $\dfrac{1}{-3} = -\dfrac{1}{3}$.

(b) The additive inverse is $-\dfrac{2}{3}$. The multiplicative inverse is $\dfrac{1}{\frac{2}{3}} = \dfrac{3}{2}$.

Exercise Set 1.3

Evaluate the absolute value expression.

1. $|3|$ 3
2. $|-4|$ 4
3. $|-6|$ 6
4. $|0.5|$ 0.5

5. $\left|-\dfrac{1}{2}\right|$ $\dfrac{1}{2}$
6. $|-7.32|$ 7.32
7. $|0|$ 0
8. $-|7|$ -7

9. $-|-7|$ -7
10. $-|-8|$ -8
11. $-\left|\dfrac{5}{9}\right|$ $-\dfrac{5}{9}$
12. $-\left|-\dfrac{5}{7}\right|$ $-\dfrac{5}{7}$

Insert <, >, or = in the shaded area to make the statement true.

13. $|6|$ ▨ $|-6|$ =
14. $|-9|$ ▨ $|3|$ >
15. -4 ▨ $|-4|$ <
16. $|-10|$ ▨ -5 >

17. -4 ▨ $-|4|$ =
18. $|-20|$ ▨ $-|24|$ >
19. $|-16|$ ▨ $-|30|$ >
20. $-|4|$ ▨ $-|8|$ >

21. $-|-31|$ ▨ $|-5|$ <
22. 6 ▨ $|-12|$ <
23. $|19|$ ▨ $|-25|$ <
24. $-|-3|$ ▨ $|-9|$ <

List the values from smallest to largest.

25. $6, 2, -1, |3|, |-5|$ $-1, 2, |3|, |-5|, 6$
26. $0, -4, -8, -|12|, |-10|$ $-|12|, -8, -4, 0, |-10|$

27. $-3, |0|, |-5|, |7|, |-12|$ $-3, |0|, |-5|, |7|, |-12|$
28. $5, -|7|, -9, |15|, |-1|$ $-9, -|7|, |-1|, 5, |15|$

29. $12, 24, |36|, |-9|, |-45|$ $|-9|, 12, 24, |36|, |-45|$
30. $-8, -12, -|9|, -|20|, -|-18|$ $-|20|, -|-18|, -12, -|9|, -$

31. $-2.1, -2, -2.4, |-2.8|, -|2.9|$
32. $-7, -7.1, -7.8, -|7.3|, -|7.4|$ $-7.8, -|7.3|, -7.1, -7, |-7.$

33. $\dfrac{1}{3}, \left|-\dfrac{1}{2}\right|, -2, \left|\dfrac{3}{5}\right|, \left|-\dfrac{3}{4}\right|$ $-2, \dfrac{1}{3}, \left|-\dfrac{1}{2}\right|, \left|\dfrac{3}{5}\right|, \left|-\dfrac{3}{4}\right|$
34. $\left|-\dfrac{5}{2}\right|, \dfrac{3}{5}, -3|, \left|-\dfrac{5}{3}\right|, \left|-\dfrac{2}{3}\right|$ $\dfrac{3}{5}, \left|-\dfrac{2}{3}\right|, \left|-\dfrac{5}{3}\right|, \left|-\dfrac{5}{2}\right|, |-3|$

Evaluate the following addition and subtraction problems.   **31.** $-|2.9|, -2.4, -2.1, -2, |-2.8|$ **49.** 11.4

35. $4 + (-3)$ 1
36. $-3 + 8$ 5
37. $-4 + 12$ 8
38. $-9 + 17$ 8

39. $-36 + 19$ -17
40. $-16 - (-5)$ -11
41. $-32 - (-14)$ -18
42. $35 - (-4)$ 39

43. $-6.28 - 3.14$ -9.42
44. $-9.5 - (-3.72)$ -5.78
45. $\dfrac{5}{6} - \dfrac{4}{5}$ $\dfrac{1}{30}$
46. $-\dfrac{3}{8} - \left(-\dfrac{5}{7}\right)$ $\dfrac{19}{56}$

47. $-3 - \dfrac{5}{12}$ $-\dfrac{41}{12}$
48. $7 - 4 - 8$ -5
49. $6.23 - 4.5 - (-9.67)$
50. $5 + (-0.43) - 6.97$ -2.4

51. $-6 - 4 - \dfrac{1}{2}$ $-\dfrac{21}{2}$
52. $(4 - \dfrac{2}{3}) + (6 - 8)$ $\dfrac{4}{3}$
53. $-2 + (4 - 6) - (3 - 8)$ 1

54. $-(-4 + 2) + (-6 + 3) + 2$ 1
55. $4 - (8 - 9) + (-6 + 8)$ 7
56. $|9 - 4| - 6$ -1

57. $|12 - 5| - |5 - 12|$ 0
58. $-|-3| - |7| + (6 + |-2|)$ -2
59. $|-4| - |-4| - |-4 - 4|$ -8

60. $\left(\dfrac{3}{5} + \dfrac{1}{6}\right) - \dfrac{1}{2}$ $\dfrac{4}{15}$
61. $\dfrac{3}{4} - \left(\dfrac{4}{5} - \dfrac{2}{3}\right)$ $\dfrac{37}{60}$
62. $\left(-\dfrac{5}{8} - \dfrac{3}{5}\right) + \dfrac{1}{3}$ $-\dfrac{107}{120}$

Evaluate the following multiplication and division problems.   **69.** 235.9192

63. $-4 \cdot 12$ -48
64. $(-8)(-9)$ 72
65. $-4\left(-\dfrac{5}{16}\right)$ $\dfrac{5}{4}$
66. $-4\left(-\dfrac{3}{4}\right)\left(-\dfrac{1}{2}\right)$ $-\dfrac{3}{2}$

67. $(-1)(-1)(-1)(2)(-3)$ 6
68. $(-2.3)(4.9)(-6.2)$ 69.874
69. $(-1.1)(3.4)(8.3)(-7.6)$
70. $-16 \div 8$ -2

71. $-80 \div (-10)$ 8
72. $36 \div \left(-\dfrac{1}{4}\right)$ -144
73. $-\dfrac{5}{9} \div \dfrac{-5}{9}$ 1
74. $\left|-\dfrac{1}{2}\right| \cdot \left|\dfrac{-3}{4}\right|$ $\dfrac{3}{8}$

75. $\left|\dfrac{-4}{7}\right| \div \dfrac{1}{14}$ 8
76. $\left|\dfrac{3}{8}\right| \div (-2)$ $-\dfrac{3}{16}$
77. $\left|\dfrac{-2}{3}\right| \div \left|\dfrac{-1}{2}\right|$ $\dfrac{4}{3}$
78. $\dfrac{-5}{9} \div |-5|$ $-\dfrac{1}{9}$

Evaluate.

79. $5 - 7$ -2
80. $-16 - 8$ -24
81. $-20 \div (-2)$ 10

82. $-\dfrac{3}{5} - \dfrac{5}{9}$ $-\dfrac{52}{45}$
83. $4\left(-\dfrac{8}{5}\right)\left(\dfrac{5}{2}\right)$ -16
84. $3 - (-4) + 6 - 3$ 10

85. $8.2 + (-4.9) - (6.8 - 9.4)$ 5.9   **86.** $(-2.7)(-12.3)(-9.6)$ -318.816 **87.** $(4.2)(-1)(-9.6)(3.8)$ 153.216

88. $-16.4 - (-9.6) - 14.8$ -21.6 **89.** $9 - (4 - 3) - (-2 - 1)$ 11   **90.** $-|4| \cdot \left| \dfrac{-1}{2} \right|$ -2

91. $-\left| \dfrac{-12}{5} \right| \cdot \left| \dfrac{3}{4} \right|$ $-\dfrac{9}{5}$ **92.** $|-1| \div \dfrac{5}{12}$ $\dfrac{12}{5}$ **93.** $\left| \dfrac{-9}{4} \right| \div \left| \dfrac{-4}{9} \right|$ $\dfrac{81}{16}$

94. $(-|3| + |5|) - (6 - |-9|)$ 5   **95.** $5 - |-2| + 3 - |-5|$ 1   **96.** $\left(\dfrac{3}{8} - \dfrac{4}{7} \right) - \left(-\dfrac{1}{2} \right)$ $\dfrac{17}{56}$

97. $\left(-\dfrac{3}{5} - \dfrac{4}{9} \right) - \left(-\dfrac{2}{3} \right)$ $-\dfrac{17}{45}$ **98.** $(|-4| - 3) - (3 \cdot |-5|)$ -14 **99.** $(25 - |36|)(-6 - 5)$ 121

100. $\left[(-2) \left| -\dfrac{1}{2} \right| \right] \div \left| -\dfrac{1}{4} \right|$ -4

Name the property.

101. $x + y = y + x$ commutative property of addition

102. $3(4 + 5) = 3 \cdot 4 + 3 \cdot 5$ distributive property

103. $3(x + 2) = 3x + 6$ distributive property

104. $3 \cdot x = x \cdot 3$ commutative property of multiplication

105. $(x + 3) + 6 = x + (3 + 6)$ associative property of addition

106. $x + 0 = x$ identity property of addition

107. $x = 1 \cdot x$ identity property of multiplication

108. $x(y + z) = xy + xz$ distributive property

109. $3y + 4 = 4 + 3y$ commutative property of addition

110. $(2x \cdot 3y) \cdot 4y = 2x \cdot (3y \cdot 4y)$

111. $3x + 2y = 2y + 3x$ commutative property of addition

112. $4(x + y + 2) = 4x + 4y + 8$ distributive property

113. $-(-1) = 1$ double negative property

114. $5 \cdot 1 = 5$ identity property of multiplication

115. $5 + 0 = 5$ identity property of addition

116. $4 \cdot \dfrac{1}{4} = 1$ inverse property of multiplication

117. $3 + (-3) = 0$ inverse property of addition

118. $6 \cdot 0 = 0$ multiplicative property of zero

119. $x \cdot \dfrac{1}{x} = 1$ inverse property of multiplication

120. $(x + y) = 1(x + y)$ identity property of multiplication

121. $(x + 2) = 1(x + 2)$ identity property of multiplication

122. $-\left(-\dfrac{1}{2} \right) = \dfrac{1}{2}$ double negative property

123. $x \cdot 0 = 0$ multiplicative property of zero

124. $-(-x) = x$ double negative property

110. associative property of multiplication

Complete the statement on the right side of the equal sign using the property indicated.

125. $x + 3 =$ Commutative property of addition $3 + x$

126. $1 \cdot x =$ Identity property of multiplication x

127. $3(x + y + 4) =$ Distributive property $3x + 3y + 12$

128. $x + 0 =$ Identity property of addition x

129. $(x \cdot 3) \cdot 4 =$ Associative property of multiplication $x \cdot (3 \cdot 4)$

130. $-(-x) =$ Double negative property x

131. $5x + (2y + 3x) =$ Associative property of addition $(5x + 2y) + 3x$

132. $a \cdot 0 =$ Multiplicative property of zero   0

133. $a + (-a) =$ Inverse property of addition   0

134. $a \cdot \dfrac{1}{a} =$ Inverse property of multiplication   1

List both the additive inverse and the multiplicative inverse for each problem.

135. 4 $-4, \dfrac{1}{4}$ **136.** -3 $3, -\dfrac{1}{3}$ **137.** $-\dfrac{2}{3}$ $\dfrac{2}{3}, -\dfrac{3}{2}$ **138.** $-\dfrac{3}{7}$ $\dfrac{3}{7}, -\dfrac{7}{3}$

139. In New York City, the temperature during a 24-hour period dropped from 46°F to -12°F. Find the change in temperature.   58°F

140. A submarine dives 412.6 feet. A short time later the submarine comes up 286.8 feet. Find the submarine's final depth from its starting point. (Consider distance in a downward direction as negative.)   125.8 ft below sea level (or -125.8 ft)

141. The Zwicks had a balance of −$42.64 in their checking account when they deposited a check for $107.38. What is their new balance? $64.74

142. On their first play, the University of Nebraska football team lost 8 yards. On their second play, they gained 17 yards. What is the gain or loss for the two plays? 9-yd gain

143. When Billy Joel signed a contract to record his latest album, he received an advance payment of $1,350,000. When the album is released and sales begin, the recording company will automatically deduct this advance from his royalties.

(a) If two weeks after the album's release the royalties for the album total $1,267,000 before the

advance is deducted, find the amount of money Billy Joel will receive from or owe to the record company. owes $83,000 (or −$83,000)

(b) If two months after the release of the album, his royalties total $2,400,000, find the amount of money he will receive from or owe to the recording company. receives $1,050,000

144. Mr. Adams had $6634 income tax withheld in 1996 by his employer. If Mr. Adams' total income tax for the 1996 year was $7496, determine the amount of Mr. Adams' income tax refund, or the balance he owed to the Internal Revenue Service. owed $862

Find the unknown number(s). Explain how you determined your answer.

145. All numbers a such that $|a| = |-a|$ all real numbers, $\mathbb{R}$

146. All numbers a such that $|a| = a$ $a \geq 0$

147. All numbers a such that $|a| = -a$ $a \leq 0$

148. All numbers a such that $|a| = -3$ { }

149. All numbers a such that $|a| = 5$ 5, −5

150. All numbers x such that $|x - 3| = |3 - x|$

151. Explain in your own words how to add two numbers with different signs.

152. Explain how the rules for multiplication and division of real numbers are similar.

153. (a) List the commutative property of addition.

(b) In your own words explain the property.

154. (a) List the associative property of multiplication.

(b) In your own words explain the property.

155. (a) List the distributive property of multiplication over addition. $a(b + c) = ab + ac$

(b) In your own words explain the property.

156. Using an example, explain why addition is not distributive over multiplication. That is, explain why $a + (b \cdot c) \neq (a + b) \cdot (a + c)$.

157. Write your own realistic word problem that involves subtracting a positive number from a negative number. Indicate the answer to your word problem.

158. Write your own realistic word problem that involves subtracting a negative number from a negative number. Indicate the answer to your problem.

150. all real numbers, $\mathbb{R}$ **153. (a)** $a + b = b + a$ **154. (a)** $(ab)c = a(bc)$ **156.** $2 + (3 \cdot 4) \neq (2 + 3) \cdot (2 + 4)$
161. (a) 3, 4, −2, 0 **(b)** 3, 4, −2, $\frac{5}{6}$, 0 **(c)** $\sqrt{3}$ **(d)** 3, 4, −2, $\frac{5}{6}$, $\sqrt{3}$, 0   **162. (a)** $A \cup B = \{1, 4, 7, 9, 12, 15\}$ **(b)** $A \cap B = \{4, 7\}$

CUMULATIVE REVIEW EXERCISES

[1.2] **159.** Answer true or false. Every irrational number is a real number. true

160. Insert either $\subseteq$ or $\nsubseteq$ in the shaded area to make a true statement. $\{3\}$ ■ $\{3, 4, 5\}$ $\subseteq$

161. Consider the set $\{3, 4, -2, \frac{5}{6}, \sqrt{3}, 0\}$. List the elements that are **(a)** integers; **(b)** rational numbers; **(c)** irrational numbers; **(d)** real numbers.

162. $A = \{4, 7, 9, 12\}$; $B = \{1, 4, 7, 15\}$. Find **(a)** $A \cup B$; **(b)** $A \cap B$.

163. Illustrate $\{x \mid -4 < x \leq 6\}$ on the number line.

Group Activity/ Challenge Problems

1. Evaluate $1 - 2 + 3 - 4 + \cdots + 99 - 100$. (*Hint:* Group in pairs of two numbers.) −50

2. Evaluate $1 + 2 - 3 + 4 + 5 - 6 + 7 + 8 - 9 + 10 + 11 - 12 + \cdots + 22 + 23 - 24$. (*Hint:* Examine in groups of three numbers.) 84

3. Evaluate $\dfrac{(1) \cdot |-2| \cdot (-3) \cdot |4| \cdot (-5)}{|-1| \cdot (-2) \cdot |-3| \cdot (4) \cdot |-5|}$. −1

4. Evaluate $\dfrac{(1)(-2)(3)(-4)(5) \cdots (97)(-98)}{(-1)(2)(-3)(4)(-5) \cdots (-97)(98)}$. 1

1.4   Order of Operations

Tape 1

1. Evaluate exponential expressions.
2. Evaluate square and higher roots.
3. Evaluate expressions using the correct order of operations.
4. Evaluate expressions for specific values of the variable.

Before we discuss the order of operations we need to speak briefly about exponents and roots. We discuss exponents in greater depth in Sections 5.1, 5.2, and 8.2, and roots (or radicals) in Chapter 8.

Exponents 1 In the expression 3^2, the 3 is called the **base** and the 2 is called the **exponent.** The expression 3^2 is read "three squared" or "three to the second power" and means

$$3^2 = \underbrace{3 \cdot 3}_{\text{2 factors of 3}}$$

The expression 5^3 is read "five cubed" or "five to the third power" and means

$$5^3 = \underbrace{5 \cdot 5 \cdot 5}_{\text{3 factors of 5}}$$

In general, the base b to the nth power, written b^n, where n is a natural number, means

$$b^n = \underbrace{b \cdot b \cdot b \cdot b \cdot \cdots \cdot b}_{n \text{ factors of } b}$$

EXAMPLE 1   Evaluate.

(a) 5^3 **(b)** $(-2)^5$ **(c)** 1^{10} **(d)** $\left(\dfrac{-3}{4}\right)^3$ **(e)** $(0.6)^3$

Solution:

(a) $5^3 = 5 \cdot 5 \cdot 5 = 125$

(b) $(-2)^5 = (-2)(-2)(-2)(-2)(-2) = -32$

(c) $1^{10} = 1$; 1 raised to any power will equal 1. Why?

(d) $\left(\dfrac{-3}{4}\right)^3 = \left(\dfrac{-3}{4}\right)\left(\dfrac{-3}{4}\right)\left(\dfrac{-3}{4}\right) = \dfrac{-27}{64}$

(e) $(0.6)^3 = (0.6)(0.6)(0.6) = 0.216$

It is not necessary to write exponents of 1. Whenever we encounter a numerical value or a variable without an exponent, we assume that it has an exponent of 1. Thus, 3 means 3^1, x means x^1, x^3y and x^3y^1, and $-xy$ means $-x^1y^1$.

Students often evaluate the expression $-x^2$ incorrectly. Study the following example and Helpful Hint carefully.

EXAMPLE 2   Evaluate $-x^2$ for each value of x. **(a)** 3   **(b)** -3

Solution: **(a)** $-x^2 = -(3)^2 = -9$ **(b)** $-x^2 = -(-3)^2 = -(9) = -9$

Helpful Hint

A negative sign directly preceding an expression that is raised to a power has the effect of making that expression negative.

$$-3^2 \quad \text{means} \quad -(3^2) \quad \text{and not} \quad (-3)^2$$
$$-x^2 \quad \text{means} \quad -(x^2) \quad \text{and not} \quad (-x)^2.$$

Example: Evaluate.

(a) -5^2 **(b)** $(-5)^2$

Solution:

(a) $-5^2 = -(5^2) = -25$ **(b)** $(-5)^2 = (-5)(-5) = 25$

Note that $-5^2 \neq (-5)^2$ since $-25 \neq 25$. Note also that $-x^2$ will always be a negative number for any nonzero value of x and that $(-x)^2$ will always be a positive number for any nonzero value of x.

EXAMPLE 3   Evaluate $-3^2 + (-2)^4 - 4^3 + (-2^4)$.

Solution: We evaluate each exponential expression. Then we add or subtract, working from left to right.

$$-3^2 + (-2)^4 - 4^3 + (-2^4) = -(3^2) + (-2)^4 - (4^3) + (-2^4)$$
$$= -9 + 16 - 64 + (-16)$$
$$= -9 + 16 - 64 - 16$$
$$= -73$$

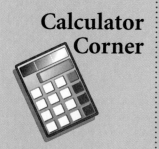

Calculator Corner

EVALUATING EXPONENTIAL EXPRESSIONS ON A CALCULATOR

Many calculators have keys that simplify finding the value of an exponential expression. The $\boxed{x^2}$ key can be used to square a number. The expression 5^2 can be evaluated on these calculators by pressing

$$5 \quad \boxed{x^2} \quad 25$$

Other calculators have a $\boxed{y^x}$ * key. These keys can be used to evaluate exponential expressions. When using these keys first enter the base, then press the $\boxed{y^x}$ key, then enter the exponent, and then press the $\boxed{=}$ key. For example, to evaluate 6^4, use the following sequence.

$$6 \quad \boxed{y^x} \quad 4 \quad \boxed{=} \quad 1296$$

---

* Some calculators use an $\boxed{x^y}$, $\boxed{a^b}$, or $\boxed{\wedge}$ key instead of the $\boxed{y^x}$ key.

Roots ② The symbol used to indicate a root, $\sqrt{}$, is called a **radical sign.** The number or expression inside the radical sign is called the **radicand.** In $\sqrt{25}$, the radicand is 25. The **principal or positive square root** of a positive number n, written $\sqrt{n}$, is the positive number that when multiplied by itself gives n. Whenever we use the words "square root" we are referring to the "principal square root."

EXAMPLE 4 Evaluate.

(a) $\sqrt{25}$ **(b)** $\sqrt{\dfrac{9}{4}}$ **(c)** $\sqrt{0.64}$

Solution:

(a) $\sqrt{25} = 5$, since $5 \cdot 5 = 25$.

(b) $\sqrt{\dfrac{9}{4}} = \dfrac{3}{2}$, since $\dfrac{3}{2} \cdot \dfrac{3}{2} = \dfrac{9}{4}$.

(c) $\sqrt{0.64} = 0.8$, since $(0.8)(0.8) = 0.64$.

The square root of 4, $\sqrt{4}$, is a rational number since it is equal to 2. The square roots of other numbers, such as $\sqrt{2}$, $\sqrt{3}$, and $\sqrt{5}$, are irrational numbers. The decimal values of such numbers can never be given exactly since irrational numbers are nonterminating, nonrepeating decimal numbers. The approximate value of $\sqrt{2}$ and other irrational numbers can be found with a calculator.

$$\sqrt{2} \approx 1.414213562 \qquad \text{from a calculator}$$

In this section we introduce square roots; cube roots, symbolized by $\sqrt[3]{}$; and higher roots. The number used to indicate the root is called the **index.**

$$\text{index} \longrightarrow \overset{\text{radical sign}}{\sqrt[m]{n}} \longleftarrow \text{radicand}$$

The index of a square root is 2. However, we generally do not show the index 2. Therefore, $\sqrt{a} = \sqrt[2]{a}$.

The concept used to explain square root can be expanded to explain cube roots and higher roots. The cube root of a number n is written $\sqrt[3]{n}$.

$$\sqrt[3]{n} = b \qquad \text{if} \qquad \underbrace{b \cdot b \cdot b}_{\text{3 factors of } b} = n$$

For example, $\sqrt[3]{8} = 2$, because $2 \cdot 2 \cdot 2 = 8$. The expression $\sqrt[m]{n}$ is read the *m*th root of *n*.

$$\sqrt[m]{n} = b \qquad \text{if} \qquad \underbrace{b \cdot b \cdot b \cdot \cdots \cdot b}_{m \text{ factors of } b} = n$$

EXAMPLE 5 Evaluate.

(a) $\sqrt[3]{27}$ **(b)** $\sqrt[3]{64}$ **(c)** $\sqrt[4]{16}$

Solution:

(a) $\sqrt[3]{27} = 3$, since $3 \cdot 3 \cdot 3 = 27$.
(b) $\sqrt[3]{64} = 4$, since $4 \cdot 4 \cdot 4 = 64$.
(c) $\sqrt[4]{16} = 2$, since $2 \cdot 2 \cdot 2 \cdot 2 = 16$.

EXAMPLE 6 Evaluate.

(a) $\sqrt[4]{81}$　　**(b)** $\sqrt[3]{\dfrac{1}{27}}$　　**(c)** $\sqrt[3]{-1}$　　**(d)** $\sqrt[3]{-8}$　　**(e)** $-\sqrt[3]{-8}$

Solution:

(a) $\sqrt[4]{81} = 3$, since $3 \cdot 3 \cdot 3 \cdot 3 = 81$.

(b) $\sqrt[3]{\dfrac{1}{27}} = \dfrac{1}{3}$ since $\left(\dfrac{1}{3}\right)\left(\dfrac{1}{3}\right)\left(\dfrac{1}{3}\right) = \dfrac{1}{27}$.

(c) $\sqrt[3]{-1} = -1$, since $(-1)(-1)(-1) = -1$.
(d) $\sqrt[3]{-8} = -2$, since $(-2)(-2)(-2) = -8$.
(e) $-\sqrt[3]{-8} = -(\sqrt[3]{-8}) = -(-2) = 2$.

Note that in Example 6(c) and (d) the cube root of a negative number is negative. Why is this so?

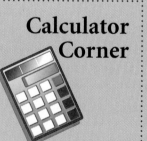

Calculator Corner

EVALUATING ROOTS ON A CALCULATOR

The square roots of numbers can be found on calculators with a square-root key, $\boxed{\sqrt{x}}$. To evaluate $\sqrt{25}$ on calculators that have this key, press

$$25 \quad \boxed{\sqrt{x}} \quad 5$$

Higher roots can be found on calculators that contain either the $\boxed{\sqrt[x]{y}}$ key or the $\boxed{y^x}$ key.* To evaluate $\sqrt[4]{625}$ on a calculator with a $\boxed{\sqrt[x]{y}}$ key, do the following:

$$625 \quad \boxed{\sqrt[x]{y}} \quad 4 \quad \boxed{=} \quad 5$$

Note that the number within the radical sign (the radicand), 625, is entered, then the $\boxed{\sqrt[x]{y}}$ key is pressed, and then the root (or index) 4 is entered. When the $\boxed{=}$ key is pressed, the answer 5 is displayed.

To evaluate $\sqrt[4]{625}$ on a calculator with a $\boxed{y^x}$ key, use the inverse key as follows:

$$625 \quad \boxed{\text{INV}} \quad \boxed{y^x} \quad 4 \quad \boxed{=} \quad 5$$

---

*Calculator keys vary. Some calculators have $\boxed{x^y}$ or $\boxed{a^b}$ instead of the $\boxed{y^x}$ key, and some calculators use a $\boxed{2^{nd}}$ or $\boxed{\text{shift}}$ key instead of the $\boxed{\text{INV}}$ key.

We discuss radicals in more detail in Chapter 8.

Order of Operations ③ You will often have to evaluate expressions containing multiple operations. To do so, follow the order of operations indicated below.

> ### Order of Operations
> To evaluate mathematical expressions use the following order:
> 1. First, evaluate the expressions within grouping symbols, including parentheses, ( ), brackets, [ ], or braces, { }. If the expression contains nested parentheses (one pair of parentheses within another pair), evaluate the expression in the innermost parentheses first.
> 2. Next, evaluate all terms containing exponents and roots.
> 3. Next, evaluate all multiplications or divisions in the order in which they occur, working from left to right.
> 4. Finally, evaluate all additions or subtractions in the order in which they occur, working from left to right.

It should be noted that a fraction bar acts as a grouping symbol. Thus, when evaluating expressions containing a fraction bar, we work separately above and below the fraction bar.

Parentheses or brackets may be used (1) to change the order of operations to be followed in evaluating an algebraic expression or (2) to help clarify the understanding of an expression.

The expression $5 + 7 \cdot 3$ means $5 + (7 \cdot 3)$ and equals 26. If we wished to have the addition performed before the multiplication, we would indicate this by placing parentheses about the $5 + 7$.

$$(5 + 7) \cdot 3 = 12 \cdot 3 = 36$$

Note that the value changed when parentheses were inserted about the $5 + 7$. Consider the expression $1 \cdot 3 + 2 \cdot 4$. According to the order of operations, multiplications are to be performed before additions. We can rewrite this expression as $(1 \cdot 3) + (2 \cdot 4)$. Notice that we did not change the order of operations. The parentheses only help clarify the order to be followed.

Brackets are sometimes used in place of parentheses to help avoid confusion. For example, the expression $7((5 \cdot 3) + 6)$ may be easier to follow when written $7[(5 \cdot 3) + 6]$.

EXAMPLE 7 Evaluate $8 + 3 \cdot 5^2 - 7$.

Solution: Colored shading will be used to indicate the order in which the operations are to be evaluated. Since there are no parentheses, we first evaluate 5^2.

$$8 + 3 \cdot 5^2 - 7 = 8 + 3 \cdot 25 - 7$$

Next, perform multiplications or divisions from left to right.

$$= 8 + 75 - 7$$

Next, perform additions or subtractions from left to right.

$$= 83 - 7$$
$$= 76$$

EXAMPLE 8   Evaluate $36 + 3\left[(12 - 4) \div \frac{1}{2}\right]$.

Solution: First, evaluate the expression in the innermost parentheses.

$$36 + 3\left[(\,12 - 4\,) \div \frac{1}{2}\right] = 36 + 3\left[\,8 \div \frac{1}{2}\,\right]$$

Next, evaluate the expression within the brackets. To do this we invert the $\frac{1}{2}$ and multiply.

$$= 36 + 3[\,8 \cdot 2\,]$$
$$= 36 + 3(16)$$

Now perform the remaining multiplication, and then add.

$$= 36 + 48$$
$$= 84$$

EXAMPLE 9   Evaluate $10 + \{6 - [4(5 - 2)]\}^2$.

Solution: First, evaluate the expression within the innermost parentheses.

$$10 + \{6 - [4(5 - 2\,)]\}^2 = 10 + \{6 - [\,4(3)\,]\}^2$$
$$= 10 + [\,6 - (12)\,]^2$$
$$= 10 + (-6)^2$$
$$= 10 + 36$$
$$= 46$$

EXAMPLE 10   Evaluate $\dfrac{6 \div 2 + 5|7 - 3|}{1 + (3 - 5) \div 2}$.

Solution: Work separately above the fraction bar and below the fraction bar.

$$\frac{6 \div 2 + 5|7 - 3|}{1 + (3 - 5) \div 2} = \frac{6 \div 2 + 5|4|}{1 + (-2) \div 2}$$
$$= \frac{3 + 20}{1 + (-1)}$$
$$= \frac{23}{0}$$

Since division by 0 is not possible, the original expression is **undefined.**

EXAMPLE 11   Evaluate $16 \div 8 \cdot 4 - \dfrac{6^2 \div 2^2 + |-12|}{|-30 - 33|}$.

Solution: Begin by simplifying the fraction.

$$16 \div 8 \cdot 4 - \frac{6^2 \div 2^2 + |-12|}{|-30 - 33|} = 16 \div 8 \cdot 4 - \frac{36 \div 4 + |-12|}{|-30 - 33|}$$

$$= 16 \div 8 \cdot 4 - \frac{9 + 12}{|-30 - 33|}$$

$$= 16 \div 8 \cdot 4 - \frac{21}{|-63|}$$

$$= 16 \div 8 \cdot 4 - \frac{21}{63}$$

$$= 16 \div 8 \cdot 4 - \frac{1}{3}$$

$$= 2 \cdot 4 - \frac{1}{3}$$

$$= 8 - \frac{1}{3} = \frac{24}{3} - \frac{1}{3} = \frac{23}{3}$$

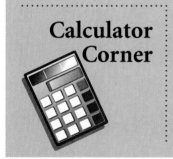

Calculator Corner

ORDER OF OPERATIONS ON A CALCULATOR

Calculators that evaluate problems using the order of operations presented in this section are said to use *algebraic logic.* If you are not sure if your calculator evaluates using the order of operations presented in this section, perform this simple test. Key in 3 $+$ 4 $\times$ 5 $=$. If the calculator displays 23, it used algebraic logic and performed the multiplication first. If it displays 35, it performed the addition first and has not followed the order of operations. *For this course you should have a calculator that uses algebraic logic.*

EXAMPLE 12   Show the sequence of keys that should be pressed to calculate each of the following on a calculator that uses algebraic logic.

(a) $\dfrac{4 + 6.2}{3.4}$ **(b)** $(8 + 3 \cdot 5 - 4)6$

(c) $(4 + 6 \cdot 3)^4$ **(d)** $\dfrac{-1.2^4 - 4^2}{5^3}$

Solution: Two sequences that may be used are illustrated for each part. The first sequence does not use parentheses, whereas the second sequence does.

(a) 4 $+$ 6.2 $=$ $\div$ 3.4 $=$ or   (4 $+$ 6.2) $\div$ 3.4 $=$

In the first display, if we had not pressed the $=$ key after the 6.2, the calculator would have evaluated 4 + (6.2 ÷ 3.4) since division is performed before addition. The answer 3 should be displayed on your calculator.

(b) 8 $+$ 3 $\times$ 5 $-$ 4 $=$ $\times$ 6 $=$ or

(8 $+$ 3 $\times$ 5 $-$ 4) $\times$ 6 $=$

If you do not use the parentheses key, it is necessary to press the $=$ key after entering the 4 to show that the entire expression, not just the 4, is to be multiplied by 6. The answer 114 should be displayed on your calculator.

(c) 4 ⊞ + ⊞ 6 ⊞ × ⊞ 3 ⊞ = ⊞ y^x ⊞ 4 ⊞ =   or

⊞ (⊞ 4 ⊞ + ⊞ 6 ⊞ × ⊞ 3 ⊞) ⊞ y^x ⊞ 4 ⊞ =

Other keys may be used in place of the ⊞ y^x key, depending on your calculator. The answer 234,256 should be displayed on your calculator.

(d) Recall that -1.2^4 means $-(1.2)^4$. The keys to press are

1.2 ⊞ y^x ⊞ 4 ⊞ = ⊞ +/− ⊞ − ⊞ 4 ⊞ x^2 ⊞ = ⊞ ÷ ⊞ 5 ⊞ y^x ⊞ 3 ⊞ =   or

⊞ (⊞ 1.2 ⊞ y^x ⊞ 4 ⊞ = ⊞ +/− ⊞ − ⊞ 4 ⊞ x^2 ⊞) ⊞ ÷ ⊞ 5 ⊞ y^x ⊞ 3 ⊞ =

The answer is -0.1445888. Why is it necessary to press the ⊞ = key after the x^2 key if parentheses are not used?

When evaluating certain expressions on a calculator it will be more convenient to use parentheses. For example, to evaluate $\dfrac{6}{4 + 11}$ use parentheses and enter 6 ⊞ ÷ ⊞ (⊞ 4 ⊞ + ⊞ 11 ⊞) ⊞ = . The answer 0.4 should be displayed on your calculator.

Evaluating Expressions Containing Variables

4 Now we will evaluate expressions for specific values of the variable.

EXAMPLE 12 Evaluate $4x^2 - 2$ when **(a)** $x = 3$ and **(b)** $x = -\frac{3}{4}$.

Solution: In part **(a)**, we substitute 3 for each x in the expression. In part **(b)**, we substitute $-\frac{3}{4}$ for each x in the expression.

(a)
$$4x^2 - 2 = 4(3)^2 - 2$$
$$= 4(9) - 2$$
$$= 36 - 2$$
$$= 34$$

(b)
$$4x^2 - 2 = 4\left(-\frac{3}{4}\right)^2 - 2$$
$$= \overset{1}{4}\left(\frac{9}{\underset{4}{\cancel{16}}}\right) - 2$$
$$= \frac{9}{4} - 2 = \frac{9}{4} - \frac{8}{4} = \frac{1}{4}$$

EXAMPLE 14 Evaluate $6 - (3x + 1) + 2x^2$ when $x = 4$.

Solution: Substitute 4 for each x in the expression; then evaluate.

$$6 - (3x + 1) + 2x^2 = 6 - [3(4) + 1] + 2(4)^2$$
$$= 6 - [12 + 1] + 2(16)$$
$$= 6 - (13) + 32$$
$$= -7 + 32$$
$$= 25$$

EXAMPLE 15   Evaluate $-x^3 - xy - y^2$ when $x = -2$ and $y = 5$.

Solution: Substitute -2 for each x and 5 for each y in the expression; then evaluate.

$$
\begin{aligned}
-x^3 - xy - y^2 &= -(-2)^3 - (-2)(5) - (5)^2 \\
&= -(-8) - (-10) - 25 \\
&= 8 + 10 - 25 \\
&= -7
\end{aligned}
$$

Calculator Corner

THE GRAPHING CALCULATOR

This is the first of a number of Calculator Corners devoted to graphing calculators. Since the keystrokes to use vary from one graphing calculator to another, in the Graphing Calculator Corners we will generally not try to cover the specific sequences of keystrokes to use to perform a task. *We suggest you read carefully the manual that came with your graphing calculator to determine the sequence of keystrokes to use.*

Some of the more popular graphing calculators at the time this book was written include Texas Instruments models TI 81, TI 82 and TI 85; Casio model FX 7700G and Sharp models EL 9200 and EL 9300; and Hewlett-Packard model 48G.

TI 82   **CASIO FX 7700G**

The purpose of the Graphing Calculator Corners is to reinforce and enhance the concepts presented in this book. No new information will be presented in the Calculator Corners. Even if you do not have or use a graphing calculator you should read these Calculator Corners for you may find they reinforce your understanding of the concepts.

(continued)

THE GRAPHING CALCULATOR   (CONTINUED)

Many graphing calculators can store an expression (or equation) and then evaluate the expression for various values of the variable or variables without having to reenter the expression each time. This is very valuable in both science and mathematics courses. For example, in Chapter 3 when we graph, we will need to evaluate an equation for various values of the variable. Below we show the screen of one graphing calculator* showing the expression $\frac{2}{3}x^2 + 2x - 4$ being evaluated for $x = 6$ and $x = -2.3$.

```
6→X:(2/3)X²+2X-4
                 32
-2.3→X:(2/3)X²+2
X-4
       -5.073333333
```

The calculator display above illustrates two important points about using graphing calculators. Since these items are very important, we will mention them again in other Graphing Calculator Corners.

1. Notice the parentheses around the 2/3. Some graphing calculators interpret $2/3x^2$ as $2/(3x^2)$. To evaluate $\frac{2}{3}x^2$ on such calculators you must use parentheses around the 2/3. You should learn how your calculator evaluates expressions like $2/3x^2$. *Whenever you are in doubt, use parentheses to prevent possible errors.*

2. In the display, you will notice that the negative sign preceding the 2.3 is slightly smaller and higher than the subtraction sign preceding the 4 in the equation. Graphing calculators generally have both a negative sign key, $(-)$, and a subtraction sign key, $-$. You must be sure to use the correct key or you will get an error. The negative sign key is used to enter a negative number. The subtraction key is used to subtract one quantity from another.

   To enter the expression $-x - 4$ on a graphing calculator, you might press

$$\underset{\substack{\uparrow \\ \text{negative} \\ \text{sign}}}{(-)} \ x \ \underset{\substack{\uparrow \\ \text{subtraction}}}{-} \ 4$$

Remember that $-x - 4$ means $-1x - 4$. By beginning with $(-)$ you are entering the coefficient -1.

---

*The screens shown in all Graphing Calculator Corners are from a TI 82 calculator.

THE GRAPHING CALCULATOR   (CONTINUED)

EXERCISES

Read the manual that comes with your graphing calculator. Use your calculator to evaluate the following. Round answers to the nearest hundredth.

1. $x^2 - 6x$ when **(a)** $x = 4$; **(b)** $x = 0.2$; **(c)** $x = \dfrac{3}{5}$.   **(a)** -8 **(b)** -1.16 **(c)** -3.24

2. $5.2x^3 - 6x + 4$ when **(a)** $x = 1$; **(b)** $x = -3$; **(c)** $x = 0.3$.

3. $x^2 y - 6x + y$ when **(a)** $x = 3$, $y = 4$; **(b)** $x = 4$, $y = 0.2$; **(c)** $x = \dfrac{1}{5}$, $y = \dfrac{5}{6}$.

The answers to odd-numbered calculator exercises are given at the back of the book.   **2.** **(a)** 3.2   **(b)** -118.4 **(c)** 2.34   **3.** **(a)** 22   **(b)** -20.6 **(c)** -0.33

Exercise Set 1.4

Evaluate without using a calculator.

1. 4^2 16

2. $(-2)^3$ -8

3. $(-4)^2$ 16

4. $(1.4)^3$ 2.744

5. -4^2 -16

6. $\left(\dfrac{5}{6}\right)^3$ $\dfrac{125}{216}$

7. $-\left(\dfrac{3}{5}\right)^4$ $-\dfrac{81}{625}$

8. $(0.3)^2$ 0.09

9. $-\left(-\dfrac{2}{3}\right)^4$ $-\dfrac{16}{81}$

10. $\sqrt{25}$ 5

11. $-\sqrt{36}$ -6

12. $\sqrt{0.81}$ 0.9

13. $\sqrt[3]{-125}$ -5

14. $\sqrt[3]{\dfrac{-27}{64}}$ $-\dfrac{3}{4}$

15. $\sqrt[3]{0.001}$ 0.1

16. $\sqrt[4]{\dfrac{1}{16}}$ $\dfrac{1}{2}$

Use a calculator to evaluate. Round answers to the nearest thousandth.

17. $(0.42)^5$ 0.013

18. $(-3.4)^{4.2}$ 170.691

19. $\left(\dfrac{5}{9}\right)^5$ 0.053

20. $\left(-\dfrac{3}{7}\right)^6$ 0.006

21. $-(2.35)^{7.4}$ -557.060

22. $\sqrt{20}$ 4.472

23. $\sqrt[3]{5}$ 1.710

24. $-\sqrt[4]{72.8}$ -2.921

25. $\sqrt[5]{1246.5}$ 4.160

26. $-\sqrt{\dfrac{8}{9}}$ -0.943

27. $-\sqrt[3]{\dfrac{20}{53}}$ -0.723

28. $\sqrt[3]{-\dfrac{20}{53}}$ -0.723

Evaluate **(a)** x^2 *and* **(b)** $-x^2$ *for the given value of x.*

29. 3   9, -9

30. 4   16, -16

31. 1   1, -1

32. -2 4, -4

33. -1 1, -1

34. -5 25, -25

35. $\dfrac{1}{3}$ $\dfrac{1}{9}, -\dfrac{1}{9}$

36. $-\dfrac{2}{5}$ $\dfrac{4}{25}, -\dfrac{4}{25}$

Evaluate **(a)** x^3 *and* **(b)** $-x^3$ *for the given value of x.*

37. 3   27, -27

38. -5 -125, 125

39. -3 -27, 27

40. -1 -1, 1

41. -2 -8, 8

42. -4 -64, 64

43. $\dfrac{2}{3}$ $\dfrac{8}{27}, -\dfrac{8}{27}$

44. $-\dfrac{3}{4}$ $-\dfrac{27}{64}, \dfrac{27}{64}$

Evaluate.

45. $4^2 + 3^2 - 2^2$ 21

46. $(-1)^3 + 1^3 + 1^{10} + (-1)^{12}$ 2

47. $-2^2 - 2^3 + 1^{10} + (-2)^3$ -19

48. $(-3)^3 - 2^2 - (-2)^2 + (4-4)^2$ -35

49. $(-2)^2 + (-3)^2 + (-3)^3 - 4^2$ -30

50. $(0.2)^2 - (1.6)^2 - (3.2)^2$ -12.76

51. $(3.7)^2 - (0.8)^2 + (2.4)^3$ 26.874

52. $\left(-\dfrac{1}{2}\right)^3 - \left(\dfrac{1}{3}\right)^2 - \left(-\dfrac{2}{3}\right)^2$ $-\dfrac{49}{72}$

53. $\left(\dfrac{3}{4}\right)^2 - \dfrac{1}{4} - \left(-\dfrac{3}{8}\right)^2 + \left(\dfrac{1}{2}\right)^3$ $\dfrac{19}{64}$

54. $\left(\dfrac{1}{2}\right)^4 - \left(\dfrac{5}{4}\right)^3 \left(-\dfrac{2}{5}\right)^3$ $\dfrac{3}{16}$

Evaluate.

55. $6 + 4 \cdot 5$ 26

56. $(6^2 - 2) \div (\sqrt{36} - 4)$ 17

57. $20 - 6 \div 3 - 4$ 14

58. $6 \div 2 + 12 - 3^2$ 6

59. $6 \div 2 + 5 \cdot \dfrac{3}{4}$ $\dfrac{27}{4}$

60. $24 \cdot 2 \div \dfrac{1}{3} \div 6$ 24

61. $\dfrac{3}{4} \div \dfrac{5}{6} + \dfrac{1}{2} \cdot \dfrac{9}{4}$ $\dfrac{81}{40}$

62. $2[1 - (4 \cdot 5)] + 6^3$ 178

63. $-2 + 5[3 - (2 - 4)]$ 23

64. $-4(5 - 2)^3 + \dfrac{5}{2}$ $-\dfrac{211}{2}$

65. $3[(4 + 6)^2 - \sqrt[3]{8}\,]$ 294

66. $\{[3(14 \div 7)]^2 - 2\}^2$ 1156

67. $\{[(12 - 15) - 3] - 2\}^2$ 64

68. $3\{6 - [(25 \div 5) - 2]\}^3$ 81

69. $4[3(2 - 6) \div (10 \div 5)^2]^2$ 36

70. $\dfrac{\frac{1}{6} - 4 \div 2}{8 - 3 + 6}$ $-\dfrac{1}{6}$

71. $\dfrac{15 \div 3 + 2 \cdot 2}{\sqrt{25} \div 5 + 8 \div 2}$ $\dfrac{9}{5}$

72. $\dfrac{\frac{1}{2} \cdot \frac{1}{3} \div 4 - 2}{3^2 - 4 \cdot 2 + 3}$ $-\dfrac{47}{96}$

73. $\dfrac{4 - (2 + 3)^2 - 6}{4(3 - 2) - 3^2}$ $\dfrac{27}{5}$

74. $\dfrac{2(-3) + 4 \cdot 5 - 3^2}{-6 + \sqrt{4}(2^2 - 1)}$ undefined

75. $\dfrac{8 - 4 \div 2 \cdot 3 - 4}{5^2 - 3^2 \cdot 2 - 7}$ undefined

76. $-2\left|-3 - \dfrac{2}{3}\right| + 4$ $-\dfrac{10}{3}$

77. $\dfrac{8 - [4 - (3 - 1)^2]}{5 - (-3)^2 + 4 \div 2}$ -4

78. $12 - 15 \div |5| + 2(|4| - 2)^2$ 17

79. $-2\,|-3| - \sqrt{36} \div |2| + 3^2$ 0

80. $\dfrac{4 - |-12| \div |3|}{2(4 - |5|) + 9}$ 0

81. $\dfrac{6 - |-3| - 4|6 - 2|}{5 - 6 \cdot 2 \div |-4|}$ $-\dfrac{13}{2}$

82. $-\dfrac{1}{4}[8 - |-6| \div 3 - 4]^2$ -1

83. $\dfrac{2}{5}\left[\sqrt[3]{27} - |-9| + 4 - 3^2\right]^2$ $\dfrac{242}{5}$

84. $\dfrac{3(5 - 2)^2}{-3^2} + \dfrac{2(3^2 - 4^2)}{4 - (-2)}$ $-\dfrac{16}{3}$

85. $\dfrac{24 - 5 - 4^2}{|-8| - 4 \div 2(3)} + \dfrac{4 - (-3)^2 - |4|}{3^2 - 4 \cdot 3 + |-7|}$ $-\dfrac{3}{4}$

86. $\dfrac{\sqrt[3]{-27} - 3^2 - 4}{|-6| - 2|-5|} \cdot \dfrac{4 - [6 \div 2 - 2]^2}{5 - [3^2 - (5 - 3)]^2}$ $-\dfrac{3}{11}$

87. $\dfrac{4(-2 - 8 \div 4^2 \cdot |8|)}{|5| - \sqrt[3]{-64}} + \dfrac{[(8 - 3)^2 - 4]^2}{-2^4 + 16}$ undefined

88. $\dfrac{\frac{1}{2}\left(4 - \frac{4}{5}\right)}{2 - \frac{1}{5}} + \dfrac{3\left(\frac{4}{3} - 3\right)^2}{9 - 3 - 3 \div |-3|}$ $\dfrac{23}{9}$

Evaluate.

89. $-3x^2 - 4$ when $x = 1$ -7

90. $2x^2 + x$ when $x = 3$ 21

91. $5x^2 - 2x + 5$ when $x = 3$ 44

92. $-3x^2 + 6x + 5$ when $x = 5$ -40

93. $3(x - 2)^2$ when $x = \dfrac{1}{4}$ $\dfrac{147}{16}$

94. $4(x + 1)^2 - 6x$ when $x = -\dfrac{5}{6}$ $\dfrac{46}{9}$

95. $-6x + 3y^2$ when $x = 2, y = 4$ 36

96. $6x + 3y - 5$ when $x = 1, y = -3$ -8

97. $4(x + y)^2 + 4x - 3y$ when $x = 2, y = -3$ 21
99. $3(a + b)^2 + 4(a + b) - 6$ when $a = 4, b = -1$ 33
101. $-3 - \{2x - [5x - (2x + 1)]\}$ when $x = 3$ -1

98. $4x^2 - 3y - 5$ when $x = 4, y = -2$ 65
100. $x^3y^2 - 6xy + 3x$ when $x = 2, y = 3$ 42
102. $-6 - \{x - [2x - (x - 3)]\}$ when $x = 4$ -3

103. $x^2y^4 - [y^3 + 3(x + y)]$ when $x = 2, y = -1$ 2

104. $\dfrac{(x - 3)^2}{9} + \dfrac{(y + 5)^2}{16}$ when $x = 4, y = 3$ $\dfrac{37}{9}$

105. $\dfrac{-b + \sqrt{b^2 - 4ac}}{2a}$ when $a = 6, b = -11, c = 3$ $\dfrac{3}{2}$

106. $\dfrac{-b - \sqrt{b^2 - 4ac}}{2a}$ when $a = 2, b = 1, c = -10$ $-\dfrac{5}{2}$

Indicate the sequence of calculator keys to press to evaluate the expression according to the order of operations on a calculator that uses algebraic logic. Use the parentheses keys only when needed. Then evaluate the sequence on your calculator and determine the answer.

107. $(4.2)^3 - (3.6)^2$

4.2 y^x 3 $-$ 3.6 x^2 $=$ 61.128

108. $\dfrac{3 - 4 \cdot 6 + 8.4}{15.2}$

3 $-$ 4 $\times$ 6 $+$ 8.4 $=$ $\div$ 15.2 $=$ -0.829

109. $\dfrac{4.4 - 3.5}{3.7 + 6.2}$

4.4 $-$ 3.5 $=$ $\div$ (3.7 $+$ 6.2) $=$ 0.090909091

110. $\dfrac{4 - 6^4 \div 3}{4.3 - 2.1}$

4 $-$ 6 y^x 4 $\div$ 3 $=$ $\div$ (4.3 $-$ 2.1) $=$ -194.5454545

111. $3.6[9.3 - (1.3)^5]$

3.6 $\times$ (9.3 $-$ 1.3 y^x 5) $=$ 20.113452

112. $\dfrac{9.6 - 4.3}{\left(\dfrac{5}{7}\right)^2}$

9.6 $-$ 4.3 $=$ $\div$ (5 $\div$ 7) x^2 $=$ 10.388

113. $\left(3 - \dfrac{5}{8}\right)^2 - \dfrac{2}{3}$

(3 $-$ 5 $\div$ 8) x^2 $-$ 2 $\div$ 3 $=$ 4.97395833

114. $\dfrac{4 - 3^2}{5^3} - 5^4$

4 $-$ 3 x^2 $=$ $\div$ 5 y^x 3 $-$ 5 y^x 4 $=$ -625.04

In Exercises 115–119 write an algebraic expression for each problem. Then evaluate the expression for the given value of the variable. **120.** The square of a real number cannot be negative. **121.** A positive number raised to an odd power is a positive number.

115. Multiply the variable x by 3. To this product add 6. Now square this sum. Find the value of this expression when $x = 3$. $(3x + 6)^2$, 225

116. Subtract 3 from x. Square this difference. Subtract 5 from this value. Now square this result. Find the value of this expression when $x = -1$. $[(x - 3)^2 - 5]^2$, 121

117. Six is added to the product of 3 and x. This expression is then multiplied by 6. Nine is then subtracted from this product. Find the value of the expression when $x = 3$. $6(3x + 6) - 9$, 81

118. The sum of x and y is multiplied by 2. Then 5 is subtracted from this product. This expression is then squared. Find the value of the expression when $x = 2$ and $y = -3$. $[2(x + y) - 5]^2$, 49

119. Three is added to x. This sum is divided by twice y. This quotient is then squared. Finally, 3 is subtracted from this expression. Find the value of the expression when $x = 5$ and $y = 2$. $[(x + 3)/2y]^2 - 3$, 1

120. Explain why $\sqrt{-4}$ cannot be a real number.

121. Explain why an odd root of a positive number will be positive.

122. Explain why an odd root of a negative number will be negative.

123. In your own words explain the order of operations you follow when you evaluate a mathematical expression.

124. (a) In your own words, explain in a step-by-step manner how you would evaluate
$$\dfrac{5 - 18 \div 3^2}{4 - 3 \cdot 2}.$$

(b) Evaluate the expression. $-\dfrac{3}{2}$

125. (a) In your own words, explain step by step how you would evaluate $\{5 - [4 - (3 - 8)]\}^2$.

(b) Evaluate the expression.   16

126. (a) In your own words, explain step by step how you would evaluate $16 \div 2^2 + 6 \cdot 4 - 24 \div 6$.

(b) Evaluate the expression.   24

122. A negative number raised to an odd power is a negative number.
123. parentheses, exponents or roots, multiplication or division left to right, addition or subtraction left to right.

[1.2] **127.** $A = \{a, b, c, d, f\}$, $B = \{b, c, f, g, h\}$. Find **(a)** $A \cap B$;
(b) $A \cup B$.

[1.3] In Exercises 128–130 the letter a represents a real number. For what values of a will the statement be true?

128. $|a| = |-a|$ all real numbers, $\mathbb{R}$

129. $|a| = a$ $a \geq 0$

130. $|a| = 4$ $4, -4$

131. List from smallest to largest: $-|6|, -4, |-5|, -|-2|, 0$.

132. Name the following property: $(2 + 3) + 5 = 2 + (3 + 5)$.   associative property of addition

---

Group Activity/ Challenge Problems

1. Evaluate $[(3 \div 6)^2 + 4]^2 + 3 \cdot 4 \div 12 \div 3$. $\dfrac{883}{48}$ or 18.396

2. Evaluate $[-2(3x^2 + 4)^2]^2 \div (3x^2 - 2)$ when $x = -2$. $131{,}072/5$ or $26{,}214.4$

3. Evaluate $\dfrac{2x + 4 - y\left(2 + \dfrac{3}{x}\right)}{\dfrac{y-2}{6} + \dfrac{3x^2}{4}}$ when $x = 2, y = 3$. $-\dfrac{15}{19}$ or ≈ -0.789

Summary

GLOSSARY

Absolute value (17): The distance of a number from 0 on the number line. The absolute value of any real number will be greater than or equal to 0.

Additive identity element (24): 0.

Additive inverse (or opposite) (17): For any number a, its additive inverse is $-a$.

Algebraic expression (or **expression**) (7): Any combination of numbers, variables, exponents, mathematical symbols, and operations.

Empty or null set (8): A set containing no elements, symbolized $\varnothing$ or $\{\ \ \}$.

Factors (22): If $a \cdot b = c$ then a and b are factors of c.

Finite set (7): A set that contains a finite number of elements.

Inequality symbols (9): $<, \leq, >, \geq, \neq$.

Infinite set (7): A set that has an infinite number of elements.

Intersection of sets (11): The intersection of set A and set B, $A \cap B$, is the set of elements that belongs to both set A and set B.

Multiplicative identity element (24): 1.

Multiplicative inverse (or reciprocal) (24): For any number a, $a \neq 0$, its multiplication inverse is $1/a$.

Principal (or positive) square root (31): The principal square root of a number n, written $\sqrt{n}$, is the positive number that when multiplied by itself gives n.

Radicand (31): The number or expression within a radical sign.

Set (7): A collection of objects or elements.

Subset (8): Set A is a subset of set B, $A \subseteq B$, if every element of set A is also an element of set B.

Union of sets (11): The union of sets A and B, $A \cup B$, is the set of elements that belongs to either set A or set B.

Variable (7): A letter used to represent a number.

Important Facts

Sets of numbers

Real numbers	$\{x \mid x$ is a point on the number line$\}$
Natural or counting numbers	$\{1, 2, 3, 4, 5, \ldots\}$
Whole numbers	$\{0, 1, 2, 3, 4, \ldots\}$
Integers	$\{\ldots, -3, -2, -1, 0, 1, 2, 3, \ldots\}$
Rational numbers	$\left\{\dfrac{p}{q} \mid p$ and q are integers, $q \neq 0\right\}$
Irrational numbers	$\{x \mid x$ is a real number that is not rational$\}$

Inequalities on the real number line

$\{x \mid x > a\}$

$\{x \mid x \leq a\}$

$\{x \mid a \leq x < b\}$

Properties of the real number system

Commutative properties	$a + b = b + a, \ ab = ba$
Associative properties	$(a + b) + c = a + (b + c), \ (ab)c = a(bc)$
Identity properties	$a + 0 = 0 + a = a, \ a \cdot 1 = 1 \cdot a = a$
Inverse properties	$a + (-a) = (-a) + a = 0, \ a \cdot \dfrac{1}{a} = \dfrac{1}{a} \cdot a = 1$
Distributive property	$a(b + c) = ab + ac$
Multiplicative property of 0	$a \cdot 0 = 0 \cdot a = 0$
Double-negative property	$-(-a) = a$

Absolute value: $|a| = \begin{cases} a, & a \geq 0 \\ -a, & a < 0 \end{cases}$

Exponents and roots

$$b^n = \underbrace{b \cdot b \cdot b \cdot \cdots \cdot b}_{n \text{ factors of } b} \qquad \sqrt[m]{n} = b \text{ if } \underbrace{b \cdot b \cdot b \cdot \cdots \cdot b}_{m \text{ factors of } b} = n$$

Order of operations

1. Parentheses or other grouping symbols
2. Exponents and roots
3. Multiplication or division from left to right
4. Addition or subtraction from left to right

Review Exercises

[1.2] *List each set in roster form.*   **1.** $\{3, 4, 5, 6\}$ **2.** $\{0, 3, 6, 9, \ldots\}$

  1. $A = \{x \mid x$ is a natural number between 2 and 7$\}$ **2.** $B = \{x \mid x$ is a whole number multiple of 3$\}$

Place either $\in$ or $\notin$ in the shaded area to make a true statement.

  3. 0 ▒ $\{0, 1, 2, 3\}$ $\in$ **4.** $\{3\}$ ▒ $\{0, 1, 2, 3\}$ $\notin$ **5.** $\{5\}$ ▒ $\{4, 5, 6\}$ $\notin$ **6.** 8 ▒ $\{1, 2, 3, 4, \ldots\}$ $\in$

Place either ⊆ or ⊄ in the shaded area to make a true statement.

7. {3} ▨ {1, 2, 3}   ⊆   **8.** 0 ▨ {0, 1, 2, 3, . . .}   ⊄   **9.** 5 ▨ {3, 4, 5, 6}   ⊄   **10.** {8} ▨ {1, 2, 3, 4, . . .}   ⊆

Let N = set of natural numbers, W = set of whole numbers, I = set of integers, Q = set of rational numbers, H = set of irrational numbers, ℝ = set of real numbers. Insert either ⊆ or ⊄ in the shaded area to make a true statement.

11. N ▨ W ⊆   **12.** Q ▨ ℝ   ⊆   **13.** H ▨ ℝ   ⊆   **14.** Q ▨ H ⊄

Consider the set of numbers $\{-3, 4, 6, \frac{1}{2}, \sqrt{5}, \sqrt{3}, 0, \frac{15}{27}, -\frac{1}{5}, 1.47\}$. List the elements of the set that are:

15. Natural numbers   4, 6   **16.** Whole numbers   4, 6, 0   **17.** Integers   −3, 4, 6, 0

18. Rational numbers   **19.** Irrational numbers $\sqrt{5}, \sqrt{3}$ **20.** Real numbers

Answer true or false.   **18.** $-3, 4, 6, \frac{1}{2}, 0, \frac{15}{27}, -\frac{1}{5}, 1.47$ **20.** $-3, 4, 6, \frac{1}{2}, \sqrt{5}, \sqrt{3}, 0, \frac{15}{27}, -\frac{1}{5}, 1.47$

21. $\dfrac{0}{1}$ is not a real number.   false   **22.** $0, \frac{3}{5}, -2$, and 4 are all rational numbers.   true

23. A real number cannot be divided by 0.   true   **24.** Every rational number and every irrational number is a real number.   true

25. $A \cup B = \{1, 2, 3, 4, 5\}, A \cap B = \{2, 3, 4, 5\}$
26. $A \cup B = \{2, 3, 4, 5, 6, 7, 8, 9\}, A \cap B = \varnothing$

Find $A \cup B$ and $A \cap B$ for each set A and B.

25. $A = \{1, 2, 3, 4, 5\}, B = \{2, 3, 4, 5\}$ **26.** $A = \{3, 5, 7, 9\}, B = \{2, 4, 6, 8\}$

27. $A = \{1, 2, 3, 4, . . .\}, B = \{2, 4, 6, . . .\}$ **28.** $A = \{4, 6, 9, 10, 11\}, B = \{3, 5, 9, 10, 12\}$

27. $A \cup B = \{1, 2, 3, 4, . . .\}, A \cap B = \{2, 4, 6, . . .\}$
28. $A \cup B = \{3, 4, 5, 6, 9, 10, 11, 12\}, A \cap B = \{9, 10\}$

Illustrate on the number line.

29. $\{x \mid x > 5\}$ **30.** $\{x \mid x \le -2\}$ **31.** $\{x \mid -1.3 < x \le 2.4\}$ **32.** $\{x \mid \frac{2}{3} \le x < 4 \text{ and } x \in N\}$

[1.3] *Insert either <, >, or = in the shaded area between the two numbers to make a true statement.*

33. −8 ▨ 0   <   **34.** −4 ▨ −3.9   <   **35.** 1.06 ▨ 1.6   <   **36.** −1.06 ▨ −1.6   >

37. |3| ▨ 3   =   **38.** |−3| ▨ 3   =   **39.** |4| ▨ |6| <   **40.** |−4| ▨ |−6| <

41. 13 ▨ |−5| >   **42.** |−12| ▨ 4   >   **43.** $\left|-\frac{2}{3}\right|$ ▨ $\frac{3}{5}$ >   **44.** $-|-2|$ ▨ -5 >

Write the numbers from smallest to largest.   **48.** $-4, -|3|, -2.1, -2$ **50.** $-3, 0, |1.6|, |-2.3|$

45. $4, -2, -5, |7|$ −5, −2, 4, |7| **46.** $0, \frac{3}{5}, 2.3, |-3|$ $0, \frac{3}{5}, 2.3, |-3|$ **47.** $|-7|, |-5|, 3, -2$ −2, 3, |−5|, |−7|

48. $-4, -2, -2.1, -|3|$ **49.** $-4, 6, -|-3|, 5$ −4, −|−3|, 5, 6   **50.** $|1.6|, |-2.3|, -3, 0$

Name the illustrated property.

51. $3(x + 2) = 3x + 6$ distributive property   **52.** $xy = yx$ commutative property of multiplication

53. $(x + 3) + 2 = x + (3 + 2)$ associative property of addition   **54.** $a + 0 = a$ identity property of addition

55. $(3x)y = 3(xy)$ associative property of multiplication   **56.** $a \cdot 1 = a$ identity property of multiplication

57. $-(-5) = 5$ double-negative property   **58.** $3(0) = 0$ multiplicative property of zero

59. $5 \cdot 1 = 5$ identity property of multiplication   **60.** $x + (-x) = 0$ inverse property of addition

61. $x \cdot \dfrac{1}{x} = 1$ inverse property of multiplication   **62.** $(x + y) = 1(x + y)$ identity property of multiplication

Complete the statement on the right side of the equal sign using the property given.

63. $x + 3 =$ commutative property $3 + x$ **64.** $3(x + 5) =$ distributive property $3x + 15$

65. $(x + 6) + (-4) =$ associative property $x + [6 + (-4)]$ **66.** $3 \cdot x =$ commutative property $x \cdot 3$

67. $(9 \cdot x) \cdot y =$ associative property $9 \cdot (x \cdot y)$ **68.** $4(x - y + 5) =$ distributive property $4x - 4y + 20$

69. $a + 0 =$ identity property a **70.** $a + (-a) =$ inverse property 0

[1.3, 1.4] *Evaluate.*

71. $4 - 2 + 3 - \dfrac{3}{5}$ $\dfrac{22}{5}$ **72.** $-4 \div (-2) + 16 \div 2^2 - \sqrt{49}$ -1 **73.** $(4 - 6) - (-3 + 5) + 12$ 8

74. $3|-2| - (4 - 3) + 2(-3)$ -1 **75.** $(6 - 9) \div (9 - 6)$ -1 **76.** $|6 - 3| \div 3 + 4 \cdot 8 - 12$ 21

77. $\sqrt[3]{27} \div 3 + |4 - 2| + 4^2$ 19 **78.** $3^2 - 6 \cdot 9 + 4 \div 2^2 - 3$ -47 **79.** $4 - (2 - 9)^0 + 3^2 \div 1 + 3$ 15

80. $4^2 - (2 - 3^2)^2 + 4^3$ 31 **81.** $-3^2 + 14 \div 2 \cdot 3 - 6$ 6 **82.** $\{[(9 \div 3)^2 - 1]^2 \div 8\}^3$ 512

83. $\dfrac{8 - 4 \div 2 + 3 \cdot 2}{\sqrt{36} \div 2^2 - 9}$ $-\dfrac{8}{5}$ **84.** $\dfrac{-(4 - 6)^2 - 3(-2) + |-6|}{18 - 9 \div 3 \cdot 5}$ $\dfrac{8}{3}$ **85.** $\dfrac{8 - [5 - (-3 + 2)] \div 2}{|5 - 3| - |5 - 8| \div 3}$ 5

86. $\dfrac{9|3 - 5| - 5|4| \div 10}{-2^2 + 2 \cdot 4 \div 2}$ undefined   **87.** $\dfrac{\dfrac{4}{5} - \dfrac{3}{4} \cdot \dfrac{1}{2}}{2 - \left(\dfrac{3}{4}\right)^2}$ $\dfrac{34}{115}$ **88.** $\dfrac{2 - \dfrac{3}{5} \div \dfrac{1}{2}}{\dfrac{3}{4} - 1} + \dfrac{\left(\dfrac{4}{9} \div \dfrac{2}{3}\right)^2}{2}$ $-\dfrac{134}{45}$

Evaluate each expression for the values given.

89. $2x^2 + 3x + 1$ when $x = 2$ 15 **90.** $4x^2 - 3y^2 + 5$ when $x = 1$ and $y = -\dfrac{1}{3}$ $\dfrac{26}{3}$

91. $(x - 2)^2 + (y - 4)^2 + 3$ when $x = 2$ and $y = 2$ 7 **92.** $-x^2y - 6xy^2 + 4y^3$ when $x = -2$ and $y = 3$ 204

Indicate the sequence of calculator keys to press to evaluate each expression according to the order of operations on a calculator that uses algebraic logic. Use parentheses keys only when needed. Then evaluate the expression on your calculator.

93. $\dfrac{1.4 - 6 \div 2.4}{3.6}$ **94.** $\dfrac{(5.3)^3 - (4.6)^2}{(1.7)^5}$ **95.** $\left(\dfrac{3}{5}\right)^2 - 4^2$ **96.** $\dfrac{4.2[1.3 - (4.6)^3]}{5.2 - 3.6}$

93. 1.4 − 6 ÷ 2.4 = ÷ 3.6 = −0.305555556   **94.** 5.3 y^x 3 − 4.6 x^2 = ÷ 1.7 y^x 5 = 8.99506077

95. (3 ÷ 5) x^2 − 4 x^2 = −15.64   **96.** 4.2 × (1.3 − 4.6 y^x 3) ÷ (5.2 − 3.6) = −252.0945

Practice Test

1. List $A = \{x \mid x$ is a natural number greater than 5$\}$ in roster form. $A = \{6, 7, 8, 9, \ldots\}$

Insert either $\subseteq$ or $\nsubseteq$ in the shaded area to make a true statement.

2. 3 $\{1, 2, 3, 4\}$ $\nsubseteq$ **3.** $\{5\}$ $\{1, 2, 3, 4, 5\}$ $\subseteq$

Answer true or false.

4. Every rational number is a real number.   true   **6.** The union of the set of rational numbers and the set

5. Every whole number is a natural number.   false   of irrational numbers is the set of real numbers.   true

Consider the set of numbers $\{-\frac{3}{5}, 2, -4, 0, \frac{19}{12}, 2.57, \sqrt{8}, \sqrt{2}, -1.92.\}$ List the elements of the set that are:

7. Rational numbers $-\frac{3}{5}, 2, -4, 0, \frac{19}{12}, 2.57, -1.92$ **8.** Real numbers $-\frac{3}{5}, 2, -4, 0, \frac{19}{12}, 2.57, \sqrt{8}, \sqrt{2}, -1.92$

Find $A \cup B$ and $A \cap B$ for sets A and B.

9. $A = \{8, 10, 11, 14\}$, $B = \{5, 7, 8, 9, 10\}$ **10.** $A = \{1, 3, 5, 7, \ldots\}$, $B = \{3, 5, 7, 9, 11\}$

9. $A \cup B = \{5, 7, 8, 9, 10, 11, 14\}$, $A \cap B = \{8, 10\}$ **10.** $A \cup B = \{1, 3, 5, 7, \ldots\}$, $A \cap B = \{3, 5, 7, 9, 11\}$

Indicate on the number line.

11. $\{x \mid -2.3 \le x < 5.2\}$ $-2.3 \quad 5.2$

12. $\left\{x \mid -\frac{5}{2} < x < \frac{6}{5} \text{ and } x \in I\right\}$ $-3 \ -2 \ -1 \ 0 \ 1 \ 2$

Insert either $>$, $<$ *or* $=$ *in the shaded area to make a true statement.*

13. $-4 \ \blacksquare \ |-9|$ $<$

14. $|-3| \ \blacksquare \ -|5|$ $>$

15. List from smallest to largest: $|3|, -|4|, -2, 6.$ $-|4|, -2, |3|, 6$

Name the property.

16. $(x + y) + 3 = x + (y + 3)$ associative property of addition

17. $3x + 4y = 4y + 3x$ commutative property of addition

18. $a + 0 = a$ identity property of addition

Evaluate.

19. $\{4 - [6 - 3(4 - 5)]\}^2 \div (-5)$ -5

20. $5^2 + 16 \div 4 - 3 \cdot 2$ 23

21. $\dfrac{-3|4 - 8| \div 2 + 4}{-\sqrt{36} + 18 \div 3^2 + 4}$ undefined

22. $\dfrac{-6^2 + 3(4 - |6|) \div 6}{4 - (-3) + 12 \div 4 \cdot 5}$ $-\frac{37}{22}$

23. $\dfrac{[4 - (2 - 5)]^2 + 6 \div 2 \cdot 5}{|4 - 6| + |-6| \div 2}$ $\frac{64}{5}$

Evaluate for the given values of x and y.

24. $-x^2 + 2xy + y^2$ when $x = 2$ and $y = 3$ 17

25. $(x - 5)^2 + 2xy^2 - 6$ when $x = 2$ and $y = -3$ 39

Chapter 2

Equations and Inequalities

See Section 2.3, Example 8

Preview and Perspective

The main emphasis of this chapter is on solving linear equations and inequalities. You will use the same procedures to solve equations in Section 2.1 that you used in elementary algebra. The equations presented in Section 2.1 will reinforce and then expand your knowledge. We will discuss solving equations containing fractions in Section 2.1. However, we will solve many more and different types of equations containing fractions in Sections 7.4, 7.5, and later sections. We will solve nonlinear equations and inequalities in later sections of the book. In Section 2.2 we will show how to evaluate formulas and how to solve for a variable in a formula. This material will help you in many mathematics and science courses.

Algebra is important because it can be used in everyday life. Sections 2.3 and 2.4 illustrate some true-life applications of algebra. To use algebra you must learn to represent real-life situations as equations. Changing real-life problems into equations is the emphasis of Sections 2.3 and 2.4. The more you practice writing equations to represent real-life situations, the better you will become at it.

Inequalities were introduced in Section 1.2. In Section 2.5 we will learn to solve linear inequalities. In Section 2.6 we will expand on the material presented in Section 2.5 when we solve inequalities containing absolute value.

2.1 Solving Linear Equations

Tape 2

1️⃣ Identify the reflexive, symmetric, and transitive properties.
2️⃣ Combine like terms.
3️⃣ Solve linear equations.
4️⃣ Solve equations containing fractions.
5️⃣ Solve proportions.
6️⃣ Identify conditional equations, inconsistent equations, and identities.

Reflexive, Symmetric, and Transitive Properties

1️⃣ In elementary algebra you learned how to solve linear equations. We review these procedures briefly in this section. Before we do so, we need to introduce three useful properties of equality: the reflexive property, the symmetric property, and the transitive property.

Properties of Equality

For all real numbers a, b, and c:

1. $a = a$.   *reflexive property*

2. If $a = b$, then $b = a$.   *symmetric property*

3. If $a = b$ and $b = c$, then $a = c$.   *transitive property*

Examples of the Reflexive Property

$$3 = 3$$
$$x + 5 = x + 5$$
$$x^2 + 2x - 3 = x^2 + 2x - 3$$

Examples of the Symmetric Property

If $x = 3$, then $3 = x$

If $y = x + 4$, then $x + 4 = y$

If $y = x^2 + 2x - 3$, then $x^2 + 2x - 3 = y$

Examples of the Transitive Property

If $x = a$ and $a = 4y$, then $x = 4y$

If $a + b = c$ and $c = 4r$, then $a + b = 4r$

If $4k + 3r = 2m$ and $2m = 5w + 3$, then $4k + 3r = 5w + 3$

In this book we will often use these properties, without referring to them by name.

Combining Terms

2 When an algebraic expression consists of several parts, the parts that are added or subtracted are called the **terms** of the expression. The expression

$$6x^2 - 3(x + y) - 4 + \frac{x + 2}{5}$$

has four terms: $6x^2$, $-3(x + y)$, -4, and $\dfrac{x + 2}{5}$.

The $+$ and $-$ signs that break up the expression into terms are a part of the term. However, when listing the terms of an expression, it is not necessary to list the $+$ sign at the beginning of a term.

Expression	*Terms*
$\frac{1}{2}x^2 - 3x - 7$	$\frac{1}{2}x^2, \quad -3x, \quad -7$
$-5x^3 + 3x^2y - 2$	$-5x^3, \quad 3x^2y, \quad -2$
$4(x + 3) + 2x + 5(x - 2) + 1$	$4(x + 3), \quad 2x, \quad 5(x - 2), \quad 1$

The numerical part of a term that precedes the variable is called its **numerical coefficient** or simply its **coefficient**. In the term $6x^2$, the 6 is the numerical coefficient. When the numerical coefficient is 1 or -1, we generally do not write the numeral 1. For example, x means $1x$, $-x^2y$ means $-1x^2y$, and $(x + y)$ means $1(x + y)$.

Term	*Numerical Coefficient*
$\dfrac{2x}{3}$	$\dfrac{2}{3}$
$-4(x + 2)$	-4
$\dfrac{x - 2}{3}$	$\dfrac{1}{3}$
$-(x + y)$	-1

Note that $\dfrac{x - 2}{3}$ means $\dfrac{1}{3}(x - 2)$ and $-(x + y)$ means $-1(x + y)$.

When a term consists of only a number, that number is often called a **constant.** For example, in the expression $x^2 - 4$, the -4 is a constant.

The **degree of a term** is the sum of the exponents on the variables in the term. For example, $3x^2$ is a second-degree term, and $-4x$ is a first-degree term ($-4x$ means $-4x^1$). The number 3 can be written as $3x^0$, so the number 3 (and every other constant) is a zero-degree term. The term $6x^2y^3$ is a fifth-degree term since $2 + 3 = 5$. The term $4xy^5$ is a sixth-degree term since the sum of the exponents is $1 + 5$ or 6.

Like terms are terms that have the same variables with the same exponents. For example, $3x$ and $5x$ are like terms, $2x^2$ and $-3x^2$ are like terms, and $3x^2y$ and $-2x^2y$ are like terms. Terms that are not like terms are said to be **unlike terms.** Some unlike terms are $2x$ and $4x^2$, and $3xy$ and $4x$.

To **simplify an expression** means to combine all like terms in the expression. To combine like terms, we can use the distributive property.

Examples of Combining Like Terms

$$5\ x - 2\ x = (5 - 2)\ x = 3x$$

$$3x^2 - 5x^2 = (3 - 5)x^2 = -2x^2$$

$$-7\ x^2y + 3\ x^2y = (-7 + 3)\ x^2y = -4x^2y$$

$$4\ (x - y) - (x - y) = 4\ (x - y) - 1\ (x - y) = (4 - 1)(x - y) = 3(x - y)$$

When simplifying expressions, we can rearrange the terms by using the commutative and associative properties discussed earlier.

EXAMPLE 1   Simplify. If an expression cannot be simplified, so state.

(a) $-2x + 5 + 3x - 7$ **(b)** $7x^2 - 2x^2 + 3x + 4$

(c) $2x - 3y + 5x - 6y + 3$

Solution:

(a) $-2x + 5 + 3x - 7 = \underbrace{-2x + 3x}_{x} \underbrace{+ 5 - 7}_{-2}$ Place like terms together.

This expression simplifies to $x - 2$.

(b) $7x^2 - 2x^2 + 3x + 4 = 5x^2 + 3x + 4$

(c) $2x - 3y + 5x - 6y + 3 = 2x + 5x - 3y - 6y + 3$ Place like terms together.

$$= 7x - 9y + 3$$

EXAMPLE 2   Simplify $7 - (2x - 5) - 3(2x + 4)$.

Solution:

$$7 - (2x - 5) - 3(2x + 4) = 7 - 1(2x - 5) - 3(2x + 4)$$

$$= 7 - 2x + 5 - 6x - 12 \qquad \text{Distributive property was used.}$$

$$= -2x - 6x + 7 + 5 - 12 \qquad \text{Rearrange terms.}$$

$$= -8x \qquad \text{Combine like terms.}$$

Solving Equations   ③ An **equation** is a mathematical statement of equality. *An equation must contain an equal sign* and a mathematical expression on each side of the equal sign.

Examples of Equations
$$x + 4 = -7$$
$$2x^2 - 4 = -3x + 5$$

The numbers that make an equation a true statement are called the **solutions** or **roots** of the equation. The solution to the equation $x + 2 = 5$ is 3. The **solution set** of an equation is the set of real numbers that make the equation true. The solution set for the equation $x + 2 = 5$ is {3}.

Two or more equations with the same solution set are called **equivalent equations.** The equations $2x + 3 = 9$, $x + 2 = 5$, and $x = 3$ are all equivalent equations since the solution set for each is {3}. Equations are generally solved by starting with the given equation and producing a series of simpler equivalent equations.

In this section we will discuss how to solve **linear equations in one variable.** A linear equation is an equation that can be written in the form $ax + b = c, a \neq 0$. Notice that the degree of the highest degree term in a linear equation is 1; for this reason, linear equations are also called **first-degree equations.**

To solve equations, we use the addition and multiplication properties to isolate the variable on one side of the equal sign.

Addition Property of Equality

If $a = b$, then $a + c = b + c$ for any a, b, and c.

The addition property of equality states that the same number can be added to both sides of an equation without changing the solution to the original equation. Since subtraction is defined in terms of addition, *the addition property of equality also allows us to subtract the same number from both sides of an equation.*

Multiplication Property of Equality

If $a = b$, then $a \cdot c = b \cdot c$ for any a, b, and c.

The multiplication property of equality states that both sides of an equation can be multiplied by the same number without changing the solution. Since division is defined in terms of multiplication, *the multiplication property of equality also allows us to divide both sides of an equation by the same nonzero number.*

To solve an equation, we will often have to use a combination of properties to isolate the variable. Our goal is to get the variable all by itself on one side of the equation. A general procedure to solve linear equations follows.

> **To Solve Linear Equations**
> 1. If the equation contains fractions, eliminate them by multiplying both sides of the equation by the least common denominator of the fractions.
> 2. Use the distributive property to remove any parentheses.
> 3. Combine like terms on each side of the equal sign.
> 4. Use the addition property of equality to rewrite the equation with all terms containing the variable on one side of the equal sign and all terms not containing the variable on the other side of the equal sign. It may be necessary to use the addition property a number of times to accomplish this. Repeated use of this property will eventually result in an equation of the form $ax = b$.
> 5. Use the multiplication property of equality to isolate the variable. This will give an answer of the form $x =$ some number.
> 6. Check the solution in the original equation by substitution.

EXAMPLE 3   Solve the equation $2x + 4 = 9$.

Solution:

$$2x + 4 = 9$$
$$2x + 4 - 4 = 9 - 4 \qquad \text{Subtract 4 from both sides of the equation.}$$
$$2x = 5$$

$$\frac{\overset{1}{\cancel{2}} x}{\cancel{2}_{1}} = \frac{5}{2} \qquad \text{Divide both sides of the equation by 2.}$$

$$x = \frac{5}{2}$$

Check:

$$2x + 4 = 9$$
$$\cancel{2}\left(\frac{5}{\cancel{2}}\right) + 4 = 9$$
$$5 + 4 = 9$$
$$9 = 9 \qquad \text{true}$$

Since the value checks, the solution is $\frac{5}{2}$.

Whenever an equation contains like terms on the same side of the equal sign, combine the like terms before using the addition or multiplication properties.

EXAMPLE 4   Solve the equation $-4.2 = 2(x - 3.5) + 3x - 1.75$.

Solution: $-4.2 = 2(x - 3.5) + 3x - 1.75$

$\qquad\qquad -4.2 = 2x - 7.0 + 3x - 1.75$ **Distributive property was used.**

$\qquad\qquad -4.2 = 5x - 8.75$ **Combine like terms.**

$\qquad -4.2 + 8.75 = 5x - 8.75 + 8.75$ **Add 8.75 to both sides of the equation.**

$\qquad\qquad\; 4.55 = 5x$

$\qquad\qquad \dfrac{4.55}{5} = \dfrac{5x}{5}$ **Divide both sides of the equation by 5.**

$\qquad\qquad\; 0.91 = x$

The solution is 0.91.

To save space, we will not always show the check of our answers. You should, however, check all your answers. When the equation contains decimal numbers, using a calculator to solve and check the equation may save you some time.

EXAMPLE 5   Solve the equation $2x + 8 - 3x = 3[-2(3x - 5) - 12]$.

Solution: First, use the distributive property and then combine like terms. Since the right side of the equation contains nested parentheses we work with the inner ones first.

$$2x + 8 - 3x = 3[-2(3x - 5) - 12]$$

$\qquad 2x + 8 - 3x = 3[-6x + 10 - 12]$ **Distributive property was used.**

$\qquad\quad -x + 8 = 3[-6x - 2]$ **Like terms were combined.**

$\qquad\quad -x + 8 = -18x - 6$ **Distributive property was used.**

$\quad 18x - x + 8 = -18x + 18x - 6$ **Add 18x to both sides of equation.**

$\qquad\quad 17x + 8 = -6$

$\quad 17x + 8 - 8 = -6 - 8$ **Subtract 8 from both sides of equation.**

$\qquad\qquad 17x = -14$

$\qquad\qquad \dfrac{17x}{17} = \dfrac{-14}{17}$ **Divide both sides of equation by 17.**

$\qquad\qquad\quad x = -\dfrac{14}{17}$

Notice the solutions to Examples 4 and 5 are not integers. You should not expect the solutions to equations to always be integer values.

In solving some of the following equations, we will omit some of the intermediate steps. Now we will illustrate how the solution may be shortened.

	Solution		*Shortened Solution*

(a) $x + 4 = 6$

$x + 4 \boxed{-4} = 6 \boxed{-4}$ ⟵ **Do this step mentally.**

$x = 2$

(b) $3x = 6$

$\dfrac{3x}{3} = \dfrac{6}{3}$ ⟵ **Do this step mentally.**

$x = 2$

(a) $x + 4 = 6$

$x = 2$

(b) $3x = 6$

$x = 2$

Equations Containing Fractions

|4| Step 1 in the procedure to solve linear equations indicates that, when an equation contains fractions, we begin by multiplying *both* sides of the equation by the least common denominator. The **least common denominator** of a set of denominators, the LCD (also called the **least common multiple, LCM**), is the smallest number that each of the denominators divides into (without remainder). For example, if the denominators of two fractions are 4 and 8, then the least common denominator is 8, since 8 is the smallest number that both 4 and 8 divide. If the denominators of two fractions are 5 and 6, then 30 is the least common denominator since 30 is the smallest number that both 5 and 6 divide.

When you multiply both sides of the equation by the LCD, *each term* in the equation will be multiplied by the least common denominator. *After this step is performed, the equation should not contain any fractions.*

EXAMPLE 6   Solve the equation $5 - \dfrac{2x}{3} = -9$.

Solution: The least common denominator is 3. Multiply both sides of the equation by 3 and then use the distributive property on the left side of the equation. *This process will eliminate all fractions from the equation.*

$$5 - \frac{2x}{3} = -9$$

$$3\left(5 - \frac{2x}{3}\right) = 3\,(-9)$$

$$3(5) - \overset{1}{\cancel{3}}\left(\frac{2x}{\cancel{3}}\right) = -27$$

$$15 - 2x = -27$$

$$-2x = -42$$

$$x = 21$$

EXAMPLE 7   Solve the equation $\dfrac{x}{4} + \dfrac{3}{5} = 2x - \dfrac{5}{3}$.

Solution: Multiply both sides of the equation by the least common denominator, 60. Then use the distributive property.

$$60\left(\frac{x}{4}+\frac{3}{5}\right)=60\left(2x-\frac{5}{3}\right)$$

$$\overset{15}{\cancel{60}}\left(\frac{x}{\cancel{4}}\right)+\overset{12}{\cancel{60}}\left(\frac{3}{\cancel{5}}\right)=60(2x)-\overset{20}{\cancel{60}}\left(\frac{5}{\cancel{3}}\right)$$

$$15x+36=120x-100$$

$$36=105x-100$$

$$136=105x$$

$$x=\frac{136}{105}$$

Calculator Corner

CHECKING SOLUTIONS BY SUBSTITUTION

Problems like Example 7, which cannot easily be checked, may be checked using a calculator. To check, substitute your solution into both sides of the equation, and see if you get the same value (there may sometimes be a slight difference in the last digits). To check if $\frac{136}{105}$ is the solution to Example 7, we use the following steps. (It is not necessary to use parentheses in the sequence of keys below, but using them may help you understand the check.)

Check Left Side of Equation

$$\frac{x}{4}+\frac{3}{5}=\frac{\frac{136}{105}}{4}+\frac{3}{5}$$

Keys to press: (136 ÷ 105) ÷ 4 + 3 ÷ 5 = 0.923809524

Check Right Side of Equation

$$2x-\frac{5}{3}=2\left(\frac{136}{105}\right)-\frac{5}{3}$$

Keys to Press: 2 × (136 ÷ 105) − 5 ÷ 3 = 0.923809524

Since each side of the equation yields the same result, 0.923809524, the solution is correct.

EXAMPLE 8   Solve the equation $-\frac{2}{5}(x+3)+4=\frac{1}{3}(x-4)$.

Solution: Multiply both sides of the equation by the LCD 15 to remove fractions. Then use the distributive property on the left side of the equation.

$$-\frac{2}{5}(x+3)+4=\frac{1}{3}(x-4)$$

$$15\left[-\frac{2}{5}(x+3)+4\right]=15\left[\frac{1}{3}(x-4)\right]\qquad\text{Multiply both sides of the equation by 15.}$$

The 15 is multiplying *two* terms on the left side of the equation. Can you identify them? Therefore, we must use the distributive property on the left side of the equation. The 15 is multiplying only *one* term on the right side of the equation.

$$\overset{3}{\cancel{15}}\left(-\frac{2}{\cancel{5}}\right)(x+3) + \boxed{15}(4) = \overset{5}{\cancel{15}}\left(\frac{1}{\cancel{3}}\right)(x-4)$$
$$\underset{1}{} \qquad\qquad\qquad\qquad \underset{1}{}$$

$$-6(x+3) + 60 = 5(x-4)$$
$$-6x - 18 + 60 = 5x - 20$$
$$-6x + 42 = 5x - 20$$
$$42 = 11x - 20$$
$$62 = 11x$$
$$\frac{62}{11} = x$$

Proportions   ⑤ Equations of the form $\dfrac{a}{b} = \dfrac{c}{d}$ are called **proportions.** Proportions can often be solved by *cross multiplication,* using the procedure that follows.

> ## Proportions
>
> If $\dfrac{a}{b} = \dfrac{c}{d}$, then $ad = bc$, $b \neq 0$, $d \neq 0$.

We will use cross multiplication to solve Example 9.

Example 9   Solve the equation $\dfrac{\frac{1}{2}x - 4}{3} = \dfrac{x + 8}{5}$.

Solution:

$$\frac{\frac{1}{2}x - 4}{3} = \frac{x + 8}{5}$$

$$5\left(\frac{1}{2}x - 4\right) = 3(x + 8) \qquad \textbf{Cross multiplication.}$$

$$\frac{5}{2}x - 20 = 3x + 24 \qquad \textbf{Distributive property.}$$

$$\boxed{2}\left(\frac{5}{2}x - 20\right) = \boxed{2}(3x + 24) \qquad \textbf{Multiply both sides of the equation by 2.}$$

$$5x - 40 = 6x + 48$$
$$-40 = x + 48$$
$$-88 = x$$

A check will show that the answer is correct.

Example 9 could also have been solved by multiplying both sides of the equation by the least common denominator, 15. Try this now to see that you obtain the same solution.

Equations that contain fractions, like those given in Examples 6 through 9, will be discussed in more detail in Section 7.4.

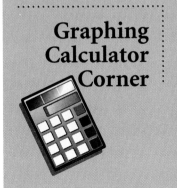

Graphing Calculator Corner

Equations in one variable may be solved using a graphing calculator. In Sections 3.2 and 4.1 we discuss the graphical interpretations of the solution to an equation. In these sections we also present techniques that can be used to solve equations in 1 variable on a graphing calculator. You may wish to review the material in those sections at this time.

Identities and Inconsistent Equations

6 All the equations discussed thus far have been **conditional equations.** They are true only under specific conditions. For example, in Example 9 the equation was true only when $x = -88$.

Consider the following equation: $2x + 1 = 5x + 1 - 3x$. Solving the equation, we obtain

$$2x + 1 = 5x + 1 - 3x$$
$$2x + 1 = 2x + 1$$
$$2x + 1 - 1 = 2x + 1 - 1 \qquad \text{Subtract 1 from both sides of the equation.}$$
$$2x = 2x$$
$$2x - 2x = 2x - 2x \qquad \text{Subtract 2x from both sides of the equation.}$$
$$0 = 0 \quad \text{true}$$

This equation, $2x + 1 = 5x + 1 - 3x$, is an example of an identity. An **identity** is an equation that is true for all real numbers. If while solving an equation you realize that both sides of the equations are identical, as in

$$2x + 1 = 2x + 1$$

the equation is an identity. The solution to $2x + 1 = 5x + 1 - 3x$ is all real numbers. **The solution to any identity is all real numbers.** If you continue to solve an equation that is an identity, you will obtain $0 = 0$, a true statement.

Now consider the equation $2(3x + 1) = 9x + 3 - 3x$.

$$2(3x + 1) = 9x + 3 - 3x$$
$$6x + 2 = 6x + 3$$
$$6x + 2 - 2 = 6x + 3 - 2 \qquad \text{Subtract 2 from both sides of the equation.}$$
$$6x = 6x + 1$$
$$6x - 6x = 6x - 6x + 1 \qquad \text{Subtract 6x from both sides of the equation.}$$
$$0 = 1 \quad \text{false}$$

Since $0 = 1$ is never a true statement, this equation has no solution. An equation that has no solution is called an **inconsistent equation.** When solving an equation that turns out to be inconsistent, do not leave the answer blank. Write "no solution" as the answer. **An inconsistent equation has no solution.**

Every linear equation is either a conditional equation with exactly one solution, an identity with an infinite number of solutions, or an inconsistent equation with no solution. Table 2.1 summarizes this information.

TABLE 2.1

Type of linear equation	Solution
Conditional equation	Has exactly one real solution
Identity	Is true for every real number; has an infinite number of solutions
Inconsistent equation	Has no solution

Exercise Set 2.1

Name the property indicated.

1. If $x = 5$, then $5 = x$ symmetric property

2. If $x + 2 = 3$, then $3 = x + 2$ symmetric property

3. If $x = 3$ and $3 = y$, then $x = y$ transitive property

4. If $x + 1 = a$ and $a = 2y$, then $x + 1 = 2y$ transitive property

5. $x + 2 = x + 2$ reflexive property

6. If $x = 4$, then $x + 3 = 4 + 3$ addition property

7. If $x = 2$, then $x - 2 = 2 - 2$ addition property

8. If $2x = 4$, then $3(2x) = 3(4)$ multiplication property

9. If $5x = 4$, then $\frac{1}{5}(5x) = \frac{1}{5}(4)$ multiplication property

10. If $x + 2 = 4$, then $x + 2 - 2 = 4 - 2$ addition property

11. If $5x = 3$, then $\frac{5x}{5} = \frac{3}{5}$ multiplication property

12. If $x - 3 = x + y$ and $x + y = z$, then $x - 3 = z$ transitive property

Give the degree of the term.

13. $4x$ first

14. $-6x^2$ second

15. $3xy$ second

16. $18x^2y^3$ fifth

17. $\frac{1}{2}x^4y$ fifth

18. 7 zero

19. -3 zero

20. $-5x$ first

21. $3x^4y^6z^3$ thirteenth

22. x^4y^6 tenth

23. $3x^5y^6z$ twelfth

24. $-2x^4y^7z^8$ nineteenth

Simplify each expression. If an expression cannot be simplified, so state.

25. $8x + 7 + 7x - 12$ $15x - 5$

26. $3x^2 + 4x + 5$ cannot be simplified

27. $5x^2 - 3x + 2x - 5$ $5x^2 - x - 5$

28. $6x^2 - 9x + 3 - 4x - 7$ $6x^2 - 13x - 4$

29. $14.6x^2 - 2.3x + 5.9x - 1.9x^2$ $12.7x^2 + 3.6x$

30. $7y + 3x - 7 + 4x - 2y$ $7x + 5y - 7$

31. $6y^2 + 6xy + 3$ cannot be simplified

32. $4x^2 - x^2 - 3x - 3x^2 + 4$ $-3x + 4$

33. $xy + 3xy + y^2 - 2$ $4xy + y^2 - 2$

34. $3x^2y + 4xy^2 - 2x^2$ cannot be simplified

35. $8.2(x - 3.4) - 1.2(9.8x + 12.4)$ $-3.56x - 42.76$

36. $6(x + 5) + 2\left(x + \frac{2}{3}\right)$ $8x + \frac{94}{3}$

37. $3\left(x + \frac{1}{2}\right) - \frac{1}{3}x + 5$ $\frac{8}{3}x + \frac{13}{2}$

38. $0.4(x - 3) + 6.5(x - 3) + 4x - 2.3$ $10.9x - 23.0$

39. $4 - [6(3x + 2) - x] + 4$ $-17x - 4$

40. $3(x + y) - 4(x + y) - 3$ $-x - y - 3$

41. $4x - [3x - (5x - 4y)] + y$ $6x - 3y$

42. $-2[3x - (2y - 1) - 5x] + y$ $4x + 5y - 2$

43. $5b - \{7[2(3b - 2) - (4b + 9)] - 2\}$ $-9b + 93$

44. $2\{[3a - (2b - 5a) - 3(2a - b) + 7] - 2\}$ $4a + 2b + 10$

Solve each equation. If an equation has no solution, so state. Use a calculator where appropriate.

45. $2x + 3 = 5$　1

46. $4 = 6 - 5x$　$\frac{2}{5}$

47. $4x + 3 = -12$　$-\frac{15}{4}$

48. $-\frac{2}{3}x = -12$　18

49. $-\frac{x}{4} = 8$　-32

50. $6 = 4 - 5x$　$-\frac{2}{5}$

51. $\frac{2.48}{3.1} = x + 6$　-5.2

52. $-2 = \frac{-4x}{5} + 9$　$\frac{55}{4}$

53. $3.2(x - 1.6) = 5.88 + 0.2x$　3.67

54. $\frac{x - 4}{3} = \frac{x + 4}{2}$　-20

55. $\frac{3x - 9}{3} = \frac{2x - 6}{6}$　3

56. $\frac{-x + 4}{3} = \frac{\frac{2}{3}x - 1}{4}$　$\frac{19}{6}$

57. $\frac{\frac{1}{4}x - 2}{5} = \frac{3(x - 2)}{4}$　$\frac{11}{7}$

58. $\frac{4x - 1.2}{0.8} = \frac{6x - 3}{1.4}$　-0.9

59. $\frac{1.5x}{5} = \frac{x - 4.2}{8}$　-3

60. $3(2x - 4) + 3(x + 1) = 9$　2

61. $\frac{1}{2}(x - 4) = 8$　20

62. $\frac{2}{3}(2x - 6) + 4 = 8$　6

63. $\frac{4x + 3}{4} = x + 6$　no solution

64. $\frac{2x - 5}{3} = -5$　-5

65. $\frac{1}{2}(3x - 5) = 6$　$\frac{17}{3}$

66. $-\frac{3}{5}(15 - 2x) = -3$　5

67. $2.5(1.6x - 3) = 4.6x$　-12.5

68. $-4.2(3.2x - 4) = 2.56x$　1.05

69. $6(x - 1) = -3(2 - x) + 3x$　all real numbers

70. $\frac{4}{3} + \frac{x}{4} = x$　$\frac{16}{9}$

71. $2(x - 3) + 2x = 4x - 5$　no solution

72. $\frac{x}{3} + 2 = x + 4$　-3

73. $\frac{11 - 3x}{4} = 2x - 4$　$\frac{27}{11}$

74. $\frac{4}{3} + 4x = 2 - x$　$\frac{2}{15}$

75. $\frac{2}{3}x = \frac{9}{4}x$　0

76. $4x - 3 = -5x - (x - 1)$　$\frac{2}{5}$

77. $3.4 - 1.2(x - 6.8) = 9.6 - 1.3x + 0.1x$　no solution

78. $0.2 - (1.76x - 3.4) = 1.3(2 - 5.2x)$　-0.2

79. $\frac{x - 25}{3} = 2x - \frac{2}{5}$　$-\frac{119}{25}$

80. $\frac{1}{2}(2x + 1) = \frac{1}{4}(x - 4)$　-2

81. $4(2 - 3x) = -[6x - (8 - 6x)]$　all real numbers

82. $4[x - (5x - 2)] = 2(x - 3)$　$\frac{7}{9}$

83. $\frac{2}{3}(x - 4) = \frac{2}{3}(4 - x)$　4

84. $-[4 - (3x - 5) + 2x] = -3(x - 2)$　$\frac{15}{4}$

85. $\frac{2(x + 2)}{5} = \frac{x}{3} - 1$　-27

86. $\frac{3x}{4} - 2 = 6 + \frac{x}{3}$　$\frac{96}{5}$

87. $\frac{x - 8}{5} + \frac{x}{3} = \frac{-8}{5}$　0

88. $\frac{x + 1}{4} = \frac{x - 4}{2} - \frac{2x - 3}{4}$　-6

89. $\frac{x - 3}{4} = \frac{2x - 1}{3} - \frac{x - 5}{6}$　-5

90. $0.04(1000) + 0.2(x + 2000) = 10{,}000$　47,800

91. $0.6(14x - 8000) = -0.4(20x + 12{,}000) + 20.6x$　0

92. $6 - \{4[x - (3x - 4) - x] + 4\} = 2(x + 3)$　2

93. $3\{[(x - 2) + 4x] - (x - 3)\} = 4 - (x - 12)$　1

94. $-3(6 - 4x) = 4 - \{5x - [6x - (4x - (3x + 2))]\}$　2

95. $-(3 - x) = 5 - \{6x - [2x - (3x - (5x - 8))]\}$　0

🖊 **96. (a)** In your own words, explain in a step-by-step manner how you would solve the equation
$$5x + 2x - 5 = 3(x - 7).$$
(b) Solve the equation.　-4

🖊 **97. (a)** In your own words, explain step by step how you would solve the equation
$$2x - \frac{2}{5} = \frac{2}{3}(x + 5).$$
(b) Solve the equation.　$\frac{14}{5}$

🖊 **98. (a)** In your own words, explain step by step how you would solve the equation
$$\frac{3}{4}(2x - 3) + \frac{1}{2} = \frac{2}{3}(x - 4).$$
(b) Solve the equation.　$-\frac{11}{10}$

🖊 **99.** What is an identity?

🖊 **100.** What is a conditional equation?

🖊 **101.** What is an inconsistent equation?

🖊 **102.** What are equivalent equations?

99. an equation that is true for all real numbers　**100.** an equation that is true only under specific conditions
101. an equation that has no solution　**102.** equations with the same solution set

103. Consider the equation $x = 4$. Give three equivalent equations. $2x = 8, \; x + 3 = 7, \; x - 2 = 2$

104. Consider the equation $2x = 5$. Give three equivalent equations. $2x + 3 = 5 + 3, \; 7(2x) = 7(5), \; \dfrac{2x}{2} = \dfrac{5}{2}$

CUMULATIVE REVIEW EXERCISES

[1.3] **105.** Write the definition of absolute value.

$$|a| = \begin{cases} a, & a \geq 0 \\ -a, & a < 0 \end{cases}$$

[1.4] *Evaluate.*

106. -3^2 -9

107. $\left(-\dfrac{3}{4}\right)^3$ $-\dfrac{27}{64}$

108. $\sqrt[3]{-64}$ -4

Group Activity/ Challenge Problems

Solve for x.

1. $-\dfrac{3}{5}(x + 2) - \dfrac{4}{3}(2x - 3) + 4 = \dfrac{1}{2}(x + 4) - 6x + 3(5 - x)$ $\dfrac{306}{157}$

2. $2\{-3[4(x + 3) + 2] + 4\} = 3[2(x + 3) + 5] + x + 6$ $\dfrac{-115}{31}$

3. $\dfrac{x}{3} + \dfrac{x - 2}{4} + \dfrac{2x - 3}{5} = \dfrac{x - 3}{6} + 4(x - 7) - x + 2$ $\dfrac{1524}{131}$

2.2   Formulas

Tape 2

1. Use subscripts and Greek letters in formulas.
2. Evaluate formulas.
3. Solve for a variable in an equation or formula.
4. Use the distributive property to solve for a variable in a formula.

Subscripts and Greek Letters

1. To give you more practice in solving equations, we will now discuss literal equations and formulas. **Literal equations** are equations that have more than one letter. **Formulas** are literal equations that are used to represent a scientific or real-life principle in mathematical terms.

Examples of Literal Equations	*Examples of Formulas*	
$5y = 2x + 3$	$A = P(1 + rt)$	from business
$x + 2y + 3z = 5$	$V = \frac{1}{2}at^2$	from physics

Often a formula contains subscripts. **Subscripts** are numbers (or other variables) placed below and to the right of variables. They are used to help clarify a formula. For example, if a formula contains two velocities, the original velocity and the final velocity, these velocities might be symbolized as V_0 and V_f, respectively. Subscripts are read using the word "sub." For example, V_f is read "V sub f" and x_2 is read "x sub 2."

Many mathematical and scientific formulas use *Greek letters*. Examples of the use of Greek letters are given in Table 2.2. *It is not necessary for you to know by name any of the Greek letters except π (pi), Δ (delta), and Σ (sigma) in this course.* Pi is used in finding the circumference and area of a circle. Delta is used when studying the slope of a line in Chapter 3, and sigma is used when discussing series in Chapter 12.

TABLE 2.2	USE OF GREEK LETTERS IN SELECTED FORMULAS	
Formula	**Greek letter**	**Greek letter represents:**
$A = \pi r^2$	π (pi)	A constant (≈ 3.14)
$S = r\theta$	θ (theta)	Angle measurement
$\bar{x} = \dfrac{\Sigma x}{n}$	Σ (capital sigma)	Summation
$z = \dfrac{x - \mu}{\sigma}$	μ (mu) σ (lower case sigma)	Mean of a set of data Standard deviation of a set of data
$m = \dfrac{\Delta y}{\Delta x}$	Δ (capital delta)	A change in value

Other Greek letters commonly used are γ (gamma), α (alpha), ϵ (epsilon), ρ (rho), λ (lambda), ω (omega), δ (lower case delta), ϕ (phi), β (beta), and χ (chi). The Greek alphabet, like our own, has both upper- and lowercase letters. For example, σ is the lowercase sigma and Σ the uppercase sigma.

Evaluating Formulas

2 Many students take algebra in preparation for another mathematics course or a science course. Evaluating formulas plays an important role in this and other courses. To **evaluate a formula** means to find the value of one of the variables when you are given the values of the other variables in the formula.

To evaluate a formula, follow the order of operations presented in Section 1.4. The order of operations is summarized and included here for you to review.

Order of Operations

1. First, evaluate any expression within parentheses, or brackets.
2. Next, evaluate all terms containing exponents and roots.
3. Next, evaluate all multiplications or divisions, working from left to right.
4. Finally, evaluate all additions or subtractions, working from left to right.

To save time, use your scientific calculator to evaluate formulas. In the following examples, *do not be concerned if you are not familiar with the formulas or symbols.*

EXAMPLE 1 Consider the compound interest formula $A = p\left(1 + \dfrac{r}{n}\right)^{nt}$. The compound interest formula is used by banks to compute the amount (or balance), A, in savings accounts that earns compound interest. In the formula, p represents the principal, r represents the interest rate, n represents the number of compounding periods (the number of times the interest is paid annually), and t represents time in years.

Mr. Johnson invests \$1000 in a savings account which earns 8% interest, compounded semiannually, for a period of 1 year. Use the compound interest formula to find the amount in the account at the end of 1 year.

Solution: The principal, p, is $1000. The rate, r, is 8% or 0.08 in decimal form. Since the interest is compounded semiannually, n is 2. The time, t, is 1 year. Substitute these values in the compound interest formula. Note that percents are always converted to decimal form before being substituted into any formula.

$$A = p\left(1 + \frac{r}{n}\right)^{nt}$$
$$= 1000\left(1 + \frac{0.08}{2}\right)^{2(1)}$$

First, we evaluate the expression within parentheses.

$$= 1000(1 + 0.04)^{2(1)}$$
$$= 1000(1.04)^{2(1)}$$

Next we raise the value within parentheses to the second power.

$$= 1000(1.04)^2$$
$$= 1000(1.0816)$$
$$= 1081.60$$

Mr. Johnson's balance is $1081.60 after 1 year. This $1081.60 represents the $1000 principal plus $81.60 in interest.

Solving for a Variable in a Formula

3 Often in science or mathematics courses we are given a formula or equation that is solved for one variable and we are asked to solve it for a different variable. To do this, treat each variable in the equation, except the one you are solving for, as if it were a constant. Then solve for the desired variable using the properties discussed previously. To solve for a given variable, it is necessary to get that variable all by itself on one side of the equal sign, that is, *isolate the variable*. The following procedure is a general procedure that may be used to isolate the variable.

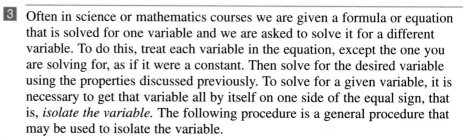

> **To Solve for a Variable in a Formula or Equation**
> 1. If the formula contains a fraction or fractions, multiply all terms by the least common denominator to remove all fractions.
> 2. Use the distributive property, if necessary, to remove parentheses.
> 3. Collect all terms containing the variable that you are solving for on one side of the equation and all terms not containing that variable on the other side of the equation.
> 4. If there is more than one term containing the variable, and the terms cannot be combined, use the distributive property to rewrite the terms as a product of two expressions where the variable you are solving for is one of the *factors*. (This process is called factoring out the variable.)
> 5. Isolate the variable you are solving for by dividing both sides of the equation by the factor that multiplies the variable.

We will use this procedure to work the examples in this section.

In Chapter 3 we will graph equations. To graph an equation, it is sometimes necessary to solve the equation for the variable *y*. Examples 2 and 3 illustrate a procedure for doing this.

EXAMPLE 2   Solve the equation $2x - 3y = 6$ for *y*.

Solution: Since this equation contains no fractions or parentheses, we begin with step 3 of the process. We will solve for the variable *y* by isolating the term containing the *y* on the left side of the equation.

$$2x - 3y = 6$$

STEP 3: $2x - 2x - 3y = -2x + 6$ **Subtract 2*x* from both sides of the equation.**

$$-3y = -2x + 6$$

STEP 5: $\dfrac{-3y}{-3} = \dfrac{-2x + 6}{-3}$ **Divide both sides of equation by −3.**

$$y = \frac{-2x + 6}{-3}$$

Since we do not want to leave the answer with a negative number in the denominator, we multiply the numerator and the denominator by −1 to get

$$y = \frac{2x - 6}{3} \quad \text{or} \quad y = \frac{2}{3}x - 2$$

EXAMPLE 3   Solve the equation $\dfrac{2}{3}(2x - y) = \dfrac{5}{7}(x - 3y) + 4$ for *y*.

Solution: Since this equation contains fractions, we begin by multiplying both sides of the equation by the least common denominator 21 to eliminate the fractions, step 1. We then use the distributive property, step 2. We then proceed to isolate the variable *y* by collecting all terms containing the variable *y* on one side of the equation and all terms not containing the variable *y* on the other side of the equation.

STEP 1: $21\left[\frac{2}{3}(2x - y)\right] = 21\left[\frac{5}{7}(x - 3y) + 4\right]$ **Multiply both sides of equation by LCD, 21.**

STEP 2: $21\left[\frac{2}{3}(2x - y)\right] = 21\left[\frac{5}{7}(x - 3y)\right] + 21 \cdot 4$ **Distributive property was used.**

$$14(2x - y) = 15(x - 3y) + 84$$

$$28x - 14y = 15x - 45y + 84$$

STEP 3: $28x - 14y + 45y = 15x - 45y + 45y + 84$ **Add 45*y* to both sides of equation.**

$$28x + 31y = 15x + 84$$

$$28x - 28x + 31y = 15x - 28x + 84$$ **Subtract 28*x* from both sides of the equation.**

$$31y = -13x + 84$$

STEP 5:
$$\frac{31y}{31} = \frac{-13x + 84}{31}$$ **Divide both sides of equation by 31.**

$$y = \frac{-13x + 84}{31} \quad \text{or} \quad y = -\frac{13}{31}x + \frac{84}{31}$$

EXAMPLE 4   The formula for the area of a triangle is $A = \frac{1}{2}bh$, where b is the length of the base and h is the height. Solve this formula for the height, h.

Solution: We are asked to express the height, h, of the triangle in terms of the triangle's area, A, and base, b. Since we are solving for h, we must isolate the h on one side of the equation. We use the appropriate properties to remove the factors $\frac{1}{2}$ and b from the right side of the equation.

$$A = \frac{1}{2}bh$$

STEP 1: $2\,A = 2\left(\frac{1}{2}\right)bh$ **Multiply both sides of the equation by 2.**

$$2A = bh$$

STEP 5: $\dfrac{2A}{b} = \dfrac{bh}{b}$ **Divide both sides of the equation by b.**

$$\frac{2A}{b} = h \quad \text{or} \quad h = \frac{2A}{b}$$

EXAMPLE 5   A formula that may be important to you now, or in the future, is the *tax-free yield formula*, $T_f = T_a(1 - F)$. This formula can be used to convert a taxable yield, T_a, into its equivalent tax-free yield, T_f, where F is the federal income tax bracket of the person.

(a) Mary is in a 28% income tax bracket. Find the equivalent tax-free yield of a 12% taxable investment.

(b) Solve this equation for T_a; that is, write an equation for taxable yield in terms of tax-free yield.

Solution:

(a) $T_f = Ta(1 - F)$
$T_f = 0.12(1 - 0.28)$
$\quad = 0.12(0.72)$
$\quad = 0.0864 \quad \text{or} \quad 8.64\%$

Thus, for Mary, or anyone else in a 28% income tax bracket, a 12% taxable yield is equivalent to an 8.64% tax-free yield.

(b) $T_f = T_a(1 - F)$

$$\frac{T_f}{1 - F} = \frac{T_a(1 - F)}{1 - F}$$ **Divide both sides of the equation by $1 - F$.**

$$\frac{T_f}{1 - F} = T_a \quad \text{or} \quad T_a = \frac{T_f}{1 - F}$$

EXAMPLE 6   A formula used in banking is $A = P(1 + rt)$. A represents the amount that must be repaid to the bank when P dollars are borrowed at simple interest rate, r, for time, t. Solve this equation for time, t.

Solution:

$$A = P(1 + rt)$$

$$A = P + Prt \qquad \text{Distributive property was used.}$$

$$A - P = P - P + Prt \qquad \text{Subtract } P \text{ from both sides of the equation to isolate the term containing the variable } t.$$

$$A - P = Prt$$

$$\frac{A - P}{Pr} = \frac{Prt}{Pr} \qquad \text{Divide both sides of equation by } Pr \text{ to isolate } t.$$

$$\frac{A - P}{Pr} = t \quad \text{or} \quad t = \frac{A - P}{Pr}$$

EXAMPLE 7   A formula used in statistics is given below. If you take a course in statistics you may use this formula often.

$$Z = \frac{\bar{x} - \mu}{\dfrac{\sigma}{\sqrt{n}}}$$

(a) Find Z when $\bar{x} = 100.4$, $\mu = 102.8$, $\sigma = 4.2$, and $n = 36$.

(b) Solve this equation for $\bar{x}$ (read "x bar").

Solution:

(a) $Z = \dfrac{\bar{x} - \mu}{\dfrac{\sigma}{\sqrt{n}}} = \dfrac{100.4 - 102.8}{\dfrac{4.2}{\sqrt{36}}} = \dfrac{-2.4}{\dfrac{4.2}{6}} = \dfrac{-2.4}{0.7} = \approx -3.43$

(b) Note that this formula contains the Greek letters μ (mu) and σ (sigma). Treat them the same as you would an English letter.

$$Z = \frac{\bar{x} - \mu}{\dfrac{\sigma}{\sqrt{n}}}$$

Begin by multiplying both sides of the equation by the denominator, $\dfrac{\sigma}{\sqrt{n}}$.

$$Z \cdot \frac{\sigma}{\sqrt{n}} = \frac{\bar{x} - \mu}{\dfrac{\sigma}{\sqrt{n}}} \cdot \frac{\sigma}{\sqrt{n}}$$

$$Z \cdot \frac{\sigma}{\sqrt{n}} = \bar{x} - \mu$$

$$\mu + Z \cdot \frac{\sigma}{\sqrt{n}} = \bar{x} - \mu + \mu \qquad \text{Add } \mu \text{ to both sides of the equation.}$$

$$\mu + Z \cdot \frac{\sigma}{\sqrt{n}} = \bar{x}$$

The formula for $\bar{x}$ may be written in a number of other forms. Other acceptable forms are $\bar{x} = Z \cdot \dfrac{\sigma}{\sqrt{n}} + \mu$ and $\bar{x} = \mu + \dfrac{Z\sigma}{\sqrt{n}}$.

Use the Distributive Property to Solve for a Variable

4 Now we will look at some examples where the variable we are solving for appears in more than one term and the terms cannot be combined.

EXAMPLE 8   Solve the equation $x(y - 3) = 2y + 4$ for y.

Solution: To solve this equation for y, we need to isolate the terms containing y. Since this equation does not contain fractions, we begin by using the distributive property, step 2 in the process on page 62.

STEP 2:
$$x(y - 3) = 2y + 4$$
$$xy - 3x = 2y + 4$$

Now collect all terms containing the variable we are solving for, y, on one side of the equation and all terms not containing y on the other side of the equation. We will choose to collect the y terms on the left side of the equation and the terms not containing y on the right side of the equation.

STEP 3:
$$xy - 2y - 3x = 2y - 2y + 4 \qquad \text{Subtract } 2y \text{ from both sides of equation.}$$
$$xy - 2y - 3x = 4$$
$$xy - 2y - 3x + 3x = 3x + 4 \qquad \text{Add } 3x \text{ to both sides of equation.}$$
$$xy - 2y = 3x + 4 \qquad \text{Terms containing } y \text{ are on one side of equation.}$$

Both terms on the left side of the equation contain y, and the terms cannot be combined. Therefore, we use the distributive property to rewrite the terms as the product of y, and another factor.

$$xy - 2y = 3x + 4$$

STEP 4:
$$y(x - 2) = 3x + 4 \qquad \text{Distributive property.}$$

STEP 5:
$$\frac{y\cancel{(x - 2)}}{\cancel{x - 2}} = \frac{3x + 4}{x - 2} \qquad \text{Divide both sides of the equation by } x - 2.$$

$$y = \frac{3x + 4}{x - 2}$$

EXAMPLE 9   A formula used for levers in physics is $d = \dfrac{fl}{f + w}$. Solve this formula for f.

Solution: We will follow the procedure given. We choose to collect the terms containing the variable f on the right side of the equation.

$$d = \frac{fl}{f + w}$$

STEP 1: $d\,(f + w) = \dfrac{fl}{\cancel{(f + w)}}\,\cancel{(f + w)}$ Multiply by the LCD, $f + w$.

STEP 2: $d(f + w) = fl$

$df + dw = fl$ Distributive property.

STEP 3: $df \;\cancel{-\; df} + dw = fl \;-\; df$ Isolate terms containing f on right side of equation.

$dw = fl - df$

STEP 4: $dw = f(l - d)$ Distributive Property

STEP 5: $\dfrac{dw}{l - d} = \dfrac{f\cancel{(l - d)}}{\cancel{l - d}}$ Isolate the f by dividing both sides of the equation by $l - d$.

Thus, $f = \dfrac{dw}{l - d}.$

EXAMPLE 10   Solve the equation $2xy + 3yz = 6xz + 3$ for z.

Solution: First, we collect all terms containing the variable we are solving for, z, on one side of the equation and all terms not containing z on the other side of the equation. The terms containing z are $3yz$ and $6xz$. We will collect these terms on the left side of the equation and the terms that do not contain the z on the right side of the equation.

STEP 3: $2xy + 3yz = 6xz + 3$

$2xy + 3yz \;-\; 6xz = 6xz \;-\; 6xz + 3$

$2xy + 3yz - 6xz = 3$

$2xy \;-\; 2xy + 3yz - 6xz = 3 \;-\; 2xy$

$3yz - 6xz = 3 - 2xy$

STEP 4: $z(3y - 6x) = 3 - 2xy$

STEP 5: $\dfrac{z\cancel{(3y - 6x)}}{\cancel{3y - 6x}} = \dfrac{3 - 2xy}{3y - 6x}$

$z = \dfrac{3 - 2xy}{3y - 6x}$

Consider the equation

$$\tfrac{1}{2}xy + \tfrac{3}{4}yz = \tfrac{3}{2}xz + \tfrac{3}{4}$$

How would you begin to solve for z? If you said multiply both sides of the equation by the least common denominator 4 to remove fractions, you answered correctly. After you multiply both sides of the equation by 4, you will end up with the equation given in Example 10. Try this now and see.

Additional examples of solving for a variable in a formula will be given in Section 7.4.

Exercise Set 2.2

Evaluate the formula for the values given. Round answers to the nearest hundredth (see Examples 1 and 7a).   **9.** 66.67

1. $P = 2l + 2w$; $l = 15, w = 6$ (mathematics)   42

2. $\bar{x} = \dfrac{x_1 + x_2 + x_3}{3}$: $x_1 = 40, x_2 = 120, x_3 = 80$
(statistics)   80

3. $A = \frac{1}{2}h(b_1 + b_2)$: $h = 10, b_1 = 20, b_2 = 30$
(mathematics)   250

4. $E = a_1p_1 + a_2p_2 + a_3p_3$: $a_1 = 10, p_1 = 0.2,$
$a_2 = 100, p_2 = 0.3, a_3 = 1000, p_3 = 0.5$
(probability)   532

5. $F = G\dfrac{m_1m_2}{r^2}$: $G = 0.5, m_1 = 100, m_2 = 200,$
$r = 4$ (physics)   625

6. $m = \dfrac{y_2 - y_1}{x_2 - x_1}$: $y_2 = 4, y_1 = -3, x_2 = -2,$
$x_1 = -6$ (mathematics) $\frac{7}{4}$

7. $z = \dfrac{\bar{x} - \mu}{\dfrac{\sigma}{\sqrt{n}}}$: $\bar{x} = 80, \mu = 70, \sigma = 15, n = 25$
(statistics)   3.33

8. $d = \sqrt{(x_2 - x_1)^2 + (y_2 - y_1)^2}$: $x_2 = 5, x_1 = -3,$
$y_2 = -6, y_1 = 3$ (mathematics) $\sqrt{145} \approx 12.04$

9. $R_T = \dfrac{R_1R_2}{R_1 + R_2}$: $R_1 = 100, R_2 = 200$ (electronics)   66.67

10. $V = \sqrt{V_x^2 + V_y^2}$: $V_x = 3, V_y = 4$ (physics)   5

11. $x = \dfrac{-b + \sqrt{b^2 - 4ac}}{2a}$: $a = 2, b = -5,$
$c = -12$ (mathematics)   4

12. $S = \pi r^2 + \pi rs$: $\pi = 3.14, r = 3, s = 4$
(mathematics)   65.94

13. $H = (0.14 + 0.47\sqrt{v})(36.5 - T)$: $v = 25,$
$T = 5$ (meteorology)   78.44

14. $\mu = f\dfrac{k(k + 1)}{n(n + 1)}$: $f = 360, n = 48, k = 16$
(banking)   41.63

15. $Z = \dfrac{p' - p}{\sqrt{\dfrac{pq}{n}}}$: $p' = 0.6, p = 0.5, q = 0.5,$
$n = 10$ (statistics) $\dfrac{0.1}{\sqrt{0.025}} \approx 0.63$

16. $\bar{x}_w = \dfrac{w_1x_1 + w_2x_2 + w_3x_3}{w_1 + w_2 + w_3}$: $x_1 = 60, x_2 = 80,$
$x_3 = 96, w_1 = 4, w_2 = 6, w_3 = 10$ (statistics)   84

17. $A = p\left(1 + \dfrac{r}{n}\right)^{nt}$: $p = 100, r = 0.06, n = 1,$
$t = 3$ (banking)   119.10

18. $r_{DD} = \dfrac{\dfrac{r_{xx} + r_{yy}}{2} - r_{xy}}{1 - r_{xy}}$: $r_{xx} = 0.7, r_{yy} = 0.9,$
$r_{xy} = 0.3$ (psychology) $\frac{5}{7}$ or ≈ 0.71

Solve each equation for y (see Examples 2 and 3).

19. $3x + y = 5$ $y = -3x + 5$

20. $3x + 2y = 8$ $y = (-3x + 8)/2$

21. $2x - y = -5$ $y = 2x + 5$

22. $2x - 4y = 6$ $y = (x - 3)/2$

23. $5x - 3y = -4$ $y = (5x + 4)/3$

24. $2y = 8x - 3$ $y = (8x - 3)/2$

25. $\frac{1}{2}x + 2y = 6$ $y = (-x + 12)/4$

26. $\frac{3}{5}x + \frac{1}{3}y = 1$ $y = (-9x + 15)/5$

27. $3(x - 2) + 3y = 6x$ $y = x + 2$

28. $2(x + 3y) = 4(x - y) + 5$
$y = (2x + 5)/10$

29. $3x - 5 = 2(3y + 6)$ $y = (3x - 17)/6$

30. $\frac{1}{5}(x - 2y) = \frac{3}{4}(y + 2) + 3$ $y = (4x - 90)$

Solve for the variable indicated (see Examples 4–7).

31. $d = rt$, for t $t = \dfrac{d}{r}$

32. $C = \pi d$, for d $d = \dfrac{C}{\pi}$

33. $A = \dfrac{1}{2}bh$, for b $b = \dfrac{2A}{h}$

34. $i = prt$, for t $t = \dfrac{i}{pr}$

35. $P = 2l + 2w$, for w $w = \dfrac{P - 2l}{2}$

36. $P = 2l + 2w$, for l $l = \dfrac{P - 2w}{2}$

47. $h = \dfrac{2A}{b_1 + b_2}$ **49.** $m = \dfrac{y - y_1}{x - x_1}$ **53.** $F = \frac{9}{5}C + 32$

37. $V = lwh$, for h $h = \dfrac{V}{lw}$

38. $V = \pi r^2 h$, for h $h = \dfrac{V}{\pi r^2}$

39. $V = \dfrac{1}{3}lwh$, for l $l = \dfrac{3V}{wh}$

40. $z = \dfrac{x - \mu}{\sigma}$, for σ $\sigma = \dfrac{x - \mu}{z}$

41. $z = \dfrac{x - \mu}{\sigma}$, for μ $\mu = x - z\sigma$

42. $A = P + Prt$, for r $r = \dfrac{A - P}{Pt}$

43. $y = mx + b$, for m $m = \dfrac{y - b}{x}$

44. $IR + Ir = E$, for R $R = \dfrac{E - Ir}{I}$

45. $y = \dfrac{kx}{z}$, for z $z = \dfrac{kx}{y}$

46. $\dfrac{P_1}{T_1} = \dfrac{P_2}{T_2}$, for T_2 $T_2 = \dfrac{T_1 P_2}{P_1}$

47. $A = \dfrac{1}{2}h(b_1 + b_2)$, for h

48. $A = \dfrac{r^2\theta}{2}$, for θ $\theta = \dfrac{2A}{r^2}$

49. $y - y_1 = m(x - x_1)$, for m

50. $R_T = \dfrac{R_1 + R_2}{2}$, for R_1 $R_1 = 2R_T - R_2$ **51.** $S = \dfrac{n}{2}(f + l)$, for n $n = \dfrac{2S}{f + l}$

52. $S = \dfrac{n}{2}(f + l)$, for l $l = \dfrac{2S - nf}{n}$

53. $C = \dfrac{5}{9}(F - 32)$, for F

54. $F = \dfrac{9}{5}C + 32$, for C $C = \frac{5}{9}(F - 32)$

55. $F = \dfrac{km_1 m_2}{d^2}$, for m_1 $m_1 = \dfrac{Fd^2}{km_2}$

56. $A = \dfrac{1}{2}h(b_1 + b_2)$, for b_1 $b_1 = \dfrac{2A - hb_2}{h}$

Solve each equation for the indicated variable (see Examples 8–10).   **57.** $y = \dfrac{3x - 4}{x - 2}$ **58.** $y = -\dfrac{x}{x - 2}$

57. $3x + 2y = xy + 4$, for y

58. $2(x + y) = x(3 + y)$, for y

59. $y = \dfrac{x}{x - 1}$, for x $x = \dfrac{y}{y - 1}$

60. $y = \dfrac{x - 2}{x + 4}$, for x $x = \dfrac{-4y - 2}{y - 1}$

61. $y = \dfrac{4 - x}{x - 2}$, for x $x = \dfrac{2y + 4}{y + 1}$

62. $3yz + 2 = 5x + z$, for z $z = \dfrac{5x - 2}{3y - 1}$

63. $2xyz + 3yz = -6xy$, for z $z = \dfrac{-6x}{2x + 3}$

64. $\dfrac{1}{3}xy - 6y = 2y + 3$, for y $y = \dfrac{9}{x - 24}$

65. $3rs - 2s = \dfrac{1}{2}(s + 2r)$, for r $r = \dfrac{5s}{6s - 2}$

66. $\dfrac{1}{2}x + \dfrac{3}{4}xy = \dfrac{1}{3}(2x - y)$, for y $y = \dfrac{2x}{9x + 4}$

Solve each formula for the indicated variable (see Examples 9 and 10).   **76.** $m = \dfrac{m_2 x_2 + m_3 x_3 - xm_2 - xm_3}{x - x_1}$

67. $A = P + Prt$, for P (banking) $P = \dfrac{A}{1 + rt}$

68. $J_0 = I\omega - I\omega_0$, for I (physics) $I = \dfrac{J_0}{\omega - \omega_0}$

69. $at_2 - at_1 + v_1 = v_2$, for a (physics) $a = \dfrac{v_2 - v_1}{t_2 - t_1}$

70. $a_n = a_1 + nd - d$, for d (mathematics) $d = \dfrac{a_n - a_1}{n - 1}$

71. $Vr - R = O - Dr$, for r (economics) $r = \dfrac{O + R}{V + D}$

72. $2P_1 - 2P_2 - P_1 P_c = P_2 P_c$, for P_c (economics) $P_c = \dfrac{2P_1 - 2P_2}{P_1 + P_2}$

73. $S_n - S_n r = a_1 - a_1 r^n$, for S_n (mathematics) $S_n = \dfrac{a_1 - a_1 r^n}{1 - r}$ **74.** $e = \dfrac{q_H + q_C}{q_H}$, for q_H (chemistry, physics) $q_H = \dfrac{q_C}{e - 1}$

75. $A = 2HW + 2LW + 2LH$, for L (mathematics) $L = \dfrac{A - 2HW}{2W + 2H}$ **76.** $xm_1 + xm_2 + xm_3 = m_1 x_1 + m_2 x_2 + m_3 x_3$, for m_1 (statistics)

77. Wok invests \$5000 in a savings account paying compound interest. Find the amount in Wok's account after 3 years if the interest is compounded quarterly and the rate is 8% (see Example 1).   \$6341.21   **78.** 5.12%

78. Phil's federal tax rate is 36%. Find the equivalent tax-free yield of an 8% taxable investment (see Example 5).

79. **(a)** In your own words, explain the procedure you would use to solve the physics formula

$$\mu = \dfrac{0.5cV^2}{Ad} \text{ for } A.$$

(b) Solve the formula for A.   **b.** $A = \dfrac{0.5cV^2}{\mu d}$

80. Consider the equation $2xy - 3x = 4 - 5y$. If we solve this equation for y we get $y = \dfrac{3x + 4}{2x + 5}$. Now consider the equation $2ab - 3a = 4 - 5b$. Notice the similarity between this equation and the first equation. Without actually solving this equation, can you give the solution for b? Give the solution and explain how you determined your answer.

$b = \dfrac{3a + 4}{2a + 5}$. Equations are same except for variables.

81. Ships at sea measure their speed in terms of knots. One knot is one nautical mile per hour. A nautical mile is about 6076 feet. When measuring a car's speed in miles per hour, a mile is 5280 ft.

   (a) Determine a formula for converting a speed in miles per hour (m) to a speed in knots (k).

   (b) Explain how you determined this formula.

   (a) $k = 1.15$ **(b)** The quotient of 6076 to 5280 is about 1.15.

SEE EXERCISE 81.

CUMULATIVE REVIEW EXERCISES

[1.4] **82.** Evaluate $-(5 - 8)^2 + |5 - 8| - 4^2$. -22

83. Evaluate $\dfrac{4 - 6 \div 3 + 5^2 - 6 \cdot 4}{5 - |6 \div (-2)|}$ $\dfrac{3}{2}$

84. Evaluate $6x^2 - 3xy + y^2$ when $x = 2, y = 3$.   15

[2.1] **85.** Solve the equation $\dfrac{1}{3}x + 4 = \dfrac{2}{5}(x - 3)$.   78

1. (a) No, an H appears on both sides of the equation. **(b)** $H = \dfrac{A - 2LW}{2W + 2L}$

Group Activity/ Challenge Problems

1. The formula used for finding the surface area of a rectangular solid is

$$A = 2HW + 2LW + 2LH$$

Consider the steps in solving the equation for H.

$$A = 2HW + 2LW + 2LH$$
$$A - 2HW = 2LW + 2LH$$
$$A - 2HW - 2LW = 2LH$$
$$\frac{A - 2HW - 2LW}{2L} = H$$

   (a) Is this formula solved correctly for H? Explain your answer.

   (b) Solve the formula correctly for H.

2. Solve the formula $r = \dfrac{s/t}{t/u}$ for **(a)** s; **(b)** u.   **(a)** $s = \dfrac{rt^2}{u}$ **(b)** $u = \dfrac{rt^2}{s}$

2.3   Applications of Algebra

Tape 2

1 Translate a verbal statement into an algebraic expression.
2 Solve application problems.

Translating to Algebra

1 The next few sections will present some of the many uses of algebra in real-life situations. Whenever possible, we include other relevant applications throughout the text.

Perhaps the most difficult part of solving a word problem is transforming the problem into an equation. Here are some examples of phrases represented as algebraic expressions.

Phrase	Algebraic Expression
a number increased by 4	$x + 4$
twice a number	$2x$
5 less than a number	$x - 5$
a number subtracted from 9	$9 - x$
6 subtracted from a number	$x - 6$
one-eighth of a number	$\frac{1}{8}x$ or $\frac{x}{8}$
2 more than 3 times a number	$3x + 2$
4 less than 6 times a number	$6x - 4$
3 times the sum of a number and 5	$3(x + 5)$

The variable x was used in the algebraic expressions, but any variable could have been used to represent the unknown quantity.

EXAMPLE 1　Express each phrase as an algebraic expression.

(a) The distance, d, increased by 15 miles

(b) 3 less than 4 times the area a

(c) 4 times a number n is decreased by 5

Solution:

(a) $d + 15$　　**(b)** $4a - 3$　　**(c)** $4n - 5$

EXAMPLE 2　Write each of the following as an algebraic expression.

(a) The cost of purchasing x shirts at $4 each

(b) The distance traveled in t hours at 55 miles per hour

(c) The number of cents in n nickels

(d) An 8% commission on sales of x dollars

Solution:

(a) We can reason like this: one shirt would cost 1(4) dollars; two shirts, 2(4) dollars; three shirts, 3(4) dollars; four shirts, 4(4) dollars, and so on. Continuing this reasoning process, we can see that x shirts would cost $x(4)$ or $4x$ dollars. We can use the same reasoning process to complete each of the following.

(b) $55t$

(c) $5n$

(d) $0.08x$ (8% is written as 0.08 in decimal form)

Helpful Hint

A percent is always a percent of some quantity. Therefore, when a percent is listed, it is *always* multiplied by a number or a variable. In the following examples we use the variable c, but any letter could be used to represent the variable.

Phrase	How Written
6% of a number	$0.06n$
the cost of an item increased by a 7% tax	$c + 0.07c$
25% off the cost of an item	$c - 0.25c$

Sometimes in a problem two numbers are related to each other. We often represent one of the numbers as a variable and the other as an expression containing that variable. We generally let the less complicated description be represented by the variable and write the second (more complex expression) in terms of the variable. In the following examples, we use x for the variable.

Phrase	One Number	Second Number
Peter's age now and Peter's age in 5 years	x	$x + 5$
one number is 3 times the other	x	$3x$
one number is 7 less than the other	x	$x - 7$
two consecutive integers	x	$x + 1$
two consecutive odd (or even) integers	x	$x + 2$
a number and the number increased by 7%	x	$x + 0.07x$
a number and the number decreased by 10%	x	$x - 0.10x$
the sum of two numbers is 10	x	$10 - x$
a 6-foot board cut in two lengths	x	$6 - x$
$10,000 shared by two people	x	$10,000 - x$

The last three examples may not be obvious. Consider "The sum of two numbers is 10." When we add x and $10 - x$ we get $x + (10 - x) = 10$. When a 6-foot board is cut in two lengths, the two lengths will be x and $6 - x$. For example, if one length is 2 feet, the other must be $6 - 2$ or 4 feet.

Example 3 For each of the following relations, select a variable to represent one quantity and express the second quantity in terms of the first.

(a) The speed of the second train is 1.2 times the speed of the first.

(b) $90 is shared by David and his brother.

(c) It takes Tom 3 hours longer than Roberta to complete the task.

(d) Hilda has $4 more than twice the amount of money Hector has.

(e) The length of a rectangle is 2 units less than 3 times its width.

Solution:

(a) Speed of first train, x; speed of second train, $1.2x$

(b) Amount David has, x; amount brother has, $90 - x$

(c) Roberta, x; Tom, $x + 3$

(d) Hector, x; Hilda, $2x + 4$

(e) Width, x; length, $3x - 2$

Solving Application Problems

2 The word **is** in a word problem often means **is equal to** and is represented by an equal sign, **=**.

Verbal Statement	*Algebraic Equation*
4 less than 3 times a number *is* 5	$3x - 4 = 5$
a number decreased by 4 *is* 3 more than twice the number	$x - 4 = 2x + 3$
the product of two consecutive integers *is* 20	$x(x + 1) = 20$
one number is 2 more than 5 times the other number; the sum of the two numbers *is* 62	$x + (5x + 2) = 62$
a number increased by 15% *is* 90	$x + 0.15x = 90$
a number decreased by 12% *is* 38	$x - 0.12x = 38$
the sum of a number and the number increased by 4% *is* 204	$x + (x + 0.04x) = 204$
the cost of renting a VCR for x days at \$15 per day *is* \$120	$15x = 120$

Although there are many types of word problems, the general procedure used to solve all word problems is basically the same.

> **To Solve a Word Problem**
> 1. Read the problem carefully.
> 2. If possible, draw a sketch to illustrate the problem.
> 3. Identify the quantity or quantities you are being asked to find.
> 4. Choose a variable to represent one quantity, *and write down exactly what it represents.* Represent any other quantities to be found in terms of this variable.
> 5. Write the word problem as an equation.
> 6. Solve the equation for the unknown quantity.
> 7. Answer the question asked. Be sure to give the proper units with your answer.
> 8. Check the solution in the *original word problem.*

Many students solve the equation for the variable, but forget to answer the question or questions asked.

EXAMPLE 4   Four subtracted from 3 times a number is 29. Find the number.

Solution: Follow the procedure on page 73 to solve the problem.

STEP 3: We are asked to find the unknown number.

STEP 4: Let x = unknown number.

STEP 5: Write the equation.

$$3x - 4 = 29$$

STEP 6: Solve the equation.

$$3x - 4 = 29$$
$$3x = 33$$
$$x = 11$$

STEP 7: Answer the question.

The number is 11.

STEP 8: Check the solution in the original word problem.
Four subtracted from 3 times a number is 29.

$$3(11) - 4 = 29$$
$$33 - 4 = 29$$
$$29 = 29 \quad \text{true}$$

EXAMPLE 5   One day Johnny Big had the following meals—breakfast: bacon, eggs, and toast with butter; lunch: a Big Mac with fries; dinner: steak, baked potato with sour cream, salad with dressing, roll and butter. After dinner he attended a movie and had a medium-size "buttered" popcorn (11 cups). The number of grams of saturated fat eaten during lunch was 2.9 grams greater than the amount eaten during breakfast. The amount of saturated fat eaten during dinner was 2.8 grams less than twice the amount eaten for breakfast. The amount of saturated fat eaten during the movie was 1 gram less than three times the amount eaten for lunch. If the total number of grams of saturated fat eaten during the day was 85.5, find the number of grams of saturated fat eaten during each meal and at the movies. (These are accurate figures. The popcorn was cooked in coconut oil. Air-popped popcorn contains no saturated fat.)

Solution:

$$\text{Let } x = \text{grams of saturated fat eaten during } \textit{breakfast}$$
$$\text{then } x + 2.9 = \text{grams of saturated fat eaten during } \textit{lunch}$$
$$2x - 2.8 = \text{grams of saturated fat eaten during } \textit{dinner}$$
$$\text{and } 3(x + 2.9) - 1 = \text{grams of saturated fat eaten during the } \textit{movie}$$

Since the total number of grams of fat eaten during the entire day is 85.5, we write

$$\left(\begin{array}{c}\text{grams during}\\\text{breakfast}\end{array}\right)+\left(\begin{array}{c}\text{grams during}\\\text{lunch}\end{array}\right)+\left(\begin{array}{c}\text{grams during}\\\text{dinner}\end{array}\right)+\left(\begin{array}{c}\text{grams during}\\\text{movie}\end{array}\right)=85.5$$

$$x + x + 2.9 + 2x - 2.8 + 3(x+2.9) - 1 = 85.5$$
$$x + x + 2.9 + 2x - 2.8 + 3x + 8.7 - 1 = 85.5$$
$$7x + 7.8 = 85.5$$
$$7x = 77.7$$
$$x = 11.1$$

Now we can answer the question asked. Grams of fat eaten during: breakfast is $x = 11.1$; lunch is $x + 2.9 = 11.1 + 2.9 = 14.0$; dinner is $2x - 2.8 = 2(11.1) - 2.8 = 19.4$; movie is $3(x + 2.9) - 1 = 3(11.1 + 2.9) - 1 = 3(14) - 1 = 41$. To check, note that $11.1 + 14.0 + 19.4 + 41 = 85.5$.

EXAMPLE 6   Mr. and Mrs. Byron live on a resort island community attached to the mainland by a toll bridge. The toll is $2 per car going to the island, but there is no toll coming from the island. Island residents can purchase a monthly pass for $20 which permits them to cross the toll bridge from the mainland for only 50 cents each time. How many times a month would one have to go to the island from the mainland for the cost of the monthly pass to equal the regular toll cost?

Solution:

Let x = number of times crossing the bridge from the mainland
then $2.00x$ = cost of x trips without pass
and $0.50x$ = cost of x trips with pass

Total cost with pass = total cost without pass
monthly fee + cost for trips = cost for trips
$$20 + 0.50x = 2.00x$$
$$20 = 1.50x$$
$$13.33 \approx x$$

The total cost would be the same in about 13 trips. Thus if the Byrons planned to make more than 13 trips monthly, the monthly pass would save them money.

EXAMPLE 7   The U.S. Bureau of Labor Statistics reported a 65% increase in the number of employed mothers from January 1970 to January 1990. If the number of employed mothers reported in 1990 was 16.8 million, how many mothers were employed in 1970?

Solution: Let x = number of employed mothers (in millions) in 1970

then $0.65x$ = increase in employed mothers from 1970 to 1990

$$\begin{pmatrix} \text{employed mothers} \\ \text{in 1970} \end{pmatrix} + \begin{pmatrix} \text{increase in} \\ \text{employed mothers} \end{pmatrix} = \begin{pmatrix} \text{employed mothers} \\ \text{in 1990} \end{pmatrix}$$

$$x \quad + \quad 0.65x \quad = 16.8$$

$$1.65x = 16.8$$

$$x \approx 10.18$$

Thus, there were about 10.2 million employed mothers in 1970.

EXAMPLE 8   On May 1, 1994, the sign outside Disneyworld's Magic Kingdom stated that the admission is $36 plus tax for adults. When paying for your ticket the actual cost you pay is $38. Determine the tax rate.

Solution: Let x = tax rate as a percent

then $0.01x$ = tax rate as a decimal

The total cost of the ticket is the ticket price plus the tax on the ticket.

$$\text{Ticket price} + \text{tax} = 38.00$$

$$36 + 36(0.01x) = 38$$

$$36 + 0.36x = 38$$

$$0.36x = 2$$

$$x = \frac{2}{0.36}$$

$$x \approx 5.555$$

The tax rate for the ticket is about 5.56%.

EXAMPLE 9   The vast majority of the population explosion has taken place in less than one-tenth of 1 percent of human history. When Columbus "discovered" the Americas 500 years ago, global population was small, numbering only 425 million. Much has changed. In just the past 40 years the world population has tripled to 5.4 billion.

   Use the information in the preceding paragraph to determine **(a)** the world's population 40 years ago. **(b)** How many years it would take the world population to double if the world population were to increase steadily at its present rate of 95 million people per year?

Solution: **(a)** We are told the world population has tripled to 5.4 billion (or 5,400,000,000) in the past 40 years.

Let x = world's population 40 years ago

then $3x$ = present world population

$$\text{Present world population} = 5.4 \text{ billion}$$

$$3x = 5.4$$

$$x = 1.8$$

Thus, 40 years ago the world population was 1.8 billion. The increase in the world population in the past 40 years is 5.4 billion $-$ 1.8 billion or 3.6 billion people.

(b) We wish to find the number of years for the population to double to 10.8 billion. The growth rate is given in *millions* per year, and the present and future population are given in *billions*. Therefore, when answering part **(b)** we will write 95 million as 95,000,000, 5.4 billion as 5,400,000,000, and 10.8 billion as 10,800,000,000.

Let x = number of years for the population to double

$95,000,000x$ = growth in population in x years

Present population + growth in population = future population

$$5,400,000,000 \quad + \quad 95,000,000x \quad = \quad 10,800,000,000$$
$$95,000,000x = 5,400,000,000$$
$$x = \frac{5,400,000,000}{95,000,000}$$
$$x \approx 56.84$$

Thus, the population, if it increased steadily at the rate of 95 million per year, would double in about 57 years. (Because the growth rate is increasing each year, the actual doubling time for the population is estimated to be about 47 years.)

In Group Activity Exercise 5, we refer to Example 9 above.

EXAMPLE 10 Beth is purchasing her first home and she is considering two banks for a $60,000 mortgage. Citicorp is charging 8.50% interest with no points for a 30-year loan. (A point is a one-time charge of 1% of the amount of the mortgage.) The monthly mortgage payments for the Citicorp mortgage would be $461.40. Citicorp is also charging a $200 application fee. BankAmerica Corporation is charging 8.00% interest with 2 points for a 30-year loan. The monthly mortgage payments for BankAmerica would be $440.04 and the cost of the points that Beth would need to pay at the time of closing is 0.02 ($60,000) = $1200. BankAmerica has waived its application fee.

(a) How long would it take for the total payments of the Citicorp mortgage to equal the total payments of the BankAmerica mortgage?

(b) If Beth plans to keep her house for 20 years, which mortgage would result in the lower total cost?

Solution:

Let x = number of months

then $461.40x$ = cost of mortgage payments for x months with the Citicorp mortgage

$440.04x$ = cost of mortgage payments for x months with the BankAmerica Corporation total cost with Citicorp = total cost with BankAmerica

mortgage payments + application fee = mortgage payments + points

461.40x   +   200   =   440.04x   +   1200

Now solve the equation.

$$461.40x + 200 = 440.04x + 1200$$
$$461.40x = 440.04x + 1000$$
$$21.36x = 1000$$
$$x \approx 46.82$$

The cost would be the same in about 46.82 months or about 3.9 years.

(b) The total cost would be the same at about 3.9 years. Prior to the 3.9 years the cost of the loan with BankAmerica would be more because of the initial $1200 charge for points. However, after the 3.9 years the cost with BankAmerica would be less because of the lower monthly payment. If you evaluate the total cost with Citicorp over 20 years (240 monthly payments), you will obtain $110,936. If you evaluate the total cost with BankAmerica over 20 years, you will obtain $106,809.60. Therefore, Beth will save $4126.40 over the 20-year period with BankAmerica.

Exercise Set 2.3

(a) Write an equation that can be used to solve the problem and (b) find the solution to the problem.

1. Kathy is 15 years older than Dawn. The sum of their ages is 41. Find Kathy's and Dawn's ages.   13, 28

2. The sum of two consecutive integers is 51. Find the two integers.   25, 26

3. The sum of two consecutive even integers is 78. Find the two integers.   38, 40

4. For two consecutive integers, the smaller plus 3 times the larger is 39. Find the integers.   9, 10

5. The larger of two numbers is 5 times the smaller. Find the two numbers if twice the smaller equals 3 less than $\frac{1}{2}$ the larger.   6, 30

6. The larger of two integers is $\frac{5}{2}$ the smaller. If twice the smaller is subtracted from twice the larger, the difference is 12. Find the two numbers.   4, 10

7. The sum of the angles of a triangle is 180°. Find the three angles of a triangle if one angle is 20° greater than the smallest angle and the third angle is twice the smallest angle.   40°, 60°, 80°

8. A number increased by 8% is 54. Find the number.   50

9. The beautiful Siberian tiger is on the endangered species list. It is estimated that only 360 are left and that their numbers are declining by about 42 a year. If this decline is not reversed, after how many years will the Siberian tiger become extinct?   8.57 yr

10. The sale price of a shirt reduced by 25% is $13.50. Find the regular price of the shirt. $18

11. Seals are migrating to Pier 39 in San Francisco. If there are presently 300 seals at the pier, and this number is increasing by an average of 32 per year, after how many years will the number of seals reach 580?   8.75 yr

SEE EXERCISE 11.

12. It cost the Macks $12.50 a week to wash and dry their clothes at the corner laundry. If a washer and drier cost a total of $940, how many weeks will it take for the laundry cost to equal the cost of a washer and drier? (Disregard energy cost.)   75.2 weeks

13. Ron Gigliotti buys a monthly bus pass, which entitles the owner to unlimited bus travel for $40 per month. Without the pass each bus ride costs $1.50. How many rides per month would Ron have to make so that it is less expensive to purchase the pass?   more than 26

14. The yearly cost for joining Sammy's Club is $35 per year. The Thompsons estimate that they save an average of 3% when compared to other discount stores that do not have a membership fee. How much would the Thompsons have to spend in a year at Sammy's to recover the yearly membership cost? $1166.67

15. The cost of renting a truck is $35 a day plus 20 cents per mile. How far can Kendra drive in 1 day if she has only $80?   225 mi

16. Mrs. Hill used her own car for a 1-day 350-mile business trip. When an employee uses their own car for business the company reimburses the employee a fixed dollar amount plus 18 cents a mile. Mrs. Hill cannot remember the fixed amount but knows she was reimbursed a total of $83.00. Find the fixed amount the company reimburses their employees. $20

17. Each week Bridget receives a flat weekly salary of $240 plus a 12% commission on the total dollar volume of all sales she makes. What must her dollar volume be in a week for her to earn $540? $2500

18. The Computer Store has reduced the price of a computer by 15%. What is the original price of the computer if the sale price is $1275? $1500

19. The amount of aluminum used for soda and beer cans in 1990 was 16,000 metric tons more than 17 times the amount used in 1970. If the difference between the amount used in 1990 and 1970 was 1,179,200 metric tons, how much aluminum was used in 1970 and in 1990?   1970: 72,700 metric tons; 1990: 1,251,900 metric tons

20. The Harts just purchased a new home and they are considering two different home appliance warrantee service plans. The Encore plan costs $480 and has a $30 deductible per visit. The BSF plan costs $360 and has a deductible of $60 per visit. How many visits would a repairperson need to make before the total cost of the two plans is equal?   4 visits

21. According to *Health* magazine the stress a bone can withstand in pounds per square inch is 6,000 pounds more than 3 times the amount that steel can withstand. If the difference between the amount of stress a bone and steel can withstand is 18,000 pounds per square inch, find the stress that both steel and a bone can withstand.   steel, 6000 lb/in²; bone, 24,000 lb/in²

22. There are 57 major sources of pollen in the United States. These pollen sources are categorized as grasses, weeds, and trees. If the number of weeds is 5 less than twice the number of grasses, and the number of trees is two more than twice the number of grasses, find the number of grasses, weeds, and trees that are major pollen sources.   grasses, 12; weeds, 19; trees, 26

SEE EXERCISE 22.

23. Caulking and weather-stripping a house are considered the easiest and most economical ways to save energy and money. An average house (12 windows and 2 doors) requires $25 worth of caulking and weather-stripping material and usually reduces heating and cooling expenses by 10% (or more). If the average house's heating and cooling bill is $74 per month, after how many months would the savings from the heating and cooling equal the cost of the caulking and weather-stripping material?   3.38 months

24. Kali is asked to write a study guide for a textbook. For her work the publishing company is giving her a choice of a one-time payment of $13,000 or $2000 plus 10% royalties per copy sold.

   (a) If her royalty rate results in a royalty of $3.20 per study guide sold, how many study guides would need to be sold for the total income received by Kali to be the same from either choice?   3437.5

   (b) If she expects 10,000 copies of the study guide to be sold, which plan should she choose?   royalty

25. Essex County has an 8% sales tax. What is the maximum price of a car if the total cost, including tax, is to be $12,000. $11,111.11

26. A security device on a car often reduces the auto insurance premium by 10%. Kathy finds it costs $260 to purchase and install a security device for her car. If the yearly insurance on the car is $460, after how many months would the security device pay for itself?   5.65 mo

27. The Midtown Tennis Club offers two payment plans for its members. Plan 1 is a monthly fee of $25 plus $10 per hour of court rental time. Plan 2 has no monthly fee, but court time costs $18.50 per hour. How many hours would Mrs. Levin have to play per month so that plan 1 becomes advantageous?   3 or more hours

SEE EXERCISE 2.7

28. Microcontrollers—tiny computers that do specific tasks in items such as cameras, appliances, or entertainment equipment—are in virtually every home. If the average number of microcontrollers increased by 110.14% from 1990 to 1994 and the average number of microcontrollers in the typical home in 1994 was 145, how many microcontrollers were in the typical home in 1990?   69

29. After Mrs. Semrau is seated in a restaurant, she realizes that she only has $10.25. If she must pay 7% sales tax and wishes to leave a 15% tip, what is the maximum price of a lunch she can order? $8.40

30. Jerry bought a bottle of perfume for a gift for his wife. The perfume cost $92 before tax. If the total price including tax was $98.90 find the tax rate.   7.5%

31. The Sanchezes are purchasing a new home and are considering a 30-year $70,000 mortgage with two different banks. Madison Savings is charging 9.0% with 0 points and First National is charging 8.5% with 2 points. First National is also charging a $200 application fee, whereas Madison is charging none. The monthly mortgage payments with Madison would be $563.50 and the monthly mortgage payments with First National would be $538.30.   **(a)** 63.49 months or 5.29 years

(a) After how many months would the total payments for the two banks be the same?

(b) If the Sanchezes plan to keep their house for 30 years, which mortgage would have the lower total cost? (See Example 10.)   First National

32. Jodi, a financial planner, is sponsoring dinner seminars. She must pay for the dinners of those attending out of her own pocket. She chooses a restaurant that seats 40 people and charges her $9.50 per person. If she earns 12% commission of sales made, how much in sales must she make from these 40 people
(a) to break even; $3166.67
(b) to make a profit of $500? $7333.33

33. The Waltons are considering refinancing their house at a lower interest rate. They have a 11.875% mortgage, are presently making monthly principal and interest payments of $510, and have 20 years left on their mortgage. Because interest rates have dropped, Countrywide Mortgage Corporation is offering them a rate of 9.5%, which would result in principal and interest payments of $420.50 for 20 years. However, to get this mortgage, their closing cost would be $2500.

(a) How many months after refinancing would the Waltons spend the same amount on their new mortgage plus closing cost as they would have on the original mortgage?   about 28 months or 2.33 yr

(b) If they plan to spend the next 20 years in the house, would they save money by refinancing?   yes

34. Mrs. Gonzalez is planning to build a sandbox for her children. She wants its length to be 3 feet more than its width. Find the length and width of the sandbox if only 22 feet of lumber are available to form the frame. Use $P = 2l + 2w$. $w = 4$ ft, $l = 7$ ft

35. The width of a rectangle is 1 meter more than $\frac{1}{2}$ its length. Find the length and width of the rectangle if its perimeter is 20 meters. $l = 6$ m, $w = 4$ m

36. Susan, a landscape architect, wishes to fence in two equal areas as illustrated in the figure on the left. If both areas are squares and the total length of fencing used is 91 meters, find the dimensions of each square.   13 m by 13 m

37. Taryn wishes to build a bookcase with four shelves as shown above. The height of the bookcase is to be 3 feet greater than the width. If only 30 feet of wood are available to build the bookcase, what will the dimensions of the bookcase be? $w = 4$ ft, $h = 7$ ft

45. exported, 12 million; incinerated, 26 million; recycled, 16 million; landfill, 186 million
48. animals, 250,000; plants, 350,000; non-beetle insects, 540,000; beetles, 360,000

38. Mike wishes to fence in three rectangular areas along a river bank as illustrated in the following figure. Each rectangle is to have the same dimensions, and the length of each rectangle is to be 1 meter greater than its width. Find the length and width of each rectangle if the total length of fencing used is 81 meters. *w* = 11 m, *l* = 12 m

39. During the first week of a going-out-of-business sale, Sam's General Store reduces all prices by 10%. The second week of the sale, Sam's reduces all items by 5 additional dollars. If Debby bought a calculator for $49 during the second week of the sale, find the original price of the calculator. $60

40. In 1996 the property tax of a community is increased by 6% over the 1995 tax. An additional surcharge of $200 is also added for a special project. If the Petersons' 1996 tax totals $2108, find their property tax for the year 1995. $1800

41. At a liquidation sale, Quality Photo Company reduces the price of all cameras by $\frac{1}{4}$, and then takes an additional $10 off. If Mark purchases a Minolta camera for $290 during this sale, find the original price of the camera. $400

42. J. P. Richardson sells each of his paintings for $50. The gallery where he displays his work charges him $810 a month plus a 10% commission on sales. How many paintings must J. P. sell in a month to break even? 18

43. A farm is divided into three regions. The area of one region is twice as large as the area of the smallest region, and the area of the third region is 4 acres less than three times the area of the smallest region. If the total acreage of the farm is 512 acres, find the area of each of the three regions. 86 acres, 172 acres, 254 acres

44. A 17 foot by 10 foot rectangular garden is divided into four parts. Two parts have the same area, the third part is twice as large as each of the first two parts, and the last part has an area 20 square feet greater than each of the first two parts. Find the area of each of the four parts of the garden. 30 sq ft, 30 sq ft, 60 sq ft, 50 sq ft

45. About 240 million tires were discarded in 1994. Discarded tires are either deposited in landfills (or stockpiled), recycled, incinerated and used as fuel, or exported. In 1994 the number incinerated was $\frac{13}{6}$ the amount exported, the amount recycled was $\frac{4}{3}$ the amount exported, and the amount placed in landfills or stockpiles was $\frac{31}{2}$ the amount exported. Determine the number of tires exported, incinerated, recycled, and placed in a landfill (or stockpiled).

46. The cost of purchasing incandescent bulbs for use over a 9750-hour period is $9.75. The energy cost for incandescent bulbs over this period is $73. The cost of one equivalent fluorescent bulb that lasts about 9750 hours is $20. By using a fluorescent bulb instead of incandescent bulbs for 9750 hours, the total savings of purchase price plus energy cost is $46.75. How much is the energy cost for using the fluorescent bulb over this period? $16

47. Ron receives a small clothing allowance from Social Security. He is also provided with a sales tax exclusion number so that when this money is spent on clothing, the 7% sales tax is waived. On all other purchases he must pay the full 7% sales tax. Last week Ron went shopping at Sears, where he used his total clothing allowance to purchase clothing, and he also purchased some other goods on which he paid tax. If he spent a total of $375 before tax and paid a total sales tax of $17.50, find his clothing allowance from Social Security. clothing, $125

48. In the world approximately 1,500,000 species have been categorized as either plants, animals, or insects. Insects are often subdivided into beetles and insects that are not beetles. There are about 100,000 more plants than animals. There are 290,000 more non-beetle insects than animals. The number of beetles is 140,000 less than twice the number of animals. Find the number of animals, plants, non-beetle insects, and beetles.

49. The five members of the Bulina family are going out to dinner with the three members of the Williams family. Before dinner, they decide that the Bulinas will pay $\frac{5}{8}$ of the bill (before tip) and the Williams will pay $\frac{3}{8}$ plus the entire 15% tip. If the total bill including the 15% tip comes to $184.60, how much will be paid by each family? Bulinas, $100.33, Williams, $84.27

50. The *FDA Consumer* indicated that the amount of caffeine in an average 5-ounce cup of coffee depends on how it is brewed. Percolator-brewed coffee contains 23.08% more caffeine than instant coffee. Drip-brewed coffee contains 76.92% more caffeine than instant coffee. If in one cup of instant coffee, one cup of percolator-brewed coffee, and one cup of drip-brewed coffee there is a total of 260 milligrams of caffeine, find the number of milligrams of caffeine in instant, percolator-brewed, and drip-brewed coffee. instant, 65 mg; percolator, 80 mg; drip, 115 mg

52. (a) $\dfrac{87 + 93 + 97 + 96 + x}{5} = 90$ **(c)** 77

51. At the time of this writing, the total number of U.S. astronauts past and present is 214. A large number of these came from six institutions. Institutions with the most astronaut graduates are: the U.S. Naval Academy, the U.S. Air Force Academy, Massachusetts Institute of Technology (MIT), U.S. Naval Postgraduate School, Purdue University, and Stanford University. The number from Purdue is 4 more than from Stanford. The number from the Postgraduate School is one more than the number from Purdue. The number from MIT is one more than the number from the Postgraduate School. The number from the Air Force Academy is 2 less than twice the number from Stanford, and the number from the Naval Academy is twice the number from Purdue. If the total number of astronauts graduating from these six sources is 133, find the number of astronauts graduating from each of these six institutions. (Pilot-astronauts must have 1000 hours of jet flying experience, meet certain physical requirements, and hold a B.S. in engineering, biology, physics, or mathematics.)

SEE EXERCISE 51.

51. Stanford, 14; Purdue, 18; Naval Post. Grad., 19; MIT, 20; Air Force, 26; Naval Academy, 36

52. To find the average of a set of test grades, we divide the sum of the test grades by the number of test grades. On her first four algebra tests, Paula's grades were 87, 93, 97, and 96.

   (a) Write an equation that can be used to determine the grade she needs to obtain on her fifth test to have a 90 average.

   (b) Explain how you determined your equation.

   (c) Solve the equation and determine the score.

53. Philip's grades on five physics hourly exams are 70, 83, 97, 84, and 74.

   (a) If the final exam will count twice as much as each hourly exam, what grade does Philip need on the final exam to have an 80 average?   76

   (b) If the highest possible grade on the final exam is 100 points, is it possible for Philip to obtain a 90 average? Explain.   No, needs score of 111

54. (a) Make up your own realistic word problem involving money. Represent this word problem as an equation.

   (b) Solve the equation and answer the word problem.

55. (a) Make up your own realistic word problem involving percents. Represent this word problem as an equation.

   (b) Solve the equation and answer the word problem.

CUMULATIVE REVIEW EXERCISES

[1.4] **56.** Evaluate $\dfrac{\{2[(5-3)-4]\}^2 \div (-8)}{-|8-5|-4^2}.$ $\dfrac{2}{19}$

[2.1] *Solve.*

57. $\frac{1}{2}x = 3(x-2)$ $\dfrac{12}{5}$

58. $\frac{1}{5}x + \frac{2}{3} = \frac{5}{4}x$ $\dfrac{40}{63}$

[2.2] **59.** Solve the equation $4x - 6y = 9$ for y.

   $y = \dfrac{4x-9}{6}$ or $y = \dfrac{2}{3}x - \dfrac{3}{2}$

1. white, 10; brown, 15; yellow, 20; red, 27; blue, 70; purple, 75; green, 100

Group Activity/ Challenge Problems

1. On a typical aircraft carrier such as the *USS America*, there are about 4700 people. However, when aircraft are taking off and landing only about 317 people are on deck and each is wearing a color-coded shirt or blouse. The colors used are: white (landing signal officers and safety observers); brown (aircraft crew); yellow (catapult officers and traffic directors); red (weapons handlers); blue (tractor devices and elevator operators); purple (fuelers); and green (aircraft maintenance and catapult crews).

SEE EXERCISE 1.

The number of brown shirts on deck is 5 more than the number of white shirts. The number of yellow shirts is twice the number of white shirts. The number of red shirts is 7 more than the number of yellow shirts. The number of blue shirts is 7 times the number of white shirts. The number of purple shirts is 5 more than the number of blue shirts. The number of green shirts is 5 times the number of yellow shirts. Can you determine the number of people wearing each of these colored shirts?

2. On Monday Linda purchased shares in a money market fund. On Tuesday the value of the shares went up 5%, and on Wednesday the value fell 5%. How much did Linda pay for the shares on Monday if she sold them on Thursday for $59.85? $60

3. The Elmers Truck Rental Agency charges $28 per day plus 15 cents a mile. If Denise rented a small truck for 3 days and the total bill was $121.68, including a 4% sales tax, how many miles did she drive?   220 mi

4. Pick any number, multiply it by 2, add 33, subtract 13, divide by 2, and subtract the number you started with. You should end with the number 10. Show that this procedure will result in the value 10 for any number, n, selected. $n, 2n, 2n + 33, 2n + 20, n + 10, 10$

 5. Read Example 9.

  (a) Explain why the equation $5.4 + 95x = 10.8$ would not result in the same solution as obtained in Example 9(b).

  (b) Insert the coefficient in the shaded area that would result in the same solution obtained in Example 9(b). Explain how you determined your answer.   0.095

$$5.4 + x = 10.8$$

2.4 Additional Application Problems

> 1 Solve motion application problems containing only one rate.
> 2 Solve motion problems containing two rates.
> 3 Solve mixture problems.

Tape 2

In this section we discuss two additional types of application problems, motion and mixture problems. We have placed motion and mixture problems in the same section because motion problems involving two rates and mixture problems are solved using similar procedures.

Motion Problems   1 A formula with many useful applications is

> amount = rate · time

The "amount" in this formula can be a measure of many different quantities, including distance or length, area, volume, or number of items produced.

When applying this formula, we must make sure that the units are consistent. For example, if the rate is given in feet per *second*, the time must be given in *seconds*. If the rate is given in gallons per *hour*, the time must be given in *hours*. If the rate is given as items produced per *minute*, the time must be given in *minutes*. Problems that can be solved using this formula are called *motion problems* because they involve movement, at a constant rate, for a certain period of time.

A nurse giving a patient an intravenous injection may use this formula to determine the drip rate of the fluid being injected. A company drilling for oil or water may use this formula to determine the amount of time needed to reach its goal. This formula may also be used to determine how fast a train or plane must travel to be at a certain place at a certain time.

EXAMPLE 1   An arborist is trying to save a white birch tree from insects called borers. The treatment is to inject 200 cubic centimeters of a lindane solution into the base of the tree over an 8-hour period. At what rate will the solution be injected into the tree?

Solution: We are given the volume and the time and are asked to find the rate.

$$\text{volume} = \text{rate} \cdot \text{time}$$
$$200 = r \cdot 8$$
$$200 = 8r$$
$$r = 25$$

Therefore, the rate of flow of the lindane solution is 25 cc/hr.

When the motion formula is used to calculate distance, the word *amount* is replaced with the word *distance* and the formula is called the *distance formula.*

> ### Distance Formula
> $$\text{Distance} = \text{rate} \cdot \text{time}$$

EXAMPLE 2   A conveyor belt at the Wonder Bread Baking Company is transporting a continuous length of dough at a rate of 1.5 feet per second. A cutting blade is activated at regular intervals to cut the dough into 10.8-inch lengths that will eventually become loaves of bread. At what time intervals should the cutting blade be activated?

Solution: Since the rate is given in feet per second and the length of the dough is given in inches, one of these quantities must be changed. Since 1 foot equals 12 inches,

$$1.5 \text{ feet per second} = (1.5)(12) = 18 \text{ inches per second}$$
$$\text{distance} = \text{rate} \cdot \text{time}$$
$$10.8 = 18 \cdot t$$
$$t = \frac{10.8}{18} = 0.6$$

The blade should cut at 0.6-second intervals.

Motion Problems with Two Rates

[2] Sometimes when a motion problem has two different rates it is helpful to put the information in a table to help analyze the problem.

EXAMPLE 3   Two trains leave San Jose at the same time traveling in the same direction on parallel tracks. One train travels at 80 miles per hour; the other travels at 60 miles per hour. In how many hours will they be 144 miles apart? (See Fig. 2.1.)

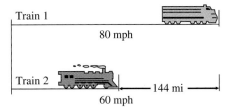

FIGURE 2.1

Solution: To solve this problem, we will use the distance formula, $d = rt$.

Let t = time when the trains are 144 miles apart

Train	Rate	Time	Distance
1	80	t	$80t$
2	60	t	$60t$

The difference in their distances is 144 miles. Thus,

$$\text{distance train 1} - \text{distance train 2} = 144$$
$$80t - 60t = 144$$
$$20t = 144$$
$$t = 7.2$$

In 7.2 hours the trains will be 144 miles apart.

EXAMPLE 4   Mrs. Sanders and her daughter Teresa jog regularly. Mrs. Sanders jogs at 5 miles per hour, Teresa at 4 miles per hour. Teresa begins jogging at noon. Mrs. Sanders begins at 12:30 P.M. and travels the same straight path.

(a) Determine the time the mother and daughter will meet.

(b) How far from the starting point will they be when they meet?

Solution:

(a) Since Mrs. Sanders is the faster jogger, she will cover the same distance in a shorter time period. When they meet they will both have traveled the same distance.

Let t = time Teresa is jogging

then $t - \dfrac{1}{2}$ = time Mrs. Sanders is jogging

Jogger	Rate	Time	Distance
Teresa	4	t	$4t$
Mrs. Sanders	5	$t - \frac{1}{2}$	$5(t - \frac{1}{2})$

When they meet they will both have covered the same distance from the starting point.

$$\text{Teresa's distance} = \text{Mrs. Sanders's distance}$$

$$4t = 5\left(t - \frac{1}{2}\right)$$

$$4t = 5t - \frac{5}{2}$$

$$-t = \frac{-5}{2}$$

$$t = \frac{5}{2}$$

They will meet $2\frac{1}{2}$ hours after Teresa begins jogging, or at 2:30 P.M.

(b) The distance can be found using either Mrs. Sanders's or Teresa's rate. We will use Teresa's.

$$d = rt$$

$$= 4\left(\frac{5}{2}\right) = \frac{20}{2} = 10$$

They will meet 10 miles from the starting point.

In Example 4, would the answer have changed if we let t represent the time Mrs. Sanders is jogging rather than the time Teresa is jogging? Try it and see.

EXAMPLE 5   A machine grinds coffee beans and then packs and seals the ground coffee in containers. When the machine runs at the slower rate it packs 400 containers per hour. At the faster rate it packs 600 containers per hour. At 9 A.M. on Tuesday the machine is turned on and runs at the slower rate. Later in the day the rate of the machine is changed to the faster rate. At 5 P.M. the machine is turned off. If the machine produced the same number of containers during the time it ran at the slower rate as it did during the time it ran at the faster rate, find **(a)** how long the machine was on at the slower rate, and **(b)** the total number of containers of coffee produced during the day.

Solution:

(a) We are told that the amounts produced at the two different rates are the same. We use this fact to help set up our equation. The machine was in operation from 9 A.M. to 5 P.M., a total of 8 hours.

Let t = time machine is on at the slower rate

then $8 - t$ = time machine is on at the faster rate.

	Rate	Time	Amount
Slower rate	400	t	$400t$
Faster rate	600	$8 - t$	$600(8 - t)$

Since the amount produced at each rate is the same,

$$\text{amount at slower rate} = \text{amount at faster rate}$$
$$400t = 600(8 - t)$$
$$400t = 4800 - 600t$$
$$1000t = 4800$$
$$t = 4.8$$

Thus, the machine was on for 4.8 hours at 400 containers per hour and $8 - t$ or $8 - 4.8 = 3.2$ hours at 600 containers per hour.

(b) The total number of containers of coffee produced can be found by adding the two amounts.

$$\begin{aligned} \text{total amount produced} &= 400t + 600(8 - t) \\ &= 400(4.8) + 600(3.2) \\ &= 1920 + 1920 \\ &= 3840 \end{aligned}$$

The answer to part **(b)**, 3840 containers, could have also been found by computing the number of containers of coffee produced at the initial rate of 400 units per hour and then doubling the value, since the same number of containers was produced at each rate.

EXAMPLE 6   A line of people 620 feet long is lined up outside a ticket window to purchase tickets to an Elton John/Billy Joel concert. During just the first $\frac{1}{2}$ hour, the line of ticket buyers moved past the ticket window at a steady speed. After the first $\frac{1}{2}$ hour a second window was opened for 3 hours. When both windows were opened, the speed of the line increased by 90 feet per hour. After this $3\frac{1}{2}$-hour period all the people in the 620-foot line received tickets. Find the speed that the line was moving, in feet per hour, during the first $\frac{1}{2}$ hour of ticket sales.

Solution: Let r = speed (or rate) during the first $\frac{1}{2}$ hour

then $r + 90$ = speed during the following 3 hours

	Rate	Time	Distance
First $\frac{1}{2}$ hour	r	$\frac{1}{2}$	$\frac{1}{2}r$
Following 3 hours	$r + 90$	3	$3(r + 90)$

Since the total distance is 620 feet,

$$\text{distance first } \tfrac{1}{2} \text{ hour} + \text{distance following 3 hours} = 620$$

$$\tfrac{1}{2}r + 3(r + 90) = 620$$

$$2\left[\tfrac{1}{2}r + 3(r + 90)\right] = 2(620)$$

$$2\left(\tfrac{1}{2}r\right) + (2)(3)(r + 90) = 2(620)$$

$$r + 6(r + 90) = 1240$$

$$r + 6r + 540 = 1240$$

$$7r + 540 = 1240$$

$$7r = 700$$

$$r = 100$$

The line moved at 100 feet per hour during the first half hour and $r + 90$ or $100 + 90 = 190$ feet per hour during the following 3 hours.

Mixture Problems

3   Any problem where two or more quantities are combined to produce a different quantity or where a single quantity is separated into two or more different quantities may be considered a **mixture problem.** As we did when working with motion problems containing two different rates, we will use tables to help organize the information.

Our first example of a mixture problem uses the **simple interest formula:**

$$\text{interest} = \text{principal} \cdot \text{rate} \cdot \text{time}$$

$$\text{or} \quad I = prt$$

This formula is used to calculate the interest earned in a savings account that gives simple interest or the interest you must pay on a simple interest loan. For example, if $2000 is placed in a savings account giving 6% simple interest for a period of 1 year, the interest earned is found as follows:

$$I = prt$$

$$= 2000(0.06)(1) = 120$$

Thus, after 1 year $120 interest is earned.

EXAMPLE 7   Mr. and Mrs. Stone invested $8000 for 1 year, part at 7% and part at $5\tfrac{1}{4}$% simple interest. If they earned $458.50 total interest, how much was invested at each rate?

Solution: Let $x = $ amount invested at 7% simple interest.

If x is the amount invested at 7% simple interest, then the amount remaining, $8000 - x$, is invested at $5\tfrac{1}{4}$% simple interest. We will construct a table to help us visualize the solution. To find the interest earned in each account, we use the simple interest formula: $I = prt$.

Account	Principal	Rate	Time	Interest
7%	x	0.07	1	$0.07x$
$5\frac{1}{4}\%$	$8000 - x$	0.0525	1	$0.0525(8000 - x)$

Since the total interest from both accounts is \$458.50, we write

interest from 7% account + interest from $5\frac{1}{4}\%$ account = total interest

$$0.07x + 0.0525(8000 - x) = 458.50$$

Now solve the equation.

$$0.07x + 420 - 0.0525x = 458.50$$
$$0.0175x + 420 = 458.50$$
$$0.0175x = 38.50$$
$$x = \frac{38.50}{0.0175} = 2200$$

Therefore, \$2200 was invested at 7% and $8000 - x$ or $8000 - 2200 = \$5800$ was invested at $5\frac{1}{4}\%$ simple interest.

EXAMPLE 8   Marni's hot dog stand in Chicago sells hot dogs for \$2.00 each and potato knishes for \$2.25 each. If the sales for the day total \$585.50 and 278 items were sold, how many of each item were sold?

Solution: Let $x =$ number of hot dogs sold

then $278 - x =$ number of knishes sold

Item	Cost of item	Number of items	Total sales
Hot dogs	2.00	x	$2.00x$
Knishes	2.25	$278 - x$	$2.25(278 - x)$

total sales of hot dogs + total sales of knishes = total sales

$$2.00x + 2.25(278 - x) = 585.50$$
$$2.00x + 625.50 - 2.25x = 585.50$$
$$-0.25x + 625.50 = 585.50$$
$$-0.25x = -40$$
$$x = \frac{-40}{-0.25} = 160$$

Therefore, 160 hot dogs and $278 - 160$ or 118 knishes were sold.

In Example 8 you could have multiplied both sides of the equation by 100 to eliminate the decimal numbers, and then solved the equation.

EXAMPLE 9   Ali, a pharmacist, has both 6% and 15% phenobarbital solutions. He receives a prescription for 0.5 liter of an 8% phenobarbital solution. How much of each solution must he mix to fill the prescription?

Solution: Let x = number of liters of 6% solution

then $0.5 - x$ = number of liters of 15% solution

The amount of phenobarbital in a solution is found by multiplying the percent strength of phenobarbital in the solution by the volume of the solution. We will draw a sketch of the problem (see Fig. 2.2) and then construct a table.

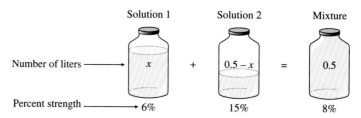

FIGURE 2.2

Solution	Strength of solution	Number of liters	Amount of phenobarbital
1	0.06	x	$0.06x$
2	0.15	$0.5 - x$	$0.15(0.5 - x)$
Mixture	0.08	0.5	$0.08(0.5)$

$$\begin{pmatrix} \text{Amount of} \\ \text{phenobarbital} \\ \text{in 6\% solution} \end{pmatrix} + \begin{pmatrix} \text{amount of} \\ \text{phenobarbital} \\ \text{in 15\% solution} \end{pmatrix} = \begin{pmatrix} \text{amount of phenobarbital} \\ \text{in mixture} \end{pmatrix}$$

$$0.06x + 0.15(0.5 - x) = 0.08(0.5)$$
$$0.06x + 0.075 - 0.15x = 0.04$$
$$0.075 - 0.09x = 0.04$$
$$-0.09x = -0.035$$
$$x = \frac{-0.035}{-0.09} = 0.39 \begin{pmatrix} \text{to nearest} \\ \text{hundredth} \end{pmatrix}$$

Ali must mix 0.39 liter of the 6% solution and $0.5 - x$ or $0.5 - 0.39 = 0.11$ liter of the 15% solution to make 0.5 liter of an 8% solution.

Exercise Set 2.4

1. A machine at a Coca-Cola bottling plant fills 82.5 bottles in 5 minutes. Find the machine's speed of filling bottles of Coke.   16.5 bottles/min

2. A patient is to receive 1200 cubic centimeters of intravenous fluid over a period of 3 hours. What should the intravenous flow rate be?   400 cc/hr

3. (a) 1.91 mph; **(b)** 1.78 mph

3. The $10\frac{1}{2}$-mile mule trip down to Phantom Ranch at the bottom of the Grand Canyon takes $5\frac{1}{2}$ hours. The return trip covers a distance of 8 miles and takes $4\frac{1}{2}$ hours. Find **(a)** the average speed going down the canyon; **(b)** the average speed coming up the canyon.

4. A laser can cut through a steel door at the rate of $\frac{1}{5}$ centimeter per minute. How thick is the door if it requires 32 minutes to cut through the steel door?   6.4 cm

5. At the dolomite quarry in Corydon, Indiana, the conveyor belt transports crushed dolomite ore to dump trucks at an average of 1800 pounds per minute. How long will it take to fill a dump truck that can hold 29,500 pounds of crushed ore?   16.39 min

6. A piece of coral is growing in the Pacific Ocean at a rate of 3 pounds per year. How long will it take for the coral to gain 24 ounces?   0.5 yr or 6 months

7. The official marathon distance, instituted at London's 1908 Olympics, is 26 miles 385 yards. The average world-class marathoner maintains an average speed of about 12.1 miles per hour over this distance. How long will it take the average marathoner to complete the race? There are 5280 feet in a mile. (In 1926, the first time women ran in a world-class marathon, the difference between the men's and women's record time was 1 hour and 20 minutes. Today the difference is only 15 minutes.)   2.17 hr

8. Under certain conditions, sound travels at 1080 feet per second. If a loud explosion occurs $2\frac{1}{2}$ miles away from you, how long will it take before you hear the noise?   12.2 sec

9. Each year Americans use 18 billion disposable diapers. If you were to lay the disposable diapers end to end, in 1 day the distance covered by the diapers would be 9167.1 miles.

(a) Determine the number of days it would take for disposable diapers placed end to end to reach from the earth to the moon, a distance of approximately 239,000 miles.   26.07 days

(b) If the diapers were laid end to end for a year, how far would they reach?   3,345,991.5 mi

10. (a) A videocassette tape is 246 meters in length. Some videocassette recorders have three tape speeds: SP for standard play, LP for long play, and SLP for super long play (some recorders refer to SLP as EP for extended play). If the tape runs for 2 hours at SP or 4 hours at LP or 6 hours at SLP, find the rate of speed of the tape at all three speeds.   123 m/hr, 61.5 m/hr, 41 m/hr

(b) If a new 8-hour tape is to be developed for the SLP rate, find the length of the new tape.   328 m

11. The moving walkway, called the "travelator," at the United Airlines Terminal at Chicago's O'Hare International Airport is like a moving conveyor belt that people walk on to increase their speed relative to the ground. If the length of one piece of the moving walkway is 275 feet, and the rate of the moving walkway is 120 feet per minute:

(a) How long will it take a briefcase placed on one end of the moving walkway to reach the other end?   2.29 min

(b) How long will it take a person walking at 150 feet per minute alongside the moving walkway to walk from one end of the moving walkway to the other end?   1.83 min

(c) How long will it take a person walking at 150 feet per minute on the moving walkway to walk from one end to the other end?   1.02min

(d) If a person walks at 150 feet per minute, how much time will he or she save by walking on the moving walkway instead of walking alongside the walkway?   0.81 min

17. 1333.33 days or 3.65 yr

18. 52 mph, 62 mph

19. freight, 50 mph; passenger, 70 mph

22. smaller hose, 200 gal/hr; larger hose, 400 gal/hr

12. Under certain conditions, sound travels at approximately 1080 feet per second. If Marsala yells down a canyon and hears her echo 1.7 seconds later, approximately how deep is the canyon?   918 feet

Write an equation that can be used to solve the motion problem. Solve the equation and answer the question asked.

13. Two planes leave Cleveland at the same time. One plane flies east at 550 miles per hour. The other flies west at 650 miles per hour. In how many hours will they be 3000 miles apart?   2.5 hr

14. Two hot-air balloons leave Albuquerque, New Mexico, going in the same direction. One balloon travels at 16 miles per hour, and the second balloon travels at 14 miles per hour. In how many hours will they be 8 miles apart?   4 hr

15. Two families on vacation leave San Diego and travel in opposite directions. After 3 hours the two cars are 330 miles apart. If one car travels at 60 miles per hour, find the speed of the other car.   50 mph

16. Two runners enter the same marathon. If the faster runner's speed is 9.2 miles per hour and the two runners are 4.8 miles apart after 3 hours, find the slower runner's speed.   7.6 mph

17. A rain forest in Brazil is 200 miles long. Bensen's logging company is working on one end, cutting down trees at a rate of 0.1 mile per day. Jenkin's logging company is working on the other end, cutting down trees at a rate of 0.05 mile per day. If both logging companies continue at their present pace, how long will it take for the two companies to meet?

18. Two cars leave from the same point at the same time, one traveling east and the other traveling west. If the car traveling west is moving 10 miles per hour faster than the car traveling east, and the two cars are 342 miles apart after 3 hours, find the speed of each car.

19. A passenger train leaves the Baltimore Depot 1.2 hours after a freight train leaves. The passenger train is traveling 20 miles per hour faster than the freight train. If the passenger train overtakes the freight train in 3 hours, find the speed of each train.

20. Two molding machines are turned on at 9 A.M. The older molding machine can produce 40 plastic buckets in 1 hour. The newer machine can produce 50 buckets in 1 hour. How long will it take the two machines to produce a total of 540 buckets?   6 hr

21. A jogger and a cyclist head for the same destination at 8 A.M. from the same point. The average speed of the cyclist is four times the speed of the jogger. In 2 hours the cyclist is 18 miles ahead of the jogger.

(a) At what rate did the cyclist ride?   12 mph

(b) How far did the cyclist ride?   24 mi

22. Two hoses are being used to fill a pool. The hose with the larger diameter supplies twice as much water as the hose with the smaller diameter. The smaller hose is on for 1.5 hours before the larger hose is turned on. If the total volume of water in the pool is 2100 gallons 3 hours after the larger hose is turned on, find the rate of flow from each hose.

23. Dr. McDonald hikes down to the bottom of Rim Rock Canyon, camps overnight, and returns the next day. Her hiking speed down averages 2.6 miles per hour and her return trip averages 1.2 miles per hour. If she spent a total of 16 hours hiking, find

(a) how long it took her to reach the bottom of the canyon;   5.05 hr

(b) the total distance traveled.   26.26 mi

SEE EXERCISE 23.

24. A Dodge has traveled 50 miles when a Chevy begins traveling in the same direction. If the Chevy travels at 65 miles per hour and the Dodge travels at 55 miles per hour, how long will it take for the Chevy to catch up to the Dodge?   5 hr

25. Two machines are packing spaghetti into boxes. The smaller machine can package 400 boxes per hour and the larger machine can package 600 boxes per hour. If the larger machine is on for 2 hours before the smaller machine is turned on, how long after the smaller machine is turned on will a total of 15,000 boxes of spaghetti be boxed?   13.8 hr

26. Mariska began driving to an out-of-town college at a speed of 45 miles per hour. One-half hour after she left, her parents realized that she had forgotten to take her wallet. They then tried to catch up to her in their car. If the parents traveled at 65 miles per hour, how long did it take them to catch Mariska?   1.125 hr

28. $4200 at 7%, $5800 at $6\frac{1}{4}$%   **29. (a)** Mobil, 67 shares; Limited, 268   **37.** 8L of 20%, 4L of 50%

Write an equation that can be used to solve the mixture problem. Solve the equation and answer the question asked.

27. Mr. Templeton invests $11,000 for 1 year, part at 9% and part at 10% simple interest. How much money was invested at each interest rate if the total interest earned from both investments is $1050? Use interest = principal · rate · time. $5000 at 9%, $6000 at 10%

28. Ms. Feldman invested $10,000 for 1 year, part at 7% and part at $6\frac{1}{4}$%. If she earned a total interest of $656.50, how much was invested at each rate?

29. Mobil Oil stock is selling at $80.75 per share and the Limited stock is selling at $17 per share. Bob Davis has a maximum of $10,000 to invest. He wishes to purchase four times as many shares of the Limited as of Mobil Oil.

  (a) How many shares of each will he purchase? Stocks can be purchased only in whole shares.

  (b) How much money will be left? $33.75

30. The price of admission at an ice hockey game is $12.50 for adults and $8.50 for children. A total of 650 tickets were sold. How many tickets were sold to children and how many to adults if a total of $7085 was collected?   390 adults, 260 children

31. Victor has a total of 33 dimes and quarters. The total value of the coins is $4.50. How many dimes and how many quarters does he have?   25 dimes, 8 quarters

32. Jim Kelly sells almonds for $6.00 per pound, and walnuts for $5.20 per pound. How many pounds of each should he mix to produce a 30-pound mixture that costs $5.50 per pound?   18.75 lb walnuts; 11.25 lb almonds

33. How many pounds of coffee costing $6.20 per pound must Larry mix with 18 pounds of coffee costing $5.80 per pound to produce a mixture that costs $6.10 per pound?   54 lb

34. How many ounces of water should a chemist mix with 16 ounces of a 25% sulfuric acid solution to reduce it to a 10% solution?   24 oz water

35. How many ounces of pure vinegar should a cook add to 40 ounces of a 10% vinegar solution to make it a 25% vinegar solution?   8 oz

36. Fifty pounds of a cement-and-sand mixture is 40% sand. How many pounds of sand must be added for the resulting mixture to be 60% sand?   25 lb sand

37. How many liters of a 20% alcohol solution and how many liters of a 50% alcohol solution must be mixed to get 12 liters of a 30% alcohol solution?

38. Two acid solutions are available to a chemist. One is a 20% sulfuric acid solution, but the label that indicates the strength of the other sulfuric acid solution is missing. Two hundred milliliters of the 20% acid solution and 100 milliliters of the solution with the unknown strength are mixed together. Upon analysis, the mixture was found to have a 25% sulfuric acid concentration. Find the strength of the solution with the missing label.   35%

39. The Bryerman Nursery sells in bulk two types of grass seeds. The lower-quality seeds have a germination rate of 76%, but the germination rate of the higher-quality seeds is not known. Twelve pounds of the higher-quality seeds are mixed with 16 pounds of the lower-quality seed. If a later analysis of the mixture finds that the mixture's germination rate was 82%, what is the germination rate of the higher-quality seed?   90%

40. The Agway nursery is selling in bulk two types of sunflower seed for bird feeding. The striped sunflower seeds cost $1.20 per pound, while the all black sunflower seeds cost $1.60 per pound. How many pounds of each should Mr. Wicker mix to get a 20-pound mixture that sells for $30?   5 lb

41. Some states allow a husband and wife to file individual state tax returns (on a single form) even though they file a joint federal return. It is usually to the taxpayer's advantage to do this when both the husband and wife work. The smallest amount of tax owed (or the largest refund) will occur when the husband's and wife's taxable incomes are the same.

Mr. Hall's 1996 taxable income was $28,200 and Mrs. Hall's income for that year was $32,450. The Halls' total tax deductions for the year were $6400. This deduction can be divided between Mr. and Mrs. Hall in any way they wish. How should the $6400 be divided between them to result in each person's having the same taxable income and therefore the greatest refund or least tax? $1075 for Mr., $5325 for Mrs

42. A certain type of engine uses a fuel mixture of 15 parts of gasoline to 1 part of oil. How much pure gasoline must be mixed with a gasoline–oil mixture, which is 75% gasoline, to make 8 quarts of the mixture to run the engine?   6 qt

Write an equation that can be used to solve the motion or mixture problem. Solve the equation and answer the question asked.

43. Two hikers visiting Yellowstone National Park start at the same point and hike in opposite directions around a thermal spring. The distance around the thermal spring is 8.2 miles. One hiker walks 0.4 mile per hour faster than the other. How fast does each hiker walk if they meet in 2 hours?   1.85 mph, 2.25 mph

SEE EXERCISE 43.

44. The Rappaports and Calters leave their homes at 8 A.M. planning to meet for a picnic at a point between them. If the Rappaports travel at 60 miles per hour and the Calters travel at 50 miles per hour, and they live 330 miles apart, at what time will they meet?   11 A.M.

45. The Bernhams decide to invest $6000 in two stocks, Apple Computer and Walmart. They wish to purchase three times as many shares of Apple as of Walmart.

   (a) If Apple shares are selling at $34 and Walmart sells at $23, how many shares of each will be purchased?   48 shares Walmart, 144 shares Apple

   (b) How much money will be left?   no money left

46. Barbara Anders invested $8000 for 1 year, part at 6% and part at 10% simple interest. How much was invested in each account if the same amount of interest was received from each account? $5000 at 6%, $3000 at 10%

47. One sump pump can remove 10 gallons of water a minute. A larger sump pump can remove 20 gallons of water a minute. How long will it take the two pumps working together to empty a 15,000-gallon swimming pool?   500 min or $8\frac{1}{3}$ hr

48. A jetliner flew from Chicago to Los Angeles at an average speed of 500 miles per hour. Then it continued on over the Pacific Ocean to Hawaii at an average speed of 550 miles per hour. If the entire trip covered $5200 miles and the part over the ocean took twice as long as the part over land, what was the time of the entire trip?   9.75 hr

49. Richard Semmler invested $4000 for 1 year in two savings accounts giving simple interest. He invested $2500 at 9% interest and the rest at a different interest rate. If the total interest received is $315, what was the rate on the second account?   6%

50. How many quarts of pure antifreeze should Mr. Alberts add to 10 quarts of a 20% antifreeze solution to make a 50% antifreeze solution?   6 qt

51. A small plane flew a round trip from Orono, Maine, to Tallahassee, Florida. The average flying speed down (with the tailwind) was 300 miles per hour. The average flying speed on the return trip (with a headwind) was 220 miles per hour. If the total flying time was 11.2 hours:

   (a) How long did it take the plane to fly from Orono to Tallahassee?   4.74 hr

   (b) Find the distance between the two airports. 1422 mi

52. A movie ticket for an adult costs $8.00 and a child's ticket costs $6.50. A total of 172 tickets are sold for a given show. If $1238 is collected for the show, how many adults and how many children attended the show?   80 adults, 92 children

53. Diedre holds two part-time jobs. One job pays $6.00 per hour and the other pays $6.50 per hour. Last week she earned a total of $114 and worked for a total of 18 hours. How many hours did she work at each job?   6 hr at $6, 12 hr at $6.50

54. Dan has a total of 12 bills in his wallet. Some are $5 bills and the rest are $10 bills. The total value of the 12 bills is $115. How many $5 bills and how many $10 bills does he have?   one $5 bill, eleven $10 bills

55. Mr. Haney mows part of his lawn in second gear and part in third gear. It took him 2 hours to mow the entire lawn and the odometer on his tractor shows that he covered 13.8 miles while cutting the grass. If he averages 4.2 miles per hour in second gear and 7.8 miles per hour in third gear, how long did he cut in second gear and in third gear?   0.5 hr in second, 1.5 hr in third

56. On a 100-mile trip to their cottage, the Ghents traveled at a steady speed for the first hour. The speed during the second hour of their trip was 16 miles per hour slower than the speed of the first hour. Find their speed during their first hour.   58 mph

57. A pint of coffee (16 ounces) containing 3% caffeine is mixed with a half-gallon of coffee (64 ounces) containing 7% caffeine. What percent of caffeine will the mixture contain?   6.2%

58. Judy has 60 ounces of water whose temperature is 92°C. How much water with a temperature of 20°C must she mix in with all the 92°C water to get a mixture whose temperature is 50°C? Neglect heat loss from the water to the surrounding air.   84 oz

59. Philip has an 80% methyl alcohol solution. He wishes to make a gallon of windshield washer solution by mixing his methyl alcohol solution with water. If 128 ounces, or a gallon, of windshield washer fluid should contain 6% methyl alcohol, how much of the 80% methyl alcohol solution and how much water must be mixed?   9.6 oz 80% sol, 118.4 oz water

60. George is making a meat loaf by combining chopped sirloin with veal. The sirloin contains 1.2 grams of fat per ounce, and the veal contains 0.3 gram of fat per ounce. If he wants his 64-ounce mixture to have only 0.8 gram of fat per ounce, find how much sirloin and how much veal he must use.   about 35.6 oz sirloin; 28.4 oz veal

61. Sundance dairy has 400 quarts of whole milk containing 5% butterfat. How many quarts of low-fat milk containing 1.5% butterfat should be added to produce milk containing 2% butterfat?   2400 qt

62. Benito can ride his bike to work in $\frac{3}{4}$ hour. If he takes his car to work, the trip takes $\frac{1}{6}$ hour. If Benito drives his car an average of 14 miles per hour faster than he rides his bike, determine the distance he travels to work.   3 mi

SEE EXERCISE 62.

63. A machine that folds and seals milk cartons can produce 50 milk cartons per minute. A new machine can produce 70 milk cartons per minute. The older machine has made 200 milk cartons when the newer machine is turned on. If both machines then continue working, how long after the new machine is turned on will the new machine produce the same total number of milk cartons as the older machine?   10 min

64. The salinity (salt content) of the Atlantic Ocean averages 37 parts per thousand. If 64 ounces of water is collected and placed in the sun, how many ounces of pure water would need to evaporate to raise the salinity to 45 parts per thousand? (Only the pure water is evaporated, the salt is left behind.)   about 11.4 oz.

65. Bert's tractor/lawn mower has no speedometer. Explain how Bert can determine the speed of his tractor without the use of a speedometer.

66. (a) Make up your own realistic motion problem that can be represented as an equation.

(b) Write the equation that represents your problem.

(c) Solve the equation, and then find the answer to your problem.

67. (a) Make up your own realistic mixture problem that can be represented as an equation.

(b) Write the equation that represents your problem.

(c) Solve the equation, and then find the answer to your problem.

68. Two rockets are launched from the Kennedy Space Center. The first rocket, launched at noon, will travel at 8000 miles per hour. The second rocket will be launched some time later and travel at 9500 miles per hour. When should the second rocket be launched if the rockets are to meet at a distance of 38,000 miles from Earth?

(a) Explain how to find the solution to this problem.

(b) Find the solution to the problem.   12:45 P.M.

SEE EXERCISE 68.

CUMULATIVE REVIEW EXERCISES

Solve.

[2.1] **69.** $0.6x + 0.22 = 0.4(x - 2.3)$ -5.7

70. $\dfrac{2}{9}x + 3 = x + \dfrac{1}{5}$ $\dfrac{126}{35} = \dfrac{18}{5}$

[2.2] **71.** Solve the equation $\dfrac{3}{5}(x - 2) = \dfrac{2}{7}(2x + 3y)$ for *y*.

71. $y = \dfrac{x - 42}{30}$

[2.3] **72.** Hertz Automobile Rental Agency charges \$30 per day plus 14 cents a mile. National Automobile Rental Agency charges \$16 per day plus 24 cents a mile for the same car. What distance would you have to drive in 1 day to make the cost of renting from Hertz equal to the cost of renting from National?   140 mi

Group Activity/ Challenge Problems

1. (a) An aircraft carrier catapults a 70,000-pound plane from 0 to 150 miles per hour in 2 seconds. How far, in feet, does the plane travel before liftoff? (Assume constant acceleration)   440 ft

(b) When landing, a tail hook on an aircraft carrier jet catches one of the four carrier steel cables, which brings the 150-mile per hour jet to a stop in 320 feet. Find the time, in seconds, it takes for the plane to come to a stop. (Assume constant deceleration)   1.45 sec

(c) When the jet is catapulted, do you believe the acceleration is constant? Explain   no

(d) When the jet lands and the tail hook catches, do you believe the deceleration is constant? Explain.   no

2. Cheryl must return her rental truck by 5 P.M. that day or pay an additional half day's rental. She is on a highway at 3 P.M. and 60 miles from the rental agency. She wishes to stop and visit some friends whose house is 5 miles off the highway. She plans to average 60 miles per hour on the highway and 30 miles per hour on route to and from her friend's house. How long can she visit with her friend? $\frac{2}{3}$ hr or 40 min

3. The Chunnel (the underwater tunnel from Folkstone, England, to Calais, France) is 31 miles long (23.5 miles under water). According to a recent newspaper article, a person will be able to board France's TGV train (the bullet train) in Paris and travel nonstop through the Chunnel and arrive in London 3 hours later. The TGV will average about 130 miles per hour from Paris to Calais. (It will travel up to 185 miles per hour during part of the trip.) It will then reduce its speed to an average of 90 miles per hour through the 31-mile Chunnel. When leaving the Chunnel in Folkstone it can travel only at an average of about 45 miles per hour for the 68-mile trip from Folkstone to London because of outdated tracks. Using the information, can you determine the distance from Paris to Calais, France?   ≈ 149 mi

SEE EXERCISE 4.

4. Two cars labeled *A* and *B* are in a 500-lap race. Each lap is 1 mile. The lead car, *A*, is averaging 125 miles per hour when it reaches the halfway point. Car *B* is exactly 6.2 laps behind.

6. 82,944 stitches require 10,368 min or 172.8 hr

(a) Find the average speed of car *B*.   121.9 mph

(b) When car A reaches the halfway point, how far behind, in seconds, is car *B* from car *A*?   0.0509 hr or 183.1 sec

5. Radar and sonar determine the distance of an object by emitting radio waves that travel at the speed of light, approximately 1000 feet per microsecond (a millionth of a second) in air. The devices determine the time it takes for its signal to travel to the object and return. If radar determines that it takes 0.6 second to receive the echo of its signal, how far is the object from the radar device?   300,000,000 ft

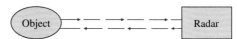

6. Lester knits at a rate of eight stitches per minute. He is planning to knit an afghan 4 feet by 6 feet. How long will it take Lester to knit the afghan if the instructions on the skein of wool indicate that four stitches equal 1 inch and six rows equal 1 inch?

7. The radiator of an automobile has a capacity of 16 quarts. It is presently filled with a 20% antifreeze solution. How many quarts must Jorge drain and replace with pure antifreeze to make the radiator contain a 50% antifreeze solution?   6 qt

2.5 Solving Linear Inequalities

Tape 3

1. Solve inequalities.
2. Graph solutions on the number line, and express solutions in interval notation and as solution sets.
3. Solve compound inequalities involving "and."
4. Solve continued inequalities using intersection of sets.
5. Solve compound inequalities involving "or."

Solve Inequalities   1. Inequalities and set builder notation were introduced in Section 1.2. You may wish to review that section now. The inequality symbols are given below.*

Inequality Symbols

$>$ is greater than

$\geq$ is greater than or equal to

$<$ is less than

$\leq$ is less than or equal to

A mathematical expression containing one or more of these symbols is called an **inequality.** The direction of the inequality symbol is sometimes called the **order** or **sense** of the inequality.

Examples of inequalities in one variable are

$$2x + 3 \leq 5 \qquad 4x > 3x - 5 \qquad 1.5 \leq -2.3x + 4.5 \qquad \frac{1}{2}x + 3 \geq 0$$

---

*$\neq$, not equal to, is also an inequality. $\neq$ means $<$ or $>$. Thus, $2 \neq 3$ means $2 < 3$ or $2 > 3$.

To solve an inequality, we must isolate the variable on one side of the inequality symbol. To isolate the variable, we use the same basic techniques used in solving equations.

Properties Used to Solve Inequalities

1. If $a > b$, then $a + c > b + c$.

2. If $a > b$, then $a - c > b - c$.

3. If $a > b$, and $c > 0$, then $ac > bc$.

4. If $a > b$ and $c > 0$, then $\dfrac{a}{c} > \dfrac{b}{c}$.

5. If $a > b$ and $c < 0$, then $ac < bc$.

6. If $a > b$ and $c < 0$, then $\dfrac{a}{c} < \dfrac{b}{c}$.

The first two properties state that the same number can be added to or subtracted from both sides of an inequality. The third and fourth properties state that both sides of an inequality can be multiplied or divided by any positive real number. The last two properties indicate that **when both sides of an inequality are multiplied or divided by a negative number, the direction of the inequality symbol reverses.**

	Example of Multiplication *by a Negative Number*	*Example of Division* *by a Negative Number*	
	$4 > -2$	$10 \geq -4$	
Multiply both sides of the inequality by −1 and reverse the direction of the inequality symbol.	$-1(4) < -1(-2)$	$\dfrac{10}{-2} \leq \dfrac{-4}{-2}$	**Divide both sides of the inequality by −2 and reverse the direction of the inequality symbol.**
	$-4 < 2$	$-5 \leq 2$	

Helpful Hint
....

Do not forget to reverse the direction of the inequality symbol when multiplying or dividing both sides of the inequality by a *negative number*.

Inequality	*Direction of Inequality* *Symbol*
$-3x < 6$	$\dfrac{-3x}{-3} > \dfrac{6}{-3}$
$-\dfrac{x}{2} > 5$	$(-2)\left(-\dfrac{x}{2}\right) < (-2)(5)$

EXAMPLE 1   Solve the inequality $2x + 6 < 12$.

Solution:

$$2x + 6 < 12$$
$$2x + 6 - 6 < 12 - 6$$
$$2x < 6$$
$$\frac{2x}{2} < \frac{6}{2}$$
$$x < 3$$

Note that the solution set is $\{x \mid x < 3\}$. Any real number less than 3 will satisfy the inequality.

Graphing Solutions   **2**   The solution to an inequality can be indicated on a number line or written as a solution set, as was explained in Section 1.2. The solution can also be written in interval notation, as explained below. Most instructors have a preferred way to indicate the solution to an inequality.

Solution of Inequality	*Solution Indicated on Number Line*	*Solution Represented in Interval Notation*
$x > a$		(a, ∞)
$x \geq a$		$[a, \infty)$
$x < a$		$(-\infty, a)$
$x \leq a$		$(-\infty, a]$
$a < x < b$		(a, b)
$a \leq x \leq b$		$[a, b]$
$a < x \leq b$		$(a, b]$
$a \leq x < b$		$[a, b)$

Recall that a solid circle on the number line indicates that the endpoint is part of the solution, and an open circle indicates that the endpoint is not part of the solution. In interval notation, brackets, [ ], are used to indicate that the endpoints are part of the solution and parentheses, ( ), indicate that the endpoints are not part of the solution. The symbol ∞ is read "infinity"; it indicates that the solution set continues indefinitely. *Whenever ∞ is used in interval notation a parenthesis must be used on the corresponding side of the interval notation.*

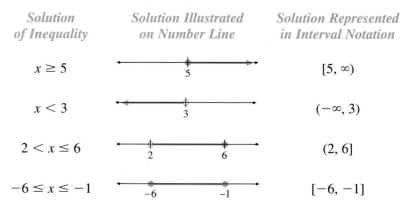

Solution of Inequality	Solution Illustrated on Number Line	Solution Represented in Interval Notation
$x \geq 5$		$[5, \infty)$
$x < 3$		$(-\infty, 3)$
$2 < x \leq 6$		$(2, 6]$
$-6 \leq x \leq -1$		$[-6, -1]$

EXAMPLE 2 Solve the following inequality and give the solution both on the number line and in interval notation.

$$3(x - 2) \leq 5x + 8$$

Solution:

$$3(x - 2) \leq 5x + 8$$

$$3x - 6 \leq 5x + 8$$

$$3x - 5x - 6 \leq 5x - 5x + 8$$

$$-2x - 6 \leq 8$$

$$-2x - 6 + 6 \leq 8 + 6$$

$$-2x \leq 14$$

$$\frac{-2x}{-2} \geq \frac{14}{-2}$$

$$x \geq -7$$

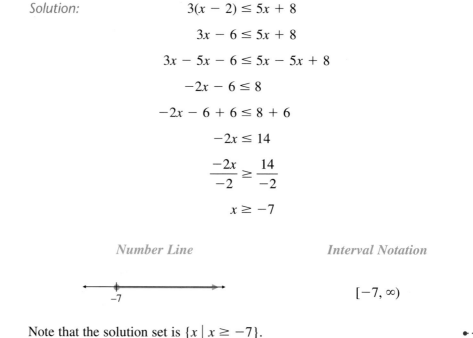

Number Line	Interval Notation
	$[-7, \infty)$

Note that the solution set is $\{x \mid x \geq -7\}$.

In Example 2 we illustrated the solution on the number line, in interval notation, and as a solution set. Your instructor may indicate which form he or she prefers.

EXAMPLE 3   Solve the inequality $\frac{1}{2}(4x + 14) \geq 5x + 4 - 3x - 10$.

Solution:

$$\frac{1}{2}(4x + 14) \geq 5x + 4 - 3x - 10$$

$$\frac{1}{2}(4x + 14) \geq 2x - 6$$

$$\left(\frac{1}{2}\right)(4x) + \left(\frac{1}{2}\right)(14) \geq 2x - 6$$

$$2x + 7 \geq 2x - 6$$

$$2x - 2x + 7 \geq 2x - 2x - 6$$

$$7 \geq -6$$

Since 7 is always greater than or equal to -6, the solution set is the set of all real numbers, $\mathbb{R}$. The solution set can also be indicated on the number line or given in interval notation.

   or $(-\infty, \infty)$

If Example 3 had resulted in the expression $7 \leq -6$, the answer would have been no solution, since 7 is never less than or equal to -6. When an inequality has no solution, its solution set is the empty or null set, $\varnothing$ or {   }.

EXAMPLE 4   Solve the inequality $\dfrac{4 - 2y}{3} \geq \dfrac{2y}{4} - 3$.

Solution: Multiply both sides of the inequality by the least common denominator, 12.

$$\overset{4}{\cancel{12}}\left(\frac{4 - 2y}{\underset{1}{\cancel{3}}}\right) \geq 12\left(\frac{2y}{4} - 3\right)$$

$$4(4) + 4(-2y) \geq \overset{3}{\cancel{12}}\left(\frac{2y}{\underset{1}{\cancel{4}}}\right) + 12(-3)$$

$$16 - 8y \geq 6y - 36$$

$$16 \geq 14y - 36$$

$$52 \geq 14y$$

$$\frac{52}{14} \geq y$$

$$\frac{26}{7} \geq y \quad \text{or} \quad y \leq \frac{26}{7}$$

$\frac{26}{7}$

In interval notation the solution is $\left(-\infty, \frac{26}{7}\right]$. The solution set is $\left\{y \mid y \leq \frac{26}{7}\right\}$.

Helpful Hint

Note that in Example 4 we indicated that $\frac{26}{7} \geq y$ can be written $y \leq \frac{26}{7}$. Generally, when writing a solution to an inequality we write the variable on the left.

For example,

$a < x$ means $x > a$ (inequality symbol points to a in both cases)

$a > x$ means $x < a$ (inequality symbol points to x in both cases)

$-6 < x$ means $x > -6$ (inequality symbol points to -6 in both cases)

$-3 > x$ means $x < -3$ (inequality symbol points to x in both cases)

EXAMPLE 5   A small single-engine airplane can carry a maximum weight of 1500 pounds. Nancy Johnson, the pilot, has to transport boxes weighing 80.4 pounds.

(a) Write an inequality that can be used to determine the maximum number of boxes that Nancy can safely place on her plane if she weighs 125 pounds.

(b) Find the maximum number of boxes that Nancy can transport.

Solution:

(a) Let n = number of boxes.

$$\text{Nancy's weight} + \text{weight of } n \text{ boxes} \leq 1500$$
$$125 \qquad + \qquad 80.4n \qquad \leq 1500$$

(b) $125 + 80.4n \leq 1500$
$80.4n \leq 1375$
$n \leq 17.1$

Therefore, Nancy can transport up to 17 boxes per trip.

EXAMPLE 6   A taxi's fare is $2.50 for the first half-mile and $1.75 for each additional half-mile. Any additional part of a half-mile will be rounded up to the next half-mile.

(a) Write an inequality that can be used to determine the maximum distance that Karen can travel if she has only $30.35.

(b) Find the maximum distance that Karen can travel.

Solution:

(a)   Let x = number of half-miles after the first

then $1.75x$ = cost of traveling x additional half-miles

$$\text{cost of first half-mile} + \text{cost of additional half-miles} \leq \text{total cost}$$
$$2.50 \qquad + \qquad 1.75x \qquad \leq 30.35$$

(b) $2.50 + 1.75x \leq 30.35$

$$1.75x \leq 27.85$$

$$x \leq \frac{27.85}{1.75}$$

$$x \leq 15.91$$

Karen can travel a distance less than or equal to 15 half-miles after the first half-mile, for a total of 16 half-miles, or 8 miles. If Karen travels for 16 half-miles after the first, she will owe $2.50 + 1.75(16) = 30.50$ dollars, which is more money than she has.

EXAMPLE 7 For a business to realize a profit, its revenue (or income), R, must be greater than its cost, C. That is, a profit will be obtained when $R > C$ (the company breaks even when $R = C$). A company that produces playing cards has a weekly cost equation of $C = 1525 + 1.7x$ and a weekly revenue equation of $R = 4.2x$, where x is the number of decks of playing cards produced and sold in a week. How many decks of cards must be produced and sold in a week for the company to make a profit?

Solution: The company will make a profit when $R > C$, or

$$4.2x > 1525 + 1.7x$$

$$2.5x > 1525$$

$$x > \frac{1525}{2.5}$$

$$x > 610$$

The company will make a profit when more than 610 decks are produced and sold in a week.

EXAMPLE 8 The 1994 Internal Revenue Tax Rate Schedule for taxpayers whose filing status is single is duplicated below.

SCHEDULE X–USE IF YOUR FILING STATUS IS SINGLE*			
If the amount on Form 1040, line 37, is: *Over—*	*But not over—*	**Enter on Form 1040, line 38**	*of the amount over—*
$0	$22,750	. . . . 15%	$0
22,750	55,100	**$3,412.50 + 28%**	22,750
55,100	115,000	**12,470.50 + 31%**	55,100
115,000	250,000	**31,039.50 + 36%**	115,000
250,000	. . . .	**79,639.50 + 39.6%**	250,000

*****Caution:** This schedule should only be used when the taxpayer meets certain requirements. In other cases, the **tax table** should be used.

(a) Write in interval notation the amounts of taxable income (form 1040, line 37) that makes up each of the five listed tax brackets, that is, the 15%, 28%, 31%, 36%, and 39.6% tax brackets.

(b) Find the tax if a single person's taxable income (line 37) is $24,800.

(c) Find the tax if a single person's taxable income is $137,600.

Solution:

(a) The words "But not over" mean "less than or equal to." The taxable incomes that make up the five tax brackets are (0, 22,750] for the 15% tax bracket, (22,750, 55,100] for the 28% tax bracket, (55,100, 115,000] for the 31% tax bracket, (115,000, 250,000] for the 36% tax bracket, and (250,000, ∞) for the 39.6% tax bracket.

(b) The tax for a single person with taxable income of $24,800 is $3412.50 plus 28% of the taxable income over $22,750. The taxable income over $22,750 is $24,800 − $22,750 = $2050.

$$\text{Tax} = 3412.50 + 0.28(2050) = 3412.50 + 574 = 3986.50$$

Therefore, the tax is $3986.50. This amount will be entered on line 38 of the 1040 tax form.

(c) A taxable income of $137,600 places a single taxpayer in the 36% tax bracket. The amount of taxable income over $115,000 is $137,600 − $115,000 = $22,600.

$$\text{Tax} = 31,039.50 + 0.36(22,600) = 31,039.50 + 8136 = 39,175.50$$

Thus, the taxable income is $39,175.50.

Graphing Calculator Corner

In Section 2.1 we indicated that the solutions to linear equations in one variable could be found or estimated using a graphing calculator. The solutions to linear inequalities in one variable such as $4x - 3 > x + 9$ may also be found or estimated using a graphing calculator. We explain how to do this in Section 3.6. You may wish to review that material now.

Compound Inequalities: And

3 A **compound inequality** is formed by joining two inequalities with the word *and* or *or*.

Examples of Compound Inequalities

$$3 < x \quad \text{and} \quad x < 5$$
$$x + 4 > 3 \quad \text{or} \quad 2x - 3 < 6$$
$$4x - 6 \geq -3 \quad \text{and} \quad x - 6 < 5$$

The solution of a compound inequality using the word *and* is all the numbers that make *both* parts of the inequality true. Consider

$$3 < x \quad \text{and} \quad x < 5$$

What are the numbers that satisfy both inequalities? The numbers that satisfy both inequalities may be easier to see if we graph the solution to each inequality on a number line (see Fig. 2.3). Note that the numbers that satisfy both inequalities are the numbers between 3 and 5. The solution set is $\{x \mid 3 < x < 5\}$.

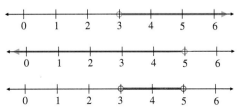

FIGURE 2.3

Recall from Chapter 1 that the intersection of two sets is the set of elements common to both sets. *To find the solution set of an inequality containing the word* **and** *take the* **intersection** *of the solution sets of the two inequalities.*

EXAMPLE 9   Solve $x + 2 \leq 5$ and $2x - 4 > -2$.

Solution: Begin by solving each inequality separately.

$$x + 2 \leq 5 \quad \text{and} \quad 2x - 4 > -2$$
$$x \leq 3 \qquad\qquad 2x > 2$$
$$\qquad\qquad\qquad x > 1$$

Now take the intersection of the sets $\{x \mid x \leq 3\}$ and $\{x \mid x > 1\}$. When we find $\{x \mid x \leq 3\} \cap \{x \mid x > 1\}$, we are finding the values of x common to both sets. Figure 2.4 illustrates that the solution set is $\{x \mid 1 < x \leq 3\}$. In interval notation, the solution is $(1, 3]$.

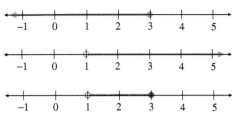

FIGURE 2.4

Continued Inequalities   ⬛4   Sometimes a compound inequality using the word *and* can be written in a shorter form. For example, $3 < x$ and $x < 5$ can be written as $3 < x < 5$. The word *and* does not appear when the inequality is written in this form, but it is implied. Inequalities written in the form $a < x < b$ are called **continued inequalities.** The compound inequality $1 < x + 5$ and $x + 5 \leq 7$ can be written $1 < x + 5 \leq 7$.

EXAMPLE 10   Solve $1 < x + 5 \leq 7$.

Solution: $1 < x + 5 \leq 7$ means $1 < x + 5$ and $x + 5 \leq 7$. Solve each inequality separately.

$$1 < x + 5 \quad \text{and} \quad x + 5 \leq 7$$
$$-4 < x \qquad\qquad\quad x \leq 2$$

Remember that $-4 < x$ means $x > -4$. Figure 2.5 illustrates that the solution set is $\{x \mid -4 < x \leq 2\}$. In interval notation, the solution is $(-4, 2]$.

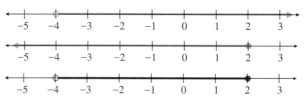

FIGURE 2.5

The inequality in Example 10, $1 < x + 5 \leq 7$, can be solved in another way. We can still use the properties discussed earlier to solve continued inequalities. However, when working with such inequalities, whatever we do to one part we must do to all three parts. In Example 10, we could have subtracted 5 from all three parts to isolate the variable in the middle and solve the inequality.

$$1 < x + 5 \leq 7$$
$$1 \boxed{-5} < x + 5 \boxed{-5} \leq 7 \boxed{-5}$$
$$-4 < x \leq 2$$

Note that this is the same solution as obtained in Example 10.

EXAMPLE 11   Solve the inequality.

$$-3 \leq 2x - 7 < 8$$

Solution: We wish to isolate the variable x. We begin by adding 7 to all three parts of the inequality.

$$-3 \leq 2x - 7 < 8$$
$$-3 \boxed{+7} \leq 2x - 7 \boxed{+7} < 8 \boxed{+7}$$
$$4 \leq 2x < 15$$

Now divide all three parts of the inequality by 2.

$$\frac{4}{2} \leq \frac{2x}{2} < \frac{15}{2}$$

$$2 \leq x < \frac{15}{2}$$

or $\left[2, \dfrac{15}{2}\right)$

The solution set is $\left\{x \mid 2 \leq x < \dfrac{15}{2}\right\}$.

EXAMPLE 12   Solve the inequality.

$$-2 < \frac{4 - 3x}{5} < 8$$

Solution: Multiply all three parts by 5 to eliminate the denominator.

$$-2 < \left(\frac{4 - 3x}{5}\right) < 8$$

$$-2\,(5) < \cancel{5}\left(\frac{4 - 3x}{\cancel{5}}\right) < 8\,(5)$$

$$-10 < 4 - 3x < 40$$

$$-10 - 4 < 4 - 4 - 3x < 40 - 4$$

$$-14 < -3x < 36$$

Now divide all three parts of the inequality by -3. Remember that when we multiply or divide an inequality by a negative number, the direction of the inequality symbol reverses.

$$\frac{-14}{-3} > \frac{-3x}{-3} > \frac{36}{-3}$$

$$\frac{14}{3} > x > -12$$

Although $\frac{14}{3} > x > -12$ is correct, we generally write continued inequalities with the lesser value on the left. We will, therefore, rewrite the solution as

$$-12 < x < \frac{14}{3}$$

The solution may also be illustrated on the number line, written in interval notation, or written as a solution set.

$$\text{or} \qquad \left(-12, \frac{14}{3}\right)$$

The solution set is $\left\{x \,|\, -12 < x < \frac{14}{3}\right\}$.

Helpful Hint

You must be very careful when writing the solution to a continued inequality. In Example 12 we can change the solution from

$$\frac{14}{3} > x > -12 \quad \text{to} \quad -12 < x < \frac{14}{3}$$

This is correct since both say that x is greater than -12 and less than $\frac{14}{3}$. Notice that the inequality symbol in both cases is pointing to the smaller number.

In Example 12, had we written the answer $\frac{14}{3} < x < -12$, we would have given the **incorrect** solution. Remember that the inequality $\frac{14}{3} < x < -12$ means

(continued on page 108)

that $\frac{14}{3} < x$ and $x < -12$. There is no number that is both greater than $\frac{14}{3}$ and less than -12. Also, by examining the inequality $\frac{14}{3} < x < -12$, it appears as if we are saying that -12 is a greater number than $\frac{14}{3}$, which is obviously *incorrect*.

It would also be **incorrect** to write the answer

$$-12 \lessgtr x \ge \tfrac{14}{3} \quad \text{or} \quad \tfrac{14}{3} \lessgtr x \ge -12$$

EXAMPLE 13 An average greater than or equal to 80 and less than 90 will result in a final grade of B in a course. Steven received grades of 85, 90, 68, and 70 on his first four exams. For Steve to receive a final grade of B in the course, between which two grades must his fifth (and last) exam fall?

Solution: Let $x =$ Steven's last exam grade.

$$80 \le \text{average of five exams} < 90$$

$$80 \le \frac{85 + 90 + 68 + 70 + x}{5} < 90$$

$$80 \le \frac{313 + x}{5} < 90$$

$$400 \le 313 + x < 450$$

$$400 - 313 \le x < 450 - 313$$

$$87 \le x < 137$$

Steven would need a minimum grade of 87 on his last exam to obtain a final grade of B. If the highest grade he could receive on the test is 100, is it possible for him to obtain a final grade of A (90 average or higher)? Explain.

Compound Inequalities: Or

⑤ The solution to a compound inequality using the word *or* is all the numbers that make *either* of the inequalities a true statement. Consider the compound inequality

$$x > 3 \quad \text{or} \quad x < 5$$

What are the numbers that satisfy the inequality? Let us graph the solution to each inequality on the number line (see Fig. 2.6). Note that every real number satisfies at least one of the two inequalities. Therefore, the solution set to the compound inequality is the set of all real numbers, $\mathbb{R}$.

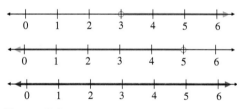

FIGURE 2.6

Recall from Chapter 1 that the *union* of two sets is the set of elements that belongs to *either* of the sets. *To find the solution set of an inequality containing the word **or**, take the **union** of the solution sets of the two inequalities that comprise the compound inequality.*

EXAMPLE 14   Solve $x + 3 \le -1$ or $-4x + 3 < -5$.

Solution: Solve each inequality separately.

$$
\begin{array}{rcl}
x + 3 \le -1 & \text{or} & -4x + 3 < -5 \\
x \le -4 & \text{or} & -4x < -8 \\
& & x > 2
\end{array}
$$

Now graph each solution on number lines and then find the union (Fig. 2.7). The union is $x \le -4$ or $x > 2$.

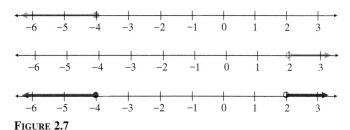

FIGURE 2.7

The solution set is $\{x \mid x \le -4\} \cup \{x \mid x > 2\}$. The union of these two sets can be written as $\{x \mid x \le -4$ or $x > 2\}$. Thus, the solution set is $\{x \mid x \le -4$ or $x > 2\}$. In interval notation, the solution is $(-\infty, -4] \cup (2, \infty)$.

We often encounter inequalities in our daily lives. For example, on a highway the minimum speed may be 45 miles per hour and a maximum speed 65 miles per hour. A restaurant may have a sign stating that maximum capacity is 300 people, and the minimum takeoff speed of an airplane may be 125 miles per hour.

Helpful Hint

There are various ways to write the solution to an inequality problem. Be sure to indicate the solution to an inequality problem in the form requested by your professor. Examples of various forms follow.

Inequality	*Number line*	*Interval Notation*	*Solution set*
$x < \dfrac{5}{3}$		$\left(-\infty, \dfrac{5}{3}\right)$	$\left\{x \mid x < \dfrac{5}{3}\right\}$
$-4 < x \le \dfrac{5}{3}$		$\left(-4, \dfrac{5}{3}\right]$	$\left\{x \mid -4 < x \le \dfrac{5}{3}\right\}$

Exercise Set 2.5   **7. (c)** $\{x \mid -6 < x \le -4\}$

Express each inequality (a) using the number line, (b) in interval notation, and (c) as a solution set (use set builder notation).

1. $x < -3$ **(a)**   **(b)** $(-\infty, -3)$ **(c)** $\{x \mid x < -3\}$ **2.** $x > \dfrac{5}{2}$ **(a)**   **(b)** $\left(\dfrac{5}{2}, \infty\right)$ **(c)** $\left\{x \mid x > \dfrac{5}{2}\right\}$

3. $x \ge 5.2$ **(a)**   **(b)** $[5.2, \infty)$ **(c)** $\{x \mid x \ge 5.2\}$ **4.** $-2 < x < 5$ **(a)**   **(b)** $(-2, 5)$ **(c)** $\{x \mid -2 < x$

5. $2 \le x < \dfrac{12}{5}$ **(a)**   **(b)** $\left[2, \dfrac{12}{5}\right)$ **(c)** $\left\{x \mid 2 \le x < \dfrac{12}{5}\right\}$ **6.** $x \ge -\dfrac{6}{5}$ **(a)**   **(b)** $\left[-\dfrac{6}{5}, \infty\right)$ **(c)** $\left\{x \mid x \ge -\dfrac{6}{5}\right\}$

7. $-6 < x \le -4$ **(a)**   **(b)** $(-6, -4]$ **8.** $-3 \le x \le 8$ **(a)**   **(b)** $[-3, 8]$ **(c)** $\{x \mid -3 \le x \le$

Solve the inequality and graph the solution on the number line.

9. $x + 3 < 8$ **10.** $2x + 3 > 4$ **11.** $3 - x < -4$

12. $4x + 3 \le -2x + 9$ **13.** $4.7x - 3.6 \ge 11.44$ **14.** $1.4x + 2.2 < 2.6x - 0.2$

15. $4(x - 2) \le 4x - 8$ **16.** $15.3 > 3(a - 1.4)$ **17.** $4b - 6 \ge 2(b + 2) + 2b$

18. $-(x - 3) + 4 \le -2x + 5$ **19.** $\dfrac{y}{3} + \dfrac{2}{5} \le 4$ **20.** $2y - 6y + 10 \le 2(-2y + 3)$

Solve the inequality and give the solution in interval notation.

21. $\dfrac{c + 3}{2} + 5 > c + 2$ $(-\infty, 9)$ **22.** $4 - 3x < 7 + 2x + 4$ $\left(-\dfrac{7}{5}, \infty\right)$ **23.** $4 + \dfrac{3x}{2} < 6$ $\left(-\infty, \dfrac{4}{3}\right)$

24. $\dfrac{3y - 6}{2} > \dfrac{2y + 5}{6}$ $\left(\dfrac{23}{7}, \infty\right)$ **25.** $\dfrac{5 - 6y}{3} \le 1 - 2y$ $\varnothing$ **26.** $\dfrac{3(x - 2)}{5} > \dfrac{5(2 - x)}{3}$ $(2, \infty)$

27. $x + 1 < 3(x + 2) - 2x$ $(-\infty, \infty)$ **28.** $\dfrac{1}{2}\left(\dfrac{3}{5}y + 4\right) \le \dfrac{1}{3}(y - 6)$ $[120, \infty)$

Solve the inequality by the method illustrated in Examples 11 and 12. Give the solution in interval notation.

29. $4 < x + 3 < 9$ $(1, 6)$ **30.** $-2 \le x - 5 < 7$ $[3, 12)$ **31.** $-3 < 5x \le 8$ $\left(-\dfrac{3}{5}, \dfrac{8}{5}\right]$
32. $-2 < -4x < 8$ $\left(-2, \dfrac{1}{2}\right)$ **33.** $4 \le 2x - 3 < 7$ $\left[\dfrac{7}{2}, 5\right)$ **34.** $-12 < 3x - 5 \le -4$ $\left(-\dfrac{7}{3}, \dfrac{1}{3}\right]$

35. $4.3 < 3.2x - 2.1 \le 16.46$ $(2, 5.8]$ **36.** $\dfrac{1}{2} < 3x + 4 < 6$ $\left(-\dfrac{7}{6}, \dfrac{2}{3}\right)$ **37.** $-7 < \dfrac{4 - 3x}{2} < 9$ $\left(-\dfrac{14}{3}, 6\right)$

38. $-6 < \dfrac{-2x - 3}{4} < 8$ $\left(-\dfrac{35}{2}, \dfrac{21}{2}\right)$

Solve the inequality and indicate the solution set.

39. $4 < \dfrac{4x - 3}{2} \le 12$ $\left\{x \mid \dfrac{11}{4} < x \le \dfrac{27}{4}\right\}$ **40.** $\dfrac{3}{5} < \dfrac{-x - 5}{3} < 6$ $\left\{x \mid -23 < x < -\dfrac{34}{5}\right\}$

41. $6 \le -3(2x - 4) < 12$ $\{x \mid 0 < x \le 1\}$ **42.** $-7 < \dfrac{4 - 2x}{3} < \dfrac{1}{3}$ $\left\{x \mid \dfrac{3}{2} < x < \dfrac{25}{2}\right\}$

43. $0 < \dfrac{2(x - 3)}{5} \le 12$ $\{x \mid 3 < x \le 33\}$ **44.** $-15 < \dfrac{3(x - 2)}{5} \le 0$ $\{x \mid -23 < x \le 2\}$

45. $-12 \le \dfrac{4 - 3x}{-5} < 2$ $\left\{x \mid -\dfrac{56}{3} \le x < \dfrac{14}{3}\right\}$ **46.** $\dfrac{1}{8} \le 4 - 2(x + 3) \le 5$ $\left\{x \mid -\dfrac{7}{2} \le x \le -\dfrac{17}{16}\right\}$

Solve the inequality and indicate the solution set.

47. $x < 4$ and $x > 2$ $\{x \mid 2 < x < 4\}$

48. $x < 4$ or $x > 2$ $\mathbb{R}$

49. $x < 2$ and $x > 4$ $\varnothing$

50. $x < 2$ or $x > 4$ $\{x \mid x < 2 \text{ or } x > 4\}$

51. $x + 2 < 3$ and $x + 1 > -2$ $\{x \mid -3 < x < 1\}$

52. $2x - 3 \le 5$ or $2x - 8 \ge 4$ $\{x \mid x \le 4 \text{ or } x \ge 6\}$

53. $5x - 3 \le 7$ or $-x + 3 < -5$ $\{x \mid x \le 2 \text{ or } x > 8\}$

54. $-2x - 3 < 2$ and $x + 6 > 4$ $\{x \mid x > -2\}$

Solve the inequality and give the solution in interval notation.

55. $3x - 6 \le 4$ or $2x - 3 < 5$ $(-\infty, 4)$

56. $-x + 6 > -3$ or $4x - 2 < 12$ $(-\infty, 9)$

57. $4x + 5 \ge 5$ and $3x - 4 \le 2$ $[0, 2]$

58. $x - 3 > -5$ and $-2x - 4 > -2$ $(-2, -1)$

59. $5x - 3 > 10$ and $4 - 3x < -2$ $\left(\frac{13}{5}, \infty\right)$

60. $x - 4 > 4$ or $3x - 5 \ge 1$ $[2, \infty)$

61. $4 - x < -2$ or $3x - 1 < -1$ $(-\infty, 0) \cup (6, \infty)$

62. $-x + 3 < 0$ or $2x - 5 \ge 3$ $(3, \infty)$

Set up an inequality that can be used to solve the problem. Solve the problem and find the desired value.

63. Cal, a janitor, must move a large shipment of books from the first floor to the fifth floor. The sign on the elevator reads "maximum weight 900 pounds." If each box of books weigh 80 pounds, find the maximum number of boxes that Cal can place in the elevator.
$80x \le 900$; 11 boxes

64. If the janitor in Exercise 63, weighing 170 pounds, must ride up with the boxes, find the maximum number of boxes of books that can be placed in the elevator. $170 + 80x \le 900$; 9 boxes

65. A telephone operator informs a customer in a phone booth that the charge for calling Denver, Colorado, is $4.25 for the first 3 minutes and 45 cents for each additional minute. Any additional part of a minute will be rounded up to the nearest minute. Find the maximum time the customer can talk if he has only $9.50.
$4.25 + 0.45x \le 9.50$; 14 min

66. A downtown parking garage in Austin, Texas charges $0.75 for the first hour and $0.50 for each additional hour. What is the maximum length of time you can park in the garage if you wish to pay no more than $3.75? $0.75 + 0.50x \le 3.75$; 7 hr

67. Miriam is considering writing and publishing her own book. She estimates her revenue equation as $R = 6.42x$, and her cost equation as $C = 10{,}025 + 1.09x$, where x is the number of books she sells. Find the minimum number of books she must sell to make a profit. See Example 7.
$10{,}025 + 1.09x < 6.42x$; 1881 books

68. Peter Collinge is opening a dry-cleaning store. He estimates his cost equation as $C = 8000 + 0.08x$ and his revenue equation as $R = 1.85x$, where x is the number of garments dry cleaned in a year. Find the minimum number of garments that must be dry cleaned in a year for Peter to make a profit.
$8000 + 0.08x < 1.85x$; 4520 garments

69. A for-profit organization can purchase a $150 bulk-mail permit and then send bulk mail at a rate of 22.8 cents per piece. Without the permit, each piece of bulk mail would cost 32 cents. Find the minimum number of pieces of bulk mail that would have to be mailed for it to be financially worthwhile for an organization to purchase the bulk-mail permit.
$150 + 0.228x < 0.32x$; more than 1630 pieces

70. The cost for mailing a package first class is 32 cents for the first ounce and 23 cents for each additional ounce. What is the maximum weight of a package that can be mailed first class for $5.00?
$0.32 + 0.23x \le 5.00$; 21 oz

71. To be eligible to continue her financial assistance for college, Nikita can earn no more than $2000 during her 8-week summer employment. She already earns $120 per week as a day-care assistant. She is considering adding an evening job at a fast-food restaurant, where she will earn $6.50 per hour. What is the maximum number of hours per week she can work at the restaurant without jeopardizing her financial assistance? $8(120) + 6.50x \le 2000$; 160 hr over 8 weeks, or 20 hr/wk

72. To receive an A in a course, Roy must obtain an average of 90 or higher on five exams. If Ray's first four exam grades are 90, 87, 96, and 95, what is the minimum grade that Ray can receive on the fifth exam to get an A in the course?
$\dfrac{90 + 87 + 96 + 95 + x}{5} \ge 90$; 82

73. To pass a course, Maria needs an average grade of 60 or more. If Maria's grades are 65, 72, 90, 47, and 62, find the minimum grade that Maria can get on her sixth and last exam and pass the course.
$\dfrac{65 + 72 + 90 + 47 + 62 + x}{6} \ge 60$; 24

74. Ms. Mahoney's grades on her first four exams are 87, 92, 70, and 75. An average greater than or equal to 80 and less than 90 will result in a final grade of B. What range of grades on Ms. Mahoney's fifth and last exam will result in a final grade of B? Assume a maximum grade of 100.
$$80 \le \frac{87 + 92 + 70 + 75 + x}{5} < 90; \ 76 \le x \le 100$$

75. For air to be considered "clean," the average of three pollutants must be less than 3.2 parts per million. If the first two pollutants are 2.7 and 3.42 ppm, what values of the third pollutant will result in clean air?
$$\frac{2.7 + 3.42 + x}{3} < 3.2; \text{ any value less than } 3.48$$

76. The water acidity in a pool is considered normal when the average pH reading of three daily measurements is greater than 7.2 and less than 7.8. If the first two pH readings are 7.48 and 7.85, find the range of pH values for the third reading that will result in the acidity level being normal.
$$7.2 < \frac{7.48 + 7.85 + x}{3} < 7.8; \ 6.27 < x < 8.07$$

77. Refer to Example 8. Find the income tax that Ms. Sharone, who is a single taxpayer, will owe if her taxable income is
(a) $28,600; $5,050.50
(b) $75,427. $18,771.89

78. What must be done when both sides of an inequality are multiplied or divided by a negative number?
reverse the direction of the inequality symbol

79. What are compound inequalities? inequalities joined with the word *and* or *or*

80. What are continued inequalities? inequalities of the form $a < x < b$

81. Explain why the inequality $4 < x < 2$ is not an acceptable answer. No real number is both greater than 4 and less than 2.

82. **(a)** Explain the step-by-step procedure to use to solve the inequality $a < bx + c < d$ for x (assume that $b > 0$).

(b) Solve the inequality for x and write the solution in interval notation. $\left(\dfrac{a - c}{b}, \dfrac{d - c}{b} \right)$

SEE EXERCISE 76.

[1.2] **83.** $A = \{1, 4, 6, 7, 9\}$, $B = \{1, 3, 4, 5, 6\}$. Find
(a) $A \cup B$;   {1, 3, 4, 5, 6, 7, 9}
(b) $A \cap B$.   {1, 4, 6}

84. $A = \left\{ -3, 4, \dfrac{5}{2}, \sqrt{7}, 0, -\dfrac{29}{80} \right\}$. List the elements that are
(a) counting numbers;   4
(b) whole numbers;   0, 4
(c) rational numbers;   −3, 4, 5/2, 0, −29/80
(d) real numbers.

[1.3] *Name the properties illustrated.* associative property
85. $(3x + 6) + 4y = 3x + (6 + 4y)$ of addition
86. $3x + y = y + 3x$ commutative property of addition
[2.2] **87.** Solve the formula $R = L + (V - D)r$ for V.
$$V = \frac{R - L + Dr}{r}$$

84. (d) $-3, 4, \dfrac{5}{2}, \sqrt{7}, 0, -\dfrac{29}{80}$

Group Activity/ Challenge Problems

1. Russell's first five grades were 82, 90, 74, 76, and 68. The final exam for the course is to count one-third in computing the final average. A final average greater than or equal to 80 and less than 90 will result in a final grade of B. What range of final-exam grades will result in Russell receiving a final grade of B in the course? Assume that a maximum grade of 100 is possible. $84 \le x \le 100$

2. Explain why the inequality $a < bx + c < d$ cannot be solved for x unless additional information is given. need to know whether $b > 0$ or $b < 0$

In Exercises 3–6 (a) explain how to solve the inequality and (b) solve the inequality and give the solution in interval notation.

3. $x < 3x - 10 < 2x$ **(b)** $(5, 10)$ **4.** $x + 5 < 3x - 8 \leq 2(x + 7)$ **(b)** $\left(\frac{13}{2}, 22\right]$

5. $x < 2x + 3 < 2x + 5$ **(b)** $(-3, \infty)$ **6.** $x + 3 < x + 1 < 2x$ **(b)** $\varnothing$

7. If $a > b$, will a^2 always be greater than b^2? Explain and give an example to support your answer.   No, $-1 > -2$ but $(-1)^2 < (-2)^2$.

2.6 Solving Equations and Inequalities Containing Absolute Values

Tape 3

[1] Understand the geometric interpretation of absolute values.
[2] Solve equations of the form $|x| = a$, $a > 0$.
[3] Solve inequalities of the form $|x| < a$, $a > 0$.
[4] Solve inequalities of the form $|x| > a$, $a > 0$.
[5] Solve inequalities of the form $|x| > a$ or $|x| < a$ when $a < 0$.
[6] Solve equations of the form $|x| = |y|$.

Geometric Interpretation of Equations and Inequalities Containing Absolute Value

[1] In Section 1.3 we introduced the concept of absolute value. We stated that the absolute value of a number may be considered the distance (without sign) from the number 0 on the number line. The absolute value of 3, written $|3|$, is 3 since it is 3 units from 0 on the number line. Similarly, the absolute value of -3, written $|-3|$, is also 3 since it is 3 units from 0 on the number line.

Consider the equation $|x| = 3$; what values of x make this equation true? We know that $|3| = 3$ and $|-3| = 3$. The solutions to $|x| = 3$ are 3 and -3. When solving the equation $|x| = 3$, we are finding the values whose distance is exactly 3 units from 0 on the number line (see Fig. 2.8a).

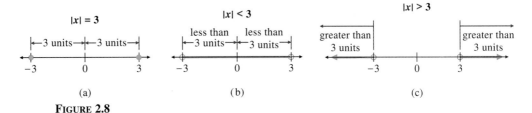

FIGURE 2.8

Now consider the inequality $|x| < 3$. To solve this inequality, we need to find the set of values whose distance is less than 3 units from 0 on the number line. These are the values of x between -3 and 3 (see Fig. 2.8b).

Consider the inequality $|x| > 3$. To solve this inequality, we need to find the set of values whose distance is greater than 3 units from 0 on the number line. These are the values that are either less than -3 or greater than 3 (see Fig. 2.8c).

In this section we will solve equations and inequalities like the following:

$$|2x - 1| = 5 \qquad |2x - 1| < 5 \qquad |2x - 1| > 5$$

The geometric interpretation of $|2x - 1| = 5$ is similar to $|x| = 3$. When solving $|2x - 1| = 5$ we are determining the set of values that result in $2x - 1$ being exactly 5 units away from 0 on the number line.

The geometric interpretation of $|2x - 1| \leq 5$ is similar to the geometrical interpretation of $|x| \leq 3$. When solving $|2x - 1| \leq 5$, we are determining the set of values that result in $2x - 1$ being less than or equal to 5 units from 0 on the number line.

The geometrical interpretation of $|2x - 1| > 5$ is similar to $|x| > 3$. When solving $|2x - 1| > 5$, we are determining the set of values that result in $2x - 1$ being more than 5 units from 0 on the number line.

We will be solving absolute value equations and inequalities algebraically in the remainder of this section. We will first solve absolute value equations, then we will solve absolute value inequalities. We will end the section by solving absolute value equations where both sides of the equation contain an absolute value, for example, $|x + 3| = |2x - 5|$.

Equations of the Form
$$|x| = a, a > 0$$

2 When solving an equation of the form $|x| = a$, $a \geq 0$, we are finding the values that are exactly a units from 0 on the number line. The following procedure may be used to solve such problems.

> **Solving Equations of the Form $|x| = a$**
> If $|x| = a$ and $a > 0$, then $x = a$ or $x = -a$.

EXAMPLE 1   Solve the equation $|x| = 4$.

Solution: Using the procedure, we get $x = 4$ or $x = -4$. The solution set is $\{-4, 4\}$.

EXAMPLE 2   Solve the equation $|x| = 0$.

Solution: The only real number whose absolute value equals 0 is 0. Thus, the solution set for $|x| = 0$ is $\{0\}$.

EXAMPLE 3   Solve the equation $|x| = -2$.

Solution: The absolute value of a number is never negative, so there are no solutions to this equation. The solution set is $\varnothing$.

EXAMPLE 4   Solve the equation $|2w - 1| = 5$.

Solution: At first this might not appear to be of the form $|x| = a$. However, if we let $2w - 1$ be x and 5 be a, you will see the equation is of this form. We are looking for the values of w such that $2w - 1$ is exactly 5 units from 0 on the number line. Thus, the quantity $2w - 1$ must be equal to 5 or -5.

$$2w - 1 = 5 \quad \text{or} \quad 2w - 1 = -5$$
$$2w = 6 \qquad\qquad 2w = -4$$
$$w = 3 \qquad\qquad\quad w = -2$$

Check:

$$w = 3 \qquad |2w - 1| = 5 \qquad w = -2 \qquad |2w - 1| = 5$$
$$|2(3) - 1| = 5 \qquad\qquad |2(-2) - 1| = 5$$
$$|6 - 1| = 5 \qquad\qquad |-4 - 1| = 5$$
$$|5| = 5 \qquad\qquad |-5| = 5$$
$$5 = 5 \quad \text{true} \qquad\qquad 5 = 5 \quad \text{true}$$

The solutions 3 and -2 each result in $2w - 1$ being 5 units from 0 on the number line. The solution set is $\{-2, 3\}$.

EXAMPLE 5   Solve the equation $\left|\frac{2}{3}z - 6\right| + 4 = 6$.

Solution: We begin by subtracting 4 from both sides of the equation to get the absolute value alone on one side of the equation.

$$\left|\frac{2}{3}z - 6\right| + 4 = 6$$

$$\left|\frac{2}{3}z - 6\right| = 2$$

Now we proceed as before. Write the two cases.

$$\frac{2}{3}z - 6 = 2 \qquad \text{or} \qquad \frac{2}{3}z - 6 = -2$$

$$\frac{2}{3}z = 8 \qquad\qquad \frac{2}{3}z = 4$$

$$2z = 24 \qquad\qquad 2z = 12$$

$$z = 12 \qquad\qquad z = 6$$

The solution set is $\{6, 12\}$.

Inequalities of the Form $|x| < a, a > 0$

③ The solution set to $|x| = 3$, $\{-3, 3\}$ contains the values that are *exactly* 3 units from 0 on the number line. The solution set to $|x| < 3$ is the set of values that are *less than 3 units* from 0 on the number line (see Fig 2.8b). This includes all real numbers between -3 and 3. The solution set is $\{x \mid -3 < x < 3\}$.

When we are asked to find the solution set to an inequality of the form $|x| < a$, we are finding the set of values that are less than a units from 0 on the number line. The solution set to $|x| \leq a$ is the set of values that are *less than or equal to* a *units from 0* on the number line.

We can use the same reasoning process to solve more complicated problems, as shown in Example 6.

EXAMPLE 6   Solve the inequality $|2x - 3| < 5$.

Solution: The solution to this inequality will be the set of values such that the distance between $2x - 3$ and 0 on the number line will be less than 5 units (see Fig. 2.9). Using Fig. 2.9, we can see that $-5 < 2x - 3 < 5$.

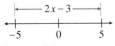

FIGURE 2.9

Solving, we get

$$-5 < 2x - 3 < 5$$
$$-2 < 2x < 8$$
$$-1 < x < 4$$

The solution set is $\{x \mid -1 < x < 4\}$. When x is any number between -1 and 4, the expression $2x - 3$ will represent a number that is less than 5 units from 0 on the number line (or a number between -5 and 5).

Using the same reasoning process, we can see that to solve inequalities of the form $|x| < a$ we use the following procedure:

> ## Solving Inequalities of the Form $|x| < a$
> If $|x| < a$ and $a > 0$, then $-a < x < a$.

EXAMPLE 7   Solve the inequality $|3x - 4| \leq 5$ and graph the solution on the number line.

Solution: Since this inequality is of the form $|x| \leq a$, we write

$$-5 \leq 3x - 4 \leq 5$$
$$-1 \leq 3x \leq 9$$
$$-\frac{1}{3} \leq x \leq 3$$

Any value of x greater than or equal to $-\frac{1}{3}$ and less than or equal to 3 would result in $3x - 4$ being less than or equal to 5 units from 0 on the number line.

EXAMPLE 8   Solve the inequality $|4.2 - x| + 1.3 < 3.6$ and graph the solution on the number line.

Solution: First isolate the absolute value by subtracting 1.3 from both sides of the inequality. Then solve as in the previous examples.

$$|4.2 - x| + 1.3 < 3.6$$
$$|4.2 - x| < 2.3$$
$$-2.3 < 4.2 - x < 2.3$$
$$-6.5 < -x < -1.9$$
$$-1(-6.5) > -1(-x) > -1(-1.9)$$
$$6.5 > x > 1.9$$
$$\text{or} \quad 1.9 < x < 6.5$$

The solution set is $\{x \mid 1.9 < x < 6.5\}$. The solution in interval notation is $(1.9, 6.5)$.

Inequalities of the Form
$|x| > a, a > 0$

4 Now we look at inequalities of the form $|x| > a$. Consider $|x| > 3$. This inequality represents the set of values that are greater than 3 units from 0 on the number line (see Fig. 2.8c). The solution set is $\{x \mid x < -3 \text{ or } x > 3\}$. The solution set to $|x| > a$ is the set of values that are *greater than* a *units from 0* on the number line.

EXAMPLE 9 Solve the inequality $|2x - 3| > 5$ and graph the solution on the number line.

Solution: The solution to $|2x - 3| > 5$ is the set of values such that the distance between $2x - 3$ and 0 on the number line will be greater than 5. The quantity $2x - 3$ must either be less than -5 or greater than 5 (see Fig. 2.10).

FIGURE 2.10

Since $2x - 3$ must be either less than -5 or greater than 5, we set up and solve the following compound inequality:

$$
\begin{array}{lcl}
2x - 3 < -5 & \text{or} & 2x - 3 > 5 \\
2x < -2 & & 2x > 8 \\
x < -1 & & x > 4
\end{array}
$$

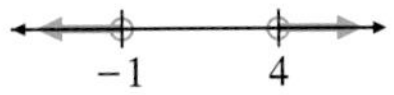

The solution set to $|2x - 3| > 5$ is $\{x \mid x < -1 \text{ or } x > 4\}$. When x is any number less than -1 or greater than 4, the expression $2x - 3$ will represent a number that is greater than 5 units from 0 on the number line (or a number less than -5 or greater than 5).

Using the same reasoning process, we can see that to solve inequalities of the form $|x| > a$ we use the following procedure.

Solving Inequalities of the Form $|x| > a$

If $|x| > a$ and $a > 0$, then $x < -a$ or $x > a$.

EXAMPLE 10 Solve the inequality $|2x - 1| \geq 7$ and graph the solution on the number line.

Solution: Since this inequality is of the form $|x| \geq a$, we use the procedure given above.

$$
\begin{array}{lcl}
2x - 1 \leq -7 & \text{or} & 2x - 1 \geq 7 \\
2x \leq -6 & & 2x \geq 8 \\
x \leq -3 & & x \geq 4
\end{array}
$$

Any value of x less than or equal to -3, or greater than or equal to 4, would result in $2x - 1$ representing a number that is greater than or equal to 7 units from 0 on the number line. The solution set is $\{x \mid x \le -3 \text{ or } x \ge 4\}$. In interval notation, the solution is $(-\infty, -3] \cup [4, \infty)$.

EXAMPLE 11   Solve the inequality $\left| \dfrac{3x - 4}{2} \right| \ge \dfrac{5}{12}$.

Solution: Since this inequality is of the form $|x| \ge a$, we write

$$\frac{3x - 4}{2} \le -\frac{5}{12} \quad \text{or} \quad \frac{3x - 4}{2} \ge \frac{5}{12}$$

Now multiply both sides of each inequality by the least common denominator, 12. Then solve each inequality.

$$\overset{6}{\cancel{12}}\left(\frac{3x - 4}{\underset{1}{\cancel{2}}}\right) \le \frac{-5}{\cancel{12}} \cdot \cancel{12} \quad \text{or} \quad \overset{6}{\cancel{12}}\left(\frac{3x - 4}{\underset{1}{\cancel{2}}}\right) \ge \frac{5}{\cancel{12}} \cdot \cancel{12}$$

$$6(3x - 4) \le -5 \qquad\qquad 6(3x - 4) \ge 5$$
$$18x - 24 \le -5 \qquad\qquad 18x - 24 \ge 5$$
$$18x \le 19 \qquad\qquad\qquad 18x \ge 29$$
$$x \le \frac{19}{18} \qquad\qquad\qquad x \ge \frac{29}{18}$$

Helpful Hint

Some general information about equations and inequalities containing absolute value follows. For real numbers a, b, and c where $a \ne 0$ and $c > 0$:

Form of Equation or Inequality	*The Solution Will Be:*	*Solution on the Number Line:*
$\lvert ax + b \rvert = c$	Two distinct numbers, p and q	
$\lvert ax + b \rvert < c$	The set of numbers between two numbers, $p < x < q$	
$\lvert ax + b \rvert > c$	The set of numbers less than one number or greater than a second number, $x < p$ or $x > q$	

Inequalities of the Form $\lvert x \rvert > a$ or $\lvert x \rvert < a$ When $a < 0$

[5] We have solved inequalities of the form $|x| < a$ where $a > 0$. Now let us consider what happens in an absolute value inequality when $a < 0$ or $a = 0$. Consider the inequality $|x| < -3$. Since $|x|$ will always have a value greater than or equal to 0 for any real number x, this inequality can never be true, and the solution is the empty set, $\varnothing$. Whenever we have an absolute value inequality of this type, the solution will be the empty set.

EXAMPLE 12   Solve the inequality $|x - 4| - 3 < -5$.

Solution: Begin by adding 3 to both sides of the inequality.

$$|x - 4| - 3 < -5$$
$$|x - 4| < -2$$

Since $|x - 4|$ will always be greater than or equal to 0 for any real number x, this inequality can never be true. Thus, the solution is the empty set, $\varnothing$.

Now consider the inequality $|x| > -3$. Since $|x|$ will always have a value greater than or equal to 0 for any real number x, this inequality will always be true. Since every value of x will make this inequality a true statement, the solution is the set of all real numbers, $\mathbb{R}$. Whenever we have an absolute value inequality of this type, the solution will be the set of all real numbers, $\mathbb{R}$.

EXAMPLE 13   Solve the inequality $|2x + 3| + 4 \geq -7$.

Solution: Begin by subtracting 4 from both sides of the inequality.

$$|2x + 3| + 4 \geq -7$$
$$|2x + 3| \geq -11$$

Since $|2x + 3|$ will always be greater than or equal to 0 for any real number x, this inequality is true for all real numbers. Thus, the solution is the set of all real numbers, $\mathbb{R}$.

Now let us discuss inequalities where one side of the inequality is 0. The only value that satisfies the equation $|x - 5| = 0$ is 5, since 5 makes the expression inside the absolute value sign 0. Now consider $|x - 5| \leq 0$. Since the absolute value can never be negative this inequality is true only when $x = 5$. The inequality $|x - 5| < 0$ has no solution. Can you explain why? What is the solution to $|x - 5| \geq 0$? Since any value of x will result in the absolute value being greater than or equal to 0, the solution is the set of all real numbers, $\mathbb{R}$. What is the solution to $|x - 5| > 0$? The solution is every real number except 5. Can you explain why 5 is excluded from the solution?

EXAMPLE 14   Solve the inequalities **(a)** $|x + 3| > 0$, and **(b)** $|3x - 4| \leq 0$.

Solution:

(a) The inequality will be true for every value of x except -3. The solution set is $\{x \mid x < -3 \text{ or } x > -3\}$.

(b) First determine the number that makes the absolute value equal to 0 by setting the expression within the absolute value equal to 0 and solving for x.

$$3x - 4 = 0$$
$$3x = 4$$
$$x = \frac{4}{3}$$

The inequality will be true only when $x = \frac{4}{3}$. The solution set is $\left\{\frac{4}{3}\right\}$.

Equations of the Form
$$|x| = |y|$$

6 Now we will discuss absolute value equations where an absolute value appears on both sides of the equation. To solve equations of the form $|x| = |y|$, use the procedure that follows.

> **Solving Equations of the Form $|x| = |y|$**
>
> If $|x| = |y|$, then $x = y$ or $x = -y$.

When solving an absolute value equation with an absolute value expression on each side of the equal sign, the two expressions must have the same absolute value. Therefore, the expressions must be equal to each other or be opposites of each other.

EXAMPLE 15   Solve the equation $|z + 3| = |2z - 7|$.

Solution: If we let $z + 3$ be x and $2z - 7$ be y, this equation is of the form $|x| = |y|$. Using the procedure given above, we obtain the two equations

$$z + 3 = 2z - 7 \quad \text{or} \quad z + 3 = -(2z - 7)$$

Now solve each equation.

$$
\begin{aligned}
z + 3 &= 2z - 7 \quad &\text{or} \quad z + 3 &= -(2z - 7) \\
3 &= z - 7 & z + 3 &= -2z + 7 \\
10 &= z & 3z + 3 &= 7 \\
& & 3z &= 4 \\
& & z &= \frac{4}{3}
\end{aligned}
$$

Check: $z = 10$

$$|z + 3| = |2z - 7|$$
$$|10 + 3| = |2(10) - 7|$$
$$|13| = |20 - 7|$$
$$|13| = |13|$$
$$13 = 13 \quad \text{true}$$

$z = \dfrac{4}{3}$

$$|z + 3| = |2z - 7|$$
$$\left|\frac{4}{3} + 3\right| = \left|2\left(\frac{4}{3}\right) - 7\right|$$
$$\left|\frac{13}{3}\right| = \left|\frac{8}{3} - \frac{21}{3}\right|$$
$$\left|\frac{13}{3}\right| = \left|-\frac{13}{3}\right|$$
$$\frac{13}{3} = \frac{13}{3} \quad \text{true}$$

The solution set is $\left\{10, \frac{4}{3}\right\}$.

EXAMPLE 16 Solve the equation $|4x - 7| = |6 - 4x|$.

Solution:

$$4x - 7 = 6 - 4x \quad \text{or} \quad 4x - 7 = -(6 - 4x)$$
$$8x - 7 = 6 \qquad\qquad 4x - 7 = -6 + 4x$$
$$8x = 13 \qquad\qquad\quad -7 = -6 \quad \text{false}$$
$$x = \frac{13}{8}$$

Since the equation $4x - 7 = -(6 - 4x)$ results in a false statement, the absolute value equation has only one solution. A check will show that the solution set is $\left\{\frac{13}{8}\right\}$.

Summary of Procedures for Solving Equations and Inequalities Containing Absolute Value

For $a > 0$.

If $|x| = a$,	then	$x = a$ or $x = -a$.
If $|x| < a$,	then	$-a < x < a$.
If $|x| > a$,	then	$x < -a$ or $x > a$.
If $|x| = |y|$,	then	$x = y$ or $x = -y$.

Exercise Set 2.6

Find the solution set for each equation.

1. $|x| = 5$ $\{-5, 5\}$

2. $|y| = 7$ $\{7, -7\}$

3. $|x| = 12$ $\{-12, 12\}$

4. $|x| = 0$ $\{0\}$

5. $|x| = -2$ $\varnothing$

6. $|x + 1| = 5$ $\{-6, 4\}$

7. $|x + 5| = 7$ $\{-12, 2\}$

8. $|3 + y| = \dfrac{3}{5}$ $\left\{-\frac{18}{5}, -\frac{12}{5}\right\}$

9. $|2.4 + 0.4x| = 4$ $\{-16, 4\}$

10. $|3x - 4| = 0$ $\left\{\frac{4}{3}\right\}$

11. $|5 - 3x| = \dfrac{1}{2}$ $\left\{\frac{3}{2}, \frac{11}{6}\right\}$

12. $|3(y + 4)| = 12$ $\{-8, 0\}$

13. $|4(x - 2)| = 18$ $\left\{-\frac{5}{2}, \frac{13}{2}\right\}$

14. $\left|\dfrac{x - 3}{4}\right| = 5$ $\{-17, 23\}$

15. $\left|\dfrac{3z + 5}{6}\right| - 3 = 6$ $\left\{-\frac{59}{3}, \frac{49}{3}\right\}$

16. $\left|\dfrac{x - 3}{4}\right| + 4 = 4$ $\{3\}$

17. $\left|\dfrac{5x - 3}{2}\right| + 2 = 6$ $\left\{-1, \frac{11}{5}\right\}$

18. $\left|\dfrac{2x + 3}{2}\right| + 1 = 4$ $\left\{-\frac{9}{2}, \frac{3}{2}\right\}$

Find the solution set for each inequality.

19. $|y| \leq 5$ $\{y \mid -5 \leq y \leq 5\}$

20. $|x| \leq 9$ $\{x \mid -9 \leq x \leq 9\}$

21. $|x - 7| \leq 9$ $\{x \mid -2 \leq x \leq 16\}$

22. $|7 - x| < 5$ $\{x \mid 2 < x < 12\}$

23. $|3z - 5| \leq 5$ $\left\{z \mid 0 \leq z \leq \frac{10}{3}\right\}$

24. $|x - 3| -2 < 3$ $\{x \mid -2 < x < 8\}$

25. $|2x + 3| - 5 \leq 10$ $\{x \mid -9 \leq x \leq 6\}$

26. $|4 - 3x| - 4 < 11$ $\left\{x \mid -\frac{11}{3} < x < \frac{19}{3}\right\}$

27. $|x - 0.4| \leq 2.3$ $\{x \mid -1.9 \leq x \leq 2.7\}$

28. $|2x - 3| < -4$ $\varnothing$

29. $|2x - 6| + 5 \leq 2$ $\varnothing$

30. $\left|\dfrac{2x - 1}{3}\right| \leq \dfrac{5}{3}$ $\{x \mid -2 \leq x \leq 3\}$

31. $\left| 5 - \dfrac{3x}{4} \right| < 8$ $\left\{ x \mid -4 < x < \dfrac{52}{3} \right\}$

32. $\left| \dfrac{x-3}{2} \right| - 4 \le -2$ $\{ x \mid -1 \le x \le 7 \}$

33. $|4x - 1| \le 0$ $\dfrac{1}{4}$

34. $|2x + 3| < 0$ $\varnothing$

Find the solution set for each inequality.

35. $|x| > 3$ $\{ x \mid x < -3 \text{ or } x > 3 \}$

36. $|y| \ge 5$ $\{ y \mid y \le -5 \text{ or } y \ge 5 \}$

37. $|x + 4| > 5$ $\{ x \mid x < -9 \text{ or } x > 1 \}$

38. $|5 - x| \ge 3$ $\{ x \mid x \le 2 \text{ or } x \ge 8 \}$

39. $|3x + 1| > 4$ $\left\{ x \mid x < -\dfrac{5}{3} \text{ or } x > 1 \right\}$

40. $|4 - 3y| \ge 8$ $\left\{ y \mid y \le -\dfrac{4}{3} \text{ or } y \ge 4 \right\}$

41. $\left| \dfrac{6 + 2z}{3} \right| > 2$ $\{ z \mid z < -6 \text{ or } z > 0 \}$

42. $\left| \dfrac{5 - 3w}{4} \right| \ge 10$ $\left\{ w \mid w \le -\dfrac{35}{3} \text{ or } w \ge 15 \right\}$

43. $|0.1x - 0.4| + 0.4 > 0.6$ $\{ x \mid x < 2 \text{ or } x > 6 \}$

44. $|2x - 1| - 4 \ge 8$ $\left\{ x \mid x \le -\dfrac{11}{2} \text{ or } x \ge \dfrac{13}{2} \right\}$

45. $\left| \dfrac{2x - 4}{3} \right| - 3 > -5$ $\mathbb{R}$

46. $\left| \dfrac{2x - 3}{4} \right| - 1 > 3$ $\left\{ x \mid x < -\dfrac{13}{2} \text{ or } x > \dfrac{19}{2} \right\}$

47. $\left| \dfrac{x}{2} + 4 \right| \ge 5$ $\{ x \mid x \le -18 \text{ or } x \ge 2 \}$

48. $\left| 4 - \dfrac{3x}{5} \right| \ge 9$ $\left\{ x \mid x \le -\dfrac{25}{3} \text{ or } x \ge \dfrac{65}{3} \right\}$

49. $|3x + 5| \ge 0$ $\mathbb{R}$

50. $|3 - 2x| \ge 0$ $\mathbb{R}$

51. $|4 - 2x| > 0$ $\{ x \mid x < 2 \text{ or } x > 2 \}$

52. $|-8y - 3| > 0$ $\left\{ y \mid y < -\dfrac{3}{8} \text{ or } y > -\dfrac{3}{8} \right\}$

Find the solution set for each equation.

53. $|2x + 1| = |4x - 9|$ $\left\{ 5, \dfrac{4}{3} \right\}$

54. $|x - 1| = |2x - 4|$ $\left\{ 3, \dfrac{5}{3} \right\}$

55. $|6x| = |3x - 9|$ $\{ -3, 1 \}$

56. $|4x - 2| = |4x - 2|$ $\mathbb{R}$

57. $\left| \dfrac{3}{4}x - 2 \right| = \left| \dfrac{1}{2}x + 5 \right|$ $\left\{ 28, -\dfrac{12}{5} \right\}$

58. $|3x - 5| = |3x + 5|$ $\{ 0 \}$

59. $\left| \dfrac{1}{2}x + \dfrac{3}{5} \right| = \left| \dfrac{1}{2}x - 1 \right|$ $\left\{ \dfrac{2}{5} \right\}$

60. $\left| \dfrac{3}{2}r + 2 \right| = \left| \dfrac{1}{2}r - 3 \right|$ $\left\{ -5, \dfrac{1}{2} \right\}$

Find the solution set for each equation or inequality.

61. $|w| = 7$ $\{ 7, -7 \}$

62. $|x - 3| = 5$ $\{ -2, 8 \}$

63. $|x - 3| < 5$ $\{ x \mid -2 < x < 8 \}$

64. $|z| \ge 2$ $\{ z \mid z \le -2 \text{ or } z \ge 2 \}$

65. $|x + 5| > 9$ $\{ x \mid x < -14 \text{ or } x > 4 \}$

66. $|3x - 4| \le -6$ $\varnothing$

67. $|2y + 4| < 1$ $\left\{ y \mid -\dfrac{5}{2} < y < -\dfrac{3}{2} \right\}$

68. $|2x - 5| + 3 \le 10$ $\{ x \mid -1 \le x \le 6 \}$

69. $|4x + 2| = 9$ $\left\{ -\dfrac{11}{4}, \dfrac{7}{4} \right\}$

70. $|2x - 4| + 2 = 10$ $\{ -2, 6 \}$

71. $|5 + 2x| > 0$ $\left\{ x \mid x < -\dfrac{5}{2} \text{ or } x > -\dfrac{5}{2} \right\}$

72. $|4 - x| = 5$ $\{ -1, 9 \}$

73. $|4 + 3x| \le 9$ $\left\{ x \mid -\dfrac{13}{3} \le x \le \dfrac{5}{3} \right\}$

74. $|2.4x + 4| + 4.9 > 1.9$ $\mathbb{R}$

75. $|3x - 5| + 4 = 2$ $\varnothing$

76. $|4 - 2x| - 5 = 5$ $\{ -3, 7 \}$

77. $\left| \dfrac{3x - 2}{4} \right| - 5 = 1$ $\left\{ -\dfrac{22}{3}, \dfrac{26}{3} \right\}$

78. $\left| \dfrac{4c - 4}{5} \right| \le 8$ $\{ c \mid -9 \le c \le 11 \}$

79. $\left| \dfrac{w + 4}{3} \right| - 1 < 3$ $\{ w \mid -16 < w < 8 \}$

80. $\left| \dfrac{3x + 4}{5} \right| > \dfrac{7}{5}$ $\left\{ x \mid x < -\dfrac{11}{3} \text{ or } x > 1 \right\}$

81. $\left| \dfrac{3x - 2}{4} \right| + 5 \ge 5$ $\mathbb{R}$

82. $\left| \dfrac{2x - 4}{5} \right| = 12$ $\{ -28, 32 \}$

83. $|2x - 8| = \left| \dfrac{1}{2}x + 3 \right|$ $\left\{ \dfrac{22}{3}, 2 \right\}$

84. $\left| \dfrac{1}{3}y + 3 \right| = \left| \dfrac{2}{3}y - 1 \right|$ $\{ 12, -2 \}$

85. $|2 - 3x| = \left| 4 - \dfrac{5}{3}x \right|$ $\left\{ -\dfrac{3}{2}, \dfrac{9}{7} \right\}$

86. $\left| \dfrac{3 - 2x}{4} \right| \ge 5$ $\left\{ x \mid x \le -\dfrac{17}{2} \text{ or } x \ge \dfrac{23}{2} \right\}$

87. $\left| 2 \left(\dfrac{3 - x}{5} \right) \right| < \dfrac{9}{5}$ $\left\{ x \mid -\dfrac{3}{2} < x < \dfrac{15}{2} \right\}$

88. $\left| 3 \left(x + \dfrac{1}{2} \right) \right| > 5$ $\left\{ x \mid x < -\dfrac{13}{6} \text{ or } x > \dfrac{7}{6} \right\}$

🖎 **89.** For what value of x will the inequality $|ax + b| \leq 0$ be true? Explain. $x = -b/a$

🖎 **90.** For what value of x will the inequality $|ax + b| > 0$ *not* be true? Explain. $x = -b/a$

95. (a) Set $ax + b = c$ or $ax + b = -c$ and solve each equation for x.   **(b)** $\left\{ x \mid x = \dfrac{c - b}{a} \text{ or } x = \dfrac{-c - b}{a} \right\}$

In Exercises 91–94, determine what values of x will make the equation true. Explain your answer.

🖎 **91.** $|x - 3| = |3 - x|$ $\mathbb{R}$

🖎 **92.** $|x - 3| = -|x - 3|$ $\{3\}$

🖎 **93.** $|x| = x$ $\{x \mid x \geq 0\}$

🖎 **94.** $|x + 2| = x + 2$ $\{x \mid x \geq -2\}$

95. (a) Explain how to find the solution to the equation $|ax + b| = c$. (Assume that $c > 0$ and $a \neq 0$.)

    (b) Solve this equation for x.

🖎 **96. (a)** Explain how to find the solution to the inequality $|ax + b| < c$. (Assume that $a > 0$ and $c > 0$.)

    (b) Solve this inequality for x.

96. (a) Write $-c < ax + b < c$ and solve the inequality for x.   **(b)** $\left\{ x \mid \dfrac{-c - b}{a} < x < \dfrac{c - b}{a} \right\}$

🖎 **97. (a)** Explain how to find the solution to the inequality $|ax + b| > c$. (Assume that $a > 0$ and $c > 0$.)

    (b) Solve the inequality for x.

🖎 **98. (a)** What is the first step in solving the inequality $-2|3x - 5| \leq -6$?

    (b) Solve the inequality and give the solution in interval notation.

97. (a) Write $ax + b < -c$ or $ax + b > c$ and solve the inequalities for x.   **(b)** $\left\{ x \mid x < \dfrac{-c - b}{a} \text{ or } x > \dfrac{c - b}{a} \right\}$

CUMULATIVE REVIEW EXERCISES

Evaluate.

[1.4] **99.** $\dfrac{1}{3} + \dfrac{1}{4} \div \dfrac{2}{5}\left(\dfrac{1}{3}\right)^2$ $\dfrac{29}{72}$

100. $4(x + 3y) - 5xy$ when $x = 1, y = 3$ 25

[2.4] **101.** Raul swims across a lake averaging 2 miles an hour. Then he turns around and swims back across the lake, averaging 1.6 miles per hour. If his total swimming time was 1.5 hours, what is the width of the lake? $2x = 1.6(1.5 - x)$; 1.33 mi

[2.5] **102.** Find the solution set to the inequality $3(x - 2) - 4(x - 3) > 2$. $\{x \mid x < 4\}$

98. (a) Divide both sides by -2 and reverse direction of the inequality

98. (b) $\left(-\infty, \dfrac{2}{3}\right] \cup \left[\dfrac{8}{3}, \infty\right)$

Group Activity/ Challenge Problems

1. Find all values of x and y such that $|x - y| = |y - x|$.   all x and all y

Solve. Explain how you determined your answer.

🖎 **2.** $|x + 1| = 2x - 1$ $\{2\}$ 🖎 **3.** $|3x + 1| = x - 3$ $\varnothing$

🖎 **4.** $|x - 2| = -(x - 2)$ $\{x \mid x \leq 2\}$

Solve by considering the possible signs for x.

5. $|x| + x = 6$ $\{3\}$ **6.** $x + |-x| = 6$ $\{3\}$

7. $|x| - x = 6$ $\{-3\}$ **8.** $x - |x| = 6$ $\varnothing$

Summary

GLOSSARY

Coefficient (or numerical coefficient) *(49):* The numerical part of a term.

Compound inequality *(104):* Two inequalities joined with the word *and* or *or*.

Conditional equation *(57):* An equation true only under specific conditions.

Constant *(49):* a term that consists of only a number.

Continued inequality *(105):* An inequality of the form $a < x < b$.

Degree of a term *(50):* The sum of the exponents on the variables in a term.

Equation *(51):* A mathematical statement of equality.

Equivalent equations *(51):* Equations with the same solution set.

Formula *(60):* An equation used to represent a scientific or real-life principle.

Identity *(57):* An equation that is true for all real numbers.

Inconsistent equation *(58):* An equation that has no solution.

Inequality *(97):* A math expression containing one or more inequality symbols.

Literal equation *(60):* An equation that has more than one variable.

Least common denominator *(54):* The smallest number divisible by a given set of denominators.

Like terms *(50):* Terms that have the same variables with the same exponents.

Linear equation *(51):* The standard form of a linear equation in one variable is $ax + b = c, a \neq 0$. A linear equation is also called a first-degree equation.

Order or sense of an inequality *(97):* The direction of the inequality symbol.

Simplify an expression *(50):* Combine like terms in the expression.

Solution set *(51):* The set of real numbers that make the equation true.

Solutions (or roots) of an equation *(51):* The number or numbers that make the equation true.

Subscript *(60):* Numbers or letters to the right of and below a variable.

Terms *(49):* The parts added or subtracted in an algebraic expression.

Unlike terms *(50):* Terms that are not "like" terms.

IMPORTANT FACTS

Properties of equality

Reflexive property: $a = a$

Symmetric property: If $a = b$, then $b = a$.

Transitive property: If $a = b$ and $b = c$, then $a = c$.

Addition property of equality: If $a = b$, then $a + c = b + c$.

Multiplication property of equality: If $a = b$, then $ac = bc$.

Proportions

If $\dfrac{a}{b} = \dfrac{c}{d}$, then $ad = bc$.

Distance formula

distance = rate · time

Properties used to solve inequalities

1. If $a > b$, then $a + c > b + c$.

2. If $a > b$, then $a - c > b - c$.

3. If $a > b$ and $c > 0$, then $ac > bc$.

4. If $a > b$ and $c > 0$, then $\dfrac{a}{c} > \dfrac{b}{c}$.

5. If $a > b$ and $c < 0$, then $ac < bc$.

6. If $a > b$ and $c < 0$, then $\dfrac{a}{c} < \dfrac{b}{c}$.

Absolute value, for $a > 0$

If $|x| = a$, then $x = a$ or $x = -a$

If $|x| < a$, then $-a < x < a$.

If $|x| > a$, then $x < -a$ or $x > a$.

If $|x| = |y|$, then $x = y$ or $x = -y$.

Review Exercises

[2.1] *State the degree of the term.*

1. $15x^4y^6$ tenth

2. $6x$ first

3. $-4xyz^5$ seventh

Simplify the expression. If an expression cannot be simplified, so state.

4. $x^2 + 3x + 6$ cannot be simplified

5. $x^2 + 2xy + 6x^2 - 4$ $7x^2 + 2xy - 4$

6. $3(x + 4) - 3x - 4$ 8

7. $2[-(x - y) + 3x] - 5y + 6$ $4x - 3y + 6$

Solve each equation. If an equation has no solution, so state.

8. $\dfrac{x - 4}{5} = 9 - x$ $\dfrac{49}{6}$

9. $3(x + 2) - 6 = 4(x - 5)$ 20

10. $3 + \dfrac{x}{2} = \dfrac{5}{6}$ $-\dfrac{13}{3}$

11. $-6 - 2x = \dfrac{1}{2}(4x + 12) + 2$ $-\dfrac{7}{2}$

12. $2\left(\dfrac{x}{2} - 4\right) = 3\left(x + \dfrac{1}{3}\right)$ $-\dfrac{9}{2}$

13. $3x - 4 = 6x + 4 - 3x$ no solution

14. $3\left[2x - \left(\dfrac{1}{2}x + 4\right)\right] = -3$ 2

[2.2] *Evaluate the formula for the values given.*

15. $P = \dfrac{nRT}{V}$: $n = 10, R = 100, T = 4, V = 20$ 200

16. $x = \dfrac{-b + \sqrt{b^2 - 4ac}}{2a}$: $a = 8, b = 10, c = -3$ $\dfrac{1}{4}$

17. $h = \dfrac{1}{2}at^2 + v_0t + h_0$: $a = -32, v_0 = 60; h_0 = 120, t = 2$ 176

18. $z = \dfrac{\bar{x} - \mu}{\dfrac{\sigma}{\sqrt{n}}}$: $\bar{x} = 60, \mu = 64, \sigma = 5, n = 25$ -4

Solve for the variable indicated.

19. $A = lw$, for l $l = A/w$

20. $A = \pi r^2 h$, for h $h = A/\pi r^2$

21. $P = 2l + 2w$, for w $w = (P - 2l)/2$

22. $d = rt$, for r $r = d/t$

23. $y = mx + b$, for m $m = (y - b)/x$

24. $2x - 3y = 5$, for y $y = (2x - 5)/3$

25. $P_1V_1 = P_2V_2$, for V_2 $V_2 = P_1V_1/P_2$

26. $S = \dfrac{3a + b}{2}$, for a $a = (2S - b)/3$

27. $K = 2(d + l)$, for l $l = (K - 2d)/2$

28. $2s - nf - nl = 0$, for n $n = \dfrac{2s}{f + l}$

29. $2(x + y) = 2 - 3xy$, for y $y = \dfrac{-2x + 2}{3x + 2}$

30. $x = \dfrac{y + 6}{y}$, for y $y = \dfrac{6}{x - 1}$

[2.3] *Write an equation that can be used to solve the problem. Solve the problem and check your answer.*

31. Four times a number increased by 12 is 32. Find the number.   5

32. One-fourth of a number plus 6 is 11. Find the number.   20

33. The sum of two consecutive odd integers is 28. Find the integers.   13, 15

34. When the price of a jacket is decreased by 60% it costs $20. Find the original price of the jacket. $50

35. A small town's population is increasing by 350 people per year. If the present population is 4750, how long will it take for the population to reach 5800?   3 yr

36. Dawn's salary is $300 per week plus 6% commission of sales. How much in sales must Dawn make to earn $650 in a week? $5833.33

37. The one-way bus fare for John to get to work is $1.65. A monthly bus pass that provides unlimited free bus travel during the month costs $27.50. How many *round trips* to and from work would John need to make in order to make the purchase of the bus pass worthwhile?   9 or more

38. At a going-out-of-business sale, furniture is selling at 40% off the regular price. In addition, green-tagged items are reduced by an additional $20. If Lalo purchased a green-tagged item and paid $120, find its regular price. $233.33

[2.4] *Solve the following motion and mixture problems.*

39. Tanya is a quality control inspector at the Eastman Kodak Company. In a typical 8-hour workday, she inspects 245 rolls of film. What is Tanya's hourly inspection rate?   30.6 rolls/hr

40. The Sampsons invest $10,000 in two accounts. One account pays 8% simple interest and the other account pays 5% simple interest. If the total interest for the year is $680, how much money was invested in each account? $6000 at 8%, $4000 at 5%

41. Two trains leave Portland at the same time traveling in opposite directions. One train travels at 60 miles per hour and the other at 90 miles per hour. In how many hours will they be 400 miles apart? $2\frac{2}{3}$ hr

42. Space Shuttle 2 takes off 0.5 hour after Shuttle 1 takes off. If Shuttle 2 travels 300 miles per hour faster than Shuttle 1 and overtakes Shuttle 1 exactly 5 hours after Shuttle 2 takes off, find **(a)** the speed of Shuttle 1; **(b)** the distance from the launch pad when Shuttle 2 overtakes Shuttle 1.   **(a)** 3000 mph; **(b)** 16,500 mi

43. Mr. Tomlins, the owner of a gourmet coffee shop, has two coffees, one selling for $6.00 per pound and the other for $6.80 per pound. How many pounds of each type of coffee should he mix to make 40 pounds of coffee to sell for $6.50 per pound?   15 lb at $6, 25 lb at $6.80

[2.3, 2.4] *Solve the following word problems.*

44. A blouse has been reduced by 12%. The sale price is $22. Find the original price. $25

45. Nicolle jogged for a distance and then turned around and walked back to her starting point. While jogging she averaged 7.2 miles per hour, and while walking she averaged 2.4 miles per hour. If the total time spent jogging and walking was 4 hours, find **(a)** how long she jogged; **(b)** the total distance she traveled.

46. Find the three angles of a triangle if one angle measures 25° greater than the smallest angle and the other angle measures 5° less than twice the smallest angle.

47. Two hoses are filling a swimming pool. The hose with the larger diameter supplies 1.5 times as much water as the hose with the smaller diameter. The larger hose is

45. **(a)** 1 hr **(b)** 14.4 mi   **46.** 40°, 65°, 75°

on for 2 hours before the smaller hose is turned on. If 5 hours after the larger hose is turned on there are 3150 gallons of water in the pool, find the rate of flow from each hose.   300 gal/hr, 450 gal/hr

48. The sum of two consecutive integers is 49. Find the integers.   24, 25

49. A clothier has two blue dye solutions, both made from the same concentrate. One solution is 6% blue dye and the other is 20% blue dye. How many ounces of the 20% solution must be mixed with 10 ounces of the 6% solution to result in the mixture being a 12% blue dye solution?   7.5 oz

50. Ken invests $12,000 in two savings accounts. One account is paying 10% simple interest and the other account is paying 6% simple interest. If the same interest is earned on each account, how much was invested at each rate? $4500 at 10%, 7500 at 6%

51. The West Ridge Fitness Center has two membership plans. The first plan is a flat $40 per month fee plus $1.00 per visit. The second plan is $25 per month plus a $4.00 per visit charge. How many visits would Mike have to make per month to make it advantageous for him to select the first plan?   more than 5

52. Two trains leave Tucson at the same time, along parallel tracks, traveling in opposite directions. The faster train travels 10 miles per hour faster than the slower train. Find the speed of the *faster* train if the trains are 510 miles apart after 3 hours.   90 mph

SEE EXERCISE 42.

[2.5] *Solve the inequality. Graph the solution on the real number line.*

53. $x - 3 \geq 4$ 7

54. $2 - x \leq 5$ −3

55. $2x + 4 > 9$ $\frac{5}{2}$

56. $16 \leq 4x - 5$ $\frac{21}{4}$

57. $\dfrac{4x + 3}{5} > -3$ $-\frac{9}{2}$

58. $2(x - 3) > 3x + 4$ −10

59. $-4(x - 2) \leq 6x + 4$ $\frac{2}{5}$

60. $\dfrac{x}{4} \geq 5 - 2x$ $\frac{20}{9}$

Write an inequality that can be used to solve the problem. Solve the inequality and answer the question.

61. A small airplane has a maximum load of 1525 pounds if it is to take off safely. If the passengers weigh 468 pounds, how many 80-pound boxes can be safely transported on the plane?   13 boxes

62. Jack, a telephone operator, informs a customer in a phone booth that the charge for calling Omaha, Nebraska, is $4.50 for the first 3 minutes and 95 cents each additional minute and any part thereof. How long can the customer talk if he has $8.65?   7 min

63. A fitness center guarantees that customers will lose a minimum of 3 pounds the first week and $1\frac{1}{2}$ pounds each additional week. Find the maximum amount of time needed to lose 27 pounds.   17 weeks

Solve the inequality. Write the solution in interval notation.

64. $1 < x - 4 < 7$ $(5, 11)$

65. $2 \leq x + 5 < 8$ $[-3, 3)$

66. $3 < 2x - 4 < 8$ $\left(\frac{7}{2}, 6\right)$

67. $-12 < 6 - 3x < -2$ $\left(\frac{8}{3}, 6\right)$

68. $-1 \leq \dfrac{2x - 3}{4} < 5$ $\left[-\frac{1}{2}, \frac{23}{2}\right)$

69. $-8 < \dfrac{4 - 2x}{3} < 0$ $(2, 14)$

70. Manuel's first four exam grades are 94, 73, 72, and 80. If a final average greater than or equal to 80 and less than 90 is needed to receive a final grade of B in the course, what range of grades on the fifth and last exam will result in Manuel receiving a B in the course? Assume a maximum grade of 100. $\{x \mid 81 \leq x \leq 100\}$

Find the solution set to each compound inequality.

71. $x < 3$ and $2x - 4 > -10$ $\{x \mid -3 < x < 3\}$

72. $2x - 1 > 5$ or $3x - 2 \leq 7$ $\mathbb{R}$

73. $3x + 5 > 2$ or $6 - x < 1$ $\{x \mid x > -1\}$

74. $4x - 3 \leq 7$ and $2x - 1 \geq 3$ $\left\{x \mid 2 \leq x \leq \frac{5}{2}\right\}$

75. $4x - 5 < 11$ and $-3x - 4 \geq 8$ $\{x \mid x \leq -4\}$

76. $\dfrac{5x - 3}{2} > 7$ or $\dfrac{2x - 1}{3} \leq -3$ $\left\{x \mid x \leq -4 \text{ or } x > \frac{17}{5}\right\}$

[2.6] *Find the solution set.*

77. $|x| = 4$ $\{-4, 4\}$

78. $|x| < 3$ $\{x \mid -3 < x < 3\}$

79. $|x| \geq 4$ $\{x \mid x \leq -4 \text{ or } x \geq 4\}$

80. $|x - 4| = 9$ $\{-5, 13\}$

81. $|x - 2| \geq 5$ $\{x \mid x \leq -3 \text{ or } x \geq 7\}$

82. $|4 - 2x| = 5$ $\left\{-\frac{1}{2}, \frac{9}{2}\right\}$

83. $|3 - 2x| < 7$ $\{x \mid -2 < x < 5\}$

84. $\left|\dfrac{2x - 3}{5}\right| = 1$ $\{-1, 4\}$

85. $\left|\dfrac{x - 4}{3}\right| < 6$ $\{x \mid -14 < x < 22\}$

86. $\left|\dfrac{4 - x}{3}\right| \geq \dfrac{1}{2}$ $\left\{x \mid x \leq \frac{5}{2} \text{ or } x \geq \frac{11}{2}\right\}$

87. $|3x - 4| = |x + 3|$ $\left\{\frac{7}{2}, \frac{1}{4}\right\}$

88. $|2x - 3| + 4 \geq -10$ $\mathbb{R}$

[2.5, 2.6] *Solve the inequality and give the solution in interval notation.*   **96.** $(-\infty, -11] \cup (14, \infty)$ **99.** $(-\infty, -12) \cup (20, \infty)$

89. $\dfrac{5x - 3}{2} > 6$ $(3, \infty)$

90. $\dfrac{x}{3} \leq 2x - 5$ $[3, \infty)$

91. $|x + 6| < -1$ $\varnothing$

92. $\left|\dfrac{x - 3}{4}\right| \leq 5$ $[-17, 23]$

93. $-6 \leq \dfrac{3 - 2x}{4} < 5$ $\left(-\frac{17}{2}, \frac{27}{2}\right]$

94. $x \leq 4$ and $4x - 6 \geq -14$ $[-2, 4]$

95. $|3(x + 2)| \leq \dfrac{9}{2}$ $\left[-\frac{7}{2}, -\frac{1}{2}\right]$

96. $\dfrac{2x - 8}{4} > 5$ or $\dfrac{3x - 2}{5} \leq -7$

97. $\dfrac{2}{3}x - 4 < \dfrac{3}{5}x + 2$ $(-\infty, 90)$

98. $|4 - 3x| \geq 5$ $\left(-\infty, -\frac{1}{3}\right] \cup [3, \infty)$

99. $\left|\dfrac{x - 4}{2}\right| - 3 > 5$

100. $\dfrac{3}{5} < \dfrac{2x - 4}{3} \leq \dfrac{9}{4}$ $\left(\frac{29}{10}, \frac{43}{8}\right]$

Practice Test

1. State the degree of the term $-6xy^2z^3$.   sixth

2. Solve the equation $3(x - 2) = 4(4 - x) + 5$. $\frac{27}{7}$

3. Solve the equation $\dfrac{3}{5} - \dfrac{x}{2} = 4$. $-\frac{34}{5}$

4. Solve the equation $\dfrac{3x}{4} - 1 = 5 + \dfrac{2x - 1}{3}$ 68

5. Find the value of S_n for the given values.

$$S_n = \dfrac{a_1(1 - r^n)}{1 - r}, \quad a_1 = 3, r = \dfrac{1}{3}, n = 3 \quad \dfrac{13}{3}$$

6. Solve $c = \dfrac{a - 3b}{2}$ for b. $b = (a - 2c)/3$

7. Solve $A = \frac{1}{2}h(b_1 + b_2)$ for b_2 $b_2 = (2A - hb_1)/h$

In Exercises 8–12, write an equation that can be used to solve the problem. Solve the equation and answer the question asked.

8. The sum of two consecutive integers is 47. Find the two integers.   23, 24

9. The cost of renting an automobile is $35 a day plus 15 cents a mile. How far can Valerie drive in 1 day on $65?   200 mi

10. Two joggers start at the same point at the same time and jog in opposite directions. Homer jogs at 4 miles per hour, while Frances jogs at $5\frac{1}{4}$ miles per hour. How far apart will they be in $1\frac{1}{4}$ hours?   11.56 mi

11. How many liters of 12% salt solution must be added to 10 liters of 25% salt solution to get a 20% salt solution?   6.25 L

12. Millie Johnson has $12,000 to invest. She places part of her money in a savings account paying 8% simple interest and the balance in a savings account paying 7% simple interest. If the total interest from the two accounts at the end of one year is $910, find the amount placed in each account. $7000 at 8%; $5000 at 7%

13. Solve the following inequality and graph the solution on the number line.
$$\frac{6 - 2x}{5} \geq -12$$
33

14. Solve the inequality and write the solution in interval notation.
$$-4 < \frac{x + 4}{2} < 8 \quad (-12, 12)$$

Find the solution set to the following equations.

15. $|x - 4| = 5$ $\{-1, 9\}$

16. $\left|2x - 3\right| = \left|\frac{1}{2}x - 10\right|$ $\left\{-\frac{14}{3}, \frac{26}{5}\right\}$

Find the solution set to the inequalities.

17. $|2x - 3| + 1 > 6$ $\{x \mid x < -1 \text{ or } x > 4\}$

18. $\left|\frac{2x - 3}{4}\right| \leq \frac{1}{2}$ $\left\{x \mid \frac{1}{2} \leq x \leq \frac{5}{2}\right\}$

----

12. Conditional: true only under specific conditions; identity true for any value; inconsistent never

Cumulative Review Test

15. (b) $\left\{x \mid -2 < x < \frac{8}{5}\right\}$ (c) $\left(-2, \frac{8}{5}\right)$ **17.** $\{x \mid x \leq -10 \text{ or } x \geq 14\}$

1. Let $A = \{1, 4, 6, 7, 9, 12\}$ and $B = \{2, 3, 4, 5, 6, 9, 10, 12\}$. Find
  (a) $A \cup B$ $\{1, 2, 3, 4, 5, 6, 7, 9, 10, 12\}$
  (b) $A \cap B$. $\{4, 6, 9, 12\}$

2. Name the indicated properties.
  (a) $4x + y = y + 4x$ commutative property of addition
  (b) $(2x)y = 2(xy)$ associative property of multiplication
  (c) $2(x + 3) = 2x + 6$ distributive property

3. Insert $<$, $>$, or $=$ in the shaded area to make the statement true: $-|-3|$ ▨ $|-5|$. $<$

Evaluate.

4. $4 - |-3| - (6 + |-3|)^2$ -80

5. $-4^2 + (-6) \div (2^3 - 2)^2$ -15

6. $x^3 - xy + y^2$ when $x = -3$ and $y = -2$ -29

7. $\dfrac{8 - \sqrt[3]{27} \cdot 3 \div 9}{|-5| - [5 - (12 \div 4)]^2}$ 7

Solve.

8. $3x - 4 = -2(x - 3) - 9$ $\frac{1}{5}$

9. $1.2(x - 3) = 2.4x - 4.98$ 1.15

10. $\dfrac{x}{4} - 5 = 3x - \dfrac{1}{3}$ $-\dfrac{56}{33}$

11. $\dfrac{\frac{1}{4}x + 2}{3} = \dfrac{x - 4}{4}$ 10

12. Explain the difference between a conditional linear equation, an identity, and an inconsistent linear equation, and give an example of each.

13. Evaluate the formula $x = \dfrac{-b + \sqrt{b^2 - 4ac}}{2a}$ for $a = 3$, $b = -8$, and $c = -3$. 3

14. Solve the formula $A = p + prt$ for t. $t = \dfrac{A - p}{pr}$

15. Solve the inequality and give the answer **(a)** on the number line; **(b)** as a solution set; **(c)** in interval notation.   **(a)** -2 $\frac{8}{5}$
$$-4 < \frac{5x - 2}{3} < 2$$

Find the solution set.   **20.** $\frac{2}{3}$ L of 50%, $1\frac{1}{3}$ L of 20%

16. $|4z + 8| = 12$ $\{1, -5\}$ **17.** $|2x - 4| - 6 \geq 18$

18. The Computer Tutor has reduced the price of a computer by 20%. Find the original price of the computer if the sale price is $1800. $2250

19. Two cars leave Caldwell, New Jersey at the same time traveling in opposite directions. The car traveling west is moving 10 miles per hour faster than the car traveling east. If the two cars are 270 miles apart after 3 hours, find the speed of each car.   40 mph, 50 mph

20. Mr. Kane has a 20% saltwater solution and a 50% saltwater solution. How much of each solution should he mix to get 2 liters of a 30% saltwater solution?

20. $\frac{2}{3}$ L of 50%, $1\frac{1}{3}$ L of 20%

Chapter *3*

Graphs and Functions

See Section 3.4, Example 9

Preview and Perspective

One of the primary goals of this book is to provide you with a good understanding of graphing and graphing techniques. Graphing is heavily used in this course and in other mathematics courses you may take. To reinforce your knowledge of this topic we introduce graphing early and discuss it repeatedly throughout the book. Many of the exercise sets have graphs taken from newspapers or magazines. The material presented in this chapter may help you understand them better.

In Section 3.4 we introduce the concept of **function.** Functions are a unifying concept used throughout all of mathematics. We also use functions throughout the book to reinforce and expand upon what you learn in this chapter.

Most of you have graphed linear equations before. However, you probably have not graphed the nonlinear equations presented in Section 3.5. Make sure you read this section carefully. We will be using the material from this section when we graph other types of nonlinear equations in Section 5.6, and when we solve nonlinear systems of equations graphically in Chapter 10.

In Section 2.5 we solved linear inequalities *in one variable.* In Section 3.6 we graph linear inequalities *in two variables.* We will use the procedures for graphing linear inequalities in two variables again when we graph systems of linear inequalities in Section 4.6.

3.1 The Cartesian Coordinate System, Distance, and Midpoint Formulas

1. Interpret bar graphs.
2. Interpret line graphs.
3. Plot points in the Cartesian coordinate system.
4. Find the distance between two points.
5. Find the midpoint of a line segment.

The emphasis of this chapter will be on graphs of equations. We will begin by introducing bar and line graphs. You see bar and line graphs daily in newspapers and magazines and being able to interpret them will help you throughout life.

Bar Graphs

1. A **bar graph** displays information using bars. Often one of the items displayed is an amount. The amount may be listed on either the horizontal or vertical axis. To determine the amount, if it is not listed next to the bar, estimate where the end of the bar would be on the axis indicating amounts.

EXAMPLE 1 The bar graph in Fig. 3.1 was constructed with information provided by the College Saving Bank. The graph indicates the cost of a four-year college education, depending on the age of children now (October 1994).

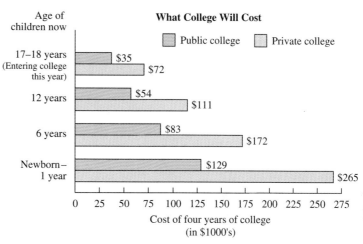

FIGURE 3.1

(a) What will be the average cost to provide a four-year college education to a newborn when he or she is old enough to attend college, at a public college and at a private college?

(b) What will be the average difference in cost of a four-year education at a public college for a student entering college now, and when a newborn enters college?

(c) Repeat part (b) for a private college.

(d) Redraw the bar graph so that the dollar amount is listed on the vertical axis.

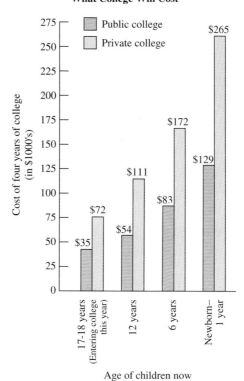

FIGURE 3.2

Solution:

(a) From the bar graph on page 131 we see the average cost for a four-year education when a newborn is old enough to enter a private college will be $265,000. The cost for a public education will be about $129,000.

(b, c) To find the difference, we subtract the present cost from the future cost.

Difference in cost at a public college = $129,000 − $35,000 = $94,000

Difference in cost at a private college = $265,000 − $72,000 = $193,000

(d) To redraw the graph we list the dollar amount on the vertical axis and the age on the horizontal axis. The bar graph is illustrated in Fig. 3.2 on page 131.

Line Graphs   ②   A **line graph** is also used to display information. On a line graph the horizontal axis is generally some unit of time, such as years, months, or days. The vertical axis displays other information. To construct a line graph, measurements are made at specific time intervals. The measurements are marked at their corresponding values along the vertical axis. Then the marks are connected by straight line segments.

EXAMPLE 2   The line graphs in Fig. 3.3 were constructed from information provided by the Cellular Telecommunications Industry Association. The monthly costs are based on the cost of phone ownership plus 250 minutes of prime-time use per month.

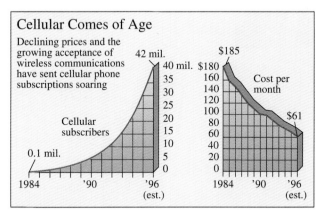

FIGURE 3.3

(a) Estimate the number of cellular subscribers, and the average cost per month, in the year 1990.

(b) Estimate the change in the number of subscribers from 1984 to 1985 and from 1995 to 1996.

(c) Describe the trend in the number of cellular subscribers from 1984 to 1996.

(d) Does there appear to be a relationship between the number of cellular subscribers and the average cost per month? Explain.

Solution:

(a) The number of subscribers in 1990 was about 5 million. To determine this, follow the vertical line up from 1990 to where the line meets the graph. Then draw a horizontal line from this point. The horizontal line meets the vertical axis at about 5 million. From the graph on the right, we see that the cost in 1990 was about $100 per month.

(b) The number of subscribers in 1985 was only slightly greater than the number of subscribers in 1984. If we assume the number of subscribers in 1985 to be 0.2 million, the difference in the number of subscribers from 1984 to 1985 is 0.2 million − 0.1million = 0.1 million. The number of subscribers in 1995 is about 30 million. Therefore, the estimated increase in the number of subscribers between 1995 and 1996 is 42 million − 30 million or 12 million.

(c) The number of cellular users has increased dramatically since 1984, with each year having more subscribers than the previous year. The growth in the number of subscribers has not been the same year after year. For example, from 1984 to 1985 the number of subscribers grew by only about 0.1 million, while from 1995 to 1996 the estimated number of subscribers grew by about 12 million. (This curve is an example of an exponential growth curve, which we will discuss in Chapter 11.)

(d) Notice that as the cost per month decreased the number of subscribers increased. As the cost decreased, more people could afford to use cellular phones. Do you believe that this trend will continue?

In Example 2(b), we had to estimate the number of subscribers. It is sometimes difficult to obtain exact values from line graphs and bar graphs. Line graphs and bar graphs are very useful in illustrating trends over a period of time.

Plotting Points in the Cartesian Coordinate System

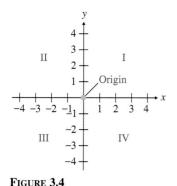

FIGURE 3.4

3 Now we will discuss graphs of equations. A graph is a picture that shows the relationship between two or more variables in an equation. Many algebraic relationships are easier to understand if we can see a visual picture of them.

Before learning how to construct a graph, you must know the **Cartesian (or rectangular) coordinate system.**

The Cartesian coordinate system, named after the French mathematician and philosopher René Descartes (1596–1650), consists of two axes (or number lines) in a plane drawn perpendicular to each other (see Fig. 3.4). Note how the two axes yield four **quadrants,** labeled I, II, III, and IV.

The horizontal axis is called the **x axis.** The vertical axis is called the **y axis.** The point of intersection of the two axes is called the **origin.** Starting from the origin and moving to the right, the numbers increase; moving to the left, the numbers decrease. Starting from the origin and moving up, the numbers increase; moving down, the numbers decrease.

To graph a point, it is necessary to know both its *x* coordinate and *y* coordinate. An **ordered pair** (*x, y*) is used to give the two coordinates of a point. If, for example, the *x* coordinate of a point is 2 and the *y* coordinate is 3, the ordered pair representing the point is (2, 3). Note that the *x* coordinate is always the first coordinate listed in the ordered pair. The point corresponding to the ordered pair (2, 3)

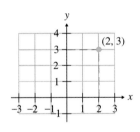

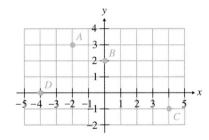

FIGURE 3.5 **FIGURE 3.6**

is plotted in Fig. 3.5. The phrase "the point corresponding to the ordered pair (2, 3)" is often abbreviated "the point (2, 3)." For example, if we write "the point (−1, 5)," it means the point corresponding to the ordered pair (−1, 5). The ordered pairs, A at (−2, 3), B at (0, 2), C at (4, −1), and D at (−4, 0) are plotted in Fig. 3.6.

EXAMPLE 3 Plot each of the following points on the same set of axes.

(a) $A(4, 2)$ **(b)** $B(0, -3)$ **(c)** $C(-3, 1)$ **(d)** $D(4, 0)$

Solution: See Fig. 3.7. Notice that when the x coordinate is 0, as in part (b), the point is on the y axis. When the y coordinate is 0, as in part (d), the point is on the x axis.

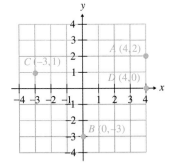

FIGURE 3.7

Distance between Two Points

4 Now we will see how to find the distance between any two points in a plane. After this, we will show how to find the midpoint of a given line segment. You need these two concepts to understand conic sections (Chapter 10).

To find the distance, d, between two points, we use the distance formula.

Distance Formula

The distance, d, between any two points (x_1, y_1) and (x_2, y_2) can be found by the distance formula

$$d = \sqrt{(x_2 - x_1)^2 + (y_2 - y_1)^2}$$

The distance between any two points will always be a positive number. Can you explain why? When finding the distance, it makes no difference which point we designate as point 1 (x_1, y_1) or point 2 (x_2, y_2). Note that the square of any real number will always be greater than or equal to zero. For example, $(5 - 2)^2 = (2 - 5)^2 = 9$.

EXAMPLE 4 Determine the distance between the points (−1, 5) and (−4, 1).

Solution: First plot the points (Fig. 3.8). Call (−1, 5) point 2 and (−4, 1) point 1. Thus, (x_2, y_2) represents (−1, 5) and (x_1, y_1) represents (−4, 1). Now use the distance formula to find the distance, d.

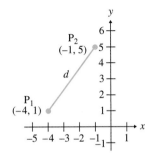

FIGURE 3.8

$$d = \sqrt{(x_2 - x_1)^2 + (y_2 - y_1)^2}$$
$$= \sqrt{[-1 - (-4)]^2 + (5 - 1)^2}$$
$$= \sqrt{(-1 + 4)^2 + 4^2}$$
$$= \sqrt{3^2 + 4^2}$$
$$= \sqrt{9 + 16} = \sqrt{25} = 5$$

Thus, the distance between the points (−1, 5) and (−4, 1) is 5 units.

If in Example 4 we had selected $(-4, 1)$ for point 2 and $(-1, 5)$ for point 1, our results would not have changed.

$$
\begin{aligned}
d &= \sqrt{(x_2 - x_1)^2 + (y_2 - y_1)^2} \\
&= \sqrt{[-4 - (-1)]^2 + (1 - 5)^2} \\
&= \sqrt{(-4 + 1)^2 + (-4)^2} \\
&= \sqrt{(-3)^2 + (-4)^2} \\
&= \sqrt{9 + 16} = \sqrt{25} = 5
\end{aligned}
$$

When using the distance formula, do not expect the value of the distance to always be a rational number. If your answer is an irrational number, such as $\sqrt{187}$, you could use a calculator to obtain an approximate decimal value. On a calculator with a square root key, we can determine that $\sqrt{187} \approx 13.674794$. Rounding this approximation to the nearest hundredth, we find that $\sqrt{187} \approx 13.67$.

Helpful Hint

Students will sometimes begin finding the distance correctly using the distance formula but will forget to take the square root of the sum $(x_2 - x_1)^2 + (y_2 - y_1)^2$ to obtain the correct answer. When taking the square root, remember that $\sqrt{a^2 + b^2} \neq a + b$.

Midpoint of a Line Segment

5 It is often necessary to find the midpoint of a line segment between two given points. To do this, we use the midpoint formula.

Midpoint Formula

Given any two points (x_1, y_1) and (x_2, y_2), the point halfway between the given points can be found by the midpoint formula:

$$
\text{midpoint} = \left(\frac{x_1 + x_2}{2}, \frac{y_1 + y_2}{2} \right)
$$

To find the midpoint, we take the average (the mean) of the two x values and of the two y values.

EXAMPLE 5 Determine the midpoint of the line segment between the points $(-3, 6)$ and $(4, 1)$.

Solution: It makes no difference which points we label (x_1, y_1) and (x_2, y_2). Let $(-3, 6)$ be (x_1, y_1) and $(4, 1)$ be (x_2, y_2) (see Fig. 3.9).

$$
\begin{aligned}
\text{Midpoint} &= \left(\frac{x_1 + x_2}{2}, \frac{y_1 + y_2}{2} \right) \\
&= \left(\frac{-3 + 4}{2}, \frac{6 + 1}{2} \right) = \left(\frac{1}{2}, \frac{7}{2} \right)
\end{aligned}
$$

The point $\left(\frac{1}{2}, \frac{7}{2} \right)$ is halfway between the points $(-3, 6)$ and $(4, 1)$.

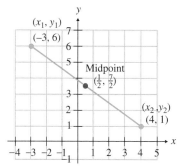

FIGURE 3.9

Graphing Calculator Corner

One of the main features that set graphing calculators apart from other calculators is their ability to display one or more graphs on their screen. We will discuss this feature throughout the book.

When graphing on a grapher you must be sure that the x and y axes are set properly on your calculator. When you set the maximum and minimum values that you wish displayed on the x axis you are setting the "*domain*," and when you set the maximum and minimum values that you wish displayed on the y *axis* you are setting the "range." You also need to provide the scale for the x and y axes. We will discuss the words *domain* and *range* further in Section 3.4 and throughout the book.

When you enter the domain and range you are defining what is to appear in the calculator's viewing **window** or display. On some graphing calculators the words *window variables* are used in place of the words domain and range. Below we show the graph of $y = x^3 - 2x + 3$ as displayed on the same calculator with two different axes instructions.

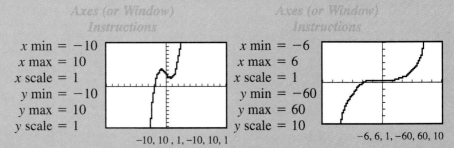

Axes (or Window) Instructions	*Axes (or Window) Instructions*
x min $= -10$	x min $= -6$
x max $= 10$	x max $= 6$
x scale $= 1$	x scale $= 1$
y min $= -10$	y min $= -60$
y max $= 10$	y max $= 60$
y scale $= 1$	y scale $= 10$
$-10, 10, 1, -10, 10, 1$	$-6, 6, 1, -60, 60, 10$

Below every calculator display illustrating a graph, we will list the values of x min, x max, x scale, y min, y max, y scale, in this order, that was used in obtaining the graph.

EXERCISES

Assume that the following axes information has been entered into your calculator. Draw a *window* (the display screen) showing the set of axes that would be displayed for your grapher.

	x min	x max	x scale	y min	y max	y scale
1.	-12	12	2	-20	20	4
2.	-50	50	5	-100	100	10
3.	-1	1	0.1	-6	9	3
4.	-10	12	2	0	600	50

1. **2.** **3.** **4.**

Exercise Set 3.1

1. $A(3, 1)$, $B(-6, 0)$, $C(2, -4)$, $D(-2, -4)$, $E(0, 3)$, $F(-8, 1)$, $G\left(\frac{3}{2}, -1\right)$

2. $A(8, 5)$ $B(14, 0)$, $C(6, -15)$, $D(-4, 20)$, $E(0, -20)$, $F(-10, -5)$, $G(-5, 10)$

List the ordered pairs corresponding to the indicated points.

1.

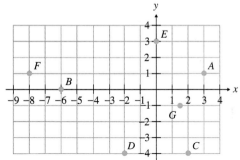

2.

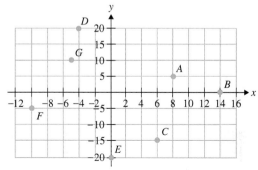

3. Graph the following points on the same axes.

    (a) $(4, 2)$ **(b)** $(-6, 2)$

    (c) $(0, -1)$ **(d)** $(-2, 0)$

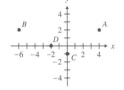

4. Graph the following points on the same axes.

    (a) $(-4, -2)$ **(b)** $(3, 2)$

    (c) $(2, -3)$ **(d)** $(-3, 3)$

Determine the distance between the points. Use a calculator with a square root key and round your answer to the nearest hundredth.

5. $(2, -2)$ and $(2, -5)$ 3

6. $(-5, 5)$ and $(-5, 1)$ 4

7. $(-4, 3)$ and $(5, 3)$ 9

8. $(-1, -1)$ and $(3, 2)$ 5

9. $(1, 4)$ and $(-3, 1)$ 5

10. $(-1, -4)$ and $(4, 8)$ 13

11. $(-3, -5)$ and $(6, -2)$ $\sqrt{90} \approx 9.49$

12. $(5, 3)$ and $(-5, -3)$ $\sqrt{136} \approx 11.66$

13. $(0, 6)$ and $(5, -1)$ $\sqrt{74} \approx 8.60$

14. $(4.2, -3.6)$ and $(-2.6, 2.3)$ $\sqrt{81.05} \approx 9.00$

15. $(-1.6, 3.5)$ and $(-4.3, -1.7)$ $\sqrt{34.33} \approx 5.86$

16. $(3, -1)$ and $\left(\frac{1}{2}, 4\right)$ $\sqrt{\frac{125}{4}} \approx 5.59$

17. $\left(\frac{3}{4}, 2\right)$ and $\left(-\frac{1}{2}, 6\right)$ $\sqrt{\frac{281}{16}} \approx 4.19$

18. $(4, 0)$ and $\left(-\frac{3}{5}, -4\right)$ $\sqrt{\frac{929}{25}} \approx 6.10$

Determine the midpoint of the line segment between the points.

19. $(5, 2)$ and $(-1, 4)$ $(2, 3)$

20. $(1, 4)$ and $(2, 6)$ $\left(\frac{3}{2}, 5\right)$

21. $(-5, 3)$ and $(5, -3)$ $(0, 0)$

22. $(0, 8)$ and $(4, -6)$ $(2, 1)$

23. $(-2, -8)$ and $(-6, -2)$ $(-4, -5)$

24. $(4, 7)$ and $(1, -3)$ $\left(\frac{5}{2}, 2\right)$

25. $(1, -6)$ and $(-8, -4)$ $\left(-\frac{7}{2}, -5\right)$

26. $(15.3, -6.2)$ and $(8.2, -12.4)$ $(11.75, -9.3)$

27. $(-9.62, 12.58)$ and $(3.52, 6.57)$ $(-3.05, 9.575)$

28. $\left(3, \frac{1}{2}\right)$ and $(2, -4)$ $\left(\frac{5}{2}, -\frac{7}{4}\right)$

29. $\left(\frac{5}{2}, 3\right)$ and $\left(2, \frac{9}{2}\right)$ $\left(\frac{9}{4}, \frac{15}{4}\right)$

30. $\left(-\frac{5}{2}, -\frac{11}{2}\right)$ and $\left(-\frac{7}{2}, \frac{3}{2}\right)$ $(-3, -2)$

Find the perimeter of the triangle determined by the three points. Use a calculator with a square root key, and round the answer to the nearest hundredth. If a calculator with a square root key is not available, leave your answer as a sum of square roots.

31. $A(7, 7)$, $B(7, 1)$, $C(-1, 1)$ $\sqrt{36} + \sqrt{64} + \sqrt{100} = 24$

32. $A(4, 3)$, $B(-2, 3)$, $C(5, 0)$ $\sqrt{36} + \sqrt{58} + \sqrt{10} \approx 16.78$

33. $A(0, -4)$, $B(2, 3)$, $C(4, 6)$ $\sqrt{53} + \sqrt{13} + \sqrt{116} \approx 21.66$

34. $A(-2, -1)$, $B(6, -8)$, $C(3, 5)$

    $\sqrt{113} + \sqrt{178} + \sqrt{61} \approx 31.78$

35. Consider the graph below.

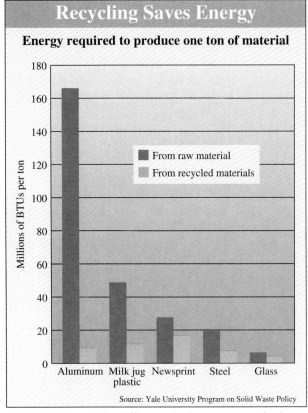

Recycling Saves Energy

Energy required to produce one ton of material

Source: Yale University Program on Solid Waste Policy

(a) Estimate the difference in the energy required to produce 1 ton of aluminum from raw materials and through recycling. ≈ 165 − 10 or 155 million Btu per ton

(b) Estimate the energy saved if 30 million tons of aluminum is produced from recycled aluminum rather than from raw materials. ≈ 4650 million Btu or 4,650,000,000 Btu

(c) Write a paragraph or two describing what this graph shows.

37. Consider the bar graph on the right.

(a) Using an outboard motor for 3 hours creates as much pollution as driving a car how many miles?
2400 mi

(b) Redraw this graph with the equivalent miles driven by car along the vertical axis and the machines along the horizontal axis.

(c) Write a paragraph explaining what this graph shows.

37. (b)

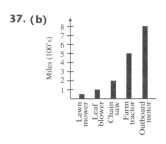

36. Consider the bar graph of grades at Stanford University.

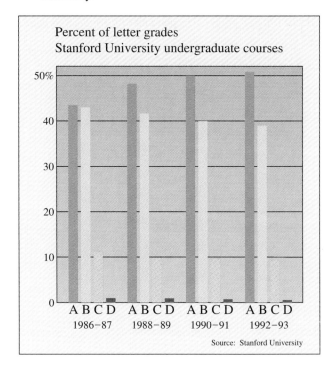

Percent of letter grades
Stanford University undergraduate courses

Source: Stanford University

(a) What percent of grades were either A or B in 1986–1987? In 1992–1993? ≈ 87%, ≈ 90%

(b) Many articles have been written on "grade inflation." Do you believe that this graph shows grade inflation? Explain.

(c) Write a paragraph describing what this graph shows.

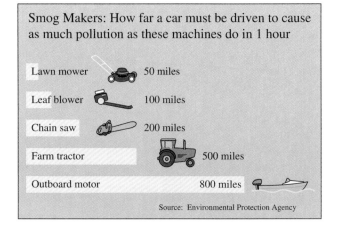

Smog Makers: How far a car must be driven to cause as much pollution as these machines do in 1 hour

Lawn mower — 50 miles

Leaf blower — 100 miles

Chain saw — 200 miles

Farm tractor — 500 miles

Outboard motor — 800 miles

Source: Environmental Protection Agency

38. Consider the bar graph below.

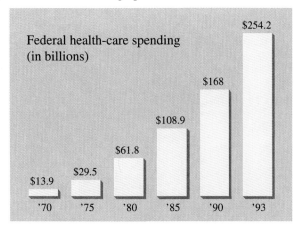

Federal health-care spending (in billions)

$254.2

$168

$108.9

$61.8

$29.5

$13.9

'70 '75 '80 '85 '90 '93

(**a**) 15.6 billion, 86.2 billion

(**a**) How much had health care spending increased from 1970 to 1975? From 1990 to 1993?

(**b**) Redraw this graph using the Cartesian coordinate system with dollars along the vertical axis and years along the horizontal axis. Your axes should be properly labeled.

(**c**) Does your graph give a different impression about federal health care spending? Explain your answer.

(**d**) Write a paragraph describing the trend in federal spending on health care from 1970 to 1993.

39. The information for the line graph that follows was provided by the U.S. Immigration and Naturalization Service.

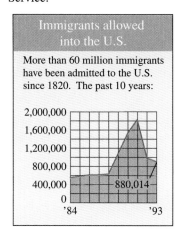

Immigrants allowed into the U.S.

More than 60 million immigrants have been admitted to the U.S. since 1820. The past 10 years:

2,000,000
1,600,000
1,200,000
800,000
400,000   880,014
0
'84   '93

(**a**) Estimate the number of immigrants admitted to the United States in 1987.   ≈ 600,000

(**b**) In what year from 1984 through 1993 were the greatest number of immigrants admitted to the United States? Estimate the number admitted that year.   1991, ≈ 1,800,000

(**c**) Estimate the total number of immigrants admitted to the United States from 1984 through 1993. Explain how you determined your answer.

(**d**) Write a paragraph describing the trend in the number of immigrants admitted to the United States from 1984 through 1993.

40. The following line graph illustrates the number of days an average upper-middle-class wage earner (earning about $50,000 per year) must work to pay for college tuition, and room and board, at an average private or public institution.

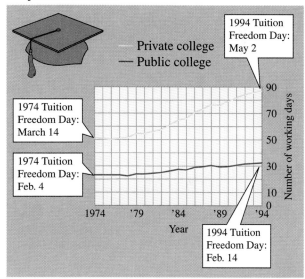

Private college
Public college

1994 Tuition Freedom Day: May 2

1974 Tuition Freedom Day: March 14

1974 Tuition Freedom Day: Feb. 4

1994 Tuition Freedom Day: Feb. 14

90
70
50
30
10
0

Number of working days

1974   '79   '84   '89   '94
Year

(**a**) 73 days, 30 days

(**a**) In 1987, estimate the number of days that had to be worked to reach "Tuition Freedom Day," for a private college and for a public college.

(**b**) What is the difference between the number of working days until tuition freedom day in a private and public college in 1974? (Assume 28 days in February.)   28 days

(**c**) What is the difference between the number of working days until tuition freedom days in a private and public college in 1994?   55 days

(**d**) Write a paragraph describing the change in the number of working days for tuition freedom for private colleges and for public colleges from 1974 to 1994.

38. (**b**)   Federal health-care spending   (**c**) ≈ 9,300,000

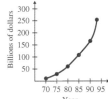

300
250
200
150
100
50

Billions of dollars

70 75 80 85 90 95
Year

41. The line graph below shows the projected world urban population.

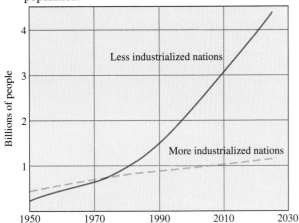

(a) The total worldwide urban population for any year may be estimated by adding together the urban population for the less industrialized nations and the more industrialized nations for that year. Estimate the *total* worldwide urban population in 1970 and in 2010.   ≈ 1.4 billion, ≈ 4.0 billion

(b) From the year 1990 estimate the time it takes for the population to double (called the *doubling time*) in the less industrialized nations.   ≈ 20 yr

(c) Draw a graph illustrating the total worldwide urban population from 1950 to 2020. The population numbers after 1995 are projections.

CUMULATIVE REVIEW EXERCISES

[2.2] **43.** Evaluate $\dfrac{-b + \sqrt{b^2 - 4ac}}{2a}$ for $a = 2$, $b = 7$, and $c = -15$. $\frac{3}{2}$

[2.3] **44.** Hertz Automobile Rental Agency charges a daily fee of $30 plus 14 cents a mile. National Automobile Rental Agency charges a daily fee of $16 plus 24 cents a mile for the same car. What distance would

44. 140 mi

(d) Write a paragraph or two describing the trend in the world urban population in the more industrialized nations and in the less industrialized nations from 1950 to 2030. Be as specific as possible in describing what you see.

SEE EXERCISE 41.

42. When the distance between two different points is found using the distance formula, why must the distance always be a positive number?

41. (c) The square of any nonzero number is positive

you have to drive in 1 day to make the cost of renting from Hertz equal to the cost of renting from National?

[2.5] **45.** Solve the inequality $-4 \le \dfrac{4 - 3x}{2} < 5$. Write the solution in set builder notation. $\{x \mid -2 < x \le 4\}$

[2.6] **46.** Find the solution set for the inequality $|3x + 2| > 5$. $\{x \mid x < -\frac{7}{3} \text{ or } x > 1\}$

Group Activity/ Challenge Problems

Plot each set of points, then find the perimeter and area of the figure whose vertices are:

1. $A(1, -1)$, $B(2, 5)$, $C(5, -1)$ $P = \sqrt{37} + \sqrt{45} + 4 \approx 16.79$, $A = 12$

2. $A(1, 1)$, $B(2, 3)$, $C(6, 3)$, $D(5, 1)$ $P = 8 + 2\sqrt{5} \approx 12.47$, $A = 8$

3. $A(1, -2)$, $B(1, 2)$, $C(6, -2)$, $D(4, 2)$
$P = 12 + \sqrt{20} \approx 16.47$, $A = 16$

3.2 Graphing Linear Equations

Tape 3

1. Write a linear equation in standard form.
2. Graph linear equations by plotting points.
3. Graph linear equations using intercepts.
4. Graph equations of the form $x = a$ and $y = b$.
5. Apply graphing to practical problems.
6. Solve linear equations in one variable graphically.

Standard Form of a Linear Equation

1. A **linear equation** is an equation whose graph is a straight line. Linear equations are also called **first-degree equations** since the degree of their highest-powered term is the first degree.

> **Standard Form of a Linear Equation**
>
> $$ax + by = c$$
>
> where, a, b, and c are real numbers, and a and b are not both 0.

Examples of Linear Equations in Standard Form

$$2x + 3y = 4 \qquad -x + 5y = -2$$

Consider the linear equation in two variables, $y = x + 1$. What is the solution? Since the equation contains two variables, its solutions must contain two numbers, one for each variable. One pair of numbers that satisfies this equation is $x = 1$ and $y = 2$. We know this because when we substitute 1 for x and 2 for y the equation checks: $2 = 1 + 1$. Therefore, one solution to the equation is the ordered pair (1, 2). However, the equation $y = x + 1$ has many other solutions. If you check the ordered pairs (2, 3), (3, 4), (−1, 0), $\left(\frac{1}{2}, \frac{3}{2}\right)$, you will see that they are all solutions. How many possible solutions does the equation have? The equation $y = x + 1$ has an unlimited or *infinite number* of solutions. Since it is not possible to list all the specific solutions to the equation, we illustrate them with a graph. **A graph of an equation is an illustration of the set of points whose coordinates satisfy the equation.**

Graphing Equations by Plotting Points

2. The graphs of all linear equations are straight lines. Since only two points are needed to draw a straight line, when graphing linear equations we need to find and plot only two ordered pairs that satisfy the equation, but it is always a good idea to use a third ordered pair as a check. If the three points are not in a straight line, you have made a mistake. A set of points that lie in a straight line are said to be **collinear.**

One method of finding ordered pairs that satisfy an equation is to solve the equation for y. Then substitute values for x and find the corresponding values of y.

EXAMPLE 1 Graph the equation $y = 3x + 6$.

Solution: This equation is already solved for y. We will find three ordered pairs that satisfy the equation by arbitrarily selecting three values for x, substituting them in the equation, and finding the corresponding values for y. In this equation we let x have values of 0, 2 and -3.

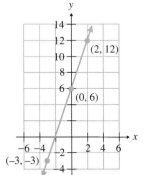

	$y = 3x + 6$	
$x = 0$	$y = 3(0) + 6 = 6$	
$x = 2$	$y = 3(2) + 6 = 12$	
$x = -3$	$y = 3(-3) + 6 = -3$	

x	y
0	6
2	12
-3	-3

FIGURE 3.10

Now plot the three ordered pairs on the same axes (Fig. 3.10). Since the three points are collinear, the line appears correct. Connect the three points with a straight line. Place arrows at the ends of the line to show that the line continues in both directions.

To plot the equation $y = 3x + 6$, we used the three values $x = 0$, $x = 2$, and $x = -3$. We could have picked three entirely different values and obtained exactly the same graph. When selecting values to substitute for x, use whatever values make the equation easy to evaluate.

The graph in Example 1 represents the set of all ordered pairs that satisfy the equation $y = 3x + 6$. If we select any point on this line, the ordered pair representing the point will be a solution to the equation $y = 3x + 6$. Similarly, any solution to the equation will be represented by a point on the line.

EXAMPLE 2 Graph the equation $-2x + 3y = -6$.

Solution: We will first solve the equation for y. This will make it easier to select values to substitute for x that can be quickly evaluated.

$$-2x + 3y = -6$$
$$3y = 2x - 6$$
$$y = \frac{2x - 6}{3}$$
$$y = \frac{2}{3}x - \frac{6}{3}$$
$$y = \frac{2}{3}x - 2$$

Now we will select values for x that make $2x/3$ integral values. $2x/3$ will have integral values when x is a multiple of 3. We will select $x = 0$, 3, and 6.

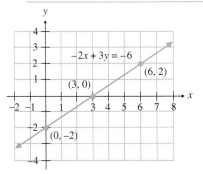

FIGURE 3.11

$$y = \frac{2}{3}x - 2$$

			x	y
$x = 0$	$y = \dfrac{2}{3}(0) - 2 = -2$		0	-2
$x = 3$	$y = \dfrac{2}{3}(3) - 2 = 0$		3	0
$x = 6$	$y = \dfrac{2}{3}(6) - 2 = 2$		6	2

Now plot the points and draw the graph (Fig. 3.11).

Graphing Calculator Corner

Graphing calculators can perform many tasks in addition to graphing. Some graphing calculators can provide tables of values that may be used in graphing. For example, the calculator window below illustrates a table of values for the equation $y = \frac{2}{3}x - 2$, which was graphed in Example 2. We can scroll up the table to find values of y when x is less than -3 and scroll down the table to find values of y when x is greater than 3. Not every calculator has this ability. Read your graphing calculator manual to determine if your graphing calculator can illustrate tables.

X	Y1
-3	-4
-2	-3.333
-1	-2.667
0	-2
1	-1.333
2	-.6667
3	0

X=-3

3.

X	Y1
0	4
1	4.5
2	5
3	5.5
4	6
5	6.5
6	7

X=0

4.

X	Y1
0	7.2
1	3.7
2	.2
3	-3.3
4	-6.8
5	-10.3
6	-13.8

X=0

1.

X	Y1
0	-2
1	2
2	6
3	10
4	14
5	18
6	22

X=0

2.

X	Y1
0	-6
1	-4.8
2	-3.6
3	-2.4
4	-1.2
5	0
6	1.2

X=0

EXERCISES

Determine if your calculator can illustrate tables. If so, use your calculator to display a table of x and y values that start at x = 0, using increments of 1, for the following graphs.

1. $y = 4x - 2$ **2.** $y = 1.2x - 6$ **3.** $y = \frac{1}{2}x + 4$ **4.** $y = -3.5x + 7.2$

EXAMPLE 3 Graph the equation $\dfrac{5}{6}x - \dfrac{1}{3}y = \dfrac{3}{2}$.

Solution: We begin by multiplying both sides of the equation by the least common denominator, 6, to eliminate fractions. Then we solve for y.

$$\frac{5}{6}x - \frac{1}{3}y = \frac{3}{2}$$

$$6\left(\frac{5}{6}x - \frac{1}{3}y\right) = 6 \cdot \frac{3}{2}$$

$$6\left(\frac{5}{6}x\right) - 6\left(\frac{1}{3}y\right) = 9$$

$$5x - 2y = 9$$

$$-2y = -5x + 9$$

$$y = \frac{-5x + 9}{-2}$$

$$y = \frac{5x - 9}{2}$$

If we wanted to we could now rewrite the equation as $y = \frac{5}{2}x - \frac{9}{2}$. However, since the constant, 9/2, is a fraction it might be easier to evaluate the equation in the form $y = \frac{5x - 9}{2}$. Notice that odd values of x will result in y being integer values and even values of x will result in y being a fractional value. Now select values for x and find the corresponding values of y.

$$y = \frac{5x - 9}{2}$$

x	y
0	$-\frac{9}{2}$
1	-2
3	3

$$x = 0 \qquad y = \frac{5(0) - 9}{2} = -\frac{9}{2}$$

$$x = 1 \qquad y = \frac{5(1) - 9}{2} = -2$$

$$x = 3 \qquad y = \frac{5(3) - 9}{2} = 3$$

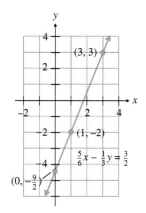

FIGURE 3.12

Plot the points and draw the straight line (Fig. 3.12).

Graphing Equations Using Intercepts

3 Let's examine two points on the graph shown in Fig. 3.11. Note that the graph crosses the x axis at the point $(3, 0)$. Therefore, $(3, 0)$ is called the **x intercept.** The graph crosses the y axis at the point $(0, -2)$. Therefore, $(0, -2)$ is called the **y intercept.** Sometimes, as in Fig. 3.12, the x and y intercepts may not be easy to identify. The x intercept in Fig. 3.12 is $\left(\frac{9}{5}, 0\right)$ and the y intercept is $\left(0, -\frac{9}{2}\right)$. Below we explain how the x and y intercepts may be determined algebraically.

> **x and y Intercepts**
>
> To find the y intercept, set $x = 0$ and solve for y.
> To find the x intercept, set $y = 0$ and solve for x.

It is often convenient to graph linear equations by finding their x and y intercepts.

EXAMPLE 4   Graph the equation $3y = 6x + 12$ by plotting the x and y intercepts.

Solution: To find the y intercept (the point where the graph crosses the y axis), set $x = 0$ and solve for y.

$$3y = 6x + 12$$
$$3y = 6(0) + 12$$
$$3y = 0 + 12$$
$$3y = 12$$
$$y = \frac{12}{3} = 4$$

The graph crosses the y axis at $y = 4$. The ordered pair representing the y intercept is $(0, 4)$.

To find the *x* intercept (the point where the graph crosses the *x* axis), set *y* = 0 and solve for *x*.

$$3y = 6x + 12$$
$$3(0) = 6x + 12$$
$$0 = 6x + 12$$
$$-12 = 6x$$
$$-\frac{12}{6} = x$$
$$-2 = x$$

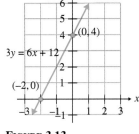

FIGURE 3.13

The graph crosses the *x* axis at *x* = −2. The ordered pair representing the *x* intercept is (−2, 0). Now plot the intercepts and draw the graph (Fig. 3.13).

When graphing equations using only the intercepts, you must be particularly careful because you have no checkpoint. If one of your intercepts is wrong, your graph will be wrong. When graphing by plotting intercepts, you may wish to plot a third point as a checkpoint.

Now return to Example 3 and make sure you can find the *x* and *y* intercepts algebraically.

Graphing Equations of the Form *x* = *a* **or** *y* = *b*

4 Examples 5 and 6 illustrate how equations of the form *x* = *a* and *y* = *b*, where *a* and *b* are constants, are graphed.

EXAMPLE 5   Graph the equation *y* = 3.

Solution: This equation can be written as *y* = 3 + 0*x*. Thus, for any value of *x* selected, *y* is 3. The graph of *y* = 3 is illustrated in Fig. 3.14.

The graph of any equation of the form *y* = *b* will always be a horizontal line for any real number *b*.

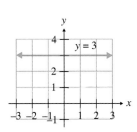

FIGURE 3.14

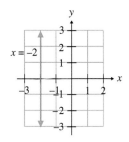

FIGURE 3.15

EXAMPLE 6   Graph the equation *x* = −2.

Solution: This equation can be written as *x* = −2 + 0*y*. Thus, for every value of *y* selected, *x* will have a value of −2 (see Fig. 3.15).

The graph of any equation of the form *x* = *a* will always be a vertical line for any real number *a*.

Applications of Graphing

⑤ Graphs are often used to show the relationship between variables. The axes of a graph do not have to be labeled x and y; they can be any designated variables. Consider the following example.

EXAMPLE 7   The yearly profit, p, of a tire store can be estimated by the formula $p = 20x - 30,000$ where x is the number of tires sold per year.

(a) Draw a graph of profit versus tires sold for up to and including 6000 tires.

(b) Estimate the number of tires that must be sold for the company to break even.

(c) Estimate the number of tires sold if the company has a $40,000 profit.

Solution: **(a)** The minimum number of tires that can be sold is 0. Therefore, negative values do not have to be indicated on the horizontal axis. We will arbitrarily select 3 values for x and find the corresponding values of p. After determining the values that are to be plotted, we construct the scales on the axes in a manner that will allow us to plot those points. The graph is illustrated in Fig. 3.16. Notice that since we were asked to draw the graph for up to 6000 tires, there is a dot on the end of the graph rather than an arrowhead.

x	p
0	$-30,000$
2000	10,000
5000	70,000

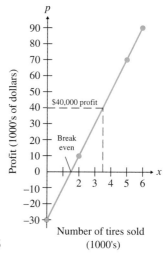

FIGURE 3.16

(b) To break even, approximately 1500 tires must be sold.

(c) To make a $40,000 profit, approximately 3500 tires must be sold.

Sometimes it is difficult to read an exact answer from a graph. To determine the exact number of tires needed to break even, substitute 0 for p in the equation $p = 20x - 30,000$ and solve for x. To determine the exact number of tires needed to obtain a $40,000 profit, substitute 40,000 for p and solve the equation for x.

EXAMPLE 8   Mr. Jordan is part owner in a newly formed toy company. His monthly salary consists of $200 plus 10% of the company's net revenue for that month.

(a) Write an equation expressing his monthly salary, s, in terms of the company's net revenue, r.

(b) Draw a graph of his monthly salary for the company's net revenue up to and including $20,000.

(c) If the company's net revenue for the month of April is $15,000, what will Mr. Jordan's salary be for April?

Solution: **(a)** His salary consists of $200 plus 10% of the net revenues, *r*. Ten percent of *r* is 0.10*r*. Thus, the equation for his salary is

$$s = 200 + 0.10r$$

(b) Select values for *r*, find the corresponding values of *s*, and then draw the graph. We can select values for *r* that are between 0 and $20,000 (see Fig. 3.17).

r	*s*
0	200
10,000	1200
20,000	2200

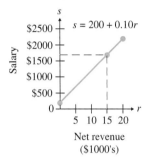

FIGURE 3.17

(c) By reading our graph carefully, we can estimate that when the company's net revenue is $15,000, Mr. Jordan's salary is about $1700.

Graphing Calculator Corner

You can often learn a lot about the general form of a graph by graphing two or more similar equations on the same axes and then comparing their graphs.

EXERCISES
   2. lines rise from left to right; greater coefficient, steeper line.
   3. lines fall from left to right; lesser coefficient, steeper line.

Graph each of the three equations on the same axes (or in the same window*) with your grapher (or by plotting points if a grapher is not available). Then answer the questions asked.*

1. $y = 2x$, $y = 2x + 1$, $y = 2x + 2$ raises or lowers it

   (a) What effect does the constant appear to have on the graph?

   (b) Do these lines appear to intersect? If not, why not?   no

2. $y = x + 2$, $y = 2x + 2$, $y = 3x + 2$

   What effect does increasing the coefficient of the *x* term from 1 to 3 appear to have on the graph?

3. $y = -x + 1$, $y = -2x + 1$, $y = -3x + 1$

   What effect does decreasing the coefficient of the *x* term from -1 to -3 appear to have on the graph?

4. $y = 1$, $y = 2$, $y = 3$ a horizontal line

   What does the graph of $y = b$, where *b* is any real number, look like?

Solving Linear Equations in One Variable Graphically

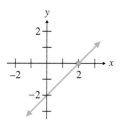

FIGURE 3.18

6 The solution to the equation $x - 2 = 0$ is 2. Now consider the equation $y = x - 2$. If 0 is substituted for y and the equation is solved, the solution is again 2. If you graph $y = x - 2$, you will see that its graph crosses the x axis at 2 (See Fig. 3.18). It is important that you realize that the x intercept of a graph is the value of x where y has a value of 0.

We can use this information to solve an equation in *one variable* graphically. To do so, rewrite the equation so that one side of the equation is equal to 0. Then replace the 0 with y and graph the equation. The x intercept of the graph will be the solution to the equation in one variable. You can use this procedure to check your solution to an equation in one variable on your graphing calculator, or you can solve the equation graphically.

In the following Graphing Calculator Corner we discuss two important features, called *trace* and *zoom*, and show an example of using the graphing calculator to solve an equation in one variable.

Graphing Calculator Corner

If you solve the equation $2(x + 3) + 3 = \frac{1}{2}x + 6$ correctly, you will obtain the solution -2. Solve this equation now and see that you obtain this value. To check this solution, or to solve the equation graphically, rewrite the equation with one side equal to 0 as follows:

$$2(x + 3) + 3 - \frac{1}{2}x - 6 = 0$$

Now replace the 0 with y and, using your calculator, graph the resulting equation.

$$y = 2(x + 3) + 3 - \frac{1}{2}x - 6$$

The x coordinate of the x intercept of the graph will be the solution to the equation $2(x + 3) + 3 = \frac{1}{2}x + 6$.

When graphing an equation on a graphing calculator it is sometimes not easy to determine the x intercept by observing the graph. Most graphing calculators have features called **trace** and **zoom**.

Using the **trace** feature you can move the cursor (the blinking box) to any point on the curve. Using the **zoom** feature you can enlarge that region of the curve at the cursor. When you use the zoom feature, your calculator adjusts the domain, range, and scale automatically. Figure 3.19 shows the graph of $y = 2(x + 3) + 3 - \frac{1}{2}x - 6$.

In Fig. 3.20 we used the trace feature to place the cursor near the x intercept. In the figure, notice the x and y values at the location of the cursor. The cursor location is at $x = -1.914894$ and $y = .12765957$. Figure 3.21 shows the graph enlarged (or zoomed in) at the cursor. In the figure the trace feature was used again (after the zoom feature was used) to move the cursor even closer to the x intercept.

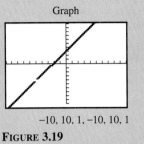

Graph

−10, 10, 1, −10, 10, 1

FIGURE 3.19

Trace feature

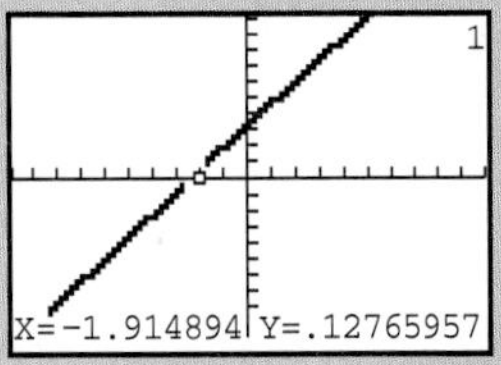

1

X=−1.914894 Y=.12765957

−10, 10, 1, −10, 10, 1

FIGURE 3.20

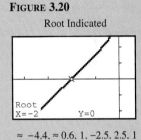

Zoom feature

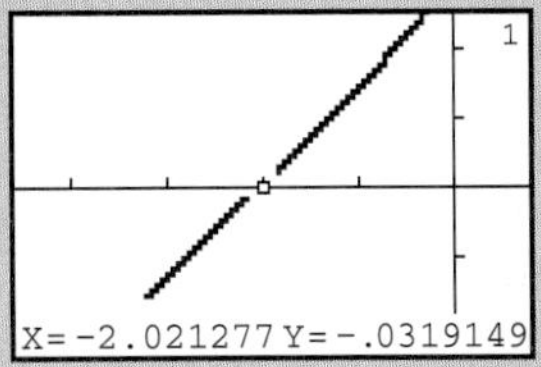

1

X=−2.021277 Y=−.0319149

≈ −4.4, ≈ 0.6, 1, −2.5, 2.5, 1

FIGURE 3.21

Root Indicated

Root
X=−2   Y=0

≈ −4.4, ≈ 0.6, 1, −2.5, 2.5, 1

FIGURE 3.22

Some graphing calculators have a feature that will actually display the root (or x intercept; see Fig. 3.22). Notice that the root is −2. The graphs displayed on your graphing calculator might look slightly different than those displayed here.

EXERCISES

Use your graphing calculator to estimate, to the nearest tenth, the solutions to the following equations.

1. $2x - 4 = \frac{1}{2}x + 8$ 8.0

2. $0.6x + 5 = 0.2x + 4$ −2.5

3. $120 - 3.42x = 5.96(x - 3)$ 14.7

4. $3.4x + 40 = (4x - 900) - 15.6$ 1592.7

Note that linear equations in one variable may also be solved graphically using systems of equations, as will be explained in Chapter 4.

Exercise Set 3.2

Graph each equation.

1. $y = 4$

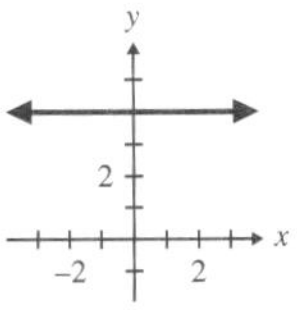

2. $x = 6$

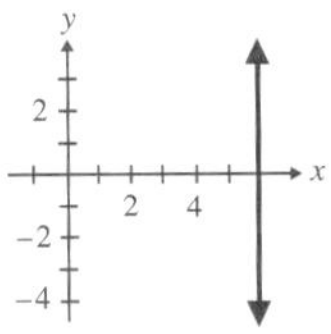

3. $x = -2$

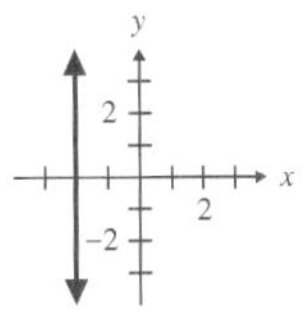

4. $y = 5$

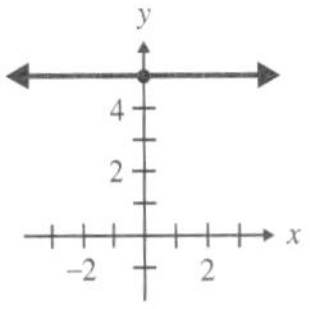

Graph each equation by solving the equation for y, selecting three values for x, and finding the corresponding values of y.

5. $y = 2x$

6. $y = -x + 3$

7. $y = x + 2$

8. $y = -\frac{1}{2}x + 5$

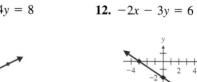

9. $2y = 2x + 4$

10. $5x - 2y = 8$

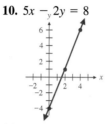

11. $-2x + 4y = 8$

12. $-2x - 3y = 6$

13. $8y - 16x = 24$

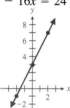

14. $y = 20x + 40$

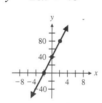

15. $2y - 50 = 100x$

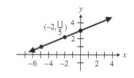

16. $-2x + 5y = 15$

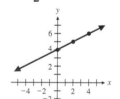

17. $5x - 2y = -7$

18. $-4x - 3y = -12$

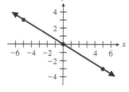

19. $y = -\frac{3}{5}x$

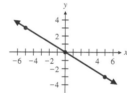

20. $y = \frac{1}{2}x + 4$

21. $y = -\frac{2}{5}x + 2$

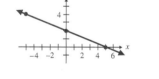

22. $x - \frac{2}{3}y = -2$

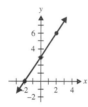

23. $\frac{2}{3}x + \frac{1}{2}y = \frac{3}{2}$

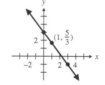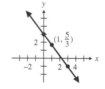

24. $\frac{4}{3}x - 2y = \frac{10}{3}$

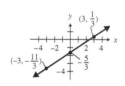

Graph each equation using x and y intercepts.

25. $y = 8x + 4$

26. $y = -2x + 6$

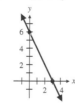

27. $y = 2x + 3$

28. $y = -6x + 5$

29. $y = 4x - 8$

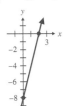

30. $4y + 3x = 12$

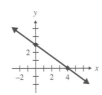

31. $4x = 3y - 9$

32. $\frac{1}{2}x + 2y = 4$

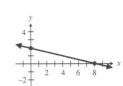

33. $30x + 25y = 50$

34. $0.6x - 1.2y = 2.4$

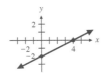

35. $0.25x + 0.50y = 1.00$

36. $-1.6y = 0.4x + 9.6$

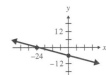

37. $\frac{1}{3}x - 2y = 6$

38. $30y + x = 45$

39. $120x - 360y = 720$

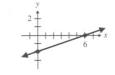

40. $20x - 240 = -60y$

41. $\frac{1}{3}x + \frac{1}{4}y = 12$

42. $-\frac{1}{2}y - \frac{1}{3}x = -1$

43. $\frac{1}{2}y = \frac{3}{8}x - \frac{3}{4}$

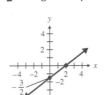

44. $\frac{1}{6}x + \frac{1}{2}y = -1$

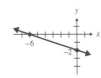

45. Using the formula distance = rate · time, or $d = rt$, draw a graph of distance versus time for a constant rate of 50 miles per hour.

46. Using the simple interest formula interest = principal · rate · time, or $i = prt$, draw a graph of interest versus time for a principal of $1000 and a rate of 8%.

47. The profit of a bicycle manufacturer can be approximated by the formula $P = 60x - 80,000$, where x is the number of bicycles produced and sold.

   (a) Draw a graph of profit versus the number of bicycles sold (for up to 5000 bicycles).

   (b) Estimate the number of bicycles that must be sold for the company to break even.   ≈1300

   (c) Estimate the number of bicycles that must be sold for the company to make $150,000 profit.   ≈3800

48. The weekly cost of operating a taxi is $50 plus 12 cents per mile.

   (a) Write an equation expressing the weekly cost, c, in terms of the number of miles, m. $c = 50 + 0.12m$

   (b) Draw a graph illustrating weekly cost versus the number of miles, up to 200, driven per week.

   (c) How many miles would Jack have to drive for the weekly cost to be $70?   ≈170 mi

   (d) If the weekly cost is $60, how many miles did Jack drive?   ≈85 mi

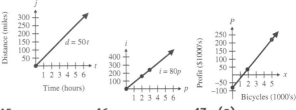

45.   **46.**   **47. (a)**

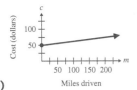

48. (b)

49. Ellen Branston's weekly salary is $200 plus 15% commission on her weekly sales. **(a)** $s = 200 + 0.15x$

   (a) Write an equation expressing Ellen's weekly salary, s, in terms of her weekly sales, x.

   (b) Draw a graph of Ellen's weekly salary versus her weekly sales, for up to $5000 in sales. Answer on page 153

   (c) What is Ellen's weekly salary if her sales were $4000? ≈$800

   (d) If her salary for the week is $400, what are her weekly sales? ≈$1300

50. Ms. Tocci, a real estate agent, makes $150 per week plus a 1% sales commission on each property she sells.

   (a) Write an equation expressing her weekly salary, s, in terms of sales, x. $s = 150 + 0.01x$

   (b) Draw a graph of her salary versus her weekly sales, for sales up to $100,000. Answer on page 153

   (c) If she sells one house per week for $80,000, what will her weekly salary be? ≈$950

Line graphs were introduced in Section 3.1. Exercises 51 and 52 will give you additional practice with this important topic.

51. Consider the graph of the Dow Jones industrial average on Tuesday, June 7, 1994.

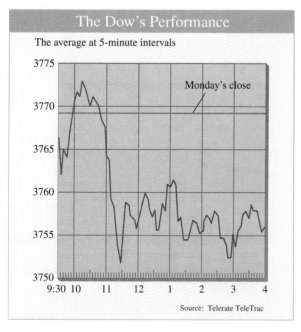

The Dow's Performance
The average at 5-minute intervals
Source: Telerate TeleTrac

   (a) Estimate the maximum and minimum values of the Dow on Tuesday, June 7.

   (b) Estimate the gain or loss of the Dow from Monday's close. loss ≈13 points

   (c) How many measurements were made to construct this graph? 78

   (d) Write a paragraph describing the performance of the Dow Jones industrial average that day.

SEE EXERCISE 51.

52. The following graph illustrates the effect of compound interest.

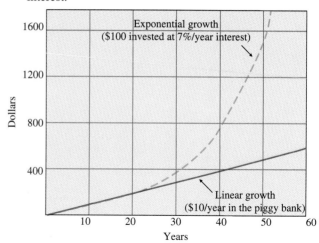

If a child puts $10 each year in a piggy bank, the savings will grow linearly, as shown by the lower curve. If, at age 10, the child invests $100 at 7% interest compounded annually, that $100 will grow exponentially.

51. maximum ≈3773, minimum ≈3752

(a) Using the linear growth curve, determine how long it would take to save $600.   ≈60 yr

(b) Using the exponential growth curve, determine how long it would take to save $600.   ≈36 yr

(c) Starting at year 20, how long would it take for the money growing at a linear rate to double?   20 yr

(d) Starting at year 20, how long would it take for the money growing exponentially to double? (Exponential growth will be discussed at length in Chapter 11.)   ≈10 yr

53. What does the graph of an equation represent?

54. What is the *standard form* of a linear equation? $ax + by = c$

53. the set of points whose coordinates that satisfy the equation
56. (a) Let $y = 0$, then solve for x.   **(b)** Let $x = 0$, then solve for y.

55. Explain the procedure to follow to graph linear equations by plotting points.

56. Explain how to find the **(a)** x intercept and **(b)** y intercept of a linear equation. Then **(c)** explain how to graph a linear equation using the intercepts.

57. Explain the procedure to use to solve a linear equation *in one variable* graphically.

58. (a) What does the graph of $y = b$, where b is any real number, look like?   a horizontal line

(b) What does the graph of $x = a$, where a is any real number, look like?   a vertical line

57. Set one side of equation equal to 0. Replace 0 with y, then graph. Solution is the x coordinate of the x intercept.

CUMULATIVE REVIEW EXERCISES

[2.1] **59.** Solve the equation $3x - 2 = \frac{1}{3}(3x - 3)$. $\frac{1}{2}$

[2.2] **60.** Solve the following formula for p_2.

$$E = a_1 p_1 + a_2 p_2 + a_3 p_3$$

[2.5] **61.** Solve the inequality $\frac{3}{5}(x - 3) > \frac{1}{4}(3 - x)$ and indicate the solution **(a)** on the number line; **(b)** in interval notation; **(c)** in set builder notation.

[2.6] **62.** Solve the equation $\left|\frac{x - 4}{3}\right| + 2 = 4$. $-2, 10$

60. $p_2 = \dfrac{E - a_1 p_1 - a_3 p_3}{a_2}$ **61. (a)** ⟵———⟶   **(b)** $(3, \infty)$ **(c)** $\{x \mid x > 3\}$

Group Activity/ Challenge Problems

In many real-life situations more than one linear equation may be needed to represent a problem. This often occurs where two or more different rates are involved. For example, when discussing federal income taxes, there are different tax rates. When two or more equations are used to represent a problem, the equations are called **piecewise equations** (or **piecewise functions**, as will be explained shortly). Following are two examples of piecewise equations and their graphs.

$$y = \begin{cases} -x + 2, & 0 \le x < 4 \\ 2x - 10, & 4 \le x < 8 \end{cases} \qquad y = \begin{cases} 2x - 1, & -2 \le x < 2 \\ x - 2, & 2 \le x < 4 \end{cases}$$

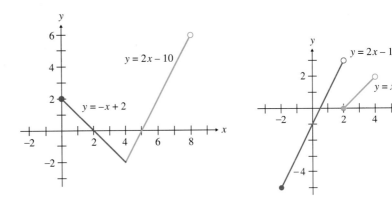

49. (b)

50. (b)

1. Denise is having some pamphlets printed at a local printer. The printer charges $200 plus $1.20 per pamphlet for an initial order of up to but not including 400 pamphlets. If the initial order is for 400 or more pamphlets the charge is $200 plus $0.80 per

1. *(a)*

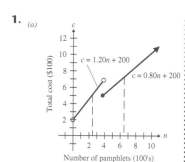

Number of pamphlets (100's)

2.
(a) $f = \begin{cases} 8, & 0 \le n \le 30 \\ 8 + 0.10(n - 30), & n > 30 \end{cases}$

(b)

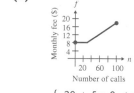

Number of calls

3. (a) $f = \begin{cases} 20 + 5n, & 0 < n < 15 \\ 20 + 10n, & n \ge 15 \end{cases}$

(b)

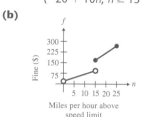

Miles per hour above
speed limit

4. (a) $y = \dfrac{x - 3}{2}$

(b)

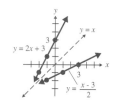

(c) The graphs that are inverse are symmetric with respect to the graph of $y = x$.

5. (a) $y = 2x - 1$

(b)

(c) The graphs that are inverse are symmetric with respect to the graph of $y = x$.

---

pamphlet. If n is the number of pamphlets initially ordered, the total cost, c, may be found by the piecewise equations.

$$c = \begin{cases} 1.20n + 200, & 0 < n < 400 \\ 0.80n + 200, & n \ge 400 \end{cases}$$

(a) Draw a graph of total cost versus the number of pamphlets printed.

(b) Estimate the total cost for orders of 250 pamphlets and 650 pamphlets, respectively. ≈$500, ≈$720

2. Under a phone company's plan, the customer pays a monthly fee of $8 and is entitled to 30 local calls at no additional charge. For more than 30 local calls, the customer must pay 10 cents on each additional call.

(a) Write a piecewise equation that represents this problem.

(b) Draw a graph showing the customer's monthly cost from 0 up to and including 100 calls per month.

3. In Washington County a person caught speeding less than 15 miles per hour over the speed limit pays a fine that includes an administrative charge of $20, plus $5 for each mile per hour he or she was traveling above the speed limit. A speeder driving 15 or more miles per hour above the limit must pay the $20 administrative charge plus $10 for each mile per hour he or she was traveling above the speed limit.

(a) Write a piecewise equation that represents this problem.

(b) Draw a graph showing the fine for a speeder going from 0 up to and including 25 miles per hour above the speed limit.

4. Consider the equation $y = 2x + 3$.   **(d)** domains of and ranges are reversed.

(a) Interchange the x and y and solve the new equation for y. The two equations are called **inverses** of each other. We will discuss inverses in detail in Chapter 9.

(b) Graph both equations on the same axes.

(c) On the same axes draw the graph of $y = x$. What is the relationship between $y = x$ and the two graphs that are inverses of each other.

(d) Complete the tables and compare the two sets of ordered pairs. Describe the relationship between the two sets of ordered pairs.

$y = 2x + 3$

x	0	1	2
y	3	5	7

(d)

$y = \dfrac{x - 3}{2}$

x	3	5	7
y	0	1	2

5. Consider the equation $y = \dfrac{x + 1}{2}$.   **(d)** domains and ranges are reversed

(a) Interchange the x and y and solve the new equation for y.

(b) Graph both equations on the same axes.

(c) On the axes draw the graph of $y = x$. What is the relationship between $y = x$ and the two graphs that are inverses of each other?

(d) Complete the tables and compare the two sets of ordered pairs. Describe the relationship between the two sets of ordered pairs.

$y = \dfrac{x + 1}{2}$

x	-1	1	3
y	0	1	2

(d)

$y = 2x - 1$

x	0	1	2
y	-1	1	3

3.3 Slope–Intercept and Point–Slope Forms of a Linear Equation

Tape 4

1. Find the slope of a line.
2. Write linear equations in slope–intercept form.
3. Graph linear equations using the slope and y intercept.
4. Use slope to recognize parallel and perpendicular lines.
5. Write linear equations in point–slope form.

Slope of a Line

1. The **slope of a line** is the ratio of the vertical change to the horizontal change between any two points on the line. As an example, consider the two points $(3, 6)$ and $(1, 2)$ on the line in Fig. 3.23a. If we draw a line parallel to the x axis through the point $(1, 2)$ and a line parallel to the y axis through the point $(3, 6)$, the two lines intersect at $(3, 2)$ (see Fig. 3.23b).

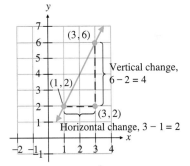

FIGURE 3.23

From Fig. 3.23b we can determine the slope of the line. The vertical change (along the y axis) is $6 - 2$, or 4 units. The horizontal change (along the x axis) is $3 - 1$, or 2 units.

$$\text{Slope} = \frac{\text{vertical change}}{\text{horizontal change}} = \frac{4}{2} = 2$$

Thus, the slope of the line through the points $(3, 6)$ and $(1, 2)$ is 2. By examining the line connecting these two points, we can see that for each two units the graph moves up the y axis it moves 1 unit to the right on the x axis (see Fig. 3.24).

Let's now determine the procedure to find the slope of a line passing through the two points (x_1, y_1) and (x_2, y_2). Consider Fig. 3.25. The vertical change can be

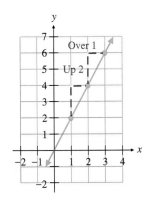

FIGURE 3.24

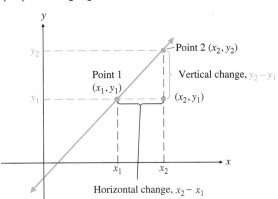

FIGURE 3.25

found by subtracting y_1 from y_2. The horizontal change can be found by subtracting x_1 from x_2.

> ### Slope
>
> The slope of the line through the distinct points (x_1, y_1) and (x_2, y_2) is
>
> $$\text{slope} = \frac{\text{change in } y \text{ (vertical change)}}{\text{change in } x \text{ (horizontal change)}} = \frac{y_2 - y_1}{x_2 - x_1}$$
>
> provided that $x_1 \neq x_2$.

It makes no difference which two points on the line are selected when finding the slope of a line. It also makes no difference which point you label (x_1, y_1) or (x_2, y_2). The letter m is used to represent the slope of a line. The Greek capital letter delta, Δ, is used to represent the words "the change in." Thus, the slope is sometimes indicated as

$$m = \frac{\Delta y}{\Delta x} = \frac{y_2 - y_1}{x_2 - x_1}$$

EXAMPLE 1   Find the slope of the line in Fig. 3.26.

Solution: Two points on the line are $(-2, 3)$ and $(1, -4)$. Let $(x_2, y_2) = (-2, 3)$ and $(x_1, y_1) = (1, -4)$. Then

$$m = \frac{y_2 - y_1}{x_2 - x_1} = \frac{3 - (-4)}{-2 - 1} = \frac{3 + 4}{-3} = -\frac{7}{3}$$

The slope of the line is $-\frac{7}{3}$. Note that if we had let $(x_1, y_1) = (-2, 3)$ and $(x_2, y_2) = (1, -4)$, the slope would still be $-\frac{7}{3}$. Try and see.

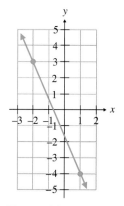

FIGURE 3.26

A line that rises going from left to right (Fig. 3.27a) has a **positive slope.** A line that neither rises nor falls going from left to right (Fig. 3.27b) has **zero slope.** And a line that falls going from left to right (Fig. 3.27c) has a **negative slope.**

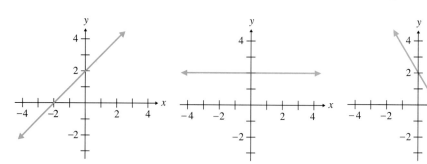

Positive slope	Zero slope	Negative slope
(a)	(b)	(c)

FIGURE 3.27

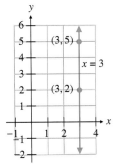

Slope is undefined

FIGURE 3.28

Consider the graph of $x = 3$ (Fig. 3.28). What is its slope? The graph is a vertical line and goes through the points (3, 2) and (3, 5). Let the point (3, 5) represent (x_2, y_2) and let (3, 2) represent (x_1, y_1). Then the slope of the line is

$$m = \frac{y_2 - y_1}{x_2 - x_1} = \frac{5 - 2}{3 - 3} = \frac{3}{0}$$

Since it is meaningless to divide by 0, we say that the slope of this line is undefined. **The slope of any vertical line is undefined.**

Helpful Hint

When students are asked to give the slope of a horizontal or a vertical line, they often answer incorrectly. When asked for the slope of a *horizontal line,* your response should be *"the slope is 0."* If you give your answer as "no slope," your instructor may well mark it wrong for these words may have various interpretations. When asked for the slope of a *vertical line,* your answer should be *"the slope is undefined."* Again, if you use the words "no slope," this may be interpreted differently by your instructor and marked wrong.

EXAMPLE 2   The following table of values, and the corresponding graph (Fig. 3.29) illustrate the U.S. public debt in billions of dollars from 1910 to 1990.

(a) Determine the slope of the line segments between 1910 and 1930 and between 1970 and 1990.

(b) Compare the two slopes found in part (a) and explain what this means in terms of the U.S. public debt.

Year	U.S. public debt (billions of dollars)
1910	1.1
1930	16.1
1950	256.1
1970	370.1
1990	3323.3

U.S. public debt

FIGURE 3.29

Solution:

(a) The slope from 1910 to 1930 is the ratio of the vertical change (public debt) to the horizontal change (year).

$$m = \frac{16.1 - 1.1}{1930 - 1910} = \frac{15}{20} = 0.75$$

Slope from 1970 to 1990:

$$m = \frac{3323.3 - 370.1}{1990 - 1970} = \frac{2953.2}{20} = 147.66$$

(b) Slope measures a rate of change. Comparing the slopes for the two 20-year periods shows that there was a much greater increase in the average rate of change in the public debt from 1970 to 1990 than from 1910 to 1930.

Look at the horizontal axis in the graph illustrated in Figure 3.29. Notice a small squiggly line near the origin. If a graph begins far from the origin (on either axis), we may use such a line to show that the axis is not uniformly marked from the origin to the starting point of the graph.

Slope–Intercept Form

2. A linear equation written in the form $y = mx + b$ is said to be in **slope–intercept form.**

Slope–Intercept Form of a Linear Equation

$$y = mx + b$$

where m **is the slope** of the line and b **is the y intercept*** of the line.

Examples of Equations in Slope-Intercept Form

$$y = 3x - 6 \qquad y = \frac{1}{2}x + \frac{3}{2}$$

This form is called the slope–intercept form because m represents the slope of the graph and b represents the y intercept.

$$\text{slope} \searrow \qquad \swarrow \; y \text{ intercept}$$
$$y = mx + b$$

Equation	Slope	y Intercept
$y = 3x - 6$	3	-6
$y = \dfrac{1}{2}x + \dfrac{3}{2}$	$\dfrac{1}{2}$	$\dfrac{3}{2}$

To write an equation in slope–intercept form, solve the equation for y.

EXAMPLE 3 Write the equation $-3x + 4y = 8$ in slope–intercept form. State the slope and y intercept.

---

*In this section, when we refer to b as the y intercept, it means the y intercept is $(0, b)$.

Solution: Solve for y.

$$-3x + 4y = 8$$
$$4y = 3x + 8$$
$$y = \frac{3x + 8}{4}$$
$$y = \frac{3}{4}x + \frac{8}{4}$$
$$y = \frac{3}{4}x + 2$$

The slope is $\frac{3}{4}$; the y intercept is 2.

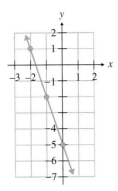

FIGURE 3.30

EXAMPLE 4   Write the equation of the line illustrated in Figure 3.30.

Solution: If we can determine the line's slope and its y intercept, we can write the equation in slope–intercept form. From the figure, we see that the y intercept is -5.

Notice that y changes 3 units for each unit change of x. Also note that the slope is negative since the line falls as it moves to the right. Therefore, the slope of the line is -3. We could also find the slope by taking two points on the line and finding $\Delta y/\Delta x$ for the two points selected.

Since the slope is -3 and the y intercept is -5, the equation of the line is $y = -3x - 5$.

Graphing Using Slope and the Y Intercept

3   One reason for studying the slope–intercept form of a line is that the information obtained from an equation in this form can be useful in drawing the graph of a linear equation. This procedure is illustrated in Example 5.

EXAMPLE 5   Graph $2y + 4x = 6$ using the y intercept and slope.

Solution: Begin by solving for y to get the equation in slope–intercept form.

$$2y + 4x = 6$$
$$2y = -4x + 6$$
$$y = -2x + 3$$

With the equation in this form, we see that the slope is -2 and the y intercept is 3. Now mark the y intercept, 3, on the y axis (Fig. 3.31). Then use the slope to obtain a second point. The slope is negative; therefore, the graph must fall as it goes from left to right. Since the slope is -2, the ratio of the vertical change to the horizontal change must be 2 to 1 $\left(\text{remember 2 means } \frac{2}{1}\right)$. Thus, if we start at $y = 3$ and move down 2 units and to the right 1 unit, we will obtain a second point on the graph.

Continue this process of moving 2 units down and 1 unit to the right to get a third point. Now draw a line through the three points to get the graph. Note that we chose to move down and to the right to get the second and third points. We could have also chosen to move up and to the left to get the second and third points.

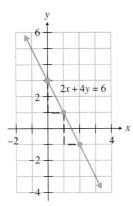

FIGURE 3.31

If you were asked to graph $y = \frac{4}{3}x - 3$ using the y intercept and slope, how would you do it? To begin, you should mark your first point at -3 on the y axis. Then you could obtain your second point by moving up 4 units and to the right 3 units.

Parallel and Perpendicular Lines

Parallel lines

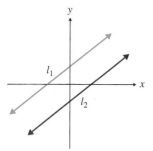

FIGURE 3.32

4 Two lines in the same plane are **parallel** when they do not intersect no matter how far they are extended. Figure 3.32 illustrates two parallel lines, l_1 and l_2. For two lines not to intersect, they must rise or fall at the same rate. That is, their slopes must be the same. **Two distinct lines are parallel if their slopes are the same.** If line l_1 has slope m_1 and line l_2 has slope m_2, and if $m_1 = m_2$, then lines l_1 and l_2 must be parallel lines.

EXAMPLE 6 Two points on l_1 are (1, 6) and (−1, 2). Two points on l_2 are (2, 3) and (−1, −3). Determine if l_1 and l_2 are parallel lines.

Solution: Determine the slopes of l_1 and l_2.

$$m_1 = \frac{6 - 2}{1 - (-1)} = \frac{4}{2} = 2 \qquad m_2 = \frac{3 - (-3)}{2 - (-1)} = \frac{6}{3} = 2$$

Since l_1 and l_2 have the same slope, 2, the two lines are parallel.

Two lines that cross at right angles (90° angles) are said to be **perpendicular lines.** Two perpendicular lines are illustrated in Figure 3.33. The square where the two lines meet is used to indicate that the two lines meet at a right angle.

Two lines will be perpendicular to each other when their slopes are negative reciprocals. For any slope a, the negative reciprocal is $\dfrac{-1}{a}$. For a slope of 2, its negative reciprocal is $\dfrac{-1}{2}$ or $-\dfrac{1}{2}$. For a slope of $-\dfrac{1}{3}$, its negative reciprocal is $\dfrac{-1}{-1/3} = (-1)(-3) = 3$.

Perpendicular lines

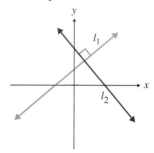

FIGURE 3.33

Slope	*Negative Reciprocal of Slope*	*Product*
2	$-\dfrac{1}{2}$	$2\left(-\dfrac{1}{2}\right) = -1$
$-\dfrac{2}{5}$	$\dfrac{5}{2}$	$-\dfrac{2}{5}\left(\dfrac{5}{2}\right) = -1$

Notice that the product of a number and its negative reciprocal equals -1. If l_1 has slope m_1 and l_2 has slope m_2, and if $m_1 m_2 = -1$, then l_1 and l_2 must be perpendicular lines.

EXAMPLE 7 Two points on l_1 are (6, 3) and (2, −3). Two points on l_2 are (0, 2) and (6, −2). Determine if l_1 and l_2 are perpendicular lines.

Solution: Determine the slopes of l_1 and l_2.

$$m_1 = \frac{3 - (-3)}{6 - 2} = \frac{6}{4} = \frac{3}{2} \qquad m_2 = \frac{2 - (-2)}{0 - 6} = \frac{4}{-6} = -\frac{2}{3}$$

Finally, determine if $m_1m_2 = -1$. If so, the lines are perpendicular.

$$m_1m_2 = \frac{3}{2}\left(-\frac{2}{3}\right) = -1$$

Since the product of the slopes equals -1, the lines are perpendicular. Note that each slope is the negative reciprocal of the other.

NOTE: Any horizontal line is perpendicular to any vertical line, although the negative reciprocal test cannot be applied. Why not?

EXAMPLE 8

(a) Determine if the graphs of the following equations are parallel lines.

$$2x - y = -4$$
$$2y = 4x - 2$$

(b) Graph both equations on the same axes.

Solution: **(a)** Two distinct lines are parallel when they have the same slope. Write each in slope–intercept form by solving each equation for y, then compare the slopes.

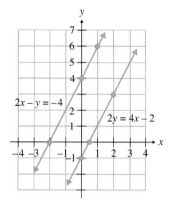

FIGURE 3.34

$$2x - y = -4 \qquad\qquad 2y = 4x - 2$$
$$-y = -2x - 4 \qquad\qquad y = \frac{4x - 2}{2}$$
$$y = \boxed{2}\,x + 4 \qquad\qquad y = \boxed{2}\,x - 1$$

Since both lines have the same slope, 2, they are parallel.

(b) Both lines are graphed in Fig. 3.34.

EXAMPLE 9   Consider the equation $2x + 4y = 8$. Determine the equation of the line that has a y intercept of 5 and is **(a)** parallel to the given line and **(b)** perpendicular to the given line.

Solution: **(a)** If we know the slope of a line and its y intercept, we can use the slope–intercept form, $y = mx + b$, to write the equation. Begin by solving the equation for y.

$$2x + 4y = 8$$
$$4y = -2x + 8$$
$$y = \frac{-2x + 8}{4}$$
$$y = -\frac{2x}{4} + \frac{8}{4}$$
$$y = -\frac{1}{2}x + 2$$

Two lines are parallel when they have the same slope. Therefore, the slope of the line parallel to the given line must be $-\frac{1}{2}$. Since its slope is $-\frac{1}{2}$ and its y intercept is 5, its equation must be

$$y = -\frac{1}{2}x + 5$$

(b) Two lines are perpendicular when their slopes are negative reciprocals of each other. We know that the slope of the given line is $-\frac{1}{2}$. Therefore, the slope of a line perpendicular to the given line must be $(-1) \div \left(-\frac{1}{2}\right)$ or 2. The line perpendicular to the given line has a y intercept of 5. Thus the equation is

$$y = 2x + 5$$

Point–Slope Form of a Linear Equation

5 When the slope of a line and a point on the line are known, we can use the **point–slope form** to determine the equation of the line. The point–slope form can be developed from the expression for the slope between any two points (x, y) and (x_1, y_1) on a line.

$$m = \frac{y - y_1}{x - x_1} \quad \text{or} \quad \frac{m}{1} = \frac{y - y_1}{x - x_1}$$

now cross multiply to obtain

$$m(x - x_1) = y - y_1 \quad \text{or} \quad y - y_1 = m(x - x_1)$$

Point–Slope Form of a Linear Equation

$$y - y_1 = m(x - x_1)$$

where **m is the slope** of the line and **(x_1, y_1) is a point on the line.**

EXAMPLE 10   Write in slope–intercept form the equation of the line that passes through the point (2, 3) and has slope 4.

Solution: Since we are given the slope of the line and a point on the line, we can write the equation in point–slope form. We can then solve the equation for y to write the equation in slope–intercept form. The slope, m, is 4. The point on the line is (2, 3); call this (x_1, y_1). Substitute 4 for m, 2 for x_1, and 3 for y_1 in the point–slope form of a line.

$$y - y_1 = m(x - x_1)$$
$$y - 3 = 4(x - 2) \qquad \text{point–slope form}$$
$$y - 3 = 4x - 8$$
$$y = 4x - 5 \qquad \text{slope–intercept form}$$

The graph of $y = 4x - 5$ has a slope of 4 and passes through the point (2, 3).

EXAMPLE 11   Determine, in slope–intercept form, the equation of the line through the points $(-1, -3)$ and $(4, 2)$.

Solution: When we are given two points on a line, we can write the equation in point–slope form. Any equation in point–slope form can be changed to slope–intercept form by solving the equation for y. We first find the slope between the two points. Let's designate $(-1, -3)$ as (x_1, y_1) and $(4, 2)$ as (x_2, y_2).

$$m = \frac{y_2 - y_1}{x_2 - x_1} = \frac{2 - (-3)}{4 - (-1)} = \frac{2 + 3}{4 + 1} = \frac{5}{5} = 1$$

Now write the equation in point–slope form using either one of the given points. This example will be worked out using the point $(-1, -3)$, as (x_1, y_1).

$$y - y_1 = m(x - x_1)$$
$$y - (-3) = 1[x - (-1)]$$
$$y + 3 = x + 1$$
$$y = x - 2 \qquad \text{slope–intercept form}$$

Rework Example 11 using the point $(4, 2)$ as (x_1, y_1). You should obtain the same answer.

EXAMPLE 12   Consider the equation $5y = -10x + 7$. Determine the equation of the line that passes through the point $\left(4, \frac{1}{3}\right)$ and is perpendicular to the given line. Write the equation in standard form $(ax + by = c)$.

Solution: First determine the slope of the given line by solving the equation for y.

$$5y = -10x + 7$$
$$y = \frac{-10x + 7}{5}$$
$$y = -2x + \frac{7}{5}$$

Since the slope of the given line is -2, the slope of the line perpendicular to it must be $\frac{1}{2}$. The line we are seeking must pass through the point $\left(4, \frac{1}{3}\right)$. Using the point–slope form, we obtain

$$y - y_1 = m(x - x_1)$$
$$y - \frac{1}{3} = \frac{1}{2}(x - 4)$$

Multiply both sides of the equation by the least common denominator, 6, to eliminate fractions.

$$6\left(y - \frac{1}{3}\right) = 6\left[\frac{1}{2}(x - 4)\right]$$
$$6y - 2 = 3(x - 4)$$
$$6y - 2 = 3x - 12$$

Now write the equation in standard form.

$$-3x + 6y - 2 = -12$$
$$-3x + 6y = -10 \qquad \text{standard form}$$

Note that $3x - 6y = 10$ is also an acceptable answer.

Helpful Hint

We have discussed three forms of a linear equation:

Standard form:	$ax + by = c$
Slope–intercept form:	$y = mx + b$
Point–slope form:	$y - y_1 = m(x - x_1)$

Consider the equation $2y = 3x + 4$. We can write this equation in all three forms as follows:

Standard form: $\qquad -3x + 2y = 4 \quad$ (or $3x - 2y = -4$)

Slope–intercept form: $\qquad y = \dfrac{3}{2}x + 2$
(solve for y)

Point–slope form: $\qquad y - 2 = \dfrac{3}{2}x \quad$ or $\quad y - 2 = \dfrac{3}{2}(x - 0)$

Note that to go from slope–intercept form to point–slope form we subtracted 2 from both sides of the equation. Also note that the point represented in point–slope form is $(0, 2)$, the y intercept.

EXAMPLE 13   The number of police officers in a city grew from 120 in 1991 to 160 in 1996. Assume constant growth since 1991.

(a) Construct a graph to illustrate this information. Place the year on the horizontal axis and number of police officers on the vertical axis. Use broken axes as in Example 2. Extend the axes far enough to plot the ordered pairs $(1991, 120)$ and $(1996, 160)$. Plot the points and draw a line through the points. Extend the line beyond 1996.

(b) Redraw the graph in part **(a)** letting 1990 be year 0 on the horizontal axis.

(c) Using the graph drawn in part **(b)**, determine an equation that can be used to predict the number of police officers in the city in the future.

(d) Using the equation found in part **(c)**, predict the number of police officers in the city in 2004.

(e) Extend the graph drawn in part **(b)** to at least 16 years on the horizontal axis and determine if the graph supports the answer found in part **(d)**.

Solution:

(a) The example indicates ordered pairs of $(1991, 120)$ and $(1996, 160)$. A graph using these values is given in Fig. 3.35.

(b) We can select any suitable year as a "reference year." Let us select the year 1990 to be our reference year and call it year 0. Then 1991 would be represented as year 1 since it is 1 year after 1990. The year 1992 would be

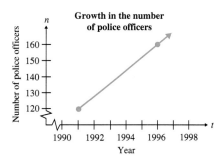

FIGURE 3.35

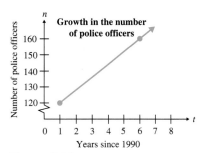

FIGURE 3.36

represented as 2, and so on. Using these values we can develop an equivalent graph (Fig. 3.36).

(c) Using the ordered pairs (1, 120) and (6, 160) in Fig. 3.36, we can find the slope of the line and then determine the equation of the line.

$$m = \frac{160 - 120}{6 - 1} = \frac{40}{5} = 8$$

Using point–slope form with the point (1, 120), we determine the equation we are seeking. We used n (number of police) for the vertical axis and t (time) for the horizontal axis.*

$$n - n_1 = m(t - t_1)$$
$$n - 120 = 8(t - 1)$$
$$n - 120 = 8t - 8$$
$$n = 8t + 112$$

As a check, if we substitute $t = 6$ (which represents the year 1996), we will obtain $n = 160$, which is what we expected.

(d) The year 2004 is 14 years after 1990. Substitute $t = 14$ in the equation to obtain the number of police officers in the year 2004.

---

*If a different reference year were selected, the equation would differ since the n-axis intercept would not be the same. However, the equation could still be used for making predictions in part (d) (see Group Activity Exercise 3).

$$n = 8t + 112$$
$$n = 8(14) + 112 = 224$$

Thus, if the growth continues at the same rate, there will be 224 police officers in the year 2004.

(e) The graph (Fig. 3.37) supports our answer to part (d). Notice the scales used on both axes were changed because of space restrictions.

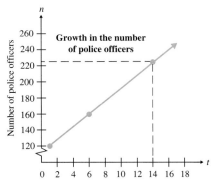

FIGURE 3.37

Exercise Set 3.3

Find the slope of the line through the given points. If the slope of the line is undefined, so state.

1. (1, 5) and (2, −3) −8
2. (3, 1) and (5, 4) $\frac{3}{2}$
3. (5, 2) and (1, 4) $-\frac{1}{2}$

4. (5, 1) and (2, 4) −1
5. (−1, 4) and (0, 3) −1
6. (2, 3) and (2, −3) undefined

7. (4, 2) and (4, −1) undefined
8. (6, −2) and (−1, −2) 0
9. (−3, 4) and (−1, −6) −5

10. (2, 5) and (−1, 5) 0
11. (−2, 3) and (7, −3) $-\frac{2}{3}$
12. (2, −4) and (−5, −3) $-\frac{1}{7}$

Solve for the given variable if the line through the two given points is to have the given slope.

13. (6, a) and (3, 4), m = 1   a = 7
14. (1, 0) and (4, y), m = 3   y = 9
15. (5, b) and (2, −4), m = 2   b = 2

16. (6, 1) and (4, d), m = 3   d = −5
17. (x, 2) and (3, −4), m = 2   x = 6
18. (−2, −3) and (x, 4), m = $\frac{1}{2}$ x = 12

19. (3, 5) and (x, 3), m = $\frac{2}{3}$ x = 0
20. (−4, −1) and (x, 2), m = $\frac{-3}{5}$ x = −9

Find the slope of the line in each of the given figures. If the slope of the line is undefined, so state. Then write an equation of the given line.

21.

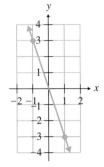

22.

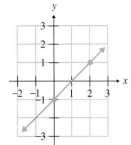

23.

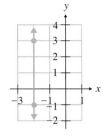

24.

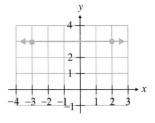

21. m = −3, y = −3x   **22.** m = 1, y = x − 1   **23.** m is undefined, x = −2   **24.** m = 0, y = 3

25.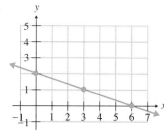

26.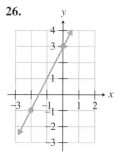

27.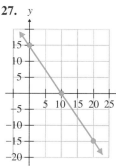

28.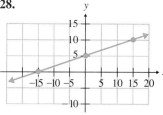

25. $m = -\frac{1}{3}$, $y = -\frac{1}{3}x + 2$ **26.** $m = 2$, $y = 2x + 3$ **27.** $m = -\frac{3}{2}$, $y = -\frac{3}{2}x + 15$ **28.** $m = \frac{1}{3}$, $y = \frac{1}{3}x + 5$

Write each equation in slope–intercept form (if not given in that form). Determine the slope and the y intercept and use them to draw the graph of the linear equation.

29. $y = -x + 2$

29.

33.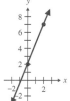

30. $2x + y = 6$
  $y = -2x + 6$

30.

32.

31. $20x - 30y = 60$
  $y = \frac{2}{3}x - 2$

32. $5y = 2x - 5$
  $y = \frac{2}{5}x - 1$

33. $-50x + 20y = 40$
  $y = \frac{5}{2}x + 2$

31.

34. $60x = -30y + 60$
  $y = -2x + 2$

34.

Two points on l_1 and two points on l_2 are given. Determine if l_1 is parallel to l_2, l_1 is perpendicular to l_2, or neither.

35. l_1: (0, 4) and (2, 8); l_2: (0, −1) and (3, 5)   parallel

36. l_1: (3, 4) and (−2, 3); l_2: (0, −3) and (2, −1)   neither

37. l_1: (3, 2) and (−1, −2); l_2: (2, 0) and (3, −1)   perpendicular

38. l_1: (0, 2) and (6, −2); l_2: (4, 0) and (6, 3)   perpendicular

39. l_1: (−1, 3) and (4, 2); l_2: (1, −3) and (4, 2)   neither

40. l_1: (1, 5) and (−2, −1); l_2: (1, −2) and (3, 2)   parallel

Determine if the two given lines are parallel, perpendicular, or neither.

41. $y = 2x - 4$
  $y = 2x + 3$ parallel

42. $2x + 3y = 6$
  $y = -\frac{2}{3}x + 5$ parallel

43. $4x + 2y = 8$
  $8x = 4 - 4y$ parallel

44. $3x - 5y = 10$
  $3y + 5x = 5$ perpendicular

45. $2x + 5y = 10$
  $-x + 3y = 9$ neither

46. $6x + 2y = 8$
  $4x - 9 = -y$ neither

47. $y = \frac{1}{2}x - 6$
  $-3y = 6x + 9$ perpendicular

48. $2y - 6 = -5x$
  $y = -\frac{5}{2}x - 2$ parallel

49. $y = 2x - 6$
  $x = -2y - 4$ perpendicular

50. $2x + y - 6 = 0$
  $6x + 3y = 12$ parallel

51. $x - 3y = -9$
  $y = 3x + 6$ neither

52. $-4x + 6y = 12$
  $2x - 3y = 6$ parallel

Use the point–slope form to find the equation of a line with the properties given. Then write the equation in slope–intercept form.

53. Slope = 4, through (2, 3) $y = 4x - 5$

54. Through (6, 3) and (5, 2) $y = x - 3$

55. Slope = −2, through (−4, 5) $y = -2x - 3$

56. Through (−4, −2) and (−2, 1) $y = \frac{3}{2}x + 4$

57. Slope = $\frac{1}{2}$, through (−1, −5) $y = \frac{1}{2}x - \frac{9}{2}$

58. Through (−4, 6) and (4, −6) $y = -\frac{3}{2}x$

59. Slope = $-\frac{2}{3}$, through (−1, −2) $y = -\frac{2}{3}x - \frac{8}{3}$

60. Slope = $\frac{3}{5}$, through (4, −2) $y = \frac{3}{5}x - \frac{22}{5}$

Find the equation of a line with the properties given. Write the equation in the form indicated.

61. Through (1, 4) parallel to $y = 2x + 4$ (slope–intercept form) $y = 2x + 2$

62. Through $\left(\frac{1}{2}, 3\right)$ parallel to $2x + 3y - 9 = 0$ (standard form) $2x + 3y = 10$

66. $3x + 2y = 6$ **67.** $y = -3x + 11$ **68.** $x - y = -7$ **69.** $y = -\frac{2}{3}x + 6$ **70.** $y = \frac{2}{5}x + \frac{1}{5}$

63. Through $\left(\frac{1}{5}, -\frac{2}{3}\right)$ parallel to $-3x = 2y + 6$ (slope–intercept form) $y = -\frac{3}{2}x - \frac{11}{30}$

64. Through $(2, 3)$ perpendicular to $y = 2x - 3$ (slope–intercept form) $y = -\frac{1}{2}x + 4$

65. Through $\left(-\frac{2}{3}, -4\right)$ perpendicular to $\frac{1}{2}x = y - 6$ (slope–intercept form) $y = -2x - \frac{16}{3}$

66. With x intercept 2 and y intercept 3 (standard form)

67. Through $(2, 5)$ and parallel to the line with x intercept 1 and y intercept 3 (slope–intercept form)

68. Through $(-3, 4)$ and perpendicular to the line with x intercept 2 and y intercept 2 (standard form)

69. Through $(6, 2)$ and perpendicular to the line with x intercept 2 and y intercept -3 (slope–intercept form)

70. Through the point $(2, 1)$ parallel to the line through the points $(3, 5)$ and $(-2, 3)$ (slope–intercept form)

71. The following graph from the National Parks Service shows the growth in attendance at our national parks from 1980 to 1993.

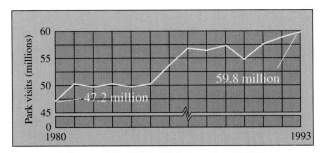

(a) Two points on the graph are $(1980, 47.2)$ and $(1993, 59.8)$. Suppose that we let 1980 be represented by year 0, then 1993 would be year 13, and the two points could be represented as $(0, 47.2)$

and $(13, 59.8)$, respectively. Find the slope of the line through these two points, rounded to the nearest hundredth. Then write the equation of the line that would go through these two points (see Example 13). $m = 0.97, n = 0.97t + 47.2$

(b) If we assume that the growth continues at the same constant rate, estimate the park visits in the year 2000.   66.6 million

(c) Write a paragraph describing the trend in park attendance from 1980 to 1993.

72. The following graph from the U.S. Department of Labor shows the number of factory workers and government employees for the years 1962 to 1992.

(b) $n = 0.32t + 10$

(a) In what year does the number of government workers outnumber the number of factory workers for the first time?   1992

(b) Assuming that the number of government workers in 1965 is 10 million and in 1990 is 18 million, write the equation of the straight line through these two points. Let 1965 be year 0 (see Example 13).

(c) If we assume that government growth continues at the same constant rate, estimate the number of government workers in the year 2000.   21.2 million

(d) Write a paragraph describing the growth for factory workers and for government workers from 1962 to 1992.

74. Select 2 points on line; find $\Delta y/\Delta x$. **78.** Change in x is zero, cannot divide by zero. **79.** It does not change.

73. The following graph appeared in the May 16, 1994 issue of *Barron's*.

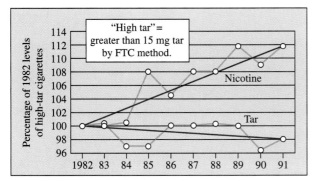

(a) Determine the equation of the straight line indicated on the nicotine graph. Assume 112 mg in 1991. Let 1982 be the reference year so that the initial point is (0, 100). $N = 1.33t + 100$

(b) Determine the equation of the straight line indicated on the tar graph. $T = -0.22t + 100$

(c) Write a paragraph describing the trend in nicotine and in tar for high-tar cigarettes from 1982 to 1991.

74. Explain how to find the slope of a line from its graph.

75. Explain what it means when the slope of a line is positive.   The line rises going from left to right.

76. Explain what it means when the slope of a line is negative.   The line falls going from left to right.

77. What is the slope of a horizontal line? Explain. $m = 0$.

78. Why is the slope of a vertical line undefined?

79. When finding the slope of a line, how does the slope change if we interchange (x_1, y_1) and (x_2, y_2)? Explain.

80. In this chapter we have discussed three forms of the equation of a line. Name the three forms, and give an example of each form.

81. Consider the equation $4 = 2y + 4x$. Write the equation in (a) standard form; (b) slope–intercept form, (c) point–slope form. Explain how you determined your answers.

82. Explain how to determine without graphing the equations if two linear equations represent (a) parallel lines; (b) perpendicular lines.

83. In Section 3.2 we explained how to graph linear equations by plotting points and by using the intercepts. In this section we discussed drawing graphs of linear equations using the slope and y intercept.

(a) Explain how to graph a linear equation by plotting points, by using the intercepts, and by using the slope and y intercept.

(b) Graph the equation $4x + 6y = 12$ using each of the three methods.

80. standard form, slope–intercept form, point–slope form
81. (a) $4x + 2y = 4$ (b) $y = -2x + 2$ (c) $y - 2 = -2(x - 0)$
82. (a) same slope (b) product of slopes $= -1$

83. (b)

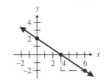

CUMULATIVE REVIEW EXERCISES

[2.3] **84.** The sum of three consecutive odd integers is 27. Find the three integers.   7, 9, 11

[2.6] **85.** $|x - a| = b$ (b) $x = a + b$ or $x = a - b$
86. $|x - a| < b$ (b) $a - b < x < a + b$
87. $|x - a| > b$ (b) $x < a - b$ or $x > a + b$

In Examples 85–87, (a) explain the procedure to solve the equation or inequality for x (assume that b > 0) and (b) solve the equation or inequality.

Group Activity/ Challenge Problems

1. The photo on page 170 is the Castle at Chichén Itza, Mexico. Each side of the castle has a stairway consisting of 91 steps. The steps of the castle are quite narrow and steep, which makes them hard to climb. The average height of the steps is 14.2 inches, and the average width is 6.4 inches.

(a) Find the total vertical distance in inches of the 91 steps.   1292.2 in.

(b) Find the total horizontal distance in inches of the 91 steps.   582.4 in.

(c) If a straight line were to be drawn connecting the tips of the steps, what would be the absolute value of the slope of this line?   2.21875

See Exercise 1.

2. A **tangent line** is a straight line that touches a curve at a single point (the tangent line may cross the curve at a different point if extended). The figure that follows shows three tangent lines to the curve at points a, b, and c.

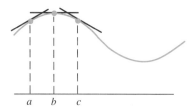

Note that the tangent line at point a has a positive slope, the tangent line at point b has a slope of 0, and the tangent line at point c has a negative slope. Now consider the curve below.

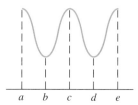

positive: (b, c) and (d, e); zero: points b, c, and d; negative: (a, b) and (c, d)

Imagine that tangent lines are drawn at all points on the curve above except at endpoints a and e. Where on the curve would the tangent lines have a positive slope, a slope of 0, a negative slope?

3. In Example 13 we replaced the year 1990 with the reference year 0 and changed all following years accordingly.   **(a)** $n = 8t - 15{,}808$, slope is the same, equation is different

(a) The original data provided is (1991, 120) and (1996, 160). Determine the equation of the line that connects these two points. Is the slope the same as that in Example 13(c)? Is the equation the same? Does this make sense? Explain.

(b) Using the equation determined in part (a), determine the number of police officers in the year 2004. Does your answer agree with the answer obtained in Example 13(d)? Explain why it does, or does not, agree.   224, answers agree

(c) Can you see any advantages in using a reference year?   smaller numbers to work with

3.4   Relations and Functions

Tape 4

|1| Identify relations.
|2| Find the domain and range of a relation.
|3| Identify functions.
|4| Use function notation.
|5| Use functions in practical applications.

Relations |1| If you plan to take additional mathematics courses, an understanding of relations and functions will be very helpful. In this section you are introduced to these important concepts. The function concept is discussed and expanded further in later sections.

A **relation** is any set of ordered pairs. A relation may be indicated by (1) a set of ordered pairs, (2) a table of values, (3) a graph, (4) a rule, or (5) an equation. For example, each of the following indicates a relation.

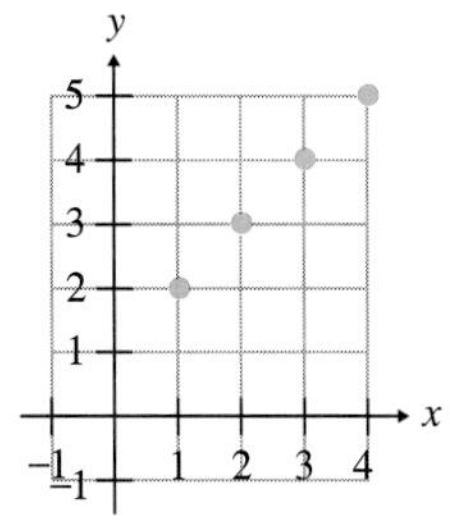

1. Set of ordered pairs $\{(1, 2), (2, 3), (3, 4), (4, 5)\}$

2. Table of values

3. Graph (see Fig. 3.38)

4. Rule: For each integer from 1 to 4 inclusive, add 1 to obtain its corresponding value.

x	1	2	3	4
y	2	3	4	5

Figure 3.38

5. Equation: $y = x + 1$, for $1 \leq x \leq 4$, $x \in N$

Note that these five examples all indicate the same relation.

Consider the following. An apple costs 30 cents. What will be your cost if you purchase 0 apples, 1 apple, 2 apples, and so on? We can indicate this using Table 3.1.

TABLE 3.1	
Number of apples, n	**Cost, c**
0	0
1	0.30
2	0.60
3	0.90
⋮	⋮
10	3.00
⋮	⋮

In general, we can see that the cost for purchasing n apples will be 30 cents times the number of apples, or $0.30n$. We can represent the cost of purchasing n apples, where n is a whole number, by the equation $c = 0.30n$. In the equation $c = 0.30n$ the cost, c, depends on the number of apples, n; thus we call c the *dependent variable* and n the *independent variable*.

Domain and Range

2 In any relation the set of values that can be used for the independent variable is called its **domain.** The set of values that represent the dependent variable is called its **range.**

Consider the equation for the cost of apples, $c = 0.30n$. What is its domain and what is its range? The domain of this relation is the set of *input values* that can be used to represent n, the number of apples. Since we cannot purchase a fractional part of an apple or a negative amount of apples, the domain is the set of numbers $\{0, 1, 2, 3, \ldots\}$. Note that the values on the left side of Table 3.1 are the elements that make up the domain. When the values in the domain, $0, 1, 2, 3, \ldots$, are substituted for n in the formula $c = 0.30n$, the values we get out are 0.00, 0.30, 0.60, 0.90, ... These values appear on the right side of Table 3.1. The range is the set of these *output values.* The range is $\{0.00, 0.30, 0.60, 0.90, \ldots\}$.

If we list the values in Table 3.1 as a set of ordered pairs, we get $\{(0, 0.00), (1, 0.30), (2, 0.60), (3, 0.90), \ldots\}$. Note that the *domain is the set of first coordinates of the ordered pairs,* and the *range is the set of second coordinates of the ordered pairs.*

When a graph is given, its domain and range can be determined by observation, as illustrated in Example 1.

EXAMPLE 1   State the domain and range of the relation shown in Fig. 3.39.

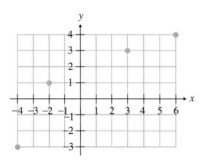   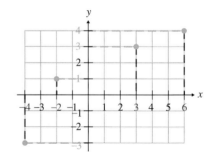

FIGURE 3.39   **FIGURE 3.40**

Solution: The domain is the set of x values (first coordinates of the ordered pairs). The values in the domain are indicated in red on Fig. 3.40.

$$\text{Domain:} \quad \{-4, -2, 3, 6\}$$

The range is the set of y values (second coordinates of the ordered pairs). The values in the range are indicated in green in Fig. 3.40.

$$\text{Range:} \quad \{-3, 1, 3, 4\}$$

The numbers in the domain and range were listed from smallest to largest; however, you may list the numbers in any order.

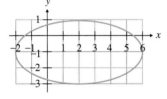

FIGURE 3.41

EXAMPLE 2   State the domain and range of the relation shown in Fig. 3.41.

Solution: The domain is the set of x values. All values of x between -2 and 6 inclusive are indicated on the graph. We can indicate this using set builder notation or interval notation.

$$\text{Domain:} \quad \{x \mid -2 \le x \le 6\} \quad \text{or} \quad [-2, 6]$$

The range is the set of y values. All values of y between -3 and 1 inclusive are indicated on the graph.

Range: $\{y \mid -3 \le y \le 1\}$ or $[-3, 1]$

EXAMPLE 3   Determine the domain and range of the relation shown in Fig. 3.42.

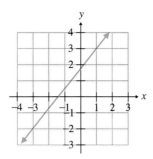

FIGURE 3.42

Solution: Since the line extends indefinitely in both directions, every value of x will be included in the domain. The domain is the set of real numbers.

Domain: $\mathbb{R}$ or $(-\infty, \infty)$

The range is also the set of real numbers since all values of y are included on the graph.

Range: $\mathbb{R}$ or $(-\infty, \infty)$

EXAMPLE 4   Determine the domain and the range of the relations in Figure 3.43.

(a)   **(b)**

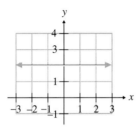   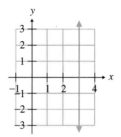

FIGURE 3.43

Solution:

(a) Domain: $\mathbb{R}$ or $(-\infty, \infty)$ **(b)** Domain: $\{3\}$
   Range: $\{2\}$ Range: $\mathbb{R}$ or $(-\infty, \infty)$

Functions   A *function* is a special type of relation. For a relation to be a function, each first coordinate in the set of ordered pairs must have a unique second coordinate. Is the relation {(1, 4), (2, 3), (3, 5), (−1, 3), (0, 6)} a function? Do any of the ordered pairs have the same first coordinate and a different second coordinate? *Since no two ordered pairs have the same first coordinate, the relation is a function.* Note that the second coordinate in the ordered pairs may repeat.

> ### Function
>
> A **function** is a relation in which no two ordered pairs have the same first coordinate and a different second coordinate.

Now consider a second relation: {(−1, 3), (4, 2), (3, 1), (2, 6), (3, 5)}. Is this relation a function? Since two ordered pairs, (3, 1) and (3, 5), have the same first coordinate and a different second coordinate, this relation is not a function.

A function may also be defined as a relation in which each element of the domain corresponds to one and only one element of the range. In other words, each *x*-value must correspond to a unique *y*-value.

Let us graph each relation above (Fig. 3.44). Note in part (a) that each *x* value has a unique *y* value. If a vertical line is drawn through any point, no other points are intersected. This relation is therefore a function. In part (b), if we draw a vertical line through the point (3, 1), it will intersect the point (3, 5). Thus, each *x* value does not have a unique *y* value and the relation is not a function.

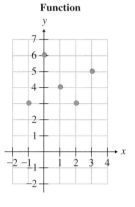

Function

(a) First set of ordered pairs

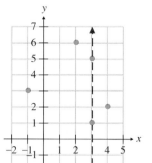

Not a Function

(b) Second set of ordered pairs

FIGURE 3.44

To determine if a graph is a function, we can use the **vertical line test.** If a vertical line can be drawn through any part of the graph and the line intersects another part of the graph, the graph is not a function. If a vertical line cannot be drawn to intersect the graph at more than one point, the graph is a function. We use the vertical line test to show that Fig. 3.45b is a function and Fig. 3.45a and c are not functions.

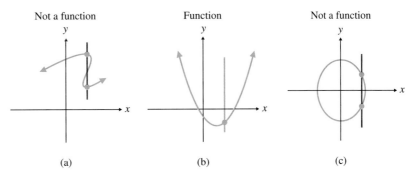

FIGURE 3.45

EXAMPLE 5   State the range and domain of the function illustrated in Fig. 3.46.

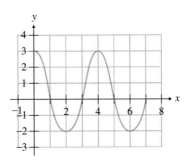

FIGURE 3.46

Solution: The domain is the set of x values, 0 through 7. Thus, the domain is $\{x \mid 0 \leq x \leq 7\}$. The range is the set of y values, -2 through 3. The range is therefore $\{y \mid -2 \leq y \leq 3\}$.

Function Notation

4   Consider the equation $y = 3x + 2$. By applying the vertical line test to its graph (Fig. 3.47), we can see that it is a function.

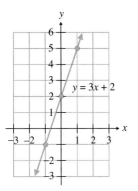

FIGURE 3.47

Since the value of y depends on the value of x and the equation is a function, we say that **y is a function of x.** The notation $y = f(x)$ is used to show that y is a function of the variable x. For this example, we can write $y = f(x) = 3x + 2$. The notation $f(x)$ is read "f of x" and *does not mean f* times x. Other letters may be used to indicate functions. For example, $g(x)$ and $h(x)$ also represent functions of x. In Section 5.5 we will use $P(x)$ to represent polynomial functions.

If y is a function of x, the notation $f(5)$, read "f of 5," means the value of y when x is 5. To evaluate a function for a specific value of x, substitute that value for x in the function. For example, if $f(x) = 3x + 2$, then $f(5)$ is found as follows:

$$f(x) = 3x + 2$$
$$f(5) = 3(5) + 2 = 17$$

Therefore, when x is 5, y is 17. The ordered pair $(5, 17)$ would appear on the graph of $y = 3x + 2$. The function $f(x) = 3x + 2$ is evaluated for other values of x below.

$$f(x) = 3x + 2$$
$$f(1) = 3(1) + 2 = 5$$
$$f(-4) = 3(-4) + 2 = -10$$
$$f(0) = 3(0) + 2 = 2$$

EXAMPLE 6   **(a)** If $f(x) = \dfrac{3}{2}x + 4$, find $f(-4)$.

(b) If $g(x) = 2x^2 + 3x - 4$, find $g\left(\dfrac{1}{4}\right)$.

Solution:

(a) $f(x) = \dfrac{3}{2}x + 4$

$$f(-4) = \frac{3}{2}(-4) + 4 = -6 + 4 = -2$$

(b) $g(x) = 2x^2 + 3x - 4$

$$g\left(\frac{1}{4}\right) = 2\left(\frac{1}{4}\right)^2 + 3\left(\frac{1}{4}\right) - 4 = 2\left(\frac{1}{16}\right) + \frac{3}{4} - 4 = \frac{1}{8} + \frac{6}{8} - \frac{32}{8} = -\frac{25}{8}$$

EXAMPLE 7   Evaluate the indicated functions at $x = 3$.

(a) $f(x) = \sqrt{x^2 + 2x + 1}$ **(b)** $g(x) = |x - 5|$ **(c)** $h(x) = \sqrt[3]{x^3 - 19}$

Solution:

(a) $f(3) = \sqrt{3^2 + 2(3) + 1} = \sqrt{16} = 4$

(b) $g(3) = |3 - 5| = |-2| = 2$

(c) $h(3) = \sqrt[3]{3^3 - 19} = \sqrt[3]{27 - 19} = \sqrt[3]{8} = 2$

Applications of Functions

⑤ Some applications of functions were discussed in Chapter 2. The examples discussed in that chapter were generally of the first degree. Now we examine applications of higher-degree functions.

EXAMPLE 8   A polygon is a closed figure whose sides are straight line segments. The number of diagonals, d, in a polygon is a function of the number of its sides n.

$$d = f(n) = \frac{1}{2}n^2 - \frac{3}{2}n$$

(a) How many diagonals has a quadrilateral (four sides)?

(b) How many diagonals has an octagon (eight sides)?

Solution:

(a) $n = 4,$ $\quad d = f(4) = \dfrac{1}{2}(4)^2 - \dfrac{3}{2}(4)$

$\qquad\qquad\qquad = \dfrac{1}{2}(16) - 6 = 2$

A quadrilateral has two diagonals.

(b) $n = 8,$ $\quad d = f(8) = \dfrac{1}{2}(8)^2 - \dfrac{3}{2}(8)$

$\qquad\qquad\qquad = \dfrac{1}{2}(64) - 12 = 20$

EXAMPLE 9   Neil Armstrong became the first person to walk on the moon on July 20, 1969. The velocity, *v*, of his spacecraft (the *Eagle*), in meters per second, was a function of time before touchdown, *t*.

$$v = f(t) = 3.2t + 0.45$$

The height, *h*, of the spacecraft above the moon's surface, in meters, was also a function of time before touchdown. Since we used $f(t)$ to represent velocity, we will use $g(t)$ to represent height.

$$h = g(t) = 1.6t^2 + 0.45t$$

What was the velocity of the spacecraft and distance from the surface of the moon at:

(a) 5 seconds before touchdown?

(b) 2 seconds before touchdown?

(c) touchdown?

Solution:

(a) To find the velocity and height at 5 seconds before touchdown, substitute $t = 5$ into the appropriate formulas.

$$v = f(t) = 3.2t + 0.45$$
$$v = f(5) = 3.2(5) + 0.45$$
$$= 16.0 + 0.45 = 16.45 \text{ meters per second}$$
$$h = g(t) = 1.6t^2 + 0.45t$$
$$h = g(5) = 1.6(5)^2 + 0.45(5)$$
$$= 1.6(25) + 2.25$$
$$= 40 + 2.25 = 42.25 \text{ meters}$$

(b) At 2 seconds before touchdown,

$$v = 3.2(2) + 0.45 = 6.4 + 0.45 = 6.85 \text{ meters per second}$$
$$h = 1.6(2)^2 + 0.45(2) = 1.6(4) + 0.45(2)$$
$$= 6.4 + 0.9 = 7.3 \text{ meters}$$

(c) At touchdown, $t = 0$:

$$v = 3.2(0) + 0.45 = 0 + 0.45 = 0.45 \text{ meter per second}$$

Thus, touchdown velocity was 0.45 meter per second.

$$h = 1.6(0^2) + 0.45(0) = 1.6(0) + 0 = 0 + 0 = 0 \text{ meters}$$

Thus, at touchdown the *Eagle* was on the moon and the distance from the moon was 0.

1. function,
D: {1, 2, 3, 4, 5},
R: {1, 2, 3, 4, 5}

2. relation,
D: {1, 2, 3, 4},
R: {1, 2, 3, 4}

3. function,
D: {1, 2, 3, 4, 5, 7},
R: {−1, 0, 2, 4, 5}

4. function,
D: {−2, −1, 0, 3, 4},
R: $\left\{-3, \dfrac{1}{2}, 1, 4, 5,\right\}$

5. relation,
D: {1, 2, 3, 5},
R: {−4, −1, 0, 1, 2}

6. function,
D: {−3, 0, 2, 3, 5, 6},
R: {2, 3, 4, 5}

7. function,
D: $\left\{-2, \dfrac{1}{2}, 0, 2, 3, 5\right\}$,
R: $\left\{-3, -1, 0, \dfrac{2}{3}, 2, 5\right\}$

8. relation,
D: $\left\{-3, \dfrac{1}{5}, \dfrac{2}{3}, 2, 5\right\}$,
R: $\left\{-3, 0, \dfrac{1}{2}, 1, 2\right\}$

9. relation,
D: {1, 2, 6},
R: {−3, 0, 2, 5}

10. relation,
D: {3},
R: {−9, −7, −3, 5}

11. relation,
D: {0, 1, 2},
R: {−7, −1, 2, 3}

12. function,
D: {−1, 0, 1, 2, 3},
R: {5}

Exercise Set 3.4

Determine which of the relations are also functions. Give the range and domain of each relation or function.

1. {(1, 4), (2, 2), (3, 5), (4, 3), (5, 1)}

2. {(1, 1), (4, 4), (3, 3), (2, 2), (4, 1)}

3. {(3, −1), (5, 0), (1, 2), (4, 4), (2, 2), (7, 5)}

4. $\left\{(-1, 1), (0, -3), (3, 4), (4, 5), \left(-2, \dfrac{1}{2}\right)\right\}$

5. {(5, 0), (3, −4), (2, −1), (5, 2) (1, 1)}

6. {(6, 3), (−3, 4), (0, 3), (5, 2), (3, 5), (2, 5)}

7. $\left\{\left(\dfrac{1}{2}, \dfrac{2}{3}\right), (3, 0), (2, -1), (5, -3), (-2, 2), (0, 5)\right\}$

8. $\left\{\left(\dfrac{1}{5}, 2\right), \left(2, \dfrac{1}{2}\right), \left(\dfrac{2}{3}, 0\right), (-3, 2), (-3, -3), (5, 1)\right\}$

9. {(6, 0), (2, −3), (1, 5), (1, 0), (1, 2)}

10. {(3, −3), (3, −7), (3, −9), (3, 5)}

11. {(0, 3), (1, 3), (2, 2), (1, −1), (2, −7)}

12. {(3, 5), (2, 5), (1, 5), (0, 5), (−1, 5)}

13.

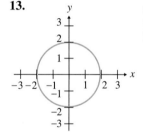

relation,
D: {x | −2 ≤ x ≤ 2},
R: {y | −2 ≤ y ≤ 2}

14.

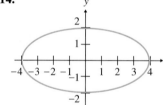

relation,
D: {x | −4 ≤ x ≤ 4},
R: {y | −2 ≤ y ≤ 2}

15.

function,
D: ℝ,
R: {y | y ≥ 0}

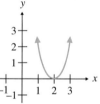

16. 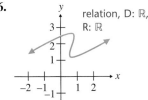 relation, D: ℝ,
R: ℝ

17. 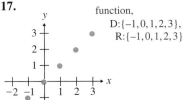 function,
D:{−1,0,1,2,3},
R:{−1,0,1,2,3}

18. 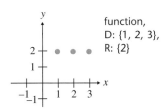 function,
D: {1, 2, 3},
R: {2}

19. 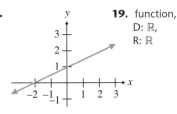 **19.** function,
D: ℝ,
R: ℝ

20. 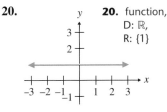 **20.** function,
D: ℝ,
R: {1}

21. 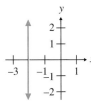 **21.** relation,
D: {−2},
R: ℝ

22. 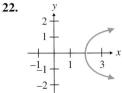 **22.** relation,
D: {x | x ≥ 2},
R: ℝ

23. 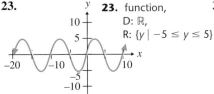 **23.** function, D: ℝ,
R: {y | −5 ≤ y ≤ 5}

24. 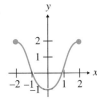 **24.** function,
D: {x | −2 ≤ x ≤ 2},
R: {y | −1 ≤ y ≤ 2}

Evaluate the functions at the values indicated.

25. $f(x) = 2x + 7$; find **(a)** $f(3)$; **(b)** $f(-2)$.   **(a)** 13 **(b)** 3

26. $f(x) = 2x + 6$; find **(a)** $f(-3)$; **(b)** $f\left(\dfrac{1}{2}\right)$.   **(a)** 0 **(b)** 7

27. $g(x) = 3 - 2x$; find **(a)** $g(2)$; **(b)** $g\left(\dfrac{1}{2}\right)$.   **(a)** −1 **(b)** 2

28. $f(x) = \dfrac{1}{2}x - 4$; find **(a)** $f(10)$; **(b)** $f(-4)$.   **(a)** 1 **(b)** −6

29. $f(x) = -\dfrac{3}{4}x + \dfrac{1}{2}$; find **(a)** $f(0)$; **(b)** $f(2)$.   **(a)** $\dfrac{1}{2}$ **(b)** −1

30. $h(x) = x^2 - x - 6$; find **(a)** $h(0)$; **(b)** $h(-1)$.   **(a)** −6 **(b)** −4

31. $g(x) = -x^2 - 2x + 3$; find **(a)** $g(-1)$; **(b)** $g\left(\dfrac{1}{2}\right)$.
(a) 4 **(b)** 7/4 or 1.75

32. $f(x) = \dfrac{1}{2}x^2 + 3x - 4$; find **(a)** $f(4)$; **(b)** $f\left(\dfrac{2}{3}\right)$.
(a) 16 **(b)** −16/9 ≈ −1.78

33. $f(x) = 2.6x^2 - 1.4x + 0.3$; find **(a)** $f(1.2)$; **(b)** $f(-3.4)$.
(a) 2.364 **(b)** 35.116

34. $f(x) = |x + 3|$; find **(a)** $f(-5)$; **(b)** $f(-12.6)$.   **(a)** 2 **(b)** 9.6

35. $g(x) = \left|\dfrac{1}{2}x + 2\right|$; find **(a)** $g(4)$; **(b)** $g\left(-\dfrac{2}{3}\right)$.   **(a)** 4 **(b)** $\dfrac{5}{3}$

36. $h(x) = -|2x - 4| + 5$; find **(a)** $h(0)$; **(b)** $h(-2)$.   **(a)** 1 **(b)** −3

37. $f(x) = |6 - 3x| - 2$; find **(a)** $f(-1)$; **(b)** $f(1.3)$.   **(a)** 7 **(b)** 0.1

38. $g(x) = x^3 - 4x^2 + 3x - 2$; find **(a)** $g(2)$; **(b)** $g(-3)$.
(a) −4 **(b)** −74

39. $f(x) = \dfrac{1}{2}x^3 + \dfrac{1}{3}x^2 + \dfrac{1}{4}x - 2$; find **(a)** $f(2)$; **(b)** $f(-3)$.
(a) 23/6 or 3.83   **(b)** −53/4 or −13.25

40. $f(x) = \sqrt{x^2 - 5}$; find **(a)** $f(3)$; **(b)** $f(4)$.   **(a)** 2   **(b)** $\sqrt{11} \approx 3.32$

41. $f(x) = -\sqrt{x^2 + 8}$; find **(a)** $f(1)$; **(b)** $f(10)$.
(a) −3   **(b)** $-\sqrt{108} \approx -10.39$

42. $g(x) = \sqrt{2x^2 - 6x - 4}$; find **(a)** $g(4)$; **(b)** $g(-2)$.   **(a)** 2 **(b)** 4

43. $f(x) = \sqrt{20 - 2x} + 5$; find **(a)** $f(8)$; **(b)** $f(2)$.   **(a)** 7 **(b)** 9

44. $f(x) = \dfrac{x^2 - 4}{x + 2}$; find **(a)** $f(2)$; **(b)** $f(-3)$.   **(a)** 0 **(b)** −5

45. $f(x) = \dfrac{x}{x^2 + 4}$; find **(a)** $f(2)$; **(b)** $f(-2)$.
(a) $\dfrac{1}{4}$ or 0.25 **(b)** $-\dfrac{1}{4}$ or −0.25

46. $h(x) = \dfrac{x^2 + 4x}{x + 6}$; find **(a)** $h(-3)$; **(b)** $h\left(\dfrac{2}{5}\right)$.
(a) −1 **(b)** $\dfrac{11}{40}$ or 0.275

47. The sum s of the first n even counting numbers is given by the function $s = f(n) = n^2 + n$. Find the sum of:

   (a) The first 10 even counting numbers   110

   (b) The first 15 even counting numbers   240

48. Use the function given in Example 8, $d = f(n) = \frac{1}{2}n^2 - \frac{3}{2}n$, to find the number of diagonals in a polygon of:

   (a) 10 sides   35   **(b)** 6 sides   9

49. Use the functions given in Example 9, $v = f(t) = 3.2t + 0.45$ and $h = g(t) = 1.6t^2 + 0.45t$, to determine the velocity and height above the surface of the moon at: **(a)** $v = 19.65$ m/s, $h = 60.3$ m

   (a) 6 seconds before touchdown **(b)** $v = 8.45$ m/s, $h = 11.125$ m

   (b) 2.5 seconds before touchdown

50. The temperature, T, in degrees Celsius, in a sauna n minutes after being turned on is given by the function $T = f(n) = -0.03n^2 + 1.5n + 14$. Find the sauna's temperature after:

   (a) 3 minutes   (18.23°C)   **(b)** 12 minutes   (27.68°C)

51. The stopping distance, d, in meters for a car traveling v kilometers per hour is given by the function $d = f(v) = 0.18v + 0.01v^2$. Find the stopping distance for the following speeds:

   (a) 50 km/hr   34 m   **(b)** 25 km/hr   10.75 m

52. The number of accidents, n, in 1 month involving drivers x years of age can be approximated by the function $n = f(x) = 2x^2 - 150x + 4000$. Find the approximate number of accidents in one month that involved:

   (a) 18-year-olds   1948   **(b)** 25-year-olds   1500

53. The profits, P, in millions of dollars, earned from constructing a building having x stories can be approximated by the function $P = f(x) = 0.02x^2 + 0.1x - 0.3$. Find the approximate profit earned from construction of an office building of:

   (a) 3 stories $0.18 million **(b)** 5 stories $0.7 million

The graphs in Ex. 57–60 are functions. Answer the questions asked.

57. Consider the graph to the right:

   (a) Explain why the graph is a function.

   (b) What is the independent variable? What is the dependent variable?   time, number of AIDS cases

   (c) Give the domain and range of the function in set builder notation.

   (d) For this graph we could state that the number of AIDS cases diagnosed, n, is a function of time, t, or $n = f(t)$. Determine n when $t = 1982$ and when $t = 1992$.   843; 47,000

   (e) Write a paragraph describing the growth in the number of AIDS cases diagnosed from 1982 to 1993.

57. (a) Each time has a unique amount.

   (c) D: $\{t \mid 1982 \le t \le 1993\}$,

      R: $\{n \mid 843 \le n \le 90{,}000\}$

54. The total number of oranges, N, in a square pyramid whose base is n by n oranges is given by the function

$$N = f(n) = \frac{1}{3}n^3 + \frac{1}{2}n^2 + \frac{1}{6}n$$

Find the number of oranges if the base is:

(a) 6 by 6 oranges   91   **(b)** 8 by 8 oranges   204

55. If the cost of a ticket to a rock concert is increased by x dollars, the estimated increase in revenue, R, in thousands of dollars is given by the function $R = f(x) = 24 + 5x - x^2$, $x < 8$. Find the increase in revenue if the cost of the ticket is increased by:

(a) $1 $28 thousand   **(b)** $4 $28 thousand

56. A roller coaster at Knott's Berry Farm, California, has a 70-foot vertical drop. The speed of the last car t seconds after it starts its descent can be approximated, in feet per second, by the function $v = f(t) = 7.2t + 4.6$. The height of the last car from the bottom of the vertical drop, in feet, can be estimated by the function $h = g(t) = -0.8t^2 - 10t + 70$, $0 \le t \le 5$. Find the last car's speed, and the height from the bottom of the drop, after:

(a) 2 seconds   Speed is 19 ft/sec, height 46.8 ft.

(b) 5 seconds   Speed is 40.6 ft/sec, height 0 ft (at 5 sec the last car is at the bottom of the drop).

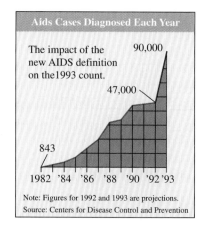

Aids Cases Diagnosed Each Year

The impact of the new AIDS definition on the 1993 count.

90,000

47,000

843

1982 '84 '86 '88 '90 '92 '93

Note: Figures for 1992 and 1993 are projections.

Source: Centers for Disease Control and Prevention

58. Consider the graph that follows:

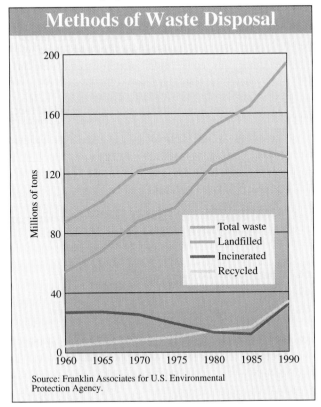

Methods of Waste Disposal

Millions of tons

Total waste
Landfilled
Incinerated
Recycled

Source: Franklin Associates for U.S. Environmental Protection Agency.

(a) Explain why each of the four curves is a function.

(b) What is the independent variable? What is the dependent variable?   time, amount of waste (or garbage)

(c) Consider the total waste curve (the green line). What is the domain indicated by the graph? Estimate the range indicated by the graph.

(d) For the total waste curve we could state that the total waste, w, is a function of time, t, or $w = f(t)$. Estimate the total waste in 1970 and in 1985.

(e) About 1985 there appears to be a change in how waste was disposed of. Explain what has happened since about 1985.

(f) Was there a greater increase in total waste from 1980 to 1985 or from 1985 to 1990? Explain.

(g) Suppose we call $f(x)$ the recycled trash function, $g(x)$ the incinerated trash function, $h(x)$ the landfill trash function, and $T(x)$ the total trash function. What is the relationship between the function $T(x)$ and the functions $f(x)$, $g(x)$, and $h(x)$?

59. This graph shows the growth of the circumference of a girl's head. The red line is the average head circumference of all girls for the given age while the green dashed lines represent the upper and lower limits of the normal range.

59. (a) Each age has a unique head circumference.
  (c) D: $\{a \mid 2 \leq a \leq 18\}$,   **(g)** $y = 0.375x + 47.25$
  R: $\{c \mid 48 \leq c \leq 55\}$

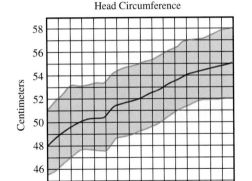

Head Circumference

Centimeters

Age

(a) Explain why the graph (the red line) is a function.

(b) What is the independent variable? What is the dependent variable?   age, head circumference

(c) What is the domain of the graph? What is the range of the average head circumference curve?

(d) What interval is considered normal for girls of age 18?   52 to 58 cm

(e) For this graph, is head circumference a function of age or is age a function of head circumference? Explain your answer.   Circumference is a function of age.

(f) Estimate the average girl's head circumference at age 10 and at age 14.   52 cm, 54 cm

(g) This graph appears to be nearly linear. Determine an equation or function that can be used to estimate the line between (2, 48) and (18, 55).

60. Consider the following graph of energy use worldwide. Primary electricity includes both hydroelectric power and nuclear power.

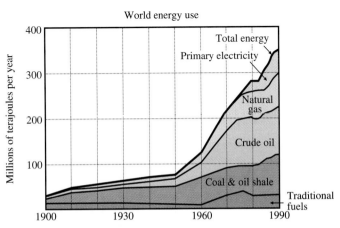

World energy use

Millions of terajoules per year

Total energy
Primary electricity
Natural gas
Crude oil
Coal & oil shale
Traditional fuels

58. (a) Each year has a unique amount of waste.
  (c) D: $\{t \mid 1960 \leq t \leq 1990\}$,   R: $\{w \mid 89 \leq w \leq 195\}$
  (d) $\approx$ 120 million tons, $\approx$ 160 million tons
  (e) less in landfill and more incinerated and recycled
  (f) 1985 to 1990   **(g)** $T(x) = f(x) + g(x) + h(x)$

62. no two ordered pairs have same first coordinate and a different second coordinate

65. If vertical line can intersect graph more than once, graph is not a function.

SEE EXERCISE 66.

(a) Explain why the total energy use is a function.

(b) What is the independent variable? What is the dependent variable?   time, energy used

Answer the following questions.

61. What is a relation?   any set of ordered pairs.

62. What is a function?

63. Are all relations also functions? Explain.   no

64. Are all functions also relations? Explain.   yes

65. Explain how to use the vertical line test to determine if a relation is a function.

60. (a) Each year has a unique amount of energy used.
  (c) D: {t | 1900 ≤ t ≤ 1990},
       R: {e | 30 ≤ e ≤ 350}
  (d) $T(t) = f(t) + g(t) + h(t) + i(t) + j(t)$

(c) What is the domain of the graph? Estimate the range of the total energy graph.

(d) Suppose that the amount of traditional fuels used is called function $f(t)$, the amount of coal and oil shale used is called function $g(t)$, the amount of crude oil used is called function $h(t)$, the amount of natural gas used is called function $i(t)$, the amount of primary electricity used is called $j(t)$, and the amount of total energy used is called function $T(t)$. Explain the relationship between $T(t)$ and the other functions for any given year. Explain how you determined your answer.

(e) Describe the total energy use from 1910 to 1950 and the total energy use from 1950 to 1990.

(f) Based on your answer to part (e) and by observing the graph, would you say that the energy use from 1910 to 1990 is linear? Explain your answer.

66. What is the domain of a relation?

67. What is the range of a relation?

68. What are the range and domain of the function $f(x) = 3x - 2$? Explain your answer.   D: ℝ,   R: ℝ

69. What are the range and domain of a function of the form $f(x) = ax + b$, $a \neq 0$. Explain your answer.
   D: ℝ,   R: ℝ

(e) appears nearly linear; appears nearly linear
(f) no, not linear
66. the set of first coordinates of the ordered pairs
67. the set of second coordinates of ordered pairs

CUMULATIVE REVIEW EXERCISES

[1.4] **70.** Evaluate $\dfrac{-6^2 - 16 \div 2 \div |-4|}{5 - 3 \cdot 2 - 4 \div 2^2}$.   19

Solve the equations.

[2.1] **71.** $\dfrac{3}{4}x + \dfrac{1}{5} = \dfrac{2}{3}(x - 2)$ $-\dfrac{92}{5}$

72. $2.6x - (-1.4x + 3.4) = 6.2$ 2.4

[2.4] **73.** Two trains leave Chicago, Illinois, traveling in the same direction along parallel tracks. The first train leaves 3 hours before the second, and its speed is 15 miles per hour faster than the second. Find the speed of each train if they are 270 miles apart 3 hours after the second train leaves Chicago.   60 mph, 75 mph

---

Group Activity/ Challenge Problems

1. If $f(x) = \dfrac{x}{x - 2}$, **(a)** what is the domain of $f(x)$? In parts (b) and (c) evaluate and explain how you determined your answer. **(b)** $f(0)$, **(c)** $f(2)$.

2. If $f(x) = \dfrac{x - 2}{x^2 - 4}$, **(a)** what is the domain of $f(x)$? In parts (b) and (c) evaluate and explain how you determined your answer. **(b)** $f(2)$, **(c)** $f(-2)$.

What are the domain and range of the following functions?

3. $f(x) = \sqrt{x}$ **4.** $f(x) = \sqrt{x + 4}$ **5.** $f(x) = \sqrt{4 - x}$ **6.** $f(x) = \sqrt[3]{x}$.

7. $g(x) = \sqrt{x^2 - 4}$ **8.** $h(x) = \sqrt{4 - x^2}$ **7.** D: {x | x ≤ −2 or x ≥ 2},   **8.** D: {x | −2 ≤ x ≤ 2},
                                                    R: {y | y ≥ 0}   R: {y | 0 ≤ y ≤ 2}

1. (a) all real numbers except 2 **(b)** 0 **(c)** undefined   **2. (a)** all real numbers except 2 and −2 **(b)** undefined **(c)** undefined
3. D: {x | x ≥ 0},   **4.** D: {x | x ≥ −4},   **5.** D: {x | x ≤ 4},   **6.** D: ℝ
    R: {y | y ≥ 0}   R: {y | y ≥ 0}   R: {y | y ≥ 0}   R: ℝ

3.5 Linear and Nonlinear Functions

Tape 4

1. Graph linear functions.
2. Graph absolute value functions.
3. Graph square root functions.

There are many different types of functions that we will discuss in this book. In this section we discuss linear functions, absolute value functions and square root functions.

The purpose of this section is (1) to give you more practice at graphing by plotting points, (2) to introduce you to both *linear* and *nonlinear* functions, (3) to reinforce the material you learned about functions, and (4) to help prepare you for the additional material on nonlinear functions that will be presented later in the book.

Remember, whenever you see the word "function" it means that value of x has a unique value of y. Also, the graph of the function will pass the vertical line test.

Linear Functions

1. All equations of the form $f(x) = ax + b$ are **linear functions.** That is, they are functions whose graphs are straight lines. We graph linear functions the same way we graph linear equations. We select values for the independent variable (often x) and find the corresponding values of the dependent variable (often y).

EXAMPLE 1   **(a)** Graph $f(x) = 2x - 3$.
(b) Give the domain and range of the function.

Solution: **(a)**   Remember that $f(x)$ is the same as y. We can make a chart of values by substituting values of x and finding the corresponding values of y [or of $f(x)$].

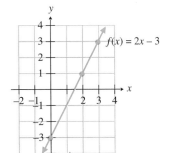

$$y = f(x) = 2x - 3$$

			x	y
$x = 0$	$y = f(0) = 2(0) - 3 = -3$		0	-3
$x = 2$	$y = f(2) = 2(2) - 3 = 1$		2	1
$x = 3$	$y = f(3) = 2(3) - 3 = 3$		3	3

The graph is illustrated in Figure 3.48. If we had wished, the vertical axis could have been labeled $f(x)$ instead of y.

(b) By observing the graph we can see that the domain is the set of all real numbers, $\mathbb{R}$. The range is also the set of all real numbers, $\mathbb{R}$.

FIGURE 3.48

An important part of the study of economics is the graphing of linear functions. The following example is a typical problem that might be found in an economics textbook.

EXAMPLE 2   It has been shown that the number of videotapes rented per week at a specific store is a function of the price of the videotape. The equation approximating the number of weekly rentals is

$$f(p) = -105p + 485, \qquad \$0.50 \le p \le \$4.50$$

where $f(p)$ is the number of rentals and p is the price per rental.

(a) What is the domain of this function?

(b) Construct a graph showing the relationship between the price of the tape and the number of rentals.

(c) Estimate the number of weekly rentals if the rental cost is $3.50.

(d) Determine the range of the function.

Solution:

(a) The domain, the set of values of p is $\{p \mid 0.50 \le p \le 4.50\}$.

(b) Construct a table of values and then graph the functions (Fig. 3.49).

p	$f(p)$
1.00	380
2.00	275
3.00	170

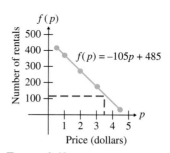

FIGURE 3.49

(c) If the rental cost is $3.50, approximately 120 tapes will be rented.

(d) The range, the set of values of $f(p)$, cannot easily be determined from the graph. Do you know how we can find them? From the graph we can see that the maximum value of $f(p)$ will occur when $p = 0.50$ and the minimum value of $f(p)$ will occur when $p = 4.50$. Evaluating $f(0.50)$ and $f(4.50)$ will give the upper and lower limits of the range.

$$f(p) = -105p + 485$$
$$f(0.50) = -105(0.50) + 485 = 432.5 \quad \text{upper limit of range}$$
$$f(4.50) = -105(4.50) + 485 = 12.5 \quad \text{lower limit of range}$$

The range is therefore $\{f(p) \mid 12.5 \le f(p) \le 432.5\}$.

Absolute Value Function

$f(x) = |x|$

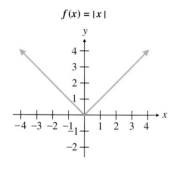

Absolute Value function
FIGURE 3.50

2 Now we introduce two **nonlinear functions,** that is, functions whose graphs are not straight lines. The first nonlinear function we will discuss is the **absolute value function.** The graphs of absolute value functions generally have a V shape as illustrated in Figure 3.50.

In Example 3 we will show how we obtain the graph of $f(x) = |x|$. In Example 4 we graph a different absolute value function.

EXAMPLE 3   **(a)** Graph the function $f(x) = |x|$.

(b) State the domain and range of the function.

Solution: **(a)**   We will substitute values for x and find the corresponding values of $f(x)$ or y, then graph the function. When an equation is not linear you must always make sure you plot enough points to get a true and accurate graph. We will select both positive and negative values of x in an organized manner and look for the V shape pattern. The value of $|x|$ is 0 when x is 0. For all other values $|x|$ is a positive number. We will therefore select $x = 0$, values of x less than 0, and values of x greater than 0, and find their corresponding values of y. For ease of reading, we list the table of values from the lowest value of x to the largest value x. The graph is illustrated in Figure 3.51.

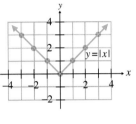

FIGURE 3.51

$$f(x) = |x|$$

		x	y
$x = -3$	$f(-3) = |-3| = 3$	−3	3
$x = -2$	$f(-2) = |-2| = 2$	−2	2
$x = -1$	$f(-1) = |-1| = 1$	−1	1
$x = 0$	$f(0) = |0| = 0$	0	0
$x = 1$	$f(1) = |1| = 1$	1	1
$x = 2$	$f(2) = |2| = 2$	2	2
$x = 3$	$f(3) = |3| = 3$	3	3

(b) By observing the graph, we can see that the domain, the x values, is the set of all real numbers $\mathbb{R}$. The range, the y values, must be greater than or equal to 0. Thus the range is $\{y \mid y \geq 0\}$.

EXAMPLE 4   **(a)** Graph $f(x) = |x + 4| - 2$.

(b) State the domain and range of the function.

Solution: **(a)** The value of $|x + 4|$ is 0 when $x = -4$. We will therefore select $x = -4$, values less than -4, and values greater than -4, and find their corresponding values of y. Make sure that you plot enough points to show the V shape. The graph is plotted in Figure 3.52.

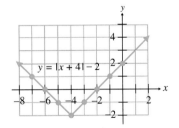

FIGURE 3.52

$$f(x) = |x + 4| - 2$$

		x	y
$x = -7$	$f(-7) = |-7 + 4| - 2 = 1$	−7	1
$x = -6$	$f(-6) = |-6 + 4| - 2 = 0$	−6	0
$x = -5$	$f(-5) = |-5 + 4| - 2 = -1$	−5	−1
$x = -4$	$f(-4) = |-4 + 4| - 2 = -2$	−4	−2
$x = -3$	$f(-3) = |-3 + 4| - 2 = -1$	−3	−1
$x = -2$	$f(-2) = |-2 + 4| - 2 = 0$	−2	0
$x = -1$	$f(-1) = |-1 + 4| - 2 = 1$	−1	1

(b) The domain is all real numbers, $\mathbb{R}$, the range is $\{y \mid y \geq -2\}$.

In Example 3, we graphed $f(x) = |x|$. Can you guess what the graph of $f(x) = -|x|$ will look like? The negative sign before the absolute value will make all nonzero values of $f(x)$ or y negative. Therefore, the graph of $f(x) = -|x|$ can have no positive y values. This means the graph cannot go above the x axis. The graph of $y = -|x|$ is illustrated in Figure 3.53. Notice that the graph of $f(x) = -|x|$ is the graph of $f(x) = |x|$ inverted.

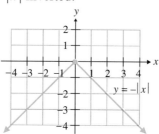

FIGURE 3.53

The Square Root Function

3 The next function we will discuss is the **square root function.** We will use the square root function in many sections of the book. The general form of the graph of the square root function is illustrated in Figure 3.54.

When graphing square root functions remember that the *radicand,* the expression inside the radical sign, must be greater than or equal to zero since the square root of a negative number is not a real number. When evaluating radical expressions your calculator should be used to save time. In Example 5, we will show how we obtain the graph of $f(x) = \sqrt{x}$. In examples 6 and 7, we graph a different square root function.

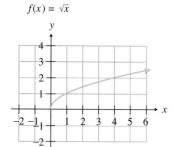

$f(x) = \sqrt{x}$

Square root function

FIGURE 3.54

EXAMPLE 5 **(a)** Consider the square root function $f(x) = \sqrt{x}$. What is its domain?

(b) Graph the function.

(c) What is the range of the function?

Solution:

(a) Since the square roots of negative numbers are not real numbers, the domain is the set of real numbers greater than or equal to 0. Domain: $\{x \mid x \geq 0\}$.

(b) When graphing we can use only values of x that are greater than or equal to 0. The graph is illustrated in Fig. 3.55.

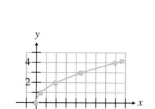

FIGURE 3.55

$$f(x) = \sqrt{x}$$

		x	y
$x = 0$	$f(0) = \sqrt{0} = 0$	0	0
$x = 1$	$f(1) = \sqrt{1} = 1$	1	1
$x = 4$	$f(4) = \sqrt{4} = 2$	4	2
$x = 9$	$f(9) = \sqrt{9} = 3$	9	3
$x = 16$	$f(16) = \sqrt{16} = 4$	16	4

We used values of x that were easy to evaluate. However, when you use a calculator, you can evaluate any nonnegative value of x. Note that the scale on the x axis is greater than the scale on the y axis because the changes in x are greater than the changes in y.

(c) By observing the graph we can see that the range is $\{y \mid y \geq 0\}$.

EXAMPLE 6   Consider the square root function $f(x) = -\sqrt{x - 3}$.

(a) What is its domain?

(b) Graph the function.

(c) What is the range of the function?

Solution: **(a)** For $\sqrt{x - 3}$ to be a real number the radicand $x - 3$ must be greater than or equal to 0. If $x = 3$ then $x - 3$ has a value of 0. For values greater than 3, the radicand $x - 3$ is positive. Therefore, the domain is all numbers greater than or equal to 3, $\{x \mid x \geq 3\}$. Only values greater than or equal to 3 may be used when graphing this equation.

(b) The graph is illustrated in Fig. 3.56.

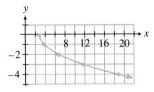

FIGURE 3.56

$$f(x) = -\sqrt{x - 3}$$

$x = 3$	$f(3) = -\sqrt{3 - 3} = -\sqrt{0} = 0$
$x = 4$	$f(4) = -\sqrt{4 - 3} = -\sqrt{1} = -1$
$x = 7$	$f(7) = -\sqrt{7 - 3} = -\sqrt{4} = -2$
$x = 12$	$f(12) = -\sqrt{12 - 3} = -\sqrt{9} = -3$
$x = 19$	$f(19) = -\sqrt{19 - 3} = -\sqrt{16} = -4$

x	y
3	0
4	-1
7	-2
12	-3
19	-4

We selected values of x that resulted in integer values of y. However, any values of x greater than or equal to 3 may be used.

(c) The range is the set of real numbers less than or equal to 0, that is, $\{y \mid y \leq 0\}$.

EXAMPLE 7   Consider the square root function $f(x) = \sqrt{16 - 4x}$.

(a) Find the domain of the function.

(b) Graph the function.

(c) Determine the range of the function.

Solution: **(a)**   To find the domain, we must find the values of x where $16 - 4x \geq 0$.

$$16 - 4x \geq 0$$
$$16 \geq 4x$$
$$\frac{16}{4} \geq \frac{4x}{4}$$
$$4 \geq x \quad \text{or} \quad x \leq 4$$

Thus, the domain is $\{x \mid x \leq 4\}$.

(b) Select values of x that are less than or equal to 4 and find the corresponding values of y. In this example we will use the calculator to find the value of y for selected values of x. The graph is illustrated in Fig. 3.57.

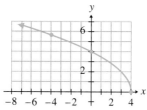

FIGURE 3.57

$$f(x) = \sqrt{16 - 4x}$$

$x = 4$	$f(4) = \sqrt{16 - 4(4)} = 0$
$x = 0$	$f(0) = \sqrt{16 - 4(0)} = 4$
$x = -4$	$f(-4) = \sqrt{16 - 4(-4)} = 5.7$
$x = -8$	$f(-8) = \sqrt{16 - 4(-8)} = 6.9$

x	y
4	0
0	4
−4	5.7
−8	6.9

(c) By observing the graph we see that the range is the set of real numbers greater than or equal to 0, $\{y \mid y \geq 0\}$.

Graphing Calculator Corner

All of the graphs illustrated in this section can be graphed on your grapher. To graph absolute value functions, use the ABS key. If your calculator does not have such a key, read your manual, which will explain how to graph absolute value functions. Use the √ key to graph square root functions.

Make sure that you use parentheses when needed when entering your function. For example, on some graphing calculators entering

abs x + 2 means $|x| + 2$ on some calculators, while on others it means $|x + 2|$ and √ x + 2 means $\sqrt{x} + 2$ on some calculators, while on others it means $\sqrt{x + 2}$.

If you are not sure whether or not parentheses are needed you should use them to avoid a possible error. Thus, if you wish to graph

$y = |x + 2|$, get $y =$, then press *abs* (x + 2)

$y = |x| + 2$, get $y =$, then press *abs* (x) + 2

$y = \sqrt{x - 7}$, get $y =$, then press √ (x − 7)

$y = \sqrt{x} - 7$, get $y =$, then press √ (x) − 7

As mentioned earlier, graphing calculators make a distinction between the negative key (−) and the subtraction key − . You must make sure you use the proper key when entering your expression. The subtraction key should not be used at the beginning of an expression or to enter a negative number. The subtraction key is used to subtract one expression from another. If you use the wrong key your calculator will give you an error message when you attempt to perform a calculation or graph an equation.

To graph $y = -\sqrt{x - 2}$, get $y =$, then press (−) √ (x − 2)

To graph $y = -|x - 3|$, get $y =$, then press (−) *abs* (x − 3)

EXERCISES

Graph on your graphing calculator. Indicate the number of x intercepts that each graph has.

1. $y = |x| - 4$ 2

2. $y = \sqrt{3x - 4}$ 1

3. $y = \left|3x - \dfrac{1}{2}\right|$ 1

4. $y = \sqrt{6 - 2x}$ 1

5. $y = -\sqrt{x^2 + 9}$ 0

6. $y = -|x - 8.64| + 2.35$ 2

7. $y = \sqrt{2x^2 - 5x - 6}$ 2

8. $y = -\sqrt{x^2 - 4x + 16}$ 0

Exercise Set 3.5

*For each function **(a)** state the domain, **(b)** graph the function, and **(c)** state the range. Use your calculator to evaluate square roots.*

1. $f(x) = 2x + 1$

D: ℝ, R: ℝ

2. $f(x) = 2x - 1$

D: ℝ, R: ℝ

3. $f(x) = \dfrac{1}{2}x - 2$

D: ℝ, R: ℝ

4. $f(x) = 3x - 5$

D: ℝ, R: ℝ

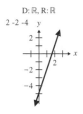

5. $f(x) = \dfrac{1}{3}x + 1$

D: ℝ, R: ℝ

6. $g(x) = \dfrac{x - 3}{2}$

D: ℝ, R: ℝ

7. $h(x) = |x - 1|$

D: ℝ, R: $\{y|y \geq 0\}$

8. $f(x) = |x + 3|$

D: ℝ, R: $\{y|y \geq 0\}$

9. $f(x) = -|x - 2|$

D: ℝ, R: $\{y|y \leq 0\}$

10. $f(x) = |2x + 1|$

D: ℝ, R: $\{y|y \geq 0\}$

11. $g(x) = |x| - 2$

D: ℝ, R: $\{y|y \geq -2\}$

12. $f(x) = |3x + 6| - 2$

D: ℝ, R: $\{y|y \geq -2\}$

13. $g(x) = \sqrt{x + 1}$

D: $\{x|x \geq -1\}$,
R: $\{y|y \geq 0\}$

14. $f(x) = \sqrt{1 - x}$

D: $\{x|x \leq 1\}$
R: $\{y|y \geq 0\}$

15. $h(x) = \sqrt{2x - 4}$

D: $\{x|x \geq 2\}$,
R: $\{y|y \geq 0\}$

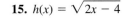

16. $f(x) = \sqrt{-x}$

D: $\{x|x \geq 0\}$,
R: $\{y|y \geq 0\}$

17. $g(x) = \sqrt{x} + 2$

D: $\{x|x \geq 0\}$,
R: $\{y|y \geq 2\}$

18. $g(x) = \sqrt{2x + 6}$

D: $\{x|x \geq -3\}$,
R: $\{y|y \geq 0\}$

19. $f(x) = |x - 4| + 2$

D: ℝ,
R: $\{y|y \geq 2\}$

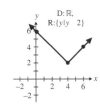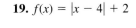

20. $g(x) = \sqrt{x^3} + 1$

D: $\{x|x \geq 0\}$,
R: $\{y|y \geq 1\}$

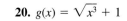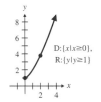

21. $f(x) = 2 - \sqrt{x}$

D: $\{x|x \geq 0\}$, R: $\{y|y \leq 2\}$

22. $h(x) = -\sqrt{2x}$

D: $\{x|x \geq 0\}$, R: $\{y|y \leq 0\}$

23. $h(x) = \sqrt{x^2}$

D: $\mathbb{R}$, R: $\{y|y \geq 0\}$

24. $g(x) = -|x + 4|$

D: $\mathbb{R}$, R: $\{y|y \leq 0\}$

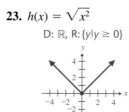

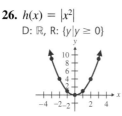

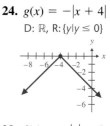

25. $g(x) = 4 - \sqrt{x^2}$

D: $\mathbb{R}$, R: $\{y|y \leq 4\}$

26. $h(x) = |x^2|$

D: $\mathbb{R}$, R: $\{y|y \geq 0\}$

27. $f(x) = |2x - 4| + 3$

D: $\mathbb{R}$, R: $\{y|y \geq 3\}$

28. $f(x) = -|x| - 3$

D: $\mathbb{R}$, R: $\{y|y \leq -3\}$

29. $h(x) = \dfrac{1}{2}x + 4$

D: $\mathbb{R}$, R: $\mathbb{R}$

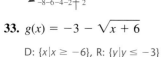

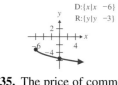

30. $g(x) = \sqrt{3x - 6}$

D: $\{x|x \geq 2\}$, R: $\{y|y \geq 0\}$

31. $f(x) = -\sqrt{2x - 6} + 4$

D: $\{x|x \geq 3\}$, R: $\{y|y \leq 4\}$

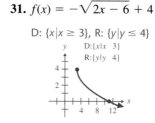

32. $g(x) = -\dfrac{1}{2}|x|$

D: $\mathbb{R}$, R: $\{y|y \leq 0\}$

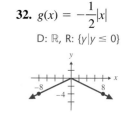

33. $g(x) = -3 - \sqrt{x + 6}$

D: $\{x|x \geq -6\}$, R: $\{y|y \leq -3\}$

34. $f(x) = 3 - |x - 1|$

D: $\mathbb{R}$, R: $\{y|y \leq 3\}$

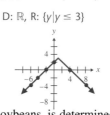

35. The price of commodities, like soybeans, is determined by **supply and demand.** If too many soybeans are produced, the supply will be greater than the demand, and the price will drop. If not enough soybeans are produced, the demand will be greater than the supply, and the price of soybeans will rise. Thus the price of soybeans is a function of the number of bushels of soybeans produced. The price of a bushel of soybeans can be estimated by the formula

$$f(Q) = -0.00004Q + 4.25$$
$$10,000 \leq Q \leq 60,000$$

In this formula, $f(Q)$ is the price of a bushel of soybeans and Q is the annual number of bushels of soybeans produced.   **(b)** approximately $2.65 per bushel

(a) Construct a graph showing the relationship between the number of bushels of soybeans produced and the price of a bushel of soybeans.

(b) Estimate the cost of a bushel of soybeans if 40,000 bushels of soybeans are produced in a given year.

36. The average annual household expenditure is a function of the average annual household income. The average expenditure can be estimated by the function

$$f(i) = 0.6i + 5000 \qquad \$3500 \leq i \leq \$50,000$$

In this formula, $f(i)$ is the average household expenditure and i is the average household income.

(a) Construct a graph showing the relationship between average household income and the average household expenditure.

(b) Estimate the average household expenditure for a family whose average household income is $30,000.   approximately $23,000

35. (a)

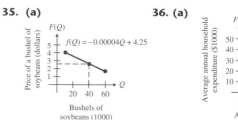

Bushels of soybeans (1000)

36. (a)

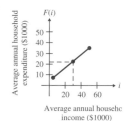

Average annual household income ($1000)

41. the square root of a negative number is not real.   **42.** finding positive square root

37. What are linear functions? functions of form $f(x) = ax + b$ **40.** What is the expression inside a radical called? the radicand

38. Name two types of nonlinear functions discussed in this section. absolute value, square root

41. Why must the radicand of a square root function always represent a number greater than or equal to 0?

39. Consider the absolute value function $f(x) = |x|$. Explain why the range of this function is the set of real numbers greater than or equal to 0. absolute value ≥ 0.

42. Consider the square root function $y = \sqrt{x}$. Why must the range always be greater than or equal to 0?

44. Reverse the direction of the inequality symbol.

46. **(a)** any set of ordered pairs **(b)** no two ordered pairs have same first coordinate and different second coordinate

CUMULATIVE REVIEW EXERCISES

[2.5] **43.** Solve the inequality $4 - \frac{1}{2}x > 2x + 3$ and indicate the solution in interval notation. $\left(-\infty, \frac{2}{5}\right)$

44. What must you do when multiplying or dividing both sides of an inequality by a negative number?

[3.1] **45.** Determine the distance and midpoint between the points $(4, -6)$ and $(-5, 3)$. $\sqrt{162} \approx 12.73$, $\left(-\frac{1}{2}, -\frac{3}{2}\right)$

[3.4] **46.** **(a)** What is a relation? **(b)** What is a function?

Group Activity/ Challenge Problems

1. Another type of nonlinear functions is *quadratic functions.* Quadratic functions will be discussed in Chapter 5. The function $f(x) = x^2 - 2x$ is a quadratic function. **(a)** Determine the domain of the function. **(b)** Graph the function. **(c)** Determine the range of the function.

2. Another type of nonlinear functions is *cubic functions.* Cubic functions will be discussed in Chapter 5. The function $f(x) = x^3 + 1$ is a cubic function. **(a)** Determine the domain of the function **(b)** Graph the function. **(c)** Determine the range of the function.

3. Another type of nonlinear function is the *cube root function.* The function $f(x) = \sqrt[3]{x}$ is a cube root function. **(a)** Determine the domain of the function. **(b)** Graph the function. **(c)** Determine the range of the function.

1. (a) D: $\mathbb{R}$
(b)

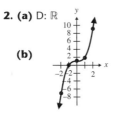

(c) $\{y | y \ge -1\}$

2. (a) D: $\mathbb{R}$

(b)

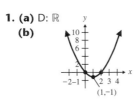

(c) R: $\mathbb{R}$

3. (a) D: $\mathbb{R}$
(b)

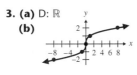

(c) R: $\mathbb{R}$

Piecewise equations were introduced in the Group Activity Exercises in Section 3.2. Piecewise equations that are functions are called **piecewise functions.** The following are piecewise functions and their graphs.

$$f(x) = \begin{cases} x + 3, & -4 \le x < -1 \\ |x - 2| - 1, & -1 \le x < 4 \end{cases}$$

$$g(x) = \begin{cases} \sqrt{x - 2}, & 2 \le x < 6 \\ |x - 2| - 2, & x < 2 \end{cases}$$

4.

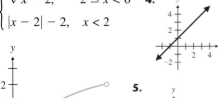

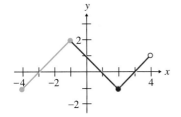

5.

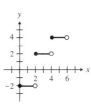

Graph the following piecewise functions.

4. $f(x) = \begin{cases} |x| + 1, & x \ge 2 \\ x + 1, & x < 2 \end{cases}$

5. $g(x) = \begin{cases} -2, & 0 \le x < 2 \\ 2, & 2 \le x < 4 \\ 4, & 4 \le x < 6 \end{cases}$

6.

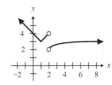

6. $g(x) = \begin{cases} \sqrt{x + 2}, & x > 2 \\ |x - 1| + 3, & x < 2 \end{cases}$

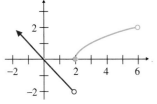

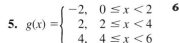

3.6 Graphing Linear Inequalities

Tape
5

1 Solve inequalities in one variable by graphing.
2 Graph linear inequalities in two variables.

**Linear Inequalities
in One Variable**

1 In Section 2.5 we explained how to solve linear inequalities *in one variable* algebraically. Examples of inequalities in one variable are $3x - 4 > 2x - 5$ and $4(x - 2) \le 2x + 3$. Now we will explain how to solve linear inequalities in *one variable* graphically, using a graphing calculator. The following Graphing Calculator Corner explains how to do this. Shortly, we will discuss how to graph linear inequalities in *two variables*.

Graphing Calculator Corner

The solutions to linear inequalities in one variable may be found or estimated using a graphing calculator. The procedure is very similar to the method used to solve linear equations in one variable as discussed on page 148. For example, consider the inequality $4x - 3 > x + 9$. Rewrite the inequality so that one side equals 0, as shown below.

$$4x - 3 > x + 9$$
$$4x - 3 - x - 9 > 0$$

Now replace 0 with the variable y and change the inequality sign to an equal sign.

$$4x - 3 - x - 9 = y \quad \text{or} \quad y = 3x - 12$$

Now graph $y = 3x - 12$. The graph is displayed in Fig. 3.58.

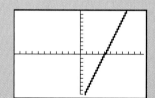

FIGURE 3.58

Since we are looking for values of x where $4x - 3 - x - 9 > 0$, the answer will be those values of x for which the y values are greater than 0, or that part of the graph above the x axis. Therefore, the solution to the inequality $4x - 3 > x + 9$ is $x > 4$. If you solve this inequality algebraically, you should obtain the same answer.

Had the inequality been $4x - 3 - (x + 9) < 0$, the answer would have been those values of x for which the graph was less than 0, or below the x axis. This procedure to find solutions for inequalities can be very helpful if the inequality is complex.

EXERCISES

Estimate, to the nearest tenth, the solution to the following inequalities using a graphing calculator.

1. $2x - 4 > \frac{1}{2}x + 6$ $x > 6.7$ **2.** $2x + 4 < -\frac{1}{5}x + 14$ $x < 4.5$

3. $2.3x - 5.6 < 9.7x - 62.4$ $x > 7.7$ **4.** $1068.4 - 20.3x > 18.4x - 246.4$

$x < 34.0$

Linear Inequalities in Two Variables

2 Now we will discuss linear inequalities in two variables.

A linear inequality results when the equal sign in a linear equation is replaced with an inequality sign. Examples of linear inequalities in two variables are:

$$2x + 3y > 2 \qquad 3y < 4x - 6$$
$$-x - 2y \leq 3 \qquad 5x \geq 2y - 3$$

> **To Graph a Linear Inequality in Two Variables**
> 1. Replace the inequality symbol with an equal sign.
> 2. Draw the graph of the equation in step 1. If the original inequality contains a $\geq$ or $\leq$ symbol, draw the graph using a solid line. If the original inequality contains a $>$ or $<$ symbol, draw the graph using a dashed line.
> 3. Select any point not on the line and determine if this point is a solution to the original inequality. If the point selected is a solution, shade the region on the side of the line containing this point. If the selected point does not satisfy the inequality, shade the region on the side of the line not containing this point.

EXAMPLE 1 Graph the inequality $y < 2x - 4$.

Solution: First graph the equation $y = 2x - 4$. Since the original inequality contains a less-than sign, $<$, use a dashed line when drawing the graph (see Fig. 3.59). The dashed line indicates that the points on this line are not solutions to the inequality $y < 2x - 4$. Select a point not on the line and determine if this point satisfies the inequality. Often the easiest point to use is the origin, (0, 0).

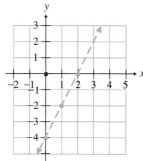

Checkpoint (0, 0)
$$y < 2x - 4$$
$$0 < 2(0) - 4$$
$$0 < 0 - 4$$
$$0 < -4 \qquad \text{false}$$

FIGURE 3.59

Since 0 is not less than -4, which is symbolized $0 \not< -4$, the point (0, 0) does not satisfy the inequality. The solution will be all the points on the opposite side of the line from the point (0, 0). Shade in this region (Fig. 3.60). Every point in the shaded area satisfies the given inequality. Let's check a few selected points: *A*, *B*, and *C*.

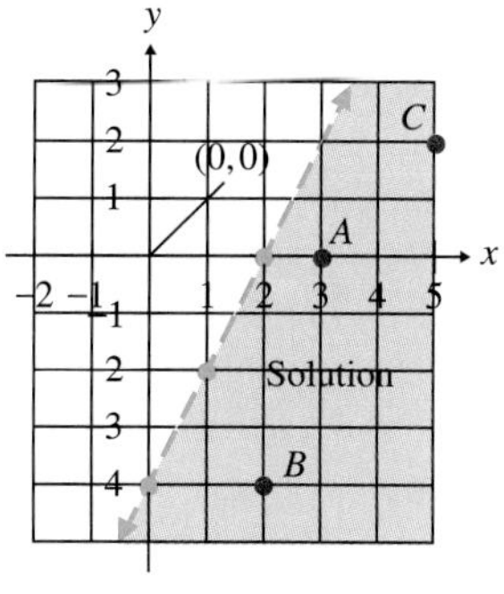

Point A	*Point B*	*Point C*
(3, 0)	(2, −4)	(5, 2)
$y < 2x - 4$	$y < 2x - 4$	$y < 2x - 4$
$0 < 2(3) - 4$	$-4 < 2(2) - 4$	$2 < 2(5) - 4$
$0 < 2$ true	$-4 < 0$ true	$2 < 6$ true

FIGURE 3.60

EXAMPLE 2 Graph the inequality $y \geq -\frac{1}{2}x$.

Solution: First graph the equation $y = -\frac{1}{2}x$. Since the inequality is $\geq$, we use a solid line to indicate that the points on the line are solutions to the inequality (Fig. 3.61). Since the point (0, 0) is on the line, we cannot select that point to find the solution. Let's arbitrarily select the point (3, 1).

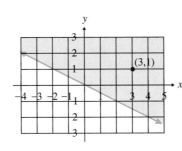

Checkpoint (3, 1)

$$y \geq -\frac{1}{2}x$$

$$1 \geq -\frac{1}{2}(3)$$

$$1 \geq -\frac{3}{2} \quad \text{true}$$

FIGURE 3.61

Since the point (3, 1) satisfies the inequality, every point on the same side of the line as (3, 1) will also satisfy the inequality $y \geq -\frac{1}{2}x$. Shade this region as indicated. Every point in the shaded region as well as every point on the line satisfies the inequality.

EXAMPLE 3 Graph the inequality $3x - 2y < -6$.

Solution: First graph the equation $3x - 2y = -6$. Since the inequality is $<$, we use a dashed line when drawing the graph (see Fig. 3.62). Substituting the checkpoint (0, 0) into the inequality results in a false statement.

Checkpoint (0, 0)

$$3x - 2y < -6$$

$$3(0) - 2(0) < -6$$

$$0 < -6 \quad \text{false}$$

The solution is, therefore, that part of the plane that does not contain the origin.

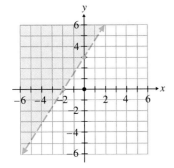

FIGURE 3.62

Exercise Set 3.6

Graph each inequality.

1. $x > 3$

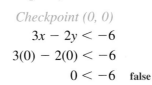

2. $x \geq \frac{5}{2}$

3. $y < -2$

4. $y < x$

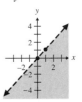

5. $y \geq 2x$

6. $y > -2x$

7. $y < 2x + 1$

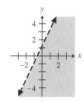

8. $y \geq 3x - 1$

9. $y < -3x + 4$

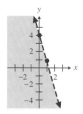

10. $y \geq 2x + 4$

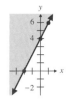

11. $y \geq \dfrac{1}{2}x - 3$

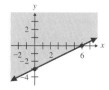

12. $y < 3x + 5$

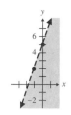

13. $y \leq \dfrac{1}{3}x + 6$

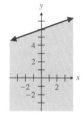

14. $y > 6x + 1$

15. $y \leq -3x + 5$

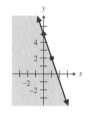

16. $y \leq \dfrac{2}{3}x + 3$

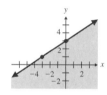

17. $y > 5x - 4$

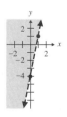

18. $y > \dfrac{2}{3}x - 1$

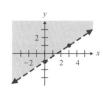

19. $2x + y < 4$

20. $3x - 4y \leq 12$

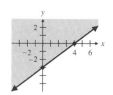

21. $2x \leq 5y + 10$

22. $\dfrac{5}{6}x - \dfrac{1}{6}y \leq \dfrac{2}{3}$

23. $-\dfrac{1}{6}x - \dfrac{1}{3}y > \dfrac{2}{3}$

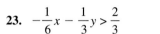
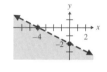

24. $-\dfrac{5}{12}x + \dfrac{1}{4}y \geq \dfrac{3}{4}$

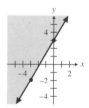

25. When graphing an inequality containing $\geq$ or $\leq$, why are points on the line solutions to the inequality?
$\geq$ means greater than or *equal to,* $\leq$ means less than or *equal to*

26. When graphing an inequality containing $>$ or $<$, why are points on the line not solutions to the inequality?
Points on the line are solutions for an equality, not $<$ or $>$.

CUMULATIVE REVIEW EXERCISES

[2.1] **27.** Solve the proportion $\dfrac{\frac{2}{3}(x-4)}{5} = \dfrac{x+8}{6}$. -56

[2.2] **28.** If $C = \bar{x} + Z\dfrac{\sigma}{\sqrt{n}}$, find C when $\bar{x} = 80$, $Z = 1.96$,
$\sigma = 3$, and $n = 25$.   81.176

[2.3] **29.** El Gigundo Department Store is going out of business. The first week all items are being reduced by 10%.

The second week all items are being reduced by an additional $2. If during the second week Sean purchases a Motley Crüe CD for $12.15, find the original cost of the CD. $15.72

[3.3] **30.** Write an equation of the line that passes through the point $(6, -2)$ and is perpendicular to the line $2x - y = 4$. $x + 2y = 2$ (other forms of the answers are possible)

Group Activity/ Challenge Problems

Graph.

1. $y < |x|$

2. $y \geq |x+4|$

3. $y \geq x^2$

4. $y < x^2 - 4$

1.

2.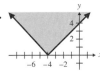

3.

4.

Summary

GLOSSARY

Cartesian coordinate system (or **rectangular coordinate system**) *(133):* Two number lines drawn perpendicular to each other, creating four quadrants.

Collinear points *(141):* Points in a straight line.

Domain *(172):* The set of first coordinates in a set of ordered pairs.

Function *(174):* A relation in which no two ordered pairs have the same first coordinate and a different second coordinate.

Graph *(141):* An illustration of the set of points whose coordinates satisfy an equation.

Linear function *(183):* A function of the form $f(x) = ax + b$.

Negative reciprocals *(160):* Two real numbers whose product is -1.

Negative slope *(156):* A line has a negative slope when it falls as it moves from left to right.

Nonlinear function *(184):* A function whose graph is not a straight line.

Ordered pair *(133):* The x and y coordinates of a point listed in parentheses, x first.

Origin *(133):* The point of intersection of the x and y axes.

Parallel lines *(160):* Two lines in the same plane that do not intersect no matter how far they are extended. Two lines are parallel when they have the same slope.

Perpendicular lines *(160):* Two lines that cross at right angles. Two lines are perpendicular when their slopes are negative reciprocals.

Positive slope *(156):* A line has a positive slope when it rises as it moves from left to right.

Range *(172):* The set of second coordinates in a set of ordered pairs.

Relation *(171):* Any set of ordered pairs.

Slope of a line *(155):* The ratio of the vertical change to the horizontal change between any two points on a line.

Trace *(148):* A feature on a graphing calculator that allows you to move the cursor along a graph.

Window *(136):* The display on the calculator.

x **Axis** *(133):* The horizontal axis in the Cartesian coordinate system.

x **Intercept** *(144):* The value of x where a graph crosses the x axis.

y **Axis** *(133):* The vertical axis in the Cartesian coordinate system.

y **Intercept** *(144):* The value of y where a graph crosses the y axis.

Zoom *(148):* A feature on a graphing calculator that allows you to enlarge the region of a curve near the cursor.

IMPORTANT FACTS

Distance formula: $d = \sqrt{(x_2 - x_1)^2 + (y_2 - y_1)^2}$

Midpoint formula: $\left(\dfrac{x_1 + x_2}{2}, \dfrac{y_1 + y_2}{2} \right)$

Slope of a line: $m = \dfrac{\Delta y}{\Delta x} = \dfrac{y_2 - y_1}{x_2 - x_1}$

Forms of a linear equation

Standard form: $ax + by = c$
Slope–intercept form: $y = mx + b$
Point–slope form: $y - y_1 = m(x - x_1)$

To find the x intercept, set $y = 0$ and solve the equation for x.

To find the y intercept, set $x = 0$ and solve the equation for y.

To write an equation in slope–intercept form, solve the equation for y.

Positive slope

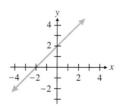

Zero slope

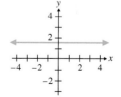

Negative slope

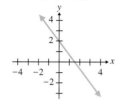

Slope is undefined

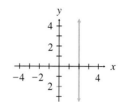

$y = |x|$

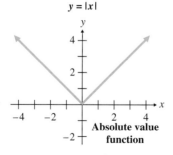

Absolute value function

$y = \sqrt{x}$

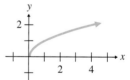

Square root function

Review Exercises

[3.1] **1.** Plot the ordered pairs on the same axes.

(a) $A\ (5, 3)$ (b) $B\ (0, 4)$ (c) $C\left(5, \dfrac{1}{2}\right)$

(d) $D\ (-4, 3)$ (e) $E\ (-6, -1)$ (f) $F\ (-2, 0)$

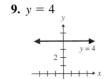

Find the distance between, and the midpoint of, the line segment between the two given points.

2. $(0, 0),\ (3, -4)$ $5,\ \left(\frac{3}{2},\ -2\right)$ **3.** $(6, 2),\ (2, -1)$ $5,\ \left(4, \frac{1}{2}\right)$ **4.** $(-2, -3),\ (3, 9)$ $13,\ \left(\frac{1}{2},\ 3\right)$

5. $(-4, 3),\ (-2, 5)$ $\sqrt{8} \approx 2.83,\ (-3, 4)$ **6.** $(3, 4),\ (5, 4)$ $2,\ (4, 4)$ **7.** $(-3, 5),\ (-3, -8)$ $13,\ \left(-3, -\frac{3}{2}\right)$

8. The graph at the right is based on information supplied by the U.S. Department of Commerce.

    (a) How many measurements were made to construct this graph?   13

    (b) In which quarter did Americans save the greatest percent of their disposable income? Estimate that percent.   fourth quarter 1992, ≈ 6.0%

    (c) In which quarter did Americans save the smallest percent of their disposable income? Estimate that percent.   first quarter 1994, ≈ 3.5%

    (d) Write a paragraph describing the changes in savings from the first quarter of 1991 to the first quarter of 1994.

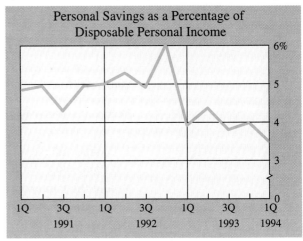

Personal Savings as a Percentage of Disposable Personal Income

[3.2, 3.3] *Graph the equation by the method of your choice.*

9. $y = 4$

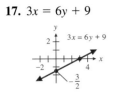

10. $x = -2$

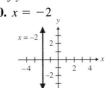

11. $y = 4x$

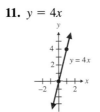

12. $y = -3x + 4$

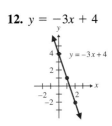

13. $y = -\dfrac{1}{2}x + 2$

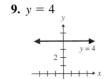

14. $2x - 3y = 12$

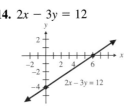

15. $2y = 3x - 6$

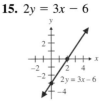

16. $5x - 2y = 10$

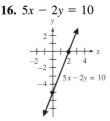

17. $3x = 6y + 9$

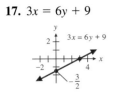

18. $25x - 50y = 200$

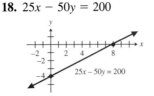

19. $3x - 2y = 150$

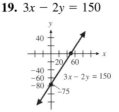

20. $\dfrac{2}{3}x = \dfrac{1}{4}y + 20$

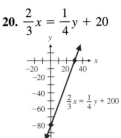

Determine the slope and y intercept of the equation.

21. $y = -x + 5$ $m = -1, b = 5$

22. $y = -4x + \dfrac{1}{2}$ $m = -4, b = \frac{1}{2}$

23. $3x + 5y = 12$ $m = -\frac{3}{5}, b = \frac{12}{5}$

24. $9x + 7y = 15$ $m = -\frac{9}{7}, b = \frac{15}{7}$

25. $x = -2$ m is undefined, no y intercept

26. $y = 6$ $m = 0, b = 6$

[3.4] *Determine the slope of the line through the two given points.*

27. $(4, 6), (5, -1)$ -7

28. $(-2, 3), (4, 1)$ $-\frac{1}{3}$

Two points on l_1 and two points on l_2 are given. Determine if l_1 is parallel to l_2, l_1 is perpendicular to l_2, or neither.

29. l_1: (4, 3) and (0, −3); l_2: (1, −1) and (2, −2)   neither

30. l_1: (3, 2) and (2, 3); l_2: (4, 1) and (1, 4)   parallel

31. l_1: (4, 0) and (1, 3); l_2: (5, 2) and (6, 3)   perpendicular

32. l_1: (−3, 5) and (2, 3); l_2: (−4, −2) and (−1, 2) neither

Solve for the given variable if the line through the two given points is to have the given slope.

33. $(5, a)$ and $(4, 2)$; $m = 1$ $a = 3$

34. $(3, 0)$ and $(5, y)$; $m = 3$ $y = 6$

35. $(-2, -1)$ and $(4, y)$; $m = -6$ $y = -37$

36. $(x, 2)$ and $(5, -2)$; $m = 2$ $x = 7$

Find the slope of each line. If the slope is undefined, so state. Then write the equation of the line.

37.   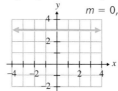 $m = 0, y = 3$ **38.** m is undefined, $x = 2$ **39.**   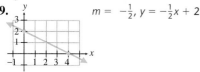 $m = -\frac{1}{2}, y = -\frac{1}{2}x + 2$

Determine if the two lines are parallel, perpendicular, or neither.

40. $y = 3x - 6$
$6y = 18x + 6$ parallel

41. $2x - 3y = 9$
$-3x - 2y = 6$ perpendicular

42. $4x - 2y = 10$
$-2x + 4y = -8$ neither

In Ex. 43–48 find the equation of the line with the properties given. Write the answer in slope–intercept form.

43. Slope $= -\frac{2}{3}$, through $(3, 2)$ $y = -\frac{2}{3}x + 4$

44. Through $(4, 3)$ and $(2, 1)$ $y = x - 1$

45. Through $(-6, 2)$ parallel to $y = 3x - 4$ $y = 3x + 20$

46. Through $(4, -2)$ parallel to $2x - 5y = 6$ $y = \frac{2}{5}x - \frac{18}{5}$

47. Through $(-3, 1)$ perpendicular to $y = \frac{3}{5}x + 5$ $y = -\frac{5}{3}x - 4$ **48.** Through $(4, 2)$ perpendicular to $4x - 2y = 8$ $y = -\frac{1}{2}x + 4$

49. The yearly profit of a bagel company can be estimated by the formula $p = 0.1x - 5000$, where x is the number of bagels sold per year.   Answer on page 200

  (a) Draw a graph of profits versus bagels sold for up to 250,000 bagels.

  (b) Estimate the number of bagels that must be sold for the company to break even.   approx. 50,000 bagels

  (c) Estimate the number of bagels sold if the company has a $20,000 profit.   approx. 250,000 bagels

50. Draw a graph illustrating the interest on a $12,000 loan for a 1-year period for various interest rates up to 20%. Use interest = principal · rate · time.   Answer on page 200

51. The graph at the right provides information on U.S. bank failures from 1980 through the first quarter of 1994. The graph at the right is based on information from the Federal Deposit Insurance Corporation (FDIC).

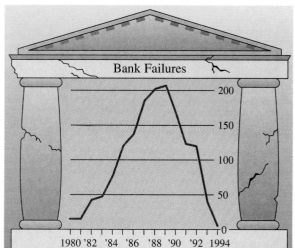

SEE EXERCISE 51.

(a) In which year did the most bank failures occur? Estimate the number of banks that failed at that time.   1989, ≈205

(b) Would you describe the growth in the number of bank failures as nearly linear from 1981 to 1989? Explain.   yes

(c) Would you describe the reduction in the number of bank failures from 1989 to 1994 as nearly linear? Explain.   yes

(d) Would you describe the activity in the number of bank failures from 1980 to 1994 as nearly linear? Explain.   no

(e) Assuming that the number of bank failures in 1980 was 12, and the number of failures in 1989 was 205, determine the equation of the straight line that connects the points on the graph at 1980 and 1989. Round the slope to two decimal places when determining the equation. Use 1980 as the reference year. $y = 21.44x + 12$

(f) Write a paragraph describing the trend in bank failures from 1980 through the first quarter of 1994.

49. (a)

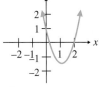

50.

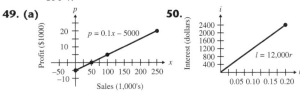

[3.5] *Give the range and domain of each relation.*   **52.** D: {−2, 0, 3, 6}, R: {−1, 4, 5, 9}   **53.** D: {½, 2, 4, 5} R: {−6, −1, 2, 3}

52. {(3, 4), (−2, 5), (0, −1), (6, 9)}

53. {(½, 2), (4, −6), (5, 3), (2, −1)}

54.

D: {x| −1 ≤ x ≤ 1}, R: {y| −1 ≤ y ≤ 1}

55.

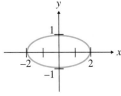

D: {x| −2 ≤ x ≤ 2}, R: {y| −1 ≤ y ≤ 1}

56.

D: ℝ, R: {y|y ≤ 0}

57.

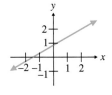

D: ℝ, R: ℝ

In Ex. 58–65, determine which of the following relations are functions.

58.

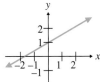

function

59.

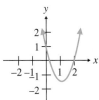

function

60.

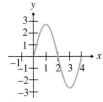

function

61.

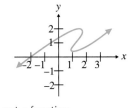

not a function

62. {(0, 4), (5, 6), (−2, 4), (1, 3)}   function

64. {(1, 4), (2, 4), (3, 4), (4, 4)}   function

63. {(6, −2), (−3, 5), (5, 2), (−1, 4)}   function

65. {(−3, 2), (4, 3), (3, 2), (4, 6)}   not a function

In Ex. 66–73, evaluate the functions at the values indicated.

66. $f(x) = -2x + 5$; find **(a)** $f(-1)$; **(b)** $f(\tfrac{1}{2})$.   **(a)** 7 **(b)** 4

67. $f(x) = -\tfrac{1}{2}x + 4$; find **(a)** $f(4)$; **(b)** $f(3)$.   **(a)** 2 **(b)** $\tfrac{5}{2}$

68. $f(x) = 3x^2 - 4x - 1$; find **(a)** $f(2)$; **(b)** $f(-5)$.   **(a)** 3 **(b)** 94

69. $f(x) = x^3 - 2x + 3$; find **(a)** $f(\tfrac{1}{2})$; **(b)** $f(2)$.   **(a)** $\tfrac{17}{8}$ **(b)** 7

70. $f(x) = (x - 2)^2 + 5$; find **(a)** $f(2)$; **(b)** $f(-3)$.   **(a)** 5 **(b)** 30

71. $g(x) = \left|\tfrac{1}{3}x - 2\right|$; find **(a)** $g(6)$; **(b)** $g(4)$.   **(a)** 0 **(b)** $\tfrac{2}{3}$ or 0.6

72. (a) $\sqrt{28} \approx 5.29$ **(b)** 7

72. $h(x) = \sqrt{x^2 + 24}$; find **(a)** $h(2)$; **(b)** $h(5)$.

73. $f(x) = \dfrac{x^2 - 9}{x - 3}$; find **(a)** $f(-3)$; **(b)** $f(4)$.   **(a)** 0 **(b)** 7

74. The number of baskets of apples, N, that are produced by x trees in a small orchard ($x \le 100$) is given by the function $N = f(x) = 40x - 0.2x^2$. How many baskets of apples are produced by:

(a) 20 trees? 720 **(b)** 50 trees? 1500

75. If a ball is dropped from the top of a 100-foot building, its height above the ground, h, at any time, t, can be found by the function $h = f(t) = -16t^2 + 100$, $t \le 2.5$. Find the height of the ball at:

(a) 1 second   84 ft **(b)** 2 seconds   36 ft

76. The graph on the right illustrates the world production of chlorofluorocarbons (CFCs) from 1940 to 1990.

  (a) Explain why the total CFC production is a function.   Each year a unique amount of CFC is produced.

  (b) What is the independent variable? What is the dependent variable?   time, amount of CFC

  (c) What is the domain of the graph? Estimate the range of the total CFC production graph.

  (d) Let the production of CFC from aerosols be called $f(t)$, and the production of CFC from nonaerosols be called $g(t)$, and total CFC production be called $T(t)$. Explain the relationship between $T(t)$ and $f(t)$ and $g(t)$. $T(t) = f(t) + g(t)$

  (e) Estimate the amount of CFC produced from aerosols and from nonaerosols in 1970. Also estimate the total CFC production in 1970.

  (f) Describe the production of CFCs from aerosols from 1940 to 1990. Describe the production of CFCs from nonaerosols from 1940 to 1990.

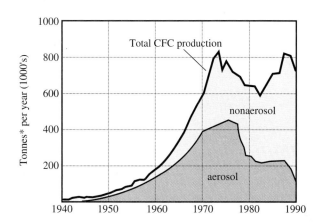

SEE EXERCISE 76.

76. **(c)** D: $\{t \mid 1940 \leq t \leq 1990\}$, R: $\{a \mid 0 \leq a \leq 815\}$
     (e)   400, 200, 600 thousand tonnes

---

*For each function (**a**) state the domain, (**b**) graph the function, and (**c**) state the range. Use your calculator to evaluate square roots.*

77. $f(x) = -3x + 2$

D: $\mathbb{R}$, R: $\mathbb{R}$

78. $g(x) = |2x - 1|$

D: $\mathbb{R}$, R: $\{y \mid y \geq 0\}$

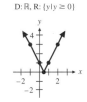

79. $f(x) = \sqrt{2x - 6}$

D: $\{x \mid x \geq 3\}$
R: $\{y \mid y \geq 0\}$

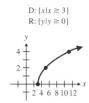

80. $f(x) = -|x - 2|$

D: $\mathbb{R}$, R: $\{y \mid y \leq 0\}$

81. $g(x) = |x - 4| + 2$

D: $\mathbb{R}$, R: $\{y \mid y \geq 2\}$

82. $h(x) = -\sqrt{4 - x} - 3$

D: $\{x \mid x \leq 4\}$
R: $\{y \mid y \leq -3\}$

83. (a)

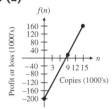

83. The profit or loss for a publishing company on a particular textbook can be estimated by the function $f(n) = 24n - 200{,}000$, where $f(n)$ is the profit or loss, and n is the number of copies of the book sold.

  (a) Construct a graph showing the relationship between the number of books sold and the profit or loss for the publishing company for up to 15,000 books.

  (b) Estimate the number of books that must be sold for the company to break even.   8300

  (c) Estimate the number of books that must be sold for the company's profit to be $100,000.   12,500

---

*A (metric) tonne is a little greater than a customary ton.

[3.6] Graph the given inequality.

84. $y \geq -3$

85. $x < 4$

86. $y \leq 4x - 3$

87. $y < \frac{1}{3}x - 2$

Practice Test

1. Find the length and midpoint of the line segment between the points $(1, 3)$ and $(-2, -1)$. $5, (-\frac{1}{2}, 1)$

2. Find the slope and y intercept of $4x - 9y = 15$.

3. Write the equation of the following graph in slope–intercept form. $y = 3x - 3$ **2.** $m = \frac{4}{9}, b = -\frac{5}{3}$

7. $y = 2x - 2$

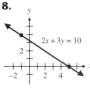

8. $2x + 3y = 10$

4. Write the equation of the line with a slope of 4 passing through the point $(-1, 3)$. Write the equation in slope–intercept form. $y = 4x + 7$

5. Write the equation of the line (in slope–intercept form) passing through the points $(3, -1)$ and $(-4, 2)$.

6. Write the equation of the line (in slope–intercept form) passing through the point $(-1, 4)$ perpendicular to $2x + 3y = 6$. $y = \frac{3}{2}x + \frac{11}{2}$ **5.** $y = -\frac{3}{7}x + \frac{2}{7}$

7. Graph $y = 2x - 2$.

8. Graph $2x + 3y = 10$.

9. The monthly profit (or loss) of Belushi's Video Store can be estimated by the function $f(c) = 2.5c - 800$,

where c is the number of tapes rented each month, and $f(c)$ is the monthly profit or loss.

(a) Construct a graph showing the relationship between the number of tapes rented each month, for up to 2000 tapes, and the monthly profit or loss.

(b) Estimate the number of tapes rented if the monthly profit is \$4000. ≈ 1900

(c) Estimate the number of tapes that need to be rented each month for the company to break even. ≈ 320

10. State the domain and range of the following relation:
$$\{(4, 0), (2, -3), (\tfrac{1}{2}, 2), (6, 9)\}$$

11. Determine which, if any, of the following are functions.

(a)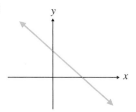

(b)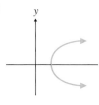

(c) $\{(1, 2), (3, 4), (-1, 2), (5, 0)\}$ (a), (c)

12. $f(x) = -2x^2 + 3x - 5$; find (a) $f(2)$; (b) $f\left(\frac{1}{2}\right)$.

13. Graph $f(x) = \sqrt{x + 3}$

14. Graph $y = |3x - 6| - 1$.

15. Graph $y \leq 4x - 2$.

10. D: $\{\frac{1}{2}, 2, 4, 6\}$, R: $\{-3, 0, 2, 9\}$ **12. (a)** -7 **(b)** -4

9. (a)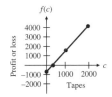

13.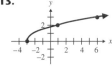

14.

15.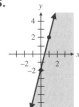

Chapter 4

Systems of Equations and Inequalities

See Section 4.3, Example 7

Preview and Perspective

In this chapter we discuss solving systems of linear equations. We solve systems of linear equations containing two equations in two variables by the following methods: by graphing, by substitution, by the addition method, by using determinants and Cramer's rule, and by using matrices. Do not be surprised if you cover only a few techniques for solving systems. Often, time does not permit covering all.

We also discuss solving systems of equations containing three equations in three variables by a variety of techniques. There are many real-life applications of systems of equations, as illustrated in Section 4.3.

In Section 4.6 we solve systems of *linear inequalities* graphically. We use the techniques that were presented in Section 3.6 to graph inequalities.

The emphasis of this chapter is on linear equations. In section 10.5 we will solve nonlinear systems of equations. A nonlinear system contains at least one equation whose graph is not a straight line. We graph nonlinear systems of inequalities in Section 10.6.

Businesses often need to consider the relationships that exist between variables. For example, a business considers items such as cost of materials, cost of labor, cost of transportation, sale price of the item manufactured, and a host of other items. Businesses relate these variables to each other as systems of equations. These systems are often solved by computer. However, individuals must first construct the system of equations to be solved, as we will learn to do in this chapter.

4.1 Solving Systems of Linear Equations

Tape 5

1. Solve systems of linear equations graphically.
2. Solve systems of linear equations by substitution.
3. Solve systems of linear equations by the addition method.
4. Understand the graphical solution to a linear equation in one variable.

It is often necessary to find a common solution to two or more linear equations. We refer to these equations as **simultaneous linear equations** or as a **system of linear equations.** For example,

$$\left. \begin{array}{l} (1)\ y = x + 5 \\ (2)\ y = 2x + 4 \end{array} \right\} \text{ system of linear equations}$$

A **solution to a system of equations** is an ordered pair or pairs that satisfy *all* equations in the system. The only solution to the system above is (1, 6).

Check in Equation (1)	*Check in Equation (2)*
(1, 6)	(1, 6)
$y = x + 5$	$y = 2x + 4$
$6 = 1 + 5$	$6 = 2(1) + 4$
$6 = 6$ true	$6 = 6$ true

The ordered pair (1, 6) satisfies *both* equations and is the solution to the system of equations.

A system of equations may consist of more than two equations. If a system consists of three equations in three variables, such as x, y, and z, the solution will be an **ordered triple** of the form (x, y, z). If the ordered triple (x, y, z) is a solution to the system, it must satisfy all three equations in the system. A system with three equations and three unknowns is referred to as a *third-order system.* Third-order systems are discussed in Section 4.2. Systems of equations may have more than three equations and three variables, but we will not discuss them in this book.

Graphing Method   **1** To solve a system of linear equations in two variables graphically, graph all equations in the system on the same axes. The solution to the system will be the ordered pair (or pairs) common to all the lines, or the point of intersection of all lines in the system.

When two lines are graphed, three situations are possible, as illustrated in Fig. 4.1. In Fig. 4.1a, lines 1 and 2 intersect at exactly one point. This system of equations has *exactly one solution.* This is an example of a **consistent** system of equations. A consistent system of equations is a system of equations that has a solution.

Lines 1 and 2 of Fig. 4.1b are different but parallel lines. The lines do not intersect, and this system of equations has *no solution.* This is an example of an **inconsistent** system of equations. An inconsistent system of equations is a system of equations that has no solution.

In Fig. 4.1c, lines 1 and 2 are actually the same line. In this case, every point on the line satisfies both equations and is a solution to the system of equations. This system has *an infinite number of solutions.* This is an example of a **dependent** system of equations. In a dependent system of linear equations both equations represent the same line. *Note that a dependent system is also a consistent system since it has solutions.*

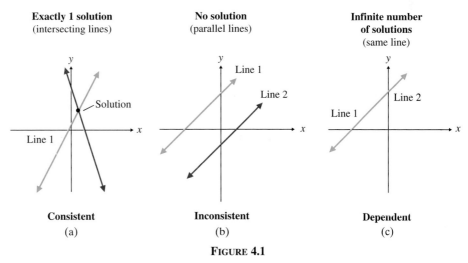

FIGURE 4.1

We can determine if a system of linear equations is consistent, inconsistent, or dependent by writing each equation in slope–intercept form and comparing the slopes and y intercepts. Note that if the slopes of the lines are different (Fig. 4.1a), the system is consistent. If the slopes are the same but the y intercepts different (Fig. 4.1b), the system is inconsistent, and if both the slopes and the y intercepts are the same (Fig. 4.1c), the system is dependent.

EXAMPLE 1   Without graphing the equations, determine whether the following system of equations is consistent, inconsistent, or dependent.

$$2x + y = 3$$
$$4x + 2y = 12$$

Solution: Write each equation in slope-intercept form.

$$2x + y = 3 \qquad\qquad 4x + 2y = 12$$
$$y = -2x + 3 \qquad\qquad 2y = -4x + 12$$
$$\qquad\qquad\qquad\qquad\qquad y = -2x + 6$$

Since both equations have the same slope, -2, and different y intercepts, the lines are parallel lines. Therefore, the system is inconsistent and has no solution.

EXAMPLE 2   Solve the following system of equations graphically.

$$y = x + 2$$
$$y = -x + 4$$

Solution: Graph both equations on the same axes (Fig. 4.2).

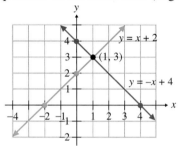

FIGURE 4.2

The solution is the point of intersection of the two lines, $(1, 3)$.

Substitution Method   **2**   Often, an exact solution to a system of equations may be difficult to find on a graph. When an exact answer is necessary, the system should be solved algebraically, either by substitution or by addition (elimination) of equations.

To Solve a Linear System of Equations by Substitution

1. Solve for a variable in either equation. (If possible, solve for a variable with a numerical coefficient of 1 to avoid working with fractions.)
2. Substitute the expression found for the variable in step 1 into the other equation. This will result in an equation containing only one variable.
3. Solve the equation obtained in step 2 to find the value of this variable.
4. Substitute the value found in step 3 into the equation from step 1. Solve the equation to find the remaining variable.
5. Check your solution in all equations in the system.

EXAMPLE 3   Solve the system of equations by substitution.

$$y = 2x + 5$$
$$y = -4x + 2$$

Solution: Since both equations are already solved for y, we can substitute $2x + 5$ for y in the second equation and then solve for the remaining variable, x.

$$2x + 5 = -4x + 2$$
$$6x + 5 = 2$$
$$6x = -3$$
$$x = -\frac{1}{2}$$

Now find y by substituting $-\frac{1}{2}$ for x in either of the original equations. We will use the first equation.

$$y = 2x + 5$$
$$= 2\left(-\frac{1}{2}\right) + 5$$
$$= -1 + 5 = 4$$

A check will show that the solution is $\left(-\frac{1}{2}, 4\right)$.

EXAMPLE 4   Solve the system of equations by substitution.

$$2x + y = 11$$
$$x + 3y = 18$$

Solution: Begin by solving for one of the variables in either of the equations. You may solve for either of the variables; however, if you solve for a variable with a numerical coefficient of 1, you may avoid working with fractions. In this system the y term in $2x + y = 11$ and the x term in $x + 3y = 18$ both have numerical coefficient 1.

Let's solve for y in $2x + y = 11$.

$$2x + y = 11$$
$$y = -2x + 11$$

Next, substitute $-2x + 11$ for y in the *other equation*, $x + 3y = 18$, and solve for the remaining variable, x.

$$x + 3y = 18$$
$$x + 3(\overbrace{-2x + 11}) = 18$$
$$x - 6x + 33 = 18$$
$$-5x + 33 = 18$$
$$-5x = -15$$
$$x = 3$$

Finally, substitute $x = 3$ in the equation $y = -2x + 11$ and solve for y.

$$y = -2x + 11$$
$$y = -2(3) + 11 = 5$$

The solution is the ordered pair (3, 5).

If, when solving a system of equations by either substitution or the addition method, you arrive at an equation that is false, such as $5 = 6$ or $0 = 3$, the system is inconsistent and has no solution. If you obtain an equation that is true, such as $6 = 6$ or $0 = 0$, the system is dependent and has an infinite number of solutions.

Helpful Hint

Students sometimes successfully solve for one of the variables and forget to solve for the other. Remember that a solution must contain a numerical value for each variable in the system.

Addition Method

3 A third and often the easiest method of solving a system of equations is the addition or elimination method. The object of this process is to obtain two equations whose sum will be an equation containing only one variable. Keep in mind that your immediate goal is to obtain one equation containing only one unknown.

EXAMPLE 5 Solve the following system of equations using the addition method.

$$x + y = 6$$
$$2x - y = 3$$

Solution: Note that one equation contains $+y$ and the other contains $-y$. By adding the equations, we can eliminate the variable y and obtain one equation containing only one unknown, x.

$$\begin{array}{r} x + y = 6 \\ 2x - y = 3 \\ \hline 3x \phantom{{}- y} = 9 \end{array}$$

Now solve for the remaining variable, x.

$$\frac{3x}{3} = \frac{9}{3}$$
$$x = 3$$

Finally, solve for y by substituting 3 for x in either of the original equations.

$$x + y = 6$$
$$3 + y = 6$$
$$y = 3$$

The solution is (3, 3).

> ## To Solve a Linear System of Equations by the Addition (or Elimination) Method
>
> 1. If necessary, rewrite each equation in standard form, that is, the terms containing variables on the left side of the equal sign and the constant on the right side of the equal sign.
> 2. If necessary, multiply one or both equations by a constant(s) so that when the equations are added, the sum will contain only one variable.
> 3. Add the respective sides of the equations. This will result in a single equation containing only one variable.
> 4. Solve for the variable in the equation obtained in step 3.
> 5. Substitute the value found in step 4 into either of the original equations. Solve that equation to find the value of the remaining variable.
> 6. Check your solution in *all* equations in the system.

In step 2 of the procedure, we indicate that it may be necessary to multiply both sides of an equation by a constant. In this text we will use brackets, [   ], to indicate multiplication of *an entire equation* by a real number. Thus, 4[   ] means to multiply the entire equation within the brackets by 4, and in general a[   ] means to multiply the entire equation within the brackets by the real number a. For example, $3[2x + 4 = -5]$ will give $6x + 12 = -15$ and $-2[5x - 6 = -8]$ will give $-10x + 12 = 16$.

EXAMPLE 6   Solve the following system of equations using the addition method.

$$2x + y = 11$$
$$x + 3y = 18$$

Solution: The object of the addition process is to obtain two equations whose sum will be an equation containing only one variable. To eliminate the variable x, we multiply the second equation by -2 and add the two equations.

$$\begin{array}{c} 2x + y = 11 \\ -2[x + 3y = 18] \end{array} \quad \text{gives} \quad \begin{array}{c} 2x + y = 11 \\ -2x - 6y = -36 \end{array}$$

Now add:

$$\begin{array}{rcr} 2x + y &=& 11 \\ -2x - 6y &=& -36 \\ \hline -5y &=& -25 \\ y &=& 5 \end{array}$$

Now solve for x by substituting 5 for y in either of the original equations.

$$2x + y = 11$$
$$2x + 5 = 11$$
$$2x = 6$$
$$x = 3$$

The solution is (3, 5). Note that we could have first eliminated the variable y by multiplying the first equation by -3 and then adding.

Sometimes both equations must be multiplied by different numbers in order for one of the variables to be eliminated. This procedure is illustrated in Examples 7 and 8.

EXAMPLE 7 Solve the following system of equations using the addition method.

$$2x + 3y = 6$$
$$5x - 4y = -8$$

Solution: The x variable can be eliminated by multiplying the first equation by -5 and the second by 2 and then adding the equations.

$$
\begin{array}{ll}
-5[2x + 3y = 6] & \quad -10x - 15y = -30 \\
2[5x - 4y = -8] & \text{gives} \quad 10x - 8y = -16
\end{array}
$$

$$
\begin{array}{r}
-10x - 15y = -30 \\
\underline{10x - 8y = -16} \\
-23y = -46 \\
y = 2
\end{array}
$$

Solve for x.

$$2x + 3y = 6$$
$$2x + 3(2) = 6$$
$$2x + 6 = 6$$
$$2x = 0$$
$$x = 0$$

The solution to this system is (0, 2).

In Example 7, the same solution could be obtained by multiplying the first equation by 5 and the second by -2 and then adding. Try it now and see.

EXAMPLE 8 Solve the following system of equations using the addition method.

$$2x + 3y = 7$$
$$5x - 7y = -3$$

Solution: We can eliminate the variable x by multiplying the first equation by -5 and the second by 2.

$$
\begin{array}{ll}
-5[2x + 3y = 7] & \quad -10x - 15y = -35 \\
2[5x - 7y = -3] & \text{gives} \quad 10x - 14y = -6
\end{array}
$$

$$-10x - 15y = -35$$
$$10x - 14y = -6$$
$$\overline{-29y = -41}$$
$$y = \frac{41}{29}$$

We can now find x by substituting $\frac{41}{29}$ for y into one of the original equations and solving for x. If you try this, you will see that, although it can be done, it gets messy. An easier method to solve for x is to go back to the original equations and eliminate the variable y.

$$\begin{array}{ll} 7[2x + 3y = 7] \\ 3[5x - 7y = -3] \end{array} \quad \text{gives} \quad \begin{array}{ll} 14x + 21y = 49 \\ 15x - 21y = -9 \end{array}$$

$$14x + 21y = 49$$
$$15x - 21y = -9$$
$$\overline{29x \qquad = 40}$$
$$x = \frac{40}{29}$$

The solution is $\left(\dfrac{40}{29}, \dfrac{41}{29}\right)$.

EXAMPLE 9   Solve the following system of equations using the addition method.

$$0.50x + 0.75y = -0.25$$
$$-0.25x - 0.125y = -0.625$$

Solution: When you have a system of equations with decimal coefficients, your first step might be to multiply each equation in the system by the power of 10 that will eliminate the decimals from the equation. Then proceed as in all other examples. Since the first equation has decimals to the hundredths place, multiply each term in the equation by 100 to eliminate the decimals. Since the second equation has decimals to the thousandths place, multiply each term in the equation by 1000 to eliminate the decimals.

$$100\,(0.50x) + 100\,(0.75y) = 100\,(-0.25) \quad \text{or} \quad 50x + 75y = -25$$

$$1000\,(-0.25x) - 1000\,(0.125y) = 1000\,(-0.625) \quad \text{or} \quad -250x - 125y = -625$$

Now solve the resulting system of equations.

$$50x + 75y = -25$$
$$-250x - 125y = -625$$

$$\begin{array}{ll} 5[50x + 75y = -25] \\ -250x - 125y = -625 \end{array} \quad \text{gives} \quad \begin{array}{ll} 250x + 375y = -125 \\ -250x - 125y = -625 \end{array}$$
$$\overline{250y = -750}$$
$$y = -3$$

Now solve for x.

$$50x + 75y = -25$$
$$50x + 75(-3) = -25$$
$$50x - 225 = -25$$
$$50x = 200$$
$$x = 4$$

Thus, the solution is $(4, -3)$.

EXAMPLE 10   Solve the system of equations using the addition method.

$$x + \frac{4}{3}y = 2$$

$$y = -\frac{2}{3}x + \frac{5}{2}$$

Solution: First, eliminate the fractions in each equation. To eliminate fractions, we multiply both sides of the first equation by 3 and both sides of the second equation by 6.

$$x + \frac{4}{3}y = 2 \qquad\qquad y = -\frac{2}{3}x + \frac{5}{2}$$

$$3\left(x + \frac{4}{3}y\right) = 3 \cdot 2 \qquad 6 \cdot y = 6\left(-\frac{2}{3}x + \frac{5}{2}\right)$$

$$3x + 4y = 6 \qquad\qquad 6y = -4x + 15$$

The new system of equations is

$$\begin{array}{cc} 3x + 4y = 6 & 3x + 4y = 6 \\ & \text{or} \\ 6y = -4x + 15 & 4x + 6y = 15 \end{array}$$

Now solve the system. We will solve for y by eliminating the terms containing x.

$$\begin{array}{ll} 4[3x + 4y = \ 6] & \\ & \text{gives} \\ -3[4x + 6y = 15] & \end{array} \qquad \begin{array}{r} 12x + 16y = \ \ 24 \\ -12x - 18y = -45 \\ \hline -2y = -21 \\ y = \dfrac{21}{2} \end{array}$$

Now solve for x by eliminating the y terms.

$$\begin{array}{ll} 3[3x + 4y = \ 6] & \\ & \text{gives} \\ -2[4x + 6y = 15] & \end{array} \qquad \begin{array}{r} 9x + 12y = \ \ 18 \\ -8x - 12y = -30 \\ \hline x \qquad\quad = -12 \end{array}$$

We could also have found x by substituting $\frac{21}{2}$ for y in either of the original equations and then solving for x. The solution to the system is $\left(-12, \frac{21}{2}\right)$.

EXAMPLE 11 Solve the following system of equations using the addition method.

$$2x + y = 3$$
$$4x + 2y = 12$$

Solution: $-2[2x + y = 3]$ gives $-4x - 2y = -6$
$$4x + 2y = 12 \qquad\qquad 4x + 2y = 12$$

$$\begin{array}{r} -4x - 2y = -6 \\ 4x + 2y = 12 \\ \hline 0 = 6 \quad \text{false} \end{array}$$

Since $0 = 6$ is a false statement, this system has no solution. The system is inconsistent and graphs of these equations are parallel lines.

EXAMPLE 12 Solve the following system of equations using the addition method.

$$x - \frac{1}{2}y = 2$$
$$y = 2x - 4$$

Solution: First, align the x and y terms on the left side of the equation.

$$x - \frac{1}{2}y = 2$$
$$2x - y = 4$$

Now proceed as in previous examples.

$$-2\left[x - \frac{1}{2}y = 2\right] \qquad\qquad -2x + y = -4$$
$$2x - y = 4 \qquad\quad \text{gives} \qquad\quad 2x - y = 4$$

$$\begin{array}{r} -2x + y = -4 \\ 2x - y = 4 \\ \hline 0 = 0 \quad \text{true} \end{array}$$

Since $0 = 0$ is a true statement, the system is dependent and has an infinite number of solutions. Both equations represent the same line. Notice that if you multiply both sides of the first equation in the solution by 2 you obtain the second equation in the solution.

We have illustrated three methods that can be used to solve a system of linear equations: graphing, substitution, and the addition method. When you are given a system of equations, which method should you use to solve the system? When you need an exact solution, graphing should not be used. Of the two algebraic methods, the addition method may be the easiest to use if there are no numerical coefficients of 1 in the system. If one or more of the equations has a coefficient of 1, you may wish to use either method. We will present a fourth method, using determinants, in Section 4.4 and a fifth method, using matrices, in section 4.5.

Graphical Interpretation to the Solution of a Linear Equation in One Variable

4 In Section 3.2, objective 6, we discussed a graphical interpretation of the solution of a linear equation in one variable. In the Graphing Calculator Corner of that section we solved the equation $2(x + 3) + 3 = \frac{1}{2}x + 6$ graphically and found the solution to be $x = -2$. The solution to a linear equation in one variable may also be determined graphically by setting each side of the equation equal to y, thus creating a system of equations. When solved graphically, the x coordinate of the point where the lines intersect will be the solution to the original equation in one variable. In Example 13 that follows we solve the same equation graphically using the system of equations. After the example, in the Graphing Calculator Corner, we show the solution to Example 13 as it might appear on a graphing calculator. This procedure may be used to solve an equation, or check a solution using a graphing calculator.

EXAMPLE 13   Solve the equation $2(x + 3) + 3 = \frac{1}{2}x + 6$ using a system of equations.

Solution: Set each side of the equation equal to y. This gives

$$y = 2(x + 3) + 3 \qquad\qquad y = 2x + 9$$

or $\qquad$ system of equations

$$y = \frac{1}{2}x + 6 \qquad\qquad y = \frac{1}{2}x + 6$$

We will solve this system graphically. The graph is shown in Fig. 4.3.

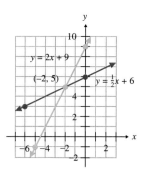

FIGURE 4.3

The solution to the original equation is the x coordinate of the point where the lines intersect. Thus, the solution to $2(x + 3) + 3 = \frac{1}{2}x + 6$ is $x = -2$, which checks with the solution obtained in Example 9 in Section 3.2.

Graphing Calculator Corner

Some graphing calculators can calculate and display the intersections of the graphs in the window. Figure 4.4 shows the graph of $y = 2(x + 3) + 3$ and $y = \frac{1}{2}x + 6$ on a graphing calculator window. Note that the x coordinate of the intersection is -2 and the y coordinate is 5. Determine how to find the point of intersection of two graphs on your calculator. Then work the following exercises.

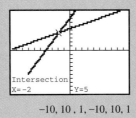

Intersection
X=-2   Y=5

$-10, 10, 1, -10, 10, 1$

FIGURE 4.4

EXERCISES

Solve each system of equations graphically. Estimate the answers to the nearest tenth.

1. $y = 4x - 2$
$y = -3x + 6$ (1.1, 2.6)

2. $y = 5x - 3$
$y = 0.3x + 5$ (1.7, 5.5)

3. $y = 2.3x + 10.4$
$y = 0.13x - 70.6$ $(-37.3, -75.5)$

4. $y = 530x + 1965$
$y = -204x - 187$ $(-2.9, 411.1)$

Solve each of the following equations in one variable graphically using systems of equations. Estimate the answers to the nearest tenth (see Example 13).

5. $5x - 6 = 2x - 8$ $(-0.7, -9.3)$

6. $3(x - 4) = 0.2(x + 5)$ (4.6, 1.9)

7. $-1.63x + 4.85 = 9.64 - 3.82x$
(2.2, 1.3)

8. $\frac{3}{4}(x - 9) = \frac{2}{3}(5x + 60)$
$(-18.1, -20.3)$

Exercise Set 4.1

Determine which, if any, of the ordered pairs or ordered triples satisfy the system of linear equations.

1. $y = 2x + 4$
$y = 2x - 1$
(a) $(0, 4)$ **(b)** $(3, 10)$ none

2. $2x - 3y = 6$
$y = \frac{2}{3}x - 2$
(a) $(3, 0)$ **(b)** $(3, -2)$ (a)

3. $0.5y = -0.5x + 2$
$2y = -2x + 8$
(a) $(2, 5)$ **(b)** $(1, 3)$ (b)

4. $3x - 4y = 8$
$2y = \frac{3}{2}x - 4$
(a) $\left(-\frac{1}{3}, -\frac{9}{4}\right)$ **(b)** $(0, -2)$ (a), (b)

5. $2x + 3y = 6$
$-2x + 5 = y$
(a) $\left(\frac{1}{2}, \frac{5}{3}\right)$ (b) $\left(\frac{9}{4}, \frac{1}{2}\right)$ (b)

6. $x + 2y - z = -5$
$2x - y + 2z = 8$
$3x + 3y + 4z = 5$
(a) $(1, 3, -2)$ (b) $(1, -2, 2)$ (b)

7. $4x + y - 3z = 1$
$2x - 2y + 6z = 11$
$-6x + 3y + 12z = -4$
(a) $(2, -1, -2)$ (b) $\left(\frac{1}{2}, 2, 1\right)$ none

8. $2x - 3y + z = 1$
$x + 2y + z = -1$
$3x - y + 3z = 4$
(a) $(1, 1, 4)$ (b) $(-3, -1, 4)$ (b)

Write each equation in slope–intercept form. Without graphing the equations, state whether the system of equations is consistent, inconsistent, or dependent. Also indicate whether the system has exactly one solution, no solution, or an infinite number of solutions.

9. $2y = -x + 5$ consistent,
$x - 2y = 1$ one

10. $3y = 2x + 3$ inconsistent,
$y = \frac{2}{3}x - 2$ no solution

11. $y = \frac{1}{2}x + 4$ dependent,
$2y = x + 8$ infinite number

12. $2x - 3y = 4$ consistent,
$3x - 2y = -2$ one

13. $x - y = 3$ inconsistent,
$2x - 2y = -2$ no solution

14. $2x = 3y + 4$ dependent,
$6x - 9y = 12$ infinite number

15. $y = \frac{3}{2}x + \frac{1}{2}$ inconsistent,
$3x - 2y = -\frac{1}{2}$ no solution

16. $x - y = 3$ consistent,
$\frac{1}{2}x - 2y = -6$ one

Determine the solution to the system of equations graphically.

17. $y = x + 4$
$y = -x + 2$ $(-1, 3)$

18. $y = 2x + 4$
$y = -3x - 6$ $(-2, 0)$

19. $y = 2x - 1$
$2y = 4x + 6$ inconsistent

20. $y = -2x - 1$
$x + 2y = 4$ $(-2, 3)$

21. $2x + 3y = 6$
$4x = -6y + 12$ dependent

22. $x + y = 1$
$3x - y = -5$ $(-1, 2)$

23. $x + 3y = 4$
$x = 1$ $(1, 1)$

24. $2x - 5y = 10$
$y = \frac{2}{5}x - 2$ dependent

25. $y = -5x + 5$
$y = 2x - 2$ $(1, 0)$

26. $4x - y = 9$
$x - 3y = 16$ $(1, -5)$

27. $2x - y = -4$
$2y = 4x - 6$ inconsistent

28. $y = -\frac{1}{3}x - 1$
$3y = 4x - 18$ $(3, -2)$

Find the solution to each system of equations by substitution.

29. $x + 2y = 9$
$x = 2y + 1$ $(5, 2)$

30. $y = x + 2$
$2y = -x - 2$ $(-2, 0)$

31. $x + y = 6$
$x = y$ $(3, 3)$

32. $2x + y = 3$ infinite number
$2y = 6 - 4x$ of solutions

33. $2x + y = 3$ no
$2x + y + 5 = 0$ solution

34. $y = 2x + 4$
$y = -\frac{3}{4}$ $\left(-\frac{19}{8}, -\frac{3}{4}\right)$

35. $x = \frac{1}{2}$
$x + \frac{1}{3}y + 6 = 0$ $\left(\frac{1}{2}, -\frac{39}{2}\right)$

36. $y = \frac{1}{3}x - 2$ infinite number
$x - 3y = 6$ of solutions

37. $x - \frac{1}{2}y = 2$ infinite number
$y = 2x - 4$ of solutions

38. $2x + 3y = 7$
$6x - y = 1$ $\left(\frac{1}{2}, 2\right)$

39. $3x + y = -1$
$y = 3x + 5$ $(-1, 2)$

40. $y = -2x + 5$
$x + 3y = 0$ $(3, -1)$

41. $y = 2x - 13$
$-4x - 7 = 9y$ $(5, -3)$

42. $x = y + 4$
$3x + 7y = -18$ $(1, -3)$

43. $5x - 2y = -7$
$5 = y - 3x$ $(-3, -4)$

44. $5x - 4y = -7$
$x - \frac{3}{5}y = -2$ $\left(-\frac{19}{5}, -3\right)$

45. $x = 3y + 5$
$y = \frac{2}{3}x + \frac{1}{2}$ $\left(-\frac{13}{2}, -\frac{23}{6}\right)$

46. $x + 2y = 4$
$x + \frac{1}{2}y = 4$ $(4, 0)$

47. $\frac{1}{2}x - \frac{1}{3}y = 2$
$\frac{1}{4}x + \frac{2}{3}y = 6$ $(8, 6)$

48. $\frac{1}{2}x + \frac{1}{3}y = 13$
$\frac{1}{5}x + \frac{1}{8}y = 5$ $(10, 24)$

Solve each system of equations using the addition method.

49. $x + y = -2$
$x - y = 4$ $(1, -3)$

50. $x - y = 12$
$x + y = 2$ $(7, -5)$

51. $3x + 2y = 15$
$x - 2y = -7$ $\left(2, \frac{9}{2}\right)$

52. $3x + 3y = 18$
$4x - y = 4$ $(2, 4)$

53. $3x + y = 6$
$-6x - 2y = 10$ no solution

54. $2x + y = 14$
$-3x + y = -2$ $\left(\frac{16}{5}, \frac{38}{5}\right)$

55. $2x + y = 6$
$3x - 2y = 16$ $(4, -2)$

56. $4x - 3y = 8$
$-2x + 5y = 14$ $\left(\frac{41}{7}, \frac{36}{7}\right)$

57. $2x - 5y = 13$
$5x + 3y = 17$ $(4, -1)$

58. $4x = 2y + 6$ infinite number
$y = 2x - 3$ of solutions

59. $3y = 2x + 4$ infinite number
$3y = 2x + 4$ of solutions

60. $5x + 4y = 10$
$-3x - 5y = 7$ $(6, -5)$

61. $4x - 3y = 8$
$-3x + 4y = 9$ $\left(\frac{59}{7}, \frac{60}{7}\right)$

62. $2x - y = 8$
$3x + y = 6$ $\left(\frac{14}{5}, -\frac{12}{5}\right)$

63. $3x + 4y = 2$
$2x = -5y - 1$ $(2, -1)$

64. $3x - 4y = 5$
$2x = 5y - 3$ $\left(\frac{37}{7}, \frac{19}{7}\right)$

65. $4x + 5y = 3$
$2x - 3y = 4$ $\left(\frac{29}{22}, -\frac{5}{11}\right)$

66. $2x + 3y = 5$
$-3x - 4y = -2$ $(-14, 11)$

67. $0.2x + 0.5y = 1.6$
$-0.3x + 0.4y = -0.1$ $(3, 2)$

68. $0.15x - 0.40y = 0.65$
$0.60x + 0.25y = -1.1$
$(-1, -2)$

69. $2.1x - 0.6y = 8.40$
$-1.5x - 0.3y = -6.00$ $(4, 0)$

70. $-0.25x + 0.10y = 1.05$
$-0.40x - 0.625y = -0.675$ $(-3, 3)$

71. $2x - \frac{1}{3}y = 6$
$5x - y = 4$ $(14, 66)$

72. $\frac{1}{2}x - \frac{1}{3}y = 1$
$\frac{1}{4}x - \frac{1}{9}y = \frac{2}{3}$ $(4, 3)$

73. $\frac{1}{3}x = 4 - \frac{1}{4}y$
$3x = 4y$ $\left(\frac{192}{25}, \frac{144}{25}\right)$

74. $\frac{1}{4}x - 3 = \frac{1}{6}y$
$y = \frac{1}{2}x + 2$ $(20, 12)$

75. $\frac{1}{5}x + \frac{1}{2}y = 4$
$\frac{2}{3}x - y = \frac{8}{3}$ $(10, 4)$

76. $\frac{2}{3}x - 4 = \frac{1}{2}y$
$x - 3y = \frac{1}{3}$ $\left(\frac{71}{9}, \frac{68}{27}\right)$

77. $2(2x + y) = -2x + y + 8$
$3(x - y) = -(y + 1)$ $(1, 2)$

78. $3(x - y) = 5(x + y) - 8$
$2x + 4(x + 3y) = 5(y - 2)$ $(-4, 2)$

79. The graph below appeared in both the *Journal of the American Medical Association* and *Scientific American*. The red line indicates the long-term trend of firearms deaths through 1992, and the purple line indicates the long-term trend in motor vehicle deaths through 1988. The black lines indicate the short-term trends in deaths from firearms and motor vehicles.

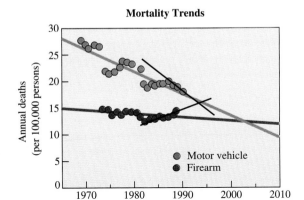

Mortality Trends

Using either the long- or short-term trends, it appears that if the trends do not change, the number of deaths from firearms will surpass the number of deaths from motor vehicle accidents as the most frequent cause of death in the United States.

(a) Discuss the long-term trend in motor vehicle deaths.

(b) Discuss the long-term trend in firearms deaths.

(c) Discuss the short-term trend in motor vehicle deaths compared with the long-term trend in motor vehicle deaths.

(d) Discuss the short-term trend in firearms deaths compared with the long-term trend in firearms deaths.

(e) Using the long-term trends, estimate when the number of deaths from firearms will equal the number of deaths from motor vehicles. ≈2003

(f) Repeat part (e) using the short-term trends. ≈1993

80. Explain how you can determine, without graphing or solving, whether a system of two linear equations is consistent, inconsistent, or dependent.

81. When solving a system of equations by addition or substitution, how will you know if the system is (a) inconsistent? (b) dependent?

82. Explain the procedure to use to solve a linear equation *in one variable* graphically using a system of equations.

80. Compare their slopes and *y* intercepts.
81. (a) Get a false statement like 6 = 0.
(b) Get a true statement like 0 = 0.
82. Set each side of equation equal to *y*. The *x* coordinate of the intersection of their graphs is the solution.

83. Rational numbers can be expressed as a quotient of two integers; irrational numbers cannot.

CUMULATIVE REVIEW EXERCISES

[1.2] **83.** Explain the difference between a rational number and an irrational number.

[1.2] **84.** (a) Are all rational numbers real numbers? yes
(b) Are all irrational numbers real numbers? yes

[2.2] **85.** Find all numbers such that $|x - 4| = |4 - x|$. ℝ

[2.2] **86.** Evaluate $A = p\left(1 + \frac{r}{n}\right)^t$, when $p = 500$, $r = 0.08$, $n = 2$, and $t = 1$. 520

[3.5] **87.** Is the relation below a function? Explain your answer. $\{(-3, 4), (7, 2), (-4, 5), (5, 0), (-3, 2)\}$ no

Group Activity/ Challenge Problems

3. (a) Infinite number; system dependent. **3. (b)** $m = -\frac{1}{2}$, $y = -\frac{1}{2}x + 5$, $b = 5$

1. The solution to the following system of equations is $(2, -3)$. Find A and B.

$$Ax + 4y = -8 \quad A = 2, B = 5$$
$$3x - By = 21$$

2. For each part explain how you determined your answer (There are many possible answers). Write a system of equations for which:

(a) $(2, 5)$ is a solution. $x + y = 7$, $x - y = -3$

(b) There is no solution. $x + y = 1$, $x + y = 2$

(c) There are infinitely many solutions. $x + y = 1$, $2x + 2y = 2$

3. The solutions of a system of linear equations are $(2, 4)$ and $(-2, 6)$.

(a) How many other solutions does the system have? Explain.

(b) Find the slope of the line containing $(2, 4)$ and $(-2, 6)$. Determine the equation of the line containing these points. Then determine the y intercept.

Solve using the addition method.

4.
$$\frac{x + 2}{2} - \frac{y + 4}{3} = 4$$
$$\frac{x + y}{2} = \frac{1}{2} + \frac{x - y}{3} \quad (8, -1)$$

5.
$$\frac{5x}{2} + 3y = \frac{9}{2} + y$$
$$\frac{1}{4}x - \frac{1}{2}y = 6x + 12 \quad \left(-\frac{105}{41}, \frac{447}{82}\right)$$

Solve the system of equations. Hint: $\dfrac{3}{a} = 3 \cdot \dfrac{1}{a} = 3x \text{ if } x = \dfrac{1}{a}.$

6.
$$\frac{3}{a} + \frac{4}{b} = -1$$
$$\frac{1}{a} + \frac{6}{b} = 2 \quad (-1, 2)$$

7.
$$\frac{6}{x} + \frac{1}{y} = -1$$
$$\frac{3}{x} - \frac{2}{y} = -3 \quad (-3, 1)$$

Solve the following system of equations, where a and b represent any nonzero constants by (a) the substitution method and (b) the addition method.

8. $4ax + 3y = 19$
$\quad -ax + \ y = \ 4 \quad (1/a, 5)$

9. $ax = 2 - by \quad (1/a, 1/b)$
$\quad -ax + 2by - 1 = 0$

10. $3ax - y = 4$
$\quad y = 7ax \quad (-1/a, \ 7)$

4.2 Third-Order Systems of Linear Equations

Tape 5

1️⃣ Solve third-order systems of equations.
2️⃣ Learn the geometrical interpretation of a third-order system.
3️⃣ Recognize inconsistent and dependent third-order systems.

Third-Order Systems

1️⃣ A third-order system consists of three equations with three unknowns. The solution of a third-order system is an ordered triple. A fourth-order system is one that consists of four equations and four unknowns. In this section we use substitution and the addition method to solve third-order systems. The procedures discussed in this section can be expanded to solve fourth-order and higher-order systems.

EXAMPLE 1 Solve the following system of equations by substitution.

$$x = 4$$
$$2x + y = 20$$
$$-x + 4y + 2z = 24$$

Solution: Since we know that $x = 4$ in the ordered triple, substitute 4 for x in the equation $2x + y = 20$, and solve for y.

$$2x + y = 20$$
$$2(4) + y = 20$$
$$8 + y = 20$$
$$y = 12$$

Now substitute $x = 4$ and $y = 12$ in the last equation and solve for z.

$$-x + 4y + 2z = 24$$
$$-(4) + 4(12) + 2z = 24$$
$$-4 + 48 + 2z = 24$$
$$44 + 2z = 24$$
$$2z = -20$$
$$z = -10$$

Check: $x = 4$, $y = 12$, $z = -10$. The solution must be checked in all three original equations.

$x = 4$	$2x + y = 20$	$-x + 4y + 2z = 24$
$4 = 4$ true	$2(4) + 12 = 20$	$-(4) + 4(12) + 2(-10) = 24$
	$20 = 20$ true	$24 = 24$ true

The solution is the ordered triple $(4, 12, -10)$. Remember that the ordered triple lists the x value first, the y value second, and the z value third.

Not every third-order system can be solved by substitution. When a third-order system cannot be solved using substitution, we can find the solution by the addition method as illustrated in Example 2.

EXAMPLE 2 Solve the system of equations.

(1) $3x + 2y + \ \ z = 4$
(2) $2x - 3y + 2z = -7$
(3) $\ \ x + 4y - \ \ z = 10$

Solution: For the sake of clarity the three equations have been labeled (1), (2), and (3). To solve this system of equations, we must first obtain two equations containing the same two variables. This is done by selecting two equations and using the addition method to eliminate one of the variables. For example, by adding equations (1) and (3) the variable z will be eliminated. Next we use a different pair of equations [either (1) and (2) or (2) and (3)] and use the addition method to eliminate the *same* variable that was eliminated previously. If we multiply equation (1) by -2 and add it to equation (2), the variable z will again be eliminated. We will then have two equations containing only two unknowns. Let us begin by adding equations (1) and (3).

$$\begin{array}{rl} (1) & 3x + 2y + z = 4 \\ (3) & \underline{x + 4y - z = 10} \\ (4) & 4x + 6y = 14 \end{array}$$

Now use a different set of equations and again eliminate the variable z.

$$\begin{array}{rl} (1) & -2[3x + 2y + z = 4] \\ (2) & 2x - 3y + 2z = -7 \end{array} \quad \text{gives} \quad \begin{array}{r} -6x - 4y - 2z = -8 \\ \underline{2x - 3y + 2z = -7} \\ (5) \quad -4x - 7y = -15 \end{array}$$

We now have a system consisting of two equations with two unknowns.

$$\begin{array}{rl} (4) & 4x + 6y = 14 \\ (5) & -4x - 7y = -15 \end{array}$$

Next we solve for one of the variables using a method presented earlier. If we add the two equations, the variable x will be eliminated.

$$\begin{array}{r} 4x + 6y = 14 \\ \underline{-4x - 7y = -15} \\ -y = -1 \\ y = 1 \end{array}$$

Next we substitute $y = 1$ in either one of the two equations containing only two variables [(4) or (5)] and solve for x.

$$\begin{array}{rl} (4) & 4x + 6y = 14 \\ & 4x + 6(1) = 14 \\ & 4x + 6 = 14 \\ & 4x = 8 \\ & x = 2 \end{array}$$

Finally, substitute $x = 2$ and $y = 1$ in any of the original equations and solve for z.

$$\begin{array}{rl} (1) & 3x + 2y + z = 4 \\ & 3(2) + 2(1) + z = 4 \\ & 6 + 2 + z = 4 \\ & 8 + z = 4 \\ & z = -4 \end{array}$$

The solution is the ordered triple $(2, 1, -4)$.

We chose first to eliminate the variable z by using equations (1) and (3) and then equations (1) and (2). We could have elected to eliminate either variable x or variable y first. For example, we could have eliminated variable x by multiplying equation (3) by -2 and then adding it to equation (2). We could also eliminate variable x by multiplying equation (3) by -3 and then adding it to equation (1). Solve the system in Example 2 by first eliminating the variable x.

EXAMPLE 3 Solve the system of equations.

$$2x - 3y + 2z = -1$$
$$x + 2y \quad\quad = 14$$
$$x \quad\quad - 3z = -5$$

Solution: The third equation does not contain y. We will therefore work to obtain another equation that does not contain y. We will use the first and second equations to do this.

$$\begin{aligned} 2[2x - 3y + 2z = -1] \\ 3[\ x + 2y \quad\quad = 14] \end{aligned} \text{ gives } \begin{aligned} 4x - 6y + 4z = -2 \\ \underline{3x + 6y \quad\quad = 42} \\ 7x \quad\quad + 4z = 40 \end{aligned}$$

We now have two equations containing only the variables x and z.

$$7x + 4z = 40$$
$$x - 3z = -5$$

Let's now eliminate the variable x.

$$\begin{aligned} 7x + 4z = 40 \\ -7[x - 3z = -5] \end{aligned} \text{ gives } \begin{aligned} 7x + \ 4z = 40 \\ \underline{-7x + 21z = 35} \\ 25z = 75 \\ z = 3 \end{aligned}$$

Now we solve for x by using one of the equations containing only the variables x and z.

$$x - 3z = -5$$
$$x - 3(3) = -5$$
$$x - 9 = -5$$
$$x = 4$$

Finally, solve for y using any of the original equations that contains y.

$$x + 2y = 14$$
$$4 + 2y = 14$$
$$2y = 10$$
$$y = 5$$

The solution is the ordered triple $(4, 5, 3)$.

Check:

$2x - 3y + 2z = -1$	$x + 2y = 14$	$x - 3z = -5$
$2(4) - 3(5) + 2(3) = -1$	$4 + 2(5) = 14$	$4 - 3(3) = -5$
$8 - 15 + 6 = -1$	$4 + 10 = 14$	$4 - 9 = -5$
$-1 = -1$	$14 = 14$	$-5 = -5$
true	true	true

> *Helpful Hint*
> ....
> If an equation in a system contains fractions, eliminate the fractions by multiplying each term in the equation by the least common denominator. Then continue to solve the system. If, for example, one equation in the system is $\frac{3}{4}x - \frac{5}{8}y + z = \frac{1}{2}$, multiply both sides of the equation by 8 to obtain the equivalent equation $6x - 5y + 8z = 4$.

Geometrical Interpretation of a Third-order System

2 When we have a system of linear equations in two variables, we can find its solution graphically using the Cartesian coordinate system. A linear equation in three variables, x, y, and z, can be graphed on a coordinate system with three axes drawn perpendicular to each other (see Fig. 4.5).

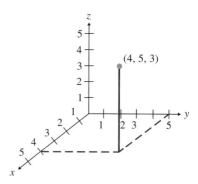

FIGURE 4.5

A point plotted in this three-dimensional system would appear to be a point in space. If we were to graph an equation such as $x + 2y + 3z = 4$, we would find that its graph would be a plane, not a line. In Example 3 we indicated the solution to be the ordered triple (4, 5, 3). This means that the three planes, one from each of the three given equations, all intersect at the point (4, 5, 3). The drawing in Exercise 34 illustrates three planes intersecting at a point.

Inconsistent and Dependent Systems

3 We discussed inconsistent and dependent systems of equations in Section 4.1. Third-order systems of equations may also be inconsistent or dependent. When solving a third-order system, if you obtain a false statement like $3 = 0$, the system is inconsistent and has no solution. This means that at least two of the planes are parallel, so the three planes cannot intersect.

When solving a third-order system, if you obtain the true statement, $0 = 0$, it indicates that the system is dependent and has an infinite number of solutions. This may happen when all three equations represent the same plane or when the intersection of the planes is a line, as in the drawing in Exercise 33. Examples 4 and 5 illustrate an inconsistent system and a dependent system, respectively.

EXAMPLE 4 Solve the system of equations.

$$x + 2y + 3z = 0$$
$$2x + 4y + 6z = 2$$
$$3x + 6y - 4z = 3$$

Solution: We will begin by eliminating x from the first two equations.

$$-2[x + 2y + 3z = 0] \quad \text{gives} \quad \begin{array}{r} -2x - 4y - 6z = 0 \\ 2x + 4y + 6z = 2 \\ \hline 0 = 2 \quad \textbf{false} \end{array}$$

Since we obtained the false statement $0 = 2$, this system is inconsistent and has no solution.

EXAMPLE 5   Solve the system of equations.

$$\begin{array}{rl} (1) & x - y + z = 1 \\ (2) & x + 2y - z = 1 \\ (3) & x - 4y + 3z = 1 \end{array}$$

Solution: We will begin by eliminating the variable x from equations (1) and (2) and then from equations (1) and (3).

$$\begin{array}{rl} (1) & -1[x - y + z = 1] \\ (2) & x + 2y - z = 1 \end{array} \quad \text{gives} \quad \begin{array}{r} -x + y - z = -1 \\ x + 2y - z = 1 \\ \hline (4) \quad 3y - 2z = 0 \end{array}$$

$$\begin{array}{rl} (1) & x - y + z = 1 \\ (3) & -1[x - 4y + 3z = 1] \end{array} \quad \text{gives} \quad \begin{array}{r} x - y + z = 1 \\ -x + 4y - 3z = -1 \\ \hline (5) \quad 3y - 2z = 0 \end{array}$$

Now eliminate the variable y using equations (4) and (5).

$$\begin{array}{rl} (4) & -1[3y - 2z = 0] \\ (5) & 3y - 2z = 0 \end{array} \quad \text{gives} \quad \begin{array}{r} -3y + 2z = 0 \\ 3y - 2z = 0 \\ \hline 0 = 0 \quad \textbf{true} \end{array}$$

Since we obtained the true statement $0 = 0$, this system is dependent and has an infinite number of solutions.

Recall from Section 4.1 that systems of equations that are dependent are also consistent since they have a solution.

Exercise Set 4.2

Solve by substitution.

1. $x = 1$
$2x + y = 4$
$-3x - y + 4z = 15$ $(1, 2, 5)$

2. $2x + 3y = 9$
$4x - 6z = 12$
$y = 5$ $(-3, 5, -4)$

3. $5x - 6z = -17$
$3x - 4y + 5z = -1$
$2z = -6$ $\left(-7, -\frac{35}{4}, -3\right)$

4. $2x - 5y = 12$
$-3y = -9$
$2x - 3y + 4z = 8$ $\left(\frac{27}{2}, 3, -\frac{5}{2}\right)$

5. $x + 2y = 6$
$3y = 9$
$x + 2z = 12$ $(0, 3, 6)$

6. $x - y + 5z = -4$
$3x - 2z = 6$
$4z = 2$ $\left(\frac{7}{3}, \frac{53}{6}, \frac{1}{2}\right)$

Solve using the addition method.

7. $\begin{aligned} x + y - z &= -3 \\ x \quad + z &= 2 \\ 2x - y + 2z &= 3 \end{aligned}$ $(-1, 1, 3)$

8. $\begin{aligned} x - 2y \quad &= 2 \\ 2x + 3y \quad &= 11 \\ -y + 4z &= 7 \end{aligned}$ $(4, 1, 2)$

9. $\begin{aligned} x \quad - 2z &= -5 \\ -y + 3z &= 3 \\ -2x \quad + z &= 4 \end{aligned}$ $(-1, 3, 2)$

10. $\begin{aligned} x - 3y \quad &= 13 \\ 2y + z &= 1 \\ y - 2z &= 11 \end{aligned}$ $\left(\frac{104}{5}, \frac{13}{5}, -\frac{21}{5} \right)$

11. $\begin{aligned} p + q + r &= 4 \\ p - 2q - r &= 1 \\ 2p - q - 2r &= -1 \end{aligned}$ $(2, -1, 3)$

12. $\begin{aligned} x - 2y + 3z &= -7 \\ 2x - y - z &= 7 \\ -x + 3y + 2z &= -8 \end{aligned}$ $(2, 0, -3)$

13. $\begin{aligned} 2x - 2y + 3z &= 5 \\ 2x + y - 2z &= -1 \\ 4x - y - 3z &= 0 \end{aligned}$ $\left(\frac{2}{3}, -\frac{1}{3}, 1 \right)$

14. $\begin{aligned} 2x - y - z &= 4 \\ 4x - 3y - 2z &= -2 \\ 8x - 2y - 3z &= 3 \end{aligned}$ $\left(-\frac{19}{2}, 10, -33 \right)$

15. $\begin{aligned} x + 2y - 3z &= 5 \\ x + y + z &= 0 \\ 3x + 4y + 2z &= -1 \end{aligned}$ $(5, -3, -2)$

16. $\begin{aligned} a + 2b + 2c &= 1 \\ 2a - b + c &= 3 \\ 4a + b + 2c &= 0 \end{aligned}$ $(-1, -2, 3)$

17. $\begin{aligned} 2a + 2b - c &= 2 \\ 3a + 4b + c &= -4 \\ 5a - 2b - 3c &= 5 \end{aligned}$ $\left(-\frac{11}{17}, \frac{7}{34}, -\frac{49}{17} \right)$

18. $\begin{aligned} x + y + z &= 0 \\ -x - y + z &= 0 \\ -x + y + z &= 0 \end{aligned}$ $(0, 0, 0)$

19. $\begin{aligned} -x + 3y + z &= 0 \\ -2x + 4y - z &= 0 \\ 3x - y + 2z &= 0 \end{aligned}$ $(0, 0, 0)$

20. $\begin{aligned} -\frac{1}{4}x + \frac{1}{2}y - \frac{1}{2}z &= -2 \\ \frac{1}{2}x + \frac{1}{3}y - \frac{1}{4}z &= 2 \\ \frac{1}{2}x - \frac{1}{2}y + \frac{1}{4}z &= 1 \end{aligned}$ $(4, 6, 8)$

21. $\begin{aligned} \frac{2}{3}x + y - \frac{1}{3}z &= \frac{1}{3} \\ \frac{1}{2}x + y + z &= \frac{5}{2} \\ \frac{1}{4}x - \frac{1}{4}y + \frac{1}{4}z &= \frac{3}{2} \end{aligned}$ $(3, -1, 2)$

22. $\begin{aligned} x - \frac{2}{3}y - \frac{2}{3}z &= -2 \\ \frac{2}{3}x + y - \frac{2}{3}z &= \frac{1}{3} \\ -\frac{1}{4}x + y - \frac{1}{4}z &= \frac{3}{4} \end{aligned}$ $\left(\frac{2}{3}, \frac{23}{15}, \frac{37}{15} \right)$

Determine whether the following systems are inconsistent, dependent, or neither.

23. $\begin{aligned} 2x + y + 2z &= 1 \\ x - 2y - z &= 0 \\ 3x - y + z &= 2 \end{aligned}$ inconsistent

24. $\begin{aligned} -x + y + z &= 0 \\ x - y + z &= 0 \\ 3x - 3y - z &= 0 \end{aligned}$ dependent

25. $\begin{aligned} 3x - 4y + z &= 4 \\ x + 2y + z &= 4 \\ -6x + 8y - 2z &= -8 \end{aligned}$ dependent

26. $\begin{aligned} 2p - 4q + 6r &= 8 \\ -p + 2q - 3r &= 6 \\ 3p + 4q + 5r &= 8 \end{aligned}$ inconsistent

27. $\begin{aligned} x + 3y + 2z &= 6 \\ x - 2y - z &= 8 \\ -3x - 9y - 6z &= -4 \end{aligned}$ inconsistent

28. $\begin{aligned} 2x - 2y + 4z &= 2 \\ -3x + y \quad &= -9 \\ 2x - y + z &= 5 \end{aligned}$ dependent

29. (a) What does the graph of an equation in two variables such as $x + 2y = 3$, represent? **(b)** What does a graph of an equation in three variables, such as $x + 2y + 3z = 6$, represent? a line, a plane

30. How many ordered triples satisfy a third-order system of equations that is **(a)** inconsistent; **(b)** dependent? **(a)** none **(b)** an infinite number

An equation in three variables, x, y, and z, represents a plane. Consider a system of equations consisting of three equations in three variables. Answer the following questions.

31. If the three planes are parallel to one another as illustrated in the figure, how many points will be common to all three planes? Is the system consistent or inconsistent? Explain your answer.

many points will be common to all three planes? Is the system consistent or inconsistent? Explain your answer.

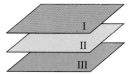

No point is common to all planes. System is inconsistent.

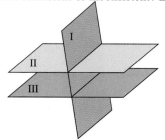

No point is common to all planes. System is inconsistent.

32. If two of the planes are parallel to each other and the third plane intersects each of the other two planes, how

33. If the three planes are as illustrated in the figure, how many points will be common to all three planes? Is the system dependent? Explain your answer.

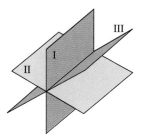

A line (infinite number of points) is common to planes. System is dependent.

34. If the three planes are as illustrated as in the figure, how many points will be common to all three planes? Is the system consistent or inconsistent? Explain your answer.

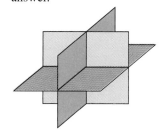

One point is common to all three planes. Therefore, the system is consistent.

35. (a) Explain, step by step, the procedure to solve the following system.
(b) Solve the system by your step-by-step procedure.

$$x + y + z = 3$$
$$x + y - z = 1$$
$$3x - 2y + z = 2 \quad \textbf{(b)} \ (1, 1, 1)$$

CUMULATIVE REVIEW EXERCISES   **36. (a)** $\frac{5}{12}$ hr (or 25 min)

[2.4] **36.** Phillipa and her son Cameron go cross-country skiing. Phillipa averages 5 miles per hour, and Cameron averages 3 miles per hour. If Cameron begins $\frac{1}{6}$ hour before his mother, **(a)** how long after Cameron starts skiing, will his mother catch up with him?
(b) How far from the starting point will they be when they meet?   1.25 mi

SEE EXERCISE 36

[2.6] *Determine the solution set.*

37. $\left| 4 - \dfrac{2x}{3} \right| > 5 \quad \left\{ x \mid x < -\dfrac{3}{2} \text{ or } x > \dfrac{27}{2} \right\}$

38. $\left| \dfrac{3x - 4}{2} \right| - 1 < 5 \quad \left\{ x \mid -\dfrac{8}{3} < x < \dfrac{16}{3} \right\}$

39. $\left| 2x - \dfrac{1}{2} \right| = -5 \quad \varnothing$

Group Activity/ Challenge Problems

1. (a) Find the values of a, b, and c such that the points $(1, -1)$, $(-1, -5)$, and $(3, 11)$ lie on the graph of $y = ax^2 + bx + c$.
(b) Find the quadratic equation whose graph passes through the three points indicated. Explain how you determined your answer.
(a) $a = 1$, $b = 2$, $c = -4$ **(b)** $y = x^2 + 2x - 4$

2. Find the quadratic equation of the form $y = ax^2 + bx + c$ whose graph passes through the points $(1, 7)$, $(-2, -5)$, and $(3, 5)$. $y = -x^2 + 3x + 5$

Find the solution to the fourth-order systems.

3.
$$3a + 2b - c \quad\quad = 0$$
$$2a \quad\quad + 2c + d = 5$$
$$a + 2b \quad\quad - d = -2$$
$$2a - b + c + d = 2 \quad (-1, 2, 1, 5)$$

4.
$$3p + 4q \quad\quad\quad = 11$$
$$2p \quad\quad + r + s = 9$$
$$q \quad - s = -2$$
$$p + 2q - r \quad = 2 \quad (1, 2, 3, 4)$$

4.3 Applications of Systems of Linear Equations

Tape 5

Applications

1 Use systems of equations to solve application problems.
2 Use third-order systems to solve application problems.

1 Many of the applications solved in earlier chapters using only one variable can now be solved using two variables. Following are some examples showing how applications can be described by systems of equations.

EXAMPLE 1   More than half of all the water used in the United States is used in livestock production. The number of gallons of water used to produce 1 pound of meat is 1250 gallons more than 50 times the number of gallons of water needed to produce 1 pound of wheat. If it requires 2525 gallons of water to produce *both* 1 pound of meat and 1 pound of wheat, find the number of gallons of water needed to produce 1 pound of meat and the number of gallons to produce 1 pound of wheat.

Solution:

Let x = number of gallons of water needed to produce 1 pound of wheat
   y = number of gallons of water needed to produce 1 pound of meat

The number of gallons of water needed to produce 1 pound of meat is 1250 more than 50 times the amount needed to produce 1 pound of wheat. Therefore,

$$y = 50x + 1250$$

Since it takes 2525 gallons of water to produce both 1 pound of meat and 1 pound of wheat,

$$x + y = 2525$$

Therefore, the system of equations is

$$y = 50x + 1250$$
$$x + y = 2525$$

We will solve this system by substitution. Substituting $50x + 1250$ for y in the second equation gives

$$x + \overbrace{y} = 2525$$
$$x + \overbrace{50x + 1250} = 2525$$
$$51x + 1250 = 2525$$
$$51x = 1275$$
$$x = 25$$

Thus, it takes 25 gallons of water to produce 1 pound of wheat and $50x + 1250 = 50(25) + 1250 = 2500$ gallons of water to produce 1 pound of meat.*

---

*According to some sources, as much as 5214 gallons of water are needed to produce 1 edible pound of beef in California.

EXAMPLE 2   A motorboat travels 30.8 miles per hour with the current and 26.4 miles per hour against the current. Find the current and the speed of the boat in still water.

Solution: Let x = speed of boat in still water
y = current

The system of equations is:

speed of boat going with the current: $x + y = 30.8$
speed of boat going against the current: $x - y = 26.4$

$$x + y = 30.8$$
$$\underline{x - y = 26.4}$$
$$2x \quad\;\; = 57.2$$
$$x = 28.6$$

The speed of the boat is 28.6 miles per hour in still water.

$$x + y = 30.8$$
$$28.6 + y = 30.8$$
$$y = 2.2$$

The current is 2.2 miles per hour.

EXAMPLE 3   Mr. Url, a salesman, receives a weekly salary plus a commission, which is a percentage of his sales. One week, on sales of $3000, his total take-home pay was $760. The next week, on sales of $4000, his total take-home pay was $880. Find his weekly salary and his commission rate.

Solution: Let x = his weekly salary,
y = his commission rate

The system of equations then becomes

$$x + 3000y = 760$$
$$x + 4000y = 880$$

$$\begin{array}{r} x + 3000y = 760 \\ -1[x + 4000y = 880] \end{array} \quad \text{gives} \quad \begin{array}{r} x + 3000y = \;\;\;760 \\ \underline{-x - 4000y = -880} \\ -\,1000y = -120 \end{array}$$

$$y = \frac{-120}{-1000} = 0.12$$

His commission is therefore 12% of his sales. Now let us find his weekly salary

$$x + 3000y = 760$$
$$x + 3000(0.12) = 760$$
$$x + 360 = 760$$
$$x = 400$$

His weekly salary is therefore $400.

EXAMPLE 4   The Channel Tunnel, called the Chunnel, which goes under the English Channel from near Folkestone, Great Britain, to near Calais, France, is about 124,088 feet (or about 23.5 miles) in length. Two teams of workers, one beginning in France and the other beginning in Great Britain, worked toward each other and met near the middle. The two teams started working on the tunnel at the same time. Due to wet soil and poor conditions in France, the team working from Great Britain completed an average of 2.2 feet per day more than the team working from France. The two teams met 1432 days after the project began. Find the rate at which each team worked.

Solution: Let x = French team's rate in feet per day
y = English team's rate in feet per day

When the two teams met, they had each worked 1432 days. We construct a table to aid us in finding the solution to the problem. Remember from Section 2.4 that distance = rate · time.

Team	Rate	Time	Distance
French	x	1432	$1432x$
English	y	1432	$1432y$

Since the rate of the English team was 2.2 feet faster than the rate of the French team,

$$y = x + 2.2$$

Since the total length (or distance) of the tunnel is 124,088 feet,

$$1432x + 1432y = 124{,}088$$

Therefore, we have the following system:

$$y = x + 2.2$$
$$1432x + 1432y = 124{,}088$$

We will solve this system by substitution:

$$1432x + 1432y = 124{,}088$$
$$1432x + 1432(x + 2.2) = 124{,}088$$
$$1432x + 1432x + 3150.4 = 124{,}088$$
$$2864x + 3150.4 = 124{,}088$$
$$2864x = 120{,}937.6$$
$$x \approx 42.2$$

Thus, the French team completed about 42.2 feet per day (about 0.008 mile per day). The English team completed 42.2 + 2.2 or about 44.4 feet per day.

Example 5   Martina, a chemist, wishes to mix a 15% sodium–iodine solution with a 40% sodium–iodine solution to get 6 liters of a 25% sodium–iodine solution. How many liters of the 15% solution and the 40% solution will she need to mix?

Solution: Let x = number of liters of the 15% solution
y = number of liters of the 40% solution

We will draw a sketch (Fig. 4.6) and then set up a table to help analyze the problem.

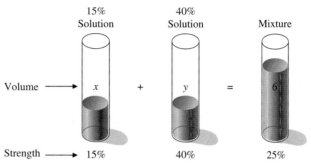

Figure 4.6

The amount of sodium–iodine in a solution is found by multiplying the percent strength of sodium–iodine in the solution by the volume of the solution.

Solution	Strength of solution	Number of liters	Amount of sodium–iodine
15% Solution	0.15	x	$0.15x$
40% Solution	0.40	y	$0.40y$
Mixture	0.25	6	0.25(6)

Since the sum of the volumes of the 15% solution and the 40% solution is 6 liters, our first equation is

$$x + y = 6$$

The second equation comes from the fact that the solutions are mixed.

$$\begin{pmatrix} \text{amount of} \\ \text{sodium–iodine} \\ \text{in 15\% solution} \end{pmatrix} + \begin{pmatrix} \text{amount of} \\ \text{sodium–iodine} \\ \text{in 40\% solution} \end{pmatrix} = \begin{pmatrix} \text{amount of} \\ \text{sodium–iodine} \\ \text{in mixture} \end{pmatrix}$$

or $0.15x + 0.40y = 0.25(6)$

The system of equations is, therefore,

$$x + y = 6$$
$$0.15x + 0.40y = 0.25(6)$$

Solving $x + y = 6$ for y we get $y = -x + 6$. Substituting $-x + 6$ for y in the second equation gives us

$$0.15x + 0.40y = 0.25(6)$$
$$0.15x + 0.40(-x + 6) = 0.25(6)$$
$$0.15x - 0.40x + 2.4 = 1.5$$
$$-0.25x + 2.4 = 1.5$$
$$-0.25x = -0.9$$
$$x = \frac{-0.9}{-0.25} = 3.6$$

Therefore, 3.6 liters of the 15% solution will be used. Since the two solutions must total 6 liters, 6 − 3.6 or 2.4 liters of the 40% solution must be used.

In Example 5, the equation $0.15x + 0.40y = 0.25(6)$ could have been simplified by multiplying both sides of the equation by 100. This would give the equation $15x + 40y = 25(6)$ or $15x + 40y = 150$. Then the system of equations would be $x + y = 6$ and $15x + 40y = 150$. If you solve this system, you should obtain the same solution. Try it and see.

Applications ② Now let us look at some applications of third-order systems.

EXAMPLE 6 Tiny Tots Toys must borrow $25,000 to pay for an expansion. They are not able to obtain a loan for the total amount from a single bank, so they take out three loans from three different banks. They borrowed some of the money at a bank that charged them 8% interest. At the second bank, they borrowed $2000 more than one-half the amount borrowed from the first bank. The interest rate at the second bank is 10%. The balance of the $25,000 is borrowed from a third bank where they paid 9% interest. The total annual interest Tiny Tots Toys pays for the three loans is $2220. How much did they borrow at each rate?

Solution: Let x = amount borrowed at first bank
y = amount borrowed at second bank
z = amount borrowed at third bank

Since the total amount borrowed is $25,000 we know that

$$x + y + z = 25{,}000$$

At the second bank, Tiny Tots Toys borrowed $2000 more than one-half the money borrowed from the first bank. Therefore, our second equation is

$$y = \frac{1}{2}x + 2000$$

Our last equation comes from the fact that the total annual interest charged by the three banks is $2220. The interest at each bank is found by multiplying the interest rate by the amount borrowed.

$$0.08x + 0.10y + 0.09z = 2220$$

Thus, our system of equation is

$$(1) \quad x + y + z = 25{,}000$$

$$(2) \quad y = \frac{1}{2}x + 2000$$

$$(3) \quad 0.08x + 0.10y + 0.09z = 2220$$

Both sides of equation (2) can be multiplied by 2 to remove fractions.

$$2(y) = 2\left(\tfrac{1}{2}x + 2000\right)$$
$$2y = x + 4000$$
$$\text{or} \qquad -x + 2y = 4000$$

The decimals in equation (3) can be removed by multiplying both sides of the equation by 100 to get

$$8x + 10y + 9z = 222{,}000$$

Our simplified system of equations is therefore

$$
\begin{aligned}
(1) \quad & x + y + z = 25{,}000 \\
(2) \quad -&x + 2y = 4000 \\
(3) \quad & 8x + 10y + 9z = 222{,}000
\end{aligned}
$$

There are various ways of solving this system. Let us start by multiplying equation (1) in the simplified system of equations by -9 and adding it to equation (3) to eliminate the variable z.

$$
\begin{array}{ll}
(1) \quad -9[x + y + z = 25{,}000] & \\
(3) \quad 8x + 10y + 9z = 222{,}000 & \text{gives}
\end{array}
\qquad
\begin{array}{r}
-9x - 9y - 9z = -225{,}000 \\
\underline{8x + 10y + 9z = 222{,}000} \\
(4) \quad -x + y = -3{,}000
\end{array}
$$

Now multiply equation (2) by -1 and add the results to equation (4) to eliminate the variable x and solve for y.

$$
\begin{array}{ll}
(4) \qquad -x + y = -3{,}000 & \\
(2) \quad -1[-x + 2y = 4000] & \text{gives}
\end{array}
\qquad
\begin{array}{r}
-x + y = -3000 \\
\underline{x - 2y = -4000} \\
-y = -7000 \\
y = 7000
\end{array}
$$

Now that we know the value of y we can solve for x.

$$
\begin{aligned}
(2) \qquad -x + 2y &= 4000 \\
-x + 2(7000) &= 4000 \\
-x + 14{,}000 &= 4000 \\
-x &= -10{,}000 \\
x &= 10{,}000
\end{aligned}
$$

Finally, solve for z.

$$(1) \qquad x + y + z = 25{,}000$$
$$10{,}000 + 7000 + z = 25{,}000$$
$$17{,}000 + z = 25{,}000$$
$$z = 8000$$

Thus, Tiny Tots Toys borrowed $10,000 at 8%, $7000 at 10%, and $8000 at 9% interest.

EXAMPLE 7   Hobson, Inc., has a small manufacturing plant that makes three types of inflatable boats: one-person, two-person, and four-person models. Each boat requires the service of three departments: cutting, assembly, and packaging. The cutting, assembly, and packaging departments are allowed to use a total of 380, 330, and 120 person-hours per week, respectively. The time requirements for each boat and department are specified in the following table. Determine how many of each type of boat Hobson's must produce each week for its plant to operate at full capacity.

Department	Time (hr)		
	One-person boat	Two-person boat	Four-person boat
Cutting	0.6	1.0	1.5
Assembly	0.6	0.9	1.2
Packaging	0.2	0.3	0.5

Solution: Let x = number of one-person boats
y = number of two-person boats
z = number of four-person boats

The total number of cutting hours for the three types of boats must equal 380 person-hours.

$$0.6x + 1.0y + 1.5z = 380$$

The total number of assembly hours must equal 330 person-hours.

$$0.6x + 0.9y + 1.2z = 330$$

The total number of packaging hours must equal 120 person-hours.

$$0.2x + 0.3y + 0.5z = 120$$

System of Equations

$$0.6x + 1.0y + 1.5z = 380$$
$$0.6x + 0.9y + 1.2z = 330$$
$$0.2x + 0.3y + 0.5z = 120$$

Multiplying each equation in the system by 10 will eliminate the decimal numbers.

Simplified System of Equations

(1) $6x + 10y + 15z = 3800$
(2) $6x + 9y + 12z = 3300$
(3) $2x + 3y + 5z = 1200$

Let's first eliminate the variable x using equations (1) and (2), and then equations (1) and (3).

(1) $6x + 10y + 15z = 3800$
(2) $-1[6x + 9y + 12z = 3300]$ gives

$$6x + 10y + 15z = 3800$$
$$-6x - 9y - 12z = -3300$$
(4) $\overline{\qquad\qquad y + 3z = 500}$

(1) $6x + 10y + 15z = 3800$
(3) $-3[2x + 3y + 5z = 1200]$ gives

$$6x + 10y + 15z = 3800$$
$$-6x - 9y - 15z = -3600$$
$$\overline{\qquad y \qquad\quad = 200}$$

Note that when we added the last two equations, both variables x and z were eliminated at the same time. Now we solve for z.

(4) $y + 3z = 500$
$$200 + 3z = 500$$
$$3z = 300$$
$$z = 100$$

Finally, find x.

(1) $6x + 10y + 15z = 3800$
$$6x + 10(200) + 15(100) = 3800$$
$$6x + 2000 + 1500 = 3800$$
$$6x + 3500 = 3800$$
$$6x = 300$$
$$x = 50$$

Thus, Hobson's should produce 50 one-person boats, 200 two-person boats, and 100 four-person boats per week.

Exercise Set 4.3

In Ex. 1–34, **(a)** *express the problem as a system of linear equations and* **(b)** *use the method of your choice to solve the problem.*

1. The sum of two numbers is 73. Find the numbers if one number is 15 less than three times the other. 22, 51

2. The sum of two consecutive odd integers is 76. Find the two numbers.   37, 39

14. $2500 at 8%, $5500 at 10%   **15.** 5.96 oz cream, 10.04 oz half and half

3. The difference of two numbers is 25. Find the two numbers if the larger is 1 less than 3 times the smaller.

4. Two angles are **complementary angles** if the sum of their measures is 90°. If the larger of two complementary angles is 15° more than 2 times the smaller angle, find the two angles.   25°, 65°

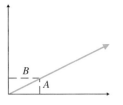

3. 13, 38

Angles *A* and *B* are complementary angles.

5. Two angles are **supplementary angles** if the sum of their measures is 180°. Find the two supplementary angles if one angle is 28° less than 3 times the other.
5. 52°, 128°

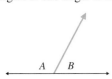

Angles *A* and *B* are supplementary angles.

6. plane, 515 mph; wind, 25 mph

6. A plane can travel 540 miles per hour with the wind and 490 miles per hour against the wind. Find the speed of the plane in still air and the speed of the wind.

7. A 50-foot length of rope is cut into two pieces. If one piece is 2 feet more than three times the other piece, find the length of the two pieces.   12 ft, 38 ft

8. Steve Trinter, an electronics salesman, earns a weekly salary plus a commission on sales. One week his salary on sales of $4000 was $660. The next week his salary on sales of $6000 was $740. Find his weekly salary and his commission rate.   salary, $500; commission, 4%

9. A car rental agency charges a daily fee plus a mileage fee. Mr. Dobson was charged $85 for 2 days and 100 miles and Mrs. Schwartz was charged $165 for 3 days and 400 miles. What is the agency's daily fee, and what is their mileage fee? $35/day and 15 cents/mile

10. The total cost of printing a political leaflet consists of a fixed charge and an additional charge for each leaflet. If the total cost for 1000 leaflets is $550 and the total cost for 2000 leaflets is $800, find the fixed charge and the charge for each leaflet.

11. Mr. Fiora, a druggist, needs 1000 milliliters of a 10% phenobarbital solution. He has only 5% and 25% phenobarbital solution available. How many milliliters of each solution should he mix to obtain the desired solution?   750 mL of 5% and 250 mL of 25%

10. fixed charge, $300; leaflet, $0.25

12. Mrs. Spinelli runs a grocery store. She wishes to mix 30 pounds of coffee to sell for a total cost of $170. To obtain the mixture, Mrs. Spinelli will mix coffee that sells for $5.20 per pound with coffee that sells for $6.30 per pound. How many pounds of each coffee should she use?   17.3 lb at $5.20, 12.7 lb at $6.30

13. Mario owns a dairy. He has milk that is 5% butterfat and skim milk, without butterfat. How much 5% milk and how much skim milk should he mix to make 100 gallons of milk that is 3.5% butterfat?

See Exercise 13   **13.** 70 gal of 5%, 30 gal of skim

14. Mr. and Mrs. McAdams invest a total of $8000 in two savings accounts. One account pays 10% interest and the other 8%. Find the amount placed in each account if they receive a total of $750 in interest after 1 year. Use interest = principal · rate · time.

15. Steve's recipe for Quiche Lorraine calls for 2 cups (16 ounces) of light cream that is 20% milk fat. It is often difficult to find light cream with 20% milk fat at the supermarket. What is commonly found is heavy cream, which is 36% milk fat, and half-and-half, which is 10.5% milk fat. How much of the heavy cream and how much of the half-and-half should Steve mix to obtain the mixture necessary for the recipe?

16. Professor Green invested $30,000, part at 12% and part at 8%. If he had invested the entire amount at 9.5%, his total annual interest would be the same as the sum of the annual interest received from the two other accounts. How much was invested in each account? $11,250 at 12%, $18,750 at 8%

17. Two cars start at the same point in Louisville and travel in opposite directions. One car travels 5 miles per hour faster than the other car. After 4 hours, the two cars are 420 miles apart. Find the speed of each car.   50 mph and 55 mph

18. The Friendly Face Fruit Juice Company sells apple juice for 8.3 cents an ounce and raspberry juice for 9.3 cents an ounce. The company wishes to market and sell 8-ounce cans of apple–raspberry juice for 8.7

20. 130 adults, 42 children

cents an ounce. How many ounces of each should be mixed?   4.8 oz apple and 3.2 oz raspberry juice

19. Blockbuster stock is selling at $35 per share and BankAmerica stock is selling at $20 per share. Geraldo has $6250 to invest. He wishes to purchase three times as many shares of Blockbuster stock as of Bank-America. How many shares of each stock should he purchase?   150 shares Blockbuster, 50 shares BankAmerica

20. A movie ticket for an adult costs $7.50 and a child's ticket costs $5.00. A total of 172 tickets are sold for a given show. If $1,185 is collected for the show, how many adults and how many children attended the show?

21. Safeway sells almonds for $6.50 per pound and walnuts for $5.90 per pound. How many pounds of each should they mix to produce a 30-pound mixture that costs $6.30 per pound?   20 lb almonds, 10 lb walnuts

22. Phong has a total of 12 bills in his wallet. Some are $5 bills and the rest are $10 bills. The total value of the bills is $115. How many $5 bills and how many $10 bills does he have?   1 $5 bill and 11 $10 bills

23. A collection of dimes and quarters has a value of $3.55. If there are a total of 25 coins, how many dimes and quarters are there?   18 dimes, 7 quarters

24. By traveling first at 40 mph and then at 50 mph, Jagat traveled 320 miles. Had he gone 10 mph faster for each period of time, he would have traveled 390 miles. How many hours did he travel at each rate?

25. Some states allow a husband and wife to file individual state tax returns (on a single form) even though they file a joint federal return. It is usually to the taxpayer's advantage to do this when both the husband and wife work. The smallest amount of tax owed (or the largest refund) will occur when the husband's and wife's taxable incomes are the same.

   Mr. Clar's 1996 taxable income was $26,200 and Mrs. Clar's income for that year was $22,450. The Clars' total tax deduction for the year was $12,400. This deduction can be divided between Mr. and Mrs. Clar in any way they wish. How should the $12,400 be divided between them to result in each person's having the same taxable income and therefore the greatest tax return? $8075 Mr., $4325 Mrs.

26. Two points on the line $y = ax + b$ are (3, 8) and $(-2, -17)$. Find the value of a and b. $a = 5$, $b = -7$

27. Two points on the line $y = ax + b$ are $(-1, 7)$ and $(\frac{1}{2}, 4)$. Find the value of a and b. $a = -2$, $b = 5$

24. 3 hr at 40 mph, 4 hr at 50 mph   **28.** faster, 8 min; slower, 5 min

28. Two photocopy machines are used to make large quantities of copies at Kinko's. The slower machine produces 75 copies per minute and the faster machine produces 120 copies per minute. The faster machine was in operation for 3 minutes before the slower machine was started. If they both continue copying together until they produced a total of 1335 copies, find the length of time both machines were in operation.

29. The sum of two times one number and one-half a second number is 35. The sum of one-half the first number and one-third the second number is 15. Find the two numbers.   10, 30

30. An automobile radiator has a capacity of 16 liters. How much pure antifreeze must be added to a mixture of water and antifreeze that is 18% antifreeze to make a mixture of 20% antifreeze that can be used to fill the radiator?   add 0.39 liter to 15.61 liters of mixture

31. Animals in an experiment are on a strict diet. Each animal is to receive, among other nutrients, 20 grams of protein and 6 grams of carbohydrates. The scientist has only two food mixes available of the following compositions.

	Protein (%)	Carbohydrate (%)
Mix A	10	6
Mix B	20	2

How many grams of each mix should be used to obtain the right diet for a single animal?   80 g A, 60 g B

32. A company that makes children's wooden chairs makes two kinds of chairs. The basic model requires 1 hour to assemble and 0.5 hour to paint, and the deluxe model requires 3.2 hours to assemble but only 0.4 hour to paint. On a particular day the company allocated 46.4 person-hours for assembling and 8.8 person-hours for painting. How many of each chair can be made?

33. By weight, one alloy of brass is 70% copper and 30% zinc. Another alloy of brass is 40% copper and 60% zinc. How many grams of each of these alloys need to be melted and combined to obtain 300 grams of a brass alloy that is 60% copper and 40% zinc?

34. A car travels 300 miles in the same amount of time that a truck travels 240 miles. If the speed of the car is 10 miles per hour faster than the speed of the truck, find both speeds.   car, 50 mph; truck, 40 mph

32. 8 basic, 12 deluxe   **33.** 200 g first alloy, 100 g second alloy

35. balcony $30, back $40, front $60
43. 10 children's, 12 standard, 8 executive

41. 4 L of 10%, 2 L of 12%, 2 L of 20%
44. 0 lb first alloy, 50 lb second alloy, 50 lb third alloy

In Ex. 35–44, **(a)** *express the problem as a third-order system of linear equations and* **(b)** *solve the problem.*

35. Three kinds of tickets are available for a Boyz II Men concert. The up-front main floor tickets are the most expensive; the seats farther back on the main floor are the second most expensive; and the balcony seats are the least expensive. The up-front main floor seats are twice as expensive as the balcony seats. The balcony seats are $10 less than the main floor seats in the back and $30 less than the up-front main floor seats. Find the price of each seat.

36. The sum of twice a number, three times a second number, and four times a third number is 40. The sum of the first and second numbers equals the third number. The third number is 2 more than twice the first number. Find the three numbers. 2, 4, 6

37. The sum of the measures of the angles of a triangle is $180°$. The smallest angle of a triangle is $\frac{2}{3}$ of the middle-sized angle. The largest angle is $30°$ less than 3 times the middle-sized angle. Find the measure of each angle. 30°, 45°, 105°

38. Find a, b, and c so that the graph of the equation $y = ax^2 + bx + c$ passes through the points $(0, -3)$, $(2, 1)$, and $(-3, -24)$. $a = -1, b = 4, c = -3$

39. Find a, b, and c so that the graph of the equation $y = ax^2 + bx + c$ passes through the points $(2, 6)$, $(3, 17)$, and $(-1, -3)$. $a = 2, b = 1, c = -4$

40. Marion received a check for $10,000. She decided to divide the money (not equally) into three different investments. She placed part of her money in a savings account paying 7% interest. The second amount, which was twice the first amount, she placed in a certificate of deposit paying 9% interest. She placed the balance in a money market fund that yielded 10% interest. If Marion's total interest over the period of 1 year was $925.00, how much was placed in each account? $1500 at 7%, $3000 at 9%, $5500 at 10%

41. A 10% solution, a 12% solution, and a 20% solution of hydrogen peroxide are to be mixed to get 8 liters of a 13% solution. How many liters of each must be mixed if the volume of the 20% solution must be 2 liters less than the volume of the 10% solution?

42. An 8% solution, a 10% solution, and a 20% solution of sulfuric acid are to be mixed to get 100 milliliters of a 12% solution. If the *volume of acid* from the 8% solution is to equal half the *volume of acid* from the

other two solutions, how much of each solution is needed? 50 mL, 8% sol; 20 mL, 10% sol; 30 mL, 20% sol

43. Donaldson Furniture Company produces three types of rocking chairs: the children's model, the standard model, and the executive model. Each chair is made in three stages: cutting, construction, and finishing. The time needed for each stage of each chair is given in the following chart. During a specific week the company has available a maximum of 154 hours for cutting, 94 hours for construction, and 76 hours for finishing. Determine how many of each chair the company should make to be operating at full capacity.

	Children's	Standard	Executive
Cutting	5 hr	4 hr	7 hr
Construction	3 hr	2 hr	5 hr
Finishing	2 hr	2 hr	4 hr

44. By volume, one alloy is 60% copper, 30% zinc, and 10% nickel. A second alloy has percentages 50, 30, and 20, respectively, of the three metals. A third alloy is 30% copper and 70% nickel. How much of each alloy must be mixed so that 100 pounds of the resulting alloy is 40% copper, 15% zinc, and 45% nickel?

45. In electronics it is necessary to analyze current flow through paths of a circuit. In three paths (A, B, and C) of a circuit, the relationships are the following:

$$I_A + I_B + I_C = 0$$
$$-8I_B + 10I_C = 0$$
$$4I_A - 8I_B = 6$$

where I_A, I_B, and I_C represent the current in paths A, B, and C, respectively. Determine the current in each path of the circuit. $\frac{27}{38}, -\frac{15}{38}, -\frac{6}{19}$

46. In physics we often study the forces acting on an object. For three forces, F_1, F_2, and F_3, acting on a beam, the following equations were obtained.

$$3F_1 + F_2 - F_3 = 2$$
$$F_1 - 2F_2 + F_3 = 0$$
$$4F_1 - F_2 + F_3 = 3$$

Find the three forces. $\left(\frac{5}{7}, \frac{6}{7}, 1\right)$

CUMULATIVE REVIEW EXERCISES

[1.4] **47.** Evaluate $\frac{1}{2}x + \frac{2}{5}xy + \frac{1}{8}y$ when $x = -2, y = 5$.

[3.1] **48.** Determine the length and midpoint of the line segment through the points $(6, -4)$ and $(2, -8)$.

47. $-\frac{35}{8}$ **48.** $\sqrt{32} \approx 5.66, (4, -6)$ **50.** Use vertical line test.

[3.3] **49.** Write an equation of the line that passes through points $(6, -4)$ and $(2, -8)$. $y = x - 10$

[3.4] **50.** Explain how to determine if a graph is a function.

1. In an article published in the *Journal of Comparative Physiology and Psychology,* J. S. Brown discusses how we often approach a situation with mixed emotions. For example, when a person is asked to give a speech, he may be a little apprehensive about his ability to do a good job. At the same time, he would like the recognition that goes along with making the speech. J. S. Brown performed an experiment on trained rats. He placed their food in a metal box. He used that same box to administer small electrical shocks to the mice. Therefore, the rats "wished" to go into the box to receive food, yet did not "wish" to go into the box for fear of receiving a small shock. Using the appropriate apparatus, Brown arrived at the following relationships:

$$\text{pull (in grams) toward food} = -\frac{1}{5}d + 70 \qquad 30 < d < 172.5$$

$$\text{pull (in grams) away from shock} = -\frac{4}{3}d + 230 \qquad 30 < d < 172.5$$

where d is the distance in centimeters from the box (and food).

(a) Using the substitution method, find the distance at which the pull toward the food equals the pull away from the shock. $\frac{2400}{17} \approx 141.2$ cm

(b) If the rat is placed 100 cm from the box (or food), what will the rat do? pull away

2. A *nonlinear system of equations* is a system of equations containing at least one equation which is not linear. (Nonlinear systems of equations will be discussed in Chapter 10.) Consider the following graph as a nonlinear system of equations. Answer the following questions.

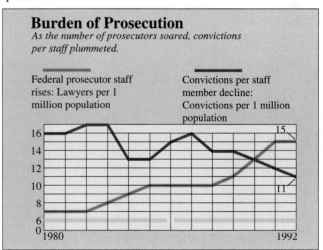

(a) How does the number of federal prosecutors change during the period 1980–1992? increasing

(b) How does the number of convictions per prosecution staff change during this period? decreasing

(c) Estimate the number of prosecutors per million population and the number of convictions per prosecution staff for 1985. 10, 13

(d) Repeat part (c) for 1992. 15, 11

(e) Do you believe there is a cause-and-effect relationship between these two graphs? Explain your answer.

Tape 6

4.4 Solving Systems of Equations Using Determinants and Cramer's Rule

1 Evaluate second-order determinants.
2 Use Cramer's rule to solve second-order systems of equations.
3 Evaluate third-order determinants.
4 Use Cramer's rule to solve third-order systems of equations.

Systems of linear equations can also be solved using determinants. Determinants are particularly useful for solving third- and higher-order systems of equations.

A **determinant** is a square array of numbers enclosed between two vertical bars. Examples of determinants are

$$\begin{vmatrix} 4 & -3 \\ 0 & 5 \end{vmatrix} \qquad \begin{vmatrix} 3 & 0 & 5 \\ 4 & -2 & 3 \\ 2 & \frac{1}{2} & -1 \end{vmatrix}$$

(a)   (b)

The numbers that make up the array are called the **elements** of the determinant. The elements of determinant (a) are 4, -3, 0, and 5.

Determinant (a) is a **second-order determinant** since it has two rows and two columns of elements. Determinant (b) is a **third-order determinant.** Determinants can be of an order greater than 3.

The **principal diagonal** of a determinant is the line of elements from the upper-left corner to the lower-right corner. The **secondary diagonal** of a determinant is the line of elements from the lower-left corner to the upper-right corner.

$$\begin{vmatrix} a_1 & b_1 \\ a_2 & b_2 \end{vmatrix} \qquad \begin{vmatrix} a_1 & b_1 \\ a_2 & b_2 \end{vmatrix}$$

principal   secondary
diagonal   diagonal

Evaluating Second-Order Determinants

1 Every determinant represents a number. The **value of a second-order determinant** is the product of the elements in its principal diagonal minus the product of the elements in its secondary diagonal.

Value of a Second-Order Determinant

$$\begin{vmatrix} a_1 & b_1 \\ a_2 & b_2 \end{vmatrix} = a_1 b_2 - a_2 b_1$$

EXAMPLE 1 Find the value of the determinants.

(a) $\begin{vmatrix} 4 & 6 \\ -3 & 2 \end{vmatrix}$ (b) $\begin{vmatrix} -3 & 4 \\ 1 & 5 \end{vmatrix}$

Solution:

(a) $a_1 = 4, a_2 = -3, b_1 = 6, b_2 = 2$ **(b)** $a_1 = -3, a_2 = 1, b_1 = 4, b_2 = 5$

$$\begin{vmatrix} 4 & 6 \\ -3 & 2 \end{vmatrix} = 4(2) - (-3)(6) \qquad \begin{vmatrix} -3 & 4 \\ 1 & 5 \end{vmatrix} = (-3)(5) - (1)(4)$$

$$= 8 + 18 \qquad\qquad\qquad\qquad = -15 - 4$$

$$= 26 \qquad\qquad\qquad\qquad\quad = -19$$

The value of determinant **(a)** is 26 and the value of determinant **(b)** is -19.

Cramer's Rule   **2** If we begin with the equations

$$a_1x + b_1y = c_1$$
$$a_2x + b_2y = c_2$$

we can use the addition method to show that

$$x = \frac{c_1b_2 - c_2b_1}{a_1b_2 - a_2b_1} \quad \text{and} \quad y = \frac{a_1c_2 - a_2c_1}{a_1b_2 - a_2b_1}$$

(see Group Activity Exercise 5). Notice that the denominators of x and y are both $a_1b_2 - a_2b_1$. Following is the determinant that yields this denominator. We have labeled this denominator D.

$$D = \begin{vmatrix} a_1 & b_1 \\ a_2 & b_2 \end{vmatrix} = a_1b_2 - a_2b_1$$

The numerators of x and y are different. Following are two determinants, labeled D_x and D_y, that yield the numerators of x and y.

$$D_x = \begin{vmatrix} c_1 & b_1 \\ c_2 & b_2 \end{vmatrix} = c_1b_2 - c_2b_1 \qquad D_y = \begin{vmatrix} a_1 & c_1 \\ a_2 & c_2 \end{vmatrix} = a_1c_2 - a_2c_1$$

We use determinants D, D_x, and D_y in Cramer's rule. Cramer's rule can be used to evaluate systems of equations.

Cramer's Rule for Second-Order Linear Systems

For a system of equations of the form

$$a_1x + b_1y = c_1$$
$$a_2x + b_2y = c_2$$

$$x = \frac{\begin{vmatrix} c_1 & b_1 \\ c_2 & b_2 \end{vmatrix}}{\begin{vmatrix} a_1 & b_1 \\ a_2 & b_2 \end{vmatrix}} = \frac{D_x}{D} \quad \text{and} \quad y = \frac{\begin{vmatrix} a_1 & c_1 \\ a_2 & c_2 \end{vmatrix}}{\begin{vmatrix} a_1 & b_1 \\ a_2 & b_2 \end{vmatrix}} = \frac{D_y}{D}, D \neq 0$$

> *Helpful Hint*
>
> The elements in determinant D are the numerical coefficients of the x and y terms in the two given equations, listed in the same order they are listed in the equations. To obtain the determinant D_x from determinant D, replace the coefficients of the x terms (the values in the first columns) with the constants of the two given equations. To obtain the determinant D_y from determinant D, replace the coefficients of the y terms (the values in the second column) with the constants of the two given equations.

EXAMPLE 2   Use Cramer's rule to evaluate the following system.

$$2x - 4y = 8$$
$$3x + 5y = -10$$

Solution: Both equations are given in the desired form, $ax + by = c$. When labeling the a, b, and c's we will refer to $2x - 4y = 8$ as equation 1 and $3x + 5y = -10$ as equation 2.

$$
\begin{array}{ccc}
a_1 & b_1 & c_1 \\
\downarrow & \downarrow & \downarrow \\
2x - & 4y = & 8 \\
3x + & 5y = & -10 \\
\uparrow & \uparrow & \uparrow \\
a_2 & b_2 & c_2
\end{array}
$$

We now find D, D_x, and D_y.

$$D = \begin{vmatrix} a_1 & b_1 \\ a_2 & b_2 \end{vmatrix} = \begin{vmatrix} 2 & -4 \\ 3 & 5 \end{vmatrix} = 2(5) - 3(-4) = 22$$

$$D_x = \begin{vmatrix} c_1 & b_1 \\ c_2 & b_2 \end{vmatrix} = \begin{vmatrix} 8 & -4 \\ -10 & 5 \end{vmatrix} = 8(5) - (-10)(-4) = 0$$

$$D_y = \begin{vmatrix} a_1 & c_1 \\ a_2 & c_2 \end{vmatrix} = \begin{vmatrix} 2 & 8 \\ 3 & -10 \end{vmatrix} = 2(-10) - (3)(8) = -44$$

Now find the value of x and y.

$$x = \frac{D_x}{D} = \frac{0}{22} = 0$$

$$y = \frac{D_y}{D} = \frac{-44}{22} = -2$$

Thus, the solution is $x = 0$, $y = -2$ or the ordered pair $(0, -2)$. A check will show that this ordered pair satisfies both equations.

When determinant $D = 0$, Cramer's rule does not apply since division by 0 is not possible. You may then use a different method to solve the system. Or, if you evaluate D_x and D_y and find they are both 0 when $D = 0$, the system is dependent and there are an infinite number of solutions. If either D_x or D_y is not 0 when $D = 0$, the system is inconsistent and there are no solutions.

Evaluating Third-Order Determinants

3 For the third-order determinant

$$\begin{vmatrix} a_1 & b_1 & c_1 \\ a_2 & b_2 & c_2 \\ a_3 & b_3 & c_3 \end{vmatrix}$$

the minor determinant of a_1 is found by crossing out the elements in the same row and column in which the element a_1 appears.

$$\begin{vmatrix} a_1 & b_1 & c_1 \\ a_2 & b_2 & c_2 \\ a_3 & b_3 & c_3 \end{vmatrix}$$

The remaining elements form the minor determinant of a_1.

$$\begin{vmatrix} b_2 & c_2 \\ b_3 & c_3 \end{vmatrix}$$

The minor determinant of a_2 is found similarly.

$$\begin{vmatrix} a_1 & b_1 & c_1 \\ a_2 & b_2 & c_2 \\ a_3 & b_3 & c_3 \end{vmatrix} \qquad \text{minor determinant of } a_2 \qquad \begin{vmatrix} b_1 & c_1 \\ b_3 & c_3 \end{vmatrix}$$

The minor determinant of a_3 is found below.

$$\begin{vmatrix} a_1 & b_1 & c_1 \\ a_2 & b_2 & c_2 \\ a_3 & b_3 & c_3 \end{vmatrix} \qquad \text{minor determinant of } a_3 \qquad \begin{vmatrix} b_1 & c_1 \\ b_2 & c_2 \end{vmatrix}$$

To evaluate third-order determinants, we use minor determinants. The following box shows how a third-order determinant may be evaluated.

Expansion of the Determinant by the Minors of the First Column

$$\begin{vmatrix} a_1 & b_1 & c_1 \\ a_2 & b_2 & c_2 \\ a_3 & b_3 & c_3 \end{vmatrix} = a_1 \begin{vmatrix} b_2 & c_2 \\ b_3 & c_3 \end{vmatrix} - a_2 \begin{vmatrix} b_1 & c_1 \\ b_3 & c_3 \end{vmatrix} + a_3 \begin{vmatrix} b_1 & c_1 \\ b_2 & c_2 \end{vmatrix}$$

where the first term uses the minor determinant of a_1, the second term uses the minor determinant of a_2, and the third term uses the minor determinant of a_3.

This method of evaluating the determinant is called **expansion of the determinant by the minors of the first column.**

EXAMPLE 3   Evaluate $\begin{vmatrix} 4 & -2 & 6 \\ 3 & 5 & 0 \\ 1 & -3 & -1 \end{vmatrix}$ using expansion of the determinant by the minors of the first column.

Solution: We will follow the procedure given in the box.

$$\begin{vmatrix} 4 & -2 & 6 \\ 3 & 5 & 0 \\ 1 & -3 & -1 \end{vmatrix} = 4\begin{vmatrix} 5 & 0 \\ -3 & -1 \end{vmatrix} - 3\begin{vmatrix} -2 & 6 \\ -3 & -1 \end{vmatrix} + 1\begin{vmatrix} -2 & 6 \\ 5 & 0 \end{vmatrix}$$

$$= 4[5(-1) - (-3)0] - 3[(-2)(-1) - (-3)6] + 1[(-2)0 - 5(6)]$$
$$= 4(-5 + 0) - 3(2 + 18) + 1(0 - 30)$$
$$= 4(-5) - 3(20) + 1(-30)$$
$$= -20 - 60 - 30$$
$$= -110$$

The determinant has a value of -110.

In Example 3 we evaluated the determinant using expansion of the determinant by the minors of the first column. To evaluate a determinant, we can use expansion of the determinant by the minors of any row or any column. To determine if the product of the element and its minor determinant is to be added or subtracted in obtaining the result, we use the following chart.

$$\begin{vmatrix} + & - & + \\ - & + & - \\ + & - & + \end{vmatrix}$$

If the element is in a position marked with $+$, the product is to be added. If the element is in a position marked with $-$, the product is to be subtracted.

EXAMPLE 4   Evaluate $\begin{vmatrix} 4 & -2 & 6 \\ 3 & 5 & 0 \\ 1 & -3 & -1 \end{vmatrix}$ using expansion of the determinant by the minors of the second row.

Solution: The signs in the second row of the chart above are $-$, $+$, $-$. Thus, the first product is to be subtracted, the second product added, and the third product subtracted.

$$\begin{vmatrix} 4 & -2 & 6 \\ 3 & 5 & 0 \\ 1 & -3 & -1 \end{vmatrix} = -3 \begin{vmatrix} -2 & 6 \\ -3 & -1 \end{vmatrix} + 5 \begin{vmatrix} 4 & 6 \\ 1 & -1 \end{vmatrix} - 0 \begin{vmatrix} 4 & -2 \\ 1 & -3 \end{vmatrix}$$

$$\qquad\qquad \underset{\text{subtract}}{\uparrow} \qquad\qquad \underset{\text{add}}{\uparrow} \qquad\qquad \underset{\text{subtract}}{\uparrow}$$

$$= -3[(-2)(-1) - (-3)6] + 5[4(-1) - 1(6)] - 0[4(-3) - 1(-2)]$$
$$= -3(2 + 18) + 5(-4 - 6) - 0(-12 + 2)$$
$$= -3(20) + 5(-10) - 0$$
$$= -60 - 50$$
$$= -110$$

Note that the same answer was obtained by evaluating the determinant using the first column or the second row. When a determinant contains one or more 0's in a particular row or column, you may wish to evaluate the determinant by expansion of the minor determinants of that row or column since it makes the calculations easier.

Cramer's Rule for a Third-Order System

[4] Cramer's rule can be extended to third-order systems of equations as follows:

Cramer's Rule for a Third-Order System

To evaluate the system

$$a_1 x + b_1 y + c_1 z = d_1$$
$$a_2 x + b_2 y + c_2 z = d_2$$
$$a_3 x + b_3 y + c_3 z = d_3$$

with

$$D = \begin{vmatrix} a_1 & b_1 & c_1 \\ a_2 & b_2 & c_2 \\ a_3 & b_3 & c_3 \end{vmatrix} \qquad D_x = \begin{vmatrix} d_1 & b_1 & c_1 \\ d_2 & b_2 & c_2 \\ d_3 & b_3 & c_3 \end{vmatrix}$$

$$D_y = \begin{vmatrix} a_1 & d_1 & c_1 \\ a_2 & d_2 & c_2 \\ a_3 & d_3 & c_3 \end{vmatrix} \qquad D_z = \begin{vmatrix} a_1 & b_1 & d_1 \\ a_2 & b_2 & d_2 \\ a_3 & b_3 & d_3 \end{vmatrix}$$

then

$$x = \frac{D_x}{D} \qquad y = \frac{D_y}{D} \qquad z = \frac{D_z}{D}, \qquad D \neq 0$$

Note that the denominators of the expressions for x, y, and z are all the same determinant, D. Note that constants, the d's, replace the a's, the numerical coefficients of the x terms, in D_x. The d's replace the b's, the numerical coefficients of the y terms in D_y. And the d's replace the c's, the numerical coefficients of the z terms in D_z.

Example 5   Solve the following system of equations using determinants.

$$3x - 2y - z = -6$$
$$2x + 3y - 2z = 1$$
$$x - 4y + z = -3$$

Solution:

$$
\begin{array}{llll}
a_1 = 3 & b_1 = -2 & c_1 = -1 & d_1 = -6 \\
a_2 = 2 & b_2 = 3 & c_2 = -2 & d_2 = 1 \\
a_3 = 1 & b_3 = -4 & c_3 = 1 & d_3 = -3
\end{array}
$$

We will use expansion of the minor determinants of the first row to evaluate D.

$$D = \begin{vmatrix} 3 & -2 & -1 \\ 2 & 3 & -2 \\ 1 & -4 & 1 \end{vmatrix} = 3\begin{vmatrix} 3 & -2 \\ -4 & 1 \end{vmatrix} - (-2)\begin{vmatrix} 2 & -2 \\ 1 & 1 \end{vmatrix} + (-1)\begin{vmatrix} 2 & 3 \\ 1 & -4 \end{vmatrix}$$

$$= 3(-5) + 2(4) - 1(-11)$$
$$= -15 + 8 + 11 = 4$$

We will evaluate D_x using expansion of the determinant by the minors of the first column.

$$D_x = \begin{vmatrix} -6 & -2 & -1 \\ 1 & 3 & -2 \\ -3 & -4 & 1 \end{vmatrix} = (-6)\begin{vmatrix} 3 & -2 \\ -4 & 1 \end{vmatrix} - (1)\begin{vmatrix} -2 & -1 \\ -4 & 1 \end{vmatrix} + (-3)\begin{vmatrix} -2 & -1 \\ 3 & -2 \end{vmatrix}$$

$$= -6(-5) - 1(-6) - 3(7)$$
$$= 30 + 6 - 21 = 15$$

We will use expansion of the determinant by the minors of the first row to evaluate D_y.

$$D_y = \begin{vmatrix} 3 & -6 & -1 \\ 2 & 1 & -2 \\ 1 & -3 & 1 \end{vmatrix} = 3\begin{vmatrix} 1 & -2 \\ -3 & 1 \end{vmatrix} - (-6)\begin{vmatrix} 2 & -2 \\ 1 & 1 \end{vmatrix} + (-1)\begin{vmatrix} 2 & 1 \\ 1 & -3 \end{vmatrix}$$

$$= 3(-5) + 6(4) - 1(-7)$$
$$= -15 + 24 + 7 = 16$$

We will evaluate D_z using expansion of the determinant by the minors of the first row.

$$D_z = \begin{vmatrix} 3 & -2 & -6 \\ 2 & 3 & 1 \\ 1 & -4 & -3 \end{vmatrix} = 3\begin{vmatrix} 3 & 1 \\ -4 & -3 \end{vmatrix} - (-2)\begin{vmatrix} 2 & 1 \\ 1 & -3 \end{vmatrix} + (-6)\begin{vmatrix} 2 & 3 \\ 1 & -4 \end{vmatrix}$$

$$= 3(-5) + 2(-7) - 6(-11)$$
$$= -15 - 14 + 66 = 37$$

We found that $D = 4$, $D_x = 15$, $D_y = 16$, and $D_z = 37$. Therefore,

$$x = \frac{D_x}{D} = \frac{15}{4} \qquad y = \frac{D_y}{D} = \frac{16}{4} = 4 \qquad z = \frac{D_z}{D} = \frac{37}{4}$$

The solution to the system is $\left(\frac{15}{4}, 4, \frac{37}{4}\right)$. Note the ordered triple lists x, y, and z in this order.

When we have a third-order system of equations in which one or more equations are missing a variable, we insert the variable with a coefficient of 0. This helps in aligning like terms. For example,

$$\begin{array}{rl} 2x - 3y + 2z = -1 \\ x + 2y = 14 \\ x - 3z = -5 \end{array} \quad \text{is written} \quad \begin{array}{rl} 2x - 3y + 2z = -1 \\ x + 2y + 0z = 14 \\ x + 0y - 3z = -5 \end{array}$$

when solving the system using determinants. Furthermore, it is very important to place the numbers in the correct column. In this example

$$D = \begin{vmatrix} 2 & -3 & 2 \\ 1 & 2 & 0 \\ 1 & 0 & -3 \end{vmatrix} \qquad D_x = \begin{vmatrix} -1 & -3 & 2 \\ 14 & 2 & 0 \\ -5 & 0 & -3 \end{vmatrix}$$

$$D_y = \begin{vmatrix} 2 & -1 & 2 \\ 1 & 14 & 0 \\ 1 & -5 & -3 \end{vmatrix} \qquad D_z = \begin{vmatrix} 2 & -3 & -1 \\ 1 & 2 & 14 \\ 1 & 0 & -5 \end{vmatrix}$$

Helpful Hint

1. When evaluating determinants, if any two rows (or columns) are identical, or identical except for opposite signs, the determinant has a value of 0. For example,

$$\begin{vmatrix} 5 & -2 \\ 5 & -2 \end{vmatrix} = 0 \quad \text{and} \quad \begin{vmatrix} 5 & -2 \\ -5 & 2 \end{vmatrix} = 0$$

$$\begin{vmatrix} 5 & -3 & 4 \\ 2 & 6 & 5 \\ 5 & -3 & 4 \end{vmatrix} = 0 \quad \text{and} \quad \begin{vmatrix} 5 & -3 & 4 \\ -5 & 3 & -4 \\ 6 & 8 & 2 \end{vmatrix} = 0$$

2. When evaluating determinants, if you use expansion of the minors about a row or column that contains one or more zeros, the determinants will usually be easier to evaluate.

As with second-order determinants, when determinant $D = 0$, Cramer's rule does not apply since division by 0 is not defined. You may then use a different method to solve the system. Or if you evaluate D_x, D_y, and D_z and they are all 0 when $D = 0$, the system is dependent and there are an infinite number of solutions. If either D_x, D_y, or D_z is not 0 when $D = 0$, the system is inconsistent and there is no solution.

Exercise Set 4.4

Evaluate the determinants.

1. $\begin{vmatrix} 2 & 3 \\ -4 & 5 \end{vmatrix}$ 22

2. $\begin{vmatrix} -1 & 3 \\ 5 & 6 \end{vmatrix}$ −21

3. $\begin{vmatrix} \frac{1}{2} & 3 \\ 2 & -4 \end{vmatrix}$ −8

4. $\begin{vmatrix} 5 & -\frac{2}{3} \\ -1 & 0 \end{vmatrix}$ −$\frac{2}{3}$

5. $\begin{vmatrix} 3 & 2 & 0 \\ 0 & 5 & 3 \\ -1 & 4 & 2 \end{vmatrix}$ −12

6. $\begin{vmatrix} 5 & -1 & 3 \\ 0 & 4 & 6 \\ 0 & 5 & -2 \end{vmatrix}$ −190

7. $\begin{vmatrix} 3 & 5 & 4 \\ -1 & -3 & -6 \\ 4 & 5 & 9 \end{vmatrix}$ −38

8. $\begin{vmatrix} 5 & -8 & 6 \\ 3 & 0 & 4 \\ -5 & -2 & 1 \end{vmatrix}$ 188

Solve the system of equations using determinants.

9. $x + 2y = 5$
$x - 2y = 1$ (3, 1)

10. $3x - 2y = 4$
$3x + y = -2$ (0, −2)

11. $x - 2y = -1$
$x + 3y = 9$ (3, 2)

12. $3x - y = 3$
$4x - 3y = 14$ (−1, −6)

13. $3x + 4y = 8$
$2x - 3y = 9$ $\left(\frac{60}{17}, -\frac{11}{17}\right)$

14. $6x + 3y = -4$
$9x + 5y = -6$ $\left(-\frac{2}{3}, 0\right)$

15. $2x = y + 5$
$6x + 2y = -5$ $\left(\frac{1}{2}, -4\right)$

16. $x + 5y = 3$
$2x + 10y = 6$ infinite number of solutions

17. $3x = -4y - 6$
$3y = -5x + 1$ (2, −3)

18. $5x - 5y = 3$
$x - y = -2$ no solution

19. $6.3x - 4.5y = -9.9$
$-9.1x + 3.2y = -2.2$ (2, 5)

20. $-1.1x + 8.3y = 36.5$
$3.5x + 1.6y = -4.1$ (−3, 4)

Solve the system of equations using determinants.

21. $x + y - z = -3$
$x + z = 2$
$2x - y + 2z = 3$ (−1, 1, 3)

22. $2x - y + 3z = 0$
$x + 2y - z = 5$
$2y + z = 1$ (2, 1, −1)

23. $-x + y = 1$
$y - z = 2$
$x + z = -2$ $\left(-\frac{1}{2}, \frac{1}{2}, -\frac{3}{2}\right)$

24. $-x + 2y + 3z = -1$
$-3x - 3y + z = 0$
$2x + 3y + z = 2$ (4, −3, 3)

25. $2x + 2y + 2z = 0$
$-x - 3y + 7z = 15$
$3x + y + 4z = 21$ $\left(\frac{165}{14}, -\frac{153}{14}, -\frac{6}{7}\right)$

26. $x - 2y + 3z = 4$
$2x - y + z = -5$
$x + y - z = -2$ $\left(-\frac{7}{3}, \frac{22}{3}, 7\right)$

27. $x - y + 2z = 3$
$x - y + z = 1$
$2x + y + 2z = 2$ (−1, 0, 2)

28. $2x + y - 2 = 0$
$3x + 2y + z = 3$
$x - 3y - 5z = 5$ (3, −4, 2)

29. $x + 2y + z = 1$
$x - y + z = 1$ infinite number
$2x + y + 2z = 2$ of solutions

30. $2x + y - 2z = -4$
$x + y + z = 1$
$x + y + 2z = 3$ (1, −2, 2)

31. $1.1x + 2.3y - 4.0z = -9.2$
$-2.3x + 4.6z = 6.9$
$-8.2y - 7.5z = -6.8$ (1, −1, 2)

32. $4.6y - 2.1z = 24.3$
$-5.6x + 1.8y = -5.8$
$2.8x - 4.7y - 3.1z = 7.0$ (2, 3, −5)

33. $x + y + z = 1$
$2x + 2y + 2z = 2$ infinite number
$3x + 3y + 3z = 3$ of solutions

34. $x - 2y + z = 2$
$4x - 6y + 2z = 1$
$2x - 3y + z = 0$ no solution

35. $4x - 3y + 8z = 12$
$2x - \frac{3}{2}y + 4z = 11$ no solution
$x - 5z = -10$

36. $0.2x - 0.1y - 0.3z = -0.1$
$0.2x - 0.1y + 0.1z = -0.9$
$0.1x + 0.2y - 0.4z = 1.7$

37. $0.6x - 0.4y + 0.5z = 3.1$
$0.5x + 0.2y + 0.2z = 1.3$
$0.1x + 0.1y + 0.1z = 0.2$ (3, −2, 1)

36. (−1, 5, −2)

38. A square array of numbers enclosed between two vertical bars; has 2 rows and 2 columns; has 3 rows and 3 columns.

38. Describe a determinant, a second-order determinant, and a third-order determinant.

39. Given a second-order determinant of the form $\begin{vmatrix} a_1 & b_1 \\ a_2 & b_2 \end{vmatrix}$, how will the value of the determinant change if the a's are switched with the b's, $\begin{vmatrix} b_1 & a_1 \\ b_2 & a_2 \end{vmatrix}$? Explain your answer.   It will have the opposite sign.

40. Given a second-order determinant of the form $\begin{vmatrix} a_1 & b_1 \\ a_2 & b_2 \end{vmatrix}$, how will the value of the determinant change if the a's are switched with each other and the b's are switched with each other, $\begin{vmatrix} a_2 & b_2 \\ a_1 & b_1 \end{vmatrix}$? Explain your answer.   It will have the opposite sign.

44.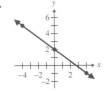

41. If $D = 0$ and D_x, D_y, and D_z also equal 0, the system is dependent.

41. Explain how you can determine if a system of equations is dependent using determinants.

42. Explain how you can determine if a system of equations is inconsistent using determinants.
If $D = 0$ and either D_x, D_y, or D_z is not equal to 0, the system is inconsistent.

CUMULATIVE REVIEW EXERCISES

[2.5] **43.** Solve the inequality $3(x - 2) < \dfrac{4}{5}(x - 4)$ and indicate the solution in interval notation. $\left(-\infty, \frac{14}{11}\right)$

Graph $3x + 4y = 8$.

[3.2] **44.** By plotting points

45. Using the x and y intercepts

[3.3] **46.** Using the slope and y intercept
Answer is below at the left.

45.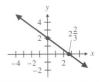

Group Activity/ Challenge Problems

46.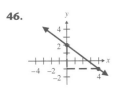

5. $x = \dfrac{c_1 b_2 - c_2 b_1}{a_1 b_2 - a_2 b_1}, \quad y = \dfrac{a_1 c_2 - a_2 c_1}{a_1 b_2 - a_2 b_1}$

Solve for the given letter.

1. $\begin{vmatrix} 4 & 6 \\ -2 & y \end{vmatrix} = 32$ 5

2. $\begin{vmatrix} b - 2 & -4 \\ b + 3 & -6 \end{vmatrix} = 14$ 5

3. $\begin{vmatrix} 3 & x & -2 \\ 0 & 5 & -6 \\ -1 & 4 & -7 \end{vmatrix} = -31$ 2

4. The evaluation of a determinant using expansion by minors can be extended to higher-order determinants. For example, when evaluating a fourth-order determinant, the pattern of $+$ and $-$ given on page 242 is extended by one row and one column. Evaluate the following determinant by using expansion by the minors of the second column.

$$\begin{vmatrix} 3 & 0 & 4 & 2 \\ 2 & 4 & 0 & 3 \\ 1 & 0 & 1 & 0 \\ 4 & 0 & 3 & 5 \end{vmatrix} \quad -28$$

5. Use the addition method to solve the following system for a) x, and b) y.

$$a_1 x + b_1 y = c_1$$
$$a_2 x + b_2 y = c_2$$

4.5 Solving Systems of Equations Using Matrices

Tape 6

1. Recognize systems of equations in triangular form.
2. Recognize matrices.
3. Solve second-order systems of linear equations.
4. Solve third-order systems of linear equations.
5. Recognize inconsistent and dependent systems.

Triangular Form ▮1▮ The following is a system of linear equations in **triangular form.**

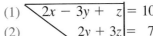

(1) $2x - 3y + z = 10$
(2) $2y + 3z = 7$
(3) $4z = 12$

A system in triangular form is easy to solve. From equation (3) we find z.

$$4z = 12$$
$$z = 3$$

By substituting $z = 3$ in equation (2), we find y.

$$2y + 3z = 7$$
$$2y + 3(3) = 7$$
$$2y = -2$$
$$y = -1$$

Now using equation (1) we find x.

$$2x - 3y + z = 10$$
$$2x - 3(-1) + 3 = 10$$
$$2x + 3 + 3 = 10$$
$$2x = 4$$
$$x = 2$$

Thus, the solution to this system is $(2, -1, 3)$.

We will use matrices to rewrite a system of equations as an *equivalent system* in triangular form.

Matrices   **2**   A **matrix** is a rectangular array of numbers within brackets. Examples of matrices are

$$\begin{bmatrix} 4 & 6 \\ 9 & -2 \end{bmatrix} \quad \begin{bmatrix} 5 & 7 & 2 \\ -1 & 3 & 4 \end{bmatrix}$$

The numbers inside the brackets are referred to as **elements** of the matrix. The plural of *matrix* is **matrices.**

The matrix on the left contains 2 rows and 2 columns and is called a 2 by 2 (2×2) matrix. The matrix on the right contains 2 rows and 3 columns and is a 2 by 3 (2×3) matrix. The number of rows is the first dimension given, and the number of columns is the second dimension given when describing the dimensions of a matrix. A **square matrix** has the same number of rows as columns. Thus, the matrix on the left is a 2×2 square matrix.

In this section we will use matrices to solve systems of linear equations. We begin by discussing solving a system with two equations and two unknowns.

The first step in solving a system of two linear equations using matrices is to write each equation in the form $ax + by = c$. The next step is to write the augmented matrix. An **augmented matrix** is made up of two smaller matrices separated by a vertical line. The numbers on the left of the vertical line are the coefficients of the variables in the system of equations, and the numbers on the right are the constants. For the system of equations

$$a_1x + b_1y = c_1$$
$$a_2x + b_2y = c_2$$

the augmented matrix is written

$$\begin{bmatrix} a_1 & b_1 & | & c_1 \\ a_2 & b_2 & | & c_2 \end{bmatrix}$$

Following is a system of equations and its augmented matrix.

System of Equations   *Augmented Matrix*

$$-x + \frac{1}{2}y = 4$$

$$-3x - 5y = -\frac{1}{2}$$

$$\begin{bmatrix} -1 & \frac{1}{2} & | & 4 \\ -3 & -5 & | & -\frac{1}{2} \end{bmatrix}$$

Notice that the bar in the augmented matrix separates the numerical coefficients from the constants. Since the matrix is just a shortened way of writing the system of equations, we can solve a linear system using matrices in a manner very similar to solving a system of equations using the addition method.

Solve Second-order Systems

3   To solve a system of equations using matrices, we use **row transformations** to rewrite the augmented matrix so that the left side of the matrix has coefficients in triangular form. We will discuss three row transformation procedures.

Procedures for Row Transformations

1. Any two rows of a matrix may be interchanged. (This is the same as interchanging any two equations in the system of equations.)

2. All the numbers in a row may be multiplied (or divided) by any nonzero real number. (This is the same as multiplying both sides of an equation by a nonzero real number.)

3. All the numbers in a row may be multiplied by any nonzero real number. These products may then be added to the corresponding numbers in any other row. (This is equivalent to eliminating a variable from a system of equations using the addition method.)

To solve a system of two linear equations using matrices, we use row transformations to obtain an augmented matrix of the form

$$\begin{bmatrix} 1 & a & | & p \\ 0 & 1 & | & q \end{bmatrix}$$

where the p and q are constants. From this type of augmented matrix we can write an equivalent system of equations in triangular form. This matrix represents the linear system

$$\begin{array}{cc} 1x + ay = p & \quad x + ay = p \\ 0x + 1y = q & \text{or} \quad\quad y = q \end{array}$$

For example,

$$\begin{bmatrix} 1 & 3 & | & 4 \\ 0 & 1 & | & 2 \end{bmatrix} \text{ represents } \begin{array}{r} x + 3y = 4 \\ y = 2 \end{array}$$

To Write a 2 × 2 Augmented Matrix in the Form $\begin{bmatrix} 1 & a & | & p \\ 0 & 1 & | & q \end{bmatrix}$

1. First use row transformations to change the element in the first column, first row to a 1.
2. Then use row transformations to change the element in the first column, second row to a 0.
3. Next use row transformations to change the element in the second column, second row to a 1.

Generally, when changing an element in the augmented matrix to 1 we use row transformation procedure 2, and when changing an element to 0 we use row transformation procedure 3. Work by columns starting from the left.

EXAMPLE 1   Solve the following system of equations using matrices:

$$2x - 3y = 10$$
$$2x + 2y = 5$$

Solution: First write the augmented matrix:

$$\begin{bmatrix} 2 & -3 & | & 10 \\ 2 & 2 & | & 5 \end{bmatrix}$$

Our goal is to obtain a matrix of the form $\begin{bmatrix} 1 & a & | & p \\ 0 & 1 & | & q \end{bmatrix}$. We begin by using row transformation procedure two to change the 2 in the first column, first row, to 1. Multiply the first row of numbers by $\frac{1}{2}$. (We abbreviate this multiplication as $\frac{1}{2}R_1$ and place it to the right of the matrix in the same row where the operation *was* performed. This may help you follow the process more clearly.)

$$\begin{bmatrix} 2\left(\frac{1}{2}\right) & -3\left(\frac{1}{2}\right) & | & 10\left(\frac{1}{2}\right) \\ 2 & 2 & | & 5 \end{bmatrix} \quad \frac{1}{2}R_1$$

This gives

$$\begin{bmatrix} 1 & -\frac{3}{2} & | & 5 \\ 2 & 2 & | & 5 \end{bmatrix}$$

The next step is to obtain 0 in the first column, second row. At present 2 is in this position. We do this by multiplying the numbers in row one by -2, and adding the products to the numbers in row 2. (This is abbreviated $-2R_1 + R_2$.)

The numbers in the first row multiplied by -2 gives

$$1(-2) \quad -\frac{3}{2}(-2) \quad 5(-2)$$

Now add these products to their respective numbers in the second row. This gives

$$\left[\begin{array}{cc|c} 1 & -\frac{3}{2} & 5 \\ 2 + 1(-2) & 2 + \left(-\frac{3}{2}\right)(-2) & 5 + 5(-2) \end{array}\right] \quad -2R_1 + R_2$$

This gives

$$\left[\begin{array}{cc|c} 1 & -\frac{3}{2} & 5 \\ 0 & 5 & -5 \end{array}\right]$$

Now obtain 1 in the second column, second row, by multiplying the second row of numbers by $\frac{1}{5}$.

$$\left[\begin{array}{cc|c} 1 & -\frac{3}{2} & 5 \\ 0\left(\frac{1}{5}\right) & 5\left(\frac{1}{5}\right) & -5\left(\frac{1}{5}\right) \end{array}\right] \quad \frac{1}{5}R_2$$

$$\left[\begin{array}{cc|c} 1 & -\frac{3}{2} & 5 \\ 0 & 1 & -1 \end{array}\right]$$

The matrix is now in the form we are seeking. The equivalent triangular system of equations is

$$x - \frac{3}{2}y = \ 5$$
$$y = -1$$

Now we can solve for x using substitution.

$$x - \frac{3}{2}y = \ 5$$
$$x - \frac{3}{2}(-1) = \ 5$$
$$x + \frac{3}{2} = \ 5$$
$$x = \frac{7}{2}$$

A check will show that the solution to the system is $\left(\frac{7}{2}, -1\right)$.

Solving Third-Order Systems

4 Now we will use matrices to solve a system of three linear equations. We use the same row transformation procedures used when solving a system of two linear equations. Our goal is to obtain an augmented matrix in the triangular form

$$\begin{bmatrix} 1 & a & b & | & p \\ 0 & 1 & c & | & q \\ 0 & 0 & 1 & | & r \end{bmatrix}$$

where p, q, and r are constants. This matrix represents the following system of equations.

$$\begin{array}{ll} 1x + ay + bz = p & \qquad x + ay + bz = p \\ 0x + 1y + cz = q \quad \text{or} & \qquad y + cz = q \\ 0x + 0y + 1z = r & \qquad \qquad z = r \end{array}$$

When constructing the augmented matrix, work by columns, from the left hand column to the right hand column. Always complete one column before moving to the next column. In each column, first obtain the 1 in the indicated position, then obtain the zeros. Example 2 illustrates this procedure.

EXAMPLE 2   Use matrices to solve the following system of equations.

$$\begin{array}{rcr} x + 2y + z & = & 0 \\ 2x - y + 2z & = & 10 \\ x + 3y - 3z & = & -14 \end{array}$$

Solution: First write the augmented matrix:

$$\begin{bmatrix} 1 & 2 & 1 & | & 0 \\ 2 & -1 & 2 & | & 10 \\ 1 & 3 & -3 & | & -14 \end{bmatrix}$$

Our first step is to use row transformations to change the first column to $\begin{smallmatrix} 1 \\ 0 \\ 0 \end{smallmatrix}$. Since the number in the first column first row is already a 1, we will work with the 2 in the first column, second row. Multiplying the numbers in the first row by -2 and adding those products to the respective numbers in the second row will result in the 2 changing to 0. The matrix is now

$$\begin{bmatrix} 1 & 2 & 1 & | & 0 \\ 0 & -5 & 0 & | & 10 \\ 1 & 3 & -3 & | & -14 \end{bmatrix} \qquad -2R_1 + R_2$$

Continuing down the first column, we now change the 1 in the third row to 0. By multiplying the numbers in the first row by -1 and adding the products to the third row, we get

$$\begin{bmatrix} 1 & 2 & 1 & | & 0 \\ 0 & -5 & 0 & | & 10 \\ 0 & 1 & -4 & | & -14 \end{bmatrix} \qquad -1R_1 + R_3$$

Now we work with the second column. We wish to change the numbers in the second column to the form $\begin{smallmatrix} a \\ 1 \\ 0 \end{smallmatrix}$, where a represents a number. Start by changing the -5 in the second row to 1 by multiplying the numbers in the second row by $-\frac{1}{5}$. This gives

$$\begin{bmatrix} 1 & 2 & 1 & | & 0 \\ 0 & 1 & 0 & | & -2 \\ 0 & 1 & -4 & | & -14 \end{bmatrix} \quad -\tfrac{1}{5}R_2$$

Continuing down the second column, we now change the 1 in the third row to 0 by multiplying the numbers in the second row by -1 and adding those products to the third row. This gives

$$\begin{bmatrix} 1 & 2 & 1 & | & 0 \\ 0 & 1 & 0 & | & -2 \\ 0 & 0 & -4 & | & -12 \end{bmatrix} \quad -1R_2 + R_3$$

Now we work with the third column. We wish to change the numbers in the third column to the form $\begin{smallmatrix} b \\ c \\ 1 \end{smallmatrix}$, where b and c represent numbers. We must change -4 in the third row to 1. We can do this by multiplying the numbers in the third row by $-\tfrac{1}{4}$. This results in the following matrix.

$$\begin{bmatrix} 1 & 2 & 1 & | & 0 \\ 0 & 1 & 0 & | & -2 \\ 0 & 0 & 1 & | & 3 \end{bmatrix} \quad -\tfrac{1}{4}R_3$$

This matrix is now in the desired form. From this matrix we obtain the following system of equations:

$$x + 2y + z = 0$$
$$y + 0z = -2$$
$$z = 3$$

From the second equation we see that $y = -2$. Now we solve for x by substituting -2 for y and 3 for z.

$$x + 2y + z = 0$$
$$x + 2(-2) + 3 = 0$$
$$x = 1$$

The solution is $(1, -2, 3)$.

Recognizing Inconsistent and Dependent Systems

5 When solving a system of two equations, if you obtain an augmented matrix in which one row of numbers on one side of the vertical line are all zeros but a zero does not appear in the same row on the other side of the vertical line, the system is inconsistent and has no solution. For example, a system of equations that yields the following augmented matrix is an inconsistent system.

$$\begin{bmatrix} 1 & 2 & | & 5 \\ 0 & 0 & | & 4 \end{bmatrix} \quad \longleftarrow \text{ inconsistent}$$

The second row of the matrix represents the equation

$$0x + 0y = 4$$

which is never true.

If you obtain a matrix in which a 0 appears across an entire row, the system of equations is dependent. For example, a system of equations that yields the following augmented matrix is a dependent system.

$$\begin{bmatrix} 1 & -3 & | & -2 \\ 0 & 0 & | & 0 \end{bmatrix} \longleftarrow \text{dependent}$$

The second row of the matrix represents the equation

$$0x + 0y = 0$$

which is always true.

Similar rules hold for systems of equations with three equations.

$$\begin{bmatrix} 1 & 2 & 4 & | & 5 \\ 0 & 0 & 0 & | & -1 \\ 0 & 1 & -2 & | & 3 \end{bmatrix} \longleftarrow \text{inconsistent system}$$

$$\begin{bmatrix} 1 & 3 & -1 & | & 2 \\ 0 & 0 & 0 & | & 0 \\ 0 & 4 & 1 & | & -3 \end{bmatrix} \longleftarrow \text{dependent system}$$

EXAMPLE 3 Solve the system of equations using matrices.

$$x + 3y + 4z = 10$$
$$5x + 10y + 5z = 12$$
$$4x + 8y + 4z = 9$$

Solution: First write the augmented matrix:

$$\begin{bmatrix} 1 & 3 & 4 & | & 10 \\ 5 & 10 & 5 & | & 12 \\ 4 & 8 & 4 & | & 9 \end{bmatrix}$$

The first column, first row, contains a 1. Change the 5 in the first column, second row, to 0 by multiplying the numbers in the first row by -5 and adding the products to the numbers in the second row.

$$\begin{bmatrix} 1 & 3 & 4 & | & 10 \\ 0 & -5 & -15 & | & -38 \\ 4 & 8 & 4 & | & 9 \end{bmatrix} \quad -5R_1 + R_2$$

Next change the 4 in the first column, third row, to 0 by multiplying the numbers in the first row by -4 and adding the products to the numbers in the third row.

$$\begin{bmatrix} 1 & 3 & 4 & | & 10 \\ 0 & -5 & -15 & | & -38 \\ 0 & -4 & -12 & | & -31 \end{bmatrix} \quad -4R_1 + R_3$$

Now we work with the second column. We change the -5 in the second column, second row, to 1 by multiplying the numbers in the second row by $-\frac{1}{5}$.

$$\begin{bmatrix} 1 & 3 & 4 & | & 10 \\ 0 & 1 & 3 & | & \frac{38}{5} \\ 0 & -4 & -12 & | & -31 \end{bmatrix} \quad -\frac{1}{5}R_2$$

Next change the -4 in the second column, third row, to 0 by multiplying the numbers in the second row by 4 and adding the products to the third row.

inconsistent system $\longrightarrow$ $\begin{bmatrix} 1 & 3 & 4 & | & 10 \\ 0 & 1 & 3 & | & \frac{38}{5} \\ 0 & 0 & 0 & | & -\frac{3}{5} \end{bmatrix}$ $4R_2 + R_3$

Since all the numbers in the third row on the left side of the augmented matrix are zeros, and the number in the third row on the right is not a zero, this system is inconsistent and has no solution.

Exercise Set 4.5

Perform the row transformation indicated and write the new matrix.

1. $\begin{bmatrix} -4 & 2 & | & 5 \\ 3 & 5 & | & -1 \end{bmatrix}$ multiplying the numbers in the first row by $-\frac{1}{4}$ $\begin{bmatrix} 1 & -\frac{1}{2} & | & -\frac{5}{4} \\ 3 & 5 & | & -1 \end{bmatrix}$

2. $\begin{bmatrix} 10 & -3 & | & 12 \\ 5 & -8 & | & -\frac{1}{2} \end{bmatrix}$ multiplying the numbers in the second row by $-\frac{1}{8}$ $\begin{bmatrix} 10 & -3 & | & 12 \\ -\frac{5}{8} & 1 & | & \frac{1}{16} \end{bmatrix}$

3. $\begin{bmatrix} 4 & 0 & 3 & | & 8 \\ 5 & -7 & 2 & | & 14 \\ -1 & 3 & 5 & | & 12 \end{bmatrix}$ multiplying the numbers in the second row by $-\frac{1}{7}$ $\begin{bmatrix} 4 & 0 & 3 & | & 8 \\ -\frac{5}{7} & 1 & -\frac{2}{7} & | & -2 \\ -1 & 3 & 5 & | & 12 \end{bmatrix}$

4. $\begin{bmatrix} 1 & 4 & -6 & | & 10 \\ 0 & 2 & 4 & | & -4 \\ 3 & -5 & 2 & | & -6 \end{bmatrix}$ multiplying the numbers in the third row by $\frac{1}{2}$ $\begin{bmatrix} 1 & 4 & -6 & | & 10 \\ 0 & 2 & 4 & | & -4 \\ \frac{3}{2} & -\frac{5}{2} & 1 & | & -3 \end{bmatrix}$

5. $\begin{bmatrix} 1 & 3 & | & 12 \\ -4 & 8 & | & -6 \end{bmatrix}$ multiplying the numbers in the first row by 4 and adding the products to the second row $\begin{bmatrix} 1 & 3 & | & 12 \\ 0 & 20 & | & 42 \end{bmatrix}$

6. $\begin{bmatrix} 1 & 5 & | & 6 \\ \frac{1}{2} & 10 & | & -4 \end{bmatrix}$ multiplying the numbers in the first row by $-\frac{1}{2}$ and adding the products to the second row $\begin{bmatrix} 1 & 5 & | & 6 \\ 0 & \frac{15}{2} & | & -7 \end{bmatrix}$

7. $\begin{bmatrix} 1 & 0 & 8 & | & \frac{1}{4} \\ 5 & 2 & 2 & | & -2 \\ 6 & -3 & 1 & | & 0 \end{bmatrix}$ multiplying the numbers in the first row by -5 and adding the products to the second row $\begin{bmatrix} 1 & 0 & 8 & | & \frac{1}{4} \\ 0 & 2 & -38 & | & -\frac{13}{4} \\ 6 & -3 & 1 & | & 0 \end{bmatrix}$

8. $\begin{bmatrix} 1 & 2 & -1 & | & 6 \\ 0 & 1 & 5 & | & 0 \\ 0 & 0 & 2 & | & 4 \end{bmatrix}$ multiplying the numbers in the second row by -2 and adding the products to the first row $\begin{bmatrix} 1 & 0 & -11 & | & 6 \\ 0 & 1 & 5 & | & 0 \\ 0 & 0 & 2 & | & 4 \end{bmatrix}$

Solve the system using matrices.

9. $\begin{aligned} x + 3y &= 3 \\ -x + y &= -3 \end{aligned}$ $(3, 0)$

10. $\begin{aligned} 2x - y &= -6 \\ 3x + y &= 1 \end{aligned}$ $(-1, 4)$

11. $\begin{aligned} 2x - 4y &= -8 \\ 3x - 2y &= 0 \end{aligned}$ $(2, 3)$

12. $4x - y = 25$
$2x - 2y = 20$ $(5, -5)$

13. $-3x + 6y = 3$
$4x - 2y = -1$ $\left(0, \frac{1}{2}\right)$

14. $-3x + 5y = -22$
$4x - 2y = 20$ $(4, -2)$

15. $2x - 5y = -6$ dependent
$-4x + 10y = 12$ system

16. $-x - 5y = -10$
$2x - 3y = 7$ $(5, 1)$

17. $12x + 10y = -14$
$4x - 3y = -11$ $(-2, 1)$

18. $4x + 2y = -10$
$-2x + y = -7$ $\left(\frac{1}{2}, -6\right)$

19. $-3x + 6y = 5$ inconsistent
$2x - 4y = 8$ system

20. $6x - 3y = 9$
$-2x + y = -3$ dependent system

21. $9x - 8y = 4$
$-3x + 4y = -1$ $\left(\frac{2}{3}, \frac{1}{4}\right)$

22. $2x - 3y = 3$
$-3x + 9y = -3$ $\left(2, \frac{1}{3}\right)$

23. $10x = 8y + 15$
$16y = -15x - 2$ $\left(\frac{4}{5}, -\frac{7}{8}\right)$

24. $8x = 9y + 4$
$24x + 6y = 1$ $\left(\frac{1}{8}, -\frac{1}{3}\right)$

Solve the system using matrices.

25. $x + y - 3z = -1$
$2x + y - z = 3$
$-x + 2y - z = -3$ $(2, 0, 1)$

26. $x - 2y + 3z = -2$
$2x - y - z = 0$
$3x + 2y - 3z = -2$ $(-1, -1, -1)$

27. $x + 2y = 5$
$y - z = -1$
$2x - 3z = 0$ $(3, 1, 2)$

28. $4x + 3y = 10$
$2x - y = 10$
$-2x + z = -9$ $(4, -2, -1)$

29. $2x + 12y + 4z = 0$
$-4x + 9y - z = 11$
$-x - 12y + 3z = -2$ $\left(-2, \frac{1}{3}, 0\right)$

30. $3x - 5y + 2z = 8$
$-x - y - z = -3$
$3x - 2y + 4z = 10$ $(2, 0, 1)$

31. $6x - 2y + 8z = 26$
$-3x + y - 4z = -13$ dependent
$x + 2y + 3z = 14$ system

32. $6x - 10y + 2z = -4$
$2x - 15y - z = -4$
$x + 5y + 3z = 6$ $\left(-2, -\frac{1}{5}, 3\right)$

33. $4x - y + z = 4$
$-6x + 3y - 2z = -5$
$2x + 5y - z = 7$ $\left(\frac{1}{2}, 2, 4\right)$

34. $-4x + 3y - 6z = 14$
$4x + 2y - 2z = -3$
$2x - 5y - 8z = -23$ $\left(-2, 3, \frac{1}{2}\right)$

35. $2x - 4y + 3z = -12$
$3x - y + 2z = -3$ inconsistent
$-4x + 8y - 6z = 10$ system

36. $3x - 2y + z = -1$
$12x - 10y - 3z = 2$
$-9x + 8y - 4z = 5$ $\left(\frac{1}{3}, \frac{1}{2}, -1\right)$

37. $5x - 3y + 4z = 22$
$-x - 15y + 10z = -15$
$-3x + 9y - 12z = -6$ $\left(5, \frac{1}{3}, -\frac{1}{2}\right)$

38. $9x - 4y + 5z = -2$
$-9x + 5y - 10z = -1$
$9x + 3y + 10z = 1$ $\left(-\frac{5}{9}, 0, \frac{3}{5}\right)$

✏ 39. If you obtain the following augmented matrix when solving a system of equations, what would your next step be in completing the process? Explain.

$$\begin{bmatrix} 1 & 3 & | & 6 \\ 0 & -1 & | & 4 \end{bmatrix}$$

✏ 41. When solving a system of equations using matrices, how will you know if the system is **(a)** dependent; **(b)** inconsistent?

39. Change the -1 to 1.

40. Change the 2 to 0.

41. **(a)** A row contains all 0's. **(b)** A row contains all zeros on left and a non zero number on right.

✏ 40. If you obtained the following augmented matrix when solving a system of equations, what would your next step be in completing the process? Explain your answer.

$$\begin{bmatrix} 1 & 3 & 7 & | & -1 \\ 0 & -1 & 5 & | & 3 \\ 2 & 4 & 6 & | & 8 \end{bmatrix}$$

CUMULATIVE REVIEW EXERCISES   **42. (a)** $\{1, 2, 3, 4, 5, 6, 9, 10\}$ **(b)** $\{4, 6\}$

[1.2] **42.** Let $A = \{1, 2, 4, 6, 9\}$ and $B = \{3, 4, 5, 6, 10\}$. Find **(a)** $A \cup B$; **(b)** $A \cap B$.

[2.5] **43.** Indicate the inequality $-2 < x \le 4$ **(a)** on the number line; **(b)** as a solution set; **(c)** in interval notation.

[3.2] **44.** What does a graph represent?

[3.4] **45.** If $f(x) = -2x^2 + 4x - 6$, find $f(-5)$. -76

43. (a) **(b)** $\{x \mid -2 < x \le 4\}$ **(c)** $(-2, 4]$

44. the set of points whose coordinates satisfy an equation

Group Activity/ Challenge Problems

A second way to use matrices to solve systems of equations is to use row transformations to write the augmented matrix in one of the forms given below. The solutions to the systems are shown below the matrices.

$$\begin{bmatrix} 1 & 0 & | & p \\ 0 & 1 & | & q \end{bmatrix}$$

$$\begin{bmatrix} 1 & 0 & 0 & | & p \\ 0 & 1 & 0 & | & q \\ 0 & 0 & 1 & | & r \end{bmatrix}$$

$$\begin{matrix} 1x + 0y = p \\ 0x + 1y = q \end{matrix} \text{ or } \begin{matrix} x = p \\ y = q \end{matrix} \qquad \begin{matrix} 1x + 0y + 0z = p \\ 0x + 1y + 0z = q \\ 0x + 0y + 1z = r \end{matrix} \text{ or } \begin{matrix} x = p \\ y = q \\ z = r \end{matrix}$$

This process is called the Gauss–Jordan elimination method.

Solve using the Gauss–Jordan method.

1. $3x + 2y = 6$
$4x - 6y = 8$ (2, 0)

2. $3x - 4y + 6z = -17$
$2x - 6y - z = 2$
$x + 2y + 2z = -4$ $\left(1, \frac{1}{2}, -3\right)$

4.6 Solving Systems of Linear Inequalities

Tape 6

 1 Solve systems of linear inequalities.
 2 Solve linear programming problems.
 3 Solve systems of linear inequalities containing absolute value.

Linear Inequalities

 1 In Section 3.6 we showed how to graph linear inequalities in two variables. In Section 4.1 we learned how to solve systems of equations graphically. In this section we show how to solve systems of linear inequalities graphically.

> ### To Solve a System of Linear Inequalities
> Graph each inequality on the same axes. The solution is the set of points whose coordinates satisfy all the inequalities in the system.

EXAMPLE 1   Determine the solution to the system of inequalities.

$$x + y \le 6$$
$$y > 2x - 3$$

Solution: First graph the inequality $x + y \le 6$ (see Fig. 4.7 on page 258). Now on the same axes graph inequality $y > 2x - 3$ (Fig. 4.8). The solution is the set of points common to the graphs of both inequalities. It is the part of the graph that contains both shadings. The dashed line is not part of the solution, but the part of the solid line that satisfies both inequalities is.

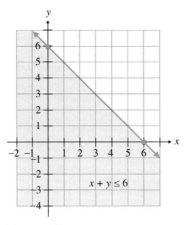

FIGURE 4.7

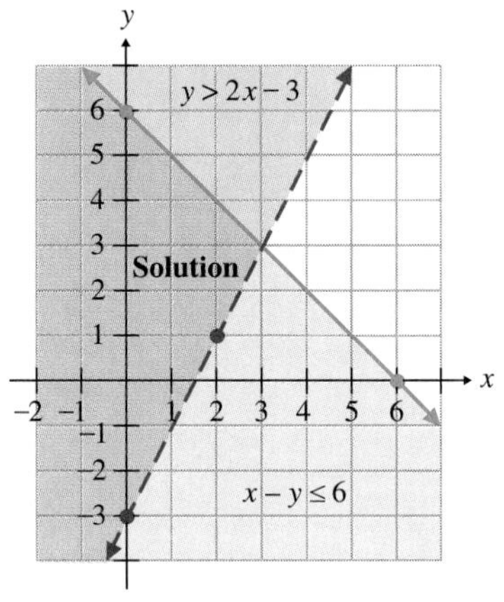

FIGURE 4.8

EXAMPLE 2 Determine the solution to the system of inequalities.

$$2x + 3y \geq 4$$
$$2x - y > -6$$

Solution: Graph $2x + 3y \geq 4$ (see Fig. 4.9). Graph $2x - y > -6$ on the same set of axes (Fig. 4.10). The solution is the part of the graph with both shadings and the part of the solid line that satisfies both inequalities.

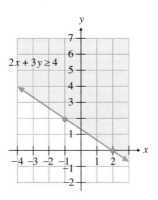

FIGURE 4.9

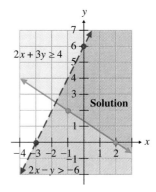

FIGURE 4.10

EXAMPLE 3 Determine the solution to the system of inequalities.

$$y < 4$$
$$x > -2$$

Solution: The solution is illustrated in Fig. 4.11.

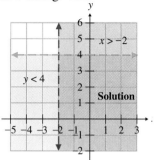

Figure 4.11

Linear Programming

2 There is a mathematical topic called **linear programming** for which you often have to graph more than two linear inequalities on the same axes. These inequalities are called **constraints.** The following two examples illustrate how to determine the solution to a system of more than two inequalities.

EXAMPLE 4   Determine the solution to the following system of inequalities.

$$x \geq 0$$
$$y \geq 0$$
$$2x + 3y \leq 12$$
$$2x + y \leq 8$$

Solution: The first two inequalities, $x \geq 0$ and $y \geq 0$, indicate that the solution must be in the first quadrant because that is the only quadrant where both x and y are positive. Figure 4.12 illustrates the graphs of the four inequalities.

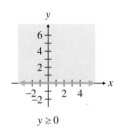 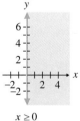   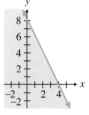

FIGURE 4.12

Figure 4.13 illustrates the graphs on the same axes and the solution to the system of inequalities. Note that every point in the shaded area and every point on the lines that form the polygonal region is part of the answer.

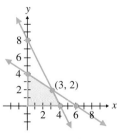

FIGURE 4.13

EXAMPLE 5   Determine the solution to the following system of inequalities.

$$x \geq 0$$
$$y \geq 0$$
$$x \leq 15$$
$$8x + 8y \leq 160$$
$$4x + 12y \leq 180$$

Solution: The first two inequalities indicate that the solution must be in the first quadrant. The third inequality indicates that x must be a value less than or equal to 15. Figure 4.14a indicates the graphs of the last three equations. Figure 4.14b indicates the solution to the system of inequalities.

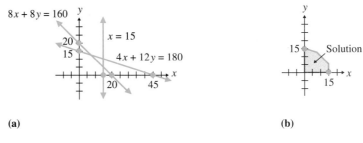

(a)   **(b)**

FIGURE 4.14

Solving Systems of Linear Inequalities with Absolute Value

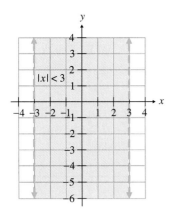

FIGURE 4.15

3 Now we will graph inequalities containing absolute value in the Cartesian coordinate system. Before we do some examples, let us recall the rules for absolute value inequalities that we learned in Section 2.6. Recall that

> If $|x| < a$ and $a > 0$, then $-a < x < a$.
> If $|x| > a$ and $a > 0$, then $x < -a$ or $x > a$.

EXAMPLE 6   Graph $|x| < 3$ in the Cartesian coordinate system.

Solution: From the rules given above we know that $|x| < 3$ means $-3 < x < 3$. Now draw dashed vertical lines through -3 and 3 and shade the area between the two (Fig. 4.15).

EXAMPLE 7   Graph $|y + 1| > 3$ in the Cartesian coordinate system.

Solution: From the rules given above we know that $|y + 1| > 3$ means $y + 1 < -3$ or $y + 1 > 3$. Solve each inequality.

$$y + 1 < -3 \quad \text{or} \quad y + 1 > 3$$
$$y < -4 \qquad\qquad y > 2$$

Now graph both inequalities and take the *union* of the two graphs. The solution is the shaded area in Fig. 4.16.

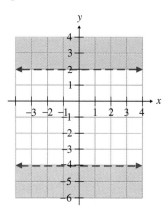

FIGURE 4.16

EXAMPLE 8 Graph the system of inequalities.

$$|x| < 3$$
$$|y + 1| > 3$$

Solution: Draw both inequalities on the same axes. Therefore, we combine the graph drawn in Example 6 with the graph drawn in Example 7 (see Fig. 4.17). The points common to both inequalities form the solution to the system.

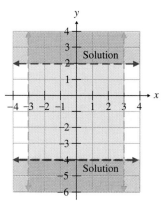

FIGURE 4.17

Exercise Set 4.6

Determine the solution to each system of inequalities.

1. $x - y > 2$
     $y < -2x + 3$

2. $y \geq 3x - 2$
     $y > -4x$

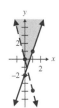

3. $y \leq x - 4$
     $y < -2x + 4$

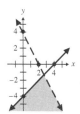

4. $2x + 3y < 6$
$4x - 2y \geq 8$

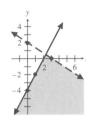

5. $y < x$
$y \geq 3x + 2$

6. $-x + 3y \geq 6$
$-2x - y > 4$

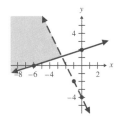

7. $4x - 2y < 6$
$y \leq -x + 4$

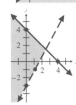

8. $y \leq 3x + 4$
$y > 2$

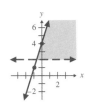

9. $-4x + 5y < 20$
$x \geq -3$

10. $3x - 4y \leq 6$
$y > -x + 4$

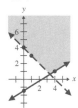

11. $x \leq 4$
$y \geq -2$

12. $x \geq 0$
$x - 3y < 6$

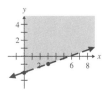

13. $5x + 2y > 10$
$3x - y > 3$

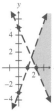

14. $3x + 2y > 8$
$x - 5y < 5$

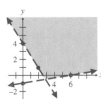

15. $-2x > y + 4$
$-x < \frac{1}{2}y - 1$

16. $y \leq 3x - 2$
$\frac{1}{3}y < x + 1$

17. $y < 3x - 4$
$6x \geq 2y + 8$

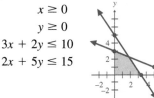

18. $\frac{1}{2}x + \frac{1}{3}y \geq 2$
$2x - 3y \leq -6$

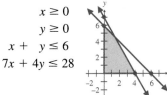

Determine the solution to each system of inequalities. Use the method discussed in Ex. 4 and 5.

19. $x \geq 0$
$y \geq 0$
$5x + 4y \leq 20$
$x + 2y \leq 6$

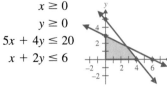

20. $x \geq 0$
$y \geq 0$
$3x + 2y \leq 10$
$2x + 5y \leq 15$

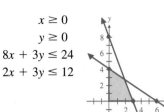

21. $x \geq 0$
$y \geq 0$
$x + y \leq 6$
$7x + 4y \leq 28$

22. $x \geq 0$
$y \geq 0$
$8x + 3y \leq 24$
$2x + 3y \leq 12$

23. $x \geq 0$
$y \geq 0$
$7x + 4y \leq 24$
$2x + 5y \leq 20$

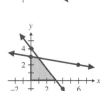

24. $x \geq 0$
$y \geq 0$
$5x + 4y \leq 16$
$x + 6y \leq 18$

25. $x \geq 0$
$y \geq 0$
$x \leq 4$
$x + y \leq 6$
$x + 2y \leq 8$

26. $x \geq 0$
$y \geq 0$
$x \leq 15$
$40x + 25y \leq 1000$
$5x + 30y \leq 900$
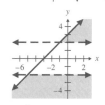

27. $x \geq 0$
$y \geq 0$
$x \leq 15$
$30x + 25y \leq 750$
$10x + 40y \leq 800$

Determine the solution to each system.

28. $|x| < 3$
$y > x$
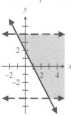

29. $|y| > 2$
$y \leq x + 3$

30. $|x| > 1$
$y \leq 3x + 2$

31. $|y| < 4$
$y \geq -2x + 2$

32. $|x| \leq 3$
$|y| > 2$

33. $|x| \geq 1$
$|y| \geq 2$
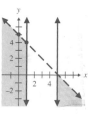

34. $|x| < 2$
$|y| \geq 3$
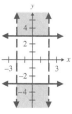

35. $|x + 2| < 3$
$|y| > 4$
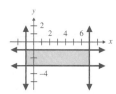

36. $|x - 3| \geq 2$
$x + y < 5$

37. $|x - 2| > 1$
$y > -2$

38. $|x - 3| \leq 4$
$|y + 2| \leq 1$
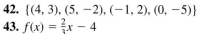

39. $|x - 3| > 4$
$|y + 1| \leq 3$

40. Is it possible for a system of linear inequalities to have no solution? Explain. Make up an example to support your answer.

40. If the boundary lines are parallel, there may be no solution.

[4.2] **CUMULATIVE REVIEW EXERCISES**   **41.** $f_2 = \dfrac{f_3 d_3 - f_1 d_1}{d_2}$

44.
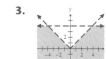

41. A formula for levers in physics is $f_1 d_1 + f_2 d_2 = f_3 d_3$. Solve this formula for f_2.

42. $\{(4, 3), (5, -2), (-1, 2), (0, -5)\}$
43. $f(x) = \frac{2}{3}x - 4$

[4.4] *In Exercises 42–44, state the range and domain of the functions:* **42.** D: $\{-1, 0, 4, 5\}$, R: $\{-5, -2, 2, 3\}$

43. D: $\mathbb{R}$, R: $\mathbb{R}$
44. D: $\mathbb{R}$, R: $\{y \mid y \geq -1\}$

Group Activity/ Challenge Problems

Determine the solution to each system.

1. $|2x - 3| - 1 > 3$
$|y - 2| < 3$

2. $\left|x + \frac{3}{2}\right| > \frac{5}{2}$
$\left|2y - \frac{1}{2}\right| \leq \frac{3}{2}$

3. $y < |x|$
$y < 4$

4. $y \geq |x - 2|$
$y \leq -|x - 2|$

1.

2.

3.

4.

Solution (2, 0)

Summary

Consistent system of equations (205): A system of equations that has a solution.

Constraints (259): The inequalities in a linear programming problem.

Dependent system of equations (205): A system of equations that has an infinite number of solutions.

Determinant (238): A square array of numbers enclosed between two vertical bars. A determinant represents a real number.

Elements of a determinant (238): The numbers in a determinant.

Inconsistent system of equations (205): A system of equations that has no solution.

Matrix (248): A rectangular array of numbers within brackets.

Second-order determinant (238): A determinant that has two rows and two columns of elements.

Solution to a system of equations (204): The ordered pair or pairs whose coordinates satisfy all equations in the system.

Square Matrix (248): A matrix that has the same number of rows and columns.

System of linear equations (204): Two or more linear equations that are solved together.

System of linear inequalities (257): Two or more linear inequalities that are solved together.

Third-order determinant (238): A determinant that has three rows and three columns of elements.

Third-order system of linear equations (218): A system of linear equations consisting of three equations with three unknowns.

IMPORTANT FACTS

Value of a second-order determinant

$$\begin{vmatrix} a_1 & b_1 \\ a_2 & b_2 \end{vmatrix} = a_1 b_2 - a_2 b_1$$

Cramer's rule: For a system of equations of the form

$$a_1 x + b_1 y = c_1$$
$$a_2 x + b_2 y = c_2$$

$$x = \frac{\begin{vmatrix} c_1 & b_1 \\ c_2 & b_2 \end{vmatrix}}{\begin{vmatrix} a_1 & b_1 \\ a_2 & b_2 \end{vmatrix}} = \frac{D_x}{D} \quad \text{and} \quad y = \frac{\begin{vmatrix} a_1 & c_1 \\ a_2 & c_2 \end{vmatrix}}{\begin{vmatrix} a_1 & b_1 \\ a_2 & b_2 \end{vmatrix}} = \frac{D_y}{D}, \quad D \neq 0$$

Value of a third-order determinant

$$
\begin{vmatrix} a_1 & b_1 & c_1 \\ a_2 & b_2 & c_2 \\ a_3 & b_3 & c_3 \end{vmatrix} = a_1 \overset{\text{Minor determinant of } a_1}{\begin{vmatrix} b_2 & c_2 \\ b_3 & c_3 \end{vmatrix}} - a_2 \overset{\text{Minor determinant of } a_2}{\begin{vmatrix} b_1 & c_1 \\ b_3 & c_3 \end{vmatrix}} + a_3 \overset{\text{Minor determinant of } a_3}{\begin{vmatrix} b_1 & c_1 \\ b_2 & c_2 \end{vmatrix}}
$$

Cramer's rule: For a system of equations of the form

$$
\begin{aligned}
a_1 x + b_1 y + c_1 z &= d_1 \\
a_2 x + b_2 y + c_2 z &= d_2 \\
a_3 x + b_3 y + c_3 z &= d_3
\end{aligned}
$$

$$
x = \frac{\begin{vmatrix} d_1 & b_1 & c_1 \\ d_2 & b_2 & c_2 \\ d_3 & b_3 & c_3 \end{vmatrix}}{\begin{vmatrix} a_1 & b_1 & c_1 \\ a_2 & b_2 & c_2 \\ a_3 & b_3 & c_3 \end{vmatrix}} = \frac{D_x}{D}, \quad y = \frac{\begin{vmatrix} a_1 & d_1 & c_1 \\ a_2 & d_2 & c_2 \\ a_3 & d_3 & c_3 \end{vmatrix}}{\begin{vmatrix} a_1 & b_1 & c_1 \\ a_2 & b_2 & c_2 \\ a_3 & b_3 & c_3 \end{vmatrix}} = \frac{D_y}{D},
$$

$$
z = \frac{\begin{vmatrix} a_1 & b_1 & d_1 \\ a_2 & b_2 & d_2 \\ a_3 & b_3 & d_3 \end{vmatrix}}{\begin{vmatrix} a_1 & b_1 & c_1 \\ a_2 & b_2 & c_2 \\ a_3 & b_3 & c_3 \end{vmatrix}} = \frac{D_z}{D}, \qquad D \neq 0
$$

Augmented matrices

The matrix $\begin{bmatrix} 1 & a & | & p \\ 0 & 1 & | & q \end{bmatrix}$ represents the system $\begin{aligned} x + a &= p \\ y &= q \end{aligned}$

The matrix $\begin{bmatrix} 1 & a & b & | & p \\ 0 & 1 & c & | & q \\ 0 & 0 & 1 & | & r \end{bmatrix}$ represents the system $\begin{aligned} x + ay + bz &= p \\ y + cz &= q \\ z &= r \end{aligned}$

Review Exercises

[4.1] *Write each equation in slope–intercept form. Without graphing or solving the system of equations, state whether the system of linear equations is consistent, inconsistent, or dependent. Also indicate whether the system has exactly one solution, no solution, or an infinite number of solutions.*

1. $x + 2y = 8$ inconsistent,
$3x + 6y = 12$ no solution

2. $y = -3x - 6$
$2x + 3y = 8$ consistent, one

3. $y = \frac{1}{2}x + 4$
$x + 2y = 8$ consistent, one

4. $6x = 4y - 8$ consistent,
$4x = 6y + 8$ one

Determine the solution to the system of equations graphically.

5. $y = x + 3$
$y = 2x + 5$ $(-2, 1)$

6. $x = -2$
$y = 3$ $(-2, 3)$

7. $2x + 2y = 8$
$2x - y = -4$ $(0, 4)$

8. $2y = 2x - 6$
$\frac{1}{2}x - \frac{1}{2}y = \frac{3}{2}$ dependent

Find the solution to the system of equations by subsitution.

9. $y = 2x + 1$
$y = 3x - 2$ $(3, 7)$

10. $y = -x + 5$
$y = 2x - 1$ $(2, 3)$

11. $y = 2x - 8$
$2x - 5y = 0$ $(5, 2)$

12. $3x + y = 17$
$\frac{1}{2}x - \frac{3}{4}y = 1$ $(5, 2)$

Find the solution to the system of equations using the addition method.

13. $x + y = 6$
$x - y = 10$ $(8, -2)$

14. $x + 2y = -3$
$2x - 2y = 6$ $(1, -2)$

15. $2x + 3y = 4$
$x + 2y = -6$ $(26, -16)$

16. $0.6x + 0.5y = 2$ $(5, -2)$
$0.25x - 0.2y = 1.65$

17. $4x - 3y = 8$
$2x + 5y = 8$ $\left(\frac{32}{13}, \frac{8}{13}\right)$

18. $-2x + 3y = 15$
$3x + 3y = 10$ $\left(-1, \frac{13}{3}\right)$

19. $x + \frac{2}{5}y = \frac{9}{5}$
$x - \frac{3}{2}y = -2$ $(1, 2)$

20. $2x + 2y = 8$
$y = 4x - 3$ $\left(\frac{7}{5}, \frac{13}{5}\right)$

21. $y = -\frac{3}{4}x + \frac{5}{2}$
$x + \frac{5}{4}y = \frac{7}{2}$ $(6, -2)$

22. $2x - 5y = 12$
$x - \frac{4}{3}y = -2$ $\left(-\frac{78}{7}, -\frac{48}{7}\right)$

23. $2x + y = 4$ infinite
$x + \frac{1}{2}y = 2$ number of solutions

24. $2x = 4y + 5$
$2y = x - 6$
no solution

[4.2] *Determine the solution to the third-order system using substitution or the addition method.*

25. $x + 2y = 12$
$4x = 8$ $\left(2, 5, \frac{34}{5}\right)$
$3x - 4y + 5z = 20$

26. $3x + 4y - 5z = 10$
$4x + 2z = 16$
$2z = -4$ $\left(5, -\frac{15}{4}, -2\right)$

27. $x + 5y + 5z = 6$
$3x + 3y - z = 10$
$x + 3y + 2z = 5$ $(1, 2, -1)$

28. $-x - y - z = -6$
$2x + 3y - z = 7$ $(3, 1, 2)$
$-3x + y + z = -6$

29. $3y - 2z = -4$
$3x - 5z = -7$ $\left(\frac{8}{3}, \frac{2}{3}, 3\right)$
$2x + y = 6$

30. $3x + 2y - 5z = 19$
$2x - 3y + 3z = -15$
$5x - 4y - 2z = -2$
$(0, 2, -3)$

31. $x - y + 3z = 1$
$-x + 2y - 2z = 1$
$x - 3y + z = 2$
no solution

32. $-2x + 2y - 3z = 6$
$4x - y + 2z = -2$
$2x + y - z = 4$
infinite number of solutions

[4.3] *In Exercises 33–38, (a) express the problem as a system of linear equations and (b) use the method of your choice to find the solution to the problem.*   **34.** 565 mph plane, 35 mph wind

33. The difference of two numbers is 18. Find the two numbers if the larger is 4 times the smaller.   6, 24

34. A plane can travel 600 miles per hour with the wind and 530 miles per hour against the wind. Find the speed of the wind and the speed of the plane in still air.

35. Curtis has a 30% acid solution and a 50% acid solution. How much of each must he mix to get 6 liters of a 40% acid solution?   3 L of each

36. The admission at an ice hockey game is $15 for adults and $11 for children. A total of 650 tickets were sold. How many tickets were sold to children and how many to adults if a total of $8790 was collected?

37. The sum of three numbers is 17. The first number is 1 more than the sum of the other two numbers, and the second number is three times the third number. Find the three numbers.   9, 6, 2

38. Mary has a total of $40,000 invested in three different savings accounts. She has some money invested in one account that gives 10% interest. The second account has $5000 less than the first account and gives 8% interest. The third account gives 6% interest. If the total annual interest that Mary receives is $3500, find the amount in each account.
$20,000 at 10%, $15,000 at 8%, $5000 at 6%
36. 410 adults, 240 children

[4.4] *Solve the system of equations using determinants.*

39. $5x + 6y = 14$
$x - 3y = 7$ $(4, -1)$

40. $3x + 5y = -2$
$5x + 3y = 2$ $(1, -1)$

41. $4x + 3y = 2$
$7x - 2y = -11$ $(-1, 2)$

42. $x + y + z = 8$
$x - y - z = 0$
$x + 2y + z = 9$ $(4, 1, 3)$

43. $x + 2y - 4z = 17$
$2x - y + z = -9$
$2x - y - 3z = -1$ $(-1, 5, -2)$

44. $y + 3z = 4$
$-x - y + 2z = 0$
$x + 2y + z = 1$ no solution

[4.5] *Solve the following systems using matrices.*

45. $-4x + 9y = 7$ $\left(-1, \frac{1}{3}\right)$
$5x + 6y = -3$

46. $2x - 3y = 4$ $\left(-\frac{5}{2}, -3\right)$
$2x = y - 2$

47. $y = 2x - 4$ infinite number
$4x = 2y + 8$ of solutions

48. $2x - y - z = 5$
  $x + 2y + 3z = -2$
  $3x - 2y + z = 2$ (2, 1, −2)

49. $3x - y + z = 2$
  $2x - 3y + 4z = 4$
  $x + 2y - 3z = -6$ no solution

50. $x + y + z = 3$
  $3x + 2y = 1$
  $y - 3z = -10$ (1, −1, 3)

[4.5] *Determine the solution to the system of inequalities.*

51. $-x + 3y > 6$
    $2x - y \leq 2$

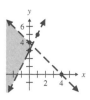

52. $5x - 2y \leq 10$
    $3x + 2y > 6$

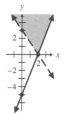

53. $y > 2x + 3$

    $y < -x + 4$

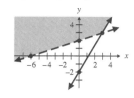

54. $x > -2y + 4$

    $y < -\frac{1}{2}x - \frac{3}{2}$

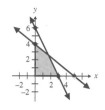

No solution

Determine the solution to the system of inequalities.

55. $x \geq 0$
      $y \geq 0$
    $x + y \leq 6$
    $4x + y \leq 8$

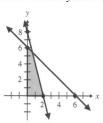

56. $x \geq 0$
      $y \geq 0$
    $2x + y \leq 6$
    $4x + 5y \leq 20$

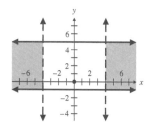

Determine the solution to the system of inequalities.

57. $|x| \leq 3$
    $|y| > 2$

58. $|x| > 4$
    $|y - 2| \leq 3$

Practice Test

Determine, without solving the system, whether the system of equations is consistent, inconsistent, or dependent. State whether the system has exactly one solution, no solution, or an infinite number of solutions.

1. $4x + 3y = -6$
  $6y = 8x + 4$ consistent, one

2. $5x + 3y = 9$
  $2y = -\frac{10}{3}x + 6$ dependent, infinite number

Solve the system of equations by the method indicated.

3. $y = 3x - 2$

  $y = -2x + 8$ (2, 4)
  graphically

4. $3x + y = 8$

  $x - y = 6$ $\left(\frac{7}{2}, -\frac{5}{2}\right)$
  substitution

5. $0.3x = 0.2y + 0.4$

  $-1.2x + 0.8y = -1.6$
  addition   dependent, infinite number

6. $\frac{3}{2}x + y = 6$
   $x - \frac{5}{2}y = -4$ $\left(\frac{44}{19}, \frac{48}{19}\right)$
   addition

7. $5x - 2y = -13$
   $2x + y = 11$
   determinants $(1, 9)$

8. $x + 5y = -2$
   $3x - y = 10$
   matrices $(3, -1)$

9. $x + y + z = 2$
   $-2x - y + z = 1$
   $x - 2y - z = 1$ $(1, -1, 2)$
   addition

10. $2x + 3y - 2z = 19$
   $x - 4y + z = -9$
   $3x - 2y - z = 7$ $(4, 3, -1)$
   any method

Express the problem as a system of linear equations and use the method of your choice to find the solution to the problem.

11. Max has cashews that sell for $7 a pound and peanuts that sell for $5.50 a pound. How much of each must he mix to get 20 pounds of a mixture that sells for $6.00 per pound? $13\frac{1}{3}$ lb peanuts, $6\frac{2}{3}$ lb cashews

13. (b)

Graph the system of inequalities and indicate its solution.

12. $3x + 2y < 9$
   $-2x + 5y \leq 10$

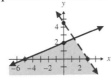

13. $|x| > 3$
   $|y| \leq 1$

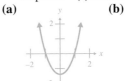

Cumulative Review Test

2. (a) $9, 1$ **(b)** $\frac{1}{2}, -4, 9, 0, -4.63, 1$ **(c)** $\frac{1}{2}, -4, 9, 0, \sqrt{3}, -4.63, 1$

3. $-|-8|, -1, \frac{5}{8}, \frac{3}{4}, |-4|, |-10|$

1. Evaluate $24 \div 4[2 - (5 - 2)]^2 - 6$. 0

2. Consider the set of numbers

$$\left\{\tfrac{1}{2}, -4, 9, 0, \sqrt{3}, -4.63, 1\right\}.$$

List the elements of the set that are (a) natural numbers, (b) rational numbers, and (c) real numbers.

3. Write the numbers from smallest to largest

$$-1, |-4|, \tfrac{3}{4}, \tfrac{5}{8}, -|-8|, |-10|,$$

Solve.

4. $-[3 - 2(x - 4)] = 3(x - 6)$ 7

5. $\frac{1}{3}x = \frac{3}{5}x + 4$ -15

6. $|4x - 3| + 2 = 10$ $\frac{11}{4}, \frac{-5}{4}$

7. Solve the formula $R = 3(a + b)$ for b. $b = \frac{R - 3a}{3}$

8. Find the solution set of the inequality below.

$$0 < \frac{3x - 2}{4} \leq 8. \quad \{x \mid \tfrac{2}{3} < x \leq \tfrac{34}{3}\}$$

9. Find the length and the midpoint of the line segment through the points $(1, 5)$ and $(-3, 2)$. $5, (-1, \tfrac{7}{2})$

10. Graph $2y = 3x - 8$

11. Write in slope–intercept form the equation of the line that is parallel to the line $2x - 3y = 8$ and passes through the point $(2, 3)$. $y = \frac{2}{3}x + \frac{5}{3}$

12. Graph the inequality $6x - 3y < 12$.

13. Determine which of the following graphs are functions. Explain.   (a) function (b) function (c) not a function

(a)

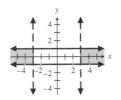

(b)

(c)

14. Graph the function $f(x) = |2x - 4| + 3$.

Solve each system of equations.

15. $3x + y = 6$
   $y = 2x + 1$ $(1, 3)$

16. $5x + 4y = 10$
   $3x + 5y = -7$ $(6, -5)$

17. $x - 2y = 0$
   $2x + z = 7$
   $y - 2z = -5$ $(2, 1, 3)$

18. If the largest angle of a triangle is nine times the measure of the smallest angle, and the middle-sized angle is $70°$ greater than the measure of the smallest angle, find the measure of the three angles. $10°, 80°, 90°$

19. Dawn speed walks at 4 miles per hour and Judy jogs at 6 miles per hour. Dawn begins walking $\frac{1}{2}$ hour before Judy starts jogging. If Judy jogs on the same path that Dawn speed walks, how long after Judy begins jogging will she catch up to Dawn?   1 hr

20. There are two different prices of seats at a rock concert. The higher-priced seats sell for $20 and the less expensive sets sell for $16. If a total of 1000 tickets are sold and the total ticket sales are $18,400, how many of each type of seat is sold?   600 at $20, 400 at $16

10.

12.

14.

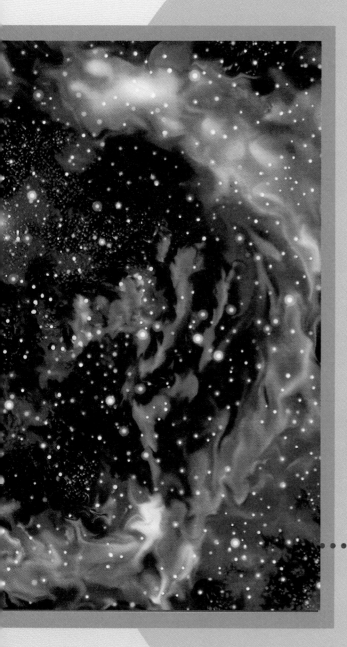

Chapter 5

Polynomials and Polynomial Functions

See Section 5.1, Exercise 110

Preview and Perspective

In this chapter we discuss the rules of exponents, polynomials, and polynomial functions. In Sections 5.1 and 5.2 we discuss exponents. These sections are important since exponents are used throughout the course. We will use these rules again with rational exponents in Section 8.2. Make sure that you understand the difference between the product and power rules and that you understand the negative exponent rule. The scientific notation presented in Section 5.1 is very important in many science courses.

Some of the concepts and material presented in this chapter you have seen before. In this chapter we expand upon your knowledge and present problems of greater difficulty. In Section 5.5 we introduce synthetic division and the remainder theorem. We will use the material presented here later in the book when we find the *x* intercepts of a graph.

Polynomial functions are introduced in Section 5.6. The material presented in this section will be very important, especially if you are planning on taking additional mathematics courses. Parabolas, that are introduced in Section 5.6, will be discussed again in Chapters 9 and 10.

5.1   Exponents and Scientific Notation

1. Learn the product rule for exponents.
2. Learn the quotient rule for exponents.
3. Learn the negative exponent rule.
4. Learn the zero exponent rule.
5. Write numbers in scientific notation.
6. Change numbers in scientific notation to decimal form.
7. Use scientific notation in calculations.

Tape 7

Before we can discuss polynomials, we need to understand exponents. In this section we will review and then expand our knowledge of exponents. Recall from Section 1.4 that

$$2^3 = \underbrace{2 \cdot 2 \cdot 2}_{\text{3 factors of 2}}$$

and

$$x^m = \underbrace{x \cdot x \cdot x \cdot x \cdot \cdots \cdot x}_{\text{\textit{m} factors of \textit{x}}}$$

The quantity 2^3 is called an **exponential expression.** In this expression, the **base** is 2 and the **exponent** is 3.

Product Rule   1   Consider the multiplication $x^3 \cdot x^5$.

$$x^3 \cdot x^5 = (x \cdot x \cdot x) \cdot (x \cdot x \cdot x \cdot x \cdot x) = x^8$$

This problem could also be simplified using the product rule for exponents.

> ## Product Rule for Exponents*
> If m and n are natural numbers and a is any real number, then
> $$a^m \cdot a^n = a^{m+n}$$

To multiply exponential expressions, maintain the common base and add the exponents.

$$x^3 \cdot x^5 = x^{3+5} = x^8$$

EXAMPLE 1   Simplify. **(a)** $3^2 \cdot 3^3$ **(b)** $x^3 \cdot x^9$ **(c)** $x \cdot x^6$

Solution:

(a) $3^2 \cdot 3^3 = 3^{2+3} = 3^5 = 243$ **(b)** $x^3 \cdot x^9 = x^{3+9} = x^{12}$

(c) $x \cdot x^6 = x^1 \cdot x^6 = x^{1+6} = x^7$

Quotient Rule   **2** Consider the division $x^7 \div x^4$.

$$\frac{x^7}{x^4} = \frac{\overset{1}{\cancel{x}} \cdot \overset{1}{\cancel{x}} \cdot \overset{1}{\cancel{x}} \cdot \overset{1}{\cancel{x}} \cdot x \cdot x \cdot x}{\underset{1}{\cancel{x}} \cdot \underset{1}{\cancel{x}} \cdot \underset{1}{\cancel{x}} \cdot \underset{1}{\cancel{x}}} = x \cdot x \cdot x = x^3$$

This problem could also be simplified using the quotient rule for exponents.

> ## Quotient Rule for Exponents
> If a is any nonzero real number and m and n are nonzero integers, then
> $$\frac{a^m}{a^n} = a^{m-n}$$

To divide expressions in exponential form, maintain the common base and subtract the exponents.

$$\frac{x^7}{x^4} = x^{7-4} = x^3$$

EXAMPLE 2   Simplify. **(a)** $\dfrac{5^4}{5^2}$ **(b)** $\dfrac{x^5}{x^2}$ **(c)** $\dfrac{y^2}{y^5}$

Solution:

(a) $\dfrac{5^4}{5^2} = 5^{4-2} = 5^2 = 25$ **(b)** $\dfrac{x^5}{x^2} = x^{5-2} = x^3$ **(c)** $\dfrac{y^2}{y^5} = y^{2-5} = y^{-3}$

Notice in Example 2(c) that the answer contains a negative exponent. Let's do part (c) again by dividing out common factors.

$$\frac{y^2}{y^5} = \frac{\overset{1}{\cancel{y}} \cdot \overset{1}{\cancel{y}}}{\underset{1}{\cancel{y}} \cdot \underset{1}{\cancel{y}} \cdot y \cdot y \cdot y} = \frac{1}{y^3}$$

By dividing out common factors from Example 2(c), we can reason that $y^{-3} = 1/y^3$.

---

* The rules given in this section and the next section also apply for rational or fractional exponents. Rational exponents will be discussed in Section 8.2. We will review these rules again at that time.

Negative Exponent Rule

③ Example 2 could also be done using the negative exponent rule.

> ### Negative Exponent Rule
> For any nonzero real number a and any whole number m,
> $$a^{-m} = \frac{1}{a^m},$$

EXAMPLE 3   Write without negative exponents.

(a) 2^{-3} (b) x^{-2} (c) $\dfrac{1}{x^{-3}}$ (d) $\dfrac{2}{y^{-5}}$

Solution:

(a) $2^{-3} = \dfrac{1}{2^3} = \dfrac{1}{8}$

(b) $x^{-2} = \dfrac{1}{x^2}$

(c) $\dfrac{1}{x^{-3}} = 1 \div x^{-3}$

$\qquad = 1 \div \dfrac{1}{x^3}$

$\qquad = \dfrac{1}{1} \cdot \dfrac{x^3}{1} = x^3$

(d) $\dfrac{2}{y^{-5}} = 2 \div y^{-5}$

$\qquad = 2 \div \dfrac{1}{y^5}$

$\qquad = \dfrac{2}{1} \cdot \dfrac{y^5}{1} = 2y^5$

Helpful Hint

Notice that **a factor can be moved from a numerator to a denominator or from a denominator to a numerator simply by changing the *sign of the exponent.***

Examples

$$\frac{1}{2^{-1}} = 2 \qquad x^{-3} = \frac{1}{x^3} \qquad 2^{-2} = \frac{1}{2^2}$$

$$6^{-1} = \frac{1}{6} \qquad \frac{1}{x^{-5}} = x^5 \qquad \frac{1}{3^{-2}} = 3^2$$

COMMON STUDENT ERROR   Study the following material carefully so that you will evaluate expressions of the form $1/a^m$ correctly.

Correct	*Incorrect*
$\dfrac{1}{3^2} = 3^{-2}$	$\dfrac{1}{3^2} = 3^2$
$\dfrac{1}{x^5} = x^{-5}$	$\dfrac{1}{x^5} = x^5$

You must understand that **an exponent applies only to the number or variable immediately preceding it unless parentheses are used.** For example, in xy^2, only the y is squared. Similarly, in the expression xy^{-1} only the y is raised to the -1 power.

$$xy^{-1} = x \cdot y^{-1} = x \cdot \frac{1}{y} = \frac{x}{y}$$

Other Examples

$$2x^{-1} = 2 \cdot \frac{1}{x} = \frac{2}{x}$$

$$-x^{-2} = -1 \cdot x^{-2} = -1 \cdot \frac{1}{x^2} = -\frac{1}{x^2}$$

$$-5x^{-4} = -5 \cdot \frac{1}{x^4} = -\frac{5}{x^4}$$

Helpful Hint

Remember that the exponent refers only to the symbol or number directly preceding it. The correct procedure for evaluating the expression, -6^2, is

$$-6^2 = -(6)^2 = -(6)(6) = -36$$

If the number -6 was to be squared, it would be written $(-6)^2$.

$$(-6)^2 = (-6)(-6) = 36$$

Notice $-6^2 \neq (-6)^2$ since $-36 \neq 36$.

Generally, we do not leave exponential expressions with negative exponents. *When we indicate that an exponential expression is to be simplified, make sure that your answer is written without negative exponents.*

EXAMPLE 4 Simplify.

(a) $-3^2x^2y^{-3}$ **(b)** $(-4)^2x^{-2}yz^{-3}$ **(c)** $4^{-2}x^{-1}y^2$ **(d)** $\dfrac{3xz^2}{y^{-4}}$

Solution:

(a) $-3^2x^2y^{-3} = -\dfrac{9x^2}{y^3}$ **(b)** $(-4)^2x^{-2}yz^{-3} = \dfrac{16y}{x^2z^3}$

(c) $4^{-2}x^{-1}y^2 = \dfrac{y^2}{4^2x^1} = \dfrac{y^2}{16x}$ **(d)** $\dfrac{3xz^2}{y^{-4}} = 3xy^4z^2$

Zero Exponent Rule   4 The last rule we will study in this section is the zero exponent rule. We introduce it now because it may be useful when we work scientific notation.

Any nonzero number divided by itself is 1. Therefore,

$$\frac{x^5}{x^5} = 1$$

By the quotient rule for exponents,

$$\frac{x^5}{x^5} = x^{5-5} = x^0$$

Since $x^0 = \dfrac{x^5}{x^5}$ and $\dfrac{x^5}{x^5} = 1$, by the transitive property of equality,

$$x^0 = 1$$

Here is the zero exponent rule.

> **Zero Exponent Rule**
>
> If a is any nonzero real number, then
>
> $$a^0 = 1$$

The zero exponent rule illustrates that any nonzero real number with an exponent of 0 equals 1. We must specify that $a \neq 0$ because 0^0 is not a real number.

EXAMPLE 5 Simplify (assume that the base is not 0).

(a) x^0 **(b)** $3x^0$ **(c)** $(5x)^0$ **(d)** $-(a + b)^0$

Solution:

(a) $x^0 = 1$ **(b)** $3x^0 = 3(1)$ **(c)** $(5x)^0 = 1$ **(d)** $-(a + b)^0 = -(1)$
$\qquad\qquad\qquad\qquad\quad = 3 \qquad\qquad\qquad\qquad\qquad\qquad\qquad\quad = -1$

Now let's look at examples that combine a number of these properties.

EXAMPLE 6 Use the rules for exponents to simplify the expressions.

(a) $4^2 \cdot 4^{-4}$ **(b)** $\dfrac{2^3}{2^{-2}}$ **(c)** $\dfrac{x^{-5}}{x^{-2}}$ **(d)** $z^3 \cdot z^{-3}$

Solution:

(a) $4^2 \cdot 4^{-4} = 4^{2 + (-4)} = 4^{-2} = \dfrac{1}{4^2} = \dfrac{1}{16}$

(b) $\dfrac{2^3}{2^{-2}} = 2^{3 - (-2)} = 2^{3 + 2} = 2^5 = 32$

(c) $\dfrac{x^{-5}}{x^{-2}} = x^{-5-(-2)} = x^{-5 + 2} = x^{-3} = \dfrac{1}{x^3}$

(d) $z^3 \cdot z^{-3} = z^{3 + (-3)} = z^0 = 1$

EXAMPLE 7 Simplify.

(a) $2^{-1} - 3 \cdot 2^{-2}$ **(b)** $-4^{-1} - 2 \cdot 3^{-1} - (4^2 \cdot 3)^0$

Solution:

(a) $2^{-1} - 3 \cdot 2^{-2} = \dfrac{1}{2} - 3 \cdot \dfrac{1}{2^2} = \dfrac{1}{2} - 3 \cdot \dfrac{1}{4}$

$$= \dfrac{1}{2} - \dfrac{3}{4} = \dfrac{2}{4} - \dfrac{3}{4} = -\dfrac{1}{4}$$

(b) $-4^{-1} - 2 \cdot 3^{-1} - (4^2 \cdot 3)^0 = -1(4^{-1}) - 2 \cdot 3^{-1} - 1$

$$= -1 \cdot \dfrac{1}{4} - 2 \cdot \dfrac{1}{3} - 1$$

$$= -\dfrac{1}{4} - \dfrac{2}{3} - 1$$

$$= -\dfrac{3}{12} - \dfrac{8}{12} - \dfrac{12}{12} = -\dfrac{23}{12}$$

EXAMPLE 8   Simplify $\dfrac{(x^{-3})(2x^6)}{x^{-5}}$.

Solution: $\dfrac{(x^{-3})(2x^6)}{x^{-5}} = \dfrac{2x^{-3+6}}{x^{-5}} = \dfrac{2x^3}{x^{-5}} = 2x^8$

The following Helpful Hint is very important. Read it carefully.

> ## Helpful Hint
>
> Notice in Example 8 that we simplifed $\dfrac{2x^3}{x^{-5}}$ to $2x^8$. This result can be obtained by rewriting x^{-5} in the denominator as x^5 in the numerator
>
> $$\dfrac{2x^3}{x^{-5}} = 2x^3 \cdot x^5 = 2x^8$$
>
> When simplifying problems of this type, we can move a factor with a negative exponent from the numerator to the denominator, or from the denominator to the numerator, *by changing the sign of the exponent.*
>
> If the same variable appears in both the numerator and denominator of an expression, we generally move the variable with the *lesser* exponent.
>
> *Examples*
>
> $\dfrac{x^{-5}}{x^{-3}} = \dfrac{1}{x^{-3} \cdot x^5} = \dfrac{1}{x^2}$ $\qquad\qquad$ $\dfrac{z^{-2}}{z^{-6}} = z^{-2} \cdot z^6 = z^4$
>
> $\dfrac{y^6}{y^{-3}} = y^6 \cdot y^3 = y^9$ $\qquad\qquad$ $\dfrac{w^{-5}}{w^4} = \dfrac{1}{w^4 \cdot w^5} = \dfrac{1}{w^9}$
>
> Now we will simplify an expression containing several variable factors.
>
> $$\dfrac{x^{-6}y^{-3}z^{-4}}{x^{-4}y^{-7}z^2} = \dfrac{y^{-3}y^7}{x^{-4}x^6z^2z^4} = \dfrac{y^4}{x^2z^6}$$

EXAMPLE 9   Simplify $\left(\dfrac{4x^3y^{-2}}{3xy^5}\right)\left(\dfrac{6x^4y^3}{4x^{-2}y^2}\right)$.

Solution: Begin by simplifying each factor and writing each factor without negative exponents. Then multiply the factors.

$$\left(\frac{4x^3y^{-2}}{3xy^5}\right)\left(\frac{6x^4y^3}{4x^{-2}y^2}\right) = \frac{4x^2}{3y^7} \cdot \frac{3x^6y}{2} = \frac{2x^8}{y^6}$$

EXAMPLE 10   Simplify $\dfrac{(6x^{-3}y^5)(5x^{-7}y^{-2})}{20x^{-4}y^5}$.

Solution: We begin by multiplying the factors in the numerator.

$$\frac{(6x^{-3}y^5)(5x^{-7}y^{-2})}{20x^{-4}y^5} = \frac{30x^{-10}y^3}{20x^{-4}y^5} = \frac{3}{2x^6y^2}$$

EXAMPLE 11   Simplify (assume that all variables used in exponents are integers).

(a) $x^{3b} \cdot x^{4b+5}$ **(b)** $\dfrac{y^{3r-2}}{y^{2r+4}}$ **(c)** $\dfrac{(x^{2p+4})(x^{p-2})}{x^{3p+1}}$

Solution:

(a)   By the product rule for exponents,
$$x^{3b} \cdot x^{4b+5} = x^{3b+(4b+5)} = x^{7b+5}$$

(b)   By the quotient rule for exponents,
$$\frac{y^{3r-2}}{y^{2r+4}} = y^{3r-2-(2r+4)} = y^{r-6}$$

(c)   Using both the product and quotient rule, we get
$$\frac{(x^{2p+4})(x^{p-2})}{x^{3p+1}} = \frac{x^{(2p+4)+(p-2)}}{x^{3p+1}}$$
$$= \frac{x^{3p+2}}{x^{3p+1}}$$
$$= x^{3p+2-(3p+1)}$$
$$= x^{3p+2-3p-1}$$
$$= x^1 \quad \text{or} \quad x$$

COMMON STUDENT ERROR A mistake sometimes made by students is to treat a

term as a *factor*. Consider the expression $\dfrac{x^6 + y^{-7}}{y^4}$.

Correct	*Incorrect*

$$\frac{x^6 + y^{-7}}{y^4} = \frac{x^6}{y^4} + \frac{y^{-7}}{y^4} = \frac{x^6}{y^4} + \frac{1}{y^{11}}$$

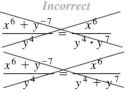

Can you explain why the procedures on the right are incorrect?

**Write Numbers
in Scientific Notation**

⑤ Scientists and engineers often deal with very large and very small numbers. For example, the frequency of an FM radio signal may be 14,200,000,000 hertz (or cycles per second) and the diameter of an atom is about 0.0000000001 meter. Because it is difficult to work with many zeros, scientists often express such numbers with exponents. For example, the number 14,200,000,000 might be written 1.42×10^{10}, and 0.0000000001 as 1×10^{-10} or simply 10^{-10}. Numbers such as 1.42×10^{10} and 1×10^{-10} (or 10^{-10}) are in a form called **scientific notation.** In scientific notation, numbers appear as a number greater than or equal to 1 and less than 10 multiplied by some power of 10. When a number written in scientific notation has no numerical coefficient showing, as in 10^5, we assume that the numerical coefficient is 1. Thus, 10^5 means 1×10^5 and 10^{-4} means 1×10^{-4}.

Examples of Numbers in Scientific Notation

$$3.2 \times 10^6 \qquad 4.176 \times 10^3 \qquad 2.64 \times 10^{-2}$$

The following will show the number 32,400 changed to scientific notation.

$$32{,}400 = 3.24 \times 10{,}000$$
$$= 3.24 \times 10^4 \quad (10{,}000 = 10^4)$$

There are four zeros in 10,000, the same number as the exponent in 10^4. The procedure for writing a number in scientific notation follows.

To Write a Number in Scientific Notation

1. Move the decimal point in the number to the right of the first nonzero digit. This gives a number greater than or equal to 1 and less than 10.

2. Count the number of places you moved the decimal point in step 1. If the original number was 10 or greater, the count is to be considered positive. If the original number was less than 1, the count is to be considered negative.

3. Multiply the number obtained in step 1 by 10 raised to the count (power) found in step 2.

EXAMPLE 12 Write the following numbers in scientific notation.

(a) 68,900   **(b)** 0.000572   **(c)** 216,000   **(d)** 0.0074

Solution:

(a) The decimal point in 68,900 is to the right of the last zero.

$$68,900. = 6.89 \times 10^4$$

The decimal point is moved four places. Since the original number is greater than 10, the exponent is positive.

(b) $0.000572 = 5.72 \times 10^{-4}$

The decimal point is moved four places. Since the original number is less than 1, the exponent is negative.

(c) $216,000. = 2.16 \times 10^5$ **(d)** $0.0074 = 7.4 \times 10^{-3}$

5 places   3 places

Change Numbers in Scientific Notation 6

> **To Convert a Number in Scientific Notation to Decimal Form**
> **1.** Observe the exponent on the base 10.
> **2. (a)** If the exponent is positive, move the decimal point in the number to the right the same number of places as the exponent. It may be necessary to add zeros to the number. This will result in a number greater than or equal to 10.
> **(b)** If the exponent is 0, the decimal point in the number does not move from its present position. Drop the factor 10^0. This will result in a number greater than or equal to 1 but less than 10.
> **(c)** If the exponent is negative, move the decimal point in the number to the left the same number of places as the exponent. It may be necessary to add zeros. This will result in a number less than 1.

EXAMPLE 13 Write each number without exponents.

(a) 2.1×10^4 **(b)** 8.73×10^{-3} **(c)** 1.45×10^8

Solution:

(a) Move the decimal point four places to the right.

$$2.1 \times \boxed{10^4} = 2.1 \times \boxed{10,000} = 21,000$$

(b) $8.73 \times 10^{-3} = 0.00873$ Move the decimal point three places to the left.

(c) $1.45 \times 10^8 = 145,000,000$ Move the decimal point eight places to the right.

Using Scientific Notation 7 We can use the rules of exponents discussed in this section when working with numbers written in scientific notation, as illustrated in the following examples.

EXAMPLE 14 Multiply (5,600,000)(0.0002) by first converting the numbers to scientific notation. Write the answer without exponents.

Solution:

$$
\begin{aligned}
(5{,}600{,}000)(0.0002) &= (5.6 \times 10^6)(2.0 \times 10^{-4}) \\
&= (5.6)(2.0) \times (10^6)(10^{-4}) \\
&= 11.2 \times 10^{6 + (-4)} \\
&= 11.2 \times 10^2 \\
&= 1120
\end{aligned}
$$

EXAMPLE 15 Divide $\dfrac{0.0000144}{0.003}$ by first converting the numbers to scientific notation. Write the answer without exponents.

Solution:

$$
\begin{aligned}
\frac{0.0000144}{0.003} &= \frac{1.44 \times 10^{-5}}{3 \times 10^{-3}} \\
&= \frac{1.44}{3} \times \frac{10^{-5}}{10^{-3}} \\
&= 0.48 \times 10^{-5 - (-3)} \\
&= 0.48 \times 10^{-5 + 3} \\
&= 0.48 \times 10^{-2} \\
&= 0.0048
\end{aligned}
$$

EXAMPLE 16 On September 30, 1992, the U.S. public debt was approximately \$4,065,000,000,000 (4 trillion 65 billion dollars). The U.S. population on that date was approximately 257,000,000.

(a) Find the average U.S. debt for every person in the United States (the per capita debt).

(b) On September 30, 1982, the U.S. debt was approximately \$1,142,000,000,000. How much larger was the debt in 1992 than 1982?

(c) How many times greater was the debt in 1992 than 1982?

Solution:

(a) To find the per capita debt, divide the public debt by the population.

$$
\frac{4{,}065{,}000{,}000{,}000}{257{,}000{,}000} = \frac{4.065 \times 10^{12}}{2.57 \times 10^{8}} \approx 1.58 \times 10^4 \approx 15{,}800
$$

Thus, the per capita debt was about \$15,800.

(b) We need to find the difference in the debt between 1992 and 1982.

$$
\begin{aligned}
4{,}065{,}000{,}000{,}000 - 1{,}142{,}000{,}000{,}000 &= 4.065 \times 10^{12} - 1.142 \times 10^{12} \\
&= (4.065 - 1.142) \times 10^{12} \\
&= 2.923 \times 10^{12} \\
&= 2{,}923{,}000{,}000{,}000
\end{aligned}
$$

(c) To find out how many times greater the 1992 public debt was, we divide as follows:

$$\frac{4{,}065{,}000{,}000{,}000}{1{,}142{,}000{,}000{,}000} = \frac{4.065 \times 10^{12}}{1.142 \times 10^{12}} \approx 3.56$$

Thus, the public debt in 1992 was about 3.56 times greater than the 1982 public debt.

Graphing Calculator Corner

SCIENTIFIC NOTATION ON CALCULATORS

Both scientific and graphing calculators display very large and very small numbers in scientific notation.

Example: On a calculator without scientific notation

$$\boxed{8000000} \ \boxed{\times} \ \boxed{600000} \ \boxed{=} \ \text{Error}$$

If your calculator can display scientific notation, it would display this calculation as follows:

Example:

$$\boxed{8000000} \ \boxed{\times} \ \boxed{600000} \ \boxed{=} \ \boxed{4.8 \quad 12}$$

In the result, $\boxed{4.8 \quad 12}$ means 4.8×10^{12}.

Example:

$$\boxed{.0000003} \ \boxed{\times} \ \boxed{.004} \ \boxed{=} \ \boxed{1.2 \quad -9}$$

In the result, $\boxed{1.2 \quad -9}$ means 1.2×10^{-9}.

To enter a number in scientific notation on a calculator you generally use either the $\boxed{\text{EE}}$ or $\boxed{\text{EXP}}$ key.

To enter 4.6×10^{-8} press $\boxed{4.6} \ \boxed{\text{EE}} \ \boxed{8} \ \boxed{^{+}/_{-}}$

The display looks like $\boxed{4.6 \quad -08}$.

On a calculator, entries in standard and scientific form may be mixed. To add $4.02 \times 10^3 + 5692$, press

$$\boxed{4.02} \ \boxed{\text{EE}} \ \boxed{3} \ \boxed{+} \ \boxed{5692} \ \boxed{=} \ \boxed{9.712 \quad 03}$$

The answer is $\boxed{9.712 \quad 03}$. Some calculators give the answer in decimal form, 9712. If a number fits on the display, some calculators convert it from scientific notation to decimal form as soon as an operation key is pressed.

EXAMPLE 17   In Example 16(a) we found that

$$\frac{4.065 \times 10^{12}}{2.57 \times 10^{8}} \approx 1.58 \times 10^{4} \approx 15{,}800$$

Show the keys to press to evaluate this quotient on a calculator.

Solution: 4.065 EE 12 ÷ 2.57 EE 8 =   1.5817   04

Some calculators will display the answer in decimal form as 15817.12062.

Exercise Set 5.1

Simplify and write the answer without negative exponents.

1. $5y^{-3}$ $\dfrac{5}{y^3}$

2. $\dfrac{1}{x^{-1}}$ x

3. $\dfrac{1}{x^{-4}}$ x^4

4. $\dfrac{3}{5y^{-2}}$ $\dfrac{3y^2}{5}$

5. $\dfrac{2x}{y^{-3}}$ $2xy^3$

6. $\dfrac{6x^4}{y^{-1}}$ $6x^4y$

7. $\dfrac{5x^{-2}y^{-3}}{2z^{-1}}$ $\dfrac{5z}{2x^2y^3}$

8. $\dfrac{4x^{-3}y}{z^4}$ $\dfrac{4y}{x^3z^4}$

9. $\dfrac{5x^{-2}y^{-3}}{z^{-4}}$ $\dfrac{5z^4}{x^2y^3}$

10. $\dfrac{10xy^5}{2z^{-3}}$ $5xy^5z^3$

11. $\dfrac{4^{-1}x^{-1}}{y}$ $\dfrac{1}{4xy}$

12. $\dfrac{5^{-1}z}{x^{-1}y^{-1}}$ $\dfrac{xyz}{5}$

Evaluate. Assume that all bases represented by variables are nonzero.

13. x^0 1

14. $5y^0$ 5

15. $-2x^0$ -2

16. $-3x^0$ -3

17. $-(a+b)^0$ -1

18. $3(a+b)^0$ 3

19. $3x^0 + 4y^0$ 7

20. $-4(x^0 - 3y^0)$ 8

Evaluate.

21. 4^{-2} 1/16

22. $(-4)^{-2}$ 1/16

23. -4^{-2} $-1/16$

24. $-(-4)^{-2}$ $-1/16$

25. 5^{-3} 1/125

26. $(-5)^{-3}$ $-1/125$

27. -5^{-3} $-1/125$

28. $-(-5^{-3})$ 1/125

29. $2 \cdot 4^{-1} - 3^{-1}$ 1/6

30. $4^{-1} - 2 \cdot 3^{-1}$ $-5/12$

31. $3 \cdot 2^{-2} + 2^{-1}$ 5/4

32. $-3^{-1} - 4 \cdot 5^{-1} - (3^2 - 5)^0$ $-32/15$

33. $2^{-2} - (-4)^{-2} \cdot 3^{-1}$ 11/48

34. $3^0 + 2^{-2} - 3^{-2} - 3 \cdot 4^{-1}$ 7/18

Simplify and write the answer without negative exponents.

51. $-10x^7z^5$ **52.** $-6y^7/xz$ **53.** $8x^7y^2/z^3$

35. $6^3 \cdot 6^{-4}$ $\dfrac{1}{6}$

36. $x^2 \cdot x^4$ x^6

37. $x^6 \cdot x^{-2}$ x^4

38. $x^{-4} \cdot x^3$ $\dfrac{1}{x}$

39. $\dfrac{3^4}{3^2}$ 9

40. $\dfrac{5^2}{5^{-2}}$ 625

41. $\dfrac{7^{-5}}{7^{-3}}$ $\dfrac{1}{49}$

42. $\dfrac{x^{-9}}{x^2}$ $\dfrac{1}{x^{11}}$

43. $\dfrac{x^{-2}}{x}$ $\dfrac{1}{x^3}$

44. $\dfrac{x^0}{x^{-3}}$ x^3

45. $\dfrac{3y^{-2}}{y^{-7}}$ $3y^5$

46. $\dfrac{x^{-3}}{x^{-5}}$ x^2

47. $2x^{-4} \cdot 6x^{-3}$ $\dfrac{12}{x^7}$

48. $(4x^2y^5)(2x^{-3}y^{-4})$ $\dfrac{8y}{x}$

49. $(-3y^{-2})(-y^3)$ $3y$

50. $(2x^{-3}y^{-4})(6x^{-4}y^7)$ $\dfrac{12y^3}{x^7}$

51. $(5x^2y^{-2}z^4)(-2x^5y^2z)$

52. $(-3x^{-4}y^6z^{-4})(2x^3yz^3)$

53. $(2x^4y^7z^9)(4x^3y^{-5}z^{-12})$

54. $\dfrac{24x^3y^2}{8xy}$ $3x^2y$

55. $\dfrac{27x^5y^{-4}}{9x^3y^2}$ $\dfrac{3x^2}{y^6}$

56. $\dfrac{6x^{-2}y^3}{2x^4y}$ $\dfrac{3y^2}{x^6}$

57. $\dfrac{9xy^{-4}}{3x^{-2}y}$ $\dfrac{3x^3}{y^5}$

58. $\dfrac{(x^{-2})(4x^2)}{x^3}$ $\dfrac{4}{x^3}$

59. $\dfrac{(2x^4)(6xy^3)}{4y^3}$ $3x^5$

60. $\dfrac{(4xy^5)(5x^4y^{-3})}{2x^5y^9}$ $\dfrac{10}{y^7}$

61. $\dfrac{(-3x^{-1}y^{-2})(2x^4y^{-3})}{6xy^4}$

62. $\dfrac{(x^4y)(3x^4y^{-3}z)}{6x^8y^2z^4}$ $\dfrac{1}{2y^4z^3}$

63. $\dfrac{(4x^{-5}y^{-2})(3x^2y^{-5})}{24x^3y^{-4}}$

64. $\left(\dfrac{5x^2y^3}{4z^3}\right)\left(\dfrac{8xy^6}{2z^3}\right)$ $\dfrac{5x^3y^9}{z^6}$

65. $\left(\dfrac{2x^5}{y^{-3}}\right)\left(\dfrac{3xy^{-2}}{z^{-3}}\right)$ $6x^6yz^3$

66. $\left(\dfrac{2x^{-2}y^{-3}}{z^3}\right)\left(\dfrac{x^3y^5}{z^{-2}}\right)$ $\dfrac{2xy^2}{z}$

67. $\left(\dfrac{3x^{-2}y^{-2}}{x^4y^{-5}}\right)\left(\dfrac{2x^3y^5}{9x^{-2}y^3}\right)$ $\dfrac{2y^5}{3x}$

61. $\dfrac{-x^2}{y^9}$ **63.** $\dfrac{1}{2x^6y^3}$

68. $\left(\dfrac{4x^{-2}y^{-2}z^3}{5x^5y^2z^5}\right)\left(\dfrac{25xy^4z^{-2}}{16x^{-3}y^5z^{-4}}\right)$ $\dfrac{5}{4x^3y^3}$

69. $\left(\dfrac{x^4y^{-3}}{x^{-4}y^{-3}z}\right)\left(\dfrac{x^3y^4z}{x^5y^{-2}z^{-3}}\right)$ $x^6y^6z^3$

70. $\dfrac{(6x^4y^{-2}z^{-1})(5x^{-5}y^4)}{9x^{-2}y^6z^{-4}}$ $\dfrac{10xz^3}{3y^4}$

Simplify. Assume that all variables represent integers.

71. $x^{4a} \cdot x^{3a+4}$ x^{7a+4} **72.** $y^{4r-2} \cdot y^{-2r+3}$ y^{2r+1} **73.** $w^{5b-2} \cdot w^{2b+3}$ w^{7b+1} **74.** $d^{5x+3} \cdot d^{-2x-3}$ d^{3x}

75. $\dfrac{x^{2w+3}}{x^{w-4}}$ x^{w+7} **76.** $\dfrac{y^{5m-1}}{y^{7m-1}}$ $\dfrac{1}{y^{2m}}$ **77.** $\dfrac{(x^{3p+5})(x^{2p-3})}{x^{4p-1}}$ x^{p+3} **78.** $\dfrac{(s^{2t-3})(s^{-t+5})}{s^{2t+4}}$ $\dfrac{1}{s^{t+2}}$

Express each number in scientific notation.

79. 3700 3.7×10^3 **80.** 900 9×10^2 **81.** 0.047 4.7×10^{-2} **82.** 0.0000462 4.62×10^{-5}

83. 19,000 1.9×10^4 **84.** 5,260,000,000 5.26×10^9 **85.** 0.00000186 1.86×10^{-6} **86.** 0.00000914 9.14×10^{-6}

Express each number without exponents.

87. 5.2×10^3 5200   **88.** 4×10^7 40,000,000   **89.** 2.13×10^{-5} 0.0000213   **90.** 9.64×10^{-7} 0.000000964

91. 3.12×10^{-1} 0.312   **92.** 4.6×10^1 46   **93.** 9×10^6 9,000,000   **94.** 7.3×10^4 73,000

Express each value without exponents.

95. $(5 \times 10^3)(3 \times 10^4)$ 150,000,000   **96.** $(1.6 \times 10^{-2})(4 \times 10^{-3})$ 0.000064   **97.** $\dfrac{8.4 \times 10^{-6}}{4 \times 10^{-4}}$ 0.021

98. $\dfrac{25 \times 10^3}{5 \times 10^{-2}}$ 500,000   **99.** $\dfrac{16,000}{0.008}$ 2,000,000   **100.** $(0.0006)(5,000,000)$ 3000

Express each value using scientific notation.

101. $(0.003)(0.00015)$ 4.5×10^{-7} **102.** $(230,000)(3000)$ 6.9×10^8 **103.** $\dfrac{1,400,000}{700}$ 2.0×10^3

104. $\dfrac{20,000}{0.0005}$ 4.0×10^7 **105.** $\dfrac{0.0000426}{200}$ 2.13×10^{-7} **106.** $\dfrac{(0.000012)(400,000)}{0.000006}$ 8.0×10^5

107. The distance to the sun is 93,000,000 miles. If a spacecraft travels at a speed of 3100 miles per hour, how long will it take for it to reach the sun?   30,000 hr

108. A computer can do one calculation in 0.0000004 second. How long would it take a computer to do a trillion (10^{12}) calculations?   400,000 sec

109. (a) Earth completes its 5.85×10^8 mile orbit around the sun in 365 days. Find the distance traveled per day.   1,602,739.7 mi

(b) Earth's speed is about 8 times faster than a bullet. Estimate the speed of a bullet in miles per hour.
$\approx$ 8347.6 mph

110. We have proof that there are at least 1 sextillion, 10^{21}, stars in the Milky Way.

(a) Write that number without exponents.

(b) How many million stars is this? Explain how you determined your answer to part (b).

111. According to U.S. Treasury, the total U.S. currency in circulation is about 3.13×10^{12} dollars. The table and circle graph that follow indicate the percent of the total circulation by denominations on March 31, 1993.

Denomination	Percent of Currency
1	1.7
5	1.9
10	3.8
20	22.5
50	11.9
100	57.7
others*	0.5

* Includes $2, $500, $1000, $5000, and $10,000 bills.

110. (a) 1,000,000,000,000,000,000,000
(b) 1,000,000,000,000,000 million

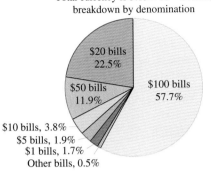

Total currency is 3.13×10^{12} dollars, breakdown by denomination

$100 bills 57.7%

$20 bills 22.5%

$50 bills 11.9%

$10 bills, 3.8%
$5 bills, 1.9%
$1 bills, 1.7%
Other bills, 0.5%

(a) Estimate the currency in circulation in $1 bills.

(b) Estimate the currency in circulation in $100 bills.

(c) Estimate the number of $1 bills in circulation.

(d) Estimate the number of $100 bills in circulation.

(e) Are there more $1 or $100 bills in circulation? Estimate the difference in the number of $1 and $100 bills in circulation.

112. The U.S. Department of Agriculture estimated the number of livestock on farms in 1890 and 1993 as follows:

	Total livestock	All cattle	Milk cows	Sheep	Hogs
1890	1.676×10^8	6.00×10^7	1.50×10^7	4.45×10^7	4.81×10^7
1993	1.8004×10^8	1.01×10^8	9.84×10^6	1.02×10^7	5.90×10^7

(a) Describe the change in the number of cattle, cows, sheep, and hogs from 1890 to 1993.

(b) How many more cattle were raised in 1993 than 1890? 4.1×10^7 or 41,000,000

(c) How many fewer milk cows were raised in 1993 than in 1890? 5.16×10^6 or 5,160,000

(d) Was there a greater change in the number of cattle or the number of milk cows from 1890 to 1993, and by how much?   cattle, 35,840,000

(e) How many times greater is the total number of livestock on farms in 1993 than in 1890? ≈ 1.07

113. If $x^{-1} = 5$, what is the value of x? Explain. $x = \frac{1}{5}$

114. If $x^{-1} = y^2$, what is x equal to? Explain. $x = \frac{1}{y^2}$

111. (a) 5.321×10^{10} or $53,210,000,000$
(b) 1.806×10^{12} or $1,806,000,000,000$ **(c)** $53,210,000,000$ **(d)** $18,060,000,000$ **(e)** $1 bills, $35,150,000,000$

115. (a) Explain the difference between the opposite of x and the reciprocal of x. For parts (b) and (c) consider

$$x^{-1} \qquad -x \qquad \frac{1}{x} \qquad \frac{1}{x^{-1}}$$

Opposite is $-x$, reciprocal is $\frac{1}{x}$

(b) Which represent (or are equal to) the *reciprocal* of x? $x^{-1}, \frac{1}{x}$

(c) Which represent the *opposite* (or *additive inverse*) of x? $-x$

116. Explain why $-2^{-2} \neq \dfrac{1}{(-2)^2}$. -2^{-2} means $-(2^{-2})$.

117. Explain how you can quickly multiply a number given in scientific notation by **(a)** 10; **(b)** 100; **(c)** 1 million. **(d)** Multiply 7.59×10^7 by 1 million. Leave your answer in scientific notation.

118. Explain how you can quickly divide a number given in scientific notation by **(a)** 10; **(b)** 100; **(c)** one million. **(d)** Divide 6.58×10^{-4} by 1 million. Leave your answer in scientific notation.

119. During a science experiment you find that the correct answer is 5.25×10^4.

(a) If you mistakenly write the answer as 4.25×10^4, by how much is your answer off?

(b) If you mistakenly write the answer as 5.25×10^5, by how much is your answer off?

(c) Which of the two errors is the more serious? Explain.   error in part (b)

117. (a) Add 1 to the exponent. **(b)** Add 2 to the exponent. **(c)** Add 6 to the exponent. **(d)** 7.59×10^{13}
118. (a) subtract 1 from the exponent. **(b)** Subtract 2 from the exponent. **(c)** subtract 6 from the exponent. **(d)** 6.58×10^{-10}
119. (a) 1.00×10^4 or 10,000   **(b)** 472,500

CUMULATIVE REVIEW EXERCISES

[1.4] **120.** Evaluate $\sqrt[3]{-125}$. -5

[2.1] **121.** Solve $-4.32 + 1.2(2x - 1.1) = 5.6x - 3.24$ -0.75

[2.3] **122.** What number when multiplied by 2 and divided by 5 equals 8?   20

[3.3] **123.** Write in slope–intercept form the equation for the line illustrated.

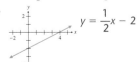

$y = \frac{1}{2}x - 2$

Group Activity/ Challenge Problems

1. **(a)** For what values of x is $x^4 > x^3$? $x < 0$ or $x > 1$
   (b) For what values of x is $x^4 < x^3$? $0 < x < 1$
   (c) For what values of x is $x^4 = x^3$? 0, 1
   (d) Why can you not say that $x^4 > x^3$? not true for $0 \le x \le 1$

2. The Richter scale measures the intensity of earthquakes. An earthquake that measures 1 on the Richter scale is barely detected by instruments. An earthquake that measures 2 is 10 times as intense as an earthquake that measures 1. An earthquake that measures 3 is 10 times as intense as one that measures 2, and $10 \cdot 10$ or 100 as intense as one that measures 1, and so on.

   (a) Use each Richter scale number as an exponent of a power of 10 and give its equivalent value. For example the Richter number 0 is equivalent to $10^0 = 1$. The Richter number 1 is equivalent to $10^1 = 10$; 2 is equivalent to $10^2 = 100$, and so on.

   (b) How many times more intense is an earthquake that measures 6 than an earthquake that measures 2 on the Richter scale? 10,000

   (c) On October 17, 1989, an earthquake measuring 6.9 on the Richter scale, with its epicenter near Santa Cruz, California, did major damage to San Francisco, California, and the surrounding area. How many times more intense was the great San Francisco earthquake than the October, 1989, earthquake? $10^{1.4} \approx 25.1$

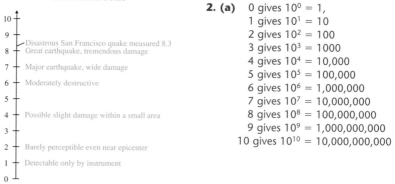

2. (a)
0 gives $10^0 = 1$,
1 gives $10^1 = 10$
2 gives $10^2 = 100$
3 gives $10^3 = 1000$
4 gives $10^4 = 10,000$
5 gives $10^5 = 100,000$
6 gives $10^6 = 1,000,000$
7 gives $10^7 = 10,000,000$
8 gives $10^8 = 100,000,000$
9 gives $10^9 = 1,000,000,000$
10 gives $10^{10} = 10,000,000,000$

3. A *light year* is the distance that light travels in 1 year.
   (a) Find the number of miles in a light year if light travels at 1.86×10^5 miles *per second*. about 5.87×10^{12} mi
   (b) If Earth is 93,000,000 miles from the sun, how long does it take for light from the sun to reach Earth? 500 sec or $8\frac{1}{3}$ min
   (c) Our galaxy, the Milky Way, is about 6.25×10^{16} miles across. If a spaceship could travel at half the speed of light, how long would it take for the craft to travel from one end of the galaxy to the other? 6.72×10^{11} sec or 21,309 yr

5.2   More on Exponents

Tape
7

1 Learn the three power rules.

**Power Rules
for Exponents**

1 In Section 5.1 we introduced a number of properties of exponents. In this section we discuss the power rules of exponents. For the sake of clarity, we will refer to the three power rules we discuss as power rules 1, 2, and 3.

Consider the problem $(x^3)^2$.

$$(x^3)^2 = x^3 \cdot x^3 = x^{3+3} = x^6$$

This problem could also be evaluated using power rule 1.

> **Power Rule for Exponents**
>
> If a is a real number and m and n are integers, then
>
> $$(a^m)^n = a^{m \cdot n} \qquad \text{Power rule 1}$$

To raise an exponential expression to a power, maintain the base and multiply the exponents.

$$(x^3)^2 = x^{3 \cdot 2} = x^6$$

EXAMPLE 1 Simplify.

(a) $(2^3)^2$ **(b)** $(x^3)^5$ **(c)** $(y^3)^{-5}$ **(d)** $(3^{-2})^3$

Solution:

(a) $(2^3)^2 = 2^{3 \cdot 2} = 2^6 = 64$

(b) $(x^3)^5 = x^{3 \cdot 5} = x^{15}$

(c) $(y^3)^{-5} = y^{3(-5)} = y^{-15} = \dfrac{1}{y^{15}}$

(d) $(3^{-2})^3 = 3^{-2(3)} = 3^{-6} = \dfrac{1}{3^6}$ or $\dfrac{1}{729}$

Helpful Hint

Students often confuse the *product rule*

$$a^m \cdot a^n = a^{m+n}$$

with the *power rule*

$$(a^m)^n = a^{m \cdot n}$$

For example, $(x^3)^2 = x^6$, not x^5.

Two additional forms of the power rule for exponents follow.

> **Power Rules for Exponents**
>
> If a and b are real numbers and m is an integer, then
>
> $$(ab)^m = a^m b^m \qquad \text{Power rule 2}$$
>
> $$\left(\frac{a}{b}\right)^m = \frac{a^m}{b^m}, \; b \neq 0 \qquad \text{Power rule 3}$$

When an expression within parentheses is raised to a power, each factor in the parentheses is raised to that power.

EXAMPLE 2   Simplify. **(a)** $(4x^3y^{-2})^3$ **(b)** $(3xy^{-3})^{-2}$ **(c)** $\left(\dfrac{3}{x^{-2}}\right)^4$

Solution:

(a) $(4x^3y^{-2})^3 = 4^3 x^9 y^{-6}$ **(b)** $(3xy^{-3})^{-2} = 3^{-2} x^{-2} y^6$ **(c)** $\left(\dfrac{3}{x^{-2}}\right)^4 = \dfrac{3^4}{x^{-8}}$

$$= 4^3 x^9 \cdot \frac{1}{y^6} \qquad\qquad = \frac{1}{3^2} \cdot \frac{1}{x^2} \cdot y^6 \qquad\qquad = \frac{81}{x^{-8}}$$

$$= \frac{64x^9}{y^6} \qquad\qquad\qquad = \frac{y^6}{9x^2} \qquad\qquad\qquad = 81x^8$$

EXAMPLE 3   Simplify. **(a)** $\left(\dfrac{2}{3}\right)^{-2}$ **(b)** $\left(\dfrac{x^2}{y}\right)^{-3}$

Solution:

(a) $\left(\dfrac{2}{3}\right)^{-2} = \dfrac{2^{-2}}{3^{-2}}$ **(b)** $\left(\dfrac{x^2}{y}\right)^{-3} = \dfrac{x^{-6}}{y^{-3}}$

$$= \frac{1}{2^2} \cdot 3^2 \qquad\qquad\qquad = \frac{1}{x^6} \cdot y^3$$

$$= \frac{9}{4} \qquad\qquad\qquad\qquad = \frac{y^3}{x^6}$$

This problem could have also been worked by using the negative exponent rule first. For example, we could have started part (a) by writing $\left(\dfrac{2}{3}\right)^{-2} = \dfrac{1}{\left(\dfrac{2}{3}\right)^2}$.

Work both parts of Example 3 now by using the negative exponent rule first.

Note that in Example 3(a), $\left(\dfrac{2}{3}\right)^{-2} = \dfrac{9}{4}$ or $\left(\dfrac{3}{2}\right)^2$. In part (b) we see that $\left(\dfrac{x^2}{y}\right)^{-3} = \dfrac{y^3}{x^6}$ or $\left(\dfrac{y}{x^2}\right)^3$. Using these examples as guides, can you guess what the simplified form of $\left(\dfrac{a}{b}\right)^{-4}$ will be without working all the steps? If you answered $\left(\dfrac{b}{a}\right)^4$ or $\dfrac{b^4}{a^4}$, you are correct.

A rational expression raised to a negative exponent can be written as a rational expression with a positive exponent by inverting the rational expression and changing the sign of the exponent.

> **Negative Exponent Rule for Fractions**
> For any a and b, $a \neq 0$, $b \neq 0$,
> $$\left(\frac{a}{b}\right)^{-m} = \left(\frac{b}{a}\right)^{m}$$

Following are some examples of the negative exponent rule for fractions.

Examples

$$\left(\frac{5}{9}\right)^{-3} = \left(\frac{9}{5}\right)^{3} \qquad \left(\frac{x^2}{y^3}\right)^{-4} = \left(\frac{y^3}{x^2}\right)^{4}$$

Whenever a fraction is raised to a negative power, you can use the property just given to rewrite the expression with a positive exponent.

EXAMPLE 4 Simplify. **(a)** $\left(\dfrac{6x^2y^4}{2x^2y}\right)^{2}$ **(b)** $\left(\dfrac{3x^4y^{-2}}{6xy^3z^{-1}}\right)^{-3}$

Solution: Exponential expressions can often be simplified in more than one order. In general, it will be easier to simplify the expression within parentheses before using the power rule.

(a) $\left(\dfrac{6x^2y^4}{2x^2y}\right)^{2} = (3y^3)^2 = 9y^6$

(b) $\left(\dfrac{3x^4y^{-2}}{6xy^3z^{-1}}\right)^{-3} = \left(\dfrac{x^3z}{2y^5}\right)^{-3} = \left(\dfrac{2y^5}{x^3z}\right)^{3} = \dfrac{8y^{15}}{x^9z^3}$

EXAMPLE 5 Simplify $\left(\dfrac{4x^{-2}y^5}{3x^5y^9}\right)^{-2}\left(\dfrac{2x^4y^{-7}}{5x^{-3}y^{-5}}\right)^{3}$.

Solution: First, simplify the expressions within parentheses and then write them without negative exponents. Use the power rule to simplify further.

$$\left(\frac{4x^{-2}y^5}{3x^5y^9}\right)^{-2}\left(\frac{2x^4y^{-7}}{5x^{-3}y^{-5}}\right)^{3} = \left(\frac{4}{3x^7y^4}\right)^{-2}\left(\frac{2x^7}{5y^2}\right)^{3}$$

$$= \left(\frac{3x^7y^4}{4}\right)^{2}\left(\frac{2x^7}{5y^2}\right)^{3}$$

$$= \left(\frac{9x^{14}y^8}{16}\right)\left(\frac{8x^{21}}{125y^6}\right)$$

$$= \frac{9x^{35}y^2}{250}$$

Example 6   Simplify $\dfrac{(2p^{-3}q^4)^{-2}(3p^{-4}q^5)^3}{(4p^{-5}q^4)^{-3}}$.

Solution: First, use the power rule; then simplify further.

$$\frac{(2p^{-3}q^4)^{-2}(3p^{-4}q^5)^3}{(4p^{-5}q^4)^{-3}} = \frac{(2^{-2}p^6q^{-8})(3^3p^{-12}q^{15})}{4^{-3}p^{15}q^{-12}}$$

$$= \frac{2^{-2}3^3p^{-6}q^7}{4^{-3}p^{15}q^{-12}}$$

$$= \frac{3^3 \cdot 4^3 q^{19}}{2^2 p^{21}}$$

$$= \frac{27 \cdot 64 q^{19}}{4 p^{21}}$$

$$= \frac{432 q^{19}}{p^{21}}$$

Common Student Error   The power rule of exponents states that

$$(ab)^m = a^m b^m \qquad \textit{Correct}$$

An error commonly made by students is to write

$$\cancel{(a + b)^m = a^m + b^m}$$
$$\cancel{(x + y)^2 = x^2 + y^2} \qquad \textit{Incorrect}$$

Select some numbers for x and y and show that $(x + y)^2 \neq x^2 + y^2$.

Summary of Rules of Exponents

For all real numbers a and b and all integers m and n:

Product rule $a^m \cdot a^n = a^{m+n}$

Quotient rule $\dfrac{a^m}{a^n} = a^{m-n}, \qquad a \neq 0$

Negative exponent rule $a^{-m} = \dfrac{1}{a^m}, \qquad a \neq 0$

Zero exponent rule $a^0 = 1, \qquad a \neq 0$

Power rules $\begin{cases} (a^m)^n = a^{m \cdot n} \\ (ab)^m = a^m b^m \\ \left(\dfrac{a}{b}\right)^m = \dfrac{a^m}{b^m}, \qquad b \neq 0 \end{cases}$

Exercise Set 5.2

Simplify and write the answers without negative exponents.

1. $(2^2)^3$ 64

2. $(3^2)^{-1}$ $\frac{1}{9}$

3. $(2^3)^{-2}$ $\frac{1}{64}$

4. $(x^3)^{-5}$ $\frac{1}{x^{15}}$

5. $(x^{-3})^{-2}$ x^6

6. $(-x)^2$ x^2

7. $(-x)^3$ $-x^3$

8. $(-x)^{-3}$ $\frac{-1}{x^3}$

9. $(-2x^{-2})^3$ $\frac{-8}{x^6}$

10. $3(x^4)^{-2}$ $\frac{3}{x^8}$

11. $\left(\frac{3}{5}\right)^2$ $\frac{9}{25}$

12. $\left(\frac{3}{4}\right)^{-2}$ $\frac{16}{9}$

13. $\left(\frac{2}{5}\right)^{-2}$ $\frac{25}{4}$

14. $\left(\frac{1}{2}\right)^{-3}$ 8

15. $\left(\frac{2x}{3}\right)^{-2}$ $\frac{9}{4x^2}$

16. $(-3x^2y)^4$ $81x^8y^4$

17. $(4x^2y^{-2})^2$ $\frac{16x^4}{y^4}$

18. $(5xy^3)^{-2}$ $\frac{1}{25x^2y^6}$

19. $(2x^3y)^{-3}$ $\frac{1}{8x^9y^3}$

20. $(3x^{-2}y)^{-2}$ $\frac{x^4}{9y^2}$

21. $(-4x^{-4}y^5)^{-3}$ $\frac{-x^{12}}{64y^{15}}$

22. $3(x^2y)^{-4}$ $\frac{3}{x^8y^4}$

23. $\left(\frac{6x}{y^2}\right)^2$ $\frac{36x^2}{y^4}$

24. $\left(\frac{3x^2y^4}{z}\right)^3$ $\frac{27x^6y^{12}}{z^3}$

25. $\left(\frac{2x^4y^5}{x^2}\right)^3$ $8x^6y^{15}$

26. $\left(\frac{3x^5y^6}{6x^4y^7}\right)^3$ $\frac{x^3}{8y^3}$

27. $\left(\frac{4xy}{y^3}\right)^{-3}$ $\frac{y^6}{64x^3}$

28. $\left(\frac{3x^{-2}}{y}\right)^{-2}$ $\frac{x^4y^2}{9}$

29. $\left(\frac{4x^{-2}y}{x^{-5}}\right)^3$ $64x^9y^3$

30. $\left(\frac{4x^2y}{x^{-5}}\right)^{-3}$ $\frac{1}{64x^{21}y^3}$

31. $\left(\frac{6x^2y}{3xz}\right)^{-3}$ $\frac{z^3}{8x^3y^3}$

32. $\left(\frac{5xy}{z^{-2}}\right)^3$ $125x^3y^3z^6$

33. $(2x^3y)(6x^4y^3)^2$ $72x^{11}y^7$

34. $(3x^4)^2(5x^3y^8)$ $45x^{11}y^8$

35. $(4x^2y^5)^2(6x^4y^3)$ $96x^8y^{13}$

36. $\frac{(2x^3y)^2}{z} \cdot \frac{xy^4}{z^3}$ $\frac{4x^7y^6}{z^4}$

37. $\left(\frac{3x^2y^{-2}}{y^3}\right)^2\left(\frac{xy^2}{3}\right)^{-2}$ $\frac{81x^2}{y^{14}}$

38. $\left(\frac{x^2y}{3}\right)^2\left(\frac{9x^2y^{-4}}{x^4y^8}\right)$ $\frac{x^2}{y^{10}}$

39. $\left(\frac{3x^2y^5}{2x^{-1}}\right)\left(\frac{2z^3}{3xy^4}\right)^4$ $\frac{8z^{13}}{27x^2y^{11}}$

40. $\frac{(2x^3y^2)^2(3xy)^{-1}}{(x^4y^3)^2}$ $\frac{4}{3x^3y^3}$

41. $\frac{(4x^2y^{-3})^2(xy^5)^{-3}}{(6x^4y^5)^3}$ $\frac{2}{27x^{11}y^{36}}$

42. $\left(\frac{x^6y^{-2}}{x^{-2}y^3}\right)^2\left(\frac{x^{-1}y^{-3}}{x^{-4}y^2}\right)^{-3}$ x^7y^5

43. $\left(\frac{x^2y^{-3}z^4}{x^{-1}y^2z^3}\right)^{-1}\left(\frac{xy^2z}{x^{-3}y^{-7}z^3}\right)^3$ $\frac{x^9y^{32}}{z^7}$

44. $\left(\frac{3x^{-4}y^{-2}}{6xy^{-4}z^2}\right)^2\left(\frac{4x^{-1}y^{-2}z^3}{2xy^2z^{-3}}\right)^{-2}$ $\frac{y^{12}}{16x^6z^{16}}$

45. $\left(\frac{6x^4y^{-6}z^4}{2xy^{-6}z^{-2}}\right)^{-2}\left(\frac{-x^4y^3}{2z^4}\right)^{-1}$ $\frac{-2}{9x^{10}y^3z^8}$

46. $\left(\frac{-x^3y^{-1}z^{-3}}{2xy^3z^{-4}}\right)^{-1}\left(\frac{2x^3y^6z^2}{4x^{-2}y^{-4}z}\right)^{-2}$ $\frac{-8}{x^{12}y^{16}z^3}$

47. $\frac{(4x^{-1}y^{-2})^{-3}(xy^3)^2}{(3x^{-1}y^3)^2}$ $\frac{x^7y^6}{576}$

48. $\frac{(3x^{-4}y^2)^3(4x^4y^3)^{-2}}{(2x^3y^5)^3}$ $\frac{27}{128x^{29}y^{15}}$

49. $\frac{(5y^3z^{-4})^{-1}(2y^{-3}z^2)^{-1}}{(3y^3z^{-2})^{-1}}$ $\frac{3y^3}{10}$

50. $\frac{(4xyz^2)^3(2xy^2z^{-3})^2}{(3x^{-1}yz^2)^{-1}}$ $768x^4y^8z^2$

Simplify. Assume that all variables used as exponents are integers.

51. $x^{-m}(x^{3m+2})^2$ x^{5m+4}

52. $y^{3b+2} \cdot (y^{2b+4})^2$ y^{7b+10}

53. $(b^{5y-2})^y \cdot b^{5y}$ b^{5y^2+3y}

54. $\frac{x^{3a} \cdot (x^{2a})^3}{x^{4a-2}}$ x^{5a+2}

55. $\frac{(m^{-5y+2})(m^{2y+3})}{m(m^{4y+1})}$ m^{3-7y}

56. $\frac{(w^{2p+4})^2(w^{-p-8})}{w^3(w^{4p-3})}$ $\frac{1}{w^p}$

What exponents must be placed in the shaded area to make the value of the expression equal to 1? Each shaded area may represent a different exponent. Explain how you determined your answer.

57. $\left(\frac{x^2y^{-2}}{x^{-3}y^{-1}}\right)^2\left(\frac{x^\blacksquare y^3}{x^7y^\blacksquare}\right)$ x^{-3}, y^1

58. $\left(\frac{x^{-2}y^3z}{x^4y^{-2}z^{-3}}\right)^3\left(\frac{x^2y^{-2}}{x^\blacksquare y^\blacksquare z^\blacksquare}\right)$ x^{-16}, y^{13}, z^{12}

59. $\left(\frac{x^{-4}y^{-2}z^7}{x^{-2}y^5z^2}\right)^{-4}\left(\frac{x^\blacksquare y^5z^{-2}}{x^4y^\blacksquare z^\blacksquare}\right)$ x^{-4}, y^{33}, z^{-22}

60. On page 287 we state that $\left(\frac{a}{b}\right)^{-m} = \left(\frac{b}{a}\right)^m$. Starting with $\left(\frac{a}{b}\right)^{-m}$, use the negative exponent rule to show that this expression is equal to $\left(\frac{b}{a}\right)^m$.

61. (a) Is $\left(-\frac{2}{3}\right)^{-2}$ equal to $\left(\frac{2}{3}\right)^{-2}$?   yes

(b) Will $(x)^{-2}$ equal $(-x)^{-2}$ for all real numbers x except 0? Explain your answer.   yes

62. (a) Is $\left(-\frac{2}{3}\right)^{-3}$ equal to $\left(\frac{2}{3}\right)^{-3}$?   no

(b) Will $(x)^{-3}$ equal $(-x)^{-3}$ for any real number x? Explain.   no

(c) What is the relationship between $(-x)^{-3}$ and $(x)^{-3}$ for nonzero real number x? opposites, $(-x)^{-3} = -(x)^{-3}$

CUMULATIVE REVIEW EXERCISES

[4.1] *In Ex. 63–65, solve the system of equations by the method indicated.*

63. $x + 3y = 10$
$2x - 4y = -10$
substitution   (1, 3)

64. $\frac{1}{2}x + \frac{3}{2}y = 7$
$x - 2y = -6$
addition   (2, 4)

[4.3]

65. $x + y = 1$
$-3x + 2y = 12$
graphing

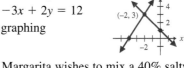
(-2, 3)

66. Margarita wishes to mix a 40% saltwater solution with a 60% saltwater solution to make 20 liters of a 45% saltwater solution. How many liters each of the 40% solution and the 60% solution should she use?
15 L of 40% solution and 5 L of 60% solution

Group Activity/ Challenge Problems

We will learn in Section 8.2 that the rules of exponents given in this and the previous section also apply when the exponents are rational numbers. Using this information and the rules of exponents evaluate each of the following.

1. $\left(\dfrac{x^{1/2}}{x^{-1}}\right)^{3/2}$ $x^{9/4}$

2. $\left(\dfrac{x^{5/8}}{x^{1/4}}\right)^{3}$ $x^{9/8}$

3. $\left(\dfrac{x^4}{x^{-1/2}}\right)^{-1}$ $\dfrac{1}{x^{9/2}}$

4. $\left(\dfrac{x^{1/2}y^{-3/2}}{x^5y^{5/3}}\right)^{2/3}$ $\dfrac{1}{x^3y^{19/9}}$

5. $\left(\dfrac{x^{1/2}y^4}{x^{-3}y^{5/2}}\right)^{2}$ x^7y^3

6. $\left(\dfrac{x^{9/5}}{y^{1/3}}\right)^{2/3}\left(\dfrac{x^{-1}y^{2/3}}{y^{-1/3}}\right)^{2}$ $\dfrac{y^{16/9}}{x^{4/5}}$

Fill in the shaded areas with the exponents that make the statement true. Each shaded area may represent a different exponent.

7. $\left(\dfrac{x^{\blacksquare}y^3}{xy^5}\right)^{2}\left(\dfrac{x^4y^3}{xy^{\blacksquare}}\right)^{3} = x^{11}y^2$ x^2, y^1

8. $\left(\dfrac{x^4y^3}{x^2y}\right)^{3}\left(\dfrac{x^{\blacksquare}y^4}{x^3y^7}\right)^{2} = \dfrac{x^{10}}{y^{12}}$ x^5, y^5

9. $\left(\dfrac{x^2y}{x^4y^3}\right)^{3}\left(\dfrac{x^{\blacksquare}y^5}{xy^7}\right)^{\blacksquare} = \dfrac{1}{y^{10}}$ $x^4, \left(\dfrac{x^4y^5}{xy^7}\right)^{2}$

10. $\left(\dfrac{x^{-2}y^3}{x^{-1}y}\right)\left(\dfrac{x^{\blacksquare}y^2}{xy^{-4}}\right)^{2} = x^5y^{12}$ x^4, y^3

5.3 Addition, Subtraction, and Multiplication of Polynomials

Tape 8

1. Identify and find the degree of a polynomial.
2. Add and subtract polynomials.
3. Multiply a monomial by a polynomial.
4. Multiply a binomial by a binomial.
5. Multiply a polynomial by a polynomial.
6. Learn two special formulas.

Degree of a Polynomial

1. A **polynomial** is a finite sum of terms in which all variables have whole-number exponents and no variable appears in a denominator. Recall that whole numbers belong to the set {0, 1, 2, 3, . . .}.

A polynomial with terms of the form ax^n is called a **polynomial in x.**

Examples of Polynomials in x	*Not Polynomials*
$5x$	$x^{1/2}$ (fractional exponent)
$6x^2 - \dfrac{1}{2}x + 4$	$2x^{-1}$ (negative exponent)

Polynomials can have more than one variable, as follows.

Examples of Polynomials in x and y

$$3xy - 6x^2y$$
$$4x^2y - 3xy^2 + 5$$

A polynomial of only one term is called a **monomial.** A **binomial** is a two-term polynomial, and a **trinomial** is a three-term polynomial. Polynomials containing more than three terms are not given special names. *Poly* is a prefix meaning *many.*

Examples of Monomials	*Examples of Binomials*	*Examples of Trinomials*
4	$x + 4$	$x^2 - 2x + 1$
$6x$	$x^2 - 6x$	$6x^2 + 3xy - 2y^2$
$\dfrac{1}{5}xyz^3$	$x^2y - y^2$	$\dfrac{1}{2}x + 3y - 6x^2y^2$

In Section 2.1 we stated that the **degree of a term** is the sum of the exponents on the variables in the term. Thus $3x^2y^3z$ is of degree 6 $(2 + 3 + 1 = 6)$.

The **degree of a polynomial** is the same as that of its highest-degree term.

Polynomial	*Degree of Polynomial*
$8x^3 + 2x^2 + 3x + 4$	third (x^3 is highest-degree term)
4	zero (4 or $4x^0$ is highest-degree term)
$6x^2 + 4x^2y^5 - 6$	seventh ($4x^2y^5$ is highest-degree term)

The polynomials $2x^3 + 4x^2 - 6x + 3$ and $4x^2 - 3xy + 5y^2$ are examples of polynomials in **descending order** of the variable x because the exponents on the variable x descend (or get lower) as the terms go from left to right. Polynomials are often written in descending order of a given variable.

EXAMPLE 1   Write each of the following polynomials in descending order of the variable x.

(a) $3x + 4x^2 - 6$ **(b)** $xy - 6x^2 + 3y^2$

Solution:

(a) $3x + 4x^2 - 6 = 4x^2 + 3x - 6$

(b) $xy - 6x^2 + 3y^2 = -6x^2 + xy + 3y^2$

Adding Polynomials ②

> **To Add Polynomials**
> Combine the like terms of the polynomials.

EXAMPLE 2   Simplify $(4x^2 - 6x + 3) + (2x^2 + 5x - 1)$.

Solution: $(4x^2 - 6x + 3) + (2x^2 + 5x - 1)$

$= 4x^2 - 6x + 3 + 2x^2 + 5x - 1$ **Remove the parentheses.**

$= \underbrace{4x^2 + 2x^2} \underbrace{- 6x + 5x} \underbrace{+ 3 - 1}$ **Rearrange terms.**

$= \quad\ 6x^2 \qquad\ -x \qquad\ +2$ **Combine like terms.**

EXAMPLE 3   Simplify $(3x^2y - 4xy + y) + (x^2y + 2xy + 3y - 2)$.

Solution: $(3x^2y - 4xy + y) + (x^2y + 2xy + 3y - 2)$

$= 3x^2y - 4xy + y + x^2y + 2xy + 3y - 2$ **Remove the parentheses.**

$= \underbrace{3x^2y + x^2y} \underbrace{- 4xy + 2xy} \underbrace{+ y + 3y} - 2$ **Rearrange terms.**

$= \quad\ 4x^2y \qquad\ -2xy \qquad\ +4y \ - 2$ **Combine like terms.**

Subtracting Polynomials

> **To Subtract Polynomials**
> 1. Remove parentheses from the polynomials being subtracted and change the sign of every term of the polynomial being subtracted.
> 2. Combine like terms.

EXAMPLE 4   Subtract $(-x^2 - 2x + 3)$ from $(x^3 + 4x + 6)$.

Solution: $(x^3 + 4x + 6) - (-x^2 - 2x + 3)$

$= x^3 + 4x + 6 + x^2 + 2x - 3$ **Remove the parentheses.**

$= x^3 + x^2 + 4x + 2x + 6 - 3$ **Rearrange terms.**

$= x^3 + x^2 + 6x + 3$ **Combine like terms.**

EXAMPLE 5   Simplify $x^2y - 4xy^2 + 5 - (2x^2y - 3y^2 + 4)$.

Solution: $x^2y - 4xy^2 + 5 - (2x^2y - 3y^2 + 4)$

$= x^2y - 4xy^2 + 5 - 2x^2y + 3y^2 - 4$ **Remove the parentheses.**

$= x^2y - 2x^2y - 4xy^2 + 3y^2 + 5 - 4$ **Rearrange terms.**

$= -x^2y - 4xy^2 + 3y^2 + 1$ **Combine like terms.**

Note that $-x^2y$ and $-4xy^2$ are not like terms since the variables have different exponents. Also, $-4xy^2$ and $3y^2$ are not like terms since $3y^2$ does not contain the variable x.

**Multiplying
a Monomial
by a Polynomial**

③ **When we multiply polynomials, each term of one polynomial must multiply each term of the other polynomial.** This results in monomials multiplying monomials. To multiply monomials, we use the rules of exponents presented earlier.

EXAMPLE 6   Multiply.   (a) $(3x^2y)(4x^5y^3)$ (b) $(-2a^4b^7)(-3a^8b^3c)$

Solution:

(a) $(3x^2y)(4x^5y^3) = 3 \cdot 4 \cdot x^2 \cdot x^5 \cdot y \cdot y^3 = 12x^{2+5}y^{1+3} = 12x^7y^4$

(b) $(-2a^4b^7)(-3a^8b^3c) = (-2)(-3)a^4 \cdot a^8 \cdot b^7 \cdot b^3 \cdot c$
$$= 6a^{4+8}b^{7+3}c$$
$$= 6a^{12}b^{10}c$$

Recall that expressions that are multiplied are called factors. In Example 6(a) both $3x^2y$ and $4x^5y^3$ are factors of the product $12x^7y^4$.

When multiplying two polynomials, when one of the polynomials is a monomial and the other is not, we can use the *expanded form of the distributive property* to perform the multiplication.

Distributive Property

$$a(b + c + d + \cdots + n) = ab + ac + ad + \cdots + an$$

EXAMPLE 7   Multiply.   (a) $3x^2\left(\dfrac{1}{6}x^3 - 5x^2\right)$ (b) $2xy(3x^2y + 6xy^2 + 4)$

Solution:

(a) $3x^2\left(\dfrac{1}{6}x^3 - 5x^2\right) = 3x^2\left(\dfrac{1}{6}x^3\right) - 3x^2(5x^2) = \dfrac{1}{2}x^5 - 15x^4$

(b) $2xy(3x^2y + 6xy^2 + 4) = (2xy)(3x^2y) + (2xy)(6xy^2) + (2xy)(4)$
$$= 6x^3y^2 + 12x^2y^3 + 8xy$$

**Multiplying a Binomial
by a Binomial**

④ Consider multiplying $(a + b)(c + d)$. Treating $(a + b)$ as a single term and using the distributive property, we get

$$(a + b)(c + d) = \boxed{(a + b)}\,c + \boxed{(a + b)}\,d$$
$$= ac + bc + ad + bd$$

When multiplying a binomial by a binomial, each term of the first binomial must be multiplied by each term of the second binomial, then all like terms are combined.

Binomials can be multiplied vertically as well as horizontally.

EXAMPLE 8   Multiply $(3x + 2)(x - 5)$

Solution: List the binomials one beneath the other. It makes no difference which one is placed on top. Then multiply each term of the top binomial by each term of the bottom binomial as shown. Remember to align like terms so that like terms can be added.

$$
\begin{array}{r}
3x + 2 \\
x - 5 \\
\hline
\end{array}
$$

$-5\,(3x + 2) \xrightarrow{} -15x - 10$ Multiply the top expression by -5.

$x\,(3x + 2) \longrightarrow 3x^2 + 2x$ Multiply the top expression by x.

$$
\begin{array}{r}
\hline
3x^2 - 13x - 10
\end{array}
$$ Add like terms in columns.

In Example 8 the binomials $3x + 2$ and $x - 5$ are *factors* of the trinomial $3x^2 - 13x - 10$.

A convenient way to multiply two binomials is called the FOIL method. To multiply two binomials using the FOIL method, list the binomials side by side. The word FOIL indicates that you multiply the **F**irst terms, **O**uter terms, **I**nner terms, and **L**ast terms of the two binomials. This procedure is illustrated in Example 9, where we multiply the same two binomials multiplied in Example 8.

EXAMPLE 9   Multiply $(3x + 2)(x - 5)$ using the FOIL method.

Solution:

$$
(3x + 2)(x - 5)
$$

$$
\underset{\text{F}}{(3x)(x)} + \underset{\text{O}}{(3x)(-5)} + \underset{\text{I}}{(2)(x)} + \underset{\text{L}}{(2)(-5)}
$$

$$
= 3x^2 - 15x + 2x - 10 = 3x^2 - 13x - 10
$$

We performed the multiplications following the FOIL order. However, any order could be followed as long as each term of one binomial is multiplied by each term of the other binomial. We used FOIL rather than OILF or any other combination of letters because it is easier to remember.

EXAMPLE 10   Multiply $(3x^2 + 6)\left(x - \dfrac{1}{2}y\right)$.

Solution:

$$
(3x^2 + 6)\left(x - \frac{1}{2}y\right) = \underset{\text{F}}{(3x^2)(x)} + \underset{\text{O}}{(3x^2)\left(-\frac{1}{2}y\right)} + \underset{\text{I}}{(6)(x)} + \underset{\text{L}}{(6)\left(-\frac{1}{2}y\right)}
$$

$$
= 3x^3 - \frac{3}{2}x^2y + 6x - 3y
$$

Multiplying a Polynomial by a Polynomial

5 When multiplying a trinomial by a binomial or a trinomial by a trinomial, we generally multiply vertically. Remember that every term of the first polynomial must be multiplied by every term of the second polynomial (see Examples 11 and 12).

EXAMPLE 11   Multiply $x^2 - 3x + 2$ by $2x^2 - 3$.

Solution: Place the longer polynomial on top; then multiply. Make sure you align like terms as you multiply so that the terms can be added more easily.

$$x^2 - 3x + 2$$
$$\underline{2x^2 \quad\quad - 3}$$

$-3\,(x^2 - 3x + 2)$ ⟶ $-3x^2 + 9x - 6$ Multiply top expression by -3.

$2x^2\,(x^2 - 3x + 2)$ → $\underline{2x^4 - 6x^3 + 4x^2}$ Multiply top expression by $2x^2$.

$\quad\quad\quad\quad\quad 2x^4 - 6x^3 + x^2 + 9x - 6$ Add like terms in columns.

EXAMPLE 12   Multiply $3x^2 + 6xy - 5y^2$ by $x + 3y$.

Solution:

$$3x^2 + 6xy - 5y^2$$
$$\underline{\quad\quad\quad\quad x + 3y}$$

$3y\,(3x^2 + 6xy - 5y^2)$ ⟶ $9x^2y + 18xy^2 - 15y^3$ Multiply top expression by $3y$.

$x\,(3x^2 + 6xy - 5y^2)$ ⟶ $\underline{3x^3 + 6x^2y - 5xy^2}$ Multiply top expression by x.

$\quad\quad\quad\quad\quad 3x^3 + 15x^2y + 13xy^2 - 15y^3$ Add like terms in columns.

Square of a Binomial   ⑥   Now we will study some special formulas. We must often square a binomial, so we have special formulas for doing so.

> ### Square of a Binomial
> $$(a + b)^2 = a^2 + 2ab + b^2$$
> $$(a - b)^2 = a^2 - 2ab + b^2$$

Examples 13 and 14 illustrate the use of the square of a binomial formula.

EXAMPLE 13   Expand. **(a)** $(3x + 5)^2$ **(b)** $(4x^2 - 3y)^2$

Solution:

(a) $(3x + 5)^2 = (3x)^2 + 2(3x)(5) + (5)^2$
$$= 9x^2 + 30x + 25$$

(b) $(4x^2 - 3y)^2 = (4x^2)^2 - 2(4x^2)(3y) + (3y)^2$
$$= 16x^4 - 24x^2y + 9y^2$$

Squaring binomials, as in Example 13, can also be done using the FOIL method.

COMMON STUDENT ERROR   Remember the middle term when squaring a binomial.

Correct	*Incorrect*
$(x + 2)^2 = (x + 2)(x + 2)$	~~$(x + 2)^2 = x^2 + 4$~~
$\quad\quad = x^2 + 4x + 4$	
$(x - 3)^2 = (x - 3)(x - 3)$	~~$(x - 3)^2 = x^2 + 9$~~
$\quad\quad = x^2 - 6x + 9$	

EXAMPLE 14 Expand $[x + (y - 1)]^2$.

Solution: This problem looks more complicated than the previous examples, but it is worked the same way as the other squares of a binomial. Treat x as the first term and $(y - 1)$ as the second term. Use the formula twice.

$$[x + (y - 1)]^2 = (x)^2 + 2(x)(y - 1) + (y - 1)^2$$
$$= x^2 + (2x)(y - 1) + y^2 - 2y + 1$$
$$= x^2 + 2xy - 2x + y^2 - 2y + 1$$

None of the six terms are like terms, so no terms can be combined. Note that $(y - 1)^2$ is also the square of a binomial and was expanded as such.

EXAMPLE 15 Use the FOIL method to multiply $(x + 6)(x - 6)$.

Solution: $(x + 6)(x - 6) = x^2 - 6x + 6x - 36 = x^2 - 36$

Difference of Two Squares

Note in Example 15 that the outer and inner terms add to 0. By examining Example 15, we see that the product of the sum and difference of the same two terms is the difference of the squares of the two terms.

> **Product of the Sum and Difference of the Same Two Terms**
>
> $$(a + b)(a - b) = a^2 - b^2$$

To multiply two binomials that differ only in the sign between their two terms, subtract the square of the second term from the square of the first term. Note that $a^2 - b^2$ represents a **difference of two squares.**

EXAMPLE 16 Multiply $\left(3x + \dfrac{2}{5}\right)\left(3x - \dfrac{2}{5}\right)$.

Solution: $\left(3x + \dfrac{2}{5}\right)\left(3x - \dfrac{2}{5}\right) = (3x)^2 - \left(\dfrac{2}{5}\right)^2 = 9x^2 - \dfrac{4}{25}$

EXAMPLE 17 Multiply $(5x + y^3)(5x - y^3)$.

Solution: $(5x + y^3)(5x - y^3) = (5x)^2 - (y^3)^2 = 25x^2 - y^6$

EXAMPLE 18 Multiply $[4x + (3y + 2)][4x - (3y + 2)]$.

Solution: Treat $4x$ as the first term and $3y + 2$ as the second term. Then we have the sum and difference of the same two terms.

$$[4x + (3y + 2)][4x - (3y + 2)] = (4x)^2 - (3y + 2)^2$$
$$= 16x^2 - (9y^2 + 12y + 4)$$
$$= 16x^2 - 9y^2 - 12y - 4$$

Exercise Set 5.3

Indicate those expressions that are polynomials. If the polynomial has a specific name, for example, "monomial" or "binomial," give the name. If the expression is not a polynomial, so state.

1. $5y$ monomial

2. $5x^2 - 6x + 9$ trinomial

3. -10 monomial

4. $5x^{-3}$ not polynomial

5. $8x^2 - 2x + 8y^2$ trinomial

6. $3x^{1/2} + 2xy$ not polynomial

7. $-2x^2 + 5x^{-1}$ not polynomial

8. $2xy + 5y^2$ binomial

Write the polynomial in descending order of the variable x. If the polynomial is already in descending order, state so. Give the degree of each polynomial.

9. $-8 - 4x - x^2$ $-x^2 - 4x - 8$, second

10. $2x + 4 - x^2$ $-x^2 + 2x + 4$, second

11. $6y^2 + 3xy + 10x^2$ $10x^2 + 3xy + 6y^2$, second

12. $-4 + x - 3x^2 + 4x^3$ $4x^3 - 3x^2 + x - 4$, third

13. $-2x^4 + 5x^2 - 4$ in descending order, fourth

14. $5xy^2 + 3x^2y - 6 - 2x^3$ $-2x^3 + 3x^2y + 5xy^2 - 6$, third

Simplify. **25.** $-x^3 + 3x2y + 4xy2$ **32.** $5x^2 - 9x - 3$ **35.** $15y^2 - 6y + 4$ **37.** $-7x^2y + 6xy^2$

15. $(6x + 3) + (x - 5)$ $7x - 2$

16. $(6x + 3) - (4x - 2)$ $2x + 5$

17. $(x^2 - 6x + 3) - (2x + 5)$ $x^2 - 8x - 2$

18. $(x - 4) - (3x^2 - 4x + 6)$ $-3x^2 + 5x - 10$

19. $(4y^2 + 6y - 3) - (2y^2 + 6)$ $2y^2 + 6y - 9$

20. $(5x - 7) + (2x^2 + 3x + 12)$ $2x^2 + 8x + 5$

21. $(-3x + 8) + (-2x^2 - 3x - 5)$ $-2x^2 - 6x + 3$

22. $(6y^2 - 6y + 4) - (-2y^2 - y + 7)$ $8y^2 - 5y - 3$

23. $(-2x^2 + 4x - 5) - (5x^2 + 3x + 7)$ $-7x^2 + x - 12$

24. $(5x^2 - x - 1) - (-3x^2 - 2x - 5)$ $8x^2 + x + 4$

25. $(-3x^3 + 4x^2y + 3xy^2) + (2x^3 - x^2y + xy^2)$

26. $(-2xy^2 + 4) - (-7xy^2 + 12)$ $5xy^2 - 8$

27. $6x^2 - 2x - [3x - (4x^2 - 6)]$ $10x^2 - 5x - 6$

28. $3xy^2 - 2x - [-(4xy^2 + 3x) - 5xy]$ $7xy^2 + 5xy + x$

29. $5w - 6w^2 - [(3w - 2w^2) - (4w + w^2)]$ $-3w^2 + 6w$

30. $-[-(5r^2 - 3r) - (2r - 3r^2) - 2r^2]$ $4r^2 - r$

31. Subtract $(4x - 6)$ from $(3x + 5)$. $-x + 11$

32. Subtract $(-x^2 + 3x + 5)$ from $(4x^2 - 6x + 2)$.

33. Add $-2x^2 + 4x - 12$ and $-x^2 - 2x$. $-3x^2 + 2x - 12$

34. Subtract $(5x^2 - 6)$ from $(2x^2 - 4x + 8)$. $-3x^2 - 4x + 14$

35. Subtract $(-6y^2 + 3y - 4)$ from $(9y^2 - 3y)$.

36. Add $6x^2 + 3xy$ and $-2x^2 + 4xy + 3y$ $4x^2 + 7xy + 3y$

37. Subtract $(5x^2y + 8)$ from $(-2x^2y + 6xy^2 + 8)$.

38. Subtract $(6x^2y + 3xy)$ from $(2x^2y + 12xy)$. $-4x^2y + 9xy$

Multiply.

39. $(4xy)(6xy^4)$ $24x^2y^5$

40. $(-2xy^4)(3x^4y^6)$ $-6x^5y^{10}$

41. $\left(\frac{5}{9}x^2y^5\right)\left(\frac{1}{5}x^5y^3z^2\right)$ $\frac{1}{9}x^7y^8z^2$

42. $2y^3(3y^2 + 2y - 6)$ $6y^5 + 4y^4 - 12y^3$

43. $-3x^2y(-2x^4y^2 + 3xy^3 + 4)$ $6x^6y^3 - 9x^3y^4 - 12x^2y$

44. $3x^4(2xy^2 + 5x^7 - 6y)$ $6x^5y^2 + 15x^{11} - 18x^4y$

45. $\frac{2}{3}yz(3x + 4y - 9y^2)$ $2xyz + \frac{8}{3}y^2z - 6y^3z$

46. $\frac{1}{2}x^2y(4x^5y^2 + 3x - 6y^2)$ $2x^7y^3 + \frac{3}{2}x^3y - 3x^2y^3$

Multiply the binomials.

47. $(4x - 6)(3x - 5)$ $12x^2 - 38x + 30$

48. $(x - y)(x + y)$ $x^2 - y^2$

49. $(4 - x)(3 + 2x^2)$ $-2x^3 + 8x^2 - 3x + 12$

50. $\left(\frac{1}{2}x + 2y\right)\left(2x - \frac{1}{3}y\right)$ $x^2 + \frac{23}{6}xy - \frac{2}{3}y^2$

51. $\left(\frac{2}{5}x - \frac{1}{5}z\right)\left(\frac{1}{3}x + z\right)$ $\frac{2}{15}x^2 + \frac{1}{3}xz - \frac{1}{5}z^2$

52. $(3xy^2 + y)(4x - 3xy)$ $-9x^2y^3 - 3xy^2 + 12x^2y^2 + 4xy$

Multiply the polynomials. **58.** $2x^5 - 7x^4 + 20x^3 - 35x^2 + 38x - 24$

53. $(x^2 - 3x + 2)(x - 4)$ $x^3 - 7x^2 + 14x - 8$

54. $(7x - 3)(-2x^2 - 4x + 1)$ $-14x^3 - 22x^2 + 19x - 3$

55. $(x - 2)(4x^2 + 9x - 2)$ $4x^3 + x^2 - 20x + 4$

56. $(5x^3 + 4x^2 - 6x + 2)(x + 5)$ $5x^4 + 29x^3 + 14x^2 - 28x + 10$

57. $(a - 3b)(2a^2 - ab + 2b^2)$ $2a^3 - 7a^2b + 5ab^2 - 6b^3$

58. $(x^3 - 2x^2 + 5x - 6)(2x^2 - 3x + 4)$

59. $(3x - 1)^3$ $27x^3 - 27x^2 + 9x - 1$

60. $(x - 2)^3$ $x^3 - 6x^2 + 12x - 8$

Multiply using either the formula for the square of a binomial or for the product of the sum and difference of the same two terms.

61. $(2x - 1)(2x + 1)$ $4x^2 - 1$

62. $(x + 2)(x + 2)$ $x^2 + 4x + 4$

63. $(2x - 3y)^2$ $4x^2 - 12xy + 9y^2$

64. $(4x - 2y)(4x + 2y)$ $16x^2 - 4y^2$

65. $(2x + 5y)^2$ $4x^2 + 20xy + 25y^2$

66. $(3x^2 - 4y)(3x^2 + 4y)$ $9x^4 - 16y^2$

67. $(5m^2 + 2n)(5m^2 - 2n)$ $25m^4 - 4n^2$

68. $[a + (b + 2)][a - (b + 2)]$ $a^2 - b^2 - 4b - 4$

69. $[y + (4 - 2x)]^2$ $y^2 + 8y - 4xy + 16 - 16x + 4x^2$

70. $[5x + (2y + 3)]^2$ $25x^2 + 20xy + 30x + 4y^2 + 12y + 9$

71. $[4 - (x - 3y)]^2$ $16 - 8x + 24y + x^2 - 6xy + 9y^2$

72. $[(x + y) + 4]^2$ $x^2 + 2xy + y^2 + 8x + 8y + 16$

73. $[(x - 2y) - 3]^2$ $x^2 - 4xy + 4y^2 - 6x + 12y + 9$

74. $[(x - 3y) - 5]^2$ $x^2 - 6xy + 9y^2 - 10x + 30y + 25$

Multiply. **89.** $\frac{2}{5}x^3y^7 - \frac{1}{6}x^6y^5 + \frac{4}{3}x^3y^7z^5$

75. $(8r^5s^4)(-3rs^9)$ $-24r^6s^{13}$

76. $\left(\frac{1}{5}a^5c^8\right)\left(\frac{2}{3}a^4b^5c^9\right)$ $\frac{2}{15}a^9b^5c^{17}$

77. $3x(x^2 + 3x - 1)$ $3x^3 + 9x^2 - 3x$

78. $\left(\frac{1}{3}x - \frac{2}{5}y\right)\left(\frac{1}{2}x - y\right)$ $\frac{1}{6}x^2 - \frac{8}{15}xy + \frac{2}{5}y^2$

79. $(3y + 4)(2y - 3)$ $6y^2 - y - 12$

80. $-\frac{3}{5}x^2y\left(-\frac{2}{3}xy^4 + \frac{1}{9}xy + 3\right)$ $\frac{2}{5}x^3y^5 - \frac{1}{15}x^3y^2 - \frac{9}{5}x^2y$

81. $\left(2x - \frac{3}{4}\right)\left(2x + \frac{3}{4}\right)$ $4x^2 - \frac{9}{16}$

82. $(4x - 5y)^2$ $16x^2 - 40xy + 25y^2$

83. $(2x^2 - 3y)(3x^2 + 2y)$ $6x^4 - 5x^2y - 6y^2$

84. $(x + 3)(2x^2 + 4x - 3)$ $2x^3 + 10x^2 + 9x - 9$

85. $(2x + 3)(4x^2 - 5x + 2)$ $8x^3 + 2x^2 - 11x + 6$

86. $(5x + 4)(x^2 - x + 4)$ $5x^3 - x^2 + 16x + 16$

87. $(2x - 3y)(3x^2 + 4xy - 2y^2)$ $6x^3 - x^2y - 16xy^2 + 6y^3$

88. $(3w^2 + 4)(3w^2 - 4)$ $9w^4 - 16$

89. $\frac{2}{3}x^2y^4\left(\frac{3}{5}xy^3 - \frac{1}{4}x^4y + 2xy^3z^5\right)$

90. $-\frac{3}{5}xy^3z^2\left(-xy^2z^5 - 5xy + \frac{1}{6}xz^7\right)$ $\frac{3}{5}x^2y^5z^7 + 3x^2y^4z^2 - \frac{1}{10}x^2y^3z$

91. $[w + (3x + 4)][w - (3x + 4)]$ $w^2 - 9x^2 - 24x - 16$

92. $[3p + (2w - 3)][3p - (2w - 3)]$ $9p^2 - 4w^2 + 12w - 9$

93. $(a + b)(a^2 - ab + b^2)$ $a^3 + b^3$

94. $(2m + n)(3m^2 - mn + 2n^2)$ $6m^3 + m^2n + 3mn^2 + 2n^3$

95. $(a + 2b)(a^2 - 2ab + 4b^2)$ $a^3 + 8b^3$

96. $(x + 3)^3$ $x^3 + 9x^2 + 27x + 27$

97. $[(3m + 2) + n][(3m + 2) - n]$ $9m^2 + 12m + 4 - n^2$

98. $[3 + (x - y)][3 - (x - y)]$ $9 - x^2 + 2xy - y^2$

In exercises 99 and 100, **(a)** *find the area of the rectangle by finding the area of the four sections and adding them.*
(b) *Multiply the two sides and compare the product with your answer to part* **(a)**.

99.

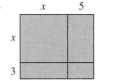

(a) and **(b)**
$x^2 + 8x + 15$

100.

(a) and **(b)**
$x^2 + 11x + 30$

Write a polynomial expression for the area of the rectangle.

101.

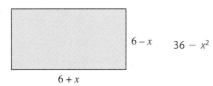

$6 - x$ $36 - x^2$

$6 + x$

102.

$5 - x$ $50 - 2x^2$

$10 + 2x$

In exercises 103 and 104, **(a)** *write a polynomial expression for the area of the shaded portion of the figure.*
(b) *The area of the shaded portion is indicated above each figure. Find the area of the larger and smaller rectangles.*

103.

Area of shaded
region = 67 sq in

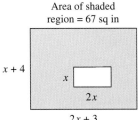

$x + 4$

x

$2x$

$2x + 3$

(a) $11x + 12$
(b) 117 sq in., 50 sq in.

104.

Area of shaded
region = 139 sq in

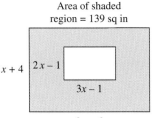

$2x + 4$ $2x - 1$

$3x - 1$

$3x + 6$

(a) $29x + 23$
(b) 216 sq in., 77 sq in.

105. What is a polynomial?   a finite sum of terms in which all variables have whole-number exponents and no variable appears in a denominator

106. When one polynomial is subtracted from another polynomial, what happens to the signs of all the terms of the polynomial being subtracted?   They all change.

107. How do you determine **(a)** the degree of a term; **(b)** the degree of a polynomial?   **107. (a)** Sum the exponents on the variables.   **(b)** It is the same as that of the highest-degree term.

108. What does it mean when a polynomial is in descending order of the variable x?   The exponents on the variable decrease from left to right.

109. In your own words, explain the formulas for the square of a binomial.

110. In your own words, explain the formula for the product of the sum and difference of the same two terms.

111. Write a fifth-degree trinomial in x in descending order that lacks fourth-, third- and second-degree terms.   One example is $x^5 + x + 1$.

112. Write a seventh-degree polynomial in y in descending order that lacks fifth-, third-, and second-degree terms.   **112.** One example is $y^7 + y^6 + y^4 + y + 1$.

CUMULATIVE REVIEW EXERCISES

[1.4] **113.** Evaluate $\dfrac{\left(\left|\frac{1}{2}\right| - \left|-\frac{1}{3}\right|\right)^2}{-\left|\frac{1}{3}\right| \cdot \left|-\frac{2}{5}\right|}$. $-\frac{15}{72} = -\frac{5}{24}$

[2.5] **114.** Solve the inequality $-4 < \dfrac{6 - 3x}{2} \le 5$ and give the solution in set builder notation.

[2.3] **115.** Ms. Jacobmeier invested a total of $10,000 in two savings accounts. One account earned 5% simple and the other account earned 6% simple interest annually. If the total interest earned

from both accounts in 1 year is $560, how much was invested in each account? Solve using only one variable.

[4.3] **116.** Solve Exercise 115 using two variables.

[3.3] **117.** Give the equation of the line that is parallel to the line $3x - 4y = 12$ and has a y intercept of -2. $3x - 4y = 8$

[4.2] **118.** Solve the system of equations:
$$-2x + 3y + 4z = 17$$
$$-5x - 3y + z = -1$$
$$-x - 2y + 3z = 18 \quad (2, -1, 6)$$

114. $\left\{x \mid -\frac{4}{3} \le x < \frac{14}{3}\right\}$ **115.** and **116.** $6000 at 6%, $4000 at 5%

1. $-4xy - 6x^2 + 4y^2 + 2x - 12$ **2.** $-6rs - 4s - 7$ **6.** $20k^{2r+4} - 15k^{2r+2} + 5k^{r+3}$
7. $12x^{3m} - 18x^m - 10x^{2m} + 15$ **8.** $x^{5n} + 2x^{3n}y^{4n} - x^{2n}y^{2n} - 2y^{6n}$

Group Activity/ Challenge Problems

Simplify.

1. $-[5xy - 2x^2 - (4y^2 - 2xy - 4) - (3xy - 5x^2 - 4) - (-3x^2 + 2x - 4)]$

2. $-3rs - \{2 - [3rs - 6s - (-2s - 5rs)] - 5\} - \{8rs - [(rs - 10) - 4rs]\}$

3. $(3r^x - 2r^{2x} - 5) + (6r^{3x} - 5r^{2x} - 8r^x - 4)$ $6r^{3x} - 7r^{2x} - 5r^x - 9$

4. $5w^{3r} - [2w^{2r} + 4w^r - w^{3r}] - [4w^{2r} - (9w^{3r} - 6w^r)]$ $15w^{3r} - 6w^{2r} - 10w^r$

Simplify.

5. $3x^t(5x^{2t-1} + 4x^{3t})$ $15x^{3t-1} + 12x^{4t}$ **6.** $5k^{r+2}(4k^{r+2} - 3k^r + k)$

7. $(6x^m - 5)(2x^{2m} - 3)$ **8.** $(x^{3n} - y^{2n})(x^{2n} + 2y^{4n})$

9. The expression $(a + b)^2$ can be represented by the following figure:

10. (a)

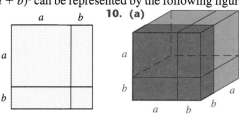

(a) Explain why this figure represents $(a + b)^2$.

(b) Find $(a + b)^2$ using the figure by finding the area of each of the four parts of the figure, then adding the areas together. $a^2 + 2ab + b^2$

10. Use Group Activity Exercise 9 as a guide.

(a) Draw a cube representing $(a + b)^3$.

(b) Find $(a + b)^3$ by adding the volume of the eight parts of the figure. $a^3 + 3a^2b + 3ab^2 + b^3$

11. If $f(x) = x^2 + 3x + 4$ find $f(a + b)$ by substituting $(a + b)$ for each x in the function.

$a^2 + 2ab + b^2 + 3a + 3b + 4$

Perform the polynomial multiplications.

12. $[(y + 1) - (x + 2)]^2$

12. $y^2 - 2y - 2xy + 2x + x^2 + 1$

13. $(x - 3y)^4$

13. $x^4 - 12x^3y + 54x^2y^2 - 108xy^3 + 81y^4$

5.4   Division of Polynomials

Tape
8

**Dividing a Polynomial
by a Monomial**

■1 Divide a polynomial by a monomial.

■2 Divide a polynomial by a binomial.

■1 In division of polynomials, division by zero is not permitted. When the denominator (or divisor) contains a variable, the variable cannot have a value that will result in the denominator being zero. (We discuss this concept further in Chapter 7.)

> **To divide a polynomial by a monomial,** divide each term of the polynomial by the monomial.

EXAMPLE 1 Divide $\dfrac{4x^2 - 8x - 3}{2x}$.

Solution:

$$\frac{4x^2 - 8x - 3}{2x} = \frac{4x^2}{2x} - \frac{8x}{2x} - \frac{3}{2x}$$

$$= 2x - 4 - \frac{3}{2x}$$

EXAMPLE 2 Divide $\dfrac{4y - 6x^4y^3 - 3x^5y^2 + 5x}{2xy^2}$.

Solution:

$$\frac{4y - 6x^4y^3 - 3x^5y^2 + 5x}{2xy^2} = \frac{4y}{2xy^2} - \frac{6x^4y^3}{2xy^2} - \frac{3x^5y^2}{2xy^2} + \frac{5x}{2xy^2}$$

$$= \frac{2}{xy} - 3x^3y - \frac{3x^4}{2} + \frac{5}{2y^2}$$

Dividing a Polynomial by a Binomial

2 We divide a polynomial by a binomial in much the same way as we perform long division.

EXAMPLE 3 Divide $\dfrac{x^2 + 7x + 10}{x + 2}$.

Solution: Rewrite the division problem as

$$x + 2 \overline{)\ x^2 + 7x + 10}$$

Divide x^2 (the first term in $x^2 + 7x + 10$) by x (the first term in $x + 2$).

$$\frac{x^2}{x} = x$$

Place the quotient, x, above the term containing x in the dividend.

$$x + 2 \overline{)\ x^2 + 7x + 10}$$

Next, multiply the x by $x + 2$ as you would do in long division and place the product under the dividend, aligning like terms.

Now subtract $x^2 + 2x$ from $x^2 + 7x$ by changing the signs of $x^2 + 2x$ and adding.

$$
\begin{array}{r}
x \\
x + 2 \overline{)\ x^2 + 7x + 10} \\
\underline{^-x^2 \mp 2x} \\
5x
\end{array}
$$

Now bring down the next term, $+10$.

$$
\begin{array}{r}
x \\
x + 2 \overline{)\ x^2 + 7x + 10} \\
\underline{x^2 + 2x} \\
5x + 10
\end{array}
$$

Divide $5x$ by x.

$$\frac{5x}{x} = +5$$

Place $+5$ above the constant in the dividend and multiply 5 by $x + 2$. Finish the problem by subtracting.

$$
\begin{array}{r}
\text{\textbf{times}} \\
x + 5 \\
x + 2 \overline{)\,x^2 + 7x + 10} \\
\underline{x^2 + 2x} \\
5x + 10 \\
\underline{5x + 10} \longleftarrow 5(x + 2) \\
0 \longleftarrow \text{remainder}
\end{array}
$$

Thus, $\dfrac{x^2 + 7x + 10}{x + 2} = x + 5$. There is no remainder.

When writing an answer in a division problem when there is a remainder, write the remainder over the divisor and add this expression to the quotient. For example, suppose that the remainder in Example 3 was 4. Then the answer would be written $x + 5 + \dfrac{4}{x + 2}$. If the remainder in Example 3 was -7, the answer would be written $x + 5 + \dfrac{-7}{x + 2}$, which we would simplify to $x + 5 - \dfrac{7}{x + 2}$.

EXAMPLE 4 Divide $\dfrac{6x^2 - 5x + 5}{2x + 3}$.

Solution: In this problem we will mentally change the signs of the terms being subtracted and then add.

$$
\begin{array}{r}
3x - 7 \\
2x + 3 \overline{)\,6x^2 - 5x + 5} \\
\underline{6x^2 + 9x} \longleftarrow 3x(2x + 3) \\
-14x + 5 \\
\underline{-14x - 21} \longleftarrow -7(2x + 3) \\
26 \longleftarrow \text{remainder}
\end{array}
$$

Thus, $\dfrac{6x^2 - 5x + 5}{2x + 3} = 3x - 7 + \dfrac{26}{2x + 3}$.

When dividing a polynomial by a binomial, the answer may be **checked** by multiplying the divisor by the quotient and then adding the remainder. You should obtain the polynomial you began with. To check Example 4 we do the following:

$$
\begin{aligned}
(2x + 3)(3x - 7) + 26 &= 6x^2 - 5x - 21 + 26 \\
&= 6x^2 - 5x + 5
\end{aligned}
$$

Since we got the polynomial we began with, our division is correct.

When you are dividing a polynomial by a binomial, you should list both the polynomial and binomial in descending order. If a term of any degree is missing, it is often helpful to include that term with a numerical coefficient of 0. For example, when dividing $(6x^2 + x^3 - 4) \div (x - 2)$, we rewrite the problem as $(x^3 + 6x^2 + 0x - 4) \div (x - 2)$ before beginning the division.

EXAMPLE 5   Divide $(4x^2 - 12x + 3x^5 - 17)$ by $(-2 + x^2)$.

Solution: Write both the dividend and divisor in descending values of the variable x. This gives $(3x^5 + 4x^2 - 12x - 17) \div (x^2 - 2)$. Where a power of x is missing, add that power of x with a coefficient of 0, then divide.

$$
\begin{array}{r}
3x^3 \qquad\quad + 6x + 4 \\
x^2 + 0x - 2 \overline{)\, 3x^5 + 0x^4 + 0x^3 + 4x^2 - 12x - 17} \\
\underline{3x^5 + 0x^4 - 6x^3} \longleftarrow\ 3x^3(x^2 + 0x - 2) \\
6x^3 + 4x^2 - 12x \\
\underline{6x^3 + 0x^2 - 12x} \longleftarrow\ 6x(x^2 + 0x - 2) \\
4x^2 + 0x - 17 \\
\underline{4x^2 + 0x - 8} \longleftarrow\ 4(x^2 + 0x - 2) \\
- 9 \longleftarrow\ \text{remainder}
\end{array}
$$

In obtaining the answer we performed the divisions

$$
\frac{3x^5}{x^2} = 3x^3 \qquad \frac{6x^3}{x^2} = 6x \qquad \frac{4x^2}{x^2} = 4
$$

The quotients $3x^3$, $6x$, and 4 were placed above their like terms in the dividend. The answer is $3x^3 + 6x + 4 - \dfrac{9}{x^2 - 2}$. You should check this answer for yourself by multiplying the divisor by the quotient and adding the remainder.

---

7. $x^3 - \dfrac{3}{2}x^2 + 3x - 2$ **12.** $\dfrac{ab}{2c} - 3 + \dfrac{5a^2b^4}{2c^2}$ **20.** $x^2 + 2x + 3 + \dfrac{1}{x + 1}$ **22.** $3x^2 - 3x + 1 + \dfrac{2}{3x + 2}$

Exercise Set 5.4

23. $2x^2 - 8x + 38 - \dfrac{156}{x + 4}$ **24.** $2x^2 + x - 2 - \dfrac{2}{2x - 1}$ **33.** $2x + 1 + \dfrac{1}{2x} + \dfrac{3}{2x^2}$

Divide.

1. $\dfrac{6x + 8}{2}$ $3x + 4$

2. $\dfrac{3x + 6}{2}$ $\dfrac{3}{2}x + 3$

3. $\dfrac{4x^2 + 2x}{2x}$ $2x + 1$

4. $\dfrac{5y^3 + 6y^2 + 3y}{3y}$ $\dfrac{5}{3}y^2 + 2y + 1$

5. $\dfrac{12x^2 - 4x - 8}{4}$ $3x^2 - x - 2$

6. $\dfrac{15y^6 + 5y^2}{5y^4}$ $3y^2 + \dfrac{1}{y^2}$

7. $\dfrac{4x^5 - 6x^4 + 12x^3 - 8x^2}{4x^2}$

8. $\dfrac{6x^2y - 9xy^2}{3xy}$ $2x - 3y$

9. $\dfrac{4x^2y^2 - 8xy^3 + 3y^4}{2y^2}$ $2x^2 - 4xy + \dfrac{3}{2}y^2$

10. $\dfrac{15x^{12} - 5x^9 + 30x^6}{5x^6}$ $3x^6 - x^3 + 6$

11. $\dfrac{6x^2y - 12x^3y^2 + 9y^3}{2xy^2}$ $\dfrac{3x}{y} - 6x^2 + \dfrac{9y}{2x}$

12. $\dfrac{a^2b^2c - 6abc^2 + 5a^3b^5}{2abc^2}$

Divide.

13. $\dfrac{x^2 + 4x + 3}{x + 1}$ $x + 3$

14. $\dfrac{x^2 + 7x + 10}{x + 5}$ $x + 2$

15. $\dfrac{2x^2 + 13x + 15}{x + 5}$ $2x + 3$

16. $\dfrac{2x^2 + x - 10}{2x + 5}$ $x - 2$

17. $\dfrac{6x^2 + x - 2}{2x - 1}$ $3x + 2$

18. $\dfrac{4r^2 - 9}{2r - 3}$ $2r + 3$

19. $\dfrac{8x^2 + 6x - 25}{4x + 9}$ $2x - 3 + \dfrac{2}{4x + 9}$

20. $\dfrac{x^3 + 3x^2 + 5x + 4}{x + 1}$

21. $\dfrac{4y^3 + 12y^2 + 7y - 3}{2y + 3}$ $2y^2 + 3y - 1$

22. $\dfrac{9x^3 - 3x^2 - 3x + 4}{3x + 2}$

23. $(2x^3 + 6x - 4) \div (x + 4)$

24. $(4x^3 - 5x) \div (2x - 1)$

25. $(3t^4 - 9t^3 + 13t^2 - 11t + 4) \div (t^2 - 2t + 1)$ $3t^2 - 3t + 4$

26. $\dfrac{3x^5 + 4x^2 - 12x - 8}{x^2 - 2}$ $3x^3 + 6x + 4$

27. $\dfrac{4x^5 - 18x^3 + 8x^2 + 18x - 12}{2x^2 - 3}$ $2x^3 - 6x + 4$

28. $\dfrac{3x^4 + 4x^3 - 32x^2 - 5x - 20}{3x^3 - 8x^2 - 5}$ $x + 4$

33. $2x + 1 + \dfrac{1}{2x} + \dfrac{3}{2x^2}$ **38.** $-x^2 - 7x - 5 - \dfrac{8}{x-1}$ **39.** $\dfrac{z}{2} + z^2 - \dfrac{3}{2}x^2y^4z^7$ **40.** $\dfrac{2}{b} - \dfrac{5abc}{3} + \dfrac{8b^3}{3c^2}$ **41.** $2x^2 - 6x + 3$ **42.** $3x^3 - 8$

Divide.

43. $x^3 + x^2 - 6$

29. $\dfrac{6x^2 + 3x + 12}{2x}$ $3x + \dfrac{3}{2} + \dfrac{6}{x}$

30. $\dfrac{6x^2 + 16x + 8}{3x + 2}$ $2x + 4$

31. $\dfrac{2x^2 + x - 10}{x - 2}$ $2x + 5$

32. $\dfrac{2x^3 - 3x^2 - 3x + 4}{x - 1}$ $2x^2 - x - 4$

33. $\dfrac{12x^3 + 6x^2 + 3x + 9}{6x^2}$

34. $\dfrac{2x^2 + 7x - 15}{2x - 3}$ $x + 5$

35. $\dfrac{-5x^3y^2 + 10xy - 6}{10x}$ $\dfrac{-x^2y^2}{2} + y - \dfrac{3}{5x}$

36. $\dfrac{2x^2 + 13x + 15}{2x + 3}$ $x + 5$

37. $\dfrac{9x^3 - x + 3}{3x - 2}$ $3x^2 + 2x + 1 + \dfrac{5}{3x -}$

38. $\dfrac{-x^3 - 6x^2 + 2x - 3}{x - 1}$

39. $\dfrac{3xyz + 6xyz^2 - 9x^3y^5z^7}{6xy}$

40. $\dfrac{6abc^3 - 5a^2b^3c^4 + 8ab^5c}{3ab^2c^3}$

41. $\dfrac{2x^4 - 8x^3 + 19x^2 - 33x + 15}{x^2 - x + 5}$

42. $\dfrac{3x^4 + 4x^3 - 32x^2 - 5x - 20}{x + 4}$

43. $\dfrac{2x^5 + 2x^4 - 3x^3 - 15x^2 + 18}{2x^2 - 3}$

44. (a) Explain how to divide a polynomial by a monomial.

   (b) Divide $\dfrac{5x^4 - 6x^3 - 4x^2 - 12x + 7}{3x}$ using the procedure given in part **(a)**.

45. (a) Explain how to divide a trinomial in x by a binomial in x.

   (b) Divide $2x^2 - 12 + 5x$ by $x + 4$ using the procedure given in part **(a)**   **(b)** $2x - 3$

44. (b) $\dfrac{5}{3}x^3 - 2x^2 - \dfrac{4}{3}x - 4 + \dfrac{7}{3x}$

46. (a) Explain how the answer may be checked when dividing a polynomial by a binomial.

   (b) Using your explanation in part **(a)** to check if the following division is correct.

   $$\dfrac{8x^2 + 2x - 15}{4x - 5} = 2x + 3$$

   (c) Check to see if the following division is correct.

   $$\dfrac{6x^2 - 23x + 14}{3x - 4} = 2x - 5 - \dfrac{6}{3x - 4}$$

46. (a) dividend = divisor · quotient + remainder
   (b), (c) correct

47. Write **(a)** the standard form of a linear equation; **(b)** the slope–intercept form of a linear equation; **(c)** the point–slope form of a linear equation.

[3.4] **48.** Is every function a relation? Is every relation a function? Explain the difference between a function and a relation.   Yes; no

49. If $f(x) = \dfrac{1}{2}x + \dfrac{3}{7}$, find $f\left(-\dfrac{2}{3}\right)$. $\dfrac{2}{21}$

[4.3] **50.** A postage meter can stamp both first-class postage, 32 cents, and bulk postage, 22.8 cents, on envelopes. If after 1 day the meter indicates that 550 envelopes were stamped and the total cost of the postage was $139.20, how many first-class and how many bulk-mail letters were stamped?   400 bulk, 150 first class

SEE EXERCISE 50.

47. (a) $ax + by = c$ **(b)** $y = mx + b$ **(c)** $y - y_1 = m(x - x_1)$

Group Activity/ Challenge Problems

Divide.

1. $\dfrac{2x^3 - x^2y - 7xy^2 + 2y^3}{x - 2y}$ $2x^2 + 3xy - y^2$

2. $\dfrac{x^3 + y^3}{x + y}$ $x^2 - xy + y^2$

Divide. The answers contain fractions.

3. $\dfrac{2x^2 + 2x - 2}{2x - 3}$ $x + \dfrac{5}{2} + \dfrac{11}{2(2x - 3)}$

4. $\dfrac{3x^3 - 5}{3x - 2}$ $x^2 + \dfrac{2}{3}x + \dfrac{4}{9} - \dfrac{37}{9(3x - 2)}$

5. The area of a rectangle is $6x^2 - 8x - 8$. If the length is $2x - 4$, find the width. $3x + 2$

Group Activity/ Challenge Problems

6. The volume of the box that follows is $2r^3 + 4r^2 + 2r$. Find w in terms of r. $w = r + 1$

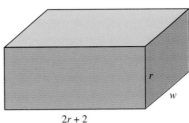

$2r + 2$

7. When a polynomial is divided by $x - 3$, the quotient is $x^2 - 3x + 4 + \dfrac{2}{x - 3}$. What is

the polynomial? Explain how you determined your answer. $x^3 - 6x^2 + 13x - 10$

Divide.

8. $\dfrac{4x^{n+1} + 2x^n - 3x^{n-1} - x^{n-2}}{2x^n}$ $2x + 1 - \dfrac{3}{2x} - \dfrac{1}{2x^2}$ **9.** $\dfrac{3x^n + 6x^{n-1} - 2x^{n-2}}{2x^{n-1}}$ $\dfrac{3}{2}x + 3 - \dfrac{1}{x}$

In the figures below, how many times greater is area or volume of the figure on the right than the figure on the left? Explain how you determined your answer.

10.

$x + 8$

$2x + 4$

$\frac{1}{2}x + 4$

$12x + 24$

3 times greater

11.

$x + 1$

$x + 2$

x

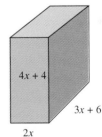

$4x + 4$

$3x + 6$

$2x$

24 times greater

5.5 Synthetic Division

Tape 8

| 1 | Divide polynomials by binomials using synthetic division. |
| 2 | Use the remainder theorem. |

1   When a polynomial is divided by a binomial of the form $x - a$, the division process can be greatly shortened by a process called **synthetic division.** Consider the following examples. In the example on the right, we use only the numerical coefficients.

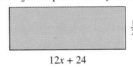

$$
\begin{array}{r}
2x^2 + 5x - 4 \\
x - 3{\overline{\smash{\big)}\,2x^3 - x^2 - 19x + 15}} \\
\underline{2x^3 - 6x^2} \\
5x^2 - 19x \\
\underline{5x^2 - 15x} \\
-4x + 15 \\
\underline{-4x + 12} \\
3
\end{array}
$$

$$
\begin{array}{r}
2 + 5 - 4 \\
1 - 3{\overline{\smash{\big)}\,2 - 1 - 1915}} \\
\underline{2 - 6} \\
5 - 19 \\
\underline{5 - 15} \\
-415 \\
\underline{-412} \\
3
\end{array}
$$

Note that the variables do not play a role in determining the numerical coefficients of the quotient. This division problem can be done more quickly and easily using synthetic division. Following is an explanation of how we use synthetic division. Consider the division

$$\frac{2x^3 - x^2 - 19x + 15}{x - 3}$$

1. Write the dividend in descending powers of x. Then list the numerical coefficients of each term in the dividend. If a term of any degree is missing, place 0 in the appropriate position to serve as a placeholder. In the problem above the numerical coefficients of the dividend are

$$2 \qquad -1 \qquad -19 \qquad 15$$

2. When dividing by a binomial of the form $x - a$, place a to the left of the line of numbers from part 1. In this problem we are dividing by $x - 3$; thus, $a = 3$. We write

$$\underline{3|} \quad 2 \qquad -1 \qquad -19 \qquad 15$$

3. Bring down the first coefficient on the left as follows:

$$\begin{array}{r|rrrr} 3 & 2 & -1 & -19 & 15 \\ \hline & 2 \end{array}$$

4. Multiply the 3 by the number brought down, the 2, to get 6. Place the 6 under the next coefficient, the -1. Then add $-1 + 6$ to get 5.

$$\begin{array}{r|rrrr} 3 & 2 & -1 & -19 & 15 \\ & & 6 & & \\ \hline & 2 & 5 \end{array}$$

5. Multiply the 3 by sum 5 to get 15. Place 15 under -19. Then add to get -4. Repeat this procedure as illustrated.

$$\begin{array}{r|rrrr} 3 & 2 & -1 & -19 & 15 \\ & & 6 & 15 & -12 \\ \hline & 2 & 5 & -4 & 3 \end{array}$$

In the last row, the first three numbers are the numerical coefficients of the quotient as shown in the long division. The last number, 3, is the remainder obtained by long division. The quotient must be one degrees less than the dividend since we are dividing by $x - 3$. The original dividend was a third-degree polynomial. Therefore, the quotient must be a second-degree polynomial. Use the first three numbers from the last row as the coefficients of a second-degree polynomial in x. This gives $2x^2 + 5x - 4$, which is the quotient. The last number, $+3$, is the remainder. Therefore,

$$\frac{2x^3 - x^2 - 19x + 15}{x - 3} = 2x^2 + 5x - 4 + \frac{3}{x - 3}.$$

EXAMPLE 1 Divide using synthetic division.

$$(6 - x^2 + x^3) \div (x + 2)$$

Solution: First, list the terms of the dividend in descending order of x.

$$(x^3 - x^2 + 6) \div (x + 2)$$

Since there is no first-degree term, insert 0 as a placeholder when listing the numerical coefficients.

Since $x + 2 = x - (-2)$, $a = -2$.

$$\begin{array}{r|rrrr} -2 & 1 & -1 & 0 & 6 \\ & & -2 & 6 & -12 \\ \hline & 1 & -3 & 6 & -6 \end{array} \longleftarrow \text{remainder}$$

Since the dividend is a third-degree equation, the quotient must be second degree. The answer is $x^2 - 3x + 6 - \dfrac{6}{x + 2}$.

EXAMPLE 2   Use synthetic division to divide.

$$(3x^4 + 11x^3 - 20x^2 + 7x + 35) \div (x + 5)$$

Solution:

$$\begin{array}{r|rrrrr} -5 & 3 & 11 & -20 & 7 & 35 \\ & & -15 & 20 & 0 & -35 \\ \hline & 3 & -4 & 0 & 7 & 0 \end{array} \longleftarrow \text{remainder}$$

Since the dividend is of the fourth degree, the quotient must be of the third degree. The quotient is $3x^3 - 4x^2 + 0x + 7$ with no remainder. This can be simplified to $3x^3 - 4x^2 + 7$.

EXAMPLE 3   Use synthetic division to divide.

$$(3x^3 - 6x^2 + 4x + 5) \div \left(x - \frac{1}{2}\right)$$

Solution:

$$\begin{array}{r|rrrr} \frac{1}{2} & 3 & -6 & 4 & 5 \\ & & \frac{3}{2} & -\frac{9}{4} & \frac{7}{8} \\ \hline & 3 & -\frac{9}{2} & \frac{7}{4} & \frac{47}{8} \end{array} \longleftarrow \text{remainder}$$

The answer is

$$3x^2 - \frac{9}{2}x + \frac{7}{4} + \frac{47}{8\left(x - \frac{1}{2}\right)} \quad \text{or} \quad 3x^2 - 4.5x + 1.75 + \frac{5.875}{x - \frac{1}{2}}$$

Remainder Theorem   **2** In Example 1, when we divided $x^3 - x^2 + 6$ by $x + 2$, we found that the remainder was -6. If we write $x + 2$ as $x - (-2)$ and evaluate the polynomial function* $P(x) = x^3 - x^2 + 6$ at -2, we obtain

$$P(x) = x^3 - x^2 + 6$$
$$P(-2) = (-2)^3 - (-2)^2 + 6 = -8 - 4 + 6 = -6$$

---

*$f(x)$ could have been used in place of $P(x)$. However, when discussing polynomial functions, $P(x)$ is generally used.

Since $P(-2) = -6$, the value of the function at $x = -2$ is -6. When we divided $x^3 - x^2 + 6$ by $x + 2$, the remainder was also -6. Is this just a coincidence? Let us try one more. In Example 3 when we divided $3x^3 - 6x^2 + 4x + 5$ by $x - \frac{1}{2}$, we obtained a remainder of $\frac{47}{8}$. Let us evaluate $P(x) = 3x^3 - 6x^2 + 4x + 5$ at $x = \frac{1}{2}$.

$$P(x) = 3x^3 - 6x^2 + 4x + 5$$

$$P\left(\frac{1}{2}\right) = 3\left(\frac{1}{2}\right)^3 - 6\left(\frac{1}{2}\right)^2 + 4\left(\frac{1}{2}\right) + 5$$

$$= 3\left(\frac{1}{8}\right) - 6\left(\frac{1}{4}\right) + 2 + 5$$

$$= \frac{3}{8} - \frac{3}{2} + 7$$

$$= \frac{3}{8} - \frac{12}{8} + \frac{56}{8} = \frac{47}{8}$$

The value of $P\left(\frac{1}{2}\right)$ is $\frac{47}{8}$, the same as the remainder obtained by synthetic division. To obtain the remainder when a polynomial $P(x)$ is divided by a binomial of the form $x - a$, we can use the Remainder Theorem.

> ### Remainder Theorem
> If the polynomial $P(x)$ is divided by $x - a$, the remainder is equal to $P(a)$.

EXAMPLE 4   Use the Remainder Theorem to find the remainder when $3x^4 + 6x^3 - 2x + 4$ is divided by $x + 4$.

Solution: First write the divisor $x + 4$ in the form $x - a$. Since $x + 4 = x - (-4)$ we evaluate $P(-4)$.

$$P(x) = 3x^4 + 6x^3 - 2x + 4$$

$$P(-4) = 3(-4)^4 + 6(-4)^3 - 2(-4) + 4$$

$$= 3(256) + 6(-64) + 8 + 4$$

$$= 768 - 384 + 8 + 4 = 396$$

Thus, when $3x^4 + 6x^3 - 2x + 4$ is divided by $x + 4$, the remainder is 396.

Using synthetic division, we will show that the remainder in Example 4 is indeed 396.

$$
\begin{array}{r|rrrrr}
-4 & 3 & 6 & 0 & -2 & 4 \\
   &   & -12 & 24 & -96 & 392 \\
\hline
   & 3 & -6 & 24 & -98 & 396
\end{array}
$$
$\longleftarrow$ remainder

In Example 2, when we divided $3x^4 + 11x^3 - 20x^2 + 7x + 35$ by $x + 5$, we found the quotient was $3x^3 - 4x^2 + 7$ and the remainder was 0. In a division problem, if the remainder is 0, both the divisor and the quotient are factors of the dividend. In Example 2 we can write

$$\frac{3x^4 + 11x^3 - 20x^2 + 7x + 35}{x + 5} = 3x^3 - 4x^2 + 7$$

$$\text{or} \quad (x + 5)(3x^3 - 4x^2 + 7) = 3x^4 + 11x^3 - 20x^2 + 7x + 35$$

Using the Remainder Theorem, we can determine if a binomial of the form $x - a$ is a factor of a polynomial $P(x)$. To do so, evaluate $P(a)$. If $P(a) = 0$, then $x - a$ divides the polynomial without remainder and $x - a$ is a factor of the polynomial.

EXAMPLE 5

(a) Show, using the Remainder Theorem, that $x + 2$ is a factor of $x^3 + 6x^2 + 11x + 6$.
(b) Find the other factor.

Solution:

(a) $x + 2 = x - (-2)$. If $P(-2) = 0$, then $x + 2$ is a factor of the polynomial.

$$P(x) = x^3 + 6x^2 + 11x + 6$$
$$P(-2) = (-2)^3 + 6(-2)^2 + 11(-2) + 6$$
$$= -8 + 24 - 22 + 6 = 0$$

Since $P(-2) = 0$, $x + 2$ is a factor of $x^3 + 6x^2 + 11x + 6$.

(b) We can find the other factor using synthetic division.

$$
\begin{array}{r|rrrr}
-2 & 1 & 6 & 11 & 6 \\
   &   & -2 & -8 & -6 \\
\hline
   & 1 & 4 & 3 & 0
\end{array}
$$

The other factor is $x^2 + 4x - 3$. Thus,
$$x^3 + 6x^2 + 11x + 6 = (x + 2)(x^2 + 4x - 3).$$

In Example 5, at $x = -2$ the value of the polynomial $x^3 + 6x^2 + 11x + 6$ is 0. Therefore, -2 is a solution to the polynomial equation $x^3 + 6x^2 + 11x + 6 = 0$. We will discuss this material further in Section 5.7 when we discuss zeros of polynomial functions.

Exercise Set 5.5

Divide using synthetic division.

1. $(x^2 + x - 6) \div (x - 2)$ $x + 3$
2. $(x^2 - 4x - 32) \div (x + 4)$ $x - 8$
3. $(x^2 + 5x - 6) \div (x + 6)$ $x - 1$
4. $(x^2 + 12x + 32) \div (x + 4)$ $x + 8$
5. $(x^2 + 5x - 12) \div (x - 3)$ $x + 8 + [12/(x - 3)]$
6. $(2x^2 - 9x + 15) \div (x - 6)$ $2x + 3 + [33/(x - 6)]$
7. $(3x^2 - 7x - 10) \div (x - 4)$ $3x + 5 + [10/(x - 4)]$
8. $(x^3 + 6x^2 + 4x - 7) \div (x + 5)$ $x^2 + x - 1 - [2/(x + 5)]$
9. $(4x^3 - 3x^2 + 2x) \div (x - 1)$ $4x^2 + x + 3 + [3/(x - 1)]$
10. $(x^3 - 7x^2 - 13x + 5) \div (x - 2)$ $x^2 - 5x - 23 - [41/(x - 2)]$

11. $3x^2 - 2x + 2 + [6/(x + 3)]$ **12.** $3x^3 + 9x^2 + 2x + 6 - [2/(x - 3)]$ **13.** $5x^2 - 11x + 14 - [20/(x + 1)]$

16. $2x^3 + 6x^2 + 17x + 56 + [156/(x - 3)]$ **18.** $z^4 + 3z^3 - 3z^2 + 3z - 3 - [7/(z + 1)]$

11. $(3x^3 + 7x^2 - 4x + 12) \div (x + 3)$ **12.** $(3x^4 - 25x^2 - 20) \div (x - 3)$

13. $(5x^3 - 6x^2 + 3x - 6) \div (x + 1)$ **14.** $(y^4 - 1) \div (y - 1)$ $y^3 + y^2 + y + 1$

15. $(x^4 + 16) \div (x + 4)$ $x^3 - 4x^2 + 16x - 64 + [272/(x + 4)]$ **16.** $(2x^4 - x^2 + 5x - 12) \div (x - 3)$

17. $(y^5 + y^4 - 10) \div (y + 1)$ $y^4 - [10/(y + 1)]$ **18.** $(z^5 + 4z^4 - 10) \div (z + 1)$

19. $(3x^3 + 2x^2 - 4x + 1) \div \left(x - \dfrac{1}{3}\right)$ $3x^2 + 3x - 3$ **20.** $(8x^3 - 6x^2 - 5x + 3) \div \left(x + \dfrac{3}{4}\right)$ $8x^2 - 12x + 4$

21. $(2x^4 - x^3 + 2x^2 - 3x + 1) \div \left(x - \dfrac{1}{2}\right)$ $2x^3 + 2x - 2$ **22.** $(9y^3 + 9y^2 - y + 2) \div \left(y + \dfrac{2}{3}\right)$ $9y^2 + 3y - 3 + \left[4/\left(y + \frac{2}{3}\right)\right]$

Determine the remainder for the following divisions using the Remainder Theorem. If the divisor is a factor of the dividend, so state.

23. $(4x^2 - 5x + 4) \div (x - 2)$ 10   **24.** $(-2x^2 + 3x - 2) \div (x + 3)$ -29

25. $(x^3 - 5x^2 + 2) \div (x + 4)$ -142 **26.** $(x^3 - 6x + 4) \div (x - 1)$ -1

27. $(x^3 - 2x^2 + 4x - 8) \div (x - 2)$ 0, factor   **28.** $(-3x^3 + 4x - 12) \div (x + 4)$ 164

29. $(-2x^3 - 6x^2 + 2x - 4) \div \left(x - \dfrac{1}{2}\right)$ $-\dfrac{19}{4}$ or -4.75 **30.** $(-5x^3 - 6) \div \left(x - \dfrac{1}{5}\right)$ $-\dfrac{151}{25}$ or -6.04

31. $(x^4 - 6x^3 + 3x^2 - 2x - 276) \div (x + 3)$ 0, factor   **32.** $(x^4 - 5x^3 - 6x + 30) \div (x - 5)$ 0, factor

Given polynomial P(x) and a value x such that P(x) = 0, find the factors of P(x) (see Example 5).

33. $P(x) = 6x^2 - x - 5$, $P(1) = 0$ **34.** $P(x) = 5x^2 + 19x + 12$, $P(-3) = 0$
$(x - 1)(6x + 5)$ $(x + 3)(5x + 4)$

35. $P(x) = x^3 + x^2 - x - 10$, $P(2) = 0$ **36.** $P(x) = x^3 + 3x^2 + x + 20$, $P(-4) = 0$
$(x - 2)(x^2 + 3x + 5)$ $(x + 4)(x^2 - x + 5)$

37. $P(x) = 2x^3 + 6x^2 - 18x + 10$, $P(-5) = 0$ **38.** $P(x) = 3x^3 - 14x^2 + 8x$, $P(4) = 0$
$(x + 5)(2x^2 - 4x + 2)$ $(x - 4)(3x^2 - 2x)$

39. $P(x) = x^3 - 3x^2 - 22x - 12$, $P(-3) = 0$ **40.** $P(x) = 6x^3 - 12x^2 + x + 5$, $P(1) = 0$
$(x + 3)(x^2 - 6x - 4)$ $(x - 1)(6x^2 - 6x - 5)$

41. $P(x) = x^3 + \dfrac{3}{2}x^2 - 5x + 2$, $P\left(\dfrac{1}{2}\right) = 0$ **42.** $P(x) = 3x^3 - 4x^2 + x + \dfrac{10}{3}$, $P\left(-\dfrac{2}{3}\right) = 0$
$\left(x - \dfrac{1}{2}\right)(x^2 + 2x - 4)$ $\left(x + \dfrac{2}{3}\right)(3x^2 - 6x + 5)$

43. (a) Describe in your own words how to divide a polynomial by $(x - a)$ using synthetic division.

   (b) Divide $x^2 + 3x - 4$ by $x - 5$ using the procedure in part **(a)**. $x + 8 + \dfrac{36}{x - 5}$

44. (a) State the Remainder Theorem in your own words.

   (b) Find the remainder when $x^2 - 6x - 4$ is divided by $x - 1$, using the procedure stated in part **(a)**. -9

45. Explain how you can determine using synthetic division if an expression of the form $x - a$ is a factor of a polynomial in x.   If the remainder is 0, $x - a$ is a factor

46. Explain how you can determine using the Remainder Theorem if an expression of the form $x - a$ is a factor of a polynomial in x.   If $P(a) = 0$, then $x - a$ is a factor.

47. Given $P(x) = ax^2 + bx + c$ and a value d such that $P(d) = 0$, explain why $x = d$ is a solution to the equation $ax^2 + bx + c = 0$.

48. Consider the trinomial $40x^2 - 313x - 56$ and the binomials $x + 8$, $x - 8$, $x + 7$. Use the Remainder Theorem to determine which of the binomials is a factor of the trinomial. Explain how you determined your answer. $x - 8$

49. If one factor of $P(x)$ is $x^2 - 5x - 12$ and if $P(-3) = 0$, find $P(x)$. Explain how you determined your answer.

50. Write a third-degree polynomial of four terms that has a factor of $x + 4$. Explain how you obtained your answer. There are many possible answers.

47. When $x = d$, $ax^2 + bx + c$ is equal to 0. Therefore, d is a solution.

49. $x^3 - 2x^2 - 27x - 36$; multiply $(x + 3)(x^2 - 5x - 12)$

50. One answer is $x^3 + 5x^2 + 5x + 4$.

CUMULATIVE REVIEW EXERCISES

[2.5] **51.** Solve the inequality and graph the solution on the number line: $-1 < \dfrac{4(3x - 2)}{3} \le 5$ $\dfrac{5}{12}$ $\dfrac{23}{12}$

[8.1] **52.** Find the perimeter of the triangle determined by the following points. $A(3, 9)$; $B(-2, 6)$; $C(8, 4)$.

[3.2] **53.** Graph the equation $20x - 60y = 120$.

[5.2] **54.** Simplify $\dfrac{(4x^{-2}y^{-3})^{-2}(2xy^{-4})^3}{(3x^{-1}y^3)^2}$. $\dfrac{x^9}{18y^{12}}$

53.

52. $\sqrt{34} + \sqrt{50} + \sqrt{104}$ (or ≈ 23.1)

---

4. $0.2x^2 - 3.92x - 1.248 - [1.1392/(x - 0.4)]$ **5.(a)** $3x^2 - 2x + 5 - [13/(3x + 5)]$
5 (b) Because the remainder is in terms of $3x + 5$ rather than $x + \frac{5}{3}$, the denominator of the remainder is altered rather than the numerator.

Group Activity/ Challenge Problems

In Exercises 1–3, explain your answer.

1. Is $x - 1$ a factor of $x^{100} + x^{99} + \cdots + x^1 + 1$?   No

2. Is $x + 1$ a factor of $x^{100} + x^{99} + \cdots + x^1 + 1$?   No

3. Is $x + 1$ a factor of $x^{99} + x^{98} + \cdots + x^1 + 1$?   Yes

4. Divide $(0.2x^3 - 4x^2 + 0.32x - 0.64)$ by $(x - 0.4)$.

5. Synthetic division can be used to divide polynomials by binomials of the form $ax - b$, $a \ne 1$. To perform this division, divide $ax - b$ by a to obtain $x - \dfrac{b}{a}$. Then place b/a to the left of the numerical coefficients of the polynomial. Work the problem as explained previously. After summing the numerical values below the line, divide all of them, except the remainder, by a. Then write the quotient of the problem using these numbers.

(a) Use this procedure to divide $(9x^3 + 9x^2 + 5x + 12)$ by $(3x + 5)$.

(b) Explain why we do not divide the remainder by a.

5.6 Polynomial Functions

Tape 8

1. Identify polynomial functions.
2. Graph quadratic functions.
3. Determine the axis of symmetry of a parabola.
4. Graph cubic functions.

Polynomial Functions

1. In Section 3.5 we discussed linear functions of the form $f(x) = ax + b$. Linear functions are a specific type of **polynomial function.** There are many other types of polynomial functions. The general form of a polynomial function is given below.

Polynomial Function of Degree n

$f(x) = a_n x^n + a_{n-1}x^{n-1} + a_{n-2}x^{n-2} + a_{n-3}x^{n-3} + \cdots + a_1 x + a_0$

where all exponents on x are whole numbers and $a_n, a_{n-1}, a_{n-2}, \ldots, a_1, a_0$ are all real numbers with $a_n \ne 0$.

Examples of Polynomial Functions

$$f(x) = 3x + 4 \qquad \text{linear function, first degree}$$

$$f(x) = 5x^2 - \frac{1}{2}x + 3 \qquad \text{quadratic function, second degree}$$

$$f(x) = 6x^3 - 4x \qquad \text{cubic function, third degree}$$

$$f(x) = \sqrt{2}x^4 - 6x \qquad \text{fourth-degree function}$$

Note that the right side of each of these functions is a polynomial since all exponents are whole numbers.

The graph of every equation of the form $y = a_n x^n + a_{n-1} x^{n-1} + a_{n-2} x^{n-2} + \cdots + a_1 x + a_0$, where all exponents are whole numbers, will pass the vertical test. Every equation of this form is therefore a function.

The domain of every polynomial function will be the set of real numbers, $\mathbb{R}$. When any real number is substituted for x in a polynomial function, the value obtained for y will be a real number. However, the range of a polynomial function is not necessarily the set of real numbers.

In Chapter 3 we graphed linear functions. Other polynomial functions can be graphed in much the same way. To graph polynomial functions, we can substitute values for x, find the corresponding values for y, and plot the points on the Cartesian coordinate system. When graphing linear functions, we had to plot only two points to draw the line. When plotting other polynomial functions, we must be sure to plot a sufficient number of points to get a true picture of the graph. The graphs of polynomial functions will be smooth curves and will pass the vertical line test for functions.

After plotting points, connect the points to get a smooth curve. When drawing the smooth curve, start with the point on the graph with the smallest x value and draw to the point having the next larger value of x. Continue this way through all of the points. Draw arrow tips on the ends of the graph to show that the graph continues in the same direction.

Quadratic Functions

2 The first function we will graph in this section is a quadratic function.

> ### Quadratic Function
> Any function of the form
> $$f(x) = ax^2 + bx + c, \quad a \neq 0$$
> where a, b, and c are real numbers, is a **quadratic function.**

In Section 3.4 we learned that $y = f(x)$. Therefore, an equation of the form $y = ax^2 + bx + c$, $a \neq 0$, is also a quadratic function.

Every quadratic function will have the shape of a **parabola** (see Fig. 5.1) when graphed. When graphing a function of the form

$$f(x) = ax^2 + bx + c \text{ (or } y = ax^2 + bx + c)$$

the sign of the numerical coefficient of the second-degree term, a, will determine whether the parabola opens upward or downward. When this coefficient is positive, the parabola will open upward, as in Fig. 5.1a. When the coefficient is negative, the parabola will open downward, as shown in Fig. 5.1b. The **vertex** is the lowest point on a parabola that opens upward and the highest point on a parabola that opens downward.

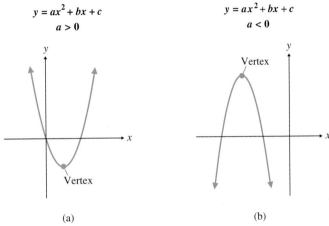

$$y = ax^2 + bx + c$$
$$a > 0$$

$$y = ax^2 + bx + c$$
$$a < 0$$

(a)   (b)

FIGURE 5.1

When graphing a quadratic equation, make sure you plot a sufficient number of points to show whether the parabola is opening upward or downward. We will graph a quadratic equation in Example 1.

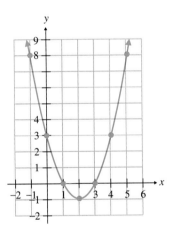

FIGURE 5.2

EXAMPLE 1 Graph $f(x) = x^2 - 4x + 3$.

Solution: Since $f(x)$ is the same as y, graphing $f(x) = x^2 - 4x + 3$ is the same as graphing $y = x^2 - 4x + 3$. Make a table of values by substituting values for x and solving for $f(x)$ or y.

$$f(x) = x^2 - 4x + 3$$

$$f(-1) = (-1)^2 - 4(-1) + 3 = 1 + 4 + 3 = 8$$
$$f(0) = 0^2 - 4(0) + 3 = 0 - 0 + 3 = 3$$
$$f(1) = 1^2 - 4(1) + 3 = 1 - 4 + 3 = 0$$
$$f(2) = 2^2 - 4(2) + 3 = 4 - 8 + 3 = -1$$
$$f(3) = 3^2 - 4(3) + 3 = 9 - 12 + 3 = 0$$
$$f(4) = 4^2 - 4(4) + 3 = 16 - 16 + 3 = 3$$
$$f(5) = 5^2 - 4(5) + 3 = 25 - 20 + 3 = 8$$

x	y
-1	8
0	3
1	0
2	-1
3	0
4	3
5	8

Now plot the points and connect them with a smooth curve (see Fig. 5.2).

Notice that the graph in Fig. 5.2 is a function since it passes the vertical line test discussed in Section 3.5. The *domain* of this function, the set of values that can be used for x, is the set of real numbers, $\mathbb{R}$. The *range,* the corresponding set of values of y, is the set of real numbers greater than or equal to -1.

Domain: $\mathbb{R}$

Range: $\{y \,|\, y \geq -1\}$

Axis of Symmetry   ③   When graphing quadratic functions, how do we decide what values to use for x? When the location of the vertex is unknown, this is a difficult question to

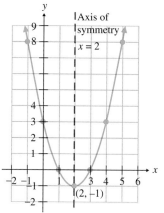

FIGURE 5.3

answer. When the location of the vertex is known, it becomes more obvious which values to use.

Let us examine the parabola in Example 1 more closely (see Fig. 5.3). Notice that the parabola is *symmetric* about a vertical line through the vertex. This means that if we folded the page along this imaginary line, called the *axis of symmetry,* the right and left sides would coincide. Every parabola that opens upward or downward will have an axis of symmetry that will be a vertical line through its vertex. If we can determine the location of the axis of symmetry, we can use it as a guide in selecting values for x. When quadratic functions of the form $y = ax^2 + bx + c$ are graphed, the axis of symmetry of the parabola will be $x = -b/2a$.

Axis of Symmetry

Given an equation of the form $y = ax^2 + bx + c$, its graph will be a parabola whose axis of symmetry is the vertical line

$$x = \frac{-b}{2a}$$

Note in Figure 5.3 that the axis of symmetry is $x = 2$. We can find this by the axis of symmetry formula as follows:

$$f(x) = x^2 - 4x + 3$$

$$a = 1, \qquad b = -4, \qquad c = 3$$

$$x = \frac{-b}{2a} = \frac{-(-4)}{2(1)} = \frac{4}{2} = 2$$

The equation of the axis of symmetry is $x = 2$. The x coordinate of the vertex of the parabola is also at 2. The y coordinate of the vertex can now be found by substituting 2 for x in the function.

$$f(x) = x^2 - 4x + 3$$
$$f(2) = 2^2 - 4(2) + 3$$
$$= 4 - 8 + 3$$
$$= -1$$

Therefore, when $x = 2$, $f(x)$ or $y = -1$. The coordinates of the vertex of the parabola are $(2, -1)$.

When graphing quadratic equations, you should first determine the axis of symmetry and the coordinates of the vertex. Use the axis of symmetry as a guide in selecting values of x to evaluate in the function. Now we will graph another quadratic function, this time making use of the axis of symmetry in selecting values for x.

EXAMPLE 2 Consider $f(x) = -x^2 + 2x + 3$.

(a) State the domain of the function.

(b) When this equation is graphed, what is the equation of the axis of symmetry of the parabola?

(c) Find the vertex of the parabola.

(d) Graph the function.

(e) Determine the range of the function.

Solution:

(a) Since any real number may be substituted for x and the corresponding value of y will be a real number, the domain is the set of real numbers, $\mathbb{R}$.

(b) $x = \dfrac{-b}{2a} = \dfrac{-(2)}{2(-1)} = \dfrac{-2}{-2} = 1$

The parabola will be symmetric about the line $x = 1$.

(c) We already know that the x coordinate of the vertex is 1 since the axis of symmetry is $x = 1$. We can find the y coordinate of the vertex by evaluating the function at $x = 1$.

$$f(x) = -x^2 + 2x + 3$$

$$f(1) = -(1)^2 + 2(1) + 3 = 4$$

Thus, the vertex is at $(1, 4)$.

(d) Since the axis of symmetry is $x = 1$, we will choose values for x of $-2, -1, 0, 1, 2, 3, 4$. Note that we selected three values less than 1 and three values greater than 1. Now we find the corresponding values of y and draw the graph as shown in Fig. 5.4.

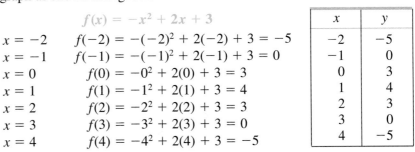

$f(x) = -x^2 + 2x + 3$			x	y
$x = -2$	$f(-2) = -(-2)^2 + 2(-2) + 3 = -5$		-2	-5
$x = -1$	$f(-1) = -(-1)^2 + 2(-1) + 3 = 0$		-1	0
$x = 0$	$f(0) = -0^2 + 2(0) + 3 = 3$		0	3
$x = 1$	$f(1) = -1^2 + 2(1) + 3 = 4$		1	4
$x = 2$	$f(2) = -2^2 + 2(2) + 3 = 3$		2	3
$x = 3$	$f(3) = -3^2 + 2(3) + 3 = 0$		3	0
$x = 4$	$f(4) = -4^2 + 2(4) + 3 = -5$		4	-5

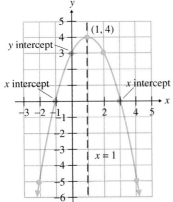

FIGURE 5.4

(e) From the graph we can see that the range of the function is $\{y \mid y \le 4\}$. We could have also determined the range by recognizing that the y coordinate of the vertex is 4, and that the graph opens downward (since $a < 0$). Therefore, the maximum value of y is 4.

The parabola in Fig. 5.4 is a smooth curve. When graphing a parabola or any polynomial function, it is often helpful to plot each point as it is determined. If the point does not appear to be part of the curve, check your calculations.

In Fig. 5.4, we see that the vertex is at the point $(1, 4)$, and the y intercept, where the graph crosses the y axis, is 3. The y intercept of any graph can be determined by substituting $x = 0$ into the equation and evaluating the equation. For the function in Example 2,

$$f(x) = -x^2 + 2x + 3$$

$$f(0) = -0^2 + 2(0) + 3 = 3$$

Can you explain why this procedure always gives the y intercept?

Also indicated in Fig. 5.4 are the x intercepts of the graph. Notice that the graph intersects the x axis at -1 and 3. These values are called the *zeros of the function* because when $x = -1$ or when $x = 3$, the value of y or $f(x)$ is 0. That is, $f(-1) = 0$ and $f(3) = 0$. Any values of the variable that result in the function having a value of 0 are called **zeros of the function**.

In Example 2 we found the x coordinate of the vertex using the formula for the axis of symmetry, $x = \dfrac{-b}{2a}$. Once we found the value of x, we substituted that value in the function to obtain the y coordinate of the vertex. Thus, the vertex of a parabola may be represented as

$$(x, f(x)) \quad \text{where} \quad x = \frac{-b}{2a}.$$

A second method of finding the y coordinate of the vertex is given in the Group Activity Exercises 1–3. You may wish to review that material now.

EXAMPLE 3   Graph $y = 2x^2 - 82x + 720$

Solution: Begin by finding the axis of symmetry and the coordinates of the vertex.

$$x = \frac{-b}{2a} = \frac{-(-82)}{2(2)} = \frac{82}{4} = 20.5$$

$$y = f(x) = 2x^2 - 82x + 720$$

$$y = f(20.5) = 2(20.5)^2 - 82(20.5) + 720 = -120.5$$

Thus, the vertex is at $(20.5, -120.5)$. We will select some values of x greater than and less than 20.5. When selecting points keep in mind that the vertex is below the x axis and the graph opens upward, so it must cross the x axis at two points. We will make sure we plot enough points to show this. In the following table, when $x = 12$, $y = 24$ and when $x = 15$, $y = -60$. Therefore, the graph must cross the x axis between $x = 12$ and $x = 15$. Can you explain why? By examining the table, can you determine between which two other x values the graph crosses the x axis?

x	y
10	100
12	24
15	-60
18	-108
20.5	-120.5
22	-116
25	-80
30	60

We selected values of x that were not necessarily equidistant from $x = 20.5$ to get a greater variety of points. We will use a scale large enough for us to plot the points $(20.5, -120.5)$ and $(10, 100)$. The y axis will extend at least from -120.5 to 100, or 220.5 units. The maximum value of x that we will plot is 30. Therefore, we will make the scale on the y axis greater than the scale on the x axis. The graph is illustrated in Fig. 5.5.

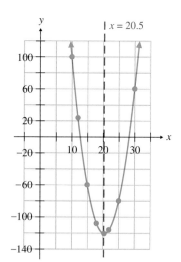

FIGURE 5.5

In a graph like Fig. 5.5, it is difficult to determine the exact x intercepts. In Chapter 9 we show how to find the x intercept using algebraic methods.

Graph Cubic Functions ▣4 Now we will graph a cubic or third-degree function by plotting points. Graphing third-degree or higher functions by this method may not result in a totally accurate graph. To graph third-degree or higher functions accurately requires a knowledge of calculus. However, this method will be sufficient for our needs.

> **Cubic Function**
>
> Any function of the form
> $$f(x) = ax^3 + bx^2 + cx + d, \quad a \neq 0$$
> where a, b, c, and d are real numbers, is a **cubic function**

Some cubic functions are illustrated in Fig. 5.6. When the coefficient of the third-degree term, a, is positive, the graph continues to increase to the right of some value of x, and continues to decrease to the left of some value of x, as in Fig. 5.6(a) and 5.6(b). When the coefficient of the third-degree term, a, is negative, the graph continues to decrease to the right of some value of x, and continues to increase to the left of some value of x, as in Fig. 5.6(c) and 5.6 (d).

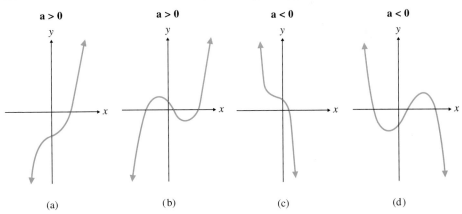

(a)   (b)   (c)   (d)

FIGURE 5.6

In any polynomial function the highest degree term is the "dominant term." The dominant term determines whether the graph will eventually increase or decrease to the right of some value of x, and eventually increase or decrease to the left of some value of x. Consider the cubic functions $f(x) = 2x^3 - 10x - 20$ and $f(x) = -2x^3 + 10x + 20$. Below we evaluate both functions at $x = 4$ and $x = -4$.

$$f(x) = 2x^3 - 10x - 20$$
$$f(4) = 2(4)^3 - 10(4) - 20$$
$$= 128 - 40 - 20 = 68$$
$$f(-4) = 2(-4)^3 - 10(-4) - 20$$
$$= -128 + 40 - 20 = -108$$

$$f(x) = -2x^3 + 10x + 20$$
$$f(4) = -2(4)^3 + 10(4) + 20$$
$$= -128 + 40 + 20 = -68$$
$$f(-4) = -2(-4)^3 + 10(-4) + 20$$
$$= 128 - 40 + 20 = 108$$

In $f(x) = 2x^3 - 10x - 20$, the coefficient of the dominant term, 2, is positive. The value of $2x^3$ will be a large positive value for large values of x. Therefore, the graph will continue to increase to the right of some value of x. The value of $2x^3$ will be a large negative number for large negative values of x. Therefore, the graph will continue to decrease to the left of some value of x.

In $f(x) = -2x^3 + 10x + 20$, the coefficient of the dominant term, -2, is negative. The value of $-2x^3$ will be a large negative number for large positive values of x. Therefore, the graph will continue to decrease to the right of some value of x. The value of $-2x^3$ will be a large positive number for large negative values of x. Therefore, the graph will continue to increase to the left of some value of x.

Now we will draw some graphs of cubic functions.

EXAMPLE 4   Graph $f(x) = x^3 - 3x + 1$.

Solution: Where do we start? Unfortunately, there is no easy way, as there was with quadratic equations, to pick a starting point. We generally start at 0, and select some values of x less than 0 and some greater than 0 and find the corresponding values of $f(x)$ or y. Then plot the points and draw a smooth curve from point to point (see Fig. 5.7).

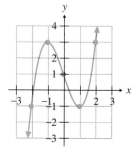

FIGURE 5.7

$f(x) = x^3 - 3x + 1$

$f(-3) = (-3)^3 - 3(-3) + 1 = -27 + 9 + 1 = -17$
$f(-2) = (-2)^3 - 3(-2) + 1 = -8 + 6 + 1 = -1$
$f(-1) = (-1)^3 - 3(-1) + 1 = -1 + 3 + 1 = 3$
$f(0) = 0^3 - 3(0) + 1 = 0 - 0 + 1 = 1$
$f(1) = 1^3 - 3(1) + 1 = 1 - 3 + 1 = -1$
$f(2) = 2^3 - 3(2) + 1 = 8 - 6 + 1 = 3$
$f(3) = 3^3 - 3(3) + 1 = 27 - 9 + 1 = 19$

x	y
-3	-17
-2	-1
-1	3
0	1
1	-1
2	3
3	19

Notice that the points $(-3, -17)$ and $(3, 19)$ listed in the table were not plotted on the graph in Fig. 5.7. Their y values are too small and too large, respectively. The arrows on the graph indicate that the graph continues in the same direction. The graph would pass through these two points if the vertical axis was extended. Unfortunately, there is no easy method to determine the vertices of the graph of a cubic equation. The graph in Fig. 5.7 is a function whose range and domain are both all real numbers, $\mathbb{R}$.

In Example 4, we graphed $f(x) = x^3 - 3x + 1$. If we multiply each term on the right sides of the function by -1, we obtain $f(x) = -x^3 + 3x - 1$. What will the graph of $f(x) = -x^3 + 3x - 1$ look like? We will graph this function in Example 5.

EXAMPLE 5   Graph $f(x) = -x^3 + 3x - 1$.

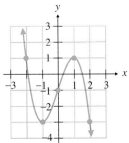

FIGURE 5.8

Solution: $f(x) = -x^3 + 3x - 1$

$$f(-3) = -(-3)^3 + 3(-3) - 1 = 17$$
$$f(-2) = -(-2)^3 + 3(-2) - 1 = 1$$
$$f(-1) = -(-1)^3 + 3(-1) - 1 = -3$$
$$f(0) = -(0)^3 + 3(0) - 1 = -1$$
$$f(1) = -(1)^3 + 3(1) - 1 = 1$$
$$f(2) = -(2)^3 + 3(2) - 1 = -3$$
$$f(3) = -(3)^3 + 3(3) - 1 = -19$$

x	y
-3	17
-2	1
-1	-3
0	-1
1	1
2	-3
3	-19

The graph is given in Fig. 5.8.

Notice for each point (x, y) on the graph of $f(x) = x^3 - 3x + 1$ in Fig. 5.7 the corresponding point on the graph of $f(x) = -x^3 + 3x - 1$ in Fig. 5.8 is $(x, -y)$. The graph in Fig. 5.7 is inverted to obtain the graph in Fig. 5.8.

EXAMPLE 6   Graph $y = x^3 + 8x^2 - 4x - 12$.

Solution: Let us start by selecting values near 0 and finding their corresponding values of y. We then plot the points and draw the graph (see Fig. 5.9).

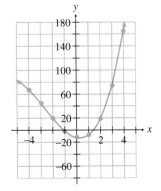

FIGURE 5.9

x	-4	-3	-2	-1	0	1	2	3	4
y	68	45	20	-1	-12	-7	20	75	164

At this point you may think you have plotted enough points to complete the graph since the values of y are large. You may believe that the graph continues upward on both sides of the y axis. However, consider the third-degree term of the equation, x^3. If x is a large positive number, x^3 will be positive. Therefore the graph will continue upwards. This agrees with the right side of the graph since the graph is increasing for positive values of x. Now consider x^3 when x is a large negative number. Since x^3 will be negative when x is negative, the graph will need to decrease to the left of some value of x. This does not agree with the left side of our graph. Therefore, we must continue selecting negative values of x until we get a sufficient number of points to illustrate that the function decreases. Continue plotting points and graphing the curve. Figure 5.10 is the complete and correct graph. Make sure that you

x	-9	-8	-7	-6	-5
y	-57	20	65	84	83

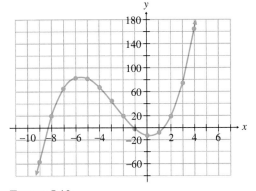

FIGURE 5.10

place arrowheads on the ends of the curve to show that it continues in the direction shown.

Determining values of y in Example 6 will take time, but you need to be persistent. Using a calculator will help save you time.

Helpful Hint

Some students start by graphing the first few points of a polynomial function correctly but stop before they complete the graph. You need to keep working until you are confident you have the entire graph completed. You might want to ask yourself, if $x = 100$, will the function be positive or negative? If $x = -100$, will the function be positive or negative? Does my graph as it is presently drawn agree with what I expect for very large or very small values of x? If not, keep working until your graph agrees with what you know about the function.

When plotting points you may wish to select values of x that differ by 3 or 5 or more units, rather than just 1 unit, so you can see more quickly where the values of y change sign. If the value of y changes sign between any two x values, the graph must cross the x axis between those x values. Can you explain why?

Graphing Calculator Corner

Sometimes it is not easy to find the proper domain and range to enter into your calculator. Let us discuss graphing two functions $y = x^3 + 60x^2 + 800$ and $y = 0.01x^2 + 0.6x + 100$. If you graph $y = x^3 + 60x^2 + 800$ using your calculator's default values (the values automatically used by your calculator unless you change them), your calculator will display an axis with no graph. This indicates that the graph is outside the calculator's window. The y intercept of this graph is 800. Therefore, you may wish to change your maximum value of the range to 1000 or 2000 or even 5000.

In Fig. 5.11 we show the graph of $y = x^3 + 60x^2 + 800$ with the domain, range, and scales used under the figure. Notice the heavy line along the y axis. This is because the range goes from -10 to 1000 with intervals of 1 unit. This creates over 1000 marks on the y axis, which all blend together, making the y axis look heavy. In Fig. 5.12 we show the graph of $y = x^3 + 60x^2 + 800$ with a different set of values for the domain and range and scales. As discussed earlier we know that this function must decrease to the left for values of x less than some value. The graph in Fig. 5.12 does not show this, so we must continue selecting lower values for the domain until the calculator displays this part of the graph. By trial and error we eventually obtain a display that accurately shows the shape of the graph (see Fig. 5.13).

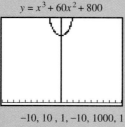

$y = x^3 + 60x^2 + 800$

$-10, 10 , 1, -10, 1000, 1$

FIGURE 5.11

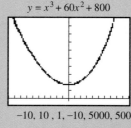

$y = x^3 + 60x^2 + 800$

$-10, 10 , 1, -10, 5000, 500$

FIGURE 5.12

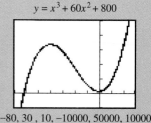

$y = x^3 + 60x^2 + 800$

$-80, 30 , 10, -10000, 50000, 10000$

FIGURE 5.13

Now consider $y = 0.01x^2 + 0.6x + 100$. If you graphed this function using the default settings, your screen would not display the graph. The y intercept is 100, so let us increase the range to a maximum of 200 and see what the graph looks like (see Fig. 5.14). We know that this graph is a parabola. However, the graph in Fig. 5.14 does not appear to be parabola. Since the coefficients of the x^2 and x terms are small, we must use large values of x to see the true picture. We will also need to increase the maximum value of the range. Figure 5.15 shows the graph with a different domain, range, and scale. In Fig. 5.15 we are starting to see the shape of the parabola. Let us increase the domain further (see Fig. 5.16). Now we see the parabola. Selecting the values to use for the domain and range is often a matter of trial and error. The more you practice, the better you will get at it. When graphing this quadratic equation, you may also find the axis of symmetry, $x = \dfrac{-b}{2a} = \dfrac{-0.6}{2(.01)} = -30$ and use this information as an aid in selecting your domain.

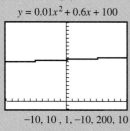

$y = 0.01x^2 + 0.6x + 100$

−10, 10 , 1, −10, 200, 10

FIGURE 5.14

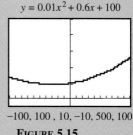

$y = 0.01x^2 + 0.6x + 100$

−100, 100 , 10, −10, 500, 100

FIGURE 5.15

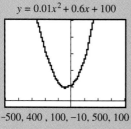

$y = 0.01x^2 + 0.6x + 100$

−500, 400 , 100, −10, 500, 100

FIGURE 5.16

EXERCISES

Graph the following functions on your calculator. Make sure that your graph illustrates all x intercepts. Indicate the number *of x intercepts that each graph has.*

1. $y = 1.6x^2 - 0.2x - 5$ 2   **2.** $y = (x - 3)^2 + 5.2$ none

3. $y = 2.1x^3 - 43.2x^2 - 6.4x + 68.9$ 3   **4.** $y = 0.02x^2 - 0.8x - 24.6$ 2

Using Polynomial Functions

5 Now we will look at an application of a polynomial function.

EXAMPLE 7   Find the maximum rectangular area that can be enclosed with 60 feet of fencing.

Solution: Begin by drawing a sketch (see Fig. 5.17).
Let x = length of the rectangle
    y = width of the rectangle
The area, A, is the product of the length and width.

$$A = xy$$

The perimeter, P, is 60 feet, thus

$$P = 2x + 2y$$
$$60 = 2x + 2y$$

Perimeter = 60 ft

y

x

FIGURE 5.17

We now have the system of equations.

$$A = xy$$
$$60 = 2x + 2y$$

Solve the second equation for y.

$$60 - 2x = 2y$$
$$30 - x = y \quad \text{or} \quad y = 30 - x$$

Now substitute $30 - x$ for y in $A = xy$.

$$A = xy$$
$$A = x(30 - x)$$
$$A = 30x - x^2 \quad \text{or} \quad A = -x^2 + 30x$$

We will now graph A as a function of x. Since area can only be positive, we will select values of x that yield nonnegative values for A. The axis of symmetry is

$$x = \frac{-b}{2a} = \frac{-30}{-2} = 15$$

We will use symmetry when drawing the graph (see Fig. 5.18).

x	A
15	225
10	200
5	125
0	0

We see from the graph that the maximum area, 225 square feet, occurs when the length, x, is 15 feet. Solving for the width using $y = 30 - x$ yields

$$y = 30 - x$$
$$= 30 - 15 = 15$$

Thus, the maximum area occurs when the shape is a square whose sides are 15 feet long.

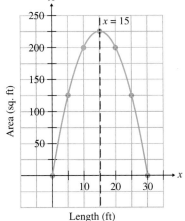

FIGURE 5.18

In Example 7 the maximum area could have been found without the graph by simply finding the vertex of the parabola (15, 225). However, the graph shows how the area varies as the length changes.

We will see numerous applications of polynomial functions in later chapters.

Exercise Set 5.6

Determine if the following are polynomial functions. If the function is not a polynomial function, explain why. If it is a polynomial function, give its degree.

1. $f(x) = x^7 - 3x^2 - 5$ yes, seventh

2. $f(x) = 2x^{-4} + x^3 + 5$ no, negative exponent

3. $f(x) = x + x^{1/2} - 2$ no, fractional exponent

4. $f(x) = \sqrt{3}x^4 - \sqrt{5}x + 7$ yes, fourth

5. $f(x) = x^4 + x^2 - x^{-1}$ no, negative exponent

6. $f(x) = 3x^5 - 7x^7 - x^9$ yes, ninth

Indicate the axis of symmetry, the coordinates of the vertex, and whether the parabola opens upward or downward.

7. $y = x^2 + 2x - 7$ $x = -1, (-1, -8),$ up

8. $y = x^2 + 4x - 9$ $x = -2, (-2, -13),$ up

9. $y = -3x^2 + 6x + 8$ $x = 1, (1, 11),$ down

10. $y = x^2 + 8x - 6$ $x = -4, (-4, -22),$ up

11. $y = -4x^2 - 8x - 12$ $x = -1, (-1, -8),$ down

12. $y = 2x^2 + 4x + 6$ $x = -1, (-1, 4),$ up

13. $y = -x^2 + x + 8$ $x = 0.5, (0.5, 8.25),$ down

14. $y = 8x^2 - 20x + 2$ $x = 1.25, (1.25, -10.5),$ up

15. $y = 4x^2 + 12x - 5$ $x = -1.5, (-1.5, -14),$ up

16. $y = -8x^2 - 12x - 5$ $x = -0.75, (-0.75, -0.5),$ down

17. $y = 0.4x^2 - 40x + 800$ $x = 50, (50, -200),$ up

18. $y = 1.6x^2 - 96x - 24.5$ $x = 30, (30, -1464.5),$ up

Determine the vertex of the parabola, then graph each quadratic function and give its domain and range.

19. $y = x^2 - 1$

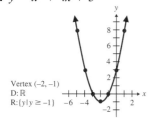

Vertex $(0, -1)$
D: $\mathbb{R}$
R: $\{y \,|\, y \geq -1\}$

20. $y = x^2 + 4$

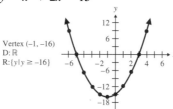

Vertex $(0, 4)$
D: $\mathbb{R}$
R: $\{y \,|\, y \geq 4\}$

21. $y = -x^2 + 3$

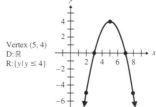

Vertex $(0, 3)$
D: $\mathbb{R}$
R: $\{y \,|\, y \leq 3\}$

22. $y = x^2 + 4x + 3$

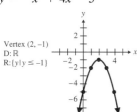

Vertex $(-2, -1)$
D: $\mathbb{R}$
R: $\{y \,|\, y \geq -1\}$

23. $y = x^2 + 2x - 15$

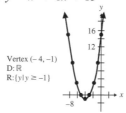

Vertex $(-1, -16)$
D: $\mathbb{R}$
R: $\{y \,|\, y \geq -16\}$

24. $y = -x^2 + 10x - 21$

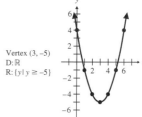

Vertex $(5, 4)$
D: $\mathbb{R}$
R: $\{y \,|\, y \leq 4\}$

25. $y = -x^2 + 4x - 5$

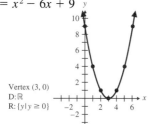

Vertex $(2, -1)$
D: $\mathbb{R}$
R: $\{y \,|\, y \leq -1\}$

26. $y = x^2 + 8x + 15$

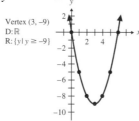

Vertex $(-4, -1)$
D: $\mathbb{R}$
R: $\{y \,|\, y \geq -1\}$

27. $y = x^2 - 6x + 4$

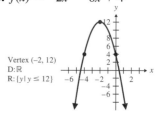

Vertex $(3, -5)$
D: $\mathbb{R}$
R: $\{y \,|\, y \geq -5\}$

28. $y = x^2 - 6x + 9$

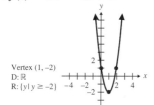

Vertex $(3, 0)$
D: $\mathbb{R}$
R: $\{y \,|\, y \geq 0\}$

29. $y = x^2 - 6x$

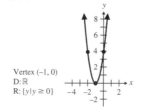

Vertex $(3, -9)$
D: $\mathbb{R}$
R: $\{y \,|\, y \geq -9\}$

30. $f(x) = -2x^2 - 8x + 4$

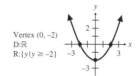

Vertex $(-2, 12)$
D: $\mathbb{R}$
R: $\{y \,|\, y \leq 12\}$

31. $f(x) = 3x^2 - 6x + 1$

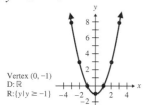

Vertex $(1, -2)$
D: $\mathbb{R}$
R: $\{y \,|\, y \geq -2\}$

32. $f(x) = 4x^2 + 8x + 4$

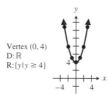

Vertex $(-1, 0)$
D: $\mathbb{R}$
R: $\{y \,|\, y \geq 0\}$

33. $f(x) = \frac{1}{2}x^2 - 2$

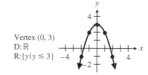

Vertex $(0, -2)$
D: $\mathbb{R}$
R: $\{y \,|\, y \geq -2\}$

34. $f(x) = \frac{1}{2}x^2 + x - 4$

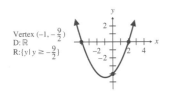

Vertex $(-1, -\frac{9}{2})$
D: $\mathbb{R}$
R: $\{y\mid y \geq -\frac{9}{2}\}$

35. $f(x) = -4x^2 + 20x + 160$

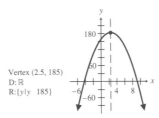

Vertex $(2.5, 185)$
D: $\mathbb{R}$
R: $\{y\mid y \quad 185\}$

36. $f(x) = 6x^2 - 48x + 60$

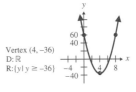

Vertex $(4, -36)$
D: $\mathbb{R}$
R: $\{y\mid y \geq -36\}$

37. $f(x) = 0.2x^2 - 0.6x - 1.2$

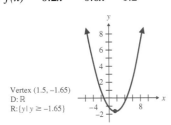

Vertex $(1.5, -1.65)$
D: $\mathbb{R}$
R: $\{y\mid y \geq -1.65\}$

38. $f(x) = -0.6x^2 - 1.2x - 2.4$

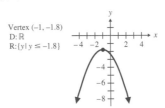

Vertex $(-1, -1.8)$
D: $\mathbb{R}$
R: $\{y\mid y \leq -1.8\}$

Graph each cubic function.

39. $y = x^3$

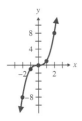

40. $y = x^3 + 1$

41. $y = x^3 + x$

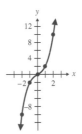

42. $y = x^3 + 2x - 1$

43. $y = x^3 + x^2 - 3x - 1$

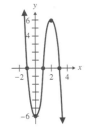

44. $f(x) = 2x^3 + x - 8$

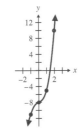

45. $f(x) = -x^3 + 3x$

46. $f(x) = -x^3 + x - 6$

47. $f(x) = -2x^3 + 6x^2 + 2x - 6$

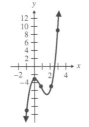

48. $y = 2x^3 - 5x^2 + x - 3$

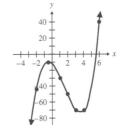

49. $f(x) = 2x^3 - 9x^2 - 10x - 12$

50. $f(x) = -x^3 + 8x^2 + 10x - 15$

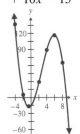

51. $f(x) = 0.8x^3 + 6x^2 - 60$ **52.** $f(x) = 0.5x^3 - 3x^2 - x$

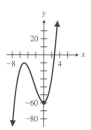   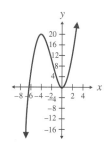

53. In Example 7, if only 48 feet of fencing were available, find the dimensions of the rectangular figure that would provide the greatest area. a square 12 ft by 12 ft

54. Suppose that a farmer wishes to place a fence near a straight riverbank to fence in cattle. The fencing will not be placed along side the riverbank, as shown in the illustration. If 120 feet of fencing is available, what should be the length and width of the fencing to obtain the maximum area? What is the maximum area?

55. The perimeter of an isosceles triangle is 24 inches. The height of the triangle is $\frac{3}{4}$ the length of the two equal sides, as shown in the figure. Find **(a)** the length of the base of the triangle that maximizes the area; **(b)** the maximum area.   **(a)** 12 in **(b)** 27 sq in.

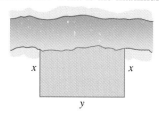

Base

56. The monthly profit, P, in thousands of dollars, of a bicycle company can be estimated by the formula $P = -2x^2 + 16x - 12$, where x is the number of bicycles, in hundreds, produced and sold per month. **(a)** Estimate the number of bicycles that must be produced and sold in a month for the company to maximize their profit. **(b)** Find the maximum profit.

57. A toy rocket is fired upward from the ground with an initial velocity of 64 feet per second. The height of the rocket from the ground at any time, t, may be found from the formula $h = -16t^2 + 64t$. Find **(a)** the time the rocket reaches its maximum height; **(b)** the maximum height obtained.   **(a)** 2 sec **(b)** 64 ft

58. An object is projected upward from a cliff 100 feet above the ocean at a velocity of 64 feet per second. The height of the object from the ocean at any time, t, can be found by the formula $h = -16t^2 + 64t + 100$.
(a) Find the time it takes for the object to reach its maximum height.   2 sec
(b) Find the maximum height achieved by the object.
(c) Estimate, to the nearest tenth second, when the object will strike the water. ≈ 5.2 sec

54. $x = 30$, $y = 60$, 1800 sq ft   **56. (a)** 400 **(b)** $20,000
58. **(b)** 164 ft

Indicate which graph or graphs cannot *be a graph of the given function. Explain your answer.*

59. $y = x^2 + bx + c$

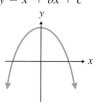

(a)   (b)   (c)

(a) and **(c)**; graph must increase to the right and left.

60. $y = x^3 + bx + c$

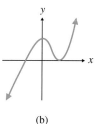   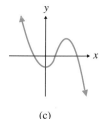

(a)   (b)   (c)

(c); graph must increase to the right and decrease to the left.

61. $a > 0$, up; $a < 0$, down **63. (a)** $x = \frac{-b}{2a}$ **(b)** substitute the value of $\frac{-b}{2a}$ into function to find y value of vertex.

64. y increases; y decreases **65.** y decreases; y increases

61. Explain how to determine if the graph of a quadratic function opens up or down.

62. What is the name given to the graph of a quadratic function? a parabola

63. For a quadratic equation, explain how to find **(a)** the axis of symmetry of its graph; **(b)** the vertex of its graph.

64. Consider the function $f(x) = x^3$. Explain what happens to y as x increases? As x decreases?

65. Consider the function $f(x) = -x^3$. Explain what happens to y as x increases? As x decreases?

66. Consider the function $f(x) = x^4$. Explain what happens to y as x increases from -3 to 3?

66. y decreases from -3 to 0; y increases from 0 to 3 (y is a minimum when $x = 0$).
67. y increases from -3 to 0; y decreases from 0 to 3 (y is a maximum when $x = 0$).
68. must continue up to the right of some value and down to the left of some value, or vice versa

67. Consider the function $f(x) = -x^4$. Explain what happens to y as x increases from -3 to 3.

68. Explain why the range of functions in one variable whose degree is odd is the set of real numbers.

69. How will the graph of the opposite (or negative) of a polynomial function compare with the graph of the polynomial function? It will be inverted.

70. (a) Make up your own quadratic function, and explain why it is a quadratic function.
(b) Graph your function.

71. (a) Make up your own cubic function, and explain why it is a cubic function.
(b) Graph your function.

CUMULATIVE REVIEW EXERCISES

Consider the system of equations $\quad x - 4y = -16$
$$2x + 3y = -10$$

[4.1] **72.** Solve using the addition method $(-8, 2)$

[4.4] **73.** Solve using determinants $(-8, 2)$

75. $15x^3 + 21x^2 - 38x + 12$

[4.3] **74.** The sum of three numbers is 12. If the sum of the two smaller numbers equals the largest number, and the largest number is four less than twice the middle-sized number, find the three numbers. (1, 5, 6)

[5.3] **75.** Multiply $(3x^2 + 6x - 4)(5x - 3)$.

Group Activity/ Challenge Problems

4. (a)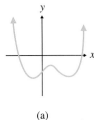

4. (b)

4. (c)

5. (a)

5. (b)

5. (c)

We will show later in the book that the y coordinate of the vertex of a parabola can also be found from the formula

$$y = \frac{4ac - b^2}{4a}$$ **2.** $(1.5, -8.5)$ **3.** $\left(\frac{2}{3}, \frac{43}{9}\right)$ or $(0.\overline{6}, 4.\overline{7})$

Therefore, the vertex of a parabola may be found from either

$$(x, f(x)) \quad \text{where } x = -\frac{b}{2a} \quad \text{or} \quad \left(-\frac{b}{2a}, \frac{4ac - b^2}{4a}\right)$$

Find the vertex of the graph of the following quadratic equations using both *methods.*

1. $f(x) = 4x^2 - 8x + 12$ (1, 8) **2.** $f(x) = 2x^2 - 6x - 4$ **3.** $f(x) = \frac{1}{2}x^2 - \frac{2}{3}x + 5$

4. Illustrate a graph of a cubic function whose third-degree term has a positive coefficient and where the graph has the following number of x intercepts: **(a)** one **(b)** two **(c)** three.

5. Illustrate a graph of a cubic function whose third-degree term has a negative coefficient and where the graph has the following number of x intercepts: **(a)** one **(b)** two **(c)** three.

6. Is it possible for the graph of a cubic function to have no x intercepts? Explain. no

By considering the dominant term determine which graph or graphs cannot *be a graph of the given function. Explain your answer.*

7. $y = x^4 + bx^3 + cx + d$ **(b)** and **(c)**; graph must increase to the right and the left.

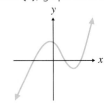

(a)

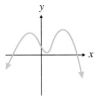

(b)

(c)

8. (b); Graph must increase to the right and decrease to the left.

9. 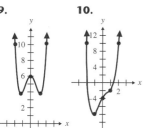 **10.**

 8. $y = x^5 + bx^4 + cx^3 + dx + e$

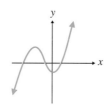   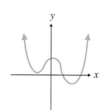   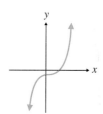

9. Graph $f(x) = x^4 - 3x^2 + 6$.   **10.** Graph $f(x) = x^4 - 2x^2 + 3x - 4$.

11. Of all numbers whose sum is 70, find the two that have the greatest product. That is, find the values of x and y that will maximize the value of P in the function $P = xy$. $x = 35, y = 35$

Summary

GLOSSARY

Binomial (291): A two-term polynomial.

Degree of a polynomial (291): The same as the highest-degree term in the polynomial.

Descending order of the variable (291): Polynomial written so that the exponents on the variable decrease as terms go from left to right.

Monomial (291): A one-term polynomial.

Polynomial (290): A finite sum of terms in which all variables have whole-number exponents and no variable appear in a denominator.

Polynomial function (311): A function of the form $f(x) = a_n x^n + a_{n-1} x^{n-1} + a_{n-2} x^{n-2} + \ldots + a_1 x + a_0$

Quadratic function (312): A function of the form $f(x) = ax^2 + bx + c, a \neq 0$.

Scientific notation (277): A form of writing large and small numbers as the product of a number greater than or equal to 1 and less than 10 and a power of 10.

Synthetic division (305): A shortened process of dividing a polynomial by a binomial of the form $x - a$.

Trinomial (291): A three-term polynomial.

IMPORTANT FACTS

Rules for exponents

1. $a^m \cdot a^n = a^{m+n}$ product rule

2. $\dfrac{a^m}{a^n} = a^{m-n}, a \neq 0$ quotient rule

3. $a^{-m} = \dfrac{1}{a^m}, a \neq 0$ negative exponent rule

4. $a^0 = 1, a \neq 0$ zero exponent rule

5. $(a^m)^n = a^{mn}$
$(ab)^m = a^m b^m$
$\left(\dfrac{a}{b}\right)^m = \dfrac{a^m}{b^m}, b \neq 0$ power rules

Parabola

Axis of symmetry: $x = \dfrac{-b}{2a}$ Vertex: $(x, f(x))$, where $x = \dfrac{-b}{2a}$

FOIL method to multiply two binomials

$(a + b)(c + d)$

F —Multiply *first* terms.
O—Multiply *outer* terms.
I —Multiply *inner* terms.
L —Multiply *last* terms.

$ac + ad + bc + bd$

Special product formulas

$(a + b)^2 = a^2 + 2ab + b^2$
$(a - b)^2 = a^2 - 2ab + b^2$ }   square of a binomial

$(a + b)(a - b) = a^2 - b^2$ product of sum and difference of same two terms (or difference of two squares)

Remainder theorem

If the polynomial $P(x)$ is divided by $x - a$, the remainder is equal to $P(a)$.

Review Exercises

[5.1, 5.2] *Simplify and write the answer without negative exponents.*

1. $4^2 \cdot 4^1$ 64

2. $x^3 \cdot x^5$ x^8

3. $\dfrac{x^6}{x^2}$ x^4

4. $\dfrac{y^{12}}{y^3}$ y^9

5. $\dfrac{x^4}{x^{-3}}$ x^7

6. $x^4 \cdot x^{-7}$ $\frac{1}{x^3}$

7. $3^{-2} \cdot 3^{-1}$ $\frac{1}{27}$

8. $3x^0$ 3

9. $4^{-1} + 2 \cdot 3^{-1} + 3^0$ $\frac{23}{12}$

10. $3^2 - 2^{-3} - 4^0$ $\frac{63}{8}$

11. $(3x^2)^2$ $9x^4$

12. $\left(\dfrac{2}{3}\right)^{-1}$ $\frac{3}{2}$

13. $\left(\dfrac{3}{4}\right)^{-2}$ $\frac{16}{9}$

14. $\left(\dfrac{x}{y^2}\right)^{-1}$ $\frac{y^2}{x}$

15. $(7x^2y^5)(-3xy^4)$ $-21x^3y^9$

16. $(4x^2y^{-3})(2x^{-4}y^2)$ $\frac{8}{x^2y}$

17. $(2x^5y^{-4})(3x^2y)(5x^{-2}y^{-3})$ $\frac{30x^5}{y^6}$

18. $\dfrac{6x^{-3}y^5}{2x^2y^{-2}}$ $\frac{3y^7}{x^5}$

19. $\dfrac{12x^{-3}y^{-4}}{4x^{-2}y^5}$ $\frac{3}{xy^9}$

20. $\dfrac{(5x^3y^2)(2xy^4z)}{20x^4y^{-2}z}$ $\frac{y^8}{2}$

21. $\left(\dfrac{3x^4y^2}{2x^{-2}y^3}\right)\left(\dfrac{4x^2y^{-3}}{6xy^4}\right)$ $\frac{x^7}{y^8}$

22. $\left(\dfrac{5x^2y}{x}\right)^3$ $125x^3y^3$

23. $\left(\dfrac{x^5y}{-3y^2}\right)^2$ $\frac{x^{10}}{9y^2}$

24. $\left(\dfrac{x^2y}{x^{-1}y^{-3}}\right)^2$ x^6y^8

25. $\left(\dfrac{-5x^{-2}y}{z^3}\right)^3$ $\frac{-125y^3}{x^6z^9}$

26. $\left(\dfrac{6xy^3}{z^2}\right)^{-2}$ $\frac{z^4}{36x^2y^6}$

27. $\left(\dfrac{9x^{-2}y}{3xy}\right)^{-3}$ $\frac{x^9}{27}$

28. $(-2x^{-3}y^2)^{-4}$ $\frac{x^{12}}{16y^8}$

29. $\left(\dfrac{2x^4y^6}{z^3}\right)^2(2x^2y^4)$ $\frac{8x^{10}y^{16}}{z^6}$

30. $\left(\dfrac{16x^4y^3z^{-2}}{4x^5y^2z^3}\right)^3$ $\frac{64y^3}{x^3z^{15}}$

31. $\left(\dfrac{2x^4y}{z^3}\right)^2\left(\dfrac{3x^{-2}y^4}{3xy^6}\right)$ $\frac{4x^5}{z^6}$

32. $\left(\dfrac{3x^4y^{-2}}{6xy^{-3}}\right)^2\left(\dfrac{2x^{-1}y^5}{3x^4y^{-2}}\right)^{-3}$ $\frac{27x^{21}}{32y^{19}}$

33. $\left(\dfrac{x^{-2}y^{-2}z}{x^4y^{-4}z^3}\right)^{-1}\left(\dfrac{2x^2y^5}{4xy^{-2}z}\right)^3$ $\frac{x^9y^{19}}{8z}$

34. $\dfrac{(4x^{-2}y^3)^{-2}(x^4y^3)^4}{(2x^{-3}y^6)^2}$ $\frac{x^{26}}{64y^6}$

Express each number in scientific notation.

35. 0.0000742 7.42×10^{-5}

36. 260,000 2.6×10^5

37. 183,000 1.83×10^5

38. 0.000001 1×10^{-6}

Chapter 6

Factoring

See Section 6.5, Exercise 70.

Preview and Perspective

The goal of this chapter is to strengthen your factoring skills. *You must have a thorough understanding of factoring to work the problems in Chapter 7.* We will also use factoring in Chapters 8, 9, and 10.

It is always a good idea to check your factoring by multiplying the factors together. We explain how to do this in various sections in the chapter. In Section 6.2 we present two methods to factor trinomials. Check with your instructor to see if he or she prefers you to use a particular method. In Section 6.2 we also introduce factoring by substitution. Pay particular attention to this, for we will use this procedure in Section 6.5 and when we discuss equations that are "quadratic in form" in Section 9.3. You should work as many exercises in Section 6.4 as possible. This exercise set contains a wide variety of problems, and will reinforce your understanding of factoring. In Section 6.5 we explain how to solve equations by factoring and also present some applications of quadratic equations. We discuss solving quadratic equations that cannot be solved by factoring in Sections 9.1 and 9.2.

6.1 Factoring a Monomial from a Polynomial and Factoring by Grouping

Tape 9

1. Find the greatest common factor.
2. Factor a monomial from a polynomial.
3. Factor by grouping.

Factoring is the opposite of multiplication. To factor an expression means to write it as a product of other expressions. For example, in Chapter 5 we learned to perform the following multiplications:

$$3x^2(6x + 3xy + 5x^3) = 18x^3 + 19x^3y + 15x^5$$

and

$$(6x + 3y)(2x - 5y) = 12x^2 - 24xy - 15y^2$$

In this chapter we learn how to determine the factors of a given expression. For example, we will learn how to perform each factoring illustrated below.

$$18x^3 + 9x^3y + 15x^5 = 3x^2(6x + 3xy + 5x^3)$$

and

$$12x^2 - 24xy - 15y^2 = (6x + 3y)(2x - 5y)$$

Finding the Greatest Common Factor

1. Recall that if $a \cdot b = c$, then a and b are said to be **factors** of c. An expression may have many factors. What are the integer factors of 12?

$$1 \cdot 12 = 12 \qquad (-1)(-12) = 12$$
$$2 \cdot 6 = 12 \qquad (-2)(-6) = 12$$
$$3 \cdot 4 = 12 \qquad (-3)(-4) = 12$$

Note that the factors of 12 are ± 1 (read "plus or minus 1"), ± 2, ± 3, ± 4, ± 6, ± 12. Generally, when asked to list the factors of a positive number, we list only the positive factors, although it should be understood the negatives of these factors are also factors. We would say the factors of 12 are 1, 2, 3, 4, 6, and 12.

What are the factors of $6x^3$?

Factors   *Factors*

$$1 \cdot 6x^3 = 6x^3 \qquad x \cdot 6x^2 = 6x^3$$
$$2 \cdot 3x^3 = 6x^3 \qquad 2x \cdot 3x^2 = 6x^3$$
$$3 \cdot 2x^3 = 6x^3 \qquad 3x \cdot 2x^2 = 6x^3$$
$$6 \cdot \ x^3 = 6x^3 \qquad 6x \cdot \ x^2 = 6x^3$$

Some factors of $6x^3$ are 1, 2, 3, 6, x, $2x$, $3x$, $6x$, x^2, $2x^2$, $3x^2$, $6x^2$, x^3, $2x^3$, $3x^3$, and $6x^3$. The opposite or negative of each of these factors is also a factor.

To factor a monomial from a polynomial, we must determine the greatest common factor (GCF) of each term in the polynomial. After the GCF is determined, we use the distributive property to factor the expression. The **greatest common factor** of two or more expressions is the greatest factor that divides into (without remainder) each expression. Consider the three numbers 12, 18, and 24. The GCF of these three numbers is 6, since 6 is the greatest number that divides into (is a factor of) each of them. If you have forgotten how to find the GCF of a set of numbers, review an arithmetic or elementary algebra text before going any further.

The GCF of a collection of terms containing variables is easily found. Consider the terms x^3, x^4, x^5, and x^6. The GCF of these terms is x^3, since x^3 is the highest power of x that divides all four terms. Note that the GCF of a collection of terms contains the *lowest power of the common variable.*

EXAMPLE 1   Find the GCF of the following terms.
(a) y^{12}, y^4, y^9, y^7 **(b)** x^3y^2, xy^4, x^4y^5 **(c)** $6x^2y^3$, $9x^3y^4$, $24x^4$

Solution:

(a) Note that y^4 is the lowest power of y that appears in any of the four terms. The GCF is, therefore, y^4.

(b) The lowest power of x that appears in any of three terms is x (or x^1). The lowest power of y that appears in any of the three terms is y^2. Thus, the GCF of the three terms is xy^2.

(c) The GCF is $3x^2$. Since y does not appear in $24x^4$, it is not part of the GCF.

EXAMPLE 2   Find the GCF of the following terms.

$$6(x - 3)^2, \quad 5(x - 3), \quad 18(x - 3)^4$$

Solution: The three numbers 6, 5, and 18 have no common factor other than 1. The lowest power of $(x - 3)$ in any of the three terms is $(x - 3)$. Thus, the GCF of the three terms is $(x - 3)$.

Factoring a Monomial from a Polynomial

2

> ### To Factor a Monomial from a Polynomial
> 1. Determine the greatest common factor of all terms in the polynomial.
> 2. Write each term as the product of the GCF and another factor.
> 3. Use the distributive property to *factor out* the GCF.

When we factor a monomial from a polynomial, we are factoring out the greatest common factor. *The first step in any factoring problem is to factor out the GCF.*

EXAMPLE 3   Factor $15x^4 - 5x^3 + 20x^2$.

Solution: The GCF is $5x^2$.

$$15x^4 - 5x^3 + 20x^2 = \boxed{5x^2} \cdot 3x^2 - \boxed{5x^2} \cdot x + \boxed{5x^2} \cdot 4$$
$$= 5x^2(3x^2 - x + 4)$$

Write each term as a product of the GCF and another factor.

Distributive property ●····

The GCF, $5x^2$, was *factored out* from each term in the polynomial. The factored form of $15x^4 - 5x^3 + 20x^2$ is $5x^2(3x^2 - x + 4)$. Thus the factors of $15x^4 - 5x^3 + 20x^2$ are $5x^2$ and $3x^2 - x + 4$.

To check the factoring process, multiply the factors using the distributive property. The product should be the expression with which you began. For instance, in Example 3,

Check: $5x^2(3x^2 - x + 4) = 5x^2(3x^2) + 5x^2(-x) + 5x^2(4)$
$$= 15x^4 - 5x^3 + 20x^2$$

EXAMPLE 4   Factor $20x^2y^3 + 6xy^4 - 12x^3y^5$.

Solution: The GCF is $2xy^3$.

$$20x^2y^3 + 6xy^4 - 12x^3y^5 = \boxed{2xy^3} \cdot 10x + \boxed{2xy^3} \cdot 3y - \boxed{2xy^3} \cdot 6x^2y^2$$
$$= 2xy^3(10x + 3y - 6x^2y^2)$$

Check: $2xy^3(10x + 3y - 6x^2y^2) = 20x^2y^3 + 6xy^4 - 12x^3y^5$ ●····

EXAMPLE 5   Factor $-36p^7q^9 + 12p^5q^{12} - 8p^4q^{15}$.

Solution: When the coefficient of the first term is negative, we can proceed in two ways. We can factor out either a positive expression or a negative expression. Both methods are shown below. The GCF is $4p^4q^9$.

Method 1: $-36p^7q^9 + 12p^5q^{12} - 8p^4q^{15} = -4p^4q^9(9p^3 - 3pq^3 + 2q^6)$

Method 2: $-36p^7q^9 + 12p^5q^{12} - 8p^4q^{15} = 4p^4q^9(-9p^3 + 3pq^3 - 2q^6)$

Both answers are correct. When we factor polynomials whose first term is negative, we will write the factors as in method 1. In Section 6.2 when we factor trinomials like $-4x^2 - 2x - 12$, the first step will be to factor out -2 to obtain $-2(2x^2 + x + 6)$.

EXAMPLE 6   Factor $3x(5x - 2) + 4(5x - 2)$.

Solution: The GCF is $(5x - 2)$. Factoring out the GCF gives

$$3x(5x - 2) + 4(5x - 2) = (3x + 4)(5x - 2)$$

We could have also used the commutative property of multiplication to write the expression as

$$(5x - 2)3x + (5x - 2)4 = (5x - 2)(3x + 4)$$

The factored forms $(3x + 4)(5x - 2)$ and $(5x - 2)(3x + 4)$ are equivalent and both are correct.

EXAMPLE 7   Factor $9(2x - 5) + 6(2x - 5)^2$.

Solution: The GCF is $3(2x - 5)$. Rewrite each term placing the factors in the GCF together.

$$9(2x - 5) + 6(2x - 5)^2 = 3(2x - 5) \cdot 3 + 3(2x - 5) \cdot 2 (2x - 5)$$
$$= 3(2x - 5)[3 + 2(2x - 5)]$$

Now simplify and combine like terms:

$$= 3(2x - 5)[3 + 4x - 10]$$
$$= 3(2x - 5)(4x - 7)$$

EXAMPLE 8   Factor $(2x - 5)(3x - 4) - (2x - 5)(x + 3)$.

Solution: Factor out the greatest common factor $(2x - 5)$ from each term; then simplify.

$$(2x - 5)(3x - 4) - (2x - 5)(x + 3) = (2x - 5)[(3x - 4) - (x + 3)]$$
$$= (2x - 5)(3x - 4 - x - 3)$$
$$= (2x - 5)(2x - 7)$$

Factoring by Grouping   ③ When a polynomial contains *four terms,* it may be possible to factor the polynomial by grouping. To factor by grouping, remove common factors from groups of terms. This procedure is illustrated in the following example. Factoring by grouping is important because you may use it to factor trinomials in Section 6.2.

EXAMPLE 9   Factor $ax + ay + bx + by$.

Solution: There is no factor (other than 1) common to all four terms. However, a is common to the first two terms and b is common to the last two terms. Factor a from the first two terms and b from the last two terms.

$$ax + ay + bx + by = a(x + y) + b(x + y)$$

Now $(x + y)$ is common to both terms. Factor out $(x + y)$.

$$a(x + y) + b(x + y) = (a + b)(x + y)$$

Thus, $ax + ay + bx + by = (a + b)(x + y)$.

> **To Factor Four Terms by Grouping**
> 1. Arrange the four terms into two groups of two terms each. Each group of two terms must have a GCF.
> 2. Factor the GCF from each group of two terms.
> 3. If the two terms formed in step 2 have a GCF, factor it out.

EXAMPLE 10 Factor $6x^2 + 9x + 8x + 12$ by grouping.

Solution: Factor $3x$ from the first two terms and 4 from the last two terms. Then factor the GCF, $2x + 3$, from the resulting two terms.

$$6x^2 + 9x + 8x + 12 = 3x(2x + 3) + 4(2x + 3)$$
$$= (3x + 4)(2x + 3)$$

Factoring by grouping problems may be checked by multiplying the factors. Check the answer to Example 10 now.

EXAMPLE 11 Factor $ax - x + a - 1$ by grouping.

Solution: $ax - x + a - 1 = x(a - 1) + 1(a - 1)$
$$= (x + 1)(a - 1)$$

Note that $a - 1$ was written as $1(a - 1)$.

EXAMPLE 12 Factor $ax - x - a + 1$ by grouping.

Solution: When x is factored from the first two terms, we get

$$ax - x - a + 1 = x(a - 1) - a + 1$$

Now factor -1 from the last two terms to get a common factor, $a - 1$.

$$= x(a - 1) - 1(a - 1)$$
$$= (x - 1)(a - 1)$$

EXAMPLE 13 Factor $2x^3 + 4x^2y - 3xy - 6y^2$.

Solution: This polynomial contains two variables, x and y. The procedure to factor is basically the same. We will factor out $2x^2$ from the first two terms and $-3y$ from the last two terms.

$$2x^3 + 4x^2y - 3xy - 6y^2 = 2x^2(x + 2y) - 3y(x + 2y)$$
$$= (2x^2 - 3y)(x + 2y)$$

EXAMPLE 14 Factor $6r^4 - 9r^3s + 8rs - 12s^2$

Solution: Factor $3r^3$ from the first two terms and $4s$ from the last two terms.

$$6r^4 - 9r^3s + 8rs - 12s^2 = 3r^3(2r - 3s) + 4s(2r - 3s)$$
$$= (3r^3 + 4s)(2r - 3s)$$

Helpful Hint
....

When factoring four terms by grouping, if the *first* term is positive and the *third* term is also positive, you must factor a positive expression from both the first two terms and the last two terms to obtain a factor common to the remaining two terms (see Example 14). If the *first* term is positive and the *third* term is negative, you must factor a positive expression from the first two terms and a negative expression from the last two terms to obtain a factor common to the remaining two terms (see Example 13).

EXAMPLE 15   Factor $x^4 - 5x^3 + 2x^2 - 10x$.

Solution: The first step in any factoring problem is to determine if all the terms have a common factor. If so, begin by factoring out the common factor. In this example, x is common to all four terms. Begin by factoring out an x. Then factor x^2 from the first two terms in parentheses and 2 from the last two terms.

$$
\begin{aligned}
x^4 - 5x^3 + 2x^2 - 10x &= x(x^3 - 5x^2 + 2x - 10) \\
&= x[x^2(x - 5) + 2(x - 5)] \\
&= x[(x^2 + 2)(x - 5)] \\
&= x(x^2 + 2)(x - 5)
\end{aligned}
$$

Exercise Set 6.1

26. $2x^2(2x + 7)(3x^3 + 2x - 1)$ **27.** $(2x + 5)(6x^3 - 2x^2 - 1)$ **30.** $(5a - 3)(4a^2 + 2a - 3)$

Factor out the greatest common factor. If an expression cannot be factored, so state.

1. $8n + 8$ $8(n + 1)$

2. $13x + 5$ cannot be factored

3. $6x^2 + 3x - 9$ $3(2x^2 + x - 3)$

4. $16x^2 - 12x - 6$ $2(8x^2 - 6x - 3)$

5. $7x^5 - 9x^4 + 3x^3$ $x^3(7x^2 - 9x + 3)$

6. $45y^{12} + 30y^{10}$ $15y^{10}(3y^2 + 2)$

7. $-24y^{15} + 9y^3 - 3y$ $-3y(8y^{14} - 3y^2 + 1)$

8. $38x^4 - 16x^5 - 9x^3$ $x^3(38x - 16x^2 - 9)$

9. $6x + 5y + 5xy$ cannot be factored

10. $-x + 3xy^2$ $-x(1 - 3y^2)$

11. $3x^2y + 6x^2y^2 + 3xy$ $3xy(x + 2xy + 1)$

12. $40x^2y^2 + 16xy^4 + 64xy^3$ $8xy^2(5x + 2y^2 + 8y)$

13. $-40x^2y^4z + 8x^6y^2z^2 + 4x^3y$ $-4x^2y(10y^3z - 2x^4yz^2 - x)$

14. $36xy^2z^3 + 36x^3y^2z + 9x^2yz$ $9xyz(4yz^2 + 4x^2y + x)$

15. $19x^4y^{12}z^{13} - 8x^5y^3z^9$ $x^4y^3z^9(19y^9z^4 - 8x)$

16. $24x^6 + 8x^4 - 4x^3y$ $4x^3(6x^3 + 2x - y)$

17. $-52x^2y^2 - 16xy^3 + 26z$ $-2(26x^2y^2 + 8xy^3 - 13z)$

18. $5x(2x - 5) + 3(2x - 5)$ $(5x + 3)(2x - 5)$

19. $3x(4x - 5)^3 + 1(4x - 5)^2$ $(4x - 5)^2(12x^2 - 15x + 1)$

20. $4x(2x + 1) + 2x + 1$ $(2x + 1)(4x + 1)$

21. $3x(2x + 5) - 6(2x + 5)^2$ $3(2x + 5)(-3x - 10)$

22. $(x - 3)(x + 1) + (x - 3)(x + 2)$ $(x - 3)(2x + 3)$

23. $(3p - q)(2p - q) + (3p - q)(p - 2q)$ $3(3p - q)(p - q)$

24. $(3r + 2)(3r - 1) - (3r + 2)(2r + 3)$ $(3r + 2)(r - 4)$

25. $(x - 2)(3x + 5) - (x - 2)(5x - 4)$ $(x - 2)(-2x + 9)$

26. $6x^5(2x + 7) + 4x^3(2x + 7) - 2x^2(2x + 7)$

27. $6x^3(2x + 5) - 2x^2(2x + 5) - (2x + 5)$

28. $5a(3x - 2)^5 + 4(3x - 2)^4$ $(3x - 2)^4(15ax - 10a + 4)$

29. $4p(2r - 3)^7 - 3(2r - 3)^6$ $(2r - 3)^6(8pr - 12p - 3)$

30. $4a^2(5a - 3) + 2a(5a - 3) - 3(5a - 3)$

Factor by grouping.

31. $x^2 + 3x - 5x - 15$ $(x - 5)(x + 3)$

32. $x^2 + 3x - 2x - 6$ $(x + 3)(x - 2)$

33. $3x^2 + 9x + x + 3$ $(3x + 1)(x + 3)$

34. $x^2 + 4x + x + 4$ $(x + 4)(x + 1)$

35. $4x^2 - 2x - 2x + 1$ $(2x - 1)^2$

36. $2x^2 + 6x - x - 3$ $(x + 3)(2x - 1)$

37. $8x^2 - 4x - 20x + 10$ $2(2x - 5)(2x - 1)$
38. $ax + ay + bx + by$ $(a + b)(x + y)$
39. $42ac + 35ad + 18bc + 15bd$ $(7a + 3b)(6c + 5d)$
40. $35x^2 - 40xy + 21xy - 24y^2$ $(5x + 3y)(7x - 8y)$
41. $x^3 - 3x^2 + 4x - 12$ $(x^2 + 4)(x - 3)$
42. $3x^2 - 18xy + 4xy - 24y^2$ $(x - 6y)(3x + 4y)$
43. $10m^2 - 12mn - 25mn + 30n^2$ $(2m - 5n)(5m - 6n)$
44. $12x^2 - 9xy + 4xy - 3y^2$ $(4x - 3y)(3x + y)$
45. $6x^3 + 18x^2 - 12x - 36$ $6(x^2 - 2)(x + 3)$
46. $2a^4 - 2a^3 - 5a^2 + 5a$ $a(2a^2 - 5)(a - 1)$
47. $2a^4b - 2ac^2 - 3a^3bc + 3c^3$ $(a^3b - c^2)(2a - 3c)$
48. $8r^2 + 6rs - 12rs - 9s^2$ $(4r + 3s)(2r - 3s)$
49. $3p^3 + 3pq^2 + 2p^2q + 2q^3$ $(3p + 2q)(p^2 + q^2)$
50. $16r^3 - 4r^2s^2 - 4rs + s^3$ $(4r - s^2)(4r^2 - s)$
66. $(4a - b)(a + 8b)$ **68.** $y(2x - 3y)(x^2y^2 + 3z^2)$ **70.** $(b - 4)(3x^2 - 2x + 5)$ **71. (a)** $(1 - 0.06)(x + 0.06x) = 0.94(1.06x)$
(b) $0.9964x$, slightly less (99.64% of the original cost)

Factor completely.

51. $20p^3 - 18p^2 + 12p$ $2p(10p^2 - 9p + 6)$
52. $6x^3 - 8x^2 - x$ $x(6x^2 - 8x - 1)$
53. $16xy^2z + 4x^3y - 8$ $4(4xy^2z + x^3y - 2)$
54. $-80x^5y^3z^4 + 36x^2yz^3$ $-4x^2yz^3(20x^3y^2z - 9)$
55. $5x^2 - 10x + 3x - 6$ $(x - 2)(5x + 3)$
56. $4x^2 + 6x - 6x - 9$ $(2x - 3)(2x + 3)$
57. $14y^3z^5 - 28y^3z^6 - 9xy^2z^2$ $y^2z^2(14yz^3 - 28yz^4 - 9x)$
58. $8x^2 - 20x - 4x + 10$ $2(2x - 1)(2x - 5)$
59. $7x^4y^9 - 21x^3y^7z^5 - 35y^8z^9$ $7y^7(x^4y^2 - 3x^3z^5 - 5yz^9)$
60. $48x^2y + 16xy^2 + 33xy$ $xy(48x + 16y + 33)$
61. $15a^2 - 18ab - 20ab + 24b^2$ $(3a - 4b)(5a - 6b)$
62. $3x(7x + 1) - 2(7x + 1)$ $(3x - 2)(7x + 1)$
63. $6x^2 - 9xy + 2xy - 3y^2$ $(3x + y)(2x - 3y)$
64. $x^2 - 3xy + 2xy - 6y^2$ $(x - 3y)(x + 2y)$
65. $5x(x + 3)^2 - 3(x + 3)$ $(x + 3)(5x^2 + 15x - 3)$
66. $(4a - b)(2a + 3b) - (4a - b)(a - 5b)$
67. $(3c - d)(c + d) - (3c - d)(c - d)$ $(3c - d)(2d)$
68. $2x^3y^3 + 6xyz^2 - 3x^2y^4 - 9y^2z^2$
69. $3x^5 - 15x^3 + 2x^3 - 10x$ $x(3x^2 + 2)(x^2 - 5)$
70. $3x^2(b - 4) - 2x(b - 4) + 5(b - 4)$

✎ **75.** What is the first step in *any* factoring problem?

✎ **76.** What is the greatest common factor of the terms of an expression?

✎ **77. (a)** Explain how to find the greatest common factor of the terms of a polynomial.

(b) Using the procedure from part **(a)**, find the greatest common factor of the polynomial $6x^2y^5 - 2x^3y + 12x^9y^3$. $2x^2y$

(c) Factor the polynomial in part **(b)**.

71. When the 1996 cars came out, the list price of one model increased by 6% over the list price of the 1995 model. Then in a special sale, the prices of all 1996 cars were reduced by 6%. The sale price can be represented by $(x + 0.06x) - 0.06(x + 0.06x)$, where x is the list price of the 1995 model.

(a) Factor out $(x + 0.06x)$ from each term.

(b) Is the sale price more or less than the price of the 1995 model?

Read Exercise 71 before working Exercises 72–74.

72. A dress is reduced by 10%, then the sale price is reduced by another 10%.

(a) Write an expression for the final price of the item.

(b) How does the final price compare with the regular price of the item? Use factoring in obtaining your answer. $0.90(0.90x) = 0.81x$

73. The price of a Toro lawn mower is increased by 15%. Then at a 4th of July sale the price is reduced by 20%.

(a) Write an expression for the final price of the item.

(b) How does the sale price compare with the regular price? Use factoring in obtaining your answer.

74. In which of the following **(a)** or **(b)**, will the final price be lower, and by how much?

(a) Decreasing the price of an item by 6%, then increasing that price by 8%.

(b) Increasing the price of an item by 6%, then decreasing that price by 8%.

SEE EXERCISE 72.

72. (a) $(x - 0.10x) - 0.10(x - 0.10x)$ **73. (a)** $(x + 0.15x) - 0.20(x + 0.15x)$ **(b)** $0.80(1.15x) = 0.92x$
74. (b) by $1.0152x - 0.9752x$ or $0.04x$ **75.** Determine if all the terms contain a GCF, and if so, factor it out.
76. the greatest factor that divides all terms in the expression   **77. (c)** $2x^2y(3y^4 - x + 6x^7y^2)$

78. When a term of a polynomial is itself the GCF, what is written in place of that term when the GCF is factored out? Explain.   the number 1

79. Consider the factoring in parts (a) and (b). Is either or both factoring correct? Explain.

   (a) $-144m^2n^3 + 44m^7n^6 - 24m^3n^8$
      $= 4m^2n^3(-36 + 11m^5n^3 - 6mn^5)$

(b) $-144m^2n^3 + 44m^7n^6 - 24m^3n^8$
   $= -4m^2n^3(36 - 11m^5n^3 + 6mn^5)$ Both are correct.

80. (a) Explain how to factor a polynomial of four terms by grouping.

   (b) Factor $6x^3 - 2xy^3 + 3x^2y^2 - y^5$ by the procedure from part **(a)**. $(2x + y^2)(3x^2 - y^3)$

CUMULATIVE REVIEW EXERCISES

[2.3] **81.** The price of a shirt is increased by 10%. This price is then decreased by \$10. If the final price of the shirt is \$17.50, find the original price of the shirt.   \$25

[4.5] **82.** Find the solution to the system of inequalities.
$$y > -3x + 4$$
$$3x - 2y \le 6$$

Divide.

[5.5] **83.** $\dfrac{3x^2 - 6xy^2 + 12xy^3}{4x^2y^2}$ $\dfrac{3}{4y^2} - \dfrac{3}{2x} + \dfrac{3y}{x}$

84. $\dfrac{6x^3 + 13x^2 - 6x - 18}{2x + 3}$ $3x^2 + 2x - 6$

82.

Group Activity/ Challenge Problems

Factor.

1. $4x^2(x - 3)^3 - 6x(x - 3)^2 + 4(x - 3)$. $2(x - 3)(2x^4 - 12x^3 + 15x^2 + 9x + 2)$

2. $12(p + 2q)^4 - 40(p + 2q)^3 + 12(p + 2q)^2$ $4(p + 2q)^2(3p + 6q - 1)(p + 2q - 3)$

3. (a) $(x + 1)^2 + (x + 1)$ $(x + 1)[(x + 1) + 1] = (x + 1)(x + 2)$

   (b) $(x + 1)^3 + (x + 1)^2$ $(x + 1)^2[(x + 1) + 1] = (x + 1)^2(x + 2)$

   (c) $(x + 1)^n + (x + 1)^{n-1}$ $(x + 1)^{n-1}[(x + 1) + 1] = (x + 1)^{n-1}(x + 2)$

   (d) $(x + 1)^{n+1} + (x + 1)^n$ $(x + 1)^n[(x + 1) + 1] = (x + 1)^n(x + 2)$

*4. $x^{3/2} + x^{1/2}$ $x^{1/2}(x + 1)$

5. $3x^{1/3} + 6x^{4/3} + 12x^{7/3}$ $3x^{1/3}(1 + 2x + 4x^2)$

6. $10x^{-2} + 25x^{-3}$ $5x^{-3}(2x + 5)$

7. $9n^{-1} + 6n^{-2} - 12n^{-3}$ $3n^{-3}(3n^2 + 2n - 4)$

8. $4x(x + 5)^{-2} + 2x(x + 5)^{-1}$ $2x(x + 5)^{-2}[2 + (x + 5)^1] = 2x(x + 5)^{-2}(x + 7)$

9. $x(2x - 3)^{-1/2} + 6(2x - 3)^{1/2}$ $(2x - 3)^{-1/2}[x + 6(2x - 3)^1] = (2x - 3)^{-1/2}(13x - 18)$

10. $x^{6m} - 2x^{4m}$ $x^{4m}(x^{2m} - 2)$

11. $x^{2mn} + x^{4mn}$ $x^{2mn}(1 + x^{2mn})$

12. $3x^{4m} - 2x^{3m} + x^{2m}$ $x^{2m}(3x^{2m} - 2x^m + 1)$

13. $r^{y+4} + r^{y+3} + r^{y+2}$ $r^{y+2}(r^2 + r + 1)$

14. $a^rb^r + c^rb^r - a^rd^r - c^rd^r$ $(b^r - d^r)(a^r + c^r)$

15. $6a^kb^k - 2a^kc^k - 9b^k + 3c^k$ $(2a^k - 3)(3b^k - c^k)$

---

*Factoring problems with fractional and negative exponents are discussed in Section 8.2.

6.2 Factoring Trinomials

Tape 9

1	Factor trinomials of the form $x^2 + bx + c$.
2	Factor trinomials of the form $ax^2 + bx + c$, $a \neq 1$, using trial and error.
3	Factor trinomals of the form $ax^2 + bx + c$, $a \neq 1$, using grouping.
4	Factor trinomials that contain a common factor.
5	Factor trinomials using substitution.

Factoring Trinomials of the Form $x^2 + bx + c$

1 In this section we learn how to factor trinomials of the form $ax^2 + bx + c$, $a \neq 0$.

Trinomials	*Coefficients*
$3x^2 + 2x - 5$	$a = 3, \quad b = 2, \quad c = -5$
$-\dfrac{1}{2}x^2 - 4x + 3$	$a = -\dfrac{1}{2}, \quad b = -4, \quad c = 3$

To Factor Trinomials of the Form $x^2 + bx + c$ (note: $a = 1$)

1. Find two numbers (or factors) whose product is c and whose sum is b.
2. The factors of the trinomial will be of the form

$$(x + \blacksquare)(x + \blacksquare)$$

$\uparrow$ $\uparrow$

one factor   other factor
determined   determined
in step 1   in step 1

If the numbers determined in step 1 are, for example, 3 and -5, the factors would be written $(x + 3)(x - 5)$. This procedure is illustrated in the following examples.

EXAMPLE 1 Factor $x^2 - x - 12$.

Solution: $a = 1, b = -1, c = -12$. We must find two numbers whose product is c, which is -12, and whose sum is b, which is -1. We begin by listing the factors of -12 and try to find a pair whose sum is -1.

Factors of -12	*Sum of Factors*
$(1)(-12)$	$1 + (-12) = -11$
$(2)(-6)$	$2 + (-6) = -4$
$(3)(-4)$	$3 + (-4) = -1$
$(4)(-3)$	$4 + (-3) = 1$
$(6)(-2)$	$6 + (-2) = 4$
$(12)(-1)$	$12 + (-1) = 11$

The numbers we are seeking are 3 and -4 because their product is -12 and their sum is -1. Now factor the trinomial using the 3 and -4.

$$x^2 - x - 12 = (x + 3)(x - 4)$$

$\uparrow$ $\uparrow$

one factor   other factor
of -12 of -12

Notice in Example 1 that we listed all the factors of -12. After the two factors whose product is c and whose sum is b are found, there is no need to go further in listing the factors. The factors were listed here to show, for example, that $(2)(-6)$ is a different set of factors than $(-2)(6)$. Note that as the positive factor increases the sum of the factors increases.

EXAMPLE 2   Factor $x^2 - 5x - 6$.

Solution: We must find two numbers whose product is -6 and whose sum is -5. The numbers are 1 and -6 because $(1)(-6) = -6$ and $1 + (-6) = -5$.

$$x^2 - 5x - 6 = (x + 1)(x - 6)$$

Since the factors may be placed in any order, $(x - 6)(x + 1)$ is also an acceptable answer.

Helpful Hint

Checking Factoring

Factoring problems can be checked by multiplying the factors obtained. If the factoring is correct, you will obtain the polynomial you started with. To check Example 2, we will multiply the factors using the FOIL method.

$$(x + 1)(x - 6) = x^2 - 6x + x - 6 = x^2 - 5x - 6$$

Since the product of the factors is the trinomial we began with, our factoring is correct. You should always check your factoring.

The procedures used to factor trinomials of the form $x^2 + bx + c$ can be used on other trinomials, as in the following example.

EXAMPLE 3   Factor $x^2 + 2xy - 15y^2$.

Solution: We must find two numbers whose product is -15 and whose sum is 2. The two numbers are 5 and -3 because $(5)(-3) = -15$ and $5 + (-3) = 2$. Since the last term of the trinomial contains y^2, the second term of each factor must contain y.

$$x^2 + 2xy - 15y^2 = (x + 5y)(x - 3y)$$

Check: $(x + 5y)(x - 3y) = x^2 - 3xy + 5xy - 15y^2$
$$= x^2 + 2xy - 15y^2$$

If each term of a trinomial has a common factor, use the distributive property to remove the common factor before following the procedure outlined earlier.

EXAMPLE 4   Factor $3x^2 - 6x - 72$.

Solution: The factor 3 is common to all three terms of the trinomial. Factor it out first.

$$3x^2 - 6x - 72 = 3(x^2 - 2x - 24)$$

The 3 that was factored out is a part of the answer but plays no further part in the factoring process. Now continue to factor $x^2 - 2x - 24$. We must find two numbers whose product is -24 and whose sum is -2. The numbers are -6 and 4.

$$3(x^2 - 2x - 24) = 3(x - 6)(x + 4)$$

Therefore, $3x^2 - 6x - 72 = 3(x - 6)(x + 4)$.

Factoring Trinomials of the Form $ax^2 + bx + c, a \neq 1$

2 Now we will look at some examples of factoring trinomials of the form

$$ax^2 + bx + c, \qquad a \neq 1$$

Two methods of factoring this type of trinomial will be illustrated. The first method, trial and error, involves trying various combinations until the correct combination is found. The second method makes use of factoring by grouping, a procedure that was presented in Section 5.1. You may use either method unless your instructor specifies that you should use a particular one.

Method 1: Trial and Error

Let us now discuss the trial-and-error method of factoring trinomials. As an aid in our explanation we will multiply $(2x + 3)(x + 1)$ using the FOIL method.

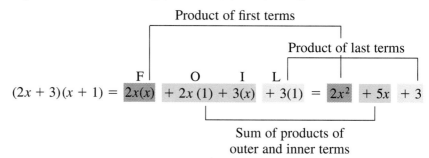

Therefore, if you are factoring the trinomial $2x^2 + 5x + 3$, you should realize that the product of the first terms of the factors must be $2x^2$, the product of the last terms must be 3, and the sum of the products of the outer and inner terms must be $5x$.

To factor $2x^2 + 5x + 3$, we begin as shown here.

$$2x^2 + 5x + 3 = (2x \qquad)(x \qquad) \qquad \text{The product of first terms is } 2x^2.$$

Now fill in the second terms using positive integers whose product is 3. Only positive integers will be considered since the product of the last terms is positive, and the sum of the products of the outer and inner terms is also positive. The two possibilities are as follows:

$$\left. \begin{array}{l} (2x + 1)(x + 3) \\ (2x + 3)(x + 1) \end{array} \right\} \quad \text{The product of the last terms is 3.}$$

To determine which factoring is correct find the sum of the product of the outer terms and the product of the inner terms. If either has a sum of $5x$, the middle term of the trinomial, that factoring is correct.

$(2x + 1)(x + 3) = 2x^2 + 6x + x + 3 = 2x^2 + 7x + 3$ **Wrong middle term**

$(2x + 3)(x + 1) = 2x^2 + 2x + 3x + 3 = 2x^2 + 5x + 3$ **Correct middle term**

Therefore, the factors of $2x^2 + 5x + 3$ are $2x + 3$ and $x + 1$. Thus,

$$2x^2 + 5x + 3 = (2x + 3)(x + 1)$$

Note that if we had begun factoring by writing

$$2x^2 + 5x + 3 = (x \qquad)(2x \qquad)$$

we could have also obtained the correct factors.

> ### To Factor Trinomials of the Form $ax^2 + bx + c$, $a \neq 1$, Using Trial and Error
> **1.** Write all pairs of factors of the coefficient of the squared term, a.
> **2.** Write all pairs of factors of the constant, c.
> **3.** Try various combinations of these factors until the correct middle term, bx, is found.

EXAMPLE 5   Factor $3x^2 - 13x + 10$.

Solution: The only factors of 3 are 1 and 3. Therefore, we write

$$3x^2 - 13x + 10 = (3x \qquad)(x \qquad)$$

The number 10 has both positive and negative factors. However, since the product of the last terms must be positive $(+10)$, and the sum of the products of the outer and inner terms must be negative (-13), the two factors of 10 must both be negative. Why? The negative factors of 10 are $(-1)(-10)$ and $(-2)(-5)$. Below is a list of the possible factors. We look for the factors that give us the correct middle term, $-13x$.

Possible Factors	*Sum of Products of Outer and Inner Terms*
$(3x - 1)(x - 10)$	$-31x$
$(3x - 10)(x - 1)$	$-13x$ ← correct middle term
$(3x - 2)(x - 5)$	$-17x$
$(3x - 5)(x - 2)$	$-11x$

Thus, $3x^2 - 13x + 10 = (3x - 10)(x - 1)$.

The following Helpful Hint is very important. Study it carefully.

Helpful Hint

Factoring by Trial and Error

When factoring a trinomial of the form $ax^2 + bx + c$, the sign of the constant term, c, is very helpful in finding the solution. If $a > 0$, then:

1. When the constant term, c, is positive, and the numerical coefficient of the x term, b, is positive, both numerical factors will be positive.

$$\textit{Example:} \quad x^2 \;\underset{\text{positive}}{+}\; \underset{\text{positive}}{7}x + 12 = (x \;\underset{\text{positive}}{+}\; 3)(x \;\underset{\text{positive}}{+}\; 4)$$

2. When c is positive and b is negative, both numerical factors will be negative.

$$\textit{Example:} \quad x^2 \;\underset{\text{negative}}{-}\; \underset{\text{positive}}{5}x + 6 = (x \;\underset{\text{negative}}{-}\; 2)(x \;\underset{\text{negative}}{-}\; 3)$$

In the first two cases the constant term, c, was positive. *Whenever the constant is positive, the signs in both factors will be the same, either both positive or both negative. Furthermore, both factors will contain the same sign as the sign of the x term in the trinomial.*

3. When c is negative, one of the numerical factors will be positive and the other will be negative.

$$\textit{Example:} \quad x^2 + x \;\underset{\text{negative}}{-}\; 6 = (x \;\underset{\text{positive}}{+}\; 3)(x \;\underset{\text{negative}}{-}\; 2)$$

Here the constant term, c, is negative. *When the constant is negative, the factors will contain opposite signs. One factor will contain a positive sign and the other factor will contain a negative sign.*

EXAMPLE 6   Factor $8x^2 - 51x + 18$.

Solution: The factors of 8 will be either $4 \cdot 2$ or $8 \cdot 1$. When there is more than one pair of factors for the first term, we generally try the medium-sized factors first. If this does not work, try the other factors. We will therefore try $4 \cdot 2$ first.

Since the constant, 18, is positive and the x term, $-51x$, is negative the numerical terms in the factors must both be negative. The possible ways to factor 18 are $(-18)(-1)$, $(-9)(-2)$, and $(-6)(-3)$. We need to find the pair of factors whose sum of the products of the outer and inner terms is $-51x$.

Possible Factors	Sum of Products of Outer and Inner Terms
$(4x - 18)(2x - 1)$	$-40x$
$(4x - 9)(2x - 2)$	$-26x$
$(4x - 6)(2x - 3)$	$-24x$
$(4x - 3)(2x - 6)$	$-30x$
$(4x - 2)(2x - 9)$	$-40x$
$(4x - 1)(2x - 18)$	$-74x$

Since the factors 4 and 2 did not yield the correct middle term, we now try 8 and 1.

$(8x - 18)(x - 1)$	$-26x$
$(8x - 9)(x - 2)$	$-25x$
$(8x - 6)(x - 3)$	$-30x$
$(8x - 3)(x - 6)$	$-51x$

The factors that give the correct term of $-51x$ are $(8x - 3)(x - 6)$. Note that once we find the factors we are seeking we can stop trying factors.

$$8x^2 - 51x + 18 = (8x - 3)(x - 6)$$

EXAMPLE 7   Factor $6x^2 - 11xy - 10y^2$.

Solution: The factors of 6 are either $6 \cdot 1$ or $2 \cdot 3$. Therefore, the factors of the trinomial may be of the form $(6x \quad)(x \quad)$ or $(2x \quad)(3x \quad)$. We begin with the middle-sized factors; thus we write

$$6x^2 - 11xy - 10y^2 = (2x \quad)(3x \quad)$$

The factors of -10 are $(-1)(10)$, $(1)(-10)$, $(-2)(5)$, and $(2)(-5)$. Since there are eight factors of -10, there will be eight pairs of possible factors to try. Can you list them? The correct factoring is

$$6x^2 - 11xy - 10y^2 = (2x - 5y)(3x + 2y)$$

In Example 7 we were fortunate to find the correct factors by using the form $(2x \quad)(3x \quad)$. If we had not found the correct factors using these, we would have tried $(6x \quad)(x \quad)$.

Method 2: Using Grouping

3   Now we will discuss the grouping method of factoring trinomials of the form $ax^2 + bx + c, a \neq 1$.

> **To Factor Trinomials of the Form $ax^2 + bx + c$, $a \neq 1$, Using Grouping**
>
> **1.** Find two numbers whose product is $a \cdot c$ and whose sum is b.
> **2.** Rewrite the middle term, bx, term using the numbers found in step 1.
> **3.** Factor by grouping.

EXAMPLE 8   Factor $2x^2 - 5x - 12$.

Solution: We see that $a = 2$, $b = -5$, and $c = -12$. We must find two numbers whose product is $a \cdot c$, or $2(-12) = -24$, and whose sum is b, -5. The two numbers are -8 and 3 because $(-8)(3) = -24$ and $-8 + 3 = -5$. Now rewrite the middle term, $-5x$, using $-8x$ and $3x$.

$$2x^2 - 5x - 12 = 2x^2 \overbrace{- 8x + 3x}^{-5x} - 12$$

Now factor by grouping as explained in Section 5.1. Factor out $2x$ from the first two terms and 3 from the last two terms.

$$2x^2 - 5x - 12 = 2x(x - 4) + 3(x - 4)$$
$$= (2x + 3)(x - 4)$$

Note in Example 8 that we wrote $-5x$ as $-8x + 3x$. As we show below, the same factors would be obtained if we wrote $-5x$ as $3x - 8x$. Therefore, it makes no difference which factor is listed first when factoring by grouping.

$$2x^2 - 5x - 12 = 2x^2 \overbrace{+ 3x - 8x}^{-5x} - 12$$
$$= x(2x + 3) - 4(2x + 3)$$
$$= (x - 4)(2x + 3)$$

EXAMPLE 9   Factor $12x^2 - 19xy + 5y^2$.

Solution: We must find two numbers whose product is $(12)(5) = 60$ and whose sum is -19. Since the product of the numbers is positive and their sum is negative, the two numbers must both be negative. Why?

The two numbers are -15 and -4 because $(-15)(-4) = 60$ and $-15 + (-4) = -19$. Now rewrite the middle term, $-19xy$, using $-15xy$ and $-4xy$. Then factor by grouping.

$$12x^2 - 19xy + 5y^2 = 12x^2 \overbrace{- 15xy - 4xy}^{-19xy} + 5y^2$$
$$= 3x(4x - 5y) - y(4x - 5y)$$
$$= (3x - y)(4x - 5y)$$

Try Example 9 again, this time writing $-19xy$ as $-4xy - 15xy$. If you do it correctly, you should get the same factors.

Factor out the Common Factor

4   The first step when factoring any trinomial whose coefficient of the squared term is not 1 is to determine if all three terms have a common factor. If so, factor out that common factor. Then factor the remaining polynomial using either the trial-and-error method or the factoring by grouping method.

Example 10 Factor $24x^3 + 60x^2 - 36x$.

Solution: By examining the three terms of the trinomial, we see that $12x$ is common to each term. We therefore begin by factoring out $12x$.

$$24x^3 + 60x^2 - 36x = 12x(2x^2 + 5x - 3)$$

Now factor $2x^2 + 5x - 3$ using either trial and error or grouping.

$$24x^3 + 60x^2 - 36x = 12x(2x - 1)(x + 3)$$

When the coefficient of the term of the highest degree is negative, we begin by factoring out a negative number or expression. Had the trinomial in Example 10 been $-24x^3 - 60x^2 + 36x$, we would have factored out $-12x$, then continued factoring to obtain

$$-24x^3 - 60x^2 + 36x = -12x(2x^2 + 5x - 3)$$
$$= -12x(2x - 1)(x + 3)$$

To factor $-3x^2 + 8x + 16$, we factor out -1 as follows:

$$-3x^2 + 8x + 16 = -1(3x^2 - 8x - 16)$$
$$= -(3x + 4)(x - 4)$$

It is important for you to realize that not every trinomial can be factored by the methods presented in this section. Consider the following example.

Example 11 Factor $2x^2 + 6x + 5$.

Solution: If you try to factor this using either trial and error or grouping, you will see it cannot be factored. Polynomials of this type are called **prime polynomials.**

Helpful Hint

We have introduced two methods of factoring trinomials of the form $ax^2 + bx + c,\ a \neq 1$. Which method should you use? If your instructor tells you to use a specific method, use it. If you may use either method, you may wish to try factoring by the trial-and-error method first, especially if the constant term has few factors. If you cannot quickly find the answer using trial and error, you may then wish to try the grouping method. After a little practice, you will be able to determine which method gives you the most success.

Factoring Using Substitution

5 Sometimes a more complicated trinomial can be factored by substituting one variable for another. The next two examples illustrate **factoring using substitution.**

EXAMPLE 12 Factor $y^4 - y^2 - 6$.

Solution: If we can rewrite this expression in the form $ax^2 + bx + c$, it will be easier to factor. Because that $(y^2)^2 = y^4$, if we substitute x for y^2, the trinomial becomes

$$y^4 - y^2 - 6 = (y^2)^2 - y^2 - 6$$
$$= x^2 - x - 6$$

Now factor $x^2 - x - 6$.

$$= (x + 2)(x - 3)$$

Finally, substitute y^2 in place of x to obtain

$$= (y^2 + 2)(y^2 - 3)$$

Thus, $y^4 - y^2 - 6 = (y^2 + 2)(y^2 - 3)$. Note that x was substituted for y^2, and then y^2 was substituted back for x.

EXAMPLE 13 Factor $3y^4 - 17y^2 - 28$.

Solution: Let $x = y^2$. Then the trinomial can be written

$$3x^2 - 17x - 28$$
$$= (3x + 4)(x - 7)$$

Now substitute y^2 for x.

$$= (3y^2 + 4)(y^2 - 7)$$

Thus, $3y^4 - 17y^2 - 28 = (3y^2 + 4)(y^2 - 7)$.

EXAMPLE 14 Factor $2(x + 5)^2 - 5(x + 5) - 12$.

Solution: We will again use a substitution, as in Examples 12 and 13. By substituting $a = x + 5$ in the equation, we obtain

$$2(x + 5)^2 - 5(x + 5) - 12 = 2a^2 - 5a - 12$$

Now factor $2a^2 - 5a - 12$.

$$= (2a + 3)(a - 4)$$

Finally, replace a with $x + 5$ to obtain

$$= [2(x + 5) + 3][(x + 5) - 4]$$
$$= [2x + 10 + 3][x + 1]$$
$$= (2x + 13)(x + 1)$$

Thus, $2(x + 5)^2 - 5(x + 5) - 12 = (2x + 13)(x + 1)$. Note that a was substituted for $x + 5$, and then $x + 5$ was substituted back for a.

In Examples 12 and 13 we used x in our substitution, whereas in Example 14, we used a. The letter selected is immaterial to the final answer.

Exercise Set 6.2

Factor each trinomial completely. If the trinomial cannot be factored, so state.

1. $x^2 + 7x + 6$ $(x + 6)(x + 1)$

2. $p^2 - 3p - 10$ $(p - 5)(p + 2)$

3. $y^2 - 12y + 11$ $(y - 1)(y - 11)$

4. $w^2 - 7w + 9$ cannot be factored

5. $x^2 - 16x + 64$ $(x - 8)^2$

6. $x^2 - 34x + 64$ $(x - 32)(x - 2)$

7. $x^2 - 11x - 30$ cannot be factored

8. $-a^2 + 18a - 45$ $-(a - 15)(a - 3)$

9. $y^2 - 9y + 15$ cannot be factored

10. $x^2 - 4xy + 3y^2$ $(x - y)(x - 3y)$

11. $x^2 - 6xy + 8y^2$ $(x - 4y)(x - 2y)$

12. $z^2 - 7yz + 10y^2$ $(z - 2y)(z - 5y)$

13. $-5x^2 - 20x - 15$ $-5(x + 1)(x + 3)$

14. $4x^2 + 12x - 16$ $4(x + 4)(x - 1)$

15. $x^2 - 12xy - 45y^2$ $(x - 15y)(x + 3y)$

16. $x^3 - 3x^2 - 18x$ $x(x - 6)(x + 3)$

17. $x^3 + 11x^2 - 42x$ $x(x + 14)(x - 3)$

18. $5p^2 - 8p + 3$ $(p - 1)(5p - 3)$

19. $4w^2 + 13w + 3$ $(4w + 1)(w + 3)$

20. $3x^2 - 11x - 6$ cannot be factored

21. $-3x^2 - 14x + 5$ $-(x + 5)(3x - 1)$

22. $5y^2 - 16y + 3$ $(5y - 1)(y - 3)$

23. $3y^2 - 2y - 5$ $(3y - 5)(y + 1)$

24. $3x^2 - 22xy + 7y^2$ $(x - 7y)(3x - y)$

25. $4x^2 + 4xy - 3y^2$ $(2x + 3y)(2x - y)$

26. $6x^3 + 5x^2 - 4x$ $x(3x + 4)(2x - 1)$

27. $8x^2 + 2x - 20$ $2(4x^2 + x - 10)$

28. $12x^3 - 12x^2 - 45x$ $3x(2x + 3)(2x - 5)$

29. $8x^2 - 8xy - 6y^2$ $2(2x - 3y)(2x + y)$

30. $18w^2 + 18wz - 8z^2$ $2(3w + 4z)(3w - z)$

31. $35x^2 + 13x - 12$ $(7x - 3)(5x + 4)$

32. $12a^2 - 11a - 5$ $(4a - 5)(3a + 1)$

33. $8x^2 - 34x + 30$ $2(4x - 5)(x - 3)$

34. $100b^2 - 90b + 20$ $10(5b - 2)(2b - 1)$

35. $x^5y - 3x^4y - 18x^3y$ $x^3y(x - 6)(x + 3)$

36. $a^3b^5 - a^2b^5 - 12ab^5$ $ab^5(a - 4)(a + 3)$

37. $a^3b + 2a^2b - 35ab$ $ab(a + 7)(a - 5)$

38. $3b^4c - 18b^3c^2 + 27b^2c^3$ $3b^2c(b - 3c)^2$

39. $6p^3q^2 - 24p^2q^3 - 30pq^4$ $6pq^2(p - 5q)(p + q)$

40. $8m^8n^3 + 4m^7n^4 - 24m^6n^5$ $4m^6n^3(m + 2n)(2m - 3n)$

41. $18x^2 + 9x - 20$ $(6x - 5)(3x + 4)$

42. $30x^2 - x - 20$ $(6x - 5)(5x + 4)$

43. $6x^2 - 43x + 20$ $(3x - 20)(2x - 1)$

44. $36x^2 - 23x - 8$ $(9x - 8)(4x + 1)$

45. $8x^4y^4 + 24x^3y^4 - 32x^2y^4$ $8x^2y^4(x + 4)(x - 1)$

46. $5a^3b^2 - 8a^2b^3 + 3ab^4$ $ab^2(5a - 3b)(a - b)$

Factor each trinomial completely.   **61.** $(2a - 5)(a - 1)(5 - a)$ **62.** $(y + 2)(2y + 3)(y + 5)$ **64.** $(3x - 1)(x + 2)(x - 2)$

47. $x^4 + x^2 - 6$ $(x^2 + 3)(x^2 - 2)$

48. $x^4 - 3x^2 - 10$ $(x^2 - 5)(x^2 + 2)$

49. $x^4 + 5x^2 + 6$ $(x^2 + 2)(x^2 + 3)$

50. $x^4 - 2x^2 - 15$ $(x^2 - 5)(x^2 + 3)$

51. $6a^4 + 5a^2 - 25$ $(2a^2 + 5)(3a^2 - 5)$

52. $(2x + 1)^2 + 2(2x + 1) - 15$ $4(x + 3)(x - 1)$

53. $4(x + 1)^2 + 8(x + 1) + 3$ $(2x + 5)(2x + 3)$

54. $(2x + 3)^2 - (2x + 3) - 6$ $2x(2x + 5)$

55. $6(a + 2)^2 - 7(a + 2) - 5$ $(3a + 1)(2a + 5)$

56. $6(p - 5)^2 + 11(p - 5) + 3$ $(2p - 7)(3p - 14)$

57. $a^2b^2 - 8ab + 15$ $(ab - 3)(ab - 5)$

58. $x^2y^2 + 10xy + 24$ $(xy + 6)(xy + 4)$

59. $3x^2y^2 - 2xy - 5$ $(3xy - 5)(xy + 1)$

60. $3p^2q^2 + 11pq + 6$ $(3pq + 2)(pq + 3)$

61. $2a^2(5 - a) - 7a(5 - a) + 5(5 - a)$

62. $2y^2(y + 2) + 13y(y + 2) + 15(y + 2)$

63. $2x^2(x - 3) + 7x(x - 3) + 6(x - 3)$ $(2x + 3)(x + 2)(x - 3)$

64. $3x^2(x - 2) + 5x(x - 2) - 2(x - 2)$

65. $y^4 - 7y^2 - 30$ $(y^2 - 10)(y^2 + 3)$

66. $3z^4 - 14z^2 - 5$ $(3z^2 + 1)(z^2 - 5)$

67. $x^2(x + 3) + 3x(x + 3) + 2(x + 3)$ $(x + 2)(x + 1)(x + 3)$

68. $x^2(x - 1) - x(x - 1) - 30(x - 1)$ $(x - 1)(x - 6)(x + 5)$

69. $5a^5b^2 - 8a^4b^3 + 3a^3b^4$ $a^3b^2(5a - 3b)(a - b)$

70. $2x^2y^6 + 3xy^5 - 9y^4$ $y^4(xy + 3)(2xy - 3)$

If we know one factor of a polynomial, we can use division or synthetic division to find a second factor (see Sections 5.4 and 5.5). In Exercises 71–78 one factor of the polynomial is given. **(a)** *Use division to find a second factor.* **(b)** *Continue factoring, if possible, until the polynomial is factored completely.*

72. (a) $3x + 8$ **(b)** $(4x - 3)(3x + 8)$

71. $2x^2 - 7x - 15; (x - 5)$ **(a)** $2x + 3$ **(b)** $(x - 5)(2x + 3)$

72. $12x^2 + 23x - 24; (4x - 3)$

73. $30x^2 - 53x + 8; (6x - 1)$

74. $x^3 + 5x^2 + 7x + 3; (x + 1)$

73. (a) $5x - 8$ **(b)** $(6x - 1)(5x - 8)$ **74. (a)** $x^2 + 4x + 3$ **(b)** $(x + 1)(x + 1)(x + 3)$ or $(x + 1)^2(x + 3)$

75. (a) $x^2 - 4x + 3$ **(b)** $(x - 2)(x - 1)(x - 3)$ **76. (a)** $10x^2 - 9x + 2$ **(b)** $(x + 3)(5x - 2)(2x - 1)$ **77. (a)** $2x^2 + 5x + 2$
(b) $(x - 4)(x + 2)(2x + 1)$ **78. (a)** $4x^2 + 17x - 15$ **(b)** $(2x - 5)(4x - 3)(x + 5)$
75. $x^3 - 6x^2 + 11x - 6; (x - 2)$ **76.** $10x^3 + 21x^2 - 25x + 6; (x + 3)$
77. $2x^3 - 3x^2 - 18x - 8; (x - 4)$ **78.** $8x^3 + 14x^2 - 115x + 75; (2x - 5)$

79. When factoring any trinomial, what should the first step always be? Factor out the GCF, if there is one.

80. If the factors of a polynomial are $(2x + 3y)$ and $(x - 4y)$, find the polynomial. Explain how you determined your answer. $2x^2 - 5xy - 12y^2$

81. If the factors of a polynomial are 3, $(4x - 5)$, and $(2x - 3)$, find the polynomial. Explain how you determined your answer. $24x^2 - 66x + 45$

82. If we know that one factor of the polynomial $x^2 + 3x - 18$ is $x - 3$, how can we find the other factor? Find the other factor. divide, $x + 6$

83. If we know that one factor of the polynomial $x^2 - xy - 6y^2$ is $x - 3y$, how can we find the other factor? Find the other factor. divide, $x + 2y$

84. On a test, Kim wrote the following factoring and did not receive full credit. Explain why Kim's factoring is not complete. 3 can be factored from $3x - 6$.
$$15x^2 - 21x - 18 = (5x + 3)(3x - 6)$$

85. (a) Explain in your own words the step-by-step procedure to factor $6x^2 - x - 12$.
(b) Factor $6x^2 - x - 12$ using the procedure you explained in part **(a)**. $(2x - 3)(3x + 4)$

86. (a) Explain in your own words the step by step procedure to factor $8x^2 - 26x + 6$.
(b) Factor $8x^2 - 26x + 6$ using the procedure you explained in part **(a)**. $2(4x - 1)(x - 3)$

CUMULATIVE REVIEW EXERCISES

[3.3] **87.** What is the slope of a horizontal line? Explain.

88. What is the slope of a vertical line? Explain.

[5.1] **89.** Write the quotient in scientific notation.
$$\frac{36{,}000{,}000}{0.0004} \quad 9.0 \times 10^{10}$$

90. $-x^2y - 8xy^2 + 6$

[5.3] **90.** Simplify $2x^2y - 6xy^2 - (3x^2y + 2xy^2 - 6)$.

[5.6] **91.** For $f(x) = x^2 - 4x + 6$, **(a)** find the axis of symmetry, and **(b)** graph the function. **(a)** $x = 2$ **(b)**

87. 0, the change in y is 0   **88.** undefined   **1.** cannot divide both sides of equation by $a - b$ since it equals 0

Group Activity/ Challenge Problems

1. Have you ever seen the "proof" that 1 is equal to 2? Here it is.

Let $a = b$	
$a^2 = b^2$	Square both sides of the equation.
$a^2 = b \cdot b$	
$a^2 = ab$	Substitute a for b.
$a^2 - b^2 = ab - b^2$	Subtract b^2 from both sides of the equation.
$(a + b)(a - b) = b(a - b)$	Factor both sides of the equation.
$\dfrac{(a + b)\cancel{(a - b)}}{\cancel{a - b}} = \dfrac{b\cancel{(a - b)}}{\cancel{a - b}}$	Divide both sides of the equation by $(a - b)$ and simplify.
$a + b = b$	
$b + b = b$	Substitute b for a.
$2b = b$	Divide both sides of the equation by b.
$\dfrac{2\cancel{b}}{\cancel{b}_1} = \dfrac{\cancel{b}}{\cancel{b}_1}$	
$2 = 1$	

But obviously, $2 \neq 1$. We must have made an error somewhere. Can you find it?

2. **(a)** Which of the following do you believe would be more difficult to factor by trial and error? Explain your answer.
$$30x^2 + 23x - 40 \quad \text{or} \quad 49x^2 - 98x + 13$$
(b) Factor both trinomials. **(b)** $(6x - 5)(5x + 8)$, $(7x - 1)(7x - 13)$

3. **(a)** If $x^2 + bx + 5$ is factorable, what are the only two possible values of b? Explain.
   (b) If $x^2 + bx + c$ is factorable and c is a prime number, what are the only two possible factors of b? Explain. $c + 1$ or $-(c + 1)$ **3(a)** 6 or -6

Consider the trinomial $ax^2 + bx + c$. We will learn later in the course that if the expression $b^2 - 4ac$, called the **discriminant**, *is not a perfect square, the trinomial cannot be factored.* **Perfect squares** *are 1, 4, 9, 16, 25, 49, and so on. The square root of a perfect square is a whole number. For the following exercises,* **(a)** *Find the value of $b^2 - 4ac$;* **(b)** *If $b^2 - 4ac$ is a perfect square, factor the polynomial; if $b^2 - 4ac$ is not a perfect square, indicate that the polynomial cannot be factored.*

4. $x^2 - 8x + 15$ **(a)** 4 **(b)** $(x - 3)(x - 5)$ **5.** $6y^2 - 5y - 6$ **(a)** 169 **(b)** $(3y + 2)(2y - 3)$
6. $x^2 - 4x + 6$ **(a)** -8 **(b)** not factorable **7.** $3t^2 - 6t + 2$ **(a)** 12 **(b)** not factorable

Factor completely.

8. $4a^{2n} - 4a^n - 15$. $(2a^n + 3)(2a^n - 5)$
9. $a^2(a + b) - 2ab(a + b) - 3b^2(a + b)$ $(a + b)^2(a - 3b)$
10. $x^2(x + y)^2 - 7xy(x + y)^2 + 12y^2(x + y)^2$ $(x + y)^2(x - 4y)(x - 3y)$
11. $3m^2(m - 2n) - 4mn(m - 2n) - 4n^2(m - 2n)$ $(m - 2n)^2(3m + 2n)$
12. $x^{2n} + 3x^n - 10$ $(x^n - 2)(x^n + 5)$
13. $12x^{2n}y^{2n} + 2x^ny^n - 2$. $2(3x^ny^n - 1)(2x^ny^n + 1)$
14. $9r^{4y} + 3r^{2y} - 2$ $(3r^{2y} + 2)(3r^{2y} - 1)$
15. $8a^{5m} - 13a^{3m} - 6a^m$ $a^m(8a^{2m} + 3)(a^{2m} - 2)$

6.3 Special Factoring Formulas

Tape 9

1 Factor the difference of two squares.
2 Factor perfect square trinomials.
3 Factor the sum and difference of two cubes.

Difference of Two Squares

1 In this section we present some special formulas for factoring the difference of two squares, perfect square trinomials, and the sum and difference of two cubes. It will be to your advantage to memorize these formulas.

The expression $x^2 - 9$ is an example of the difference of two squares.

$$x^2 - 9 = (x)^2 - (3)^2$$

To factor the difference of two squares, it is convenient to use the difference-of-two-squares formula. This formula was first presented in Section 5.3.

> **Difference of Two Squares**
> $$a^2 - b^2 = (a + b)(a - b)$$

EXAMPLE 1 Factor using the difference-of-two-squares formula.

(a) $x^2 - 16$ **(b)** $16x^2 - 9y^2$

Solution: Rewrite each expression as a difference of two squares.

(a) $x^2 - 16 = (x)^2 - (4)^2$
$\qquad\qquad = (x + 4)(x - 4)$

(b) $16x^2 - 9y^2 = (4x)^2 - (3y)^2$
$\qquad\qquad\qquad = (4x + 3y)(4x - 3y)$ •

EXAMPLE 2   Factor each of the following differences of squares.

(a) $x^6 - y^4$ **(b)** $2z^4 - 162x^6$

Solution: Rewrite each expression as a difference of two squares and then use the formula to factor the expression.

(a) $x^6 - y^4 = (x^3)^2 - (y^2)^2$ **(b)** $2z^4 - 162x^6 = 2(z^4 - 81x^6)$
$$= (x^3 + y^2)(x^3 - y^2)$$
$$= 2[(z^2)^2 - (9x^3)^2]$$
$$= 2(z^2 + 9x^3)(z^2 - 9x^3)$$

EXAMPLE 3   Factor $x^4 - 16y^4$.

Solution: $x^4 - 16y^4 = (x^2)^2 - (4y^2)^2$
$$= (x^2 + 4y^2)(x^2 - 4y^2)$$

Note that $(x^2 - 4y^2)$ is also a difference of two squares. We use the difference-of-two-squares formula a second time to obtain

$$= (x^2 + 4y^2)[(x)^2 - (2y)^2]$$
$$= (x^2 + 4y^2)(x + 2y)(x - 2y)$$

EXAMPLE 4   Factor $(x - 5)^2 - 9$ using the formula for the difference of two squares.

Solution: We can express $(x - 5)^2 - 9$ as a difference of two squares.

$$(x - 5)^2 - 9 = (x - 5)^2 - 3^2$$
$$= [(x - 5) + 3][(x - 5) - 3]$$
$$= (x - 2)(x - 8)$$

Thus, $(x - 5)^2 - 9$ factors into $(x - 2)(x - 8)$.

Note: **It is not possible to factor the sum of two squares of the form $a^2 + b^2$ over the set of real numbers.**

For example, it is not possible to factor $x^2 + 4$ since $x^2 + 4 = x^2 + 2^2$, which is a sum of two squares.

Perfect Square Trinomials

2   In Section 5.3 we saw that

$$(a + b)^2 = a^2 + 2ab + b^2$$
$$(a - b)^2 = a^2 - 2ab + b^2$$

If we reverse the left and right sides of these two formulas, we obtain two special factoring formulas.

> **Perfect Square Trinomials**
> $$a^2 + 2ab + b^2 = (a + b)^2$$
> $$a^2 - 2ab + b^2 = (a - b)^2$$

These two trinomials are called **perfect square trinomials** since each is the square of a binomial. *To be a perfect square trinomial, the first and last terms must be the squares of some expression and the middle term must be twice the product of the first and last terms.* When you are given a trinomial to factor, determine if it is a perfect square trinomial before you attempt to factor it by the procedures explained in Section 6.2. If it is a perfect square trinomial, you can factor it using the formulas given above.

Perfect Square Trinomials

$$y^2 + 6y + 9 \qquad \text{or} \qquad y^2 + 2(y)(3) + 3^2$$
$$9a^2b^2 - 24ab + 16 \qquad \text{or} \qquad (3ab)^2 - 2(3ab)(4) + 4^2$$
$$(r + s)^2 + 6(r + s) + 9 \qquad \text{or} \qquad (r + s)^2 + 2(r + s)(3) + 3^2$$

Now let us factor some perfect square trinomials.

EXAMPLE 5 Factor $x^2 - 8x + 16$.

Solution: Since the first and last terms x^2 and 4^2, are squares, this trinomial might be a perfect square trinomial. To determine if it is, take twice the product of x and 4 to see if you obtain $8x$.

$$2(x)(4) = 8x$$

Since $8x$ is the middle term and since the sign of the middle term is negative, we factor as follows:

$$x^2 - 8x + 16 = (x - 4)^2$$

EXAMPLE 6 Factor $9x^4 - 12x^2 + 4$.

Solution: The first term is a square, $(3x^2)^2$, as is the last term, 2^2. Since $2(3x^2)(2) = 12x^2$, we factor as follows:

$$9x^4 - 12x^2 + 4 = (3x^2 - 2)^2$$

EXAMPLE 7 Factor $(a + b)^2 + 6(a + b) + 9$.

Solution: The first term, $(a + b)^2$, is a square. The last term, 9 or 3^2, is a square. The middle term is $2(a + b)(3) = 6(a + b)$. Therefore, this is a perfect square trinomial. Thus,

$$(a + b)^2 + 6(a + b) + 9 = [(a + b) + 3]^2 = (a + b + 3)^2$$

EXAMPLE 8 Factor $x^2 - 6x + 9 - y^2$.

Solution: Since $x^2 - 6x + 9$ is a perfect square trinomial, we write

$$(x - 3)^2 - y^2$$

Now $(x - 3)^2 - y^2$ is a difference of two squares; therefore

$$(x - 3)^2 - y^2 = [(x - 3) + y][(x - 3) - y]$$
$$= (x - 3 + y)(x - 3 - y)$$

Thus, $x^2 - 6x + 9 - y^2 = (x - 3 + y)(x - 3 - y)$.

The polynomial in Example 8 has four terms. In Section 6.1 we learned to factor polynomials with four terms by grouping. If you study Example 9, you will see that no matter how you arrange the four terms they cannot be arranged so that the first two terms have a common factor and the last two terms have a common factor. Whenever a polynomial of four terms cannot be factored by grouping, try to rewrite three of the terms as the square of a binomial and then factor using the difference-of-two-squares formula.

Example 9 Factor $4a^2 + 12ab + 9b^2 - 25$.

Solution: We first notice that this polynomial of four terms cannot be factored by grouping. We next look to see if three terms of the polynomial can be expressed as the square of a binomial. Since this can be done, we write the three terms as the square of a binomial. We complete our factoring using the difference-of-two-squares formula.

$$
\begin{aligned}
4a^2 + 12ab + 9b^2 - 25 &= (2a + 3b)^2 - 5^2 \\
&= [(2a + 3b) + 5][(2a + 3b) - 5] \\
&= (2a + 3b + 5)(2a + 3b - 5)
\end{aligned}
$$

Sum and Difference of Two Cubes

3 Earlier in this section we factored the difference of two squares. Now we will factor the sum and difference of two cubes. Consider the product of $(a + b)(a^2 - ab + b^2)$.

$$
\begin{array}{r}
a^2 - ab + b^2 \\
a + b \\
\hline
a^2b - ab^2 + b^3 \\
a^3 - a^2b + ab^2 \\
\hline
a^3 \qquad\qquad\qquad + b^3
\end{array}
$$

Thus, $a^3 + b^3 = (a + b)(a^2 - ab + b^2)$. Using multiplication, we can also show that $a^3 - b^3 = (a - b)(a^2 + ab + b^2)$. Formulas for factoring the sum and the difference of two cubes appear in the following box.

> **Sum of Two Cubes**
> $$a^3 + b^3 = (a + b)(a^2 - ab + b^2)$$
>
> **Difference of Two Cubes**
> $$a^3 - b^3 = (a - b)(a^2 + ab + b^2)$$

Example 10 Factor $x^3 + 27$.

Solution: Rewrite $x^3 + 27$ as a sum of two cubes, $x^3 + 3^3$. Let x correspond to a and 3 to b. Then factor using the sum-of-two cubes formula.

$$
\begin{aligned}
a^3 + b^3 &= (a + b)(a^2 - ab + b^2) \\
x^3 + 3^3 &= (x + 3)[x^2 - x(3) + 3^2] \\
&= (x + 3)(x^2 - 3x + 9)
\end{aligned}
$$

Thus, $x^3 + 27 = (x + 3)(x^2 - 3x + 9)$.

EXAMPLE 11   Factor $27x^3 - 8y^6$.

Solution: We first observe that $27x^3$ and $8y^6$ have no common factors other than 1. Since we can express both $27x^3$ and $8y^6$ as cubes, we can factor using the difference-of-two-cubes formula.

$$27x^3 - 8y^6 = (3x)^3 - (2y^2)^3$$
$$= (3x - 2y^2)[(3x)^2 + (3x)(2y^2) + (2y^2)^2]$$
$$= (3x - 2y^2)(9x^2 + 6xy^2 + 4y^4)$$

Thus, $27x^3 - 8y^6 = (3x - 2y^2)(9x^2 + 6xy^2 + 4y^4)$.

EXAMPLE 12   Factor $8y^3 - 64x^6$.

Solution: First factor out 8, which is common to both terms.

$$8y^3 - 64x^6 = 8(y^3 - 8x^6)$$

Next factor $y^3 - 8x^6$ by writing it as a difference of two cubes.

$$8(y^3 - 8x^6) = 8[(y)^3 - (2x^2)^3]$$
$$= 8(y - 2x^2)[y^2 + y(2x^2) + (2x^2)^2]$$
$$= 8(y - 2x^2)(y^2 + 2x^2y + 4x^4)$$

Thus, $8y^3 - 64x^6 = 8(y - 2x^2)(y^2 + 2x^2y + 4x^4)$.

EXAMPLE 13   Factor $(x - 2)^3 + 64$.

Solution: Write $(x - 2)^3 + 64$ as a sum of two cubes, then use the sum-of-two-cubes formula to factor.

$$(x - 2)^3 + (4)^3 = [(x - 2) + 4][(x - 2)^2 - (x - 2)(4) + (4)^2]$$
$$= (x - 2 + 4)(x^2 - 4x + 4 - 4x + 8 + 16)$$
$$= (x + 2)(x^2 - 8x + 28)$$

Helpful Hint

The square of binomial has a 2 as part of the middle term of the trinomial.

$$(a + b)^2 = a^2 + 2ab + b^2$$
$$(a - b)^2 = a^2 - 2ab + b^2$$

The sum or the difference of two cubes has a factor which is similar to the trinomial in the square of the binomial. However, the middle term does not contain a 2.

$$a^3 + b^3 = (a + b)(a^2 - ab + b^2)$$
$$a^3 - b^3 = (a - b)(a^2 + ab + b^2)$$
$$\underbrace{\qquad}_{\text{not } 2ab}$$

Exercise Set 6.3   **40.** $(6a + 2b)(2a − 8b) = 4(3a + b)(a − 4b)$

Use the difference-of-two-squares formula or the perfect square trinomial formula to factor. If the polynomial cannot be factored, so state.

1. $x^2 − 81$ $(x + 9)(x − 9)$

2. $x^2 − 9$ $(x + 3)(x − 3)$

3. $x^2 + 9$ cannot be factored

4. $1 − 4x^2$ $(1 + 2x)(1 − 2x)$

5. $1 − 9a^2$ $(1 + 3a)(1 − 3a)$

6. $x^2 − 36y^2$ $(x + 6y)(x − 6y)$

7. $25 − 16y^4$ $(5 + 4y^2)(5 − 4y^2)$

8. $x^6 − 144y^4$ $(x^3 + 12y^2)(x^3 − 12y^2)$

9. $81a^4 − 16b^2$ $(9a^2 + 4b)(9a^2 − 4b)$

10. $x^2y^2 − 1$ $(xy + 1)(xy − 1)$

11. $a^2b^2 − 49c^2$ $(ab + 7c)(ab − 7c)$

12. $4a^2c^2 − 16x^2y^2$ $4(ac + 2xy)(ac − 2xy)$

13. $9x^2y^2 − 4x^2$ $x^2(3y + 2)(3y − 2)$

14. $100 − (x + y)^2$ $(10 + x + y)(10 − x − y)$

15. $36 − (x − 6)^2$ $x(12 − x)$

16. $(2x + 3)^2 − 9$ $4x(x + 3)$

17. $a^2 − (3b + 2)^2$ $(a + 3b + 2)(a − 3b − 2)$

18. $a^2 + 2ab + b^2$ $(a + b)^2$

19. $x^2 + 10x + 25$ $(x + 5)^2$

20. $25 − 10t + t^2$ $(5 − t)^2$

21. $4 + 4x + x^2$ $(2 + x)^2$

22. $y^2 − 8y + 16$ $(y − 4)^2$

23. $4x^2 − 20xy + 25y^2$ $(2x − 5y)^2$

24. $9y^2 + 6yz + z^2$ $(3y + z)^2$

25. $9a^2 + 12a + 4$ $(3a + 2)^2$

26. $25a^2b^2 − 20ab + 4$ $(5ab − 2)^2$

27. $w^4 + 16w^2 + 64$ $(w^2 + 8)^2$

28. $x^2 − 5xy + 25y^2$ cannot be factored

29. $(x + y)^2 + 2(x + y) + 1$ $(x + y + 1)^2$

30. $(x + 1)^2 + 6(x + 1) + 9$ $(x + 4)^2$

31. $a^4 − 2a^2b^2 + b^4$ $(a + b)^2(a − b)^2$

32. $(w − 3)^2 + 8(w − 3) + 16$ $(w + 1)^2$

33. $x^2 + 6x + 9 − y^2$ $(x + 3 + y)(x + 3 − y)$

34. $9 − (x^2 − 8x + 16)$ $(x − 1)(7 − x)$

35. $25 − (x^2 + 4x + 4)$ $(x + 7)(−x + 3)$

36. $a^2 + 2ab + b^2 − 16c^2$ $(a + b + 4c)(a + b − 4c)$

37. $9a^2 − 12ab + 4b^2 − 9$ $(3a − 2b + 3)(3a − 2b − 3)$

38. $x^4 − 6x^2 + 9$ $(x^2 − 3)^2$

39. $(x + y)^2 − (x − y)^2$ $4xy$

40. $(4a − 3b)^2 − (2a + 5b)^2$

Factor using the sum- or difference-of-two-cubes formula.   **56.** $(2x + y − 4)(4x^2 + 4xy + y^2 + 8x + 4y + 16)$

41. $x^3 − 27$ $(x − 3)(x^2 + 3x + 9)$

42. $y^3 + 125$ $(y + 5)(y^2 − 5y + 25)$

43. $x^3 + y^3$ $(x + y)(x^2 − xy + y^2)$

44. $x^3 − 8a^3$ $(x − 2a)(x^2 + 2ax + 4a^2)$

45. $64 − a^3$ $(4 − a)(16 + 4a + a^2)$

46. $w^3 − 216$ $(w − 6)(w^2 + 6w + 36)$

47. $27y^3 − 8x^3$ $(3y − 2x)(9y^2 + 6xy + 4x^2)$

48. $5x^3 + 40y^3$ $5(x + 2y)(x^2 − 2xy + 4y^2)$

49. $24x^3 − 81y^3$ $3(2x − 3y)(4x^2 + 6xy + 9y^2)$

50. $y^6 + x^9$ $(y^2 + x^3)(y^4 − x^3y^2 + x^6)$

51. $5x^3 − 625y^3$ $5(x − 5y)(x^2 + 5xy + 25y^2)$

52. $16y^6 − 250x^3$ $2(2y^2 − 5x)(4y^4 + 10xy^2 + 25x^2)$

53. $(x + 1)^3 + 1$ $(x + 2)(x^2 + x + 1)$

54. $(x − 3)^3 + 8$ $(x − 1)(x^2 − 8x + 19)$

55. $(x − y)^3 − 27$ $(x − y − 3)(x^2 − 2xy + y^2 + 3x − 3y + 9)$

56. $(2x + y)^3 − 64$

57. $b^3 − (b + 3)^3$ $−9(b^2 + 3b + 3)$

58. $(m − n)^3 − (m + n)^3$ $−2n(3m^2 + n^2)$

Factor using a special factoring formula.

59. $121y^4 − 49x^2$ $(11y^2 + 7x)(11y^2 − 7x)$

60. $a^4 − 4b^4$ $(a^2 + 2b^2)(a2 − 2b^2)$

61. $16y^2 − 81x^2$ $(4y + 9x)(4y − 9x)$

62. $49 − 64x^2y^2$ $(7 + 8xy)(7 − 8xy)$

63. $25x^4 − 81y^6$ $(5x^2 + 9y^3)(5x^2 − 9y^3)$

64. $(x + y)^2 − 16$ $(x + y + 4)(x + y − 4)$

65. $x^3 − 64$ $(x − 4)(x^2 + 4x + 16)$

66. $2a^2 − 24a + 72$ $2(a − 6)^2$

67. $9x^2y^2 + 24xy + 16$ $(3xy + 4)^2$

68. $a^4 + 12a^2 + 36$ $(a^2 + 6)^2$

69. $a^4 + 2a^2b^2 + b^4$ $(a^2 + b^2)^2$

70. $8y^3 − 125x^6$ $(2y − 5x^2)(4y^2 + 10x^2y + 25x^4)$

71. $x^2 − 2x + 1 − y^2$ $(x − 1 + y)(x − 1 − y)$

72. $4r^2 + 4rs + s^2 − 9$ $(2r + s + 3)(2r + s − 3)$

73. $(x + y)^3 + 1$ $(x + y + 1)(x^2 + 2xy + y^2 − x − y + 1)$

74. $9x^2 − 6xy + y^2 − 4$ $(3x − y + 2)(3x − y − 2)$

75. $(m + n)^2 − (2m − n)^2$ $3m(−m + 2n)$

76. $(r + p)^3 + (r − p)^3$ $2r(r^2 + 3p^2)$

87. For the second term to be positive, the factors must be of the form $(a + b)(a + b)$ or $(a - b)(a - b)$. Neither of these have a product of $a^2 + b^2$.

*In Exercises 77–81, **(a)** Find the area or volume of the shaded figure by subtracting the smaller area or volume from the larger. The formula to find the area or volume is given under the figure. **(b)** Write the expression obtained in part (a) in factored form. Part of the GCF in Exercises 78, 80, and 81 is π.*

77.

Squares

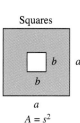

$A = s^2$

(a) $a^2 - b^2$
(b) $(a + b)(a - b)$

78.

Circles

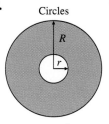

$A = \pi r^2$

(a) $\pi R^2 - \pi r^2$
(b) $\pi(R + r)(R - r)$

79.

Rectangular solid

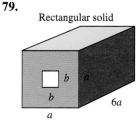

$V = lwh$

(a) $6a^3 - 6ab^2$
(b) $6a(a + b)(a - b)$

80.

Cylinder

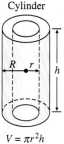

$V = \pi R^2 h$

(a) $\pi R^2 h - \pi r^2 h$
(b) $\pi h(R + r)(R - r)$

81.

Sphere

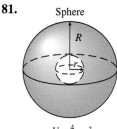

$V = \frac{4}{3}\pi r^3$

(a) $\frac{4}{3}\pi R^3 - \frac{4}{3}\pi r^3$
(b) $\frac{4}{3}\pi(R - r)(R^2 + Rr + r^2)$

82. Explain in your own words how to determine whether or not a trinomial is a perfect square trinomial.

83. Find two values of b that will make $4x^2 + bx + 9$ a perfect square trinomial. Explain how you determined your answer. $-12, +12$

84. Find two values of c that will make $16x^2 + cx + 4$ a perfect square trinomial. Explain how you determined your answer. $16, -16$

85. Find the value of c that will make $25x^2 + 20x + c$ a perfect square trinomial. Explain how you determined your answer. $c = 4$

86. Find the value of d that will make $49x^2 - 42x + d$ a perfect square trinomial. Explain how you determined your answer. 9

87. Explain why a sum of two squares, $a^2 + b^2$, cannot be factored over the set of real numbers.

CUMULATIVE REVIEW EXERCISES

[1.2] **88.** Consider the set of elements
$\{-2, \frac{5}{9}, -1.67, 0, \sqrt{3}, -\sqrt{6}, 3, 6\}$.
List the elements that are:

(a) Counting numbers $3, 6$
(b) Rational numbers $-2, \frac{5}{9}, -1.67, 0, 3, 6$
(c) Irrational numbers $\sqrt{3}, -\sqrt{6}$
(d) Real numbers $-2, \frac{5}{9}, -1.67, 0, \sqrt{3}, -\sqrt{6}, 3, 6$

Place the proper symbol, either ∈ or ⊆, in the shaded area to make the statement true.

89. 6 ▨ $\{3, 4, 5, 6, 7\}$ ∈
90. $\{b\}$ ▨ $\{a, b, c, d\}$ ⊆

[2.2] **91.** Given the formula $z = \dfrac{p' - p}{\sqrt{\dfrac{pq}{n}}}$, find the value of

z when $p' = 0.4$, $p = 0.3$, $q = 0.7$, and $n = 4$. ≈ 0.44

[2.3] **92.** The length of a rectangular hallway is 2 feet greater than twice its width. Find the length and width of the hallway if its perimeter is 22 feet. $w = 3$ ft, $l = 8$ ft

---

Group Activity/ Challenge Problems

Factor. Note that $\sqrt{a} \cdot \sqrt{a} = a, a \geq 0$.

1. $x^2 - 7$ $(x + \sqrt{7})(x - \sqrt{7})$
2. $2x^2 - 15$ $(x\sqrt{2} + \sqrt{15})(x\sqrt{2} - \sqrt{15})$

*Now we will construct **perfect square trinomials**, which are trinomials that may be written as the square of a binomial. For example, $x^2 + 10x + 25$ is a perfect square trinomial since $x^2 + 10x + 25 = (x + 5)^2$. Also, $x^2 - 8x + 16$ is a perfect square trinomial since it may be written $(x - 4)^2$. There is a relationship between the coefficient b in the middle term, bx, and the constant term, c, in perfect square trinomials. In Exercises 3–6, **(a)** find the value of b or c to make a perfect trinomial and **(b)** replace b or c with the value* *obtained in part (a), then write the perfect square trinomial as the square of a binomial. (We will explain how to construct perfect square trinomials in more detail in Chapter 9.)*

3. $x^2 + bx + 9$ **4.** $x^2 - bx + 25$ **5.** $x^2 + 4x + c$ **6.** $x^2 - 8x + c$

$b = 6, (x + 3)^2$ $b = 10, (x - 5)^2$ $c = 4, (x + 2)^2$ $c = 16, (x - 4)^2$

7. Explain how b and c are related if the trinomial $x^2 + bx + c$ is a perfect square trinomial. $(\frac{b}{2})^2 = c$

8. The figure to the left shows how we *complete the square*. The sum of the areas of the 3 parts of the square that are shaded in blue is

$$x^2 + 4x + 4x \quad \text{or} \quad x^2 + 8x$$

(a) Find the area of the fourth part (in red) to complete the square.   16

(b) Find the sum of the areas of the four parts of the square. $x^2 + 8x + 16$

(c) This process has resulted in a perfect square trinomial in part **(b)**. Write this perfect square trinomial as the square of a binomial. $(x + 4)^2$

9. Factor $(m - n)^3 - (9 - n)^3$. $(m - 9)(m^2 - 3mn + 3n^2 + 9m - 27n + 81)$

Factor completely.

10. $64x^{4a} - 9y^{6a}$ **11.** $16p^{8w} - 49p^{6w}$ **12.** $a^{2n} - 16a^n + 64$

$(8x^{2a} + 3y^{3a})(8x^{2a} - 3y^{3a})$ $p^{6w}(4p^w + 7)(4p^w - 7)$ $(a^n - 8)^2$

13. $144r^{8k} + 48r^{4k} + 4$ **14.** $x^{3n} - 8$ **15.** $27x^{3m} + 64x^{6m}$

$4(6r^{4k} + 1)^2$ $(x^n - 2)(x^{2n} + 2x^n + 4)$ $x^{3m}(3 + 4x^m)(9 - 12x^m + 16x^{2m})$

16. The expression $x^6 - 1$ can be factored using either the difference of two squares or the difference of two cubes. At first the factors do not appear the same. But with a little algebraic manipulation they can be shown to be equal. Factor $x^6 - 1$ using **(a)** the difference of two squares and **(b)** the difference of two cubes. **(c)** Show these two answers are equal by factoring the answer received in part **(a)** completely. Then multiply the two binomials by the two trinomials.   **(a)** $(x^3 + 1)(x^3 - 1)$ **(b)** $(x^2 - 1)(x^4 + x^2 + 1)$

6.4   General Review of Factoring

Tape 9

A Review of Factoring

1️⃣ Factor polynomials using a combination of factoring techniques.

1️⃣ We have presented a number of factoring methods. Now we will combine problems and techniques from the previous sections.

A general procedure to factor any polynomial follows.

> ### To Factor a Polynomial
>
> **1.** Determine if all the terms in the polynomial have a greatest common factor other than 1. If so, factor out the GCF.
>
> **2.** If the polynomial has two terms, determine if it is a difference of two squares or a sum or difference of two cubes. If so, factor using the appropriate formula.
>
> **3.** If the polynomial has three terms, determine if it is a perfect square trinomial. If so, factor accordingly. If it is not, factor the trinomial using trial and error, grouping, or substitution, as explained in Section 6.2.

> **4.** If the polynomial has more than three terms, try factoring by grouping. If that does not work, see if three of the terms are the square of a binomial.
>
> **5.** As a final step, examine your factored polynomial to see if any factors listed have a common factor and can be factored further. If you find a common factor, factor it out at this point.

The following examples illustrate how to use the procedure.

EXAMPLE 1   Factor $3x^4 - 27x^2$.

Solution: First, check for a greatest common factor other than 1. Since $3x^2$ is common to both terms, factor it out.

$$3x^4 - 27x^2 = 3x^2(x^2 - 9) = 3x^2(x + 3)(x - 3)$$

Note that $x^2 - 9$ is factored as a difference of two squares.

EXAMPLE 2   Factor $3x^2y^2 - 24xy^2 + 48y^2$.

Solution: Begin by factoring the GCF, $3y^2$, from each term.

$$3x^2y^2 - 24xy^2 + 48y^2 = 3y^2(x^2 - 8x + 16) = 3y^2(x - 4)^2$$

Note that $x^2 - 8x + 16$ is a perfect square trinomial. If you did not recognize this, you would still obtain the correct answer by factoring the trinomial into $(x - 4)(x - 4)$.

EXAMPLE 3   Factor $24x^2 - 6xy + 16xy - 4y^2$.

Solution: Always begin by determining if all the terms in the polynomial have a common factor. In this example the number 2 is common to all terms. Factor out the 2; then factor the remaining four-term polynomial by grouping.

$$
\begin{aligned}
24x^2 - 6xy + 16xy - 4y^2 &= 2[12x^2 - 3xy + 8xy - 2y^2] \\
&= 2[3x(4x - y) + 2y(4x - y)] \\
&= 2(3x + 2y)(4x - y)
\end{aligned}
$$

EXAMPLE 4   Factor $10a^2b - 15ab + 20b$.

Solution: $10a^2b - 15ab + 20b = 5b(2a^2 - 3a + 4)$

Since $2a^2 - 3a + 4$ cannot be factored, we stop here.

EXAMPLE 5   Factor $2x^4y + 54xy$.

Solution:
$$
\begin{aligned}
2x^4y + 54xy &= 2xy(x^3 + 27) \\
&= 2xy(x + 3)(x^2 - 3x + 9).
\end{aligned}
$$

Note that $x^3 + 27$ was factored as a sum of two cubes.

EXAMPLE 6 Factor $6x^2 - 3x + 6y^2 - 9$.

Solution: First, factor 3 from all four terms.

$$6x^2 - 3x + 6y^2 - 9 = 3(2x^2 - x + 2y^2 - 3)$$

Now see if the four terms within parentheses can be factored by grouping. Since these four terms cannot be factored by grouping, see if three of the terms can be written as the square of a binomial. No matter how we rearrange the terms this cannot be done. We conclude that this expression cannot be factored further. Thus,

$$6x^2 - 3x + 6y^2 - 9 = 3(2x^2 - x + 2y^2 - 3)$$

EXAMPLE 7 Factor $3x^2 - 18x + 27 - 3y^2$.

Solution: Factor out 3 from all four terms.

$$3x^2 - 18x + 27 - 3y^2 = 3(x^2 - 6x + 9 - y^2)$$

Now try factoring by grouping. Since the four terms within parentheses cannot be factored by grouping, see if three of the terms can be written as the square of a binomial. Since this can be done, we express $x^2 - 6x + 9$ as $(x - 3)^2$ and then use the difference-of-two-squares formula. Thus,

$$3x^2 - 18x + 27 - 3y^2 = 3[(x - 3)^2 - y^2]$$
$$= 3[(x - 3 + y)(x - 3 - y)]$$
$$= 3(x - 3 + y)(x - 3 - y)$$

Exercise Set 6.4

Factor each of the following completely.

1. $3x^2 + 3x - 36$ $\quad 3(x + 4)(x - 3)$
2. $2x^2 - 16x + 32$ $\quad 2(x - 4)^2$
3. $10s^2 + 19s - 15$ $\quad (5s - 3)(2s + 5)$
4. $6x^3y^2 + 10x^2y^3 + 8x^2y^2$ $\quad 2x^2y^2(3x + 5y + 4)$
5. $-8r^2 + 26r - 15$ $\quad -(4r - 3)(2r - 5)$
6. $3x^3 - 12x^2 - 36x$ $\quad 3x(x + 2)(x - 6)$
7. $2x^2 - 72$ $\quad 2(x + 6)(x - 6)$
8. $4x^2 - 4y^2$ $\quad 4(x - y)(x + y)$
9. $5x^5 - 45x$ $\quad 5x(x^2 + 3)(x^2 - 3)$
10. $6x^2y^2z^2 - 24x^2y^2$ $\quad 6x^2y^2(z + 2)(z - 2)$
11. $3x^6 - 3x^5 - 12x^5 + 12x^4$ $\quad 3x^4(x - 4)(x - 1)$
12. $2x^2y^2 + 6xy^2 - 10xy^2 - 30y^2$ $\quad 2y^2(x - 5)(x + 3)$
13. $5x^4y^2 + 20x^3y^2 - 15x^3y^2 - 60x^2y^2$ $\quad 5x^2y^2(x - 3)(x + 4)$
14. $6x^2 - 15x - 9$ $\quad 3(2x + 1)(x - 3)$
15. $x^4 - x^2y^2$ $\quad x^2(x + y)(x - y)$
16. $4x^3 + 108$ $\quad 4(x + 3)(x^2 - 3x + 9)$
17. $x^7y^2 - x^4y^2$ $\quad x^4y^2(x - 1)(x^2 + x + 1)$
18. $x^4 - 16$ $\quad (x - 2)(x + 2)(x^2 + 4)$
19. $x^5 - 16x$ $\quad x(x^2 + 4)(x + 2)(x - 2)$
20. $12x^2y^2 + 33xy^2 - 9y^2$ $\quad 3y^2(4x - 1)(x + 3)$
21. $4x^6 + 32y^3$ $\quad 4(x^2 + 2y)(x^4 - 2x^2y + 4y^2)$
22. $12x^4 - 6x^3 - 6x^3 + 3x^2$ $\quad 3x^2(2x - 1)^2$
23. $2(a + b)^2 - 18$ $\quad 2(a + b + 3)(a + b - 3)$
24. $12x^3y^2 + 4x^2y^2 - 40xy^2$ $\quad 4xy^2(3x - 5)(x + 2)$
25. $6x^2 + 36xy + 54y^2$ $\quad 6(x + 3y)^2$
26. $3x^2 - 30x + 75$ $\quad 3(x - 5)^2$
27. $(x + 2)^2 - 4$ $\quad x(x + 4)$
28. $4y^4 - 36x^6$ $\quad 4(y^2 + 3x^3)(y^2 - 3x^3)$
29. $(2a + b)(2a - 3b) - (2a + b)(a - b)$ $\quad (2a + b)(a - 2b)$
30. $pq + 6q + pr + 6r$ $\quad (q + r)(p + 6)$
31. $(y + 3)^2 + 4(y + 3) + 4$ $\quad (y + 5)^2$
32. $b^4 + 2b^2 + 1$ $\quad (b^2 + 1)^2$
33. $45a^4 - 30a^3 + 5a^2$ $\quad 5a^2(3a - 1)^2$
34. $(x + 1)^2 - (x + 1) - 6$ $\quad (x + 3)(x - 2)$

35. $x^3 + \dfrac{1}{27}$ $\left(x + \frac{1}{3}\right)\left(x^2 - \frac{1}{3}x + \frac{1}{9}\right)$

36. $8y^3 - \dfrac{1}{8}$ $\left(2y - \frac{1}{2}\right)\left(4y^2 + y + \frac{1}{4}\right)$

37. $3x^3 + 2x^2 - 27x - 18$ $(x + 3)(x - 3)(3x + 2)$

38. $6y^3 + 14y^2 + 4y$ $2y(3y + 1)(y + 2)$

39. $a^3b - 16ab^3$ $ab(a + 4b)(a - 4b)$

40. $x^6 + y^6$ $(x^2 + y^2)(x^4 - x^2y^2 + y^4)$

41. $81 - (x^2 + 2xy + y^2)$ $(9 + x + y)(9 - x - y)$

42. $x^2 - 2xy + y^2 - 25$ $(x - y + 5)(x - y - 5)$

43. $24x^2 - 34x + 12$ $2(3x - 2)(4x - 3)$

44. $40x^2 + 52x - 12$ $4(2x + 3)(5x - 1)$

45. $16x^2 - 34x - 15$ $(8x + 3)(2x - 5)$

46. $7(a - b)^2 + 4(a - b) - 3$ $(7a - 7b - 3)(a - b + 1)$

47. $x^4 - 81$ $(x^2 + 9)(x + 3)(x - 3)$

48. $(x + 2)^2 - 12(x + 2) + 36$ $(x - 4)^2$

49. $5bc - 10cx - 6by + 12xy$ $(5c - 6y)(b - 2x)$

50. $16y^4 - 9y^2$ $y^2(4y + 3)(4y - 3)$

51. $3x^4 - x^2 - 4$ $(3x^2 - 4)(x^2 + 1)$

52. $x^2 + 16x + 64 - 100y^2$ $(x + 8 + 10y)(x + 8 - 10y)$

53. $y^2 - (x^2 - 8x + 16)$ $(y + x - 4)(y - x + 4)$

54. $4a^3 + 32$ $4(a + 2)(a^2 - 2a + 4)$

55. $24ax + 18x + 36ay + 27y$ $3(2x + 3y)(4a + 3)$

56. $2(y + 4)^2 + 5(y + 4) - 12$ $(2y + 5)(y + 8)$

57. $x^6 - 11x^3 + 30$ $(x^3 - 6)(x^3 - 5)$

58. $a^2 + 12ab + 36b^2 - 16c^2$ $(a + 6b + 4c)(a + 6b - 4c)$

59. $y - y^3$ $y(1 + y)(1 - y)$

60. $6x^4y + 15x^3y - 9x^2y$ $3x^2y(x + 3)(2x - 1)$

61. $4x^2y^2 + 12xy + 9$ $(2xy + 3)^2$

62. $x^4 - 2x^2y^2 + y^4$ $(x + y)^2(x - y)^2$

63. $6r^2s^2 + rs - 1$ $(3rs - 1)(2rs + 1)$

64. $4x^4 + 12x^2 + 9$ $(2x^2 + 3)^2$

In Exercises 65–68, (a) write an expression for the shaded area and (b) write the expression in factored form.

65.

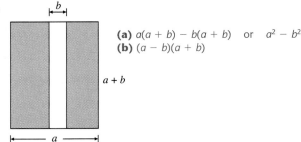

(a) $a(a + b) - b(a + b)$ or $a^2 - b^2$
(b) $(a - b)(a + b)$

66.

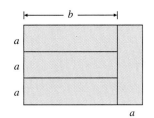

(a) $3ab + 3a^2$
(b) $3a(b + a)$

67.

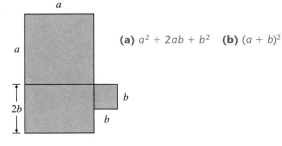

(a) $a^2 + 2ab + b^2$ **(b)** $(a + b)^2$

68.

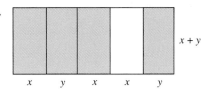

(a) $x(x + y) + y(x + y) + x(x + y) + y(x + y)$ **(b)** $2(x + y)^2$

69. (a) Write an expression for the surface area of the four sides of the box below (omit top and bottom).

   (b) Write the expression in factored form.

(a) $a(a - b) + a(a - b) + b(a - b) + b(a - b)$
(b) $2(a + b)(a - b)$

70. Explain the possible procedures that may be used to factor a polynomial of **(a)** two terms; **(b)** three terms; **(c)** four terms.

[2.1] *Solve.*

71. $4(x - 2) = 3(x - 4) - 4$ -8

72. $-5(x - 2) + 3 = -5x - 6$ no solution

Find the solution set for each inequality.

[2.5] **73.** $4(x - 3) < 6(x - 4)$. $\{x \mid x > 6\}$

[2.6] **74.** $|2x - 3| > -4$. $\mathbb{R}$

1. (a) $x^{-4}(x^2 - 5x + 6)$ **(b)** $x^{-4}(x - 3)(x - 2)$ **2. (a)** $x^{-1}(x^4 - 2x^2 - 3)$ **(b)** $x^{-1}(x^2 - 3)(x^2 +$

Group Activity/ Challenge Problems

*We have worked only with integer exponents in this chapter (except Exercise 4 in Group Activity of Section 6.1). However, fractional exponents may also be factored out of an expression. The expressions below are not polynomials. (**a**) Factor out the variable with the lowest (or most negative) exponent from each of the following. (Fractional exponents are discussed in Section 8.2.) (**b**) Factor further if possible.*

1. $x^{-2} - 5x^{-3} + 6x^{-4}$, factor out x^{-4} **2.** $x^3 - 2x^1 - 3x^{-1}$ factor out x^{-1}

3. $4x^{3/2} + x^1 - 2x^{1/2}$, factor out $x^{1/2}$ **4.** $x^{5/2} - 3x^{3/2} - 2x^{1/2}$, factor out $x^{1/2}$

5. $5x^{1/2} + 2x^{-1/2} - 3x^{-3/2}$, factor out $x^{-3/2}$

3. (a) $x^{1/2}(4x + x^{1/2} - 2)$ **(b)** not possible   **4. (a)** $x^{1/2}(x^2 - 3x - 2)$ **(b)** not possible

5. (a) $x^{-3/2}(5x^2 + 2x - 3)$ **(b)** $x^{-3/2}(5x - 3)(x + 1)$

6.5 Solving Equations Using Factoring

Tape 10

1. Use the zero-factor property to solve equations.
2. Use factoring to solve equations.
3. Solve equations quadratic in form.
4. Use factoring to solve application problems.
5. Use factoring to find the x intercepts of a graph.

Quadratic equations in two variables, x and y, were introduced and graphed in Section 5.6. In this section we explain how to use factoring to solve quadratic equations in one variable. Every quadratic equation has a second-degree term as its highest term.

Examples of Quadratic Equations

$$3x^2 + 6x - 4 = 0$$
$$5x = 2x^2 - 4$$
$$(x + 4)(x - 3) = 0$$

Any quadratic equation can be written in standard form.

> **Standard Form of a Quadratic Equation**
> $$ax^2 + bx + c = 0, \quad a \neq 0$$
> where a, b, and c are real numbers.

Before going any further, make sure that you can convert each of the three quadratic equations given above to standard form, with $a > 0$.

Zero-Factor Property   **1** To solve equations using factoring, we use the **zero-factor property.**

> ### Zero-Factor Property
> For all real numbers a and b, if $a \cdot b = 0$, then either $a = 0$ or $b = 0$, or both a and $b = 0$.

The zero-factor property indicates that, if the product of two factors equals zero, one (or both) of the factors must be zero.

EXAMPLE 1   Solve the equation $(x + 5)(x - 3) = 0$.

Solution: Since the product of the factors equals 0, according to the zero-factor property, one or both factors must equal zero. Set each factor equal to 0 and solve each equation separately.

$$x + 5 = 0 \qquad \text{or} \qquad x - 3 = 0$$
$$x = -5 \qquad\qquad\qquad x = 3$$

Thus, if x is either -5 or 3, the product of the factors is 0.

Check:

$x = -5$	$x = 3$
$(x + 5)(x - 3) = 0$	$(x + 5)(x - 3) = 0$
$(-5 + 5)(-5 - 3) = 0$	$(3 + 5)(3 - 3) = 0$
$0(-8) = 0$	$8(0) = 0$
$0 = 0$ true	$0 = 0$ true

Solving Equations   **2**
by Factoring

> ### To Solve an Equation by Factoring
> 1. Use the addition property to remove all terms from one side of the equation. This will result in one side of the equation being equal to 0.
> 2. Combine like terms in the equation and then factor.
> 3. Set each factor *containing a variable* equal to zero, solve the equations, and find the solutions.
> 4. Check the solutions in the original equation.

EXAMPLE 2   Solve the equation $2x^2 = 12x$.

Solution: First, make the right side of the equation equal to 0 by subtracting $12x$ from both sides of the equation. Then factor the left side of the equation.

$$2x^2 - 12x = 0$$
$$2x(x - 6) = 0$$

Now set each factor equal to zero.

$$2x = 0 \quad \text{or} \quad x - 6 = 0$$
$$x = 0 \qquad\qquad x = 6$$

A check will show that the numbers 0 and 6 both satisfy the equation $2x^2 = 12x$.

COMMON STUDENT ERROR The zero-factor property can be used only when one side of the equation is equal to 0.

Correct	*Incorrect*
$(x - 4)(x + 3) = 0$	$(x - 4)(x + 3) = 2$
$x - 4 = 0 \quad \text{or} \quad x + 3 = 0$	$x - 4 = 2 \quad \text{or} \quad x + 3 = 2$
$x = 4 \qquad\qquad x = -3$	$x = 6 \qquad\qquad x = -1$

In the incorrect process illustrated on the right, the zero-factor property cannot be used since the right side of the equation is not equal to 0. Example 3 shows how to solve such problems correctly.

EXAMPLE 3 Solve the equation $(x - 1)(3x + 2) = 4x$.

Solution: Since the right side of the equation is not 0, we cannot use the zero-factor property yet. Begin by multiplying the factors on the left side of the equation. Then subtract $4x$ from both sides of the equation to obtain 0 on the right side. Then factor and solve the equation.

$$(x - 1)(3x + 2) = 4x$$
$$3x^2 - x - 2 = 4x$$
$$3x^2 - 5x - 2 = 0$$
$$(3x + 1)(x - 2) = 0$$

$$3x + 1 = 0 \qquad \text{or} \qquad x - 2 = 0$$
$$3x = -1 \qquad\qquad\qquad x = 2$$
$$x = -\frac{1}{3}$$

The solutions are $-\frac{1}{3}$ and 2.

EXAMPLE 4 Solve the equation $3x^2 + 2x - 12 = -7x$.

Solution:

$3x^2 + 2x - 12 = -7x$	
$3x^2 + 9x - 12 = 0$	$7x$ was added to both sides of equation.
$3(x^2 + 3x - 4) = 0$	Factor out 3.
$3(x + 4)(x - 1) = 0$	Factor the trinomial.
$x + 4 = 0 \qquad \text{or} \qquad x - 1 = 0$	Solve for x.
$x = -4 \qquad \text{or} \qquad x = 1$	

Since the factor 3 does not contain a variable, we do not have to set it equal to zero. Only the numbers -4 and 1 satisfy the equation $3x^2 + 2x - 12 = -7x$.

EXAMPLE 5   Solve the equation $2x(x + 2) = x(x - 3) - 12$.

Solution:
$$2x(x + 2) = x(x - 3) - 12$$
$$2x^2 + 4x = x^2 - 3x - 12$$
$$x^2 + 7x + 12 = 0$$
$$(x + 4)(x + 3) = 0$$

$x + 4 = 0$ or $x + 3 = 0$

$x = -4$ or $x = -3$

Helpful Hint

When solving an equation whose highest powered term has a *negative coefficient,* we generally make it positive by multiplying both sides of the equation by -1. This makes the factoring process easier, as in the following example.

$$-x^2 + 5x + 6 = 0$$
$$-1(-x^2 + 5x + 6) = -1 \cdot 0$$
$$x^2 - 5x - 6 = 0$$

Now solve the equation $x^2 - 5x - 6 = 0$ by factoring.

$$(x - 6)(x + 1) = 0$$

$x - 6 = 0$ or $x + 1 = 0$

$x = 6$ $x = -1$

The numbers 6 and -1 both satisfy the original equation, $-x^2 + 5x + 6 = 0$.

The equations in Examples 1 through 5 were all quadratic equations that were rewritten in the form $ax^2 + bx + c = 0$ and solved by factoring. Other methods that can be used to solve quadratic equations include completing the square and the quadratic formula; we discuss these methods in Chapter 9.

The zero-factor property can be extended to three or more factors as illustrated in Example 6.

EXAMPLE 6   Solve the equation $2x^3 + 5x^2 - 3x = 0$.

Solution: First factor, then set each factor containing an x equal to 0.

$$2x^3 + 5x^2 - 3x = 0$$
$$x(2x^2 + 5x - 3) = 0$$
$$x(2x - 1)(x + 3) = 0$$

$x = 0$ or $2x - 1 = 0$ or $x + 3 = 0$

$2x = 1$ $x = -3$

$$x = \frac{1}{2}$$

The numbers 0, $\frac{1}{2}$, and -3 are all solutions to the equation.

Note that the equation in Example 6 is not a quadratic equation because its highest degree term is 3, not 2. This is a *cubic,* or *third-degree, equation,* as was discussed in Chapter 5.

Equations Quadratic in Form

③ In Section 6.2, in Examples 12 and 14, we factored the expressions $y^4 - y^2 - 6$ and $2(x + 5)^2 - 5(x + 5) - 12$. If these expressions were set equal to 0, we would have $y^4 - y^2 - 6 = 0$ and $2(x + 5)^2 - 5(x + 5) - 12 = 0$. These two equations are examples of *equations quadratic in form.* Any equation that can be written in the form $au^2 + bu + c = 0$, $a \neq 0$ is an **equation quadratic in form.** In $y^4 - y^2 - 6 = 0$, if we let $u = y^2$, this equation becomes $u^2 - u - 6 = 0$. In $2(x + 5)^2 - 5(x + 5) - 12 = 0$, if we let $u = x + 5$ the equation becomes $2u^2 - 5u - 12 = 0$. To solve equations that are quadratic in form, rewrite the equation in the form $au^2 + bu + c = 0$ by making a substitution. Then factor the left side of the equation and set each factor equal to 0. Example 7 illustrates how to solve equations like $2(x + 5)^2 - 5(x + 5) - 12 = 0$. In Section 9.3 we will show how to solve equations like $y^4 - y^2 - 6 = 0$, and equations that contain fractional exponents, such as $6y^{2/5} + 5y^{1/5} - 6 = 0$, that are quadratic in form.

EXAMPLE 7 Solve the equation $4(2w + 1)^2 - 16(2w + 1) + 15 = 0$.

Solution: If we let $u = (2w + 1)$, the equation becomes

$$4(2w + 1)^2 - 16(2w + 1) + 15 = 0$$
$$4u^2 - 16u + 15 = 0$$

Now factor and solve.

$$(2u - 3)(2u - 5) = 0$$

$$2u - 3 = 0 \quad \text{or} \quad 2u - 5 = 0$$
$$2u = 3 \qquad\qquad 2u = 5$$
$$u = \frac{3}{2} \qquad\qquad u = \frac{5}{2}$$

We are not finished. Since the variable in the original equation is w we must solve for w, not u. Therefore, we substitute back $2w + 1$ for u and solve for w.

$$u = \frac{3}{2} \qquad\qquad u = \frac{5}{2}$$
$$2w + 1 = \frac{3}{2} \qquad\qquad 2w + 1 = \frac{5}{2}$$
$$2w = \frac{1}{2} \qquad\qquad 2w = \frac{3}{2}$$
$$w = \frac{1}{4} \qquad\qquad w = \frac{3}{4}$$

Check: $\qquad w = \dfrac{1}{4}$ $\qquad\qquad\qquad\qquad\qquad w = \dfrac{3}{4}$

$$4(2w + 1)^2 - 16(2w + 1) + 15 = 0 \qquad\qquad 4(2w + 1)^2 - 16(2w + 1) + 15 = 0$$

$$4\left[2\left(\frac{1}{4}\right) + 1\right]^2 - 16\left[2\left(\frac{1}{4}\right) + 1\right] + 15 = 0 \qquad 4\left[2\left(\frac{3}{4}\right) + 1\right]^2 - 16\left[2\left(\frac{3}{4}\right) + 1\right] + 15 = 0$$

$$4\left(\frac{3}{2}\right)^2 - 16\left(\frac{3}{2}\right) + 15 = 0 \qquad\qquad 4\left(\frac{5}{2}\right)^2 - 16\left(\frac{5}{2}\right) + 15 = 0$$

$$4\left(\frac{9}{4}\right) - 24 + 15 = 0 \qquad\qquad\qquad 4\left(\frac{25}{4}\right) - 40 + 15 = 0$$

$$9 - 24 + 15 = 0 \qquad\qquad\qquad\qquad 25 - 40 + 15 = 0$$

$$0 = 0, \quad \text{true} \qquad\qquad\qquad\qquad\qquad 0 = 0, \quad \text{true}$$

Therefore, the solutions are $\dfrac{1}{4}$ and $\dfrac{3}{4}$.

Applications   **4**   Now let us look at some application problems that use factoring in their solution.

Example 8   The area of a triangle is 27 square inches. Find the base and height if its height is 3 inches less than twice its base.

Solution: Let $x =$ base, then $2x - 3 =$ height (Fig. 6.1).

$$\text{Area} = \frac{1}{2}(\text{base})(\text{height})$$

$$27 = \frac{1}{2}(x)(2x - 3) \qquad \textbf{Multiply both sides of equation}$$
$$\textbf{by 2 to remove fractions.}$$

$$2(27) = 2\left[\frac{1}{2}(x)(2x - 3)\right]$$

$$54 = x(2x - 3)$$

$$54 = 2x^2 - 3x$$

or $\qquad 2x^2 - 3x - 54 = 0$

$$(2x + 9)(x - 6) = 0$$

$$2x + 9 = 0 \qquad \text{or} \qquad x - 6 = 0$$

$$2x = -9 \qquad\qquad\qquad x = 6$$

$$x = -\frac{9}{2}$$

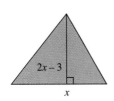

Figure 6.1

Since the dimensions of a geometric figure cannot be negative, we can eliminate $x = -\frac{9}{2}$ as an answer to our problem.

$$\text{Base} = x = 6 \text{ inches}$$

$$\text{Height} = 2x - 3 = 2(6) - 3 = 9 \text{ inches}$$

EXAMPLE 9 A projectile on top of a building 384 feet high is fired upward with a velocity of 32 feet per second. The projectile's distance, s, above the ground at any time, t, is given by the formula $s = -16t^2 + 32t + 384$. Find the time that it takes for the object to strike the ground.

Solution: When the object strikes the ground, its distance from the ground is 0. Substituting $s = 0$ into the equation gives

$$0 = -16t^2 + 32t + 384$$

$$\text{or} \quad -16t^2 + 32t + 384 = 0$$

$$-16(t^2 - 2t - 24) = 0$$

$$-16(t + 4)(t - 6) = 0$$

$$t + 4 = 0 \quad \text{or} \quad t - 6 = 0$$

$$t = -4 \qquad\qquad t = 6$$

Since t is the number of seconds, $t = -4$ is not a possible answer. The projectile will strike the ground in 6 seconds.

Finding the x Intercepts

⑤ Consider the graph of the function $y = x^2 - 2x - 8$ (Fig. 6.2).

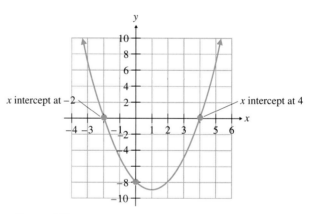

x intercept at -2 *x* intercept at 4

FIGURE 6.2

The x intercepts of the graph are at 4 and -2. The value of y at the x intercepts is 0. Therefore, to find the x intercepts of a function we can replace y with 0 and solve the resulting equation. In the function above, when y is replaced with 0, we get $x^2 - 2x - 8 = 0$. The solution to the equation $x^2 - 2x - 8 = 0$ is illustrated below.

$$x^2 - 2x - 8 = 0$$

$$(x - 4)(x + 2) = 0$$

$$x - 4 = 0 \quad \text{or} \quad x + 2 = 0$$

$$x = 4 \qquad\qquad x = -2$$

Notice that the solutions of $x^2 - 2x - 8 = 0$ are 4 and -2. These values are also the x intercepts of the graph of the function $y = x^2 - 2x - 8$.

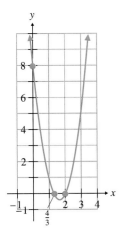

FIGURE 6.3

EXAMPLE 10   Use factoring to determine the x intercepts of the graph of the equation $y = 3x^2 - 10x + 8$.

Solution: At the x intercepts the value of y is 0. Therefore, we set $y = 0$ and solve the resulting equation $3x^2 - 10x + 8 = 0$.

$$3x^2 - 10x + 8 = 0$$
$$(3x - 4)(x - 2) = 0$$

$$3x - 4 = 0 \quad \text{or} \quad x - 2 = 0$$
$$3x = 4 \qquad\qquad x = 2$$
$$x = \frac{4}{3}$$

The x intercepts are at 4/3 and 2. We have illustrated the graph of the function $y = 3x^2 - 10x + 8$ in Fig. 6.3.

EXAMPLE 11   Use factoring to find the x intercepts of the graph of the function $y = 2x^3 + x^2 - 6x$.

Solution: Set y equal to 0 and solve the resulting equation.

$$2x^3 + x^2 - 6x = 0$$
$$x(2x^2 + x - 6) = 0$$
$$x(2x - 3)(x + 2) = 0$$

$$x = 0 \quad \text{or} \quad 2x - 3 = 0 \quad \text{or} \quad x + 2 = 0$$
$$2x = 3 \qquad\qquad x = -2$$
$$x = \frac{3}{2}$$

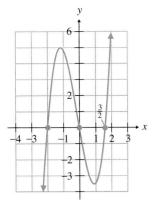

FIGURE 6.4

Thus, the solutions to the equation $2x^3 + x^2 - 6x = 0$ are 0, 3/2, and -2. The y intercepts of the graph of the function $y = 2x^3 + x^2 - 6x$ are also 0, 3/2, and -2 (see Fig. 6.4).

We will use the information presented here when we find the x intercepts of certain graphs in later chapters. Remember that to find the x intercepts of a graph, or the zeros of a function, we can replace y or $f(x)$ with 0 and solve the resulting equation.

Helpful Hint

If the function $P(x) = ax^2 + bx + c$ has *zeros* at m and n, that is, $P(m) = 0$ and $P(n) = 0$, then m and n are *x intercepts* of the graph of $y = ax^2 + bx + c$. The values of m and n are also *solutions* or *roots* of the equation $ax^2 + bx + c = 0$. In some books and graphing calculator manuals the words *zeros*, *x intercepts*, *roots*, and *solutions* are all used synonymously.

The graphing calculator can be used as an aid in factoring expressions. Suppose that you wish to factor the trinomial $5x^2 + 46x + 48$. If you write the equation $y = 5x^2 + 46x + 48$ and then graph it, you can see where the graph crosses the x axis (see Fig. 6.5).

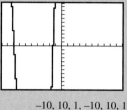

$-10, 10, 1, -10, 10, 1$

FIGURE 6.5

The graph appears to cross the x axis at $x = -8$ and somewhere between $x = -1$ and $x = -2$. To determine if -8 is indeed an x intercept, replace y with $P(x)$, then evaluate $P(-8)$ to see if we get 0.

$$P(x) = 5x^2 + 46x + 48$$
$$P(-8) = 5(-8)^2 + 46(-8) + 48 = 320 - 368 + 48 = 0$$

Since $P(-8) = 0$, when x is -8 then y is 0. Therefore, the graph crosses the x axis at -8.

Since $P(-8) = 0$, by the remainder theorem, we know that when $5x^2 + 46x + 48$ is divided by $x + 8$, the remainder is 0. Thus $x + 8$ must be factor of $5x^2 + 46x + 48$. We can now use trial and error (or division) to obtain the second factor.

$$(x + 8)(\quad\quad) = 5x^2 + 46x + 48$$
$$(x + 8)(5x + 6) = 5x^2 + 46x + 48$$

Thus, $5x^2 + 46x + 48 = (x + 8)(5x + 6)$.

When graphing a function, if the x intercepts are not integer values, it does not necessarily mean that the polynomial is not factorable. It may mean that the factors do not contain integer values. Consider the function $f(x) = 4x^2 - 4x - 15$. The x intercepts of its graph are at $-3/2$ and $5/2$. The trinomial factors as follows:

$$4x^2 - 4x - 15 = (2x + 3)(2x - 5)$$

EXERCISES

Use your graphing calculator to find at least one factor, and then write the expression in factored form.

1. $x^2 - 16x + 48$ $(x - 4)(x - 12)$ **2.** $x^2 - 45x + 200$ $(x - 5)(x - 40)$

3. $x^2 - 51x - 540$ $(x + 9)(x - 60)$ **4.** $x^2 - 25x + 126$ $(x - 7)(x - 18)$

Solve the following equations using your graphing calculator by replacing the 0 with y, and determining the x intercepts of the graphs of the function. Estimate the solutions to the nearest tenth. If the equation has no real solutions, so state.

5. $x^2 + 3x - 54 = 0$ $-9, 6$ **6.** $2x^2 + 33x - 164 = 0$ $-20.5, 4$

7. $0.6x^2 + 7.1x - 1.2 = 0$ $0.2, -12$ **8.** $-2.3x^2 - 6.9x + 21.6$ $-4.9, 1.9$

Exercise Set 6.5

Solve.

1. $x(x + 5) = 0$ 0, −5

4. $2(x + 3)(x - 5) = 0$ −3, 5

7. $4x - 12 = 0$ 3

10. $x^2 + 4x = 0$ 0, −4

13. $x^2 + x - 12 = 0$ −4, 3

16. $3y^2 - 2 = -y$ $\frac{2}{3}$, −1

19. $3x^2 - 6x - 72 = 0$ −4, 6

22. $3x^2 - 9x - 30 = 0$ −2, 5

25. $-16x - 3 = -12x^2$ $-\frac{1}{6}, \frac{3}{2}$

28. $-2y^2 + 24y - 22 = 0$ 1, 11

31. $3p^2 = 22p - 7$ $\frac{1}{3}$, 7

34. $3x^2 = 7x + 20$ $-\frac{5}{3}$, 4

37. $6x^2 = 16x$ 0, $\frac{8}{3}$

40. $2x^4 - 32x^2 = 0$ 0, 4, −4

43. $(x - 7)(x + 5) = -20$ −3, 5

46. $(x - 4)^2 - 4 = 0$ 2, 6

49. $2(x + 2)(x - 2) = (x - 2)(x + 3) - 2$ 0, 1

51. $2x^3 + 16x^2 + 30x = 0$ 0, −3, −5

5. $-\frac{5}{2}$, 3, −2

2. $3x(x - 5) = 0$ 0, 5

5. $(2x + 5)(x - 3)(3x + 6) = 0$

8. $9x - 27 = 0$ 3

11. $9x^2 = -18x$ 0, −2

14. $x(x + 6) = -9$ −3

17. $-z^2 - 3z = -18$ 3, −6

20. $x^3 = 3x^2 + 18x$ 0, 6, −3

23. $2y^2 + 22y + 60 = 0$ −5, −6

26. $-7x - 10 = -6x^2$ 2, $-\frac{5}{6}$

29. $3x^3 - 8x^2 - 3x = 0$ $0, -\frac{1}{3}$, 3

32. $5w^2 - 16w = -3$ $\frac{1}{5}$, 3

35. $4x^3 + 4x^2 - 48x = 0$ 0, −4, 3

38. $4x^2 = 9$ $-\frac{3}{2}, \frac{3}{2}$

41. $(x + 4)^2 - 16 = 0$ 0, −8

44. $(x + 1)^2 = 3x + 7$ −2, 3

47. $(b - 1)(3b + 2) = 4b$ $-\frac{1}{3}$, 2

50. $2(a + 3)(a - 5) = 2(a - 1) + 8$ −3, 6

52. $18x^3 - 15x^2 = 12x$ $0, \frac{4}{3}, -\frac{1}{2}$

3. $5x(x + 9) = 0$ 0, −9

6. $x(2x + 3)(x - 5) = 0$ $0, -\frac{3}{2}$, 5

9. $-x^2 + 12x = 0$ 0, 12

12. $x^2 + 6x + 5 = 0$ −1, −5

15. $x(x - 12) = -20$ 2, 10

18. $3x^2 = -21x - 18$ −6, −1

21. $x^3 + 19x^2 = 42x$ 0, 2, −21

24. $8x^2 + 14x - 15 = 0$ $\frac{3}{4}, -\frac{5}{2}$

27. $-28x^2 + 15x - 2 = 0$ $\frac{1}{4}, \frac{2}{7}$

30. $z^3 + 16z^2 = -64z$ 0, −8

33. $3r^2 + r = 2$ $\frac{2}{3}$, −1

36. $x^2 - 25 = 0$ 5, −5

39. $25x^3 - 16x = 0$ $0, \frac{4}{5}, -\frac{4}{5}$

42. $(2x + 5)^2 - 9 = 0$ −1, −4

45. $6a^2 - 12 - 4a = 19a - 32$ $\frac{5}{2}, \frac{4}{3}$

48. $2(a^2 - 3) - 3a = 2(a + 3)$ $-\frac{3}{2}$, 4

Use factoring to find the x intercepts of the graphs of the equations (see Examples 10 and 11).

53. $y = x^2 + 2x - 24$ 4, −6

55. $y = x^2 + 14x + 49$ −7

57. $y = 15x^2 - 14x - 8$ $-\frac{2}{5}, \frac{4}{3}$

59. $y = 6x^3 - 23x^2 + 20x$ $0, \frac{4}{3}, \frac{5}{2}$

54. $y = x^2 - 14x + 48$ 6, 8

56. $y = 2x^2 - 10x + 12$ 2, 3

58. $y = 6x^2 + 7x - 24$ $\frac{3}{2}, -\frac{8}{3}$

60. $y = 12x^3 - 39x^2 + 30x$ $0, \frac{5}{4}$, 2

Write the problem as an equation. Solve the equation and answer the question.   **65.** $w = 3$ ft, $l = 12$ ft

61. The product of two consecutive positive integers is 72. Find the two integers.   8, 9

62. The product of two consecutive positive odd integers is 99. Find the two integers.   9, 11

63. The product of two positive numbers is 108. Find the two numbers if one is 3 more than the other.   9, 12

64. The product of two positive numbers is 35. Find the two numbers if one number is 3 less than twice the other.   5, 7

65. The area of a rectangle is 36 square feet. Find the length and width if the length is 4 times the width.

66. The area of a rectangle is 54 square inches. Find the length and width if the length is 3 inches less than twice the width. $w = 6$ in., $l = 9$ in.

67. The base of a triangle is 6 centimeters greater than its height. Find the base and height if the area is 80 square centimeters. $b = 16$ cm, $h = 10$ cm

68. The height of a triangle is 1 centimeter less than twice its base. Find the base and height if the triangle's area is 33 square centimeters. $b = 6$ cm, $h = 11$ cm

69. A model rocket will be launched from a hill 80 feet above sea level. The launch site is next to the ocean (sea level), and the rocket will fall into the ocean. The rocket's distance, s, above sea level at any time, t, is found by the equation $s = -16t^2 + 64t + 80$. Find the time it takes for the rocket to strike the ocean.   5 sec

77. (b) One possible answer is $x^2 + 7x + 10 = 0$. **(c)** An infinite number; $f(x) = a(x^2 + 7x + 10)$, where a is any real number except 0, will have x intercepts of -2 and -5. **(d)** An infinite number; $a(x^2 + 7x + 10) = 0$, where a is any real number except 0 will have solutions -2 and -5.

70. A rubber ball is at the top of a 96-foot waterfall. The ball's distance from the pool of water at the bottom of the falls t seconds after it goes over the falls can be found by using the formula $d = -16t^2 + 96$. Find the ball's distance above the pool after **(a)** 1 second; **(b)** 1.5 seconds. **(c)** When will the ball hit the pool of water at the bottom of the falls? **72.** 4 sec **73.** 5 sec

71. If the sides of a square are increased by 4 meters, the area becomes 121 square meters. Find the length of a side of the original square. 7 m

72. An egg is dropped from a helicopter 256 feet above the ground. The distance of the egg from the ground at the time t is given by the equation $s = -16t^2 + 256$. Find the time it takes for the egg to strike the ground.

73. A car key is thrown downward from the top of a 640-foot-tall building with a velocity of 48 feet per second. The key's distance, s, from the ground at any time, t, is found by the equation $s = -16t^2 - 48t + 640$. Find the time it takes for the key to strike the ground.

74. (a) Explain the zero-factor property in your own words.
(b) Solve the equation $(3x - 7)(2x + 3) = 0$ using the zero-factor property. $\frac{7}{3}, -\frac{3}{2}$

75. (a) Explain why the equation $(x + 3)(x + 4) = 2$ **cannot** be solved by writing $x + 3 = 2$ or $x + 4 = 2$.
(b) Solve the equation $(x + 3)(x + 4) = 2$. $-2, -5$

76. (a) Explain in your own words how to solve an equation using factoring.
(b) Solve the equation $-x - 20 = -12x^2$ using the procedure in part (a). $\frac{4}{3}, -\frac{5}{4}$

77. Consider the following graph of a quadratic function:

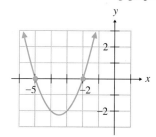

(a) Write a quadratic function that has the x intercepts indicated. One possible answer is $f(x) = x^2 + 7x + 10$.
(b) Write a quadratic equation in one variable that has solutions of -2 and -5.
(c) How many different quadratic functions can have x intercepts of -2 and -5? Explain.
(d) How many different quadratic equations in one variable can have solutions of -2 and -5? Explain.

78. When a constant is factored out of an equation, why is it not necessary to set that constant equal to 0 when solving the equation? constant term does not contain a varia

79. Consider the equation $x^2 - 6x + 9 = 0$. Solving the equation, we obtain

$$(x - 3)(x - 3) = 0$$
$$x - 3 = 0 \quad \text{or} \quad x - 3 = 0$$
$$x = 3 \quad \text{or} \quad x = 3$$

This equation has only one real solution, $x = 3$. Since we obtained 3 twice when solving the equation, we say that 3 is a **root of multiplicity two.**

(a) Graph $y = (x - 3)(x - 3)$ and determine where and how many times the graph crosses the x axis.
(b) How many times will the graph of $y = (x - r)(x - r)$ intersect the x axis, and where will it intersect? Explain.
(c) Write an equation in one variable that has 2 as a root of multiplicity two. $x^2 - 4x + 4 = 0$
(d) Write an equation in two variables that has only one x intercept at $x = 2$. $y = x^2 - 4x + 4$

80. Refer to Exercise 79 before working this problem.
(a) Solve the equation $x^2 + 4x + 4 = 0$ by factoring. What solution does each factor give when you use the zero-factor property? -2
(b) Graph the equation $y = x^2 + 4x + 4$. How many x intercepts does it have? one, at $x = 2$
(c) How many solutions will an equation of the form $(x - a)(x - a) = 0$ have? What is the solution?
(d) When an equation of the form $y = (x - a)(x - a)$ is graphed, how many x intercepts will the graph have? Where will the intercept or intercepts be?

81. Consider the polynomial function $P(x) = ax^2 + bx + c, a > 0$.
(a) How many possible intercepts may the graph have? Sketch each of the possibilities.
(b) How many possible real solutions may the equation $ax^2 + bx + c = 0, a > 0$ have? Explain your answer to part (b) by using the sketches in part (a). none, one or two

82. The graph of the equation $y = x^2 + 2x + 4$ is illustrated below. **(a)** none

81. (a) None

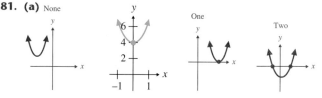

(a) How many x intercepts does the graph have?
(b) How many real solutions does the equation $x^2 + 2x + 4 = 0$ have? Explain your answer. none

70. (a) 80 feet **(b)** 60 feet **(c)** $\sqrt{6} \approx 2.45$ seconds.
75. (a) Zero-factor property holds only when one side of the equation is 0.
79. (a) One; the x intercept is at $(3, 0)$.
(b) One; it will have its x intercept at $(r, 0)$.

80. (c) One; the solution is $x = a$.
(d) One; at $(a, 0)$

83. (a) Carmen, 4.8 mph, Bob, 6 mph; **(b)** 24 mi

CUMULATIVE REVIEW EXERCISES

[2.3] **83.** Two distance runners, Carmen and Bob, run the same course with Bob running 1.2 miles per hours faster than Carmen. If Bob finishes in 4 hours and Carmen finishes in 5 hours, **(a)** what is the rate of each; **(b)** how long is the course?

[3.5] **84.** Graph $f(x) = -|2x - 4| + 1$.

[4.1] **85.** Solve the system of equations.   (3, 0)

$$3x + 5y = 9$$
$$2x - y = 6$$

[4.6] **86.** Find the solution to the system of inequalities.   ∅

$$2y > 6x + 12$$
$$\tfrac{1}{2}y < \tfrac{3}{2}x + 2$$

[5.4] **87.** Divide $(6x^2 - x - 12) \div (2x - 3)$. $3x + 4$

84.

---

1. $x^2 + 3x - 10 = 0$ **2.** $6x^2 - 7x + 2 = 0$ **3.** $x^3 - 2x^2 - 3x = 0$ **4.** $x^4 - 4x^3 - 7x^2 + 22x + 24 = 0$

Group Activity/ Challenge Problems

Write an equation that has the solutions indicated.

1. $2, -5$ **2.** $\dfrac{1}{2}, \dfrac{2}{3}$ **3.** $-1, 0, 3$ **4.** $-2, -1, 3, 4,$

Solve.

5. $(x + 3)^2 + 2(x + 3) = 24$ $1, -9$ **6.** $2(x + 1)^2 - 5(x + 1) - 3 = 0$ $2, -\tfrac{3}{2}$

7. $6(x - 2)^2 = -19(x - 2) - 10$ $\tfrac{4}{3}, -\tfrac{1}{2}$ **8.** $(3x + 2)^2 = (x - 4)^2$ $\tfrac{1}{2}, -3$

9. $x^4 - 5x^2 + 4 = 0$ $\pm 1, \pm 2$ **10.** $x^4 - 13x^2 = -36$ $\pm 2, \pm 3$

11. $x^6 - 9x^3 + 8 = 0$ $2, 1$

In more advanced mathematics courses you may need to solve an equation for y′ (read y prime). When doing so, treat the y′ as a different variable from y. Solve each of the following for y′.

12. $xy' + yy' = 1$ $y' = \dfrac{1}{x + y}$ **13.** $xy - xy' = 3y' + 2$ $y' = \dfrac{xy - 2}{x + 3}$ **14.** $2xyy' - xy = x - 3y'$

$$y' = \dfrac{x + xy}{2xy + 3}$$

Summary

GLOSSARY

Equation quadratic in form (366): An equation of the form $au^2 + bu + c = 0$, $a \neq 0$.

Factoring (332): Writing an expression as the product of other expressions.

Greatest common factor (333): The greatest common factor of two or more expressions is the greatest factor that divides each expression.

Standard form of a quadratic equation (362): $ax^2 + bx + c = 0$, $a \neq 0$

Zero-factor property (363): If $a \cdot b = 0$, then either $a = 0$ or $b = 0$ or both a and $b = 0$.

IMPORTANT FACTS

Special factoring formulas

$a^2 - b^2 = (a + b)(a - b)$	difference of two squares
$a^2 + 2ab + b^2 = (a + b)^2$	
$a^2 - 2ab + b^2 = (a - b)^2$	perfect square trinomials
$a^3 + b^3 = (a + b)(a^2 - ab + b^2)$	sum of two cubes
$a^3 - b^3 = (a - b)(a^2 + ab + b^2)$	difference of two cubes

NOTE: The sum of two squares, $a^2 + b^2$, cannot be factored over the set of real numbers.

Review Exercises

[6.1] *Find the greatest common factor for each set of terms.*

1. $40x^2$, $36x^3$, $16x^5$ $4x^2$

2. $12xy$, $36xy^2$, $18x^2y$ $6xy$

3. $15x^3y^2z^5$, $-6x^2y^3$, $30xy^4z$ $3xy^2$

4. $x(2x - 5)$, $3(2x - 5)^2$, $5(2x - 5)^3$ $2x - 5$

5. $x(x + 5)$, $x + 5$, $2(x + 5)^2$ $x + 5$

6. $2x$, $(x - 2)$, $(x - 2)^2$ 1

Factor out the greatest common factor.

7. $12x^2 + 4x + 8$ $4(3x^2 + x + 2)$

8. $60x^4 + 6x^9 - 18x^5y^2$ $6x^4(10 + x^5 - 3xy^2)$

9. $2x(4x - 3) + 4x - 3$ $(2x + 1)(4x - 3)$

10. $3x(x - 1)^2 - 2(x - 1)$ $(x - 1)(3x^2 - 3x - 2)$

11. $4x(2x - 1) + 3(2x - 1)^2$ $(2x - 1)(10x - 3)$

12. $12xy^4z^3 + 6x^2y^3z^2 - 15x^3y^2z^3$ $3xy^2z^2(4y^2z + 2xy - 5x^2z)$

Factor by grouping.

13. $x^2 + 3x + 2x + 6$ $(x + 2)(x + 3)$

14. $5x^2 + 20x - x - 4$ $(5x - 1)(x + 4)$

15. $5x^2 - xy + 20xy - 4y^2$ $(x + 4y)(5x - y)$

16. $12x^2 - 8xy + 15xy - 10y^2$ $(4x + 5y)(3x - 2y)$

17. $(3x - y)(x + 2y) - (3x - y)(5x - 7y)$ $(3x - y)(-4x + 9y)$

18. $3a^4 - 12a^2b + 9a^2b - 36b^2$ $3(a^2 + 3b)(a^2 - 4b)$

32. $(x + 9)(x + 11)$ **33.** $(3x + 8)(x - 4)$

[6.2] *Factor each trinomial.*

19. $x^2 + 8x + 15$ $(x + 5)(x + 3)$

20. $x^2 - 8x + 15$ $(x - 5)(x - 3)$

21. $-x^2 + 12x + 45$ $-(x - 15)(x + 3)$

22. $x^2 - 5xy - 50y^2$ $(x - 10y)(x + 5y)$

23. $x^2 - 15xy - 54y^2$ $(x - 18y)(x + 3y)$

24. $2x^2 + 16x + 32$ $2(x + 4)^2$

25. $6x^2 - 19x - 20$ $(6x + 5)(x - 4)$

26. $8x^3 + 10x^2 - 25x$ $x(4x - 5)(2x + 5)$

27. $16x^2 + 8xy - 35y^2$ $(4x - 5y)(4x + 7y)$

28. $4x^3 - 9x^2 + 5x$ $x(4x - 5)(x - 1)$

29. $12x^3 + 61x^2 + 5x$ $x(12x + 1)(x + 5)$

30. $x^4 - 3x^2 - 10$ $(x^2 - 5)(x^2 + 2)$

31. $x^4 - x^2 - 20$ $(x^2 + 4)(x^2 - 5)$

32. $(x + 5)^2 + 10(x + 5) + 24$

33. $3(x + 2)^2 - 16(x + 2) - 12$

In Exercises 34–36 one factor of the polynomial is given. **(a)** *Use division to find a second factor.* **(b)** *Continue factoring until the polynomial is completely factored.*

35. (a) $9x^2 + 15x - 50$ **(b)** $(x - 3)(3x - 5)(3x + 10)$

34. $24x^2 + 65x + 21$, $(3x + 7)$

35. $9x^3 - 12x^2 - 95x + 150$, $(x - 3)$

36. $6x^3 - 49x^2 + 125x - 100$, $(2x - 5)$ **36. (a)** $3x^2 - 17x + 20$ **(b)** $(2x - 5)(x - 4)(3x - 5)$

34. (a) $8x + 3$ **(b)** $(3x + 7)(8x + 3)$

[6.3] *Use a special factoring formula to factor.*

37. $x^2 - 36$ $(x + 6)(x - 6)$

38. $4x^2 - 16y^4$ $4(x + 2y^2)(x - 2y^2)$

39. $x^4 - 81$ $(x^2 + 9)(x - 3)(x + 3)$

40. $(x + 2)^2 - 9$ $(x - 1)(x + 5)$

41. $(x - 3)^2 - 4$ $(x - 1)(x - 5)$

42. $4x^2 - 12x + 9$ $(2x - 3)^2$

43. $9y^2 + 24y + 16$ $(3y + 4)^2$

44. $w^4 - 16w^2 + 64$ $(w^2 - 8)^2$

45. $a^2 + 6ab + 9b^2 - 4c^2$ $(a + 3b + 2c)(a + 3b - 2c)$

46. $x^3 - 8$ $(x - 2)(x^2 + 2x + 4)$

47. $8x^3 + 27$ $(2x + 3)(4x^2 - 6x + 9)$

48. $27x^3 - 8y^3$ $(3x - 2y)(9x^2 + 6xy + 4y^2)$

49. $8y^6 - 125x^3$ $(2y^2 - 5x)(4y^4 + 10xy^2 + 25x^2)$

50. $(x + 1)^3 - 8$ $(x - 1)(x^2 + 4x + 7)$

[6.1–6.4] *Factor completely.*

51. $x^2y^2 - 2xy^2 - 15y^2$ $y^2(x + 3)(x - 5)$

52. $3x^3 - 18x^2 + 24x$ $3x(x - 4)(x - 2)$

53. $3x^3y^4 + 18x^2y^4 - 6x^2y^4 - 36xy^4$ $3xy^4(x - 2)(x + 6)$

54. $3y^5 - 27y$ $3y(y^2 + 3)(y^2 - 3)$

55. $2x^3y + 16y$ $2y(x + 2)(x^2 - 2x + 4)$

56. $5x^4y + 20x^3y + 20x^2y$ $5x^2y(x + 2)^2$

57. $6x^3 - 21x^2 - 12x$ $3x(2x + 1)(x - 4)$

58. $x^2 + 10x + 25 - y^2$ $(x + 5 + y)(x + 5 - y)$

59. $3x^3 + 24y^3$ $3(x + 2y)(x^2 - 2xy + 4y^2)$

60. $x^2(x + 4) + 3x(x + 4) - 4(x + 4)$ $(x + 4)^2(x - 1)$

61. $4(2x + 3)^2 - 12(2x + 3) + 5$ $(4x + 1)(4x + 5)$

62. $4x^4 + 4x^2 - 3$ $(2x^2 - 1)(2x^2 + 3)$

63. $(x - 1)x^2 - (x - 1)x - 2(x - 1)$ $(x + 1)(x - 2)(x - 1)$

64. $9ax - 3bx + 12ay - 4by$ $(3x + 4y)(3a - b)$

65. $6p^2q^2 - 5pq - 6$ $(2pq - 3)(3pq + 2)$

66. $9x^4 - 12x^2 + 4$ $(3x^2 - 2)^2$

67. $4y^2 - (x^2 + 4x + 4)$ $(2y + x + 2)(2y - x - 2)$
69. $6x^4y^4 + 9x^3y^4 - 27x^2y^4$ $3x^2y^4(x + 3)(2x - 3)$

68. $6(2a + 3)^2 - 7(2a + 3) - 3$ $2(3a + 5)(4a + 3)$
70. $x^3 - \frac{8}{27}y^6$ $\left(x - \frac{2}{3}y^2\right)\left(x^2 + \frac{2}{3}xy^2 + \frac{4}{9}y^4\right)$

[6.5] *In Exercises 71–74, (a) write an expression for the shaded area of the figure; (b) write the expression in part (a) in factored form.* **71. (a)** $a^2 - 4b^2$ **(b)** $(a + 2b)(a - 2b)$ **72. (a)** $2ab + 2b^2$ **(b)** $2b(a + b)$

71.

72.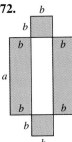

73.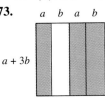

$a + 3b$

74.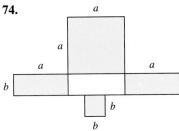

73. (a) $2a(a + 3b) + b(a + 3b)$ **(b)** $(2a + b)(a + 3b)$ **74. (a)** $a^2 + 2ab + b^2$ **(b)** $(a + b)^2$

Solve.
75. $(x - 5)(3x + 2) = 0$ $5, -\frac{2}{3}$
78. $x^2 - 2x - 24 = 0$ $6, -4$
81. $3x^2 + 21x + 30 = 0$ $-2, -5$
84. $8x^2 - 3 = -10x$ $\frac{1}{4}, -\frac{3}{2}$

76. $2x^2 = 3x$ $0, \frac{3}{2}$
79. $x^2 + 8x + 15 = 0$ $-3, -5$
82. $x^3 - 6x^2 + 8x = 0$ $0, 2, 4$
85. $4x^2 = 16$ $2, -2$

77. $15x^2 + 20x = 0$ $0, -\frac{4}{3}$
80. $x^2 = -2x + 8$ $-4, 2$
83. $12x^3 - 13x^2 - 4x = 0$ $0, \frac{4}{3}, -\frac{1}{4}$
86. $x(x + 3) = 2(x + 4) - 2$ $2, -3$

Use factoring to find the x intercepts of the graphs of the following equations.
87. $y = 2x^2 - 2x - 60$ $-5, 6$
88. $y = 20x^2 - 49x + 30$ $\frac{6}{5}, \frac{5}{4}$
89. $y = 14x^2 - 41x - 28$ $\frac{7}{2}, -\frac{4}{7}$

Write the problem as an equation. Solve the equation and answer the question. **93.** 5 in., 9 in.

90. The product of two positive integers is 30. Find the integers if the larger is 4 less than twice the smaller.   5, 6

91. The area of a rectangle is 63 square feet. Find the length and width of the rectangle if the length is 2 feet greater than the width. $w = 7$ ft, $l = 9$ ft

92. The base of a triangle is 3 more than twice the height. Find the base and height if the area of the triangle is 22 square meters. $h = 4$ m, $b = 11$ m

93. One square has a side 4 inches longer than the side of a second square. If the area of the larger square is 81 square inches, find the length of a side of each square.

94. A rocket is projected upward from the top of a 144-foot-tall building with a velocity of 128 feet per second. The rocket's distance from the ground, s, at any time, t, is given by the formula $s = -16t^2 + 128t + 144$. Find the time it takes for the rocket to strike the ground.   9 sec

Practice Test

Factor completely.
1. $4x^2y - 4x$ $4x(xy - 1)$
3. $9x^3y^2 + 12x^2y^5 - 27xy^4$ $3xy^2(3x^2 + 4xy^3 - 9y^2)$
5. $2x^2 + 4xy + 3xy + 6y^2$ $(2x + 3y)(x + 2y)$
7. $3x^3 - 6x^2 - 9x$ $3x(x - 3)(x + 1)$
9. $5x^2 + 17x + 6$ $(5x + 2)(x + 3)$
11. $27x^3y^6 - 8y^6$ $y^6(3x - 2)(9x^2 + 6x + 4)$
13. $2x^4 + 5x^2 - 18$ $(2x^2 + 9)(x^2 - 2)$

2. $3x^2 + 12x + 2x + 8$ $(3x + 2)(x + 4)$
4. $5(x - 2)^2 + 15(x - 2)$ $5(x - 2)(x + 1)$
6. $x^2 - 7xy + 12y^2$ $(x - 4y)(x - 3y)$
8. $6x^2 - 7x + 2$ $(3x - 2)(2x - 1)$
10. $81x^2 - 16y^4$ $(9x + 4y^2)(9x - 4y^2)$
12. $(x + 3)^2 + 2(x + 3) - 3$ $(x + 2)(x + 6)$

14. One factor of $4x^3 - 5x^2 - 38x - 24$ is $x + 2$.
(a) Use division to find a second factor. $4x^2 - 13x - 12$
(b) Factor the polynomial completely. $(x + 2)(x - 4)(4x + 3)$

Solve.

15. $2(x - 5)(3x + 2) = 0$ $5, -\frac{2}{3}$

16. $4x^2 - 18 = 21x$ $-\frac{3}{4}, 6$

17. $x^3 + 4x^2 - 5x = 0$ $0, -5, 1$

18. Use factoring to find the *x* intercepts of the graph of the equation $y = 8x^2 + 10x - 3$. $\frac{1}{4}, -\frac{3}{2}$

19. The area of a triangle is 28 square meters. If the base of the triangle is 2 meters greater than 3 times the height, find the base and height of the triangle.
$h = 4$ m, $b = 14$ m

20. A baseball is projected upward from the top of a 448-foot-tall building with an initial velocity of 48 feet per second. The distance *s*, of the baseball from the ground at any time, *t*, is given by the equation $s = -16t^2 + 48t + 448$. Find the time after which the baseball strikes the ground.   7 sec

---

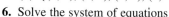

Cumulative Review Test

1. Evaluate $\dfrac{\sqrt[3]{27} - \sqrt[3]{-8} + |-4|}{3^0 - 12 \div 3 \div 4 - 8}$. $-\dfrac{9}{8}$

2. Solve the equations $\frac{1}{3}(x - 6) = \frac{3}{4}(2x - 1)$. $-\dfrac{15}{14}$

3. Solve the formula $3P = \dfrac{2L - W}{4}$ for *L*. $L = \dfrac{12P + W}{2}$

4. Graph the inequality $2x - y \le 6$.

5. Indicate whether the following sets of ordered pairs are functions. Explain your answer.
(a) $\{(0, 1), (3, -2), (-2, 6), (5, 6)\}$
(b) $\{(1, 2), (3, 4), (5, 6), (1, 0)\}$

6. Solve the system of equations
$\left(\dfrac{20}{11}, -\dfrac{14}{11}\right)$
$3x - 2y = 8$
$2x - 5y = 10$.

7. Graph the equation.
$y = |x + 4| - 2$

8. Simplify $\left(\dfrac{8x^{-2}y^3}{4xy^{-1}}\right)\left(\dfrac{2xy^5}{x^{-3}}\right)$ $4xy^9$

9. Simplify $\dfrac{(2p^4q^3)(3pq^4)^3}{(4p^{-2}q^3)^2}$ $\dfrac{27p^{11}q^9}{8}$

10. Simplify $3x^2 - 4x - 6 - (5x - 4x^2 - 6)$. Write the answer in descending powers of the variable. $7x^2 - 9x$

11. Multiply $(x^2 - 3x - 6)(2x - 5)$. $2x^3 - 11x^2 + 3x + 30$

12. Divide $\dfrac{9x^3y^5 - 8x^2y^4 - 12xy}{3x^2y}$. $3xy^4 - \dfrac{8y^3}{3} - \dfrac{4}{x}$

13. If $f(x) = 3x^3 - 6x^2 - 4x + 3$, find $f(2)$. -5

14. Graph $f(x) = x^2 - 6x + 8$.

5. (a) Yes, each *x* has a unique *y*. **(b)** No, (1, 2) and (1, 0) have the same *x* coordinate.

Factor exercises 15–18.

15. $x^4 - 3x^3 + 2x^2 - 6x$ $x(x^2 + 2)(x - 3)$

16. $12x^2y - 27xy + 6y$ $3y(x - 2)(4x - 1)$

17. $y^4 + 2y^2 - 24$ $(y + 2)(y - 2)(y^2 + 6)$

18. $8x^3 - 27y^6$ $(2x - 3y^2)(4x^2 + 6xy^2 + 9y^4)$

19. Copy World charges 15 cents a page for making a master copy from material that must be hand fed into the copier. After the master copy is made, they can make additional copies from the master copy for 5 cents a page. John has a manuscript to be copied, but since the pages must be hand fed into the copier, John has a master copy made and six additional copies of the manuscript made from the master copy. If his total bill before tax is $279, how many pages are in the manuscript?   620 pages

20. Santo's first four test grades are 68, 72, 90, and 86. What range of grades on his fifth test will result in an average greater than or equal to 70 and less than 80? $34 \le x < 84$

4.

7.

14.

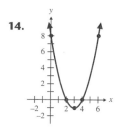

Chapter **7**

Rational Expressions and Equations

See Section 7.5, Exercise 32.

In this chapter we explain how to add, subtract, multiply, and divide rational expressions. Rational expressions are expressions that contain fractions. We also explain how to solve equations containing rational expressions. Equations containing rational expressions are sometimes referred to as rational equations.

In Section 7.1 we introduce rational expression and discuss the domains of rational expressions and rational functions. If you take a higher mathematics course, you will probably discuss rational functions in much more detail and graph them. If you own a graphing calculator, this is a good time to experiment with your grapher.

To succeed in Sections 7.1 and 7.2, you must use the factoring techniques presented in Chapter 6.

In Chapter 2 we solved some equations containing fractions. In Section 7.4 we present and solve many types of equations containing rational expressions. Applications of equations containing rational expressions are included in Sections 7.4 and 7.5.

A knowledge of variation is important in many science courses. In Section 7.6 we discuss various types of variations, including direct, inverse, joint, and combined variations.

7.1 The Domain of Rational Functions and Multiplication and Division of Rational Expressions

Tape 10

> 1 Find the domain of a rational expression or rational function.
> 2 Reduce a rational expression to its lowest terms.
> 3 Multiply rational expressions.
> 4 Divide rational expressions.

The Domain of Rational Functions

1 *To understand rational expressions, you must have a thorough understanding of the factoring techniques discussed in Chapter 6.* A **rational expression** (also called an **algebraic fraction**) is an expression of the form p/q, where p and q are polynomials and $q \neq 0$. Examples of rational expressions are

$$\frac{2}{3}, \quad \frac{x+3}{x}, \quad \frac{x^2+4x}{x-3}, \quad \frac{x}{x^2-4}$$

Note that the denominator of a rational expression cannot equal 0 because division by 0 is undefined. In the expression $(x+3)/x$, x cannot equal 0, since the denominator would then equal 0. In $(x^2+4x)/(x-3)$, x cannot equal 3 because that would result in the denominator having a value of 0. What values of x cannot be used in the expression $x/(x^2-4)$? If you answered 2 and -2, you answered correctly.

Whenever we write a rational expression containing a variable in the denominator, we always assume that the value or values of the variable that make the denominator 0 are excluded.

EXAMPLE 1   Find the values of the variable that must be excluded in the rational expression $\dfrac{x-2}{3x-8}$.

Solution: The values that make the denominator equal to 0 must be excluded. What values of x will make the denominator equal to 0? We can determine this by setting the denominator equal to 0 and solving the equation for x.

$$3x - 8 = 0$$
$$3x = 8$$
$$x = \frac{8}{3}$$

If x were $\frac{8}{3}$, the denominator would be 0. Therefore, the number $\frac{8}{3}$ must be excluded when considering the rational expression.

EXAMPLE 2   Find the values of the variable that must be excluded in the rational expression $\dfrac{3}{y^2-2y-15}$.

Solution: We must determine which values of y will make the denominator equal to 0. To do this, set the denominator equal to 0 and solve for y.

$$y^2 - 2y - 15 = 0$$
$$(y-5)(y+3) = 0$$
$$y - 5 = 0 \quad \text{or} \quad y + 3 = 0$$
$$y = 5 \qquad\qquad y = -3$$

Therefore, the numbers -3 and 5 must be excluded when considering the rational expression.

Rational Functions   In Section 3.5 we introduced certain types of functions, including absolute value functions and square root functions. In Section 5.6 we discussed polynomial functions. Now we briefly introduce rational functions. A **rational function** is a function of the form $y = \dfrac{p}{q}\left[\text{or } f(x) = \dfrac{p}{q}\right]$, where p and q are polynomials and $q \neq 0$. Examples of rational functions are:

$$y = \frac{1}{x} \qquad y = \frac{x^2+9}{x+3} \qquad \text{and} \qquad y = \frac{x^2}{x^2-4}$$

In Section 3.5 we discussed the domain of relations and functions. When discussing rational functions, the **domain** will be the set of values that can be used to replace the variable. For example, in the rational function $f(x) = \dfrac{x+2}{x-3}$, the domain will be all real numbers except 3, written $\{x \mid x \text{ is a real number and } x \neq 3\}$. If x were 3, the denominator would be 0, and division by 0 is undefined. We will not discuss the range or the graphs of rational functions except in the Graphing Calculator Corner and the Group Activity exercises at the end of this section. If you take another mathematics course, you may spend time graphing rational functions.

EXAMPLE 3   Find the domain of the rational functions.

(a) $y = \dfrac{2}{x}$ **(b)** $f(x) = \dfrac{x^2}{x^2 - 4}$ **(c)** $f(x) = \dfrac{x}{2x^2 - 7x - 15}$

Solution:

(a) Since the denominator cannot have a value of 0, the domain is all real numbers except 0, $\{x \mid x \text{ is a real number and } x \neq 0\}$.

(b) The denominator cannot have a value of 0. To determine the values of x that make the denominator 0, we solve the equation $x^2 - 4 = 0$.

$$x^2 - 4 = 0$$
$$(x + 2)(x - 2) = 0$$
$$x + 2 = 0 \quad \text{or} \quad x - 2 = 0$$
$$x = -2 \qquad\qquad x = 2$$

Therefore, the domain of the function is $\{x \mid x \text{ is a real number and } x \neq -2 \text{ and } x \neq 2\}$.

(c) The denominator cannot have a value of 0. To determine the values of x that make the denominator 0, we solve the equation $2x^2 - 7x - 15 = 0$.

$$2x^2 - 7x - 15 = 0$$
$$(2x + 3)(x - 5) = 0$$
$$2x + 3 = 0 \quad \text{or} \quad x - 5 = 0$$
$$2x = -3 \qquad\qquad x = 5$$
$$x = -\frac{3}{2}$$

Therefore, the domain of the function is $\{x \mid x \text{ is a real number and } x \neq -\dfrac{3}{2} \text{ and } x \neq 5\}$.

Graphing Calculator Corner

If you have a graphing calculator, you may wish to experiment by graphing some rational functions. This will give you some idea of the wide variety of graphs of rational functions.

If you graph $y = \dfrac{x^2}{x^2 - 4}$ on a graphing calculator, the display might look like that in Fig. 7.1.

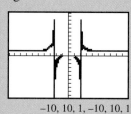

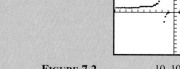

FIGURE 7.1   −10, 10, 1, −10, 10, 1   **FIGURE 7.2**   −10, 10, 1, −10, 10, 1

Notice what appear to be vertical lines at $x = -2$ and $x = 2$. This calculator is in a mode called *connected mode*. When a calculator is in connected mode, it connects all points it plots, going from the point with the smallest x coordinate to the next larger one. Just to the left of -2, the value of y is a very large positive number, and just to the right of -2 the value of y is a very large negative number. The vertical line is the calculator's attempt to connect the point with this very large positive y value with the point with this very large negative y value. A similar situation occurs at $x = 2$.

You may wish to sometimes place your calculator in *dot mode*. When the calculator is in dot mode it displays unconnected points that have been calculated. Read the manual that comes with your calculator to learn how to change from connected to dot mode, or vice versa. The graph in Fig. 7.2 shows the same graph as in Fig. 7.1 except that this time the calculator is in dot mode.

EXERCISES

For each function (a) determine its domain, and (b) graph the function in connected mode, then change your calculator to dot mode and observe the difference. Describe in your own words the appearance of the graph.

1. $y = \dfrac{1}{x}$ **(a)** $\{x \mid x \neq 0\}$ **2.** $y = \dfrac{3}{x - 2}$ **(a)** $\{x \mid x \neq 2\}$

3. $y = \dfrac{x + 4}{x^2}$ **(a)** $\{x \mid x \neq 0\}$ **4.** $f(x) = \dfrac{x^2}{x + 2}$ **(a)** $\{x \mid x \neq -2\}$

5. $f(x) = \dfrac{x}{x^2 - 4}$ **(a)** $\{x \mid x \neq -2 \text{ or } x \neq 2\}$

6. (a) *Make up a rational function that contains a variable in both the numerator and denominator. Write down the function, then graph the function using your calculator.* **(b)** *Describe the appearance of the graph in your own words.*

Reducing Rational Expressions

2 When we work problems containing rational expressions, we must make sure that we write the answer in lowest terms. A rational expression is **reduced to its lowest terms** when the numerator and denominator have no common factors other than 1. You know that the fraction $\frac{6}{9}$ is not in reduced form because the 6 and 9 both contain the common factor of 3. When the 3 is factored out, the reduced fraction is $\frac{2}{3}$.

$$\frac{6}{9} = \frac{\overset{1}{\cancel{3}} \cdot 2}{\underset{1}{\cancel{3}} \cdot 3} = \frac{2}{3}$$

The rational expression $\dfrac{ab - b^2}{2b}$ is not in reduced form because both the numerator and denominator have a common factor, b. To reduce this expression, factor b from each term in the numerator; then divide it out.

$$\frac{ab - b^2}{2b} = \frac{\cancel{b}(a - b)}{2\cancel{b}} = \frac{a - b}{2}$$

$\dfrac{ab - b^2}{2b}$ becomes $\dfrac{a - b}{2}$ when reduced to lowest terms.

> **To Reduce Rational Expressions**
> **1.** Factor both numerator and denominator as completely as possible.
> **2.** Divide both the numerator and the denominator by any common factors.

EXAMPLE 4   Reduce $\dfrac{x^2 + 2x - 3}{x + 3}$ to its lowest terms.

Solution: Factor the numerator; then divide out the common factor.

$$\frac{x^2 + 2x - 3}{x + 3} = \frac{\cancel{(x + 3)}(x - 1)}{\cancel{x + 3}} = x - 1$$

The rational expression reduces to $x - 1$.

When the terms in a numerator differ only in sign from the terms in a denominator, we can factor out -1 from either the numerator or denominator. **When -1 is factored from a polynomial, the sign of each term in the polynomial changes**. For example,

$$-2x + 3 = -1(2x - 3) = -(2x - 3)$$
$$6 - 5x = -1(-6 + 5x) = -(5x - 6)$$
$$-3x^2 + 5x - 6 = -1(3x^2 - 5x + 6) = -(3x^2 - 5x + 6)$$

EXAMPLE 5   Reduce $\dfrac{3x^2 + 19x - 14}{2 - 3x}$.

Solution:

$$\frac{3x^2 + 19x - 14}{2 - 3x} = \frac{(3x - 2)(x + 7)}{2 - 3x}$$

$$= \frac{\cancel{(3x - 2)}(x + 7)}{-1\cancel{(3x - 2)}} = -(x + 7)$$

In Example 5, $3x - 2$ appeared in the numerator and $2 - 3x$ appeared in the denominator. Since the expressions differ only in sign, we factored out -1 from $2 - 3x$. Expressions that differ only in sign are said to be *opposites*. Thus $3x - 2$ and $2 - 3x$ are opposites.

EXAMPLE 6 Reduce $\dfrac{5x^2y + 10xy^2 - 25x^2y^3}{5x^2y}$.

Solution: Factor the numerator. The greatest common factor of each term in the numerator is $5xy$. Then divide out the common factors.

$$\frac{5x^2y + 10xy^2 - 25x^2y^3}{5x^2y} = \frac{5xy(x + 2y - 5xy^2)}{5x^2y}$$

$$= \frac{x + 2y - 5xy^2}{x}$$

EXAMPLE 7 Reduce $\dfrac{x^2 - x - 12}{(x + 2)(x - 4) + x(x - 4)}$.

Solution: Factor the numerator and denominator. Each term in the denominator has $x - 4$ as a factor. Then divide out the common factor $x - 4$.

$$\frac{x^2 - x - 12}{(x + 2)(x - 4) + x(x - 4)} = \frac{(x + 3)(x - 4)}{(x - 4)[(x + 2) + x]}$$

$$= \frac{(x + 3)\cancel{(x - 4)}}{\cancel{(x - 4)}(2x + 2)}$$

$$= \frac{x + 3}{2x + 2}$$

COMMON STUDENT ERROR

Incorrect   *Incorrect*

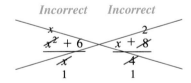

Remember that you can divide out only common **factors.** Only when expressions are **multiplied** can they be factors. Neither of the expressions above can be simplified from their original form.

Correct   *Incorrect*

$$\frac{x^2 - 4}{x - 2} = \frac{(x + 2)\cancel{(x - 2)}}{\cancel{x - 2}}$$

$$= x + 2$$

When we add rational expressions in Section 7.2 we will need to rewrite expressions so that they have a common denominator. In Example 8 we will illustrate how this may be done.

EXAMPLE 8 Write each rational expression with the indicated denominator.

(a) $\dfrac{3}{xy^2}, x^5y^4z^6$ **(b)** $\dfrac{p-1}{p+3}, 2p^2 + 5p - 3$

Solution:

(a) To determine the expression that the numerator and denominator of the fraction $\dfrac{3}{xy^2}$ must be multiplied by to obtain a denominator of $x^5y^4z^6$, we can divide the desired denominator by the present denominator, as follows:

$$\frac{x^5y^4z^6}{xy^2} = x^4y^2z^6$$

Thus, both the numerator and denominator of $\dfrac{3}{xy^2}$ must be multiplied by $x^4y^2z^6$ to obtain an equivalent fraction with the denominator $x^5y^4z^6$.

$$\frac{3}{xy^2} \cdot \boxed{\frac{x^4y^2z^6}{x^4y^2z^6}} = \frac{3x^4y^2z^6}{x^5y^4z^6}$$

(b) To determine the expression that the numerator and denominator of the fraction $\dfrac{p-1}{p+3}$ must be multiplied by to obtain the denominator $2p^2 + 5p - 3$, we can divide as follows:

$$\frac{2p^2 + 5p - 3}{p+3} = \frac{(2p-1)(p+3)}{(p+3)} = 2p - 1$$

Now multiply both the numerator and denominator of $\dfrac{p-1}{p+3}$ by $2p - 1$.

$$\frac{p-1}{p+3} \cdot \boxed{\frac{2p-1}{2p-1}} = \frac{2p^2 - 3p + 1}{2p^2 + 5p - 3}$$

Multiplying Rational Expressions

③ Now that we know how to reduce a rational expression we can discuss multiplication of rational expressions. To multiply fractions, we divide out their common factors, then multiply the numerators and multiply the denominators.

> **Multiplication**
>
> $$\frac{a}{b} \cdot \frac{c}{d} = \frac{a \cdot c}{b \cdot d}, \qquad b \neq 0, \quad d \neq 0$$

We follow the same basic procedure to multiply rational expressions as illustrated in Example 9.

EXAMPLE 9 Multiply $\dfrac{3x^2}{z^2} \cdot \dfrac{2z^5}{9x}$.

Solution:
$$\dfrac{\overset{1}{\cancel{3x^2}}\,\overset{x}{}}{\underset{1}{\cancel{z^2}}} \cdot \dfrac{\overset{z^3}{\cancel{2z^5}}}{\underset{3\ 1}{\cancel{9}\,\cancel{x}}} = \dfrac{x \cdot 2z^3}{1 \cdot 3} = \dfrac{2xz^3}{3}$$

To Multiply Rational Expressions

1. Factor all numerators and denominators as far as possible.
2. Divide out any common factors.
3. Multiply the numerators and multiply the denominators.
4. Reduce the answer when possible.

If all common factors were factored out in step 2, you will be unable to reduce the answer in step 4. However, if you missed a common factor in step 2, you can factor it out in step 4.

EXAMPLE 10 Multiply $\dfrac{x - 5}{4x} \cdot \dfrac{x^2 - 2x}{x^2 - 7x + 10}$.

Solution: $\dfrac{x - 5}{4x} \cdot \dfrac{x^2 - 2x}{x^2 - 7x + 10} = \dfrac{\cancel{x - 5}}{4\cancel{x}} \cdot \dfrac{\cancel{x}(x - 2)}{(x - 2)(x - 5)} = \dfrac{1}{4}$

EXAMPLE 11 Multiply $\dfrac{2x - 5}{x - 4} \cdot \dfrac{x^2 - 8x + 16}{5 - 2x}$.

Solution:

$$\dfrac{2x - 5}{x - 4} \cdot \dfrac{x^2 - 8x + 16}{5 - 2x} = \dfrac{2x - 5}{x - 4} \cdot \dfrac{(x - 4)(x - 4)}{5 - 2x}$$

$$= \dfrac{\cancel{2x - 5}}{\cancel{x - 4}} \cdot \dfrac{(\cancel{x - 4})(x - 4)}{-1(\cancel{2x - 5})}$$

$$= \dfrac{x - 4}{-1} = -(x - 4) \quad \text{or} \quad -x + 4$$

EXAMPLE 12 Multiply $\dfrac{x^2 - y^2}{x + y} \cdot \dfrac{x + 2y}{2x^2 - xy - y^2}$.

Solution:

$$\dfrac{x^2 - y^2}{x + y} \cdot \dfrac{x + 2y}{2x^2 - xy - y^2} = \dfrac{(x + y)(x - y)}{x + y} \cdot \dfrac{x + 2y}{(2x + y)(x - y)} = \dfrac{x + 2y}{2x + y}$$

EXAMPLE 13   Multiply $\dfrac{ab - ac + bd - cd}{ab + ac + bd + cd} \cdot \dfrac{b^2 + bc + bd + cd}{b^2 + bd - bc - cd}$.

Solution: Factor both numerators and denominators by grouping; then divide out common factors.

$$\frac{ab - ac + bd - cd}{ab + ac + bd + cd} \cdot \frac{b^2 + bc + bd + cd}{b^2 + bd - bc - cd} = \frac{a(b - c) + d(b - c)}{a(b + c) + d(b + c)} \cdot \frac{b(b + c) + d(b + c)}{b(b + d) - c(b + d)}$$

$$= \frac{(a + d)(b - c)}{(a + d)(b + c)} \cdot \frac{(b + d)(b + c)}{(b - c)(b + d)} = 1$$

Dividing Rational Expressions

4 To divide numerical fractions, we invert the divisor and multiply.

$$\frac{a}{b} \div \frac{c}{d} = \frac{a}{b} \cdot \frac{d}{c} = \frac{a \cdot d}{b \cdot c}, \qquad b \neq 0, \quad c \neq 0, \quad d \neq 0$$

We follow the same basic procedure to divide rational expressions, as illustrated in Example 14.

EXAMPLE 14   Divide $\dfrac{12x^4}{5y^3} \div \dfrac{3x^5}{10y}$.

Solution: $\dfrac{12x^4}{5y^3} \div \dfrac{3x^5}{10y} = \dfrac{\overset{4}{\cancel{12}}x^4}{\underset{1}{\cancel{5}}y^3} \cdot \dfrac{\overset{2}{\cancel{10}}y}{\underset{1}{\cancel{3}}x^5} = \dfrac{4 \cdot 2}{y^2x} = \dfrac{8}{xy^2}$

To Divide Rational Expressions
Invert the divisor (the second or bottom fraction) and then multiply the resulting rational expressions.

EXAMPLE 15   Divide $\dfrac{x^2 - 9}{x + 4} \div \dfrac{x - 3}{x + 4}$.

Solution:

$$\frac{x^2 - 9}{x + 4} \div \frac{x - 3}{x + 4} = \frac{x^2 - 9}{x + 4} \cdot \frac{x + 4}{x - 3}$$

$$= \frac{(x + 3)(x - 3)}{x + 4} \cdot \frac{x + 4}{x - 3} = x + 3$$

EXAMPLE 16   Divide $\dfrac{12x^2 - 22x + 8}{3x} \div \dfrac{3x^2 + 2x - 8}{2x^2 + 4x}$.

Solution:

$$\frac{12x^2 - 22x + 8}{3x} \div \frac{3x^2 + 2x - 8}{2x^2 + 4x} = \frac{12x^2 - 22x + 8}{3x} \cdot \frac{2x^2 + 4x}{3x^2 + 2x - 8}$$

$$= \frac{2(6x^2 - 11x + 4)}{3x} \cdot \frac{2x(x + 2)}{(3x - 4)(x + 2)}$$

$$= \frac{2(3x - 4)(2x - 1)}{3x} \cdot \frac{2x(x + 2)}{(3x - 4)(x + 2)}$$

$$= \frac{4(2x - 1)}{3}$$

EXAMPLE 17   Divide $\dfrac{x^4 - y^4}{x - y} \div \dfrac{x^2 + xy}{x^2 - 2xy + y^2}$.

Solution:

$$\frac{x^4 - y^4}{x - y} \div \frac{x^2 + xy}{x^2 - 2xy + y^2} = \frac{x^4 - y^4}{x - y} \cdot \frac{x^2 - 2xy + y^2}{x^2 + xy}$$

$$= \frac{(x^2 + y^2)(x^2 - y^2)}{x - y} \cdot \frac{(x - y)(x - y)}{x(x + y)}$$

$$= \frac{(x^2 + y^2)(x + y)(x - y)}{x - y} \cdot \frac{(x - y)(x - y)}{x(x + y)}$$

$$= \frac{(x^2 + y^2)(x - y)^2}{x}$$

EXAMPLE 18   Perform the indicated operations.

$$\frac{x^4 - y^4}{x - y} \div \frac{x^2 + xy}{x^2 - 2xy + y^2} \cdot \frac{x^2}{x^2 - 2xy + y^2}$$

Solution: When a problem contains more than one operation, we follow the priority of operations given in Section 1.4. Since multiplication and division have the same priority, and there are no parentheses, we work from left to right. In Example 17 we found that

$$\frac{x^4 - y^4}{x - y} \div \frac{x^2 + xy}{x^2 - 2xy + y^2} = \frac{(x^2 + y^2)(x - y)^2}{x}$$

Using this result we can write

$$\frac{x^4 - y^4}{x - y} \div \frac{x^2 + xy}{x^2 - 2xy + y^2} \cdot \frac{x^2}{x^2 - 2xy + y^2} = \frac{(x^2 + y^2)(x - y)^2}{x} \cdot \frac{x^2}{x^2 - 2xy + y^2}$$

$$= \frac{(x^2 + y^2)(x - y)^2}{x} \cdot \frac{x^2}{(x - y)^2}$$

$$= x(x^2 + y^2)$$

Exercise Set 7.1

Determine the values that are excluded in the following functions.

1. $\dfrac{6}{x}$ 0

2. $\dfrac{2x}{2x - 6}$ 3

3. $\dfrac{5}{x^2 - 9}$ 3, −3

4. $\dfrac{3}{x^2 - 64}$ 8, −8

5. $\dfrac{4}{2x^2 - 15x + 25}$ 5, $\frac{5}{2}$

6. $\dfrac{2}{(x - 2)^2}$ 2

7. $\dfrac{x - 3}{x^2 + 4}$ none

8. $\dfrac{-2}{16 - r^2}$ 4, −4

9. $\{x \mid x \neq -3 \text{ and } x \neq 2\}$ **10.** $\{x \mid x \neq -1 \text{ and } x \neq -6\}$ **11.** $\{x \mid x \neq 3 \text{ and } x \neq -7\}$

Determine the domain of each function. **12.** $\{p \mid p \neq 1\}$ **13.** $\{z \mid z \neq \frac{15}{8}\}$ **14.** $\{x \mid x \neq -3\}$ **15.** $\{a \mid a \neq \frac{1}{2} \text{ and } a \neq -2\}$ **16.** $\{x \mid x \neq 0\}$

9. $y = \dfrac{5}{(x + 3)(x - 2)}$

10. $y = \dfrac{3}{x^2 + 7x + 6}$

11. $f(x) = \dfrac{x + 3}{x^2 + 4x - 21}$

12. $f(p) = \dfrac{p + 1}{p - 1}$

13. $f(z) = \dfrac{-2}{-8z + 15}$

14. $y = \dfrac{4x - 6}{x^2 + 6x + 9}$

15. $f(a) = \dfrac{3a^2 - 6a + 4}{2a^2 + 3a - 2}$

16. $f(x) = \dfrac{4 - 2x}{x^3 + 9x}$

Write each rational expression in reduced form. **21.** $\dfrac{2(x^2 + 2xy - 3y^2)}{3}$

17. $\dfrac{x - xy}{x}$ $1 - y$

18. $\dfrac{5x^2 - 25x}{10}$ $\dfrac{x(x - 5)}{2}$

19. $\dfrac{3x + xy}{y + 3}$ x

20. $\dfrac{5x^2 - 10xy}{25x}$ $\dfrac{x - 2y}{5}$

21. $\dfrac{6x^2 + 12xy - 18y^2}{9}$

22. $\dfrac{4x^2y + 12xy + 18x^3y^3}{8xy^2}$ $\dfrac{2x + 6 + 9x^2y^2}{4y}$

23. $\dfrac{4r - 2}{2 - 4r}$ -1

24. $\dfrac{4x^2 - 9}{2x^2 - x - 3}$ $\dfrac{2x + 3}{x + 1}$

25. $\dfrac{x^2 - 2x - 24}{6 - x}$ $-(x + 4)$

26. $\dfrac{4x^2 - 16x^4 + 6x^5y}{8x^3y}$ $\dfrac{2 - 8x^2 + 3x^3y}{4xy}$

27. $\dfrac{x^2 + 5x + 6}{x^2 - 3x - 10}$ $\dfrac{x + 3}{x - 5}$

28. $\dfrac{y^2 - 10yz + 24z^2}{y^2 - 5yz + 4z^2}$ $\dfrac{y - 6z}{y - z}$

29. $\dfrac{8x^3 - 125y^3}{2x - 5y}$ $4x^2 + 10xy + 25y^2$

30. $\dfrac{x(x - 1) + x(x - 4)}{2x - 5}$ x

31. $\dfrac{(x + 1)(x - 3) + (x + 1)(x - 2)}{2(x + 1)}$ $\dfrac{2x - 5}{2}$

32. $\dfrac{(2x - 5)(x + 4) - (2x - 5)(x + 1)}{3(2x - 5)}$ 1

33. $\dfrac{x^2 - 8x + 16}{x^2 + 3x - 4x - 12}$ $\dfrac{x - 4}{x + 3}$

34. $\dfrac{xy - yw + xz - zw}{xy + yw + xz + zw}$ $\dfrac{x - w}{x + w}$

35. $\dfrac{a^2 + 3a - ab - 3b}{a^2 - ab + 5a - 5b}$ $\dfrac{a + 3}{a + 5}$

36. $\dfrac{a^3 - b^3}{a^2 - b^2}$ $\dfrac{a^2 + ab + b^2}{a + b}$

37. $\dfrac{x^2 + 2x - 3}{x^3 + 27}$ $\dfrac{x - 1}{x^2 - 3x + 9}$

38. $\dfrac{x^3 + 3x^2 - 4x - 12}{x^2 + 5x + 6}$ $x - 2$

Write each rational expression with the denominator indicated.

39. $\dfrac{5}{x}, x^2$ $\dfrac{5x}{x^2}$

40. $\dfrac{3r}{r + 2}, 5r + 10$ $\dfrac{15r}{5r + 10}$

41. $\dfrac{5y}{y + 3}, y^2 - 3y - 18$ $\dfrac{5y(y - 6)}{y^2 - 3y - 18}$

42. $\dfrac{1}{x^2y^3}, 2x^3y^5z^2$ $\dfrac{2xy^2z^2}{2x^3y^5z^2}$

43. $\dfrac{4}{p}, p^2(p - 2)$ $\dfrac{4p(p - 2)}{p^2(p - 2)}$

44. $\dfrac{x + 1}{2(x - 4)}, 8(x - 4)(x + 2)$

45. $\dfrac{p}{2p - 1}, 6p^2 + p - 2$ $\dfrac{p(3p + 2)}{6p^2 + p - 2}$

46. $\dfrac{z}{z + 2y}, z^3 + 8y^3$ $\dfrac{z(z^2 - 2zy + 4y^2)}{z^3 + 8y^3}$

44. $\dfrac{4(x + 1)(x + 2)}{8(x - 4)(x + 2)}$

Multiply or divide as indicated. Write all answers in lowest terms.

47. $\dfrac{3x}{2y} \cdot \dfrac{y^3}{6}$ $\dfrac{xy^2}{4}$

48. $\dfrac{16x^2}{y^4} \cdot \dfrac{5x^2}{4y^2}$ $\dfrac{20x^4}{y^6}$

49. $\dfrac{9x^3}{4} \div \dfrac{3}{16y^2}$ $12x^3y^2$

50. $\dfrac{80m^4}{49x^5y^7} \cdot \dfrac{14x^{12}y^5}{25m^5}$ $\dfrac{32x^7}{35my^2}$

84. $\dfrac{(x + 2)(3x + 1)}{(2x - 3)(x + 1)}$ **85.** $\dfrac{(a + b)^2}{ab}$ **86.** $\dfrac{(x + y)(3x - 1)}{(2x - 3)(x - 4)}$

51. $\dfrac{12a^2}{4bc} \div \dfrac{3a^2}{bc}$ 1

52. $\dfrac{-25a^3b^5}{6a^4b^2} \div \dfrac{-5a^4b^7}{12ab^4}$ $\dfrac{10}{a^4}$

53. $\dfrac{6x^5y^3 \cdot 6x^4y^2}{5xz^3} \div \dfrac{12y^4}{(2x^4z^2)^2}$ $\dfrac{12x^{16}yz}{5}$

54. $\dfrac{(4x^3z)^2}{(6x^5y^3)(3x^4y^6)} \div \dfrac{4(x^4y^5)^2}{8x^4y^6}$ $\dfrac{16z^2}{9x^7y^{13}}$

55. $\dfrac{(-3x^2y^4)^2 \cdot (2x^3y^5)^3}{(6x^4y^2)^2} \cdot \dfrac{(xy)^3}{(4xy^2)^2}$ $\dfrac{x^6y^{18}}{8}$

56. $\dfrac{(5x^2y^3)^3 \cdot (x^2y)^4}{(3xy^4)^3} \cdot \dfrac{9(2x^5y^6)^2}{(25x^4y^6)(2x^4y)^3}$ $\dfrac{5x^5y^4}{6}$

57. $\dfrac{4 - x}{x - 4} \cdot \dfrac{x - 3}{3 - x}$ 1

58. $\dfrac{2a + 2b}{3} \div \dfrac{a^2 - b^2}{a - b}$ $\dfrac{2}{3}$

59. $\dfrac{x^2 + 7x + 12}{x + 4} \cdot \dfrac{1}{x + 3}$ 1

60. $\dfrac{x^2 + 3x - 10}{2x} \cdot \dfrac{x^2 - 3x}{x^2 - 5x + 6}$ $\dfrac{x + 5}{2}$

61. $\dfrac{x^2 + 10x + 21}{x + 7} \div (x^2 - 5x - 24)$ $\dfrac{1}{x - 8}$

62. $(x - 3) \div \dfrac{x^2 + 3x - 18}{x}$ $\dfrac{x}{x + 6}$

63. $\dfrac{x^2 - 9x + 14}{x^2 - 5x + 6} \div \dfrac{x^2 - 5x - 14}{x + 2}$ $\dfrac{1}{x - 3}$

64. $\dfrac{1}{x^2 - 17x + 30} \div \dfrac{1}{x^2 + 7x - 18}$ $\dfrac{x + 9}{x - 15}$

65. $\dfrac{a - b}{9a + 9b} \div \dfrac{a^2 - b^2}{a^2 + 2a + 1}$ $\dfrac{(a + 1)^2}{9(a + b)^2}$

66. $\dfrac{2x + 4y}{x^2 + 4xy + 4y^2} \cdot \dfrac{2x^2 + 7xy + 6y^2}{4x^2 + 14xy + 12y^2}$ $\dfrac{1}{x + 2y}$

67. $\dfrac{2x^2 - 5x - 12}{6x^2 + x - 12} \cdot \dfrac{3x^2 - x - 4}{4x^2 + 5x + 1}$ $\dfrac{x - 4}{4x + 1}$

68. $\dfrac{6x^3 - x^2 - x}{2x^2 + x - 1} \cdot \dfrac{x^2 - 1}{x^3 - 2x^2 + x}$ $\dfrac{3x + 1}{x - 1}$

69. $\dfrac{x + 2}{x^3 - 8} \cdot \dfrac{(x - 2)^2}{x^2 + 4}$ $\dfrac{(x + 2)(x - 2)}{(x^2 + 2x + 4)(x^2 + 4)}$

70. $\dfrac{x^2 - y^2}{x^2 - 2xy + y^2} \div \dfrac{x + y}{(x - y)^2}$ $x - y$

71. $\dfrac{x^4 - y^8}{x^2 + y^4} \div \dfrac{x^2 - y^4}{3x^2}$ $3x^2$

72. $\dfrac{(x^2 - y^2)^2}{(x^2 - y^2)^3} \div \dfrac{x^2 + y^2}{x^4 - y^4}$ 1

73. $\dfrac{2x^4 + 4x^2}{6x^2 + 14x + 4} \div \dfrac{x^2 + 2}{3x^2 + x}$ $\dfrac{x^3}{x + 2}$

74. $\dfrac{8a^3 - 1}{4a^2 + 2a + 1} \div \dfrac{a - 1}{(a - 1)^2}$ $(2a - 1)(a - 1)$

75. $\dfrac{2x^3 - 7x^2 + 3x}{x^2 + 2x - 3} \cdot \dfrac{x^2 + 3x}{(x - 3)^2}$ $\dfrac{x^2(2x - 1)}{(x - 1)(x - 3)}$

76. $\dfrac{4x + y}{5x + 2y} \cdot \dfrac{25x^2 - 5xy - 6y^2}{20x^2 - 7xy - 3y^2}$ 1

77. $\dfrac{3r^2 + 17rs + 10s^2}{6r^2 + 13rs - 5s^2} \div \dfrac{6r^2 + rs - 2s^2}{6r^2 - 5rs + s^2}$ $\dfrac{r + 5s}{2r + 5s}$

78. $\dfrac{ac - ad + bc - bd}{ac + ad + bc + bd} \cdot \dfrac{pc + pd - qc - qd}{pc - pd + qc - qd}$ $\dfrac{p - q}{p + q}$

79. $\dfrac{2p^2 + 2pq - pq^2 - q^3}{p^3 + p^2 + pq^2 + q^2} \div \dfrac{p^3 + p + p^2q + q}{p^3 + p + p^2 + 1}$ $\dfrac{2p - q^2}{p^2 + q^2}$

80. $\dfrac{x^3 - 4x^2 + x - 4}{x^4 - x^3 + x^2 - x} \cdot \dfrac{2x^3 + 2x^2 + x + 1}{2x^3 - 8x^2 + x - 4}$ $\dfrac{x + 1}{x(x - 1)}$

81. $\dfrac{x^2 + 5x + 6}{x^2 - x - 20} \cdot \dfrac{2x^2 + 6x - 8}{x^2 - 9} \cdot \dfrac{x^2 - 3x}{x - 1}$ $\dfrac{2x(x + 2)}{x - 5}$

82. $\dfrac{x^2 - 1}{x^2 + x} \cdot \dfrac{2x + 2}{1 - x^2} \cdot \dfrac{x^2 + x - 2}{x^2 - x}$ $\dfrac{-2(x + 2)}{x^2}$

83. $\dfrac{x^3 + 64}{x - 2} \cdot \dfrac{x^2 - 4}{x + 4} \cdot \dfrac{x}{x + 2}$ $x(x^2 - 4x + 16)$

84. $\dfrac{2x^2 - 3x - 14}{2x^2 - 9x + 7} \div \dfrac{6x^2 + x - 15}{3x^2 + 2x - 5} \cdot \dfrac{6x^2 - 7x - 3}{2x^2 - x - 3}$

85. $\dfrac{a^2 - b^2}{2a^2 - 3ab + b^2} \cdot \dfrac{2a^2 - 7ab + 3b^2}{a^2 + ab} \div \dfrac{ab - 3b^2}{a^2 + 2ab + b^2}$

86. $\dfrac{10x^2 - 17x + 3}{15x^2 - 8x + 1} \div \dfrac{4x^2 - 12x + 9}{3x^2 + 3xy - x - y} \cdot \dfrac{6x^2 - 11x + 3}{2x^2 - 11x + 12}$

87. $\dfrac{5x^2(x - 1) - 3x(x - 1) - 2(x - 1)}{10x^2(x - 1) + 9x(x - 1) + 2(x - 1)} \cdot \dfrac{2x + 1}{x + 3}$ $\dfrac{x - 1}{x + 3}$

88. $\dfrac{x^2(3x - y) - 5x(3x - y) - 24(3x - y)}{x^2(3x - y) - 9x(3x - y) + 8(3x - y)} \cdot \dfrac{x - 1}{x + 3}$ 1

Determine the polynomial to be placed in the shaded area that will result in a true statement. Explain how you determined your answer.

89. $\dfrac{\rule{2cm}{0.4cm}}{x^2 + 2x - 15} = \dfrac{1}{x - 3}$ $x + 5$

90. $\dfrac{\rule{2cm}{0.4cm}}{3x + 4} = x - 3$ $3x^2 - 5x - 12$

91. $\dfrac{y^2 - y - 20}{\rule{2cm}{0.4cm}} = \dfrac{y + 4}{y + 1}$ $y^2 - 4y - 5$

92. $\dfrac{\rule{2cm}{0.4cm}}{6p^2 + p - 15} = \dfrac{2p - 1}{2p - 3}$ $6p^2 + 7p - 5$

93. (a) $x^2 + x - 2$ **(b)** Factors must be $(x - 1)(x + 2)$.
95. (a) $2x^2 + x - 6$ **(b)** Factors must be $(x + 2)(2x - 3)$.

94. (a) $4x^2 + 4x - 15$ **(b)** Factors must be $(2x - 3)(2x + 5)$.
96. (a) $2r^2 + 3r - 2$ **(b)** Factors must be $(r + 2)(2r - 1)$.

*In Exercises 93 through 96, **(a)** determine the polynomial to be placed in the shaded area that will result in a true statement and **(b)** explain how you determined your answer.*

93. $\dfrac{x^2 - x - 12}{x^2 + 2x - 3} \cdot \dfrac{\rule{1.2cm}{0.3cm}}{x^2 - 2x - 8} = 1$

94. $\dfrac{x^2 - 4}{(x + 2)^2} \cdot \dfrac{2x^2 + x - 6}{\rule{1.2cm}{0.3cm}} = \dfrac{x - 2}{2x + 5}$

95. $\dfrac{x^2 - 9}{2x^2 + 3x - 2} \div \dfrac{2x^2 - 9x + 9}{\rule{1.2cm}{0.3cm}} = \dfrac{x + 3}{2x - 1}$

96. $\dfrac{4r^2 - r - 18}{\rule{1.2cm}{0.3cm}} \div \dfrac{4r^3 - 9r^2}{6r^2 - 9r + 3} = \dfrac{3(r - 1)}{r^2}$

97. Make up a rational expression that is undefined at $x = 2$ and $x = -3$. Explain how you determined your answer.   One possible answer is $\dfrac{1}{(x - 2)(x + 3)}$.

98. Consider the rational function $f(x) = \dfrac{1}{x}$. Explain why this function can never equal 0. Numerator is never 0.

99. Consider the rational function $f(x) = \dfrac{x - 4}{x^2 - 4}$. For what value of x, if any, will this function **(a)** equal 0; **(b)** be undefined? Explain.   **(a)** 4 **(b)** 2 and -2

100. Consider the function $f(x) = \dfrac{x^2 - 9}{x^2 - 4}$. For what values of x, if any, will this function **(a)** equal zero; **(b)** be undefined? Explain.   **(a)** 3 and -3 **(b)** 2 and -2

101. Explain why $\dfrac{\sqrt{x}}{x + 1}$ is not a rational expression.

102. Explain why $\dfrac{2}{\sqrt{y} + 3}$ is not a rational expression.

101. $\sqrt{x}$ is not a polynomial. **102.** $\sqrt{y} + 3$ is not a polynomial.
104. (a) Factor out -1 from either the numerator or denominator.

103. (a) Explain how to reduce a rational expression.
   (b) Using the procedure stated in part **(a)**, reduce
   $\dfrac{6x^2 + 7x - 20}{4x^2 - 25}$ **(b)** $\dfrac{3x - 4}{2x - 5}$

104. (a) Explain how to reduce a rational expression where the numerator and denominator differ only in sign.
   (b) By the procedure explained in part **(a)**, reduce the expression $\dfrac{3x^2 - 2x - 8}{-3x^2 + 2x + 8}$ **(b)** -1

105. (a) Explain how to multiply rational expressions.
   (b) Multiply using the procedure given in part **(a)**.
   $\dfrac{6a^2 + a - 1}{3a^2 + 2a - 1} \cdot \dfrac{3a^2 + 4a + 1}{6a^2 + 5a + 1}$ 1

106. (a) Explain how to divide rational expressions.
   (b) Divide using the procedure given in part **(a)**.
   $\dfrac{r + 2}{r^2 + 7r + 12} \div \dfrac{(r + 2)^2}{r^2 + 5r + 6}$ $\dfrac{1}{r + 4}$

CUMULATIVE REVIEW EXERCISES

[2.2] **107.** Solve the formula $V = \frac{4}{3}\pi r^2 h$ for h. $h = \dfrac{3V}{4\pi r^2}$

[2.5] **108.** Solve the inequality $-4 < 3x - 4 < 8$. Write the solution in interval notation. $(0, 4)$

[3.4] **109.** Find the slope and y intercept of the equation $3(y - 4) = -(x - 2)$. $m = -\frac{1}{3}$, y intercept $= \frac{14}{3}$

[4.2] **110.** Solve the system of equations using the addition method.
   $x + 2y = 4$
   $2y = 6x + 6$ $\left(-\frac{2}{7}, \frac{15}{7}\right)$

[5.1] **111.** Simplify $\dfrac{(4x^{-3}y^4)(2x^2y^{-1})}{12x^{-2}y^3}$. $\dfrac{2x}{3}$

[5.3] **112.** Simplify $3x^2y - 4xy + 2y^2 - (3xy + 6y^2 + 2x)$.
   $3x^2y - 7xy - 4y^2 - 2x$

1. (a) $\{x \mid x \text{ is a real number and } x \neq 0\}$

Group Activity/ Challenge Problems

1. Consider the rational function $y = \dfrac{1}{x}$.

   (a) What is the domain of the function?
   (b) Complete the table by finding the values of y for the given values of x.

(b)

x	-10	-1	-0.5	-0.1	0.1	0.5	1	10
y	-0.1	-1	-2	-10	10	2	1	0.1

(c)

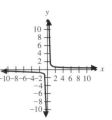

(c) Use the information from parts (a) and (b) to draw the graph of the function. *Hint:* The function is not defined at $x = 0$. This means the graph cannot pass through an imaginary vertical line at $x = 0$.

2. Consider the rational function $y = \dfrac{x^2 - 4}{x - 2}$.   **2. (a)** $\{x \,|\, x$ is a real number and $x \neq 2\}$

(a) What is the domain of the function?

(b) Complete the table by finding the values of y for the given values of x.

(c)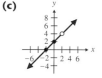

x	-2	-1	0	1	1.9	2.1	3	4	5	6
(b) y	0	1	2	3	3.9	4.1	5	6	7	8

(c) Use the information from parts (a) and (b) to draw the graph of the function.

Write in lowest terms.

3. $\dfrac{x^{5y} + 3x^{4y}}{3x^{3y} + x^{4y}}$ x^y

4. $\dfrac{m^{2x} - m^x - 2}{m^{2x} - 4}$ $\dfrac{m^x + 1}{m^x + 2}$

5. Consider the triangle below. If the area is $a^2 + 2ab - 3b^2$ and the base is $a + 3b$, find the height h. Use area $= \frac{1}{2}$(base)(height). $2(a - b)$

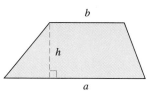

$a + 3b$

6. Consider the trapezoid below. If the area of the trapezoid is $a^2 - b^2$ find the height. Use area $= \frac{1}{2}$(height)(base 1 + base 2). $2(a - b)$

b

h

a

Simplify.

7. $\dfrac{x + 2}{x^2 + 5x + 6} \cdot \dfrac{2x^2 + 7x + 3}{4x^2 + 4x + 1} \cdot \dfrac{6x^2 + 5x + 1}{3x^2 + x} \cdot \dfrac{x^2 - 4}{x^2 + 2x}$ $\dfrac{x - 2}{x^2}$

8. $\dfrac{8x^2(x - 3y) + 6x(x - 3y) + 1(x - 3y)}{4x^2(x - 3y) - 4x(x - 3y) - 3(x - 3y)} \div \dfrac{4x + 1}{2x^2 - 3x}$ x

9. $\dfrac{(a + b)^2}{a - b} \cdot \dfrac{a^3 - b^3}{a^2 - b^2} \div \dfrac{a^2 + 2ab + b^2}{(a - b)^2} \cdot \dfrac{1}{a - b}$ $\dfrac{a^2 + ab + b^2}{a + b}$

10. $\dfrac{(x - p)^n}{x^{-2}} \div \dfrac{(x - p)^{2n}}{x^{-4}}$ $\dfrac{1}{x^2(x - p)^n}$

7.2 Addition and Subtraction of Rational Expressions

1. Add or subtract rational expressions with a common denominator.
2. Find the least common denominator (LCD).
3. Add or subtract rational expressions with unlike denominators.

**Adding
and Subtracting
Expressions
with a Common
Denominator**

1 When adding (or subtracting) two rational expressions with a common denominator we add (or subtract) the numerators while keeping the common denominator.

Addition	Subtraction
$\dfrac{a}{c} + \dfrac{b}{c} = \dfrac{a+b}{c}, \quad c \neq 0$	$\dfrac{a}{c} - \dfrac{b}{c} = \dfrac{a-b}{c}, \quad c \neq 0$

EXAMPLE 1 Add $\dfrac{3}{x+2} + \dfrac{x-4}{x+2}$.

Solution: Since the denominators are the same, we add the numerators and keep the common denominator.

$$\frac{3}{x+2} + \frac{x-4}{x+2} = \frac{3+(x-4)}{x+2} = \frac{x-1}{x+2}$$

To Add or Subtract Rational Expressions with a Common Denominator

1. Add or subtract the numerators.
2. Place the sum or difference of the numerators found in step 1 over the common denominator.
3. Reduce the expression, if possible.

EXAMPLE 2 Add $\dfrac{x^2 + 3x - 2}{(x+5)(x-2)} + \dfrac{4x+12}{(x+5)(x-2)}$.

Solution:

$$\frac{x^2 + 3x - 2}{(x+5)(x-2)} + \frac{4x+12}{(x+5)(x-2)} = \frac{x^2 + 3x - 2 + (4x+12)}{(x+5)(x-2)}$$

$$= \frac{x^2 + 7x + 10}{(x+5)(x-2)}$$

$$= \frac{(x+5)(x+2)}{(x+5)(x-2)} = \frac{x+2}{x-2}$$

When subtracting rational expressions, be sure to subtract the entire numerator of the fraction being subtracted. Study the common student error that follows very carefully.

COMMON STUDENT ERROR   The error presented here is sometimes made by students. Study the information presented so that you will not make this error.

How do you simplify this problem?

$$\frac{4x}{x - 2} - \frac{2x + 1}{x - 2}$$

Correct

$$\frac{4x}{x - 2} - \frac{2x + 1}{x - 2} = \frac{4x - (2x + 1)}{x - 2}$$

$$= \frac{4x - 2x - 1}{x - 2}$$

$$= \frac{2x - 1}{x - 2}$$

Incorrect

$$\frac{4x}{x - 2} - \frac{2x + 1}{x - 2} = \frac{4x - 2x + 1}{x - 2}$$

$$= \frac{2x + 1}{x - 2}$$

The procedure on the right side is incorrect because the *entire numerator*, $2x + 1$, must be subtracted from $4x$. Instead, only $2x$ was subtracted. Note that **the sign of each term** (not just the first term) **in the numerator of the fraction being subtracted must change.** Note that $-(2x + 1) = -2x - 1$ by the distributive property discussed in Section 1.3.

EXAMPLE 3   Subtract $\dfrac{3x}{x - 6} - \dfrac{x^2 - 4x + 6}{x - 6}$.

Solution:

$$\frac{3x}{x - 6} - \frac{x^2 - 4x + 6}{x - 6} = \frac{3x - (x^2 - 4x + 6)}{x - 6}$$

$$= \frac{3x - x^2 + 4x - 6}{x - 6}$$

$$= \frac{-x^2 + 7x - 6}{x - 6}$$

$$= \frac{-(x^2 - 7x + 6)}{x - 6}$$

$$= \frac{-(x - 6)(x - 1)}{x - 6} = -(x - 1)$$

Finding the Least Common Denominator

2   To add two numerical fractions with *unlike denominators*, we must first obtain a common denominator.

EXAMPLE 4   Add $\dfrac{3}{5} + \dfrac{4}{7}$.

Solution: The least common denominator (LCD) of 5 and 7 is 35. Thirty-five is the smallest number divisible by both 5 and 7. Now rewrite each fraction so that it has a denominator equal to the LCD.

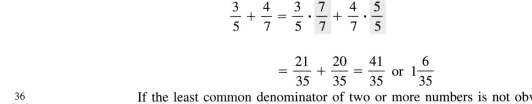

$$\frac{3}{5} + \frac{4}{7} = \frac{3}{5} \cdot \frac{7}{7} + \frac{4}{7} \cdot \frac{5}{5}$$

$$= \frac{21}{35} + \frac{20}{35} = \frac{41}{35} \text{ or } 1\frac{6}{35}$$

Prime factorization: $2 \cdot 2 \cdot 3 \cdot 3 = 2^2 \cdot 3^2$

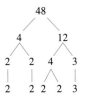

$2 \cdot 2 \cdot 2 \cdot 2 \cdot 3 = 2^4 \cdot 3$

FIGURE 7.3

If the least common denominator of two or more numbers is not obvious, it can be found by writing each denominator as a product of prime numbers. **Prime numbers** are natural numbers greater than 1 that are divisible by only two numbers, themselves and 1. The first 10 prime numbers are 2, 3, 5, 7, 11, 13, 17, 19, 23, and 29. Writing a number as a product of prime numbers is called the *prime factorization* of the number. Consider the addition problem $\frac{1}{36} + \frac{5}{48}$. To find the least common denominator, we determine the prime factorization of each denominator using *tree diagrams* as shown in Fig. 7.3. In this figure we find the prime factorizations of the numbers 36 and 48.

The tree diagrams in Fig. 7.3 can be formed in many ways. For example, we could have started the diagram for 36 as 9 · 4 or 12 · 3 or 18 · 2 or any other numbers whose product is 36. Regardless of how you start you should obtain the same prime factorization. For practice, determine the prime factorizations for 36 and 48 by starting the tree diagrams with the products 9 · 4 and 6 · 8, respectively. You should obtain the same prime factorizations. When writing the prime factorization of a number, list the prime factors that appear more than once in exponential form.

To determine the least common denominator, multiply each prime number that appears in either prime factorization. If a prime number is common to both factorizations, use the prime with the highest power. The prime numbers that appear in either of the factorizations are 2 and 3. The greatest number of twos that appear is 4 (in 48), and the greatest number of threes that appear is 2 (in 36). Thus, the least common denominator is found as follows:

$$\text{LCD} = 2^4 \cdot 3^2 = 16 \cdot 9 = 144$$

Therefore, the least common denominator of the two fractions is 144. Below we add the fractions.

$$\frac{1}{36} + \frac{5}{48} = \frac{4}{4} \cdot \frac{1}{36} + \frac{5}{48} \cdot \frac{3}{3} = \frac{4}{144} + \frac{15}{144} = \frac{19}{144}$$

To add or subtract rational expressions with unlike denominators, we must first write each expression with a common denominator.

> **To Find the Least Common Denominator of Rational Expressions**
>
> 1. Factor each denominator completely. Any factors that occur more than once should be expressed as powers. For example, $(x + 5)(x + 5)$ should be expressed as $(x + 5)^2$.
>
> *(continued on next page)*

> **2.** List all different factors (other than 1) that appear in any of the denominators. When the same factor appears in more than one denominator, write the factor with the *highest* power that appears.
>
> **3.** The least common denominator is the product of all the factors found in step 2.

EXAMPLE 5   Find the LCD.

(a) $\dfrac{3}{5x} - \dfrac{2}{x^2}$ **(b)** $\dfrac{1}{18x^3y} + \dfrac{5}{27x^2y^3}$ **(c)** $\dfrac{3}{x} - \dfrac{2y}{x+5}$

Solution:

(a) The factors that appear in the denominators are 5 and x. List each factor with its highest power. The LCD is the product of these factors.

$$\overset{\text{highest power of } x}{\text{LCD} = 5 \cdot x^2 = 5x^2}$$

(b) Using a tree diagram, we can determine that $18 = 2 \cdot 3^2$ and $27 = 3^3$. The variable factors that appear are x and y. Using the highest powers of the factors, we obtain the LCD.

$$\text{LCD} = 2 \cdot 3^3 \cdot x^3 y^3 = 54x^3y^3$$

(c) The factors that appear are x and $(x + 5)$. Note that the x in the second denominator, $x + 5$, is not a factor of that denominator since the operation is addition rather than multiplication.

$$\text{LCD} = x(x + 5)$$

EXAMPLE 6   Find the LCD.

(a) $\dfrac{3}{2x^2 - 4x} + \dfrac{x^2}{x^2 - 4x + 4}$ **(b)** $\dfrac{5x}{x^2 - x - 12} - \dfrac{6x^2}{x^2 - 7x + 12}$

Solution:

(a) Factor both denominators.

$$\frac{3}{2x^2 - 4x} + \frac{x^2}{x^2 - 4x + 4} = \frac{3}{2x(x - 2)} + \frac{x^2}{(x - 2)^2}$$

The factors that appear are 2, x, and $x - 2$. List the highest powers of each of these factors that appear.

$$\text{LCD} = 2 \cdot x \cdot (x - 2)^2 = 2x(x - 2)^2$$

(b) Factor both denominators.

$$\frac{5x}{x^2 - x - 12} - \frac{6x^2}{x^2 - 7x + 12} = \frac{5x}{(x + 3)(x - 4)} - \frac{6x^2}{(x - 3)(x - 4)}$$

$$\text{LCD} = (x + 3)(x - 4)(x - 3)$$

Note that although $(x - 4)$ is a common factor of each denominator, the highest power of that factor that appears in either denominator is 1.

Adding and Subtracting Rational Expressions with Unlike Denominators

3 The method used to add or subtract rational expressions with unlike denominators is illustrated in Example 7.

EXAMPLE 7 Add $\dfrac{3}{x} + \dfrac{5}{y}$.

Solution: First, determine the LCD.

$$LCD = xy$$

Now write each fraction with the LCD. We do this by multiplying **both** numerator and denominator of each fraction by any factors needed to obtain the LCD.

In this problem the fraction on the left must be multiplied by y/y and the fraction on the right must be multiplied by x/x.

$$\frac{3}{x} + \frac{5}{y} = \frac{y}{y} \cdot \frac{3}{x} + \frac{5}{y} \cdot \frac{x}{x} = \frac{3y}{xy} + \frac{5x}{xy}$$

By multiplying both the numerator and denominator by the same factor, we are in effect multiplying by 1, which does not change the value of the fraction, only its appearance. Thus, the new fraction is equivalent to the original fraction.

Now add the numerators while leaving the LCD alone.

$$\frac{3y}{xy} + \frac{5x}{xy} = \frac{3y + 5x}{xy} \text{ or } \frac{5x + 3y}{xy}$$

Therefore, $\dfrac{3}{x} + \dfrac{5}{y} = \dfrac{5x + 3y}{xy}$.

To Add or Subtract Rational Expressions with Unlike Denominators

1. Determine the LCD.
2. Rewrite each fraction as an equivalent fraction with the LCD. This is done by multiplying both the numerator and denominator of each fraction by any factors needed to obtain the LCD.
3. Leave the denominator in factored form, but multiply out the numerator.
4. Add or subtract the numerators while maintaining the LCD.
5. When possible, factor the numerator and reduce fractions.

EXAMPLE 8 Add $\dfrac{5}{4x^2y} + \dfrac{3}{14xy^3}$.

Solution: The LCD is $28x^2y^3$. We must write each fraction with the denominator $28x^2y^3$. To do this, multiply the fraction on the left by $7y^2/7y^2$ and the fraction on the right by $2x/2x$.

$$\frac{5}{4x^2y} + \frac{3}{14xy^3} = \frac{7y^2}{7y^2} \cdot \frac{5}{4x^2y} + \frac{3}{14xy^3} \cdot \frac{2x}{2x}$$

$$= \frac{35y^2}{28x^2y^3} + \frac{6x}{28x^2y^3}$$

$$= \frac{35y^2 + 6x}{28x^2y^3}$$

EXAMPLE 9   Subtract $\dfrac{x+2}{x-4} - \dfrac{x+3}{x+4}$.

Solution: The LCD is $(x-4)(x+4)$. Write each fraction with the denominator $(x-4)(x+4)$.

$$\frac{x+2}{x-4} - \frac{x+3}{x+4} = \frac{x+4}{x+4} \cdot \frac{x+2}{x-4} - \frac{x+3}{x+4} \cdot \frac{x-4}{x-4}$$

$$= \frac{(x+4)(x+2)}{(x+4)(x-4)} - \frac{(x+3)(x-4)}{(x+4)(x-4)}$$

Use the FOIL method to multiply each numerator.

$$= \frac{x^2 + 6x + 8}{(x+4)(x-4)} - \frac{x^2 - x - 12}{(x+4)(x-4)}$$

$$= \frac{x^2 + 6x + 8 - (x^2 - x - 12)}{(x+4)(x-4)}$$

$$= \frac{x^2 + 6x + 8 - x^2 + x + 12}{(x+4)(x-4)}$$

$$= \frac{7x + 20}{(x+4)(x-4)}$$

EXAMPLE 10   Add $\dfrac{4}{x-3} + \dfrac{x+5}{3-x}$.

Solution: Note that each denominator is the opposite, or additive inverse, of the other. (The terms of one denominator differ only in sign from the terms of the other denominator.) When this special situation arises, we can multiply the numerator and denominator of either one of the fractions by -1 to obtain the LCD.

$$\frac{4}{x-3} + \frac{x+5}{3-x} = \frac{4}{x-3} + \frac{-1}{-1} \cdot \frac{(x+5)}{(3-x)}$$

$$= \frac{4}{x-3} + \frac{-x-5}{x-3} = \frac{-x-1}{x-3}$$

EXAMPLE 11 Subtract $\dfrac{3x + 4}{2x^2 - 5x - 12} - \dfrac{2x - 3}{5x^2 - 18x - 8}$.

Solution: Factor the denominator of each expression.

$$\frac{3x + 4}{2x^2 - 5x - 12} - \frac{2x - 3}{5x^2 - 18x - 8} = \frac{3x + 4}{(2x + 3)(x - 4)} - \frac{2x - 3}{(5x + 2)(x - 4)}$$

The LCD is $(2x + 3)(x - 4)(5x + 2)$.

$$\frac{3x + 4}{(2x + 3)(x - 4)} - \frac{2x - 3}{(5x + 2)(x - 4)} = \frac{5x + 2}{5x + 2} \cdot \frac{3x + 4}{(2x + 3)(x - 4)} - \frac{2x - 3}{(5x + 2)(x - 4)} \cdot \frac{2x + 3}{2x + 3}$$

$$= \frac{15x^2 + 26x + 8}{(5x + 2)(2x + 3)(x - 4)} - \frac{4x^2 - 9}{(5x + 2)(2x + 3)(x - 4)}$$

$$= \frac{15x^2 + 26x + 8 - (4x^2 - 9)}{(5x + 2)(2x + 3)(x - 4)}$$

$$= \frac{15x^2 + 26x + 8 - 4x^2 + 9}{(5x + 2)(2x + 3)(x - 4)}$$

$$= \frac{11x^2 + 26x + 17}{(5x + 2)(2x + 3)(x - 4)}$$

EXAMPLE 12 $\dfrac{x - 1}{x - 2} - \dfrac{x + 1}{x + 2} + \dfrac{x - 6}{x^2 - 4}$.

Solution: First, factor $x^2 - 4$. The LCD of the three fractions is $(x + 2)(x - 2)$.

$$\frac{x - 1}{x - 2} - \frac{x + 1}{x + 2} + \frac{x - 6}{x^2 - 4} = \frac{x - 1}{x - 2} - \frac{x + 1}{x + 2} + \frac{x - 6}{(x + 2)(x - 2)}$$

$$= \frac{x + 2}{x + 2} \cdot \frac{x - 1}{x - 2} - \frac{x + 1}{x + 2} \cdot \frac{x - 2}{x - 2} + \frac{x - 6}{(x + 2)(x - 2)}$$

$$= \frac{x^2 + x - 2}{(x + 2)(x - 2)} - \frac{x^2 - x - 2}{(x + 2)(x - 2)} + \frac{x - 6}{(x + 2)(x - 2)}$$

$$= \frac{x^2 + x - 2 - (x^2 - x - 2) + (x - 6)}{(x + 2)(x - 2)}$$

$$= \frac{x^2 + x - 2 - x^2 + x + 2 + x - 6}{(x + 2)(x - 2)}$$

$$= \frac{3x - 6}{(x + 2)(x - 2)}$$

$$= \frac{3(x - 2)}{(x + 2)(x - 2)} = \frac{3}{x + 2}$$

Exercises 69 through 76 that follow involve more than one operation. When working these exercises, work each part of the exercise separately, following the priority of operations given in Section 1.4. For example, if asked to simplify

$$\left(4 - \frac{1}{x-2}\right) \cdot \frac{x^2-4}{4x^2-5x-9}$$

you would begin by performing the operation within the parentheses. For $\left(4 - \dfrac{1}{x-2}\right)$ you should obtain $\dfrac{4x-9}{x-2}$. Then you would multiply

$$\frac{4x-9}{x-2} \cdot \frac{x^2-4}{4x^2-5x-9}$$

to obtain the answer of $\dfrac{x+2}{x+1}$. Work the entire problem now, starting from the beginning, to see if you obtain this answer.

Exercise Set 7.2

9. $\dfrac{x+5}{x+3}$ **10.** $\dfrac{x+4}{x-4}$

Add or subtract.

1. $\dfrac{2x-7}{3} - \dfrac{4}{3}$ $\dfrac{2x-11}{3}$

2. $\dfrac{-2x+6}{x^2+x-6} + \dfrac{3x-3}{x^2+x-6}$ $\dfrac{1}{x-2}$

3. $\dfrac{2x+4}{(x+2)(x-3)} - \dfrac{x+7}{(x+2)(x-3)}$ $\dfrac{1}{x+2}$

4. $\dfrac{4r+12}{3-r} - \dfrac{3r+15}{3-r}$ -1

5. $\dfrac{x^2-2}{x^2+6x-7} - \dfrac{-4x+19}{x^2+6x-7}$ $\dfrac{x-3}{x-1}$

6. $\dfrac{-x^2}{x^2+5xy-14y^2} + \dfrac{x^2+xy+7y^2}{x^2+5xy-14y^2}$ $\dfrac{y}{x-2y}$

7. $\dfrac{3r^2+15r}{r^3+2r^2-8r} + \dfrac{2r^2+5r}{r^3+2r^2-8r}$ $\dfrac{5}{r-2}$

8. $\dfrac{x^3-10x^2+35x}{x(x-6)} - \dfrac{x^2+5x}{x(x-6)}$ $x-5$

9. $\dfrac{3x^2-x}{2x^2-x-21} + \dfrac{3x-8}{2x^2-x-21} - \dfrac{x^2-x+27}{2x^2-x-21}$

10. $\dfrac{2x^2+8x-15}{2x^2-13x+20} - \dfrac{2x+10}{2x^2-13x+20} - \dfrac{3x-5}{2x^2-13x+20}$

Find the least common denominator.   **17.** $(a-8)(a+3)(a+8)$ **20.** $(x+y)(3x+2y)(2x+3y)$

11. $\dfrac{5x}{x+1} + \dfrac{6}{x+2}$ $(x+1)(x+2)$

12. $\dfrac{-4}{8x^2y^2} + \dfrac{7}{5x^4y^5}$ $40x^4y^5$

13. $\dfrac{x+3}{16x^2y} - \dfrac{x^2}{3x^3}$ $48x^3y$

14. $\dfrac{9}{(x-4)(x+3)} - \dfrac{x+8}{x-4}$ $(x-4)(x+3)$

15. $6z^2 + \dfrac{9z}{z-3}$ $z-3$

16. $\dfrac{b^2+3}{18b} - \dfrac{b-7}{12(b+5)}$ $36b(b+5)$

17. $\dfrac{a-2}{a^2-5a-24} + \dfrac{3}{a^2+11a+24}$

18. $\dfrac{6x+5}{x^2-4} - \dfrac{3x}{x^2-5x-14}$ $(x-2)(x+2)(x-7)$

19. $\dfrac{6}{x+3} - \dfrac{x+5}{x^2-4x+3}$ $(x+3)(x-3)(x-1)$

20. $\dfrac{3x-5}{6x^2+13xy+6y^2} + \dfrac{3}{3x^2+5xy+2y^2}$

21. $\dfrac{3}{x^2+3x-4} - \dfrac{4}{4x^2+5x-9} + \dfrac{x+2}{4x^2+25x+36}$
$(x-1)(x+4)(4x+9)$

22. $\dfrac{x}{2x^2-7x+3} + \dfrac{x-3}{4x^2+4x-3} - \dfrac{x^2+1}{2x^2-3x-9}$
$(x-3)(2x-1)(2x+3)$

40. $\dfrac{-2x-7}{(x-4)(x+4)(2x-3)}$ **50.** $\dfrac{x^2-3y^2}{(x-2y)(x+y)(x-y)}$ **60.** $\dfrac{11x-37}{(x-9)(x-4)(2x+1)}$ **61.** $\dfrac{8m^2+5mn}{(2m+3n)(3m+2n)(2m+n)}$

Add or subtract.

23. $\dfrac{4}{3x}+\dfrac{2}{x}$ $\dfrac{10}{3x}$

24. $\dfrac{6}{x^2}+\dfrac{3}{2x}$ $\dfrac{3x+12}{2x^2}$

25. $\dfrac{5}{6y}+\dfrac{3}{4y^2}$ $\dfrac{10y+9}{12y^2}$

26. $\dfrac{3x}{4y}+\dfrac{5}{6xy}$ $\dfrac{9x^2+10}{12xy}$

27. $\dfrac{5}{12x^4y}-\dfrac{1}{5x^2y^3}$ $\dfrac{25y^2-12x^2}{60x^4y^3}$

28. $\dfrac{3}{4xy^3}+\dfrac{1}{6x^2y}$ $\dfrac{9x+2y^2}{12x^2y^3}$

29. $\dfrac{4x}{3xy}+2$ $\dfrac{4+6y}{3y}$

30. $\dfrac{5}{b-2}+\dfrac{3x}{2-b}$ $\dfrac{5-3x}{b-2}$

31. $\dfrac{x}{x-y}-\dfrac{x}{y-x}$ $\dfrac{2x}{x-y}$

32. $\dfrac{b}{a-b}+\dfrac{a+b}{b}$ $\dfrac{a^2}{b(a-b)}$

33. $\dfrac{z+5}{z-5}-\dfrac{z-5}{z+5}$ $\dfrac{20z}{(z+5)(z-5)}$

34. $\dfrac{x+7}{x+3}-\dfrac{x-3}{x+7}$ $\dfrac{14x+58}{(x+3)(x+7)}$

35. $\dfrac{x}{x^2-9}-\dfrac{4(x-3)}{x+3}$ $\dfrac{-4x^2+25x-36}{(x+3)(x-3)}$

36. $\dfrac{4x}{x-4}+\dfrac{x+4}{x+1}$ $\dfrac{5x^2+4x-16}{(x-4)(x+1)}$

37. $\dfrac{2m+1}{m-5}-\dfrac{4}{m^2-3m-10}$ $\dfrac{2m^2+5m-2}{(m-5)(m+2)}$

38. $\dfrac{x}{x+1}+\dfrac{1}{x^2+2x+1}$ $\dfrac{x^2+x+1}{(x+1)^2}$

39. $\dfrac{-x^2+5x}{(x-5)^2}+\dfrac{x+1}{x-5}$ $\dfrac{1}{x-5}$

40. $\dfrac{4}{(2x-3)(x+4)}-\dfrac{3}{(x+4)(x-4)}$

41. $\dfrac{x}{x^2+2x-8}+\dfrac{x+2}{x^2-3x+2}$ $\dfrac{2x^2+5x+8}{(x-1)(x+4)(x-2)}$

42. $\dfrac{5x}{x^2-9x+8}-\dfrac{3(x+2)}{x^2-6x-16}$ $\dfrac{2x+3}{(x-8)(x-1)}$

43. $5-\dfrac{x-1}{x^2+3x-10}$ $\dfrac{5x^2+14x-49}{(x+5)(x-2)}$

44. $\dfrac{3x}{2x-3}+\dfrac{3x+6}{2x^2+x-6}$ $\dfrac{3(x+1)}{2x-3}$

45. $\dfrac{3a-4}{4a+1}+\dfrac{3a+6}{4a^2+9a+2}$ $\dfrac{3a-1}{4a+1}$

46. $\dfrac{7}{3q^2+q-4}+\dfrac{9q+2}{3q^2-2q-8}$ $\dfrac{3q-4}{(q-1)(q-2)}$

47. $\dfrac{x+3}{3x^2+6x-9}+\dfrac{x-3}{6x^2-15x+9}$ $\dfrac{x-2}{(x-1)(2x-3)}$

48. $\dfrac{1}{x^2-y^2}+\dfrac{4}{x^2-2xy-3y^2}$ $\dfrac{5x-7y}{(x+y)(x-y)(x-3y)}$

49. $\dfrac{x-y}{x^2-4xy+4y^2}+\dfrac{x-3y}{x^2-4y^2}$ $\dfrac{2x^2-4xy+4y^2}{(x-2y)^2(x+2y)}$

50. $\dfrac{x+2y}{x^2-xy-2y^2}-\dfrac{y}{x^2-3xy+2y^2}$

51. $\dfrac{2x}{x-3}-\dfrac{2x}{x+3}+\dfrac{36}{x^2-9}$ $\dfrac{12}{x-3}$

52. $\dfrac{3}{p-1}+\dfrac{4}{p+1}+\dfrac{p+2}{p^2-1}$ $\dfrac{8p+1}{(p+1)(p-1)}$

53. $\dfrac{x^2+2}{x^2-x-2}+\dfrac{1}{x+1}-\dfrac{x}{x-2}$ 0

54. $\dfrac{2}{x^2-16}+\dfrac{x+1}{x^2+8x+16}+\dfrac{3}{x-4}$ $\dfrac{4x^2+23x+52}{(x+4)^2(x-4)}$

55. $\dfrac{3x+2}{x-5}+\dfrac{x}{3x+4}-\dfrac{7x^2+24x+28}{3x^2-11x-20}$ 1

56. $\dfrac{4}{3x-2}-\dfrac{1}{x-4}+5$ $\dfrac{15x^2-69x+26}{(3x-2)(x-4)}$

57. $\dfrac{x}{x^2-10x+24}-\dfrac{3}{x-6}+1$ $\dfrac{x-6}{x-4}$

58. $3-\dfrac{4}{8r^2+2r-15}+\dfrac{r+2}{4r-5}$ $\dfrac{26r^2+13r-43}{(4r-5)(2r+3)}$

59. $\dfrac{3}{5x+6}+\dfrac{x^2-x}{5x^2-4x-12}-\dfrac{4}{x-2}$ $\dfrac{x^2-18x-30}{(5x+6)(x-2)}$

60. $\dfrac{3}{x^2-13x+36}+\dfrac{4}{2x^2-7x-4}+\dfrac{1}{2x^2-17x-9}$

61. $\dfrac{m}{6m^2+13mn+6n^2}+\dfrac{2m}{4m^2+8mn+3n^2}$

62. $\dfrac{(x-y)^2}{x^3-y^3}+\dfrac{1}{x^2+xy+y^2}$ $\dfrac{x-y+1}{x^2+xy+y^2}$

63. $\dfrac{5r-2s}{25r^2-4s^2}-\dfrac{2r-s}{10r^2-rs-2s^2}$ 0

64. $\dfrac{6}{(2r-1)^2}+\dfrac{2}{2r-1}-4$ $\dfrac{-16r^2+20r}{(2r-1)^2}$

65. $\dfrac{2}{2m^2-5mn+2n^2}-\dfrac{3}{6m^2+mn-2n^2}+\dfrac{4}{3m^2-4mn-4n^2}$ $\dfrac{11m+6n}{(2m-n)(m-2n)(3m+2n)}$

66. $\dfrac{5}{6p^2 + pq - 15q^2} + \dfrac{2}{2p^2 - 7pq + 6q^2} - \dfrac{1}{3p^2 - pq - 10q^2}$ $\quad \dfrac{9p + 3q}{(2p - 3q)(3p + 5q)(p - 2q)}$

67. $\dfrac{2}{2x + 3y} - \dfrac{4x^2 - 6xy + 9y^2}{8x^3 + 27y^3} \quad \dfrac{1}{2x + 3y}$

68. $\dfrac{4}{4x - 5y} - \dfrac{3x^2 + 2y^2}{64x^3 - 125y^3}$

Perform the indicated operations. **68.** $\dfrac{61x^2 + 80xy + 98y^2}{(4x - 5y)(16x^2 + 20xy + 25y^2)}$ **75.** $\dfrac{4x^2 - 13x + 5}{(2x - 1)(x - 3)}$

69. $\left(3 + \dfrac{1}{x + 3}\right)\left(\dfrac{x + 3}{x - 2}\right) \quad \dfrac{3x + 10}{x - 2}$

70. $\left(\dfrac{3}{r + 1} - \dfrac{4}{r - 2}\right)\left(\dfrac{r - 2}{r + 10}\right) \quad -\dfrac{1}{r + 1}$

71. $\left(\dfrac{5}{a - 5} - \dfrac{2}{a + 3}\right) \div (3a + 25) \quad \dfrac{1}{(a - 5)(a + 3)}$

72. $(x + 3)\left(\dfrac{x - 4}{x^2 + x - 6}\right) + \dfrac{2}{x + 1} \quad \dfrac{x^2 - x - 8}{(x - 2)(x + 1)}$

73. $\left(\dfrac{x^2 + 4x - 5}{2x^2 + x - 3} \cdot \dfrac{2x + 3}{x + 1}\right) - \dfrac{2}{x + 2} \quad \dfrac{x^2 + 5x + 8}{(x + 1)(x + 2)}$

74. $\left(\dfrac{x + 5}{x - 3} - x\right) \div \dfrac{1}{x - 3} \quad -x^2 + 4x + 5$

75. $\dfrac{4x^2 - 17x + 4}{x^2 - 2x - 8} \div \dfrac{4x^2 + 4x - 3}{2x^2 + 7x + 6} + \dfrac{2}{2x^2 - 7x + 3}$

76. $\left(\dfrac{x + 5}{x^2 - 25} + \dfrac{1}{x + 5}\right)\left(\dfrac{2x^2 - 13x + 15}{4x^2 - 6x}\right) \quad \dfrac{1}{x + 5}$

*In Exercises 77–79, (**a**) determine the polynomial to be placed in the shaded area that will result in a true statement; (**b**)*
explain how you determined your answer. **77.** $-5x^2 + x + 6$ **78.** $7x^2 - 6x + 6$ **80. (c)** $(8x + 11)(8x - 11)(x - 2)$

 **77.** $\dfrac{2x^2 - 5x - 12}{x^2 + 3x - 10} + \dfrac{\rule{1.5cm}{0.3cm}}{x^2 + 3x - 10} = \dfrac{-3x^2 - 4x - 6}{x^2 + 3x - 10}$

81. (a) Explain how to add or subtract two rational expressions.

78. $\dfrac{5x^2 - 6}{x^2 - x - 1} - \dfrac{\rule{1.5cm}{0.3cm}}{x^2 - x - 1} = \dfrac{-2x^2 + 6x - 12}{x^2 - x - 1}$

(b) Add $\dfrac{4}{x + 2} + \dfrac{x}{3x^2 - 4x - 20}$ following the procedure given in part **(a)**. $\quad \dfrac{13x - 40}{(x + 2)(3x - 10)}$

79. $\dfrac{r^2 - 6}{r^2 - 5r + 6} - \dfrac{\rule{1.5cm}{0.3cm}}{r^2 - 5r + 6} = \dfrac{1}{r - 2} \quad r^2 - r - 3$

80. (a) What is the least common denominator of two or more rational expressions?

(b) Explain how to find the LCD.

(c) Using the procedure given in part **(b)**, find the LCD of

$$\dfrac{5}{64x^2 - 121} \quad \text{and} \quad \dfrac{1}{8x^2 - 27x + 22}$$

82. Show that $\dfrac{a}{b} + \dfrac{c}{d} = \dfrac{ad + bc}{bd}$.

83. Show that $x^{-1} + y^{-1} = \dfrac{x + y}{xy}$.

*Consider the rectangles given below. Find (**a**) the perimeter; (**b**) the area.*

84.

$\dfrac{a + b}{a}$

$\dfrac{a - b}{a}$

(a) 4 **(b)** $\dfrac{a^2 - b^2}{a^2}$

85.

$\dfrac{a + 2b}{b}$

$\dfrac{-a + 2b}{b}$

$\dfrac{4b^2 - a^2}{b^2}$

(a) 8 **(b)**

CUMULATIVE REVIEW EXERCISES

[2.4] **86.** A bottling machine fills and caps bottles at a rate of 80 per minute. Then the machine is slowed down and fills and caps bottles at a rate of 60 per minute. If the sum of the two time periods was 14 minutes and the number of bottles filled and capped at the higher rate was the same as the number filled and capped at the lower rate, determine **(a)** how long the machine was used at the faster rate, and **(b)** the total number of bottles filled and capped over the 14-minute period.
86. (a) 6 min **(b)** 960 bottles

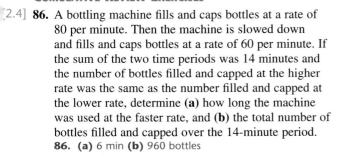

[2.3] **87.** The price of a suit is increased by 20%. The price is then decreased by $20. If the suit sells for $196, find the original price of the suit. $180

[5.4] **88.** Divide $\dfrac{9x^4y^6 - 3x^3y^2 + 5xy^5}{3xy^4}$. $3x^3y^2 - \dfrac{x^2}{y^2} + \dfrac{5y}{3}$

89. Divide $\dfrac{6x^2 + 5x - 4}{3x + 4}$. $2x - 1$

Group Activity/ Challenge Problems

Tape 11

1. The weighted average of two values a and b is given by $a\left(\dfrac{x}{n}\right) + b\left(\dfrac{n - x}{n}\right)$, where $\dfrac{x}{n}$ is the weight given to a and $\dfrac{n - x}{n}$ is the weight given to b.

(a) Express this sum as a single fraction. $\dfrac{ax + bn - bx}{n}$

(b) On exam a you received a grade of 60 and on exam b you received a grade of 92. If exam a counts $\frac{2}{5}$ of your final grade and exam b counts $\frac{3}{5}$, determine your weighted average.   79.2

2. Show that $\left(\dfrac{x}{y}\right)^{-1} + \left(\dfrac{y}{x}\right)^{-1} + (xy)^{-1} = \dfrac{x^2 + y^2 + 1}{xy}$.

3. Express each sum as a single fraction.

(a) $1 + \dfrac{1}{x}$ $\dfrac{x + 1}{x}$

(b) $1 + \dfrac{1}{x} + \dfrac{1}{x^2}$ $\dfrac{x^2 + x + 1}{x^2}$

(c) $1 + \dfrac{1}{x} + \dfrac{1}{x^2} + \cdots + \dfrac{1}{x^5}$ $\dfrac{x^5 + x^4 + x^3 + x^2 + x + 1}{x^5}$

(d) $1 + \dfrac{1}{x} + \dfrac{1}{x^2} + \cdots + \dfrac{1}{x^n}$ $\dfrac{x^n + x^{n-1} + x^{n-2} + \cdots + 1}{x^n}$

Add or subtract as indicated.

4. $\dfrac{3}{x^2 - x - 6} - \dfrac{2}{x^2 + x - 6}$ **4.** $\dfrac{(x + 6)(x - 1)}{(x + 2)(x - 3)(x - 2)(x + 3)}$

5. $\dfrac{3}{x^3 + 27} + \dfrac{4}{x^2 - 9}$ **5.** $\dfrac{4x^2 - 9x + 27}{(x + 3)(x - 3)(x^2 - 3x + 9)}$

6. $(a - b)^{-1} + (a - b)^{-2}$ **6.** $\dfrac{a - b + 1}{(a - b)^2}$

7. $(2x - 4)^{-1} + (2x + 3)^{-1}$ **7.** $\dfrac{4x - 1}{(2x - 4)(2x + 3)}$

8. $\left(\dfrac{a - b}{a}\right)^{-1} - \left(\dfrac{a + b}{a}\right)^{-1}$ **8.** $\dfrac{2ab}{(a - b)(a + b)}$

7.3   Complex Fractions

Tape 11

1️⃣ Recognize complex fractions.
2️⃣ Simplify complex fractions by multiplying by a common denominator.
3️⃣ Simplify complex fractions by simplifying the numerator and denominator.

Recognize Complex Fractions

1️⃣ A **complex fraction** is one that has a fractional expression in its numerator or its denominator or both its numerator and denominator. Examples of complex fractions include

$$\dfrac{\frac{2}{3}}{\frac{5}{}}, \quad \dfrac{\frac{x + 1}{x}}{3x}, \quad \dfrac{x}{\frac{y}{x + 1}}, \quad \dfrac{\frac{a + b}{a}}{\frac{a - b}{b}}, \quad \dfrac{3 + \frac{1}{x}}{\frac{1}{x^2} + \frac{3}{x}}$$

The expression above the main fraction line is the numerator, and the expression below the main fraction line is the denominator of the complex fraction.

We will explain two methods that can be used to simplify complex fractions.

secondary fraction $\longrightarrow$ $\dfrac{\dfrac{a+b}{a}}{\dfrac{a-b}{b}}$ $\longleftarrow$ numerator of complex fraction

$\longleftarrow$ main fraction line

secondary fraction $\longrightarrow$ $\longleftarrow$ denominator of complex fraction

Method 1: Multiplying by a Common Denominator

2 The first method involves multiplying both the numerator and denominator of the complex fraction by a common denominator.

> **To Simplify a Complex Fraction by Multiplying by a Common Denominator**
> 1. Find the least common denominator of each of the two secondary fractions.
> 2. Next find the LCD of the complex fraction. The LCD of the complex fraction will be the LCD of the two expressions found in step 1.
> 3. Multiply both secondary fractions by the LCD of the complex fraction found in step 2.
> 4. Simplify when possible.

EXAMPLE 1   Simplify $\dfrac{\dfrac{2}{3}+\dfrac{3}{4}}{\dfrac{3}{4}-\dfrac{1}{2}}$.

Solution:

Step 1:   The LCD of the numerator of the complex fraction is 12. The LCD of the denominator is 4.

Step 2:   The LCD of the complex fraction is the LCD of 12 and 4, which is 12.

Step 3:   Multiply both secondary fractions by 12.

$$\frac{\dfrac{2}{3}+\dfrac{3}{4}}{\dfrac{3}{4}-\dfrac{1}{2}}=\frac{12\left(\dfrac{2}{3}+\dfrac{3}{4}\right)}{12\left(\dfrac{3}{4}-\dfrac{1}{2}\right)}=\frac{12\left(\dfrac{2}{3}\right)+12\left(\dfrac{3}{4}\right)}{12\left(\dfrac{3}{4}\right)-12\left(\dfrac{1}{2}\right)}$$

Step 4:   Simplify.

$$=\frac{\overset{4}{\cancel{12}}\left(\dfrac{2}{\cancel{3}}\right)+\overset{3}{\cancel{12}}\left(\dfrac{3}{\cancel{4}}\right)}{\overset{3}{\cancel{12}}\left(\dfrac{3}{\cancel{4}}\right)-\overset{6}{\cancel{12}}\left(\dfrac{1}{\cancel{2}}\right)}$$

$$=\frac{8+9}{9-6}=\frac{17}{3}$$

EXAMPLE 2   Simplify $\dfrac{\dfrac{2}{x^2} - \dfrac{3}{x}}{\dfrac{x}{5}}$.

Solution: The LCD of the numerator of the complex fraction is x^2. The LCD of the denominator is 5. Therefore, the LCD of the complex fraction is $5x^2$. Multiply the numerator and denominator by $5x^2$.

$$\frac{\dfrac{2}{x^2} - \dfrac{3}{x}}{\dfrac{x}{5}} = \frac{5x^2\left(\dfrac{2}{x^2} - \dfrac{3}{x}\right)}{5x^2\left(\dfrac{x}{5}\right)} = \frac{5x^2\left(\dfrac{2}{x^2}\right) - 5x^2\left(\dfrac{3}{x}\right)}{5x^2\left(\dfrac{x}{5}\right)}$$

$$= \frac{10 - 15x}{x^3}$$

$$= \frac{5(2 - 3x)}{x^3}$$

EXAMPLE 3   Simplify $\dfrac{a + \dfrac{1}{b}}{b + \dfrac{1}{a}}$.

Solution: Multiply the numerator and denominator of the complex fraction by its LCD, ab.

$$\frac{a + \dfrac{1}{b}}{b + \dfrac{1}{a}} = \frac{ab\left(a + \dfrac{1}{b}\right)}{ab\left(b + \dfrac{1}{a}\right)} = \frac{a^2b + a}{ab^2 + b}$$

$$= \frac{a(ab + 1)}{b(ab + 1)} = \frac{a}{b}$$

EXAMPLE 4   Simplify $\dfrac{a^{-1} + ab^{-2}}{ab^{-2} - a^{-2}b^{-1}}$.

Solution: Rewrite each expression without negative exponents.

$$\frac{a^{-1} + ab^{-2}}{ab^{-2} - a^{-2}b^{-1}} = \frac{\dfrac{1}{a} + \dfrac{a}{b^2}}{\dfrac{a}{b^2} - \dfrac{1}{a^2b}}$$

The LCD of the numerator is ab^2. The LCD of the denominator is a^2b^2. The LCD of the complex fraction is a^2b^2.

$$= \frac{a^2b^2\left(\dfrac{1}{a} + \dfrac{a}{b^2}\right)}{a^2b^2\left(\dfrac{a}{b^2} - \dfrac{1}{a^2b}\right)} = \frac{ab^2 + a^3}{a^3 - b}$$

Method 2: Simplifying Numerator and Denominator

③ Complex fractions can also be simplified as follows:

> **To Simplify a Complex Fraction by Simplifying the Numerator and the Denominator**
> 1. Add or subtract each secondary fraction as indicated.
> 2. Invert and multiply the denominator of the complex fraction by the numerator of the complex fraction.
> 3. Simplify when possible.

Example 5 will show how Example 4 can be completed by simplifying the numerator and denominator.

EXAMPLE 5   Simplify $\dfrac{a^{-1} + ab^{-2}}{ab^{-2} - a^{-2}b^{-1}}$.

Solution: $\dfrac{a^{-1} + ab^{-2}}{ab^{-2} - a^{-2}b^{-1}} = \dfrac{\dfrac{1}{a} + \dfrac{a}{b^2}}{\dfrac{a}{b^2} - \dfrac{1}{a^2b}}$

Now add the fractions in the numerator and subtract the fractions in the denominator. The lowest common denominator of the numerator of the complex fraction is ab^2. The LCD of the denominator of the complex fraction is a^2b^2.

$$\frac{\dfrac{b^2}{b^2} \cdot \dfrac{1}{a} + \dfrac{a}{b^2} \cdot \dfrac{a}{a}}{\dfrac{a^2}{a^2} \cdot \dfrac{a}{b^2} - \dfrac{1}{a^2b} \cdot \dfrac{b}{b}} = \frac{\dfrac{b^2}{ab^2} + \dfrac{a^2}{ab^2}}{\dfrac{a^3}{a^2b^2} - \dfrac{b}{a^2b^2}} = \frac{\dfrac{a^2 + b^2}{ab^2}}{\dfrac{a^3 - b}{a^2b^2}}$$

Now invert the denominator of the complex fraction and multiply it by the numerator.

$$\frac{a^2 + b^2}{ab^2} \cdot \frac{a^2b^2}{a^3 - b} = \frac{a(a^2 + b^2)}{a^3 - b} \quad \text{or} \quad \frac{a^3 + ab^2}{a^3 - b}$$

Compare Examples 4 and 5 and note that we obtain the same answer.

When doing the exercises, unless a particular method is specified, you may use either.

Exercise Set 7.3

Simplify.

1. $\dfrac{1 + \dfrac{3}{5}}{2 + \dfrac{1}{5}}$ $\dfrac{8}{11}$

2. $\dfrac{1 - \dfrac{9}{16}}{3 + \dfrac{4}{5}}$ $\dfrac{35}{304}$

3. $\dfrac{2 + \dfrac{3}{8}}{1 + \dfrac{1}{3}}$ $\dfrac{57}{32}$

4. $\dfrac{\dfrac{3}{5} + \dfrac{2}{7}}{\dfrac{1}{5} + \dfrac{5}{6}}$ $\dfrac{6}{7}$

5. $\dfrac{\dfrac{4}{9} - \dfrac{3}{8}}{4 - \dfrac{3}{5}}$ $\dfrac{25}{1224}$

6. $\dfrac{1 - \dfrac{x}{y}}{x}$ $\dfrac{y - x}{xy}$

7. $\dfrac{\dfrac{x^2 y}{4}}{\dfrac{2}{x}}$ $\dfrac{x^3 y}{8}$

8. $\dfrac{\dfrac{15a}{b^2}}{\dfrac{b^3}{5}}$ $\dfrac{75a}{b^5}$

9. $\dfrac{\dfrac{8x^2 y}{3z^3}}{\dfrac{4xy}{9z^5}}$ $6xz^2$

10. $\dfrac{\dfrac{36x^4}{5y^4 z^5}}{\dfrac{9xy^2}{15z^5}}$ $\dfrac{12x^3}{y^6}$

11. $\dfrac{x + \dfrac{1}{y}}{\dfrac{x}{y}}$ $\dfrac{xy + 1}{x}$

12. $\dfrac{x - \dfrac{x}{y}}{\dfrac{1 + x}{y}}$ $\dfrac{x(y - 1)}{1 + x}$

13. $\dfrac{\dfrac{9}{x} + \dfrac{3}{x^2}}{3 + \dfrac{1}{x}}$ $\dfrac{3}{x}$

14. $\dfrac{\dfrac{2}{a} + \dfrac{1}{2a}}{a + \dfrac{a}{2}}$ $\dfrac{5}{3a^2}$

15. $\dfrac{3 - \dfrac{1}{y}}{2 - \dfrac{1}{y}}$ $\dfrac{3y - 1}{2y - 1}$

16. $\dfrac{\dfrac{x}{y} - \dfrac{y}{x}}{\dfrac{x + y}{x}}$ $\dfrac{x - y}{y}$

17. $\dfrac{\dfrac{a^2}{b} - b}{\dfrac{b^2}{a} - a}$ $\dfrac{-a}{b}$

18. $\dfrac{\dfrac{1}{x} + \dfrac{2}{x^2}}{2 + \dfrac{1}{x^2}}$ $\dfrac{x + 2}{2x^2 + 1}$

19. $\dfrac{\dfrac{a}{b} - 2}{\dfrac{-a}{b} + 2}$ -1

20. $\dfrac{\dfrac{x^2 - y^2}{x}}{\dfrac{x + y}{x^3}}$ $x^2(x - y)$

21. $\dfrac{\dfrac{4x + 8}{3x^2}}{\dfrac{4x}{6}}$ $\dfrac{2(x + 2)}{x^3}$

22. $\dfrac{\dfrac{a}{a + 1} - 1}{\dfrac{2a + 1}{a - 1}}$ $\dfrac{-a + 1}{(a + 1)(2a + 1)}$

23. $\dfrac{\dfrac{x}{4} - \dfrac{1}{x}}{1 + \dfrac{x + 4}{x}}$ $\dfrac{x - 2}{8}$

24. $\dfrac{1 + \dfrac{x}{x + 1}}{\dfrac{2x + 1}{x - 1}}$ $\dfrac{x - 1}{x + 1}$

25. $\dfrac{\dfrac{1}{x - 1} + 1}{\dfrac{1}{x + 1} - 1}$ $\dfrac{x + 1}{1 - x}$

26. $\dfrac{\dfrac{a + 1}{a - 1} + \dfrac{a - 1}{a + 1}}{\dfrac{a + 1}{a - 1} - \dfrac{a - 1}{a + 1}}$ $\dfrac{a^2 + 1}{2a}$

27. $\dfrac{\dfrac{a - 2}{a + 2} - \dfrac{a + 2}{a - 2}}{\dfrac{a - 2}{a + 2} + \dfrac{a + 2}{a - 2}}$ $\dfrac{-4a}{a^2 + 4}$

28. $\dfrac{\dfrac{5}{5 - x} + \dfrac{6}{x - 5}}{\dfrac{3}{x} + \dfrac{2}{x - 5}}$ $\dfrac{x}{5(x - 3)}$

29. $\dfrac{\dfrac{2}{m} + \dfrac{1}{m^2} + \dfrac{3}{m - 1}}{\dfrac{2}{m - 1}}$ $\dfrac{5m^2 - m - 1}{2m^2}$

30. $\dfrac{\dfrac{3}{x^2} - \dfrac{1}{x} + \dfrac{2}{x - 2}}{\dfrac{1}{x}}$ $\dfrac{x^2 + 5x - 6}{x(x - 2)}$

Simplify.

31. $2a^{-2} + b$ $\dfrac{2 + a^2 b}{a^2}$

32. $3a^{-2} + b^{-1}$ $\dfrac{3b + a^2}{a^2 b}$

33. $(a^{-1} + b^{-1})^{-1}$ $\dfrac{ab}{b + a}$

34. $\dfrac{a^{-1} + b^{-1}}{\dfrac{1}{ab}}$ $a + b$

35. $\dfrac{a^{-1} + 1}{b^{-1} - 1}$ $\dfrac{b(1 + a)}{a(1 - b)}$

36. $\dfrac{\dfrac{a}{b} + a^{-1}}{\dfrac{b}{a} + a^{-1}}$ $\dfrac{a^2 + b}{b(b + 1)}$

37. $\dfrac{x^{-1} - y^{-1}}{x^{-1} + y^{-1}}$ $\dfrac{y - x}{x + y}$

38. $\dfrac{x^{-2} + \dfrac{1}{x}}{x^{-1} + x^{-2}}$ 1

39. $\dfrac{a^{-1} + b^{-1}}{(a + b)^{-1}} \qquad \dfrac{(a + b)^2}{ab}$ **40.** $\dfrac{3a^{-1} - b^{-1}}{(a - b)^{-1}} \qquad \dfrac{(3b - a)(a - b)}{ab}$ **41.** $2x^{-1} - (3y)^{-1} \qquad \dfrac{6y - x}{3xy}$ **42.** $\dfrac{\dfrac{5}{x} + \dfrac{1}{y}}{(x - y)^{-1}} \qquad \dfrac{(5y + x)(x - y)}{xy}$

43. $\dfrac{\dfrac{2}{xy} - \dfrac{3}{y} + \dfrac{5}{x}}{3x^{-1} - 2y^{-2}} \qquad \dfrac{2y - 3xy + 5y^2}{3y^2 - 2x}$

44. $\dfrac{4m^{-1} + 3n^{-1} + (2mn)^{-1}}{\dfrac{5}{m} + \dfrac{3}{n}} \qquad \dfrac{8n + 6m + 1}{10n + 6m}$

45. The efficiency of a jack, E, is given by the formula

$$E = \dfrac{\dfrac{1}{2}h}{h + \dfrac{1}{2}}$$

where h is determined by the pitch of the jack's thread.

Pitch

Determine the efficiency of a jack whose value of h is:

(a) $\dfrac{2}{3} \quad \dfrac{2}{7}$ **(b)** $\dfrac{4}{5} \quad \dfrac{4}{13}$

46. If two resistors with resistances R_1 and R_2 are connected in parallel, their combined resistance, R_T, can be found from the formula

$$R_T = \dfrac{1}{\dfrac{1}{R_1} + \dfrac{1}{R_2}}$$

Simplify the right side of the formula. $\dfrac{R_1 R_2}{R_1 + R_2}$

47. If three resistors with resistances R_1, R_2, and R_3 are connected in parallel, their combined resistance can be found by the following formula:

$$R_T = \dfrac{1}{\dfrac{1}{R_1} + \dfrac{1}{R_2} + \dfrac{1}{R_3}} \qquad R_T = \dfrac{R_1 R_2 R_3}{R_2 R_3 + R_1 R_3 + R_1 R_2}$$

Simplify the right side of this formula.

48. A formula used in the study of optics is

$$f = (p^{-1} + q^{-1})^{-1}$$

where p is the object's distance from a lens, q is the image distance from the lens, and f is the focal length of the lens. Express the right side of the formula without any negative exponents. $pq/(p + q)$

49. What is a complex fraction?

50. We have indicated two procedures for evaluating complex fractions. Which procedure do you prefer? Briefly explain why.

49. one that has a fractional expression in its numerator or denominator or both

CUMULATIVE REVIEW EXERCISES

[1.4] **51.** Evaluate $\dfrac{\left|-\dfrac{3}{9}\right| - \left(-\dfrac{5}{9}\right) \cdot \left|-\dfrac{3}{8}\right|}{|-5 - (-3)|} \qquad \dfrac{13}{48}$

[2.6] **52.** Find the solution set to the inequality

$\left|\dfrac{4 - 2x}{3}\right| \geq 3$. $\left\{x \,\middle|\, x \leq -\dfrac{5}{2} \text{ or } x \geq \dfrac{13}{2}\right\}$

[3.6] **53.** Graph the inequality $6y - 3x < 12$.

[5.5] **54.** Divide using synthetic division:

$x^3 - 7x^2 - 13x + 9 \div (x - 2) \qquad x^2 - 5x - 23 - \dfrac{37}{x - 2}$

53.

Group Activity/ Challenge Problems

Simplify.

1. $\dfrac{\dfrac{3}{(x-2)^2} + \dfrac{1}{(x-2)}}{\dfrac{2}{(x-2)^2}}$ $\dfrac{x+1}{2}$

2. $\dfrac{\dfrac{5}{x^2-4} + \dfrac{3}{x-2} - \dfrac{2}{x+2}}{\dfrac{x}{x^2-4}}$ $\dfrac{x+15}{x}$

3. $\dfrac{\dfrac{3}{x^2-x-6} + \dfrac{4}{x^2+6x+8}}{\dfrac{5}{x^2+x-12}}$ $\dfrac{7x}{5(x+2)}$

4. $\dfrac{1}{2a + \dfrac{1}{2a + \dfrac{1}{2a}}}$ $\dfrac{4a^2+1}{4a(2a^2+1)}$

5. $\dfrac{1}{x + \dfrac{1}{x + \dfrac{1}{x+1}}}$ $\dfrac{x^2+x+1}{x^3+x^2+2x+1}$

6. $\dfrac{1}{2 + \dfrac{1}{2 + \dfrac{1}{2}}}$ $\dfrac{5}{12}$

7.4 Solving Rational Equations

Tape 11

1. Solve equations containing rational expressions.
2. Check solutions.
3. Solve proportions by cross multiplication.
4. Interpret solutions of equations in one variable graphically.
5. Solve application problems using rational expressions.
6. Solve for a variable in a formula containing rational expressions.

Solving Rational Equations

1. In Sections 7.1 through 7.3 we presented techniques to add, subtract, multiply, and divide rational expressions. In this section we present a method for solving rational equations. A **rational equation** is an equation that contains at least one rational expression.

> ### To Solve Rational Equations
> 1. Determine the LCD of all rational expressions in the equation.
> 2. Multiply **both** sides of the equation by the LCD. This will result in every term in the equation being multiplied by the LCD.
> 3. Remove any parentheses and combine like terms on each side of the equation.
> 4. Solve the equation using the properties discussed in earlier sections.
> 5. Check the solution in the original equation.

In step 2, we multiply both sides of the equation by the LCD to eliminate fractions. In some examples we will not show the check to save space.

EXAMPLE 1   Solve $\dfrac{x}{4} + \dfrac{1}{2} = \dfrac{x-1}{2}$.

Solution: Multiply both sides of the equation by the LCD, 4. Then use the distributive property, which results in each term in the equation being multiplied by the LCD.

$$4\left(\frac{x}{4} + \frac{1}{2}\right) = \frac{x-1}{2} \cdot 4$$

$$4\left(\frac{x}{4}\right) + 4\left(\frac{1}{2}\right) = 2(x-1)$$

$$x + 2 = 2x - 2$$

$$2 = x - 2$$

$$4 = x$$

Checking Solutions   2   **Whenever a variable appears in any denominator, you must check your apparent solution in the original equation. When checking, if an apparent solution makes any denominator equal to zero, that value is not a solution to the equation.** Such values are called **extraneous roots** or **extraneous solutions.** An extraneous root is a number obtained when solving an equation that is not a solution to the original equation.

EXAMPLE 2   Solve $2 - \dfrac{4}{x} = \dfrac{1}{3}$.

Solution: Multiply both sides of the equation by the LCD, $3x$.

$$3x\left(2 - \frac{4}{x}\right) = \left(\frac{1}{3}\right) \cdot 3x$$

$$3x(2) - 3x\left(\frac{4}{x}\right) = \left(\frac{1}{3}\right)3x$$

$$6x - 12 = x$$

$$5x - 12 = 0$$

$$5x = 12$$

$$x = \frac{12}{5}$$

Check: $2 - \dfrac{4}{x} = \dfrac{1}{3}$

$x = \dfrac{12}{5}$ $2 - \dfrac{4}{(12/5)} = \dfrac{1}{3}$

$2 - \dfrac{20}{12} = \dfrac{1}{3}$

$\dfrac{1}{3} = \dfrac{1}{3}$ true

Example 3   Solve $x + \dfrac{12}{x} = -7$.

Solution:

$$x \cdot \left(x + \frac{12}{x}\right) = -7 \cdot x \qquad \text{Multiply both sides of the equation by the LCD, } x.$$

$$x(x) + x\left(\frac{12}{x}\right) = -7x$$

$$x^2 + 12 = -7x$$

$$x^2 + 7x + 12 = 0$$

$$(x + 3)(x + 4) = 0$$

$$x + 3 = 0 \quad \text{or} \quad x + 4 = 0$$

$$x = -3 \qquad\qquad x = -4$$

Checks of -3 and -4 will show that they are solutions to the equation.

Example 4   Solve $\dfrac{2x}{x^2 - 4} + \dfrac{1}{x - 2} = \dfrac{2}{x + 2}$.

Solution: First factor the denominator $x^2 - 4$, then find the LCD.

$$\frac{2x}{(x + 2)(x - 2)} + \frac{1}{x - 2} = \frac{2}{x + 2}$$

The LCD is $(x + 2)(x - 2)$. Multiply both sides of the equation by the LCD, and then use the distributive property. This process will eliminate the fractions from the equation.

$$(x + 2)(x - 2) \cdot \left[\frac{2x}{(x + 2)(x - 2)} + \frac{1}{x - 2}\right] = \frac{2}{x + 2} \cdot (x + 2)(x - 2)$$

$$(x + 2)(x - 2) \cdot \frac{2x}{(x + 2)(x - 2)} + (x + 2)(x - 2) \cdot \frac{1}{x - 2} = \frac{2}{x + 2} \cdot (x + 2)(x - 2)$$

$$2x + (x + 2) = 2(x - 2)$$

$$3x + 2 = 2x - 4$$

$$x + 2 = -4$$

$$x = -6$$

A check will show that -6 is the solution.

Example 5   Solve the equation $\dfrac{22}{2p^2 - 9p - 5} - \dfrac{3}{2p + 1} = \dfrac{2}{p - 5}$.

Solution: Factor the denominator, then determine the LCD.

$$\frac{22}{(2p + 1)(p - 5)} - \frac{3}{2p + 1} = \frac{2}{p - 5}$$

Multiply both sides of the equation by the LCD, $(2p + 1)(p - 5)$.

$$(2p+1)(p-5) \cdot \frac{22}{(2p+1)(p-5)} - (2p+1)(p-5) \cdot \frac{3}{2p+1} = (2p+1)(p-5) \cdot \frac{2}{p-5}$$

$$22 - 3(p - 5) = 2(2p + 1)$$
$$22 - 3p + 15 = 4p + 2$$
$$37 - 3p = 4p + 2$$
$$35 = 7p$$
$$5 = p$$

The solution appears to be 5. However, since a variable appears in a denominator, this solution must be checked.

Check: $p = 5$

$$\frac{22}{2p^2 - 9p - 5} - \frac{3}{2p + 1} = \frac{2}{p - 5}$$

$$\frac{22}{2(5)^2 - 9(5) - 5} - \frac{3}{2(5) + 1} = \frac{2}{5 - 5}$$

$$\text{undefined} \longrightarrow \frac{22}{0} - \frac{3}{11} = \frac{2}{0} \longleftarrow \text{undefined}$$

Since 5 makes a denominator 0 and division by 0 is undefined, 5 is an extraneous solution. Therefore, you should write **"no solution"** as your answer.

Helpful Hint

Remember, whenever you solve an equation, where a variable appears in any denominator you must check the apparent solution to make sure it is not an extraneous solution. If the apparent solution makes any denominator 0, then it is an extraneous solution and not a true solution to the equation.

Solving Proportions ③ Proportions, equations of the form $\frac{a}{b} = \frac{c}{d}$, were introduced in Section 2.1. Proportions are one type of rational equation. Recall that to solve a proportion we often use *cross multiplication*.

$$\text{If } \frac{a}{b} = \frac{c}{d}, \text{ then } ad = bc, \quad b \neq 0, \quad d \neq 0$$

When we perform cross multiplication we are actually multiplying both sides of the equation by the lowest common denominators of the fractions in the equation. In the proportion $\dfrac{a}{b} = \dfrac{c}{d}$, the least common denominator is bd.

$$\frac{a}{b} = \frac{c}{d}$$

$$bd \cdot \frac{a}{b} = \frac{c}{d} \cdot bd$$

$$ad = bc$$

Notice we obtain the same results. Sometimes we cross multiply as a first step in solving a proportion to save a step in the process.

When solving a proportion where the denominator of one or more of the ratios contains a variable, you must check to make sure that your solution is not extraneous.

Proportions are often used when working with similar figures. **Similar figures** are figures whose corresponding angles are equal and whose corresponding sides are in proportion. Figure 7.4 illustrates two sets of similar figures.

In Fig. 7.4a, the ratio of the length of side AB to the length of side BC is the same as the ratio of the length of side $A'B'$ to the length of side $B'C'$. That is,

$$\frac{AB}{BC} = \frac{A'B'}{B'C'}$$

In a pair of similar figures, if the length of a side is unknown, it can often be found by using proportions, as illustrated in Example 6.

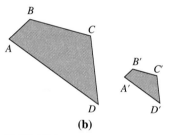

(a)

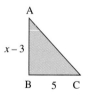

(b)

FIGURE 7.4

EXAMPLE 6 Triangles ABC and $A'B'C'$ in Fig. 7.5 are similar figures. Find the length of sides AB and $B'C'$.

Solution: We can set up a proportion and then solve for x. Then we can find the lengths.

$$\frac{AB}{BC} = \frac{A'B'}{B'C'}$$

$$\frac{x-3}{5} = \frac{8}{x}$$

$$x(x-3) = 8 \cdot 5$$

$$x^2 - 3x = 40$$

$$x^2 - 3x - 40 = 0$$

$$(x-8)(x+5) = 0$$

$$x - 8 = 0 \quad \text{or} \quad x + 5 = 0$$

$$x = 8 \qquad\qquad x = -5$$

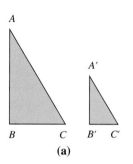

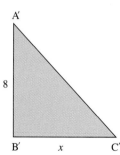

FIGURE 7.5

Since the length of the side of a triangle cannot be a negative number, -5 is not a possible answer. Substituting 8 for x, we see that the length of side $B'C'$ is 8 and the length of side AB is $8 - 3$ or 5.

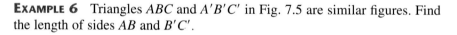

Check: $\dfrac{AB}{BC} = \dfrac{A'B'}{B'C'}$

$\dfrac{5}{5} = \dfrac{8}{8}$

$1 = 1$ **true**

In Example 7 that follows we begin solving the proportion by multiplying both sides of the equation by the LCD, rather than showing the cross multiplication first.

EXAMPLE 7   Solve $\dfrac{x^2}{x-4} = \dfrac{16}{x-4}$.

Solution: This equation is a proportion. We will solve this equation by multiplying both sides of the equation by the LCD, $x - 4$.

$$(x-4) \cdot \dfrac{x^2}{x-4} = \dfrac{16}{x-4} \cdot (x-4)$$

$$x^2 = 16$$

$$x^2 - 16 = 0 \qquad \textbf{Factor the difference of squares.}$$

$$(x + 4)(x - 4) = 0$$

$$x + 4 = 0 \quad \text{or} \quad x - 4 = 0$$

$$x = -4 \qquad\qquad x = 4$$

Check:

$$x = -4 \qquad\qquad\qquad x = 4$$

$$\dfrac{x^2}{x-4} = \dfrac{16}{x-4} \qquad\qquad \dfrac{x^2}{x-4} = \dfrac{16}{x-4}$$

$$\dfrac{(-4)^2}{-4-4} = \dfrac{16}{-4-4} \qquad\qquad \dfrac{(4)^2}{4-4} = \dfrac{16}{4-4}$$

$$\dfrac{16}{-8} = \dfrac{16}{-8} \qquad\qquad \dfrac{16}{0} = \dfrac{16}{0} \longleftarrow \textbf{undefined}$$

$$-2 = -2 \qquad \text{true}$$

Since $x = 4$ results in a denominator of 0, 4 is *not* a solution to the equation. It is an extraneous root. The only solution to the equation is -4.

In Example 7, what would you obtain if you began by cross multiplying? Solve Example 7 now by beginning with cross multiplication.

Graphical Interpretation of Solutions to Rational Equations

4   In Example 1 on page 409 we solved the equation $\dfrac{x}{4} + \dfrac{1}{2} = \dfrac{x-1}{2}$ and found the solution to be $x = 4$. Now we will discuss what the solution to an equation means from a graphical view point. We will use the same procedure used in Example 13 on page 214. You may wish to review that example and the Graphing Calculator Corner that follows that example now.

We will consider each side of the equation $\dfrac{x}{4} + \dfrac{1}{2} = \dfrac{x-1}{2}$ as separate functions. The functions are

$$y = \frac{x}{4} + \frac{1}{2} \quad \text{and} \quad y = \frac{x-1}{2}.$$

We can consider these two equations as a system of equations. If we graph these two equations on the same axis the graphs intersect at the point whose coordinates are $\left(4, \frac{3}{2}\right)$ (see Fig. 7.6). The x coordinate, the 4, is the solution to the equation $\dfrac{x}{4} + \dfrac{1}{2} = \dfrac{x-1}{2}$. When $x = 4$, both sides of the equation have the same value, $\frac{3}{2}$. The solution, 4, is the same solution obtained in Example 1.

$y = \dfrac{x}{4} + \dfrac{1}{2}$ $\qquad\qquad$ $y = \dfrac{x-1}{2}$

x	y
0	$\dfrac{1}{2}$
2	1
4	$\dfrac{3}{2}$
6	2

x	y
0	$-\dfrac{1}{2}$
1	0
4	$\dfrac{3}{2}$
5	2

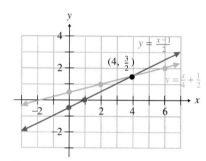

FIGURE 7.6

EXERCISES

Study objective 4, then solve the following equations graphically by setting each side of the equation equal to y *and finding the* x *coordinate of the intersection of the system of equations. Estimate the solutions to the nearest tenth.*

1. $\dfrac{x}{6} + \dfrac{1}{3} = \dfrac{x}{9}$ -6.0

2. $\dfrac{1.5x}{5} = \dfrac{10.2 - x}{8}$ 3.0

3. $\dfrac{2}{3}x - 3 = \dfrac{1}{3}x - 1$ 6.0

4. $\dfrac{4x - 1.2}{0.8} = \dfrac{6x - 3}{1.4}$ -0.9

Applications   **5**   Now let us look at some applications of rational equations.

EXAMPLE 8   A formula frequently used in optics is

$$\frac{1}{p} + \frac{1}{q} = \frac{1}{f}$$

where p represents the distance of the object from a mirror (or lens), q represents the distance of the image from the mirror (or lens), and f represents the focal length of the mirror (or lens). If the focal length of a curved mirror is 10 centimeters, how far from the mirror will the image appear when the object is 30 centimeters from the mirror?

Solution: The object distance, p, is 30 centimeters and the focal length, f, is 10 centimeters. We are asked to find the image distance, q.

$$\frac{1}{p} + \frac{1}{q} = \frac{1}{f}$$

$$\frac{1}{30} + \frac{1}{q} = \frac{1}{10}$$

Multiply both sides of the equation by the LCD, $30q$.

$$30q\left(\frac{1}{30} + \frac{1}{q}\right) = 30q\left(\frac{1}{10}\right)$$

$$\cancel{30}q\left(\frac{1}{\cancel{30}}\right) + 30\cancel{q}\left(\frac{1}{\cancel{q}}\right) = \overset{3}{\cancel{30}}q\left(\frac{1}{\cancel{10}}\right)$$

$$q + 30 = 3q$$

$$30 = 2q$$

$$15 = q$$

Thus, the image will appear 15 centimeters from the mirror.

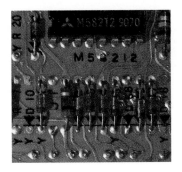

EXAMPLE 9   In electronics the total resistance, R_T, of resistors connected in a parallel circuit is determined by the formula

$$\frac{1}{R_T} = \frac{1}{R_1} + \frac{1}{R_2} + \frac{1}{R_3} + \cdots + \frac{1}{R_n}$$

where $R_1, R_2, R_3, \ldots, R_n$ are the resistances of the individual resistors (measured in ohms) in the circuit. Find the total resistance if two resistors, one of 200 ohms and the other of 300 ohms, are connected in a parallel circuit.

Solution: Since there are only two resistances, we use the formula

$$\frac{1}{R_T} = \frac{1}{R_1} + \frac{1}{R_2}$$

Let $R_1 = 200$ ohms and $R_2 = 300$ ohms; then

$$\frac{1}{R_T} = \frac{1}{200} + \frac{1}{300}$$

Multiply both sides of the equation by the LCD, $600R_T$.

$$600R_T \cdot \frac{1}{R_T} = 600R_T\left(\frac{1}{200} + \frac{1}{300}\right)$$

$$600\cancel{R_T} \cdot \frac{1}{\cancel{R_T}} = \overset{3}{\cancel{600}}R_T\left(\frac{1}{\cancel{200}}\right) + \overset{2}{\cancel{600}}R_T\left(\frac{1}{\cancel{300}}\right)$$

$$600 = 3R_T + 2R_T$$

$$600 = 5R_T$$

$$R_T = \frac{600}{5} = 120$$

Thus, the total resistance of the parallel circuit is 120 ohms. Notice that the resistors actually have less resistance when connected in a parallel circuit than separately.

---

6 Now we will do an example where we solve for a variable in a formula that contains rational expressions.

EXAMPLE 10   In Example 8 we used the formula $\frac{1}{p} + \frac{1}{q} = \frac{1}{f}$. Solve this formula for f.

Solution: Our goal is to isolate the variable f. We begin by multiplying both sides of the equation by the least common denominator, pqf, to eliminate fractions.

$$\frac{1}{p} + \frac{1}{q} = \frac{1}{f}$$

$$pqf\left(\frac{1}{p} + \frac{1}{q}\right) = pqf\left(\frac{1}{f}\right)$$

$$pqf\left(\frac{1}{p}\right) + pqf\left(\frac{1}{q}\right) = pqf\left(\frac{1}{f}\right)$$

$$qf + pf = pq$$

$$f(q + p) = pq$$

$$\frac{f\cancel{(q + p)}}{\cancel{(q + p)}} = \frac{pq}{q + p}$$

$$f = \frac{pq}{q + p} \quad \text{or} \quad f = \frac{pq}{p + q}$$

Exercise Set 7.4

Solve each equation and check your solution.

1. $\dfrac{2}{5} = \dfrac{x}{10}$ 4

2. $\dfrac{3}{k} = \dfrac{9}{6}$ 2

3. $\dfrac{x}{8} = \dfrac{-15}{4}$ −30

4. $\dfrac{a}{25} = \dfrac{12}{10}$ 30

5. $\dfrac{9}{3b} = \dfrac{-6}{2}$ -1

6. $\dfrac{1}{4} = \dfrac{z+1}{8}$ 1

7. $\dfrac{4x+5}{6} = \dfrac{7}{2}$ 4

8. $\dfrac{a}{5} = \dfrac{a-3}{2}$ 5

9. $\dfrac{6x+7}{10} = \dfrac{2x+9}{6}$ 3

10. $\dfrac{n}{10} = 9 - \dfrac{n}{5}$ 30

11. $\dfrac{x}{3} - \dfrac{3x}{4} = -\dfrac{5x}{12}$ all real numbers

12. $\dfrac{2}{8} + \dfrac{3}{4} = \dfrac{w}{5}$ 5

13. $\dfrac{3}{4} - x = 2x$ $\dfrac{1}{4}$

14. $\dfrac{2}{y} + \dfrac{1}{2} = \dfrac{5}{2y}$ 1

15. $\dfrac{5}{3x} + \dfrac{3}{x} = 1$ $\dfrac{14}{3}$

16. $\dfrac{x}{4} - \dfrac{x}{6} = \dfrac{1}{4}$ 3

17. $\dfrac{x-1}{x-5} = \dfrac{4}{x-5}$ no solution

18. $\dfrac{2x+3}{x+1} = \dfrac{3}{2}$ -3

19. $\dfrac{5y-2}{7} = \dfrac{15y-2}{28}$ $\dfrac{6}{5}$

20. $\dfrac{2}{x+1} = \dfrac{1}{x-2}$ 5

21. $\dfrac{5.6}{-x-6.2} = \dfrac{2}{x}$ ≈ -1.63

22. $\dfrac{4.5}{y-3} = \dfrac{6.9}{y+3}$ 14.25

23. $\dfrac{x-2}{x+4} = \dfrac{x+1}{x+10}$ 8

24. $\dfrac{x-3}{x+1} = \dfrac{x-6}{x+5}$ $\dfrac{9}{7}$

25. $x - \dfrac{4}{3x} = -\dfrac{1}{3}$ $-\dfrac{4}{3}, 1$

26. $\dfrac{b}{2} - \dfrac{4}{b} = -\dfrac{7}{2}$ $1, -8$

27. $\dfrac{2x-1}{3} - \dfrac{x}{4} = \dfrac{7.4}{6}$ 3.76

28. $x + \dfrac{3}{x} = \dfrac{12}{x}$ $3, -3$

29. $x + \dfrac{6}{x} = -5$ $-2, -3$

30. $\dfrac{15}{x} + \dfrac{9x-7}{x+2} = 9$ 3

31. $\dfrac{3y-2}{y+1} = 4 - \dfrac{y+2}{y-1}$ 4

32. $\dfrac{2b}{b+1} = 2 - \dfrac{5}{2b}$ -5

33. $\dfrac{1}{x+3} + \dfrac{1}{x-3} = \dfrac{-5}{x^2-9}$ $-\dfrac{5}{2}$

34. $c - \dfrac{c}{3} + \dfrac{c}{5} = 26$ 30

35. $\dfrac{2}{x-3} - \dfrac{4}{x+3} = \dfrac{8}{x^2-9}$ 5

36. $\dfrac{2}{w-5} = \dfrac{22}{2w^2-9w-5} - \dfrac{3}{2w+1}$ no solution

37. $\dfrac{y}{2y+2} + \dfrac{2y-16}{4y+4} = \dfrac{2y-3}{y+1}$ no solution

38. $\dfrac{3}{x+3} + \dfrac{5}{x+4} = \dfrac{12x+19}{x^2+7x+12}$ 2

39. $\dfrac{1}{x+2} + \dfrac{1}{x-2} = \dfrac{4}{x^2-4}$ no solution

40. $\dfrac{4r-1}{r^2+5r-14} = \dfrac{1}{r-2} - \dfrac{2}{r+7}$ $\dfrac{12}{5}$

41. $\dfrac{5}{x^2+4x+3} + \dfrac{2}{x^2+x-6} = \dfrac{3}{x^2-x-2}$ $\dfrac{17}{4}$

42. $\dfrac{2}{x^2+2x-8} - \dfrac{1}{x^2+9x+20} = \dfrac{4}{x^2+3x-10}$ $-\dfrac{4}{3}$

For each pair of similar figures, find the length of the two unknown sides (that is, those two sides indicated with the variable x).

43. 12, 2

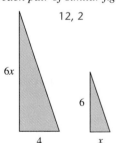

44. 10, 7

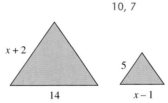

45. 12, 4

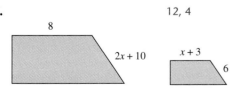

46. 30, 4

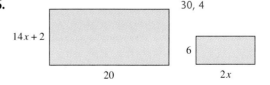

Solve the formula for the variable indicated.

47. $d = \dfrac{fl}{f+w}$, for w (physics) $w = \dfrac{fl-df}{d}$

48. $d = \dfrac{fl}{f+w}$, for f (physics) $f = \dfrac{dw}{l-d}$

49. $\dfrac{1}{p} + \dfrac{1}{q} = \dfrac{1}{f}$, for p (optics) $p = \dfrac{qf}{q-f}$

50. $\dfrac{1}{R_T} = \dfrac{1}{R_1} + \dfrac{1}{R_2}$, for R_T (electronics) $R_T = \dfrac{R_1 R_2}{R_1 + R_2}$

51. $\dfrac{1}{R_T} = \dfrac{1}{R_1} + \dfrac{1}{R_2}$, for R_1 (electronics) $R_1 = \dfrac{R_T R_2}{R_2 - R_T}$

52. $\dfrac{1}{p} + \dfrac{1}{q} = \dfrac{1}{f}$, for q (optics) $q = \dfrac{pf}{p - f}$

53. $z = \dfrac{\bar{x} - \mu}{\dfrac{\sigma}{\sqrt{n}}}$, for $\bar{x}$ (statistics) $\bar{x} = \mu + \dfrac{z\sigma}{\sqrt{n}}$

54. $z = \dfrac{\bar{x} - \mu}{\dfrac{\sigma}{\sqrt{n}}}$, for μ (statistics) $\mu = \bar{x} - \dfrac{z\sigma}{\sqrt{n}}$

55. $E = \dfrac{q}{\epsilon_0 A} - \dfrac{q'}{\epsilon_0 A}$, for q (physics) $q = q' + E\epsilon_0 A$

56. $S_f - S_i = -\dfrac{Q}{T_1} + \dfrac{Q}{T_2}$, for Q (chemistry) $Q = \dfrac{S_i T_1 T_2 - S_f T_1 T_2}{T_1 - T_2}$

57. $\dfrac{1}{C_T} = \dfrac{1}{C_1} + \dfrac{1}{C_2} + \dfrac{1}{C_3}$, for C_T (electronics)

58. $\dfrac{n_1}{o} + \dfrac{n_2}{i} = \dfrac{n_1 - n_2}{r}$, for n_2 (optics) $n_2 = \dfrac{n_1 oi - n_1 ir}{or + oi}$

Simplify the expression in (a) and solve the equation in (b).

57. $C_T = \dfrac{C_1 C_2 C_3}{C_2 C_3 + C_1 C_3 + C_1 C_2}$

59. (a) $\dfrac{2}{x - 2} + \dfrac{3}{x^2 - 4}$ **(b)** $\dfrac{2}{x - 2} + \dfrac{3}{x^2 - 4} = 0$ **(a)** $\dfrac{2x + 7}{(x - 2)(x + 2)}$ **(b)** $-\dfrac{7}{2}$

60. (a) $\dfrac{4}{x - 3} + \dfrac{5}{2x - 6} + \dfrac{1}{2}$ **(b)** $\dfrac{4}{x - 3} + \dfrac{5}{2x - 6} = \dfrac{1}{2}$ **(a)** $\dfrac{x + 10}{2(x - 3)}$ **(b)** 16

61. (a) $\dfrac{b + 3}{b} - \dfrac{b + 4}{b + 5} - \dfrac{15}{b^2 + 5b}$ **(b)** $\dfrac{b + 3}{b} - \dfrac{b + 4}{b + 5} = \dfrac{15}{b^2 + 5b}$ **(a)** $\dfrac{4}{b + 5}$ **(b)** no solution

62. (a) $\dfrac{4x + 3}{x^2 + 11x + 30} - \dfrac{3}{x + 6} + \dfrac{2}{x + 5}$ **(b)** $\dfrac{4x + 3}{x^2 + 11x + 30} - \dfrac{3}{x + 6} = \dfrac{2}{x + 5}$ **(a)** $\dfrac{3x}{(x + 6)(x + 5)}$ **(b)** -24

Use the formula $\dfrac{1}{p} + \dfrac{1}{q} = \dfrac{1}{f}$ *for Exercises 63 through 68 (see Example 8).* **67.** Object: 16 cm, image: 48 cm

63. Find the distance of the image from the mirror if the object is 12 inches from the mirror and the focal length is 6 inches. 12 in.

64. If the object distance is 12 inches and the image distance is 20 inches, find the focal length of the mirror. 7.5 in

65. If the focal length of the mirror is 8 centimeters and the image distance from the mirror is 15 centimeters, find the object's distance from the mirror. 17.14 cm

69. 155.6 ohms **70.** 300 ohms

66. If the object distance is 15 centimeters and the image distance is 9 centimeters, find the focal length of the mirror. 5.625 cm

67. The focal length of a mirror is 12 centimeters. Find the object's distance and image distance if the image distance is 3 times the object distance.

68. The focal length of a mirror is 2 inches. Find the object's distance and image distance if the object's distance is 3 inches more than the image distance. 3 in, 6 in

Refer to Example 9 for Exercises 69 through 72.

73. a number obtained when solving an equation that is not a true solution to the original equation

69. What is the total resistance in the circuit if resistors of 200 ohms and 700 ohms are connected in parallel?

70. What is the total resistance in the circuit if resistors of 500 ohms and 750 ohms are connected in parallel?

71. What is the total resistance in the circuit if resistors of 300 ohms, 500 ohms, and 3000 ohms are connected in parallel? 176.47 ohms

72. Three resistors of identical resistance are to be connected in parallel. What should be the resistance of each resistor if the circuit is to have a total resistance of 700 ohms? 2100 ohms

73. What is an extraneous root?

74. Under what circumstances is it necessary to check your answers for extraneous roots?

74. when there is a variable in any denominator

75. Consider the equation $\dfrac{x}{4} - \dfrac{x}{3} = 2$ and the expression $\dfrac{x}{4} - \dfrac{x}{3} + 2$.

(a) What is the first step in solving the equation? Explain what effect the first step will have on the equation.

(b) Solve the equation. -24.

(c) What is the first step in simplifying the expression? Explain what effect the first step has when simplifying the expression.

(d) Simplify the expression. $\dfrac{-x + 24}{12}$.

75. (a) Multiply both sides of the equation by the LCD, 12. This removes fractions from equation.

(c) Write each term with common denominator 12. Then fractions can be added and subtracted.

76. figures whose corresponding angles are the same and whose corresponding sides are in proportion

76. What are similar figures?

77. Explain why you **cannot** rewrite the expression
$$\frac{4}{x-2} - \frac{3}{x+4} \text{ as either } \frac{4}{x-2} - \frac{3}{x+4} = 0$$
or $\dfrac{4}{x-2} = \dfrac{3}{x+4}$.

77. You cannot just change an expression to an equation by adding 0.

78. (a) Explain in your own words how to solve a rational equation.

(b) Solve $\dfrac{3}{x-4} + \dfrac{1}{x+4} = \dfrac{4}{x^2-16}$ following your procedure in part (a). -1

80. D: $\{x \mid x \geq 4\}$
R: $\{y \mid y \geq 2\}$

CUMULATIVE REVIEW EXERCISES

[3.4] **79.** $f(x) = \frac{1}{2}x^2 - 3x + 4$ find $f(5)$. $\frac{3}{2}$

[3.5] **80.** Graph the function $f(x) = \sqrt{x-4} + 2$ and give the domain and range.

[5.6] **81.** Graph the function $f(x) = x^2 - 4x - 6$ and give the domain and range.

D: $\mathbb{R}$
R: $\{y \mid y \geq -10\}$

[6.3] **82.** Factor $8x^3 - 64y^6$. $(2x - 4y^2)(4x^2 + 8xy^2 + 16y^4)$

---

1. There are many answers. One answer is $\dfrac{1}{x-4} + \dfrac{1}{x+2} = 0$. **2.** There are many answers. One answer is $\dfrac{x}{2} + \dfrac{x}{2} = x$.

3. There are many answers. One answer is $\dfrac{1}{x} + \dfrac{1}{x} = \dfrac{2}{x}$.

Group Activity/ Challenge Problems

1. Make up an equation that cannot have 4 or -2 as a solution. Explain how you determined your answer.

2. Make up an equation containing the sum of two rational expressions in the variable x whose solution is the set of *real numbers*. Explain how you determined your answer.

3. Make up an equation in the variable x containing the sum of two rational expressions whose solution is the set of real numbers except 0. Explain how you determined your answer.

Solve. **6. (b)** Tax-Free Money Market portfolio since 9.71% > 7.68%

4. $5p^{-1} + 2p^{-2} - 3 = 0$ $-\frac{1}{3}, 2$ **5.** $x(x-2)^{-1} + 4 = 2(x-2)^{-1}$ no solution

6. In Example 5 on page 64 we discussed the formula for converting a tax-free investment, T_f, to a taxable investment, T_a. The formula we used was

$$T_a = \frac{T_f}{1-f}$$

In the formula, f represents your federal tax bracket. Some investments, such as certain municipal bonds and municipal bond funds, are not only federally tax free but are also state and county or city tax free. When you wish to compare a taxable investment with an investment which is federal, state, and county tax free, you must replace f in the previous formula with your combined *true tax bracket,* which is $[f + (s + c)(1 - f)]$, where s is your state tax bracket and c is your county or local tax bracket. If you are investing in a federal, state, and county tax-free investment, the formula above then becomes

$$T_a = \frac{T_f}{1 - [f + (s + c)(1 - f)]}$$

Mr. Levy, who lives in Detroit, Michigan, is in a 4.6% state tax bracket, a 3% city tax bracket, and a 33% federal tax bracket. He is choosing between the Fidelity Michigan Triple *Tax-Free* Money Market Portfolio yielding 6.01% and the Fidelity *Taxable* Cash Reserve Money Market Fund yielding 7.68%.

(a) Using his tax brackets, determine the taxable equivalent of the 6.01% tax free yield. 9.71%

(b) Which investment should Mr. Levy make? Explain your answer.

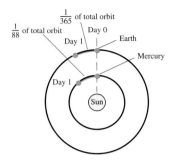

7. The synodic period of Mercury is the time required for swiftly moving Mercury to gain one lap on Earth in their orbits around the sun. If the orbital periods (in Earth days) of the two planets are designated P_m and P_e, Mercury will be seen on the average to move $1/P_m$ of a revolution per day, while Earth moves $1/P_e$ of a revolution per day in pursuit. Mercury's daily gain on Earth is $(1/P_m) - (1/P_e)$ of a revolution, so that the time for Mercury to gain one complete revolution on Earth, the synodic period s, may be found by the formula

$$\frac{1}{s} = \frac{1}{P_m} - \frac{1}{P_e}$$

If $P_e = 365$ days and P_m is 88 days, find the synodic period in Earth days. 115.96 days

7.5 Applications of Rational Equations

1 Solve work problems.
2 Solve number problems.
3 Solve motion problems.

Tape 11

Some applications of equations with rational expressions were illustrated in Section 7.4. In this section we examine some additional applications. We study work problems first.

Work Problems

1 Problems where two or more machines or people work together to complete a certain task are sometimes referred to as **work problems.** Work problems often involve rational equations. Generally, work problems are based on the fact that the part of the work done by person 1 (or machine 1) plus the part of the work done by person 2 (or machine 2) is equal to the total amount of work done by both people (or both machines).

Part of task done by first person or machine	+	Part of task done by second person or machine	=	1 (one whole task completed)

When we do work problems, we designate the total task completed as 1 (for 1 whole task completed). To determine the part of the task done by each person or machine we use the formula

part of task completed = rate · time

This formula is very similar to the formula *amount = rate · time* that was discussed in Section 2.4.

Let us now discuss how to determine the rate. If, for example, John can do a particular task in 5 hours, he could complete 1/5 of the task in 1 hour. Thus, his rate is 1/5 of the task per hour. If Kishi can do a job in 6 hours, her rate is 1/6 of the job per hour. Similarly, if Maria can do a job in x minutes, her rate is $1/x$ of the job per minute. *In general, if a person or machine can complete a task in x units of time, the rate is $1/x$.*

EXAMPLE 1   Carmine can mow Mr. Polyo's lawn in 3 hours. Justin can mow Mr. Polyo's lawn in 4 hours. How long will it take to mow the lawn if both Carmine and Justin work together?

Solution: Let $x =$ time, in hours, for both boys together to mow the lawn. We construct a table to assist us in finding the part of the task completed by Carmine and Justin.

	Rate of work	Time worked	Part of task completed
Carmine	$\dfrac{1}{3}$	x	$\dfrac{x}{3}$
Justin	$\dfrac{1}{4}$	x	$\dfrac{x}{4}$

$$\left(\begin{array}{c}\text{part of lawn mowed}\\\text{by Carmine in } x \text{ hours}\end{array}\right) + \left(\begin{array}{c}\text{part of lawn mowed}\\\text{by Justin in } x \text{ hours}\end{array}\right) = 1 \text{ (entire lawn mowed)}$$

$$\frac{x}{3} \qquad + \qquad \frac{x}{4} \qquad = \qquad 1$$

Multiply both sides of the equation by the LCD, 12; then solve for x, the number of hours.

$$12\left(\frac{x}{3} + \frac{x}{4}\right) = 12 \cdot 1$$

$$12\left(\frac{x}{3}\right) + 12\left(\frac{x}{4}\right) = 12$$

$$4x + 3x = 12$$

$$7x = 12$$

$$x = \frac{12}{7} \quad \textbf{1.71 hours (to the nearest hundredth)}$$

The two boys together can mow the lawn in about 1.71 hours. This answer is reasonable because this time is less than it takes either boy to mow the lawn by himself.

EXAMPLE 2   Mr. and Mrs. Love just moved into a home that has a Jacuzzi bathtub. When they turn on the tap water to fill the tub, the water is cloudy from lack of use by the previous owner. They wish to run as much water through the tub as possible until the water clears. To accomplish this they turn on the cold water tap (to save energy, they don't use the hot water tap) and they open the drain of the tub. The cold water tap can fill the tub in 7.6 minutes and the drain can empty the tub in 10.3 minutes. If the drain is open and the cold water faucet is turned on, how long will it take before the water fills the tub and begins to overflow?

Solution: As the water from the faucet is filling the tub, water going down the drain is emptying the tub. Thus, the faucet and drain are working against each other. Let x = amount of time needed to fill the tub.

	Rate of work	Time worked	Part of tub filled or emptied
Faucet filling tub	$\dfrac{1}{7.6}$	x	$\dfrac{x}{7.6}$
Drain emptying tub	$\dfrac{1}{10.3}$	x	$\dfrac{x}{10.3}$

Since the faucet and drain are working against each other, we will *subtract* the part of the water being emptied from the part of water being added to the tub.

$$\left(\begin{array}{c}\text{part of tub filled}\\ \text{in } x \text{ hours}\end{array}\right) - \left(\begin{array}{c}\text{part of tub emptied}\\ \text{in } x \text{ hours}\end{array}\right) = 1 \text{ (whole tub filled)}$$

$$\frac{x}{7.6} \qquad - \qquad \frac{x}{10.3} \qquad = 1$$

Using a calculator, we can determine that the LCD is $(7.6)(10.3) = 78.28$. Now multiply both sides of the equation by 78.28 to remove fractions.

$$\overset{10.3}{\cancel{78.28}}\left(\frac{x}{\cancel{7.6}}\right) - \overset{7.6}{\cancel{78.28}}\left(\frac{x}{\cancel{10.3}}\right) = 78.28(1)$$

$$10.3x - 7.6x = 78.28$$

$$2.7x = 78.28$$

$$x \approx 28.99$$

Thus, the tub will fill in about 29 minutes.

EXAMPLE 3   Patty and Mike work in the sporting goods department at Sears, where they assemble bicycles. When Patty and Mike work together, they can assemble a bicycle in 20 minutes. When Patty assembles a bike by herself, it takes her 36 minutes. How long would it take Mike to assemble the bike by himself?

Solution: Let x = amount of time for Mike to assemble the bike by himself. We know that when working together they can assemble the bike in 20 minutes. We organize this information in the table that follows.

	Rate of work	Time worked	Part of bicycle completed
Patty	$\dfrac{1}{36}$	20	$\dfrac{20}{36}$
Mike	$\dfrac{1}{x}$	20	$\dfrac{20}{x}$

$$\left(\begin{array}{c}\text{part of bicycle} \\ \text{assembled by Patty}\end{array}\right) + \left(\begin{array}{c}\text{part of bicycle} \\ \text{assembled by Mike}\end{array}\right) = 1$$

$$\frac{20}{36} + \frac{20}{x} = 1$$

$$36x\left(\frac{20}{36} + \frac{20}{x}\right) = 36x \cdot 1$$

$$36x\left(\frac{20}{36}\right) + 36x\left(\frac{20}{x}\right) = 36x$$

$$20x + 720 = 36x$$

$$720 = 16x$$

$$45 = x$$

Thus, Mike can assemble a bike by himself in 45 minutes.

Number Problems

2 Now let us look at a problem where we must find a number.

EXAMPLE 4   What number multiplied by the numerator and added to the denominator of the fraction $\frac{4}{7}$ makes the resulting fraction equal to $\frac{5}{3}$?

Solution: Let x = unknown number.

$$\frac{4x}{7 + x} = \frac{5}{3} \qquad \text{\textbf{Now cross multiply.}}$$

$$3(4x) = 5(7 + x)$$

$$12x = 35 + 5x$$

$$7x = 35$$

$$x = 5$$

The number is 5.

Check: $\dfrac{4 \cdot 5}{7 + 5} = \dfrac{20}{12} = \dfrac{5}{3}.$

EXAMPLE 5   When the reciprocal of 3 times a number is subtracted from 1, the result is the reciprocal of twice the number. Find the number.

Solution: Let x = unknown number. Then $3x$ is 3 times the number, and $\dfrac{1}{3x}$ is the reciprocal of 3 times the number. Twice the number is $2x$, and $\dfrac{1}{2x}$ is the reciprocal of twice the number.

$$1 - \frac{1}{3x} = \frac{1}{2x} \qquad \text{The LCD is } 6x.$$

$$6x\left(1 - \frac{1}{3x}\right) = \frac{1}{2x} \cdot 6x$$

$$6x(1) - 6x\left(\frac{1}{3x}\right) = 6x\left(\frac{1}{2x}\right)$$

$$6x - 2 = 3$$

$$6x = 5$$

$$x = \frac{5}{6}$$

A check will show that the number is $\frac{5}{6}$.

Motion Problems   ❸ The last type of problem we will look at is motion problems. Recall that we discussed motion problems earlier in Section 2.4. In that section we learned that distance = rate · time. Sometimes it is convenient to solve for the time when solving motion problems.

$$\text{Time} = \frac{\text{distance}}{\text{rate}}$$

EXAMPLE 6   Amy Schumacher can fly her plane 300 miles against the wind in the same time it takes her to fly 400 miles with the wind. If the wind blows at 20 miles per hour, find the speed of the plane in still air.

Solution: Let x = speed of plane in still air. Let us set up a table to help analyze the problem.

Plane	Distance	Rate	Time
Against wind	300	$x - 20$	$\dfrac{300}{x - 20}$
With wind	400	$x + 20$	$\dfrac{400}{x + 20}$

Since the times are the same, we set up and solve the following equation:

$$\frac{300}{x - 20} = \frac{400}{x + 20}$$

$$300(x + 20) = 400(x - 20)$$

$$300x + 6000 = 400x - 8000$$

$$6000 = 100x - 8000$$

$$14{,}000 = 100x$$

$$140 = x$$

The speed of the plane in still air is 140 miles per hour.

EXAMPLE 7   Mary Kay rides her bike to and from her home to San Francisco City College. Going to school, she rides mostly downhill and averages 15 miles per hour. Coming home, mostly uphill, she averages only 6 miles per hour. If it takes her $\frac{1}{2}$ hour longer for her to get home than to ride to school, how far is the college from her home?

Solution: Let $x =$ the distance from her home to the college. Note that in this problem the times are not equal. Her time returning is $\frac{1}{2}$ hour longer than going. Therefore, to make the times equal, we must add $\frac{1}{2}$ hour to her time going (or subtract $\frac{1}{2}$ hour from her time returning).

	Distance	Rate	Time
Going	x	15	$\dfrac{x}{15}$
Returning	x	6	$\dfrac{x}{6}$

$$\text{time going} + \frac{1}{2} \text{ hour} = \text{time returning}$$

$$\frac{x}{15} + \frac{1}{2} = \frac{x}{6}$$

$$30\left(\frac{x}{15}\right) + 30\left(\frac{1}{2}\right) = 30\left(\frac{x}{6}\right)$$

$$2x + 15 = 5x$$

$$15 = 3x$$

$$5 = x$$

Therefore, Mary Kay lives 5 miles from San Francisco City College.

EXAMPLE 8   The number 4 train in the New York City subway system goes from Woodlawn/Jerome Avenue in the Bronx to Flatbush Avenue/Brooklyn College in Brooklyn. The one-way distance between these two stops is 24.2 miles. On this route, two tracks run parallel to each other, one for the local train and the other for the express train. The local train stops at every station (48 stations), while the express stops at only certain stations (33 stations). The local and express trains leave Woodlawn/Jerome Avenue at the same time. When the express reaches the end of the line at Flatbush Avenue/Brooklyn College, the local is at Wall Street, 7.8 miles from Flatbush. If the express averages 5.2 miles per hour faster than the local, find the speed of the two trains.

Solution:

$$\text{Let } x = \text{speed of local}$$
$$\text{then } x + 5.2 = \text{speed of express}$$

In the same time that the express reaches the end of the line, 24.2 miles, the local will have traveled $24.2 - 7.8 = 16.4$ miles.

Train	Distance	Rate	Time
Local	16.4	x	$\dfrac{16.4}{x}$
Express	24.2	$x + 5.2$	$\dfrac{24.2}{x + 5.2}$

$$\frac{16.4}{x} = \frac{24.2}{x + 5.2}$$

$$16.4(x + 5.2) = 24.2x$$

$$16.4x + 85.28 = 24.2x$$

$$85.28 = 7.8x$$

$$10.9 \approx x$$

The local averages 10.9 miles per hour and the express averages $10.9 + 5.2 = 16.1$ miles per hour.

EXAMPLE 9 Dawn, who lives in Buffalo, New York, travels to college in South Bend, Indiana. The speed limit on some of the expressways is 55 miles per hour, while on others it is 65 miles per hour. The total distance traveled by Dawn is 490 miles. If Dawn follows the speed limits, and the total trip takes 8 hours, how long did she drive at 55 miles per hour, and how long did she drive at 65 miles per hour?

Solution: Let x = number of miles driven at 55 mph
then $490 - x$ = number of miles driven at 65 mph

Speed Limit	Distance	Rate	Time
55 mph	x	55	$\dfrac{x}{55}$
65 mph	$490 - x$	65	$\dfrac{490 - x}{65}$

Since the total time is 8 hours, we write

$$\frac{x}{55} + \frac{490 - x}{65} = 8$$

The LCD of 55 and 65 is 715.

$$715\left(\frac{x}{55} + \frac{490 - x}{65}\right) = 715 \cdot 8$$

$$715\left(\frac{x}{55}\right) + 715\left(\frac{490 - x}{65}\right) = 5720$$

$$13x + 11(490 - x) = 5720$$

$$13x + 5390 - 11x = 5720$$

$$2x + 5390 = 5720$$

$$2x = 330$$

$$x = 165$$

The number of miles driven at 55 mph is 165 miles. Then the time driven at 55 mph is 165/55 = 3 hours, and the time driven at 65 mph is (490 − 165) / 65 = 325 / 65 = 5 hours.

Notice that in Example 9 the answer to the question was not the value obtained for *x*. The value obtained was a distance, and the question asked us to find the time. *When working word problems, you must read and work the problems very carefully and make sure you answer the question that was asked.*

A number of examples discussed in this section can be solved in other ways. In this section our emphasis is on solving rational equations, so we worked these examples using fractions. It is important for you to realize that problems can often be solved in more than one way. Try solving Examples 6 and 9 in a different way now.

Exercise Set 7.5

Solve each problem.

1. At the Boeing Corporation it takes one computer 4 hours to print checks for its employees and a second computer 5 hours to complete the same job. How long will it take the two computers together to complete the job? $\dfrac{x}{4} + \dfrac{x}{5} = 1$; $\dfrac{20}{9}$ or 2.22 hr

2. Ramon can mow a lawn on a rider lawn mower in 4 hours. Donna can mow the lawn in 6 hours with a push lawn mower. How long will it take them to mow the lawn together? $\dfrac{x}{4} + \dfrac{x}{6} = 1$; $\dfrac{12}{5}$ or 2.4 hr

3. A $\frac{1}{2}$-inch-diameter hose can fill a swimming pool in 8 hours. A $\frac{4}{5}$-inch-diameter hose can fill the same pool in 5 hours. How long will it take to fill the pool when both hoses are used? $\dfrac{x}{8} + \dfrac{x}{5} = 1$; $\dfrac{40}{13}$ or 3.08 hr

4. A conveyor belt operating at full speed can fill a tank with topsoil in 3 hours. When a valve at the bottom of the tank is opened, the tank will empty in 4 hours. If the conveyor belt is operating at full speed and the valve at the bottom of the tank is open, how long will it take to fill the tank? $\dfrac{x}{3} - \dfrac{x}{4} = 1$; 12 hr

5. A factory making antifreeze has vats to hold the antifreeze. Each vat has an inlet valve and an outlet valve. The vat can be filled with antifreeze in 20 hours when the inlet valve is wide open and the outlet valve is closed. That vat can be emptied in 25 hours when the outlet valve is wide open and the inlet valve is closed. If a new vat is placed in operation and both the inlet valve and outlet valve are wide open, how long will it take to fill the vat? $\dfrac{x}{20} - \dfrac{x}{25} = 1$; 100 hr

6. When Mrs. Dellaquila rides the power lawn mower and Mr. Dellaquila uses the push lawn mower, they can mow their large lawn together in 2 hours. Mrs. Dellaquila can mow the entire lawn by herself on the rider lawn mower in 3 hours. How long would it take Mr. Dellaquila to mow the entire lawn by himself using the push mower? $\dfrac{2}{3} + \dfrac{2}{x} = 1$; 6 hr

7. Dr. Indiana Jones and his father, Dr. Henry Jones, are archeologists working on a dig near the Forum in Rome. Indiana and his father working together can unearth a specific plot of land in 2.6 months. Indiana can unearth the entire area by himself in 3.9 months. How long would it take the father to unearth the entire area by himself? $\dfrac{2.6}{3.9} + \dfrac{2.6}{x} = 1$; 7.8 mo

See Exercise 7.

9. $\dfrac{3.2}{5.7} + \dfrac{3.2}{x} = 1$; 7.30 hr   **12.** $\dfrac{x}{6} + \dfrac{x}{5} + \dfrac{x}{4} = 1$; $\dfrac{60}{37}$ or 1.62 hr   **13.** $\dfrac{12}{x} + \dfrac{12}{2x} = 1$; 18 hr

8. Franki can plant a garden by herself in 4 hours. When her young son Garyn helps her, the total time it takes them to plant the garden is 3 hours. How long would it take Garyn to plant the garden himself? $\dfrac{3}{4} + \dfrac{3}{x} = 1$; 12 hr

9. When a professor and a graduate student work together, they can mark a set of mathematics final exams in 3.2 hours. When the professor marks the papers alone, it takes 5.7 hours. How long would it take the graduate student to mark the papers alone?

10. When only the cold water valve is opened, a washtub will fill in 8 minutes. When only the hot water valve is opened, the washtub will fill in 12 minutes. When the drain of the washtub is open, it will drain completely in 7 minutes. If both the hot and cold water valves are open and the drain is open, how long will it take for the washtub to fill? $\dfrac{x}{8} + \dfrac{x}{12} - \dfrac{x}{7} = 1$; $\dfrac{168}{11}$ or 15.27 min

11. A large tank is being used on the Donovan's farm to irrigate the crops. The tank has two inlet pipes and one outlet pipe. The two inlet pipes can fill the tank in 10 and 12 hours, respectively. The outlet pipe can empty the tank in 15 hours. If the tank is empty, how long would it take to fill the tank when all three valves are open? $\dfrac{x}{10} + \dfrac{x}{12} - \dfrac{x}{15} = 1$; $\dfrac{60}{7}$ or 8.57 hr

12. A fire department uses 3 pumps to remove water from flooded basements. One pump can remove all the water from a flooded basement in 6 hours. The second pump can remove the same amount of water in 5 hours, and the third pump requires only 4 hours to remove the water. If all 3 pumps work together to remove the water from the flooded basement, how long will it take to empty the basement?

13. It takes Lee twice as long as Nancy to knit an afghan. If together they knit an afghan in 12 hours, how long would it take Nancy to knit the afghan by herself?

14. A roofer requires 15 hours to put a new roof on a house. Anna, the roofer's apprentice, can reroof the house by herself in 20 hours. After working alone on a roof for 6 hours, the roofer leaves for another job. Anna takes over and completes the job. How long will it take Anna to complete the job? $\dfrac{6}{15} + \dfrac{x}{20} = 1$; 12 hr

15. Two pipes are used to fill an oil tanker. When the larger pipe is used alone, it takes 60 hours to fill the tanker. When the smaller pipe is used alone, it takes 80 hours to fill the tanker. If the large pipe begins filling the tanker, and after 20 hours the large pipe is closed down and the smaller pipe is opened, how much longer will it take to finish filling the tanker using only the smaller pipe? $\dfrac{20}{60} + \dfrac{x}{80} = 1$; $\dfrac{160}{3}$ or 53.33 hr

16. What number multiplied by the numerator and added to the denominator of the fraction $\frac{4}{3}$ makes the resulting fraction $\frac{5}{2}$? $\dfrac{4 \cdot x}{3 + x} = \dfrac{5}{2}$; 5

17. What number added to the numerator and multiplied by the denominator of the fraction $\frac{3}{2}$ makes the resulting fraction $\frac{1}{8}$? $\dfrac{3 + x}{2 \cdot x} = \dfrac{1}{8}$; -4

18. One number is twice another. The sum of their reciprocals is $\frac{3}{4}$. Find the numbers. $\dfrac{1}{x} + \dfrac{1}{2x} = \dfrac{3}{4}$; 2, 4

19. The sum of the reciprocals of two consecutive integers is $\frac{11}{30}$. Find the two integers. $\dfrac{1}{x} + \dfrac{1}{x + 1} = \dfrac{11}{30}$; 5, 6

20. The sum of the reciprocals of two consecutive even integers is $\frac{5}{12}$. Find the two integers.

21. When a number is added to both the numerator and denominator of the fraction $\frac{5}{7}$, the resulting fraction is $\frac{4}{5}$. Find the number added. $\dfrac{5 + x}{7 + x} = \dfrac{4}{5}$; 3

22. When 3 is added to twice the reciprocal of a number, the sum is $\frac{31}{10}$. Find the number. $\dfrac{2}{x} + 3 = \dfrac{31}{10}$; 20

23. The reciprocal of 3 less than a certain number is twice the reciprocal of 6 less than twice the number. Find the number(s).

24. If 3 times a number is added to twice the reciprocal of the number, the answer is 5. Find the number(s).

25. If 3 times the reciprocal of a number is subtracted from twice the reciprocal of the square of the number, the difference is -1. Find the number(s).

26. A Greyhound bus can travel 400 kilometers in the same time that an Amtrak train can travel 600 kilometers. If the speed of the train is 40 kilometers per hour greater than that of the bus, find the speeds of the bus and train. $\dfrac{400}{x} = \dfrac{600}{x + 40}$; 80 kph, 120 kph

27. The speed of a boat in still water is 20 miles per hour. It takes the same amount of time for the boat to travel 3 miles downstream (with the current) as it does to travel 2 miles upstream (against the current). Find the speed of the current. $\dfrac{3}{20 + x} = \dfrac{2}{20 - x}$; 4 mph

20. $\dfrac{1}{x} + \dfrac{1}{x + 2} = \dfrac{5}{12}$; 4, 6   **23.** $\dfrac{1}{x - 3} = 2\left(\dfrac{1}{2x - 6}\right)$; all real numbers except 3   **24.** $3x + \dfrac{2}{x} = 5$; $\dfrac{2}{3}$ or 1   **25.** $2\left(\dfrac{1}{x^2}\right) - 3\left(\dfrac{1}{x}\right) = -1$; 1 or 2

28. The rate of a bicyclist is 8 miles per hour faster than that of a jogger. If the bicyclist travels 10 miles in the same amount of time that the jogger travels 5 miles, find the rate of the jogger. $\dfrac{10}{x+8} = \dfrac{5}{x};$ 8 mph

29. Two cross-country skiers ski along the same path. One skier averages 8 miles per hour, while the other averages 6 miles per hour. If it takes the slower skier $\frac{1}{2}$ hour longer than the faster skier to reach the designated resting point, how far is the resting point from the starting point? $\dfrac{x}{8} + \dfrac{1}{2} = \dfrac{x}{6};$ 12 mi

30. The current of a river is 3 miles per hour. It takes a motorboat 3 hours to travel 12 miles upstream and return 12 miles downstream. What is the speed of the boat in still water? $\dfrac{12}{x+3} + \dfrac{12}{x-3} = 3;$ 9 mph

31. Ray starts out on a boating trip at 8 A.M. Ray's boat can go 20 miles per hour in still water.

  (a) How far downstream can Ray go if the current is 5 miles per hour and he wishes to go down and back in 4 hours? $\dfrac{x}{25} + \dfrac{x}{15} = 4;$ $\dfrac{300}{8}$ or 37.5 mi

  (b) At what time must Ray turn back? 9:30 AM

32. Anne drove in one day from Front Royal, Virginia, to Ashville, NC, along the scenic Skyline drive and Blue Ridge Parkway, a distance of 492 miles. For part of the trip she drove at a steady rate of 50 miles per hour, but in some of the more scenic areas she drove at a steady rate of 35 miles per hour. If the total time of the trip was 11.13 hours, how far did she travel at each speed?

33. $\dfrac{5.4}{4.2x} + \dfrac{2.3}{x} = 1.5;$ ≈ 2.39 mph

33. Each morning, Ron takes his horse, Beauty, for a walk on Pfeiffer Beach in Big Sur, California. He typically rides Beauty for 5.4 miles, then walks Beauty 2.3 miles. His speed when riding is 4.2 times his speed when walking. If his total outing takes 1.5 hours, find the rate at which he walks Beauty.

34. A train and car leave from a railroad station at the same time headed for the State Fair on the other side of the state. The car averages 50 miles per hour and the train averages 70 miles per hour. If the train arrives at the fair 2 hours ahead of the car, find the distance from the railroad station to the State Fair.

See Exercise 33.

35. A train and a plane leave from Boston at the same time for a destination 900 miles away. If the speed of the plane is 5 times the speed of the train, and the plane arrives 12 hours before the train, find the speeds of the train and the plane.

36. Two brothers are long-distance swimmers. Jim averages 3.6 miles per hour and Pete averages 2.4 miles per hour. The brothers start swimming at the same time across Lake Mead to a point on the other side of the lake. If Jim arrives 0.2 hour ahead of Pete, find the distance they swam. (Of course each swimmer was accompanied by a boat for safety reasons.)

37. Two rockets are to be launched at the same time from NASA headquarters in Houston, Texas, and are to meet at a space station many miles from earth. The first rocket is to travel at 20,000 miles per hour and the second rocket will travel at 18,000 miles per hour. If the first rocket is scheduled to reach the space station 0.6 hour before the second rocket, how far is the space station from NASA headquarters?

32. $\dfrac{x}{50} + \dfrac{492-x}{35} = 11.13;$ 341.5 mi at 50 mph, 150.5 mi at 35 mph   **34.** $\dfrac{x}{70} + 2 = \dfrac{x}{50};$ 350 mi   **35.** $\dfrac{900}{5x} + 12 = \dfrac{900}{x};$ train,

60 mi, plane, 300 mi   **36.** $\dfrac{x}{3.6} + 0.2 = \dfrac{x}{2.4};$ 1.44 mi   **37.** $\dfrac{x}{20,000} + 0.6 = \dfrac{x}{18,000};$ 108,000 mi   **38.** $-\dfrac{5}{2}x^2 + \dfrac{7}{5}xy - 6y^2$

39. $12x^3 - 34x^2 + 21x + 4$ **40.** $4x + 5 + \dfrac{22}{3x-2}$

CUMULATIVE REVIEW EXERCISES

[5.3] **38.** Subtract $\frac{1}{2}x^2 - 3x^2 + 2xy - (\frac{3}{5}xy + 6y^2)$.

39. Multiply $(4x^2 - 6x - 1)(3x - 4)$.

[5.4] **40.** Divide $(12x^2 + 7x + 12) \div (3x - 2)$.

[6.2] **41.** Factor $8x^2 + 26x + 15$. $(4x+3)(2x+5)$

Group Activity/ Challenge Problems

1. An officer flying a California Highway Patrol aircraft determines that a car 10 miles ahead of her is speeding at 90 miles per hour.

    (a) If the aircraft is traveling 450 miles per hour, how far will the car have traveled in the time it takes the aircraft to reach it? $\dfrac{x}{90} = \dfrac{x + 10}{450}$; 2.5 mi

    (b) How long, in minutes, will it take for the aircraft to reach the car? $0.02\overline{7}$ hr = $1.\overline{6}$ min

    (c) If the pilot wishes to reach the car in exactly 1 minute, how fast must the airplane fly?   690 mph

    (d) If this problem had not indicated specific speeds but stated that the plane's speed was 4 times the car's speed, could you have answered parts **(a)** and **(b)**? If your answer is yes, solve parts **(a)** and **(b)** using this new information. If your answer is no, explain why.   no

2. A trip up Mt. Pilatus, near Lucerne, Switzerland, involves riding an inclined railroad up to the top of the mountain, spending time at the top, then coming down the opposite side of the mountain in an aerial tram. A van or bus then returns you to your starting point 18 kilometers away, where your vehicle is parked.

    The distance traveled up the mountain is 7.5 kilometers and the distance traveled down the mountain is 8.7 kilometers. The speed coming down the mountain is 1.2 times the speed going up. It takes 36 minutes for the van to return you to your vehicle. If the Lieblichs stayed at the top of the mountain for 3 hours and the total time of their outing (from when they started going up the mountain until they were returned to their car) took 9 hours, find the speed, in kilometers per hour, of the inclined railroad. $\dfrac{7.5}{r} + \dfrac{8.7}{1.2r} + 3 + 0.6 = 9.0$; $2.73\overline{148}$ km/hr

7.6 Variation

Tape 12

1. Solve direct variation problems.
2. Solve inverse variation problems.
3. Solve joint variation problems.
4. Solve combination of variation problems.

In Sections 7.4 and 7.5 we saw many applications of equations containing rational expressions. In this section we see still more applications.

Direct Variation

1. Many scientific formulas are expressed as variations. A **variation** is an equation that relates one variable to one or more other variables using the operations of multiplication or division (or both operations). There are essentially three types of variation problems: direct, inverse, and joint variation.

In **direct variation** the two related variables will both increase together or both decrease together; that is, as one increases so does the other, and as one decreases so does the other.

Consider a car traveling at 30 miles an hour. The car travels 30 miles in 1 hour, 60 miles in 2 hours, and 90 miles in 3 hours. Notice that as the time increases, the distance traveled increases, and as the time decreases, the distance traveled decreases.

The formula used to calculate distance traveled is

$$\text{distance} = \text{rate} \cdot \text{time}$$

Since the rate is a constant, 30 miles per hour, the formula can be written

$$d = 30t$$

We say that distance varies directly as time or that distance is directly proportional to time.

The equation above is an example of a direct variation.

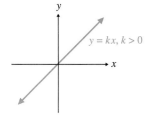

FIGURE 7.7

> **Direct Variation**
>
> If a variable y varies directly with a variable x, then
> $$y = kx$$
> where k is the constant of proportionality (or the variation constant).

The graph of $y = kx$, $k > 0$, is always a straight line that goes through the origin (see Fig. 7.7). The slope of the line depends on the value of k. The greater the value of k, the greater the slope.

EXAMPLE 1   The circumference of a circle, C, is directly proportional to (or varies directly as) its radius, r. Write the equation for the circumference of a circle if the constant of proportionality, k, is 2π.

Solution:

$$C = kr \ (C \text{ varies directly as } r)$$
$$C = 2\pi r \ (\text{constant of proportionality is } 2\pi)$$

EXAMPLE 2 The resistance, R, of a wire varies directly as its length, L.
(a) Write this variation as an equation.
(b) Find the resistance (measured in ohms) of a 20-foot length of wire assuming that the constant of proportionality for the wire is 0.007.

Solution:

(a) $R = kL$
(b) $R = 0.007(20) = 0.14$
The resistance of the wire is 0.14 ohm.

In certain variation problems the constant of proportionality, k, may not be
known. In such cases it can often be found by substituting given values in the variation formula and solving for k.

EXAMPLE 3 The gravitational force of attraction, F, between an object and
the earth is directly proportional to the mass, m, of the object. If the force of
attraction is 640 when the object's mass is 20, find the constant of proportionality.

Solution:

$$F = km$$
$$640 = k20$$
$$\frac{640}{20} = \frac{20k}{20}$$
$$32 = k$$

Thus, the constant of proportionality is 32.

EXAMPLE 4 y varies directly as the square of z. If y is 80 when z is 20, find
y when z is 90.

Solution: Since y varies directly as the *square of z,* we begin with the formula $y = kz^2$. Since the constant of proportionality is not given, we must first
find k using the given information.

$$y = kz^2$$
$$80 = k(20)^2$$
$$80 = 400k$$
$$\frac{80}{400} = \frac{400k}{400}$$
$$0.2 = k$$

We now use $k = 0.2$ to find y when z is 90.

$$y = kz^2$$
$$y = 0.2(90)^2$$
$$y = 1620$$

Thus, when z equals 90, y equals 1620.

Inverse Variation

2   A second type of variation is **inverse variation.** When two quantities vary inversely, it means that as one quantity increases, the other quantity decreases, and vice versa.

To explain inverse variation, we use the formula, distance = rate · time. If we solve for time, we get time = distance/rate. Assume that the distance is fixed at 120 miles; then

$$\text{time} = \frac{120}{\text{rate}}$$

At 120 miles per hour it would take 1 hour to cover this distance. At 60 miles an hour, it would take 2 hours. At 30 miles an hour, it would take 4 hours. Note that as the rate (or speed) decreases the time increases, and vice versa.

The equation above can be written

$$t = \frac{120}{r}$$

This equation is an example of an inverse variation. The time and rate are inversely proportional. The constant of proportionality is 120.

> **Inverse Variation**
>
> If a variable y varies inversely with a variable x, then
>
> $$y = \frac{k}{x} \qquad (\text{or} \quad xy = k)$$
>
> where k is the constant of proportionality.

Two quantities vary inversely, or are inversely proportional, when as one quantity increases the other quantity decreases. The graph of $y = k/x$, for $k > 0$ and $x > 0$, will have the shape illustrated in Fig. 7.8. The graph of an inverse variation is not defined at $x = 0$ for 0 is not in the domain of the function $y = k/x$. We will discuss graphs of this type further in Chapter 10.

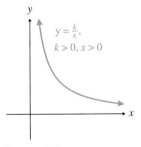

$y = \frac{k}{x}$,
$k > 0, x > 0$

FIGURE 7.8

EXAMPLE 5   The illuminance, I, of a light source varies inversely as the square of the distance, d, from the source. Assuming that the illuminance is 75 units at a distance of 6 meters, find the equation that expresses the relationship between the illuminance and the distance.

Solution: Since the illuminance varies inversely as the *square* of the distance the general form of the equation is

$$I = \frac{k}{d^2} \quad (\text{or} \quad Id^2 = k)$$

To find k, we substitute the given values for I and d.

$$75 = \frac{k}{6^2}$$

$$75 = \frac{k}{36}$$

$$(75)(36) = k$$

$$2700 = k$$

Thus, the formula is $I = \dfrac{2700}{d^2}$.

EXAMPLE 6 y varies inversely as x. If $y = 8$ when $x = 15$, find y when $x = 18$.

Solution: First write the inverse variation, then solve for k.

$$y = \frac{k}{x}$$

$$8 = \frac{k}{15}$$

$$120 = k$$

Now substitute 120 for k in $y = \dfrac{k}{x}$ and find y when $x = 18$.

$$y = \frac{120}{x} = \frac{120}{18} = 6.7 \qquad \text{(to the nearest tenth)}$$

Joint Variation ③ One quantity may vary directly as a product of two or more other quantities. This type of variation is called **joint variation.**

> **Joint Variation**
> The general form of a joint variation, where y varies directly as x and z, is
> $$y = kxz$$
> where k is the constant of proportionality.

EXAMPLE 7 The area, A, of a triangle varies jointly as its base, b, and height, h. If the area of a triangle is 48 square inches when its base is 12 inches and its height is 8 inches, find the area of a triangle whose base is 15 inches and height is 20 inches.

Solution: First write the joint variation; then substitute the known values and solve for *k*.

$$A = kbh$$
$$48 = k(12)(8)$$
$$48 = k(96)$$
$$\frac{48}{96} = k$$
$$k = \frac{1}{2}$$

Now solve for the area of the given triangle.

$$A = kbh$$
$$= \frac{1}{2}(15)(20)$$
$$= 150 \text{ square inches}$$

Summary of Variations

Direct	Inverse	Joint
$y = kx$	$y = \dfrac{k}{x}$	$y = kxz$

Combined Variation

4   Often in real-life situations one variable varies as a combination of variables. The following examples illustrate the use of **combined variations.**

EXAMPLE 8   The owners of the Freeport Pretzel Shop find that their weekly sales of pretzels, *S*, varies directly with their advertising budget, *A*, and inversely with their pretzel price, *P*. When their advertising budget is $400, and the price is $1, they sell 6200 pretzels.
(a) Write an equation of variation expressing *S* in terms of *A* and *P*. Include the value of the constant.
(b) Find the expected sales if the advertising budget is $600 and the price is $1.20.

Solution:
(a) We begin with the equation

$$S = \frac{kA}{P}$$

Now find *k* using the known values.

$$6200 = \frac{k(400)}{1}$$
$$6200 = 400k$$
$$15.5 = k$$

Therefore, the equation for the sales of pretzels is $S = \dfrac{15.5A}{P}$.

(b) $S = \dfrac{15.5A}{P}$

$= \dfrac{15.5(600)}{1.20} = 7750$

They can expect to sell 7750 pretzels.

EXAMPLE 9 The electrostatic force, F, of repulsion between two positive electrical charges is jointly proportional to the two charges, q_1 and q_2, and inversely proportional to the square of the distance, d, between the two charges. Express F in terms of q_1, q_2, and d.

Solution: $F = \dfrac{kq_1q_2}{d^2}$

EXAMPLE 10 A varies jointly as B and C and inversely as the square of D. If $A = 1$ when $B = 9$, $C = 4$, and $D = 6$, find A when $B = 8$, $C = 12$, and $D = 5$.

Solution: $A = \dfrac{kBC}{D^2}$

We must first find the constant of proportionality, k, by substituting the known values for A, B, C, and D and solving for k.

$$1 = \frac{k(9)(4)}{6^2}$$

$$1 = \frac{36k}{36}$$

$$1 = k$$

Thus, the constant of proportionality equals 1. Now we find A for the corresponding values of B, C, and D.

$$A = \frac{(1)(8)(12)}{5^2} = \frac{96}{25} = 3.84$$

Exercise Set 7.6

Use your intuition to determine if the variation between the indicated quantities is direct or inverse.

1. The speed and distance traveled by a car in a specified time period.   direct

2. The distance between two cities on a map and the actual distance between the two cities.   direct

3. The diameter of a hose and volume of water coming from the hose.   direct

4. A weight and the force needed to lift that weight.   direct

5. The cubic-inch displacement in liters and the horsepower of the engine.   direct

6. The volume of a balloon and its radius.   direct

7. The light illuminating an object and the distance the light is from the object.   inverse

8. The length of a board and the force needed to break the board at the center.   inverse

16. $x = ky$; 72   **19.** $x = k/y$; 0.2   **24.** $T = kD^2/F$; 51.2   **32.** $F = kq_1q_2/d^2$; 672   **33.** $S = kIT^2$; 0.2   **36.** $I = k/d^2$; 31.25 footcandles

9. The shutter opening of a camera and the amount of sunlight that reaches the film.   direct

10. A person's weight (due to the earth's gravity) and his distance from the earth.   inverse

11. The number of pages a person can read in a fixed period of time and his reading speed.   direct

12. The time it takes an ice cube to melt in water and the temperature of the water.   inverse

13. The time needed to properly expose a film and the size of the opening of the camera lens.   inverse

14. The time to reach a certain point for a plane flying with the wind and the speed of the wind.   inverse

15. The number of calories eaten and the amount of exercise required to burn off those calories.   direct

*For Exercises 16 through 42, **(a)** write the variation and **(b)** find the quantity indicated.*

16. x varies directly as y. Find x when $y = 12$ and $k = 6$.

17. C varies directly as the square of Z. Find C when $Z = 9$ and $k = \frac{3}{4}$. $C = kZ^2$; 60.75

18. y varies directly as R. Find y when $R = 180$ and $k = 1.7$. $y = kR$; 306

19. x varies inversely as y. Find x when $y = 25$ and $k = 5$.

20. R varies inversely as W. Find R when $W = 160$ and $k = 8$. $R = k/W$; 1/20

21. L varies inversely as the square of P. Find L when $P = 4$ and $k = 100$. $L = k/P^2$; 6.25

22. A varies directly as B and inversely as C. Find A when $B = 12$, $C = 4$, and $k = 3$. $A = kB/C$; 9

23. A varies jointly as R_1 and R_2 and inversely as the square of L. Find A when $R_1 = 120$, $R_2 = 8$, $L = 5$, and $k = \frac{3}{2}$. $A = kR_1R_2/L^2$; 57.6

24. T varies directly as the square of D and inversely as F. Find T when $D = 8$, $F = 15$, and $k = 12$.

25. x varies directly as y. If x is 9 when y is 18, find x when y is 36. $x = ky$; 18

26. Z varies directly as W. If Z is 7 when W is 28, find Z when W is 140. $Z = kW$; 35

27. y varies directly as the square of R. If y is 5 when $R = 5$, find y when R is 10. $y = kR^2$; 20

28. S varies inversely as G. If S is 12 when G is 0.4, find S when G is 5. $S = k/G$; 0.96

29. C varies inversely as J. If C is 7 when J is 0.7, find C when J is 12. $C = k/J$; 0.41

30. x varies inversely as the square of P. If $x = 10$ when P is 6, find x when $P = 20$. $x = k/P^2$; 0.9

31. F varies jointly as M_1 and M_2 and inversely as d. If F is 20 when $M_1 = 5$, $M_2 = 10$, and $d = 0.2$, find F when $M_1 = 10$, $M_2 = 20$, and $d = 0.4$. $F = kM_1M_2/d$; 40

32. F varies jointly as q_1 and q_2 and inversely as the square of d. If F is 8 when $q_1 = 2$, $q_2 = 8$, and $d = 4$, find F when $q_1 = 28$, $q_2 = 12$, and $d = 2$.

33. S varies jointly as I and the square of T. If S is 8 when $I = 20$ and $T = 4$, find S when $I = 2$ and $T = 2$.

34. The volume of a gas, V, varies inversely as its pressure, P. If the volume, V, is 800 cc when the pressure is 200 millimeters (mm) of mercury, find the volume when the pressure is 25 mm of mercury. $V = k/P$; 6400 cc

35. The length a spring will stretch, S, varies directly with the force (or weight), F, attached to the spring. If a spring stretches 1.4 inches when 20 pounds is attached, how far will it stretch when 10 pounds is attached? $S = kF$; 0.7 in

36. The intensity, I, of light received at a source varies inversely as the square of the distance, d, from the source. If the light intensity is 20 footcandles at 15 feet, find the light intensity at 12 feet.

37. The weekly videotape rentals, R, at Busterblock Video vary directly with their advertising budget, A, and inversely with the daily rental price, P. When their advertising budget is $400 and the rental price is $2 per day, they rent 4600 tapes per week. How many tapes would they rent per week if they increased their advertising budget to $500 and raised their rental price to $2.50? $R = \dfrac{kA}{P}$; 4600

38. On earth the weight of an object varies directly with its mass. If an object with a weight of 256 pounds has a mass of 8 slugs, find the mass of an object weighing 120 pounds. $W = km$; 3.75 slugs

See Exercise 34.

39. The weight, W, of an object in the earth's atmosphere varies inversely with the square of the distance, d, between the object and the center of the earth. A 140-pound person standing on earth is approximately 4000 miles from the earth's center. Find the weight (or gravitational force of attraction) of this person at a distance 100 miles from the earth's surface.

40. The wattage rating of an appliance, W, varies jointly as the square of the current, I, and the resistance, R. If the wattage is 1 watt when the current is 0.1 ampere and the resistance is 100 ohms, find the wattage when the current is 0.4 ampere and the resistance is 250 ohms.

41. The electrical resistance of a wire, R, varies directly as its length, L, and inversely as its cross-sectional area, A. If the resistance of a wire is 0.2 ohm when the length is 200 feet and its cross-sectional area is 0.05 square inch, find the resistance of a wire whose length is 5000 feet with a cross-sectional area of 0.01 square inch. $R = kL/A$; 25 ohms

42. The number of phone calls between two cities during a given time period, N, varies directly as the populations p_1 and p_2 of the two cities and inversely to the distance, d, between them. If 100,000 calls are made between two cities 300 miles apart and the populations of the cities are 60,000 and 200,000, how many calls are made between two cities with populations of 125,000 and 175,000 that are 450 miles apart?

43. Write a paragraph explaining the various types of variations. Include in your discussion the terms direct, inverse, joint, and combined variation. Give your own example of each type of variation.

44. (a) If y varies directly as x and the constant of proportionality is 2, does x vary directly or inversely as y? Explain.   directly

   (b) Give the new constant of proportionality for x as a variation of y. $\frac{1}{2}$ or 0.5

45. (a) If y varies inversely as x and the constant of proportionality is 0.3, does x vary directly or inversely as y? Explain.   inversely

   (b) Give the new constant of proportionality for x as a variation of y.   stays 0.3

39. $W = k/d^2$; 133.25 lb   **40.** $W = kI^2R$; 40 watts   **42.** $N = \dfrac{Kp_1p_2}{d}$; 121,528

CUMULATIVE REVIEW EXERCISES

[3.3] **46.** Find the variable d if the line through the two given points is to have the given slope:
(5, 1) and $(-4, d)$, $m = \frac{2}{3}$. $d = -5$

[4.3] **47.** Mr. Wilcox, a salesman, earns a base weekly salary plus a commission on his sales. The first week in February, on sales of $5000, his total income was $550. The second week in February, on sales of $8000, his total income was $640. Find his weekly salary and his commission rate.

47. weekly salary, $400; commission rate, 3%

[4.6] **48.** Graph the inequality $|x - 2| < 4$ in the cartesian coordinate system.
[5.6]

49. Graph $f(x) = -x^3 + 4x - 6$.

48.   **49.**

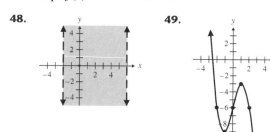

Group Activity/ Challenge Problems

1. An article in the magazine *Outdoor and Travel Photography* states, "If a surface is illuminated by a point-source of light, the intensity of illumination produced is inversely proportional to the square of the distance separating them. In practical terms, this means that foreground objects will be grossly overexposed if your background subject is properly exposed with a flash. Thus direct flash will not offer pleasing results if there are any intervening objects between the foreground and the subject."

   If the subject you are photographing is 4 feet from the flash, and the illumination on this subject is 1/16 of the light of the flash, what is the intensity of illumination on an intervening object that is 3 feet from the flash? $\frac{1}{9}$

2. In a specific region of the country, the amount of a customer's water bill, W, is directly proportional to the average daily temperature for the month, T, the lawn area, A, and the square root of F, where F is the family size, and inversely proportional to the number of inches of rain, R.

2. $W = \dfrac{kTA\sqrt{F}}{R}$; $124.92   **3. (b)** Force is 24 times greater than original force.

In one month, the average daily temperature is 78° and the number of inches of rain is 5.6. If the average family of 4 who have 1000 square feet of lawn pay $68 for water, estimate the water bill in the same month for the average family of six who have 1500 square feet of lawn.

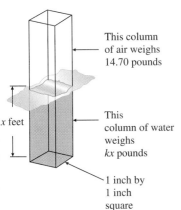

This column of air weighs 14.70 pounds

x feet

This column of water weighs kx pounds

1 inch by 1 inch square

3. One of Newton's laws states that the force of attraction, F, between two masses is directly proportional to the masses of the two objects, m_1 and m_2, and inversely proportional to the square of the distance, d, between the two masses.

   (a) Write the formula that represents Newton's law. $F = \dfrac{km_1m_2}{d^2}$

   (b) What happens to the force of attraction if one mass is doubled, the other mass is tripled, and the distance between the objects is halved?

4. The pressure, P, in pounds per square inch (psi) on an object x feet below the sea is 14.70 psi plus the product of a constant of proportionality, k, and the number of feet, x, the object is below sea level (see the figure at the left). The 14.70 represents the weight, in pounds, of the column of air (from sea level to the top of the atmosphere) standing over a 1 inch by 1 inch square of seawater. The kx represents the weight, in pounds, of a column of water 1 inch by 1 inch by x feet.

   (a) Write a formula for the pressure on an object x feet below sea level. $P = 14.7 + kx$

   (b) If the pressure gauge in a submarine 60 feet deep registers 40.5 psi, find the constant k.   0.43

   (c) A submarine is built to withstand a pressure of 160 psi. How deep can the submarine go? ≈ 337.9 ft

Summary

GLOSSARY

Combined variation (435): A variation problem that involves two or more types of variations.

Complex fraction (402): A fractional expression that has a fraction in its numerator or its denominator or both its numerator and denominator.

Constant of proportionality (431): The constant in a variation problem.

Domain of a rational expression (379): The set of values that can replace the variable in the expression.

Rational expression (378): An expression of the form p/q, where p and q are polynomials, $q \neq 0$.

Rational function (379): Functions of the form $f(x) = p/q$, where p and q are polynomials, $q \neq 0$.

Reduced to lowest terms (381): A rational expression is reduced to its lowest terms when the numerator and denominator have no common factor other than 1.

Variation (431): An equation that relates one variable to one or more other variables using the operations of multiplication or division (or both operations).

> **IMPORTANT FACTS**
>
> *Types of Variations*
>
> | Direct | Inverse | Joint |
> |--------|---------|-------|
> | $y = kx$ | $y = \dfrac{k}{x}$ | $y = kxz$ |

Review Exercises

[7.1] *Determine the values of the variable that must be excluded in the rational expression.*

1. $\dfrac{3}{x - 4}$ 4

2. $\dfrac{x}{x + 1}$ -1

3. $\dfrac{-2x}{x^2 + 5}$ none

Determine the domain of the rational function.

4. $y = \dfrac{0}{(x + 3)^2}$
$\{x \mid x \text{ is a real number and } x \neq -3\}$

5. $f(x) = \dfrac{x + 6}{x^2}$
$\{x \mid x \text{ is a real number and } x \neq 0\}$

6. $f(x) = \dfrac{x^2 - 2}{x^2 - 3x - 10}$
$\{x \mid x \text{ is a real number and } x \neq 5 \text{ and } x \neq -2\}$

Write each expression in reduced form.

7. $\dfrac{x^2 + xy}{x + y}$ x

8. $\dfrac{x^2 - 9}{x + 3}$ $x - 3$

9. $\dfrac{4 - 5x}{5x - 4}$ -1

10. $\dfrac{x^2 + 2x - 3}{x^2 + x - 6}$ $\dfrac{x - 1}{x - 2}$

11. $\dfrac{2x^2 - 6x + 5x - 15}{2x^2 + 7x + 5}$ $\dfrac{x - 3}{x + 1}$

12. $\dfrac{a^3 - 8}{a^2 - 4}$ $\dfrac{a^2 + 2a + 4}{a + 2}$

13. $x(x + 1)$ **14.** $(x + y)(x - y)$ **15.** $(x + 7)(x - 5)(x + 2)$ **16.** $(x + 2)^2(x - 2)(x + 1)$

[7.2] *Find the least common denominator.*

13. $\dfrac{6x}{x + 1} - \dfrac{3}{x}$

14. $\dfrac{9x - 3}{x + y} - \dfrac{4x + 7}{x^2 - y^2}$

15. $\dfrac{19x - 5}{x^2 + 2x - 35} + \dfrac{3x - 2}{x^2 + 9x + 14}$

16. $\dfrac{3}{(x + 2)^2} - \dfrac{6(x + 3)}{x^2 - 4} - \dfrac{4x}{x + 1}$

[7.1, 7.2] *Perform the indicated operations.*   **24.** $\dfrac{-3.2x^2 - 8.2x + 7.3}{1.7x^2 - 2.4x + 1.5}$

17. $\dfrac{15x^2y^3}{3z} \cdot \dfrac{6z^3}{5xy^3}$ $6xz^2$

18. $\dfrac{1}{x - 2} \cdot \dfrac{2 - x}{2}$ $-\dfrac{1}{2}$

19. $\dfrac{(8xy^2)(xy^2)}{(z)(xz^4)} \div \dfrac{2x^2y^4}{(x^2z^5)^2}$ $4x^3z^5$

20. $\dfrac{4}{2x} + \dfrac{x}{x^2}$ $\dfrac{3}{x}$

21. $\dfrac{4x + 4y}{x^2y} \cdot \dfrac{y^3}{8x}$ $\dfrac{(x + y)y^2}{2x^3}$

22. $\dfrac{4x^2 - 11x + 4}{x - 3} - \dfrac{x^2 - 4x + 10}{x - 3}$ $3x + 2$

23. $\dfrac{a - 2}{a + 3} \cdot \dfrac{a^2 + 4a + 3}{a^2 - a - 2}$ 1   **24.** $\dfrac{2.6x^2 - 3.5x + 0.4}{1.7x^2 - 2.4x + 1.5} - \dfrac{5.8x^2 + 4.7x - 6.9}{1.7x^2 - 2.4x + 1.5}$

25. $\dfrac{5x}{3xy} - \dfrac{4}{x^2}$ $\dfrac{5x^2 - 12y}{3x^2y}$

26. $6 + \dfrac{x}{x + 2}$ $\dfrac{7x + 12}{x + 2}$

27. $5 - \dfrac{3}{x + 3}$ $\dfrac{5x + 12}{x + 3}$

28. $\dfrac{x^2 - y^2}{x - y} \cdot \dfrac{x + y}{xy + x^2}$ $\dfrac{x + y}{x}$

29. $\dfrac{1}{a^2 + 8a + 15} \div \dfrac{3}{a + 5}$ $\dfrac{1}{3(a + 3)}$

30. $\dfrac{a + c}{c} - \dfrac{a - c}{a}$ $\dfrac{a^2 + c^2}{ac}$

31. $\dfrac{4x^2 + 8x - 5}{2x + 5} \cdot \dfrac{x + 1}{4x^2 - 4x + 1}$ $\dfrac{x + 1}{2x - 1}$

32. $(x + 3) \div \dfrac{x^2 - 4x - 21}{x - 7}$ 1

33. $\dfrac{x^2 - 3xy - 10y^2}{6x} \div \dfrac{x + 2y}{12x^2}$ $2x(x - 5y)$

34. $\dfrac{2}{3x} - \dfrac{3x}{3x - 6}$ $\dfrac{-3x^2 + 2x - 4}{3x(x - 2)}$

35. $\dfrac{x - 4}{x - 5} - \dfrac{3}{x + 5}$ $\dfrac{x^2 - 2x - 5}{(x + 5)(x - 5)}$

36. $\dfrac{x + 3}{x^2 - 9} + \dfrac{2}{x + 3}$ $\dfrac{3(x - 1)}{(x + 3)(x - 3)}$

37. $\dfrac{1}{a - 3} \cdot \dfrac{a^2 - 2a - 3}{a^2 + 3a + 2}$ $\dfrac{1}{a + 2}$

38. $\dfrac{4x^2 - 16y^2}{9} \div \dfrac{(x + 2y)^2}{12}$ $\dfrac{16(x - 2y)}{3(x + 2y)}$

39. $\dfrac{4}{(x+2)(x-3)} - \dfrac{4}{(x-2)(x+2)}$ $\dfrac{4}{(x+2)(x-3)(x-2)}$

40. $\dfrac{2x^2 + 10x + 12}{(x+2)^2} \cdot \dfrac{x+2}{x^3 + 5x^2 + 6x}$ $\dfrac{2}{x(x+2)}$

41. $\dfrac{x+2}{x^2 - x - 6} + \dfrac{x-3}{x^2 - 8x + 15}$ $\dfrac{2(x-4)}{(x-3)(x-5)}$

42. $\dfrac{x+5}{x^2 - 15x + 50} - \dfrac{x-2}{x^2 - 25}$ $\dfrac{22x+5}{(x-5)(x-10)(x+5)}$

43. $\dfrac{y^4 - x^6}{x^3 - y^2} \div (y^2 - x^3)$ $\dfrac{x^3 + y^2}{x^3 - y^2}$

44. $\dfrac{1}{x+3} - \dfrac{2}{x-3} + \dfrac{6}{x^2 - 9}$ $\dfrac{-1}{x-3}$

45. $\dfrac{x^3 + 27}{4x^2 - 4} \div \dfrac{x^2 - 3x + 9}{(x-1)^2}$ $\dfrac{(x+3)(x-1)}{4(x+1)}$

46. $\dfrac{x-4}{x-5} - \dfrac{3}{x+5} - \dfrac{10}{x^2 - 25}$ $\dfrac{x+3}{x+5}$

47. $\dfrac{x^2 - 8x + 16}{2x^2 - x - 6} \cdot \dfrac{2x^2 - 7x - 15}{x^2 - 2x - 24} \div \dfrac{x^2 - 9x + 20}{x^2 + 2x - 8}$ $\dfrac{x-4}{x-6}$

48. $\dfrac{x^2 - x - 56}{x^2 + 14x + 49} \cdot \dfrac{x^2 + 4x - 21}{x^2 - 9x + 8} + \dfrac{3}{x^2 + 8x - 9}$

48. $\dfrac{x^2 + 6x - 24}{(x-1)(x+9)}$

[7.3] *Simplify the complex fractions.*

49. $\dfrac{4 - \dfrac{9}{16}}{1 + \dfrac{5}{8}}$ $\dfrac{55}{26}$

50. $\dfrac{\dfrac{15xy}{6z}}{\dfrac{3x}{z^2}}$ $\dfrac{5yz}{6}$

51. $\dfrac{x + \dfrac{1}{y}}{y^2}$ $\dfrac{xy+1}{y^3}$

52. $\dfrac{\dfrac{4}{x} + \dfrac{2}{x^2}}{6 - \dfrac{1}{x}}$ $\dfrac{4x+2}{6x^2 - x}$

53. $\dfrac{a^{-1} + 2}{a^{-1} + \dfrac{1}{a}}$ $\dfrac{2a+1}{2}$

54. $\dfrac{x^{-2} + \dfrac{1}{x}}{\dfrac{1}{x^2} - \dfrac{1}{x}}$ $\dfrac{x+1}{-x+1}$

[7.4] *Solve the equation.*

55. $\dfrac{3}{x} = \dfrac{8}{24}$ 9

56. $\dfrac{5.6}{a} = \dfrac{14.6}{7.3}$ 2.8

57. $\dfrac{x+3}{5} = \dfrac{9}{5}$ 6

58. $\dfrac{x}{3.4} = \dfrac{x-4}{5.2}$ $-7.\overline{5}$

59. $\dfrac{3x+4}{5} = \dfrac{2x-8}{3}$ 52

60. $\dfrac{x}{4.8} + \dfrac{x}{2} = 1.7$ 2.4

61. $\dfrac{4}{x} - \dfrac{1}{6} = \dfrac{1}{x}$ 18

62. $\dfrac{1}{x-2} + \dfrac{1}{x+2} = \dfrac{1}{x^2 - 4}$ $\dfrac{1}{2}$

63. $\dfrac{x-3}{x-2} + \dfrac{x+1}{x+3} = \dfrac{2x^2 + x + 1}{x^2 + x - 6}$ -6

64. $\dfrac{x}{x^2 - 9} + \dfrac{2}{x+3} = \dfrac{4}{x-3}$ -18

Solve each problem.

65. Three resistors of 200, 400, and 1200 ohms, respectively, are connected in parallel. Find the total resistance of the circuit.   120 ohms

66. Two resistors are to be connected in parallel. One is to contain twice the resistance of the other. What should be the resistance of each resistor if the circuit's total resistance is to be 600 ohms?   900 ohms, 1800 ohms

67. What is the focal length of a curved mirror if the object distance is 12 centimeters and the image distance is 4 centimeters?   3 cm

68. The focal length of a curved mirror is 10 centimeters. Find the object's distance from the lens if the image distance is twice the object's distance.   15 cm

[7.5] *Solve each problem.*   **69.** $\dfrac{x}{3} + \dfrac{x}{4} = 1$; $\dfrac{12}{7}$ or 1.71 hr   **70.** $\dfrac{40}{75} + \dfrac{40}{x} = 1$; $\dfrac{3000}{35}$ or 85.71 hr   **71.** $\dfrac{5 \cdot x}{8 + x} = 1$; 2

69. It takes Dan 3 hours to mow Mr. Lee's lawn. It takes Kim 4 hours to mow the same lawn. How long will it take them working together to mow Mr. Lee's lawn?

70. Annette and Pete are both copy editors for a publishing company. Together they can edit a 500-page manuscript in 40 hours. If Annette by herself can edit the manuscript in 75 hours, how long will it take Pete to edit the manuscript by himself?

71. What number multiplied by the numerator and added to the denominator of the fraction $\frac{5}{8}$ makes the result equal to 1?

72. When the reciprocal of twice a number is subtracted from 1, the result is the reciprocal of 3 times the number. Find the number.

73. Kit motorboat can travel 15 miles per hour in still water. Traveling with the current of a river, the boat can travel 20 miles in the same time it takes to go 10 miles against the current. Find the current.

74. A small plane and a car start from the same location, at the same time, heading toward the same town 450 miles away. The speed of the plane is 3 times the speed of the car. The plane arrives at the town 6 hours ahead of the car. Find the speeds of the car and the plane.

72. $1 - \dfrac{1}{2x} = \dfrac{1}{3x}$; $\dfrac{5}{6}$

73. $\dfrac{20}{15+x} = \dfrac{10}{15-x}$; 5 mph

74. $\dfrac{450}{3x} + 6 = \dfrac{450}{x}$; 50 mph, 150 mph

[7.6]
Find the quantity indicated.

75. *A* is directly proportional to *B*. If *A* is 120 when *B* = 80, find *A* when *B* = 50.   75

76. *A* is directly proportional to the square of *C*. If *A* is 5 when *C* is 5, find *A* when *C* = 10.   20

77. *x* is inversely proportional to *y*. If *x* is 20 when *y* = 5, find *x* when *y* = 100.   1

78. *W* is directly proportional to *L* and inversely proportional to *A*. If *W* = 80 when *L* = 100 and *A* = 20, find *W* when *L* = 50 and *A* = 40.   20

79. *z* is jointly proportional to *x* and *y* and inversely proportional to the square of *r*. If *z* is 12 when *x* is 20, *y* = 8, and *r* = 8, find *z* when *x* = 10, *y* = 80, and *r* = 3.   426.7

80. The scale of a map is 1 inch to 60 miles. How large a distance on the map represents 300 miles?   5 in

81. An electric company charges $0.162 per kilowatthour. What is the electric bill if 740 kilowatthours are used in a month? $119.88

82. The distance, *d*, an object drops in free fall is directly proportional to the square of the time, *t*. If an object falls 16 feet in 1 second, how far will an object fall in 5 seconds?   400 ft

83. The area, *A*, of a circle varies directly with the square of its radius, *r*. If the area is 78.5 when the radius is 5, find the area when the radius is 8.   200.96

84. The time, *t*, for an ice cube to melt is inversely proportional to the temperature of the water it is in. If it takes an ice cube 1.7 minutes to melt in 70°F water temperature, how long will it take an ice cube of the same size to melt in 50°F water?   2.38 min

Practice Test

1. (a) Find the domain of $\dfrac{x+4}{x^2-3x-28}$ then **(b)** reduce to lowest terms.   **(a)** $\{x \mid x \neq 7, x \neq -4\}$ **(b)** $\dfrac{1}{x-7}$

Perform the operations indicated.

2. $\dfrac{6x^2y^4 \cdot 3xy^4}{4z^2 \cdot 6x^2y^3} \cdot \dfrac{(2x^2y)^4}{x^5y^7}$ $\dfrac{12x^4y^2}{z^2}$

3. $\dfrac{a^2-9a+14}{a-2} \cdot \dfrac{a^2-4a-21}{(a-7)^2}$ $a+3$

4. $\dfrac{x^2-9y^2}{3x+6y} \div \dfrac{x+3y}{x+2y}$ $\dfrac{x-3y}{3}$

5. $\dfrac{x^3+y^3}{x+y} \div \dfrac{x^2-xy+y^2}{x^2+y^2}$ x^2+y^2

6. $\dfrac{5}{x} + \dfrac{3}{2x^2}$ $\dfrac{10x+3}{2x^2}$

7. $\dfrac{x-5}{x^2-16} - \dfrac{x-2}{x^2+2x-8}$ $\dfrac{-1}{(x+4)(x-4)}$

8. $\dfrac{x+1}{4x^2-4x+1} + \dfrac{3}{2x^2+5x-3}$ $\dfrac{x(x+10)}{(2x-1)^2(x+3)}$

Simplify.

9. $\dfrac{\dfrac{1}{x}+\dfrac{1}{y}}{\dfrac{1}{x}-\dfrac{1}{y}}$ $\dfrac{y+x}{y-x}$

10. $\dfrac{x+\dfrac{x}{y}}{x^{-1}+y^{-1}}$ $\dfrac{x^2(y+1)}{y+x}$

In Exercises 11 and 12, solve the equation.

11. $\dfrac{x}{3} - \dfrac{x}{4} = 5$ 60

12. $\dfrac{x}{x-8} + \dfrac{6}{x-2} = \dfrac{x^2}{x^2-10x+16}$ 12

13. *P* varies directly as *Q* and inversely as *R*. If *P* = 8 when *Q* = 4 and *R* = 10, find *P* when *Q* = 10 and *R* = 20.   10

14. *W* varies jointly as *P* and *Q* and inversely as the square of *T*. If *W* = 6 when *P* = 20, *Q* = 8, and *T* = 4, find *W* if *P* = 30, *Q* = 4, and *T* = 8.   1.125

15. Kris can level a 1-acre field in 8 hours on his tractor. Heather can level a 1-acre field in 5 hours on her tractor. How long will it take them to level a 1-acre field if they work together?

15. $\dfrac{x}{8} + \dfrac{x}{5} = 1$; $\dfrac{40}{13}$ or 3.08 hr

Chapter **8**

Roots, Radicals, and Complex Numbers

See Section 8.6, Exercise 79.

Preview and Perspective

We graphed equations containing square roots in Section 3.5. In this chapter we discuss roots and radicals further. We explain how to add, subtract, multiply, and divide radical expressions. We also introduce imaginary numbers and complex numbers.

Section 8.1 gives some basic concepts and definitions of roots. In Section 8.2 we change expressions from radical form to exponential form, and vice versa. We also work problems using rational exponents. The rules of exponents discussed in Sections 5.1 and 5.2 still apply to rational exponents and we use those rules again here.

In Section 8.4 we discuss rationalizing the denominator, which removes radicals from a denominator. Make sure that you understand the three requirements for a radical expression to be simplified, as discussed in Section 8.4. We learn to add and subtract radicals in Section 8.5. We also discuss rationalizing denominators further in this section.

In Section 8.6 we discuss how to solve equations that contain radical expressions. We will use these procedures again in Chapters 9 and 10. Section 8.6 also illustrates some applications of radical equations.

Imaginary numbers and complex numbers are introduced in Section 8.7. These numbers play a very important role in higher mathematics courses. We will be using imaginary and complex numbers throughout Chapter 9.

8.1 Roots and Radicals

1 Find principal square roots.
2 Find cube and higher roots.
3 Evaluate radical expressions using absolute value.

Tape 12

In this chapter we expand on the concept of radicals introduced in Chapter 1, 3, and 4. So far we have discussed only square and cube roots. In this chapter we will also discuss higher roots.

In the expression $\sqrt{x}$, the $\sqrt{}$ is called the **radical sign.** The number or expression within the radical sign is called the **radicand.**

radical sign

$$\sqrt{x}$$

radicand

The entire expression, including the radical sign and radicand, is called the **radical expression.** Another part of the radical expression is its index. The **index** tells the "root" of the expression. Square roots have an index of 2. The index of square roots is generally not written.

$$\sqrt{x} \quad \text{means} \quad \sqrt[2]{x}$$

Principal Square Root

1 Every positive number has two square roots, a principal or positive square root and a negative square root. For any positive number x, the positive square root is written $\sqrt{x}$, and the negative square root is written $-\sqrt{x}$.

Number	Principal or Positive Square Root	Negative Square Root
25	$\sqrt{25}$	$-\sqrt{25}$
36	$\sqrt{36}$	$-\sqrt{36}$

> The **principal** or **positive square root** of a positive real number x, written $\sqrt{x}$, is that *positive* number whose square equals x.

Examples

$$\sqrt{25} = 5 \qquad \text{since } 5^2 = 5 \cdot 5 = 25$$
$$\sqrt{36} = 6 \qquad \text{since } 6^2 = 6 \cdot 6 = 36$$
$$\sqrt{\frac{4}{9}} = \frac{2}{3} \qquad \text{since } \left(\frac{2}{3}\right)^2 = \left(\frac{2}{3}\right)\left(\frac{2}{3}\right) = \frac{4}{9}$$

In this book whenever we use the words *square root* we will be referring to the principal or positive square root. Thus if you are asked to find the value of $\sqrt{25}$, your answer will be 5. Note that both $\sqrt{25}$ and $\sqrt{36}$ have square roots that are integers. Thus, $\sqrt{25}$ and $\sqrt{36}$ are rational numbers. However, not every square root is a rational number. For example, $\sqrt{10}$ is not a rational number but is an irrational number. There is no integer whose square is 10.

When we defined the principal square root, we indicated that we were taking the square root "of a positive real number x." Do you know why we had to specify that we were taking the square root of a *positive* real number? Consider the square root of -25, written $\sqrt{-25}$. What is its value? Is $\sqrt{-25}$ equal to 5? Is it equal to -5? The answer to both questions is no.

$$\sqrt{-25} \neq 5 \qquad \text{since } 5^2 = 25 \quad (\text{not } -25)$$
$$\sqrt{-25} \neq -5 \qquad \text{since } (-5)^2 = 25 \quad (\text{not } -25)$$

Since the square of any real number will always be greater than or equal to 0, there is no real number that equals -25 when squared. For this reason, $\sqrt{-25}$ is *not a real number*. We will discuss numbers like $\sqrt{-25}$ later in this chapter. Since the square of any real number cannot be negative, *the square root of a negative number is not a real number.*

Other types of radical expressions have different indexes. For example, $\sqrt[3]{x}$ is the third or cube root of x. The index of cube roots is 3. In the expression $\sqrt[5]{xy}$, read "the fifth root of xy," the index is 5 and the radicand is xy.

Even Indexes 2 Radical expressions that have indexes of 2, 4, 6, . . . or any even integer are **even roots.** Square roots are even roots since their index is 2. Radical expressions that have indexes of 3, 5, 7, . . . or any odd integer are **odd roots.**

Examples of Even Roots	Examples of Odd Roots
$\sqrt{9}, \ \sqrt[4]{x}, \ \sqrt[12]{9x^5}$	$\sqrt[3]{27}, \ \sqrt[5]{x}, \ \sqrt[17]{6x^4}$

> The nth root of x, $\sqrt[n]{x}$, where n is an *even index* and x is a *positive* real number, is that *positive* real number c such that $c^n = x$.

Examples of Even Roots

$$\sqrt{9} = 3 \qquad \text{since } 3^2 = 3 \cdot 3 = 9$$
$$\sqrt[4]{16} = 2 \qquad \text{since } 2^4 = 2 \cdot 2 \cdot 2 \cdot 2 = 16$$
$$\sqrt[4]{81} = 3 \qquad \text{since } 3^4 = 3 \cdot 3 \cdot 3 \cdot 3 = 81$$

Note that

$$-\sqrt{9} = -3$$
$$-\sqrt[4]{16} = -2$$
$$-\sqrt[4]{81} = -3$$

When considering radical expressions with even indexes, the radicand must be a positive value or 0 if the number is to be real. For example, is $\sqrt[4]{-16}$ equal to any real number? Is there a real number that when raised to the fourth power equals -16? Since any real number raised to an even power cannot be negative, there is no real number that equals $\sqrt[4]{-16}$. Thus, *when the index is even, the radicand must be nonnegative for the radical to be a real number.* The *n*th root of 0, $\sqrt[n]{0}$, is 0.

Odd Indexes

> The *n*th root of x, $\sqrt[n]{x}$, where n is an *odd index* and x is *any real number,* is that real number c such that $c^n = x$.

Examples of Odd Roots

$$\sqrt[3]{8} = 2 \qquad \text{since } 2^3 = 2 \cdot 2 \cdot 2 = 8$$
$$\sqrt[3]{-8} = -2 \qquad \text{since } (-2)^3 = (-2)(-2)(-2) = -8$$
$$\sqrt[5]{243} = 3 \qquad \text{since } 3^5 = 3 \cdot 3 \cdot 3 \cdot 3 \cdot 3 = 243$$
$$\sqrt[5]{-243} = -3 \qquad \text{since } (-3)^5 = (-3)(-3)(-3)(-3)(-3) = -243$$

Note that

$$\sqrt[3]{8} = 2, \qquad \sqrt[3]{-8} = -2$$
$$-\sqrt[3]{8} = -2, \qquad -\sqrt[3]{-8} = -(-2) = 2$$

An odd root of a positive number is a positive number, and an odd root of a negative number is a negative number.

Table 8.1 summarizes the information about even and odd roots.

TABLE 8.1		
	n is even	**n is odd**
$x > 0$	$\sqrt[n]{x}$ is a positive real number	$\sqrt[n]{x}$ is a positive real number
$x < 0$	$\sqrt[n]{x}$ is not a real number	$\sqrt[n]{x}$ is a negative real number
$x = 0$	$\sqrt[n]{0} = 0$	$\sqrt[n]{0} = 0$

Helpful Hint

There is an important difference between $-\sqrt[4]{16}$ and $\sqrt[4]{-16}$. The number $-\sqrt[4]{16}$ is the opposite of $\sqrt[4]{16}$. Since $\sqrt[4]{16} = 2$, $-\sqrt[4]{16} = -2$. However, $\sqrt[4]{-16}$ is not a real number since no real number when raised to the fourth power equals -16.

$$-\sqrt[4]{16} = -(\sqrt[4]{16}) = -2$$
$$\sqrt[4]{-16} \quad \text{is not a real number.}$$

EXAMPLE 1   Indicate whether or not the radical expression is a real number. If the number is a real number, find its value.

(a) $\sqrt[4]{-81}$ **(b)** $-\sqrt[4]{81}$ **(c)** $\sqrt[3]{-64}$ **(d)** $-\sqrt[3]{-64}$

Solution:

(a) Not a real number. Even roots of negative numbers are not real numbers.
(b) Real number, $-\sqrt[4]{81} = -(\sqrt[4]{81}) = -(3) = -3$.
(c) Real number, $\sqrt[3]{-64} = -4$ since $(-4)^3 = -64$.
(d) Real number, $-\sqrt[3]{-64} = -(-4) = 4$

Evaluating Radicals Using Absolute Value

3   When $\sqrt{a^2}$ is evaluated for any nonzero real number a, what is its sign? Let us substitute values for a, one positive and one negative, and then examine the results.

$$a = 2: \quad \sqrt{a^2} = \sqrt{2^2} = \sqrt{4} = 2$$
$$a = -2: \quad \sqrt{a^2} = \sqrt{(-2)^2} = \sqrt{4} = 2$$

By examining these examples and other examples we can make up, we can reason that $\sqrt{a^2}$ *will always be a positive real number* for any nonzero real number a. Recall from Section 1.3 that the *absolute value* of any real number a, or $|a|$, is also a positive number for any nonzero number. We use these facts to reason that

For any real number a,
$$\sqrt{a^2} = |a|$$

Examples

$$\sqrt{2^2} = |2| = 2 \qquad\qquad \sqrt{7^2} = |7| = 7$$
$$\sqrt{(-2)^2} = |-2| = 2 \qquad \sqrt{(-7)^2} = |-7| = 7$$

EXAMPLE 2   Use absolute value to evaluate.

(a) $\sqrt{5^2}$ **(b)** $\sqrt{(-5)^2}$ **(c)** $\sqrt{(-71)^2}$

Solution:

(a) $\sqrt{5^2} = |5| = 5$ **(b)** $\sqrt{(-5)^2} = |-5| = 5$ **(c)** $\sqrt{(-71)^2} = |-71| = 71$

You may be tempted to generalize that $\sqrt{a^2} = a$, *but this is true only when* $a \geq 0$. When $a < 0$, then $\sqrt{a^2} = -a$. This concept is illustrated using values of $a = 2$ and $a = -2$.

$$a = 2$$
$$\sqrt{a^2} = \sqrt{2^2} = \sqrt{4} = 2$$
Since $a = 2$, $\sqrt{a^2} = a$
That is, $\sqrt{2^2} = 2$

$$a = -2$$
$$\sqrt{a^2} = \sqrt{(-2)^2} = \sqrt{4} = 2$$
Since $\sqrt{a^2} = 2$ when $a = -2$, $\sqrt{a^2} \neq a$.
When $a < 0$, $\sqrt{a^2} = -a$.
Note that $\sqrt{(-2)^2} = -(-2) = 2$

If you do not know whether or not the a in the expression $\sqrt{a^2}$ is a number greater than or equal to zero, then you must simplify $\sqrt{a^2}$ as $|a|$.

To what is $\sqrt{(x+1)^2}$ equal? Since you do not know whether $x + 1$ represents a positive or negative number, you must write $\sqrt{(x+1)^2} = |x + 1|$. We will now show that $\sqrt{(x+1)^2} = |x + 1|$ for values of $x = -3$ and $x = 5$.

$$x = -3$$
$$\sqrt{(x+1)^2} = |x + 1|$$
$$\sqrt{(-3+1)^2} = |-3 + 1|$$
$$\sqrt{(-2)^2} = |-2|$$
$$\sqrt{4} = 2$$
$$2 = 2 \quad \text{true}$$

$$x = 5$$
$$\sqrt{(x+1)^2} = |x + 1|$$
$$\sqrt{(5+1)^2} = |5 + 1|$$
$$\sqrt{6^2} = |6|$$
$$\sqrt{36} = 6$$
$$6 = 6 \quad \text{true}$$

Try a few other values for x and show that $\sqrt{(x+1)^2} = |x + 1|$ for those values.

When you are asked to find the square root of an expression that is squared, and you do not know whether the expression represents a positive or a negative number, you must write the answer as the absolute value of the expression.

EXAMPLE 3   Write as an absolute value.

(a) $\sqrt{(x+2)^2}$ **(b)** $\sqrt{(y-7)^2}$ **(c)** $\sqrt{(x^2 - 5x + 6)^2}$

Solution:

(a) $\sqrt{(x+2)^2} = |x + 2|$
(b) $\sqrt{(y-7)^2} = |y - 7|$
(c) $\sqrt{(x^2 - 5x + 6)^2} = |x^2 - 5x + 6|$

When you are asked to evaluate $\sqrt{a^2}$, and you do not know whether a represents a positive or negative number, you must write the answer as $|a|$. However, if you are asked to evaluate $\sqrt{a^2}$ and you know that a represents a positive number, you can write the answer as a.

Exercise Set 8.1

Evaluate the radical expression if it is a real number. Approximate irrational numbers to the nearest hundredth. If the expression is not a real number, indicate so. The procedure for evaluating radicals on the calculator is given on page 32.

1. $\sqrt{25}$ 5

2. $\sqrt[3]{27}$ 3

3. $\sqrt[3]{-27}$ -3

4. $\sqrt[5]{32}$ 2

5. $\sqrt[3]{125}$ 5

6. $\sqrt[4]{81}$ 3

7. $\sqrt{-9}$ not real

8. $\sqrt[6]{64}$ 2

9. $\sqrt[3]{-8}$ -2 | **10.** $\sqrt[3]{216}$ 6 | **11.** $\sqrt{529}$ 23 | **12.** $\sqrt[4]{256}$ 4

13. $\sqrt[5]{-1}$ -1 | **14.** $\sqrt[3]{-343}$ -7 | **15.** $\sqrt[3]{343}$ 7 | **16.** $\sqrt[5]{-32}$ -2

17. $\sqrt[4]{-16}$ not real | **18.** $\sqrt[4]{16}$ 2 | **19.** $-\sqrt{-25}$ not real | **20.** $\sqrt[3]{-64}$ -4

21. $-\sqrt[3]{102.4}$ -4.68 | **22.** $\sqrt{1600}$ 40 | **23.** $\sqrt{\dfrac{25}{9}}$ $\frac{5}{3}$ | **24.** $\sqrt[3]{\dfrac{1}{8}}$ $\frac{1}{2}$

25. $\sqrt{-36}$ not real | **26.** $\sqrt[4]{-50}$ not real | **27.** $\sqrt[5]{16.2}$ 1.75 | **28.** $-\sqrt{92.6}$ -9.62

Use absolute value to evaluate.

29. $\sqrt{6^2}$ 6 | **30.** $\sqrt{(-6)^2}$ 6 | **31.** $\sqrt{(-1)^2}$ 1 | **32.** $\sqrt{(-17)^2}$ 17

33. $\sqrt{(43)^2}$ 43 | **34.** $\sqrt{(-96)^2}$ 96 | **35.** $\sqrt{(147.23)^2}$ 147.23 | **36.** $\sqrt{(-147.23)^2}$ 147.23

37. $\sqrt{(-0.03)^2}$ 0.03 | **38.** $\sqrt{(-89)^2}$ 89 | **39.** $\sqrt{\left(-\dfrac{156}{5}\right)^2}$ $\frac{156}{5}$ or 31.2 | **40.** $\sqrt{\left(\dfrac{40}{9}\right)^2}$ $\frac{40}{9}$ or $4.\overline{4}$

Write as an absolute value. **46.** $|x^2 - 3x + 4|$ **49.** $|y^2 - 4y + 3|$ **52.** $|3w^4 - 4w|$

41. $\sqrt{(y-8)^2}$ $|y-8|$ | **42.** $\sqrt{(x-7)^2}$ $|x-7|$ | **43.** $\sqrt{(x-3)^2}$ $|x-3|$ | **44.** $\sqrt{(3x^2-y)^2}$ $|3x^2-y|$

45. $\sqrt{(3x+5)^2}$ $|3x+5|$ | **46.** $\sqrt{(x^2-3x+4)^2}$ | **47.** $\sqrt{(6-3x)^2}$ $|6-3x|$ | **48.** $\sqrt{(4-5x^2)^2}$ $|4-5x^2|$

49. $\sqrt{(y^2-4y+3)^2}$ | **50.** $\sqrt{(x^2-3x)^2}$ $|x^2-3x|$ | **51.** $\sqrt{(8a-b)^2}$ $|8a-b|$ | **52.** $\sqrt{(3w^4-4w)^2}$

Write each radicand as the square of an expression. Then use absolute value to simplify.

53. $\sqrt{a^8}$ $|a^4|$ | **54.** $\sqrt{r^{12}}$ $|r^6|$ | **55.** $\sqrt{a^2+2ab+b^2}$ $|a+b|$ **56.** $\sqrt{a^2-6a+9}$ $|a-3|$

57. (a) How many square roots does every positive real number have? Name them. two, positive and negative

(b) Find all square roots of the number 36. $6, -6$

(c) When we refer to "the square root," which square root are we referring to? positive

(d) Find the square root of 36. 6

58. (a) What are even roots? Give an example of an even root. roots with even indexes, $\sqrt{6}$

(b) What are odd roots? Give an example of an odd root. roots with odd indexes, $\sqrt[3]{8}$

59. Explain why $\sqrt{-49}$ is not a real number.

60. Will a radical expression with an odd index and a real number as the radicand always be a real number? Explain your answer. yes

61. Will a radical expression with an even index and a real number as the radicand always be a real number? Explain your answer. no

62. (a) To what is $\sqrt{a^2}$ equal? $|a|$

(b) To what is $\sqrt{a^2}$ equal if we know $a \geq 0$? a

63. If $\sqrt[n]{x} = a$, then to what is x equal? $x = a^n$

64. Select a value for x and show that $\sqrt{(2x-1)^2} \neq 2x - 1$. select a value less than $\frac{1}{2}$

65. Select a value for x and show that $\sqrt{(5x-3)^2} \neq 5x - 3$. select a value less than $\frac{3}{5}$

66. For what values of x will $\sqrt{(x-1)^2} = x - 1$? Explain how you determined your answer. $x \geq 1$

67. For what values of x will $\sqrt{(x+4)^2} = x + 4$? Explain how you determined your answer. $x \geq -4$

68. For what values of x will $\sqrt{(2x-6)^2} = 2x - 6$? Explain how you determined your answer. $x \geq 3$

69. For what values of x will $\sqrt{(4x-4)^2} = 4x - 4$? Explain how you determined your answer. $x \geq 1$

70. (a) For what values of a is $\sqrt{a^2} = |a|$? all real numbers

(b) For what values of a is $\sqrt{a^2} = a$? $a \geq 0$

71. Under what circumstances is the expression $\sqrt[n]{x}$ not a real number? when n is an even integer and $x < 0$

72. Explain why the expression $\sqrt[n]{x^n}$ is a real number for any real number x. If n is even, finding even root of positive number. If n is odd, expression is real.

73. Under what circumstances is the expression $\sqrt[n]{x^m}$ not a real number? When n is even, m is odd, and $x < 0$.

59. no real number when squared gives -49.

CUMULATIVE REVIEW EXERCISES

[6.1–6.5] *Factor.*

74. $3y^2 - 18y + 27 - 3z^2$ $3(y - 3 + z)(y - 3 - z)$ | **76.** $(x + 2)^2 - (x + 2) - 12$ $(x - 2)(x + 5)$

75. $x^3 + \dfrac{1}{27}$ $(x + \frac{1}{3})(x^2 - \frac{1}{3}x + \frac{1}{9})$ | **77.** $2x^4 - 3x^3 - 6x^2 + 9x$ $x(x^2 - 3)(2x - 3)$

Group Activity/ Challenge Problems

By substituting values for a and b, determine if the following are true. We will discuss some of these statements later in the chapter.

1. Is $\sqrt{a} \cdot \sqrt{b} = \sqrt{ab}$?   yes

2. Is $\dfrac{\sqrt{a}}{\sqrt{b}} = \sqrt{\dfrac{a}{b}}$?   yes

3. Is $\sqrt{a} + \sqrt{b} = \sqrt{a+b}$?   no

4. Is $\sqrt{a} - \sqrt{b} = \sqrt{a-b}$?   no

5. Is $\sqrt[3]{a} \cdot \sqrt[3]{b} = \sqrt[3]{a \cdot b}$?   yes

6. Is $\dfrac{\sqrt[3]{a}}{\sqrt[3]{b}} = \sqrt[3]{\dfrac{a}{b}}$?   yes

Use absolute value to simplify.

7. $\sqrt{9m^2 - 30mn + 25n^2}$ $|3m - 5n|$

8. $\sqrt{p^4 + 4p^2q + 4q^2}$ $|p^2 + 2q|$

Evaluate.

9. $\sqrt{(\sqrt{5})^2}$ $\sqrt{5} \approx 2.24$

10. $\sqrt{\sqrt{\sqrt{6}}}$ $\sqrt[8]{6} \approx 1.25$

Simplify.

11. $\sqrt{(\sqrt{x})^2}, x \geq 0$ $\sqrt{x}$

12. $\sqrt{\sqrt{\sqrt{x}}}, x \geq 0$ $\sqrt[8]{x}$

8.2   Rational Exponents

Tape 12

1. Change a radical expression to an exponential expression.
2. Change an exponential expression to a radical expression.
3. Apply the rules of exponents to rational and negative exponents.
4. Factor expressions with rational and negative exponents.

In this section we discuss changing radical expressions to exponential expressions, and vice versa. When you see a rational (or fractional) exponent, you should realize that the expression can be written as a radical expression.

In Section 8.1 we indicated that

$$\sqrt{a^2} = |a|$$

If we know that a in the expression $\sqrt{a^2}$ represents a value greater than or equal to 0, then we can write

$$\sqrt{a^2} = a, \quad a \geq 0$$

> **For the remainder of this chapter we assume that all variables represent positive real numbers.**

We make this assumption so that we can write many answers without absolute value signs. With this assumption, when we evaluate a radical like $\sqrt{y^2}$, we can write the answer as y rather than $|y|$.

Changing from Radical to Exponential Form

1. A radical expression of the form $\sqrt[n]{a}$ can be written as an exponential expression using the following rule.

> For any nonnegative number a and positive integer n
> $$\sqrt[n]{a} = a^{1/n}$$

Examples

$$\sqrt{6} = \sqrt[2]{6} = 6^{1/2} \qquad \sqrt[3]{x} = x^{1/3}$$
$$\sqrt{x} = \sqrt[2]{x} = x^{1/2} \qquad \sqrt[4]{y} = y^{1/4}$$
$$\sqrt[3]{9} = 9^{1/3} \qquad \sqrt[5]{3} = 3^{1/5}$$

We can expand the rule so that radicals of the form $\sqrt[n]{a^m}$ can be written as exponential expressions.

> For any positive number a, and intergers m and n, $n \geq 2$,
>
> $$\sqrt[n]{a^m} = (\sqrt[n]{a})^m = a^{m/n} \quad \longleftarrow \quad \text{index}$$
>
> power $\nearrow$

This rule can be used to change an expression from radical form to exponential form, and vice versa. When changing a radical expression to exponential form, the *power* is placed in the *numerator,* and the *index or root* is placed in the *denominator* of the rational exponent. Thus, for example, $\sqrt[3]{x^4}$ can be written $x^{4/3}$. Also $(\sqrt[5]{y})^2$ can be written $y^{2/5}$. Additional examples follow.

Examples

$$\sqrt{y^3} = y^{3/2} \qquad \sqrt[3]{z^2} = z^{2/3} \qquad \sqrt[5]{2^8} = 2^{8/5}$$
$$(\sqrt{z})^3 = z^{3/2} \qquad (\sqrt[4]{x})^3 = x^{3/4} \qquad (\sqrt[4]{6})^3 = 6^{3/4}$$

By this rule, for nonnegative values of the variable we can write

$$\sqrt[3]{x^4} = (\sqrt[3]{x})^4 \qquad (\sqrt[5]{y})^2 = \sqrt[5]{y^2}$$

Changing from Exponential to Radical Form

2 Exponential expressions with rational exponents can be converted to radical expressions by reversing the procedure. The *numerator* of the rational exponent is the *power,* and the *denominator* of the rational exponent is the *index or root* of the radical expression. Here are some examples.

Examples

$$x^{1/2} = \sqrt{x} \qquad\qquad 5^{1/3} = \sqrt[3]{5}$$
$$6^{2/3} = \sqrt[3]{6^2} \text{ or } (\sqrt[3]{6})^2 \qquad y^{3/10} = \sqrt[10]{y^3} \text{ or } (\sqrt[10]{y})^3$$
$$x^{9/5} = \sqrt[5]{x^9} \text{ or } (\sqrt[5]{x})^9 \qquad y^{10/3} = \sqrt[3]{y^{10}} \text{ or } (\sqrt[3]{y})^{10}$$

You may choose, for example to write $6^{2/3}$ as either $\sqrt[3]{6^2}$ or $(\sqrt[3]{6})^2$.

When you rewrite a radical expression in exponential form, sometimes the exponential expression can be simplified. You may be able to simplify a radical expression by changing it to exponential form, as in Example 1.

EXAMPLE 1 Write in exponential form and then simplify.
(a) $\sqrt[6]{(16)^3}$ **(b)** $(\sqrt[3]{x})^9$ **(c)** $\sqrt[12]{y^4}$ **(d)** $(\sqrt[10]{z})^5$

Solution:

(a) $\sqrt[6]{(16)^3} = 16^{3/6} = 16^{1/2} = \sqrt{16} = 4$ **(b)** $(\sqrt[3]{x})^9 = x^{9/3} = x^3$

(c) $\sqrt[12]{y^4} = y^{4/12} = y^{1/3} = \sqrt[3]{y}$ **(d)** $(\sqrt[10]{z})^5 = z^{5/10} = z^{1/2} = \sqrt{z}$

Now consider $\sqrt[5]{x^5}$. When written in exponential form, this is $x^{5/5} = x^1 = x$. This leads to the following rule.

> For any nonnegative number a,
>
> $$\sqrt[n]{a^n} = (\sqrt[n]{a})^n = a^{n/n} = a$$

Thus if the index and the exponent are the same in a radical expression, the expression simplifies to a (if a is assumed to be positive).

Examples

$\sqrt{5^2} = 5 \qquad \sqrt[4]{y^4} = y$

$\sqrt{x^2} = x \qquad (\sqrt[5]{z})^5 = z$

Remember, we are assuming that all the variables in this chapter represent positive numbers. This allows us to write the answers to $\sqrt{x^2}$ as x and $(\sqrt[4]{z})^4$ as z.

Apply the Rules of Exponents

3 In Sections 5.1 and 5.2 we introduced and discussed the rules of exponents. In Chapter 5 we used only exponents that were whole numbers. The rules still apply when the exponents are rational numbers. Let us review those rules now.

> ## Rules of Exponents
>
> For all real numbers a and b and all rational numbers m and n,
>
> Product rule $\qquad a^m \cdot a^n = a^{m+n}$
>
> Quotient rule $\qquad \dfrac{a^m}{a^n} = a^{m-n}, \quad a \neq 0$
>
> Negative exponent rule $\qquad a^{-m} = \dfrac{1}{a^m}, \quad a \neq 0$
>
> Zero exponent rule $\qquad a^0 = 1, \quad a \neq 0$
>
> Power rules $\qquad \begin{cases} (a^m)^n = a^{m \cdot n} \\ (ab)^m = a^m b^m \\ \left(\dfrac{a}{b}\right)^m = \dfrac{a^m}{b^m}, \quad b \neq 0 \end{cases}$

Using these rules, we will now work some problems in which the exponents are rational numbers.

Example 2 Evaluate. **(a)** $8^{2/3}$ **(b)** $8^{-2/3}$

Solution:

(a) First rewrite the expression in radical form, and then evaluate.

$$8^{2/3} = (\sqrt[3]{8})^2 = 2^2 = 4$$

(b) Begin by using the negative exponent rule.

$$8^{-2/3} = \frac{1}{8^{2/3}} = \frac{1}{4}$$

Example 2**(a)** could also have been evaluated as follows:

$$8^{2/3} = \sqrt[3]{8^2} = \sqrt[3]{64} = 4$$

However, it is generally easier to evaluate the root before the power.

EXAMPLE 3   Evaluate $\left(\dfrac{4}{25}\right)^{-1/2}$.

Solution: Begin by using the negative exponent rule, and then write the expression in radical form.

$$\left(\frac{4}{25}\right)^{-1/2} = \frac{1}{\left(\dfrac{4}{25}\right)^{1/2}} = \frac{1}{\sqrt{\dfrac{4}{25}}} = \frac{1}{\dfrac{2}{5}} = \frac{5}{2}$$

Example 3 could also have been evaluated by using the rule $\left(\dfrac{x}{y}\right)^{-m} = \left(\dfrac{y}{x}\right)^{m}$.

This rule was discussed in Section 5.2. The example could be evaluated as follows:

$$\left(\frac{4}{25}\right)^{-1/2} = \left(\frac{25}{4}\right)^{1/2} = \sqrt{\frac{25}{4}} = \frac{5}{2}$$

EXAMPLE 4   Write each expression as a square root, and then simplify if possible.   **(a)** $-25^{1/2}$ **(b)** $(-25)^{1/2}$

Solution:

(a) Recall that $-x^2$ means $-(x^2)$. The same principle applies here.

$$-25^{1/2} = -(25^{1/2}) = -\sqrt{25} = -5$$

(b) Since the negative sign is within the parentheses, we write $(-25)^{1/2} = \sqrt{-25}$, which is not a real number.

EXAMPLE 5   Simplify $x^{1/2} \cdot x^{2/3}$ and write the result in exponential form.

Solution: Use the product rule, and then simplify.

$$x^{1/2} \cdot x^{2/3} = x^{(1/2)+(2/3)} = x^{(3/6)+(4/6)} = x^{7/6}$$

EXAMPLE 6   Simplify $\dfrac{y^{3/5}}{y^{1/3}}$ and write the result in exponential form.

Solution: Using the quotient rule, we write

$$\frac{y^{3/5}}{y^{1/3}} = y^{(3/5)-(1/3)} = y^{(9/15)-(5/15)} = y^{4/15}$$

EXAMPLE 7 Simplify $(x^{3/5})^{-1/4}$ and write the result in radical form.

Solution: First, use the power rule. Then use the negative exponent rule.

$$(x^{3/5})^{-1/4} = x^{(3/5)(-1/4)} = x^{-3/20} = \frac{1}{x^{3/20}} = \frac{1}{\sqrt[20]{x^3}}$$

In Example 7, we could have first applied the negative exponent rule and then applied the power rule. Try working Example 7 by using the negative exponent rule first.

EXAMPLE 8 Simplify $\left(\dfrac{8x^4y^{1/3}}{16x^{1/2}y}\right)^2$ and write the result in exponential form.

Solution: We will use the quotient rule to simplify the expression within parentheses first.

$$\left(\frac{8x^4y^{1/3}}{16x^{1/2}y}\right)^2 = \left(\frac{x^{4-(1/2)}y^{(1/3)-1}}{2}\right)^2 = \left(\frac{x^{7/2}y^{-2/3}}{2}\right)^2 = \frac{x^7y^{-4/3}}{4} = \frac{x^7}{4y^{4/3}}$$

Now work Example 8 by using the power rule first. Which method do you prefer? Why?

EXAMPLE 9 Simplify $\dfrac{p^{-2/3}p^{1/3}q^{-3/2}}{p^{-1/5}(q^{1/3})^{-2}}$.

Solution:
$$\frac{p^{-2/3}p^{1/3}q^{-3/2}}{p^{-1/5}(q^{1/3})^{-2}} = \frac{p^{(-2/3)+(1/3)}q^{-3/2}}{p^{-1/5}q^{-2/3}}$$
$$= \frac{p^{-1/3}q^{-3/2}}{p^{-1/5}q^{-2/3}}$$
$$= p^{(-1/3)-(-1/5)}q^{-3/2-(-2/3)}$$
$$= p^{(-1/3)+(1/5)}q^{(-3/2)+(2/3)}$$
$$= p^{-2/15}q^{-5/6} = \frac{1}{p^{2/15}q^{5/6}}$$

EXAMPLE 10 Simplify $2y^{-5/11}(y^{2/3} - 4y^{-2})$.

Solution:
$$2y^{-5/11}(y^{2/3} - 4y^{-2}) = 2y^{-5/11} \cdot y^{2/3} - 2y^{-5/11} \cdot 4y^{-2}$$
$$= 2y^{(-5/11)+(2/3)} - 8y^{(-5/11)-2}$$
$$= 2y^{(-15/33)+(22/33)} - 8y^{(-5/11)-(22/11)}$$
$$= 2y^{7/33} - 8y^{-27/11} = 2y^{7/33} - \frac{8}{y^{27/11}}$$

Factoring Expressions Containing Rational and Negative Exponents

4 One reason we introduce both rational and negative exponents is that you may need to use and understand them in a later mathematics course. In Section 6.1 we discussed factoring a monomial from a polynomial. Recall that the terms in polynomials must have nonnegative integer exponents. Now we will factor a common factor from expressions that are not polynomials. We begin by reviewing a factoring problem like those discussed in Section 6.1.

EXAMPLE 11   Factor $x^2 + x^7$.

Solution: Recall that we factor out any variables common to each term in the polynomial. In this problem the variable x is common to each term. We factor out each variable with the *lowest* (or lesser) exponent. The lowest power of x is x^2.

$$x^2 + x^7 = x^2(1 + x^5)\overset{7 - 2 = 5}{}$$

Notice that the exponent of the x in the parentheses can be determined by *subtracting* the exponent on the x being factored out, the 2, from the exponent on the x it is being factored from, the 7 ($7 - 2 = 5$). The factoring can be checked using the laws of exponents. Notice that $x^2 \cdot x^5 = x^{2+5} = x^7$.

In Example 11 we used subtraction to determine the exponent on the variable left in parentheses after the common factor was factored out. The exponent was obtained by *subtracting the lesser exponent from the greater exponent.* We can use the same procedure when factoring out factors that have rational and negative exponents.

EXAMPLE 12   Factor $x^{1/5} + x^{6/5}$.

Solution:
The lesser exponent is 1/5. Thus, we will factor out $x^{1/5}$ from both terms. When we factor out $x^{1/5}$ from $x^{6/5}$, what is left? We can determine the exponent on the remaining term by subtracting the smaller exponent (the exponent on the variable being factored out, the 1/5) from the larger exponent (the exponent on the variable it is being factored from, the 6/5).

$$x^{1/5} + x^{6/5} = x^{1/5}(1 + x^1)\overset{\frac{6}{5} - \frac{1}{5} = \frac{5}{5} = 1}{}$$

Since it is not necessary to write an exponent of 1, the expression factors as follows.

$$x^{1/5} + x^{6/5} = x^{1/5}(1 + x)$$

Check: We check the factoring process using both the distributive property and the laws of exponents.

$$x^{1/5}(1 + x) = x^{1/5} \cdot 1 + x^{1/5} \cdot x$$
$$= x^{1/5} + x^{(1/5) + 1}$$
$$= x^{1/5} + x^{(1/5) + (5/5)}$$
$$= x^{1/5} + x^{6/5}$$

Since we obtained the expression we started with, our factoring is correct.

Now let us look at some examples that contain negative exponents. We can subtract the exponent on the variable being factored out from the exponent on the variable it is being factored from to obtain the exponent on the remaining variable. As with the other examples, we subtract the lesser exponent from the greater exponent to obtain the remaining exponent.

Example 13   Factor $y^{-3} + y^{-5}$, and then write the result without negative exponents.

Solution: We must factor out the y with the lesser exponent. Since -5 is less than -3, we factor out y^{-5} from both terms.

$$\overset{-3 - (-5) = -3 + 5 = 2}{y^{-3} + y^{-5} = y^{-5}(y^2 + 1)}$$

Remember that for an expression to be simplified the answer should be written without negative exponents. Thus, we write

$$y^{-5}(y^2 + 1) = \frac{1}{y^5}(y^2 + 1) = \frac{y^2 + 1}{y^5}$$

Therefore,

$$y^{-3} + y^{-5} = \frac{y^2 + 1}{y^5}$$

In Example 13, we can check the factoring by multiplying the factors, as follows:

$$y^{-5}(y^2 + 1) = y^{-5} \cdot y^2 + y^{-5} \cdot 1 = y^{-3} + y^{-5}$$

Since the product results in the expression we began with, our factoring is correct.

Example 14   Factor $5z^{-3/4} - 2z^{-3/8}$.

Solution: Since $-\dfrac{3}{4}$ is less than $-\dfrac{3}{8}$ we factor $z^{-3/4}$ from each term.

$$\overset{-\frac{3}{8} - (-\frac{3}{4}) = -\frac{3}{8} + \frac{3}{4} = \frac{3}{8}}{5z^{-3/4} - 2z^{-3/8} = z^{-3/4}(5 - 2z^{3/8})}$$

Check: $z^{-3/4}(5 - 2z^{3/8}) = 5 \cdot z^{-3/4} - z^{-3/4} \cdot 2z^{3/8} = 5z^{-3/4} - 2z^{-3/8}$

Now we will factor expressions of three-term containing rational exponents as if they were trinomials. To factor the following expressions, we will use the substitution procedure that was introduced in Section 6.2.

EXAMPLE 15   Factor $x^{2/3} - 7x^{1/3} + 10$.

Solution: We first notice that $x^{2/3}$ is the square of $x^{1/3}$, $(x^{1/3})^2 = x^{2/3}$. Therefore, this expression can be changed to a trinomial using a substitution. We will substitute y for $x^{1/3}$ to obtain a trinomial in y.

$$x^{2/3} - 7x^{1/3} + 10 = (x^{1/3})^2 - 7x^{1/3} + 10$$
$$= y^2 - 7y + 10$$

We now factor the trinomial $y^2 - 7y + 10$ and then substitute $x^{1/3}$ back for each y.

$$y^2 - 7y + 10 = (y - 5)(y - 2)$$
$$= (x^{1/3} - 5)(x^{1/3} - 2)$$

Check:

$$(x^{1/3} - 5)(x^{1/3} - 2) = (x^{1/3})(x^{1/3}) + (x^{1/3})(-2) + (-5)(x^{1/3}) + (-5)(-2)$$
$$= x^{2/3} - 2x^{1/3} - 5x^{1/3} + 10$$
$$= x^{2/3} - 7x^{1/3} + 10$$

Since the check results in the original expression, our factoring is correct. Thus, $x^{2/3} - 7x^{1/3} + 10 = (x^{1/3} - 5)(x^{1/3} - 2)$.

EXAMPLE 16   Factor $6y^{2/5} + 5y^{1/5} - 6$.

Solution: We first notice that $y^{2/5}$ is the square of $y^{1/5}$, $(y^{1/5})^2 = y^{2/5}$. If we substitute z for $y^{1/5}$, we will obtain a trinomial in z.

$$6y^{2/5} + 5y^{1/5} - 6 = 6(y^{1/5})^2 + 5y^{1/5} - 6$$
$$= 6z^2 + 5z - 6$$

Now factor the trinomial and then substitute $y^{1/5}$ back for each z.

$$6z^2 + 5z - 6 = (3z - 2)(2z + 3)$$
$$= (3y^{1/5} - 2)(2y^{1/5} + 3)$$

Thus, $6y^{2/5} + 5y^{1/5} - 6 = (3y^{1/5} - 2)(2y^{1/5} + 3)$.

In Example 15 we used the variable y in our substitution. In Example 16 we used the variable z. The final result will be independent of the variable selected in the substitution, and the final result will always be written with the variable in the original expression. We will see factoring problems like these again in Section 9.3.

Calculator Corner

On page 32 we discussed finding roots on a calculator. Now that we have discussed rational exponents, let us discuss how to evaluate expressions with rational exponents. There are many ways to evaluate an expression like $845^{3/5}$. The easiest may be to use the y^x key with parentheses keys as shown below.

To evaluate $845^{3/5}$, press

$$845 \boxed{y^x} \boxed{(} \boxed{3} \boxed{\div} \boxed{5} \boxed{)} \boxed{=} 57.03139903$$

To evaluate $845^{-3/5}$, which means $\dfrac{1}{845^{3/5}}$, press

$$845 \boxed{y^x} \boxed{(} \boxed{3} \boxed{+/_-} \boxed{\div} \boxed{5} \boxed{)} \boxed{=} 0.017534201$$

Note that $-845^{3/5} = -(845)^{3/5} = -57.03139903$
and $-845^{-3/5} = -(845)^{-3/5} = -0.017534201$

EXAMPLE 17   Show the keys to press to evaluate $\sqrt[4]{192}$ on a calculator.

Solution: We will show two methods that may be used.

Method 1: $192 \boxed{\text{inv}} \boxed{y^x} \boxed{4} \boxed{=} 3.722419436$

Method 2:   Treat $\sqrt[4]{192}$ as $192^{1/4}$.

$$192 \boxed{y^x} \boxed{(} \boxed{1} \boxed{\div} \boxed{4} \boxed{)} \boxed{=} 3.722419436$$

Exercise Set 8.2

In this exercise set assume all variables represent positive real numbers. Write in exponential form.

23. $\dfrac{1}{(\sqrt[3]{a^2 - 4b^2})^2}$

1. $\sqrt{x^3}$ $x^{3/2}$

2. $\sqrt{y^5}$ $y^{5/2}$

3. $\sqrt{4^5}$ $4^{5/2}$

4. $\sqrt[3]{z^2}$ $z^{2/3}$

5. $\sqrt[5]{x^4}$ $x^{4/5}$

6. $\sqrt[3]{z^5}$ $z^{5/3}$

7. $(\sqrt{x})^3$ $x^{3/2}$

8. $(\sqrt[3]{y})^2$ $y^{2/3}$

9. $(\sqrt[4]{5})^{13}$ $5^{13/4}$

10. $\sqrt[4]{x^7}$ $x^{7/4}$

11. $\sqrt[8]{y^{23}}$ $y^{23/8}$

12. $\sqrt[17]{a^{12}}$ $a^{12/17}$

Write in radical form.

13. $x^{1/2}$ $\sqrt{x}$

14. $y^{2/3}$ $\sqrt[3]{y^2}$

15. $z^{3/2}$ $\sqrt{z^3}$

16. $5^{1/2}$ $\sqrt{5}$

17. $(24y^2)^{1/2}$ $\sqrt{24y^2}$

18. $(35c^2)^{5/2}$ $(\sqrt{35c^2})^5$

19. $(19x^2y^4)^{-1/2}$ $\dfrac{1}{\sqrt{19x^2y^4}}$

20. $(24xy^2)^{1/2}$ $\sqrt{24xy^2}$

21. $(2m^2n^3)^{2/5}$ $(\sqrt[5]{2m^2n^3})^2$

22. $(5r + s^2)^{1/4}$ $\sqrt[4]{5r + s^2}$

23. $(a^2 - 4b^2)^{-2/3}$

24. $(3r^2 + 2m)^{-1/3}$ $\dfrac{1}{\sqrt[3]{3r^2 + 2m}}$

Simplify each radical expression by changing the expression to exponential form. Write the answer in radical form (see Example 1).

25. $\sqrt{y^6}$ y^3

26. $\sqrt{x^{12}}$ x^6

27. $\sqrt{z^8}$ z^4

28. $\sqrt[3]{x^6}$ x^2

29. $\sqrt[3]{x^9}$ x^3

30. $\sqrt[6]{y^2}$ $\sqrt[3]{y}$

31. $\sqrt[10]{z^5}$ $\sqrt{z}$

32. $\sqrt{2^4}$ 4

33. $(\sqrt{5.1})^2$ 5.1   **34.** $\sqrt[4]{(6.83)^4}$ 6.83   **35.** $\sqrt[6]{y^6}$ y **36.** $(\sqrt[5]{x})^5$ x

37. $(\sqrt[8]{x})^2$ $\sqrt[4]{x}$ **38.** $\sqrt[3]{4^6}$ 16   **39.** $(\sqrt[3]{x})^{15}$ x^5 **40.** $(\sqrt[4]{y})^{40}$ y^{10}

41. $(\sqrt[18]{8})^{12}$ 4   **42.** $\sqrt[10]{8^5}$ $\sqrt{8}$ **43.** $\sqrt[18]{y^6}$ $\sqrt[3]{y}$ **44.** $(\sqrt[10]{y})^5$ $\sqrt{y}$

Evaluate if possible. If the expression is not a real number, so state (see Examples 2–4).

45. $4^{1/2}$ 2   **46.** $8^{2/3}$ 4   **47.** $27^{2/3}$ 9   **48.** $-8^{1/3}$ -2

49. $(-4)^{1/2}$ not a real number **50.** $\left(\frac{4}{9}\right)^{1/2}$ $\frac{2}{3}$ **51.** $\left(\frac{9}{25}\right)^{1/2}$ $\frac{3}{5}$ **52.** $\left(\frac{1}{8}\right)^{1/3}$ $\frac{1}{2}$

53. $-16^{1/2}$ -4 **54.** $(-16)^{1/2}$ not a real number   **55.** $-27^{1/3}$ -3 **56.** $4^{-1/2}$ $\frac{1}{2}$

57. $27^{-1/3}$ $\frac{1}{3}$ **58.** $16^{-3/2}$ $\frac{1}{64}$ **59.** $4^{-3/2}$ $\frac{1}{8}$ **60.** $81^{-3/4}$ $\frac{1}{27}$

61. $-\left(\frac{4}{49}\right)^{-1/2}$ $-\frac{7}{2}$ **62.** $\left(\frac{25}{121}\right)^{-1/2}$ $\frac{11}{5}$ **63.** $\left(\frac{64}{27}\right)^{-4/3}$ $\frac{81}{256}$ **64.** $\left(\frac{81}{16}\right)^{-3/4}$ $\frac{8}{27}$

65. $25^{1/2} + 169^{1/2}$ 18   **66.** $25^{-1/2} + 36^{-1/2}$ $\frac{11}{30}$ **67.** $343^{-1/3} + 9^{-1/2}$ $\frac{10}{21}$ **68.** $16^{-1/2} - 625^{-3/4}$ $\frac{121}{500}$

Simplify. Write the answer in exponential form (see Examples 5–9).

69. $x^5 \cdot x^{1/2}$ $x^{11/2}$ **70.** $x^{1/3} \cdot x^{3/8}$ $x^{17/24}$ **71.** $\dfrac{x^{1/2}}{x^{1/3}}$ $x^{1/6}$ **72.** $(x^{2/3})^3$ x^2

73. $(x^{1/5})^{2/3}$ $x^{2/15}$ **74.** $x^{-3/5}$ $\dfrac{1}{x^{3/5}}$ **75.** $(x^{1/2})^{-2}$ $\dfrac{1}{x}$ **76.** $(z^{-1/4})^{-1/2}$ $z^{1/8}$

77. $(6^{-1/3})^0$ 1   **78.** $\dfrac{x^4}{x^{-1/2}}$ $x^{9/2}$ **79.** $\dfrac{5y^{-1/3}}{60y^{-2}}$ $\dfrac{y^{5/3}}{12}$ **80.** $x^{-1/2} \cdot x^{-3/5}$ $\dfrac{1}{x^{11/10}}$

81. $4x^{5/3} \cdot 2x^{-7/2}$ $\dfrac{8}{x^{11/6}}$ **82.** $(x^{-2/5})^{1/3}$ $\dfrac{1}{x^{2/15}}$ **83.** $\left(\dfrac{8}{64x}\right)^{1/3}$ $\dfrac{1}{2x^{1/3}}$ **84.** $\left(\dfrac{81}{3y^4}\right)^{1/3}$ $\dfrac{3}{y^{4/3}}$

85. $\left(\dfrac{22x^{3/7}}{2x^{1/2}}\right)^2$ $\dfrac{121}{x^{1/7}}$ **86.** $\left(\dfrac{x^{-1/3}}{x^{-2}}\right)^{1/2}$ $x^{5/6}$ **87.** $\left(\dfrac{y^4}{4y^{-2/5}}\right)^{-3}$ $\dfrac{64}{y^{66/5}}$ **88.** $\left(\dfrac{81z^{1/2}y^3}{9z^{1/2}}\right)^{1/2}$ $3y^{3/2}$

89. $\left(\dfrac{x^{3/4}y^{-2}}{x^{1/2}y^2}\right)^4$ $\dfrac{x}{y^{16}}$ **90.** $\left(\dfrac{x^{1/2}y^3}{z^{-3}}\right)\left(\dfrac{xy^{2/3}}{z^2}\right)$ $x^{3/2}y^{11/3}z$ **91.** $\left(\dfrac{250a^{-3/4}b^{-5}}{2a^{-2}b^{-2}}\right)^{2/3}$ $\dfrac{25a^{5/6}}{b^2}$

92. $\left(\dfrac{x^{1/3}y^{2/3}}{16z^5}\right)\left(\dfrac{x^2y}{z^6}\right)^{3/4}$ $\dfrac{x^{11/6}y^{17/12}}{16z^{19/2}}$ **93.** $\dfrac{r^{1/2}r^{-2/3}r^{1/4}}{(r^{2/3})^2}$ $\dfrac{1}{r^{5/4}}$ **94.** $\left(\dfrac{a^{-2}a^{-3/4}b^{1/2}}{(a^4)^{-2/3}b^{-3/5}}\right)^2$ $\dfrac{b^{11/5}}{a^{1/6}}$

Multiply (see Example 10).

95. $3z^{-1/2}(2z^4 - z^{1/2})$ $6z^{7/2} - 3$ **96.** $-2x^{-4/9}(2x^{1/9} - x^2)$ $-\dfrac{4}{x^{1/3}} + 2x^{14/9}$ **97.** $5x^{-1}(x^{-4} + 2x^{-1/2})$ $\dfrac{5}{x^5} + \dfrac{10}{x^{3/2}}$

98. $-4a^{3/2}(a^{3/2} - a^{-3/2})$ $-4a^3 + 4$ **99.** $-4x^{5/3}(-2x^{1/2} + x^{1/3})$ $8x^{13/6} - 4x^2$ **100.** $\dfrac{1}{2}x^{-2}(6x^{4/3} - 8x^{-1/2})$ $\dfrac{3}{x^{2/3}} - \dfrac{4}{x^{5/2}}$

Factor. Write the answer without negative exponents (see Examples 11–14).

101. $x^{3/2} + x^{1/2}$ $x^{1/2}(x + 1)$ **102.** $x^{1/4} - x^{5/4}$ $x^{1/4}(1 - x)$ **103.** $y^{1/3} - y^{4/3}$ $y^{1/3}(1 - y)$

104. $x^{-1/2} + x^{1/2}$ $\dfrac{1 + x}{x^{1/2}}$ **105.** $y^{-3/5} + y^{2/5}$ $\dfrac{1 + y}{y^{3/5}}$ **106.** $x^2 - x^{-1}$ $\dfrac{x^3 - 1}{x}$

107. $y^{-1} - y$ $\dfrac{1 - y^2}{y}$ **108.** $y^2 - y^{-2}$ $\dfrac{y^4 - 1}{y^2}$ **109.** $x^{-7} + x^{-5}$ $\dfrac{1 + x^2}{x^7}$

110. $x^{-12} + x^{-7}$ $\dfrac{1 + x^5}{x^{12}}$ **111.** $x^{-1/2} + x^{-5/2}$ $\dfrac{x^2 + 1}{x^{5/2}}$ **112.** $y^{-9/5} - y^{-4/5}$ $\dfrac{1 - y}{y^{9/5}}$

113. $2x^{-4} - 6x^{-5}$ $\dfrac{2(x - 3)}{x^5}$ **114.** $3x^{1/2} - 6x^{-1/2}$ $\dfrac{3(x - 2)}{x^{1/2}}$

Factor.

115. $x^{2/3} + 2x^{1/3} - 3$ $(x^{1/3} + 3)(x^{1/3} - 1)$ **116.** $x^{2/5} + 4x^{1/5} - 5$ $(x^{1/5} + 5)(x^{1/5} - 1)$

117. $x + 6x^{1/2} + 9$ $(x^{1/2} + 3)(x^{1/2} + 3)$ **118.** $x^{1/2} + x^{1/4} - 20$ $(x^{1/4} + 5)(x^{1/4} - 4)$

119. $2x^{2/7} + x^{1/7} - 3$ $(2x^{1/7} + 3)(x^{1/7} - 1)$ **120.** $3x^{1/2} - 10x^{1/4} - 8$ $(3x^{1/4} + 2)(x^{1/4} - 4)$

121. $4x^{4/5} + 8x^{2/5} + 3$ $(2x^{2/5} + 3)(2x^{2/5} + 1)$ **122.** $6x^{2/3} - 5x^{1/3} + 1$ $(3x^{1/3} - 1)(2x^{1/3} - 1)$

123. $15x^{1/3} - 14x^{1/6} + 3$ $(5x^{1/6} - 3)(3x^{1/6} - 1)$ **124.** $8x + 2x^{1/2} - 1$ $(4x^{1/2} - 1)(2x^{1/2} + 1)$

Use a calculator to evaluate each of the following. If the number is irrational, give the answer to the nearest hundredth.

125. $\sqrt{120}$ 10.95
126. $\sqrt[3]{168}$ 5.52
127. $\sqrt[5]{402.83}$ 3.32
128. $\sqrt[4]{1096}$ 5.75

129. $45^{2/3}$ 12.65
130. $697.2^{3/2}$ 18409.25
131. $1000^{-1/2}$ 0.03
132. $8060^{-3/2}$ 0.00

133. $247^{4/5}$ 82.06
134. $-62^{-1/2}$ -0.13
135. $-807.52^{3/4}$ -151.48 **136.** $-916^{-2/3}$ -0.01

137. Under what conditions will $\sqrt[n]{a^n} = (\sqrt[n]{a})^n = a$?

138. By selecting values for a and b, show that $(a^2 + b^2)^{1/2}$ *is not equal to* $a + b$.

139. By selecting values for a and b, show that $(a^{1/2} + b^{1/2})^2$ *is not equal to* $a + b$.

140. By writing $\sqrt{\sqrt{x}}$ in exponential form, show that $\sqrt{\sqrt{x}} = \sqrt[4]{x}, x \geq 0.$ $(x^{1/2})^{1/2} = x^{1/4} = \sqrt[4]{x}$

141. Determine if $\sqrt[3]{\sqrt{x}} = \sqrt{\sqrt[3]{x}}, x \geq 0.$ Explain how you determined your answer.

137. n is odd, or n is even and $a \geq 0$ **138.** $(1^2 + 1^2)^{1/2} \neq 1 + 1;\ \sqrt{2} \neq 2$
139. $(4^{1/2} + 9^{1/2})^2 \neq 4 + 9;\ 25 \neq 13$ **141.** They are equal; both equal $x^{1/6}$.

CUMULATIVE REVIEW EXERCISES

[3.4] **142.** Determine which of the following graphs are functions and which are relations.

(a) 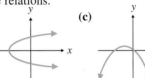   **(b)**   **(c)**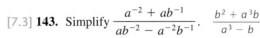

c is a function; a, b, and c are relations.

[7.3] **143.** Simplify $\dfrac{a^{-2} + ab^{-1}}{ab^{-2} - a^{-2}b^{-1}} \cdot \dfrac{b^2 + a^3b}{a^3 - b}$

[7.4] **144.** Solve the equation $\dfrac{3x - 2}{x + 4} = \dfrac{2x + 1}{3x - 2}.$ 0, 3

[7.5] **145.** Amy can fly her plane 500 miles against the wind in the same time it takes her to fly 560 miles with the wind. If the wind blows at 25 miles per hour, find the speed of the plane in still air.   441.67 mph

8. $\dfrac{2(3x - 2)}{(6x - 5)^3}$ **10. (a)** $x - y$ **(b)** $(x^{1/2} + y^{1/2})(x^{1/2} - y^{1/2})$

Group Activity/ Challenge Problems

Tape 13

1. Evaluate $(3^{\sqrt{2}})^{\sqrt{2}}$. Explain how you determined your answer.   9

2. (a) On your calculator evaluate 3^{π}.   31.5442807

(b) Explain why your value from part **(a)** does or does not make sense.

Simplify by writing as exponential expressions with rational exponents. Write the answer in exponential form.

3. $\sqrt{xy^5} \cdot \sqrt{x^2y^3}$ $x^{3/2}y^4$ **4.** $\sqrt[4]{xy^5} \cdot \sqrt[3]{x^2y^4}$ $x^{11/12}y^{31/12}$ **5.** $\sqrt{\sqrt[3]{\sqrt[4]{x}}}$ $x^{1/24}$

Determine the index to be placed in the shaded area to make the statement true. Explain how you determined your answer.

6. $\sqrt[4]{\sqrt{\sqrt{x}}} = x^{1/24}$ 3   **7.** $\sqrt[4]{\sqrt[5]{\sqrt{\sqrt[3]{z^{1/20}}}}} = z^{1/20}$ 2

Factor. Write the answers without negative exponents.

8. $(6x - 5)^{-3} + (6x - 5)^{-2}$ **9.** $(2x + 3)^{-1/3} + (2x + 3)^{2/3}$ $\dfrac{2x + 4}{(2x + 3)^{1/3}}$

10. (a) Multiply $(x^{1/2} + y^{1/2})(x^{1/2} - y^{1/2})$.   **(b)** Factor $x - y$.

11. Work Exercise 10, then factor $x + 2x^{1/2}y^{1/2} + y$. $(x^{1/2} + y^{1/2})(x^{1/2} + y^{1/2})$

12. Refer to Example 17. Explain why both methods yield the same result. The mth root of a positive expression is equal to the expression to the $\frac{1}{m}$ th power.

8.3   Multiplying and Simplifying Radicals

1. Apply the product rule for radicals.
2. Simplify radicals whose radicands are natural numbers.
3. Simplify radicals whose radicands are variables.
4. Simplify radicals.
5. Simplify products of two radicals.

Product Rule for Radicals

1 We will first simplify radicals using the product rule.

> ### Product Rule for Radicals
> For nonnegative real numbers a and b,
> $$\sqrt[n]{a} \cdot \sqrt[n]{b} = \sqrt[n]{ab}$$

Examples of the Product Rule

$$\sqrt{60} = \begin{cases} \sqrt{1} \cdot \sqrt{60} \\ \sqrt{2} \cdot \sqrt{30} \\ \sqrt{3} \cdot \sqrt{20} \\ \sqrt{4} \cdot \sqrt{15} \\ \sqrt{5} \cdot \sqrt{12} \\ \sqrt{6} \cdot \sqrt{10} \end{cases}$$

$\sqrt{60}$ can be factored into any of these forms

$$\sqrt[3]{60} = \begin{cases} \sqrt[3]{1} \cdot \sqrt[3]{60} \\ \sqrt[3]{2} \cdot \sqrt[3]{30} \\ \sqrt[3]{3} \cdot \sqrt[3]{20} \\ \sqrt[3]{4} \cdot \sqrt[3]{15} \\ \sqrt[3]{5} \cdot \sqrt[3]{12} \\ \sqrt[3]{6} \cdot \sqrt[3]{10} \end{cases}$$

$\sqrt[3]{60}$ can be factored into any of these forms

$$\sqrt{x^7} = \begin{cases} \sqrt{x} \cdot \sqrt{x^6} \\ \sqrt{x^2} \cdot \sqrt{x^5} \\ \sqrt{x^3} \cdot \sqrt{x^4} \end{cases}$$

$\sqrt{x^7}$ can be factored into any of these forms

$$\sqrt[3]{x^7} = \begin{cases} \sqrt[3]{x} \cdot \sqrt[3]{x^6} \\ \sqrt[3]{x^2} \cdot \sqrt[3]{x^5} \\ \sqrt[3]{x^3} \cdot \sqrt[3]{x^4} \end{cases}$$

$\sqrt[3]{x^7}$ can be factored into any of these forms

Perfect Powers of Natural Numbers

2 To help clarify our explanations, we will introduce **perfect powers**. A number is a **perfect square** if it is the square of a natural number. A number is a **perfect cube** if it is a cube of a natural number. A number is a **perfect fourth power** if it is the fourth power of a natural number, and so on.

Some perfect squares and perfect cubes numbers follow.

Squares of natural numbers: 1^2, 2^2, 3^2, 4^2, 5^2, 6^2, 7^2, 8^2, 9^2, ...

Perfect squares: 1, 4, 9, 16, 25, 36, 49, 64, 81, ...

Cubes of natural numbers: 1^3, 2^3, 3^3, 4^3, 5^3, 6^3, 7^3, 8^3, 9^3, ...

Perfect cubes: 1, 8, 27, 64, 125, 216, 343, 512, 729, ...

Note that the square root of any perfect square is a whole number. For example,
$$\sqrt{36} = \sqrt{6^2} = 6^{2/2} = 6$$

Similarly, the cube root of any perfect cube is a whole number. For example,
$$\sqrt[3]{125} = \sqrt[3]{5^3} = 5^{3/3} = 5$$

The fourth root of any perfect fourth-power is a whole number, etc.

Now we will discuss how to simplify radicals whose radicands are natural numbers.

> **To Simplify Radicals Whose Radicands Are Natural Numbers**
> **1.** Write the radicand as the product of two numbers, one of which is the largest perfect power number for the given index.
> **2.** Use the product rule to write the expression as a product of roots.
> **3.** Find the roots of any perfect power numbers.

If we are simplifying a *square* root, we will write the radicand as the product of the largest *perfect square* and another number. If we are simplifying a *cube* root, we will write the radicand as the product of the largest *perfect cube* and another number, and so on.

EXAMPLE 1   Simplify $\sqrt{32}$.

Solution: Since we are evaluating a square root, we look for the largest perfect square that divides 32. The largest perfect square that divides, or is a factor of, 32 is 16.

$$\sqrt{32} = \sqrt{16 \cdot 2} = \sqrt{16}\sqrt{2} = 4\sqrt{2}$$

In Example 1, if you first believed that 4 was the largest perfect square that divided 32, you could proceed as follows:

$$\sqrt{32} = \sqrt{4 \cdot 8} = \sqrt{4}\sqrt{8} = 2\sqrt{8}$$
$$= 2\sqrt{4}\sqrt{2} = 2 \cdot 2\sqrt{2} = 4\sqrt{2}$$

Note that the final result is the same, but you must perform more steps.

The chart on page 461 can help you determine the largest perfect square or perfect cube that divides the radicand when simplifying square or cube root problems.

EXAMPLE 2   Simplify $\sqrt{60}$.

Solution: $\sqrt{60} = \sqrt{4}\sqrt{15} = 2\sqrt{15}$

In Example 2, $\sqrt{15}$ can be factored into $\sqrt{5}\sqrt{3}$; however, since neither 5 nor 3 is a perfect square $\sqrt{15}$ cannot be simplified. The simplified form is $2\sqrt{15}$.

EXAMPLE 3   Simplify **(a)** $\sqrt[3]{54}$ **(b)** $\sqrt[3]{375}$

Solution: **(a)** The largest perfect cube that is a factor of 54 is 27.

$$\sqrt[3]{54} = \sqrt[3]{27}\sqrt[3]{2} = 3\sqrt[3]{2}$$

(b) The largest perfect cube that is a factor of 375 is 125.

$$\sqrt[3]{375} = \sqrt[3]{125}\sqrt[3]{3} = 5\sqrt[3]{3}$$

Simplifying Radicals Whose Radicands Contain Variables

3 Now we will discuss perfect powers of variables for an index. The radicand x^n is a **perfect square** when n is a multiple of 2 (an even natural number). The radicand x^n is a **perfect cube** when n is a multiple of 3. In general, the radicand x^n is a **perfect power** when n is a **multiple of the index** of the radical (or when n is divisible by the index).

Following are some perfect square, perfect cube, and perfect fourth powers of the variable x.

Perfect squares: $\quad x^2, \ x^4, \ x^6, \ x^8, x^{10}, \ldots$

Perfect cubes: $\quad x^3, \ x^6, \ x^9, x^{12}, x^{15}, \ldots$

Perfect fourth powers of x: $\quad x^4, \ x^8, x^{12}, x^{16}, x^{20}, \ldots$

Perfect powers of x for index n: $\quad x^n, x^{2n}, x^{3n}, x^{4n}, x^{5n}, \ldots$

Helpful Hint

A quick way to determine if a radicand x^n is a perfect power for an index is to determine if the exponent n is divisible by the index of the radical. For example, consider $\sqrt[5]{x^{20}}$. Since the exponent, 20, is divisible by the index, 5, x^{20} is a perfect fifth power. Now consider $\sqrt[6]{x^{20}}$. Since the exponent, 20, is not divisible by the index, 6, x^{20} is not a perfect sixth power. However, x^{18} and x^{24} are both perfect sixth powers since 6 divides both 18 and 24.

A radical can often be simplified by writing the radical in exponential form, as in Example 4.

EXAMPLE 4 Simplify. (a) $\sqrt{x^4}$ (b) $\sqrt[3]{x^{12}}$ (c) $\sqrt[6]{y^{24}}$

Solution:

(a) $\sqrt{x^4} = x^{4/2} = x^2$ (b) $\sqrt[3]{x^{12}} = x^{12/3} = x^4$ (c) $\sqrt[6]{y^{24}} = y^{24/6} = y^4$

To Simplify Radicals Whose Radicands Are Variables

1. Write each variable as the product of two factors, one of which is the largest perfect power of the variable for the index.
2. Use the product rule to write the radical expression as a product of radicals. Place all perfect powers under the same radical.
3. Find the roots of any perfect powers.

EXAMPLE 5 Simplify. (a) $\sqrt{x^9}$ (b) $\sqrt[3]{x^{14}}$ (c) $\sqrt[5]{x^{23}}$

Solution: (a) The largest perfect square less than or equal to x^9 is x^8.

$$\sqrt{x^9} = \sqrt{x^8 \cdot x} = \sqrt{x^8} \cdot \sqrt{x} = x^{8/2}\sqrt{x} = x^4\sqrt{x}$$

(b) The largest perfect cube less than or equal to x^{14} is x^{12}.

$$\sqrt[3]{x^{14}} = \sqrt[3]{x^{12} \cdot x^2} = \sqrt[3]{x^{12}} \sqrt[3]{x^2} = x^{12/3} \sqrt[3]{x^2} = x^4 \sqrt[3]{x^2}$$

(c) The largest perfect fifth power less than or equal to x^{23} is x^{20}.

$$\sqrt[5]{x^{23}} = \sqrt[5]{x^{20} \cdot x^3} = \sqrt[5]{x^{20}}\sqrt[5]{x^3} = x^{20/5}\sqrt[5]{x^3} = x^4\sqrt[5]{x^3}$$

EXAMPLE 6   Evaluate $\sqrt{x^{12}y^{17}}$.

Solution: x^{12} is a perfect square for the index 2. The highest perfect square that is a factor of y^{17} is y^{16}. Write y^{17} as $y^{16} \cdot y^1$.

$$\sqrt{x^{12}y^{17}} = \sqrt{x^{12} \cdot y^{16} \cdot y} = \sqrt{x^{12}y^{16}}\sqrt{y}$$
$$= \sqrt{x^{12}}\sqrt{y^{16}}\sqrt{y}$$
$$= x^{12/2}y^{16/2}\sqrt{y}$$
$$= x^6 y^8 \sqrt{y}$$

Often the steps where we change the radical expression to exponential form are done mentally, <u>and those steps</u> are not illustrated. For instance, in Example 6, we might change $\sqrt{x^{12}y^{16}}$ to x^6y^8 mentally and not show the intermediate steps.

EXAMPLE 7   Simplify $\sqrt[4]{x^6 y^{23}}$.

Solution: We begin by finding the highest perfect fourth powers of x^6 and y^{23}. For an index of 4, the highest perfect power that is a factor of x^6 is x^4. The highest perfect power that is a factor of y^{23} is y^{20}.

$$\sqrt[4]{x^6 y^{23}} = \sqrt[4]{x^4 \cdot x^2 \cdot y^{20} \cdot y^3}$$
$$= \sqrt[4]{x^4 y^{20} \cdot x^2 y^3}$$
$$= \sqrt[4]{x^4 y^{20}}\sqrt[4]{x^2 y^3}$$
$$= xy^5 \sqrt[4]{x^2 y^3}$$

Simplifying Radicals   **4**   Now we give a general procedure for simplifying radicals.

> ### To Simplify Radicals
> 1. If the radicand contains a numerical factor, write it as a product of two numbers, one of which is the largest perfect power for the index.
> 2. Write each variable factor as a product of two factors, one of which is the largest perfect power of the variable for the index.
> 3. Use the product rule to write the radical expression as a product of radicals. Place all the perfect powers (numbers and variables) under the same radical.
> 4. Simplify the radical containing the perfect powers.

EXAMPLE 8   Simplify $\sqrt{80x^5 y^{12} z^3}$.

Solution: The highest perfect square that is a factor of 80 is 16. You should recognize that $80 = 16 \cdot 5$. The highest perfect square that is a factor of x^5 is x^4. Note that $x^5 = x^4 \cdot x$. The expression y^{12} is a perfect square. Finally, the highest perfect square that is a factor of z^3 is z^2. We know that $z^3 = z^2 \cdot z$. Place all the perfect squares under the same radical, and then simplify.

$$\sqrt{80x^5y^{12}z^3} = \sqrt{16 \cdot 5 \cdot x^4 \cdot x \cdot y^{12} \cdot z^2 \cdot z}$$
$$= \sqrt{16x^4y^{12}z^2 \cdot 5xz}$$
$$= \sqrt{16x^4y^{12}z^2} \cdot \sqrt{5xz}$$
$$= 4x^2y^6z \sqrt{5xz}$$

EXAMPLE 9 Simplify $\sqrt[3]{54x^{17}y^{25}}$.

Solution: The highest perfect cube that is a factor of 54 is 27. The highest perfect cube that is a factor of x^{17} is x^{15}. The highest perfect cube that is a factor of y^{25} is y^{24}.

$$\sqrt[3]{54x^{17}y^{25}} = \sqrt[3]{27 \cdot 2 \cdot x^{15} \cdot x^2 \cdot y^{24} \cdot y}$$
$$= \sqrt[3]{27x^{15}y^{24} \cdot 2x^2y}$$
$$= \sqrt[3]{27x^{15}y^{24}} \cdot \sqrt[3]{2x^2y}$$
$$= 3x^5y^8 \sqrt[3]{2x^2y}$$

Helpful Hint

In Example 7 we showed that

$$\sqrt[4]{x^6y^{23}} = xy^5\sqrt[4]{x^2y^3}$$

This radical can also be simplified by dividing the exponents on the variables in the radicand, 6 and 23, by the index, 4, and observing the quotients and remainders.

quotient	quotient	remainder	remainder
$6 \div 4$	$23 \div 4$	$6 \div 4$	$23 \div 4$

$$\sqrt[4]{x^6y^{23}} = x^1y^5\sqrt[4]{x^2y^3}$$

Can you explain why this procedure works? You may wish to use this procedure to work or check certain problems.

Multiplying Radicals ⑤ To multiply radicals, we use the product rule given earlier. After multiplying we can often simplify the new radical (see Examples 10 and 11).

EXAMPLE 10 Multiply and simplify. **(a)** $\sqrt{2}\sqrt{8}$ **(b)** $\sqrt[3]{2x}\sqrt[3]{4x^2}$

Solution:

(a) $\sqrt{2}\sqrt{8} = \sqrt{2 \cdot 8}$ **(b)** $\sqrt[3]{2x}\sqrt[3]{4x^2} = \sqrt[3]{2x \cdot 4x^2}$
$\qquad\qquad = \sqrt{16} = 4$ $\qquad\qquad\qquad = \sqrt[3]{8x^3} = 2x$

EXAMPLE 11 Multiply and simplify.

(a) $\sqrt[4]{8x^3y}\sqrt[4]{8x^6y^2}$ **(b)** $\sqrt[3]{5xy^4}\sqrt[3]{50x^2y^{18}}$

Solution:

(a) $\sqrt[4]{8x^3y}\,\sqrt[4]{8x^6y^2} = \sqrt[4]{8x^3y \cdot 8x^6y^2}$ **(b)** $\sqrt[3]{5xy^4}\,\sqrt[3]{50x^2y^{18}} = \sqrt[3]{5xy^4 \cdot 50x^2y^{18}}$

$$= \sqrt[4]{64x^9y^3} \qquad\qquad\qquad\qquad = \sqrt[3]{250x^3y^{22}}$$

$$= \sqrt[4]{16x^8}\,\sqrt[4]{4xy^3} \qquad\qquad\qquad = \sqrt[3]{125x^3y^{21}}\,\sqrt[3]{2y}$$

$$= 2x^2\sqrt[4]{4xy^3} \qquad\qquad\qquad\quad = 5xy^7\sqrt[3]{2y}$$

When a radical is simplified, the radicand does not have any variable with an exponent greater than or equal to the index.

EXAMPLE 12   Multiply and simplify $\sqrt{2x}(\sqrt{8x} - \sqrt{32})$.

Solution: Begin by using the distributive property.

$$\sqrt{2x}(\sqrt{8x} - \sqrt{32}) = (\sqrt{2x})(\sqrt{8x}) + (\sqrt{2x})(-\sqrt{32})$$

$$= \sqrt{16x^2} - \sqrt{64x}$$

$$= 4x - \sqrt{64}\sqrt{x}$$

$$= 4x - 8\sqrt{x}$$

Note in Example 12 that the same result could be obtained by first simplifying $\sqrt{8x}$ and $\sqrt{32}$ and then multiplying. You may wish to try this now.

EXAMPLE 13   Multiply and simplify $\sqrt[3]{3x^2y}(\sqrt[3]{9xy^5} + \sqrt[3]{18x^8y^{10}})$.

Solution:

$$\sqrt[3]{3x^2y}(\sqrt[3]{9xy^5} + \sqrt[3]{18x^8y^{10}}) = (\sqrt[3]{3x^2y})(\sqrt[3]{9xy^5}) + (\sqrt[3]{3x^2y})(\sqrt[3]{18x^8y^{10}})$$

$$= \sqrt[3]{27x^3y^6} + \sqrt[3]{54x^{10}y^{11}}$$

$$= 3xy^2 + \sqrt[3]{27x^9y^9} \cdot \sqrt[3]{2xy^2}$$

$$= 3xy^2 + 3x^3y^3\sqrt[3]{2xy^2}$$

We will do additional multiplication of radicals in Sections 8.4 and 8.5.

Exercise Set 8.3

In this exercise set, assume that all variables represent positive real numbers. Simplify.

1. $\sqrt{50}$ $5\sqrt{2}$ **2.** $\sqrt{40}$ $2\sqrt{10}$ **3.** $\sqrt{32}$ $4\sqrt{2}$ **4.** $\sqrt{72}$ $6\sqrt{2}$

5. $\sqrt[3]{16}$ $2\sqrt[3]{2}$ **6.** $\sqrt[3]{24}$ $2\sqrt[3]{3}$ **7.** $\sqrt[3]{54}$ $3\sqrt[3]{2}$ **8.** $\sqrt[4]{80}$ $2\sqrt[4]{5}$

9. $-\sqrt{x^3}$ $-x\sqrt{x}$ **10.** $\sqrt{y^5}$ $y^2\sqrt{y}$ **11.** $7\sqrt{x^{11}}$ $7x^5\sqrt{x}$ **12.** $\sqrt{a^{30}}$ a^{15}

13. $\sqrt{b^{27}}$ $b^{13}\sqrt{b}$ **14.** $\sqrt[3]{y^7}$ $y^2\sqrt[3]{y}$ **15.** $\sqrt[3]{y^9}$ $y^2\sqrt[3]{y}$ **16.** $\sqrt[4]{b^{23}}$ $b^5\sqrt[4]{b^3}$

17. $\sqrt{24x^3}$ $2x\sqrt{6x}$ **18.** $\sqrt{20x^7}$ $2x^3\sqrt{5x}$ **19.** $3\sqrt[3]{24y^7}$ $6y^2\sqrt[3]{3y}$ **20.** $-3\sqrt[4]{16x^{10}}$ $-6x^2\sqrt[4]{x^2}$

21. $\sqrt{x^3y^7}$ $xy^3\sqrt{xy}$ **22.** $2\sqrt{50xy^4}$ $10y^2\sqrt{2x}$ **23.** $\sqrt[3]{81x^6y^8}$ $3x^2y^2\sqrt[3]{3y^2}$ **24.** $\sqrt[3]{16x^3y^6}$ $2xy^2\sqrt[3]{2}$

25. $4\sqrt[3]{54x^{12}y^{13}}$ $12x^4y^4\sqrt[3]{2y}$ **26.** $\sqrt[4]{x^9y^{12}z^{15}}$ $x^2y^3z^3\sqrt[4]{xz^3}$ **27.** $-\sqrt[5]{64x^{12}y^7}$ $-2x^2y\sqrt[5]{2x^2y^2}$ **28.** $\sqrt[3]{18w^{12}v^9r^{31}}$ $w^4v^3r^{10}\sqrt[3]{18r}$

29. $\sqrt[3]{32c^4w^9z}$ $2cw^3\sqrt[3]{4cz}$ **30.** $\sqrt[4]{32x^8y^9z^{19}}$ $2x^2y^2z^4\sqrt[4]{2yz^3}$ **31.** $\sqrt[3]{81x^7y^{21}z^{50}}$ $3x^2y^7z^{16}\sqrt[3]{3xz^2}$ **32.** $\sqrt[3]{18x^4y^7z^{15}}$ $xy^2z^5\sqrt[3]{18xy}$

Simplify.

33. $\sqrt{5}\sqrt{5}$ 5

34. $\sqrt{60}\sqrt{5}$ $10\sqrt{3}$

35. $\sqrt[3]{2}\sqrt[3]{4}$ 2

36. $\sqrt[3]{2}\sqrt[3]{28}$ $2\sqrt[3]{7}$

37. $\sqrt[3]{3}\sqrt[3]{54}$ $3\sqrt[3]{6}$

38. $\sqrt{5x^2}\sqrt{8x^3}$ $2x^2\sqrt{10x}$

39. $\sqrt{15xy^4}\sqrt{6xy^3}$ $3xy^3\sqrt{10y}$

40. $(\sqrt{6xy^2})^2$ $6xy^2$

41. $(\sqrt{4x^3y^2})^2$ $4x^3y^2$

42. $\sqrt{9x^3y^7}\sqrt{3xy^4}$ $3x^2y^5\sqrt{3y}$

43. $\sqrt[3]{5xy^2}\sqrt[3]{25x^4y^{12}}$ $5xy^4\sqrt[3]{x^2y^2}$

44. $\sqrt[3]{9x^7y^{12}}\sqrt[3]{6x^4y}$ $3x^3y^4\sqrt[3]{2x^2y}$

45. $(\sqrt[3]{2x^3y^4})^2$ $x^2y^2\sqrt[3]{4y^2}$

46. $(\sqrt[3]{5x^2y^6})^2$ $xy^4\sqrt[3]{25x}$

47. $\sqrt[4]{12xy^4}\sqrt[4]{2x^3y^9z^7}$ $xy^3z\sqrt[4]{24yz^3}$

48. $\sqrt[4]{3x^9y^{12}}\sqrt[4]{54x^4y^7}$ $3x^3y^4\sqrt[4]{2xy^3}$

49. $\sqrt[5]{x^{24}y^{30}z^9}\sqrt[5]{x^{13}y^8z^7}$ $x^7y^7z^3\sqrt[5]{x^2y^3z}$

50. $\sqrt[4]{8x^4yz^3}\sqrt[4]{2x^2y^3z^7}$ $2xyz^2\sqrt[4]{x^2z^2}$

60. $2x^2y^2\sqrt[4]{y} - x^2y^2\sqrt[4]{6}$ **61.** $4x^5y^3\sqrt[3]{x} + 4xy^4\sqrt[3]{2x^2y^2}$

Simplify.

51. $\sqrt{2}(\sqrt{6} + \sqrt{2})$ $2\sqrt{3} + 2$

52. $\sqrt{5}(\sqrt{5} + \sqrt{3})$ $5 + \sqrt{15}$

53. $\sqrt{3}(\sqrt{12} - \sqrt{6})$ $6 - 3\sqrt{2}$

54. $2(2\sqrt{8} - 3\sqrt{2})$ $2\sqrt{2}$

55. $\sqrt{2}(\sqrt{18} + \sqrt{8})$ 10

56. $\sqrt{2x}(\sqrt{8x} - \sqrt{32})$ $4x - 8\sqrt{x}$

57. $\sqrt{3y}(\sqrt{27y^2} - \sqrt{y})$ $9y\sqrt{y} - y\sqrt{3}$

58. $\sqrt[3]{x}(\sqrt[3]{x^2} + \sqrt[3]{x^5})$ $x + x^2$

59. $\sqrt[3]{2x^2y}(\sqrt[3]{4xy^5} + \sqrt[3]{12x^{10}y})$

60. $\sqrt[4]{2x^3y^2}(\sqrt[4]{8x^5y^7} - \sqrt[4]{3x^5y^6})$

61. $2\sqrt[3]{x^4y^5}(\sqrt[3]{8x^{12}y^4} + \sqrt[3]{16xy^9})$

62. $\sqrt[3]{4x^2y^6}(\sqrt[3]{9x^8y^5} - \sqrt[3]{7x^9y})$

63. $3\sqrt{2xy^4}(\sqrt{20x^4y^8} - 2\sqrt{6xy^9})$

64. $\sqrt[5]{8x^4y^6}(\sqrt[5]{4x^6y^9} - \sqrt[5]{10xy^7})$

59. $2xy^2 + 2x^4\sqrt[3]{3y^2}$

62. $x^3y^3\sqrt[3]{36xy^2} - x^3y^2\sqrt[3]{28x^2y}$

63. $6x^2y^6\sqrt{10x} - 12xy^6\sqrt{3y}$ **64.** $2x^2y^3 - xy^2\sqrt[5]{80y^3}$

Simplify. These exercises are a combination of the exercises presented earlier in this exercise set.   **89.** $xy\sqrt[3]{12x^2y^2} - 2x^2y^2\sqrt[3]{3}$

65. $\sqrt{24}$ $2\sqrt{6}$

66. $\sqrt{200}$ $10\sqrt{2}$

67. $\sqrt[3]{32}$ $2\sqrt[3]{4}$

68. $\sqrt[4]{162}$ $3\sqrt[4]{2}$

69. $\sqrt[3]{x^5}$ $x\sqrt[3]{x^2}$

70. $\sqrt[4]{y^{13}}$ $y^4\sqrt[4]{y}$

71. $\sqrt{36x^5}$ $6x^2\sqrt{x}$

72. $\sqrt[3]{80x^{11}}$ $2x^3\sqrt[3]{10x^2}$

73. $\sqrt{x^5y^{12}}$ $x^2y^6\sqrt{x}$

74. $\sqrt[3]{x^9y^{11}z}$ $x^3y^3\sqrt[3]{y^2z}$

75. $\sqrt[4]{16ab^{17}c^9}$ $2b^4c^2\sqrt[4]{abc}$

76. $\sqrt[5]{32a^2b^5}$ $2b\sqrt[5]{a^2}$

77. $\sqrt{75}\sqrt{6}$ $15\sqrt{2}$

78. $\sqrt[4]{8}\sqrt[4]{10}$ $2\sqrt[4]{5}$

79. $\sqrt{15x^2}\sqrt{6x^5}$ $3x^3\sqrt{10x}$

80. $\sqrt{14xy^2}\sqrt{3xy^3}$ $xy^2\sqrt{42y}$

81. $\sqrt{20xy^4}\sqrt{6x^5y^7}$ $2x^3y^5\sqrt{30y}$

82. $\sqrt{6}(4 - \sqrt{2})$ $4\sqrt{6} - 2\sqrt{3}$

83. $\sqrt{x}(\sqrt{x} + 3)$ $x + 3\sqrt{x}$

84. $\sqrt{y}(\sqrt{y^3} - 2)$ $y^2 - 2\sqrt{y}$

85. $\sqrt[3]{4xy^2}\sqrt[3]{4xy^4}$ $2y^2\sqrt[3]{2x^2}$

86. $(\sqrt[3]{4x^5y^2})^2$ $2x^3y\sqrt[3]{2xy}$

87. $\sqrt[3]{y}(2\sqrt[3]{y} - \sqrt[3]{y^8})$ $2\sqrt[3]{y^2} - y^3$

88. $\sqrt[3]{2x^9y^6z}\sqrt[3]{12xy^4z^3}$ $2x^3y^3z\sqrt[3]{3xyz}$

89. $\sqrt[3]{3xy^2}(\sqrt[3]{4x^4y^3} - \sqrt[3]{8x^5y^4})$

90. $\sqrt[4]{4xy^2}(\sqrt[4]{2x^5y^6} + \sqrt[4]{5x^9y^2})$ $xy^2\sqrt[4]{8x^2} + x^2y\sqrt[4]{20x^2}$

✎ **91. (a)** How do you obtain the numbers that are perfect squares?   **(a)** square the natural numbers
   (b) List the first six perfect squares.   **(b)** 1, 4, 9, 16, 25, 36

✎ **92. (a)** How do you obtain the numbers that are perfect cubes?   **(a)** cube the natural numbers
   (b) List the first six perfect cube numbers.

✎ **93. (a)** How do you obtain numbers that are perfect fifth powers? **(a)** Raise the natural numbers to the fifth power.
   (b) List the first five perfect fifth-power numbers.

✎ **94.** In your own words state the product rule for radicals.

✎ **95.** We stated that for nonnegative real numbers a and b that $\sqrt[n]{a} \cdot \sqrt[n]{b} = \sqrt[n]{ab}$. Why is it necessary to specify that both a and b are nonnegative real numbers?

✎ **96. (a)** In your own words, explain how to simplify radicals.
   (b) Simplify $\sqrt{32x^5y^4}$ using the procedure given in part **(a)**. $4x^2y^2\sqrt{2x}$

92. (b) 1, 8, 27, 64, 125, 216

93. (b) 1, 32, 243, 1024, 3125

95. If n is even and a or b is negative, numbers are not real.

97. a quotient of two integers, denominator not zero

98. a number that can be represented on the real number line

99. a real number that cannot be expressed as a quotient of two integers

102. (a)   **(b)** $(-\frac{1}{2}, 4]$ **(c)** $\{x \mid -\frac{1}{2} < x \le 4\}$

CUMULATIVE REVIEW EXERCISES

[1.2] ✎ **97.** What is a rational number?

[1.3] ✎ **98.** What is a real number?

✎ **99.** What is an irrational number?

100. What is the definition of $|a|$? $|a| = \begin{cases} a, & a \ge 0 \\ -a, & a < 0 \end{cases}$

[2.2] **101.** Solve the formula $E = \frac{1}{2}mv^2$ for m. $m = \dfrac{2E}{v^2}$

[2.5] **102.** Solve the inequality $-4 < 2x - 3 \le 5$ and indicate the solution **(a)** on the number line; **(b)** in internal notation; **(c)** in set builder notation.

Group Activity/ Challenge Problems

1. The formula for the period of a pendulum (the time required for the pendulum to make one complete swing back and forth) is $T = 2\pi \sqrt{\dfrac{l}{g}}$ where T is the period in seconds, l is its length in feet, and g is the acceleration of gravity. On Earth, gravity is 32 ft/sec². The formula when used on Earth becomes

$$T = 2\pi \sqrt{\dfrac{l}{32}}.$$

(a) Find the period of a pendulum whose length is 6 feet.   ≈ 2.72 sec

(b) If the length of a pendulum is doubled, what effect will this have on the period? Explain. $\sqrt{2} \cdot T$

(c) The gravity on the moon is 1/6 of that on Earth. If a pendulum has a period of 2 seconds on Earth, what will be the period of the same pendulum on the moon?   The period is $2\sqrt{6}$ or ≈4.90 sec.

2. A formula for the length of a diagonal from the upper corner of a box to the opposite lower corner is

$$d = \sqrt{L^2 + W^2 + H^2}$$

where L, W, and H are the length, width and height respectively.

(a) Find the diagonal of a box with length of 2.4 feet, width of 2 feet and height 18 inches.   **(a)** $\sqrt{12.01} \approx 3.47$ ft   **(b)** The length of the diagonal is doubled.

(b) If the length, width, and height are all doubled, how will the diagonal change?

3. Consider $F = \sqrt{ab^2}$. How will the value of F change if:

(a) Both a and b are doubled?   **(a)** $\sqrt{8} \cdot F$

(b) a is doubled and b is halved?   **(b)** $\sqrt{\dfrac{1}{2}} \cdot F$

(c) Both a and b are halved?   **(c)** $\sqrt{\dfrac{1}{8}} \cdot F$

Multiply each of the following using the FOIL method. We will discuss problems like these in Section 8.4.

4. $(\sqrt{3} + \sqrt{2})(\sqrt{3} - \sqrt{2})$ 1   **5.** $(2\sqrt{x} - 3\sqrt{y})(2\sqrt{x} + 3\sqrt{y})$ $4x - 9y$

6. $(5\sqrt{a} - 4\sqrt{b})(2\sqrt{a} + 3\sqrt{b})$ $10a + 7\sqrt{ab} - 12b$

7. Prove $\sqrt{a \cdot b} = \sqrt{a}\,\sqrt{b}$ by converting $\sqrt{a \cdot b}$ to exponential form.
$(ab)^{1/2} = a^{1/2}b^{1/2} = \sqrt{a}\,\sqrt{b}$

8.4 Dividing and Simplifying Radicals

Tape 13

1. Apply the quotient rule for radicals.
2. Know when a radical is simplified.
3. Rationalize denominators.
4. Rationalize denominators using the conjugate.

Quotient Rule for Radicals

1. In mathematics we sometimes need to divide one radical expression by another. To divide radicals, or to simplify radicals containing fractions, we use the quotient rule for radicals.

> ### Quotient Rule for Radicals
> For nonnegative real numbers a and b,
> $$\frac{\sqrt[n]{a}}{\sqrt[n]{b}} = \sqrt[n]{\frac{a}{b}} \qquad b \neq 0$$

Examples 1 through 3 illustrate how to use the quotient rule to simplify radical expressions.

EXAMPLE 1 Simplify. **(a)** $\dfrac{\sqrt{75}}{\sqrt{3}}$ **(b)** $\dfrac{\sqrt[3]{24}}{\sqrt[3]{3}}$

Solution: **(a)** $\dfrac{\sqrt{75}}{\sqrt{3}} = \sqrt{\dfrac{75}{3}} = \sqrt{25} = 5$ **(b)** $\dfrac{\sqrt[3]{24}}{\sqrt[3]{3}} = \sqrt[3]{\dfrac{24}{3}} = \sqrt[3]{8} = 2$

EXAMPLE 2 Simplify. **(a)** $\sqrt{\dfrac{9}{4}}$ **(b)** $\sqrt[3]{\dfrac{8}{27}}$

Solution: **(a)** $\sqrt{\dfrac{9}{4}} = \dfrac{\sqrt{9}}{\sqrt{4}} = \dfrac{3}{2}$ **(b)** $\sqrt[3]{\dfrac{8}{27}} = \dfrac{\sqrt[3]{8}}{\sqrt[3]{27}} = \dfrac{2}{3}$

EXAMPLE 3 Simplify.   **(a)** $\sqrt{\dfrac{16x^2}{8}}$ **(b)** $\sqrt{\dfrac{4x^5y^7}{16x^3y^{13}}}$ **(c)** $\sqrt[4]{\dfrac{15xy^5}{3x^9y}}$

Solution:

(a) $\sqrt{\dfrac{16x^2}{8}} = \sqrt{2x^2} = \sqrt{2}\sqrt{x^2} = \sqrt{2}\,x$ or $x\sqrt{2}$

(b) $\sqrt{\dfrac{4x^5y^7}{16x^3y^{13}}} = \sqrt{\dfrac{x^2}{4y^6}} = \dfrac{\sqrt{x^2}}{\sqrt{4y^6}} = \dfrac{x}{2y^3}$

(c) $\sqrt[4]{\dfrac{15xy^5}{3x^9y}} = \sqrt[4]{\dfrac{5y^4}{x^8}} = \dfrac{\sqrt[4]{5y^4}}{\sqrt[4]{x^8}} = \dfrac{\sqrt[4]{y^4}\,\sqrt[4]{5}}{x^2} = \dfrac{y\sqrt[4]{5}}{x^2}$

Simplified Radicals   **2**   After you have simplified a radical expression, you should check it to make sure that it is simplified as far as possible.

> ### A Radical Expression Is Simplified When the Following Are All True
> **1.** No perfect powers are factors of the radicand.
> **2.** No radicand contains a fraction.
> **3.** No denominator contains a radical.

Rationalizing Denominators

3 When the denominator of a fraction contains a radical, we generally simplify the expression by **rationalizing the denominator.** To rationalize a denominator is to remove all radicals from the denominator. Denominators are rationalized because, without a calculator, it is often easier to evaluate a fraction with a whole-number denominator than one with a radical.

To rationalize a denominator, multiply both the numerator and the denominator of the fraction by the denominator, or by a radical that will result in the radicand in the denominator becoming a perfect power. The following examples illustrate the procedure.

EXAMPLE 4 Simplify $\dfrac{1}{\sqrt{5}}$.

Solution: To simplify this expression, we must rationalize the denominator.

$$\frac{1}{\sqrt{5}} = \frac{1}{\sqrt{5}} \cdot \frac{\sqrt{5}}{\sqrt{5}} = \frac{\sqrt{5}}{\sqrt{25}} = \frac{\sqrt{5}}{5}$$

In Example 4, multiplying both the numerator and denominator by $\sqrt{5}$ is equivalent to multiplying the fraction by 1. This does not change the value of the fraction, but it makes the radicand in the denominator a perfect square.

EXAMPLE 5 Simplify.

(a) $\sqrt{\dfrac{2}{3}}$ **(b)** $\dfrac{x}{3\sqrt{2}}$ **(c)** $\sqrt{\dfrac{x}{y}}$

Solution:

(a) $\sqrt{\dfrac{2}{3}} = \dfrac{\sqrt{2}}{\sqrt{3}} = \dfrac{\sqrt{2}}{\sqrt{3}} \cdot \dfrac{\sqrt{3}}{\sqrt{3}} = \dfrac{\sqrt{6}}{3}$

(b) $\dfrac{x}{3\sqrt{2}} = \dfrac{x}{3\sqrt{2}} \cdot \dfrac{\sqrt{2}}{\sqrt{2}} = \dfrac{x\sqrt{2}}{3 \cdot 2} = \dfrac{x\sqrt{2}}{6}$

(c) $\sqrt{\dfrac{x}{y}} = \dfrac{\sqrt{x}}{\sqrt{y}} = \dfrac{\sqrt{x}}{\sqrt{y}} \cdot \dfrac{\sqrt{y}}{\sqrt{y}} = \dfrac{\sqrt{xy}}{\sqrt{y^2}} = \dfrac{\sqrt{xy}}{y}$

EXAMPLE 6 Simplify $\sqrt[3]{\dfrac{3}{5}}$.

Solution: $\sqrt[3]{\dfrac{3}{5}} = \dfrac{\sqrt[3]{3}}{\sqrt[3]{5}}$. Since the denominator is a cube root, we must multiply the numerator and denominator by the cube root of an expression that will result in the product of the radicands in the denominator being a perfect cube. Multiply both the numerator and denominator by $\sqrt[3]{5^2}$.

$$\frac{\sqrt[3]{3}}{\sqrt[3]{5}} = \frac{\sqrt[3]{3}}{\sqrt[3]{5}} \cdot \frac{\sqrt[3]{5^2}}{\sqrt[3]{5^2}}$$

$$= \frac{\sqrt[3]{3} \cdot \sqrt[3]{5^2}}{\sqrt[3]{5^3}}$$

$$= \frac{\sqrt[3]{3}\sqrt[3]{25}}{5}$$

$$= \frac{\sqrt[3]{75}}{5}$$

EXAMPLE 7 Simplify $\sqrt[3]{\dfrac{x}{2y^2}}$.

Solution: First use the quotient rule to rewrite the cube root as the quotient of two cube roots.

$$\sqrt[3]{\frac{x}{2y^2}} = \frac{\sqrt[3]{x}}{\sqrt[3]{2y^2}}$$

Then multiply both the numerator and denominator by the cube root of an expression that will make the product of the radicands in the denominator a perfect cube.

The resulting exponents in the radicand in the denominator must be divisible by 3. Since the denominator is $\sqrt[3]{2y^2}$, we must multiply the expression by $\sqrt[3]{2^2 y}$. Note that $2 \cdot 2^2 = 2^3$ and $y^2 \cdot y = y^3$. Multiply both the numerator and denominator by $\sqrt[3]{2^2 y}$ and then simplify.

$$\frac{\sqrt[3]{x}}{\sqrt[3]{2y^2}} = \frac{\sqrt[3]{x}}{\sqrt[3]{2y^2}} \cdot \frac{\sqrt[3]{2^2 y}}{\sqrt[3]{2^2 y}}$$

$$= \frac{\sqrt[3]{x}\sqrt[3]{4y}}{\sqrt[3]{2^3 y^3}}$$

$$= \frac{\sqrt[3]{4xy}}{2y}$$

EXAMPLE 8 Simplify $\sqrt{\dfrac{12x^3 y^5}{5z}}$.

Solution: $\sqrt{\dfrac{12x^3 y^5}{5z}} = \dfrac{\sqrt{12x^3 y^5}}{\sqrt{5z}}$

Now simplify the numerator.

$$\frac{\sqrt{12x^3 y^5}}{\sqrt{5z}} = \frac{\sqrt{4x^2 y^4}\sqrt{3xy}}{\sqrt{5z}} = \frac{2xy^2 \sqrt{3xy}}{\sqrt{5z}}$$

Now rationalize the denominator.

$$\frac{2xy^2\sqrt{3xy}}{\sqrt{5z}} = \frac{2xy^2\sqrt{3xy}}{\sqrt{5z}} \cdot \frac{\sqrt{5z}}{\sqrt{5z}}$$

$$= \frac{2xy^2\sqrt{15xyz}}{\sqrt{5^2z^2}}$$

$$= \frac{2xy^2\sqrt{15xyz}}{5z}$$

Conjugates

4 We will discuss adding and subtracting radicals in detail in Section 8.5. In this section we give only a brief introduction to this topic. Only radicals that have the same radicands and index may be added or subtracted. We add such radicals, called *like radicals*, by adding or subtracting their coefficients and multiplying this sum or difference by the like radical.

Examples

$$3\sqrt{3} - 3\sqrt{3} = 0 \qquad 3\sqrt{x} + \sqrt{x} = 3\sqrt{x} + 1\sqrt{x} = 4\sqrt{x}$$

$$2\sqrt{5} - 9\sqrt{5} = -7\sqrt{5} \qquad 3\sqrt{y} - 3\sqrt{y} = 0$$

The terms $\sqrt{2} + \sqrt{3}$ cannot be added in radical form since the radicands are different.

When the denominator of a rational expression is a binomial that contains a radical, we rationalize the denominator. We do this by multiplying both the numerator and the denominator of the fraction by the **conjugate** of the denominator. The conjugate of a binomial is a binomial having the same two terms with the sign of the second term changed.

Expression	*Conjugate*
$3 + \sqrt{2}$	$3 - \sqrt{2}$
$2\sqrt{3} - \sqrt{5}$	$2\sqrt{3} + \sqrt{5}$
$\sqrt{x} + \sqrt{y}$	$\sqrt{x} - \sqrt{y}$
$a + \sqrt{b}$	$a - \sqrt{b}$

When a binomial is multiplied by its conjugate using the FOIL method, the outer and inner products will sum to zero.

EXAMPLE 9 Multiply $(2 + \sqrt{3})(2 - \sqrt{3})$.

Solution: Multiply using the FOIL method.

$$\overset{\text{F}}{2(2)} + \overset{\text{O}}{2(-\sqrt{3})} + \overset{\text{I}}{2(\sqrt{3})} + \overset{\text{L}}{\sqrt{3}(-\sqrt{3})} = 4 - 2\sqrt{3} + 2\sqrt{3} - \sqrt{9}$$

$$= 4 - \sqrt{9}$$

$$= 4 - 3 = 1$$

In Example 9, we would get the same result using the formula for the difference-of-two-squares, $(a + b)(a - b) = a^2 - b^2$.

$$(2 + \sqrt{3})(2 - \sqrt{3}) = 2^2 - (\sqrt{3})^2$$
$$= 4 - 3 = 1$$

EXAMPLE 10 Multiply $(\sqrt{3} - \sqrt{5})(\sqrt{3} + \sqrt{5})$.

Solution: $(\sqrt{3} - \sqrt{5})(\sqrt{3} + \sqrt{5}) = (\sqrt{3})^2 - (\sqrt{5})^2$
$$= 3 - 5 = -2$$

EXAMPLE 11 Simplify $\dfrac{5}{2 + \sqrt{3}}$.

Solution: To simplify this expression, we must rationalize the denominator. Multiply the numerator and denominator by $2 - \sqrt{3}$, the conjugate of $2 + \sqrt{3}$.

$$\frac{5}{2 + \sqrt{3}} \cdot \frac{2 - \sqrt{3}}{2 - \sqrt{3}} = \frac{5(2 - \sqrt{3})}{(2 + \sqrt{3})(2 - \sqrt{3})}$$

$$= \frac{5(2 - \sqrt{3})}{4 - 3}$$

$$= 5(2 - \sqrt{3}) \quad \text{or} \quad 10 - 5\sqrt{3}$$

EXAMPLE 12 Simplify $\dfrac{6}{\sqrt{5} - \sqrt{2}}$.

Solution:

$$\frac{6}{\sqrt{5} - \sqrt{2}} \cdot \frac{\sqrt{5} + \sqrt{2}}{\sqrt{5} + \sqrt{2}} = \frac{6(\sqrt{5} + \sqrt{2})}{5 - 2}$$

$$= \frac{\overset{2}{\cancel{6}}(\sqrt{5} + \sqrt{2})}{\underset{1}{\cancel{3}}}$$

$$= 2(\sqrt{5} + \sqrt{2}) \quad \text{or} \quad 2\sqrt{5} + 2\sqrt{2}$$

EXAMPLE 13 Simplify $\dfrac{x - \sqrt{y}}{x + \sqrt{y}}$.

Solution: Multiply both the numerator and denominator of the fraction by the conjugate of the denominator, $x - \sqrt{y}$.

$$\frac{x - \sqrt{y}}{x + \sqrt{y}} \cdot \frac{x - \sqrt{y}}{x - \sqrt{y}} = \frac{x^2 - x\sqrt{y} - x\sqrt{y} + \sqrt{y^2}}{x^2 - y}$$

$$= \frac{x^2 - 2x\sqrt{y} + y}{x^2 - y}$$

Remember that you cannot divide out x^2 or y because they are terms, not factors.

COMMON STUDENT ERROR   The following simplifications are correct because the numbers and variables divided out are not within square roots.

Correct   *Correct*

$$\frac{\overset{2}{\cancel{6}}\sqrt{2}}{\underset{1}{\cancel{3}}} = 2\sqrt{2} \qquad \frac{\cancel{x}\sqrt{2}}{\cancel{x}} = \sqrt{2}$$

An expression within a square root cannot be divided by an expression not within the square root.

Correct   *Incorrect*

$$\frac{\sqrt{2}}{2} \quad \text{cannot be simplified further} \qquad \frac{\sqrt{\cancel{2}^1}}{\underset{1}{\cancel{2}}} = \sqrt{1} = 1$$

$$\frac{\sqrt{x^3}}{x} = \frac{\sqrt{x^2}\sqrt{x}}{x} = \frac{\cancel{x}\sqrt{x}}{\cancel{x}} = \sqrt{x} \qquad \frac{\sqrt{x\cancel{^3}^2}}{\cancel{x}} = \sqrt{x^2} = x$$

Exercise Set 8.4

In this exercise set assume that all variables represent positive real numbers.

Simplify.

1. $\sqrt{\dfrac{27}{3}}$ 3

2. $\sqrt{\dfrac{4}{25}}$ $\dfrac{2}{5}$

3. $\dfrac{\sqrt{3}}{\sqrt{27}}$ $\dfrac{1}{3}$

4. $\sqrt{\dfrac{16}{25}}$ $\dfrac{4}{5}$

5. $\sqrt[3]{\dfrac{2}{16}}$ $\dfrac{1}{2}$

6. $\dfrac{\sqrt[3]{108}}{\sqrt[3]{3}}$ $\sqrt[3]{36}$

7. $\dfrac{-\sqrt{24}}{\sqrt{3}}$ $-2\sqrt{2}$

8. $\sqrt[3]{\dfrac{c^3}{27}}$ $\dfrac{c}{3}$

9. $\sqrt{\dfrac{r^4}{25}}$ $\dfrac{r^2}{5}$

10. $\dfrac{\sqrt[3]{2x^6}}{\sqrt[3]{16x^3}}$ $\dfrac{x}{2}$

11. $-\sqrt{\dfrac{27x^6}{3x^2}}$ $-3x^2$

12. $-\sqrt{\dfrac{72x^2y^5}{8x^2y^7}}$ $-\dfrac{3}{y}$

13. $\sqrt[3]{\dfrac{7xy}{8x^{13}}}$ $\dfrac{\sqrt[3]{7y}}{2x^4}$

14. $\sqrt[3]{\dfrac{25x^2y^5}{5x^8y}}$ $\dfrac{y\sqrt[3]{5y}}{x^2}$

15. $\sqrt[4]{\dfrac{20x^4}{81}}$ $\dfrac{x\sqrt[4]{20}}{3}$

16. $\sqrt[4]{\dfrac{3a^6b^5}{16a^2b^{13}}}$ $\dfrac{a\sqrt[4]{3}}{2b^2}$

Simplify.

17. $\dfrac{1}{\sqrt{3}}$ $\dfrac{\sqrt{3}}{3}$

18. $\dfrac{3}{\sqrt{3}}$ $\sqrt{3}$

19. $\dfrac{\sqrt{m}}{\sqrt{2}}$ $\dfrac{\sqrt{2m}}{2}$

20. $\dfrac{5\sqrt{3}}{\sqrt{5}}$ $\sqrt{15}$

21. $\dfrac{2x}{\sqrt{6}}$ $\dfrac{x\sqrt{6}}{3}$

22. $-\dfrac{\sqrt{x}}{\sqrt{y}}$ $-\dfrac{\sqrt{xy}}{y}$

23. $\dfrac{2\sqrt{3}}{\sqrt{w}}$ $\dfrac{2\sqrt{3w}}{w}$

24. $\sqrt{\dfrac{4}{5x}}$ $\dfrac{2\sqrt{5x}}{5x}$

25. $\sqrt{\dfrac{5m}{8}}$ $\dfrac{\sqrt{10m}}{4}$

26. $-\sqrt{\dfrac{x}{2y^2}}$ $-\dfrac{\sqrt{2x}}{2y}$

27. $\dfrac{2\sqrt{3}}{\sqrt{y^3}}$ $\dfrac{2\sqrt{3y}}{y^2}$

28. $\dfrac{2n}{\sqrt{18n}}$ $\dfrac{\sqrt{2n}}{3}$

29. $-\dfrac{4\sqrt{5}}{\sqrt{32a}}$ $-\dfrac{\sqrt{10a}}{2a}$

30. $\sqrt{\dfrac{120x}{4y^3}}$ $\dfrac{\sqrt{30xy}}{y^2}$

31. $\sqrt{\dfrac{8x^5y}{2z}}$ $\dfrac{2x^2\sqrt{xyz}}{z}$

32. $\sqrt{\dfrac{18x^4y^6}{3z}}$ $\dfrac{x^2y^3\sqrt{6z}}{z}$

33. $\sqrt{\dfrac{5pq^4}{2r}}$ $\dfrac{q^2\sqrt{10pr}}{2r}$

34. $-\sqrt{\dfrac{20y^4z^3}{3x}}$ $-\dfrac{2y^2z\sqrt{15xz}}{3x}$

35. $\sqrt{\dfrac{5xy^6}{6z}}$ $\dfrac{y^3\sqrt{30xz}}{6z}$

36. $\sqrt{\dfrac{15x^5z^7}{2y}}$ $\dfrac{x^2z^3\sqrt{30xyz}}{2y}$

37. $-\sqrt{\dfrac{18x^4y^3}{2z^3}}$ $-\dfrac{3x^2y\sqrt{yz}}{z^2}$

38. $\sqrt{\dfrac{45y^{12}z^{10}}{2x}}$ $\dfrac{3y^6z^5\sqrt{10x}}{2x}$

Simplify.

39. $\dfrac{1}{\sqrt[3]{2}}$ $\dfrac{\sqrt[3]{4}}{2}$

40. $\dfrac{2}{\sqrt[3]{4}}$ $\sqrt[3]{2}$

41. $\dfrac{1}{\sqrt[3]{3}}$ $\dfrac{\sqrt[3]{9}}{3}$

42. $\dfrac{5}{\sqrt[3]{x}}$ $\dfrac{5\sqrt[3]{x^2}}{x}$

43. $\sqrt[3]{\dfrac{5x}{y}}$ $\dfrac{\sqrt[3]{5xy^2}}{y}$

44. $\sqrt[3]{\dfrac{1}{4x}}$ $\dfrac{\sqrt[3]{2x^2}}{2x}$

45. $-\sqrt[3]{\dfrac{5c}{9y^2}}$ $-\dfrac{\sqrt[3]{15cy}}{3y}$

46. $\dfrac{3}{\sqrt[4]{a}}$ $\dfrac{3\sqrt[4]{a^3}}{a}$

47. $\dfrac{5m}{\sqrt[4]{2}}$ $\dfrac{5m\sqrt[4]{8}}{2}$

48. $\sqrt[4]{\dfrac{5}{3x^3}}$ $\dfrac{\sqrt[4]{135x}}{3x}$

49. $\sqrt[4]{\dfrac{2x^3}{4y^2}}$ $\dfrac{\sqrt[4]{8x^3y^2}}{2y}$

50. $\sqrt[3]{\dfrac{3x^2}{2y^2}}$ $\dfrac{\sqrt[3]{12x^2y}}{2y}$

51. $\sqrt[3]{\dfrac{15x^6y^7}{2z^2}}$ $\dfrac{x^2y^2\sqrt[3]{60yz}}{2z}$

52. $\sqrt[3]{\dfrac{8xy^2}{2z^2}}$ $\dfrac{\sqrt[3]{4xy^2z}}{z}$

53. $\sqrt[3]{\dfrac{32r^4s^9}{4r^5}}$ $\dfrac{2s^3\sqrt[3]{r^2}}{r}$

54. $\sqrt[4]{\dfrac{5x^4y^5z}{2x^7}}$ $\dfrac{y\sqrt[4]{40xyz}}{2x}$

Simplify.

55. $(3 - \sqrt{3})(3 + \sqrt{3})$ 6

56. $(4 + \sqrt{2})(4 - \sqrt{2})$ 14

57. $(6 - \sqrt{5})(6 + \sqrt{5})$ 31

58. $(\sqrt{8} - 3)(\sqrt{8} + 3)$ −1

59. $(\sqrt{x} + 5)(\sqrt{x} - 5)$ $x - 25$

60. $(\sqrt{6} + x)(\sqrt{6} - x)$ $6 - x^2$

61. $(\sqrt{x} + y)(\sqrt{x} - y)$ $x - y^2$

62. $(\sqrt{x} + \sqrt{y})(\sqrt{x} - \sqrt{y})$ $x - y$

63. $(x + \sqrt{y})(x - \sqrt{y})$ $x^2 - y$

64. $(\sqrt{7} + \sqrt{3})(\sqrt{7} - \sqrt{3})$ 4

65. $(5 - \sqrt{y})(5 + \sqrt{y})$ $25 - y$

66. $(\sqrt{3} - \sqrt{5})(\sqrt{3} + \sqrt{5})$ −2

67. $(2\sqrt{3} - \sqrt{2})(2\sqrt{3} + \sqrt{2})$ 10

68. $(3\sqrt{a} - 7\sqrt{b})(3\sqrt{a} + 7\sqrt{b})$ $9a - 49b$

Simplify.   **85.** $\dfrac{c + \sqrt{cd} - \sqrt{2cd} - d\sqrt{2}}{c - d}$ **87.** $\dfrac{x\sqrt{y} - y\sqrt{x}}{x - y}$

69. $\dfrac{3}{1 + \sqrt{2}}$ $3\sqrt{2} - 3$

70. $\dfrac{1}{2 + \sqrt{3}}$ $2 - \sqrt{3}$

71. $\dfrac{3}{\sqrt{6} - 5}$ $\dfrac{-3\sqrt{6} - 15}{19}$

72. $\dfrac{3}{\sqrt{2} + 5}$ $\dfrac{-3\sqrt{2} + 15}{23}$

73. $\dfrac{4}{\sqrt{2} - 7}$ $\dfrac{-4\sqrt{2} - 28}{47}$

74. $\dfrac{2}{\sqrt{2} + \sqrt{3}}$ $2\sqrt{3} - 2\sqrt{2}$

75. $\dfrac{\sqrt{5}}{2\sqrt{5} - \sqrt{6}}$ $\dfrac{10 + \sqrt{30}}{14}$

76. $\dfrac{8}{\sqrt{5} - \sqrt{8}}$ $\dfrac{-8\sqrt{5} - 16\sqrt{2}}{3}$

77. $\dfrac{1}{\sqrt{17} - \sqrt{8}}$ $\dfrac{\sqrt{17} + 2\sqrt{2}}{9}$

78. $\dfrac{2}{6 + \sqrt{x}}$ $\dfrac{12 - 2\sqrt{x}}{36 - x}$

79. $\dfrac{3\sqrt{5}}{\sqrt{a} - 3}$ $\dfrac{3\sqrt{5a} + 9\sqrt{5}}{a - 9}$

80. $\dfrac{5}{3 + \sqrt{x}}$ $\dfrac{15 - 5\sqrt{x}}{9 - x}$

81. $\dfrac{4\sqrt{x}}{\sqrt{x} - y}$ $\dfrac{4x + 4y\sqrt{x}}{x - y^2}$

82. $\dfrac{\sqrt{8x}}{x + \sqrt{y}}$ $\dfrac{2x\sqrt{2x} - 2\sqrt{2xy}}{x^2 - y}$

83. $\dfrac{\sqrt{2} - 2\sqrt{3}}{\sqrt{2} + 4\sqrt{3}}$ $\dfrac{-13 + 3\sqrt{6}}{23}$

84. $\dfrac{\sqrt{x} - 2}{\sqrt{x} + 4}$ $\dfrac{x - 6\sqrt{x} + 8}{x - 16}$

85. $\dfrac{\sqrt{c} - \sqrt{2d}}{\sqrt{c} - \sqrt{d}}$

86. $\dfrac{\sqrt{a^3} + \sqrt{a^7}}{\sqrt{a}}$ $a + a^3$

87. $\dfrac{2\sqrt{xy} - \sqrt{xy}}{\sqrt{x} + \sqrt{y}}$

88. $\dfrac{2}{\sqrt{x + 2} - 3}$ $\dfrac{2\sqrt{x + 2} + 6}{x - 7}$

Simplify. These exercises are a combination of the exercises presented earlier in this exercise set.   **93.** −1

89. $\sqrt{\dfrac{x}{9}}$ $\dfrac{\sqrt{x}}{3}$

90. $\sqrt[4]{\dfrac{x^4}{16}}$ $\dfrac{x}{2}$

91. $\sqrt{\dfrac{2}{5}}$ $\dfrac{\sqrt{10}}{5}$

92. $\sqrt{\dfrac{x}{y}}$ $\dfrac{\sqrt{xy}}{y}$

93. $(\sqrt{5} + \sqrt{6})(\sqrt{5} - \sqrt{6})$ **94.** $\sqrt[3]{\dfrac{1}{3}}$ $\dfrac{\sqrt[3]{9}}{3}$

95. $\sqrt{\dfrac{24x^3y^6}{5z}}$ $\dfrac{2xy^3\sqrt{30xz}}{5z}$

96. $\dfrac{6}{4 - \sqrt{y}}$ $\dfrac{24 + 6\sqrt{y}}{16 - y}$

97. $\sqrt{\dfrac{12xy^4}{2x^3y^4}}$ $\dfrac{\sqrt{6}}{x}$

98. $\dfrac{4x}{\sqrt[3]{5y}}$ $\dfrac{4x\sqrt[3]{25y^2}}{5y}$

99. $(\sqrt{x} + 3)(\sqrt{x} - 3)$ $x - 9$ **100.** $\dfrac{\sqrt{x}}{\sqrt{x} + 5\sqrt{y}}$ $\dfrac{x - 5\sqrt{xy}}{x - 25y}$

101. $-\dfrac{7\sqrt{x}}{\sqrt{98}}$ $-\dfrac{\sqrt{2x}}{2}$

102. $\sqrt{\dfrac{2xy^4}{18xy^2}}$ $\dfrac{y}{3}$

103. $\sqrt[4]{\dfrac{3y^2}{2x}}$ $\dfrac{\sqrt[4]{24x^3y^2}}{2x}$

104. $\sqrt{\dfrac{25x^2y^5}{3z}}$ $\dfrac{5xy^2\sqrt{3yz}}{3z}$

105. $\sqrt[3]{\dfrac{32y^{12}z^{10}}{2x}}$

106. $\dfrac{\sqrt{3} + \sqrt{4}}{\sqrt{2} + \sqrt{3}}$

107. $\dfrac{\sqrt{ar}}{\sqrt{a} - 2\sqrt{r}}$

108. $\sqrt[4]{\dfrac{2}{9x}}$

109. $\dfrac{\sqrt[3]{6x}}{\sqrt[3]{5xy}}$

110. $(3\sqrt{y} - 2\sqrt{x})(5\sqrt{y} + \sqrt{x})$

111. $\sqrt[4]{\dfrac{2x^7y^{12}z^4}{3x^9}}$

112. $\dfrac{3}{\sqrt{y + 3} - \sqrt{y}}$

105. $\dfrac{2y^4z^3\sqrt[3]{2x^2z}}{x}$ **106.** $-\sqrt{6} + 3 - 2\sqrt{2} + 2\sqrt{3}$ **107.** $\dfrac{a\sqrt{r} + 2r\sqrt{a}}{a - 4r}$ **108.** $\dfrac{\sqrt[4]{18x^3}}{3x}$

109. $\dfrac{\sqrt[3]{150y^2}}{5y}$ **110.** $15y - 7\sqrt{xy} - 2x$ **111.** $\dfrac{y^3z\sqrt[4]{54x^2}}{3x}$ **112.** $\sqrt{y + 3} + \sqrt{y}$

115. to remove radicals from a denominator   **117. (b)** $-\dfrac{7 + 2\sqrt{10}}{3}$ **118.** both equal 0.447213596   **120.** both equal 0.843432665.

113. In your own words state the quotient rule for radicals.

114. (a) What is the conjugate of a binomial?
(b) What is the conjugate of $x - \sqrt{3}$? $x + \sqrt{3}$

115. What does it mean to rationalize a denominator?

116. (a) Explain how to rationalize a denominator that contains a radical expression of one term.
(b) Rationalize $\dfrac{4}{\sqrt{3y}}$ using part (a). $\dfrac{4\sqrt{3y}}{3y}$

117. (a) Explain how to rationalize a denominator that contains a binomial in which one or both terms is a radical expression.
(b) Rationalize $\dfrac{\sqrt{2} + \sqrt{5}}{\sqrt{2} - \sqrt{5}}$ using part **(a)**.

118. In Example 4 we showed $\dfrac{1}{\sqrt{5}} = \dfrac{\sqrt{5}}{5}$. Using your calculator, show that this is true.

119. Consider the expression $\dfrac{1}{\sqrt{18}}$. Rationalize the denominator by:

(a) First simplifying $\sqrt{18}$ and then rationalizing
(b) Multiplying both numerator and denominator by
(b) $\dfrac{\sqrt{2}}{\sqrt{2}}$ **(c)** $\dfrac{\sqrt{18}}{\sqrt{18}}$ **(a), (b)**, and **(c)** $\dfrac{\sqrt{2}}{6}$

120. In Example 6 we showed that $\sqrt[3]{\dfrac{3}{5}} = \dfrac{\sqrt[3]{75}}{5}$. Using your calculator, show that this is true.

121. Which is greater, $\dfrac{2}{\sqrt{2}}$ or $\dfrac{3}{\sqrt{3}}$? Explain. $\dfrac{3}{\sqrt{3}}$

122. Which is greater, $\dfrac{\sqrt{3}}{2}$ or $\dfrac{2}{\sqrt{3}}$? Explain. $\dfrac{2}{\sqrt{3}}$

123. Use a calculator to determine if $\sqrt[3]{\dfrac{2}{3}}$ is equal to $\dfrac{\sqrt[3]{18}}{3}$.   equal

124. Use a calculator to determine if $\sqrt[4]{\dfrac{5}{9}}$ is equal to $\dfrac{\sqrt[4]{30}}{3}$.   not equal

125. What are the three conditions that must be met for a radical expression to be simplified?

126. We stated that for nonnegative real numbers a and b, $b \neq 0$, $\dfrac{\sqrt[n]{a}}{\sqrt[n]{b}} = \sqrt[n]{\dfrac{a}{b}}$. Why is it necessary to specify that both a and b are nonnegative real numbers?

128. (a)
Solve equation for y. Select values for x and find corresponding values for y.

128. (b)
To find y intercept, set $x = 0$ and solve for y. Set $y = 0$ and solve for x.

$\left(\dfrac{4}{3}, 0\right)$

128. (c)
Mark y intercept on axes; then use slope to determine a second point.

CUMULATIVE REVIEW EXERCISES

[2.4] **127.** Two cars leave from West Point at the same time traveling in opposite directions. One travels 10 miles per hour faster than the other. If the two cars are 270 miles apart after 3 hours, find the speed of each car.   40 mph, 50 mph

[3.2–3.3] **128.** Explain how to graph a linear equation by **(a)** plotting points; **(b)** using the intercepts; **(c)** using the slope and y intercepts. Plot $y = 3x - 4$ using each method.

[5.3] **129.** Multiply $(4x^2 - 3x - 2)(2x - 3)$.

[6.5] **130.** Solve $(2x - 3)(x - 2) = 4x - 6$. $4, \frac{3}{2}$

126. If n is even and a or b is negative the numbers are not real.   **129.** $8x^3 - 18x^2 + 5x + 6$

125. (1) No perfect powers are factors of any radicand. (2) No radicand contains fractions. (3) No radicals in any denominator.

Group Activity/ Challenge Problems

Rationalize the denominator.

1. $\dfrac{1}{\sqrt{a + b}}$ $\dfrac{\sqrt{a + b}}{a + b}$ **2.** $\dfrac{3}{\sqrt{2a - 3b}}$ $\dfrac{3\sqrt{2a - 3b}}{2a - 3b}$

In higher math courses it may be necessary to rationalize the numerator of radical expressions. Rationalize the numerator of each of the following. (Your answers will contain radicals in the denominators.)

3. $\dfrac{\sqrt{6}}{3}$ $\dfrac{2}{\sqrt{6}}$ **4.** $\dfrac{5 - \sqrt{5}}{6}$ $\dfrac{10}{15 + 3\sqrt{5}}$

5. $\dfrac{4\sqrt{x} - \sqrt{3}}{x}$ $\dfrac{16x - 3}{4x\sqrt{x} + x\sqrt{3}}$ **6.** $\dfrac{\sqrt{x + h} - \sqrt{x}}{h}$, $h \neq 0$ $\dfrac{1}{\sqrt{x + h} + \sqrt{x}}$

7. Prove $\sqrt[n]{\dfrac{a}{b}} = \dfrac{\sqrt[n]{a}}{\sqrt[n]{b}}$ by converting $\sqrt[n]{\dfrac{a}{b}}$ to exponential form. $\left(\dfrac{a}{b}\right)^{1/n} = \dfrac{a^{1/n}}{b^{1/n}} = \dfrac{\sqrt[n]{a}}{\sqrt[n]{b}}$

8. (a) Multiply $(\sqrt[3]{3} - \sqrt[3]{2})(\sqrt[3]{9} + \sqrt[3]{6} + \sqrt[3]{4})$ and show that the answer is 1.

   (b) Consider $3 - 2$ as $3^{3/3} - 2^{3/3}$, which may be written $(3^{1/3})^3 - (2^{1/3})^3$.
   Factor $(3^{1/3})^3 - (2^{1/3})^3$ using the formula for the difference of two cubes and show that you get the product in part **(a)**.

   (c) Using the information from parts **(a)** and **(b)** factor $x + y$ using the formula for the sum of two cubes. $(\sqrt[3]{x} + \sqrt[3]{y})(\sqrt[3]{x^2} - \sqrt[3]{xy} + \sqrt[3]{y^2})$

(b) $(\sqrt[3]{3} - \sqrt[3]{2})(\sqrt[3]{3^2} + \sqrt[3]{3}\sqrt[3]{2} + \sqrt[3]{2^2})$. This equals the answer in part **(a)**.

8.5 Adding and Subtracting Radicals

Tape 13

1 Add and subtract radicals.

Adding and Subtracting Radicals

1 **Like radicals** are radicals having the same radicand and index. **Unlike radicals** are radicals differing in either the radicand or the index.

Examples of Like Radicals	*Examples of Unlike Radicals*	
$\sqrt{5}, 3\sqrt{5}$	$\sqrt{5}, \sqrt[3]{5}$	indexes differ
$5\sqrt{7}, -2\sqrt{7}$	$\sqrt{5}, \sqrt{7}$	radicands differ
$\sqrt{x}, 5\sqrt{x}$	$\sqrt{x}, \sqrt{2x}$	radicands differ
$\sqrt[3]{2x}, -4\sqrt[3]{2x}$	$\sqrt{x}, \sqrt[3]{x}$	indexes differ
$\sqrt[4]{xy}, -\sqrt[4]{xy}$	$\sqrt[3]{xy}, \sqrt[3]{x^2y}$	radicands differ

Like radicals are added and subtracted in much the same way that like terms are added or subtracted. To add or subtract like radicals, add or subtract their numerical coefficients and multiply this sum or difference by the like radical.

Examples of Adding and Subtracting Like Radicals

$$3\sqrt{5} + 2\sqrt{5} = (3 + 2)\sqrt{5} = 5\sqrt{5}$$
$$5\sqrt{x} - 7\sqrt{x} = (5 - 7)\sqrt{x} = -2\sqrt{x}$$
$$\sqrt[3]{4x} + 5\sqrt[3]{4x} = (1 + 5)\sqrt[3]{4x} = 6\sqrt[3]{4x}$$
$$4\sqrt{5x} - y\sqrt{5x} = (4 - y)\sqrt{5x}$$

EXAMPLE 1 Simplify.

(a) $6 + 4\sqrt{2} - \sqrt{2} + 3$ **(b)** $2\sqrt[3]{x} + 5x + 4\sqrt[3]{x} - 3$

Solution:

(a) $6 + 4\sqrt{2} - \sqrt{2} + 3 = 3\sqrt{2} + 9$ (or $9 + 3\sqrt{2}$)

(b) $2\sqrt[3]{x} + 5x + 4\sqrt[3]{x} - 3 = 6\sqrt[3]{x} + 5x - 3$

It is sometimes possible to convert unlike radicals into like radicals by simplifying one or more of the radicals.

EXAMPLE 2 Simplify $\sqrt{3} + \sqrt{27}$.

Solution: Since $\sqrt{3}$ and $\sqrt{27}$ are unlike radicals, they cannot be added in their present form. We can simplify $\sqrt{27}$ to obtain like radicals.

$$\sqrt{3} + \sqrt{27} = \sqrt{3} + \sqrt{9}\sqrt{3}$$
$$= \sqrt{3} + 3\sqrt{3} = 4\sqrt{3}$$

To Add or Subtract Radicals
1. Simplify each radical expression.
2. Combine like radicals (if there are any).

EXAMPLE 3 Simplify $4\sqrt{24} + \sqrt{54}$.

Solution:
$$4\sqrt{24} + \sqrt{54} = 4 \cdot \sqrt{4} \cdot \sqrt{6} + \sqrt{9} \cdot \sqrt{6}$$
$$= 4 \cdot 2\sqrt{6} + 3\sqrt{6}$$
$$= 8\sqrt{6} + 3\sqrt{6} = 11\sqrt{6}$$

EXAMPLE 4 Simplify $2\sqrt{45} - \sqrt{80} + \sqrt{20}$.

Solution:
$$2\sqrt{45} - \sqrt{80} + \sqrt{20} = 2 \cdot \sqrt{9} \cdot \sqrt{5} - \sqrt{16} \cdot \sqrt{5} + \sqrt{4} \cdot \sqrt{5}$$
$$= 2 \cdot 3\sqrt{5} - 4\sqrt{5} + 2\sqrt{5}$$
$$= 6\sqrt{5} - 4\sqrt{5} + 2\sqrt{5} = 4\sqrt{5}$$

EXAMPLE 5 Simplify $\sqrt[3]{27} + \sqrt[3]{81} - 4\sqrt[3]{3}$.

Solution: $\sqrt[3]{27} + \sqrt[3]{81} - 4\sqrt[3]{3} = 3 + \sqrt[3]{27}\sqrt[3]{3} - 4\sqrt[3]{3}$
$$= 3 + 3\sqrt[3]{3} - 4\sqrt[3]{3} = 3 - \sqrt[3]{3}$$

EXAMPLE 6 Simplify $\sqrt{x^2} - \sqrt{x^2 y} + x\sqrt{y}$.

Solution: $\sqrt{x^2} - \sqrt{x^2 y} + x\sqrt{y} = x - \sqrt{x^2}\sqrt{y} + x\sqrt{y}$
$$= x - x\sqrt{y} + x\sqrt{y} = x$$

EXAMPLE 7 Simplify $\sqrt[3]{x^{10}y^2} - \sqrt[3]{x^4 y^8}$.

Solution: $\sqrt[3]{x^{10}y^2} - \sqrt[3]{x^4 y^8} = \sqrt[3]{x^9} \cdot \sqrt[3]{xy^2} - \sqrt[3]{x^3 y^6} \cdot \sqrt[3]{xy^2}$
$$= x^3 \sqrt[3]{xy^2} - xy^2 \sqrt[3]{xy^2}$$

Now factor out the common factor $\sqrt[3]{xy^2}$.

$$= (x^3 - xy^2)\sqrt[3]{xy^2}$$

EXAMPLE 8   Simplify $4\sqrt{2} - \dfrac{1}{\sqrt{8}} + \sqrt{32}$.

Solution: $4\sqrt{2} - \dfrac{1}{\sqrt{8}} + \sqrt{32} = 4\sqrt{2} - \dfrac{1}{\sqrt{8}} \cdot \dfrac{\sqrt{2}}{\sqrt{2}} + \sqrt{16}\sqrt{2}$

$$= 4\sqrt{2} - \dfrac{\sqrt{2}}{\sqrt{16}} + 4\sqrt{2}$$

$$= 4\sqrt{2} - \dfrac{\sqrt{2}}{4} + 4\sqrt{2}$$

$$= \left(4 - \dfrac{1}{4} + 4\right)\sqrt{2} = \dfrac{31\sqrt{2}}{4}$$

Now that we have discussed adding and subtracting radical expressions in more depth, let us multiply a few more radicals.

EXAMPLE 9   Simplify $(3\sqrt{6} - \sqrt{3})^2$.

Solution: $(3\sqrt{6} - \sqrt{3})^2 = (3\sqrt{6} - \sqrt{3})(3\sqrt{6} - \sqrt{3})$

Now multiply the factors using the FOIL method.

$(3\sqrt{6})(3\sqrt{6}) + (3\sqrt{6})(-\sqrt{3}) + (-\sqrt{3})(3\sqrt{6}) + (-\sqrt{3})(-\sqrt{3})$
$= 9(6) - 3\sqrt{18} - 3\sqrt{18} + 3$
$= 54 - 3\sqrt{18} - 3\sqrt{18} + 3$
$= 57 - 6\sqrt{18}$
$= 57 - 6\sqrt{9}\sqrt{2}$
$= 57 - 18\sqrt{2}$

EXAMPLE 10   Simplify $(\sqrt[3]{x} - \sqrt[3]{2y^2})(\sqrt[3]{x^2} - \sqrt[3]{8y})$.

Solution: Multiply the factors using the FOIL method.

$(\sqrt[3]{x})(\sqrt[3]{x^2}) + (\sqrt[3]{x})(-\sqrt[3]{8y}) + (-\sqrt[3]{2y^2})(\sqrt[3]{x^2}) + (-\sqrt[3]{2y^2})(-\sqrt[3]{8y})$
$= \sqrt[3]{x^3} - \sqrt[3]{8xy} - \sqrt[3]{2x^2y^2} + \sqrt[3]{16y^3}$
$= x - 2\sqrt[3]{xy} - \sqrt[3]{2x^2y^2} + 2y\sqrt[3]{2}$

Helpful Hint

The product rule and quotient rule for radicals presented in Sections 8.3 and 8.4 are

$$\sqrt[n]{a} \cdot \sqrt[n]{b} = \sqrt[n]{ab} \qquad \dfrac{\sqrt[n]{a}}{\sqrt[n]{b}} = \sqrt[n]{\dfrac{a}{b}}$$

Students often incorrectly assume similar properties exist for addition and subtraction. They do not. To illustrate this, let n be a square root (index 2), $a = 9$, and $b = 16$.

$$\sqrt[n]{a} + \sqrt[n]{b} \neq \sqrt[n]{a + b}$$
$$\sqrt{9} + \sqrt{16} \neq \sqrt{9 + 16}$$
$$3 + 4 \neq \sqrt{25}$$
$$7 \neq 5$$

70. $5 + \sqrt[3]{36} + \sqrt[3]{6}$ **71.** $8 - 2\sqrt[3]{18} - \sqrt[3]{12}$ **72.** $2\sqrt[3]{2x^2} + 2\sqrt[3]{5x} - 2\sqrt[3]{xy} - \sqrt[3]{20y}$

Exercise Set 8.5   **3.** $9\sqrt{10} + 2$ **4.** $5 - 4\sqrt{3}$ **9.** $7\sqrt{5} + 2\sqrt[3]{x}$ **10.** $x + \sqrt{x} + 4\sqrt{y}$ **12.** $7\sqrt{x} + 4 + 2x$

In this exercise set assume that all variables represent positive real numbers.

Simplify.

1. $4\sqrt{3} - 2\sqrt{3}$ $2\sqrt{3}$ **2.** $6\sqrt[3]{7} - 8\sqrt[3]{7}$ $-2\sqrt[3]{7}$ **3.** $4\sqrt{10} + 6\sqrt{10} - \sqrt{10} + 2$

4. $2\sqrt{3} - 2\sqrt{3} - 4\sqrt{3} + 5$ **5.** $12\sqrt[3]{15} + 5\sqrt[3]{15} - 8\sqrt[3]{15}$ $9\sqrt[3]{15}$ **6.** $4\sqrt{x} + \sqrt{x}$ $5\sqrt{x}$

7. $3\sqrt{y} - 6\sqrt{y}$ $-3\sqrt{y}$ **8.** $3\sqrt{y} - \sqrt{y} + 3$ $3 + 2\sqrt{y}$ **9.** $3\sqrt{5} - \sqrt[3]{x} + 4\sqrt{5} + 3\sqrt[3]{x}$

10. $\sqrt{x} + \sqrt{y} + x + 3\sqrt{y}$ **11.** $5 + 4\sqrt[3]{x} - 8\sqrt[3]{x}$ $5 - 4\sqrt[3]{x}$ **12.** $5\sqrt{x} + 4 + 3\sqrt{x} + 2x - \sqrt{x}$

Simplify.   **23.** $-16\sqrt{5x}$ **24.** $-7x\sqrt{3}$ **25.** $-27x\sqrt{2}$ **35.** $(6xy - x^2y)\sqrt[4]{3x}$ **36.** $3x^3y^2\sqrt{xy}$ **37.** $4x^2\sqrt[3]{x^2y^2}$

13. $\sqrt{8} - \sqrt{12}$ $2(\sqrt{2} - \sqrt{3})$ **14.** $\sqrt{75} + \sqrt{108}$ $11\sqrt{3}$ **15.** $-6\sqrt{75} + 4\sqrt{125}$ $-30\sqrt{3} + 20\sqrt{5}$

16. $3\sqrt{250} + 5\sqrt{160}$ $35\sqrt{10}$ **17.** $-4\sqrt{90} + 3\sqrt{40} + 2\sqrt{10}$ $-4\sqrt{10}$ **18.** $8\sqrt{45} + 7\sqrt{20} + 2\sqrt{5}$ $40\sqrt{5}$

19. $3\sqrt{40x^2y} + 2x\sqrt{490y}$ $20x\sqrt{10y}$ **20.** $\sqrt{500xy^2} + y\sqrt{320x}$ $18y\sqrt{5x}$ **21.** $4\sqrt{32} - \sqrt{18} + 2\sqrt{128}$ $29\sqrt{2}$

22. $5\sqrt{8} + 2\sqrt{50} - 3\sqrt{72}$ $2\sqrt{2}$ **23.** $2\sqrt{5x} - 3\sqrt{20x} - 4\sqrt{45x}$ **24.** $3\sqrt{27x^2} - 2\sqrt{108x^2} - \sqrt{48x^2}$

25. $3\sqrt{50x^2} - 3\sqrt{72x^2} - 8x\sqrt{18}$ **26.** $\sqrt[3]{54} - \sqrt[3]{16}$ $\sqrt[3]{2}$ **27.** $4\sqrt[3]{5} - 5\sqrt[3]{40}$ $-6\sqrt[3]{5}$

28. $\sqrt[3]{108} + 2\sqrt[3]{32}$ $7\sqrt[3]{4}$ **29.** $2\sqrt[3]{16} + \sqrt[3]{54}$ $7\sqrt[3]{2}$ **30.** $\sqrt[3]{27} - 5\sqrt[3]{8}$ -7

31. $3\sqrt{45x^3} + \sqrt{5x}$ $(9x + 1)\sqrt{5x}$ **32.** $2\sqrt[3]{x^4y^2} + 3x\sqrt[3]{xy^2}$ $5x\sqrt[3]{xy^2}$ **33.** $2a\sqrt{20a^3b^2} + 2b\sqrt{45a^5}$ $10a^2b\sqrt{5a}$

34. $x\sqrt[3]{x^2y} - \sqrt[3]{8x^5y}$ $-x\sqrt[3]{x^2y}$ **35.** $3y\sqrt[4]{48x^5} - x\sqrt[4]{3x^5y^4}$ **36.** $\sqrt{4x^7y^5} + 3x^2\sqrt{x^3y^5} - 2xy\sqrt{x^5y^3}$

37. $x\sqrt[3]{27x^5y^2} - x^2\sqrt[3]{x^2y^2} + 2\sqrt[3]{x^8y^2}$ **38.** $2\sqrt[3]{x^7y^7} - 3x\sqrt[3]{x^4y^7}$ $-x^2y^2\sqrt[3]{xy}$ **39.** $\sqrt[3]{128x^9y^{10}} - 2x^2y\sqrt[3]{16x^3y^7}$ 0

40. $5\sqrt[3]{320x^5y^8} + 2x\sqrt[3]{135x^2y^8}$ $26xy^2\sqrt[3]{5x^2y^2}$

Simplify.

41. $\dfrac{1}{\sqrt{2}} + \dfrac{\sqrt{2}}{2}$ $\sqrt{2}$ **42.** $\dfrac{1}{\sqrt{3}} + \dfrac{\sqrt{3}}{3}$ $\dfrac{2\sqrt{3}}{3}$ **43.** $\sqrt{3} - \dfrac{1}{\sqrt{3}}$ $\dfrac{2\sqrt{3}}{3}$

44. $\sqrt{6} - \sqrt{\dfrac{2}{3}}$ $\dfrac{2\sqrt{6}}{3}$ **45.** $\sqrt{\dfrac{1}{6}} + \sqrt{24}$ $\dfrac{13\sqrt{6}}{6}$ **46.** $3\sqrt{2} - \dfrac{2}{\sqrt{8}} + \sqrt{50}$ $\dfrac{15\sqrt{2}}{2}$

47. $\sqrt{\dfrac{1}{2}} + 3\sqrt{2} + \sqrt{18}$ $\dfrac{13\sqrt{2}}{2}$ **48.** $\dfrac{3}{\sqrt{18}} - 2\sqrt{18} + \sqrt{\dfrac{5}{8}}$ **49.** $4\sqrt{x} + \dfrac{1}{\sqrt{x}} + \sqrt{\dfrac{1}{x}}$ $2\sqrt{x}\left(2 + \dfrac{1}{x}\right)$

50. $\dfrac{1}{3} + \dfrac{1}{\sqrt{3}} + \sqrt{75}$ $\dfrac{1}{3} + \dfrac{16\sqrt{3}}{3}$ **51.** $\dfrac{1}{2}\sqrt{18} - \dfrac{3}{\sqrt{2}} - 3\sqrt{50}$ $-15\sqrt{2}$ **52.** $\dfrac{\sqrt{3}}{3} + 2\sqrt{\dfrac{1}{3}} + \sqrt{12}$ $3\sqrt{3}$

53. $\sqrt{\dfrac{3}{8}} + \sqrt{\dfrac{3}{2}}$ $\dfrac{3\sqrt{6}}{4}$ **54.** $-2\sqrt{\dfrac{x}{y}} + 3\sqrt{\dfrac{y}{x}}$ $\left(\dfrac{-2}{y} + \dfrac{3}{x}\right)\sqrt{xy}$ **48.** $\dfrac{-22\sqrt{2} + \sqrt{10}}{4}$

Simplify.   **61.** $8 + 2\sqrt{15}$ **62.** $10 - 3\sqrt{6}$ **67.** $x - \sqrt{3x} - 6$ **68.** $\sqrt{2yz} - 2y\sqrt{2} + 2z\sqrt{3} - 4\sqrt{3yz}$ **69.** $18x - \sqrt{3xy} - y$

55. $(\sqrt{3} + 4)(\sqrt{3} + 5)$ $23 + 9\sqrt{3}$ **56.** $(\sqrt{3} + 1)(\sqrt{3} - 6)$ $-3 - 5\sqrt{3}$ **57.** $(1 + \sqrt{5})(6 + \sqrt{5})$ $11 + 7\sqrt{5}$

58. $(3 - \sqrt{2})(4 - \sqrt{8})$ $16 - 10\sqrt{2}$ **59.** $(4 - \sqrt{2})(5 + \sqrt{2})$ $18 - \sqrt{2}$ **60.** $(5\sqrt{6} + 3)(4\sqrt{6} - 2)$ $114 + 2\sqrt{6}$

61. $(\sqrt{5} + \sqrt{3})(\sqrt{5} + \sqrt{3})$ **62.** $(4\sqrt{3} + \sqrt{2})(\sqrt{3} - \sqrt{2})$ **63.** $(\sqrt{2} - \sqrt{3})(\sqrt{3} + \sqrt{8})$ $1 - \sqrt{6}$

64. $(\sqrt{3} + 4)^2$ $19 + 8\sqrt{3}$ **65.** $(2 - \sqrt{3})^2$ $7 - 4\sqrt{3}$ **66.** $(2\sqrt{5} - 3)^2$ $29 - 12\sqrt{5}$

67. $(\sqrt{x} + \sqrt{3})(\sqrt{x} - \sqrt{12})$ **68.** $(\sqrt{y} + \sqrt{6z})(\sqrt{2z} - \sqrt{8y})$ **69.** $(2\sqrt{3x} - \sqrt{y})(3\sqrt{3x} + \sqrt{y})$

70. $(\sqrt[3]{9} + \sqrt[3]{2})(\sqrt[3]{3} + \sqrt[3]{4})$ **71.** $(\sqrt[3]{4} - \sqrt[3]{6})(\sqrt[3]{2} - \sqrt[3]{36})$ **72.** $(\sqrt[3]{4x} - \sqrt[3]{2y})(\sqrt[3]{4x} + \sqrt[3]{10})$

Simplify. These exercises are a combination of the exercises presented earlier in this exercise set.

73. $\sqrt{5} + 2\sqrt{5}$ $3\sqrt{5}$ **74.** $-2\sqrt{x} - 3\sqrt{x}$ $-5\sqrt{x}$ **75.** $\sqrt{125} + \sqrt{20}$ $7\sqrt{5}$

76. $3\sqrt{7} + 2\sqrt{63} - 2\sqrt{28}$ $5\sqrt{7}$ **77.** $\dfrac{\sqrt{6}}{2} + \dfrac{1}{\sqrt{6}}$ $\dfrac{2\sqrt{6}}{3}$ **78.** $(\sqrt{5} + 2)(7 + \sqrt{5})$ $19 + 9\sqrt{5}$

82. $3\sqrt{3} + 6\sqrt{2}$ **88.** $x - 2\sqrt{5xy} + 5y$ **90.** $x - 2\sqrt[3]{x^2y^2} - \sqrt[3]{xy} + 2y$ **91.** $(5xy^2 + 2x)\sqrt[3]{xy}$ **92.** $a - 3\sqrt[3]{a} + 5\sqrt[3]{a^2} - 15$ **94.** 0

79. $-\sqrt[4]{x} + 6\sqrt[4]{x} - 2\sqrt[4]{x}$ $3\sqrt[4]{x}$ **80.** $2\sqrt[3]{81} + 4\sqrt[3]{24}$ $14\sqrt[3]{3}$ **81.** $2 + 3\sqrt{y} - 6\sqrt{y} + 5$ $7 - 3\sqrt{y}$

82. $4\sqrt{3} - \dfrac{3}{\sqrt{3}} + 2\sqrt{18}$ **83.** $(3\sqrt{2} - 4)(\sqrt{2} + 5)$ $-14 + 11\sqrt{2}$ **84.** $(\sqrt{5} + \sqrt{2})(\sqrt{2} + \sqrt{20})$ $12 + 3\sqrt{10}$

85. $4\sqrt{3x^3} - \sqrt{12x}$ $(4x - 2)\sqrt{3x}$ **86.** $2b\sqrt[4]{a^4b} + ab\sqrt[4]{16b}$ $4ab\sqrt[4]{b}$ **87.** $\dfrac{3}{\sqrt{y}} - \sqrt{\dfrac{9}{y}} + \sqrt{y}$ $\sqrt{y}$

88. $(\sqrt{x} - \sqrt{5y})(\sqrt{x} - \sqrt{5y})$ **89.** $2\sqrt[3]{24a^3y^4} + 4a\sqrt[3]{81y^4}$ $16ay\sqrt[3]{3y}$ **90.** $(\sqrt[3]{x^2} - \sqrt[3]{y})(\sqrt[3]{x} - 2\sqrt[3]{y^2})$

91. $2x\sqrt[3]{xy} + 5y\sqrt[3]{x^4y^4}$ **92.** $(\sqrt[3]{a} + 5)(\sqrt[3]{a^2} - 3)$ **93.** $\dfrac{2}{\sqrt{50}} - 3\sqrt{50} - \dfrac{1}{\sqrt{8}}$ $\dfrac{-301\sqrt{2}}{20}$

94. $\sqrt{48} + 2\sqrt{75} - 3\sqrt{27} - 5\sqrt{3}$ **95.** $2\sqrt{\dfrac{8}{3}} - 4\sqrt{\dfrac{100}{6}}$ $\dfrac{-16\sqrt{6}}{3}$ **96.** $-5\sqrt{\dfrac{x^2}{y^3}} + \dfrac{2x}{y}\sqrt{\dfrac{1}{y}}$ $\dfrac{-3x\sqrt{y}}{y^2}$

Find the perimeter and area of the following figures. Write the area and perimeter in radical form with the radicals simplified.

97.

$P = 14\sqrt{5}$ $A = 60$

98.
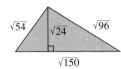
$P = 12\sqrt{6}$ $A = 30$

 99. What are like radicals?

 100. (a) Explain how to add like radicals.
   (b) Using the procedure in part **(a)** add $\dfrac{3}{5}\sqrt{5} + \dfrac{5}{4}\sqrt{5}$.

101. Use a calculator to estimate $\sqrt{3} + 3\sqrt{2}$. ≈ 5.97
102. Use a calculator to estimate $2\sqrt{3} + \sqrt{5}$. ≈ 5.70

 103. Which is greater, $\dfrac{1}{\sqrt{3} + 2}$ or $2 + \sqrt{3}$? (Do not use a calculator.) Explain how you determined your answer. $2 + \sqrt{3}$

 104. Which is greater,
   $\dfrac{1}{\sqrt{3}} + \sqrt{75}$ or $\dfrac{2}{\sqrt{12}} + \sqrt{48} + 2\sqrt{3}$? (Do not use a calculator.) Explain how you determined your answer.

 105. Does $\sqrt{a} + \sqrt{b} = \sqrt{a + b}$? Explain your answer and give examples supporting your answer.   no

 106. Since $64 + 36 = 100$, does $\sqrt{64} + \sqrt{36} = \sqrt{100}$? Explain your answer.   no

 107. If $\sqrt{a + b} = \sqrt{c + d}$, must $a + b = c + d$? Explain your answer and give examples supporting your answer.   yes

99. radicals with the same radicands and index   **100. (b)** $\dfrac{37\sqrt{5}}{20}$ **104.** $\dfrac{2}{\sqrt{12}} + \sqrt{48} + 2\sqrt{3}$ **111.** $3x^2y\sqrt[3]{y} - x^4y^2\sqrt[3]{3y^2}$

CUMULATIVE REVIEW EXERCISES

[5.2] **108.** Simplify $\dfrac{(2x^{-2}y^3)^2(x^{-1}y^{-3})}{(xy^2)^{-2}}$ $\dfrac{4y^7}{x^3}$

[8.2] **110.** Simplify $\left(\dfrac{x^{3/4}y^{2/3}}{x^{1/2}y}\right)^2$ $\dfrac{x^{1/2}}{y^{2/3}}$

[6.5] **109.** Solve the equation $20x^2 + 3x - 9 = 0$. $\dfrac{3}{5}, -\dfrac{3}{4}$

[8.3] **111.** Simplify $\sqrt[3]{3x^2y}\,(\sqrt[3]{9x^4y^3} - \sqrt[3]{x^{10}y^7})$.

1. $\dfrac{4\sqrt{3} + 9}{3}$ **2.** $\dfrac{8\sqrt{5} + 20}{5}$

If the figures below are similar, find the length of the side x. Write the answer in radical form with a rationalized denominator.

1.
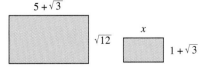

2.
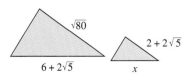

Group Activity/ Challenge Problems

Factor each of the following by factoring out a radical expression.

3. $\sqrt{10} + \sqrt{5}$ $\sqrt{5}(\sqrt{2} + 1)$ **4.** $\sqrt{6} - \sqrt{18}$ $\sqrt{6}(1 - \sqrt{3})$

Write each term as the square of an expression, then factor using the formula for the difference of two squares.

5. $x^2 - 3$ $(x + \sqrt{3})(x - \sqrt{3})$
6. $2x^2 - 7$ $(\sqrt{2}x + \sqrt{7})(\sqrt{2}x - \sqrt{7})$

Simplify.

7. $\dfrac{1}{\sqrt[5]{3x^7y^9}} + \dfrac{2\sqrt[5]{81x^3y}}{3x^2y^2}$ $\dfrac{\sqrt[5]{81x^3y}}{x^2y^2}$
8. $\dfrac{1}{\sqrt[4]{3x^5y^6z^{13}}}$ $\dfrac{\sqrt[4]{27x^3y^2z^3}}{3x^2y^2z^4}$

8.6   Solving Radical Equations

1 Solve radical equations containing one radical.
2 Solve radical equations containing two radicals.
3 Understand the graphical interpretation of a solution to a radical equation.
4 Solve applications problems using radical equations.
5 Solve for a variable in a radicand.

Tape 14

Solving Radical Equations Containing One Radical

1 A **radical equation** is an equation that contains a variable in a radicand. Some examples of radical equations are

$$\sqrt{x} = 4, \qquad \sqrt[3]{y + 4} = 9, \qquad \sqrt{x - 2} = 4 + \sqrt{x + 8}$$

> ## To Solve Radical Equations
> **1.** Rewrite the equation so that one radical containing a variable is by itself on one side of the equation.
> **2.** Raise each side of the equation to a power equal to the index of the radical.
> **3.** Collect and combine like terms.
> **4.** If the equation still contains a term with a variable in a radicand, repeat steps 1 through 3.
> **5.** Solve the resulting equation for the variable.
> **6.** Check all solutions in the original equations for extraneous solutions.

Recall from Section 7.4 that an extraneous solution is a number obtained when solving an equation that is not a solution to the original equation.

The following examples illustrate the procedure for solving radical equations.

EXAMPLE 1   Solve the equation $\sqrt{x} = 6$.

Solution: The square root containing the variable is already by itself on one side of the equation. Square both sides of the equation.

$$\sqrt{x} = 6 \qquad\qquad Check: \quad \sqrt{x} = 6$$
$$(\sqrt{x})^2 = (6)^2 \qquad\qquad\qquad \sqrt{36} = 6$$
$$x = 36 \qquad\qquad\qquad\qquad 6 = 6 \quad \text{true}$$

EXAMPLE 2 Solve the equation $\sqrt{x + 4} - 6 = 0$.

Solution:

$$\sqrt{x + 4} - 6 = 0$$

$$\sqrt{x + 4} = 6 \qquad \text{Isolate the radical containing the variable.}$$

$$(\sqrt{x + 4})^2 = 6^2 \qquad \text{Square both sides of the equation.}$$

$$x + 4 = 36 \qquad \text{Now solve for the variable.}$$

$$x = 32$$

A check will show that 32 is the solution.

EXAMPLE 3 Solve the equation $\sqrt[3]{x} + 4 = 6$.

Solution: Since the 4 is outside the radical, we first subtract 4 from both sides of the equation to isolate the radical.

$$\sqrt[3]{x} + 4 = 6$$

$$\sqrt[3]{x} = 2$$

Now cube both sides of the equation.

$$(\sqrt[3]{x})^3 = 2^3$$

$$x = 8$$

A check will show that 8 is the solution.

EXAMPLE 4 Solve the equation $\sqrt{2x - 3} = x - 3$.

Solution: Since the radical is already isolated we square both sides of the equation.

$$(\sqrt{2x - 3})^2 = (x - 3)^2$$

$$2x - 3 = (x - 3)(x - 3)$$

$$2x - 3 = x^2 - 6x + 9$$

$$0 = x^2 - 8x + 12$$

Now factor.

$$x^2 - 8x + 12 = 0$$

$$(x - 6)(x - 2) = 0$$

$$x - 6 = 0 \qquad \text{or} \qquad x - 2 = 0$$

$$x = 6 \qquad\qquad\qquad x = 2$$

Check:

$x = 6$	$x = 2$
$\sqrt{2x - 3} = x - 3$	$\sqrt{2x - 3} = x - 3$
$\sqrt{2(6) - 3} = 6 - 3$	$\sqrt{2(2) - 3} = 2 - 3$
$\sqrt{9} = 3$	$\sqrt{1} = -1$
$3 = 3$ true	$1 = -1$ false

Thus, 6 is a solution, but 2 is not a solution to the equation. The 2 is an extraneous solution because 2 satisfies the equation $(\sqrt{2x - 3})^2 = (x - 3)^2$, but not the original equation, $\sqrt{2x - 3} = x - 3$.

Helpful Hint

Don't forget to check your solutions in the original equation. Remember that when you raise both sides of an equation to a power you may introduce extraneous solutions.

Consider the equation $x = 2$. Note what happens when you square both sides of the equation.

$$x = 2$$
$$x^2 = 2^2$$
$$x^2 = 4$$

Note that the equation $x^2 = 4$ has two solutions, $+2$ and -2. Since the original equation $x = 2$ has only one solution, 2, we introduced the extraneous solution, -2.

EXAMPLE 5   Solve the equation $2x - 5\sqrt{x} - 3 = 0$.

Solution: First, write the equation with the square root containing the variable by itself on one side of the equation.

$$2x - 5\sqrt{x} - 3 = 0$$
$$-5\sqrt{x} = -2x + 3$$
$$\text{or } 5\sqrt{x} = 2x - 3$$

Now square both sides of the equation.

$$(5\sqrt{x})^2 = (2x - 3)^2$$
$$25x = (2x - 3)(2x - 3)$$
$$25x = 4x^2 - 12x + 9$$
$$0 = 4x^2 - 37x + 9$$
$$0 = (4x - 1)(x - 9)$$
$$4x - 1 = 0 \quad \text{or} \quad x - 9 = 0$$
$$4x = 1 \qquad\qquad x = 9$$
$$x = \frac{1}{4}$$

Check: $x = \dfrac{1}{4}$ $\qquad\qquad\qquad\qquad\qquad\qquad x = 9$

$$2x - 5\sqrt{x} - 3 = 0 \qquad\qquad 2x - 5\sqrt{x} - 3 = 0$$
$$2\left(\frac{1}{4}\right) - 5\sqrt{\frac{1}{4}} - 3 = 0 \qquad\qquad 2(9) - 5\sqrt{9} - 3 = 0$$
$$\frac{1}{2} - 5\left(\frac{1}{2}\right) - 3 = 0 \qquad\qquad 18 - 5(3) - 3 = 0$$
$$\qquad\qquad\qquad\qquad\qquad 18 \overset{?}{=} 15 \quad 3 = 0$$
$$-5 = 0 \quad \textbf{false} \qquad\qquad\qquad 0 = 0 \quad \textbf{true}$$

The solution is 9. The value $\frac{1}{4}$ is an extraneous solution.

Solving Equations Containing Two Radical Expressions

2 Now we will look at some equations that contain two radical expressions.

Example 6   Solve the equation $\sqrt{4x^2 + 16} = 2\sqrt{x^2 + 3x - 2}$.

Solution: Since the two radicals appear on different sides of the equation, we square both sides of the equation.

$$(\sqrt{4x^2 + 16})^2 = (2\sqrt{x^2 + 3x - 2})^2$$
$$4x^2 + 16 = 4(x^2 + 3x - 2)$$
$$4x^2 + 16 = 4x^2 + 12x - 8$$
$$16 = 12x - 8$$
$$24 = 12x$$
$$2 = x$$

A check will show that 2 is the solution.

Example 7   Solve the equation $3\sqrt[3]{x - 2} = \sqrt[3]{17x - 14}$.

Solution:

$$3\sqrt[3]{x - 2} = \sqrt[3]{17x - 14}$$
$$(3\sqrt[3]{x - 2})^3 = (\sqrt[3]{17x - 14})^3 \quad \text{Cube both sides of the equation.}$$
$$27(x - 2) = 17x - 14$$
$$27x - 54 = 17x - 14$$
$$10x - 54 = -14$$
$$10x = 40$$
$$x = 4$$

A check will show that the solution is 4.

Since radical expressions may be represented using rational exponents, radical equations may also be given with rational exponents. For instance, Example 7 may be written as $3(x - 2)^{1/3} = (17x - 14)^{1/3}$. To solve this equation, we cube both sides of the equation, just as we did in Example 7.

$$[3(x - 2)^{1/3}]^3 = [(17x - 14)^{1/3}]^3$$
$$3^3(x - 2)^1 = (17x - 14)^1$$
$$27(x - 2) = 17x - 14$$

This is the third step of the solution in Example 7. If you continue solving, you will find that $x = 4$.

When a radical equation contains two radical terms and a third nonradical term, you will sometimes need to raise both sides of the equation to a given power twice to obtain the solution. First, isolate one radical term. Then raise both sides of the equation to a given power. This will eliminate one of the radicals. Next, isolate the remaining radical on one side of the equation. Then raise both sides of the equation to the given power a second time. This procedure is illustrated in Example 8.

EXAMPLE 8 Solve the equation $\sqrt{5x-1} - \sqrt{3x-2} = 1$.

Solution: We must isolate one radical term on one side of the equation. We will begin by adding $\sqrt{3x-2}$ to both sides of the equation to isolate $\sqrt{5x-1}$. Then we will square both sides of the equation and combine like terms.

$$\sqrt{5x-1} = 1 + \sqrt{3x-2}$$
$$(\sqrt{5x-1})^2 = (1 + \sqrt{3x-2})^2$$
$$5x - 1 = (1 + \sqrt{3x-2})(1 + \sqrt{3x-2})$$
$$5x - 1 = 1 + \sqrt{3x-2} + \sqrt{3x-2} + (\sqrt{3x-2})^2$$
$$5x - 1 = 1 + 2\sqrt{3x-2} + 3x - 2$$
$$5x - 1 = 3x - 1 + 2\sqrt{3x-2}$$

Now we isolate the remaining radical term. We then square both sides of the equation again and solve for x.

$$2x = 2\sqrt{3x-2}$$
$$(2x)^2 = (2\sqrt{3x-2})^2$$
$$4x^2 = 4(3x-2)$$
$$4x^2 = 12x - 8$$
$$4x^2 - 12x + 8 = 0$$
$$4(x^2 - 3x + 2) = 0$$
$$4(x - 2)(x - 1) = 0$$
$$x - 2 = 0 \quad \text{or} \quad x - 1 = 0$$
$$x = 2 \qquad\qquad x = 1$$

A check will show that both 2 and 1 are solutions to the equation.

Graphical Interpretation to the Solution of a Radical Equation

3 In Section 3.5 we introduced and graphed selected square root functions. We will use the graphs of square root functions to help explain the solution of a radical equation in one variable. Let us consider the equation $\sqrt{2x-3} = x - 3$, which was solved in Example 4. The solution was 6. Suppose we graph each side of the equation separately. To do so we write the two functions $y = \sqrt{2x-3}$ and $y = x - 3$. Now we have a system of equations. We will graph both equations on the same axes and determine the point of intersection.

Let us first graph $y = \sqrt{2x-3}$. We can find the domain by determining where $2x - 3 \geq 0$ (remember the radicand of a square root must be greater than or equal to 0 for the expression to represent a real number).

$$2x - 3 \geq 0$$
$$2x \geq 3$$
$$x \geq \frac{3}{2}$$

Therefore, the domain of $y = \sqrt{2x - 3}$ is $\{x \mid x \geq \frac{3}{2}\}$. Below we select some values for x and find the corresponding values for y. The graph of $y = \sqrt{2x - 3}$ is illustrated in blue in Fig. 8.1. The graph of $y = x - 3$ is a straight line. The domain of the function is the set of real numbers. Below we select some values for x and find the corresponding values of y. The graph of $y = x - 3$ is illustrated in red in Fig. 8.1.

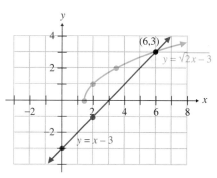

$y = \sqrt{2x - 3}$			$y = x - 3$	
x	y		x	y
$\frac{3}{2}$	0		0	-3
2	1		2	-1
$\frac{7}{2}$	2		6	3
6	3			

FIGURE 8.1

Notice that the graphs intersect at (6, 3). The x coordinate of the ordered pair, 6, is the solution to the original equation. This checks with the solution obtained in Example 4.

Graphing Calculator Corner

EXERCISES

Use your graphing calculator to solve the equations. Round solutions to the nearest tenth.

1. $\sqrt{x + 8} = \sqrt{3x + 5}$ 1.5

2. $\sqrt{10x - 16} - 15 = 0$ 24.1

3. $\sqrt[3]{5x^2 - 6} - 4 = 0$ $-3.7, 3.7$

4. $\sqrt[3]{5x^2 - 10} = \sqrt[3]{4x + 95}$ $-4.2, 5$

Applications of Radical Equations

4 Now we will look at a few of the many applications of radicals.

EXAMPLE 9   A telephone pole is at a right, or 90°, angle with the ground (see Fig. 8.2). The length, l, of a wire from height a on the pole to a point at a distance b from the pole's base can be found by the formula $l = \sqrt{a^2 + b^2}$.

FIGURE 8.2

Find the length of the wire that connects to the pole 40 feet above the ground and is anchored to the ground 20 feet from the base of the pole.

Solution: If we substitute 40 for a and 20 for b in the formula, we get

$$l = \sqrt{a^2 + b^2}$$
$$= \sqrt{(40)^2 + (20)^2}$$
$$= \sqrt{1600 + 400}$$
$$= \sqrt{2000}$$
$$\approx 44.7$$

Thus, the wire is about 44.7 feet long.

The formula used in Example 9 is a special case of the Pythagorean theorem that we will discuss in the next chapter. The formula above can be adapted to many situations involving right triangles.

EXAMPLE 10   The length of time it takes for a pendulum to make one complete swing back and forth is called the period of the pendulum. The period of a pendulum, T, in seconds, can be found by the formula $T = 2\pi\sqrt{\dfrac{L}{32}}$, where L is the length of the pendulum in feet. Find the period of a pendulum if its length is 4 feet.

Solution: Substitute 4 for L and 3.14 for π in the formula. If you have a calculator that has a $\boxed{\pi}$ key, use it to enter π.

$$T = 2\pi\sqrt{\frac{L}{32}}$$
$$= 2(3.14)\sqrt{\frac{4}{32}}$$
$$= 2(3.14)\sqrt{0.125} \approx 2.22$$

Thus, the period is about 2.22 seconds. If you have a grandfather clock with a 4-foot pendulum, it will take about 2.22 seconds for it to swing back and forth.

EXAMPLE 11   The area of a triangle is $A = \frac{1}{2}bh$. If the height is not known, but we know the length of each of the three sides, we can use *Heron's formula* to find the area.

$$A = \sqrt{S(S - a)(S - b)(S - c)}$$

where *a, b,* and *c* are the lengths of the three sides and

$$S = \frac{a + b + c}{2}$$

Use Heron's formula to find the area of a triangle whose sides are 3 inches, 4 inches, and 5 inches.

Solution: The triangle is illustrated in Fig. 8.3. Let $a = 3$, $b = 4$, and $c = 5$. First, find the value of *S*.

$$S = \frac{3 + 4 + 5}{2} = \frac{12}{2} = 6$$

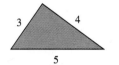

FIGURE 8.3

Now find the area.

$$\begin{aligned} A &= \sqrt{S(S - a)(S - b)(S - c)} \\ &= \sqrt{6(6 - 3)(6 - 4)(6 - 5)} \\ &= \sqrt{6(3)(2)(1)} \\ &= \sqrt{36} = 6 \end{aligned}$$

The area of the triangle is 6 square inches.

Solve for a Variable   **5** You may be given a formula and be asked to solve for a variable in a radicand. To do so, follow the same general procedure used to solve a radical equation. Begin by isolating the radical expression. Then raise both sides of the equation to the same power as the index of the radical. This procedure is illustrated in Example 12(b).

EXAMPLE 12   A formula in statistics for finding the maximum error of estimation is $E = Z\dfrac{\sigma}{\sqrt{n}}$.

(a) Find *E* if $Z = 1.28$, $\sigma = 5$, and $n = 36$.
(b) Solve this equation for *n*.

Solution:

(a) $E = Z\dfrac{\sigma}{\sqrt{n}} = 1.28\left(\dfrac{5}{\sqrt{36}}\right) = 1.28\left(\dfrac{5}{6}\right) \approx 1.07$

(b) First multiply both sides of the equation by $\sqrt{n}$ to eliminate fractions.

Then isolate $\sqrt{n}$. Finally, solve for n by squaring both sides of the equation.

$$E = Z\frac{\sigma}{\sqrt{n}}$$

$$\sqrt{n}(E) = \left(Z\frac{\sigma}{\sqrt{n}}\right)\sqrt{n}$$

$$\sqrt{n}(E) = Z\sigma$$

$$\sqrt{n} = \frac{Z\sigma}{E}$$

$$(\sqrt{n})^2 = \left(\frac{Z\sigma}{E}\right)^2$$

$$n = \left(\frac{Z\sigma}{E}\right)^2 \quad \text{or} \quad n = \frac{Z^2\sigma^2}{E^2}$$

Exercise Set 8.6   **13, 15, 20, 36, 44, 49** no solution   **29.** -7 **33.** 10   **35.** -3 **37.** 4

Solve and check your solution(s). If the equation has no real solution, so state.

1. $\sqrt{x} = 5$ 25

2. $\sqrt{x} = 9$ 81

3. $\sqrt[3]{x} = 2$ 8

4. $\sqrt[3]{x} = 4$ 64

5. $\sqrt[4]{x} = 3$ 81

6. $\sqrt{x-3} + 5 = 6$ 4

7. $-\sqrt{2x+4} = -6$ 16

8. $\sqrt{x+3} = 5$ 4

9. $\sqrt[3]{2x+11} = 3$ 8

10. $\sqrt[3]{6x-3} = 3$ 5

11. $\sqrt[3]{3x+4} = 7$ 9

12. $2\sqrt{4x-3} = 10$ 7

13. $\sqrt{2x-3} = 2\sqrt{3x-2}$

14. $\sqrt{8x-4} = \sqrt{7x+2}$ 6

15. $\sqrt{5x+10} = -\sqrt{3x+8}$

16. $\sqrt[4]{x+8} = \sqrt[4]{2x}$ 8

17. $\sqrt[3]{6x+1} = \sqrt[3]{2x+5}$ 1

18. $\sqrt[4]{3x+1} = 2$ 5

19. $\sqrt{x^2+9x+3} = -x$ $-\frac{1}{3}$

20. $\sqrt{x^2+3x+9} = x$

21. $\sqrt{m^2+4m-20} = m$ 5

22. $\sqrt{5a+1} - 11 = 0$ 24

23. $\sqrt{x^2-2} = x+4$ $-\frac{9}{4}$

24. $\sqrt{x^2+3} = x+1$ 1

25. $-\sqrt{x} = 2x - 1$ $\frac{1}{4}$

26. $\sqrt{3x+4} = x - 2$ 7

27. $\sqrt{x+7} = 2x - 1$ 2

28. $\sqrt[3]{3x-1} + 4 = 0$ -21

29. $\sqrt[3]{x-12} = \sqrt[3]{5x+16}$

30. $\sqrt{6x-1} = 3x$ $\frac{1}{3}$

31. $\sqrt{8b-15} + b = 10$ 5

32. $\sqrt[3]{4x-3} - 3 = 0$ $\frac{15}{2}$

33. $(x+15)^{1/2} - x + 5 = 0$

34. $(2x^2+4x+6)^{1/2} = \sqrt{2x^2+6}$ 0

35. $(r+2)^{1/3} = (3r+8)^{1/3}$

36. $\sqrt[4]{x+5} = -3$

37. $(5x+18)^{1/4} = (9x+2)^{1/4}$

38. $(3x+6)^{1/3} + 3 = 0$ -11

39. $(x^2+4x+4)^{1/2} - x - 3 = 0$ $-\frac{5}{2}$

40. $(5a+2)^{1/4} = (2a+16)^{1/4}$ $\frac{14}{3}$

Solve. You will have to square both sides of the equation twice to eliminate all radicals (see Example 8).

41. $\sqrt{2a-3} = \sqrt{2a} - 1$ 2

42. $\sqrt{x+2} = \sqrt{x+16}$ 9

43. $\sqrt{x+1} = 2 - \sqrt{x}$ $\frac{9}{16}$

44. $\sqrt{x+3} = \sqrt{x} - 3$

45. $\sqrt{x+7} = 5 - \sqrt{x-8}$ 9

46. $\sqrt{y+2} = 1 + \sqrt{y-3}$ 7

47. $\sqrt{b-3} = 4 - \sqrt{b+5}$ 4

48. $\sqrt{4x-3} = 2 + \sqrt{2x-5}$ 3, 7

49. $\sqrt{r+10} + 3 + \sqrt{r-5} = 0$

50. $\sqrt{y+1} = \sqrt{y+2} - 1$ -1

51. $\sqrt{2x+4} - \sqrt{x+3} - 1 = 0$ 6

52. $2 + \sqrt{x+8} = \sqrt{3x+12}$ 8

Solve each formula for the variable indicated.

53. $p = \sqrt{2v}$, for v $v = \frac{p^2}{2}$

54. $l = \sqrt{4r}$, for r $r = \frac{l^2}{4}$

55. $v = \sqrt{2gh}$, for g $g = \frac{v^2}{2h}$

56. $v = \sqrt{\frac{2E}{m}}$, for E $E = \frac{v^2 m}{2}$

57. $v = \sqrt{\frac{FR}{m}}$, for F $F = \frac{v^2 m}{R}$

58. $\omega = \sqrt{\frac{a_0}{x_0}}$, for x_0 $x_0 = \frac{a_0}{\omega^2}$

59. $x = \sqrt{\frac{m}{k}}\, V_0$, for m $m = \frac{x^2 k}{v_0^2}$

60. $T = 2\pi\sqrt{\frac{L}{32}}$, for L $L = \frac{8T^2}{\pi^2}$

63. $\sqrt{18.25} \approx 4.27$m   **64.** $\sqrt{40} \approx 6.32$ m   **68.** $\sqrt{6.37} \approx 2.52$ in   **72.** $6.28\sqrt{1.25} \approx 7.02$ sec   **73.** $\sqrt{576} = 24$ sq in

Use the formula given in Example 9 to find the length of side x.

61.   **62.**

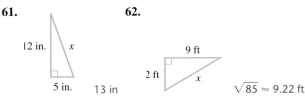

12 in.   x

5 in.   13 in

9 ft

2 ft   x

$\sqrt{85} \approx 9.22$ ft

Use the formula given in Example 9 to answer Exercises 63–65.

63. How long a wire does a phone company worker need to reach from the top of a 4-meter telephone pole to a point 1.5 meters from the base of the pole?

64. Ms. Song Tran places an extension ladder against her house. The base of the ladder is 2 meters from the house and the ladder rests against the house 6 meters above the ground. How far is her ladder extended?

65. A regulation baseball diamond is a square with 90 feet between bases. How far is second base from home plate? $\sqrt{16,200} \approx 127.28$ ft

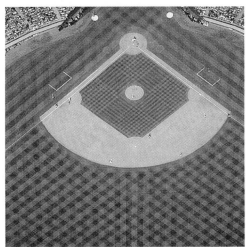

66. When you are given the area of a square, the length of a side can be found by the formula $s = \sqrt{A}$. Find the side of a square that has an area of 64 inches.   8 in.

67. Find the side of a square that has an area of 60 square meters. $\sqrt{60} \approx 7.75$ m

68. When you are given the area of a circle, its radius can be found by the formula $r = \sqrt{A/\pi}$. Find the radius of a circle of an area of 20 square inches.

69. On Earth, the velocity of an object, in feet per second, after freely falling h feet may be found by the formula $V = \sqrt{64h}$. Find the velocity of a shoe after it has fallen 80 feet. $\sqrt{5120} \approx 71.55$ ft/sec

70. Find the velocity of an object after it has fallen 50 feet.

70. $\sqrt{3200} \approx 56.57$ ft/sec   **78.** $\sqrt{124,215,000} \approx 11,145.18$ m/s

71. Find the period of the pendulum if its length is 8 feet. Use $T = 2\pi\sqrt{L/32}$. Refer to Example 10.   3.14 sec

72. Find the period of a 40-foot pendulum.

73. Find the area of a triangle if its three sides are 6 inches, 8 inches, and 10 inches. Use $A = \sqrt{S(S-a)(S-b)(S-c)}$. Refer to Example 11.

74. Find the area of a triangle if its three sides are 4 inches, 10 inches, and 12 inches. $\sqrt{351} \approx 18.73$ sq in.

75. For any planet, its "year" is the time it takes for the planet to revolve once around the sun. The number of Earth days in a given planet's year, N, is approximated by the formula $N = 0.2(\sqrt{R})^3$ where R is the mean distance of the planet to the sun in millions of kilometers. Find the number of Earth days in the year of the planet Earth whose mean distance to the sun is 149.4 million kilometers. $0.2(\sqrt{149.4})^3 \approx 365.2$ days

76. Find the number of earth days in the year of the planet Mercury, whose mean distance to the sun is 58 million kilometers. $0.2(\sqrt{58})^3 \approx 88.3$ days

77. When two forces, F_1 and F_2, pull at right angles to each other as illustrated below, the resultant, or the effective force, R, can be found by the formula $R = \sqrt{F_1^2 + F_2^2}$. Two cars are trying to pull a third out of the mud, as illustrated. If car A is exerting a force of 600 pounds and car B is exerting a force of 800 pounds, find the resulting force on the car stuck in the mud. $\sqrt{1,000,000} = 1000$ lb

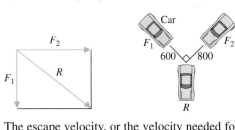

F_2

F_1 R

Car

F_1 F_2

600   800

R

78. The escape velocity, or the velocity needed for a spacecraft to escape a planet's gravitational field, is

82. $\sqrt{x+3}$ cannot equal a negative number. **85.** $\sqrt{x-3}$ cannot equal a negative number, and it must equal -3.

found by the formula $v_e = \sqrt{2gR}$, where g is the force of gravity of the planet and R is the radius of the planet. Find the escape velocity for Earth, in meters per second, where $g = 9.75$ meters per second squared and $R = 6{,}370{,}000$ meters.

79. A formula used in the study of shallow-water wave motion is $c = \sqrt{gH}$, in which c is wave velocity, H is water depth, and g is the acceleration due to gravity. Find the wave velocity if the water's depth is 10 feet. (Use $g = 32$ ft/sec².) $\sqrt{320} \approx 17.89$ ft/sec

80. The length of the diagonal of a rectangular solid is given by $d = \sqrt{a^2 + b^2 + c^2}$. Find the length of the diagonal of a suitcase of length 37 inches, width 15 inches, and depth 9 inches. $\sqrt{1675} \approx 40.93$ in

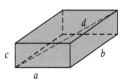

81. A formula we have already mentioned and will be discussing in more detail shortly is the quadratic formula

$$x = \frac{-b \pm \sqrt{b^2 - 4ac}}{2a}$$

(a) Find x when $a = 1, b = 0, c = -4$. $2, -2$
(b) Find x when $a = 1, b = 1, c = -12$. $3, -4$
(c) Find x when $a = 2, b = 5, c = -12$. $-4, \frac{3}{2}$
(d) Find x when $a = -1, b = 4, c = 5$ $5, -1$

82. Consider the equation $\sqrt{x+3} = -\sqrt{2x-1}$. Explain why this equation can have no real solution.

83. Consider the equation $-\sqrt{x^2} = \sqrt{(-x)^2}$. By studying the equation, can you determine its solution? Explain your answer. 0

84. Consider the equation $\sqrt[3]{x^2} = -\sqrt[3]{x^2}$. By studying the equation, can you determine its solution? Explain. 0

85. Explain without solving the equation how you can tell that $\sqrt{x-3} + 3 = 0$ has no solution.

86. Why is it necessary to check solutions to radical equations? They may be extraneous solutions.

87. (a) Solve the equation $\sqrt{2x+12} = 4$. 2
(b) Graph $y = \sqrt{2x+12}$ and $y = 4$ and determine the x coordinate of the point of intersection.

(c) Does the x coordinate of the inersection agree with your answer to part (a)? yes

88. (a) Solve the equation $\sqrt{2x-3} = x - 3$. 6
(b) Graph $y = \sqrt{2x-3}$ and $y = x - 3$ and determine the x coordinate of the intersection.
(c) Does the x coordinate of the intersection agree with your answer to part (a)? yes

89. (a) Consider the equation $\sqrt{4x-12} = x - 3$. Setting each side of the equation equal to y yields the following system of equations.

$$y = \sqrt{4x-12} \quad \textbf{(a)}\ 3, 7 \quad \textbf{(b)}\ \text{yes}$$
$$y = x - 3 \quad \textbf{(c)}\ 3, 7$$

The graphs of the equations in the system are illustrated below. From the graph determine the values that appear to be solutions to the equation $\sqrt{4x-12} = x - 3$. Explain how you determined your answer.
(b) Substitute the values found in part (a) into the original equation and determine if they are the solutions to the equation.
(c) Solve the equation $\sqrt{4x-12} = x - 3$ algebraically and see if your solution agrees with the values obtained in part (a).

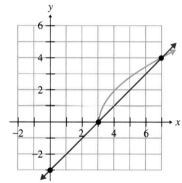

90. The graph of the equation $y = \sqrt{x-3} + 2$ is illustrated below. **90. (a)** $\{x \mid x \geq 3\}$ **(b)** no real solutions
(a) What is the domain of the function?
(b) How many real solutions does the equation $\sqrt{x-3} + 2 = 0$ have? List all the real solutions. Explain how you determined your answer.

87. (b)

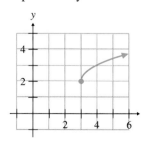

88. (b)

91. Consider the equation solved in Example 5,
$2x - 5\sqrt{x - 3} = 0$.

(a) Refer back to the example and determine how many real solutions and how many extraneous solutions the equation has.   one real, 9; one extraneous

(b) If we graph $y = 2x - 5\sqrt{x - 3}$, how many x intercepts will the graph have, and where will they occur? Explain your answer.   one, at $x = 9$

(c) What is the domain of the function? $\{x \mid x \geq 0\}$

(d) Graph $y = 2x - 5\sqrt{x - 3}$ by plotting points or using a grapher, and determine if your answer to part **(b)** was correct.

Yes

CUMULATIVE REVIEW EXERCISES

[2.2] **92.** Solve the formula $P_1P_2 - P_1P_3 = P_2P_3$ for P_2

[7.1] **93.** Reduce $\dfrac{(2x + 3)(3x - 4) - (2x + 3)(5x - 1)}{(2x + 3)}$

94. Find the domain of $\dfrac{x + 2}{3x^2 - 10x - 8}$.

Perform the operations indicated.

95. $\dfrac{4a^2 - 9b^2}{4a^2 + 12ab + 9b^2} \cdot \dfrac{6a^2b}{8a^2b^2 - 12ab^3}$ $\dfrac{3a}{2b(2a + 3b)}$

96. $(t^2 - t - 12) \div \dfrac{t^2 - 9}{t^2 - 3t}$ $t(t - 4)$

[7.2] **97.** $\dfrac{2}{x + 3} - \dfrac{1}{x - 3} + \dfrac{2x}{x^2 - 9}$ $\dfrac{3}{x + 3}$

[7.4] **98.** Solve $2 + \dfrac{3x}{x - 1} = \dfrac{8}{x - 1}$.   2

92. $P_2 = \dfrac{P_1P_3}{P_1 - P_3}$ **93.** $-2x - 3$

94. $\{x \mid x \neq 4 \text{ and } x \neq -\frac{2}{3}\}$

Group Activity/ Challenge Problems

Solve.

1. $(x^2 + 4x + 4)^{1/2} - x - 2 = 0$ $\mathbb{R}$

2. $\sqrt{x^2 - 4} = (x^2 - 4)^{1/2}$ $x \leq -2$ or $x \geq 2$

3. $\sqrt{4x + 1} - \sqrt{3x - 2} = \sqrt{x - 5}$ 6

4. $\dfrac{x + \sqrt{x + 3}}{x - \sqrt{x + 3}} = 3$ 6

5. $\sqrt{\sqrt{x + 25} - \sqrt{x}} = 5$ no solution

6. $\sqrt{\sqrt{x + 9} + \sqrt{x}} = 3$ 16

7. $(3p - 1)^{2/3} = (5p^2 - p)^{1/3}$ $\frac{1}{4}$, 1

8. (a) Solve the equation $\sqrt{x - 4} = \sqrt{2x - 3}$.   no solution

(b) Show, using graphing, that the equation has no real solution.

8. (b)

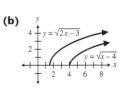

Solve each of the following equations for n.

9. $z = \dfrac{\bar{x} - \mu}{\dfrac{\sigma}{\sqrt{n}}}$ $n = \dfrac{z^2\sigma^2}{(\bar{x} - \mu)^2}$

10. $z = \dfrac{p' - p}{\sqrt{\dfrac{pq}{n}}}$ $n = \dfrac{z^2pq}{(p' - p)^2}$

11. A formula used to determine the frequency of a vibrating spring is $f = \dfrac{1}{2\pi}\sqrt{\dfrac{k}{m}}$,

where f is the frequency of oscillation in cycles per second (also called hertz), k is the spring stiffness constant, and m is the mass of the spring. Find the resulting frequency of a spring with a stiffness constant of 10^5 dynes/cm and a mass of 1000 grams.   1.59 cps

8.7   Complex Numbers

Tape 14

1. Recognize an imaginary number.
2. Recognize a complex number.
3. Add and subtract complex numbers.
4. Multiply complex numbers.
5. Find the conjugate of a complex number.
6. Divide complex numbers.

Imaginary Numbers   1. In Section 8.1 we stated that the square roots of negative numbers, such as $\sqrt{-4}$, are not real numbers. Numbers like $\sqrt{-4}$ are called **imaginary numbers.** There is no real number that when multiplied by itself is -4.

$$\sqrt{-4} \neq 2 \qquad \text{since} \qquad 2 \cdot 2 = 4$$
$$\sqrt{-4} \neq -2 \qquad \text{since} \qquad (-2)(-2) = 4$$

Numbers such as $\sqrt{-4}$ are called imaginary because when they were introduced many mathematicians refused to believe that they existed. Although they do not belong to the set of real numbers, the imaginary numbers do exist and are very useful in mathematics and science.

Every imaginary number has $\sqrt{-1}$ as a factor. For example,

$$\sqrt{-4} = \sqrt{4}\sqrt{-1}$$
$$\sqrt{-9} = \sqrt{9}\sqrt{-1}$$
$$\sqrt{-7} = \sqrt{7}\sqrt{-1}$$

The $\sqrt{-1}$, called the **imaginary unit,** is often denoted by the letter i.

$$i = \sqrt{-1}$$

We can therefore write

$$\sqrt{-4} = \sqrt{4}\sqrt{-1} = 2\sqrt{-1} = 2i$$
$$\sqrt{-9} = \sqrt{9}\sqrt{-1} = 3\sqrt{-1} = 3i$$
$$\sqrt{-7} = \sqrt{7}\sqrt{-1} = \sqrt{7}i \ \text{ or } \ i\sqrt{7}$$

In this book we will generally write $i\sqrt{7}$ rather than $\sqrt{7}i$ to avoid confusion with $\sqrt{7i}$.

To help in writing square roots of negative numbers using i, we give the following rule.

For any positive real number n,

$$\sqrt{-n} = i\sqrt{n}$$

Examples

$$\sqrt{-4} = i\sqrt{4} = 2i \qquad\qquad \sqrt{-3} = i\sqrt{3}$$
$$\sqrt{-25} = i\sqrt{25} = 5i \qquad\qquad \sqrt{-10} = i\sqrt{10}$$

Any number that can be expressed in the form $bi,$ where b is any nonzero real number and $i = \sqrt{-1}$, is an **imaginary number.** For example, $3i$ and $i\sqrt{7}$ are imaginary numbers. Since the numbers $\sqrt{-4}$ and $\sqrt{-15}$ can also be placed in the bi form, these numbers are also imaginary numbers.

Complex Numbers   2️⃣ Now we are prepared to discuss complex numbers.

> **Complex Number**
> Every number of the form
> $$a + bi$$
> where a and b are real numbers, is a complex number.

Every real number and every imaginary number are also complex numbers. A complex number has two parts: a real part, a, and an imaginary part, b.

real part ⟶   ⟵ imaginary part
$$a + bi$$

If $b = 0$, the complex number is a real number. If $a = 0$, the complex number is a *pure imaginary number.*

Examples of Complex Numbers

$3 + 4i$	$a = 3, b = 4$	
$5 - i\sqrt{3}$	$a = 5, b = -\sqrt{3}$	
5	$a = 5, b = 0$	(real number, $b = 0$)
$2i$	$a = 0, b = 2$	(imaginary number, $a = 0$)
$-i\sqrt{7}$	$a = 0, b = -\sqrt{7}$	(imaginary number, $a = 0$)

We stated that all real numbers and imaginary numbers are also complex numbers. The relationship between the various sets of numbers is illustrated in Fig. 8.4.

Complex Numbers

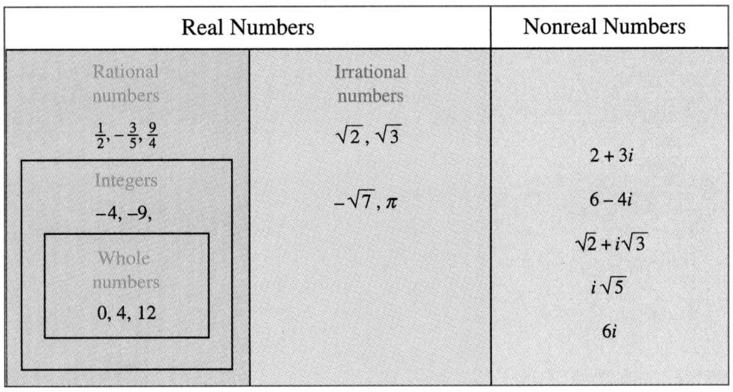

Real Numbers		Nonreal Numbers
Rational numbers $\frac{1}{2}, -\frac{3}{5}, \frac{9}{4}$ Integers $-4, -9,$ Whole numbers $0, 4, 12$	Irrational numbers $\sqrt{2}, \sqrt{3}$ $-\sqrt{7}, \pi$	$2 + 3i$ $6 - 4i$ $\sqrt{2} + i\sqrt{3}$ $i\sqrt{5}$ $6i$

FIGURE 8.4

EXAMPLE 1 Write each of the following complex numbers in the form $a + bi$.

(a) $3 + \sqrt{-16}$ **(b)** $5 - \sqrt{-12}$ **(c)** 4 **(d)** $\sqrt{-18}$ **(e)** $6 + \sqrt{5}$

Solution:

(a) $3 + \sqrt{-16} = 3 + \sqrt{16}\sqrt{-1}.$
$$= 3 + 4i$$

(b) $5 - \sqrt{-12} = 5 - \sqrt{12}\sqrt{-1}$
$$= 5 - \sqrt{4}\sqrt{3}\sqrt{-1}$$
$$= 5 - 2\sqrt{3}\,i \text{ or } 5 - 2i\sqrt{3}$$

(c) $4 = 4 + 0i$

(d) $\sqrt{-18} = 0 + \sqrt{-18}$
$$= 0 + \sqrt{9}\sqrt{2}\sqrt{-1}$$
$$= 0 + 3\sqrt{2}\,i \text{ or } 0 + 3i\sqrt{2}$$

(e) Both 6 and $\sqrt{5}$ are real numbers. $(6 + \sqrt{5}) + 0i.$

Complex numbers can be added, subtracted, multiplied, and divided. To perform these operations, we use the definitions that $i = \sqrt{-1}$ and

$$i^2 = -1$$

Adding and Subtracting Complex Numbers

3 We will first explain how to add or subtract complex numbers. The procedures to multiply and divide complex numbers will be explained shortly.

> ## To Add or Subtract Complex Numbers
> 1. Change all imaginary numbers to bi form.
> 2. Add (or subtract) the real parts of the complex numbers.
> 3. Add (or subtract) the imaginary parts of the complex numbers.
> 4. Write the answer in the form $a + bi$.

EXAMPLE 2 Add $(4 + 13i) + (-6 - 8i)$.

Solution:

$$(4 + 13i) + (-6 - 8i) = 4 + 13i - 6 - 8i$$
$$= 4 - 6 + 13i - 8i$$
$$= -2 + 5i$$

EXAMPLE 3 Subtract $\left(-6 - \frac{1}{3}i\right) - \left(\frac{5}{2} - 4i\right)$.

Solution: $\left(-6 - \frac{1}{3}i\right) - \left(\frac{5}{2} - 4i\right) = -6 - \frac{1}{3}i - \frac{5}{2} + 4i$

$$= -6 - \frac{5}{2} - \frac{1}{3}i + 4i$$

$$= -\frac{12}{2} - \frac{5}{2} - \frac{1}{3}i + \frac{12}{3}i$$

$$= -\frac{17}{2} + \frac{11}{3}i$$

EXAMPLE 4 Add $(6 - \sqrt{-8}) + (4 + \sqrt{-18})$.

Solution:

$(6 - \sqrt{-8}) + (4 + \sqrt{-18}) = (6 - \sqrt{8}\sqrt{-1}) + (4 + \sqrt{18}\sqrt{-1})$
$= (6 - \sqrt{4}\sqrt{2}\sqrt{-1}) + (4 + \sqrt{9}\sqrt{2}\sqrt{-1})$
$= (6 - 2i\sqrt{2}) + (4 + 3i\sqrt{2})$
$= 6 + 4 - 2i\sqrt{2} + 3i\sqrt{2}$
$= 10 + i\sqrt{2}$

Multiplying Complex Numbers | 4 |

To Multiply Complex Numbers

1. Change all imaginary numbers to bi form.
2. Multiply the complex numbers as you would multiply polynomials.
3. Subtitute -1 for each i^2.
4. Combine the real parts and the imaginary parts. Write the answer in $a + bi$ form.

EXAMPLE 5 Multiply. **(a)** $3i(5 - 2i)$. **(b)** $\sqrt{-4}(\sqrt{-2} + 7)$.

Solution:
(a) $3i(5 - 2i) = 3i(5) + (3i)(-2i)$
$= 15i - 6i^2$
$= 15i - 6(-1)$
$= 15i + 6$ or $6 + 15i$

(b) $\sqrt{-4}(\sqrt{-2} + 7) = 2i(i\sqrt{2} + 7)$
$= (2i)(i\sqrt{2}) + (2i)(7)$
$= 2i^2\sqrt{2} + 14i$
$= 2(-1)\sqrt{2} + 14i$
$= -2\sqrt{2} + 14i$

COMMON STUDENT ERROR   What is $\sqrt{-4} \cdot \sqrt{-2}$?

Correct	*Incorrect*

$$\sqrt{-4} \cdot \sqrt{-2} = 2i \cdot i\sqrt{2}$$
$$= 2i^2\sqrt{2}$$
$$= 2(-1)\sqrt{2}$$
$$= -2\sqrt{2}$$

$$\cancel{\sqrt{-4} \cdot \sqrt{-2} = \sqrt{8}}$$
$$\cancel{= \sqrt{4} \cdot \sqrt{2}}$$
$$\cancel{= 2\sqrt{2}}$$

Recall that $\sqrt{a} \cdot \sqrt{b} = \sqrt{ab}$ for *nonnegative* real numbers a and b.

EXAMPLE 6   Multiply $(3 - \sqrt{-8})(\sqrt{-2} + 5)$.

Solution: $(3 - \sqrt{-8})(\sqrt{-2} + 5) = (3 - \sqrt{8}\sqrt{-1})(\sqrt{2}\sqrt{-1} + 5)$

$$= (3 - 2i\sqrt{2})(i\sqrt{2} + 5)$$

Now use the FOIL method to multiply.

$$= (3)(i\sqrt{2}) + (3)(5) + (-2i\sqrt{2})(i\sqrt{2}) + (-2i\sqrt{2})(5)$$
$$= 3i\sqrt{2} + 15 - 2i^2(2) - 10i\sqrt{2}$$
$$= 3i\sqrt{2} + 15 - 2(-1)(2) - 10i\sqrt{2}$$
$$= 3i\sqrt{2} + 15 + 4 - 10i\sqrt{2}$$
$$= 19 - 7i\sqrt{2}$$

EXAMPLE 7   Show that $x = 1 + i\sqrt{3}$ is a solution to the equation $x^2 - 2x + 4 = 0$.

Solution: Substitute $1 + i\sqrt{3}$ for each x and determine if the resulting statement is true.

$$x^2 - 2x + 4 = 0$$
$$(1 + i\sqrt{3})^2 - 2(1 + i\sqrt{3}) + 4 = 0$$
$$(1 + i\sqrt{3})(1 + i\sqrt{3}) - 2 - 2i\sqrt{3} + 4 = 0$$
$$1 + 2i\sqrt{3} + 3i^2 - 2 - 2i\sqrt{3} + 4 = 0$$
$$1 + 2i\sqrt{3} - 3 - 2 - 2i\sqrt{3} + 4 = 0$$
$$0 = 0 \qquad \text{true}$$

Since we obtain a true statement, $1 + i\sqrt{3}$ is a solution.

A second solution to $x^2 - 2x + 4 = 0$ is $1 - i\sqrt{3}$. Show now that this is also a solution to the equation.

Conjugates   **5**   The **conjugate** of a complex number $a + bi$ is $a - bi$. For example,

Complex Number	*Conjugate*
$3 + 4i$	$3 - 4i$
$1 - i\sqrt{3}$	$1 + i\sqrt{3}$
$2i$ (or $0 + 2i$)	$-2i$ (or $0 - 2i$)

When a complex number is multiplied by its conjugate using the FOIL method, the inner and outer products will sum to zero. For example,

$$(5 + 2i)(5 - 2i) = 25 - 10i + 10i - 4i^2$$
$$= 25 - 4i^2$$
$$= 25 - 4(-1) = 25 + 4 = 29$$

Divide Complex 6
Numbers

> ## To Divide Complex Numbers
> 1. Change all imaginary numbers to bi form.
> 2. Write the division problem as a fraction.
> 3. Rationalize the denominator by multiplying both the numerator and denominator by the conjugate of the denominator.
> 4. Write the answer in $a + bi$ form.

EXAMPLE 8   Simplify (a) $\dfrac{4 + i}{i}$ (b) $\dfrac{\sqrt{-108}}{\sqrt{-3}}$

Solution: Begin part **(a)** by multiplying both numerator and denominator by $-i$, the conjugate of i. Begin part **(b)** by writing the radicals in terms of i.

(a) $\dfrac{4 + i}{i} \cdot \dfrac{-i}{-i} = \dfrac{(4 + i)(-i)}{-i^2}$

$$= \frac{-4i - i^2}{-i^2}$$

$$= \frac{-4i - (-1)}{-(-1)}$$

$$= \frac{-4i + 1}{1}$$

$$= 1 - 4i$$

(b) $\dfrac{\sqrt{-108}}{\sqrt{-3}} = \dfrac{i\sqrt{108}}{i\sqrt{3}}$

$$= \frac{\sqrt{108}}{\sqrt{3}}$$

$$= \sqrt{\frac{108}{3}}$$

$$= \sqrt{36}$$

$$= 6$$

EXAMPLE 9   Divide $\dfrac{6 - 5i}{2 - i}$.

Solution: Multiply both numerator and denominator by $2 + i$, the conjugate of $2 - i$.

$$\frac{6 - 5i}{2 - i} \cdot \frac{2 + i}{2 + i} = \frac{12 + 6i - 10i - 5i^2}{4 - i^2}$$

$$= \frac{12 - 4i - 5(-1)}{4 - (-1)}$$

$$= \frac{17 - 4i}{5} \quad \text{or} \quad \frac{17}{5} - \frac{4}{5}i$$

EXAMPLE 10   A concept needed for the study of electronics is *impedance*. Impedance affects the current in a circuit. The impedance, Z, in a circuit is found by the formula $Z = \dfrac{V}{I}$, where V is voltage and I is current. Find Z when $V = 1.6 - 0.3i$ and $I = -0.2i$, where $i = \sqrt{-1}$.

Solution: $Z = \dfrac{V}{I} = \dfrac{1.6 - 0.3i}{-0.2i}$. Now multiply both the numerator and denominator by $0.2i$.

$$Z = \frac{1.6 - 0.3i}{-0.2i} \cdot \frac{0.2i}{0.2i} = \frac{0.32i - 0.06i^2}{-0.04i^2}$$

$$= \frac{0.32i + 0.06}{0.04}$$

$$= \frac{0.32i}{0.04} + \frac{0.06}{0.04}$$

$$= 8i + 1.5 \quad \text{or} \quad 1.5 + 8i$$

Most algebra books use i as the imaginary unit. However, most electronics books use j as the imaginary unit because i is often used to represent current.

Using $i = \sqrt{-1}$ and $i^2 = -1$, we can find other powers of i. For example,

$$i^3 = i^2 \cdot i = -1 \cdot i = -i \qquad i^6 = i^4 \cdot i^2 = 1(-1) = -1$$
$$i^4 = i^2 \cdot i^2 = (-1)(-1) = 1 \qquad i^7 = i^4 \cdot i^3 = 1(-i) = -i$$
$$i^5 = i^4 \cdot i^1 = 1 \cdot i = i \qquad i^8 = i^4 \cdot i^4 = (1)(1) = 1$$

Note that successive powers of i rotate through the four values i, -1, $-i$, 1 (see Fig. 8.5).

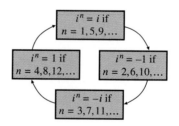

FIGURE 8.5

EXAMPLE 11   Evaluate.   **(a)** i^{35} **(b)** i^{81}

Solution: Write each expression as a product of factors such that the exponent of one factor is the largest multiple of 4 less than or equal to the given exponent. Then write this factor as i^4 raised to some power. Since i^4 has a value of 1, the expression i^4 raised to a power will also have a value of 1.

(a) $i^{35} = i^{32} \cdot i^3 = (i^4)^8 \cdot i^3 = 1 \cdot i^3 = 1\,(-i) = -i$

(b) $i^{81} = i^{80} \cdot i^1 = (i^4)^{20} \cdot i = 1 \cdot i = i$

Helpful Hint

A quick way of evaluating i^n is to divide the exponent by 4 and observe the remainder.

If the remainder is 0, the value is 1.   If the remainder is 2, the value is -1.
If the remainder is 1, the value is i.   If the remainder is 3, the value is $-i$.

For Example 11(**a**) $\quad 4\overline{)35} \atop 32$

$3 \longleftarrow$ answer is $-i$

For 11(**b**) $\quad 4\overline{)81} \atop 80$

$1 \longleftarrow$ answer is i

Exercise Set 8.7

60. $\dfrac{5\sqrt{2} - 2i\sqrt{6}}{37}$ **61.** $\dfrac{(5\sqrt{10} - 2\sqrt{15}) + (10\sqrt{2} + 5\sqrt{3})i}{45}$ **62.** $\dfrac{(6\sqrt{3} - \sqrt{15}) - (6\sqrt{5} + 3)i}{4}$

Write each expression as a complex number in the form $a + bi$.

1. $3 \quad 3 + 0i$ **2.** $\sqrt{9} \quad 3 + 0i$ **3.** $3 + \sqrt{-4} \quad 3 + 2i$ **4.** $-\sqrt{5} \quad -\sqrt{5} + 0i$

5. $6 + \sqrt{3} \quad (6 + \sqrt{3}) + 0i$ **6.** $\sqrt{-8} \quad 0 + 2i\sqrt{2}$ **7.** $\sqrt{-25} \quad 0 + 5i$ **8.** $2 + \sqrt{-5} \quad 2 + i\sqrt{5}$

9. $4 + \sqrt{-12} \quad 4 + 2i\sqrt{3}$ **10.** $\sqrt{-4} + \sqrt{-16} \quad 0 + 6i$ **11.** $\sqrt{-25} - 2i \quad 0 + 3i$ **12.** $3 + \sqrt{-72} \quad 3 + 6i\sqrt{2}$

13. $9 - \sqrt{-9} \quad 9 - 3i$ **14.** $\sqrt{75} - \sqrt{-128}$ $\quad 5\sqrt{3} - 8i\sqrt{2}$ **15.** $2i - \sqrt{-80}$ $\quad 0 + (2 - 4\sqrt{5})i$ **16.** $\sqrt{288} - \sqrt{-96}$ $\quad 12\sqrt{2} - 4i\sqrt{6}$

Add or subtract.

17. $(12 - 6i) + (3 + 2i) \quad 15 - 4i$

18. $(6 - 3i) - 2(2 - 4i) \quad 2 + 5i$

19. $(12 + \frac{5}{9}i) - (4 - \frac{3}{4}i) \quad 8 + \frac{47}{36}i$

20. $(\frac{5}{8} + \sqrt{-4}) + (\frac{2}{3} + 7i) \quad \frac{31}{24} + 9i$

21. $(13 - \sqrt{-4}) - (-5 + \sqrt{-9}) \quad 18 - 5i$

22. $(7 + \sqrt{5}) + (2\sqrt{5} + \sqrt{-5}) \quad (7 + 3\sqrt{5}) + i\sqrt{5}$

23. $(\sqrt{3} + \sqrt{2}) + (3\sqrt{2} - \sqrt{-8}) \quad (4\sqrt{2} + \sqrt{3}) - 2i\sqrt{2}$

24. $(3 - \sqrt{-72}) + (4 - \sqrt{-32}) \quad 7 - 10i\sqrt{2}$

25. $(19 + \sqrt{-147}) + (\sqrt{-75}) \quad 19 + 12i\sqrt{3}$

26. $(13 + \sqrt{-108}) - (\sqrt{49} - \sqrt{-48}) \quad 6 + 10i\sqrt{3}$

27. $(\sqrt{12} + \sqrt{-49}) - (\sqrt{49} - \sqrt{-12})$ $\quad (2\sqrt{3} - 7) + (7 + 2\sqrt{3})i$

28. $(\sqrt{20} - \sqrt{-12}) + (2\sqrt{5} + \sqrt{-75}) \quad 4\sqrt{5} + 3i\sqrt{3}$

Multiply.

29. $2(-3 - 2i) \quad -6 - 4i$ **30.** $-3(\sqrt{5} + 2i) \quad -3\sqrt{5} - 6i$ **31.** $i(6 + i) \quad -1 + 6i$

32. $2i(2 - 5i) \quad 10 + 4i$ **33.** $-3.5i(6.4 - 1.8i) \quad -6.3 - 22.4i$ **34.** $\sqrt{-5}(2 + 3i) \quad -3\sqrt{5} + 2i\sqrt{5}$

35. $\sqrt{-4}(\sqrt{3} + 2i) \quad -4 + 2i\sqrt{3}$ **36.** $\sqrt{-8}(\sqrt{2} - \sqrt{-2}) \quad 4 + 4i$ **37.** $\sqrt{-6}(\sqrt{3} + \sqrt{-6}) \quad -6 + 3i\sqrt{2}$

38. $-\sqrt{-2}(3 - \sqrt{-8}) \quad -4 - 3i\sqrt{2}$ **39.** $(3 + 2i)(1 + i) \quad 1 + 5i$ **40.** $(3 - 4i)(6 + 5i) \quad 38 - 9i$

41. $(4 - 6i)(3 - i) \quad 6 - 22i$ **42.** $(3i + 4)(2i - 3) \quad -18 - i$ **43.** $(\frac{1}{4} + \sqrt{-3})(2 + \sqrt{3})$

44. $(2 - 3i)(4 + \sqrt{-4}) \quad 14 - 8i$ **45.** $(5 - \sqrt{-8})(\frac{1}{4} + \sqrt{-2}) \quad \frac{21}{4} + \frac{9}{2}i\sqrt{2}$ **46.** $(\frac{3}{5} - \frac{1}{4}i)(\frac{2}{3} + \frac{2}{5}i) \quad \frac{1}{2} + \frac{11}{150}i$

43. $\left(\frac{1}{2} + \frac{\sqrt{3}}{4}\right) + (2\sqrt{3} + 3)i$

Divide.

47. $\dfrac{-5}{-3i} \quad -\dfrac{5i}{3}$ **48.** $\dfrac{2}{5i} \quad -\dfrac{2i}{5}$ **49.** $\dfrac{2 + 3i}{2i} \quad \dfrac{3 - 2i}{2}$ **50.** $\dfrac{1 + i}{-3i} \quad \dfrac{-1 + i}{3}$

51. $\dfrac{2 + 5i}{5i} \quad \dfrac{5 - 2i}{5}$ **52.** $\dfrac{6}{2 + i} \quad \dfrac{12 - 6i}{5}$ **53.** $\dfrac{7}{7 - 2i} \quad \dfrac{49 + 14i}{53}$ **54.** $\dfrac{4 + 2i}{1 + 3i} \quad 1 - i$

55. $\dfrac{6 - 3i}{4 + 2i} \quad \dfrac{9 - 12i}{10}$ **56.** $\dfrac{4 - 3i}{4 + 3i} \quad \dfrac{7 - 24i}{25}$ **57.** $\dfrac{4}{6 - \sqrt{-4}} \quad \dfrac{3 + i}{5}$ **58.** $\dfrac{5}{3 + \sqrt{-5}} \quad \dfrac{15 - 5i\sqrt{5}}{14}$

69. $6 + i\sqrt{6}$ **70.** $\sqrt{2} + 2i\sqrt{3}$ **73.** $2\sqrt{15} + 3i\sqrt{2}$ **87.** $-3.33 - 9.01i$ **88.** $(-3\sqrt{10} + 15) + (5\sqrt{6} + 3\sqrt{15})i$

59. $\dfrac{\sqrt{6}}{\sqrt{3} - \sqrt{-9}}$ $\dfrac{\sqrt{2} + i\sqrt{6}}{4}$ **60.** $\dfrac{\sqrt{2}}{5 + \sqrt{-12}}$ **61.** $\dfrac{\sqrt{10} + \sqrt{-3}}{5 - \sqrt{-20}}$ **62.** $\dfrac{12 - \sqrt{-12}}{\sqrt{3} + \sqrt{-5}}$

63. $\dfrac{\sqrt{-60}}{\sqrt{-2}}$ $\sqrt{30}$ **64.** $\dfrac{\sqrt{-150}}{\sqrt{6}}$ $5i$ **65.** $\dfrac{\sqrt{-80}\sqrt{-5}}{\sqrt{-2}}$ $10i\sqrt{2}$ **66.** $\dfrac{\sqrt{-32}}{\sqrt{-18}\sqrt{2}}$ $\dfrac{2\sqrt{2}}{3}$

Perform the operations indicated. These exercises are a combination of the exercises presented earlier in this exercise set.

67. $(4 - 2i) + (3 - 5i)$ $7 - 7i$ **68.** $(\frac{1}{2} + i) - (\frac{3}{5} - \frac{2}{3}i)$ $-\frac{1}{10} + \frac{5}{3}i$ **69.** $(8 - \sqrt{-6}) - (2 - \sqrt{-24})$

70. $(\sqrt{8} - \sqrt{2}) - (\sqrt{-12} - \sqrt{-48})$ **71.** $5.2(4 - 3.2i)$ $20.8 - 16.64i$ **72.** $-0.6i(3 + 5i)$ $3 - 1.8i$

73. $\sqrt{-6}(\sqrt{3} - \sqrt{-10})$ **74.** $(5 + 2i)(3 - 5i)$ $25 - 19i$ **75.** $(\sqrt{3} + 2i)(\sqrt{6} - \sqrt{-8})$ $7\sqrt{2}$

76. $\dfrac{6}{2i}$ $-3i$ **77.** $\dfrac{5 - 4i}{2i}$ $\dfrac{-4 - 5i}{2}$ **78.** $\dfrac{1}{2 - 3i}$ $\dfrac{2 + 3i}{13}$

79. $\dfrac{4}{\sqrt{3} - \sqrt{-4}}$ $\dfrac{4\sqrt{3} + 8i}{7}$ **80.** $\dfrac{5 - 2i}{3 + 2i}$ $\dfrac{11 - 16i}{13}$ **81.** $\left(5 - \dfrac{5}{9}i\right) - \left(2 - \dfrac{3}{5}i\right)$ $3 + \dfrac{2}{45}i$

82. $\dfrac{4}{7}\left(4 - \dfrac{3}{5}i\right)$ $\dfrac{16}{7} - \dfrac{12}{35}i$ **83.** $\left(\dfrac{2}{3} - \dfrac{1}{5}i\right)\left(\dfrac{3}{5} - \dfrac{3}{4}i\right)$ $\dfrac{1}{4} - \dfrac{31}{50}i$ **84.** $\sqrt{\dfrac{4}{9}}\left(\sqrt{\dfrac{25}{36}} - \sqrt{-\dfrac{4}{25}}\right)$ $\dfrac{5}{9} - \dfrac{4}{15}i$

85. $\dfrac{\sqrt{-96}}{\sqrt{-24}}$ 2 **86.** $\dfrac{-1 - 2i}{2 + \sqrt{-5}}$ $\dfrac{(-2 - 2\sqrt{5}) + (\sqrt{5} - 4)i}{9}$

87. $(5.23 - 6.41i) - (8.56 - 4.5i)$ $-7.1i$ **88.** $(\sqrt{-6} + 3)(\sqrt{-15} + 5)$

Evaluate for the given value of x.

89. $x^2 - 2x + 5, x = 1 - 2i$ 0 **90.** $x^2 - 2x + 5, x = 1 + 2i$ 0

91. $x^2 + 2x + 8, x = -1 + i\sqrt{5}$ 2 **92.** $x^2 + 2x + 8, x = -1 - i\sqrt{5}$ 2

Determine if the given value of x is a solution to the equation.

93. $x^2 - 4x + 5 = 0, x = 2 - i$ yes   **94.** $x^2 - 4x + 5 = 0, x = 2 + i$ yes

95. $x^2 - 6x + 12 = 0, x = -3 + i\sqrt{3}$ no   **96.** $x^2 - 6x + 12 = 0, x = 3 - i\sqrt{3}$ yes

Indicate whether the value is i, −1, −i, or 1.

97. i^{10} -1 **98.** i^{43} $-i$ **99.** i^{200} 1 **100.** i^{211} $-i$ **101.** i^{93} i **102.** i^{103} $-i$ **103.** i^{907} $-i$ **104.** i^{1113} i

105. Find the impedance, Z, using the formula $Z = \dfrac{V}{I}$ when $V = 1.8 + 0.5i$ and $I = 0.6i$. See Example 10.

106. Refer to Exercise 105. Find the impedance when $V = 2.4 - 0.6i$ and $I = -0.4i$ $1.5 + 6i$

107. Under certain conditions, the total impedance, Z_T, of a circuit is given by the formula

$$Z_T = \dfrac{Z_1 Z_2}{Z_1 + Z_2}$$

Find Z_T when $Z_1 = 2 - i$ and $Z_2 = 4 + i$.

108. Determine whether i^{-1} is equal to i, -1, $-i$, or 1. Show your work. $-i$

109. Determine whether i^{-2} is equal to i, -1, $-i$ or 1. Show your work. -1

105. $0.83 - 3i$ **107.** $1.5 - 0.33i$

Answer true or false.

110. Every real and every imaginary number is a complex number.   true

111. Every complex number is a real number.   false

112. The product of two pure imaginary numbers is always a real number.   true

113. The sum of two pure imaginary numbers is always an imaginary number.   false

114. The product of two complex numbers is always a real number.   false

115. The sum of two complex numbers is always a complex number.   true

116. List, if possible, a number that is *not*
   (a) a rational number; $\sqrt{2}$
   (b) an irrational number;   1
   (c) a real number; $\sqrt{-3}$ or $2i$
   (d) an imaginary number;   6
   (e) a complex number.

117. Write a paragraph or two explaining the relationship between the real numbers, imaginary numbers, and complex numbers. Include in your discussion how the various sets of numbers relate to each other.
116.(e) every number we have studied is a complex number

Cumulative Review Exercises

[4.3] **118.** Mr. Tomlins, a grocer, has two coffees, one selling for $5.00 per pound and the other for $5.80 per pound. How many pounds of each type of coffee should he mix to make 40 pounds of coffee to sell for $5.50 per pound?   15 lb at $5.00, 25 lb at $5.80

[5.6] **119.** Find the axis of symmetry, the coordinates of the vertex, and whether the graph opens up or down. Then graph the function, $f(x) = x^2 - 2x + 1$, and give the domain and range.
$x = 1$, $(1, 0)$, up, D: $\mathbb{R}$, R: $\{y | y \geq 0\}$

Factor.

[6.2] **120.** $4x^4 + 12x^2 + 9$. $(2x^2 + 3)^2$

121. $15r^2s^2 + rs - 6$. $(3rs + 2)(5rs - 3)$

Group Activity/ Challenge Problems

*In Chapter 10 we will use the **quadratic formula*** $x = \dfrac{-b \pm \sqrt{b^2 - 4ac}}{2a}$ *to solve equations of the form* $ax^2 + bx + c = 0$. *(a) Use the quadratic formula to solve the following quadratic equations. (b) Check each of the two solutions by substituting the values found for x (one at a time) back in the original equation. In these exercises, the* $\pm$, *read "plus or minus," results in two distinct complex answers.*

1. $x^2 - 4x + 6 = 0$ $2 \pm i\sqrt{2}$

2. $x^2 - 2x + 6 = 0$ $1 \pm i\sqrt{5}$

Given the complex numbers $a = 3 + 2i\sqrt{3}$, $b = 1 + i\sqrt{3}$, *evaluate each of the following.*

3. $a + b$ $4 + 3i\sqrt{3}$ **4.** $a - b$ $2 + i\sqrt{3}$ **5.** ab $-3 + 5i\sqrt{3}$ **6.** $\dfrac{a}{b}$ $\dfrac{9 - i\sqrt{3}}{4}$

7. For any two complex numbers, a and b, will
   (a) $a + b = b + a$;   yes
   (b) $a - b = b - a$;   no
   (c) $ab = ba$;   yes
   (d) $a \div b = b \div a$?   no
   Explain your answer and give an example to support your answer.

Perform the operations indicated.

10. $\dfrac{-1 + 21i}{221}$

8. $\dfrac{2}{2 - 3i} + \dfrac{1}{1 + i}$ $\dfrac{21 - i}{26}$

9. $\dfrac{3i}{3 + i} + \dfrac{i}{1 - i}$ $\dfrac{-1 + 7i}{5}$

10. $(3 - 2i)^{-1} - (4 - i)^{-1}$

Summary

Glossary

Complex number (495): A number of the form $a + bi$. Every real number and every imaginary number are complex numbers.

Conjugate (472): The conjugate of $a + b$ is $a - b$.

Even roots (445): Radicals with indexes of 2, 4, 6, . . . are said to be even roots.

Imaginary number (495): A number that can be expressed in the form bi, $b \neq 0$.

Imaginary unit (494): $\sqrt{-1}$. *Note:* $\sqrt{-1} = i$.

Index (444): The root of an expression.

Like radicals (477): Radicals that have the same index and radicand.

Odd roots (445): Radicals with indexes of 3, 5, 7, . . . are said to be odd roots.

Perfect power (463): For $\sqrt[m]{x^n}$, x^n is a perfect power if n is a multiple of m.

Principle (or positive) square root (445): The principal square root of a positive real number x, written $\sqrt{x}$, is the positive number whose square equals x.

Radical equation (482): An equation that contains a variable in a radicand.

Radical expression (444): The radical sign and radicand together are a radical expression.

Radical sign (444): $\sqrt{}$

Rationalizing the denominator (470): To write a rational expression in a form that has no radical in its denominator.

IMPORTANT FACTS

$$\sqrt[n]{x} = a \text{ if } a^n = x$$
$$\sqrt{a^2} = |a|$$
$$\sqrt{a^2} = a, \quad a \geq 0$$
$$\sqrt[n]{a^n} = a, \quad a \geq 0$$
$$\sqrt[n]{x} = x^{1/n}, \quad x \geq 0$$
$$\sqrt[n]{x^m} = (\sqrt[n]{x})^m = x^{m/n}, \quad x > 0$$
$$\sqrt[n]{a}\sqrt[n]{b} = \sqrt[n]{ab}, \quad a \geq 0, b \geq 0$$
$$\frac{\sqrt[n]{a}}{\sqrt[n]{b}} = \sqrt[n]{\frac{a}{b}}, \quad a \geq 0, b > 0$$

A radical is simplified when the following are all true:

1. No perfect powers are factors of any radicand.
2. No radicand contains a fraction.
3. No denominator contains a radical.

Powers of i

$$i = \sqrt{-1}, i^2 = -1, i^3 = -i, i^4 = 1$$

Review Exercises

[8.1] *Evaluate.*

1. $\sqrt{9}$ 3

2. $\sqrt{25}$ 5

3. $\sqrt[3]{-8}$ -2

4. $\sqrt[4]{256}$ 4

5. $\sqrt[3]{27}$ 3

6. $\sqrt[3]{-27}$ -3

7. $-\sqrt{144}$ -12

8. $-\sqrt{256}$ -16

Use absolute value to evaluate.

9. $\sqrt{(-7)^2}$ 7

10. $\sqrt{(-93.4)^2}$ 93.4

Write as an absolute value.   **14.** $|x^2 - 4x + 12|$

11. $\sqrt{x^2}$ $|x|$

12. $\sqrt{(x-2)^2}$ $|x-2|$

13. $\sqrt{(x-y)^2}$ $|x-y|$

14. $\sqrt{(x^2 - 4x + 12)^2}$

For the remainder of these review exercises, assume that all variables represent positive real numbers.

[8.2] *Write in exponential form.*

15. $\sqrt{x^5}$ $x^{5/2}$

16. $\sqrt[3]{x^5}$ $x^{5/3}$

17. $(\sqrt[4]{y})^{15}$ $y^{15/4}$

18. $\sqrt[7]{5^2}$ $5^{2/7}$

Write in radical form.

19. $a^{1/2}$ $\sqrt{a}$

20. $y^{3/5}$ $\sqrt[5]{y^3}$

21. $(2m^2n)^{9/5}$ $(\sqrt[5]{2m^2n})^9$

22. $(2a + 3b)^{-3/4}$ $\dfrac{1}{(\sqrt[4]{2a+3b})}$

46. $\dfrac{2(2+3x^2)}{x^6}$ **47.** $(x^{1/4}-1)(x^{1/4}-4)$ **48.** $(x^{1/5}+3)(x^{1/5}-5)$ **49.** $(2x^{1/2}-3)(3x^{1/2}+2)$ **50.** $(4x^{1/3}-1)(2x^{1/3}+3)$

Simplify each radical expression by changing the expression to exponential form. Write the answer in radical form.

23. $\sqrt{5^6}$ 125   **24.** $\sqrt[3]{3^6}$ 9   **25.** $\sqrt[6]{y^2}$ $\sqrt[3]{y}$ **26.** $\sqrt{x^{10}}$ x^5

27. $(\sqrt[3]{4})^6$ 16   **28.** $\sqrt[5]{9^{10}}$ 81   **29.** $\sqrt[20]{x^4}$ $\sqrt[5]{x}$ **30.** $\sqrt[9]{x^9}$ x

Evaluate if possible. If the expression is not a real number, so state.

31. $-25^{1/2}$ -5 **32.** $(-25)^{1/2}$ not a real number **33.** $\left(\dfrac{8}{27}\right)^{-1/3}$ $\dfrac{3}{2}$ **34.** $36^{-1/2}-8^{-2/3}$ $-\dfrac{1}{12}$

Simplify. Write the answer without negative exponents.

35. $x^{3/5}\cdot x^{-1/3}$ $x^{4/15}$ **36.** $\left(\dfrac{64}{y^6}\right)^{1/3}$ $\dfrac{4}{y^2}$ **37.** $\left(\dfrac{y^{-3/5}}{y^{1/5}}\right)^{2/3}$ $\dfrac{1}{y^{8/15}}$ **38.** $\left(\dfrac{30x^4y^{-2}}{5z^{1/2}}\right)^2\left(\dfrac{xy^{1/2}}{4y^3z^2}\right)^2$ $\dfrac{9x^{10}}{4y^9z^5}$

Multiply.   **39.** $2z^2-4z^{4/3}$ **40.** $2x^{-4}-8$ **41.** $-6a^{-3}-2a^{-3/2}$ **42.** $3r^{-13/6}+1$

39. $z^{1/3}(2z^{5/3}-4z)$ **40.** $2x^{-1}(x^{-3}-4x)$ **41.** $-2a^{-2}(3a^{-1}+a^{1/2})$ **42.** $\dfrac{3}{4}r^{-2/3}\left(4r^{-3/2}+\dfrac{4}{3}r^{2/3}\right)$

Factor each of the following. Write the answer without negative exponents.

43. $x^{1/5}+x^{6/5}$ $x^{1/5}(1+x)$ **44.** $x^{-3}-x^{-4}$ $\dfrac{x-1}{x^4}$ **45.** $x^{-1/2}+x^{-2/3}$ $\dfrac{x^{1/6}+1}{x^{2/3}}$ **46.** $4x^{-6}+6x^{-4}$

Factor.

47. $x^{1/2}-5x^{1/4}+4$ **48.** $x^{2/5}-2x^{1/5}-15$ **49.** $6x-5x^{1/2}-6$ **50.** $8x^{2/3}+10x^{1/3}-3$

Use a calculator to evaluate. If the answer is irrational, round to the nearest hundredth.

51. $\sqrt{260}$ 16.12   **52.** $\sqrt[3]{5060.8}$ 17.17   **53.** $512^{3/5}$ 42.22   **54.** $-2162^{3/4}$ -317.06

Simplify.   **58.** $\dfrac{15x^{1/2}y^{4/3}z^{1/3}}{2}$ **67.** $2x^2yz^3\sqrt[5]{x^2y^2z^2}$ **70.** $2x^2y^3\sqrt[3]{4x^2}$ **72.** $2x^2y^4\sqrt[4]{x}$ **73.** $6x-2\sqrt{15x}$ **74.** $4y-y^3\sqrt[3]{y^2}$

[2-8.5]   **55.** $\left(\dfrac{3r^2p^{1/3}}{r^{1/2}p^{4/3}}\right)^3$ $\dfrac{27r^{9/2}}{p^3}$ **56.** $\left(\dfrac{x^{3/5}\cdot 2x^{1/3}}{y^{1/4}\cdot 3y^{1/4}}\right)^4$ $\dfrac{16x^{56/15}}{81y^2}$ **57.** $\left(\dfrac{4y^{2/5}\cdot z^{1/3}}{x^{-1}y^{3/5}}\right)^{-1}$ $\dfrac{y^{1/5}}{4xz^{1/3}}$ **58.** $\dfrac{5x^{-1}y^{-2}z^{1/3}}{2x^{-2}y^{-1/3}}\cdot\dfrac{3x^{1/2}y^2}{xy^{-1}}$

59. $\sqrt{24}$ $2\sqrt{6}$ **60.** $\sqrt{80}$ $4\sqrt{5}$ **61.** $\sqrt[3]{16}$ $2\sqrt[3]{2}$ **62.** $\sqrt[3]{54}$ $3\sqrt[3]{2}$

63. $\sqrt{50x^3y^7}$ $5xy^3\sqrt{2xy}$ **64.** $\sqrt[3]{9x^6y^5}$ $x^2y\sqrt[3]{9y^2}$ **65.** $\sqrt[4]{16x^9y^{12}}$ $2x^2y^3\sqrt[4]{x}$ **66.** $\sqrt[3]{125x^7y^{10}}$ $5x^2y^3\sqrt[3]{xy}$

67. $\sqrt[5]{32x^{12}y^7z^{17}}$ **68.** $\sqrt{20}\sqrt{5}$ 10   **69.** $\sqrt{5x}\sqrt{8x^5}$ $2x^3\sqrt{10}$ **70.** $\sqrt[3]{2x^4y^5}\sqrt[3]{16x^4y^4}$

71. $(\sqrt[3]{5x^2y^3})^2$ $xy^2\sqrt[3]{25x}$ **72.** $\sqrt[4]{8x^4y^7}\sqrt[4]{2x^5y^9}$ **73.** $\sqrt{3x}(\sqrt{12x}-\sqrt{20})$ **74.** $\sqrt[3]{y}(4\sqrt[3]{y^2}-\sqrt[3]{y^{10}})$

75. $\sqrt[3]{2x^2y^3}(\sqrt[3]{4x^4y^7}+\sqrt[3]{9xy^{12}})$ $2x^2y^3\sqrt[3]{y}+xy^5\sqrt[3]{18}$ **76.** $\sqrt[4]{3x^3y^2}(\sqrt[4]{2x^5y^9}+\sqrt[4]{27x^9y^3})$ $x^2y^2\sqrt[4]{6y^3}+3x^3y\sqrt[4]{y}$

77. $\sqrt{\dfrac{1}{4}}$ $\dfrac{1}{2}$ **78.** $\sqrt{\dfrac{36}{25}}$ $\dfrac{6}{5}$ **79.** $\sqrt[3]{\dfrac{x^3}{8}}$ $\dfrac{x}{2}$ **80.** $\dfrac{\sqrt[3]{2x^9}}{\sqrt[3]{16x^6}}$ $\dfrac{x}{2}$

81. $\sqrt{\dfrac{32x^2y^5}{2x^4y}}$ $\dfrac{4y^2}{x}$ **82.** $\sqrt[3]{\dfrac{108x^3y^6}{2y^3}}$ $3xy\sqrt[3]{2}$ **83.** $\sqrt{\dfrac{75x^2y^5}{3x^4y^7}}$ $\dfrac{5}{xy}$ **84.** $\dfrac{1}{\sqrt{2}}$ $\dfrac{\sqrt{2}}{2}$

85. $\dfrac{x}{\sqrt{7}}$ $\dfrac{x\sqrt{7}}{7}$ **86.** $\sqrt{\dfrac{2}{5}}$ $\dfrac{\sqrt{10}}{5}$ **87.** $\sqrt{\dfrac{12x}{5y}}$ $\dfrac{2\sqrt{15xy}}{5y}$ **88.** $\dfrac{2}{\sqrt[3]{x}}$ $\dfrac{2\sqrt[3]{x^2}}{x}$

89. $\sqrt[3]{\dfrac{3x}{5y}}$ $\dfrac{\sqrt[3]{75xy^2}}{5y}$ **90.** $\sqrt{\dfrac{3x^2}{y}}$ $\dfrac{x\sqrt{3y}}{y}$ **91.** $\sqrt{\dfrac{18x^4y^5}{3z}}$ $\dfrac{x^2y^2\sqrt{6yz}}{z}$ **92.** $\sqrt{\dfrac{125x^2y^5}{3z}}$ $\dfrac{5xy^2\sqrt{15yz}}{3z}$

93. $\sqrt[4]{\dfrac{2x^2y^6}{8x^3}}$ $\dfrac{y\sqrt[4]{4x^3y^2}}{2x}$ **94.** $\sqrt{\dfrac{20y^6z^9}{6x^3}}$ $\dfrac{y^3z^4\sqrt{30xz}}{3x^2}$ **95.** $\sqrt[3]{\dfrac{4x^5y^3}{x^6}}$ $\dfrac{y\sqrt[3]{4x^2}}{x}$ **96.** $\sqrt[3]{\dfrac{y^6}{2x^2}}$ $\dfrac{y^2\sqrt[3]{4x}}{2x}$

97. $(3-\sqrt{2})(3+\sqrt{2})$ 7   **98.** $(\sqrt{3}-\sqrt{5})(\sqrt{3}+\sqrt{5})$ -2 **99.** $(\sqrt{x}+y)(\sqrt{x}-y)$ $x-y^2$

100. $(x-\sqrt{y})(x+\sqrt{y})$ x^2-y **101.** $(\sqrt{3}+5)^2$ $28+10\sqrt{3}$ **102.** $(\sqrt{5}-\sqrt{20})^2$ 5

103. $(\sqrt{x}-\sqrt{3y})(\sqrt{x}+\sqrt{5y})$ **104.** $(2\sqrt{x}+3\sqrt{y})(\sqrt{x}-\sqrt{y})$ **105.** $(\sqrt[3]{2x}-\sqrt[3]{3y})(\sqrt[3]{3x}-\sqrt[3]{2y})$

106. $\dfrac{5}{2+\sqrt{5}}$ $-10+5\sqrt{5}$ **107.** $\dfrac{x}{3+\sqrt{x}}$ $\dfrac{3x-x\sqrt{x}}{9-x}$ **108.** $\dfrac{\sqrt{x}}{\sqrt{x}+\sqrt{y}}$ $\dfrac{x-\sqrt{xy}}{x-y}$

103. $x+\sqrt{5xy}-\sqrt{3xy}-y\sqrt{15}$ **104.** $2x+\sqrt{xy}-3y$ **105.** $\sqrt[3]{6x^2}-\sqrt[3]{4xy}-\sqrt[3]{9xy}+\sqrt[3]{6y^2}$

109. $\dfrac{\sqrt{30} + \sqrt{15} + 2\sqrt{3} + \sqrt{6}}{3}$ **118.** $(3x^2y^3 - 4x^3y^4)\sqrt{x}$

109. $\dfrac{\sqrt{5} + \sqrt{2}}{\sqrt{6} - \sqrt{3}}$

110. $\dfrac{\sqrt{x} - 2\sqrt{y}}{\sqrt{x} - \sqrt{y}}$ $\dfrac{x - \sqrt{xy} - 2y}{x - y}$

111. $\dfrac{4}{\sqrt{y+2} - 3}$ $\dfrac{4\sqrt{y+2} + 12}{y - 7}$

112. $2\sqrt{\dfrac{3}{8}} + 4\sqrt{\dfrac{2}{3}}$ $\dfrac{11\sqrt{6}}{6}$

113. $\sqrt[3]{x} + 3\sqrt[3]{x} - 2\sqrt[3]{x}$ $2\sqrt[3]{x}$

114. $\sqrt[3]{16} - \sqrt[3]{54}$ $-\sqrt[3]{2}$

115. $\sqrt{3} + \sqrt{27} - \sqrt{192}$ $-4\sqrt{3}$

116. $\sqrt[3]{16} - 5\sqrt[3]{54} + 2\sqrt[3]{64}$ $8 - 13\sqrt[3]{2}$

117. $4\sqrt{2} - \dfrac{3}{\sqrt{32}} + \sqrt{50}$ $\dfrac{69\sqrt{2}}{8}$

118. $3\sqrt{x^5y^6} - \sqrt{16x^7y^8}$

119. $2\sqrt[3]{x^7y^8} - \sqrt[3]{x^4y^2} + 3\sqrt[3]{x^{10}y^2}$ $(2x^2y^2 - x + 3x^3)\sqrt[3]{xy^2}$

[8.6] *Solve each equation and check your solutions.*

120. $\sqrt[3]{x} = 4$ 64

121. $\sqrt{3x+4} = \sqrt{5x+12}$

122. $2 + \sqrt[3]{x} = 4$ 8

123. $\sqrt{x^2 + 2x - 4} = x$ 2

124. $\sqrt[3]{x-9} = \sqrt[3]{5x+3}$ -3

125. $(x^2 + 5)^{1/2} = x + 1$ 2

126. $\sqrt{x+3} = \sqrt{3x+9}$ 0, 9

127. $\sqrt{6x-5} - \sqrt{2x+6} - 1 = 0$ 5   **121.** no solution

Solve for the variable indicated.

128. $V = \sqrt{\dfrac{2L}{w}}$, for L $L = \dfrac{V^2w}{2}$

129. $r = \sqrt{\dfrac{A}{\pi}}$, for A $A = \pi r^2$

Solve.

130. How long a wire does a phone company need to reach the top of a 5-meter telephone pole from a point 2 meters from the base of the pole? Use $l = \sqrt{a^2 + b^2}$. $\sqrt{29} \approx 5.39$ m

131. Use the formula $v = \sqrt{2gh}$ to find the velocity of an object after it has fallen 20 feet ($g = 32$).

132. Use the formula $T = 2\pi\sqrt{L/32}$ to find the period of a pendulum, T, if its length L is 64 feet. You may leave your answer in simplified radical form.

131. $\sqrt{1280} \approx 35.78$ ft/sec

132. $2\pi\sqrt{2} \approx 2.83\pi \approx 8.89$ sec

In Exercises 133 and 134, (a) solve the system of equations. (b) Solve the equation graphically, using a system of equations. (c) Does the answer obtained in part (b) agree with the answer obtained in part (a)?

133. $\sqrt{x+4} = 3$ **(a)** 5 **(c)** yes **(b)**

134. $\sqrt{x+4} = x + 2$ **(a)** 0 **(c)** yes **(b)**
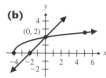

[8.7] *Write each expression as a complex number in the form $a + bi$.*

135. 5 $5 + 0i$

136. -6 $-6 + 0i$

137. $2 - \sqrt{-256}$ $2 - 16i$

138. $3 + \sqrt{-16}$ $3 + 4i$

142. $3\sqrt{3} + (\sqrt{5} - \sqrt{7})i$ **148.** $(24 + 3\sqrt{5}) + (4\sqrt{3} - 6\sqrt{15})i$ **153.** $\dfrac{(\sqrt{10} - 6\sqrt{2}) + (3\sqrt{2} + 2\sqrt{10})i}{10}$

Perform the operations.

139. $(3 + 2i) + (4 - i)$ $7 + i$

140. $(4 - 6i) - (3 - 4i)$ $1 - 2i$

141. $(5 + \sqrt{-9}) - (3 - \sqrt{-4})$ $2 + 5i$

142. $(\sqrt{3} + \sqrt{-5}) + (2\sqrt{3} - \sqrt{-7})$

143. $4(3 + 2i)$ $12 + 8i$

144. $-2(\sqrt{3} - i)$ $-2\sqrt{3} + 2i$

145. $\sqrt{8}(\sqrt{-2} + 3)$ $6\sqrt{2} + 4i$

146. $\sqrt{-6}(\sqrt{6} + \sqrt{-6})$ $-6 + 6i$

147. $(4 + 3i)(2 - 3i)$ $17 - 6i$

148. $(6 + \sqrt{-3})(4 - \sqrt{-15})$

149. $\dfrac{2}{3i}$ $\dfrac{-2i}{3}$

150. $\dfrac{2 + \sqrt{3}}{2i}$ $\dfrac{(-2 - \sqrt{3})i}{2}$

151. $\dfrac{5}{3 + 2i}$ $\dfrac{15 - 10i}{13}$

152. $\dfrac{\sqrt{3}}{5 - \sqrt{-6}}$ $\dfrac{5\sqrt{3} + 3i\sqrt{2}}{31}$

153. $\dfrac{\sqrt{5} + 3i}{\sqrt{2} - \sqrt{-8}}$

Evaluate the equation for the given value of x.

154. $x^2 - 2x + 9$, $x = 1 + 2i\sqrt{2}$ 0

155. $x^2 - 2x + 12$, $x = 1 - 2i$ 7

Indicate whether the value is i, −1, −i, or 1.

156. i^{53} i

157. i^{19} −i

158. i^{404} 1

159. i^{5326} −1

Practice Test

11. $(2xy + 2x^2y^2)\sqrt[3]{y^2}$

1. Use absolute value to evaluate $\sqrt{(-26)^2}$.   26

2. Write as an absolute value $\sqrt{(3x - 4)^2}$. $|3x - 4|$

3. Simplify $\left(\dfrac{y^{2/3} \cdot y^{-1}}{y^{1/4}}\right)^2$. $\dfrac{1}{y^{7/6}}$

4. Factor $2x^{2/3} + x^{1/3} - 10$. $(2x^{1/3} + 5)(x^{1/3} - 2)$

Simplify. Assume that all variables represent positive real numbers.

5. $\sqrt{50x^5y^8}$ $5x^2y^4\sqrt{2x}$

6. $\sqrt[3]{4x^5y^2}\,\sqrt[3]{10x^6y^8}$ $2x^3y^3\sqrt[3]{5x^2y}$

7. $\sqrt{\dfrac{2x^4y^5}{8z}}$ $\dfrac{x^2y^2\sqrt{yz}}{2z}$

8. $\sqrt[3]{\dfrac{1}{x}}$ $\dfrac{\sqrt[3]{x^2}}{x}$

9. $\dfrac{\sqrt{2}}{2 + \sqrt{8}}$ $\dfrac{2 - \sqrt{2}}{2}$

10. $\sqrt{27} + 2\sqrt{3} - 5\sqrt{75}$ $-20\sqrt{3}$

11. $\sqrt[3]{8x^3y^5} + 2\sqrt[3]{x^6y^8}$

12. $(\sqrt{5} - 3)(2 - \sqrt{8})$ $2\sqrt{5} - 2\sqrt{10} - 6 + 6\sqrt{2}$

Solve.

13. $\sqrt{4x - 3} = 7$ 13

14. $\sqrt{x^2 - x - 12} = x + 3$ −3

15. $\sqrt{x - 15} = \sqrt{x} - 3$ 16

16. Solve the formula $w = \dfrac{\sqrt{2gh}}{4}$ for h. $h = \dfrac{8w^2}{g}$

17. A ladder is placed against a house. If the base of the ladder is 5 feet from the house and the ladder rests on the house 12 feet above the ground, find the length of the ladder.   13 ft

18. Multiply $(6 - \sqrt{-4})(3 + \sqrt{-2})$.

19. Divide $\dfrac{\sqrt{5}}{2 - \sqrt{-8}}$. $\dfrac{\sqrt{5} + i\sqrt{10}}{6}$

20. Evaluate $x^2 + 6x + 12$ for $x = -3 + i$.   2

18. $(18 + 2\sqrt{2}) + (6\sqrt{2} - 6)i$

---

2. (a) a set of ordered pairs **(b)** a set of ordered pairs no two of which have the same first coordinate

Cumulative Review Test

1. Solve $\frac{1}{5}(x - 3) = \frac{3}{4}(x + 3) - x$. $\frac{57}{9}$

2. Define **(a)** a relation; **(b)** a function.

3. (a) State the domain of $f(x) = \sqrt{x} + 2$.
   (b) Graph $f(x)$ and state the range.   answers on p. 508

4. Solve the system of equations.
   $\left(\frac{2}{11}, -\frac{16}{11}\right)$ $x - 4y = 6$
   $3x - y = 2$

5. Simplify $\left(\dfrac{3x^2y^{-2}}{x^4y^{-5}}\right)\left(\dfrac{2xy^3}{x^2y^{-3}}\right)^{-1}$. $\dfrac{3}{2xy^3}$

6. Multiply $(3x^2 - 4x - 6)(2x - 5)$. $6x^3 - 23x^2 + 8x + 30$

7. Divide $\dfrac{3x^2 + 10x + 10}{x + 2}$. $3x + 4 + \dfrac{2}{x + 2}$

8. Graph $f(x) = x^3 + 4x - 1$.   answer on p. 508

9. Factor $2x^2 - 12x + 18 - 2y^2$. $2(x - 3 + y)(x - 3 - y)$

10. Find the domain of $\dfrac{x - 4}{5x - 3}$. $x \neq \frac{3}{5}$

11. Reduce $\dfrac{(x + 2)(x - 4) + (x - 1)(x - 4)}{3(x - 4)}$. $\dfrac{2x + 1}{3}$

12. Multiply $\dfrac{4x^2 + 8x + 3}{2x^2 - x - 1} \cdot \dfrac{x^2 - 1}{4x^2 + 12x + 9}$. $\dfrac{x + 1}{2x + 3}$

13. Subtract $\dfrac{x + 1}{x^2 + 2x - 3} - \dfrac{x}{2x^2 + 11x + 15}$.

14. Solve $4 - \dfrac{5}{y} = \dfrac{4y}{y + 1}$. -5

15. Simplify $\left(\dfrac{x^2 y^{1/2}}{x^{1/4}}\right)^2$. $x^{7/2}y$

16. Simplify $\sqrt[3]{4x^{10}y^{20}} \cdot \sqrt[3]{4x^3y^9}$. $2x^4y^9\sqrt[3]{2xy^2}$

17. Solve $\sqrt{2x^2 + 7} + 3 = 8$. $3, -3$

18. Divide $\dfrac{2}{3 + \sqrt{-6}}$ $\dfrac{6 - 2i\sqrt{6}}{15}$

19. Jim by himself can paint the living room in his house in 2 hours. Jim's son Mike can paint the same room by himself in 3 hours. How long will it take them to paint the room if they work together? $1\frac{1}{5}$hr

20. A wire reaches from the top of a 30-foot telephone pole to a point on the ground 20 feet from the base of the pole. What is the length of the wire? $\sqrt{1300} \approx 36.1$ ft

13. $\dfrac{x^2 + 8x + 5}{(x + 3)(x - 1)(2x + 5)}$

- - -

3. (a) D: $\{x \mid x \geq -2\}$,   **(b)**   **(c)** R: $\{y \mid y \geq 0\}$

8.

Chapter 9

Quadratic Functions and the Algebra of Functions

See Section 9.2, Exercise 97

Preview and Perspective

We discussed and graphed quadratic functions in Section 5.6. Now we expand upon the concepts. Section 9.1 introduces completing the square and Section 9.2 discusses the quadratic formula. After studying these sections, you will know three techniques for solving quadratic equations: factoring (when possible), completing the square, and the quadratic formula.

In Section 9.3 we introduce equations that can be expressed and solved as if they were quadratic equations. Such equations are called equations quadratic in form.

We solved linear inequalities in one variable in Section 2.5. In Section 9.4 we solve quadratic and other nonlinear inequalities in one variable.

In Sections 9.5 and 9.6 we expand upon our knowledge of functions when we discuss operations that can be performed on functions and inverse functions.

9.1 Solving Quadratic Equations by Completing the Square

Tape 14

1. Use the square root property to solve equations.
2. Use the square root property to solve for a variable in a formula.
3. Write perfect square trinomials.
4. Solve quadratic equations by completing the square.

In this section we introduce two concepts, the square root property and completing the square. The square root property will be used in several sections in this book.

In Section 6.5, we solved quadratic, or second degree, equations by factoring. Quadratic equations that cannot be solved by factoring can be solved by completing the square, or by the quadratic formula, which is presented in Section 9.2.

The Square Root Property

1   In Section 8.1 we stated that every positive number has two square roots. Thus far, we have been using only the positive square root. In this section we use both the positive and negative square roots of a number.
The positive square root of 25 is 5.

$$\sqrt{25} = 5$$

The negative square root of 25 is -5.

$$-\sqrt{25} = -5$$

Notice that $5 \cdot 5 = 25$ and $(-5)(-5) = 25$. The two square roots of 25 are $+5$ and -5. A convenient way to indicate the two square roots of a number is to use the plus or minus symbol, $\pm$. For example, the square roots of 25 can be indicated ± 5, read "plus or minus 5." The equation $x^2 = 25$ has two solutions, the two square roots of 25, they are ± 5. If you check each root, you will see that each value satisfies the equation. The square root property can be used to find the solutions to equations of the form $x^2 = a$.

Square Root Property
If $x^2 = a$, where a is a real number, then $x = \pm\sqrt{a}$.

EXAMPLE 1   Solve the equation $x^2 - 9 = 0$.

Solution: Add 9 to both sides of the equation to get the variable by itself on one side.

$$x^2 = 9$$

Use the square root property,

$$x = \pm\sqrt{9}$$
$$x = \pm 3$$

Check the solutions in the original equation.

$x = 3$	$x = -3$
$x^2 - 9 = 0$	$x^2 - 9 = 0$
$3^2 - 9 = 0$	$(-3)^2 - 9 = 0$
$0 = 0$ true	$0 = 0$ true

In both cases the check is true, which means that both 3 and −3 are solutions to the equation.

EXAMPLE 2   Solve the equation $x^2 + 5 = 65$.

Solution: Begin by subtracting 5 from both sides of the equation.

$$x^2 = 60$$
$$x = \pm\sqrt{60} = \pm\sqrt{4}\sqrt{15} = \pm 2\sqrt{15}$$

The solutions are $2\sqrt{15}$ and $-2\sqrt{15}$.

Not all quadratic equations have real solutions, as is illustrated in Example 3.

EXAMPLE 3   Solve the equation $x^2 + 7 = 0$.

Solution:
$$x^2 + 7 = 0$$
$$x^2 = -7$$
$$x = \pm\sqrt{-7} = \pm i\sqrt{7}$$

The solutions are $i\sqrt{7}$ and $-i\sqrt{7}$.

EXAMPLE 4   Solve the equation $(x - 4)^2 = 32$.

Solution: Begin by taking the square root of both sides of the equation.

$$(x - 4)^2 = 32$$
$$x - 4 = \pm\sqrt{32}$$
$$x = 4 \pm \sqrt{32}$$
$$x = 4 \pm \sqrt{16}\sqrt{2}$$
$$x = 4 \pm 4\sqrt{2}$$

The solutions are $4 + 4\sqrt{2}$ and $4 - 4\sqrt{2}$.

Solve for a Variable in a Formula

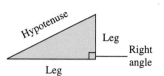

FIGURE 9.1

2 When the square of a variable appears in a *formula,* you may need to use the square root property to solve for the variable. However, *when you use the square root property in most formulas, you will use only the principal or positive root,* because you are generally solving for a quantity that cannot be negative.

Consider a right triangle (see Fig. 9.1). The two shorter sides of a right triangle are called the **legs** and the side opposite the right angle is called the **hypotenuse.** The Pythagorean theorem expresses the relationship between the legs of the triangle and its hypotenuse.

> **Pythagorean Theorem**
>
> The square of the hypotenuse of a right triangle is equal to the sum of the squares of the two legs. If a and b represent the legs and c represents the hypotenuse, then
> $$a^2 + b^2 = c^2$$

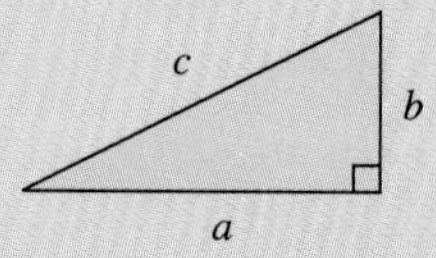

When we solve the Pythagorean theorem for a, b, or c we use the square root property. However, when we use the square root property we take only the positive square root since a length cannot be a negative number.

EXAMPLE 5 Find the hypotenuse of the right triangle whose legs are 3 feet and 4 feet.

Solution: It is often helpful when using the Pythagorean theorem to draw a picture of the problem before using the formula (Fig. 9.2). When drawing the picture, it makes no difference which leg is called a and which leg is called b.

$$a^2 + b^2 = c^2$$
$$3^2 + 4^2 = c^2$$
$$9 + 16 = c^2$$
$$25 = c^2$$
$$\sqrt{25} = c \qquad \text{square root property using only the positive root}$$
$$5 = c$$

FIGURE 9.2

The hypotenuse is 5 feet.

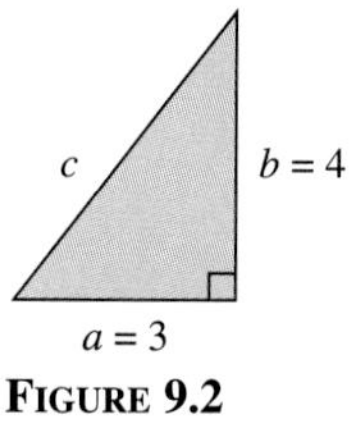

Perfect Square Trinomials

3 Now that we know the square root property we can focus our attention on completing the square. To understand this procedure you need to know how to form perfect square trinomials. Perfect square trinomials were introduced in Section 6.3.

A perfect square trinomial is a trinomial that can be expressed as the square of a binomial. Some examples follow.

Perfect Square Trinomials	*Factors*	*Square of a Binomial*
$x^2 + 8x + 16$	$= (x + 4)(x + 4)$	$= (x + 4)^2$
$x^2 - 8x + 16$	$= (x - 4)(x - 4)$	$= (x - 4)^2$
$x^2 + 10x + 25$	$= (x + 5)(x + 5)$	$= (x + 5)^2$
$x^2 - 10x + 25$	$= (x - 5)(x - 5)$	$= (x - 5)^2$

In each of the squared terms in the trinomials above the numerical coefficient is 1. Whenever the coefficient of the squared term is 1, there is a relationship between the coefficient of the first-degree term and the constant term. In every perfect square trinomial the constant term is the square of one-half the coefficient of the first-degree term.

Let us examine some perfect square trinomials for which the coefficient of the squared term is 1.

$$x^2 + 8x + 16 = (x + 4)^2$$
$$\left[\tfrac{1}{2}(8)\right]^2 = (4)^2$$

$$x^2 - 10x + 25 = (x - 5)^2$$
$$\left[\tfrac{1}{2}(-10)\right]^2 = (-5)^2$$

When such a perfect square trinomial is written as the square of a binomial the constant in the binomial is one-half the value of the coefficient of the first-degree term in the perfect square trinomial. For example,

$$x^2 + 8x + 16 = (x + 4)^2$$
$$\tfrac{1}{2}(8)$$

$$x^2 - 10x + 25 = (x - 5)^2$$
$$\tfrac{1}{2}(-10)$$

Completing the Square   4   Now we introduce completing the square. To solve a quadratic equation by completing the square, we add (or subtract) a constant to (or from) both sides of the equation so that the remaining trinomial is a perfect square trinomial. Then we use the square root property to solve the resulting equation. We will now summarize the procedure.

> ### To Solve a Quadratic Equation by Completing the Square
>
> **1.** Use the multiplication (or division) property of equality if necessary to make the numerical coefficient of the squared term equal to 1.
>
> **2.** Rewrite the equation with the constant by itself on the right side of the equation.
>
> **3.** Take one-half the numerical coefficient of the first-degree term, square it, and add this quantity to both sides of the equation.
>
> *(continued on page 514)*

> **4.** Replace the trinomial with the square of a binomial.
> **5.** Use the square root property to take the square root of both sides of the equation.
> **6.** Solve for the variable.
> **7.** Check your solutions in the original equation.

EXAMPLE 6   Solve the equation $x^2 + 6x + 5 = 0$ by completing the square.

Solution: Since the numerical coefficient of the squared term is 1, step 1 is not necessary.

STEP 2:   Move the constant, 5, to the right side of the equation by subtracting 5 from both sides of the equation.

$$x^2 + 6x + 5 = 0$$
$$x^2 + 6x = -5$$

STEP 3:   Determine the square of one-half the numerical coefficient of the first-degree term.

$$\frac{1}{2}(6) = 3, \qquad 3^2 = \boxed{9}$$

Add this value to both sides of the equation.

$$x^2 + 6x \boxed{+\ 9} = -5 \boxed{+\ 9}$$

$$x^2 + 6x + 9 = 4$$

STEP 4:   By following this procedure we produce a perfect square trinomial on the left side of the equation. The expression $x^2 + 6x + 9$ is a perfect square trinomial that can be expressed as $(x + 3)^2$.

$$x^2 + 6x + 9 = 4$$

$\frac{1}{2}$ the numerical coefficient of the
first-degree term is $\frac{1}{2}(6) = +3$

$$(x + 3)^2 = 4$$

STEP 5:   Take the square root of both sides of the equation.

$$x + 3 = \pm\sqrt{4}$$
$$x + 3 = \pm 2$$

STEP 6:   Finally, solve for x by subtracting 3 from both sides of the equation.

$$x + 3 \boxed{-\ 3} = \boxed{-3} \pm 2$$

$$x = -3 \pm 2$$

$$x = -3 + 2 \qquad \text{or} \qquad x = -3 - 2$$

$$x = -1 \qquad\qquad\qquad x = -5$$

STEP 7: Check both solutions in the original equation.

$$x = -1$$
$$x^2 + 6x + 5 = 0$$
$$(-1)^2 + 6(-1) + 5 = 0$$
$$1 - 6 + 5 = 0$$
$$0 = 0 \quad \text{true}$$

$$x = -5$$
$$x^2 + 6x + 5 = 0$$
$$(-5)^2 + 6(-5) + 5 = 0$$
$$25 - 30 + 5 = 0$$
$$0 = 0 \quad \text{true}$$

Since each number checks, both -1 and -5 are solutions to the original equation.

Helpful Hint

When solving the equation $x^2 + bx + c = 0$, by completing the square, we first obtain the expression $x^2 + bx$ on the left side of the equation by moving the constant to the right side of the equation (step 2 in the procedure). We then add $\left(\dfrac{b}{2}\right)^2$ to both sides of the equation (step 3) to complete the square. This gives $x^2 + bx + \left(\dfrac{b}{2}\right)^2$ on the left side of the equation and a constant on the right side of the equation. In step 4 of the procedure we replace the trinomial $x^2 + bx + \left(\dfrac{b}{2}\right)^2$ with the square of its equivalent binomial, which is $\left(x + \dfrac{b}{2}\right)^2$. In the figure that follows we show why $x^2 + bx + \left(\dfrac{b}{2}\right)^2 = \left(x + \dfrac{b}{2}\right)^2$. The figure is a square with sides of length $x + \dfrac{b}{2}$. The area is therefore $\left(x + \dfrac{b}{2}\right)^2$. The area of each of the four sections that comprise the square is found by multiplying its length by its width. The area of the square can also be determined by adding the areas of the four sections as follows:

$$x^2 + \frac{b}{2}x + \frac{b}{2}x + \left(\frac{b}{2}\right)^2 = x^2 + bx + \left(\frac{b}{2}\right)^2$$

By comparing the areas we see that $x^2 + bx + \left(\dfrac{b}{2}\right)^2 = \left(x + \dfrac{b}{2}\right)^2$.

The area of this piece represents the term we add to each side of the equation when we complete the square.

EXAMPLE 7   Solve the equation $-x^2 = -3x - 18$ by completing the square.

Solution: The numerical coefficient of the squared term must be 1, not -1. Therefore, we begin by multiplying both sides of the equation by -1 to make the coefficient of the squared term equal to 1.

$$-x^2 = -3x - 18$$
$$-1(-x^2) = -1(-3x - 18)$$
$$x^2 = 3x + 18$$

Now move all terms except the constant to the left side of the equation.

$$x^2 - 3x = 18$$

Take half the numerical coefficient of the x term, square it, and add this product to both sides of the equation. Then rewrite the left side of the equation as the square of a binomial.

$$\frac{1}{2}(-3) = -\frac{3}{2} \qquad \left(-\frac{3}{2}\right)^2 = \frac{9}{4}$$

$$x^2 - 3x + \frac{9}{4} = 18 + \frac{9}{4}$$

$$\left(x - \frac{3}{2}\right)^2 = 18 + \frac{9}{4}$$

$$\left(x - \frac{3}{2}\right)^2 = \frac{72}{4} + \frac{9}{4}$$

$$\left(x - \frac{3}{2}\right)^2 = \frac{81}{4}$$

$$x - \frac{3}{2} = \pm\sqrt{\frac{81}{4}}$$

$$x - \frac{3}{2} = \pm\frac{9}{2}$$

$$x = \frac{3}{2} \pm \frac{9}{2}$$

$$x = \frac{3}{2} + \frac{9}{2} \quad \text{or} \quad x = \frac{3}{2} - \frac{9}{2}$$

$$x = \frac{12}{2} = 6 \qquad\qquad x = -\frac{6}{2} = -3$$

In the following examples we will not illustrate some of the intermediate steps.

EXAMPLE 8   Solve the equation $x^2 - 6x + 17 = 0$.

Solution:

$$x^2 - 6x + 17 = 0$$
$$x^2 - 6x = -17$$
$$x^2 - 6x + 9 = -17 + 9$$
$$(x - 3)^2 = -8$$
$$x - 3 = \pm\sqrt{-8}$$
$$x - 3 = \pm 2i\sqrt{2}$$
$$x = 3 \pm 2i\sqrt{2}$$

The solutions are $3 + 2i\sqrt{2}$ and $3 - 2i\sqrt{2}$. Note that the solutions to the equation $x^2 - 6x + 17 = 0$ are not real. The solutions are complex numbers.

EXAMPLE 9 Solve the equation $-3m^2 + 6m + 24 = 0$ by completing the square.

Solution: To solve an equation by completing the square, the numerical coefficient of the squared term should be 1. Since the numerical coefficient of the squared term is -3, we multiply both sides of the equation by $-1/3$ to make the numerical coefficient of the squared term equal to 1.

$$-3m^2 + 6m + 24 = 0$$

$$-\frac{1}{3}(-3m^2 + 6m + 24) = -\frac{1}{3}(0)$$

$$m^2 - 2m - 8 = 0$$

Now proceed as before.

$$m^2 - 2m = 8$$
$$m^2 - 2m + 1 = 8 + 1$$
$$(m - 1)^2 = 9$$
$$m - 1 = \pm 3$$
$$m = 1 \pm 3$$
$$m = 1 + 3 \qquad \text{or} \qquad m = 1 - 3$$
$$m = 4 \qquad\qquad\qquad m = -2$$

If you were asked to solve the equation $-\frac{1}{4}x^2 + 2x - 8 = 0$ by completing the square, what would you do first? If you answered "multiply both sides of the equation by -4 to make the coefficient of the squared term 1," you answered correctly. To solve the equation $\frac{2}{3}x^2 + 3x - 5 = 0$, you could multiply both sides of the equation by $\frac{3}{2}$ to obtain a squared term with a coefficient of 1.

In examples 6 through 9 we solved quadratic equations by completing the square. Generally, quadratic equations that cannot be solved by factoring will be solved by the *quadratic formula*, which will be presented in the next section. We introduced completing the square because we use it to derive the quadratic formula in Section 9.2. We will use completing the square in Chapter 10.

Exercise Set 9.1

Use the square root property to solve each equation.

1. $x^2 = 25$ ± 5

2. $x^2 = 18$ $\pm 3\sqrt{2}$

3. $y^2 = 75$ $\pm 5\sqrt{3}$

4. $x^2 - 7 = 19$ $\pm\sqrt{26}$

5. $z^2 + 12 = 40$ $\pm 2\sqrt{7}$

6. $y^2 + 15 = 80$ $\pm\sqrt{65}$

7. $(x - 4)^2 = 16$ $8, 0$

8. $(y - 3)^2 = 45$ $3 \pm 3\sqrt{5}$

9. $\left(z + \frac{1}{3}\right)^2 = \frac{4}{9}$ $\frac{1}{3}, -1$

10. $(x - 0.2)^2 = 0.64$ $1, -0.6$

11. $(x + 1.8)^2 = 0.81$ $-0.9, -2.7$

12. $\left(x + \frac{1}{2}\right)^2 = \frac{4}{9}$ $\frac{1}{6}, -\frac{7}{6}$

13. $(2x - 5)^2 = 12$ $\dfrac{5 \pm 2\sqrt{3}}{2}$

14. $(4y + 1)^2 = 8$ $\dfrac{-1 \pm 2\sqrt{2}}{4}$

15. $\left(2y + \frac{1}{2}\right)^2 = \frac{4}{25}$ $-\frac{1}{20}, -\frac{9}{20}$

16. $\left(3x - \frac{1}{4}\right)^2 = \frac{9}{25}$ $\frac{17}{60}, -\frac{7}{60}$

Use the square root property to solve each formula for the indicated variable. Use only the positive root.

17. $A = s^2$, for s $s = \sqrt{A}$

18. $r = \frac{1}{2}t^2$, for t $t = \sqrt{2r}$

19. $A = \pi r^2$, for r $r = \sqrt{\dfrac{A}{\pi}}$

20. $k = mv^2$, for v $v = \sqrt{\dfrac{k}{m}}$

21. $F_x^2 + F_y^2 = F^2$, for F $F = \sqrt{F_x^2 + F_y^2}$

22. $a^2 + b^2 = c^2$, for b $b = \sqrt{c^2 - a^2}$

23. $V^2 = V_x^2 + V_y^2$, for V_x $V_x = \sqrt{V^2 - V_y^2}$

24. $E = \frac{1}{9}rp^2$, for p $p = \sqrt{\dfrac{9E}{r}}$

25. $L = a^2 - b^2$, for b $b = \sqrt{a^2 - L}$

26. $H = i^2R$, for i $i = \sqrt{\dfrac{H}{R}}$

27. $v = m + nt^2$, for t $t = \sqrt{\dfrac{v - m}{n}}$

28. $E = \frac{1}{2}mv^2$, for v $v = \sqrt{\dfrac{2E}{m}}$

29. $d = b^2 - 4ac$, for b $b = \sqrt{d + 4ac}$

30. $V = \pi r^2h$, for r $r = \sqrt{\dfrac{V}{\pi h}}$

31. $w = 3l + 2d^2$, for d $d = \sqrt{\dfrac{w - 3l}{2}}$

32. $A = P(1 + r)^2$, for r $r = \sqrt{\dfrac{A}{P}} - 1$

Use the Pythagorean theorem to find the unknown length.

33.

34.
$\sqrt{52} \approx 7.21$

35.
$\sqrt{175} \approx 13.23$

36.
$\sqrt{67} \approx 8.19$

37.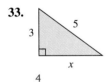
$\sqrt{41} \approx 6.40$

38.
$\sqrt{6} \approx 2.45$

39.
$\sqrt{128} \approx 11.31$

40.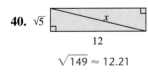
$\sqrt{149} \approx 12.21$

41.
10

42.
$\sqrt{77} \approx 8.77$

Solve each equation by completing the square.

43. $x^2 + 2x - 3 = 0$ 1, −3
44. $x^2 - 6x + 8 = 0$ 2, 4
45. $x^2 - 4x - 5 = 0$ 5, −1

46. $x^2 + 8x + 12 = 0$ −2, −6
47. $x^2 + 3x + 2 = 0$ −2, −1
48. $x^2 + 4x - 32 = 0$ 4, −8

49. $x^2 - 8x + 15 = 0$ 5, 3
50. $x^2 - 9x + 14 = 0$ 2, 7
51. $x^2 + 2x + 15 = 0$ $-1 \pm i\sqrt{14}$

52. $x^2 + 5x + 4 = 0$ −1, −4
53. $x^2 = -5x - 6$ −2, −3
54. $x^2 - 2x + 4 = 0$ $1 \pm i\sqrt{3}$

55. $x^2 + 9x + 18 = 0$ −3, −6
56. $x^2 - 9x + 18 = 0$ 3, 6
57. $x^2 = 15x - 56$ 7, 8

58. $x^2 = 3x + 28$ 7, −4
59. $-4x = -x^2 + 12$ 6, −2
60. $x^2 + 3x + 6 = 0$ $(-3 \pm i\sqrt{15})/2$

61. $\frac{1}{2}x^2 + x - 3 = 0$ $-1 \pm \sqrt{7}$
62. $x^2 - 4x + 2 = 0$ $2 \pm \sqrt{2}$
63. $6x + 6 = -x^2$ $-3 \pm \sqrt{3}$

64. $x^2 - x - 3 = 0$ $(1 \pm \sqrt{13})/2$
65. $-x^2 + 5x = -8$ $(5 \pm \sqrt{57})/2$
66. $\frac{1}{4}x^2 + \frac{3}{4}x - \frac{3}{2} = 0$ $(-3 \pm \sqrt{33})/2$

67. $-\frac{1}{4}x^2 - \frac{1}{2}x = 0$ 0, −2
68. $2x^2 - 6x = 0$ 0, 3
69. $12x^2 - 4x = 0$ $0, \frac{1}{3}$

70. $6x^2 = 9x$ $0, \frac{3}{2}$
71. $-\frac{1}{2}x^2 - x + \frac{3}{2} = 0$ 1, −3
72. $2x^2 + 2x - 24 = 0$ 3, −4

73. $2x^2 + 18x + 4 = 0$ $(-9 \pm \sqrt{73})/2$
74. $2x^2 = 8x + 90$ 9, −5
75. $3x^2 + 33x + 72 = 0$ −8, −3

76. $3x^2 + 2x - 1 = 0$ $\frac{1}{3}$, −1
77. $\frac{2}{3}x^2 + \frac{4}{3}x + 1 = 0$ $(-2 \pm i\sqrt{2})/2$
78. $3x^2 - 8x + 4 = 0$ $2, \frac{2}{3}$

79. $-3x^2 + 6x = 6$ $1 \pm i$
80. $2x^2 - x = -5$ $(1 \pm i\sqrt{39})/4$
81. $\frac{5}{2}x^2 + \frac{3}{2}x - \frac{5}{4} = 0$ $(-3 \pm \sqrt{59})/10$

82. $\frac{3}{4}x^2 - \frac{1}{2}x - \frac{3}{20} = 0$ $(5 \pm \sqrt{70})/15$
83. $3x^2 + \frac{1}{2}x = -4$ $(-1 \pm i\sqrt{191})/12$
84. $2x^2 - \frac{1}{3}x = -2$ $(1 \pm i\sqrt{143})/12$

85. Solve by completing the square: $x^2 + 2ax + a^2 = k$.

86. Solve by completing the square: $(x \pm a)^2 - k = 0$.

87. The product of two consecutive positive odd integers is 63. Find the two odd integers.   7, 9

88. The larger of two integers is 2 more than twice the smaller. Find the two numbers if their product is 12.

89. The length of a rectangle is 2 feet more than twice its width. Find the dimensions of the rectangle if the area is 60 square feet.   5 ft, 12 ft

90. Find the length of the side of a square whose diagonal is 10 feet longer than the length of its side.

91. Find the length of the side of a square whose diagonal is 12 feet longer than the length of a side.

92. The length of a rectangle is 3 inches more than its width. If the length of the diagonal is 15 inches, find the dimensions of the rectangle.   9 in by 12 in

93. What is the first step in completing the square?

94. Explain how to determine if a trinomial is a perfect square trinomial. $(b/2)^2$ must equal c

95. Write a paragraph explaining in your own words how to construct a perfect square trinomial.

85. $-a \pm \sqrt{k}$ **86.** $a \pm \sqrt{k}, -a \pm \sqrt{k}$ **88.** 2, 6

90. $10 + 10\sqrt{2} \approx 24.14$ ft

91. $12 + 12\sqrt{2} \approx 28.97$ ft   **93.** make $a = 1$

CUMULATIVE REVIEW EXERCISES

[2.2] **96.** Solve the equation for z: $2xy - 3yz = -xy + z$.

[8.1] **97.** Express $\sqrt{(x^2 - 4x)^2}$ as an absolute value. $|x^2 - 4x|$

[8.2] **98.** Evaluate $25^{-1/2}$ $\frac{1}{5}$

99. Simplify $\dfrac{x^{3/4} y^{1/2}}{x^{1/4} y^2}$ $\dfrac{x^{1/2}}{y^{3/2}}$

96. $z = \dfrac{3xy}{3y + 1}$

Group Activity/ Challenge Problems

1. The surface area S and volume V of a right circular cylinder of radius r and height h are given by the formulas   **(b)** $r = 4\sqrt{\pi}/\pi \approx 2.26$ in   **(c)** $r \approx 2.1$ in

$$S = 2\pi r^2 + 2\pi rh, \qquad V = \pi r^2 h$$

(a) Find the surface area of the cylinder if its height is 10 inches and its volume is 160 cubic inches. $S = 32 + 80\sqrt{\pi} \approx 173.80$ sq in

(b) Find the radius if the height is 10 inches and the volume is 160 cubic inches.

(c) Find the radius if the height is 10 inches and the surface area is 160 square inches.

2. The distance formula, $d = \sqrt{(x_2 - x_1)^2 + (y_2 - y_1)^2}$ was discussed in Section 3.1. Derive the distance formula using the Pythagorean theorem and the square root property.

Write each equation in the form $a(x - h)^2 + b(y - k)^2 = c$ by completing the square twice: once for the x terms and once for the y terms. (We will do this type of problem in Chapter 10.)

3. $x^2 + 4x + y^2 - 6y = 3$ $(x + 2)^2 + (y - 3)^2 = 16$

4. $4x^2 + 9y^2 - 48x + 72y + 144 = 0$ $4(x - 6)^2 + 9(y + 4)^2 = 144$

5. $x^2 - 4y^2 + 4x - 16y - 28 = 0$ $(x + 2)^2 - 4(y + 2)^2 = 16$

6. Find all four solutions to the equation $x^4 = 16$. $\pm 2, \pm 2i$

9.2   Solving Quadratic Equations by the Quadratic Formula

Tape 15

1️⃣ Derive the quadratic formula.

2️⃣ Use the quadratic formula to solve equations.

3️⃣ Write a quadratic equation, given its solutions.

4️⃣ Use the discriminant to determine the number of solutions to a quadratic equation.

5️⃣ Sketch quadratic functions.

6️⃣ Use quadratic equations to solve applied problems.

The Quadratic Formula

1️⃣ The quadratic formula can be used to solve any quadratic equation. *It is the most useful and most versatile method of solving quadratic equations.* It is generally used in place of completing the square because of its efficiency.

The standard form of a quadratic equation is $ax^2 + bx + c = 0$, where a is the numerical coefficient of the squared term, b is the numerical coefficient of the first-degree term, and c is the constant.

Quadratic Equation in Standard Form	*Values of Coefficients*
$x^2 - 3x + 4 = 0$	$a = 1, \quad b = -3, \quad c = 4$
$3x^2 - 4 = 0$	$a = 3, \quad b = 0, \quad c = -4$
$-5x^2 + 3x = 0$	$a = -5, \quad b = 3, \quad c = 0$

We can derive the quadratic formula by starting with a quadratic equation in standard form and completing the square as discussed in the preceding section.

$$ax^2 + bx + c = 0$$

$$\frac{ax^2}{a^2} + \frac{b}{a}x + \frac{c}{a} = 0 \qquad \text{Divide both sides of the equation by } a.$$

$$x^2 + \frac{b}{a}x = \frac{-c}{a} \qquad \text{Subtract } c/a \text{ from both sides of the equation.}$$

$$x^2 + \frac{b}{a}x + \frac{b^2}{4a^2} = \frac{-c}{a} + \frac{b^2}{4a^2} \qquad \text{Take } 1/2 \text{ of } b/a \text{ that is, } b/2a, \text{ and square it to get } b^2/4a^2. \text{ Then add this expression to both sides of the equation.}$$

$$\left(x + \frac{b}{2a}\right)^2 = \frac{-c}{a} + \frac{b^2}{4a^2} \qquad \text{Rewrite the left side of the equation as the square of a binomial.}$$

$$\left(x + \frac{b}{2a}\right)^2 = \frac{4a}{4a} \cdot \frac{-c}{a} + \frac{b^2}{4a^2} \qquad \text{Obtain a common denominator so that the fractions can be added.}$$

$$\left(x + \frac{b}{2a}\right)^2 = \frac{-4ac + b^2}{4a^2}$$

$$\left(x + \frac{b}{2a}\right)^2 = \frac{b^2 - 4ac}{4a^2}$$

$$x + \frac{b}{2a} = \pm \sqrt{\frac{b^2 - 4ac}{4a^2}} \qquad \text{Use the square root property.}$$

$$x + \frac{b}{2a} = \pm \frac{\sqrt{b^2 - 4ac}}{2a}$$

$$x = \frac{-b}{2a} \pm \frac{\sqrt{b^2 - 4ac}}{2a} \qquad \text{Subtract } b/2a \text{ from both sides of the equations.}$$

$$x = \frac{-b \pm \sqrt{b^2 - 4ac}}{2a} \qquad \text{The quadratic formula}$$

Use the Quadratic Formula ②

To Solve a Quadratic Equation by the Quadratic Formula

1. Write the quadratic equation in standard form, $ax^2 + bx + c = 0$, and determine the numerical values for a, b, and c.

2. Substitute the values for a, b, and c in the quadratic formula and then evaluate the formula to obtain the solution.

The Quadratic Formula

$$x = \frac{-b \pm \sqrt{b^2 - 4ac}}{2a}$$

EXAMPLE 1 Solve the equation $x^2 + 2x - 8 = 0$ using the quadratic formula.

Solution: $a = 1, b = 2, c = -8$.

$$x = \frac{-b \pm \sqrt{b^2 - 4ac}}{2a}$$

$$x = \frac{-(2) \pm \sqrt{(2)^2 - 4(1)(-8)}}{2(1)}$$

$$= \frac{-2 \pm \sqrt{4 + 32}}{2}$$

$$= \frac{-2 \pm \sqrt{36}}{2}$$

$$= \frac{-2 \pm 6}{2}$$

$$x = \frac{-2 + 6}{2} \qquad \text{or} \qquad x = \frac{-2 - 6}{2}$$

$$= \frac{4}{2} = 2 \qquad\qquad\qquad = \frac{-8}{2} = -4$$

A check will show that both 2 and -4 are solutions to the equation.

The solution to Example 1 could also be obtained by factoring as illustrated below.

$$x^2 + 2x - 8 = 0$$
$$(x + 4)(x - 2) = 0$$
$$x + 4 = 0 \quad \text{or} \quad x - 2 = 0$$
$$x = -4 \qquad\qquad x = 2$$

When you are given a quadratic equation to solve, and the method to solve it has not been specified, you may try solving by factoring first (as we discussed in Section 6.5). If the equation can not be easily factored, use the quadratic formula.

COMMON STUDENT ERROR   The entire numerator of the quadratic formula must be divided by $2a$.

Correct

$$x = \frac{-b \pm \sqrt{b^2 - 4ac}}{2a}$$

Incorrect

$$x = -b \pm \frac{\sqrt{b^2 - 4ac}}{2a}$$

$$x = \frac{-b}{2a} \pm \sqrt{b^2 - 4ac}$$

EXAMPLE 2   Solve the equation $2x^2 + 4x - 5 = 0$ using the quadratic formula.

Solution: $a = 2, \quad b = 4, \quad c = -5.$

$$x = \frac{-b \pm \sqrt{b^2 - 4ac}}{2a}$$

$$x = \frac{-4 \pm \sqrt{(4)^2 - 4(2)(-5)}}{2(2)} = \frac{-4 \pm \sqrt{56}}{4} = \frac{-4 \pm 2\sqrt{14}}{4}$$

Now factor out 2 from both terms in the numerator; then divide out the common factor.

$$x = \frac{\overset{1}{2}(-2 \pm \sqrt{14})}{\underset{2}{4}} = \frac{-2 \pm \sqrt{14}}{2}*$$

Thus, the solutions are $\dfrac{-2 + \sqrt{14}}{2}$ and $\dfrac{-2 - \sqrt{14}}{2}$.

If you try solving the equation $2x^2 + 4x - 5 = 0$ by factoring, you will see that $2x^2 + 4x - 5$ is not factorable. Therefore, Example 2 could not be solved by factoring.

---

*Solutions will be given in this form in the Answer Section.

COMMON STUDENT ERROR   Many students use the quadratic formula correctly until the last step, where they make an error. Below are illustrated both the correct and incorrect procedures for simplifying an answer.

When *both* terms in the numerator *and* the denominator have a common factor, that common factor may be divided out, as follows:

Correct

$$\frac{2 + 4\sqrt{3}}{2} = \frac{\overset{1}{2}(1 + 2\sqrt{3})}{\underset{1}{2}} = 1 + 2\sqrt{3}$$

$$\frac{6 + 3\sqrt{3}}{6} = \frac{\overset{1}{3}(2 + \sqrt{3})}{\underset{2}{6}} = \frac{2 + \sqrt{3}}{2}$$

Below are some common errors. Study them carefully so you will not make them. Can you explain why each of the following procedures is incorrect?

Incorrect

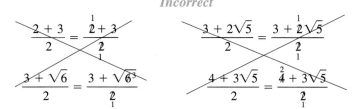

Note that $(2 + 3)/2$ simplifies to $5/2$. However, $(3 + 2\sqrt{5})/2$, $(3 + \sqrt{6})/2$, and $(4 + 3\sqrt{5})/2$ cannot be simplified any further.

EXAMPLE 3   Solve the quadratic equation $-2p^2 - 5p = 6$.

Solution: Do not let the change in variable worry you. The quadratic formula is used exactly the same way as when x is the variable.

$$-2p^2 - 5p - 6 = 0$$
$$a = -2, \qquad b = -5 \qquad c = -6$$
$$p = \frac{-b \pm \sqrt{b^2 - 4ac}}{2a}$$
$$= \frac{-(-5) \pm \sqrt{(-5)^2 - 4(-2)(-6)}}{2(-2)}$$
$$= \frac{5 \pm \sqrt{25 - 48}}{-4}$$
$$= \frac{5 \pm \sqrt{-23}}{-4} = \frac{5 \pm i\sqrt{23}}{-4}$$

$$p = \frac{5 + i\sqrt{23}}{-4} \qquad \text{or} \qquad p = \frac{5 - i\sqrt{23}}{-4}$$
$$= \frac{-5 - i\sqrt{23}}{4} \qquad\qquad\qquad = \frac{-5 + i\sqrt{23}}{4}$$

The solutions are $\dfrac{-5 + i\sqrt{23}}{4}$ and $\dfrac{-5 - i\sqrt{23}}{4}$. Note that neither solution is real.

In Section 8.7 we explained how to check equations whose solutions are complex numbers. If you check the solutions to Example 3 you will see that both check. Check the solutions now.

EXAMPLE 4 Solve the equation $x^2 + \dfrac{2}{5}x - \dfrac{1}{3} = 0$ using the quadratic formula.

Solution: We could solve this equation using the quadratic formula with $a = 1$, $b = \frac{2}{5}$, and $c = -\frac{1}{3}$.

When a quadratic equation contains fractions, it is generally easier to begin by multiplying both sides of the equation by the least common denominator. In this example, the least common denominator is 15.

$$15\left(x^2 + \frac{2}{5}x - \frac{1}{3}\right) = 15(0)$$

$$15x^2 + 6x - 5 = 0$$

Now use the quadratic formula with $a = 15$, $b = 6$, $c = -5$.

$$x = \frac{-b \pm \sqrt{b^2 - 4ac}}{2a}$$

$$= \frac{-6 \pm \sqrt{6^2 - 4(15)(-5)}}{2(15)}$$

$$= \frac{-6 \pm \sqrt{336}}{30}$$

$$= \frac{-6 \pm \sqrt{16}\sqrt{21}}{30}$$

$$= \frac{-6 \pm 4\sqrt{21}}{30}$$

$$= \frac{\overset{1}{\cancel{2}}\,(-3 \pm 2\sqrt{21})}{\underset{15}{\cancel{30}}}$$

$$= \frac{-3 \pm 2\sqrt{21}}{15}$$

The solutions are $\dfrac{-3 + 2\sqrt{21}}{15}$ and $\dfrac{-3 - 2\sqrt{21}}{15}$. The same solution could be obtained using the equivalent equation with fractional coefficients.

If all the numerical coefficients in a quadratic equation have a common factor, you should factor it out before using the quadratic formula. For example, consider the equation $3x^2 + 12x + 3 = 0$. Here $a = 3$, $b = 12$, and $c = 3$. If we use the quadratic formula, we would eventually obtain $x = -2 \pm \sqrt{3}$ as solutions. By factoring the equation before using the formula, we get

$$3x^2 + 12x + 3 = 0$$
$$3(x^2 + 4x + 1) = 0$$

If we consider $x^2 + 4x + 1 = 0$, then $a = 1$, $b = 4$, and $c = 1$. If we use these new values of a, b, and c in the quadratic formula, we will obtain the identical solution, $x = -2 \pm \sqrt{3}$. However, the calculations with these smaller values of a, b, and c are simplified. Solve both equations now using the quadratic formula to convince yourself.

Writing Quadratic Equations

3️⃣ If we are given the solutions of an equation, we can find the equation by working backwards. This procedure is illustrated in Example 5.

EXAMPLE 5

(a) Write an equation whose solutions are -3 and 2.

(b) Write a function whose x intercepts are -3 and 2.

Solution:

(a) If the solutions are -3 and 2, the factors must be $(x + 3)(x - 2)$. Therefore, the equation is

$$(x + 3)(x - 2) = 0$$
$$\text{or} \quad x^2 + x - 6 = 0$$

Many other equations have roots -3 and 2. In fact, any equation of the form $a(x + 3)(x - 2) = 0$, for any real number a, has the roots -3 and 2. Can you explain why this is true?

(b) Since the roots of the equation $x^2 + x - 6 = 0$ are -3 and 2, the x intercepts of the graph of the function $y = x^2 + x - 6$ are -3 and 2. Remember, to find the x intercepts we replace y with 0 and solve for x.

The Discriminant

4️⃣ The expression under the radical sign in the quadratic formula is called the **discriminant**.

$$\underbrace{b^2 - 4ac}_{\text{discriminant}}$$

The discriminant gives the number and nature of solutions of a quadratic equation.

Solutions of a Quadratic Equation

For a quadratic equation of the form $ax^2 + bx + c = 0$, $a \neq 0$:

If $b^2 - 4ac > 0$, the quadratic equation has two distinct real number solutions.

If $b^2 - 4ac = 0$, the quadratic equation has a single real number solution.

If $b^2 - 4ac < 0$, the quadratic equation has no real number solution.

EXAMPLE 6

(a) Find the discriminant of the equation $x^2 - 8x + 16 = 0$.

(b) How many real number solutions does the given equation have?

(c) Use the quadratic formula to find the solution(s).

Solution:

(a) $a = 1, \quad b = -8, \quad c = 16.$

$$b^2 - 4ac = (-8)^2 - 4(1)(16)$$
$$= 64 - 64 = 0$$

(b) Since the discriminant equals zero, there is a single real number solution.

(c)
$$x = \frac{-b \pm \sqrt{b^2 - 4ac}}{2a}$$

$$= \frac{-(-8) \pm \sqrt{0}}{2(1)} = \frac{8 \pm 0}{2} = \frac{8}{2} = 4$$

The only solution is 4.

EXAMPLE 7   Without actually finding the solutions, determine if the following equations have two distinct real number solutions, a single real number solution, or no real number solution.

(a) $2x^2 - 4x + 6 = 0$ **(b)** $x^2 - 5x - 8 = 0$ **(c)** $4x^2 - 12x = -9$

Solution: We use the discriminant of the quadratic formula to answer these questions.

(a) $b^2 - 4ac = (-4)^2 - 4(2)(6) = 16 - 48 = -32$
Since the discriminant is negative, this equation has no real number solution.

(b) $b^2 - 4ac = (-5)^2 - 4(1)(-8) = 25 + 32 = 57$
Since the discriminant is positive, this equation has two distinct real number solutions.

(c) First, rewrite $4x^2 - 12x = -9$ as $4x^2 - 12x + 9 = 0$.

$$b^2 - 4ac = (-12)^2 - 4(4)(9) = 144 - 144 = 0$$

Since the discriminant is zero, this equation has a single real number solution.

In this section we study quadratic functions in more depth. Recall from Section 5.8 that quadratic functions are of the form $f(x) = ax^2 + bx + c, a \neq 0$. Since y may replace $f(x)$, equations of the form $y = ax^2 + bx + c, a \neq 0$, are also quadratic functions. The graph of every quadratic equation of this form is a *parabola.*

When graphing quadratic functions you can use two methods to determine if the graph will have no, exactly one, or two distinct x intercepts. One method is to use the discriminant. The value of y at the x intercepts is 0. If we substitute 0 for y in the equation $y = ax^2 + bx + c$, we obtain

$$ax^2 + bx + c = 0$$

The solutions to this equation, if they are real, will be the x intercepts of the graph of $y = ax^2 + bx + c$. The solutions to the equation $ax^2 + bx + c = 0$ may be found by factoring, by the quadratic formula, or by completing the square. To determine the *number of x intercepts* all we need to do is evaluate the discriminant, $b^2 - 4ac$. If the discriminant is greater than 0, then $ax^2 + bx + c = 0$ has two distinct real number solutions and the graph of $y = ax^2 + bx + c$ will have two distinct x intercepts (see Fig. 9.3a).

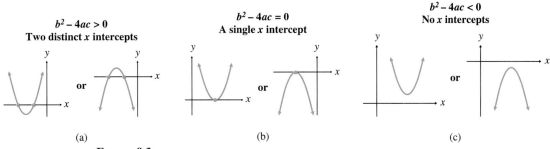

(a)

(b)

(c)

FIGURE 9.3

If $b^2 - 4ac = 0$, the equation $ax^2 + bx + c = 0$ has a single real number solution, and the graph of $y = ax^2 + bx + c$ has only one x intercept. This can happen only when the vertex of the graph is on the x axis (see Fig. 9.3b). If $b^2 - 4ac < 0$, the equation $ax^2 + bx + c = 0$ has no real number solutions and the graph of $y = ax^2 + bx + c$ has no x intercept (Fig. 9.3c).

A second method to determine the number of x intercepts of the function is to find the vertex and determine whether the graph opens upward or downward. If the y coordinate of the vertex is negative and the graph opens upward ($a > 0$), or the y coordinate of the vertex is positive and the graph opens downward ($a < 0$), the graph must also have two distinct x intercepts (see Fig. 9.3a). If the y coordinate of the vertex is 0, it means the vertex of the graph is on the x axis and the graph can only have one x intercept (see Fig. 9.3b). If the y coordinate of the vertex is positive and the graph opens upward or if the y coordinate of the vertex is negative and the graph opens downward, the graph can have no x intercept (see Fig. 9.3c). We can find the y coordinate of the vertex in one of two ways. We can find the x value of the vertex using $x = \dfrac{-b}{2a}$ and then substitute this value in the function to find the y coordinate of the vertex. A second method to find the y coordinate of the vertex is to use the formula

$$y = \frac{4ac - b^2}{4a}$$

We will derive this formula in Section 10.3. This formula may be easier to use to find the y coordinate if the x coordinate is a fraction.

Sketching Quadratic Functions

⑤ We can *sketch* the graph of a quadratic equation by noticing whether the parabola opens upward or downward, and finding the y intercept, the vertex, and the x intercepts. Recall that, to find the y intercept, set $x = 0$ and solve for y. Example 8 illustrates how a quadratic function may be sketched.

EXAMPLE 8   Consider the equation $y = -x^2 + 8x - 12$.

(a) Determine whether the parabola opens upward or downward.
(b) Find the y intercept.
(c) Find the vertex.
(d) Find the x intercepts (if any).
(e) Sketch the graph.

Solution:

(a) Since a is -1, which is less than 0, the parabola opens downward.
(b) To find the y intercept, set $x = 0$ and solve for y.

$$y = -(0)^2 + 8(0) - 12 = -12$$

The y intercept is -12.

(c)
$$x = \frac{-b}{2a} = \frac{-8}{2(-1)} = 4$$

$$y = \frac{4ac - b^2}{4a} = \frac{4(-1)(-12) - 8^2}{4(-1)} = \frac{48 - 64}{-4} = 4$$

The vertex is at $(4, 4)$. The y coordinate of the vertex could also be found by substituting 4 for x in the function and finding the corresponding value of y, which is 4.

(d) To find the x intercepts, we set $y = 0$.

$$0 = -x^2 + 8x - 12$$
or $\quad -x^2 + 8x - 12 = 0$

We can multiply both sides of the equation by -1 and then factor.

$$-1(-x^2 + 8x - 12) = -1(0)$$
$$x^2 - 8x + 12 = 0$$
$$(x - 6)(x - 2) = 0$$
$$x - 6 = 0 \quad \text{or} \quad x - 2 = 0$$
$$x = 6 \quad \text{or} \quad x = 2$$

Thus, the x intercepts are 2 and 6. These values could also be found by the quadratic formula (or by completing the square).

(e) Now we use all this information to sketch the graph (Fig. 9.4).

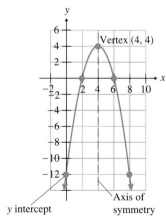

FIGURE 9.4

If you obtain irrational values when finding x intercepts by the quadratic formula, use your calculator to estimate these values, then plot these decimal values. For example, if you obtain $x = (2 \pm \sqrt{10})/2$ you would evaluate $(2 + \sqrt{10})/2$ and $(2 - \sqrt{10})/2$ on your calculator and obtain 2.58 and -0.58 respectively, to the nearest hundredth. The x intercepts would therefore be $(2.58, 0)$ and $(-0.58, 0)$.

EXAMPLE 9   Consider the equation $y = 2x^2 + 3x + 4$.

(a) Determine whether the parabola opens upward or downward.

(b) Find the y intercept.

(c) Find the vertex.

(d) Find the x intercepts, if any.

(e) Sketch the graph.

Solution:

(a) Since a is 2, which is greater than 0, the parabola opens upward.

(b) $y = 2(0)^2 + 3(0) + 4 = 4$. The y intercept is 4.

(c) $x = \dfrac{-b}{2a} = \dfrac{-3}{2(2)} = \dfrac{-3}{4}$

$y = \dfrac{4ac - b^2}{4a} = \dfrac{4(2)(4) - 3^2}{4(2)} = \dfrac{32 - 9}{8} = \dfrac{23}{8}$

The vertex is $\left(-\frac{3}{4}, \frac{23}{8}\right)$.

(d) To find the x intercepts, set $y = 0$.

$$0 = 2x^2 + 3x + 4$$

This trinomial cannot be factored. To determine if this equation has any real solutions, we will evaluate the discriminant.

$$b^2 - 4ac = 3^2 - 4(2)(4) = 9 - 32 = -23$$

Since the discriminant is less than 0, this equation has no real solutions. We should have expected this answer because the y coordinate of the vertex is a positive number and therefore above the x axis. Since the parabola opens upward, it cannot intersect the x axis.

(e) The graph is sketched in Fig. 9.5.

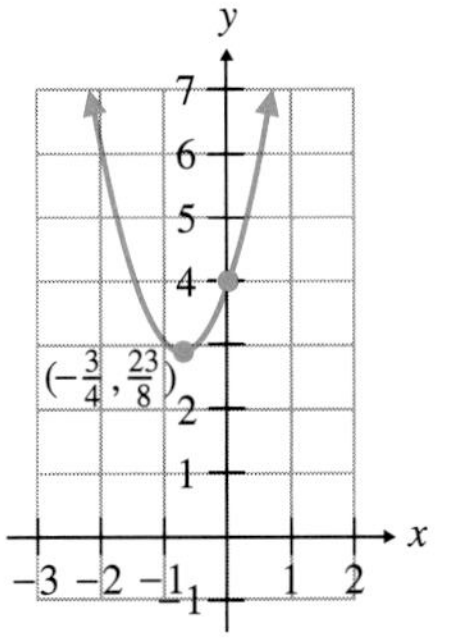

FIGURE 9.5

When sketching a parabola, as in Examples 8 and 9, the exact curve of the graph may be slightly inaccurate, for we are not plotting point by point. However, for our needs a sketch is generally sufficient.

Applications of Quadratic Equations

6 We will now look at some applications of quadratic equations.

EXAMPLE 10   Laserox, a startup company, projects that its annual profits, in thousands of dollars, over the first 8 years of operation can be approximated by the function $p(n) = 1.2n^2 + 4n - 8$, where n is the number of years completed.

(a) Estimate the profit (or loss) of the company after the first year.

(b) Estimate the profit (or loss) of the company after 8 years.

(c) Estimate the time needed for the company to break even.

Solution:

(a) To estimate the profit after 1 year, evaluate the function at 1.

$$p(n) = 1.2n^2 + 4n - 8$$
$$p(1) = 1.2(1)^2 + 4(1) - 8 = -2.8$$

Thus, at the end of the first year the company projects a loss of $2800.

(b) $p(8) = 1.2(8)^2 + 4(8) - 8 = 100.8$

Thus, at the end of the eighth year the projected profit is $100,800.

(c) The company will break even when the profit is 0. Thus, to find the break-even point we solve the equation

$$1.2n^2 + 4n - 8 = 0$$

We can use the quadratic formula to solve this equation.

$$n = \frac{-b \pm \sqrt{b^2 - 4ac}}{2a}$$

$$= \frac{-4 \pm \sqrt{4^2 - 4(1.2)(-8)}}{2(1.2)}$$

$$= \frac{-4 \pm \sqrt{16 + 38.4}}{2.4}$$

$$= \frac{-4 \pm \sqrt{54.4}}{2.4}$$

$$\approx \frac{-4 \pm 7.376}{2.4}$$

$$n = \frac{-4 + 7.376}{2.4} \approx 1.4 \quad \text{or} \quad n = \frac{-4 - 7.376}{2.4} \approx -4.74$$

Since time cannot be negative, the break-even time is about 1.4 years.

An important formula in physics is $h = \frac{1}{2}at^2 + v_0 t + h_0$. When an object is projected upward from an initial height h_0, with initial velocity of v_0, this formula can be used to find the height h of the object above the ground at any time, t. The a in the formula is the acceleration of gravity. Since the acceleration of Earth's gravity is -32 ft/sec^2, we use -32 for a in the formula when discussing Earth. This formula can also be used in describing projectiles on the moon and other planets, but the value of a in the formula will need to change for each planetary body. We will use this formula in Example 11.

EXAMPLE 11   Jennifer is standing on top of a building and throws a ball upward from a height of 60 feet with an initial velocity of 30 feet per second. Use the formula $h = \frac{1}{2}at^2 + v_0 t + h_0$ to answer these questions:

(a) How long after the ball is thrown, to the nearest tenth of a second, will the ball be 25 feet from the ground?

(b) How long after the ball is thrown will the ball strike the ground?

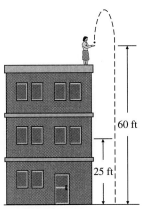

FIGURE 9.6

Solution:

(a) We will illustrate this problem with a diagram (see Fig. 9.6). Here $a = -32$, $v_0 = 30$, and $h_0 = 60$. We are asked to find the time, t, it takes for the ball to reach a height, h, of 25 feet above the ground. We substitute these values into the formula and then solve for t.

$$h = \frac{1}{2} at^2 + v_0 t + h_0$$

$$25 = \frac{1}{2}(-32)t^2 + 30t + 60$$

Now write the quadratic equation in standard form and solve for t by using the quadratic formula.

$$0 = -16t^2 + 30t + 35$$

$$\text{or} \quad -16t^2 + 30t + 35 = 0$$

$$a = -16, \quad b = 30, \quad c = 35$$

$$t = \frac{-b \pm \sqrt{b^2 - 4ac}}{2a} = \frac{-30 \pm \sqrt{(30)^2 - 4(-16)(35)}}{2(-16)} = \frac{-30 \pm \sqrt{3140}}{-32}$$

$$t = \frac{-30 + \sqrt{3140}}{-32} \quad \text{or} \quad t = \frac{-30 - \sqrt{3140}}{-32}$$

$$\approx -0.8 \qquad\qquad \approx 2.7$$

Since time cannot be negative, the only acceptable solution is 2.7 seconds. Thus, about 2.7 seconds after the ball is thrown, it will be 25 feet above the ground.

(b) We wish to find the time at which the ball strikes the ground. When the ball strikes the ground, its distance above the ground is 0. We substitute $h = 0$ into the formula and solve.

$$h = \frac{1}{2} at^2 + v_0 t + h_0$$

$$0 = \frac{1}{2}(-32)t^2 + 30t + 60$$

$$0 = -16t^2 + 30t + 60$$

$$a = -16, \quad b = 30, \quad c = 60$$

$$t = \frac{-b \pm \sqrt{b^2 - 4ac}}{2a} = \frac{-30 \pm \sqrt{(30)^2 - 4(-16)(60)}}{2(-16)} = \frac{-30 \pm \sqrt{4740}}{-32}$$

$$t = \frac{-30 + \sqrt{4740}}{-32} \quad \text{or} \quad t = \frac{-30 - \sqrt{4740}}{-32}$$

$$\approx -1.2 \qquad\qquad \approx 3.1$$

Since time cannot be negative, the solution is 3.1 seconds. Thus, the ball will strike the ground approximately 3.1 seconds after it is thrown.

Exercise Set 9.2

no real solution: **2, 3, 4, 7, 10**   one real solution: **6, 9, 11**   two real solutions: **1, 5, 8, 12**

49. $\dfrac{-0.94 \pm \sqrt{32.3116}}{3.24}$ or $\dfrac{-47 \pm \sqrt{80,779}}{162}$ **50.** $\dfrac{2.04 \pm i\sqrt{38.9904}}{3.48}$ or $\dfrac{51 \pm i\sqrt{24,369}}{87}$

Use the discriminant to determine whether each equation has two distinct real solutions, a single real solution, or no real solution.

1. $x^2 + 4x - 3 = 0$

2. $3x^2 + x + 3 = 0$

3. $2x^2 - 4x + 7 = 0$

4. $-2x^2 + x - 8 = 0$

5. $5x^2 + 3x - 7 = 0$

6. $2x^2 = 16x - 32$

7. $-3x^2 + 5x - 8 = 0$

8. $4.1x^2 - 3.1x - 5.2 = 0$

9. $x^2 + 10.2x + 26.01 = 0$

10. $\dfrac{1}{2}x^2 + 3x + 12 = 0$

11. $x^2 = -3x - \dfrac{9}{4}$

12. $\dfrac{x^2}{3} = \dfrac{2x}{5}$

Use the quadratic formula to solve the equation.   **47.** $\dfrac{-0.6 \pm \sqrt{0.84}}{0.2}$ or $-3 \pm \sqrt{21}$ **48.** $\dfrac{5.6 \pm \sqrt{35.04}}{4.6}$ or $\dfrac{28 \pm 2\sqrt{219}}{23}$

13. $x^2 - 3x + 2 = 0$ 1, 2

14. $x^2 + 6x + 8 = 0$ $-2, -4$

15. $x^2 - 9x + 20 = 0$ 4, 5

16. $x^2 - 3x - 10 = 0$ 5, -2

17. $x^2 - 6x = -5$ 1, 5

18. $x^2 = 13x - 36$ 4, 9

19. $x^2 - 36 = 0$ ± 6

20. $x^2 - 25 = 0$ ± 5

21. $x^2 - 6x = 0$ 0, 6

22. $x^2 - 3x = 0$ 0, 3

23. $z^2 + 17z + 72 = 0$ $-8, -9$

24. $2x^2 - 3x + 2 = 0$ $(3 \pm i\sqrt{7})/4$

25. $2y^2 - 7y + 4 = 0$ $(7 \pm \sqrt{17})/4$

26. $2x^2 - 7x = -5$ $1, \frac{5}{2}$

27. $6x^2 = -x + 1$ $\frac{1}{3}, -\frac{1}{2}$

28. $2x^2 = 4x + 1$ $(2 \pm \sqrt{6})/2$

29. $3w^2 - 4w + 5 = 0$ $(2 \pm i\sqrt{11})/3$

30. $4s^2 - 8s + 6 = 0$ $(2 \pm i\sqrt{2})/2$

31. $x^2 + 7x = -3$ $(-7 \pm \sqrt{37})/2$

32. $x^2 - 2x - 1 = 0$ $1 \pm \sqrt{2}$

33. $-6x^2 + 21x = -27$ $\frac{9}{2}, -1$

34. $-x^2 + 2x + 15 = 0$ 5, -3

35. $(2a + 3)(3a - 1) = 2$ $\frac{1}{2}, -\frac{5}{3}$

36. $(2w - 6)(3w + 4) = -20$ $2, -\frac{1}{3}$

37. $-2x^2 = x + 3$ $(-1 \pm i\sqrt{23})/4$

38. $9x^2 + 6x + 1 = 0$ $-\frac{1}{3}$

39. $2x^2 + 6x = 0$ 0, -3

40. $3x^2 - 5x = 0$ $0, \frac{5}{3}$

41. $m = \dfrac{-m + 6}{m - 4}$ $\dfrac{3 \pm \sqrt{33}}{2}$

42. $3p = \dfrac{5p + 6}{2p + 3}$ $\dfrac{-1 \pm \sqrt{10}}{3}$

43. $\dfrac{1}{2}x^2 + 2x + \dfrac{2}{3} = 0$ $\dfrac{-6 \pm 2\sqrt{6}}{3}$

44. $x^2 - \dfrac{x}{5} - \dfrac{1}{3} = 0$ $\dfrac{3 \pm \sqrt{309}}{30}$

45. $-x^2 + \dfrac{11}{3}x + \dfrac{10}{3} = 0$ $\dfrac{11 \pm \sqrt{241}}{6}$

46. $x^2 - \dfrac{7}{6}x + \dfrac{2}{3} = 0$ $\dfrac{7 \pm i\sqrt{47}}{12}$

47. $0.1x^2 + 0.6x - 1.2 = 0$

48. $-2.3x^2 + 5.6x + 0.4 = 0$

49. $-1.62x^2 - 0.94x + 4.85 = 0$

50. $1.74x^2 - 2.04x + 6.2 = 0$

Write an equation in two variables that has the x intercepts given.

51. 4, 6 $y = x^2 - 10x + 24$

52. $-2, 5$ $y = x^2 - 3x - 10$

53. 3, -4 $y = x^2 + x - 12$

54. 0, 4 $y = x^2 - 4x$

55. 2, -3 $y = x^2 + x - 6$

56. $-1, -6$ $y = x^2 + 7x + 6$

57. 2, 2 $y = x^2 - 4x + 4$

58. $\dfrac{1}{2}, 3$ $y = 2x^2 - 7x + 3$

59. $-2, \dfrac{2}{3}$ $y = 3x^2 + 4x - 4$

60. $-\dfrac{3}{5}, \dfrac{2}{3}$ $y = 15x^2 - x - 6$

61. $-\dfrac{1}{2}, \dfrac{2}{3}$ $y = 6x^2 - x - 2$

62. $\dfrac{3}{5}, \dfrac{1}{4}$ $y = 20x^2 - 17x + 3$

(a) Determine whether the parabola opens upward or downward. (b) Find the y intercept. (c) Find the vertex. (d) Find the x intercepts (if any). Use a calculator to find an approximate value for the x intercepts if they are irrational. (e) Sketch the graph.

63. $y = x^2 + 8x + 15$
(a) upward (b) 15   (e)
(c) $(-4, -1)$
(d) $-3, -5$

64. $y = x^2 + 2x - 3$
(a) upward (b) -3 (e)
(c) $(-1, -4)$
(d) 1, -3

65. $f(x) = x^2 - 6x + 4$
(a) upward (b) 4   (e)
(c) $(3, -5)$
(d) $3 \pm \sqrt{5}$

66. $y = -x^2 + 4x - 5$
(a) downward (b) -5 (e)
(c) $(2, -1)$
(d) no x intercepts

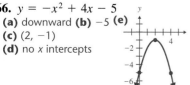

67. $y = x^2 + 6x + 9$
(a) upward (b) 9   (e)
(c) $(-3, 0)$ (d) -3

68. $y = -2x^2 + 4x - 8$
(a) downward (b) -8 (e)
(c) $(1, -6)$,
(d) no x intercepts

69. $y = 2x^2 - x - 6$
(a) upward (b) -6 (e)
(c) $(\frac{1}{4}, -\frac{49}{8})$ (d) $-\frac{3}{2}, 2$

70. $y = -3x^2 + 6x - 9$
(a) downward (b) -9 (e)
(c) $(1, -6)$
(d) no x intercepts

71. $y = 3x^2 + 4x + 3$
(a) upward (b) 3 (e)
(c) $(-\frac{2}{3}, \frac{5}{3})$
(d) no x intercepts
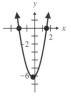

72. $f(x) = -3x^2 - 2x - 6$
(a) downward (b) -6 (e)
(c) $(-\frac{1}{3}, -\frac{17}{3})$
(d) no x intercepts

73. $f(x) = -2x^2 - 6x + 4$
(a) downward (b) 4 (e)
(c) $(-\frac{3}{2}, \frac{17}{2})$
(d) $(-3 \pm \sqrt{17})/2$

74. $f(x) = 2x^2 + x - 6$
(a) upward (b) -6 (e)
(c) $(-\frac{1}{4}, -\frac{49}{8})$
(d) $\frac{3}{2}, -2$
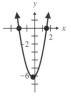

75. $y = x^2 + 4$
(a) upward (b) 4 (e)
(c) $(0, 4)$
(d) no x intercepts
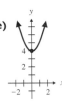

76. $y = x^2 + 4x$
(a) upward (b) 0 (e)
(c) $(-2, -4)$
(d) $0, -4$

77. $y = -x^2 + 6x$
(a) downward (b) 0 (e)
(c) $(3, 9)$
(d) $0, 6$

78. $f(x) = 3x^2 + 10x$
(a) upward (b) 0 (e)
(c) $(-\frac{5}{3}, -\frac{25}{3})$ (d) $0, -\frac{10}{3}$
(e)

79. $f(x) = -5x^2 + 5$
(a) downward (b) 5 (e)
(c) $(0, 5)$ (d) $-1, 1$

80. $f(x) = 2x^2 - 6x + 4$
(a) upward (b) 4 (e)
(c) $(\frac{3}{2}, -\frac{1}{2})$ (d) $1, 2$

81. $y = 3x^2 + 4x - 6$
(a) upward (b) -6 (e)
(c) $(-\frac{2}{3}, -\frac{22}{3})$
(d) $(-2 \pm \sqrt{22})/3$

82. $f(x) = -x^2 + 3x - 5$
(a) downward (b) -5
(c) $(\frac{3}{2}, -\frac{11}{4})$ (e)
(d) no x intercepts

83. $y = -x^2 + 3x + 6$
(a) downward (b) 6 (e)
(c) $(\frac{3}{2}, \frac{33}{4})$
(d) $(3 \pm \sqrt{33})/2$
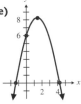

84. $f(x) = -2x^2 - 6x + 5$
(a) downward (b) 5 (e)
(c) $(-\frac{3}{2}, \frac{19}{2})$
(d) $(-3 \pm \sqrt{19})/2$

85. $f(x) = -4x^2 + 6x - 9$
(a) downward (b) -9 (e)
(c) $(\frac{3}{4}, -\frac{27}{4})$
(d) no x intercepts

86. $f(x) = -2x^2 + 5x + 4$
(a) downward (b) 4 (e)
(c) $(\frac{5}{4}, \frac{57}{8})$
(d) $(5 \pm \sqrt{57})/4$

In Exercises 87–101, use a calculator as needed to give the solution in decimal form. Round irrational numbers to the nearest hundredth. **90.** 50 ft by 300 ft or 150 ft by 100 ft

87. Twice the square of a positive number increased by 3 times the number is 14. Find the number.   2

88. Three times the square of a positive number decreased by twice the number is 21. Find the number.   3

89. The length of a rectangular garden is 2 feet less than 3 times its width. Find the length and width if the area of the garden is 21 square feet. $w = 3$ ft, $l = 7$ ft

90. Lora wishes to form a rectangular region along a river bank by constructing fencing as illustrated in the diagram below. If she has only 400 feet of fencing and wishes to enclose an area of 15,000 square feet, find the dimensions of the rectangular region.

River

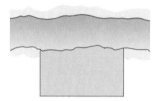

Fence

91. John Williams, a professional photographer, has a 6-inch by 8-inch photo. He wishes to reduce the photo by the same amount on each side so that the resulting photo will have half the area of the original photo. By how much will he have to reduce the length of each side?   2 in

92. The temperature, T, in degrees Fahrenheit in a car's radiator during the first 4 minutes of driving is a function of time, t. The temperature can be found by formula $T = 6.2t^2 + 12t + 32, 0 \le t \le 4$.

(a) What is the car's radiator temperature at the instant the car is turned on?   32°F

(b) What is the car's radiator temperature after the car has been driven for 1 minute?   50.2°F

(c) How long after the car has begun operating will the car's radiator temperature reach 120°F?   2.92 min

93. A video store's weekly profit, P, in thousands of dollars, is a function of the rental price of the tapes, t. The profit equation is $P = 0.2t^2 + 1.5t - 1.2$, $0 \le t \le 5$.

(a) What is the store's weekly proft or loss if they charge $1 per tape?   0.5 thousand

(b) What is the weekly profit if they charge $5 per tape?   11.3 thousand

(c) At what tape rental price will their weekly profit be 1.6 thousand dollars? $1.55

94. The cost, C, in thousands of dollars of a ranch house in Norfolk, Virginia, is a function of the number of square feet of the house, s. The cost of a house can be approximated by the formula

$$C = -0.01s^2 + 80s + 20,000, \qquad 1200 \le s \le 4000$$

(a) Find the cost of a 1500-square-foot house. $117,500

(b) How large a house can Mr. Dodge purchase if he has $150,000 to spend on a house?   2268 sq ft

95. At a college, records show that the average person's grade point average, G, is a function of the number of hours he or she studies and does homework per week, h. The grade point average can be estimated by the equation $G = 0.01h^2 + 0.2h + 1.2, 0 \le h \le 8$.

(a) What is the average GPA of the average student who studies for 0 hours a week?   1.2

(b) What is the average GPA of a student who studies 4 hours per week?   2.16

(c) To obtain a 3.2 GPA, how many hours per week would the average student need to study?   7.3 hr

96. The roller coaster at Busch Gardens in Williamsburg, Virginia, has a number of steep drops. One of its drops has a vertical distance of 62 feet. The speed of the last car, in feet per second, t seconds after it has begun its drop can be calculated by the formula $s = 6.74t + 2.3, 0 \le t \le 4$. The height of the last car from the bottom of the drop t seconds after it has begun this drop can be found by the formula $h = -3.3t^2 - 2.3t + 62, 0 \le t \le 4$.

(a) Find the time it takes for the last car to travel from the top of the drop to the bottom of the drop.   4 sec

(b) Find the speed of the last car when it reaches the bottom of the drop.   29.26 ft/sec

97. The San Francisco Philharmonic is trying to set the price of its concert tickets. If the price is too low, they will not make enough money to cover expenses, and if their price is too high, not enough people will wish to pay the price of a ticket. They estimate that their total income, I, in hundreds of dollars, per concert can be approximated by the formula $I = -x^2 + 24x - 44, 0 \le x \le 24$, where x is the cost of a ticket.

(a) Draw a graph of income versus the cost of a ticket.

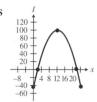

100. (a) Courtney's rock (5.2 sec vs. 6.8 sec)   **104.** No x intercepts if vertex above x axis and $a > 0$, or below x axis and $a < 0$; one intercept if vertex is on x axis. All other cases have two x intercepts.

(b) Determine the minimum cost of a ticket for the Philharmonic to break even. $2

(c) Determine the maximum cost of a ticket that the Philharmonic can charge and break even. $22

(d) How much should they charge to receive the maximum income? $12

(e) Find the maximum income. $10,000

98. An object is projected upward with an initial velocity of 192 feet per second. The object's distance above the ground, d, after t seconds may be found by the formula $d = -16t^2 + 192t$.

(a) Find the object's distance from the ground in 3 seconds.   432 ft

(b) Make a graph of distance versus time.

(c) What is the maximum height the object will reach?   576 ft   **(d)** 6 sec   **(e)** 12 sec

(d) At what time will it reach its maximum height?

(e) At what time will the object strike the ground?

In Exercises 99 through 101 use the equation $h = \frac{1}{2}at^2 + v_o t + h_o$ *(Refer to Example 11).*

99. A horseshoe is thrown upward from an initial height of 80 feet with an initial velocity of 60 feet per second. How long after the horseshoe is projected upward

(a) will it be 20 feet from the ground? $\approx$ 4.57 sec

(b) will it strike the ground? $\approx$ 4.79 sec

100. Jim is on the fourth floor of an eight-story building and Courtney is on the roof. Jim is 60 feet above the ground while Courtney is 120 feet above the ground.

(a) If Jim throws a rock upward with an initial velocity of 100 feet per second at the same time that Courtney throws a rock upward at 60 feet per second, whose rock will hit the ground first?

(b) Will the rocks ever be at the same distance above the ground? If so, at what times?   yes, at 1.5 sec

101. Gravity on the moon is about one-sixth of that on Earth. Suppose Neil Armstrong is standing on a hill on the moon 60 feet high. If he jumps upward with a velocity of 40 feet per second, how long will it take for him to land on the ground below the hill? $\approx$ 16.37 sec

🖉 **102.** Consider the two equations $-6x^2 + \frac{1}{2}x - 5 = 0$ and $6x^2 - \frac{1}{2}x + 5 = 0$? Must the solutions to these two equations be the same? Explain your answer.   yes

🖉 **103.** To find the x intercepts of the graph of an equation of the form $y = ax^2 + bx + c$, we set $y = 0$ and solve the resulting equation. Explain why **(a)** if the discriminant is negative, the graph of $y = ax^2 + bx + c$ has no x intercept; **(b)** if the discriminant is 0, the

SEE EXERCISE 101.

graph has one x intercept; and **(c)** if discriminant is positive, the graph has two distinct x intercepts.

🖉 **104.** By observing the value of the coefficient of the squared term in a quadratic function, and by determining the coordinates of vertex of its graph, explain how you can determine the number of x intercepts the parabola has.

🖉 **105.** Consider the equations $y = x^2 - 8x + 12$ and $y = -x^2 + 8x - 12$.

(a) Without graphing, can you explain how the graphs of the two equations compare?

(b) Will the graphs have the same x intercepts? Explain.   yes

(c) Will the graphs have the same vertex? Explain.   no

(d) Graph both equations on the same axes.

🖉 **106. (a)** What is the discriminant of a quadratic equation of the form $ax^2 + bx + c = 0$? **(b)** Write a paragraph or two explaining the relationship between the value of the discriminant and the number of real solutions to a quadratic equation. In your paragraph, explain *why* the value of the discriminant determines the number of real solutions.   **(a)** $b^2 - 4ac$

98. (b)

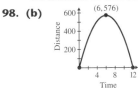

105. (d)

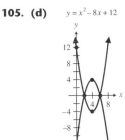

CUMULATIVE REVIEW EXERCISES   **108.** $2x^2y^4\sqrt[3]{x} + 2xy^3\sqrt[3]{3}$

[8.3] **107.** Simplify $\sqrt[5]{64x^9y^{12}z^{20}}$. $2xy^2z^4\sqrt[5]{2x^4y^2}$

108. Simplify $\sqrt[3]{4x^2y^8}\,(\sqrt[3]{2x^5y^4} + \sqrt[3]{6xy})$.

[8.4] **109.** Simplify $\dfrac{x + \sqrt{y}}{x - \sqrt{y}}$. $\dfrac{x^2 + 2x\sqrt{y} + y}{x^2 - y}$

[8.6] **110.** Solve the equation $\sqrt{2x + 4} - 1 = \sqrt{x + 3}$.   6

---

1. $2\sqrt{5}, -\sqrt{5}$ **2.** $-2\sqrt{6}, -3\sqrt{6}$ **3.** $1, -1, 2, -2$ **4.** $x = \dfrac{-1 \pm \sqrt{1 + 4y}}{2}$ **5.** $x = \dfrac{-3y \pm \sqrt{9y^2 - 16}}{2}$ **6.** $x = \dfrac{-3 \pm \sqrt{9 + 20}}{2y}$

Group Activity/ Challenge Problems

Solve using the quadratic formula.

1. $x^2 - \sqrt{5}x - 10 = 0$.   **2.** $x^2 + 5\sqrt{6}x + 36 = 0$.   **3.** $a^4 - 5a^2 + 4 = 0$.

Use the quadratic formula to solve for x.

4. $x^2 + x - y = 0$ **5.** $x^2 + 3yx + 4 = 0$ **6.** $yx^2 + 3x - 5 = 0$

7. Explain why the graph of the function $f(x) = ax^2 + bx + c, a \neq 0$, has its vertex at the ordered pair $\left(\dfrac{-b}{2a}, f\left(\dfrac{-b}{2a}\right)\right)$.   The x coordinate is $\frac{-b}{2a}$. The y coordinate is found by evaluating $f(\frac{-b}{2a})$.

8. A metal cube expands when heated. If each edge increases 0.20 millimeter after being heated and the total volume increases by 6 cubic millimeters, find the original length of a side of the cube. $(-0.12 + \sqrt{14.3952})/1.2 \approx 3.0618$mm

9. The "golden ratio," which occurs often in mathematics, was introduced by the ancient Greeks. To obtain the golden ratio, divide a line segment such that the ratio of the larger part to the whole is equal to the ratio of the smaller part to the larger part. Consider the following diagram.

Using the diagram, the golden ratio will be $\dfrac{a}{a + b} = \dfrac{b}{a}$.

(a) If we let the smaller part be 1, determine the value of a, the larger part. $\dfrac{1 + \sqrt{5}}{2}$

(b) Then substitute the values for a and b and use cross multiplication to show that $\dfrac{a}{a + b} = \dfrac{b}{a}$.

10. The equation $x^n = 1$ has n solutions (including the complex solutions). Find the six solutions to $x^6 = 1$. *Hint:* Rewrite the equation as $x^6 - 1 = 0$, then factor using the formula for the difference of two squares. $\pm 1, \dfrac{1 \pm i\sqrt{3}}{2}, \dfrac{-1 \pm i\sqrt{3}}{2}$

9.3   Writing Equations in Quadratic Form

Tape 15

1. Apply quadratic equations to real-life situations.
2. Solve equations that are quadratic in form.
3. Solve equations with rational exponents.

In this section we will rewrite equations that are not given in quadratic form into quadratic form. Then we can solve them by factoring, completing the square, or using the quadratic formula. We have already done similar problems in other sections of the text. For example, in Section 8.6 we started with $\sqrt{2x - 3} = x - 3$. We squared both sides of the equation and simplified to obtain $x^2 - 8x + 12 = 0$, which we solved by factoring. We will also study some application problems that can be expressed as quadratic equations.

Applications   **1** We have already seen many applications of quadratic equations. In this section we will look at additional applications. First we will discuss motion problems. Motion problems were introduced and discussed in Section 2.4. They were also discussed in Sections 4.3 and 7.6.

MOTION PROBLEMS   **EXAMPLE 1**   In 5 hours the Dechs traveled 12 miles downriver in their motorboat and then returned. If the river's current is 2 miles per hour, find the rate (or speed) of the motorboat in still water.

Solution: We are asked to find the rate of the boat in still water. Let $r =$ the rate of the boat in still water. We know that the total time of the trip is 5 hours. Thus, the time downriver plus the time upriver must sum to 5 hours. Since distance = rate × time, we can find the time by dividing the distance by the rate.

	Distance	Rate	Time
Downriver (with current)	12	$r+2$	$\dfrac{12}{r+2}$
Upriver (against current)	12	$r-2$	$\dfrac{12}{r-2}$

$$\text{time downriver} + \text{time upriver} = \text{total time}$$

$$\frac{12}{r+2} + \frac{12}{r-2} = 5$$

$$(r+2)(r-2)\left(\frac{12}{r+2} + \frac{12}{r-2}\right) = (r+2)(r-2)\,(5) \qquad \text{Multiply by the LCD to remove fractions.}$$

$$(r+2)(r-2)\left(\frac{12}{r+2}\right) + (r+2)(r-2)\left(\frac{12}{r-2}\right) = (r+2)(r-2)(5) \qquad \text{Distributive property.}$$

$$12(r-2) + 12(r+2) = 5(r^2 - 4)$$
$$12r - 24 + 12r + 24 = 5r^2 - 20$$
$$24r = 5r^2 - 20$$
$$\text{or} \quad 5r^2 - 24r - 20 = 0$$

Using the quadratic formula with $a = 5$, $b = -24$, and $c = -20$, we obtain

$$r = \frac{24 \pm \sqrt{976}}{10}$$

$$r \approx 5.5 \quad \text{or} \quad r \approx -0.7$$

Since the rate is not negative, the rate or speed of the boat in still water is about 5.5 miles per hour.

Notice that in real-life situations most answers are not integral values.

WORK PROBLEMS   Let us do one example involving a work problem. Work problems were discussed in Section 7.5. You may wish to review that section before studying the next example.

EXAMPLE 2 Two sump pumps are emptying a flooded basement. Together they can empty the basement in 6 hours. One of the pumps has a slightly larger horsepower and could do the job by itself in 2 hours less time that the other pump could if it were working alone. How long would it take each pump to complete the job alone?

Solution: Recall from Section 7.5 that the rate of work multiplied by the time worked gives the part of the task completed.

Let t = number of hours for slower pump to complete the job by itself, then $t - 2$ = number of hours for the faster pump to complete the job by itself.

	Rate of Work	Time Worked	Part of Task Completed
Slower pump	$\dfrac{1}{t}$	6	$\dfrac{6}{t}$
Faster pump	$\dfrac{1}{t-2}$	6	$\dfrac{6}{t-2}$

$$\begin{pmatrix} \text{part of task} \\ \text{by slower pump} \end{pmatrix} + \begin{pmatrix} \text{part of task} \\ \text{by faster pump} \end{pmatrix} = 1$$

$$\frac{6}{t} + \frac{6}{t-2} = 1$$

$$t(t-2)\left(\frac{6}{t} + \frac{6}{t-2}\right) = t(t-2)(1)$$
Multiply both sides of equation by the LCD $t(t-2)$.

$$t(t-2)\left(\frac{6}{t}\right) + t(t-2)\left(\frac{6}{t-2}\right) = t^2 - 2t$$
Distributive property.

$$6(t-2) + 6t = t^2 - 2t$$

$$6t - 12 + 6t = t^2 - 2t$$

$$t^2 - 14t + 12 = 0$$

Using the quadratic formula, we obtain:

$$t = \frac{14 \pm \sqrt{148}}{2}$$

$$t \approx 13.1 \quad \text{or} \quad t \approx 0.9$$

Both 13.1 and 0.9 satisfy the equation $\dfrac{6}{x} + \dfrac{6}{x-2} = 1$ (with some round-off error). However, if we accept 0.9 as a solution, then the faster pump could complete the task in a negative time ($x - 2 = 0.9 - 2 = -1.1$ hours), which is not possible. Therefore, 0.9 hour is not an acceptable solution. The only solution is 13.1 hours. The slower pump takes approximately 13.1 hours by itself, and the faster pump takes approximately $13.1 - 2$ or 11.1 hours by itself to empty the basement.

Equations Quadratic in Form

2 In Section 6.5 we introduced solving equations quadratic in form. Recall that an *equation is quadratic in form* if it can be expressed as

$$au^2 + bu + c = 0, \quad a \neq 0$$

Now we will solve additional equations that are quadratic in form. To solve equations quadratic in form we use the following procedure.

To Solve Equations Quadratic in Form

1. If necessary, rewrite the equation in descending order of the variable with one side of the equation equal to 0.
2. Rewrite the variable in the highest-degree term as the square of the variable in the middle term.
3. Make a substitution that will result in an equation of the form $au^2 + bu + c = 0$, $a \neq 0$ where u is a function of the original variable.
4. Solve the equation $au^2 + bu + c = 0$ for u by factoring, by the quadratic formula, or by completing the square.
5. Replace u with the function of the original variable from step 3.
6. Solve the resulting equation for the original variable.
7. Check for extraneous solutions by substituting the apparent solutions in the original equation.

We will illustrate this procedure in Example 3.

EXAMPLE 3   Solve the equation $p^4 + 2p^2 = 8$.

Solution:

STEP 1: $\qquad p^4 + 2p^2 - 8 = 0$ **Set equation equal to 0.**

STEP 2: $\qquad (p^2)^2 + 2p^2 - 8 = 0$ **Write p^4 as $(p^2)^2$ to obtain equation in desired form.**

STEP 3:   Now let $u = p^2$. This gives an equation quadratic in form.

$$u^2 + 2u - 8 = 0$$

STEP 4: $\qquad (u + 4)(u - 2) = 0$ **Solve the equation.**

$$u + 4 = 0 \qquad \text{or} \qquad u - 2 = 0$$
$$u = -4 \qquad\qquad\quad u = 2$$

STEP 5: $\qquad p^2 = -4 \qquad\qquad p^2 = 2$ **Replace u with p^2. Solve for p.**

STEP 6: $\qquad p = \pm\sqrt{-4} \qquad\quad p = \pm\sqrt{2}$

$$p = \pm 2i$$

STEP 7:   Check the four possible solutions in the original equation.

$$p = 2i$$
$$p^4 + 2p^2 = 8$$
$$(2i)^4 + 2(2i)^2 = 8$$
$$2^4 i^4 + 2(2^2)(i^2) = 8$$
$$16(1) + 8(-1) = 8$$
$$16 - 8 = 8$$
true

$$p = -2i$$
$$p^4 + 2p^2 = 8$$
$$(-2i)^4 + 2(-2i)^2 = 8$$
$$(-2)^4 i^4 + 2(-2)^2 i^2 = 8$$
$$16(1) + 8(-1) = 8$$
$$16 - 8 = 8$$
true

$$p = \sqrt{2}$$
$$p^4 + 2p^2 = 8$$
$$(\sqrt{2})^4 + 2(\sqrt{2})^2 = 8$$
$$4 + 2(2) = 8$$
$$8 = 8$$
true

$$p = 2\sqrt{2}$$
$$p^4 + 2p^2 = 8$$
$$(-\sqrt{2})^4 + 2(-\sqrt{2})^2 = 8$$
$$4 + 2(2) = 8$$
$$8 = 8$$
true

Thus, the solutions are $2i$, $-2i$, $\sqrt{2}$ and $-\sqrt{2}$.

The solutions to equations like $p^4 + 2p^2 = 8$ will always check unless a mistake has been made. In equations like this, extraneous solutions will not be introduced. However, extraneous solutions may be introduced when working with rational exponents, as will be explained shortly.

> ### Helpful Hint
>
> Students sometimes solve the equation for u but then forget to complete the problem by solving for the original variable. Remember that if the original equation is in x you must obtain values for x. If the original equation is in p (as in Example 3) you must obtain values for p.

EXAMPLE 4   Solve the equation $6x^4 + 19x^2 - 7 = 0$.

Solution:
$$6x^4 + 19x^2 - 7 = 0$$
$$6(x^2)^2 + 19x^2 - 7 = 0$$

If we let $u = x^2$, the equation becomes

$$6u^2 + 19u - 7 = 0$$
$$(3u - 1)(2u + 7) = 0$$
$$3u - 1 = 0 \quad \text{or} \quad 2u + 7 = 0$$
$$3u = 1 \qquad\qquad 2u = -7$$
$$u = \frac{1}{3} \qquad\qquad u = 2\frac{7}{2}$$

Now substitute back x^2 for u and use the square root property.

$$x^2 = \frac{1}{3} \qquad\qquad x^2 = -\frac{7}{2}$$

$$x = \pm\sqrt{\frac{1}{3}} \qquad\qquad x = \pm\sqrt{2\frac{7}{2}}$$

$$x = \pm\frac{\sqrt{1}}{\sqrt{3}} \cdot \frac{\sqrt{3}}{\sqrt{3}} \qquad x = \pm\frac{\sqrt{-7}}{\sqrt{2}} \cdot \frac{\sqrt{2}}{\sqrt{2}}$$

$$= \pm\frac{\sqrt{3}}{3} \qquad\qquad = \pm\frac{\sqrt{-14}}{2} = \pm\frac{i\sqrt{14}}{2}$$

The solutions are $\pm\dfrac{\sqrt{3}}{3}$ and $\pm\dfrac{i\sqrt{14}}{2}$.

EXAMPLE 5   Solve the equation $2x^{-2} + x^{-1} - 1 = 0$.

Solution: This equation can be expressed as

$$2(x^{-1})^2 + x^{-1} - 1 = 0$$

Let $u = x^{-1}$, then the equation becomes

$$2u^2 + u - 1 = 0$$
$$(2u - 1)(u + 1) = 0$$

$$2u - 1 = 0 \qquad \text{or} \qquad u + 1 = 0$$

$$u = \frac{1}{2} \qquad\qquad\qquad u = -1$$

Now substitute x^{-1} for u.

$$x^{-1} = \frac{1}{2} \qquad \text{or} \qquad x^{-1} = -1$$

$$\frac{1}{x} = \frac{1}{2} \qquad\qquad \frac{1}{x} = -1$$

$$x = 2 \qquad\qquad\qquad x = -1$$

A check will show that both 2 and -1 are solutions to the original equation.

The equation in Example 5 could also be expressed as

$$\frac{2}{x^2} + \frac{1}{x} - 1 = 0$$

A second method to solve this equation is to multiply both sides of the equation by the least common denominator, x^2, then simplify.

$$x^2\left(\frac{2}{x^2} + \frac{1}{x} - 1\right) = x^2 \cdot 0$$
$$2 + x - x^2 = 0$$
$$x^2 - x - 2 = 0$$
$$(x - 2)(x + 1) = 0$$
$$x - 2 = 0 \qquad \text{or} \qquad x + 1 = 0$$
$$x = 2 \qquad\qquad\qquad x = -1$$

A third method to solve the equation $2x^{-2} + x^{-1} - 1 = 0$ is to factor the equation as $(2x^{-1} - 1)(x^{-1} + 1) = 0$ and setting each factor equal to 0. Try this now. Many of the equations solved in this section may be solved in more than one way.

Equations with Rational Exponents

3 To solve equations quadratic in form with rational exponents, we raise both sides of the equation to some power to eliminate the rational exponents. Recall that we did this in Section 8.6 when we solved radical equations. Whenever you raise both sides of an equation to a power, you may introduce extraneous solutions. **Therefore, whenever you raise both sides of an**

equation to a power, you must check all apparent solutions in the original equation to make sure that none are extraneous. We will now work two examples showing how to solve equations that contain rational exponents. We use the same procedure as used earlier.

EXAMPLE 6   Solve the equation $x^{2/5} + x^{1/5} - 6 = 0$.

Solution: This equation can be rewritten as

$$(x^{1/5})^2 + x^{1/5} - 6 = 0$$

Let $u = x^{1/5}$. Then the equation becomes

$$u^2 + u - 6 = 0$$
$$(u + 3)(u - 2) = 0$$
$$u + 3 = 0 \quad \text{or} \quad u - 2 = 0$$
$$u = -3 \qquad\qquad u = 2$$

Now substitute back $x^{1/5}$ for u and raise both sides of the equation to the fifth power to remove the rational exponents.

$$x^{1/5} = -3 \quad \text{or} \quad x^{1/5} = 2$$
$$(x^{1/5})^5 = (-3)^5 \qquad (x^{1/5})^5 = 2^5$$
$$x = -243 \qquad\qquad x = 32$$

The two *possible* solutions are -243 and 32. Remember that whenever we raise both sides of an equation to a power, as we did here, we need to check for extraneous solutions.

Check: $x = -243$ $\qquad\qquad\qquad\qquad$ $x = 32$

$$x^{2/5} + x^{1/5} - 6 = 0 \qquad\qquad x^{2/5} + x^{1/5} - 6 = 0$$
$$(-243)^{2/5} + (-243)^{1/5} - 6 = 0 \qquad (32)^{2/5} + (32)^{1/5} - 6 = 0$$
$$\left(\sqrt[5]{-243}\right)^2 + \sqrt[5]{-243} - 6 = 0 \qquad \left(\sqrt[5]{32}\right)^2 + \sqrt[5]{32} - 6 = 0$$
$$(-3)^2 - 3 - 6 = 0 \qquad\qquad 4 + 2 - 6 = 0$$
$$9 - 9 = 0 \qquad\qquad\qquad\qquad \textbf{true}$$
$$\textbf{true}$$

Since both values check the solutions are -243 and 32.

EXAMPLE 7   Solve the equation $2p - \sqrt{p} - 10 = 0$.

Solution: We can express this equation as

$$2p - p^{1/2} - 10 = 0$$
$$2(p^{1/2})^2 - p^{1/2} - 10 = 0$$

If we let $u = p^{1/2}$, this equation is quadratic in form.

$$2u^2 - u - 10 = 0$$
$$(2u - 5)(u + 2) = 0$$
$$2u - 5 = 0 \quad \text{or} \quad u + 2 = 0$$
$$2u = 5 \qquad\qquad u = -2$$
$$u = \frac{5}{2}$$

However, since our original equation is in the variable p we must solve for p. Now substitute $p^{1/2}$ for u.

$$p^{1/2} = \frac{5}{2} \qquad\qquad p^{1/2} = -2$$

Now square both sides of the equation.

$$(p^{1/2})^2 = \left(\frac{5}{2}\right)^2 \qquad\qquad (p^{1/2})^2 = (-2)^2$$

$$p = \frac{25}{4} \qquad\qquad p = 4$$

We must now check both apparent solutions in the original equation.

$$Check: \quad p = \frac{25}{4} \qquad\qquad\qquad p = 4$$

$$2p - \sqrt{p} - 10 = 0 \qquad\qquad 2p - \sqrt{p} - 10 = 0$$

$$2\left(\frac{25}{4}\right) - \sqrt{\frac{25}{4}} - 10 = 0 \qquad 2(4) - \sqrt{4} - 10 = 0$$

$$\frac{25}{2} - \frac{5}{2} - 10 = 0 \qquad\qquad 8 - 2 - 10 = 0$$

$$0 = 0 \quad \text{true} \qquad\qquad\qquad -4 = 0 \quad \text{false}$$

Since 4 does not check, it is an extraneous solution. The only solution is 25/4.

Example 7 could also be solved by writing the equation as $\sqrt{p} = 2p - 10$ and squaring both sides of the equation. Try this now.

Exercise Set 9.3

Solve.

1. $1 - \dfrac{5}{x} - \dfrac{6}{x^2} = 0$ 6, -1

2. $4 + \dfrac{9}{b} + \dfrac{2}{b^2} = 0$ $-2, -\dfrac{1}{4}$

3. $1 - \dfrac{3}{x} + \dfrac{2}{x^2} = 0$ 1, 2

4. $\dfrac{8}{x^2} - \dfrac{5}{x} + 1 = 0$ $\dfrac{5 \pm i\sqrt{7}}{2}$

5. $\dfrac{3}{x-2} + \dfrac{4}{(x-2)^2} = 1$ 1, 6

6. $\dfrac{2}{a} + \dfrac{a}{a-2} = 4$ $\dfrac{5 \pm \sqrt{13}}{3}$

7. $2(p+3) - 4 = \dfrac{3}{p+3}$ $\dfrac{-4 \pm \sqrt{10}}{2}$

8. $\dfrac{6}{r} = \dfrac{r}{r+1} + 1$ $\dfrac{5 \pm \sqrt{73}}{4}$

9. $2 = \dfrac{-2}{x} - \dfrac{2}{x^2}$ $\dfrac{-1 \pm i\sqrt{3}}{2}$

10. $\dfrac{3}{b^2 - 4} = \dfrac{1}{b-2} + 2$ $\dfrac{-1 \pm \sqrt{73}}{4}$

11. $\dfrac{2(x-2)}{x+2} = \dfrac{x-3}{x}$ $\dfrac{3 \pm i\sqrt{15}}{2}$

12. $\dfrac{5(x-3)}{x} + 2 = x - 1$ 3, 5

Solve by substitution.

13. $x^4 + 2x^2 - 8 = 0$ $\pm\sqrt{2}, \pm 2i$

14. $p^4 + 4p^2 - 5 = 0$ $\pm 1, \pm i\sqrt{5}$

15. $y^4 - 7y^2 = -12$ $\pm 2, \pm\sqrt{3}$

16. $2x^4 = 3x^2 + 20$ $\pm 2, \pm\dfrac{i\sqrt{10}}{2}$

17. $-3a^4 - 14a^2 = -5$ $\pm\dfrac{\sqrt{3}}{3}, \pm i\sqrt{5}$

18. $x^{-2} + 6x^{-1} - 16 = 0$ $\dfrac{1}{2}, -\dfrac{1}{8}$

19. $r^{-2} + 6r^{-1} + 9 = 0$ $-\dfrac{1}{3}$

20. $x^{-2} - x^{-1} = 20$ $\dfrac{1}{5}, -\dfrac{1}{4}$

21. $2b^{-2} = -7b^{-1} - 3$ $-2, -\dfrac{1}{3}$

22. $6m^{-2} - m^{-1} - 12 = 0$ $-\dfrac{3}{4}, \dfrac{2}{3}$

Solve by substitution. Be sure to check for extraneous solutions.

23. $x^{2/3} - 5x^{1/3} + 6 = 0$ 8, 27

24. $x^{2/3} + 11x^{1/3} + 28 = 0$ $-64, -343$

25. $p - 13\sqrt{p} + 40 = 0$ 25, 64

26. $2a - 7\sqrt{a} - 30 = 0$ 36

27. $3b + b^{1/2} = 4$ 1

28. $x = -4\sqrt{x} + 12$ 4

43. If a vertical line can be drawn to intersect a graph at more than one point, the graph is not a function.
45. the set of values that can be used for the independent variable   **46.** the set of values obtained for the dependent variable
29. $c^{2/5} - 9c^{1/5} = -18$ 243, 7776   **30.** $y^{2/5} + 8y^{1/5} + 16 = 0$ -1024 **31.** $-2a - 5a^{1/2} + 3 = 0$ $\frac{1}{4}$
32. $r^{2/3} - 7r^{1/3} + 6 = 0$ 1, 216

33. Mr. Winkle jogs 6 miles, and then turns around and jogs back to his starting point. The first part of his jog he was going mostly uphill, so his speed was 2 miles per hour slower than his speed returning. If the total time he spent jogging was $1\frac{3}{4}$ hours, find his rate going and his rate returning.   going, 6 mph; returning, 8 mph

34. A truck driver is transporting a heavy load from Detroit, Michigan, to Chicago, Illnios. On his return trip to Detroit, since his truck is lighter, he averages 10 miles per hour faster than his trip down. If the distance traveled each way is 300 miles and the total time spent driving is 11 hours, find his speed going and coming.   50 mph going, 60 mph returning

35. Amit Patel flew his single-engine Cesna airplane 80 miles with the wind, then turned around and returned against the wind. If the wind speed was a constant 30 miles per hour and the total time going and coming was 1.3 hours, find the speed of the plane in still air.

36. Two mechanics take 6 hours to rebuild an engine when they work together. If they worked alone, the more experienced mechanic could complete the job 1 hour faster than the less experienced mechanic. How long would it take each of them to rebuild the engine working alone?   faster, 11.52 hr; slower, 12.52 hr

37. After a small oil spill, two cleanup ships were sent to siphon off the oil floating in Baffin Bay. The newer ship can clean up the entire spill by itself in 3 hours less time than the older ship takes by itself. Working together the two ships can clean up the oil spill in 8 hours. How long will it take the newer ship by itself to clean up the spill? ≈ 14.64 hr

SEE EXERCISE 37.

38. A small electric heater requires 6 minutes longer to raise the temperature in an unheated garage to a comfortable level than does a larger electric heater. Together the two heaters can raise the garage temperature to a comfortable level in 42 minutes. How long would it take each heater by itself to raise the temperature in the garage to a comfortable level?

The amount of money, A, in an account in which P dollars was invested at r percent interest compounded annually for 2 years may be found by the formula

$$A = P(1 + r)^2$$

39. If $600 grows to $660 in 2 years, find the interest rate. $\approx 4.88\%$

40. If $1000 grows to $1180 in 2 years, find the interest rate. $\approx 8.63\%$

35. 130 mph   **38.** larger, 81.1 min; smaller, 87.1 min

CUMULATIVE REVIEW EXERCISES

[2.2] **41.** Solve $S = 2\pi rh + 2\pi r^2$ for h. $h = \dfrac{S - 2\pi r^2}{2\pi r}$

[3.4] **42.** Is the set of ordered pairs $\{(2, 3), (4, 5), (6, 0), (2, 1)\}$ a function? Explain.   No

43. Explain how the vertical line test may be used to determine if a graph is a function.

44. If $f(x) = \dfrac{3}{2}x^3 - \dfrac{1}{2}x^2 + 3$, find $f\left(\dfrac{1}{3}\right)$.   3

45. What is the domain of a function?

46. What is the range of a function?

47. Is $f(x) = \dfrac{1}{x}$ a polynomial function? Explain.   No

Group Activity/ Challenge Problems

1. Solve the following equation (**a**) by multiplying both sides of the equation by the LCD and (**b**) by writing the equation with negative exponents and using substitution.

$$1 = \frac{2}{x} - \frac{2}{x^2}$$ (**a**) and (**b**) $1 \pm i$

Solve.

2. $4 - (x - 2)^{-1} = 3(x - 2)^{-2}$ $3, \frac{5}{4}$ **3.** $(x^2 + x)^2 - 18(x^2 + x) + 72 = 0$ $2, -3, 3, -4$

4. Sumatra drove from Lubbock, Texas, to Plainview, Texas, a distance of 60 miles. She then stopped for 2.5 hours to see a friend in Plainview before continuing her journey from Plainview to Amarillo, Texas, a distance of 100 miles. If she drove 10 miles per hour faster from Lubbock to Plainview and the total time of the trip was 5.5 hours, find her average speed from Lubbock to Plainview.   60 mph

9.4   Quadratic and Other Inequalities in One Variable

Tape 15

1 Solve quadratic inequalities.
2 Solve other inequalities.
3 Solve or check inequalities graphically.

In Section 2.5 we discussed linear inequalities in one variable. Now we discuss quadratic inequalities in one variable. This section will also review both interval notation and set builder notation, which were introduced in Sections 1.2 and 2.5.

When the equal sign in a quadratic equation of the form $ax^2 + bx + c = 0$ is replaced by an inequality sign, we get a quadratic inequality.

Examples of Quadratic Inequalities

$$x^2 + x - 12 > 0, \qquad 2x^2 - 9x - 5 \leq 0$$

The **solution to a quadratic inequality** is the set of all values that make the inequality a true statement. For example, if we substitute 5 for x in $x^2 + x - 12 > 0$, we obtain

$$x^2 + x - 12 > 0$$
$$5^2 + 5 - 12 > 0$$
$$25 + 5 - 12 > 0$$
$$18 > 0 \qquad \text{true}$$

The inequality is true when x is 5, therefore 5 satisfies the inequality. However, 5 is not the only solution, for there are other values that satisfy (or are solutions to) the inequality. Does 4 satisfy the inequality? Does 2 satisfy the inequality?

Solving Quadratic Inequalities

1 A number of methods can be used to find the solution to quadratic inequalities. We will begin by introducing a **sign graph.** Consider the inequality $x^2 + x - 12 > 0$. The first step in obtaining the solution is to factor the left side of the inequality

$$x^2 + x - 12 > 0$$
$$(x + 4)(x - 3) > 0$$

We now determine the values of x that make the product of the factors $x + 4$ and $x - 3$ positive. The product of two positive numbers is a positive number, and the product of two negative numbers is also a positive number. Therefore, we must find the values of x that make both factors positive numbers or both factors negative numbers. One way to do this is to use number lines. Consider the factor $x - 3$. For values of x less than 3, the value of $x - 3$ will be negative, and for values of x greater than 3, the value of $x - 3$ will be positive. This is illustrated in Fig. 9.7.

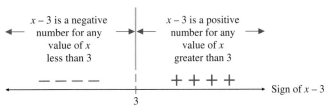

FIGURE 9.7

Now consider the factor $x + 4$. This factor will be negative when x is less than -4 and positive when x is greater than -4 (see Fig. 9.8).

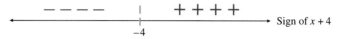

FIGURE 9.8

If we draw the two number lines together and draw a dashed vertical line through the values -4 and 3, we get the sign graph in Fig. 9.9. Note that when we place the two lines together the -4 is marked to the left of 3, since -4 is less than 3. Also note that the two vertical lines divide the sign graph into three vertical regions labeled *A, B,* and *C.*

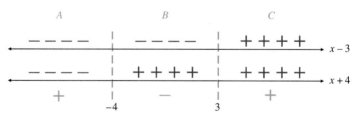

FIGURE 9.9

Next determine the sign of the product for each region. If the two signs in a vertical region are the same, $(-)(-)$ or $(+)(+)$, the product will be positive. If the signs in a vertical region are different, $(+)(-)$ or $(-)(+)$, the product will be negative (see Fig. 9.10).

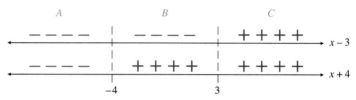

FIGURE 9.10

Since we are solving the inequality $x^2 + x - 12 > 0$, the solutions are the numbers in regions A and C. Any number less than -4 or greater than 3 is a solution to the inequality. Since the inequality symbol is $>$, the numbers -4 and 3, called the **boundary values,** do not satisfy the inequality. Thus, the solution in interval notation is $(-\infty, -4) \cup (3, \infty)$. The solution in set builder notation is $\{x \mid x < -4 \text{ or } x > 3\}$. The solution on the number line is illustrated in Fig. 9.11.

FIGURE 9.11

An alternative method to using the sign graph is to use a single number line. For example, consider again the inequality $x^2 + x - 12 > 0$. This inequality could also have been solved by setting $x^2 + x - 12$ equal to 0 and solving the resulting equation.

$$x^2 + x - 12 = 0$$
$$(x + 4)(x - 3) = 0$$
$$x + 4 = 0 \quad \text{or} \quad x - 3 = 0$$
$$x = -4 \qquad\qquad x = 3$$

As we saw with the sign graph in Fig. 9.10, the boundary values -4 and 3, break the number line into three regions, $(-\infty, -4)$, $(-4, 3)$, and $(3, -\infty)$, labeled A, B, and C, respectively.

Rather than drawing a sign graph we can draw a single number line containing these three regions, labeled A, B, and C, as illustrated in Fig. 9.12.

Next, select one test value in *each* region. Then substitute each of those numbers, one at a time, in either $x^2 + x - 12 > 0$ or $(x + 4)(x - 3) > 0$ and determine if they result in a true statement. If the test value results in a true statement, all values in that region will also satisfy the inequality. If the test value results in a false statement, no numbers in that region will satisfy the inequality.

In this example we will use the test values of -5 in region A, 0 in region B, and 4 in region C (see Fig. 9.13).

FIGURE 9.12

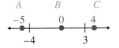

FIGURE 9.13

Region A, x = −5	**Region B, x = 0**	**Region C, x = 4**
Is $x^2 + x - 12 > 0$?	Is $x^2 + x - 12 > 0$?	Is $x^2 + x - 12 > 0$?
$(-5)^2 - 5 - 12 > 0$	$0 + 0 - 12 > 0$	$4^2 + 4 - 12 > 0$
$8 > 0$	$-12 > 0$	$8 > 0$
true	false	true

Since the test values in both regions A and C satisfy the inequality, the solution is all real numbers in regions A or C. The inequality symbol is $>$. The values -4 and 3 are not included in the solution since they make the inequality equal to 0. The solution is $(-\infty, -4) \cup (3, \infty)$, the same as we found earlier.

To Solve Quadratic and Other Inequalities

1. Write the inequality as an equation and solve the equation.

2. If the inequality contains a variable in any denominator, determine the value or values that make the denominator equal to 0.

3. Construct a number line. Mark each solution that is determined in step 1, and values obtained in step 2 on the number line. Make sure that you mark these values from the lowest value on the left to the greatest value on the right.

4. Select a test value in each region of the number line.

5. Test each value in step 4 in the inequality to determine if it satisfies the inequality.

6. Test each boundary value to determine if it is a solution to the inequality. Remember that division by 0 is not permitted.

7. Write the solution in the form requested by your instructor.

EXAMPLE 1   Graph the solution to $x^2 + 9x \leq -18$ on the number line.

Solution: First, change the inequality to the equation $x^2 + 9x = -18$, then solve.

$$x^2 + 9x + 18 = 0$$
$$(x + 6)(x + 3) = 0$$
$$x + 6 = 0 \quad \text{or} \quad x + 3 = 0$$
$$x = -6 \qquad\qquad x = -3$$

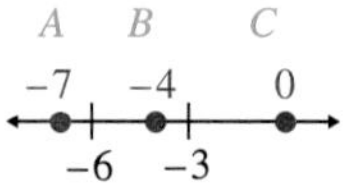

FIGURE 9.14

Now indicate the regions and the test values on the number line (see Fig. 9.14). In the figure we have illustrated test values at -7, -4, and 0.

Region A, x = −7	Region B, x = −4	Region C, x = 0
$x^2 + 9x \leq -18$	$x^2 + 9x \leq -18$	$x^2 + 9x \leq -18$
$49 - 63 \leq -18$	$16 - 36 \leq -18$	$0 + 0 \leq -18$
$-14 \leq -18$ false	$-20 \leq -18$ true	$0 \leq -18$ false

Only the values in region B satisfy the inequality. Since the original inequality symbol is $\leq$, the boundary values -6 and -3 satisfy the inequality. Therefore, the solution is $[-6, -3]$. The solution is illustrated on the number line in Fig. 9.15.

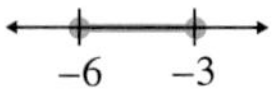

FIGURE 9.15

EXAMPLE 2 Solve the inequality $x^2 - 10x + 25 \geq 0$. Indicate the solution **(a)** on the number line, **(b)** in interval notation, and **(c)** in set builder notation.

Solution:
$$x^2 - 10x + 25 = 0$$
$$(x - 5)(x - 5) = 0$$
$$x - 5 = 0 \quad \text{or} \quad x - 5 = 0$$
$$x = 5 \qquad\qquad x = 5$$

FIGURE 9.16

Since both factors are the same there is only one boundary value (see Fig. 9.16). Both test values, 4 and 6, result in true statements.

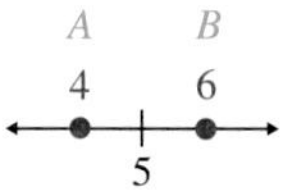

Region A, x = 4	Region B, x = 6
$x^2 - 10x + 25 \geq 0$	$x^2 - 10x + 25 \geq 0$
$16 - 40 + 25 \geq 0$	$36 - 60 + 25 \geq 0$
$1 \geq 0$ true	$1 \geq 0$ true

In this special case, both factors are the same. The solution is the set of real numbers, $\mathbb{R}$.

(a) 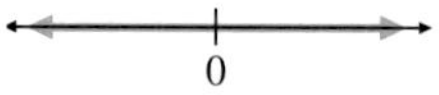 **(b)** $(-\infty, \infty)$ **(c)** $\{x \mid -\infty < x < \infty\}$

In Example 2, if we consider the inequality $x^2 - 10x + 25 \geq 0$ as $(x - 5)^2 \geq 0$ we can see that the solution must be the set of real numbers, since $(x - 5)^2$ must be greater than or equal to 0 for any real number x.

EXAMPLE 3 Solve the inequality $x^2 - 2x - 4 \geq 0$. Express the solution in interval notation.

Solution: First we need to solve the equation $x^2 - 2x - 4 = 0$. Since this equation is not factorable, we use the quadratic formula to solve.

$$x = \frac{-b \pm \sqrt{b^2 - 4ac}}{2a}$$

$$= \frac{2 \pm \sqrt{4 - 4(1)(-4)}}{2(1)} = \frac{2 \pm \sqrt{20}}{2} = \frac{2 \pm 2\sqrt{5}}{2} = 1 \pm \sqrt{5}$$

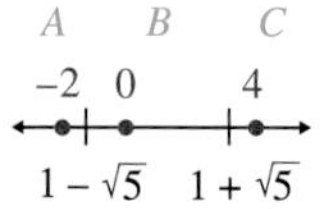

The value of $1 - \sqrt{5}$ is about -1.24 and the value of $1 + \sqrt{5}$ is about 3.24. We will select test values of -2, 0, and 4 (see Fig. 9.17).

Region A, $x = -2$	Region B, $x = 0$	Region C, $x = 4$
$x^2 - 2x - 4 \geq 0$	$x^2 - 2x - 4 \geq 0$	$x^2 - 2x - 4 \geq 0$
$4 + 4 - 4 \geq 0$	$0 - 0 - 4 \geq 0$	$16 - 8 - 4 \geq 0$
$4 \geq 0$	$-4 \geq 0$	$4 \geq 0$
true	false	true

$1 - \sqrt{5}$ $1 + \sqrt{5}$

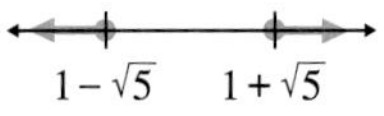

FIGURE 9.18

The boundary values are part of the solution because the inequality symbol is $\geq$ and the boundary values make the inequality equal to 0. Thus, the solution in interval notation is $\left(-\infty, 1 - \sqrt{5}\right] \cup \left[1 + \sqrt{5}, \infty\right)$. The solution is illustrated on the number line in Fig. 9.18.

Helpful Hint

Notice in Example 1, which may be written $x^2 + 9x + 18 \leq 0$, the solution was the middle region of the number line, region B. In Example 3, $x^2 - 2x - 4 \geq 0$, the solution was the union of the two outer regions of the number line, region $A \cup$ region C. All quadratic inequalities, of the form $ax^2 + bx + c \leq 0$, where $a > 0$, *with two distinct real solutions,* will have the middle region of the number line as its solution. All quadratic inequalities of the form $ax^2 + bx + c \geq 0$, where $a > 0$, *with two distinct real solutions* will have the union of the two outer regions of the number line as its solutions. Can you explain why this must be true?

Example 2 does not have two *distinct* real solutions. Therefore, this Helpful Hint does not apply to that example.

Solving Other Inequalities

2 A procedure similar to the one used earlier can be used to solve other inequalities, as illustrated in the following examples.

EXAMPLE 4   Solve the inequality $(3x - 2)(x + 3)(x + 5) < 0$. Illustrate the solution on the number line and write the solution in set builder notation.

Solution: Use the zero factor property to solve the equation $(3x - 2)(x + 3)(x + 5) = 0$.

$$3x - 2 = 0 \quad \text{or} \quad x + 3 = 0 \quad \text{or} \quad x + 5 = 0$$

$$x = \frac{2}{3} \qquad\qquad x = -3 \qquad\qquad x = -5$$

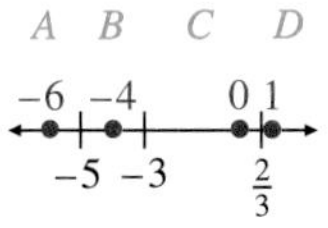

A   B   C   D

-6 -4 0 1

-5 -3 $\frac{2}{3}$

FIGURE 9.19

The solutions $\frac{2}{3}$, -3, and -5 break the number line into four regions (see Fig. 9.19). The test values we will use are -6, -4, 0 and 1.

Region A, $x = -6$	Region B, $x = -4$
$(3x - 2)(x + 3)(x + 5) < 0$	$(3x - 2)(x + 3)(x + 5) < 0$
$(-20)(-3)(-1) < 0$	$(-14)(-1)(1) < 0$
$-60 < 0$ true	$14 < 0$ false

Region C, $x = 0$	Region D, $x = 1$
$(3x - 2)(x + 3)(x + 5) < 0$	$(3x - 2)(x + 3)(x + 5) < 0$
$(-2)(3)(5) < 0$	$(1)(4)(6) < 0$
$-30 < 0$ true	$24 < 0$ false

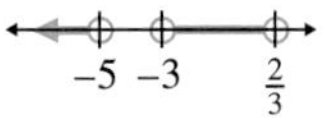

FIGURE 9.20

Since the original inequality symbol is $<$, the boundary values are not part of the solution. The solution, regions A and C, is illustrated on the number line in Fig. 9.20. The solution in set builder notation is $\left\{x \mid x < -5 \text{ or } -3 < x < \frac{2}{3}\right\}$.

In Examples 5 through 8 we solve **rational inequalities,** which are inequalities that contain a rational expression.

EXAMPLE 5 Solve the inequality $\dfrac{x + 3}{x - 4} \le 0$ and graph the solution on the number line.

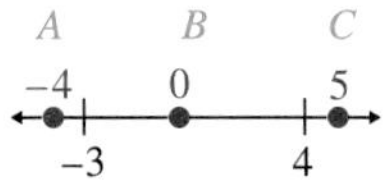

FIGURE 9.21

Solution: The solution to the equation $\dfrac{x + 3}{x - 4} = 0$ is -3 since -3 makes the numerator 0. The number 4 makes the denominator equal to 0. We use these two numbers, -3 and 4, when setting up our number line (see Fig. 9.21). The test values we selected are -4, 0, and 5.

Region A, x = −4	*Region B, x = 0*	*Region C, x = 5*
$\dfrac{x + 3}{x - 4} \le 0$	$\dfrac{x + 3}{x - 4} \le 0$	$\dfrac{x + 3}{x - 4} \le 0$
$\dfrac{1}{8} \le 0$ **false**	$-\dfrac{3}{4} \le 0$ true	$8 \le 0$ **false**

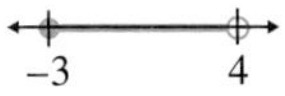

The values in region B satisfy the inequality. Now check the boundary values -3 and 4. Since -3 results in the inequality being true, -3 is a solution. Since division by 0 is not permitted, 4 is not a solution. Thus, the solution is $[-3, 4)$. The solution is illustrated on the number line in Fig. 9.22.

FIGURE 9.22

EXAMPLE 6 Solve the inequality $\dfrac{6}{x - 2} > 4$ and write the solution in interval notation.

Solution: First solve the equation $\dfrac{6}{x - 2} = 4$.

$$\frac{6}{x - 2} = 4$$
$$6 = 4(x - 2)$$
$$6 = 4x - 8$$
$$14 = 4x$$
$$\frac{7}{2} = x$$

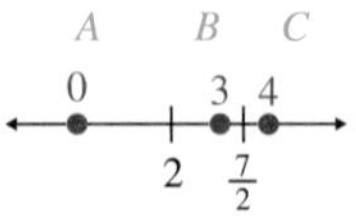

The equation $\dfrac{6}{x - 2} = 4$ is not defined at 2. Therefore, we mark the values 7/2 and 2 on the number line, which creates 3 regions (see Fig. 9.23). Our test values will be 0, 3, and 4.

Region A, x = 0	*Region B, x = 3*	*Region C, x = 4*
$\dfrac{6}{x - 2} > 4$	$\dfrac{6}{x - 2} > 4$	$\dfrac{6}{x - 2} > 4$
$-3 > 4$ **false**	$6 > 4$ true	$3 > 4$ **false**

FIGURE 9.23

The values between 2 and 7/2 satisfy the inequality. Since the inequality symbol is $>$ the number 7/2 is not a part of the solution. When $x = 7/2$ the value of $6/(x - 2)$ is 4. The denominator is 0 when $x = 2$, therefore 2 is not a part of the solution. The solution in interval notation is $(2, 7/2)$.

Helpful Hint

When solving the inequality in Example 6, $\dfrac{6}{x - 2} > 4$, your first reaction might be to multiply both sides of the inequality by $x - 2$. This is wrong. Do you know why? Remember that if we multiply both sides of an inequality by a negative number we must change the direction of the equality symbol. Since we do not know the value of x, we do not know whether $x - 2$ represents a positive or negative number. Therefore, it would be incorrect to multiply both sides of the inequality by $x - 2$.

EXAMPLE 7 Solve the inequality $\dfrac{x}{x + 4} < 1$.

Solution: If we solve this equation we see that there is no solution.

$$\frac{x}{x + 4} = 1$$
$$x = 1(x + 4)$$
$$x = x + 4$$
$$0 = 4 \quad \text{false}$$

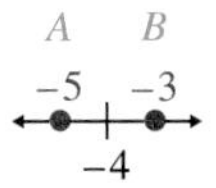

FIGURE 9.24

Since the number -4 makes the denominator 0 the number line will contain two regions (see Fig. 9.24). Our test points will be -5 and -3.

Region A, $x = -5$	*Region B, $x = -3$*
$\dfrac{x}{x + 4} < 1$	$\dfrac{x}{x + 4} < 1$
$5 < 1$ **false**	$-3 < 1$ **true**

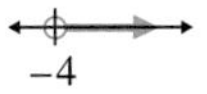

FIGURE 9.25

The solution, the values greater than -4, region B, is illustrated on the number line in Fig. 9.25. The solution in interval notation is $(-4, \infty)$.

EXAMPLE 8 Solve the inequality $\dfrac{(x - 3)(x + 4)}{x + 1} \geq 0$. Illustrate the solution on the number line and give the solution in interval notation.

Solution: The solution to the equation $\dfrac{(x - 3)(x + 4)}{x + 1} = 0$ is 3 and -4 since these are the values that make the numerator equal to 0. The equation is not defined at -1. We, therefore, use the values 3, -4, and -1 to determine the regions on the number line (see Fig. 9.26). Checking test values at $-5, -2, 0$ and 4 we find that the values in regions B and D, $-4 < x < -1$

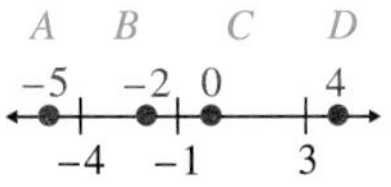

FIGURE 9.26

FIGURE 9.27

and $x > 3$, satisfy the inequality. Check the test values yourself to verify this. The values 3 and -4 make the inequality equal to 0 and are part of the solution. The inequality is not defined at -1, so -1 is not part of the solution. The solution is $[-4, -1) \cup [3, \infty)$. The solution is illustrated on the number line in Fig. 9.27.

Helpful Hint

When checking test values it is generally not necessary to find the exact values. When one side of the inequality is > 0, ≥ 0, < 0, or ≤ 0 we need only to determine the sign of the other side of the inequality at the test value. For example, if one factor of an expression is $x - 3$ and the test value is -5, this factor would have a negative value since $-5 - 3$ is negative. We may test the test values in Example 8 as follows.

Region A, $x = -5$	*Region B, $x = -2$*	*Region C, $x = 0$*	*Region D, $x = 4$*
$\dfrac{(x-3)(x+4)}{x+1} \geq 0$	$\dfrac{(x-3)(x+4)}{x+1} \geq 0$	$\dfrac{(x-3)(x+4)}{x+1} \geq 0$	$\dfrac{(x-3)(x+4)}{x+1} \geq 0$
$\dfrac{(-)(-)}{(-)} \geq 0$	$\dfrac{(-)(+)}{(-)} \geq 0$	$\dfrac{(-)(+)}{(+)} \geq 0$	$\dfrac{(+)(+)}{(+)} \geq 0$
$\dfrac{(+)}{(-)} \geq 0$	$\dfrac{(-)}{(-)} \geq 0$	$\dfrac{(-)}{(+)} \geq 0$	$\dfrac{(+)}{(+)} \geq 0$
$(-) \geq 0$	$(+) \geq 0$	$(-) \geq 0$	$(+) \geq 0$
false	**true**	**false**	**true**

We see that the values in regions B and D satisfy the inequality, as we found in Example 8.

Sometimes it may be convenient to factor an expression and use the factored form to check the test values. Check the test values in Example 1 by factoring the left side of the inequality $x^2 + 9x + 18 \leq 0$, substituting the test values, and noting the signs.

Graphical Solutions to Inequalities

3 Consider the inequality solved at the beginning of this section, $x^2 + x - 12 > 0$. We found the solution $(-\infty, -4) \cup (3, \infty)$, that is, $x < -4$ or $x > 3$. Let us graph the related function $y = x^2 + x - 12$ (see Fig. 9.28).

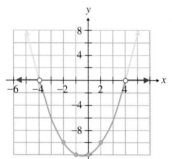

FIGURE 9.28

If you study the graph in Fig. 9.28 you will see that the graph lies above the x axis when $x < -4$ or $x > 3$. Recall that these are the values of x that satisfy the inequality. When the graph in Fig. 9.28 lies above the x axis, $y > 0$, so $x^2 + x - 12 > 0$. Therefore, the solution to the inequality $x^2 + x - 12 > 0$ may be found graphing the function $y = x^2 + x - 12$ and observing the values of x for which the graph lies above the x axis.

In Example 8 we solved the inequality $\dfrac{(x - 3)(x + 4)}{x + 1} \geq 0$. The solution is

$-4 \leq x < -1$ or $x > 3$. The graph of the function $y = \dfrac{(x - 3)(x + 4)}{x + 1}$ is

illustrated in Fig. 9.29. This graph can be obtained quickly with a graphing calculator.

The graph is greater than or equal to 0 where the graph intersects or rises

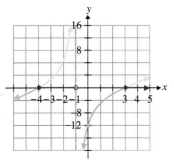

FIGURE 9.29

above the x axis. From the graph we see that $y > 0$, or $\dfrac{(x - 3)(x + 4)}{x + 1} \geq 0$, in the intervals $-4 \leq x < -1$ and $x \geq 3$. This is consistent with the answer obtained in Example 8.

The solution to the inequality $\dfrac{(x - 3)(x + 4)}{x + 1} < 0$ would be the values of x where the graph in Fig. 9.29 is below the x axis, $x < -4$ or $-1 < x < 3$. The domain of the function $y = \dfrac{(x - 3)(x + 4)}{x + 1}$ is the set of all real numbers except -1.

Observe the dashed line at $x = -1$ on the graph in Fig. 9.29. This line, called a *vertical asymptote,* is not a part of the graph. We often add the vertical asymptote at values of x where the graph is undefined as an aid in drawing and interpreting the graph. We see that one part of the graph decreases as x gets close to -1 while the other part of the graph increases as x gets close to -1. Note that there is no value of y for $x = -1$, which is what we expect since the function is undefined at $x = -1$.

When solving or checking an inequality graphically pay particular attention to the boundary values. You need to refer back to the original inequality symbol to determine which boundary values are part of the solution.

Using the information presented in this objective, you can use a graphing calculator to (1) check your solutions to inequalities, or (2) to solve inequalities in one variable graphically.

Graphing Calculator Corner

EXERCISES

Using the information presented in objective 3, solve the inequalities graphically. Use the trace and zoom features, and round solutions to the nearest hundredth.

1. $1.3x^2 - 2.5x - 6.2 > 0$
   $x < -1.42$ or $x > 3.35$

2. $(2x - 3.4)(x - 1.7) < 0$
   no solution

3. $\dfrac{x - 1}{(x - 2)(x + 3)} \geq 0$
   $-3 < x \leq 1$ or $x > 2$

4. $\dfrac{(3 - 4x)(1.2x - 4)}{2.7x - 5.4} > 0$
   $x < 0.75$ or $2.00 < x < 3.33$

Exercise Set 9.4

1. [number line: -2, 5]   **2.** [number line: -7, -1]

Solve each inequality and graph the solution on the number line.

1. $x^2 - 3x - 10 \geq 0$

2. $x^2 + 8x + 7 < 0$

3. $x^2 + 4x > 0$ [number line: -4, 0]

4. $x^2 - 5x \geq 0$ [number line: 0, 5]

5. $x^2 - 16 < 0$ [number line: -4, 4]

6. $y^2 - 25 \leq 0$ [number line: -5, 5]

7. $2x^2 + 5x - 3 \geq 0$ [number line: -3, $\frac{1}{2}$]

8. $2x^2 > 13x - 6$ [number line: $\frac{1}{2}$, 6]

9. $3n^2 - 7n \leq 6$ [number line: $-\frac{2}{3}$, 3]

10. $a^2 \geq -9a - 14$ [number line: -7, -2]

11. $x^2 \leq 2x + 35$ [number line: -5, 7]

12. $x^2 \geq -4x$ [number line: -4, 0]

13. $6r^2 < 36$ $-\sqrt{6} \quad \sqrt{6}$

17. $\dfrac{6-3\sqrt{2}}{2} \quad \dfrac{6+3\sqrt{2}}{2}$

14. $5x^2 + 19x \le 4$ $-4 \quad \dfrac{1}{5}$

15. $x^2 - x > 5$ $\dfrac{1-\sqrt{21}}{2} \quad \dfrac{1+\sqrt{21}}{2}$

16. $3x^2 + 5x - 3 \le 0$ $\dfrac{-5-\sqrt{61}}{6} \quad \dfrac{-5+\sqrt{61}}{6}$

17. $2x^2 - 12x + 9 \le 0$

18. $6w^2 > 5w + 6$ $-\dfrac{2}{3} \quad \dfrac{3}{2}$

19. $4x^2 - 11x \le 20$ $-\dfrac{5}{4} \quad 4$

20. $5x^2 \le -20x - 4$

Solve each inequality and give the solution in interval notation.

21. $(x - 1)(x + 1)(x + 4) > 0$ $(-4, -1) \cup (1, \infty)$

22. $(x - 3)(x + 2)(x + 5) \le 0$ $(-\infty, -5] \cup [-2, 3]$

23. $(b - 4)(b - 1)(b + 3) \le 0$ $(-\infty, -3] \cup [1, 4]$

24. $(2x - 4)(x + 3)(x + 6) > 0$ $(-6, -3) \cup (2, \infty)$

25. $x(x - 3)(2x + 6) \ge 0$ $[-3, 0] \cup [3, \infty)$

26. $(x - 3)(x + 4)(x - 2) \le 0$ $(-\infty, -4] \cup [2, 3]$

27. $(2c + 5)(3c - 6)(c + 6) > 0$ $(-6, -\frac{5}{2}) \cup (2, \infty)$

28. $(2p - 1)(p + 5)(3p + 6) \ge 0$ $[-5, -2] \cup [\frac{1}{2}, \infty)$

29. $(x + 2)(x + 2)(3x - 8) \ge 0$ $[\frac{8}{3}, \infty)$

30. $(x + 3)^2 (4x - 5) \le 0$ $(-\infty, \frac{5}{4}]$

31. $(5x + 3)(5x + 3)(x - 4) < 0$ $(-\infty, -\frac{3}{5}) \cup (-\frac{3}{5}, 4)$

32. $x(x - 5)(x + 5) > 0$ $(-5, 0) \cup (5, \infty)$

Solve each inequality and give the solution in set builder notation.   **33.** $\{x \mid x < -3 \text{ or } x > 1\}$ **36.** $\{x \mid x \le -6 \text{ or } x > -2\}$

33. $\dfrac{x + 3}{x - 1} > 0$

34. $\dfrac{x - 5}{x + 2} < 0$ $\{x \mid -2 < x < 5\}$

35. $\dfrac{y - 4}{y - 1} \le 0$ $\{y \mid 1 < y \le 4\}$

36. $\dfrac{x + 6}{x + 2} \ge 0$

37. $\dfrac{2x - 4}{x - 1} < 0$ $\{x \mid 1 < x < 2\}$

38. $\dfrac{3x + 6}{x + 4} \ge 0$

39. $\dfrac{3a + 6}{2a - 1} \ge 0$

40. $\dfrac{3x + 4}{2x - 1} < 0$

41. $\dfrac{x + 4}{x - 4} \le 0$ $\{x \mid -4 \le x < 4\}$

42. $\dfrac{k + 3}{k} \ge 0$

43. $\dfrac{4x - 2}{2x - 4} > 0$

44. $\dfrac{3x + 5}{x - 2} \le 0$

38. $\{x \mid x < -4 \text{ or } x \ge -2\}$ **39.** $\{a \mid a \le -2 \text{ or } a > \frac{1}{2}\}$ **40.** $\{x \mid -\frac{4}{3} < x < \frac{1}{2}\}$ **42.** $\{k \mid k \le -3 \text{ or } k > 0\}$ **43.** $\{x \mid x < \frac{1}{2} \text{ or } x > 2\}$

Solve each inequality and give the solution in interval notation.   **44.** $\{x \mid -\frac{5}{3} \le x < 2\}$ **45.** $(-\infty, -6) \cup (-2, 4)$

45. $\dfrac{(x + 2)(x - 4)}{x + 6} < 0$

46. $\dfrac{(x - 3)(x - 6)}{x + 4} \ge 0$ $(-4, 3] \cup [6, \infty)$

47. $\dfrac{(w - 6)(w - 1)}{w - 3} \ge 0$ $[1, 3) \cup [6, \infty)$

48. $\dfrac{x + 6}{(x - 2)(x + 4)} > 0$ $(-6, -4) \cup (2, \infty)$

49. $\dfrac{x - 6}{(x + 4)(x - 1)} \le 0$ $(-\infty, -4) \cup (1, 6]$

50. $\dfrac{x}{(x + 3)(x - 3)} \le 0$ $(-\infty, -3) \cup [0, 3)$

51. $\dfrac{(x - 3)(2x + 5)}{(x - 6)} > 0$ $(-\frac{5}{2}, 3) \cup (6, \infty)$

52. $\dfrac{r(r - 3)}{2r + 6} < 0$ $(-\infty, -3) \cup (0, 3)$

53. $\dfrac{(z + 2)(2z - 3)}{z} \ge 0$ $[-2, 0) \cup [\frac{3}{2}, \infty)$

54. $\dfrac{(x - 4)(3x - 2)}{x + 2} \le 0$ $(-\infty, \ 2) \cup [\frac{2}{3}, 4]$

Solve each inequality and graph the solution on the number line.

55. $\dfrac{2}{x - 3} \ge -1$ $1 \quad 3$

56. $\dfrac{3}{m - 1} \le -1$ $-2 \quad 1$

57. $\dfrac{4}{x - 2} \ge 2$ $2 \quad 4$

58. $\dfrac{2}{2a - 1} > 2$ $\dfrac{1}{2} \quad 1$

59. $\dfrac{2p - 5}{p - 4} \le 1$ $1 \quad 4$

60. $\dfrac{2x}{x + 1} > 1$ $-1 \quad 1$

61. $\dfrac{w}{3w - 2} > -2$ $\dfrac{2}{3} \quad$

62. $\dfrac{x - 1}{2x + 6} \le -3$ $-3 \quad -\dfrac{17}{7}$

63. $\dfrac{x}{3x - 1} < -1$ $\dfrac{1}{4} \quad \dfrac{1}{3}$

64. $\dfrac{4x - 5}{x + 3} < 3$ $-3 \quad 14$

65. $\dfrac{x + 8}{x + 2} > 1$ -2

66. $\dfrac{4x + 2}{2x - 3} \ge 2$ $\dfrac{3}{2}$

67. The graph of $y = \dfrac{x^2 - 4x + 4}{x - 4}$ is sketched below. Determine the solution to the following inequalities.

68. The graph of $y = \dfrac{x^2 + x - 6}{x - 4}$ is illustrated below. Determine the solution to the following inequalities.

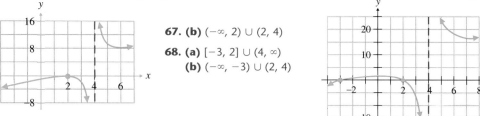

67. (b) $(-\infty, 2) \cup (2, 4)$

68. (a) $[-3, 2] \cup (4, \infty)$
     (b) $(-\infty, -3) \cup (2, 4)$

(a) $\dfrac{x^2 - 4x + 4}{x - 4} > 0$ $(4, \infty)$ **(b)** $\dfrac{x^2 - 4x + 4}{x - 4} < 0$

Explain how you determined your answer.

(a) $\dfrac{x^2 + x - 6}{x - 4} \ge 0$ **(b)** $\dfrac{x^2 + x - 6}{x - 4} < 0$

Explain how you determined your answer.

69. The graph of $y = \dfrac{(x-2)(x-2)(x+4)}{x+2}$ is illustrated below. Determine the solution to the following inequalities.

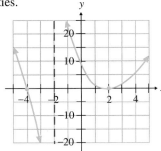

(a) $\dfrac{(x-2)(x-2)(x+4)}{x+2} \geq 0$ $(-\infty, -4] \cup (-2, \infty)$

(b) $\dfrac{(x-2)(x-2)(x+4)}{x+2} \leq 0$ $[-4, -2)$
Explain how you determined your solution.

70. What is the solution to the inequality   all real numbers $(x+3)^2(x-1)^2 \geq 0$? Explain your answer.

71. What is the solution to the inequality $0, 3, -4$ $x^2(x-3)^2(x+4)^2 \leq 0$? Explain your answer.

72. What is the solution to the inequality   no real solution $x^2(x-3)^2(x+4)^2 < 0$? Explain your answer.

73. What is the solution to the inequality $\dfrac{x^2}{(x+1)^2} \geq 0$? Explain your answer.

74. What is the solution to the inequality $\dfrac{x^2}{(x-3)^2} > 0$? Explain your answer.   all real numbers except 0 and 3

73. all real numbers except -1

CUMULATIVE REVIEW EXERCISES

In exercises 75–76, find the domain of the function.

[3.5] **75.** $f(x) = \sqrt{4-x}$. $\{x \mid x \leq 4\}$

[7.1] **76.** $f(x) = \dfrac{3}{x^2 - 4}$. $\{x \mid x \neq 2 \text{ and } x \neq -2\}$

[7.3] **77.** Simplify $\dfrac{ab^{-2} - a^{-1}b}{a^{-2} + ab^{-1}} \cdot \dfrac{a^3 - ab^3}{b^2 + a^3b}$

[8.7] **78.** Multiply $(\sqrt{-8} + \sqrt{2})(\sqrt{-2} - \sqrt{8})$. $-8 - 6i$

Group Activity/ Challenge Problems

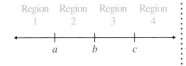

4(a) $x^2 - 8x + 15 \leq 0$ **(b)** $x^2 - 3x > 0$ **(c)** $x^2 - 8x + 16 \leq 0$ **(d)** $x^2 < 0$

Solve the following inequalities. Graph the solution on the number line.

1. $(x+1)(x-3)(x+5)(x+9) \geq 0$ 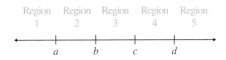   **2.** $\dfrac{(x-4)(x+2)}{x(x+6)} \geq 0$

3. $x^2 - x + 1 > 0$

4. Write a quadratic inequality with the following solutions. Many answers are possible. Explain how you determined your answers.
   (a) $[3, 5]$ **(b)** $(-\infty, 0) \cup (3, \infty)$ **(c)** $\{4\}$ **(d)** $\varnothing$ **(e)** $\mathbb{R}$ $x^2 \geq 0$

5. Consider the number line on the left, where a, b, and c are distinct real numbers.
   (a) In which regions will the real numbers satisfy the inequality $(x-a)(x-b)(x-c) > 0$? Explain.   regions 2 and 4
   (b) In which regions will the real numbers satisfy the inequality $(x-a)(x-b)(x-c) < 0$? Explain.   regions 1 and 3

6. Consider the number line below where a, b, c, and d are distinct real numbers.

| Region 1 | Region 2 | Region 3 | Region 4 | Region 5 |

$a \qquad b \qquad c \qquad d$

   (a) In which regions do the real numbers satisfy the inequality $(x-a)(x-b)(x-c)(x-d) > 0$? Explain.   regions 1, 3, and 5
   (b) In which regions do the real numbers satisfy the inequality $(x-a)(x-b)(x-c)(x-d) < 0$? Explain.   regions 2 and 4

Region 1   Region 2   Region 3   Region 4

$a \qquad b \qquad c$

9.5 The Algebra of Functions

Tape 16

1. Find the sum, difference, product, and quotient of functions.
2. Find the composition of functions.

Now that we have introduced a wide variety of functions we can discuss some operations performed on functions.

Sum, Difference, Product and Quotient of Functions

1. We can add, subtract, multiply, and divide functions. We can also perform another type of operation called finding the composite function. We will first discuss addition, subtraction, multiplication, and division of functions. Then we will discuss how to determine composite functions.

> If $f(x)$ represents one function and $g(x)$ represents a second function, then the following operations on functions may be performed:
>
> **Sum of functions:** $(f + g)(x) = f(x) + g(x)$
>
> **Difference of functions:** $(f - g)(x) = f(x) - g(x)$
>
> **Product of functions:** $(fg)(x) = f(x) \cdot g(x)$
>
> **Quotient of functions:** $\left(\dfrac{f}{g}\right)(x) = \dfrac{f(x)}{g(x)}, \quad g(x) \neq 0$

EXAMPLE 1 If $f(x) = x^2 + x - 6$ and $g(x) = x - 2$, find **(a)** $(f + g)(x)$, **(b)** $(f - g)(x)$, and **(c)** $(g - f)(x)$.

Solution:

(a) $(f + g)(x) = f(x) + g(x) = (x^2 + x - 6) + (x - 2)$
$$= x^2 + x - 6 + x - 2$$
$$= x^2 + 2x - 8$$

(b) $(f - g)(x) = f(x) - g(x) = (x^2 + x - 6) - (x - 2)$
$$= x^2 + x - 6 - x + 2$$
$$= x^2 - 4$$

(c) $(g - f)(x) = g(x) - f(x) = (x - 2) - (x^2 + x - 6)$
$$= x - 2 - x^2 - x + 6$$
$$= -x^2 + 4$$

EXAMPLE 2 If $f(x) = x^2 + x - 6$ and $g(x) = x - 2$, find **(a)** $(f + g)(2)$ and **(b)** $f(2) + g(2)$.

Solution:

(a) In Example 1 (a), we found that for these particular functions $(f + g)(x) = x^2 + 2x - 8$. Using this information, we can evaluate $(f + g)(2)$ by substituting 2 for each x in $(f + g)(x)$.

$$(f + g)(x) = x^2 + 2x - 8$$
$$(f + g)(2) = 2^2 + 2(2) - 8 = 4 + 4 - 8 = 0$$

Thus, $(f + g)(2) = 0$.

(b) We will evaluate $f(2) + g(2)$ by substituting 2 for x in $f(x)$ and 2 for x in $g(x)$. We will then add the values to obtain the sum of the functions when x is 2.

$$f(x) + g(x) = (x^2 + x - 6) + (x - 2)$$
$$f(2) + g(2) = (2^2 + 2 - 6) + (2 - 2)$$
$$= (4 + 2 - 6) + (2 - 2) = 0 + 0 = 0$$

We see by comparing parts **(a)** and **(b)** that $(f + g)(2) = f(2) + g(2)$.   •····

EXAMPLE 3 If $f(x) = x^2 + x - 6$ and $g(x) = x - 2$, find (a) $(fg)(x)$ and **(b)** $\left(\dfrac{f}{g}\right)(x)$.

Solution:

(a) $(fg)(x) = f(x)g(x) = (x^2 + x - 6)(x - 2)$
$$= x^3 - x^2 - 8x + 12$$

If you have forgotten how to multiply polynomials, see Section 5.3.

(b) $\left(\dfrac{f}{g}\right)(x) = \dfrac{f(x)}{g(x)} = \dfrac{x^2 + x - 6}{x - 2}$

$$= \dfrac{(x + 3)(x - 2)}{x - 2} = x + 3$$

In the quotient, $\dfrac{x^2 + x - 6}{x - 2}$, x may not have a value of 2. Therefore, 2 cannot be in the domain of the simplified expression $x + 3$. We write the answer as

$$\left(\dfrac{f}{g}\right)(x) = x + 3, \quad x \neq 2$$ •····

EXAMPLE 4 Consider $f(x) = x - 3$ and $g(x) = 2x - 6$. The sum of these functions is $(f + g)(x) = f(x) + g(x) = (x - 3) + (2x - 6) = 3x - 9$. In Fig. 9.30 we illustrate the graphs of $f(x)$, $g(x)$, and $(f + g)(x)$ on the same axis. Explain how to graph $(f + g)(x)$ using the graphs of $f(x)$ and $g(x)$.

Solution: For each value of x add the values of $f(x)$ to that of $g(x)$ to obtain $(f + g)(x)$. For example, at $x = 0$, $f(0) = -3$, $g(0) = -6$, and $(f + g)(0) = -3 + (-6) = -9$. If you study the graph of $(f + g)(x)$, you will see that $(f + g)(0) = -9$. At $x = 4$, $f(4) = 1$ and $g(4) = 2$, thus $(f + g)(4) = 1 + 2 = 3$ (see Fig. 9.30).   •····

In Example 4, we could have obtained the graphs of $(f - g)(x)$, $(f \cdot g)(x)$, and $\left(\dfrac{f}{g}\right)(x)$ using similar procedures with the appropriate operations. What would be the values of $(f - g)(5)$, $(f \cdot g)(5)$, and $\left(\dfrac{f}{g}\right)(5)$? The answers are $2 - 4$ or -2, $2 \cdot 4$ or 8, and $\dfrac{2}{4}$ or $\dfrac{1}{2}$, respectively.

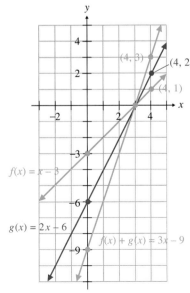

FIGURE 9.30

Graphing Calculator Corner

Graphing calculators can graph the sums, differences, products, and quotients of functions. One way to do this is to enter the individual functions. Then, following instructions that come with your calculator, you can add, subtract, multiply, or divide the functions. For example, the screen below on the left shows one type of calculator ready to graph $y_1 = x - 3$, $y_2 = 2x + 4$, and the sum of the functions, $y_3 = y_1 + y_2$. The graph on the right shows the two functions and the sum of the functions.

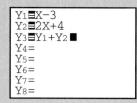

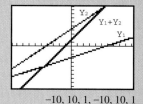

$-10, 10, 1, -10, 10, 1$

1. straight line, x intercept $-1.\overline{3}$, y intercept 4

2. straight line, x intercept 6, y intercept 6

3. straight line, x intercept 6, y intercept -6

4. parabola, vertex at $(-2.25, -15.125)$ x intercepts at -5 and 0.5

EXERCISES

Enter the following functions as $y_1 = x + 5$ and $y_2 = 2x - 1$. Then have your calculator graph y_1 and y_2 and the following. Describe the graphs in Exercises 1–4.

1. $y_1 + y_2$ **2.** $y_1 - y_2$ **3.** $y_2 - y_1$ **4.** $y_1 \cdot y_2$

Composition of Functions

2 Now let us consider another operation performed on functions, the **composition operation.** When we use the composition operation on two functions, the result is called the **composite function.** The composition operation is indicated by a small open circle, ∘, or by using parentheses and brackets. If $f(x)$ and $g(x)$ are two functions, *the composite function of f with g* may be indicated by

$$(f \circ g)(x) \quad \text{or} \quad f[g(x)]$$

The composite function of g with f may be indicated by

$$(g \circ f)(x) \quad \text{or} \quad g[f(x)]$$

To explain how to determine a composite function, we will use the notation $f[g(x)]$ to represent the composite function of f with g.

Consider the functions $f(x) = x^2 - 2x + 3$ and $g(x) = x - 5$. How would you find $f(4)$? To find $f(4)$ you substitute 4 for each x in $f(x)$.

$$f(x) = x^2 - 2x + 3$$
$$f(4) = 4^2 - 2(4) + 3 = 16 - 8 + 3 = 11$$

How would you find $f(a)$? To find $f(a)$, you substitute a for each x in $f(x)$.

$$f(x) = x^2 - 2x + 3$$
$$f(a) = a^2 - 2a + 3$$

Using the illustrations just presented, how do you think you would find $f[g(x)]$? If you answered "substitute function $g(x)$ for each x in function $f(x)$," you answered correctly.

$$f(x) = x^2 - 2x + 3, \qquad g(x) = x - 5$$

$$f[g(x)] = (x - 5)^2 - 2(x - 5) + 3$$
$$= (x - 5)(x - 5) - 2x + 10 + 3$$
$$= x^2 - 10x + 25 - 2x + 13$$
$$= x^2 - 12x + 38$$

Therefore, the composite function of f with g is $x^2 - 12x + 38$.

$$(f \circ g)(x) = f[g(x)] = x^2 - 12x + 38$$

How do you think we would determine $g[f(x)]$ or $(g \circ f)(x)$? If you answered, "substitute $f(x)$ for each x in $g(x)$" you answered correctly.

$$g(x) = x - 5, \qquad f(x) = x^2 - 2x + 3$$

$$g[f(x)] = (x^2 - 2x + 3) - 5$$
$$= x^2 - 2x + 3 - 5$$
$$= x^2 - 2x - 2$$

Therefore, the composite function of g with f is $x^2 - 2x - 2$.

$$(g \circ f)(x) = g[f(x)] = x^2 - 2x - 2$$

By comparing the illustrations above we see that $f[g(x)] \neq g[f(x)]$.

EXAMPLE 5   Given $f(x) = x^2 + 4$ and $g(x) = \sqrt{x - 2}$, find **(a)** $(f \circ g)(x)$ and **(b)** $(g \circ f)(x)$.

Solution:

(a) To find $(f \circ g)(x)$, we substitute $g(x)$, which is $\sqrt{x - 2}$, for each x in $f(x)$. You should realize that $\sqrt{x - 2}$ is a real number only when $x \geq 2$.

$$f(x) = x^2 + 4$$
$$(f \circ g)(x) = f[g(x)] = (\sqrt{x - 2})^2 + 4 = x - 2 + 4 = x + 2, \ x \geq 2.$$

Since values of $x < 2$ are not in the domain of $g(x)$, values of $x < 2$ are not in the domain of $(f \circ g)(x)$.

(b) To find $(g \circ f)(x)$, we substitute $f(x)$, which is $x^2 + 4$, for each x in $g(x)$.

$$g(x) = \sqrt{x - 2}$$
$$(g \circ f)(x) = g[f(x)] = \sqrt{(x^2 + 4) - 2} = \sqrt{x^2 + 2}$$

EXAMPLE 6   Given $f(x) = x - 1$ and $g(x) = x + 7$, find **(a)** $(f \circ g)(x)$, **(b)** $(f \circ g)(2)$,   **(c)** $(g \circ f)(x)$, and   **(d)** $(g \circ f)(2)$.

Solution:

(a) $f(x) = x - 1$

$$(f \circ g)(x) = f[g(x)] = (x + 7) - 1 = x + 6$$

(b) $(f \circ g)(2)$ can be found by substituting 2 for each x in $(f \circ g)(x)$.
$$(f \circ g)(x) = x + 6$$
$$(f \circ g)(2) = 2 + 6 = 8$$

(c) $g(x) = \boxed{x} + 7$

$$(g \circ f)(x) = g[f(x)] = (x - 1) + 7 = x + 6$$

(d) Since $(g \circ f)(x) = x + 6$, $(g \circ f)(2) = 2 + 6 = 8$.

In general $(f \circ g)(x) \neq (g \circ f)(x)$. In Example 6, $(f \circ g)(x) = (g \circ f)(x)$, but this is only due to the specific functions used.

Helpful Hint

Do not confuse finding the product of two functions with finding a composite function.

Product of functions f and g:	$(fg)(x)$ or $(f \cdot g)(x)$
Composite function of f with g:	$(f \circ g)(x)$

When multiplying functions f and g, we can use a dot between the f and g. When finding the composite function of f with g, we use a small *open* circle.

Exercise Set 9.5

Given $f(x) = x^2 + 2x - 8$ and $g(x) = x - 2$, find the following.

1. **(a)** $(f + g)(x)$ **(b)** $(f + g)(2)$ **(a)** $x^2 + 3x - 10$ **(b)** 0

2. **(a)** $(f - g)(x)$ **(b)** $(f - g)(4)$ **(a)** $x^2 + x - 6$ **(b)** 14

3. **(a)** $(fg)(x)$ **(b)** $(fg)(-1)$ **(a)** $x^3 - 12x + 16$ **(b)** 27

4. **(a)** $\left(\dfrac{f}{g}\right)(x)$ **(b)** $\left(\dfrac{f}{g}\right)(0)$ **(a)** $x + 4$, $x \neq 2$ **(b)** 4

5. **(a)** $(f \circ g)(x)$ **(b)** $(f \circ g)(3)$ **(a)** $x^2 - 2x - 8$ **(b)** -5

6. **(a)** $(g \circ f)(x)$ **(b)** $(g \circ f)(3)$ **(a)** $x^2 + 2x - 10$ **(b)** 5

Given $f(x) = x^2 - 4$ and $g(x) = x + 2$, find the following.

7. **(a)** $(g + f)(x)$ **(b)** $(g + f)(-2)$ **(a)** $x^2 + x - 2$ **(b)** 0

8. **(a)** $(g - f)(x)$ **(b)** $(g - f)(5)$ **(a)** $-x^2 + x + 6$ **(b)** -14

9. **(a)** $(gf)(x)$ **(b)** $(gf)(-4)$ **(a)** $x^3 + 2x^2 - 4x - 8$ **(b)** -24

10. **(a)** $\left(\dfrac{f}{g}\right)(x)$ **(b)** $\left(\dfrac{f}{g}\right)(4)$ **(a)** $x - 2$, $x \neq -2$ **(b)** 2

11. **(a)** $(g \circ f)(x)$ **(b)** $(g \circ f)(4)$ **(a)** $x^2 - 2$ **(b)** 14

12. **(a)** $(f \circ g)(x)$ **(b)** $(f \circ g)(4)$ **(a)** $x^2 + 4x$ **(b)** 32

Given $f(x) = x - 4$ and $g(x) = \sqrt{x + 6}$, $x \geq -6$, find the following.

13. **(a)** $(f + g)(x)$ **(b)** $(f + g)(3)$ **(a)** $x - 4 + \sqrt{x + 6}$ **(b)** 2

14. **(a)** $(f - g)(x)$ **(b)** $(f - g)(3)$ **(a)** $x - 4 - \sqrt{x + 6}$ **(b)**

15. **(a)** $(fg)(x)$ **(b)** $(fg)(10)$ **(a)** $(x - 4)\sqrt{x + 6}$ **(b)** 24

16. **(a)** $\left(\dfrac{g}{f}\right)(x)$ **(b)** $\left(\dfrac{g}{f}\right)(10)$ **(a)** $\dfrac{\sqrt{x + 6}}{x - 4}$, $x \neq 4$ **(b)** $\frac{2}{3}$

17. **(a)** $(f \circ g)(x)$ **(b)** $(f \circ g)(7)$ **(a)** $\sqrt{x + 6} - 4$ **(b)** $\sqrt{13} - 4$

18. **(a)** $(g \circ f)(x)$ **(b)** $(g \circ f)(8)$ **(a)** $\sqrt{x + 2}$ **(b)** $\sqrt{10}$

Using the graph to the right, find the value of the following.

19. $(f + g)(2)$ 3

20. $(f - g)(2)$ 5

21. $(fg)(2)$ -4

22. $\left(\dfrac{f}{g}\right)(2)$ -4

23. $(f - g)(-2)$ -3

24. $(g - f)(0)$ -1

25. $\left(\dfrac{g}{f}\right)(-2)$ undefined

26. $(gf)(0)$ 2

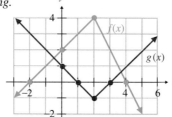

27. Using the graphs of $f(x)$ and $g(x)$ as illustrated in Exercises 19–26, construct the graph of $(f + g)(x)$.

28. Consider the following illustration.

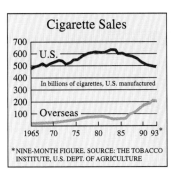

Cigarette Sales

700
600 — U.S.
500
400
300   In billions of cigarettes, U.S. manufactured
200
100 — Overseas

1965 70 75 80 85 90 93*

*NINE-MONTH FIGURE. SOURCE: THE TOBACCO INSTITUTE, U.S. DEPT. OF AGRICULTURE

27.

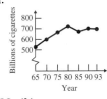

(a) Explain why each of the two graphs may be considered functions.

(b) Considering the red graph as $f(x)$ and the green graph as $g(x)$, construct the graph of $(f + g)(x)$ using intervals of 5 years starting with 1965.

(c) What does $(f + g)(x)$ represent?

29. Consider the following illustration.

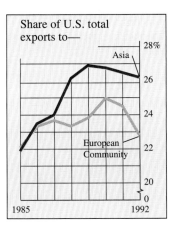

Share of U.S. total exports to—

28%
Asia
26
24
European
Community 22
20
0

1985   1992

(a) Explain why each of the two graphs may be considered functions.

(b) Considering the red graph as $f(x)$ and the green graph as $g(x)$, construct the graph of $(f + g)(x)$.

(c) What does $(f + g)(x)$ represent?

30. Is $(f + g)(x) = (g + f)(x)$ for all values of x? Explain and give an example to support your answer.

31. Is $(f - g)(x) = (g - f)(x)$ for all values of x? Explain and give an example to support your answer.

32. Is $\left(\dfrac{f}{g}\right)(x) = \left(\dfrac{g}{f}\right)(x)$ for all values of x? Explain and give an example to support your answer.

33. Is $(fg)(x) = (gf)(x)$ for all values of x? Explain and give an example to support your answer.

34. **(a)** Make up your own polynomial function, $f(x)$. Does $f(x) + f(x) = 2 \cdot f(x)$ for your function?

(b) Do you believe that $f(x) + f(x) = 2 \cdot f(x)$ for all functions? Explain.   **(b)** yes

35. Is $(f \circ g)(x) = (g \circ f)(x)$ for all values of x? Explain and give an example to support your answer.

36. Consider the functions $f(x) = \sqrt{x + 5}, x \geq -5$, and $g(x) = x^2 - 5, x \geq 0$.

(a) Show that for $x \geq 0$ $(f \circ g)(x) = (g \circ f)(x)$.

(b) Explain why we need to stipulate that $x \geq 0$ for part **(a)** to be true.

37. Consider the functions $f(x) = x^3 + 2$ and $g(x) = \sqrt[3]{x - 2}$. $(f \circ g)(x) = (g \circ f)(x) = x$

(a) Show that $(f \circ g)(x) = (g \circ f)(x)$.

(b) What is the domain of $f(x), g(x), (f \circ g)(x)$ and $(g \circ f)(x)$? Explain. $\mathbb{R}$ for all of them

Cigarette sales in the U.S. and overseas

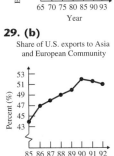

800
700
600
500

65 70 75 80 85 90 93
Year

29. (b)

Share of U.S. exports to Asia and European Community

53
51
49
47
45
43

85 86 87 88 89 90 91 92
Year

28. (c) sales of cigarettes in the United States and overseas   **29. (c)** share of us exports to Asia and the European community
30. Yes, addition is commutative.   **31.** No, subtraction is not commutative.   **32.** No, division is not commutative.   **33.** Yes, multiplication is commutative.   **35.** No composition of functions is not commutative.   **36. (a)** $(f \circ g)(x) = x, (g \circ f)(x) = x$,
(b) without specifying that $x \geq 0$, $(f \circ g)(x) = \sqrt{x^2} = |x|$, not x

CUMULATIVE REVIEW EXERCISES

[3.4] **38.** Are all functions relations? Are all relations functions? Explain.   yes, no

[5.6] **39.** Are all linear functions polynomial functions? Are all quadratic functions polynomial functions? Explain.

40. Give an example of a polynomial function that is not a linear or quadratic function. $f(x) = x^3$

41. Are any polynomial functions square root functions? Are any square root functions polynomial functions? Explain.   no, no

[7.1] **42.** Are all polynomial functions rational functions? Are all rational functions polynomial functions? Explain.
39. yes, yes   **42.** yes, no

2. (c) 51.84 π or $\approx$ 162.86 sq in   **3. (a)** {(1, 7), (2, −1), (3, 10)}   **(b)**{(1, −3), (2, 7), (3, −2)}
(c) {(1, 10), (2, −12), (3, 24)}   **(d)** {(1, $\frac{2}{5}$), (2, $-\frac{3}{4}$), (3, $\frac{2}{3}$)}

Group Activity/ Challenge Problems

1. When a pebble is thrown into a pond of water the circle formed by the pebble hitting the water expands with time. The surface area of the expanding circle may be found by the formula $A = \pi r^2$. The radius, r, of the circle, in feet, is a function of time, t. Suppose that the function is $r(t) = 2t$.

(a) Find the surface area of the circle at 3 seconds. $36\pi \approx 113.10$ sq ft

(b) Express the surface area as a function of time by finding $A \circ r$. $A = 4\pi t^2$

(c) Using the function found in part (b) find the surface area of the circle after 4 seconds. $64\pi \approx 201.06$ sq ft

2. The surface area of a spherical balloon of radius r, in inches, is found by $S(r) = 4\pi r^2$. If the balloon is being blown up at a constant rate by a machine, then the radius of the balloon is a function of time. Suppose that this function is $r(t) = 1.2t$.

(a) Find the surface area at 2 seconds. $23.04\pi \approx 72.38$ sq in

(b) Express the surface area as a function of time by finding $S \circ r$. $S = 5.76\pi t^2$

(c) Using the function found in part (b), find the surface area after 3 seconds.

3. Consider $f(x) = \{(1, 2), (2, 3), (3, 4)\}$ and $g(x) = \{(1, 5), (2, −4), (3, 6)\}$. Find

(a) $(f + g)(x)$;   (b) $(f − g)(x)$;   (c) $(fg)(x)$;   (d) $\left(\dfrac{f}{g}\right)(x)$.

9.6   Inverse Functions

Tape
16

1 Identify one-to-one functions.
2 Find the inverse function of a set of ordered pairs.
3 Find inverse functions.
4 Show that $(f \circ f^{-1})(x) = (f^{-1} \circ f)(x) = x$.

One-to-One Functions

1 Inverse functions are an important concept that must be understood before we can discuss exponential and logarithmic functions in Chapter 11. However, before we discuss inverse functions we need to explain what is meant by *one-to-one functions*.

Consider the function $f(x) = x^2$ (see Fig. 9.31). Note that it is a function since it passes the vertical line test. For each value of x, there is a unique value of y. Does each value of y also have a unique value of x? The answer is no, as illustrated in Fig. 9.32. Note that for the indicated value of y there are two values of x, namely x_1 and x_2. If we limit the domain of $f(x) = x^2$ to values of x greater than or equal to 0, then each *x* value has a unique *y* value and each *y* value also has a unique *x* value (see Fig. 9.33).

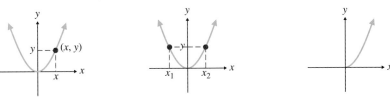

FIGURE 9.31   **FIGURE 9.32**   **FIGURE 9.33**

The function $f(x) = x^2$, $x \geq 0$, Fig. 9.33, is an example of a one-to-one function. In a **one-to-one function,** each value in the range has a unique value in the domain. Thus, if *y* is a one-to-one function of *x*, in addition to each *x* value having a unique *y* value (the definition of a function), each *y* value must also have a unique *x* value. For a function to be a one-to-one function, it must pass not only a **vertical line test** (the test to ensure that it is a function) but also a **horizontal line test** (to test the one-to-one criteria).

In Fig. 9.34 all the graphs are functions since they all pass the vertical line test. However, only the graphs in parts (a), (d), and (e) are one-to-one functions since they also pass the horizontal line test.

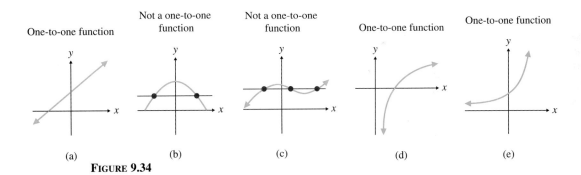

FIGURE 9.34

EXAMPLE 1   Determine which of the following functions are one-to-one functions.

(a)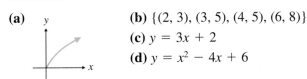

(b) {(2, 3), (3, 5), (4, 5), (6, 8)}

(c) $y = 3x + 2$

(d) $y = x^2 - 4x + 6$

Solution: **(a)** Yes, the graph is a one-to-one function because it passes the horizontal line test.

(b) No, the set of ordered pairs is not a one-to-one function. Note that the *y* value 5 is associated with two *x* values, 3 and 4.

(c) Yes, the graph of this function is a straight line, and straight lines (except for horizontal lines) pass the horizontal line test.

(d) No, the graph of this function is a parabola, and parabolas do not pass the horizontal line test.

Inverse Functions   **2** Now that we have discussed one-to-one functions we can introduce inverse functions. You must be aware that **only one-to-one functions have inverse functions.** If a function is one-to-one, its **inverse function** may be obtained by interchanging the first and second coordinates in each ordered pair of the function. Thus, for each ordered pair (x, y) in the function, the ordered pair (y, x) will be in the inverse function. For example,

Function: $\{(1, 4), (2, 0), (3, 7), (-2, 1), (-1, -5)\}$
Inverse function: $\{(4, 1), (0, 2), (7, 3), (1, -2), (-5, -1)\}$

Note that the domain of the function becomes the range of the inverse function, and the range of the function is the domain of the inverse function.

If we graph the points in the function and the points in the inverse function (Fig. 9.35), we see that the points are symmetric about the line $y = x$.

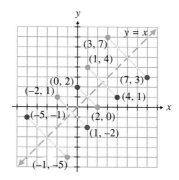

● Ordered pair in function
● Ordered pair in inverse function

FIGURE 9.35

For a function $f(x)$, the notation $f^{-1}(x)$ represents its inverse function. Note that the -1 in the notation is *not* an exponent.

> **Inverse Function**
>
> If $f(x)$ is a one-to-one function with ordered pairs of the form (x, y), its inverse function, $f^{-1}(x)$, is a one-to-one function with ordered pairs of the form (y, x).

When a function $f(x)$ and its inverse function $f^{-1}(x)$ are graphed on the same axes, $f(x)$ *and* $f^{-1}(x)$ are *symmetric about the line* $y = x$.

Finding Inverse   **3** When a one-to-one function is given as an equation, its inverse function can
Functions   be found by following this procedure:

> **To Find the Inverse Function of a One-to-One Function of the Form $y = f(x)$**
>
> 1. Interchange the two variables x and y.
> 2. Solve the equation for y. The resulting equation is the inverse function.

The following example will illustrate the procedure.

EXAMPLE 2 (a) Find the inverse function of $y = f(x) = 4x + 2$.
(b) On the same axes, graph both $f(x)$ and $f^{-1}(x)$.

Solution:

(a) $y = 4x + 2$.

First interchange x and y.

$$x = 4y + 2$$

Now solve for y.

$$x - 2 = 4y$$

$$\frac{x - 2}{4} = y$$

$$y = f^{-1}(x) = \frac{x - 2}{4}$$

(b)

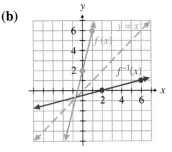

Note the symmetry of $f(x)$ and $f^{-1}(x)$ about the line $y = x$. Also note that both the domain and range of both $f(x)$ and $f^{-1}(x)$ are the set of real numbers, $\mathbb{R}$.

In Chapter 8 when we solved radical equations we raised each side of the equation to the same power. We can use that same procedure to solve cubic equations. For example, to solve the equation $x^3 = 8$ we can raise each side of the equation to the one-third power, which is equivalent to taking the cubic root of each side of the equation. Recall from Chapter 8 that $\sqrt[3]{a^3} = a$ for any real number a. Below we solve two cubic equations. The equation on the left is solved for x, while the one on the right is solved for b.

$$
\begin{aligned}
x^3 &= 2 & a &= b^3 + 5 \\
\sqrt[3]{x^3} &= \sqrt[3]{8} & a - 5 &= b^3 \\
x &= \sqrt[3]{8} & \sqrt[3]{a - 5} &= \sqrt[3]{b^3} \\
x &= 2 & \sqrt[3]{a - 5} &= b
\end{aligned}
$$

EXAMPLE 3 (a) Find the inverse function of $y = f(x) = x^3 + 2$.
(b) On the same axes, graph both $f(x)$ and $f^{-1}(x)$.

Solution:

(a) Interchange x and y; then solve for y.   **(b)**

$$x = y^3 + 2$$
$$x - 2 = y^3$$
$$\sqrt[3]{x - 2} = \sqrt[3]{y^3}$$
$$\sqrt[3]{x - 2} = y$$
$$y = f^{-1}(x) = \sqrt[3]{x - 2}$$

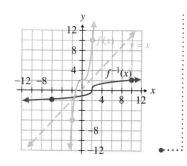

In Example 3, if we evaluate $f(x) = x^3 + 2$ at 3 we obtain $f(3) = 3^3 + 2 = 29$. Suppose that we evaluate the inverse function $f^{-1}(x) = \sqrt[3]{x - 2}$ at 29, what will we obtain? Remember for each ordered pair (x, y) in $f(x)$, the ordered pair (y, x) is in $f^{-1}(x)$. Since $(3, 29)$ is an ordered pair of $f(x)$, then $(29, 3)$ is an ordered pair of $f^{-1}(x)$. Thus, $f^{-1}(29) = 3$.

$$f^{-1}(29) = \sqrt[3]{29 - 2} = \sqrt[3]{27} = 3$$

Graphing Calculator Corner

If you graph the function $y = x^3 + 2$ and its inverse function $y = \sqrt[3]{x - 2}$ on a grapher, you would expect the two graphs to be symmetric with respect to the line $y = x$. Although this is true, it may not appear this way on a grapher. Figure 9.36 illustrates the two graphs graphed on a grapher.

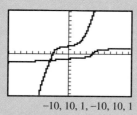

−10, 10, 1, −10, 10, 1

FIGURE 9.36

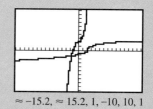

≈ −15.2, ≈ 15.2, 1, −10, 10, 1

FIGURE 9.37

The graphs do not appear to be symmetric about the line $y = x$. Since the horizontal axis is longer than the vertical axis, the distance between marks on the horizontal axis is generally greater than the distance between marks on the vertical axis. This appears to distort the graph. Many calculators have a feature to "square" the axes. When this feature is used, the window is still rectangular, but the distance between marks on each axis is equalized, so the graphs appear more symmetric about the line $y = x$. Figure 9.37 shows the same two equations graphed using this feature.

You may not be able to tell conclusively by looking at two graphs if they are inverses of each other. However, you should be able to tell that two graphs are not inverses if their graphs are not symmetric to $y = x$.

EXERCISES

Use your grapher to determine if the following equations appear to be inverses.

1. $y = 3x - 4$, $y = \dfrac{x}{3} + \dfrac{4}{3}$ yes

2. $y = \sqrt{4 - x^2}$, $y = \sqrt{4 - 2x}$ no

3. $y = x^3 - 12$, $y = \sqrt[3]{x + 12}$ yes

4. $y = x^5 + 5$, $y = \sqrt[5]{x - 5}$ yes

Composite Functions and Their Inverses

In Section 9.5 we discussed the composition of functions. If two functions $f(x)$ and $f^{-1}(x)$ are inverses of each other, $(f \circ f^{-1})(x) = x$ and $(f^{-1} \circ f)(x) = x$.

EXAMPLE 4   In Example 3 we determined that $f(x) = x^3 + 2$ and $f^{-1}(x) = \sqrt[3]{x - 2}$ are inverse functions. Show that **(a)** $(f \circ f^{-1})(x) = x$ and **(b)** $(f^{-1} \circ f)(x) = x$.

Solution:

(a) To determine $(f \circ f^{-1})(x)$, substitute $f^{-1}(x)$ for each x in $f(x)$.

$$f(x) = x^3 + 2$$
$$(f \circ f^{-1})(x) = (\sqrt[3]{x - 2})^3 + 2$$
$$= x - 2 + 2 = x$$

(b) To determine $(f^{-1} \circ f)(x)$, substitute $f(x)$ for each x in $f^{-1}(x)$.

$$f^{-1}(x) = \sqrt[3]{x - 2}$$
$$(f^{-1} \circ f)(x) = \sqrt[3]{(x^3 + 2) - 2}$$
$$= \sqrt[3]{x^3} = x$$

By parts **(a)** and **(b)**, $(f \circ f^{-1})(x) = (f^{-1} \circ f)(x) = x$.

Because a function and its inverse "undo" each other, the composite of a function with its inverse results in the given value from the domain. For example, for any function $f(x)$ and its inverse $f^{-1}(x)$, $(f^{-1} \circ f)(3) = 3$, and $(f \circ f^{-1})\left(-\frac{1}{2}\right) = -\frac{1}{2}$.

Exercise Set 9.6

Determine whether each function is a one-to-one function.

1.

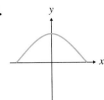

no

2.

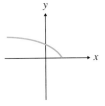

yes

3.

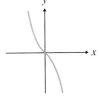

yes

4.

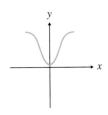

no

5.

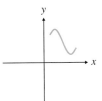

no

6.

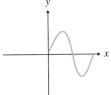

no

20. D: $f(x)$, R: $f^{-1}(x)$: $\{-4, -2, 2, 5, 6\}$, R: $f(x)$, D: $f^{-1}(x)$: $\{-3, 0, \frac{1}{2}, 2, 3\}$

22. D: $f(x)$, R: $f^{-1}(x)$: $\{-\frac{5}{3}, 0, \sqrt{3}, 2.3\}$, R: $f(x)$, D: $f^{-1}(x)$: $\{-2, -\frac{1}{2}, \frac{3}{8}, 4\}$

7. $\{(-2, 4), (3, -7), (5, 3), (-6, 0)\}$ yes   **8.** $\{(-4, 2), (2, 3), (4, 1), (0, 4)\}$ yes   **9.** $\{(-4, 2), (5, 3), (0, 2), (3, 7)\}$ no

10. $\{(-4, 5), (1, 4), (-3, 5), (4, 2)\}$ no   **11.** $y = x - 4$ yes   **12.** $y = 3x - 6$ yes

13. $y = x^2 - 4$ no   **14.** $y = x^2 - 2x + 4$ no   **15.** $y = x^2 - 4, x \geq 0$ yes

16. $y = x^2 - 4, x \leq 0$ yes   **17.** $y = x^3$ yes   **18.** $y = \sqrt{x}$. yes

23. D: $f(x)$, R: $f^{-1}(x)$: $\{-1, 1, 2, 4\}$, R: $f(x)$, D: $f^{-1}(x)$: $\{-3, -1, 0, 2\}$ **24.** $f(x)$, D: $\{x | x \geq 0\}$, R: $\{y | y \geq 0\}$, $f^{-1}(x)$, D: $\{x | x \geq 0\}$, R: $\{y | y \geq 0\}$

25. $f(x)$, D: $\{x | x \geq 2\}$, R: $\{y | y \geq 0\}$, $f^{-1}(x)$, D: $\{x | x \geq 0\}$, R: $\{y | y \geq 2\}$ **26.** $f(x)$, D: $\mathbb{R}$, R: $\mathbb{R}$, $f^{-1}(x)$, D: $\mathbb{R}$, R: $\mathbb{R}$

For the given function in Exercises 19–26 find the domain and range of both $f(x)$ and $f^{-1}(x)$.

19. $\{(4, 0), (9, 3), (2, 7), (-1, 6), (-2, 4)\}$

20. $\left\{(-2, -3), (-4, 0), (5, 3), (6, 2), \left(2, \frac{1}{2}\right)\right\}$

21. $\{(1.7, 3), (-2.9, 4), (5.7, -3.4), (0, 9.76)\}$

22. $\left\{(2.3, -2), \left(0, \frac{3}{8}\right), \left(-\frac{5}{3}, -\frac{1}{2}\right), (\sqrt{3}, 4)\right\}$

23.

24.

25.

26.

Find $f^{-1}(x)$ and graph $f(x)$ and $f^{-1}(x)$ on the same axes.

27. $y = f(x) = 2x + 8$

$f^{-1}(x) = (x - 8)/2$

28. $y = f(x) = -3x + 6$

$f^{-1}(x) = (-x/3) + 2$

29. $y = f(x) = -3x - 10$

$f^{-1}(x) = -(x + 10)/3$

30. $y = f(x) = \frac{1}{2}x + 3$

$f^{-1}(x) = 2x - 6$

31. $y = f(x) = 2x - \frac{3}{5}$

$f^{-1}(x) = (5x + 3)/10$

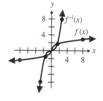

32. $y = f(x) = \frac{x + 3}{6}$

$f^{-1}(x) = 6x - 3$

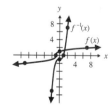

33. $y = f(x) = \sqrt{x}, x \geq 0$

$f^{-1}(x) = x^2, x \geq 0$

34. $y = f(x) = \sqrt{x + 1}, x \geq -1$

$f^{-1}(x) = x^2 - 1, x \geq 0$

35. $y = f(x) = \sqrt[3]{x}$

$f^{-1}(x) = x^3$

36. $y = f(x) = \sqrt[3]{x + 1}$

$f^{-1}(x) = x^3 - 1$

37. $y = f(x) = \frac{1}{x}, x > 0$

$f^{-1}(x) = \frac{1}{x}, x > 0$

38. $y = f(x) = \frac{1}{x}$

$f^{-1}(x) = \frac{1}{x}$

19. D: $f(x)$, R: $f^{-1}(x)$: $\{-2, -1, 2, 4, 9\}$, R: $f(x)$, D: $f^{-1}(x)$: $\{0, 3, 4, 6, 7\}$

21. D: $f(x)$, R: $f^{-1}(x)$: $\{-2.9, 0, 1.7, 5.7\}$, R: $f(x)$, D: $f^{-1}(x)$: $\{-3.4, 3, 4, 9.76\}$

48. $y = f^{-1}(x) = \frac{15}{22}x$, x is ft/sec, y is mi/hr   **49.** $y = f^{-1}(x) = \frac{9}{5}x + 32$, x is degrees Celsius, y is degrees Fahrenheit

For each pair of inverse functions, show that $(f \circ f^{-1})(x) = x$ and $(f^{-1} \circ f)(x) = x$.

39. $f(x) = x - 4$, $f^{-1}(x) = x + 4$

40. $f(x) = \dfrac{2x - 3}{4}$, $f^{-1}(x) = \dfrac{4x + 3}{2}$

41. $f(x) = 3x + 2$, $f^{-1}(x) = \dfrac{x - 2}{3}$

42. $f(x) = \sqrt[3]{x - 2}$, $f^{-1}(x) = x^3 + 2$

43. $f(x) = \dfrac{x - 4}{5}$, $f^{-1}(x) = 5x + 4$

44. $f(x) = \sqrt[3]{x + 8}$, $f^{-1}(x) = x^3 - 8$

45. $f(x) = \dfrac{2}{x}$, $f^{-1}(x) = \dfrac{2}{x}$

46. $f(x) = \sqrt{x + 1}$, $f^{-1}(x) = x^2 - 1, x \geq 0$

47. The function $y = f(x) = 12x$ converts feet, x, into inches, y. Find the inverse function that converts inches into feet. In the inverse function, what do x and y now represent? $y = f^{-1}(x) = \frac{x}{12}$, x is inches, y is feet

48. The function $y = f(x) = \frac{22}{15}x$ converts miles per hour, x, into feet per second, y. Find the inverse function that converts feet per second to miles per hour. In the inverse function, what do x and y now represent?

49. The function $y = f(x) = \frac{5}{9}(x - 32)$ converts degrees Fahrenheit, x, to degrees Celsius, y. Find the inverse function that changes degrees Celsius into degrees Fahrenheit. In the inverse function, what do x and y now represent?

50. The function $y = f(x) = \pi x^2$ may be used to find the area, y, of a circle of radius, x. Find the inverse function, which gives the radius when the area of a circle is known. In the inverse function, what do x and y now represent? $y = f^{-1}(x) = \sqrt{\frac{x}{\pi}}$, x is the area, y is the radius

51. What are one-to-one functions?

52. Which functions may have inverse functions?

53. What is the relationship between the domain and range of a function and the domain and range of its inverse function?

54. Explain why the function $y = x^2$ has no inverse function.   It is not a one-to-one function.

55. For the function $f(x) = x^3$, $f(2) = 2^3 = 8$. Explain why $f^{-1}(8) = 2$.   The range of $f^{-1}(x)$ is the domain of $f(x)$.

56. For the function $f(x) = x^4, x > 0, f(2) = 16$. Explain why $f^{-1}(16) = 2$. The range of $f^{-1}(x)$ is the domain of $f(x)$.

57. What is the value of $(f \circ f^{-1})(6)$? Explain.

58. Write a paragraph explaining why, if $f(x)$ and $g(x)$ are inverse functions, then $(f \circ g)(x) = x$ and $(g \circ f)(x) = x$.

51. functions for which each value in the range has a unique value in the domain.   **52.** one-to-one functions   **53.** Domain of function is the range of the inverse function. Range of function is the domain of inverse function.   **57.** 6, $(f \circ f^{-1})(x) = x$

CUMULATIVE REVIEW EXERCISES

[4.2] **59.** Solve the system of equations.

$$2x + 3y - 4z = 18$$
$$x - y - z = 3$$
$$x - 2y - 2z = 2 \quad (4, 2, -1)$$

[5.5] **60.** Divide using synthetic division.

$$(x^3 + 6x^2 + 6x - 8) \div (x + 2) \quad x^2 + 4x - 2 - \frac{4}{x + 2}$$

[8.4] **61.** Simplify $\sqrt{\dfrac{24x^3y^2}{3xy^3}}$. $\dfrac{2x\sqrt{2y}}{y}$

[9.1] **62.** Solve the equation $x^2 + 2x - 6 = 0$ by completing the square. $x = -1 \pm \sqrt{7}$

2. (a), (c)

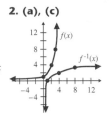

Group Activity/ Challenge Problems

1. (a) no, not one-to-one function   **(b)** yes

1. (a) Does the function $f(x) = |x|$ have an inverse? Explain why.

(b) If the domain is limited to $x \geq 0$, does the function have an inverse? Explain.

(c) Find the inverse function of $f(x) = |x|, x \geq 0$. $f^{-1}(x) = x, x \geq 0$

2. Consider the function $f(x) = 2^x$. This is an example of an exponential function that we will discuss in Chapter 11.

(a) Graph this function by substituting values for x and finding the corresponding values of y.

(b) Do you believe this function has an inverse? Explain your answer.   yes

(c) Using the graph in part (*a*), draw the inverse function, $f^{-1}(x)$ on the same axes.

3.

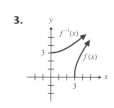

Find $f^{-1}(x)$ and graph $f(x)$ and $f^{-1}(x)$ on the same axes. Use the fact that the domain of $f(x)$ is the range of $f^{-1}(x)$ and the range of $f(x)$ is the domain of $f^{-1}(x)$ to graph $f^{-1}(x)$.

3. $y = \sqrt{x^2 - 9}, x \geq 3$ $f^{-1}(x) = \sqrt{x^2 + 9}; x \geq 0$ **4.**

4. $y = \sqrt{x^2 - 9}, x \leq -3$ $f^{-1}(x) = -\sqrt{x^2 + 9}; x \geq 0$

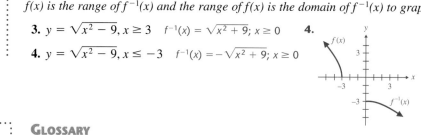

Summary

GLOSSARY

Boundary values *(546):* The values that make an inequality equal to zero or undefined.

Discriminant *(525):* For a quadratic equation of the form $ax^2 + bx + c = 0$, the discriminant is $b^2 - 4ac$.

Equation quadratic in form *(539):* An equation that may be expressed as $au^2 + bu + c = 0$, where u is a function of the variable in the given equation.

Horizontal line test *(563):* A test used to determine if a function is a one-to-one function. If a horizontal line cannot be drawn to intersect the graph of a function at more than one point, the function is a one-to-one function.

Inverse function *(564):* If $f(x)$ is a one-to-one function with ordered pairs of the form (x, y), its inverse function, $f^{-1}(x)$, will be a one-to-one function with ordered pairs of the form (y, x).

One-to-one function *(563):* A function where each value in the range has a unique value in the domain.

Perfect square trinomial *(512):* A trinomial that can be expressed as the square of a binomial.

Quadratic function *(526):* Functions of the form $f(x) = ax^2 + bx + c, a \neq 0$.

Standard form of a quadratic equation *(520):* $ax^2 + bx + c = 0, a \neq 0$.

IMPORTANT FACTS

Square root property: If $x^2 = a$, then $x = \pm\sqrt{a}$.

Pythagorean theorem: $a^2 + b^2 = c^2$

Quadratic formula: $x = \dfrac{-b \pm \sqrt{b^2 - 4ac}}{2a}$

Discriminant: $b^2 - 4ac$

If $b^2 - 4ac > 0$, the quadratic equation has two distinct real solutions.

If $b^2 - 4ac = 0$, the quadratic equation has one real solution.

If $b^2 - 4ac < 0$, the quadratic equation has no real solution.

Vertex of a parabola is at $\left(\dfrac{-b}{2a}, \dfrac{4ac - b^2}{4a}\right)$.

Algebra of functions

Sum of functions: $(f + g)(x) = f(x) + g(x)$

Difference of functions: $(f - g)(x) = f(x) - g(x)$

Product of functions: $(fg)(x) = f(x) \cdot g(x)$

Quotient of functions: $\left(\dfrac{f}{g}\right)(x) = \dfrac{f(x)}{g(x)}$, $g(x) \neq 0$

Composite function of f with g: $(f \circ g)(x) = f[g(x)]$

Composite function of g with f: $(g \circ f)(x) = g[f(x)]$

If $f(x)$ and $g(x)$ are inverse functions then $(f \circ g)(x) = (g \circ f)(x) = x$.

Review Exercises

[9.1] *Use the square root property to solve each equation for x.*

1. $(x - 4)^2 = 20$ $4 \pm 2\sqrt{5}$

2. $(3x - 4)^2 = 60$ $\dfrac{4 \pm 2\sqrt{15}}{3}$

3. $\left(x - \frac{2}{3}\right)^2 = \frac{1}{9}$ $1, \frac{1}{3}$

4. $\left(2x - \frac{1}{2}\right)^2 = 4$ $\frac{5}{4}, -\frac{3}{4}$

Use the square root property to solve each formula for the indicated variable. Use only the positive square root.

5. $F_T^2 = F_a^2 + F_b^2$, for F_b $F_b = \sqrt{F_T^2 - F_a^2}$

6. $L = 3r + 2s^2$, for s $s = \sqrt{\dfrac{L - 3r}{2}}$

Use the Pythagorean Theorem to solve for x.

7.

$\sqrt{128} \approx 11.31$

8.

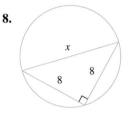

$\sqrt{128} \approx 11.31$

Solve each equation by completing the square.

9. $x^2 - 8x + 15 = 0$ $3, 5$

10. $x^2 - 3x - 54 = 0$ $9, -6$

11. $x^2 = -5x + 6$ $1, -6$

12. $x^2 + 2x - 5 = 0$ $-1 \pm \sqrt{6}$

13. $x^2 - 3x + 8 = 0$ $(3 \pm i\sqrt{23})/2$

14. $2x^2 - 8x = -64$ $2 \pm 2i\sqrt{7}$

15. $4a^2 - 2a + 12 = 0$ $(1 \pm i\sqrt{47})/4$

16. $-x^2 - 6x + 10 = 0$ $-3 \pm \sqrt{19}$

[9.2] *Determine whether the equation has two distinct real solutions, a single real solution, or no real solution.*

17. $3x^2 - 4x - 20 = 0$

18. $-3x^2 + 4x = 9$

19. $2x^2 + 6x + 7 = 0$

20. $p^2 - p + 8 = 0$

21. $n^2 - 12n = -36$

22. $3x^2 - 4x + 5 = 0$

23. $-3x^2 - 4x + 8 = 0$

24. $x^2 - 9x + 6 = 0$

no real solutions 18, 19, 20, 22

one real solution: 21

two real solutions: 17, 23, 24

Solve by the quadratic formula.

25. $x^2 - 9x + 14 = 0$ $2, 7$

26. $x^2 + 7x - 30 = 0$ $3, -10$

27. $x^2 = 7x - 10$ $2, 5$

28. $5x^2 - 7x = 6$ $2, -\frac{3}{5}$

29. $x^2 - 18 = 7x$ $-2, 9$

30. $x^2 - x + 30 = 0$ $(1 \pm i\sqrt{119})/2$

31. $6d^2 + d - 15 = 0$ $\frac{3}{2}, -\frac{5}{3}$

32. $2x^2 + 4x - 3 = 0$ $(-2 \pm \sqrt{10})/2$

33. $-2x^2 + 3x + 6 = 0$ $(3 \pm \sqrt{57})/4$

34. $x^2 - 6x + 7 = 0$ $3 \pm \sqrt{2}$

35. $3x^2 - 6x - 8 = 0$ $(3 \pm \sqrt{33})/3$

36. $2x^2 - 5x = 0$ $0, \frac{5}{2}$

[9.2] *Find the solution to the quadratic equation by the method of your choice.*

37. $x^2 - 11x + 24 = 0$ $3, 8$

38. $x^2 - 16x + 63 = 0$ $7, 9$

39. $x^2 = -3x + 40$ $5, -8$

40. $x^2 + 6x = 27$ $3, -9$

41. $k^2 - 4k - 60 = 0$ $10, -6$

42. $x^2 - x + 42 = 0$ $(1 \pm i\sqrt{167})/2$

43. $x^2 + 11x + 12 = 0$ $(-11 \pm \sqrt{73})/2$

44. $x^2 = 25$ $5, -5$

45. $x^2 + 6x = 0$ $0, -6$

46. $2x^2 + 5x = 3$ $\frac{1}{2}, -3$

47. $(x - 2)(5x + 3) = -4$

48. $6x^2 + 5x = 6$ $\frac{2}{3}, -\frac{3}{2}$

49. $x^2 + 3x - 6 = 0$ $(-3 \pm \sqrt{33})/2$

50. $3x^2 - 11x + 10 = 0$ $2, \frac{5}{3}$

51. $1.2r^2 + 5.7r = 2.3$ $\dfrac{-5.7 \pm \sqrt{43.53}}{2.4}$

52. $-2x^2 + 6x = -9$ $(3 \pm 3\sqrt{3})/2$

53. $2x^2 - 5x = 0$ $0, \frac{5}{2}$

54. $3y = \dfrac{2y + 3}{y - 2}$ $3, -\frac{1}{3}$

55. $x^2 + \dfrac{5}{4}x = \dfrac{3}{8}$ $\frac{1}{4}, -\frac{3}{2}$

56. $x^2 = \dfrac{5}{6}x + \dfrac{25}{6}$ $\frac{5}{2}, -\frac{5}{3}$

47. $(7 \pm \sqrt{89})/10$

79. b. $\sqrt{112.5} \approx 10.61$ sec **82.** larger 23.51 hr; smaller 24.51 hr

(a) *In exercises 57–62, determine whether the parabola opens upward or downard,* **(b)** *find the y intercept,* **(c)** *find the vertex,* **(d)** *find the x intercepts if they exist, and* **(e)** *sketch the graph.*

57. $y = x^2 + 6x$ **(e)**

(a) upward
(b) 0, **(c)** $(-3, -9)$
(d) 0, −6

58. $y = x^2 + 2x - 8$ **(e)**

(a) upward **(b)** −8
(c) $(-1, -9)$
(d) −4, 2

59. $y = 2x^2 + 4x - 16$ **(e)**

(a) upward **(b)** −16
(c) $(-1, -18)$
(d) −4, 2

60. $y = -x^2 - 9$ **(e)**
(a) downward
(b) −9 **(c)** $(0, -9)$
(d) no x intercepts

61. $y = -2x^2 - x + 15$ **(e)**
(a) downward
(b) 15 **(c)** $\left(-\frac{1}{4}, \frac{121}{8}\right)$
(d) $\frac{5}{2}$, −3

62. $y = x^2 + 3x + 8$ **(e)**
(a) upward
(b) 8 **(c)** $\left(-\frac{3}{2}, \frac{23}{4}\right)$
(d) no x intercepts

Write an equation in two variables whose graph has the given x intercepts.

63. 3, −2 $y = x^2 - x - 6$ **64.** $\frac{2}{3}$, −3 $y = 3x^2 + 7x - 6$ **65.** −3, −3 $y = x^2 + 6x + 9$ **66.** $\frac{1}{2}, \frac{2}{3}$ $y = 6x^2 - 7x + 2$

[9.3] *Solve.*   **69.** $\dfrac{-1 \pm 3\sqrt{5}}{2}$ **71.** $\pm 2\sqrt{2}, \pm i\sqrt{3}$ **72.** $\frac{3}{2}, -\frac{2}{5}$ **74.** $\frac{27}{8}$, 8

67. $1 - \dfrac{12}{x} + \dfrac{35}{x^2} = 0$ 5, 7   **68.** $\dfrac{3}{a-1} + \dfrac{2}{a} = 4$ 2, $\frac{1}{4}$ **69.** $\dfrac{5}{r^2-9} = \dfrac{1}{r-3} + 1$ **70.** $\dfrac{5(x-1)}{x+1} = \dfrac{x-1}{x}$ 1, $\frac{1}{4}$

71. $p^4 - 5p^2 = 24$ **72.** $6m^{-2} + 11m^{-1} - 10 = 0$ **73.** $4x + 23\sqrt{x} - 6 = 0$ $\frac{1}{16}$ **74.** $2m^{2/3} - 7m^{1/3} = -6$

[9.1–9.3]
75. The product of two consecutive positive integers is 90. Find the integers.   9, 10

76. The larger of two positive numbers is 4 greater than the smaller. Find the two numbers if their product is 45.   5, 9

77. The length of a rectangle is 1 inch less than twice its width. Find the sides of the rectangle if its area is 66 square inches. $w = 6$ in, $l = 11$ in

78. The value, V, in dollars per acre of a wheat crop d days after planting is given by the formula $V = 12d - 0.05d^2$, $20 < d < 80$. Find the value of an acre of wheat after it has been planted 50 days. $475

79. The distance, d, in feet, that an object is from the ground t seconds after being dropped from an airplane is given by the formula $d = -16t^2 + 1800$.
(a) Find the distance the object is from the ground 3 seconds after it has been dropped.   **(a)** 1656 ft
(b) When will the object hit the ground?

80. If an object is thrown upward from the top of a 100-foot-tall building, its height above the ground, h, at any time, t, can be found by the formula $h = -16t^2 + 16t + 100$.   **(b)** $\frac{4 + \sqrt{416}}{8} \approx 3.05$ sec
(a) Find the height of the object at 2 seconds.   68 ft
(b) When will the object hit the ground?

81. A tractor has an oil leak. The amount of oil, L, in milliliters per hour that leaks out is a function of the tractor's operating temperature, t, in degrees Celsius. The formula to determine the amount of oil that leaks out is $L = 0.0004t^2 + 0.16t + 20$, $100°C \le t \le 160°C$.
(a) How many milliliters of oil will leak out in one hour if the operating temperature of the tractor is 100°C?   40 mi
(b) If oil is leaking out at 53 milliliters per hour, what is the operating termperature of the tractor?   150°C

82. Two molding machines can complete an order in 12 hours. The larger machine could complete the order by itself in 1 hour less time than the smaller machine could by itself. How long will it take each machine to complete the order working by itself?

83. Ben drove 20 miles at a constant speed, then increased his speed by 10 miles per hour for the next 30 miles. If the time required to travel the 50 miles was 0.9 hours, find the speed he drove during the first 20 miles.   50 mph

84. Carmen canoed downstream going with the current for 3 miles, then turned around and canoed upstream against the current to her starting point. If the total time she spent canoeing was 4 hours and the current was 0.4 miles per hour, what is the speed she canoes in still water?   1.6 mph

[9.4] *Graph the solution to each inequality on the number line.*

85. $x^2 + 6x + 5 \ge 0$
86. $x^2 + 2x - 15 \le 0$
87. $x^2 \le 11x - 20$
88. $2x^2 + 6x > 0$
89. $3x^2 + 8x > 16$
90. $4x^2 - 9 \le 0$
91. $5x^2 - 25 > 0$
92. $9x^2 > 25$

Solve each inequality. Give the solutions to Exercises 93–98 in set builder notation and the solutions to Exercises 99–104 in interval notation. **97.** $\{x \mid -3 < x < -1 \text{ or } x > 2\}$ **98.** $\{x \mid x \le 0 \text{ or } 3 \le x \le 5\}$ **99.** $\left[-\frac{4}{3}, 1\right] \cup [3, \infty)$ **100.** $(-\infty, -5) \cup (-2, 0)$

102. $(-\infty, -3) \cup (2, 5)$

93. $\dfrac{x+2}{x-3} > 0$ $\{x \mid x < -2 \text{ or } x > 3\}$

94. $\dfrac{x-5}{x+2} \le 0$ $\{x \mid -2 < x \le 5\}$

95. $\dfrac{2x-4}{x+1} \ge 0$ $\{x \mid x < -1 \text{ or } x \ge 2\}$

96. $\dfrac{3x+5}{x-6} < 0$ $\{x \mid -\frac{5}{3} < x < 6\}$

97. $(x+3)(x+1)(x-2) > 0$

98. $x(x-3)(x-5) \le 0$

99. $(3x+4)(x-1)(x-3) \ge 0$

100. $2x(x+2)(x+5) < 0$

101. $\dfrac{x(x-4)}{x+2} > 0$ $(-2, 0) \cup (4, \infty)$

102. $\dfrac{(x-2)(x-5)}{x+3} < 0$

103. $\dfrac{x-3}{(x+2)(x-5)} \ge 0$ $(-2, 3] \cup (5, \infty)$

104. $\dfrac{x(x-5)}{x+3} \le 0$ $(-\infty, -3) \cup [0, 5]$

Solve each inequality and graph the solution on the number line.

105. $\dfrac{3}{x+4} \ge -1$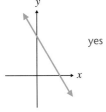

106. $\dfrac{2x}{x-2} \le 1$

107. $\dfrac{2x+3}{3x-5} < 4$

108. $\dfrac{4x}{x+4} \ge 9$

9.5] *Given $f(x) = x^2 - 3x + 4$ and $g(x) = 2x - 5$, find the following.*

109. $(f+g)(x)$ $x^2 - x - 1$

110. $(f+g)(3)$ 5

111. $(g-f)(x)$ $-x^2 + 5x - 9$

112. $(g-f)(-1)$ -15

113. $(fg)(x)$

114. $(fg)(5)$ 70

115. $\left(\dfrac{f}{g}\right)(x)$

116. $\left(\dfrac{f}{g}\right)(2)$ -2

117. $(f \circ g)(x)$

118. $(f \circ g)(2)$ 8

119. $(g \circ f)(x)$ $2x^2 - 6x + 3$

120. $(g \circ f)(-3)$ 39

Given $f(x) = 3x + 2$ and $g(x) = \sqrt{x-4}$, $x \ge 4$, find the following.

121. $(f+g)(x)$

122. $(f-g)(x)$

123. $(fg)x$ $(3x+2)\sqrt{x-4}$

124. $\left(\dfrac{g}{f}\right)(x)$ $\dfrac{\sqrt{x-4}}{3x+2}$, $x \ne -\frac{2}{3}$

125. $(f \circ g)(x)$ $3\sqrt{x-4} + 2$

126. $(g \circ f)(x)$ $\sqrt{3x-2}$, $x \ge \frac{2}{3}$

113. $2x^3 - 11x^2 + 23x - 20$

117. $4x^2 - 26x + 44$

115. $\dfrac{x^2 - 3x + 4}{2x - 5}$, $x \ne \frac{5}{2}$

121. $3x + 2 + \sqrt{x-4}$

122. $3x + 2 - \sqrt{x-4}$

Using the following graph, find each of the following.

127. $(f+g)(1)$ -1

128. $(f-g)(3)$ 2

129. $(gf)(-1)$ -9

130. $\left(\dfrac{g}{f}\right)(0)$ -3

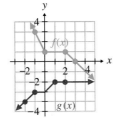

131. Using the graphs of $f(x)$ and $g(x)$ as illustrated in Exercises 127–130, graph $(f+g)(x)$.

131.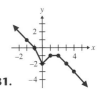

9.6] *Determine whether each function is a one-to-one function.*

132. yes

133. yes

134. no

135. $\{(2, 3), (4, 0), (-5, 7), (3, 8)\}$ yes

136. $\left\{(0, -2), (5, 6), (3, -2), \left(\dfrac{1}{2}, 4\right)\right\}$ no

137. $y = \sqrt{x+1}$, $x \ge -1$ yes

138. $y = x^2 - 9$ no

139. $f(x)$, D: $\{-4, 0, 5, 6\}$, R: $\{-3, 2, 3, 7\}$ **140.** $f(x)$, D: $\{-3, -1, \frac{1}{2}, \sqrt{5}\}$, R: $\{2, \sqrt{7}, 3, 8\}$
$f^{-1}(x)$, D: $\{-3, 2, 3, 7\}$, R: $\{-4, 0, 5, 6\}$ $f^{-1}(x)$, D: $\{2, \sqrt{7}, 3, 8\}$, R: $\{-3, -1, \frac{1}{2}, \sqrt{5}\}$

For the function find the domain and range of both $f(x)$ and $f^{-1}(x)$.

139. $\{(5, 3), (6, 2), (-4, -3), (0, 7)\}$ **140.** $\left\{\left(\frac{1}{2}, 2\right), (-3, 8), (-1, 3), (\sqrt{5}, \sqrt{7})\right\}$

141. 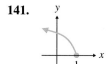 $f(x)$, D: $\{x \mid x \le 1\}$, R: $\{y \mid y \ge 0\}$
$f^{-1}(x)$, D: $\{x \mid x \ge 0\}$, R: $\{y \mid y \le 1\}$

142. $f(x)$, D: $\{x \mid x \ge 0\}$ R: $\{y \mid y \ge 2\}$
$f^{-1}(x)$, D: $\{x \mid x \ge 2\}$, R: $\{y \mid y \ge 0\}$

Find $f^{-1}(x)$ and graph $f(x)$ and $f^{-1}(x)$ on the same axes.

143. $y = f(x) = 4x - 2$ $f^{-1}(x) = (x + 2)/4$

144. $y = f(x) = -3x - 5$
$f^{-1}(x) = -\frac{1}{3}(x + 5)$

145. $y = f(x) = \sqrt[3]{x - 1}$ **143.**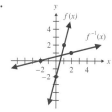

146. $y = f(x) = \sqrt{x - 1}, x \ge 1$
$f^{-1}(x) = x^2 + 1, x \ge 0$

$f^{-1}(x) = x^3 + 1$

144.   **145.**   **146.**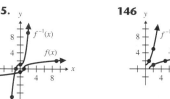

Practice Test

Solve by completing the square.

1. $x^2 = -x + 12$ $3, -4$

2. $4x^2 + 8x = -12$ $-1 \pm i\sqrt{2}$

Solve by the quadratic formula.

3. $x^2 - 5x - 6 = 0$ $-1, 6$

4. $x^2 + 5 = -8x$ $-4 \pm \sqrt{11}$

Solve by the method of your choice.

5. $3x^2 - 5x = 0$ $0, \frac{5}{3}$

6. $-2x^2 = 9x - 5$ $\frac{1}{2}, -5$

7. Solve the formula $P = 3a - b^2$ for b.
Use only the positive root. $b = \sqrt{3a - P}$

 8. Determine whether the following equation has two distinct real solutions, a single unique solution, or no real solution: $5x^2 = 4x + 2$. Explain your answer.
two real solutions

Solve.

9. $10m^4 + 21m^2 = 10$ $\pm\frac{\sqrt{10}}{5}, \pm\frac{i\sqrt{10}}{2}$

10. $3r^{2/3} + 11r^{1/3} - 42 = 0$ $\frac{343}{27}, -216$

Graph the solution to the inequality on the number line.

11. $x^2 - x \ge 42$

12. $\dfrac{(x + 3)(x - 4)}{x + 1} \ge 0$

Solve the inequality. Write the answer in (a) interval notation and (b) set builder notation.

13. $\dfrac{x + 3}{x + 2} \le -1$ **(a)** $[-\frac{5}{2}, -2)$ **(b)** $\{x \mid -\frac{5}{2} \le x < -2\}\}$

14. Consider the quadratic equation $y = x^2 - 2x - 8$.
  (a) Determine whether the parabola opens upward or downward.
  (b) Find the y intercept.   **(c)** Find the vertex.
  (d) Find the x intercepts if they exist. **(a)** upward **(b)** -8
  (e) Sketch the graph. **(c)** $(1, -9)$ **(d)** $4, -2,$

e.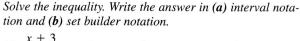

15. Write an equation in two variables whose x intercepts are $-6, \frac{1}{2}$. $y = 2x^2 + 11x - 6$

16. The length of a rectangle is 4 feet greater than twice its width. Find the length and width of the rectangle if the area of the rectangle is 48 square feet. $w = 4$ ft, $l = 12$ ft

17. Kerry throws a ball upward from the top of a building. The distance, d, of the ball from the ground at any time t is $d = -16t^2 + 64t + 80$. How long will it take for the ball to strike the ground?   5 sec

Given $f(x) = x^2 - x + 8$ and $g(x) = 2x - 4$, find the following.

18. $(g - f)(x)$ $-x^2 + 3x - 12$

19. $(f \circ g)(x)$ $4x^2 - 18x + 28$

20. Find the inverse function of $f(x)$.
$y = f(x) = 2x + 4$ $f^{-1}(x) = (x - 4)/2$

Chapter **10**

Conic Sections

See Section 10.1, Group Activity 1.

Preview and Perspective

The focus of this chapter is on graphing conic sections. These include the circle, ellipse, parabola, and hyperbola. We have already discussed parabolas. In this chapter we will learn more about them.

We solved linear systems of equations graphically in Section 4.1. In Section 10.5 we solve nonlinear systems of equations. The systems we solve will contain the equation of at least one conic section. We will provide both graphical and algebraic solutions to the systems.

In Section 4.6 we graphed linear systems of inequalities in two variables. In Section 10.6 we graph nonlinear systems of inequalities, where at least one inequality is a conic section.

Conic sections are important in higher mathematics courses and in certain science and engineering courses. If you take additional mathematics courses, you will probably cover this material again, but in more depth.

10.1 The Circle

Tape 16

1 Identify and describe the conic sections.
2 Graph circles with the center at the origin.
3 Graph circles with the center at (h, k).

Conic Sections

1 In previous chapters we discussed parabolas. A parabola is one type of conic section. Parabolas will be discussed further in Section 10.3. Other conic sections are circles, ellipses, and hyperbolas. Each of these shapes is called a conic section because each can be made by slicing a cone and observing the shape of the slice. The methods used to slice the cone to obtain each conic section are illustrated in Fig. 10.1.

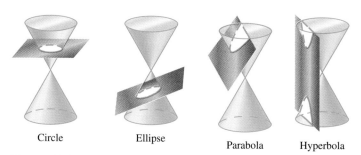

Circle   Ellipse   Parabola   Hyperbola

FIGURE 10.1

Circle at the Origin

2 A **circle** may be defined as the set of points in a plane that are the same distance from a fixed point called its **center**.

The formula for the **standard form** of a circle whose center is at the origin may be derived using the distance formula discussed in Section 3.1. Let (x, y) be a point on a circle of radius r with center at $(0, 0)$ (see Fig. 10.2). Using the distance formula, we have

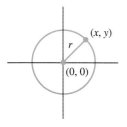

FIGURE 10.2

$$d = \sqrt{(x_2 - x_1)^2 + (y_2 - y_1)^2}$$
$$\text{or} \quad r = \sqrt{(x - 0)^2 + (y - 0)^2}$$
$$r = \sqrt{x^2 + y^2}$$
$$r^2 = x^2 + y^2 \quad \text{or} \quad x^2 + y^2 = r^2$$

> **Circle with Its Center at the Origin and Radius r**
>
> $$x^2 + y^2 = r^2$$

For example, $x^2 + y^2 = 16$ is a circle with its center at the origin and radius 4, and $x^2 + y^2 = 7$ is a circle with its center at the origin and radius $\sqrt{7}$. Note that $4^2 = 16$ and $(\sqrt{7})^2 = 7$.

EXAMPLE 1 Graph

(a) $x^2 + y^2 = 64$ **(b)** $y = \sqrt{64 - x^2}$ **(c)** $y = -\sqrt{64 - x^2}$

Solution:

(a) If we rewrite the equation as

$$x^2 + y^2 = 8^2$$

we see that the radius is 8. The graph is illustrated in Fig. 10.3.

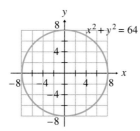

FIGURE 10.3

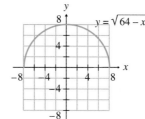

FIGURE 10.4

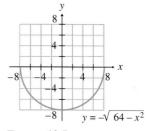

FIGURE 10.5

(b) If we solve the equation $x^2 + y^2 = 64$ for y, we obtain

$$y^2 = 64 - x^2$$
$$y = \pm\sqrt{64 - x^2}$$

The graph of $y = \sqrt{64 - x^2}$, where y represents the principal square root, lies above and on the x axis. For any value of x in the domain of the function, the value of y must be greater than or equal to 0. Why? The graph is the semicircle shown in Fig. 10.4.

(c) The graph of $y = -\sqrt{64 - x^2}$ is also a semicircle. However, this graph lies below and on the x axis. For any value of x in the domain of the function, the value of y must be less than or equal to 0. Why? The graph is shown in Fig. 10.5.

Consider the equations $y = \sqrt{64 - x^2}$ and $y = -\sqrt{64 - x^2}$ in Example 1(**b**) and (**c**). If you square both sides of the equations and rearrange the terms, you will obtain $x^2 + y^2 = 64$. Try this now and see.

Graphing Calculator Corner

When using your calculator to obtain a graph, you insert the equation you wish to graph to the right of $y =$. The equation you insert must be a function. Circles are not functions since they do not pass the vertical line test. To graph the equation $x^2 + y^2 = 64$, which from Example 1 we know is a circle of radius 8, we solve the equation for y to obtain $y = \pm\sqrt{64 - x^2}$. We then graph the two functions $y = \sqrt{64 - x^2}$ and $y = -\sqrt{64 - x^2}$ on the same axes to obtain the circle. These graphs are illustrated for one type of calculator in Fig. 10.6. Because of the distortion (described in the Calculator Corner in Section 9.6), the graph does not appear to be a circle. When you use the *square feature* of the calculator, the figure appears as a circle. (see Fig. 10.7).

3.

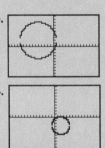

4.

1.

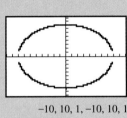

$-10, 10, 1, -10, 10, 1$

FIGURE 10.6

2.

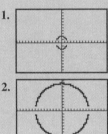

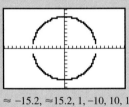

$\approx -15.2, \approx 15.2, 1, -10, 10, 1$

FIGURE 10.7

EXERCISES

Using the square feature, graph the equations on your graphing calculator.

1. $x^2 + y^2 = 5$ **2.** $4x^2 + 4y^2 = 324$
3. $(x + 5)^2 + (y - 2)^2 = 36$ **4.** $(x - 2)^2 + (y + 3)^2 = 9$

Circle at (h, k)

FIGURE 10.8

3 The standard form of a circle with center at (h, k) and radius r can be derived using the distance formula. Let (h, k) be the center of the circle and let (x, y) be any point on the circle (see Fig. 10.8). If the radius r represents the distance between points (x, y) and (h, k), then by the distance formula

$$r = \sqrt{(x - h)^2 + (y - k)^2}$$

We now square both sides of the equation to obtain the standard form of a circle with center at (h, k) and radius r.

$$r^2 = (x - h)^2 + (y - k)^2$$

> **Circle with Its Center at (h, k) and Radius r**
>
> $$(x - h)^2 + (y - k)^2 = r^2$$

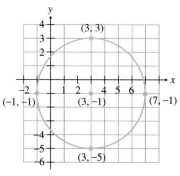

FIGURE 10.9

EXAMPLE 2

(a) Determine the equation of the circle with center at $(3, -1)$ and radius 4.

(b) Sketch the circle.

Solution:

(a) The center is $(3, -1)$. Thus, h is 3 and k is -1. The radius, r, is 4.

$$(x - h)^2 + (y - k)^2 = r^2$$
$$(x - 3)^2 + [y - (-1)]^2 = 4^2$$
$$(x - 3)^2 + (y + 1)^2 = 16$$

(b) See Fig. 10.9.

EXAMPLE 3 Determine the equation of the circle shown in Fig. 10.10.

Solution: The center is $(-3, 2)$ and the radius is 3. The equation is therefore

$$[x - (-3)]^2 + (y - 2)^2 = 3^2$$
$$(x + 3)^2 + (y - 2)^2 = 9$$

In Example 4, and other examples in this chapter, we complete the square. If you have forgotten the procedure to complete the square, review Section 9.1 now.

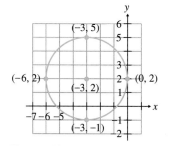

FIGURE 10.10

EXAMPLE 4

(a) Show that the graph of the equation $x^2 + y^2 + 6x - 2y + 6 = 0$ is a circle by using the procedure for completing the square to rewrite this equation in standard form.

(b) Determine the center and radius of the circle and then sketch the circle.

Solution: **(a)** First rewrite the equation, placing all the terms containing like variables together.

$$x^2 + 6x + y^2 - 2y + 6 = 0$$

Then move the constant to the right side of the equation.

$$x^2 + 6x + y^2 - 2y = -6$$

Now we complete the square twice, once for each variable. We will first work with the variable x.

$$x^2 + 6x + 9 + y^2 - 2y = -6 + 9$$

Now work with the variable y.

$$x^2 + 6x + 9 + y^2 - 2y + 1 = -6 + 9 + 1$$

or

$$x^2 + 6x + 9 + y^2 - 2y + 1 = 4$$
$$(x + 3)^2 + (y - 1)^2 = 4$$
$$(x + 3)^2 + (y - 1)^2 = 2^2$$

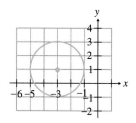

FIGURE 10.11

(b) The center of the circle is at $(-3, 1)$ and the radius is 2. The circle is sketched in Fig. 10.11.

Exercise Set 10.1

Write the equation of the circle with the given center and radius; then sketch the graph of the equation.

1. Center $(0, 0)$, radius 3

$x^2 + y^2 = 9$

2. Center $(0, 0)$, radius 5

$x^2 + y^2 = 25$

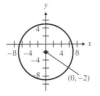

3. Center $(3, 0)$, radius 1

$(x - 3)^2 + y^2 = 1$

7.

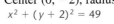

4. Center $(0, -2)$, radius 7

$x^2 + (y + 2)^2 = 49$

5. Center $(-6, 5)$, radius 5

$(x + 6)^2 + (y - 5)^2 = 25$

8.

6. Center $(-4, -1)$, radius 4

$(x + 4)^2 + (y + 1)^2 = 16$

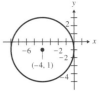

7. Center $(4, 7)$, radius $\sqrt{8}$ $(x - 4)^2 + (y - 7)^2 = 8$

8. Center $(0, -2)$, radius $\sqrt{12}$ $x^2 + (y + 2)^2 = 12$

Sketch the graph of each equation.

9. $x^2 + y^2 = 16$

10. $x^2 + y^2 = 9$

11. $x^2 + y^2 = 3$

12. $x^2 + y^2 = 10$

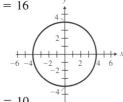

13. $x^2 + (y - 3)^2 = 4$

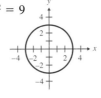

14. $(x + 4)^2 + y^2 = 25$

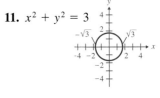

15. $(x - 2)^2 + (y + 3)^2 = 16$

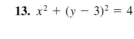

16. $(x + 8)^2 + (y + 2)^2 = 9$

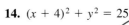

17. $(x + 1)^2 + (y - 4)^2 = 36$

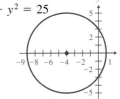

18. $y = \sqrt{16 - x^2}$

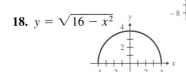

19. $y = \sqrt{25 - x^2}$

20. $y = -\sqrt{4 - x^2}$

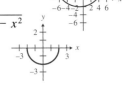

21. $y = -\sqrt{49 - x^2}$

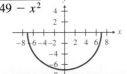

22. $y = -\sqrt{81 - x^2}$

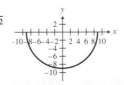

Write the equation of the circle.

23.

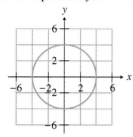

$x^2 + y^2 = 16$

24.

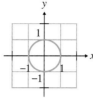

$x^2 + y^2 = 1$

25.

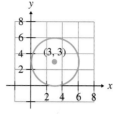

$(x - 3)^2 + (y - 3)^2 = 9$

26.

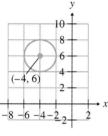

$(x + 4)^2 + (y - 6)^2 = 4$

27.

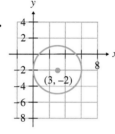

$(x - 3)^2 + (y + 2)^2 = 9$

28.

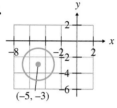

$(x + 5)^2 + (y + 3)^2 = 4$

Use the procedure for completing the square to write the equation in standard form; then sketch the graph (see Example 4).

29. $x^2 + y^2 + 10y - 75 = 0$
$x^2 + (y + 5)^2 = 10^2$

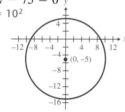

30. $x^2 + y^2 - 4y = 0$
$x^2 + (y - 2)^2 = 2^2$

31. $x^2 + 8x - 9 + y^2 = 0$
$(x + 4)^2 + y^2 = 5^2$

32. $x^2 + y^2 + 6x - 4y + 9 = 0$
$(x + 3)^2 + (y - 2)^2 = 2^2$

33. $x^2 + y^2 + 2x - 4y - 4 = 0$
$(x + 1)^2 + (y - 2)^2 = 3^2$

34. $x^2 + y^2 + 4x - 6y - 3 = 0$
$(x + 2)^2 + (y - 3)^2 = 4^2$

35. $x^2 + y^2 + 6x - 2y + 6 = 0$
$(x + 3)^2 + (y - 1)^2 = 2^2$

36. $x^2 + y^2 + 8x - 4y + 4 = 0$
$(x + 4)^2 + (y - 2)^2 = 4^2$

37. $x^2 + y^2 - 8x + 2y + 13 = 0$
$(x - 4)^2 + (y + 1)^2 = 2^2$

38. $x^2 - x + y^2 + 3y - \dfrac{3}{2} = 0$

$\left(x - \dfrac{1}{2}\right)^2 + \left(y + \dfrac{3}{2}\right)^2 = 2^2$

39. What is the definition of a circle?

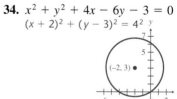

a set of points in a plane that are the same distance from a fixed point

[6.3] **40.** Factor $8x^3 - 64$. $8(x - 2)(x^2 + 2x + 4)$

Solve the formula for the indicated variable.

[7.4] **41.** $E = 1 - \dfrac{T_1}{T_2}$ for T_2. $T_2 = \dfrac{T_1}{1 - E}$

[9.1] **42.** $E = \dfrac{V + P^2}{2}$ for P. $P = \sqrt{2E - V}$

[9.2] **43.** Solving using the quadratic formula:

$$3x^2 - 5x + 5 = 0 \qquad \dfrac{5 \pm i\sqrt{35}}{6}$$

Group Activity/ Challenge Problems

1. A highway department is planning to construct a semicircular tunnel through a mountain. The tunnel is to be large enough so that a truck 8 feet wide and 10 feet tall will pass through the center of the tunnel with 1 foot to spare when driving down the center of the tunnel (as shown in the figure below.) Determine the minimum radius of the tunnel.

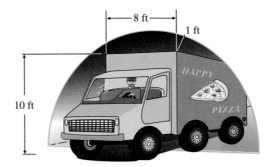

$1 + \sqrt{116} \approx 11.77$ ft

2. Consider the figure below. Write an equation for **(a)** the blue circle; **(b)** the red circle; **(c)** the green circle. **(d)** Find the shaded area.

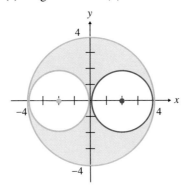

(a) $x^2 + y^2 = 16$
(b) $(x - 2)^2 + y^2 = 4$
(c) $(x + 2)^2 + y^2 = 4$
(d) 8π

3. Find the shaded area of the square in the figure on the left. The equation of the circle is $x^2 + y^2 = 9$.   4.5 sq units

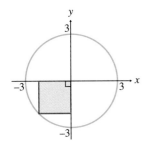

4. Consider the equations $x^2 + y^2 = 16$ and $(x - 2)^2 + (y - 2)^2 = 16$. By considering the center and radius of each circle, determine the number of points of intersection of the two circles.   2

5. Find the area between the two concentric circles whose equations are $(x - 2)^2 + (y + 4)^2 = 16$ and $(x - 2)^2 + (y + 4)^2 = 64$. *Concentric circles* are circles that have the same center. 48π

10.2 The Ellipse

1️⃣ Graph ellipses.

Graph Ellipses

1️⃣ An **ellipse** may be defined as a set of points in a plane, the sum of whose distances from two fixed points is a constant. The two fixed points are called the **foci** (each is a focus) of the ellipse (see Fig. 10.12).

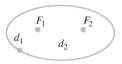

FIGURE 10.12

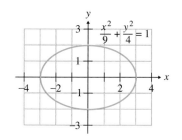

FIGURE 10.13

We can construct an ellipse using a length of string and two thumbtacks. Place the two thumbtacks fairly close together (Fig. 10.13). Then tie the ends of the string to the thumbtacks. With a pencil or pen pull the string taut, and, while keeping the string taut, draw the ellipse by moving the pencil around the thumbtacks.

The standard form of an ellipse with its center at the origin (Fig. 10.14) follows.

FIGURE 10.14

> ### Ellipse with Its Center at the Origin
>
> $$\frac{x^2}{a^2} + \frac{y^2}{b^2} = 1$$
>
> where a and $-a$ are the x intercepts and b and $-b$ are the y intercepts.

In Fig. 10.14, the line segment from $-a$ to a on the x axis is the *longer* or **major axis** and the line segment from $-b$ to b is the *shorter* or **minor axis** of the ellipse.

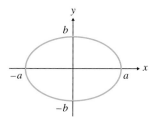

FIGURE 10.15

EXAMPLE 1 Sketch the graph of $\dfrac{x^2}{9} + \dfrac{y^2}{4} = 1$.

Solution: We can rewrite the equation as

$$\frac{x^2}{3^2} + \frac{y^2}{2^2} = 1$$

Thus, $a = 3$ and the x intercepts are ± 3. Since $b = 2$, the y intercepts are ± 2. The ellipse is illustrated in Fig. 10.15.

An equation may be camouflaged so that it may not be obvious that its graph is an ellipse. This is illustrated in Example 2.

EXAMPLE 2 Sketch the graph of $20x^2 + 9y^2 = 180$.

Solution: If we divide both sides of the equation by 180 to make the right side of the equation equal to 1, we obtain an equation that we can recognize as an ellipse.

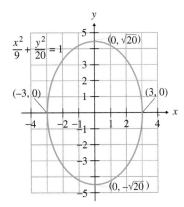

FIGURE 10.16

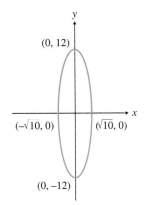

FIGURE 10.17

$$\frac{20x^2 + 9y^2}{180} = \frac{180}{180}$$

$$\frac{20x^2}{180} + \frac{9y^2}{180} = 1$$

$$\frac{x^2}{9} + \frac{y^2}{20} = 1$$

The equation can now be recognized as an ellipse in standard form.

$$\frac{x^2}{a^2} + \frac{y^2}{b^2} = 1$$

Since $a^2 = 9$, $a = 3$. We know that $b^2 = 20$; thus $b = \sqrt{20}$ (or approximately 4.47).

$$\frac{x^2}{3^2} + \frac{y^2}{(\sqrt{20})^2} = 1$$

The x intercepts are ± 3. The y intercepts are $\pm\sqrt{20}$. The graph is illustrated in Fig. 10.16. Note that the major axis lies along the y axis instead of along the x axis.

EXAMPLE 3 Write the equation of the ellipse illustrated in Fig. 10.17.

Solution: The x intercepts are $\pm\sqrt{10}$; thus $a = \sqrt{10}$ and $a^2 = 10$. The y intercepts are ± 12; thus, $b = 12$ and $b^2 = 144$.

$$\frac{x^2}{a^2} + \frac{y^2}{b^2} = 1$$

$$\frac{x^2}{10} + \frac{y^2}{144} = 1$$

Exercise Set 10.2

Sketch the graph of each equation.

1. $\dfrac{x^2}{4} + \dfrac{y^2}{1} = 1$

2. $\dfrac{x^2}{9} + \dfrac{y^2}{4} = 1$

3. $\dfrac{x^2}{4} + \dfrac{y^2}{9} = 1$

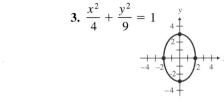

4. $\dfrac{x^2}{25} + \dfrac{y^2}{9} = 1$

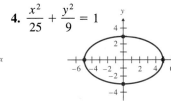

5. $\dfrac{x^2}{16} + \dfrac{y^2}{25} = 1$

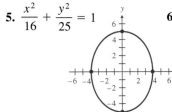

6. $\dfrac{x^2}{9} + \dfrac{y^2}{121} = 1$

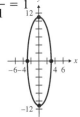

7. $9x^2 + 12y^2 = 108$

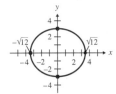

8. $9x^2 + 4y^2 = 36$

13. The set of points in a plane, the sum of whose distances from two fixed points is a constant.   **14.** $a > b$, ellipse, major axis is along the x axis; $a < b$, ellipse, major axis is along the y axis; $a = b$, circle   **16.** If $a = b$, the formula for a circle is obtained.

9. $100x^2 + 25y^2 = 400$ **10.** $x^2 + 36y^2 = 36$ **11.** $9x^2 + 25y^2 = 225$ **12.** $x^2 + 2y^2 = 8$

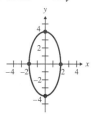   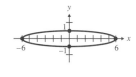   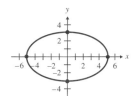

13. What is the definition of an ellipse?

14. In your own words discuss the graphs of $\dfrac{x^2}{a^2} + \dfrac{y^2}{b^2} = 1$ when $a > b$, $a < b$, and $a = b$.

15. How many points are on the graph of $9x^2 + 36y^2 = 0$? Explain.   one, at (0, 0).

16. Explain why the circle is a special case of the ellipse.

17. Consider the graph of the equation $\dfrac{x^2}{a^2} + \dfrac{y^2}{b^2} = 1$, where $a > b$. Explain what will happen to the shape of the graph as the value of b gets closer to the value of a. What is the shape of the graph when $a = b$?   becomes more circular. When $a = b$, the graph is a circle.

CUMULATIVE REVIEW EXERCISES

Solve the following equations or inequalities. Indicate the solution on the number line.

[2.5] **18.** $-3 \le 4 - \frac{1}{2}x < 6$

[2.6] **19.** $|2x - 4| = 8$

20. $|2x - 4| \le 8$

21. $|2x - 4| > 8$

. . . .

Group Activity/ Challenge Problems

The standard form of an ellipse with center at (h, k) is

$$\frac{(x - h)^2}{a^2} + \frac{(y - k)^2}{b^2} = 1$$

where a and b are distances from the center to the vertices, as shown.

3.

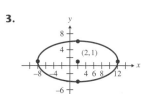   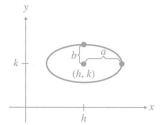

1.

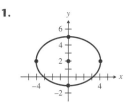

2.

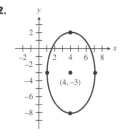

Sketch the graph of the equations.

1. $\dfrac{x^2}{16} + \dfrac{(y - 2)^2}{9} = 1$ **2.** $\dfrac{(x - 4)^2}{9} + \dfrac{(y + 3)^2}{25} = 1$

3. Write the following equation in standard form. Determine the center of the ellipse and then sketch the ellipse.

$$x^2 + 4y^2 - 4x - 8y - 92 = 0 \qquad \frac{(x - 2)^2}{100} + \frac{(y - 1)^2}{25} = 1$$

4. How many points of intersection will the graphs of the equations $x^2 + y^2 = 16$ and $\dfrac{x^2}{4} + \dfrac{y^2}{9} = 1$ have? Explain.   none

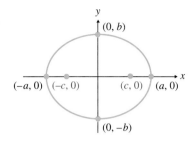

5. The equation of an ellipse is $\dfrac{x^2}{a^2} + \dfrac{y^2}{b^2} = 1$. In an ellipse with foci $(-c, 0)$ and $(c, 0)$ (see figure at left) it can be shown that $a^2 = b^2 + c^2$ (the Pythagorean relationship).

  (a) If the center of the ellipse is at the origin, the x intercepts are at $(-5, 0)$ and $(5, 0)$ and the y intercepts are at $(0, 3)$ and $(0, -3)$, find the location of the foci.
  $(4, 0), (-4, 0)$

  (b) If the x intercepts are $(-13, 0)$ and $(13, 0)$ and the foci are at $(-5, 0)$ and $(5, 0)$, find the y intercepts. $(0, 12)\ (0, -12)$

6. In Exercise 5, we explained that $a^2 = b^2 + c^2$. The ratio c/a is called the *eccentricity* of the ellipse. The greater the eccentricity, the greater the ellipse is "out of roundness." A circle, which is a special form of an ellipse, has eccentricity 0. The eccentricity of the orbits of Mercury, Venus, Earth, and Mars about the sun are 0.206, 0.007, 0.017, and 0.093, respectively. The eccentricity of the moon's orbit about Earth is 0.053. Of the four planets mentioned which orbit is **(a)** the most circular; **(b)** the least circular? **(c)** Use $a^2 = b^2 + c^2$ to find the eccentricity of an ellipse whose x intercepts are $(13, 0)$ and $(-13, 0)$ and whose y intercepts are $(0, 12)$ and $(0, -12)$.   **(a)** Venus **(b)** Mercury
(c) $\frac{5}{13} \approx 0.38$

7. Ultrasound waves are used to shatter kidney stones. The procedure makes use of the fact that sound waves emitted from one foci of an ellipse reflect off the surface of the ellipse to the other foci (see the figure below). Do research and write a report describing the procedure used to shatter kidney stones. Make sure that you explain how the waves are directed on the stone.

8. The rotunda of the Capital Building in Washington, D.C. is a "whispering gallery." Do research and explain why one person standing at a certain point can whisper something and someone standing a considerable distance away can hear it.

10.3   The Parabola

Tape
17

1. Derive the equation $y = a(x - h)^2 + k$.
2. Graph parabolas of the forms $y = a(x - h)^2 + k$ and $x = a(y - k)^2 + h$.
3. Convert equations from the form $y = ax^2 + bx + c$ to the form $y = a(x - h)^2 + k$.

Equation of a Parabola

1. We have discussed the parabola in Chapters 5 and 9. In this section we will discuss the parabola further. Below we start with a quadratic equation of the form $y = ax^2 + bx + c$. Completing the square on the first two terms, we obtain another form of the quadratic equation.

$$y = ax^2 + bx + c$$

$$y = a\left(x^2 + \frac{b}{a}x\right) + c$$

$$y = a\left[x^2 + \frac{b}{a}x + \left(\frac{b}{2a}\right)^2\right] + c - a\left(\frac{b}{2a}\right)^2$$

$$y = a\left(x + \frac{b}{2a}\right)^2 + c - a\left(\frac{b^2}{4a^2}\right)$$

$$y = a\left(x + \frac{b}{2a}\right)^2 + c\left(\frac{4a}{4a}\right) - \frac{b^2}{4a}$$

$$y = a\left(x + \frac{b}{2a}\right)^2 + \frac{4ac}{4a} - \frac{b^2}{4a}$$

$$y = a\left(x + \frac{b}{2a}\right)^2 + \frac{4ac - b^2}{4a}$$

$$y = a\left[x - \left(\frac{-b}{2a}\right)\right]^2 + \frac{4ac - b^2}{4a}$$

We know from earlier sections that $\dfrac{-b}{2a}$ is the x coordinate of the vertex of the parabola. If x equals $\dfrac{-b}{2a}$, the term $a\left[x - \left(\dfrac{-b}{2a}\right)\right]^2$ equals 0, and y equals $\dfrac{4ac - b^2}{4a}$. Thus, the y coordinate at the vertex must be $\dfrac{4ac - b^2}{4a}$. Therefore, the coordinates of the vertex of a parabola are

$$\left(\frac{-b}{2a}, \frac{4ac - b^2}{4a}\right)$$

Let us simplify the equation by substituting $h = \dfrac{-b}{2a}$ and $k = \dfrac{4ac - b^2}{4a}$. The equation then becomes

$$y = a(x - h)^2 + k$$

We can therefore reason that the graph of an equation of the form $y = a(x - h)^2 + k$ is a parabola with vertex at (h, k). If a in the equation $y = a(x - h)^2 + k$ is a positive number, the parabola will open upward, and if a is a negative number, the parabola will open downward.

Parabolas can also open to the right or left. The graph of an equation of the form $x = a(y - k)^2 + h$ will be a parabola whose vertex is at the point (h, k). If a is a positive number, the parabola will open to the right, and if a is a negative number, the parabola will open to the left. The four different forms of a parabola are shown in Fig. 10.18.

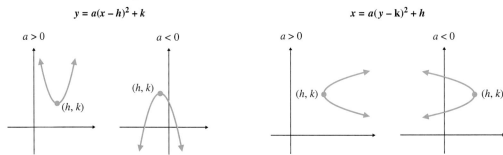

FIGURE 10.18

Parabola with Vertex at (h, k)

1. $y = a(x - h)^2 + k, a > 0$ (opens upward)
2. $y = a(x - h)^2 + k, a < 0$ (opens downward)
3. $x = a(y - k)^2 + h, a > 0$ (opens to the right)
4. $x = a(y - k)^2 + h, a < 0$ (opens to the left)

Note that equations of the form $y = a(x - h)^2 + k$ are functions since their graphs pass the vertical line test. However, equations of the form $x = a(y - k)^2 + h$ are not functions since their graphs do not pass the vertical line test.

Sketching Parabolas   **2**   Now we will sketch some parabolas.

EXAMPLE 1   Sketch the graph of $y = (x - 2)^2 + 3$.

Solution: The graph opens upward since the equation is of the form $y = a(x - h)^2 + k$ and $a = 1$, which is greater than 0. The vertex is at $(2, 3)$ (see Fig. 10.19). Let $x = 0$ and note that the y intercept is $(0 - 2)^2 + 3 = 4 + 3$ or 7.

EXAMPLE 2   Sketch the graph of $x = -2(y + 4)^2 - 1$.

Solution: The graph opens to the left since the equation is of the form $x = a(y - k)^2 + h$ and $a = -2$, which is less than 0 (see Fig. 10.20). The equation can be expressed as $x = -2[y - (-4)]^2 - 1$. Thus, $h = -1$ and $k = -4$. The vertex of the graph is $(-1, -4)$. If we set $y = 0$, we see that the x intercept is at $-2(0 + 4)^2 - 1 = -2(16) - 1$ or -33.

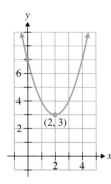

FIGURE 10.19

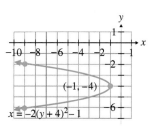

Figure 10.20

Changing the Form of an Equation

☒3 In Examples 3 and 4, we will convert an equation from the form $y = ax^2 + bx + c$ to the form $y = a(x - h)^2 + k$ before graphing.

Example 3

(**a**) Write the equation $y = x^2 - 6x + 8$ in the form $y = a(x - h)^2 + k$.

(**b**) Sketch the graph of $y = x^2 - 6x + 8$.

Solution:

(**a**) We convert $y = x^2 - 6x + 8$ to the form $y = a(x - h)^2 + k$ by completing the square.

$$y = x^2 - 6x + 8$$

Take one-half the coefficient of the x term; then square it.

$$\frac{-6}{2} = -3, \qquad (-3)^2 = 9$$

Now add $+9$ and -9 to the right side of the equation to obtain

$$y = x^2 - 6x \boxed{+ 9} \boxed{- 9} + 8$$

By doing this, we have created a perfect square trinomial plus a constant on the right side.

$$y = \underbrace{x^2 - 6x + 9}_{(x - 3)^2} \underbrace{- 9 + 8}_{-1}$$

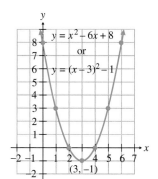

Figure 10.21

(**b**) The vertex of the parabola is at $(3, -1)$ and the parabola opens upward since $a = 1$. The y intercept is $(0 - 3)^2 - 1 = 9 - 1$ or 8. The x intercepts, 2 and 4, can be found by solving the equation $x^2 - 6x + 8 = 0$. Can you explain why? The graph is sketched in Fig. 10.21.

Example 4

(**a**) Write the equation $y = 2x^2 + 4x - 6$ in the form $y = a(x - h)^2 + k$.

(**b**) Sketch the graph of $y = 2x^2 + 4x - 6$.

Solution:

(a) First, factor 2 from the two terms containing the variable to make the coefficient of the squared term 1. Do not factor 2 from the constant, -6.

$$y = 2x^2 + 4x - 6$$
$$y = 2(x^2 + 2x) - 6$$

Now complete the square on the expression in parentheses by taking one-half the coefficient of the x term and squaring it.

$$\frac{2}{2} = 1, \qquad 1^2 = 1$$

Now add $+1$ inside the parentheses. Since the terms inside the parentheses are multiplied by 2, we are really adding 2(1) or 2. Therefore, we must also add -2 outside the parentheses. In doing this, we are not changing the equation, since we are adding 2 and -2, which sums to zero.

$$y = 2(x^2 + 2x + 1) - 2 - 6$$
$$y = 2(x + 1)^2 - 8$$

(b) This parabola opens upward since $a = 2$, which is greater than 0. Its vertex is at $(-1, -8)$. The y intercept is at $2(0 + 1)^2 - 8 = 2(1) - 8$ or -6 (see Fig. 10.22).

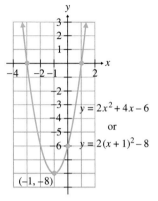

$y = 2x^2 + 4x - 6$
or
$y = 2(x + 1)^2 - 8$
$(-1, -8)$

FIGURE 10.22

EXAMPLE 5

(a) Write the equation $x = -2y^2 + 4y + 5$ in the form $x = a(y - k)^2 + h$.

(b) Sketch the graph of $x = -2y^2 + 4y + 5$.

Solution:

(a) First factor -2 from the first two terms. Then complete the square on the expression within parentheses.

$$x = -2y^2 + 4y + 5$$
$$x = -2(y^2 - 2y) + 5$$
$$x = -2(y^2 - 2y + 1) + 2 + 5$$
$$x = -2(y - 1)^2 + 7$$

(b) Since $a < 0$, this parabola opens to the left. Note that when $y = 0$, $x = 5$. Therefore, the x intercept is 5. The vertex of the parabola is $(7, 1)$. The graph is shown in Fig. 10.23.

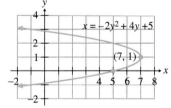

$x = -2y^2 + 4y + 5$
$(7, 1)$

FIGURE 10.23

Exercise Set 10.3

Sketch the graph of each equation.

1. $y = (x - 2)^2 + 3$

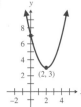

$(2, 3)$

2. $y = (x + 1)^2 - 2$

$(-1, -2)$

3. $y = -(x - 3)^2 - 6$

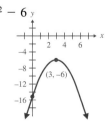

$(3, -6)$

4. $x = (y + 6)^2 + 1$

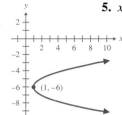

5. $x = (y - 4)^2 - 3$

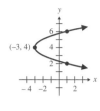

6. $x = -(y - 5)^2 + 4$

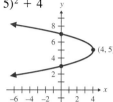

7. $y = 2(x + 6)^2 - 4$

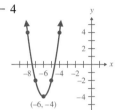

8. $y = -3(x - 5)^2 + 3$

9. $x = -5(y + 3)^2 - 6$

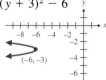

10. $x = -(y - 7)^2 + 8$

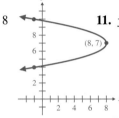

11. $y = -2\left(x + \dfrac{1}{2}\right)^2 + 6$

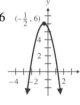

12. $y = -\left(x - \dfrac{5}{2}\right)^2 + \dfrac{1}{2}$

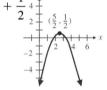

Write each equation in the form $y = a(x - h)^2 + k$ or $x = a(y - k)^2 + h$, and then sketch the graph of the equation.

13. $y = x^2 + 2x$
$y = (x + 1)^2 - 1$

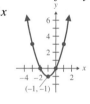

14. $y = x^2 - 4x$
$y = (x - 2)^2 - 4$

15. $x = y^2 + 6y$
$x = (y + 3)^2 - 9$

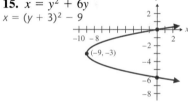

16. $y = x^2 - 6x + 8$
$y = (x - 3)^2 - 1$

17. $y = x^2 + 2x - 15$
$y = (x + 1)^2 - 16$

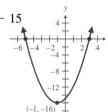

18. $x = y^2 + 8y + 7$
$x = (y + 4)^2 - 9$

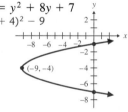

19. $x = -y^2 + 6y - 9$
$x = -(y - 3)^2$

20. $y = -x^2 + 4x - 4$
$y = -(x - 2)^2$

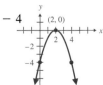

21. $y = x^2 + 7x + 10$
$y = \left(x + \dfrac{7}{2}\right)^2 - \dfrac{9}{4}$

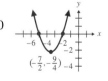

22. $x = -y^2 + 3y - 4$
$x = -\left(y - \dfrac{3}{2}\right)^2 - \dfrac{7}{4}$

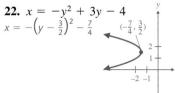

23. $y = 2x^2 - 4x - 4$
$y = 2(x - 1)^2 - 6$

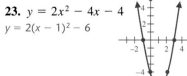

24. $x = 3y^2 - 12y - 36$
$x = 3(y - 2)^2 - 48$

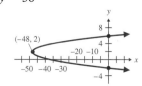

25. $y = a(x - h)^2 + k$ opens up if $a > 0$, down if $a < 0$. $x = a(y - k)^2 + h$ opens right if $a > 0$, left if $a < 0$.
26. Maximum or minimum value, k, will occur when $x = h$. Therefore, vertex is at (h, k). **28.** No, will not pass the vertical line test.
D: $\{x \mid x \geq h\}$; R: $\mathbb{R}$.

25. Explain how to determine the direction the parabola will open by observing the equation.

26. When functions of the form $y = a(x - h)^2 + k$ are graphed, their graphs will be parabolas with vertex at the point (h, k). By studying the equation, explain why the vertex occurs at (h, k).

27. Will all parabolas of the form $y = a(x - h)^2 + k$, $a > 0$, be functions? Explain your answer. What will be the domain and range of $y = a(x - h)^2 + k$, $a > 0$?

27. Yes, D; $\mathbb{R}$; R: $\{y \mid y \geq k\}$.

28. Will all parabolas of the form $x = a(y - k)^2 + h$, $a > 0$, be functions? Explain your answer. What will be the domain and range of $x = a(y - k)^2 + h$, $a > 0$?

29. How will the graphs of $y = 2(x - 3)^2 + 4$ and $y = -2(x - 3)^2 + 4$ compare? Explain your answer.
Both have vertex (3, 4). $y = 2(x - 3)^2 + 4$ opens up and has no x intercepts; $y = -2(x - 3)^2 + 4$ opens down and has two x intercepts.

CUMULATIVE REVIEW EXERCISES   **32.(a)** $\frac{13}{9}$ **(b)** $-2a^2 - 4ab - 2b^2 - a - b + 2$

[3.3] **30.** Write the equation, in slope–intercept form, of the graph that passes through the points $(-6, 4)$ and $(-2, 2)$. $y = -\frac{1}{2}x + 1$

[4.4] **31.** Evaluate the determinant: $\begin{vmatrix} 4 & 0 & 3 \\ 5 & 2 & -1 \\ 3 & 6 & 4 \end{vmatrix}$.   128

[5.6] **32.** $f(x) = -2x^2 - x + 2$; find
(a) $f\left(\frac{1}{3}\right)$;   **(b)** $f(a + b)$.

[7.6] **33.** T varies jointly as m_1 and m_2 and inversely as the square of R. If $T = \frac{3}{2}$ when $m_1 = 6$, $m_2 = 4$, and $R = 4$, find T when $m_1 = 6$, $m_2 = 10$, and $R = 2$.   15

Group Activity/ Challenge Problems

1. The St. Louis Arch, pictured here, appears to be a parabola, but it is really not. Do research and write a report on the name of this curve.

2. The photo shows the bridge on Route 255 which crosses the Mississippi River near Jefferson Barracks, Missouri. The bridge is a parabola with equation $x^2 = 1135y$. The horizontal road on the bridge is 909 feet, and the highest point of the bridge is 182 feet above the center of the road below.

(a) What is the *horizontal distance* of the road from the center of the bridge to either end of the bridge?   454.5 ft

(b) The equation $x^2 = 1135y$ may be written $y = x^2/1135$. Evaluate y at the following values of x: 0, 100, 200, and 454.5.   0, ≈8.8, ≈35.2, ≈182

(c) Consider the highest point on the bridge to be at the origin. Explain what x and y represent in the equation in part (**b**) and how this equation is used in the design and construction of the bridge.   *x:* horizontal distance from the vertex, *y:* vertical distance from the vertex

3. If you take the ends of a piece of rope in your hands and let the rope hang in the middle, is the shape formed by the rope a parabola? Do research to obtain your answer.

4. What is the maximum number and the minimum number of points of intersection possible for the graphs of $y = a(x - h)^2 + k$ and $x = a(y - k)^2 + h$? Explain.   4, 0

5.(c)

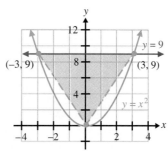

5. In the Group Activity exercises in Section 4.2 we found that the equation of a parabola could be determined from three points on the parabola. A parabola intersects the axes at $(2, 0)$, $(-4, 0)$, and $(0, -8)$.

(a) Find the equation of the parabola in the form $y = ax^2 + bx + c$. $y = x^2 + 2x - 8$

(b) Find the equation of the parabola in the form $y = a(x - h)^2 + k$. $y = (x + 1)^2 - 9$

(c) Sketch the parabola.

6. Consider the figure below.

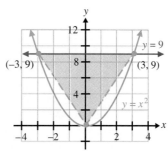

(a) Find the area of the triangle outlined in green.   27 sq units

(b) When a triangle is inscribed within a parabola, as in the figure, the area within the parabola from the base of the triangle is $\frac{4}{3}$ the area of the triangle. Find the area within the parabola from $x = -3$ to $x = 3$.   36 sq units

7. We see parabolic reflectors all around us. They are in TV satellite antennas, flashlights, headlights, and mirrors in telescopes, to name just a few. Do research and explain how a parabolic antenna is used to capture signals reflected from space.

10.4   The Hyperbola

1 Graph hyperbolas in standard form using asymptotes.
2 Graph hyperbolas in nonstandard form.
3 Review conic sections.
4 Identify nonstandard forms of conic sections.

Tape 17

Graphing Hyperbolas in Standard Form

1 A **hyperbola** is the set of points in a plane the difference of whose distances from two fixed points (called foci) is a constant. A hyperbola may look like a pair of parabolas (Fig. 10.24). However, the shapes are actually quite different. A hyperbola has two **vertices.** The point halfway between the two vertices is the **center** of the hyperbola. The dashed lines in the figure are called **asymptotes.** The asymptotes are not a part of the hyperbola, but are used as an aid in graphing the hyperbola. We will discuss asymptotes shortly.

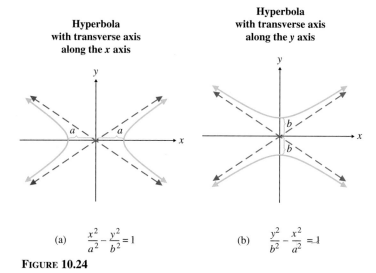

Hyperbola with transverse axis along the x axis

Hyperbola with transverse axis along the y axis

(a) $\dfrac{x^2}{a^2} - \dfrac{y^2}{b^2} = 1$ (b) $\dfrac{y^2}{b^2} - \dfrac{x^2}{a^2} = 1$

FIGURE 10.24

Also given in Fig. 10.24 is the standard form of the equation for each hyperbola. The axis through the center that intersects the vertices is called the **transverse axis.** The axis perpendicular to the transverse axis is the **conjugate axis.** In Fig. 10.24a, the transverse axis lies along the x axis and the conjugate axis lies along the y axis. Both vertices are a units from the origin. In Fig. 10.24b, the transverse axis lies along the y axis and the conjugate axis lies along the x axis. Both vertices are b units from the origin. Note that in the standard form of the equation the denominator of the x^2 term is always a^2 and the denominator of the y^2 term is always b^2.

A hyperbola whose transverse and conjugate axes are parallel to the coordinate axes has either x intercepts (Fig. 10.24a) or y intercepts (Fig. 10.24b), but not both. When written in standard form, the intercepts will be on the axis indicated by the variable with the positive coefficient. For example, the graph of $\dfrac{x^2}{25} - \dfrac{y^2}{9} = 1$ will intersect the x axis, and the graph of $\dfrac{y^2}{9} - \dfrac{x^2}{25} = 1$ will intersect the y axis. In either case the intercepts will be the positive and the negative square root of the denominator of the positive term. Thus, the graph of $\dfrac{x^2}{25} - \dfrac{y^2}{9} = 1$ has x intercepts ± 5 and the graph of $\dfrac{y^2}{9} - \dfrac{x^2}{25} = 1$ has y intercepts ± 3. Note that the intercepts are the vertices of the hyperbola.

Asymptotes can help you graph hyperbolas. The asymptotes are two straight lines that go through the center of the hyperbola. As the values of x and y get larger, the graph of the hyperbola approaches the asymptotes. The equations of the asymptotes of a hyperbola whose center is the origin are

$$y = \frac{b}{a}x \quad \text{and} \quad y = -\frac{b}{a}x$$

The asymptote can be drawn quickly by plotting the four points (a, b), $(-a, b)$, $(a, -b)$, and $(-a, -b)$, then connecting these points with dashed lines to form a rectangle. Next draw dashed lines through the opposite corners of the rectangle to obtain the graphs of the asymptote.

Hyperbola with Its Center at the Origin

Transverse Axis along x Axis *(opens to the right and left)*	*Transverse Axis along y Axis* *(opens upward and downward)*
$$\dfrac{x^2}{a^2} - \dfrac{y^2}{b^2} = 1$$	$$\dfrac{y^2}{b^2} - \dfrac{x^2}{a^2} = 1$$

Asymptotes

$$y = \frac{b}{a}x \quad \text{and} \quad y = -\frac{b}{a}x$$

EXAMPLE 1

(a) Determine the equations of the asymptotes of the hyperbola with equation

$$\frac{x^2}{9} - \frac{y^2}{16} = 1$$

(b) Sketch the hyperbola using the asymptotes.

Solution:

(a) The value of a^2 is 9; the positive square root of 9 is 3. The value of b^2 is 16; the positive square root of 16 is 4. The asymptotes are

$$y = \frac{b}{a}x \quad \text{and} \quad y = -\frac{b}{a}x$$

or

$$y = \frac{4}{3}x \quad \text{and} \quad y = -\frac{4}{3}x$$

(b) To graph the hyperbola, we first graph the asymptotes. To graph the asymptotes we can plot the points (3, 4), (−3, 4), (3, −4), and (−3, −4) and draw the rectangle as illustrated in Fig. 10.25. The asymptotes are the dashed lines through the opposite corners of the rectangle.

  Since the x term in the original equation is positive, the graph intersects the x axis. Since the denominator of the positive term is 9, the vertices are at 3 and −3. Now draw the hyperbola by letting the hyperbola approach its asymptotes (Fig. 10.26). Note that the asymptotes are drawn using dashed lines since they are not part of the hyperbola. They are used merely to help sketch the graph.

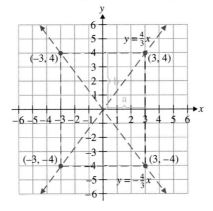

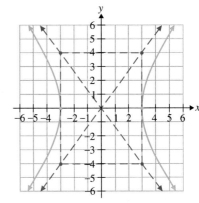

FIGURE 10.25   **FIGURE 10.26** $\dfrac{x^2}{9} - \dfrac{y^2}{16} = 1$

EXAMPLE 2

(a) Show that the equation $-25x^2 + 4y^2 = 100$ is a hyperbola by expressing the equation in standard form.

(b) Determine the equation of the asymptotes of the graph.

(c) Sketch the graph.

Solution:

(a) Divide both sides of the equation by 100 to obtain 1 on the right side of the equation.

$$\frac{-25x^2 + 4y^2}{100} = \frac{100}{100}$$

$$\frac{-25x^2}{100} + \frac{4y^2}{100} = 1$$

$$\frac{-x^2}{4} + \frac{y^2}{25} = 1$$

Rewriting the equation in standard form (positive term first), we get

$$\frac{y^2}{25} - \frac{x^2}{4} = 1$$

(b) The equations of the asymptotes are

$$y = \frac{5}{2}x \quad \text{and} \quad y = -\frac{5}{2}x$$

(c) The graph intersects the y axis at 5 and -5. Figure 10.27a illustrates the asymptotes, and Fig. 10.27b illustrates the hyperbola.

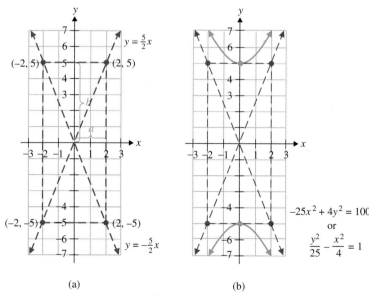

(a)   (b)

FIGURE 10.27

Nonstandard Form of the Hyperbola

2 Another form of the hyperbola is $xy = c$, where c is a nonzero constant. The following equations are examples of hyperbolas in nonstandard form:

$$xy = 4, \qquad xy = -6, \qquad x = \frac{1}{y}, \qquad y = -\frac{3}{x}$$

Note that $x = \dfrac{1}{y}$ is equivalent to $xy = 1$ and $y = -\dfrac{3}{x}$ is equivalent to $xy = -3$. The graphs of equations of the form $xy = c$ are hyperbolas whose asymptotes are the x and y axes.

EXAMPLE 3

(a) Sketch the graph of $xy = 6$.

(b) Sketch the graph of $xy = -6$.

Solution: **(a)** $xy = 6$ or $y = 6/x$. We will graph the equation $y = 6/x$ in two parts. It is important that you realize that the equation is not defined when x is 0 since division by 0 is not permitted. First we will make a table of values for x less than 0. Then we will make a table of values for x greater than 0. This will allow us to see what happens to the graph as the values of x approach 0 from the left and from the right.

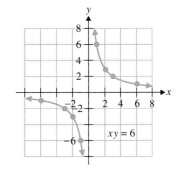

FIGURE 10.28

x	y	x	y
-6	-1	6	1
-3	-2	3	2
-2	-3	2	3
-1	-6	1	6
$-\frac{1}{2}$	-12	$\frac{1}{2}$	12
$-\frac{1}{6}$	-36	$\frac{1}{6}$	36

By looking at the tables and the graphs (Fig. 10.28), we can see that as x gets closer and closer to 0 coming from the left (values of x less than 0), the values of y decrease. As x gets closer and closer to zero coming from the right (values of x greater than 0), the values of y increase. The graph does not cross the y axis, which is what we expect since the equation $y = 6/x$ is not defined when $x = 0$.

(b) $xy = -6$ or $y = -6/x$. We will follow the same procedure as in part (a) to graph $y = -6/x$ (see Fig. 10.29).

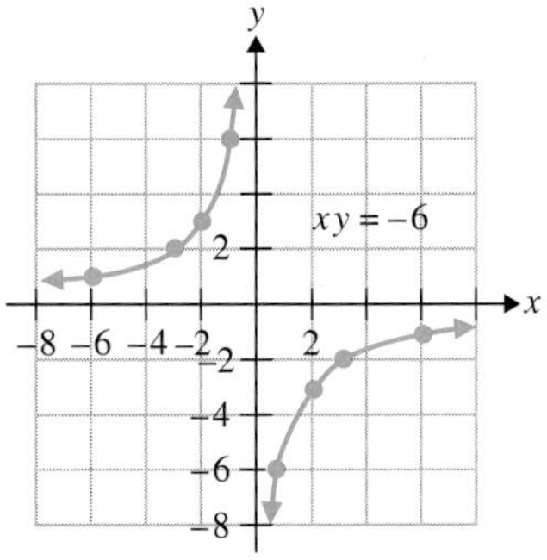

x	y	x	y
-6	1	6	-1
-3	2	3	-2
-2	3	2	-3
-1	6	1	-6
$-\frac{1}{2}$	12	$\frac{1}{2}$	-12
$-\frac{1}{6}$	36	$\frac{1}{6}$	-36

FIGURE 10.29

Summary of Conic Sections

3 The following chart summarizes conic sections.

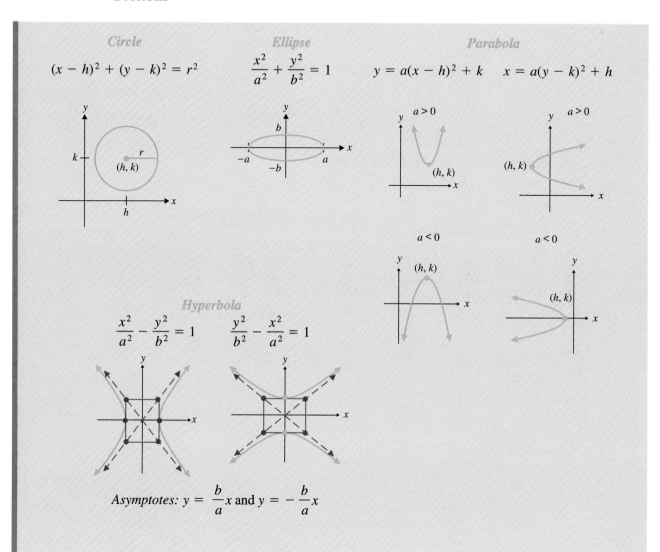

Circle

$$(x - h)^2 + (y - k)^2 = r^2$$

Ellipse

$$\frac{x^2}{a^2} + \frac{y^2}{b^2} = 1$$

Parabola

$$y = a(x - h)^2 + k \qquad x = a(y - k)^2 + h$$

Hyperbola

$$\frac{x^2}{a^2} - \frac{y^2}{b^2} = 1 \qquad \frac{y^2}{b^2} - \frac{x^2}{a^2} = 1$$

$$\text{Asymptotes: } y = \frac{b}{a}x \text{ and } y = -\frac{b}{a}x$$

Nonstandard Forms of Conic Sections

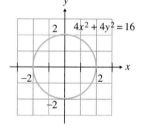

FIGURE 10.30

4 Often, a conic section is given in a nonstandard form. Conic sections given in nonstandard form can be recognized with a little practice. The circle, ellipse, and hyperbola can be expressed by an equation of the form

$$ax^2 + by^2 = c^2$$

If a and b are both positive and $a = b$, the equation is a circle whose center is the origin. For example, $4x^2 + 4y^2 = 16$ is a circle of radius 2 (see Fig. 10.30).

If a and b are both positive and $a \neq b$, the equation is an ellipse whose center is at the origin. If $a < b$, the major axis is along the x axis (see Fig. 10.31a). If $a > b$, the major axis is along the y axis (see Fig. 10.31b).

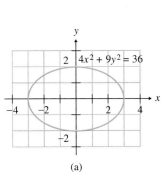

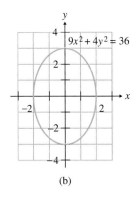

FIGURE 10.31

If a and b have opposite signs, the equation is a hyperbola whose center is at the origin. If $a > 0$, the transverse axis is along the x axis (see Fig. 10.32a). If $b > 0$, its transverse axis is along the y axis (see Fig. 10.32b).

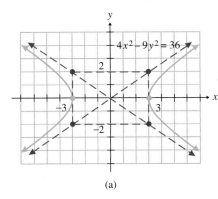

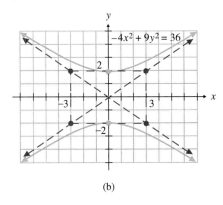

FIGURE 10.32

Nonstandard forms of parabolas are $y = ax^2 + bx + c$ (parabola opens upward

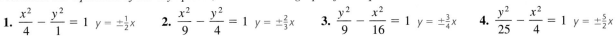

Exercise Set 10.4

Determine the equations of the asymptotes, and sketch the graph of the equation.

1. $\dfrac{x^2}{4} - \dfrac{y^2}{1} = 1$ $y = \pm\frac{1}{2}x$ **2.** $\dfrac{x^2}{9} - \dfrac{y^2}{4} = 1$ $y = \pm\frac{2}{3}x$ **3.** $\dfrac{y^2}{9} - \dfrac{x^2}{16} = 1$ $y = \pm\frac{3}{4}x$ **4.** $\dfrac{y^2}{25} - \dfrac{x^2}{4} = 1$ $y = \pm\frac{5}{2}x$

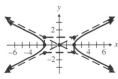

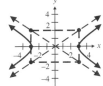

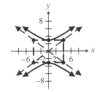

5. $\dfrac{y^2}{25} - \dfrac{x^2}{36} = 1$ $y = \pm\frac{5}{6}x$ **6.** $\dfrac{x^2}{9} - \dfrac{y^2}{25} = 1$ $y = \pm\frac{5}{3}x$ **7.** $\dfrac{x^2}{4} - \dfrac{y^2}{4} = 1$ $y = \pm x$ **8.** $\dfrac{y^2}{49} - \dfrac{x^2}{100} = 1$ $y = \pm\frac{7}{10}x$

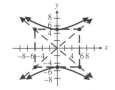

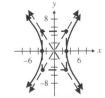

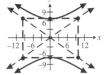

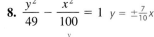

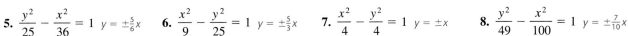

9. $\dfrac{y^2}{16} - \dfrac{x^2}{81} = 1$ $y = \pm\frac{4}{9}x$ **10.** $\dfrac{x^2}{25} - \dfrac{y^2}{16} = 1$ $y = \pm\frac{4}{5}x$ **11.** $\dfrac{y^2}{25} - \dfrac{x^2}{16} = 1$ $y = \pm\frac{5}{4}x$ **12.** $\dfrac{y^2}{4} - \dfrac{x^2}{36} = 1$ $y = \pm\frac{1}{3}x$

   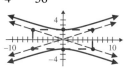

Write each equation in standard form, determine the equation of the asymptotes, and sketch the graph.

13. $16x^2 - 4y^2 = 64$
$(x^2/4) - (y^2/16) = 1,$
$y = \pm 2x$

14. $25x^2 - 16y^2 = 400$
$(x^2/16) - (y^2/25) = 1,$
$y = \pm\frac{5}{4}x$

15. $9y^2 - x^2 = 9$
$(y^2/1) - (x^2/9) = 1,$
$y = \pm\frac{1}{3}x$

16. $4y^2 - 25x^2 = 100$
$(y^2/25) - (x^2/4) = 1,$
$y = \pm\frac{5}{2}x$

 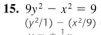 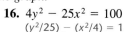 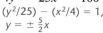 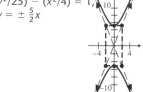

17. $4y^2 - 36x^2 = 144$
$(y^2/36) - (x^2/4) = 1,$
$y = \pm 3x$

18. $x^2 - 25y^2 = 25$
$(x^2/25) - (y^2/1) = 1,$
$y = \pm\frac{1}{5}x$

19. $25x^2 - 4y^2 = 100$
$(x^2/4) - (y^2/25) = 1,$
$y = \pm\frac{5}{2}x$

20. $25x^2 - 9y^2 = 225$
$(x^2/9) - (y^2/25) = 1,$
$y = \pm\frac{5}{3}x$

 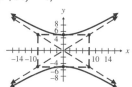 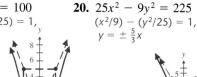

21. $81x^2 - 9y^2 = 729$
$(x^2/9) - (y^2/81) = 1,$
$y = \pm 3x$

22. $64y^2 - 25x^2 = 1600$
$(y^2/25) - (x^2/64) = 1,$
$y = \pm\frac{5}{8}x$

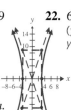

Graph each hyperbola.

23. $xy = 10$ **24.** $xy = 1$ **25.** $xy = -8$ **26.** $xy = -4$ **27.** $y = \dfrac{12}{x}$ **28.** $y = -\dfrac{3}{x}$

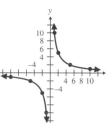  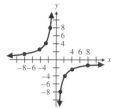 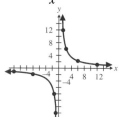 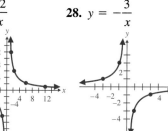

Indicate whether the graph of the equation is a parabola, a circle, an ellipse, or a hyperbola.

29. $y = 6x^2 + 4x + 3$ parabola

30. $4x^2 + 4y^2 = 16$ circle

31. $5x^2 - 5y^2 = 25$ hyperbola

32. $9x^2 - 16y^2 = 36$ hyperbola

33. $x = y^2 + 6y - 7$ parabola

34. $x^2 - 4y^2 = 36$ hyperbola

35. $-2x^2 + 4y^2 = 16$ hyperbola

36. $3x^2 + 3y^2 = 12$ circle

37. $5x^2 + 10y^2 = 12$ ellipse

38. $9x^2 + 16y^2 = 144$ ellipse

39. $x = 3y^2 - y + 7$ parabola

40. $6x^2 - 9y^2 = 36$ hyperbola

41. $6x^2 + 6y^2 = 36$ circle

42. $-y^2 + 4x^2 = 16$ hyperbola

43. $-3x^2 - 3y^2 = -27$ circle

44. $-6x^2 + 2y^2 = -6$ hyperbola

45. $-6y^2 + x^2 = -9$ hyperbola

46. $4x^2 - 9y^2 = 36$ hyperbola

47. What are asymptotes? How do you find the equations of the aymptotes of a hyperbola?

48. If you are given only the equations of the two asymptotes of a hyperbola, can you determine if the transverse axis lies along the x axis or y axis? Explain.

49. Discuss the graph of $\dfrac{x^2}{a^2} - \dfrac{y^2}{b^2} = 1$ for nonzero real numbers a and b. Include the transverse and conjugate axes, vertices, and asymptotes.

50. Discuss the graph of $\dfrac{y^2}{b^2} - \dfrac{x^2}{a^2} = 1$ for nonzero real numbers a and b. Include transverse and conjugate axes, vertices, and asymptotes.

51. Consider the equation $ax^2 + by^2 = c^2$, where a, b, and c represent real numbers. Write a report explaining under what conditions the graph of this equation will be **(a)** a circle, **(b)** an ellipse with major axis along the x axis, **(c)** an ellipse with major axis along the y axis, **(d)** a hyperbola with transverse axis along the x axis, and **(e)** a hyperbola with transverse axis along the y axis. In your report, discuss the characteristics of each graph using a, b, and c to represent real numbers.

47. lines that the hyperbola approaches; $y = \pm \frac{b}{a}x$ **48.** no
49. Hyperbola with vertices at $(a, 0)$ and $(-a, 0)$. Transverse axis along x axis. Asymptotes, $y = \pm \frac{b}{a}x$.   **50.** Hyperbola with vertices at $(0, b)$ and $(0, -b)$. Transverse axis along y axis. Asymptotes, $y = \pm \frac{b}{a}x$.
53. Domain: values of the independent variable. Range: values obtained for the dependent variable.

CUMULATIVE REVIEW EXERCISES

[2.1] **52.** Solve the equation $\dfrac{x}{2} + \dfrac{2}{3}(x - 6) = x + 4$.   48

[3.4] **53.** What are the range and domain of a function?

[4.6] **54.** Determine the solution to the system of inequalities graphically.

$$6x - 2y < 12$$
$$y \geq -2x + 3$$

[9.2] **55.** Solve the quadratic equation
$$-2x^2 + 6x - 5 = 0.$$

55. $\dfrac{3 \pm i}{2}$

54.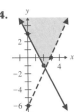

Group Activity/ Challenge Problems

1. If the equation $\dfrac{x^2}{a^2} - \dfrac{y^2}{b^2} = 1$, where $a > b$, is graphed, and then the values of a and b are interchanged, and the new equation is graphed, how will the two graphs compare? Explain your answer.

2. If the equation $\dfrac{x^2}{a^2} - \dfrac{y^2}{b^2} = 1$, where $a > b$, is graphed, and then the signs of each term on the left side of the equation are changed, and the new equation is graphed, how will the two graphs compare? Explain your answer.

The standard forms of a hyperbola with center at (h, k) their graphs follow

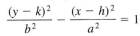

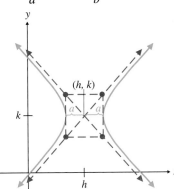

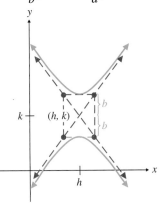

1. The transverse axis of both graphs is along the x axis. The vertices of the second graph will be closer to the origin, and the second graph will open wider.

2. The transverse axis of the first graph is along the x axis. The transverse axis of the second is along the y axis. The vertices of the second graph will be closer to the origin, and the second graph will be wider.

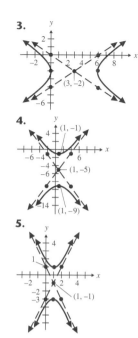

3. $y + 2 = \pm\frac{2}{3}(x - 3)$ **4.** $y + 5 = \pm\frac{4}{3}(x - 1)$ **5.** $[(y + 1)^2/4] - [(x - 1)^2/1] = 1$, $y + 1 = \pm 2(x - 1)$

The asymptotes of both graphs are

$$y - k = \pm\frac{b}{a}(x - h)$$

Write the equations of the asymptotes, then use the asymptotes to sketch the graph.

3. $\dfrac{(x - 3)^2}{9} - \dfrac{(y + 2)^2}{4} = 1$ **4.** $\dfrac{(y + 5)^2}{16} - \dfrac{(x - 1)^2}{9} = 1$

5. Write the equation in standard form, then write the equation of the asymptotes and sketch the graph.

$$y^2 - 4x^2 + 2y + 8x - 7 = 0$$

6. The roof of the Saddledome, a building in Calgary, Canada, where concerts and sporting events are held, is a *hyperbolic paraboloid*. We have discussed both parabolas and hyperbolas.

(a) Can you guess the shape of a hyperbolic paraboloid?

(b) Make a sketch of what you believe the roof looks like?

10.5   Nonlinear Systems of Equations and Their Applications

Tape 17

Solving Nonlinear Systems Using Substitution

1️⃣ Solve a nonlinear system using substitution.
2️⃣ Solve a nonlinear system using addition.
3️⃣ Use nonlinear systems to solve applied problems.

1️⃣ In Chapter 4 we discussed systems of linear equations. Here we discuss nonlinear systems of equations. A **nonlinear system of equations** is a system of equations in which at least one equation is not linear (that is, one whose graph is not a straight line).

The solution to a system of equations is the point or points that satisfy all equations in the system. Consider the system of equations

$$x^2 + y^2 = 25$$
$$3x + 4y = 0$$

Both equations are graphed on the same axes in Fig. 10.33. Note that the graphs appear to intersect at the points $(-4, 3)$ and $(4, -3)$. The check shows these points satisfy all equations in the system and are therefore solutions to the system.

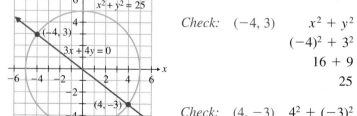

FIGURE 10.33

Check: $(-4, 3)$ $x^2 + y^2 = 25$ $3x + 4y = 0$
                                  $(-4)^2 + 3^2 = 25$ $3(-4) + 4(3) = 0$
                                  $16 + 9 = 25$ $-12 + 12 = 0$
                                  $25 = 25$ true $0 = 0$ true

Check: $(4, -3)$ $4^2 + (-3)^2 = 25$ $3(4) + 4(-3) = 0$
                                  $16 + 9 = 25$ $12 - 12 = 0$
                                  $25 = 25$ true $0 = 0$ true

The graphical procedure for solving a system of equations may be inaccurate since you have to estimate the point or points of intersection. An exact answer may be obtained algebraically.

To solve a system of equations algebraically, we often solve one or more of the equations for one of the variables and then substitute. This procedure is illustrated in Examples 1 and 2.

EXAMPLE 1   Solve the previous system of equations algebraically using substitution.

$$x^2 + y^2 = 25$$
$$3x + 4y = 0$$

Solution: Solve the linear equation $3x + 4y = 0$ for either x or y. We will select to solve for y.

$$3x + 4y = 0$$
$$4y = -3x$$
$$y = -\frac{3x}{4}$$

Now substitute $-\dfrac{3x}{4}$ for y in the equation $x^2 + y^2 = 25$ and solve for the remaining variable, x.

$$x^2 + y^2 = 25$$
$$x^2 + \left(-\frac{3x}{4}\right)^2 = 25$$
$$x^2 + \frac{9x^2}{16} = 25$$
$$16\left(x^2 + \frac{9x^2}{16}\right) = 16(25)$$
$$16x^2 + 9x^2 = 400$$
$$25x^2 = 400$$
$$x^2 = \frac{400}{25} = 16$$
$$x = \pm\sqrt{16} = \pm 4$$

Next, find the corresponding value of y for each value of x by substituting each value of x (one at a time) in the equation solved for y.

$x = 4$	$x = -4$
$y = -\dfrac{3x}{4}$	$y = -\dfrac{3x}{4}$
$y = \dfrac{-3(4)}{4}$	$y = \dfrac{-3(-4)}{4}$
$y = -3$	$y = 3$

The solutions are $(4, -3)$ and $(-4, 3)$. This checks with the solution obtained graphically in Fig. 10.33.

Note that our objective in using substitution is to obtaining a single equation containing only one variable.

EXAMPLE 2   Solve the following system of equations using substitution.

$$y = x^2 - 3$$
$$x^2 + y^2 = 9$$

Solution: Since both equations contain x^2, we will solve one of the equations for x^2. We will choose to solve $y = x^2 - 3$ for x^2.

$$y = x^2 - 3$$
$$y + 3 = x^2$$

Now substitute $y + 3$ for x^2 in the equation $x^2 + y^2 = 9$.

$$x^2 + y^2 = 9$$
$$(y + 3) + y^2 = 9$$
$$y^2 + y + 3 = 9$$
$$y^2 + y - 6 = 0$$
$$(y + 3)(y - 2) = 0$$
$$y + 3 = 0 \quad \text{or} \quad y - 2 = 0$$
$$y = -3 \qquad\qquad y = 2$$

Now find the corresponding values of x.

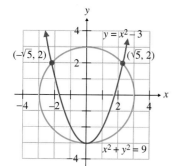

FIGURE 10.34

$y = -3$	$y = 2$
$y = x^2 - 3$	$y = x^2 - 3$
$-3 = x^2 - 3$	$2 = x^2 - 3$
$0 = x^2$	$5 = x^2$
$0 = x$	$\pm\sqrt{5} = x$

This system has three solutions: $(0, -3)$, $(\sqrt{5}, 2)$, and $(-\sqrt{5}, 2)$.

Note that in Example 2 the graph of the equation $y = x^2 - 3$ is a parabola and the graph of the equation $x^2 + y^2 = 9$ is a circle. The graphs of both equations are illustrated in Fig. 10.34.

Helpful Hint

Students will sometimes solve for one variable and assume that they have the solution. Remember that the solution to a system in two variables, if one exists, consists of one or more *ordered pairs*.

Solving Nonlinear Systems Using Addition

2 We can often solve systems of equations more easily using the addition method that was discussed in Section 4.1. As with the substitution method, our objective is to obtain a single equation containing only one variable.

EXAMPLE 3   Solve the system of equations by the addition method.

$$x^2 + y^2 = 9$$
$$2x^2 - y^2 = -6$$

Solution: If we add the two equations, we will obtain one equation containing only one variable.

$$
\begin{aligned}
x^2 + y^2 &= 9 \\
2x^2 - y^2 &= -6 \\
\hline
3x^2 &= 3 \\
x^2 &= 1 \\
x &= \pm 1
\end{aligned}
$$

Now solve for the variable y by substituting $x = \pm 1$ in either the original equations.

$x = 1$	$x = -1$
$x^2 + y^2 = 9$	$x^2 + y^2 = 9$
$1^2 + y^2 = 9$	$(-1)^2 + y^2 = 9$
$1 + y^2 = 9$	$1 + y^2 = 9$
$y^2 = 8$	$y^2 = 8$
$y = \pm\sqrt{8} = \pm 2\sqrt{2}$	$y = \pm\sqrt{8} = \pm 2\sqrt{2}$

There are four solutions to this system of equations:

$$(1, 2\sqrt{2}), (1, -2\sqrt{2}), (-1, 2\sqrt{2}), \text{ and } (-1, -2\sqrt{2})$$

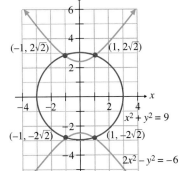

FIGURE 10.35

The graphs of the equations in the system solved in Example 3 are given in Fig. 10.35. Notice the four points of intersection of the two graphs.

It is possible that a system of equations has no real solution (therefore, the graphs do not intersect). Example 4 illustrates such a case.

EXAMPLE 4   Solve the system of equations using the addition method.

$$x^2 + 4y^2 = 16$$
$$x^2 + \ y^2 = 1$$

Solution: Recall from Section 4.1 that in this text we place an equation within brackets, [ ], to indicate that the entire equation is to be multiplied by the value to the left of the brackets.

$$
\begin{array}{c}
\begin{aligned}
x^2 + 4y^2 &= 16 \\
-1[x^2 + y^2 &= 1]
\end{aligned}
\quad \text{gives} \quad
\begin{aligned}
x^2 + 4y^2 &= 16 \\
-x^2 - y^2 &= -1 \\
\hline
3y^2 &= 15 \\
y^2 &= 5 \\
y &= \pm\sqrt{5}
\end{aligned}
\end{array}
$$

Now solve for x.

$$y = \sqrt{5}$$
$$x^2 + y^2 = 1$$
$$x^2 + (\sqrt{5})^2 = 1$$
$$x^2 + 5 = 1$$
$$x^2 = -4$$
$$x = \pm\sqrt{-4}$$
$$x = \pm 2i$$

$$y = -\sqrt{5}$$
$$x^2 + y^2 = 1$$
$$x^2 + (-\sqrt{5})^2 = 1$$
$$x^2 + 5 = 1$$
$$x^2 = -4$$
$$x = \pm\sqrt{-4}$$
$$x = \pm 2i$$

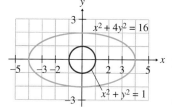

FIGURE 10.36

Since x is an imaginary number for both values of y, this system of equations has no real solution.

The graphs of the equations in Example 4 are shown in Fig. 10.36. Notice that the two graphs do not intersect; therefore, there is no real solution. This agrees with the answer we obtained in Example 4.

EXAMPLE 5 Solve the system of equations by the addition method.

$$6x^2 + y^2 = 10$$
$$2x^2 + 4y^2 = 40$$

Solution: We can choose to eliminate either the variable x or the variable y. Here we will eliminate the variable x.

$$
\begin{array}{l}
6x^2 + y^2 = 10 \\
-3[2x^2 + 4y^2 = 40]
\end{array}
\quad \text{gives} \quad
\begin{array}{l}
6x^2 + y^2 = 10 \\
-6x^2 - 12y^2 = -120 \\
\hline
-11y^2 = -110
\end{array}
$$

$$y^2 = \frac{-110}{-11} = 10$$
$$y = \pm\sqrt{10}$$

Now find the corresponding values of x by substituting $y = \pm\sqrt{10}$ in either of the original equations.

$$y = \sqrt{10}$$
$$6x^2 + y^2 = 10$$
$$6x^2 + (\sqrt{10})^2 = 10$$
$$6x^2 + 10 = 10$$
$$6x^2 = 0$$
$$x^2 = 0$$
$$x = 0$$

$$y = -\sqrt{10}$$
$$6x^2 + y^2 = 10$$
$$6x^2 + (-\sqrt{10})^2 = 10$$
$$6x^2 + 10 = 10$$
$$6x^2 = 0$$
$$x^2 = 0$$
$$x = 0$$

The solutions are $(0, \sqrt{10})$ and $(0, -\sqrt{10})$.

Applications of Nonlinear Systems **3** Now we will study some applications of nonlinear systems.

EXAMPLE 6 The area, A, of a rectangle is 80 square feet and its perimeter, P, is 36 feet. Find the length and width of the rectangle.

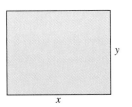

FIGURE 10.37

Solution: Begin by drawing a rectangle (see Fig. 10.37).

$$\text{Let } x = \text{length}$$
$$y = \text{width}$$

Since $A = xy$ and $P = 2x + 2y$, the system of equations is

$$xy = 80$$
$$2x + 2y = 36$$

We will solve this system using substitution. Since $2x + 2y = 36$ is a linear equation, we solve this equation for y (we could also have solved for x).

$$2x + 2y = 36$$
$$2y = 36 - 2x$$
$$y = \frac{36 - 2x}{2} = \frac{36}{2} - \frac{2x}{2} = 18 - x$$

Now substitute $18 - x$ for y in $xy = 80$.

$$xy = 80$$
$$x(18 - x) = 80$$
$$18x - x^2 = 80$$
$$x^2 - 18x + 80 = 0$$
$$(x - 10)(x - 8) = 0$$
$$x - 10 = 0 \quad \text{or} \quad x - 8 = 0$$
$$x = 10 \qquad\qquad x = 8$$

If x is 10, then $y = 18 - x = 18 - 10 = 8$. And if $x = 8$, then $y = 18 - 8 = 10$. Thus, the dimensions of the rectangle are 8 feet by 10 feet.

In Example 6 the graph of $xy = 80$ is a hyperbola, and the graph of $2x + 2y = 36$ is a straight line. If you graph these two equations, the graphs will intersect at (8, 10) and (10, 8). Try solving the system given in Example 6 graphically now.

EXAMPLE 7   A company breaks even when its cost equals its revenue. When its cost is greater than its revenue, the company has a loss. When its revenue exceeds its cost, the company makes a profit.

Hike 'n' Bike Company produces and sells bicycles. Its weekly cost equation is $C = 50x + 400$, $0 \le x \le 160$, and its weekly revenue equation is $R = 100x - 0.3x^2$, $0 \le x \le 160$, where x is the number of bicycles produced and sold each week. Find the number of bicycles that must be produced and sold for Hike 'n' Bike to break even.

Solution: The system of equations is

$$C = 50x + 400$$
$$R = 100x - 0.3x^2$$

6. $(2, \sqrt{2})$, $(2, -\sqrt{2})$, $(-1, \sqrt{5})$, $(-1, -\sqrt{5})$ **8.** $(2, 2\sqrt{2})$, $(2, -2\sqrt{2})$, $(-2, 2\sqrt{2})$, $(-2, -2\sqrt{2})$

11. $(0, -3)$, $(\sqrt{5}, 2)$, $(-\sqrt{5}, 2)$ **13.** $(2, 0)$, $(-2, 0)$ **14.** $(\sqrt{19}, \sqrt{6})$, $(\sqrt{19}, -\sqrt{6})$, $(-\sqrt{19}, \sqrt{6})$, $(-\sqrt{19}, -\sqrt{6})$

For Hike 'n' Bike to break even, its cost must equal its revenue. Thus, we write

$$C = R$$
$$50x + 400 = 100x - 0.3x^2$$

Writing this quadratic equation in standard form, we obtain

$$0.3x^2 - 50x + 400 = 0, \qquad 0 \le x \le 160$$

We will solve this equation using the quadratic formula.

$$a = 0.3, \qquad b = -50, \qquad c = 400$$

$$x = \frac{-b \pm \sqrt{b^2 - 4ac}}{2a}$$

$$= \frac{-(-50) \pm \sqrt{(-50)^2 - 4(0.3)(400)}}{2(0.3)}$$

$$= \frac{50 \pm \sqrt{2020}}{0.6}$$

$$x = \frac{50 + \sqrt{2020}}{0.6} \approx 158.2 \qquad \text{or} \qquad x = \frac{50 - \sqrt{2020}}{0.6} \approx 8.4$$

The cost will equal the revenue and the company will break even when approximately 8 bicycles are sold. The cost will also equal the revenue when approximately 158 bicycles are sold. The company will make a profit when between 9 and 158 bicycles are sold. When fewer than 9 or more than 158 bicycles are sold, the company will take a loss (see Fig. 10.38).

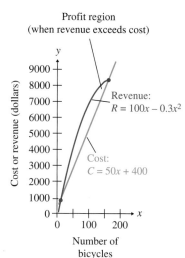

FIGURE 10.38

15. $(3, 2)$, $(3, -2)$, $(-3, 2)$, $(-3, -2)$ **16.** $(\sqrt{2}, \sqrt{2})$, $(\sqrt{2}, -\sqrt{2})$, $(-\sqrt{2}, \sqrt{2})$, $(-\sqrt{2}, -\sqrt{2})$,

17. $(3, 0)$, $(-3, 0)$ **18.** $(1, 3)$, $(1, -3)$, $(-1, 3)$, $(-1, -3)$

Exercise Set 10.5

19. $(\sqrt{6}, \sqrt{3})$, $(\sqrt{6}, -\sqrt{3})$, $(-\sqrt{6}, \sqrt{3})$, $(-\sqrt{6}, -\sqrt{3})$, **20.** $(1, 3)$, $(1, -3)$, $(-1, 3)$, $(-1, -3)$

Find all real solutions to the system of equations using the substitution method.

1. $x^2 + y^2 = 4$
$\quad x - 2y = 4$ $(0, -2)$, $(\frac{8}{5}, -\frac{6}{5})$

2. $x^2 + y^2 = 9$ $(3, 0)$, $(-\frac{9}{5}, \frac{12}{5})$
$\quad x + 2y - 3 = 0$

3. $y = x^2 - 5$ $(-4, 11)$, $(\frac{5}{2}, \frac{5}{4})$
$\quad 3x + 2y = 10$

4. $x + y = 4$ $(\frac{5}{2}, \frac{3}{2})$
$\quad x^2 - y^2 = 4$

5. $2x^2 - y^2 = -8$
$\quad x - y = 6$ $(2, -4)$, $(-14, -20)$

6. $y^2 = -x + 4$
$\quad x^2 + y^2 = 6$

7. $x^2 - 4y^2 = 16$
$\quad x^2 + y^2 = 1$ no real solution

8. $2x^2 + y^2 = 16$
$\quad x^2 - y^2 = -4$

9. $xy = 4$ $(\frac{4}{3}, 3)$, $(-\frac{1}{2}, -8)$
$\quad 6x - y = 5$

10. $xy = 4$ $(4, 1)$, $(1, 4)$
$\quad y = 5 - x$

11. $y = x^2 - 3$
$\quad x^2 + y^2 = 9$

12. $x^2 + y^2 = 25$ $(-5, 0)$, $(4, 3)$
$\quad x - 3y = -5$

21. no real solution **22.** $(5, 3)$, $(5, -3)$, $(-5, 3)$, $(-5, -3)$ **23.** $(\sqrt{5}, 2)$, $(\sqrt{5}, -2)$, $(-\sqrt{5}, 2)$, $(-\sqrt{5}, -2)$,

Find all real solutions to the system of equations using the addition method. **24.** $(1, 2\sqrt{2})$, $(1, -2\sqrt{2})$, $(-1, 2\sqrt{2})$, $(-1, -2\sqrt{2})$

13. $x^2 - y^2 = 4$
$\quad x^2 + y^2 = 4$

14. $x^2 + y^2 = 25$
$\quad x^2 - 2y^2 = 7$

15. $x^2 + y^2 = 13$
$\quad 2x^2 + 3y^2 = 30$

16. $3x^2 - y^2 = 4$
$\quad x^2 + 4y^2 = 10$

17. $4x^2 + 9y^2 = 36$
$\quad 2x^2 - 9y^2 = 18$

18. $5x^2 - 2y^2 = -13$
$\quad 3x^2 + 4y^2 = 39$

19. $2x^2 + 3y^2 = 21$
$\quad x^2 + 2y^2 = 12$

20. $2x^2 + y^2 = 11$
$\quad x^2 + 3y^2 = 28$

21. $-x^2 - 2y^2 = 6$
$\quad 5x^2 + 15y^2 = 20$

22. $x^2 - 2y^2 = 7$
$\quad x^2 + y^2 = 34$

23. $x^2 + y^2 = 9$
$\quad 16x^2 - 4y^2 = 64$

24. $3x^2 + 4y^2 = 35$
$\quad 2x^2 + 5y^2 = 42$

Solve.

25. The sum of two numbers is 12 and the sum of their squares is 74. Find the two numbers. 5, 7

26. The sum of two numbers is 14 and the sum of their squares is 106. Find the numbers. 5, 9

27. The sum of the squares of two numbers is 34. The difference of their squares is 16. Find the two numbers.   5, 3 or 5, −3 or −5, 3 or −5, −3

28. The difference of the squares of two numbers is 80. The sum of their squares is 208. Find the two numbers.   12, 8 or 12, −8 or −12, 8 or −12, −8

29. The square of one number is 11 less than the square of another. The sum of their squares is 61. Find the numbers.   5, 6 or 5, −6 or −5, 6 or −5, −6

30. The product of two numbers is 135. The sum of the two numbers is 24. Find the numbers.   9, 15

31. The area of a rectangle is 84 square feet and its perimeter is 38 feet. Find the dimensions of the rectangle.   7 ft by 12 ft

32. The area of a rectangle is 80 square feet and its perimeter is 36 feet. Find the dimensions of the rectangle.   8 ft by 10 ft

33. A garden is shaped like a right triangle with a perimeter of 30 feet and a hypotenuse of 13 feet. Find the length of the legs of the triangle.   5 ft, 12 ft

34. A sail on a sailboat is shaped like a right triangle with a perimeter of 36 meters and a hypotenuse 15 meters. Find the length of the legs of the triangle.   9m, 12m

SEE EXERCISE 34.

41. $r = 12\%$, $p = \$600$

35. The area of a rectangle is 300 square meters and its length is 5 meters longer than its width. Find the length and width of the rectangle. $w = 15m$, $l = 20m$

36. The area of a rectangle is 72 square yards and its length is twice its width. Find the length and width of the rectangle. $w = 6$ yd, $l = 12$ yd

37. A rectangular area is to be fenced in along a river bank as illustrated. If 20 feet of fencing encloses an area of 48 square feet, find the dimensions of the enclosed area.   6 ft by 8 ft or 4 ft by 12 ft

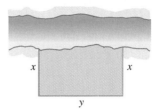

38. A football is thrown up from the ground. Its distance from the ground at any time t is given by the formula $d = -16t^2 + 64t$. At the same time that the football is thrown, a baseball is thrown up from the top of an 80-foot building. Its distance from the ground at any time t is given by the formula $d = -16t^2 + 16t + 80$. Find the time at which the two balls will be the same distance from the ground.   1.67 sec

39. A quarter is thrown down from an airplane flying at a height of 2000 feet. The distance of the quarter above the ground at any time t is found by the formula $d = -16t^2 - 10t + 2000$. At the instant the quarter is thrown from the plane, a snowball is thrown upward from the top of a 100-foot-tall building. The distance from the ground of the snowball at any time t is found by the formula $d = -16t^2 + 800t + 100$. At what time will the quarter and snowball pass each other? 2.35 sec

40. Simple interest is calculated using the simple interest formula, interest = principal · rate · time or $i = prt$. When a certain principal is invested at a certain interest rate for 1 year, the interest obtained is \$7.50. If the principal is increased by \$25 and the interest rate is decreased by 1%, the interest remains the same. Find the principal and the interest rate. $r = 6\%$, $p = \$125$

41. When a certain principal is invested at a certain interest rate for 1 year, the interest obtained is \$72. If the principal is increased by \$120 and the interest rate is decreased by 2%, the interest remains the same. Find the principal and the interest rate. Use $i = prt$.

Cost and revenue equations are given below. Find the break-even point(s) (see Example 7).

42. $C = 10x + 300$, $R = 30x - 0.1x^2$ ≈ 16 and ≈ 184

43. $C = 80x + 900$, $R = 120x - 0.2x^2$ ≈ 26 and ≈ 174

44. $C = 12.6x + 150$, $R = 42.8x - 0.3x^2$ ≈ 5 and ≈ 95

45. $C = 0.6x^2 + 9$, $R = 12x - 0.2x^2$ ≈ 1 and ≈ 14

46. a system of equations where at least one equation is not a linear or first-degree equation
50. parentheses, exponents, multiplication or division from left to right, addition or subtraction from left to right

46. What is a system of nonlinear equations?

47. Make up your own nonlinear system of equations whose solution is the empty set. Explain how you know the system has no solution.

48. If a system of equations consists of an ellipse and a hyperbola, what is the maximum number of points of intersection? Make a sketch to illustrate this.

49. If a system of equations consists of two hyperbolas, what is the maximum number of points of intersection? Make a sketch to illustrate this.

48. 4   **49.** 8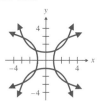

CUMULATIVE REVIEW EXERCISES

[1.4] **50.** List the order of operations we follow when evaluating an expression.

[2.2] **51.** Use the compound interest formula $A = P\left(1 + \dfrac{r}{n}\right)^{nt}$ to find the amount, A, when the principal P is \$5000, the rate r is 8%, the number of compounding periods, n, is 2, and the number of years, t, is 2.   \$5849.29

52. Solve $\dfrac{3}{5}(2x - y) = \dfrac{3}{4}(2x - 3y) + 6$ for y.

[2.5] **53.** Solve the inequality $\dfrac{3 - 4y}{3} \geq \dfrac{2y - 6}{4} - \dfrac{7}{6}$. $y \leq 2$

52. $y = \dfrac{2x + 40}{11}$

5. $(\sqrt{2}, \sqrt{2}), (-\sqrt{2}, -\sqrt{2}), \left(\sqrt{3}, \dfrac{2\sqrt{3}}{3}\right), \left(-\sqrt{3}, -\dfrac{2\sqrt{3}}{3}\right)$

Group Activity/ Challenge Problems

1. Find the length of the legs of a right triangle if the hypotenuse is 13 meters and the area is 30 square meters.   5 m, 12 m

2. Find the length of the legs of a right triangle if the hypotenuse is 7 ft and the area is 11.76 square feet.   4.2 ft, 5.6 ft

3. In the figure below, R represents the radius of the larger circle and r represents the radius of the smaller circle. If $R = 2r$ and if the shaded area is 122.5π, find r and R.

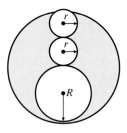

$r = 3.5$ $R = 7$

Find all the solutions to each system of equations. You may need to review earlier material about equations quadratic in form.

4. $x^2 + 2xy - y^2 = 7$ (2, 1), (−2, −1)
    $x^2 - y^2 = 3$ (−i, 2i), (i, −2i)

5. $2x^2 - 3xy + 3y^2 = 4$
    $2x^2 - 7xy + 3y^2 = -4$

10.6   Nonlinear Systems of Inequalities

Tape 18

[1] Graph second-degree inequalities.
[2] Solve a system of nonlinear inequalities graphically.

Graph Second Degree Inequalities

[1] In Sections 10.1 through 10.5, we graphed second-degree equations, such as $x^2 + y^2 = 25$. Now we will graph second-degree inequalities, such as $x^2 + y^2 \leq 25$. To graph second-degree inequalities, we use a procedure

similar to the one we used to graph linear inequalities. That is, graph the equation using a solid line if the inequality symbol is ≤ or ≥, and a dashed line if the inequality symbol is < or >, and then shade in the region that satisfies the inequality. Examples 1 and 2 illustrate this procedure.

EXAMPLE 1 Graph the inequality $\dfrac{x^2}{4} + \dfrac{y^2}{9} \geq 1$.

Solution: First, graph the equation $\dfrac{x^2}{4} + \dfrac{y^2}{9} = 1$. Use a solid line when drawing the ellipse since the inequality symbol is ≥ (Fig. 10.39a). Next select a test point not on the graph and test it in the original inequality. We will select the test point (0, 0).

Check: (0, 0)

$$\frac{x^2}{4} + \frac{y^2}{9} \geq 1$$

$$\frac{0^2}{4} + \frac{0^2}{9} \geq 1$$

$$0 + 0 \geq 1$$

$$0 \geq 1 \qquad \textbf{false}$$

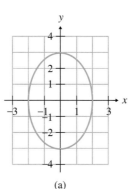 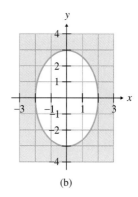

(a)   (b)

FIGURE 10.39

The point (0, 0), which is within the ellipse, is not a solution. Thus, all the points outside the ellipse satisfy the inequality. Shade in this outer area (Fig. 10.39b). The shaded area, and the ellipse itself, represent the solution to the inequality.

EXAMPLE 2 Graph the inequality. $\dfrac{y^2}{25} - \dfrac{x^2}{9} < 1$

Solution: Graph the equation $\dfrac{y^2}{25} - \dfrac{x^2}{9} = 1$. Use a dashed line when drawing

the hyperbola since the inequality symbol is $<$ (Fig. 10.40a). Next select the test point. We will use $(0, 0)$.

$$\frac{y^2}{25} - \frac{x^2}{9} < 1$$

$$\frac{0^2}{25} - \frac{0^2}{9} < 1$$

$$0 - 0 < 1$$

$$0 < 1 \quad \text{true}$$

Since $(0, 0)$ is a solution, we shade in the region containing the point $(0, 0)$. The solution is graphed in Fig. 10.40b.

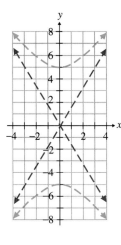 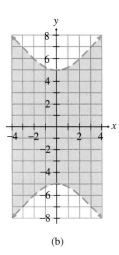

(b)

FIGURE 10.40

Systems of Inequalities

2 Now we will find the graphical solution to a system of inequalities in which at least one inequality is not linear. The solution to a system of inequalities is the set of ordered pairs that satisfy all inequalities in the system.

> **To Solve a System of Inequalities Graphically**
> 1. Graph one inequality.
> 2. On the same axes, graph the second inequality. Use different shading than was used in the first inequality.
> 3. The solution is the area where the shaded areas from both inequalities overlap.

EXAMPLE 3 Solve the system of inequalities graphically.

$$x^2 + y^2 < 25$$
$$2x + y \geq 4$$

Solution: First, graph the inequality, $x^2 + y^2 < 25$. The inner region of the circle satisfies the inequality (Fig. 10.41). On the same axes, graph the second inequality, $2x + y \geq 4$. The solution to the system is the area containing both types of shading and that part of the straight line within the boundaries of the circle (see Fig. 10.42).

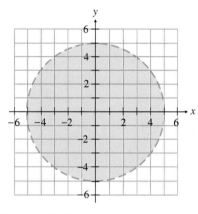

FIGURE 10.41

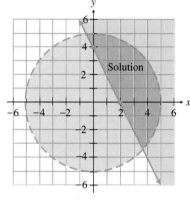

FIGURE 10.42

EXAMPLE 4 Solve the system of inequalities graphically.

$$\frac{x^2}{4} - \frac{y^2}{9} \leq 1$$
$$y > (x + 2)^2 - 4$$

Solution: Graph each inequality on the same axes (Fig. 10.43). The solution includes the area where the two types of shadings overlap and the solid line within the parabola.

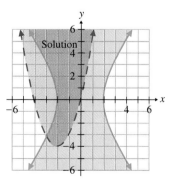

FIGURE 10.43

Exercise Set 10.6

Solve the system of inequalities graphically.

1. $x^2 + y^2 \geq 16$
$x + y < 5$

2. $\dfrac{x^2}{25} - \dfrac{y^2}{9} < 1$
$2x + 3y > 6$
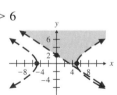

3. $4x^2 + y^2 > 16$
$y \geq 2x + 2$
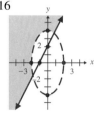

4. $x^2 - y^2 > 1$
$\dfrac{x^2}{9} + \dfrac{y^2}{4} \leq 1$
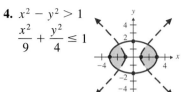

5. $x^2 + y^2 \leq 36$
$y < (x + 1)^2 - 5$
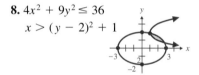

6. $y \geq x^2 - 2x + 1$
$x^2 + y^2 > 4$

7. $\dfrac{y^2}{16} - \dfrac{x^2}{4} \geq 1$
$\dfrac{x^2}{4} - \dfrac{y^2}{1} < 1$
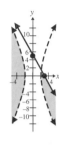

8. $4x^2 + 9y^2 \leq 36$
$x > (y - 2)^2 + 1$
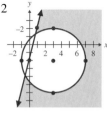

9. $xy \leq 6$
$2x - y \leq 8$
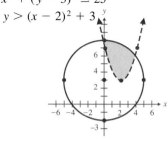

10. $25x^2 - 4y^2 > 100$
$5x + 3y \leq 15$

11. $(x - 3)^2 + (y + 2)^2 \geq 16$
$y \leq 4x - 2$

12. $x^2 + (y - 3)^2 \leq 25$
$y > (x - 2)^2 + 3$

CUMULATIVE REVIEW EXERCISES

[5.2] **13.** Simplify $\left(\dfrac{4x^{-2}y^3}{2xy^{-4}}\right)^2 \left(\dfrac{3xy^{-1}}{6x^4y^{-3}}\right)^{-2}$ $16y^{10}$

16.

Solve the equation $x^2 - 2x - 4 = 0$:

[9.1] **14.** By completing the square $1 \pm \sqrt{5}$

[9.2] **15.** By using the quadratic formula $1 \pm \sqrt{5}$

[9.3] **16.** Graph the solution to the inequality
$(x + 4)(x - 2)(x - 4) \leq 0$ on the number line.

Group Activity/ Challenge Problems

Graph the following systems of inequalities

1. $y > 4x - 6$
$x^2 + y^2 \geq 36$
$2x + y \leq 8$

1.

2. $x^2 + y^2 > 25$
$\dfrac{x^2}{25} + \dfrac{y^2}{9} > 1$
$2x - 3y < 12$

2.

Summary

GLOSSARY

Asymptotes of a hyperbola *(594):* Two lines through the center of the hyperbola that help in graphing the hyperbola.

Circle *(576):* The set of points in a plane that are the same distance from a fixed point called the center.

Conic sections *(576):* Circles, ellipses, hyperbolas, and parabolas.

Conjugate axis of a hyperbola *(594):* The axis perpendicular to the transverse axis.

Ellipse *(583):* The set of points in a plane the sum of whose distance from two fixed points, called foci, is a constant.

Hyperbola *(593):* The set of points in a plane the difference of whose distance from two fixed points, called foci, is a constant.

Major axis of ellipse *(583):* The longer axis through the center of the ellipse.

Minor axis of ellipse *(583):* The shorter axis through the center of the ellipse.

Nonlinear system of equations *(602):* A system of equations in which at least one is not linear.

Parabola *(587):* The graph of an equation of the form $y = a(x - h)^2 + k$ or $x = a(y - k)^2 + h$.

Transverse axis of a hyperbola *(594):* The axis through the center of the hyperbola that intersects the vertices.

IMPORTANT FACTS

Circle
$$(x - h)^2 + (y - k)^2 = r^2$$
center at (h, k)
radius r

Ellipse
$$\frac{x^2}{a^2} + \frac{y^2}{b^2} = 1$$
center at $(0, 0)$

Parabola
$$y = a(x - h)^2 + k$$
vertex at (h, k)
opens upward when $a > 0$
opens downward when $a < 0$

$$x = a(y - k)^2 + h$$
vertex at (h, k)
opens right when $a > 0$
opens left when $a < 0$

Hyperbola
$$\frac{x^2}{a^2} - \frac{y^2}{b^2} = 1$$
x axis transverse axis

$$\frac{y^2}{b^2} - \frac{x^2}{a^2} = 1$$
y axis transverse axis

asymptotes:
$$y = \frac{b}{a}x \text{ or } y = -\frac{b}{a}x$$

Review Exercises

[10.1]   *Write the equation of the circle in standard form and sketch the graph.*

1. Center (0, 0), radius 5
$x^2 + y^2 = 5^2$

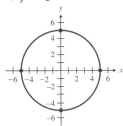

2. Center (−3, 4), radius 3
$(x + 3)^2 + (y − 4)^2 = 3^2$

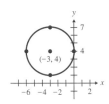

3. Center (4, 2), radius $\sqrt{8}$
$(x − 4)^2 + (y − 2)^2 = (\sqrt{8})^2$

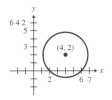

Sketch the graph of each equation.

4. $x^2 + y^2 = 16$

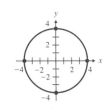

5. $y = −\sqrt{25 − x^2}$

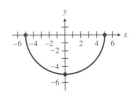

6. $(x − 2)^2 + (y + 3)^2 = 25$

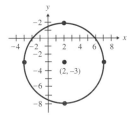

Determine the equation of the circle.

7.

$(x + 1)^2 + (y − 1)^2 = 4$

8.

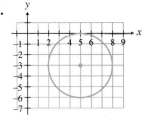

$(x − 5)^2 + (y + 3)^2 = 9$

Write the equation in standard form and sketch the graph.

9. $x^2 + y^2 − 4y = 0$
$x^2 + (y − 2)^2 = 2^2$

10. $x^2 + y^2 − 2x + 6y + 1 = 0$
$(x − 1)^2 + (y + 3)^2 = 3^2$

11. $x^2 − 8x + y^2 − 10y + 40 = 0$
$(x − 4)^2 + (y − 5)^2 = 1^2$

12. $x^2 + y^2 − 4x + 10y + 17 = 0$
$(x − 2)^2 + (y + 5)^2 = (\sqrt{12})^2$

[10.2] *Sketch the graph of each equation.*

13. $\dfrac{x^2}{9} + \dfrac{y^2}{4} = 1$

14. $\dfrac{x^2}{16} + \dfrac{y^2}{1} = 1$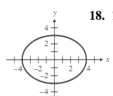

15. $\dfrac{x^2}{9} + \dfrac{y^2}{64} = 1$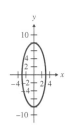

16. $4x^2 + 9y^2 = 36$

17. $9x^2 + 16y^2 = 144$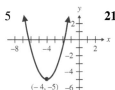

18. $16x^2 + y^2 = 16$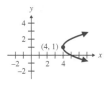

[10.3] *Graph each equation.*

19. $y = (x - 3)^2 + 4$

20. $y = (x + 4)^2 - 5$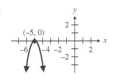

21. $x = (y - 1)^2 + 4$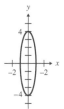

22. $x = -2(y + 4)^2 - 3$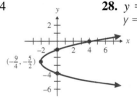

23. $y = -3(x + 5)^2$

Write each equation in the form $y = a(x - h)^2 + k$ or $x = a(y - k)^2 + h$ and graph the equation.

24. $y = x^2 - 6x$
$y = (x - 3)^2 - 9$

25. $y = x^2 - 2x - 3$
$y = (x - 1)^2 - 4$

26. $x = -y^2 - 2y + 8$
$x = -(y + 1)^2 + 9$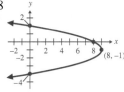

27. $x = y^2 + 5y + 4$
$x = \left(y + \dfrac{5}{2}\right)^2 - \dfrac{9}{4}$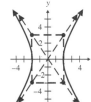

28. $y = 2x^2 - 8x - 24$
$y = 2(x - 2)^2 - 32$

[10.4] *Determine the equations of the asymptotes and sketch the graph.*

29. $\dfrac{x^2}{4} - \dfrac{y^2}{9} = 1$
$y = \pm\dfrac{3}{2}x$

30. $\dfrac{y^2}{16} - \dfrac{x^2}{4} = 1$
$y = \pm 2x$

31. $\dfrac{y^2}{9} - \dfrac{x^2}{25} = 1$
$y = \pm\dfrac{3}{5}x$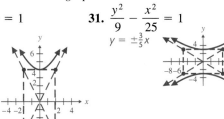

32. $\dfrac{x^2}{4} - \dfrac{y^2}{36} = 1$
$y = \pm 3x$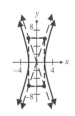

Write each equation in standard form, determine the equations of the asymptotes, and sketch the graph.

33. $9y^2 - 4x^2 = 36$
$(y^2/4) - (x^2/9) = 1,$
$y = \pm\frac{2}{3}x$

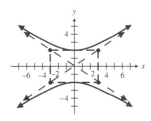

34. $x^2 - 16y^2 = 16$
$(x^2/16) - (y^2/1) = 1,$
$y = \pm\frac{1}{4}x$

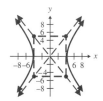

35. $25x^2 - 16y^2 = 400$
$(x^2/16) - (y^2/25) = 1,$
$y = \pm\frac{5}{4}x$

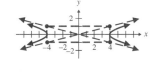

36. $49y^2 - 9x^2 = 441$
$(y^2/9) - (x^2/49) = 1,$
$y = \pm\frac{3}{7}x$

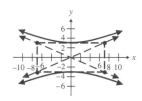

Graph each equation.

37. $xy = 6$

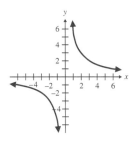

38. $xy = -8$

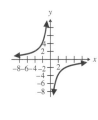

39. $y = \dfrac{3}{x}$

40. $y = -\dfrac{10}{x}$

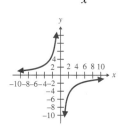

[10.1–10.4]

Identify the graph as a circle, ellipse, parabola, or hyperbola, and sketch the graph.

41. $\dfrac{x^2}{4} - \dfrac{y^2}{16} = 1$

hyperbola

42. $16x^2 + 9y^2 = 144$

ellipse

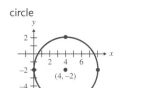

43. $(x - 4)^2 + (y + 2)^2 = 16$

circle

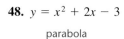

44. $x^2 - 25y^2 = 25$

hyperbola

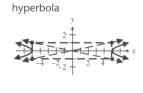

45. $\dfrac{x^2}{64} + \dfrac{y^2}{9} = 1$

ellipse

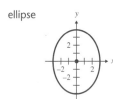

46. $y = (x - 2)^2 - 5$

parabola

$(2, -5)$

47. $4x^2 + 9y^2 = 36$

ellipse

48. $y = x^2 + 2x - 3$

parabola

$(-1, -4)$

49. $x = -2(y + 3)^2 - 1$ **50.** $25y^2 - 9x^2 = 225$ **51.** $x^2 - 4x + y^2 + 6y = -4$ **52.** $x = -y^2 - 6y - 7$
parabola   hyperbola   circle   parabola

          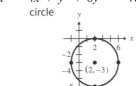

0.5] *Find all real solutions to the systems of equations using the substitution method.*

53. $x^2 + y^2 = 9$ $(-3, 0)$,   **54.** $xy = 5$ $(\frac{1}{3}, 15)$,   **55.** $x^2 + y^2 = 4$ $(2, 0)$,   **56.** $x^2 + 4y^2 = 4$ no real
$y = 3x + 9$ $(-\frac{12}{5}, \frac{9}{5})$ $3x + y = 16$ $(5, 1)$ $x^2 - y^2 = 4$ $(-2, 0)$ $x^2 - 6y^2 = 12$ solution

Find all real solutions to the systems of equations using the addition method.

57. $x^2 + y^2 = 16$ $(4, 0)$,   **58.** $x^2 + y^2 = 25$ **59.** $-4x^2 + y^2 = -12$ **60.** $-2x^2 - 3y^2 = -6$
$x^2 - y^2 = 16$ $(-4, 0)$ $x^2 - 2y^2 = -2$ $8x^2 + 2y^2 = -8$ $5x^2 + 4y^2 = 15$
                                 $(4, 3), (4, -3), (-4, 3), (-4, -3)$ no real solution $(\sqrt{3}, 0), (-\sqrt{3}, 0)$

61. The square of one number is 9 less than the square of another. The sum of their squares is 41. Find the numbers.   4, 5 or 4, −5 or −4, 5, or −4, −5

62. A right triangle with a hypotenuse of 13 meters has a perimeter of 30 meters. Find the length of the legs of the triangle.   5 m, 12 m

0.6] *Graph each system of nonlinear inequalities.*

63. $2x + y \geq 6$ **64.** $xy > 5$ **65.** $4x^2 + 9y^2 \leq 36$ **66.** $\dfrac{x^2}{4} - \dfrac{y^2}{9} > 1$

$x^2 + y^2 < 9$ $y < 3x + 4$ $x^2 + y^2 > 25$ $y \geq (x - 3)^2 - 4$

                                                    no real solution

Practice Test   **7.** $(\sqrt{6}, \sqrt{10}), (\sqrt{6}, -\sqrt{10}), (-\sqrt{6}, \sqrt{10}), (-\sqrt{6}, -\sqrt{10})$.

1. Write the equation of the circle with center at $(-3, -1)$ and radius 9. Then sketch the circle.

2. Write the equation in standard form; then sketch the graph.
$$x^2 + y^2 - 2x - 6y + 1 = 0$$

3. Sketch the graph of $9x^2 + 16y^2 = 144$.

4. Sketch the graph of $y = -2(x - 3)^2 - 9$.

5. Sketch the graph of $\dfrac{y^2}{25} - \dfrac{x^2}{1} = 1$

6. Sketch the graph of $y = \dfrac{8}{x}$.

7. Solve the system of equations by the method of your choice.
$$x^2 + y^2 = 16$$
$$2x^2 - y^2 = 2$$

8. Sketch the system of inequalities.
$$\dfrac{x^2}{9} - \dfrac{y^2}{25} < 1$$
$$x^2 + y^2 \leq 4$$

9. The product of two numbers is 54. Their sum is 29. Find the numbers.   2, 27

1.

1. $(x + 3)^2 + (y + 1)^2 = 9^2$ **2.** $(x - 1)^2 + (y - 3)^2 = 3^2$

3.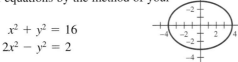

4.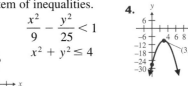

5.   **6.**   **8.**

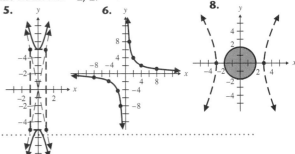

Cumulative Review Test

1. Solve the system of equations algebraically:
$$2x - y = 6$$
$$3x + 2y = 4 \quad \left(\frac{16}{7}, -\frac{10}{7}\right)$$

2. Simplify $\left(\dfrac{2x^0y^{2/3}}{x^2}\right)^2\left(\dfrac{3x^{1/3}y^{1/2}}{y^{-3/2}}\right)$. $\dfrac{12y^{10/3}}{x^{11/3}}$

3. If $f(x) = x^2 + 2x + 5$, find $f(a + 3)$. $a^2 + 8a + 20$

4. Multiply $\dfrac{6x^2 + 5x - 4}{2x^2 - 3x + 1} \cdot \dfrac{4x^2 - 1}{8x^3 + 1}$. $\dfrac{(3x + 4)(2x - 1)}{(x - 1)(4x^2 - 2x + 1)}$

5. Subtract $\dfrac{x}{x + 3} - \dfrac{x + 1}{2x^2 - 2x - 24}$. $\dfrac{2x^2 - 9x - 1}{2(x + 3)(x - 4)}$

6. Solve $\dfrac{y + 1}{y + 3} + \dfrac{y - 3}{y - 2} = \dfrac{2y^2 - 15}{y^2 + y - 6}$.   4

7. Simplify $\sqrt{\dfrac{12x^5y^3}{8z}}$. $\dfrac{x^2y\sqrt{6xyz}}{2z}$

8. Simplify $\dfrac{6}{\sqrt{3} - \sqrt{5}}$. $-3(\sqrt{3} + \sqrt{5})$

9. Solve $3\sqrt[3]{2x + 2} = \sqrt[3]{80x - 24}$.   3

10. Evaluate $(5 - 4i)(5 + 4i)$.   41

11. Solve $(x - 3)^2 = 28$. $3 \pm 2\sqrt{7}$

12. Solve $3x^2 - 4x - 8 = 0$ by the quadratic formula.

13. Solve $3p^{2/3} + 14p^{1/3} - 24 = 0$. $\dfrac{64}{27}, -216$

14. Solve the inequality $\dfrac{3x - 2}{x + 4} \geq 0$ and graph the solution on the number line.

15. Consider the equation $y = x^2 - 4x + 4$.
  (a) Determine whether the parabola opens upward or downward.   upward
  (b) Find the y intercept.   4
  (c) Find the vertex.   (2, 0)
  (d) Find the x intercepts if they exist.   (2, 0)
  (e) Sketch the graph.

16. If $f(x) = x^2 + 6x$ and $g(x) = 2x - 3$, find
  (a) $(f \circ g)(x)$;   **(b)** $(g \circ f)(x)$.

17. Graph $9x^2 + 4y^2 = 36$.

18. Graph $\dfrac{y^2}{25} - \dfrac{x^2}{16} = 1$.

19. The price of a suit is reduced by 20% and then reduced an additional $25. If the sale price of the suit is $155, find the original cost of the suit. $225

20. The Nut Shop sells cashews for $7 per pound and peanuts for $5 per pound. If a customer wants a 4-pound mixture of these two nuts and wants the mixture to cost $25 before tax, how many pounds of each nut should be mixed? $2\frac{1}{2}$ lb cashews, $1\frac{1}{2}$ lb peanuts

12. $\dfrac{2 \pm 2\sqrt{7}}{3}$

16. (a) $4x^2 - 9$ **(b)** $2x^2 + 12x - 3$

---

14.    **15 (e)** 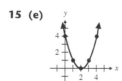   **17.** 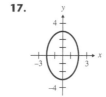   **18.**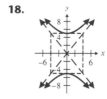

Chapter **11**

Exponential and Logarithmic Functions

See Section 11.3, Exercise 74

Preview and Perspective

Exponential and logarithmic functions have a wide variety of uses, some of which you will see as you read through this chapter. These topics are discussed further in higher-level mathematics courses and most science courses. You often read in newspaper and magazine articles that health care spending, the federal deficit, the number of cases of AIDS, and the world population, to list just a few, are growing exponentially. By the time you finish this chapter you should have a clear understanding of just what this means.

Exponential and logarithmic functions are introduced in Section 11.1 and are used throughout the chapter. The properties of logarithms are discussed in Section 11.2. Common logarithms, that is, logarithms to the base 10, are discussed in Section 11.3. In Section 11.4 we explain how to solve equations that contain logarithmic and exponential functions. We use the properties discussed in earlier sections of the chapter to solve these equations.

In Section 11.5 we introduce very special functions, the natural exponential function and the natural logarithmic function. In both of these functions the base is *e*, an irrational number whose value is approximately 2.7183. Many natural phenomena, such as carbon dating, radioactive decay, and the growth of savings invested in an account compounding interest continuously, can be described by natural exponential functions.

11.1 Exponential and Logarithmic Functions

Tape 18

1. Graph exponential functions.
2. Study applications of exponential functions.
3. Evaluate exponential equations.
4. Change an exponential equation to logarithmic form.
5. Graph logarithmic functions.
6. Graph inverse functions of the form $y = a^x$ and $y = \log_a x$ on the same axes.
7. Solve logarithmic equations.

In this section we introduce the exponential function and the logarithmic function. We present both functions in the same section because the two are closely related. We will show that the exponential function and the logarithmic function are inverse functions as defined in Section 9.6.

Exponential Functions

1. An **exponential function** is a function of the form $y = f(x) = a^x$, where a is a positive real number not equal to 1. Examples of exponential functions are

$$y = 2^x, \qquad y = 5^x, \qquad y = \left(\frac{1}{2}\right)^x$$

> ### Exponential Function
> For any real number $a > 0$ and $a \neq 1$,
> $$f(x) = a^x$$
> is an exponential function.

Exponential functions can be graphed by selecting values for x, finding the corresponding values of y [or $f(x)$], and plotting the points.

EXAMPLE 1   Graph the exponential function $y = 2^x$. State the domain and range of the function.

Solution: First, construct a table of values.

x	-4	-3	-2	-1	0	1	2	3	4
y	$\frac{1}{16}$	$\frac{1}{8}$	$\frac{1}{4}$	$\frac{1}{2}$	1	2	4	8	16

Now plot these points and connect them with a smooth curve (Fig. 11.1).

Domain: $\mathbb{R}$

Range: $\{y|y > 0\}$

The domain of this function is the set of real numbers, $\mathbb{R}$. The range is the set of values greater than 0. If you study the equation $y = 2^x$, you should realize that y must always be positive because 2 is positive.

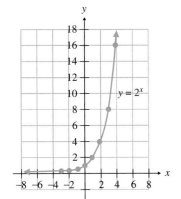

FIGURE 11.1

EXAMPLE 2   Graph $y = \left(\frac{1}{2}\right)^x$. State the domain and range of the function.

Solution: We will construct a table of values and plot the curve (Fig. 11.2).

x	-4	-3	-2	-1	0	1	2	3	4
y	16	8	4	2	1	$\frac{1}{2}$	$\frac{1}{4}$	$\frac{1}{8}$	$\frac{1}{16}$

The domain is the set of real numbers $\mathbb{R}$. The range is $\{y|y > 0\}$.

What will the graph of $y = 2^{-x}$ look like? Remember that 2^{-x} means $\frac{1}{2^x}$ or $\left(\frac{1}{2}\right)^x$. Thus, the graph of $y = 2^{-x}$ will be identical to the graph in Fig. 11.2. Now consider the equation $y = \left(\frac{1}{2}\right)^{-x}$. This equation may be rewritten as $y = 2^x$ since $\left(\frac{1}{2}\right)^{-x} = 2^x$. Thus, the graph of $y = \left(\frac{1}{2}\right)^{-x}$ will be identical to the graph in Fig. 11.1.

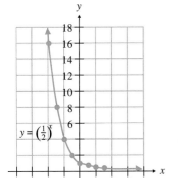

FIGURE 11.2

Note that the graphs in Fig. 11.1 and 11.2 are both one-to-one functions. Both graphs appear to be horizontal near the x axis, but in fact they are not.

In Fig. 11.1, when $x = -2$, $y = 1/4$; when $x = -3$, $y = 1/8$; when $x = -10$, $y = 1/1024$ (about 0.001); and so on. As the values of x decrease, the values of y get closer to 0 and the curve gets closer the x axis.

In Fig. 11.2, when $x = 2$, $y = 1/4$; when $x = 3$, $y = 1/8$; when $x = 10$, $y = 1/1024$; and so on. As the values of x increase, the values of y get closer to 0 and the closer the curve gets to the x axis.

The graphs of exponential functions of the form $y = a^x$ are similar to

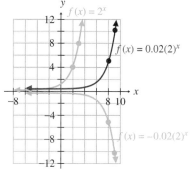

FIGURE 11.3

Fig. 11.1 when a > 1 and similar to Fig. 11.2 when 0 < a < 1. Note that $y = 1^x$ is not a one-to-one function, so we exclude it from our discussion of exponential functions.

Not all exponential functions go through the point (0, 1). For example, $f(x) = 2^x + 5$ is an exponential function. Its graph resembles the graph of $f(x) = 2^x$ in Fig. 11.1 but is raised 5 units. Thus, the graph of $f(x) = 2^x + 5$ has the same shape as $f(x) = 2^x$ but its y intercept is (0, 6). Exponential functions can also have numerical coefficients preceding the exponential part of the function. For example, $f(x) = 0.02(2)^x$ is also an exponential function, as is $f(x) = -0.02(2)^x$. The graphs of $f(x) = 2^x$, $f(x) = 0.02(2)^x$ and $f(x) = -0.02(2)^x$ are illustrated in Fig. 11.3.

The y intercepts of the graphs $f(x) = 2^x$, $f(x) = 0.02(2)^x$, and $f(x) = -0.02(2)^x$, are 1, 0.02, and −0.02, respectively.

In exponential functions, the variable may have a coefficient other than 1. For example, $f(x) = 2.35(4)^{0.3x}$ and $f(x) = -4.6(3)^{-2.16x}$ are exponential functions.

Graphing Calculator Corner

Since the y values of exponential functions may become very large very quickly, make sure the range in your window is sufficient. If you graph $y = 2000(1.08)^x$ on a standard calculator window, you will not see any of the graph. Can you explain why?

1.

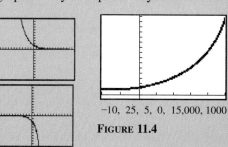

−10, 25, 5, 0, 15,000, 1000

FIGURE 11.4

2.

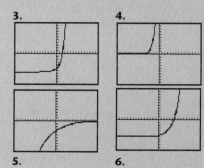

3.   **4.**

5.   **6.**

In Fig. 11.4 we show the graph of the function for the domain from −10 to 25 and the range from 0 to 15,000.

EXERCISES

Use your calculator to graph the following exponential functions.

1. $y = 0.6^x$ **2.** $y = -(3.2)^x$ **3.** $y = 2.5^x - 6.9$

4. $y = 3.9^x + 3$ **5.** $y = -3.7(1.2)^{-x}$ **6.** $y = 0.1(2.1)^x - 5.7$

Applications   **2** Many real-life situations can be described by graphs that resemble exponential functions. Books discussing world situations often use such graphs. The graph of the world population (Fig. 11.5) and the corresponding text are taken from *Beyond the Limits* by Donella Meadows, Dennis Meadows, and Jorgen Randers.

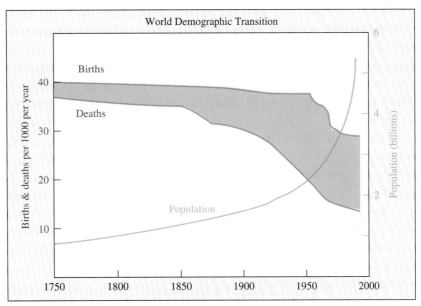

FIGURE 11.5

The shaded gap between births and deaths shows the rate at which the population grows. Until about 1970 the average human death rate was dropping faster than the birth rate, and the population growth rate was increasing. Since 1970 the average birth rate has dropped slightly faster than the death rate. Therefore, the rate of population growth has decreased somewhat—though the growth continues to be exponential. (*Source: United Nations.*)

EXAMPLE 3   Consider the graph in Fig. 11.5.

(a) Estimate the world population and the birth and death rates in 1750.

(b) Estimate the world population and the birth and death rates in 1900.

(c) From 1750 to 1900, would you describe the growth in population as linear or exponential? Explain.

(d) From 1900 to present day, would you describe the growth in population as linear or exponential? Explain.

(e) Estimate how long it took the 1750 world population to double.

(f) Estimate how long it will take for the 1900 world population to double.

Solution:

(a) The world population in 1750 was about 0.8 billion people. The birth rate was about 40 per 1000 people. The death rate was about 37 per 1000 people.

(b) The world population in 1900 was about 1.6 billion people. The birth rate was about 38 per 1000 people. The death rate was about 30 per 1000 people.

(c) The growth in world population from 1750 to 1900 was close to linear growth. The population grew by about the same amount each year from 1750 to 1900.

(d) The growth in world population from 1900 to present day is exponential. Each year the population increases more than it did in the preceding year.

(e) The world population in 1750 was about 0.8 billion people. The world population grew to about 1.6 billion people in 1900. Therefore, it took about 150 years, 1750–1900, for the world population to double from its 1750 population.

(f) The world population was about 1.6 billion people in 1900. The world population grew to about 3.2 billion people about 1965. Therefore, it took only about 65 years for the world population to double from its 1900 population.

The United Nations estimates that the world's population will continue to grow exponentially and reach about 8 billion people in the year 2016 (see Group Activity Exercise 2).

A second interesting graph and its corresponding text from *Beyond the Limits,* shows how quickly all remaining rain forests on Earth could be depleted (see Fig. 11.6).

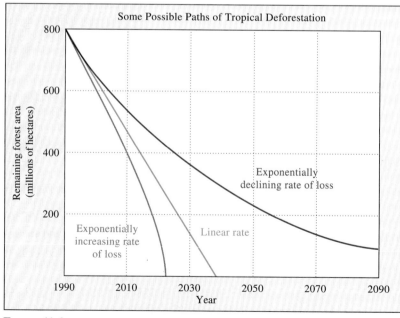

FIGURE 11.6

Estimates of the future loss of tropical forests depend upon assumptions about demographic, legal, and economic trends. The three scenarios, shown in this plot, indicate a range of possible futures if there is no concerted and effective agreement to protect the remaining forests. If the initial loss of 17 million hectares per year increases with population at about 2.3% per year, the forest will be gone by 2020. If the rate of loss is constant at 17 million hectares per year, the forest will be gone by about 2040. If the rate of loss is 2.1% of the remaining area each year, the forest will decline gradually toward zero over 100 years or more.

EXAMPLE 4   Study Fig. 11.6 and read the explanation above (about 41% of the remaining rain forest in 1990 was in Brazil).

(a) Estimate the year in which only 200 million hectares of rain forest will remain, assuming that the rate of loss of rain forest increases by 2.3% per year (the present annual growth rate in world population) over the present rate of loss.

(b) Estimate the year in which only 200 million hectares of rain forest will remain, assuming the present rate of loss of rain forests (17 million hectares per year).

(c) Estimate the year in which 200 million hectares of rain forest will remain if the clearing rate remains a constant percent of the remaining forest (2.1% per year). Each year the area cut will be slightly less than the year before because the area of the remaining rain forest decreases with each cut.

Solution:

(a) Using the green line, we see that in the year 2016 about 200 hectares of rain forest will remain.

(b) Using the blue line, we see that in the year 2026 about 200 hectares of rain forest will remain.

(c) Using the red line, we see that in the year 2054 about 200 hectares will remain.

The loss of tropical rain forests and many other items that we discuss daily, including the growth of the U.S. debt, are exponential in form (see Group Activity Exercise 3).

Exponential Equations

3   An **exponential equation** is one that has a variable as an exponent. Exponential equations are often used to describe the growth and decay of some quantities. Example 5 illustrates an exponential equation used in genetics.

EXAMPLE 5   The number of gametes, g, in a certain species of plant is determined by the equation $g = 2^n$, where n is the number of cells individuals of the species commonly have. Determine the number of gametes if the individual has 12 cells.

Solution: By evaluating 2^{12} on a calculator, we can determine that an individual of this species, with 12 cells, has 4096 gametes.

EXAMPLE 6   The formula $A = P(1 + r)^n$ is called the **compound interest formula.** When interest is compounded periodically (yearly, monthly, daily), this formula can be used to find the amount, A, in the account after n periods. In the formula, P is the principal, or the original amount invested, r is the interest rate per compounding period in decimal form, and n is the number of compounding periods. Suppose that \$10,000 is invested at 8% interest compounded quarterly for 6 years. Then $P = 10,000$; the interest rate per compounding period, r, is $\dfrac{8\%}{4}$ or 2%, and the number of compounding periods, n, is $6 \cdot 4$ or 24.

Find the amount in the account after 6 years.

Solution:

$$A = P(1 + r)^n$$
$$= 10,000(1 + 0.02)^{24}$$
$$= 10,000(1.02)^{24}$$
$$= 10,000(1.608437) \text{ from a calculator}$$
$$= 16,084.37$$

Therefore, the original \$10,000 investment has grown to \$16,084.37.

EXAMPLE 7   Carbon 14 dating is used by scientists to find the age of fossils and other artifacts. The formula used in carbon dating is

$$A = A_0 \cdot 2^{-t/5600}$$

where A_0 represents the amount of carbon 14 present when the fossil was formed and A represents the amount of carbon 14 present after t years. If 500 grams of carbon 14 were present when an organism died, how many grams will be found in the fossil 2000 years later?

Solution:

$$A = A_0 \cdot 2^{-t/5600}$$
$$= 500(2)^{-2000/5600}$$
$$= 500(0.7791646)$$
$$\approx 390.35 \text{ grams}$$

Logarithms   ④   Consider the exponential function $y = 2^x$. Recall from Section 9.6 that to find the inverse function, we interchange x and y and solve the equation for y. Interchanging x and y gives the equation $x = 2^y$, which is the inverse of $y = 2^x$. But at this time we have no way of solving the equation $x = 2^y$ for y. To solve this equation for y, we introduce a new definition.

Definition of Logarithm

For all positive numbers a, where $a \neq 1$,

$$y = \log_a x \quad \text{means} \quad x = a^y$$

In the expression $y = \log_a x$, the word *log* is an abbreviation for the word *logarithm*. $y = \log_a x$ is read "y is the logarithm of x to the base a." The letter y represents the logarithm, the letter a represents the base, and the letter x represents the number.

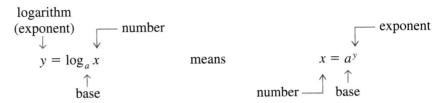

In words, the logarithm of the number x to the base a is the exponent to which the base a must be raised to equal the number x. For example,

$$2 = \log_{10} 100 \quad \text{means} \quad 100 = 10^2$$

In $\log_{10} 100 = 2$, the logarithm is 2, the base is 10, and the number is 100. The logarithm, 2, is the exponent to which the base, 10, must be raised to equal the number, 100. Note $10^2 = 100$.

The following are some examples of how an exponential expression can be converted to a logarithmic expression.

Exponential Form	*Logarithmic Form*
$10^0 = 1$	$\log_{10} 1 = 0$
$4^2 = 16$	$\log_4 16 = 2$
$\left(\dfrac{1}{2}\right)^5 = \dfrac{1}{32}$	$\log_{1/2} \dfrac{1}{32} = 5$
$5^{-2} = \dfrac{1}{25}$	$\log_5 \dfrac{1}{25} = -2$

Now let us do a few examples involving conversion from exponential form to logarithmic form, and vice versa.

EXAMPLE 8   Write in logarithmic form.

(a) $3^4 = 81$ **(b)** $\left(\dfrac{1}{5}\right)^3 = \dfrac{1}{125}$ **(c)** $2^{-4} = \dfrac{1}{16}$

Solution: **(a)** $\log_3 81 = 4$ **(b)** $\log_{1/5} \dfrac{1}{125} = 3$ **(c)** $\log_2 \dfrac{1}{16} = -4$

EXAMPLE 9   Write in exponential form.

(a) $\log_6 36 = 2$ **(b)** $\log_3 9 = 2$ **(c)** $\log_{1/3} \dfrac{1}{81} = 4$

Solution: **(a)** $6^2 = 36$ **(b)** $3^2 = 9$ **(c)** $\left(\dfrac{1}{3}\right)^4 = \dfrac{1}{81}$

EXAMPLE 10   Write in exponential form; then find the unknown value.

(a) $y = \log_5 25$ **(b)** $2 = \log_a 16$ **(c)** $3 = \log_{1/2} x$

Solution:

(a) $5^y = 25$ Since $5^2 = 25$, $y = 2$.   **(b)** $a^2 = 16$. Since $4^2 = 16$, $a = 4$.

(c) $\left(\dfrac{1}{2}\right)^3 = x$. Since $\left(\dfrac{1}{2}\right)^3 = \dfrac{1}{8}$, $x = \dfrac{1}{8}$.

Graphs of Logarithmic Functions

⑤ Now that we know how to convert from exponential form to logarithmic form, and vice versa, we can graph logarithmic functions. Equations of the form $y = \log_a x$, $a > 0$, $a \neq 1$, and $x > 0$, are called **logarithmic functions** since their graphs pass the vertical line test. To graph a logarithmic function, change it to exponential form then plot points. This procedure is illustrated in Examples 11 and 12.

EXAMPLE 11 Graph $y = \log_2 x$. State the domain and range of the function.

Solution: $y = \log_2 x$ means $x = 2^y$. Using $x = 2^y$, construct a table of values. The table will be easier to develop by selecting values for y and finding the corresponding values for x.

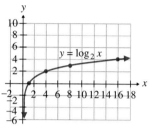

FIGURE 11.7

x	$\dfrac{1}{16}$	$\dfrac{1}{8}$	$\dfrac{1}{4}$	$\dfrac{1}{2}$	1	2	4	8	16
y	-4	-3	-2	-1	0	1	2	3	4

Now draw the graph (Fig. 11.7). The domain, the set of x values, is $\{x \mid x > 0\}$. The range, the set of y values, is all real numbers, $\mathbb{R}$.

EXAMPLE 12 Graph $y = \log_{1/2} x$. State the domain and range of the function.

Solution: $y = \log_{1/2} x$ means $x = \left(\frac{1}{2}\right)^y$. Construct a table of values by selecting values for y and finding the corresponding values of x.

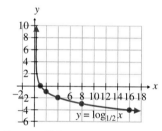

FIGURE 11.8

x	16	8	4	2	1	$\dfrac{1}{2}$	$\dfrac{1}{4}$	$\dfrac{1}{8}$	$\dfrac{1}{16}$
y	-4	-3	-2	-1	0	1	2	3	4

The graph is illustrated in Fig. 11.8. The domain is $\{x \mid x > 0\}$. The range is the set of real numbers, $\mathbb{R}$.

If we study the domains in Examples 11 and 12, we see that the domains of both $y = \log_2 x$ and $y = \log_{1/2} x$ are $\{x \mid x > 0\}$. In fact, **for any logarithmic function $y = \log_a x$, the domain is $\{x \mid x > 0\}$.** Also note that the graphs in Examples 11 and 12 are both one-to-one functions.

Graphs of Inverse Functions

⑥ Earlier we stated that functions of the form $y = 2^x$ and $x = 2^y$ are inverse functions. Since $x = 2^y$ means $y = \log_2 x$, the functions $y = 2^x$ and $y = \log_2 x$ are inverse functions. In fact, for any $a > 0$, $a \neq 1$, the functions $y = a^x$ and $y = \log_a x$ are inverse functions.

The graphs of $y = 2^x$ and $y = \log_2 x$ are illustrated in Fig. 11.9. Note that they are symmetric with respect to the line $y = x$. Since these graphs are inverses of each other, the symmetry is expected.

The graphs of $y = \left(\frac{1}{2}\right)^x$ and $y = \log_{1/2} x$ are illustrated in Fig. 11.10. They are also inverses of each other and are symmetric with respect to the line $y = x$.

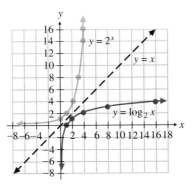

FIGURE 11.9

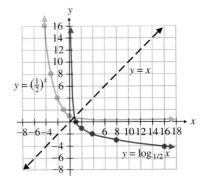

FIGURE 11.10

Solving Logarithmic Equations

7 When finding the logarithm of an expression, the expression is called the **argument** of the logarithm. For example, in $\log_{10} 3$ the 3 is the argument, and in $\log_{10}(2x + 4)$ the $(2x + 4)$ is the argument. When the argument contains a variable, we assume that the argument represents a positive value. *Remember, only logarithms of positive numbers exist.*

A **logarithmic equation** is one in which a variable appears in the argument of a logarithm. Exponential and logarithmic equations are discussed in more detail in Sections 11.4 and 11.5. One application of logarithms is to measure the intensity of earthquakes. Example 13 illustrates this use of logarithms.

EXAMPLE 13   The magnitude, R, of an earthquake on the Richter scale is given by the formula

$$R = \log_{10} I$$

The I represents the number of times greater (or more intense) the earthquake is than the smallest measurable activity that can be measured on the seismograph.

(a) If an earthquake measures 4 on the Richter scale, how many times more intense is it than the smallest measurable activity?

(b) How many times more intense is an earthquake that measures 5 than an earthquake that measures 4?

Solution: **(a)** $R = \log_{10} I$

$$4 = \log_{10} I$$

or $\qquad 10^4 = I$

$$10,000 = I$$

Therefore, an earthquake that measures 4 is 10,000 times more intense than the smallest measurable activity.

(b) $\qquad\qquad 5 = \log_{10} I$

$$10^5 = I$$

$$100,000 = I$$

Since $(10,000)(10) = 100,000$, an earthquake that measures 5 is 10 times more intense than an earthquake that measures 4.

Exercise Set 11.1

Graph the exponential function (see Examples 1 and 2).

1. $y = 2^x$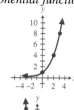

2. $y = 3^x$

3. $y = \left(\dfrac{1}{2}\right)^x$

4. $y = \left(\dfrac{1}{3}\right)^x$

5. $y = 4^x$

6. $y = 5^x$

7. $y = 3^{-x}$ (same as 4)

8. $y = \left(\dfrac{1}{3}\right)^{-x}$ (same as 2)

Graph the logarithmic function (see Examples 11 and 12).

9. $y = \log_2 x$

10. $y = \log_3 x$

11. $y = \log_{1/2} x$

12. $y = \log_{1/3} x$

13. $y = \log_5 x$

14. $y = \log_{1/4} x$

Graph each pair of functions on the same axes.

15. $y = 2^x, y = \log_{1/2} x$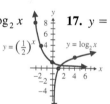

16. $y = \left(\dfrac{1}{2}\right)^x, y = \log_2 x$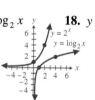

17. $y = 2^x, y = \log_2 x$

18. $y = \left(\dfrac{1}{2}\right)^x, y = \log_{1/2} x$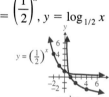

Write each expression in logarithmic form.

19. $2^3 = 8$ $\log_2 8 = 3$ **20.** $5^2 = 25$ $\log_5 25 = 2$ **21.** $3^5 = 243$ $\log_3 243 = 5$ **22.** $9^{1/2} = 3$ $\log_9 3 = \frac{1}{2}$

23. $8^{1/3} = 2$ $\log_8 2 = \frac{1}{3}$ **24.** $\left(\dfrac{1}{2}\right)^5 = \dfrac{1}{32}$ $\log_{1/2} \frac{1}{32} = 5$ **25.** $\left(\dfrac{1}{4}\right)^2 = \dfrac{1}{16}$ $\log_{1/4} \frac{1}{16} = 2$ **26.** $2^{-3} = \dfrac{1}{8}$ $\log_2 \frac{1}{8} = -3$

27. $5^{-2} = \dfrac{1}{25}$ $\log_5 \frac{1}{25} = -2$ **28.** $4^3 = 64$ $\log_4 64 = 3$ **29.** $4^{-3} = \dfrac{1}{64}$ $\log_4 \frac{1}{64} = -3$ **30.** $64^{1/3} = 4$ $\log_{64} 4 = \frac{1}{3}$

31. $16^{-1/2} = \dfrac{1}{4}$ $\log_{16} \frac{1}{4} = -\frac{1}{2}$ **32.** $36^{1/2} = 6$ $\log_{36} 6 = \frac{1}{2}$ **33.** $8^{-1/3} = \dfrac{1}{2}$ $\log_8 \frac{1}{2} = -\frac{1}{3}$ **34.** $81^{-1/4} = \dfrac{1}{3}$ $\log_{81} \frac{1}{3} = -\frac{1}{4}$

Write each expression in exponential form.   **47.** $10^3 = 1000$

35. $\log_2 8 = 3$ $2^3 = 8$ **36.** $\log_3 9 = 2$ $3^2 = 9$ **37.** $\log_4 64 = 3$ $4^3 = 64$ **38.** $\log_{1/3} \dfrac{1}{9} = 2$ $(\frac{1}{3})^2 = \frac{1}{9}$

39. $\log_{1/2} \dfrac{1}{16} = 4$ $(\frac{1}{2})^4 = \frac{1}{16}$ **40.** $\log_4 \dfrac{1}{16} = -2$ $4^{-2} = \frac{1}{16}$ **41.** $\log_5 \dfrac{1}{125} = -3$ $5^{-3} = \frac{1}{125}$ **42.** $\log_9 3 = \dfrac{1}{2}$ $9^{1/2} = 3$

43. $\log_{125} 5 = \dfrac{1}{3}$ $125^{1/3} = 5$ **44.** $\log_8 \dfrac{1}{64} = -2$ $8^{-2} = \frac{1}{64}$ **45.** $\log_{27} \dfrac{1}{3} = -\dfrac{1}{3}$ $27^{-1/3} = \frac{1}{3}$ **46.** $\log_{10} 100 = 2$ $10^2 = 100$

47. $\log_{10} 1000 = 3$ **48.** $\log_6 216 = 3$ $6^3 = 216$

49. $6y = 36$, $y = 2$ **50.** $3y = 27$, $y = 3$ **51.** $2^3 = x$, $x = 8$ **52.** $3^5 = x$, $x = 243$ **54.** $100^{1/2} = x$, $x = 10$ **55.** $a^3 = 64$, $a = 4$
56. $a^2 = 16$, $a = 4$ **57.** $\left(\frac{1}{2}\right)^4 = x$, $x = \frac{1}{16}$ **58.** $\left(\frac{1}{4}\right)^3 = x$, $x = \frac{1}{64}$ **59.** $a^5 = 32$, $a = 2$ **60.** $2y = \frac{1}{4}$, $y = -2$

Write each logarithm in exponential form. Then find the unknown value.

49. $y = \log_6 36$ **50.** $y = \log_3 27$ **51.** $3 = \log_2 x$ **52.** $5 = \log_3 x$

53. $y = \log_{64} 8$ $64^y = 8$, $y = \frac{1}{2}$ **54.** $\frac{1}{2} = \log_{100} x$ **55.** $3 = \log_a 64$ **56.** $2 = \log_a 16$

57. $4 = \log_{1/2} x$ **58.** $3 = \log_{1/4} x$ **59.** $5 = \log_a 32$ **60.** $y = \log_2 \frac{1}{4}$

61. $\frac{1}{3} = \log_8 x$ $8^{1/3} = x$, $x = 2$ **62.** $-\frac{1}{4} = \log_{16} x$ $16^{-1/4} = x$, $x = \frac{1}{2}$ **65.** 256   **68.** 10.6g

63. The graph below shows world population growth through history. The world population has been growing exponentially since the beginning of the industrial revolution. The following table indicates the population in 1650 (about 0.5 billion people) and each later year for which the world population doubled.

World Population Growth through History

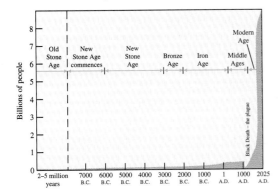

Year	Population (billions)
1650	0.5
1850	1
1930	2
1976	4
2016	8

63. (a)

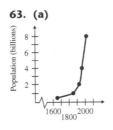

(a) Draw a graph of the data provided in the table. Place the year on the horizontal axis.

(b) How long did it take for the population to double (called the *doubling time*) from 0.5 billion to 1 billion people?   200 yr

(c) What is the doubling time for the population to increase from 4 billion to the projected 8 billion people?   40 yr

(d) From your graph in part (a) estimate the world population in 1950.   ≈ 2.7 billion

64. The following graph indicates linear growth of $100 invested at 7% simple interest and exponential growth at 7% interest compounded annually. In the formulas,

A represents the amount in dollars and *t* represents the time in years.

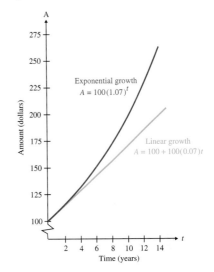

(a) Use the graph to estimate the doubling time for $100 invested at 7% simple interest.   ≈ 14 yr

(b) What is the doubling time for $100 invested at 7% interest compounded annually?   ≈ 10 yr

(c) Estimate the difference in amounts after 10 years for $100 invested by each method.   ≈ $25

(d) Most banks compound interest daily instead of annually. What effect does this have on the total amount? Explain.   increases it

65. Use the formula $g = 2^n$ to determine the number of gametes a plant has if it has 8 cells (see Example 5).

66. If José invests $5000 at 6% interest compounded quarterly, find the amount after 4 years. Use $A = P(1 + r)^n$ (see Example 6). $6344.93

67. If Marsha invests $8000 at 8% interest compounded quarterly, find the amount after 5 years. $11,887.58

68. If 12 grams of carbon 14 is originally present in a certain animal bone, how much will remain at the end of 1000 years? Use $A = A_0 \cdot 2^{-t/5600}$ (see Example 7).

69. If 60 grams of carbon 14 are originally present in the fossil Jonas found at the archeological site, how much will remain after 10,000 years?   17.4 g

80. $(\frac{1}{27}, -3)$, $(\frac{1}{9}, -2)$, $(\frac{1}{3}, -1)$, $(1, 0)$, $(3, 1)$, $(9, 2)$, $(27, 3)$
81. Graphs are symmetric about $y = x$. For each ordered pair (x, y) on $y = a^x$, (y, x) is on $y = \log_a x$.

70. The expected future population of Ackworth, which presently has 2000 residents, can be approximated by the formula $y = 2000 (1.2)^{0.1x}$, where x is the number of years in the future. Find the expected population of the town in **(a)** 10 years; **(b)** 100 years.   **(a.)** 2400   **(b.)** 12,383

71. The amount of a radioactive substance present, in grams, at time t in years is given by the formula $y = 80(2)^{-0.4t}$. Find the number of grams present in **(a)** 10 years; **(b)** 100 years.   **(a)** 5 g **(b)** 7.28×10^{-11} g

72. The number of a certain type of bacteria present in a culture is determined by the equation $y = 5000(3)^x$, where x is the number of days the culture has been growing. Find the number of bacteria in **(a)** 5 days; **(b)** 7 days.   **(a)** 1,215,000 **(b)** 10,935,000

73. If the magnitude of an earthquake is 7 on the Richter scale, how many times greater is the earthquake than the smallest measurable activity? Use $R = \log_{10} I$ (see Example 13).   10,000,000

74. How many times greater is an earthquake that measures 3 on the Richter scale than an earthquake that measures 1?   100

75. The average U.S. resident used 110,000 gallons of water in 1995. Suppose that each year after 1995 the average resident is able to reduce the amount of water used by 5%. The amount of water used by the average resident t years after 1995 could then be found by the formula $A = 110,000(0.95)^t$.
  (a) Explain why this formula may be used to find the amount of water used.
  (b) What would be the average amount of water used in the year 2000?   85,115.9 gal

76. Presently about $\frac{2}{3}$ of all aluminum cans are recycled each year, while about $\frac{1}{3}$ are disposed of in landfills. The recycled aluminum is used to make new cans. Americans used about 170,000,000 aluminum cans in

1995. The number of new cans made each year from recycled 1995 aluminum cans n years later can be estimated by the formula $A = 170,000,000 \left(\frac{2}{3}\right)^n$.

  (a) Explain why the formula may be used to estimate the number of the cans made from recycled aluminum cans n years after 1995.
  (b) How many cans will be made from 1995 recycled aluminum cans in 1998?   ≈ 5.04 million

77. For the logarithmic function $y = \log_a x$:
  (a) What are the restrictions on a? $a > 0$, $a \neq 1$
  (b) What is the domain of the function? $\{x \mid x > 0\}$
  (c) What is the range of the function? $\mathbb{R}$

78. For the logarithmic function $y = \log_a(x - 3)$, what must be true about x? Explain. $x > 3$

79. We stated earlier that, for exponential functions $f(x) = a^x$, the value of a cannot equal 1.
  (a) What does the graph of $f(x) = a^x$ look like when $a = 1$?   a horizontal line through 1
  (b) Is $f(x) = a^x$ a function when $a = 1$?   yes
  (c) Does $f(x) = a^x$ have an inverse function when $a = 1$? Explain your answer.   no

80. If some points on the graph of the exponential function $f(x) = a^x$ are $(-3, \frac{1}{27})$, $(-2, \frac{1}{9})$, $(-1, \frac{1}{3})$, $(0, 1)$, $(1, 3)$, $(2, 9)$, and $(3, 27)$, list some points on the graph of the logarithmic function $f(x) = \log_a x$. Explain.

81. Discuss the relation between the graphs $y = a^x$ and $y = \log_a x$ for $a > 0$ and $a \neq 1$.

82. Discuss the graphs of $y = a^x$ **(a)** for $0 < a < 1$ and **(b)** for $a > 1$.

83. How will the graphs of $y = a^x$ and $y = a^x + k$ compare? Explain.

84. Explain why $y = \log_a x$, $a > 0$, $a \neq 1$, is not defined for $x \leq 0$. *Hint:* Change the equation to exponential form. $a^y = x$; if $a > 0$, x must be greater than 0

82. (a) y intercept is $(0, 1)$, y gets closer to 0 as x increases. **(b)** y intercept is at $(0, 1)$, y gets closer to 0 as x decreases.   **83.** same basic shapes, however, $y = a^x + k$ will be k units higher or lower than $y = a^x$. The y intercept of $y = a^x + k$ is $(0, 1 + k)$.

CUMULATIVE REVIEW EXERCISES

[6.1–6.4] *Factor.*

85. $24x^2 - 6xy + 16xy - 4y^2$ $2(3x + 2y)(4x - y)$

86. $2(a - 3)^2 + 7(a - 3) - 15$ $(2a - 9)(a + 2)$

87. $4x^4 - 36x^2$ $4x^2(x + 3)(x - 3)$

88. $8x^3 + \dfrac{1}{27}$ $\left(2x + \frac{1}{3}\right)\left(4x^2 - \frac{2}{3}x + \frac{1}{9}\right)$

Group Activity/ Challenge Problems

1. Do you believe that the world's population can continue to grow exponentially forever? Explain.

2. Refer to Fig. 11.5. Explain how the world population curve may be constructed using the birth and death curves.

3. We see exponential functions daily. For example, the U.S. debt, U.S. spending on health care, and the growing number of cases of AIDS are all increasing exponentially.

    (a) Use an almanac or other reference source and construct a graph of the U.S. debt (the budget deficit) from 1970 to 1995.

    (b) Explain why this graph is exponential.

4. In Exercise 64 we graph the amount for various years when $100 is invested at 7% simple interest and at 7% interest compounded annually.   **4. (a)** $201.36

    (a) Use compound interest formula given in Example 6 to determine the amount if $100 is compounded daily at 7% for 10 years (assume 365 days per year).

    (b) Estimate the difference in the amount in 10 years for the $100 invested at 7% simple interest versus the 7% interest compounded daily. $131.36

Graph.

5.

5. $y = 2^{-x + 1}$

6. $y = 1000(2)^{0.6t}$, $0 \le t \le 10$

Solve the following equations. We will discuss problems like this in Section 11.4.

7. $2^{3x + 2} = 16$ $\frac{2}{3}$

8. $16^x = 64$ $\frac{3}{2}$

6.

9. Bob gives Larry $1 on day 1, $2 on day 2, $4 on day 3, $8 on day 4, and continues this doubling process for 30 days.   **(a)** 2^{n-1}

    (a) Express the amount, using exponential form, that Bob gives Larry on day n.

    (b) How much, in dollars, will Bob give Larry on day 30? Write the amount in exponential form. Then use a calculator to evaluate. $2^{29} = $536,870,912

    (c) Express the total amount Bob gives Larry over the 30 days as a sum of exponential terms. (Do not find the actual value.) $2^0 + 2^1 + 2^2 + \cdots + 2^{29}$

11.2 Properties of Logarithms

    1 Use the product rule for logarithms.
    2 Use the quotient rule for logarithms.
    3 Use the power rule for logarithms.

Product Rule   **1** To be able to do calculations using logarithms, you must understand their properties. We use these properties in Section 11.4 when we solve exponential and logarithmic equations. The first property we discuss is the product rule for logarithms.

> **Product Rule for Logarithms**
> For positive real numbers x, y, and a, $a \neq 1$,
> $$\log_a xy = \log_a x + \log_a y \qquad \text{Property 1}$$

To prove this property, let $\log_a x = m$ and $\log_a y = n$. Now write each logarithm in exponential form.

$$\log_a x = m \quad \text{means} \quad a^m = x$$
$$\log_a y = n \quad \text{means} \quad a^n = y$$

By substitution and using the rules of exponents, we see that

$$xy = a^m \cdot a^n = a^{m+n}$$

We can now convert $xy = a^{m+n}$ to logarithmic form.

$$xy = a^{m+n} \quad \text{means} \quad \log_a xy = m + n$$

Finally, substituting $\log_a x$ for m and $\log_a y$ for n, we obtain

$$\log_a xy = \log_a x + \log_a y$$

which is property 1.

Examples of Property 1

$$\log_3 5 \cdot 7 = \log_3 5 + \log_3 7$$
$$\log_4 3x = \log_4 3 + \log_4 x$$
$$\log_8 x^2 = \log_8 (x \cdot x) = \log_8 x + \log_8 x \quad \text{or} \quad 2 \log_8 x$$

Quotient Rule **2** Now we give the quotient rule for logarithms, which we refer to as property 2.

> **Quotient Rule for Logarithms**
> For positive real numbers x, y, and a, $a \neq 1$,
> $$\log_a \frac{x}{y} = \log_a x - \log_a y \qquad \text{Property 2}$$

Examples of Property 2

$$\log_3 \frac{12}{4} = \log_3 12 - \log_3 4$$

$$\log_6 \frac{x}{3} = \log_6 x - \log_6 3$$

$$\log_5 \frac{x}{x+2} = \log_5 x - \log_5 (x+2)$$

Power Rule   3   The last property we discuss in this section is the power rule for logarithms.

> **Power Rule for Logarithms**
> If x and a are positive real numbers, $a \neq 1$, and n is any real number, then
> $$\log_a x^n = n \log_a x \qquad\qquad \text{Property 3}$$

Examples of Property 3

$$\log_2 4^3 = 3 \log_2 4$$
$$\log_{10} x^2 = 2 \log_{10} x$$
$$\log_5 \sqrt{12} = \log_5 (12)^{1/2} = \frac{1}{2} \log_5 12$$
$$\log_8 \sqrt[5]{x+3} = \log_8 (x+3)^{1/5} = \frac{1}{5} \log_8 (x+3)$$

Properties 2 and 3 can be proved in a manner similar to that given for property 1 (see Group Activity Exercises 6 and 7).

EXAMPLE 1   Use properties 1 through 3 to expand.

(a) $\log_8 \dfrac{27}{43}$
(b) $\log_4 (64)(180)$
(c) $\log_{10} (32)^{1/5}$

Solution:

(a) $\log_8 \dfrac{27}{43} = \log_8 27 - \log_8 43$

(b) $\log_4 (64)(180) = \log_4 64 + \log_4 180.$

(c) $\log_{10} (32)^{1/5} = \dfrac{1}{5} \log_{10} 32$

Often we will have to use two or more of these properties in the same problem.

EXAMPLE 2   Expand.

(a) $\log_{10} 4(x+2)^3$
(b) $\log_5 \dfrac{(4-x)^2}{3}$

(c) $\log_5 \left(\dfrac{4-x}{3}\right)^2$
(d) $\log_5 \dfrac{[x(x+4)]^3}{2}$

Solution:

(a) $\log_{10} 4(x+2)^3 = \log_{10} 4 + \log_{10} (x+2)^3$
$$= \log_{10} 4 + 3 \log_{10} (x+2)$$

(b) $\log_5 \dfrac{(4-x)^2}{3} = \log_5 (4-x)^2 - \log_5 3$
$$= 2 \log_5 (4-x) - \log_5 3$$

(c) $\log_5\left(\dfrac{4-x}{3}\right)^2 = 2\log_5\left(\dfrac{4-x}{3}\right)$

$$= 2\,[\log_5(4-x) - \log_5 3]$$
$$= 2\log_5(4-x) - 2\log_5 3$$

(d) $\log_5\dfrac{[x(x+4)]^3}{2} = \log_5[x(x+4)]^3 - \log_5 2$

$$= 3\log_5 x(x+4) - \log_5 2$$
$$= 3\,[\log_5 x + \log_5(x+4)] - \log_5 2$$
$$= 3\log_5 x + 3\log_5(x+4) - \log_5 2$$

Note that property 1 can be expanded to evaluate the product of 3 or more quantities. For example, $\log_5 xyz = \log_5 x + \log_5 y + \log_5 z$.

Helpful Hints

In Example 2(b), when we expanded $\log_5\dfrac{(4-x)^2}{3}$, we first used the quotient rule. In Example 2(c), when we expanded $\log_5\left(\dfrac{4-x}{3}\right)^2$, we first used the power rule. Do you see the difference in the two problems? In $\log_5\dfrac{(4-x)^2}{3}$ just the numerator of the argument is squared; therefore, we use the quotient rule first. In $\log_5\left(\dfrac{4-x}{3}\right)^2$ the entire argument is squared, and so we use the power rule first. If we wished, we could rewrite $\log_5\left(\dfrac{4-x}{3}\right)^2$ as $\log_5\dfrac{(4-x)^2}{3^2}$. Then we could use the quotient rule to expand the expression.

Show for yourself that $\log_5\dfrac{(4-x)^2}{3^2}$ gives the same answer as $\log_5\left(\dfrac{4-x}{3}\right)^2$.

EXAMPLE 3 Write each of the following as the logarithm of a single expression.
(a) $3\log_8(x+2) - \log_8 x$
(b) $\log_7(x+1) + 2\log_7(x+4) - 3\log_7(x-5)$

Solution:
(a) $3\log_8(x+2) - \log_8 x = \log_8(x+2)^3 - \log_8 x$

$$= \log_8\dfrac{(x+2)^3}{x}$$

(b) $\log_7(x+1) + 2\log_7(x+4) - 3\log_7(x-5)$
$$= \log_7(x+1) + \log_7(x+4)^2 - \log_7(x-5)^3$$
$$= \log_7(x+1)(x+4)^2 - \log_7(x-5)^3$$
$$= \log_7\dfrac{(x+1)(x+4)^2}{(x-5)^3}$$

COMMON STUDENT ERROR

The Correct Rules Are

$$\log_a (x \cdot y) = \log x + \log y$$

$$\log_a \frac{x}{y} = \log_a x - \log_a y$$

A mistake sometimes made by students is to use the following *incorrect procedures.*

Incorrect

$$\log_a (x + y) = \log_a x + \log_a y$$

$$\log_a (x - y) = \log_a x - \log_a y$$

$$\log_a (x \cdot y) = (\log_a x)(\log_a y)$$

Incorrect

$$\frac{\log_a x}{\log_a y} = \log_a x - \log_a y$$

$$\frac{\log_a x}{\log_a y} = \log_a \frac{x}{y}$$

For example,

$$\log_a (x + 2) \neq \log_a x + \log_a 2$$

$$\log_a (3x) \neq (\log_a 3)(\log_a x)$$

$$\frac{\log_a x}{\log_a 2} \neq \log_a x - \log_a 2$$

$$\frac{\log_a 10}{\log_a x} \neq \log_a \frac{10}{x}$$

The last properties we will discuss in this section will be used to solve equations in Section 11.4.

Additional Properties of Logarithms

If $a > 0$, and $a \neq 1$, then

$$\log_a a^x = x \qquad\qquad \text{Property 4}$$

and $\qquad a^{\log_a x} = x \quad (x > 0) \qquad$ Property 5

Examples of Property 4

$$\log_6 6^5 = 5$$
$$\log_6 6^x = x$$

Examples of Property 5

$$3^{\log_3 7} = 7$$
$$5^{\log_5 x} = x \ (x > 0)$$

EXAMPLE 4   Evaluate.   **(a)** $\log_5 25$ **(b)** $\sqrt{16}^{\log_4 9}$

Solution:

(a) $\log_5 25$ may be written as $\log_5 5^2$. By property 4

$$\log_5 25 = \log_5 5^2 = 2$$

(b) $\sqrt{16}^{\log_4 9}$ may be written $4^{\log_4 9}$. Then use property 5.

$$\sqrt{16}^{\log_4 9} = 4^{\log_4 9} = 9$$

Exercise Set 11.2

Use properties 1–3 to expand.

1. $\log_3 7 \cdot 12$ $\log_3 7 + \log_3 12$

2. $\log_5 8 \cdot 29$ $\log_5 8 + \log_5 29$

3. $\log_8 7(x + 3)$ $\log_8 7 + \log_8 (x + 3)$

4. $\log_{10} x(x - 2)$ $\log_{10} x + \log_{10} (x - 2)$

5. $\log_4 \dfrac{15}{7}$ $\log_4 15 - \log_4 7$

6. $\log_9 \dfrac{\sqrt{x}}{12}$ $\frac{1}{2} \log_9 x - \log_9 12$

7. $\frac{1}{2} \log_{10} x - \log_{10} (x - 3)$ **12.** $2 \log_8 x + \log_8 (x - 2)$ **13.** $\frac{1}{2} [5 \log_4 x - \log_4 (x + 4)]$ **14.** $2 \log_{10} (x - 3) + 3 \log_{10} x$
15. $4 \log_{10} x - 3 \log_{10} (a + 2)$ **17.** $-2 \log_8 x + \log_8 (x - 6)$ **19.** $\log_{10} 2 + \log_{10} m - \log_{10} 3 - \log_{10} n$

7. $\log_{10} \dfrac{\sqrt{x}}{x - 3}$

8. $\log_5 3^{12}$ $12 \log_5 3$

9. $\log_8 x^4$ $4 \log_8 x$

10. $\log_5 (x + 4)^3$ $3 \log_5 (x + 4)$

11. $\log_{10} 3(8^2)$ $\log_{10} 3 + 2 \log_{10} 8$

12. $\log_8 x^2(x - 2)$

13. $\log_4 \sqrt{\dfrac{x^5}{x + 4}}$

14. $\log_{10} (x - 3)^2 x^3$

15. $\log_{10} \dfrac{x^4}{(a + 2)^3}$

16. $\log_7 x^2(x - 2)$ $2 \log_7 x + \log_7 (x - 2)$ **17.** $\log_8 \dfrac{x(x - 6)}{x^3}$

18. $\log_{10} \left(\dfrac{x}{6}\right)^2$ $2 (\log_{10} x - \log_{10} 6)$

19. $\log_{10} \dfrac{2m}{3n}$

20. $\log_5 \dfrac{\sqrt{a}\sqrt[3]{b}}{\sqrt[4]{c}}$ $\frac{1}{2} \log_5 a + \frac{1}{3} \log_5 b - \frac{1}{4} \log_5 c$

Write as a logarithm of a single expression.   **25.** $\log_{10} [x(x - 4)/(x + 1)]$ **28.** $\log_7 [(x + 3)^5 (x - 1)^2/ \sqrt{x}]$
21. $2 \log_{10} x - \log_{10} (x - 2)$ $\log_{10} [x^2/(x - 2)]$

22. $3 \log_8 x + 2 \log_8 (x + 1)$ $\log_8 x^3(x + 1)^2$

23. $2 (\log_5 x - \log_5 4)$ $\log_5 (x/4)^2$

24. $\dfrac{1}{2} [\log_6 (x - 1) - \log_6 3]$ $\log_6 \sqrt{\frac{x - 1}{3}}$

25. $\log_{10} x + \log_{10} (x - 4) - \log_{10} (x + 1)$

26. $2 \log_5 x + \log_5 (x - 4) + \log_5 (x - 2)$ $\log_5 x^2(x - 4)(x - 2)$

27. $\dfrac{1}{2} [\log_7 (x - 2) - \log_7 x]$ $\log_7 \sqrt{\frac{x - 2}{x}}$

28. $5 \log_7 (x + 3) + 2 \log_7 (x - 1) - \dfrac{1}{2} \log_7 x$

29. $2 \log_9 5 + \dfrac{1}{3} \log_9 (x + 7) - \dfrac{1}{2} \log_9 x$ $\log_9 \frac{25 \sqrt[3]{x + 7}}{\sqrt{x}}$

30. $5 \log_6 (x + 3) - [2 \log_6 (x - 4) + 3 \log_6 x]$

31. $4 \log_6 3 - [2 \log_6 (x + 3) + 4 \log_6 x]$

32. $2 \log_7 (x - 6) + 3 \log_7 (x + 3) - [5 \log_7 2 + 3 \log_7 (x - 2)]$

Find the value by writing each argument using the numbers 2 and/or 5 and using the values $\log_{10} 2 = 0.3010$ *and*
$\log_{10} 5 = 0.6990$.   **30.** $\log_6 [(x + 3)^5/((x - 4)^2 x^3)]$ **31.** $\log_6 [3^4/((x + 3)^2 \cdot x^4)]$ **32.** $\log_7 [(x - 6)^2(x + 3)^3/(2^5(x - 2)^3)]$
33. $\log_{10} 10$ 1

34. $\log_{10} 0.4$ -0.3980

35. $\log_{10} 2.5$ 0.3980

36. $\log_{10} 4$ 0.6020

37. $\log_{10} 25$ 1.3980

38. $\log_{10} 8$ 0.9030

Evaluate (see Example 4).
39. $6^{\log_6 12}$ 12

40. $\log_5 5$ 1

41. $(2^3)^{\log_8 5}$ 5

42. $\log_8 64$ 2

43. $\log_3 27$ 3

44. $2 \log_9 \sqrt{9}$ 1

45. $5 \left(\sqrt[3]{27}\right)^{\log_3 5}$ 25

46. $\dfrac{1}{2} \log_6 \sqrt[3]{6}$ $\frac{1}{6}$

Determine whether each statement is true or false.   **47.** false   **52.** false
47. $\log_{10} 3 + \log_{10} 4 = \log_{10} (3 + 4)$ **48.** $(\log_4 5)(\log_4 8) = \log_4 (5 \cdot 8)$ false   **49.** $\log_3 8 - \log_3 4 = \log_3 \dfrac{8}{4}$ true

50. $\dfrac{\log_6 9}{\log_6 2} = \log_6 9 - \log_6 2$ false

51. $\dfrac{\log_5 8}{\log_5 2} = \log_5 \dfrac{8}{2}$ false

52. $(\log_4 5)(\log_4 8) = \log_4 5 + \log_4 8$

53. $2 \log_{10} 10 = (\log_{10} 10)^2$ false

54. $3 \log_{10} 10 = \log_{10} 10^3$ true

55. $\log_6 a^6 = a$ false

56. $\log_6 6^a = a$ true

57. $10^{\log_{10} x} = x$ $(x > 0)$ true

58. $3^{\log_3 3} = 3$ true

59. Explain why $\log_a a = 1$ for any base a $(a > 0)$. $\log_a a^1 = 1$ by property 5.
60. In your own words, explain the product rule for logarithms.   **64.** $\dfrac{3x^2 + 28x + 9}{(x - 4)(x - 3)(2x + 5)}$
61. In your own words, explain the quotient rule for logarithms.
62. In your own words, explain the power rule for logarithms.   **65.** $2\frac{2}{9}$ days   **66.** $2x^3y^5 \sqrt[3]{6x^2y^2}$

CUMULATIVE REVIEW EXERCISES

Perform the indicated operations.

[7.1] **63.** $\dfrac{2x + 5}{x^2 - 7x + 12} \div \dfrac{x - 4}{2x^2 - x - 15}$. $\dfrac{(2x + 5)^2}{(x - 4)^2}$

[7.2] **64.** $\dfrac{2x + 5}{x^2 - 7x + 12} - \dfrac{x - 4}{2x^2 - x - 15}$.

[7.5] **65.** Mike can paint a house by himself in 4 days and Jill can paint the same house by herself in 5 days. How long would it take them to paint the house together?

[8.3] **66.** Multiply and then simplify $\sqrt[3]{4x^4y^7} \cdot \sqrt[3]{12x^7y^{10}}$

1. $\frac{1}{4} \log_2 x + \frac{1}{4} \log_2 y + \frac{1}{3} \log_2 a - \frac{1}{5} \log_2 (a - b)$ 2. $2[\log_3 (a^2 + b^2) + 2 \log_3 c - \log_3 (a - b) - \log_3 (b + c) - \log_3 (c + d)]$
3. (a) 0.864 (b) 0.216 (c) 0.108 (d) 4.32 4. (a) 0.7000 (b) 0.3000 (c) no

Group Activity/ Challenge Problems

Use properties 1–3 to expand.

1. $\log_2 \dfrac{\sqrt[4]{xy}\,\sqrt[3]{a}}{\sqrt[5]{a - b}}$

2. $\log_3 \left[\dfrac{(a^2 + b^2)(c^2)}{(a - b)(b + c)(c + d)} \right]^2$

3. If $\log_{10} x = 0.4320$, find (a) $\log_{10} x^2$; (b) $\log_{10} \sqrt{x}$; (c) $\log_{10} \sqrt[4]{x}$; (d) $\log_{10} x^{10}$.

4. If $\log_{10} x = 0.5000$ and $\log_{10} y = 0.2000$, find (a) $\log_{10} xy$; (b) $\log_{10} \left(\dfrac{x}{y} \right)$. (c) Using just the information provided, is it possible to find $\log_{10} (x + y)$? Explain.

5. Are the graphs of $y = \log_b x^2$ and $y = 2 \log_b x$ the same? Explain your answer by discussing the domains of each equation.

6. Prove the quotient rule for logarithms.

7. Prove the power rule for logarithms.

5. No, the domain of $y = \log_b x^2$ is all real numbers except 0; the domain of $y = 2 \log_b x$ is $\{x \mid x > 0\}$.

11.3 Common Logarithms

Tape 18

1   Find common logarithms of powers of 10.
2   Find common logarithms on a calculator.
3   Find antilogarithms on a calculator.

Find Common Logarithms of Powers of 10

1   The properties discussed in Section 11.2 can be used with any valid base (a real number greater than 0 and not equal to 1). However, since we are used to working in base 10, we will often use the base 10 when computing with logarithms. Base 10 logarithms are called **common logarithms.** When we are working with common logarithms, it is not necessary to list the base. Thus, $\log x$ means $\log_{10} x$.

The properties of logarithms written as common logarithms follow. For positive real numbers x and y, and any real number n

1. $\log xy = \log x + \log y$

2. $\log \dfrac{x}{y} = \log x - \log y$

3. $\log x^n = n \log x$.

In Chapter 5 we learned that 1 can be expressed as 10^0 and the 10 can be expressed as 10^1. Since, for example, the 5 is between 1 and 10, it must also be between 10^0 and 10^1.

$$1 < 5 < 10$$
$$10^0 < 5 < 10^1$$

The number 5 can be expressed as the base 10 raised to an exponent between 0 and 1. The number 5 is equal to $10^{0.69897}$. The common logarithm of 5 is 0.69897.

$$\log 5 = 0.69897$$

> ## Common Logarithm
> The **common logarithm** of a positive real number is the **exponent** to which the base 10 is raised to obtain the number.
>
> $$\text{If } \log N = L, \quad \text{then} \quad 10^L = N.$$

For example, if $\log 5 = 0.69897$, then $10^{0.69897} = 5$.
Now consider the number 50.

$$10 < 50 < 100$$
$$10^1 < 50 < 10^2$$

The number 50 can be expressed as the base 10 raised to an exponent between 1 and 2. The number $50 = 10^{1.69897}$; thus $\log 50 = 1.69897$.

Find Logarithms on a Calculator

2 To find common logarithms of numbers, we can use a calculator that has a logarithm key, $\boxed{\log}$.

The logarithms of most numbers are irrational numbers. Even the values given by calculators are usually only approximations of the actual values. Even though we are working with approximations when evaluating most logarithms, we generally write the logarithm with an equal sign. Thus, rather than writing $\log 6 \approx 0.7781513$, we will write $\log 6 = 0.7781513$.

Calculator Corner

FINDING COMMON LOGARITHMS USING A CALCULATOR

Common logarithms can be found using a calculator with a $\boxed{\log}$ key.
To find the common logarithm of a real number greater than zero, enter the number, then press the logarithm key. The answer will then be displayed.

Example	Keys to Press	Answer Displayed
Find log 400	400 $\boxed{\log}$	2.60206
Find log 0.0538	0.0538 $\boxed{\log}$	-1.2692177

EXAMPLE 1 Indicate the keys you would press to find the following logarithms using a calculator and then give the value obtained.

(a) log 962   **(b)** log 1   **(c)** log 0.00046   **(d)** log -5.2

Solution: The expression following $\boxed{\log}$ is the answer displayed on the calculator.

(a) 962 $\boxed{\log}$ 2.9831751   **(b)** 1 $\boxed{\log}$ 0
(c) 0.00046 $\boxed{\log}$ -3.3372422 **(d)** 5.2 $\boxed{+/-}$ $\boxed{\log}$ Error

Recall that the logarithms of negative numbers do not exist.

EXAMPLE 2 Find the exponent to which the base 10 must be raised to obtain the number 43,600.

Solution: We are asked to find the exponent, which is a logarithm. We need to determine log 43,600.

$$43,600 \;\boxed{\text{log}}\; 4.6394865$$

Thus, the exponent is 4.6394865. Note that $10^{4.6394865} = 43,600$.

The question that should now be asked is, "if we know the common logarithm of a number, how do we find the number?" For example, if log $N = 3.406$, what is N? When we find the value of the number from the logarithm, we say we are finding the **antilogarithm** or **inverse logarithm.**

If the logarithm of N is L, then N is the antilogarithm or inverse logarithm of L.

Antilogarithm

$$\text{If } \log N = L, \quad \text{then} \quad N = \text{antilog } L.$$

Finding Antilogarithms

3   When we are given the common logarithm, which is the exponent on the base 10, the *antilog is the number* obtained when the base 10 is raised to that exponent. In Example 1 we found log $962 = 2.9831751$. If we are given the logarithm, or exponent, 2.9831751, the antilog is the value of $10^{2.9831751}$, or 962. Thus, if log $962 = 2.9831751$, then antilog $2.9831751 = 962$. In Example 1 we showed that log $0.00046 = -3.3372422$. Thus, antilog $-3.3372422 = 0.00046$. When finding an antilog, we are converting from a logarithm, or exponent, to a number.

Calculator Corner

FINDING ANTILOGARITHMS ON A CALCULATOR

To find antilogarithms on a calculator we first enter the logarithm for which we are finding the antilog. Then press the second function key, $\boxed{\text{2nd}}$, the inverse key, $\boxed{\text{inv}}$, or the shift key, $\boxed{\text{SHIFT}}$, depending on which of these keys your calculator has. Next press the log key, $\boxed{\text{log}}$. After the log key is pressed the antilog will be displayed.

On some calculators 10^x is listed directly above the $\boxed{\text{log}}$ key. As explained earlier, the antilog of the logarithm (or exponent) x is the number obtained when the base 10 is raised to the logarithm (or exponent) x. Thus, the antilog is actually the value of 10^x, where x is the logarithm (or exponent).

In the following examples we will use the $\boxed{\text{inv}}$ key. If your calculator used a $\boxed{\text{2nd}}$ key or a $\boxed{\text{SHIFT}}$ key, you would use that in place of the $\boxed{\text{inv}}$ key.

Example	*Keys to Press*	*Answer Displayed*
Find antilog 2.9831751	2.9831751 $\boxed{\text{inv}}$ $\boxed{\text{log}}$	962.00006*
Find antilog -3.3372422	3.3372422 $\boxed{^+/_-}$ $\boxed{\text{inv}}$ $\boxed{\text{log}}$	0.00046**

When you are finding the antilog of a negative value, enter the value and then press the $\boxed{^+/_-}$ key before pressing the inverse and logarithm keys.

---

* Some calculators give slightly different answers, depending upon their electronics.
** Some calculators may display answers in scientific notation form.

EXAMPLE 3 Indicate the keys you would press to find the following antilogarithms on a calculator and give the value obtained.

(a) antilog 6.827   **(b)** antilog 0

(c) antilog -2.35 **(d)** antilog -5.822

Solution:

(a) 6.827 `inv` `log` 6714288.4   **(b)** 0 `inv` `log` 1

(c) 2.35 `+/−` `inv` `log` 0.0044668   **(d)** 5.822 `+/−` `inv` `log` 0.0000015

EXAMPLE 4 Find the value obtained when the base 10 is raised to the -1.052 power.

Solution: We are asked to find the value of $10^{-1.052}$. Since we are given the exponent, or logarithm, we can find the value by taking the antilog of -1.052.

$$\text{antilog } -1.052 = 0.0887156$$

Thus, $10^{-1.052} = 0.0887156$.

EXAMPLE 5 Find N if $\log N = 3.742$.

Solution: We are given the logarithm and asked to find the antilog, or the number N.

$$\text{antilog } 3.742 = 5520.7743$$

Thus, $N = 5520.7743$.

Since we generally do not need the eight-place accuracy given by most calculators, in the exercise set that follows we will round logarithms to four decimal places and antilogarithms to three *significant digits*. In a number written in decimal form, any zeros preceding the first nonzero digit are not significant digits. The first nonzero digit in a number, moving from left to right, is the first significant digit.

	Example	
0.006 3402	first significant digit is shaded
3.04 24080	first three significant digits are shaded
0.0000 138 483	first three significant digits are shaded
206,4 35.05	first four significant digits are shaded

EXAMPLE 6 Find the following antilogs and round to three significant digits.

(a) antilog 6.827   **(b)** antilog -2.35

Solution:

(a) In Example 3(a) we found antilog $6.827 = 6{,}714{,}288.4$. Rounding to three significant digits, we get antilog $6.827 = 6{,}71\,0{,}000$.

(b) In Example 3(c) we found antilog $-2.35 = 0.0044668$. Rounding to three significant digits, we get antilog $-2.35 = 0.00447$.

Exercise Set 11.3

Find the common logarithm of the number. Round the answer to four decimal places.

1. 870   2.9395
2. 36   1.5563
3. 8   0.9031
4. 19,200   4.2833

5. 1000   3.0000
6. 0.00152   −2.8182
7. 0.0000857   −4.0671
8. 27,700   4.4425

9. 100   2.0000
10. 0.000835   −3.0783
11. 1.74   0.2405
12. 3.75   0.5740

13. 0.375   −0.4260
14. 0.0000375   −4.4260
15. 0.00872   −2.0595
16. 960   2.9823

Find the antilog of the logarithm. Round the answer to three significant digits.

17. 0.5416   3.48
18. 2.6464   443
19. 2.3201   209
20. 5.8149   653,000

21. −1.0585   0.0874
22. −2.3382   0.00459
23. 0.0000   1.00
24. 5.5922   391,000

25. 2.5011   317
26. −4.4306   0.0000371
27. −0.1543   0.701
28. −1.2549   0.0556

Find the number N. Round N to three significant digits.

29. $\log N = 2.0000$ 100
30. $\log N = 1.6730$ 47.10
31. $\log N = -3.104$ 0.000787
32. $\log N = 1.9330$ 85.7

33. $\log N = 3.8202$ 6610
34. $\log N = 2.7404$ 550
35. $\log N = -1.06$ 0.0871
36. $\log N = -1.1469$ 0.0713

37. $\log N = -0.3686$ 0.428
38. $\log N = 1.5159$ 32.8
39. $\log N = -0.3936$ 0.404
40. $\log N = -1.3206$ 0.0478

To what exponent must the base 10 be raised to obtain each of the following values? Round your answer to four decimal places.

41. 2370   3.3747
42. 817,000   5.9122
43. 0.0410   −1.3872
44. 0.00612   −2.2132

45. 102   2.0086
46. 8.92   0.9504
47. 0.00128   −2.8928
48. 73,700,000   7.8675

Find the value obtained when 10 is raised to the following exponents. Round to three significant digits.

49. 2.5866   386.0
50. 3.7118   5150
51. −0.158   0.695
52. −2.2351   0.00582

53. −1.6091   0.0246
54. 4.8537   71,400
55. 1.1903   15.5
56. −2.1918   0.00643

By changing the logarithm to exponential form, evaluate the common logarithm without the use of a calculator.

57. $\log 1$ 0
58. $\log 100$ 2
59. $\log 0.1$ −1
60. $\log 1000$ 3

61. $\log 0.01$ −2
62. $\log 10$ 1
63. $\log 0.001$ −3
64. $\log 10,000$ 4

In Section 11.2 we stated that for a > 0, and a ≠ 1, $\log_a a^x = x$ and $a^{\log_a x} = x \ (x > 0)$. Rewriting these properties using common logarithms (a = 10), we obtain $\log 10^x = x$ and $10^{\log x} = x \ (x > 0)$, respectively. Use these properties to evaluate each of the following.

65. $\log 10^5$ 5
66. $\log 10^{6.7}$ 6.7
67. $10^{\log 7}$ 7
68. $10^{\log 8.3}$ 8.3

69. $6 \log 10^{5.2}$ 31.2
70. $8 \log 10^{4.6}$ 36.8
71. $5(10^{\log 9.4})$ 47
72. $2.3(10^{\log 5.2})$ 11.96

Find the solution. Round answers to the nearest hundredth.

73. The magnitude of an earthquake on the Richter scale is given by the formula $R = \log I$, where I is the number of times more intense the quake is than the minimum level for comparison.   **(b)** 19,500

(a) Find the Richter scale number for an earthquake that is 12,000 times more intense than the minimum level for comparison.   4.08

(b) If the Richter scale number of an earthquake is 4.29, how many times more intense is the earthquake than the minimum level for comparison?

74. In astronomy, a formula used to find the diameter, in kilometers, of minor planets (also called asteroids) is $\log d = 3.7 - 0.2g$, where g is a quantity called the

74. **(a)** 31.62 km **(b)** 0.50 km **(c)** 14.68

absolute magnitude of the minor planet. Find the diameter of a minor planet if its absolute magnitude is **(a)** 11; **(b)** 20. **(c)** Find the absolute magnitude of the minor planet whose diameter is 5.8 kilometers.

75. A formula sometimes used to estimate the seismic energy released by an earthquake is $\log E = 11.8 + 1.5m_s$, where E is the seismic energy and m_s is the surface wave magnitude.

(a) Find the energy released in an earthquake whose surface wave magnitude is 6. 6.31×10^{20}

(b) If the energy released during an earthquake is 1.2×10^{15}, what is its magnitude?   2.19

76. The sound pressure level, s_p, is given by the formula

$$s_p = 20 \log \frac{p_r}{0.0002},$$ where p_r is the sound pressure in

dynes/cm².

  (a) Find the sound pressure level if the sound pressure is 0.0036 dynes/cm².   25.1

  (b) If the sound pressure level is 10.0, find the sound pressure.   0.000632 dynes/cm²

77. The pH is a measure of the acidity or alkalinity of a solution. The pH of water, for example, is 7. In general, acids have pH numbers less than 7 while alkaline solutions have pH numbers greater than 7. The pH of a solution is defined as

$$\text{pH} = -\log[\text{H}_3\text{O}^+]$$

where H_3O^+ represents the hydronium ion concentration of the solution. Find the pH of a solution whose hydronium ion concentration is 2.8×10^{-3}.   2.55

78. On your calculator, you find log 462 and obtain the value 1.6646. Can this value be correct? Explain.

79. On your calculator, you find log 6250 and obtain the value 2.7589. Can this value be correct? Explain.

80. On your calculator, you find log 0.163 and obtain the value −2.7878. Can this value be correct? Explain.

81. On your calculator, you find log 0.0024 and obtain the value −1.6198. Can this value be correct? Explain.

82. On your calculator, you find log −1.23 and obtain the value 0.08991. Can this value be correct? Explain.

78. No; since $10^2 = 100$, log 462 must be greater than 2.

79. No; since 10^3 is 1000, log 6250 must be greater than 3.

80. No; since $10^0 = 1$ and $10^{-1} = 0.1$, log 0.163 must be between 0 and −1.

81. No; since $10^{-2} = 0.01$ and $10^{-3} = 0.001$, log 0.0024 must be between −2 and −3.

82. No; the log of a negative number does not exist.

86.

85.

CUMULATIVE REVIEW EXERCISES

[9.2] **83.** Solve the quadratic equation using the quadratic formula. $\dfrac{-2 \pm 2i\sqrt{5}}{3}$

$$-3x^2 - 4x - 8 = 0$$

84. In 4 hours the Simpsons traveled 15 miles down river in their motorboat, and then turned around and returned home. If the river current is 5 miles per hour, find the speed of their boat in still water.   10 mph

[9.4] **85.** Graph the solution to $\dfrac{2x - 3}{5x + 10} < 0$ on the number line.

[10.3] **86.** Sketch the graph of $x = 3(y - 2)^2 + 1$.

Group Activity/ Challenge Problems

3.

In Section 11.5 we introduce the formula $\log_a x = \dfrac{\log_b x}{\log_b a}$ when $a > 0$, $a \neq 1$, $b > 0$, $b \neq 1$,

and $x > 0$. Use this formula, called the change of base formula, to evaluate the following logarithms. (Hint: Let $b = 10$.) Round values to hundredths.

1. $\log_3 45$ 3.46   **2.** $\log_5 30$ 2.11

3. Use the fact that $\log_a x = \dfrac{\log_b x}{\log_b a}$, where $b = 10$, to graph the equation $y = \log_2 x$ for $x > 0$. Use a grapher if available.

4. The Richter scale for measuring earthquakes was developed by Charles R. Richter. The Richter scale relates the magnitude, M, of the earthquake to the release of energy, E (in ergs) by the formula

$$M = \frac{\log E - 11.8}{1.5}$$

An earthquake releases 15^{18} ergs of energy. What is the magnitude of such an earthquake on the Richter scale?   6.2

11.4   Exponential and Logarithmic Equations

1. Solve exponential and logarithmic equations.
2. Solve some practical problems using exponential and logarithmic equations.

**Solving Exponential
and Logarithmic
Equations**

1. In Section 11.1 we introduced exponential and logarithmic equations. In this section we give more examples of their use and discuss further procedures for solving such equations.

    To solve exponential and logarithmic equations, we often use properties 6a through 6d.

Properties for Solving Exponential and Logarithmic Equations

a. If $x = y$, then $a^x = a^y$.

b. If $a^x = a^y$, then $x = y$.

c. If $x = y$, then $\log x = \log y$ ($x > 0, y > 0$).

d. If $\log x = \log y$, then $x = y$ ($x > 0, y > 0$).   Properties 6a–6d

---

We will be referring to these properties when explaining the solutions to the examples in this section.

EXAMPLE 1   Solve the equation $8^x = \dfrac{1}{2}$.

Solution:
$$8^x = \frac{1}{2}$$
$$(2^3)^x = \frac{1}{2}$$
$$2^{3x} = 2^{-1}$$

Using property 6b, we can write
$$3x = -1$$
$$x = -\frac{1}{3}$$

    When both sides of the exponential equation cannot be written as a power of the same base, we often begin by taking the logarithm of both sides of the equation, as in Example 2. In the following examples, we will round logarithms to the nearest ten-thousandth.

EXAMPLE 2   Solve the equation $5^n = 20$.

Solution: Take the logarithm of both sides of the equation and solve for n.
$$\log 5^n = \log 20$$
$$n \log 5 = \log 20$$
$$n = \frac{\log 20}{\log 5} \approx \frac{1.3010}{0.6990} \approx 1.861$$

Some logarithmic equations can be solved by expressing the equation in exponential form. **It is necessary to check logarithmic equations for extraneous solutions.** When checking a solution, if you obtain the logarithm of a nonpositive number, the solution is extraneous.

EXAMPLE 3 Solve the equation $\log_2 (x + 1)^3 = 4$.

Solution: Write the equation in exponential form.

$$(x + 1)^3 = 2^4$$
$$(x + 1)^3 = 16$$
$$x + 1 = \sqrt[3]{16}$$
$$x = -1 + \sqrt[3]{16}$$

Check:

$$\log_2 (x + 1)^3 = 4$$
$$\log_2 [(-1 + \sqrt[3]{16}) + 1]^3 = 4$$
$$\log_2 (\sqrt[3]{16})^3 = 4$$
$$\log_2 16 = 4$$
$$2^4 = 16 \qquad \text{Write in exponential form.}$$
$$16 = 16 \qquad \text{true}$$

Other logarithmic equations can be solved using the properties of logarithms given in earlier sections.

EXAMPLE 4 Solve the equation $\log (3x + 2) + \log 9 = \log (x + 5)$.

Solution:

$$\log (3x + 2) + \log 9 = \log (x + 5)$$
$$\log (3x + 2)(9) = \log (x + 5)$$
$$(3x + 2)(9) = (x + 5) \qquad \text{property 6d}$$
$$27x + 18 = x + 5$$
$$26x + 18 = 5$$
$$26x = -13$$
$$x = -\frac{1}{2}$$

Check for yourself that the solution is $-\frac{1}{2}$.

EXAMPLE 5 Solve the equation $\log x + \log (x + 1) = \log 12$.

Solution:

$$\log x + \log (x + 1) = \log 12$$
$$\log x(x + 1) = \log 12$$
$$x(x + 1) = 12$$
$$x^2 + x = 12$$
$$x^2 + x - 12 = 0$$
$$(x + 4)(x - 3) = 0$$
$$x + 4 = 0 \qquad \text{or} \qquad x - 3 = 0$$
$$x = -4 \qquad\qquad x = 3$$

Check:

$$x = 3$$

$$\log x + \log (x + 1) = \log 12$$
$$\log 3 + \log 4 = \log 12$$
$$\log (3)(4) = \log 12$$
$$\log 12 = \log 12 \quad \text{true}$$

$$x = -4$$

$$\log x + \log (x + 1) = \log 12$$
$$\log (-4) + \log (-3) = \log 12$$

Stop. ↑   ↑
Logarithms of negative numbers are not real numbers.

Thus, -4 is an extraneous solution. The only solution is 3.

Graphing Calculator Corner

We have indicated how equations in one variable may be solved graphically. Logarithmic and exponential equations may also be solved graphically by graphing each side of the equation and finding the x coordinate of point of intersection of the two graphs. In Example 5 we found that the solution to the equation $\log x + \log(x + 1) = \log 12$ was $x = 3$. Figure 11.11 shows the graphical solution to this equation. Can you determine which side of the equation is illustrated by the horizontal line? Why? The horizontal line is the graph of $y = \log 12$ since $\log 12$ is a constant. Notice that the x coordinate of the point of intersection of the two graphs, 3, is the solution to the equation.

EXERCISES

Using your calculator estimate the solutions to the nearest tenth. If a real solution does not exist, state so.

1. $\log (x + 3) + \log x = \log 16$ 2.8
2. $\log (3x + 5) = 2.3x - 6.4$ 3.3
3. $5.6 \log (5x - 12) = 2.3 \log (x - 5.4)$ φ
4. $5.6 \log (x + 12.2) - 1.6 \log (x - 4) = 20.3 \log (2x - 6)$ 4.2

−2, 10 , 1, −1, 2, 1

FIGURE 11.11

EXAMPLE 6   Solve the equation $\log (3x - 5) - \log 5x = 1.23$.

Solution:

$$\log (3x - 5) - \log 5x = 1.23$$

$$\log \frac{3x - 5}{5x} = 1.23 \qquad \text{property 2}$$

$$\frac{3x - 5}{5x} = \text{antilog } 1.23$$

$$\frac{3x - 5}{5x} = 17.0$$

$$3x - 5 = 5x(17.0)$$

$$3x - 5 = 85x$$

$$-5 = 82x$$

$$x = -\frac{5}{82} \approx -0.061$$

Check:
$$\log (3x - 5) - \log 5x = 1.23$$
$$\log [3(-0.061) - 5] - \log [(5)(-0.061)] = 1.23$$
$$\log (-5.183) - \log (-0.305) = 1.23$$

Since we have the logarithms of negative numbers, -0.061 is an extraneous solution. Thus, this equation is no solution, that is, its solution is the empty set, $\{\quad\}$.

Applications of Exponential and Logarithmic Equations

2 In Section 11.1 we introduced the compound interest formula, $A = P(1 + r)^n$. We found A at that time by using a calculator. The value of A can also be found by using logarithms, as in Example 7.

EXAMPLE 7   The amount of money, A, accumulated in a savings account for a given principal P, interest rate r, and number of compounding periods n can be found by the formula

$$A = P(1 + r)^n$$

Suppose $1000 is invested in a savings account at 8% interest compounded annually for 5 years. The amount accumulated at the end of 5 years is

$$A = 1000(1 + 0.08)^5$$

Use logarithms to find the amount accumulated.

Solution: Begin by taking the logarithms of both sides of the equation,

$$
\begin{aligned}
\log A &= \log [1000(1 + 0.08)^5] \\
&= \log 1000 + \log (1 + 0.08)^5 &&\textbf{Property 1} \\
&= \log 1000 + \log (1.08)^5 \\
&= \log 1000 + 5 \log 1.08 &&\textbf{Property 3} \\
&= 3.00 + 5(0.0334) \\
&= 3.00 + 0.167 \\
&= 3.167 \\
A &= \text{antilog } 3.167 \\
A &\approx 1469
\end{aligned}
$$

In 5 years $1000 would grow to about $1469. This amount includes the $1000 principal and about $469 interest.

EXAMPLE 8   If there are initially 1000 bacteria in a culture, and the number of bacteria doubles each hour, the number of bacteria after t hours can be found by the formula

$$N = 1000(2)^t$$

How long will it take for the culture to grow to 30,000 bacteria?

Solution:

$$N = 1000(2)^t$$
$$30{,}000 = 1000(2)^t$$

Now take the logarithm of both sides of the equation.

$$\log 30{,}000 = \log [1000(2)^t]$$
$$\log 30{,}000 = \log 1000 + \log 2^t$$
$$\log 30{,}000 = \log 1000 + t(\log 2)$$
$$4.4771 = 3.000 + t(0.3010)$$
$$4.4771 = 3.000 + 0.3010t$$
$$1.4771 = 0.3010t$$
$$\frac{1.4771}{0.3010} = t$$
$$4.91 \approx t$$

In approximately 4.91 hours there will be 30,000 bacteria in the culture.

Exercise Set 11.4

Solve the exponential equation without using a calculator.

1. $3^x = 243$ 5

2. $16^x = \dfrac{1}{4}$ $-\dfrac{1}{2}$

3. $5^x = 125$ 3

4. $5^{-x} = \dfrac{1}{25}$ 2

5. $2^{3x-2} = 16$ 2

6. $64^x = 4^{4x+1}$ -1

7. $27^x = 3^{2x+3}$ 3

8. $\left(\dfrac{1}{2}\right)^x = 8$ -3

Use a calculator to solve. Round your answers to hundredths.

9. $8^x = 60$ 1.97

10. $1.05^x = 15$ 55.50

11. $4^{x-1} = 20$ 3.16

12. $2.3^{x-1} = 5.6$ 3.07

13. $1.63^{x+1} = 25$ 5.59

14. $4^x = 9^{x-2}$ 5.42

15. $3^{x+4} = 6^x$ 6.34

16. $5^x = 2^{x+5}$ 3.78

Use a calculator to solve the logarithmic equation. Round answers to the nearest hundredth. If the equation has no real solution, so state.

17. $\log_4 (x+1)^3 = 3$ 3

18. $\log_3 (x-2)^2 = 2$ 5, −1

19. $\log_2 (x+4)^2 = 4$ 0, −8

20. $\log_2 x + \log_2 3 = 1$ $\frac{2}{3}$

21. $\log (2x-3)^3 = 3$ $\frac{13}{2}$

22. $\log_3 3 + \log_3 x = 3$ 9

23. $\log (x+2) = \log (3x-1)$ $\frac{3}{2}$

24. $\log x = \log (1-x)$ $\frac{1}{2}$

25. $\log (3x-1) + \log 4 = \log (9x+2)$ 2

26. $\log (x+3) + \log x = \log 4$ 1

27. $\log x + \log (3x-5) = \log 2$ 2

28. $\log (x+4) - \log x = \log (x+1)$ 2

29. $\log x + \log 4 = 0.56$ 0.91

30. $\log (x+4) - \log x = 1.22$ 0.26

31. $2 \log x - \log 4 = 2$ 20

32. $\log 6000 - \log (x+2) = 3.15$ 2.25

33. $\log x + \log (x-3) = 1$ 5

34. $2 \log_2 x = 2$ 2

35. $\log x = \frac{1}{3} \log 27$ 3

36. $\log_7 x = \frac{3}{2} \log_7 64$ 512

37. $\log_8 x = 3 \log_8 2 - \log_8 4$ 2

38. $\log_4 x + \log_4(6x-7) = \log_4 5$ $\frac{5}{3}$

Find the solution. Round answers to the nearest hundredth.

39. Find the amount accumulated if Marlina puts $1200 in a savings account offering 10% interest compounded annually for 5 years. Use $A = P(1+r)^n$ (see Example 7). $1932.61

40. If the initial amount of bacteria in the culture in Example 8 is 4500, when will the number of bacteria in the culture reach 50,000? Use $N = 4500(2)^t$.   3.47 hr

41. If after 4 hours the culture in Example 8 contains 2224 bacteria, how many bacteria were present initially?   139

42. The amount, A, of 200 grams of a certain radioactive material remaining after t years can be found by the equation $A = 200(0.800)^t$. When will 40 grams remain?   7.21 yr

43. A machine purchased for business can be depreciated to reduce income tax. The value of the machine at the end of its useful life is called its *scrap value*. When the machine depreciates by a constant percentage annually, its scrap value, S, is

$$S = c(1 - r)^n$$

where c is the original cost, r is the annual rate of depreciation as a decimal, and n is the useful life in years. Find the scrap value of a machine that costs $50,000, has a useful life of 12 years, and has an annual depreciation rate of 15%. $7112.10

44. If the machine in Exercise 43 costs $100,000, has a useful life of 15 years, and has an annual depreciation rate of 8%, find its scrap value. $28,629.74

45. The power gain, P, of an amplifier is defined as

$$P = \log\left(10\,\frac{P_{out}}{P_{in}}\right)$$

where P_{out} is the power output in watts and P_{in} is the power input in watts. If an amplifier has an output power of 12.6 watts and an input power of 0.146 watts, find the power gain.   2.94

46. Measured on the Richter scale, the magnitude, R, of an earthquake of intensity I is defined by

$$R = \log I$$

where I is the number of times greater (or more intense) the earthquake is than the minimum level for comparison.

(a) How many times more intense was the 1906 San Francisco earthquake, which measured 8.25 on the Richter scale, than the minimum level for comparison?   177,827,941

(b) How many times more intense is an earthquake that measures 6.4 on the Richter scale than one that measures 4.7?   50.12

47. The decibel scale is used to measure the magnitude of sound. The magnitude d, in decibels, of a sound is defined to be

$$d = 10 \log I$$

where I is the number of times greater (or more intense) the sound is than the minimum intensity of audible sound.

(a) An airplane engine (nearby) measures 120 decibels. How many times greater than the minimum level of audible sound is the airplane engine?   1,000,000,000,000

(b) The intensity of the noise in a busy city street is 70 decibels. How many times greater is the intensity of the sound of the airplane engine than the sound of the city street?   100,000

See Exercise 45.

48. After solving a logarithmic equation, what must you do?   check for extraneous solutions

---

[2.5] **49.** Solve the inequality $\dfrac{x - 4}{2} - \dfrac{2x - 5}{5} > 3$ and indicate the solution in **(a)** set builder notation and **(b)** interval notation.   **(a)** $\{x \mid x > 40\}$, **(b)** $(40, \infty)$

[3.5, **50.** Indicate which of the graphs on the right are functions.
11.1] If the graph is a function, is it a one-to-one function?
   (b) and **(c)** are functions; only **(b)** is one-to-one

51. $2x^3 - 11x^2 + 18x - 9$

[5.3] **51.** Multiply $(x^2 - 4x + 3)(2x - 3)$.

[5.4] **52.** Divide $2x^2 + 11x + 15$ by $x + 4$. $2x + 3 + \dfrac{3}{x + 4}$

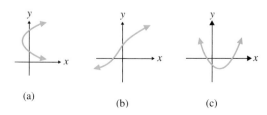

(a)   (b)   (c)

1. log (0.1) is < 0, so must change direction of inequality symbol in step 2.

1. In the following procedure, we begin with a true statement and end with a false statement. Can you find the error?

$2 < 3$	true
$2 \log(0.1) < 3 \log(0.1)$	Multiply both sides by log (0.1).
$\log (0.1)^2 < \log (0.1)^3$	property 3
$(0.1)^2 < (0.1)^3$	property 6d
$0.01 < 0.001$	false

Solve the systems of equations.

2. $2^x = 8^y$
  $x + y = 4$ (3, 1)

3. $3^{2x} = 9^{y+1}$
  $x - 2y = -3$ (5, 4)

4. $\log (x + y) = 2$
  $x - y = 8$ (54, 46)

5. $\log (x + y) = 3$
  $2x - y = 5$ (335, 665)

6. Use equations that are quadratic in form to solve the following equation.
$$2^{2x} - 6(2^x) + 8 = 0 \qquad x = 1 \text{ or } x = 2$$

11.5 The Natural Exponential Function and Natural Logarithms

Tape 19

1. Identify the natural exponential function.
2. Identify the natural logarithmic function.
3. Show that the natural exponential function and the natural logarithmic functions are inverse functions.
4. Use a calculator to find natural logarithms and natural exponential values.
5. Find natural logarithms using the change of base formula.
6. Solve natural logarithmic and natural exponential equations.
7. Study applications of natural exponential functions and natural logarithms.

The natural exponential function and *its inverse*, the natural logarithmic function which we present in this section are exponential functions and logarithmic functions of the type presented in the previous sections. They share all the properties of exponential functions and logarithmic functions discussed earler. The importance of these special functions lies in the many varied applications in real life of a unique irrational number designated by the letter *e*.

The Natural Exponential Function

1. In Section 11.1 we indicated that exponential functions were of the form $f(x) = a^x$, $a > 0$ and $a \neq 1$. Now we introduce a very special exponential function. It is called the **natural exponential function** and it uses the number *e*. Like the irrational number π, the number *e* is an irrational number whose value can only be approximated by a decimal number. The number *e* plays a very important role in higher-level mathematics courses. The value of *e* is approximately 2.7183. Now we define the natural exponential function.

> ## The Natural Exponential Function
>
> $$f(x) = e^x$$
>
> where $e \approx 2.7183$.

The Natural Logarithmic Function

2 We discussed common logarithms in Section 11.3. Now we will discuss natural logarithms. **Natural logarithms** are logarithms to the base e. Natural logarithms are indicated by the letters ln. The notation *ln x* is read the "natural log of *x*."

> ## Natural Logarithms
>
> $$\log_e x = \ln x$$

You must remember that the base of the natural logarithm is e. Thus, when you change a natural logarithm to exponential form, the base of the exponential expression will be e.

> For $x > 0$ if $y = \ln x$, then $e^y = x$.

EXAMPLE 1 Find the value by changing the natural logarithm to exponential form. **(a)** ln 1   **(b)** ln e

Solution:

(a) Let $y = \ln 1$; then $e^y = 1$. Since any nonzero value to the 0th power equals 1, y must equal 0. Thus, $\ln 1 = 0$.

(b) Let $y = \ln e$; then $e^y = e$. For e^y to equal e, y must equal 1. Thus, $\ln e = 1$. •

Inverse Functions

3 The functions $y = a^x$ and $y = \log_a x$ are inverse functions. Similarly, the functions $y = e^x$ and $y = \ln x$ are inverse functions. (Remember, $y = \ln x$ means $y = \log_e x$.) The graphs of $y = e^x$ and $y = \ln x$ are illustrated in Fig. 11.12. Notice that the graphs are symmetric about the line $y = x$, which is what we expect of inverse functions.

Note that the graph of $y = e^x$ is similar to graphs of the form $y = a^x$, $a > 1$, and that the graph of $y = \ln x$ is similar to graphs of the form $y = \log_a x$, $a > 1$.

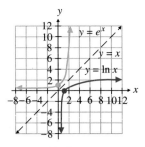

FIGURE 11.12

Finding Values   **4**   Now we will learn how to find natural logarithms on a calculator.

Calculator Corner

FIND NATURAL LOGARITHMS ON A CALCULATOR

Natural logarithms can be found using a calculator that has a ln key. Natural logarithms are found in the same manner that we found common logarithms on a calculator, except we use the natural log key, ln , instead of the common log key, log .

Example	Keys to Press	Answer Displayed
Find ln 242	242 ln	5.4889377
Find ln 0.85	0.85 ln	−0.1625189

When finding the natural logarithm of a number, we are finding an exponent. The natural logarithm of a number is the exponent to which the base e must be raised to obtain that number. For example,

$$\text{if } \ln 242 = 5.4889377, \text{ then } e^{5.4889377} = 242$$
$$\text{if } \ln 0.85 = -0.1625189, \text{ then } e^{-0.1625189} = 0.85$$

EXAMPLE 2   Indicate the keys to press to evaluate each of the natural logarithms. Then give the value obtained on a calculator.

(a) ln 1462   **(b)** ln 0.000381

Solution:

(a) 1462 ln 7.2875606   **(b)** 0.000381 ln −7.8727112

Since $y = \ln x$ and $y = e^x$ are inverse functions of one another, we can use the inverse key inv in combination with the natural log key ln to obtain values for e^x.

Calculator Corner

FINDING VALUES OF e^x ON A CALCULATOR

To find values of e^x on a calculator, first enter the exponent on e. Then press either the shift key, SHIFT , the second function key, 2nd , or the inverse key, inv , depending on your calculator. Then press the natural log key, ln . After the ln key is pressed, the value of e^x will be displayed. Since e^x is the inverse function of ln x, many calculators list e^x directly above the ln key. In the examples below we will use the inv key. If your calculator uses a SHIFT or 2nd key, use that in place of the inv key.

Example	Keys to Press	Answer Displayed
Find $e^{5.24}$	5.24 inv ln	188.6701
Find $e^{-1.639}$	1.639 +/− inv ln	0.1941741

Since e is about 2.7183, if we evaluated $(2.7183)^{5.24}$ we would obtain a value close to 188.6701. Also, if we found ln 188.6701 on a calculator, we would obtain 5.24. What do you think we would get if we evaluated ln 0.1941741 on a calculator? If you answered -1.639, you answered correctly.

EXAMPLE 3 Indicate the keys you would press to evaluate each of the following on a calculator. Then indicate the value obtained.

(a) $e^{7.214}$ **(b)** $e^{-1.245}$

Solution:

(a) 7.214 inv ln 1358.3147   **(b)** 1.245 +/- inv ln 0.2879409

EXAMPLE 4 Find N if **(a)** ln $N = 4.92$ and **(b)** ln $N = -0.0253$.

Solution:

(a) If we write ln $N = 4.92$ in exponential form, we get $e^{4.92} = N$. Thus, we simply need to evaluate $e^{4.92}$ to determine N. We do this by pressing the following keys on the calculator.

$$\text{4.92} \quad \text{inv} \quad \text{ln} \quad \text{137.00261}$$

Thus, $N = 137.00261$. Note that $e^{4.92} = 137.00261$.

(b) 0.0253 +/- inv ln 0.9750174.
Thus, $N = 0.9750174$.

Changing Bases ⑤ If you are given a logarithm in a base other than 10 or e you will not be able to evaluate it on your calculator directly. When this occurs you can use the change of base formula.

> **Change of Base Formula**
> If $a > 0$, $a \neq 1$, $b > 0$, $b \neq 1$, and $x > 0$, then
> $$\log_a x = \frac{\log_b x}{\log_b a}$$

In the change of base formula, 10 is often used in place of base b because we can find common logarithms on a calculator. Replacing base b with 10, we get

$$\log_a x = \frac{\log_{10} x}{\log_{10} a} \qquad \text{or} \qquad \log_a x = \frac{\log x}{\log a}$$

EXAMPLE 5 Use the change of base formula to find $\log_3 24$.

Solution: If we substitute 3 for a and 24 for x in $\log_a x = \dfrac{\log x}{\log a}$, we obtain

$$\log_3 24 = \frac{\log 24}{\log 3} \approx \frac{1.3802}{0.4771} \approx 2.8929$$

Note that $3^{2.8929} \approx 24$.

We can use the same procedure as in Example 5 to find natural logarithms using the change of base formula. For example, to evaluate $\ln 20$ (or $\log_e 20$), we can substitute e for a and 20 for x in the formula $\log_a x = \dfrac{\log x}{\log a}$.

$$\log_e 20 = \frac{\log 20}{\log e} \approx \frac{1.3010}{0.4343} \approx 2.9956$$

Thus, $\ln 20 \approx 2.9956$. If you find $\ln 20$ on a calculator, you will obtain a very close value.

Since $\log e \approx 0.4343$, to evaluate natural logarithms using common logarithms, we use the formula

$$\ln x \approx \frac{\log x}{0.4343}$$

EXAMPLE 6   Use the change of base formula to find $\ln 95$.

Solution: $\ln 95 = \dfrac{\log 95}{\log e} \approx \dfrac{1.9778}{0.4343} \approx 4.5540$

Solving Natural Logarithmic and Natural Exponential Equations

6 The properties of logarithms discussed in Section 11.2 still hold true for natural logarithms. Following is a summary of these properties in the notation of natural logarithms.

Properties for Natural Logarithms

$$\ln xy = \ln x + \ln y \qquad (x > 0 \text{ and } y > 0)$$

$$\ln \frac{x}{y} = \ln x - \ln y \qquad (x > 0 \text{ and } y > 0)$$

$$\ln x^n = n \ln x \qquad (x > 0)$$

Consider the expression $\ln e^x$, which means $\log_e e^x$. From property 4 on page 639, $\log_e e^x = x$. Thus $\ln e^x = x$. Similarly, $e^{\ln x} = e^{\log x} = x$ by property 5. Although $\ln e^x = x$ and $e^{\ln x} = x$ are just special cases of properties 4 and 5, respectively, we will call these properties 7 and 8 so that we can make reference to them.

Additional Properties for Natural Logarithms and Natural Exponential Expressions

$$\ln e^x = x \qquad\qquad\qquad \text{Property 7}$$

$$e^{\ln x} = x, \qquad x > 0 \qquad\qquad \text{Property 8}$$

Using property 7, $\ln e^x = x$, we can state, for example, that $\ln e^{kt} = kt$, and $\ln e^{-2.06t} = -2.06t$. Using property 8, $e^{\ln x} = x$, we can state, for example, that $e^{\ln(t+2)} = t + 2$ and $e^{\ln kt} = kt$.

EXAMPLE 7   Solve the equation $\ln y - \ln (x + 6) = t$ for y.

Solution: $\ln y - \ln(x + 6) = t.$

$$\ln \frac{y}{x + 6} = t$$

To eliminate the natural logarithm on the left side of the equation, we rewrite both sides of the equation in exponential form, with the base e. Since $\ln [y/(x + 6)] = t$ we are permitted to write $e^{\ln [y/(x + 6)]} = e^t$ by property 6a given on page 647 (if $x = y$, then $a^x = a^y$).

$$e^{\ln[y/(x + 6)]} = e^t$$

By Property 8 we know that $e^{\ln[y/(x + 6)]} = y/(x + 6)$. Replacing $e^{\ln[y/(x + 6)]}$ with $y/(x + 6)$, we obtain

$$\frac{y}{x + 6} = e^t$$

Now solve the equation for y.

$$y = (x + 6)e^t$$

In Example 7 we could have changed from $\ln \dfrac{y}{x + 6} = t$ to $\dfrac{y}{x + 6} = e^t$ using the definition of logarithm as shown below.

$$\ln \frac{y}{x + 6} = t \quad \text{means} \quad \log_e \frac{y}{x + 6} = t$$

Now change the logarithm on the right to exponential form.

$$\log_e \frac{y}{x + 6} = t \quad \text{means} \quad e^t = \frac{y}{x + 6}$$

The procedure used in Example 7 reinforces some properties of logarithms.

EXAMPLE 8   Solve the equation $450 = 225e^{-0.4t}$ for t

Solution: Begin by dividing both sides of the equation by 225 to isolate $e^{-0.4t}$.

$$\frac{450}{225} = \frac{\cancel{225}e^{-0.4t}}{\cancel{225}}$$
$$2 = e^{-0.4t}$$

Now take the natural log of both sides of the equation to eliminate the exponential expression on the right side of the equation.

$$\ln 2 = \ln e^{-0.4t}$$
$$\ln 2 = -0.4t \qquad \text{(ln } e^{-0.4t} = -0.4t \text{ by property 7)}$$
$$0.6931472 = -0.4t$$
$$\frac{0.6931472}{-0.4} = t$$
$$-1.732868 = t$$

EXAMPLE 9   Solve the equation $P = P_0 e^{kt}$ for t.

Solution: We can follow the same procedure as given in Example 8.

$$P = P_0 e^{kt}$$

$$\frac{P}{P_0} = \frac{\cancel{P_0} e^{kt}}{\cancel{P_0}}$$

$$\frac{P}{P_0} = e^{kt}$$

$$\ln \frac{P}{P_0} = \ln e^{kt}$$

$$\ln P - \ln P_0 = \ln e^{kt}$$

$$\ln P - \ln P_0 = kt$$

$$\frac{\ln P - \ln P_0}{k} = t$$

Applications   **7** Now let us look at some applications that involve the natural exponential function and natural logarithms.

When a quantity increases or decreases at an *exponential rate,* a formula often used to find the value of P after time t is

$$P = P_0 e^{kt}$$

where P_0 is the initial or starting value, and k is the constant rate of growth or decay.

EXAMPLE 10   Banks often credit compound interest on a continuous basis. When interest is compounded continuously, the balance P, in the account, at any time t can be calculated by the exponential growth formula $P = P_0 e^{kt}$, where P_0 is the principal initially invested, and k is the interest rate.

(a) Suppose the interest rate is 10% compounded continuously and $1000 is initially invested. Determine the balance in the account after 1 year.

(b) How long will it take the account to double in value?

Solution:

(a) $P = P_0 e^{kt}$
$P = 1000 e^{(0.10)(1)} = 1000 e^{0.10} = 1000(1.1051709) = 1105.1709.$

We found the value of $e^{0.10}$ by using a calculator. Thus, after 1 year, the balance in the account is $1105.17. Since the initial amount invested was $1000, the interest that has accumulated in 1 year is $105.17.

(b) For the value of the account to double, the balance in the account would have to reach $2000. Therefore, we substitute 2000 for P and solve for t.

$$P = P_0 e^{kt}$$
$$2000 = 1000 e^{0.10t}$$
$$2 = e^{0.10t}$$
$$\ln 2 = \ln e^{0.10t}$$
$$\ln 2 = 0.10t$$
$$\frac{\ln 2}{0.10} = t$$
$$\frac{0.6931472}{0.10} = t$$
$$6.931472 = t$$

Thus, with an interest rate of 10% compounded continuously, the account would double in slightly over 6.9 years.

EXAMPLE 11 Assume that the value of the island of Manhattan has grown at an exponential rate of 8% per year since 1626, when Peter Minuit of the Dutch West India Company purchased the island for $24. What is the value of the island of Manhattan in 1996, 370 years later?

Solution: Since the value is increasing exponentially, we can use the exponential growth formula. We will use V_0 to represent initial value, and V to represent the value in 1996.

$$V = V_0 e^{kt}$$
$$= 24 e^{(0.08)(370)}$$
$$= 24 e^{29.6}$$
$$\approx 24(7.1634 \times 10^{12}) \qquad \text{from a calculator}$$
$$\approx 1.7192 \times 10^{14}$$
$$\$171{,}920{,}000{,}000{,}000$$

Thus, if the value grows exponentially at 8%, the value of Manhattan after 370 years is $171,920,000,000,000.

EXAMPLE 12 Strontium 90 is a radioactive isotope that decays exponentially at 2.8% per year. The amount of strontium 90 left after t years can be found by the formula $P = P_0 e^{-0.028t}$. Suppose there are initially 1000 grams of strontium 90:

(a) Find the number of grams of strontium 90 left after 50 years.

(b) Find the half-life of strontium 90.

Solution:

(a) We use the same formula as in Example 10 except that the exponent is negative because 2.8% is a rate of decay.

$$P = P_0 e^{-0.028t}$$
$$= 1000 e^{-0.028(50)} = 1000 e^{-1.4} = 1000(0.246597) = 246.597$$

Thus, after 50 years, 246.597 grams of strontium 90 remains.

(b) To find the half-life, we need to determine when 500 grams of strontium 90 is left.

$$P = P_0 e^{-0.028t}$$
$$500 = 1000\, e^{-0.028t}$$
$$0.5 = e^{-0.028t}$$
$$\ln 0.5 = \ln e^{-0.028t}$$
$$-0.6931472 = -0.028t$$
$$\frac{-0.6931472}{-0.028} = t$$
$$24.755257 = t$$

Thus, the half-life of strontium 90 is about 24.8 years.

EXAMPLE 13 The formula for estimating the number, N, of a particular toy sold is $N = 400 + 250 \ln a$, where a is the amount of money spent on advertising the toy.

(a) If $2000 is spent on advertising, what are the expected sales of the toy?

(b) How much money should be spent on advertising to sell 1500 toys?

Solution:

(a) $N = 400 + 250 \ln a$
$= 400 + 250 \ln 2000$
$= 400 + 250(7.6009025) = 400 + 1900.2256 = 2300.2256$

Thus, approximately 2300 toys are expected to be sold.

(b)
$$N = 400 + 250 \ln a$$
$$1500 = 400 + 250 \ln a$$
$$1100 = 250 \ln a$$
$$\frac{1100}{250} = \ln a$$
$$4.4 = \ln a$$
$$e^{4.4} = e^{\ln a}$$
$$81.45 = a$$

Thus, about $81 should be spent on advertising to sell 1500 toys.

Exercise Set 11.5

Find the following values. Round values to four decimal places.

1. $\ln 50$ 3.9120   **2.** $\ln 0.432$ -0.8393 **3.** $\ln 302$ 5.7104   **4.** $\ln 0.0038$ -5.5728

Find the value of N. Round values to three significant digits.

5. $\ln N = 1.6$ 4.95   **6.** $\ln N = 4.96$ 143   **7.** $\ln N = -2.63$ 0.0721   **8.** $\ln N = 0.632$ 1.88

Use the change of base formula to find the value of the following logarithms.

9. $\ln 40$ 3.6889   **10.** $\ln 562$ 6.3315   **11.** $\ln 0.046$ -3.0791 **12.** $\ln 198$ 5.2883

13. $\log_3 25$ 2.9300   **14.** $\log_7 96$ 2.3456   **15.** $\log_2 20$ 4.3219   **16.** $\log_5 0.463$ -0.4784

17. $\ln 2700$ 7.9010   **18.** $\log_6 4000$ 4.6290   **19.** $\log_3 0.0049$ -4.8411 **20.** $\ln 8462$ 9.0433

Solve the following logarithmic equations.

21. $\ln x + \ln (x - 1) = \ln 12$ 4   **22.** $\ln (x + 3) + \ln (x + 2) = \ln 6$ 0   **23.** $\ln x = 5 \ln 2 - \ln 8$ 4

24. $\ln x + \ln (x - 1) = \ln 2$ 2   **25.** $\ln (x^2 - 4) - \ln (x + 2) = \ln 1$ 3   **26.** $\ln x = \dfrac{3}{2} \ln 4$ 8

Each of the following equations is in the form $P = P_0 e^{kt}$. Solve each equation for the remaining variable. Remember, e is a constant.

27. $P = 500 e^{1.6(1.2)}$ $P = 3410.48$ **28.** $1000 = V_0 e^{0.6(2)}$ $V_0 = 301.19$ **29.** $50 = P_0 e^{-0.05(3)}$ $P_0 = 58.09$

30. $120 = 60 e^{2t}$ $t = 0.3466$ **31.** $90 = 30 e^{1.4t}$ $t = 0.7847$ **32.** $20 = 40 e^{-0.5t}$ $t = 1.3863$

33. $100 = 50 e^{k(3)}$ $k = 0.2310$ **34.** $10 = 50 e^{k(4)}$ $k = -0.4024$ **35.** $20 = 40 e^{k(2.4)}$ $k = -0.2888$

36. $100 = A_0 e^{-0.02(3)}$ $A_0 = 106.18$ **37.** $A = 6000 e^{-0.08(3)}$ $A = 4719.77$ **38.** $75 = 100 e^{-0.04t}$ $t = 7.1921$

Solve for the variable indicated.

39. $V = V_0 e^{kt}$ for V_0 $V_0 = \dfrac{V}{e^{kt}}$ **40.** $P = P_0 e^{kt}$ for P_0 $P_0 = \dfrac{P}{e^{kt}}$ **41.** $P = 150 e^{4t}$ for t $t = \dfrac{\ln P - \ln 150}{4}$

42. $200 = P_0 e^{kt}$ for t $t = \dfrac{\ln 200 - \ln P_0}{k}$ **43.** $A = A_0 e^{kt}$ for k $k = \dfrac{\ln A - \ln A_0}{t}$ **44.** $140 = R_0 e^{kt}$ for k $k = \dfrac{\ln 140 - \ln R_0}{t}$

45. $\ln y - \ln x = 2.3$ for y $y = xe^{2.3}$ **46.** $\ln y + 5 \ln x = \ln 2$ for y $y = \dfrac{2}{x^5}$ **47.** $\ln y - \ln (x + 3) = 6$ for y $y = (x + 3)e^6$

48. $\ln (x + 2) - \ln (y - 1) = \ln 5$ for y $y = \dfrac{x + 7}{5}$

49. The intensity of light as it passes through a certain medium is found by the formula $x = k(\ln I_0 - \ln I)$. Solve this equation for I_0. $I_0 = Ie^{x/k}$

50. The distance traveled by a train initially moving at velocity v_0 after the engine is shut off can be calculated by the formula $x = \dfrac{1}{k} \ln (kv_0 t + 1)$. Solve this equation for v_0. $v_0 = \dfrac{e^{xk} - 1}{kt}$

53. (a) $5867.55,   **(b)** 8.66 yr

Use a calculator if available to solve.

53. If $5000 is invested at 8% compounded continuously, **(a)** determine the balance in the account after 2 years, and **(b)** how long would it take the value of the account to double? (See Example 10.)

54. Refer to Example 12. Determine the volume of strontium 90 remaining after 20 years if there were originally 70 grams.   39.98 g

55. For a certain soft drink, the percentage of a target market, $f(t)$, that buys the soft drink is a function of the number of days, t, that the soft drink is advertised. The function that describes this relationship is $f(t) = 1 - e^{-0.04t}$.
   (a) What percentage of the target market buys the soft drink after 50 days of advertising?   86.47%
   (b) How many days of advertising are needed if 75% of the target market is to buy the soft drink?   34.66 days

51. A formula used in studying the action of a protein molecule is $\ln M = \ln Q - \ln (1 - Q)$. Solve this equation for Q. $Q = M/(1 + M)$

52. An equation relating the current and time in an electric circuit is $\ln i - \ln I = \dfrac{-t}{RC}$. Solve this equation for i.
   52. $i = Ie^{-t/RC}$

56. For a certain type of tie, the number of ties, $N(a)$, sold is a function of the dollar amount spent on advertising, a (in thousands of dollars). The function that describes this relationship is $N(a) = 800 + 300 \ln a$.
   (a) How many ties were sold after $1500 (or 1.5 thousand) was spent on advertising?   922
   (b) How much money must be spent on advertising to sell 1000 ties?   1.95 thousand (or $1950)

57. It was found in a psychological study that the average walking speed, $f(P)$, of a person living in a city is a function of the population of the city. For a city of population P, the average walking speed in feet per second is given by $f(P) = 0.37 \ln P + 0.05$. The population of Nashville, Tennessee, is 972,000.
   (a) What is the average walking speed of a person living in Nashville?   **(a)** 5.15 ft/sec   **(b)** 5.96 ft/sec

(b) What is the average walking speed of a person living in New York City, population 8,567,000?

(c) If the average walking speed of the people in a certain city is 5.0 feet per second, what is the population of the city?   646,000

58. The percentage of doctors who accept and prescribe a new drug is given by the formula $P(t) = 1 - e^{-0.22t}$, where t is the time in months since the drug has been placed on the market. What percentage of doctors accept a new drug 2 months after it is placed on the market?   35.6%

59. The world population in 1994 is estimated at 5.66 billion people. It is estimated that the world's population continues to grow exponentially at the rate of 2.0% per year. The world's expected population, in billions, in t years is given by the formula $P(t) = 5.66e^{0.02t}$.

(a) Find the expected world's population in the year 2000.   6.38 billion

(b) In how many years will the world's population double?   34.66 yr

60. Plutonium, which is commonly used in nuclear reactors, decays at a rate of 0.003% per year. The formula $A = A_0e^{kt}$ can be used to find the amount of plutonium remaining from an initial amount, A_0, after t years. In the formula the k is replaced with -0.00003.

(a) If 1000 grams of plutonium is present in 1990, how many grams of plutonium will remain in the year 2090, 100 years later?   997 g

(b) Find the half-life of plutonium.   23,104.91 yr

61. Carbon dating is used to estimate the age of ancient plants and objects. The radioactive element, carbon 14, is most often used for this purpose. Carbon 14 decays at a rate of 0.01205% per year. The amount of carbon 14 remaining in an object after t years can be found by the function $f(t) = v_0e^{-0.0001205t}$, where v_0 is the initial amount present.

(a) If an ancient animal bone originally had 20 grams of carbon 14, and when found it had 9 grams of carbon 14, how old is the bone?   6626.62 yr

(b) How old is an item that has 50% of its original carbon 14 remaining?   5752.26 yr

62. Determine the age of a fossil if it contains 80% of its amount of carbon 14 (see Exercise 61).   1851.8 yr

63. At what rate, compounded continuously, must a sum of money be invested if it is to double in 6 years?

64. How much money must be deposited today to become $20,000 in 18 years if invested at 6% compounded continuously? $6791.91

65. The power supply of a satellite is a radioisotope. The power P, in watts, remaining in the power supply is a function of the time the satellite is in space.

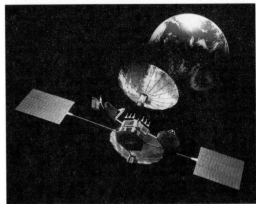

(a) If there are 50 grams of the isotope originally, the power remaining after t days is $P = 50e^{-0.002t}$. Find the power remaining after 100 days.   40.94 watts

(b) When will the power remaining drop to 10 watts?

66. What is the approximate value of e?   2.718

67. If $e^x = 12.183$, find the value of x. Explain how you obtained your answer.   2.5000; find ln 12.183

68. To what exponent must the base e be raised to obtain the value 184.93? Explain how you obtained your answer.   5.2200; find ln 184.93

69. Find the value of $e^{4.32}$. Explain how you obtained your answer.   75.1886; press 4.32 inv ln

70. Find the value of e raised to the exponent -1.73. Explain how you obtained your answer.

63. ≈ 11.55%
65. (b) 804.72 days
70. 0.1773; press 1.73 +/- inv ln

CUMULATIVE REVIEW EXERCISES

[5.2] **71.** Simplify $\dfrac{(x^2y^{-2})^{-1}(4xy^3)^2}{(x^{-2}y^3)^{-2}}$. $\dfrac{16y^{14}}{x^4}$

[7.4] **72.** Simplify $\dfrac{\dfrac{3}{x^2} - \dfrac{2}{x}}{\dfrac{x}{4}}$. $\dfrac{12 - 8x}{x^3}$

[7.5] **73.** Solve the formula $\dfrac{1}{f} = \dfrac{1}{p} + \dfrac{1}{q}$ for q. $q = \dfrac{fp}{p - f}$

[8.3] **74.** Simplify $\sqrt[3]{128x^7y^9z^{13}}$. $4x^2y^3z^4\sqrt[3]{2xz}$

Group Activity/ Challenge Problems

1. During the nuclear accident at Chernobyl in the Ukraine in 1986, two of the radioactive materials that escaped into the atmosphere were cesium 137, with a decay rate of 2.3%, and strontium 90, with a decay rate of 2.8%.   **(b)** ≈ 31.66% of original amount

(a) Which material will decompose more quickly? Explain.   Strontium 90

(b) What percentage of the cesium will remain in 2036, 50 years after the accident?

2. Sometimes exponential equations of the form $y = ab^{kt}$ are rewritten in the form $y = ae^{k't}$, where e is used as the base rather than b. Rewrite the function $y = 3(2)^{0.5t}$ in the form $y = ae^{k't}$ $3(e^{\ln 2})^{0.5t} \approx 3e^{0.347t}$

3. (a) The inverse function of $y = e^x$ is $y = \ln x$. Find the inverse function of $y = e^{-x}$ by using procedures discussed in Section 11.1 and in this section. $y = -\ln x$

(b) Graph $y = e^{-x}$ and $y = -\ln x$ on the same axes.

(c) Do $y = e^{-x}$ and $y = -\ln x$ appear to be inverse functions? Explain.   yes

(d) What are the domain and range of $y = e^{-x}$?   D: $\mathbb{R}$, R: $\{y \mid y > 0\}$

(e) What are the domain and range of $y = -\ln x$?   D: $\{x \mid x > 0\}$, R: $\mathbb{R}$

3. (b)

4. When a drug is given to a patient, the rate at which the drug dissipates from the body is proportional to the amount, A, of drug given. The elimination of a drug from the body as a function of time can be given by the formula $f(t) = Ae^{kt}$, where $k = -(\ln 2)/H$, and H is the time for $\frac{1}{2}$ of the drug to be eliminated. Doctors must often prescribe repeated doses of a drug at proper time intervals and amounts so that the drug neither wears off completely nor builds up to toxic levels in the person's body. Do research and write a report on how the medical profession determines what amount of a drug to give a patient and the proper time intervals in which to give the drug.

Summary

GLOSSARY

Antilogarithm *(643):* If $\log N = L$, then $N = $ antilog L.

Argument of a logarithm *(631):* When finding the logarithm of an expression, the expression is called the argument. In $y = \log (x + 3)$, the argument is $x + 3$.

Common logarithm *(641):* Logarithms to the base 10 are common logarithms.

Exponential equation *(627):* An equation that has a variable as an exponent.

Exponential function *(622):* A function of the form $f(x) = a^x$, $a > 0$, $a \neq 1$.

Logarithm *(628):* $y = \log_a x$ means $x = a^y$, $a > 0$, $a \neq 1$.

Logarithmic equation *(631):* An equation in which a variable appears in the argument of the logarithm.

Logarithmic functions *(630):* Equations of the form $y = \log_a x$.

Natural exponential function *(653):* A function of the form $y = e^x$.

Natural logarithm *(654):* A logarithm to the base e, $\log_e x = \ln x$.

IMPORTANT FACTS

$y = a^x$ and $y = \log_a x$ are inverse functions.

$y = e^x$ and $y = \ln x$ are inverse functions.

The domain of a logarithm function of the form $y = \log_a x$ is $x > 0$.

Properties of logarithms

1. $\log_a xy = \log_a x + \log_a y$

2. $\log_a \dfrac{x}{y} = \log_a x - \log_a y$

3. $\log_a x^n = n \log_a x$

4. $\log_a a^x = x$

5. $a^{\log_a x} = x \ (x > 0)$

Change of base formula

$$\log_a x = \frac{\log_b x}{\log_b a}$$

5.

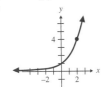

To solve exponential and logarithmic equations, we also use these properties:

6. (a) If $x = y$, then $a^x = a^y$.

6. (b) If $a^x = a^y$, then $x = y$.

6. (c) If $x = y$, then $\log x = \log y$ $(x > 0, y > 0)$.

6. (d) If $\log x = \log y$, then $x = y$ $(x > 0, y > 0)$.

Natural logarithms

7. $\ln e^x = x$

8. $e^{\ln x} = x, x > 0$

Review Exercises

[11.1] *Graph the following functions.*

1. $y = 2^x$

2. $y = \left(\dfrac{1}{2}\right)^x$

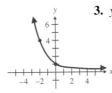

3. $y = \log_2 x$

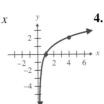

4. $y = \log_{1/2} x$

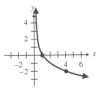

5. On the same axes, graph $y = 3^x$ and $y = \log_3 x$.

Write the logarithmic form.

6. $4^2 = 16 \quad \log_4 16 = 2$

7. $8^{1/3} = 2 \quad \log_8 2 = \frac{1}{3}$

8. $6^{-2} = \frac{1}{36} \quad \log_6 \frac{1}{36} = -2$

9. $25^{1/2} = 5 \quad \log_{25} 5 = \frac{1}{2}$

Write in exponential form.

10. $\log_5 25 = 2 \quad 5^2 = 25$

11. $\log_{1/3} \dfrac{1}{9} = 2 \quad \left(\frac{1}{3}\right)^2 = \frac{1}{9}$

12. $\log_3 \dfrac{1}{9} = -2 \quad 3^{-2} = \frac{1}{9}$

13. $\log_2 32 = 5 \quad 2^5 = 32$

Write in exponential form and find the missing value.

14. $3 = \log_4 x \quad 4^3 = x, 64$

15. $2 = \log_4 x \quad 4^2 = x, 16$

16. $3 = \log_a 8 \quad a^3 = 8, 2$

17. $-3 = \log_{1/4} x \quad \left(\frac{1}{4}\right)^{-3} = x, 64$

18. Consider the graph below, which shows the worldwide growth in the use of computer CD-ROM drives.

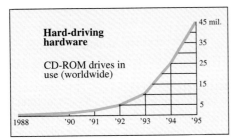

(a) Was the growth more nearly linear or exponential from 1988 to 1990? Explain. linear

(b) In the period 1988–1995, is the graph more nearly linear or exponential? Explain. exponential

(c) How many measurements were used in constructing this graph? 8

(d) Estimate the time for the number of drives to double from the number used in 1992. 1 yr

(e) Estimate the time for the number of drives to double from the number used in 1993. less than 1 yr, $\approx$ 9 months

(f) Estimate when 35 million drives were in use worldwide. mid-1994

20. $5 \log (x - 8)$ **21.** $\log 2 + \log (x - 3) - \log x$

22. $4 \log x - \log 39 - \log (2x + 8)$

[11.2] *Use the properties of logarithms to expand each expression.*

19. $\log_8 \sqrt{12} \quad \frac{1}{2} \log_8 12$

20. $\log (x - 8)^5$

21. $\log \dfrac{2(x - 3)}{x}$

22. $\log \dfrac{x^4}{39(2x + 8)}$

Write as the logarithm of a single expression.

23. $\log_5 (x - 2) - 2 \log_5 x$ $\log_5[(x - 2)/x^2]$

25. $3[\log 2 + \log x] - \log y$ $\log [(2x)^3/y]$

27. $\frac{1}{2}[\ln x - \ln (x + 2)] - \ln 2$

26. $\log_8 [(x + 3)^2(x - 1)^4/\sqrt{x}]$ **27.** $\ln \dfrac{\sqrt{\dfrac{x}{x + 2}}}{2}$ **28.** $\ln \dfrac{x^3 \sqrt{x + 1}}{(x + 4)^3}$

24. $2 \log x - 3 \log (x + 1)$ $\log [x^2/(x + 1)^3]$

26. $2 \log_8 (x + 3) + 4 \log_8(x - 1) - \frac{1}{2} \log_8 x$

28. $3 \ln x + \frac{1}{2} \ln (x + 1) - 3 \ln (x + 4)$

Evaluate.

29. $8^{\log_8 9}$ 9

30. $\log_4 4^5$ 5

31. $3 \log_7 49$ 6

32. $4^{\log_8 \sqrt{8}}$ 2

[11.3] *Use a calculator to find the common logarithm. Round your answer to four decimal places.*

33. 8200   3.9138

34. 0.000716 -3.1451

35. 0.00189 -2.7235

36. 17,600   4.2455

Use a calculator to find the antilog. Give the antilog to three significant digits.

37. 1.7528   56.6

38. 2.9186   829

39. -1.0991 0.0796

40. -1.3747 0.0422

Use a calculator to find N. Round your answer to three significant digits.

41. $\log N = 2.3304$ 214

42. $\log N = -1.2262$ 0.0594

Evaluate.

43. $\log 10^4$ 4

44. $10^{\log 3}$ 3

45. $7.5 \log 10^{4.2}$ 31.5

46. $3(10^{\log 1.7})$ 5.1

[11.4] *Solve without using a calculator.*

47. $9 = 3^x$ 2

48. $49^x = \dfrac{1}{7}$ $-\frac{1}{2}$

49. $2^{2x + 3} = 32$ 1

50. $27^x = 3^{2x + 5}$ 5

Solving using a calculator. Round solutions to thousandths.

51. $4^x = 37$ 2.605

52. $(3.2)^x = 187$ 4.497

53. $(10.9)^{x + 1} = 492$ 1.595

54. $3^{x + 2} = 8^x$ 2.24

Solve the logarithmic equation.

55. $\log_5(x + 2) = 3$ 123

56. $\log (3x + 2) = \log 300$ $\frac{298}{3}$

57. $\log x - \log(3x - 5) = \log 2$ 2

58. $\log_3 x + \log_3 (2x + 1) = 1$ 1

59. $\ln x = \frac{2}{3} \ln 8$ 4

60. $\ln (x + 1) - \ln (x - 2) = \ln 4$ 3

[11.5] *Solve each exponential equation for the remaining variable.*

61. $40 = 20e^{0.6t}$ $t = 1.1552$

62. $P_o = 80e^{-0.02(10)}$ $P_o = 65.50$

63. $100 = A_o e^{-0.42(3)}$ $A_o = 352.54$

Solve for the indicated variable.

64. $w = w_o e^{kt}$ for w_o $w_o = \dfrac{w}{e^{kt}}$

65. $A = A_o e^{kt}$ for t $t = \dfrac{\ln A - \ln A_o}{k}$

66. $150 = 600e^{kt}$ for k $k = \dfrac{\ln 0.25}{t}$

67. $\ln y - \ln x = 2$ for y $y = xe^2$

68. $\ln (y + 3) - \ln (x + 1) = \ln 5$ for y **68.** $y = 5x + 2$

69. $\ln (x - 3) - \ln y = 1.4$ for y **69.** $y = \dfrac{x - 3}{e^{1.4}}$

Use the change of vase formula to evaluate.

70. $\ln 450$ 6.1092

71. $\log_3 50$ 3.5609

72. $\log_5 0.0862$ -1.5229

[11.1–11.5]   **73.** $25,723.08

73. Find the amount of money accumulated if Mrs. Elwood puts $12,000 in a savings account yielding 10% interest per year for 8 years. Use $A = P(1 + r)^n$.

74. Plutonium is a radioactive element that decays over time. If there are originally 1000 mg of plutonium, the amount remaining, R, after t years is

$$R = 1000(0.5)^{0.000041t}$$

Calculate the amount of plutonium present after 20,000 years.   566.5 mg

75. The bacteria Escherichia coli are commonly found in the bladders of humans. Suppose that 2000 bacteria are present at time 0. Then the number of bacteria present t minutes later may be found by the formula

$$N(t) = 2000(2)^{0.05t}$$

(a) When will 50,000 bacteria be present?   92.88 min

(b) Suppose that a human bladder infection is classified as a condition with 120,000 bacteria. When would a person develop a bladder infection if he started with 2000 bacteria?   118.14 min

76. A class of history students is given a final exam at the end of the course. As part of a research project the students are also given equivalent forms of the exam each month for n months. The average grade of the class after n months may be found by the formula

$$A(n) = 72 - 18 \log (n + 1), \qquad n \geq 0$$

(a) What was the class average when the students took the original exam ($n = 0$)?   72

(b) What was the class average for the exam given 2 months later? ≈ 63.4

(c) After how many months was the class average 59.4? ≈ 4 months

77. The atmospheric pressure, P, in pounds per square inch at an elevation of x feet above sea level can be found by the formula $P = 14.7e^{-0.00004x}$. Find the atmospheric pressure at an elevation of 2000 feet.

78. If \$10,000 is placed in a savings account paying 7% interest compounded continuously, find the time needed for the account to double in value.   9.90 yr

77. 13.57 lb/in²

Practice Test

1. Graph $y = 2^x$.

2. Graph $y = \log_2 x$.

3. Write $4^{-3} = \frac{1}{64}$ in logarithmic form. $\log_4 \frac{1}{64} = -3$

4. Write $\log_3 243 = 5$ in exponential form. $3^5 = 243$

Write in exponential form and find the missing value.

5. $4 = \log_2 x$ $2^4 = x, 16$ **6.** $y = \log_{27} 3$ $27^y = 3, \frac{1}{3}$

7. Expand $\log_3 \dfrac{x(x - 4)}{x^2}$ $\log_3 x + \log_3(x - 4) - 2 \log_3 x$

8. Write as the logarithm of a single expression.

$$3 \log_8 (x - 4) + 2 \log_8 (x + 1) - \frac{1}{2} \log_8 x$$

8. $\log_8 [(x - 4)^3 (x + 1)^2 / \sqrt{x}]$

9. Find log 4620.   3.6646

10. Find log 0.000638. -3.1952

11. If $\log N = -2.3002$, find N.   0.00501

12. Solve for x: $3^x = 123$.   4.38

13. Solve for x: $\log 4x = \log (x + 3) + \log 2$.   3

14. What amount of money accumulates if Say-Chun puts \$3500 in a savings account yielding 6% interest compounded quarterly for 10 years? Use $A = P(1 + r)^n$.

14. \$6349.06

15. The amount of carbon 14 remaining after t years is found by the formula $v = v_o e^{-0.0001205t}$, where v_o is the original amount of carbon 14. If a fossil originally contained 60 grams of carbon 14, and now contains 40 grams of carbon 14, how old is the fossil?   3364.86 yr

1.
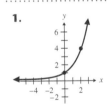

2.
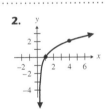

Chapter *12*

Sequences, Series, and the Binomial Theorem

See Section 12.2, Exercise 50

Sequences and series are discussed in this chapter. A sequence is a list of numbers in a specific order and a series is the sum of the numbers in a sequence. In this book we discuss two types of sequences and series: arithmetic and geometric. Series can be used to solve many real-life problems as illustrated in this chapter. If you take higher-level mathematics courses, you may use sequences to approximate some irrational numbers. Other types of sequences not discussed in this book, including the Fibonacci sequence, may be discussed in other mathematics courses.

In Section 12.1 we use the summation symbol, Σ. If you take a course in statistics you will use this symbol often.

We introduce the binomial theorem for expanding an expression of the form $(a + b)^n$ in Section 12.5

12.1   Sequences and Series

Tape 19

1️⃣ Find the terms of a sequence given the general term.
2️⃣ Write a series.
3️⃣ Find partial sums of a series.
4️⃣ Use summation notation, Σ.

**Find the Terms of a
Sequence**

1️⃣ Consider the following two lists of numbers:

$$5, 10, 15, 20, 25, 30, \ldots$$
$$2, 4, 8, 16, 32, \ldots$$

The three dots at the ends of the lists indicate that the lists continue in the same manner. Can you determine the next two numbers in the first list? In the second list? The next numbers in the first list would seem to be 35 and 40. In the second list of numbers they would seem to be 64 and 128. The two lists of numbers above are examples of sequences. A **sequence** (or **progression**) of numbers is a list of numbers arranged in a specified order. Consider the sequence 5, 10, 15, 20, 25, 30, ... The first term is 5. We indicate this by writing $a_1 = 5$. Since the second term is 10, $a_2 = 10$, and so on.

> An **infinite sequence** is a function whose domain is the set of natural numbers.

Consider the infinite sequence 5, 10, 15, 20, 25, 30, 35, . . .

$$
\begin{array}{llllllllll}
\text{Domain:} & \{1, & 2, & 3, & 4, & 5, & 6, & 7, & \ldots, & n, & \ldots\} \\
& \downarrow & \downarrow & \downarrow & \downarrow & \downarrow & \downarrow & \downarrow & & \downarrow \\
\text{Range:} & \{5, & 10, & 15, & 20, & 25, & 30, & 35, & \ldots, & 5n, & \ldots\}
\end{array}
$$

Note that the terms of the sequence 5, 10, 15, 20, . . . are found by multiplying each natural number by 5. For any natural number n, the corresponding term in the sequence is $5 \cdot n$ or $5n$. The **general term of the sequence,** a_n, which defines the sequence, is $a_n = 5n$.

$$a_n = f(n) = 5n$$

To find the twelfth term of the sequence, substitute 12 for n in the general term of the sequence, $a_{12} = 5 \cdot 12 = 60$. Thus, the twelfth term of the sequence is 60. Note that the terms in the sequence are the function values, or the numbers in the range of the function. When writing the sequence, we do not use set braces. The general form of a sequence is

$$a_1, a_2, a_3, a_4, \ldots, a_n, \ldots$$

For the infinite sequence $2, 4, 8, 16, 32, \ldots, 2^n, \ldots$ we can write

$$a_n = f(n) = 2^n$$

Notice that when $n = 1$, $a_1 = 2^1 = 2$; when $n = 2$, $a_2 = 2^2 = 4$; when $n = 3$, $a_3 = 2^3 = 8$; when $n = 4$, $a_4 = 2^4 = 16$; and so on. What is the seventh term of this sequence? The answer is $a_7 = 2^7 = 128$.

A sequence may also be finite.

> A **finite sequence** is a function whose domain includes only the first n natural numbers.

A finite sequence has only a finite number of terms.

Examples of Finite Sequences

5, 10, 15, 20	domain is $\{1, 2, 3, 4\}$
2, 4, 8, 16, 32	domain is $\{1, 2, 3, 4, 5\}$

EXAMPLE 1   Write the finite sequence defined by $a_n = 2n + 1$, for $n = 1, 2, 3, 4$.

Solution:

$$a_1 = 2n + 1$$
$$a_1 = 2(1) + 1 = 3$$
$$a_2 = 2(2) + 1 = 5$$
$$a_3 = 2(3) + 1 = 7$$
$$a_4 = 2(4) + 1 = 9$$

Thus, the sequence is 3, 5, 7, 9.

Since each term of the sequence in Example 1 is larger than the preceding term, it is called an **increasing sequence.**

EXAMPLE 2   Given $a_n = (2n + 3)/n^2$, find

(a) The first term in the sequence

(b) The third term in the sequence

(c) The fifth term in the sequence

Solution:

(a) When $n = 1$, $a_1 = \dfrac{2(1) + 3}{1^2} = \dfrac{5}{1} = 5$.

(b) When $n = 3$, $a_3 = \dfrac{2(3) + 3}{3^2} = \dfrac{9}{9} = 1$.

(c) When $n = 5$, $a_5 = \dfrac{2(5) + 3}{5^2} = \dfrac{13}{25}$.

Note in Example 2 that since there is no restriction on n, a_n is the general term of an infinite sequence.

In Example 2, since each term of the sequence generated by $a_n = (2n + 3)/n^2$ will be smaller than the preceding term, the sequence is called a **decreasing sequence.**

EXAMPLE 3   Find the first four terms of the sequence whose general term is $a_n = (-1)^n(n)$.

Solution:
$$a_n = (-1)^n(n)$$
$$a_1 = (-1)^1(1) = -1$$
$$a_2 = (-1)^2(2) = 2$$
$$a_3 = (-1)^3(3) = -3$$
$$a_4 = (-1)^4(4) = 4$$

If we write the sequence in Example 3, we get $-1, 2, -3, 4, \ldots, (-1)^n(n)$. Notice that each term alternates in sign. We call this an **alternating sequence.**

Write a Series   **2**   A **series** is the sum of a sequence. A series may be finite or infinite depending on whether the sequence it is based on is finite or infinite. Examples of a finite sequence and a finite series are

$$a_1, a_2, a_3, a_4, a_5 \qquad \text{finite sequence}$$
$$a_1 + a_2 + a_3 + a_4 + a_5 \qquad \text{finite series}$$

Examples of an infinite sequence and an infinite series are

$$a_1, a_2, a_3, a_4, a_5, \ldots, a_n, \ldots \qquad \text{infinite sequence}$$
$$a_1 + a_2 + a_3 + a_4 + a_5 + \cdots + a_n + \cdots \qquad \text{infinite series}$$

EXAMPLE 4   Write the first eight terms of the sequence; then write the series that represents the sum of that sequence if

(a) $a_n = \left(\dfrac{1}{2}\right)^n$ **(b)** $a_n = \left(\dfrac{1}{2}\right)^{n-1}$

Solution:

(a) We begin with $n = 1$; thus, the first eight terms of the sequence whose general term is $a_n = \left(\frac{1}{2}\right)^n$ are

$$\left(\frac{1}{2}\right)^1, \left(\frac{1}{2}\right)^2, \left(\frac{1}{2}\right)^3, \left(\frac{1}{2}\right)^4, \left(\frac{1}{2}\right)^5, \left(\frac{1}{2}\right)^6, \left(\frac{1}{2}\right)^7, \left(\frac{1}{2}\right)^8$$

or

$$\frac{1}{2}, \frac{1}{4}, \frac{1}{8}, \frac{1}{16}, \frac{1}{32}, \frac{1}{64}, \frac{1}{128}, \frac{1}{256}$$

The series that represents the sum of the sequence is

$$\frac{1}{2} + \frac{1}{4} + \frac{1}{8} + \frac{1}{16} + \frac{1}{32} + \frac{1}{64} + \frac{1}{128} + \frac{1}{256}$$

(b) We again begin with $n = 1$; thus, the first eight terms of the sequence whose general term is $a_n = \left(\frac{1}{2}\right)^{n-1}$ are

$$\left(\frac{1}{2}\right)^{1-1}, \left(\frac{1}{2}\right)^{2-1}, \left(\frac{1}{2}\right)^{3-1}, \left(\frac{1}{2}\right)^{4-1}, \left(\frac{1}{2}\right)^{5-1}, \left(\frac{1}{2}\right)^{6-1}, \left(\frac{1}{2}\right)^{7-1}, \left(\frac{1}{2}\right)^{8-1}$$

or

$$\left(\frac{1}{2}\right)^{0}, \left(\frac{1}{2}\right)^{1}, \left(\frac{1}{2}\right)^{2}, \left(\frac{1}{2}\right)^{3}, \left(\frac{1}{2}\right)^{4}, \left(\frac{1}{2}\right)^{5}, \left(\frac{1}{2}\right)^{6}, \left(\frac{1}{2}\right)^{7}$$

or

$$1, \frac{1}{2}, \frac{1}{4}, \frac{1}{8}, \frac{1}{16}, \frac{1}{32}, \frac{1}{64}, \frac{1}{128}$$

The series that represents the sum of this sequence is

$$1 + \frac{1}{2} + \frac{1}{4} + \frac{1}{8} + \frac{1}{16} + \frac{1}{32} + \frac{1}{64} + \frac{1}{128}$$

Partial Sums

3 A **partial sum of a series** is the sum of a finite number of consecutive terms of the series, beginning with the first term.

$$
\begin{array}{ll}
s_1 = a_1 & \text{first partial sum} \\
s_2 = a_1 + a_2 & \text{second partial sum} \\
s_3 = a_1 + a_2 + a_3 & \text{third partial sum} \\
\quad\vdots & \\
s_n = a_1 + a_2 + a_3 + \cdots + a_n & n\text{th partial sum}
\end{array}
$$

EXAMPLE 5 Given the infinite sequence defined by $a_n = (1 + n^2)/n$, find:
(a) The first partial sum   **(b)** The third partial sum

Solution:

(a) $s_1 = a_1 = \dfrac{1 + 1^2}{1} = \dfrac{1 + 1}{1} = 2$

(b) $s_3 = a_1 + a_2 + a_3$

$$= \frac{1 + 1^2}{1} + \frac{1 + 2^2}{2} + \frac{1 + 3^2}{3}$$

$$= 2 + \frac{5}{2} + \frac{10}{3} = \frac{47}{6} \text{ or } 7\frac{5}{6}$$

Summation Notation

4 In Section 2.2 we introduced the **summation symbol, Σ,** which is the Greek capital letter *sigma*. This symbol plays a very important role in mathematics and statistics. The summation notation gives a compact way to write a series from the general term of the corresponding sequence. For example, the first five terms of the sequence with general term $a_n = 2n + 3$, where $n = 1, 2, 3, 4, 5$, are

$$5, 7, 9, 11, 13$$

The series representing the sum of this sequence is

$$5 + 7 + 9 + 11 + 13$$

The series representing the sum of this five-term sequence can be indicated using the summation notation:

$$\sum_{n=1}^{5} (2n + 3)$$

To list and evaluate the series represented by $\sum_{n=1}^{5} (2n + 3)$, we first substitute 1 for n in $2n + 3$ and list the value obtained. Then we substitute 2 for n in $2n + 3$ and list the value. We follow this procedure for the values 1 through 5. We then sum these values to obtain the series value.

$$\sum_{n=1}^{5} (2n + 3) = (2 \cdot 1 + 3) + (2 \cdot 2 + 3) + (2 \cdot 3 + 3) + (2 \cdot 4 + 3) + (2 \cdot 5 + 3)$$

$$= 5 + 7 + 9 + 11 + 13 = 45$$

Notice that the series generated by $\sum_{n=1}^{5} (2n + 3)$ is $5 + 7 + 9 + 11 + 13$, and the sum of the series is 45.

The letter n used in the summation notation is called the **index of summation** or simply the *index*. Any letter can be used for the index. Thus, for example, $\sum_{n=1}^{5} (2n + 3) = \sum_{i=1}^{5} (2i + 3)$. The 1 below the summation symbol is called the **lower limit,** and the 5 above the summation symbol is called the **upper limit** of the summation. The expression $\sum_{n=1}^{5} (2n + 3)$ is read "the sum of the terms $2n + 3$ as n varies from 1 to 5."

EXAMPLE 6 Write out the series $\sum_{k=1}^{6} (k^2 + 1)$ and find the sum of the series.

Solution:

$$\sum_{k=1}^{6} (k^2 + 1) = (1^2 + 1) + (2^2 + 1) + (3^2 + 1) + (4^2 + 1) + (5^2 + 1) + (6^2 + 1)$$

$$= 2 + 5 + 10 + 17 + 26 + 37 = 97$$

EXAMPLE 7 Consider the general term of a sequence $a_n = 2n^2 - 4$. Represent the third partial sum in summation notation.

Solution: The third partial sum will be the sum of the first three terms,

$a_1 + a_2 + a_3$. We can represent the third partial sum as $\sum\limits_{n=1}^{3} (2n^2 - 4)$.

EXAMPLE 8 For the following set of values $x_1 = 3$, $x_2 = 4$, $x_3 = 5$, $x_4 = 6$,

and $x_5 = 7$, does $\sum\limits_{i=1}^{5} x_i^2 = \left(\sum\limits_{i=1}^{5} x_i\right)^2$?

Solution:
$$\sum_{i=1}^{5} x_i^2 = x_1^2 + x_2^2 + x_3^2 + x_4^2 + x_5^2$$
$$= 3^2 + 4^2 + 5^2 + 6^2 + 7^2$$
$$= 9 + 16 + 25 + 36 + 49 = 135$$
$$\left(\sum_{i=1}^{5} x_i\right)^2 = (x_1 + x_2 + x_3 + x_4 + x_5)^2$$
$$= (3 + 4 + 5 + 6 + 7)^2 = (25)^2 = 625$$

Since $135 \neq 625$, $\sum\limits_{i=1}^{5} x_i^2 \neq \left(\sum\limits_{i=1}^{5} x_i\right)^2$.

When a summation symbol is written without any upper and lower limits, it means that all the given data are to be summed.

EXAMPLE 9 A formula used to find the arithmetic mean, $\bar{x}$ (read x bar), of a set of data is $\bar{x} = \dfrac{\sum x}{n}$, where n is the number of pieces of data.

Sally's five test grades are 70, 90, 83, 74, and 92. Find the arithmetic mean of her grades.

Solution: $\bar{x} = \dfrac{\sum x}{n} = \dfrac{70 + 90 + 83 + 74 + 92}{5} = \dfrac{409}{5} = 81.8$

Exercise Set 12.1

2. 5, 7, 9, 11, 13 **6.** 0, 2, 6, 12, 20 **7.** $\dfrac{3}{2}, \dfrac{4}{3}, \dfrac{5}{4}, \dfrac{6}{5}, \dfrac{7}{6}$ **8.** $\dfrac{3}{4}, \dfrac{4}{5}, \dfrac{5}{6}, \dfrac{6}{7}, \dfrac{7}{8}$ **9.** $-1, 1, -1, 1, -1$

11. 4, -8, 16, -32, 64 **12.** 3, 8, 15, 24, 35

Write the first five terms of the sequence whose nth term is shown.

1. $a_n = 2n$ 2, 4, 6, 8, 10   **2.** $a_n = 2n + 3$ **3.** $a_n = \dfrac{n+5}{n}$ $6, \dfrac{7}{2}, \dfrac{8}{3}, \dfrac{9}{4}, 2$ **4.** $a_n = \dfrac{n}{n^2}$ $1, \dfrac{1}{2}, \dfrac{1}{3}, \dfrac{1}{4}, \dfrac{1}{5}$

5. $a_n = \dfrac{1}{n}$ $1, \dfrac{1}{2}, \dfrac{1}{3}, \dfrac{1}{4}, \dfrac{1}{5}$ **6.** $a_n = n^2 - n$ **7.** $a_n = \dfrac{n+2}{n+1}$ **8.** $a_n = \dfrac{n+2}{n+3}$

9. $a_n = (-1)^n$ **10.** $a_n = (-1)^{2n}$ 1, 1, 1, 1, 1 **11.** $a_n = (-2)^{n+1}$ **12.** $a_n = n(n+2)$

Find the indicated term of the sequence whose nth term is shown.

13. $a_n = 2n + 3$, twelfth term 27   **14.** $a_n = 2^n$, seventh term 128   **15.** $a_n = 2n - 4$, fifth term 6

21. $s_1 = 7$, $s_3 = 27$ **22.** $s_1 = 1$, $s_3 = \frac{43}{10}$ **23.** $s_1 = 3$, $s_3 = 17$ **24.** $s_1 = 0$, $s_3 = \frac{5}{6}$ **26.** $s_1 = 1$, $s_3 = \frac{49}{6}$
16. $a_n = (-1)^n$, eighth term 1 **17.** $a_n = (-2)^n$, fourth term 16 **18.** $a_n = n(n + 5)$, eighth term 104

19. $a_n = \dfrac{n^2}{(2n + 1)}$, ninth term $\dfrac{81}{19}$ **20.** $a_n = \dfrac{n(n + 1)}{n^2}$, tenth term $\dfrac{11}{10}$

Find the first and third partial sums, s_1 and s_3, for the sequences. **27.** $s_1 = \dfrac{1}{2}$, $s_3 = 7$ **28.** $s_1 = 2$, $s_3 = \dfrac{17}{4}$

21. $a_n = 2n + 5$ **22.** $a_n = \dfrac{3n}{n + 2}$ **23.** $a_n = 2^n + 1$ **24.** $a_n = \dfrac{n - 1}{n + 1}$

25. $a_n = (-1)^{2n}$ $s_1 = 1$, $s_3 = 3$ **26.** $a_n = \dfrac{2n^2}{n + 1}$ **27.** $a_n = \dfrac{n^2}{2}$ **28.** $a_n = \dfrac{n + 3}{2n}$

Write the next three terms of each sequence. **36.** $-50, -60, -70$ **41.** $-25, -33, -41$

29. $2, 4, 8, 16, 32, \ldots$ 64, 128, 256 **30.** $\dfrac{1}{2}, \dfrac{1}{3}, \dfrac{1}{4}, \dfrac{1}{5}, \ldots$ $\dfrac{1}{6}, \dfrac{1}{7}, \dfrac{1}{8}$ **31.** $3, 5, 7, 9, 11, 13, \ldots$ 15, 17, 19

32. $5, 10, 15, 20, 25, \ldots$ 30, 35, 40 **33.** $1, \dfrac{1}{2}, \dfrac{1}{3}, \dfrac{1}{4}, \dfrac{1}{5}, \ldots$ $\dfrac{1}{6}, \dfrac{1}{7}, \dfrac{1}{8}$ **34.** $\dfrac{2}{3}, \dfrac{3}{4}, \dfrac{4}{5}, \dfrac{5}{6}, \dfrac{6}{7}, \ldots$ $\dfrac{7}{8}, \dfrac{8}{9}, \dfrac{9}{10}$

35. $-1, 1, -1, 1, -1, \ldots$ 1, -1, 1 **36.** $-10, -20, -30, -40, \ldots$ **37.** $1, \dfrac{1}{3}, \dfrac{1}{9}, \dfrac{1}{27}, \ldots$ $\dfrac{1}{81}, \dfrac{1}{243}, \dfrac{1}{729}$

38. $\dfrac{1}{3}, \dfrac{2}{3}, \dfrac{3}{3}, \dfrac{4}{3}, \ldots$ $\dfrac{5}{3}, \dfrac{6}{3}, \dfrac{7}{3}$ **39.** $1, -\dfrac{1}{2}, \dfrac{1}{4}, -\dfrac{1}{8}, \ldots$ $\dfrac{1}{16}, -\dfrac{1}{32}, \dfrac{1}{64}$ **40.** $\dfrac{2}{3}, \dfrac{1}{3}, \dfrac{1}{6}, \dfrac{1}{12}, \ldots$ $\dfrac{1}{24}, \dfrac{1}{48}, \dfrac{1}{96}$

41. $7, -1, -9, -17, \ldots$ **42.** $37, 32, 27, 22, \ldots$ 17, 12, 7

Write out the series, and then find the sum of the series. **43.** $2 + 5 + 8 + 11 + 14$; 40 **45.** $-1 + 5 + 15 + 29 + 47 + 69$; 164

43. $\displaystyle\sum_{n=1}^{5} (3n - 1)$ **44.** $\displaystyle\sum_{k=1}^{4} (k^2 - 1)$ $0 + 3 + 8 + 15$; 26 **45.** $\displaystyle\sum_{k=1}^{6} (2k^2 - 3)$

46. $\displaystyle\sum_{i=1}^{4} \dfrac{i^2}{2}$ $\dfrac{1}{2} + 2 + \dfrac{9}{2} + 8$; 15 **47.** $\displaystyle\sum_{n=2}^{4} \dfrac{n^2 + n}{n + 1}$ $2 + 3 + 4$; 9 **48.** $\displaystyle\sum_{n=2}^{5} \dfrac{n^3}{n + 1}$ $\dfrac{8}{3} + \dfrac{27}{4} + \dfrac{64}{5} + \dfrac{125}{6}$; $\dfrac{2583}{60}$

For the general term a_n, write an expression using Σ to represent the indicated partial sum.

49. $a_n = n + 3$, fifth partial sum $\displaystyle\sum_{n=1}^{5} (n + 3)$ **50.** $a_n = n^2 + 1$, fourth partial sum $\displaystyle\sum_{n=1}^{4} (n^2 + 1)$

51. $a_n = \dfrac{n^2}{4}$, third partial sum $\displaystyle\sum_{n=1}^{3} \dfrac{n^2}{4}$ **52.** $a_n = \dfrac{n^2 + 1}{n + 1}$, third partial sum $\displaystyle\sum_{n=1}^{3} \dfrac{n^2 + 1}{n + 1}$

For the set of values $x_1 = 2$, $x_2 = 3$, $x_3 = 5$, $x_4 = -1$, and $x_5 = 4$, find each of the following.

53. $\displaystyle\sum_{i=1}^{5} x_i$ 13 **54.** $\displaystyle\sum_{i=1}^{5} x_i^2$ 55 **55.** $\left(\displaystyle\sum_{i=1}^{5} x_i\right)^2$ 169 **56.** $\displaystyle\sum_{i=1}^{4} (x_i^2 + 3)$ 51

57. $\displaystyle\sum_{i=3}^{5} \dfrac{2x_i}{3}$ $\dfrac{16}{3}$ **58.** $\displaystyle\sum_{i=2}^{4} (x_i^2 + x_i)$ 42

Find the arithmetic mean, $\bar{x}$, of the following sets of data (see Example 9).
59. $15, 20, 25, 30, 35$ 25 **60.** $16, 20, 96, 18, 25$ 35
61. $72, 83, 4, 60, 18, 20$ 42.83 **62.** $5, 12, 9, 12, 17, 36, 70$ 23

63. Write $\displaystyle\sum_{i=1}^{n} x_i$ as a sum of terms. $x_1 + x_2 + x_3 + \ldots + x_n$ **64.** Solve the formula $\bar{x} = \dfrac{\Sigma x}{n}$ for **(a)** Σx; **(b)** n.

🔖 **65.** What is a sequence? 🔖 **66.** What is a series? the sum of the terms of a sequence

🔖 **67.** What is the nth partial sum of a series? 🔖 **68.** How is the following notation read: $\displaystyle\sum_{i=1}^{5} (x_i + 2)$

64. (a) $\Sigma x = n\bar{x}$, **(b)** $n = \dfrac{\Sigma x}{\bar{x}}$ **65.** a list of numbers arranged in a specific order **67.** the sum of the first consecutive n terms of a series
68. the sum of the terms $(x_i + 2)$ as i varies from 1 to 5 **69.** $5, \dfrac{3}{2}$

[6.5] **69.** Solve the equation $2x^2 + 15 = 13x$ using factoring.

[9.2] **70.** How many real solutions does the equation
$6x^2 - 3x - 4 = 2$ have? Explain how you obtained your answer.   *two; $b^2 - 4ac > 0$*

[10.2] **71.** Sketch the graph of $\dfrac{x^2}{4} + \dfrac{y^2}{1} = 1$.

[10.5] **72.** Solve the following system of equations
$$x^2 + y^2 = 5$$
$$x = 2y.$$
(2, 1), (−2, −1)

Group Activity/ Challenge Problems

When no upper and lower limits are placed on a summation symbol, it indicates that the sum of all values is to be found. Consider the following values:

$$x_1 = 1, \quad x_2 = 3, \quad x_3 = 5, \quad x_4 = 7, \quad x_5 = 9$$
$$f_1 = 3, \quad f_2 = 4, \quad f_3 = 5, \quad f_4 = 0, \quad f_5 = 2$$

Then

$$\Sigma x = x_1 + x_2 + x_3 + x_4 + x_5 = 1 + 3 + 5 + 7 + 9 = 25$$
$$\Sigma f = f_1 + f_2 + f_3 + f_4 + f_5 = 3 + 4 + 5 + 0 + 2 = 14$$
$$\Sigma xf = x_1 f_1 + x_2 f_2 + x_3 f_3 + x_4 f_4 + x_5 f_5$$
$$= 1(3) + 3(4) + 5(5) + 7(0) + 9(2) = 58$$

A 10 point quiz is given to a statistics class of 20 students. The results of the quiz are indicated in the table. The grade on the quiz is represented by x and the frequency of the grade (or the number of students who obtained that grade) is represented by f. For example, from the table we see that two students received a grade of 10.

x	6	7	8	9	10
f	1	5	7	5	2

For the values in the table, find:

1. Σf 20   **2.** Σxf 162   **3.** $\Sigma x^2 f$ 1334   **4.** $(\Sigma xf)^2$ 26,244

5. The mean, $\bar{x}$, of a set of data may be found by the formula $\bar{x} = \dfrac{\Sigma xf}{n}$, where $n = \Sigma f$.
Use this formula to find the mean of this set of data.   8.1

6. A formula used to calculate **standard deviation** in statistics is

$$s \approx \sqrt{1.147} \approx 1.07 \qquad s = \sqrt{\dfrac{n(\Sigma x^2 f) - (\Sigma xf)^2}{n(n-1)}}$$

where $n = \Sigma f$. Find the standard deviation of the set of quiz grades.

12.2   Arithmetic Sequences and Series

Tape 19

1 Find the common difference in an arithmetic sequence.
2 Find the nth term of an arithmetic sequence.
3 Find the nth partial sum of an arithmetic series.

Common Difference

1 A sequence in which each term after the first differs from the preceding term by a constant amount is called an **arithmetic sequence** or **arithmetic progression.** The constant amount by which each pair of successive terms differs is called the **common difference,** d. The common difference can be found by subtracting any term from the term that directly follows it.

Arithmetic Sequence	*Common Difference*
$1, 4, 7, 10, 13, 16, \ldots$	$d = 4 - 1 = 3$
$-7, -2, 3, 8, 13, \ldots$	$d = -2 - (-7) = -2 + 7 = 5$
$\dfrac{7}{2}, \dfrac{2}{2}, -\dfrac{3}{2}, -\dfrac{8}{2}, -\dfrac{13}{2}, -\dfrac{18}{2}, \ldots$	$d = \dfrac{2}{2} - \dfrac{7}{2} = -\dfrac{5}{2}$

EXAMPLE 1   Write the first five terms of the arithmetic sequence with **(a)** first term 6 and common difference 3 and **(b)** first term 3 and common difference -2.

Solution:

(a) Start with 6 and keep adding 3. The sequence is 6, 9, 12, 15, 18.

(b) $3, 1, -1, -3, -5$.

---

nth Term of an Arithmetic Series

2 In general, an arithmetic sequence with first term a_1 and common difference d has the following terms:

$$a_1 = a_1, \quad a_2 = a_1 + d, \quad a_3 = a_1 + 2d, \quad a_4 = a_1 + 3d, \quad \text{and so on.}$$

If we continue this process, we can see that the nth term, a_n, can be found by the following formula:

> **nth Term of an Arithmetic Sequence**
> $$a_n = a_1 + (n - 1)d$$

EXAMPLE 2

(a) Write an expression for the general (or nth) term, a_n, of the arithmetic sequence whose first term is -3 and whose common difference is 4.

(b) Find the twelfth term of the sequence.

Solution:

(a) The nth term of the sequence is $a_n = a_1 + (n - 1)d$. Substituting $a_1 = -3$ and $d = 4$, we obtain

$$\begin{aligned}
a_n &= a_1 + (n - 1)d \\
&= -3 + (n - 1)4 \\
&= -3 + 4(n - 1) \\
&= -3 + 4n - 4 \\
&= 4n - 7
\end{aligned}$$

Thus, $a_n = 4n - 7$.

(b) $a_n = 4n - 7$

$a_{12} = 4(12) - 7 = 48 - 7 = 41.$

The twelfth term in the sequence is 41.

EXAMPLE 3 Find the number of terms in the arithmetic sequence $5, 9, 13, 17, \ldots, 41$.

Solution: The first term, a_1, is 5; the nth term is 41, and the common difference, d, is 4.

$$a_n = a_1 + (n - 1)d$$
$$41 = 5 + (n - 1)4$$
$$41 = 5 + 4n - 4$$
$$41 = 4n + 1$$
$$40 = 4n$$
$$10 = n$$

The sequence has 10 terms.

---

nth Partial Sum of an Arithmetic Series

3 An **arithmetic series** is the sum of the terms of an arithmetic sequence. A finite arithmetic series can be written

$$s_n = a_1 + (a_1 + d) + (a_1 + 2d) + (a_1 + 3d) + \cdots + (a_n - 2d) + (a_n - d) + a_n$$

If we consider the last term as a_n, the term before the last term will be $a_n - d$, the second before the last term will be $a_n - 2d$, and so on.

A formula for the nth partial sum, s_n, can be obtained by adding the reverse of s_n to itself.

$$s_n = a_1 + (a_1 + d) + (a_1 + 2d) + \cdots + (a_n - 2d) + (a_n - d) + a_n$$
$$s_n = a_n + (a_n - d) + (a_n - 2d) + \cdots + (a_1 + 2d) + (a_1 + d) + a_1$$
$$\overline{2s_n = (a_1 + a_n) + (a_1 + a_n) + (a_1 + a_n) + \cdots + (a_1 + a_n) + (a_1 + a_n) + (a_1 + a_n)}$$

Since the right side of the equation contains n terms of $(a_1 + a_n)$, we can write

$$2s_n = n(a_1 + a_n)$$

Therefore,

> **nth Partial Sum of an Arithmetic Series**
> $$s_n = \frac{n(a_1 + a_n)}{2}$$

EXAMPLE 4 Find the sum of the first 25 natural numbers.

Solution: The arithmetic sequence is $1, 2, 3, 4, 5, 6, \ldots, 25$. The first term, a_1, is 1; the last term, a_n, is 25. There are 25 terms; thus, $n = 25$.

$$s_n = \frac{n(a_1 + a_n)}{2} = \frac{25(1 + 25)}{2} = \frac{25(26)}{2} = 25(13) = 325$$

The sum of the first 25 natural numbers is 325. Thus, $s_{25} = 325$.

EXAMPLE 5   The first term of an arithmetic sequence is 4, and the last term is 31. If $s_n = 175$, find the number of terms in the sequence and the common difference.

Solution: $a_1 = 4$, $a_n = 31$, and $s_n = 175$.

$$s_n = \frac{n(a_1 + a_n)}{2}$$

$$175 = \frac{n(4 + 31)}{2}$$

$$175 = \frac{35n}{2}$$

$$350 = 35n$$

$$10 = n$$

There are 10 terms in the sequence. We can now find the common difference.

$$a_n = a_1 + (n - 1)d$$
$$31 = 4 + (10 - 1)d$$
$$31 = 4 + 9d$$
$$27 = 9d$$
$$3 = d$$

The common difference is 3. The sequence is 4, 7, 10, 13, 16, 19, 22, 25, 28, 31.

Examples 6 and 7 illustrate some applications of arithmetic sequences and series.

EXAMPLE 6   Donna Stansell is given a starting salary of $25,000 and is promised a $1200 raise after each of the next 8 years. Find her salary during her eighth year of work.

Solution: Her salaries after the first few years would be

$$\$25{,}000, \ \$26{,}200, \ \$27{,}400, \ \$28{,}600, \ldots$$

Since we are adding a constant amount each year, this is an arithmetic sequence. The general term of an arithmetic sequence is $a_n = a_1 + (n - 1)d$. In this example, $a_1 = 25{,}000$ and $d = 1200$. Thus, for $n = 8$, Donna's salary would be

$$a_8 = 25{,}000 + (8 - 1)1200$$
$$= 25{,}000 + 7(1200)$$
$$= 25{,}000 + 8400$$
$$= 33{,}400$$

If we listed all the salaries for the 8-year period, they would be
$25,000, $26,200, $27,400, $28,600, $29,800, $31,000, $32,200, $33,400.

EXAMPLE 7 Each swing of a pendulum is 3 inches shorter than its preceding swing. The first swing is 8 feet.

(a) Find the length of the twelfth swing.

(b) Determine the total distance traveled by the pendulum during the first 12 swings.

Solution:

(a) Since each swing is decreasing by a constant amount, this problem can be represented as an arithmetic series. Since the first swing is given in feet and the decrease in swing in inches, we will change 3 inches to 0.25 feet ($3 \div 12 = 0.25$). The twelfth swing can be considered a_{12}. The difference, d, is negative since the distance is decreasing with each swing.

$$a_n = a_1 + (n - 1)d$$
$$a_{12} = 8 + (12 - 1)(-0.25)$$
$$= 8 + 11(-0.25)$$
$$= 8 - 2.75$$
$$= 5.25 \text{ feet}$$

The twelfth swing is 5.25 feet long.

(b) The total distance traveled during the first 12 swings can be found using the formula

$$s_n = \frac{n(a_1 + a_n)}{2}$$

$$s_{12} = \frac{12(a_1 + a_{12})}{2}$$

$$= \frac{12(8 + 5.25)}{2} = \frac{12(13.25)}{2} = 6(13.25) = 79.5 \text{ feet}$$

The pendulum travels 79.5 feet during its first 12 swings.

Exercise Set 12.2

4. $-8, -11, -14, -17, -20; a_n = -8 - 3(n-1)$ **6.** $-\frac{5}{3}, -2, -\frac{7}{3}, -\frac{8}{3}, -3; a_n = -\frac{5}{3} - \frac{1}{3}(n-1)$

Write the first five terms of the arithmetic sequence with the given first term and common difference. Write the expression for the general (or nth) term, a_n, of the arithmetic sequence.

1. $a_1 = 3, d = 4$ $3, 7, 11, 15, 19; a_n = 3 + 4(n-1)$ **2.** $a_1 = 8, d = 2$ $8, 10, 12, 14, 16; a_n = 8 + 2(n-1)$

3. $a_1 = -5, d = 2$ $-5, -3, -1, 1, 3; a_n = -5 + 2(n-1)$ **4.** $a_1 = -8, d = -3$

5. $a_1 = \frac{1}{2}, d = \frac{3}{2}$ $\frac{1}{2}, 2, \frac{7}{2}, 5, \frac{13}{2}; a_n = \frac{1}{2} + \frac{3}{2}(n-1)$ **6.** $a_1 = -\frac{5}{3}, d = -\frac{1}{3}$

7. $a_1 = 100, d = -5$ $100, 95, 90, 85, 80; a_n = 100 - 5(n-1)$ **8.** $a_1 = \frac{5}{4}, d = -\frac{3}{4}$ $\frac{5}{4}, \frac{2}{4}, -\frac{1}{4}, -\frac{4}{4}, -\frac{7}{4}; a_n = \frac{5}{4} - \frac{3}{4}(n-1)$

Find the desired quantity of the arithmetic sequence.

9. $a_1 = 4, d = 3$; find a_7 22 **10.** $a_1 = 8, d = -2$; find a_6 -2

11. $a_1 = -6, d = -1$; find a_{18} -23 **12.** $a_1 = -15, d = 3$; find a_{20} 42

13. $a_1 = -2, d = \frac{5}{3}$; find a_{10} 13 **14.** $a_1 = 5, a_8 = -21$; find d $-\frac{26}{7}$

15 $a_1 = 3, a_9 = 19$; find d 2 **16.** $a_1 = \frac{1}{2}, a_7 = \frac{19}{2}$; find d $\frac{3}{2}$

17. $a_1 = 4, a_n = 28, d = 3$; find n 9 **18.** $a_1 = -2, a_n = -20, d = -3$; find n 7

19. $a_1 = -\frac{7}{3}, a_n = -\frac{17}{3}, d = -\frac{2}{3}$; find n 6 **20.** $a_1 = 100, a_n = 60, d = -8$; find n 6

Find the sum, s_n, and common difference, d.

21. $a_1 = 1, a_n = 19, n = 10$ $s_{10} = 100, d = 2$ **22.** $a_1 = -5, a_n = 13, n = 7$ $s_7 = 28, d = 3$

23. $a_1 = \frac{3}{5}, a_n = 2, n = 8$ $s_8 = \frac{52}{5}, d = \frac{1}{5}$ **24.** $a_1 = 12, a_n = -23, n = 8$ $s_8 = -44, d = -5$

25. $a_1 = \frac{12}{5}, a_n = \frac{28}{5}, n = 5$ $s_5 = 20, d = \frac{4}{5}$ **26.** $a_1 = -3, a_n = 15.5, n = 6$ $s_6 = 37.5, d = 3.7$

27. $a_1 = 7, a_n = 67, n = 11$ $s_{11} = 407, d = 6$ **28.** $a_1 = 14.25, a_n = 18.75, n = 31$ $s_{31} = 511.5, d = 0.15$

Write the first four terms of each sequence; then find a_{10} and s_{10}.

29. $a_1 = 5, d = 3$ $5, 8, 11, 14; a_{10} = 32; s_{10} = 185$ **30.** $a_1 = -4, d = -2$ $-4, -6, -8, -10; a_{10} = -22, s_{10} = -130$

31. $a_1 = -8, d = -5$ $-8, -13, -18, -23; a_{10} = -53; s_{10} = -305$ **32.** $a_1 = \frac{7}{2}, d = \frac{5}{2}$ $\frac{7}{2}, \frac{12}{2}, \frac{17}{2}, \frac{22}{2}; a_{10} = 26, s_{10} = 147.5$

33. $a_1 = 100, d = -7$ $100, 93, 86, 79; a_{10} = 37, s_{10} = 685$ **34.** $a_1 = -15, d = 4$ $-15, -11, -7, -3; a_{10} = 21, s_{10} = 30$

35. $a_1 = \frac{9}{5}, d = \frac{3}{5}$ $\frac{9}{5}, \frac{12}{5}, \frac{15}{5}, \frac{18}{5}; a_{10} = \frac{36}{5}, s_{10} = 45$ **36.** $a_1 = 35, d = 3$ $35, 38, 41, 44; a_{10} = 62, s_{10} = 485$

Find the number of terms in each sequence and find s_n.

37. $1, 4, 7, 10, \ldots, 43$ $n = 15, s_n = 330$ **38.** $-8, -6, -4, -2, \ldots, 42$ $n = 26, s_n = 442$

39. $-9, -5, -1, 3, \ldots, 27$ $n = 10, s_n = 90$ **40.** $\frac{1}{2}, \frac{2}{2}, \frac{3}{2}, \frac{4}{2}, \frac{5}{2}, \ldots, \frac{17}{2}$ $n = 17, s_n = \frac{153}{2}$

41. $-\frac{5}{6}, -\frac{7}{6}, -\frac{9}{6}, -\frac{11}{6}, \ldots, -\frac{21}{6}$ $n = 9, s_n = -\frac{39}{2}$ **42.** $7, 14, 21, 28, \ldots, 63$ $n = 9, s_n = 315$

43. $-12, -16, -20, \ldots, -52$ $n = 11, s_n = -352$ **44.** $9, 12, 15, 18, \ldots, 93$ $n = 29, s_n = 1479$

45. Mr. Baudean is given a starting salary of $20,000 and is told he will receive a $1000 raise at the end of each year for the next 5 years. **(b)** $a_n = 20,000 + 1000(n - 1)$

   (a) Write a sequence showing his salaries for the next 5 years.   20,000; 21,000; 22,000; 23,000; 24,000

   (b) What is the general term for this sequence?

   (c) If this procedure is extended, find his salary after 12 years. $a_{13} = 32,000$

46. Find the sum of the first 1000 positive integers. 500,500

47. Find the sum of the numbers between 50 and 200 inclusive.   18,875

48. Determine how many numbers between 7 and 1610 are divisible by 6.   267

49. An object falls 16.0 feet during the first second, 48.0 feet during the second second, 80.0 feet during the third second, and so on.

   (a) How far will it fall in its tenth second?   304 ft

   (b) How far will the object fall, in total, during the first 10 seconds?   1600 ft

50. Each time a ball bounces, the height attained is 6 inches less than the previous bounce. If its first bounce reaches a height of 6 feet, find the height attained on the eleventh bounce.   1 foot

51. If you are given $1 on January 1, $2 on January 2, $3 on the 3rd, and so on, how much money will you have received in total by January 31? $496

52. Jack piles logs so that there are 20 logs in the bottom layer, and each layer contains one log less than the layer below it. How many logs are on the pile?

53. What is an arithmetic sequence?

54. What is an arithmetic series?

55. How can the common difference in an arithmetic sequence be found?

52. 210 logs

53. a sequence in which each term differs from the preceding term by a constant amount

54. the sum of the terms of an arithmetic sequence.

55. by subtracting any term from its following term

CUMULATIVE REVIEW EXERCISES

[4.1] **56.** Consider the system of equations

$$2x + 3y = -4$$
$$-x - y = -1$$

Without solving the system, determine how many solutions the system has. Explain.

[4.1] **57.** Solve the system of equations in Problem 56. $(7, -6)$

56. one; slopes of lines are different.

[3.6] **58.** Graph $|x - 2| < 4$ in the Cartesian coordinate system.

[9.4] **59.** Graph the inequality $y \geq 2x^2 - 4x + 6$.

58.   **59.**

<table>
<tr><td rowspan="7">

Group Activity/ Challenge Problems

</td><td>

1. Another formula that may be used to find the nth partial sum of an arithmetic series is

$$s_n = \frac{n}{2}[2a_1 + (n - 1)d]$$

Derive this formula using the two formulas presented in this section.

2. The sum of the interior angles of a triangle, as quadrilateral, a pentagon, and a sextagon is 180°, 360°, 540°, and 720°, respectively. Use the pattern here to find the formula for the sum of the interior angles of a polygon with n sides. $s_n = 180° (n - 2)$

In calculus, a topic of importance is limits. Consider $a_n = 1/x$. Since the value of $1/x$ gets closer and closer to 0 as x gets larger and larger, we say that the limit of a_n as x approaches infinity is 0. Notice that $1/x$ can never equal 0, but its value approaches 0 as x gets larger and larger. Find the limit of each of the following expressions as x approaches infinity.

3. $a_n = \dfrac{1}{x - 2}$ 0

4. $a_n = \dfrac{x}{x + 1}$ 1

5. $a_n = \dfrac{1}{x^2 + 2}$ 0

6. $a_n = \dfrac{2x + 1}{x}$ 2

7. $a_n = \dfrac{4x - 3}{3x + 1}$ $\frac{4}{3}$

</td></tr>
</table>

12.3 Geometric Sequences and Series

Tape 20

1. Find the common ratio in a geometric sequence.
2. Find the nth term of a geometric sequence.
3. Find the nth partial sum of a geometric series.

Common Ratio

1. A **geometric sequence** (or **geometric progression**) is a sequence in which each term after the first is the same multiple of the preceding term. The common multiple is called the **common ratio.** The common ratio, r, in any geometric sequence can be found by dividing any term, except the first, by the preceding term. Consider the geometric sequence

$$1, 3, 9, 27, 81, \ldots, 3^{n-1}, \ldots$$

The common ratio is 3 since $3 \div 1 = 3$ (or $9 \div 3 = 3$, and so on).

The common ratio of the geometric sequence

$$4, 8, 16, 32, 64, \ldots, 4(2^{n-1}), \ldots \text{ is } 2.$$

The common ratio of the geometric sequence

$$3, 12, 48, 192, 768, \ldots, 3(4^{n-1}), \ldots \text{ is } 4.$$

The common ratio of the geometric sequence

$$7, \frac{7}{2}, \frac{7}{4}, \frac{7}{8}, \frac{7}{16}, \ldots, 7\left(\frac{1}{2}\right)^{n-1}, \ldots \text{ is } \frac{1}{2}.$$

EXAMPLE 1 Determine the first five terms of the geometric sequence if $a_1 = 4$ and $r = \frac{1}{2}$.

Solution:

$$a_1 = 4, \quad a_2 = 4 \cdot \frac{1}{2} = 2, \quad a_3 = 2 \cdot \frac{1}{2} = 1, \quad a_4 = 1 \cdot \frac{1}{2} = \frac{1}{2}, \quad a_5 = \frac{1}{2} \cdot \frac{1}{2} = \frac{1}{4}$$

Thus, the first five terms of the geometric sequence are

$$4, 2, 1, \frac{1}{2}, \frac{1}{4}$$

nth Term of a Geometric Series

2. In general, a geometric sequence with first term a_1 and common ratio r has the following terms:

$$a_1, \quad a_1 r, \quad a_1 r^2, \quad a_1 r^3, \quad a_1 r^4, \ldots, a_1 r^{n-1}, \ldots$$

↑	↑	↑	↑	↑	↑
1st term	2nd term	3rd term	4th term	5th term	nth term

Thus, we can see that the nth term of a geometric sequence is given by the following formula:

nth Term of a Geometric Sequence

$$a_n = a_1 r^{n-1}$$

EXAMPLE 2

(a) Write an expression for the general (or nth) term, a_n, of the geometric sequence with $a_1 = 3$ and $r = -2$.

(b) Find the twelfth term of this sequence.

Solution:

(a) The nth term of the sequence is $a_n = a_1 r^{n-1}$. Substituting $a_1 = 3$ and $r = -2$, we obtain

$$a_n = a_1 r^{n-1} = 3(-2)^{n-1}$$

Thus, $a_n = 3(-2)^{n-1}$.

(b) $a_n = 3(-2)^{n-1}$

$a_{12} = 3(-2)^{12-1} = 3(-2)^{11} = 3(-2048) = -6144$

The twelfth term of the sequence is -6144. The first twelve terms of the sequence are 3, -6, 12, -24, 48, -96, 192, -384, 768, -1536, 3072, -6144.

EXAMPLE 3   Find r and a_1 for the geometric sequence with $a_2 = 24$ and $a_5 = 648$.

Solution: The sequence can be represented with blanks for the missing terms.

$$\underline{}, \; 24, \; \underline{}, \; \underline{}, \; 648$$
$$\qquad \uparrow \qquad\qquad\quad \uparrow$$
$$\qquad a_2 \qquad\qquad\quad a_5$$

If we assume that a_2 is the first term of a sequence with the same common ratio, we obtain

$$24, \; \underline{}, \; \underline{}, \; 648$$
$$\uparrow \qquad\qquad\quad \uparrow$$
$$\text{1st} \qquad\qquad \text{4th}$$
$$\text{term} \qquad\qquad \text{term}$$

Now use the formula. Let the first term, a_1, be 24 and the number of terms, n, be 4.

$$a_n = a_1 r^{n-1}$$
$$648 = 24r^{4-1}$$
$$648 = 24r^3$$
$$\frac{648}{24} = r^3$$
$$27 = r^3$$
$$3 = r$$

Thus, the common ratio is 3.

The first term of the original sequence must be $24 \div 3$ or 8. Thus, $a_1 = 8$. The first term could also be found using the formula with $n = 5$.

$$a_n = a_1 r^{n-1}$$
$$648 = a_1 3^{5-1}$$
$$648 = a_1 3^4$$
$$648 = a_1(81)$$
$$\frac{648}{81} = a_1$$
$$8 = a_1$$

nth Partial Sum of a Geometric Series

3 A **geometric series** is the sum of the terms of a geometric sequence. The sum of the first n terms, s_n, of a geometric sequence can be expressed as

$$s_n = a_1 + a_1 r + a_1 r^2 + a_1 r^3 + \cdots + a_1 r^{n-1}$$

If we multiply both sides of the equation by r, we obtain

$$r s_n = a_1 r + a_1 r^2 + a_1 r^3 + \cdots + a_1 r^n$$

Now subtract the second equation from the first.

$$s_n = a_1 + a_1 r + a_1 r^2 + a_1 r^3 + \cdots + a_1 r^{n-1}$$
$$r s_n = \qquad a_1 r + a_1 r^2 + a_1 r^3 + \cdots + a_1 r^{n-1} + a_1 r^n$$
$$\overline{s_n - r s_n = a_1 \qquad\qquad\qquad\qquad\qquad\qquad - a_1 r^n}$$

or

$$s_n - r s_n = a_1 - a_1 r^n$$
$$s_n(1 - r) = a_1 - a_1 r^n$$
$$s_n = \frac{a_1 - a_1 r^n}{1 - r}$$

Thus, we have the following formula for the nth partial sum of a geometric series.

> **nth Partial Sum of a Geometric Series**
>
> $$s_n = \frac{a_1(1 - r^n)}{1 - r}, \qquad r \neq 1$$

EXAMPLE 4   Find the seventh partial sum of a geometric series whose first term is 8 and whose common ratio is $\frac{1}{2}$.

Solution: $s_n = \dfrac{a_1(1 - r^n)}{1 - r}$

$$s_7 = \frac{8\left[1 - \left(\frac{1}{2}\right)^7\right]}{1 - \frac{1}{2}} = \frac{8\left(1 - \frac{1}{128}\right)}{\frac{1}{2}} = \frac{8\left(\frac{127}{128}\right)}{\frac{1}{2}} = \frac{127}{16} \cdot \frac{2}{1} = \frac{127}{8}$$

Thus, $s_7 = \frac{127}{8}$.

EXAMPLE 5   Given $s_n = 93$, $a_1 = 3$, and $r = 2$, find n.

Solution:

$$s_n = \frac{a_1(1 - r^n)}{1 - r}$$

$$93 = \frac{3(1 - 2^n)}{1 - 2}$$

$$93 = \frac{3(1 - 2^n)}{-1}$$

$$-93 = 3(1 - 2^n)$$

$$-31 = 1 - 2^n$$

$$-32 = -2^n$$

$$32 = 2^n$$

$$2^5 = 2^n$$

Therefore, $n = 5$.

Examples 6 and 7 illustrate applications of geometric sequences and series.

EXAMPLE 6   A certain substance decomposes and loses 20% of its weight each hour. If the original quantity of the substance is 300 grams, how much remains after 7 hours?

Solution: This problem can be considered as a geometric sequence since the substance is decreasing by a certain rate (or percent) each hour. Often, when working with a sequence, it is helpful to write out the first few terms. In this problem, since we are asked for the amount of the substance *remaining,* the rate, r, is $100\% - 20\% = 80\%$ or 0.80. To obtain the terms in the sequence, each term is multiplied by 0.80, giving the amount of the substance left in each succeeding hour.

Amount remaining at beginning of:

1st hour	2nd hour	3rd hour	4th hour
$a_1 = 300$	$a_2 = 300(0.80)$	$a_3 = 300(0.80)^2$	$a_4 = 300(0.80)^3$

In this geometric sequence, $a_1 = 300$ and $r = 0.8$. In general, the amount of substance remaining *at the beginning* of the nth hour is $a_n = 300(0.80)^{n-1}$.

We are asked to find the amount remaining after 7 hours. Thus, we must find a_8, since the amount remaining after 7 hours is the same as the amount remaining at the beginning of the eighth hour.

$$a_8 = 300(0.8)^{8-1}$$
$$= 300(0.8)^7$$
$$= 300(0.2097)$$
$$= 62.91 \text{ grams}$$

Note that $a_n = 300(0.80)^n$ could also be used to find the amount *remaining after n hours.* Thus, after 7 hours, $a_7 = 300(0.80)^7$. In the solution to this example, we used the form $a_n = a_1 r^{n-1}$ rather than $a_n = a_1 r^n$, since the first form was presented in the section.

EXAMPLE 7 Mary Foster invests $1000 at 8% interest compounded annually in a savings account. Determine the amount in her account at the end of 6 years.

Solution: At the beginning of the second year, the amount is $1000 + 0.08(1000) = 1000(1 + 0.8) = 1000(1.08)$. At the beginning of the third year, this new amount increases by 8% to give $1000(1.08)^2$, and so on.

Amount in account at beginning of:

1st year	2nd year	3rd year	4th year
$a_1 = 1000$	$a_2 = 1000(1.08)$	$a_3 = 1000(1.08)^2$	$a_4 = 1000(1.08)^3$

This is a geometric sequence with $a_1 = 1000$ and $r = 1.08$. Since we are seeking the amount at the end of 6 years, which is the same as the beginning of the seventh year, we will find a_7.

$$a_n = 1000(1.08)^{n-1}$$
$$a_7 = 1000(1.08)^{7-1}$$
$$= 1000(1.08)^6$$
$$= 1000(1.58687)$$
$$= 1586.87$$

After 6 years the amount in the account is $1586.87. The amount of interest earned is $1586.87 − $1000 = $586.87.

Exercise Set 12.3
1. 5, 15, 45, 135, 405   **2.** 6, 3, $\frac{3}{2}, \frac{3}{4}, \frac{3}{8}$ **3.** 90, 30, 10, $\frac{10}{3}, \frac{10}{9}$ **4.** −12, 12, −12, 12, −12

Determine the first five terms of the geometric sequence.   **5.** −15, 30, −60, 120, −240   **6.** 1, $-\frac{1}{2}, \frac{1}{4}, -\frac{1}{8}, \frac{1}{16}$ **7.** 3, $\frac{9}{2}, \frac{27}{4}, \frac{81}{8}, \frac{243}{16}$

1. $a_1 = 5, r = 3$ **2.** $a_1 = 6, r = \dfrac{1}{2}$ **3.** $a_1 = 90, r = \dfrac{1}{3}$ **4.** $a_1 = -12, r = -1$

5. $a_1 = -15, r = -2$ **6.** $a_1 = 1, r = -\dfrac{1}{2}$ **7.** $a_1 = 3, r = \dfrac{3}{2}$ **8.** $a_1 = 60, r = \dfrac{1}{3}$

Find the indicated term of the geometric sequence.   60, 20, $\frac{20}{3}, \frac{20}{9}, \frac{20}{27}$

9. $a_1 = 5, r = 2$; find a_6 160   **10.** $a_1 = -12, r = \dfrac{1}{2}$; find a_{10} $-\frac{3}{128}$ **11.** $a_1 = 18, r = 3$; find a_7 13,122

12. $a_1 = -20, r = -2$; find a_{10} 10,240   **13.** $a_1 = 2, r = \dfrac{1}{2}$; find a_8 $\frac{1}{64}$ **14.** $a_1 = 5, r = \dfrac{2}{3}$; find a_9 $\frac{1280}{6561}$

15. $a_1 = -3, r = -2$; find a_{12} 6144   **16.** $a_1 = 80, r = \dfrac{1}{3}$; find a_{12} $\frac{80}{177,147}$

Find the sum.

17. $a_1 = 3, r = 4$; find s_5 1023   **18.** $a_1 = 9, r = \dfrac{1}{2}$; find s_6 $\frac{567}{32}$ **19.** $a_1 = 80, r = 2$; find s_7 10,160

20. $a_1 = 1, r = -2$; find s_{12} −1365   **21.** $a_1 = -30, r = -\dfrac{1}{2}$; find s_9 $-\frac{2565}{128}$ **22.** $a_1 = \dfrac{3}{5}, r = 3$; find s_7 $\frac{3279}{5}$

23. $a_1 = -9, r = \dfrac{2}{5}$; find s_5 $-\frac{9279}{625}$ **24.** $a_1 = 35, r = \dfrac{1}{5}$; find s_{12} 43.75

Find the common ratio, r; then write an expression for the general (or nth) term, a_n, for the geometric sequence.

25. $5, \dfrac{5}{2}, \dfrac{5}{4}, \dfrac{5}{8}, \ldots$ $r = \frac{1}{2}, a_n = 5\left(\frac{1}{2}\right)^{n-1}$

26. $3, 9, 27, 81, \ldots$
$r = 3, a_n = 3(3)^{n-1} = 3^n$

27. $2, -6, 18, -54, \ldots$ $r = -3, a_n = 2(-3)^{n-1}$

28. $\dfrac{3}{4}, \dfrac{6}{12}, \dfrac{12}{36}, \dfrac{24}{108}$ $r = \frac{2}{3}, a_n = \frac{3}{4}\left(\frac{2}{3}\right)^{n-1}$

29. $-1, -3, -9, -18, \ldots$
$r = 3, a_n = -1(3)^{n-1}$

30. $\dfrac{4}{3}, \dfrac{8}{3}, \dfrac{16}{3}, \dfrac{32}{3}, \ldots$ $r = 2, a_n = \frac{4}{3}(2)^{n-1}$

31. In a geometric series, $a_3 = 28$ and $a_5 = 112$; find r and a_1. $r = 2, a_1 = 7$, or $r = -2, a_1 = 7$

32. In a geometric series, $a_2 = 27$ and $a_5 = 1$; find r and a_1. $r = \frac{1}{3}, a_1 = 81$

33. In a geometric series, $a_2 = 15$ and $a_5 = 405$; find r and a_1. $r = 3, a_1 = 5$

34. In a geometric series, $a_2 = 12$ and $a_5 = -324$; find r and a_1. $r = -3, a_1 = -4$

35. Your salary increases by 15% per year. If your initial salary is $20,000, what will your salary be at the start of your twenty-fifth year? $572,503.52

36. A ball, when dropped, rebounds to four-fifths of its original height. How high will the ball rebound after the fourth bounce if it is dropped from a height of 30 feet?   12.288 ft

37. A substance loses half its mass each day. If there is initially 300 grams of the substance, find:
  (a) The number of days after which only 37.5 grams of the substance remain.   3 days
  (b) The amount of the substance remaining after 8 days.   1.172 g

38. The number of a certain type of bacteria doubles every hour. If there are initially 1000 bacteria, how long will it take for the number of bacteria to reach 64,000?   6 hr

39. The population of the United States is 260 million. If the population grows at a rate of 5.5% per year, find:
  (a) The population in 12 years. 494.31 million **(b)** 13.9 yr
  (b) The number of years for the population to double.

40. A piece of farm equipment that costs $75,000 decreases in value by 15% per year. Find the value of the equipment after 4 years. $39,150.47

41. The amount of light filtering through a lake diminishes by one-half for each meter of depth.
  (a) Write a sequence indicating the amount of light remaining at depths of 1, 2, 3, 4, and 5 meters.
  (b) What is the general term for this sequence?
  (c) What is the remaining light at a depth of 7 meters?

42. You invest $10,000 in a savings account paying 6% interest annually. Find the amount in your account at the end of 8 years. $15,938.48

43. A tracer dye is injected into Mark for medical reasons. After each hour, two-thirds of the previous hour's dye remains. How much dye remains in Mark's system after 10 hours? $\left(\frac{2}{3}\right)^{10}$ or 1.7%

44. If you start with $1 and double your money each day, how many days will it take to surpass $1,000,000?   21 days

45. One method of depreciating an item on an income tax return is the declining balance method. With this method, a given percentage of the cost of the item is depreciated each year. Suppose that an item has a 5-year life and is depreciated using the declining balance method. Then at the end of its first year it loses $\frac{1}{5}$ of its value and $\frac{4}{5}$ of its value remains. At the end of the second year it loses $\frac{1}{5}$ of the remaining $\frac{4}{5}$ of its value, and so on. A car has a 5-year life expectancy and costs $9800.   **(a)** $7840, $6272, $5017.60
  (a) Write a sequence showing the value of the car remaining for each of the first 3 years.
  (b) What is the general term of this sequence?
  (c) Find the value of the car at the end of 5 years.

46. What is a geometric sequence?

47. What is a geometric series?

48. Explain how to find the common ratio in a geometric sequence.

SEE EXERCISE 41.

45. (b) $a_n = 7840\left(\frac{4}{5}\right)^{n-1}$ **(c)** $3211.26

46. Each term after the first is the same multiple of the preceding term.

47. The sum of the terms of a geometric sequence

48. Divide any term by the term that precedes it.

41. (a) $\frac{1}{2}, \frac{1}{4}, \frac{1}{8}, \frac{1}{16}, \frac{1}{32}$ **(b)** $a_n = a_1 r^{n-1} = \frac{1}{2}\left(\frac{1}{2}\right)^{n-1} = \left(\frac{1}{2}\right)^n$ **(c)** $\frac{1}{128}$ or 0.78%

[7.5] **49.** It takes Mrs. Donovan twice as long to load a truck as Mr. Donovan. If together they can load the truck in 8 hours, how long would it take Mr. Donovan to load the truck by himself?   12 hr

51. $3x^2y^2\sqrt[3]{y} - 2xy\sqrt[3]{9y^2}$

[8.2] **50.** Simplify $\left(\dfrac{x^{1/2}y^{1/3}}{4x^{-3/2}y^{3/5}}\right)^{1/2}$ $\dfrac{x}{2y^{2/15}}$

[8.3] **51.** Simplify $\sqrt[3]{9x^2y}\,(\sqrt[3]{3x^4y^6} - \sqrt[3]{8xy^4})$.

[8.5] **52.** Simplify $x\sqrt{y} - 2\sqrt{x^2y} + \sqrt{4x^2y}$. $x\sqrt{y}$

Group Activity/ Challenge Problems

1. In Exercise Set 11.4, number 43, a formula for scrap value was given. The scrap value, S, is found by $S = c(1 - r)^n$ where c is the original cost, r is the annual depreciation rate, and n is the number of years the object is depreciated.

  (a) If you have not already done so, do Exercise 45 of this section to find the value of the car remaining at the end of 5 years. $3211.26

  (b) Use the formula above to find the scrap value of the car at the end of 5 years and compare this answer with the answer found in part (**a**). $3211.26

2. Find the sum of the sequence 1, 2, 4, 8, . . . , 1,048,576 and the number of terms in the sequence. $n = 21$, $s = 2,097,151$

12.4 Infinite Geometric Series

Tape 20

1 Identify infinite geometric series.
2 Find the sum of an infinite geometric series.

Infinite Geometric Series

1 All the geometric sequences that we have examined thus far have been finite since they have had a last term. The following sequence is an example of an infinite geometric sequence.

$$1, \frac{1}{2}, \frac{1}{4}, \frac{1}{8}, \frac{1}{16}, \ldots, \left(\frac{1}{2}\right)^{n-1}, \ldots$$

Note that the three dots at the end of the sequence indicate that the sequence continues indefinitely in the same manner. The sum of the terms in an infinite geometric sequence form an **infinite geometric series.** For example,

$$1 + \frac{1}{2} + \frac{1}{4} + \frac{1}{8} + \frac{1}{16} + \cdots + \left(\frac{1}{2}\right)^{n-1} + \cdots$$

is an infinite geometric series. Let's find some partial sums.

Sum of first two terms, s_2: $1 + \dfrac{1}{2} = \dfrac{3}{2} = 1.5$

Sum of first three terms, s_3: $1 + \dfrac{1}{2} + \dfrac{1}{4} = \dfrac{7}{4} = 1.75$

Sum of first four terms, s_4: $1 + \dfrac{1}{2} + \dfrac{1}{4} + \dfrac{1}{8} = \dfrac{15}{8} = 1.875$

Sum of first five terms, s_5: $\qquad 1 + \dfrac{1}{2} + \dfrac{1}{4} + \dfrac{1}{8} + \dfrac{1}{16} = \dfrac{31}{16} = 1.9375$

Sum of first six terms, s_6: $\qquad 1 + \dfrac{1}{2} + \dfrac{1}{4} + \dfrac{1}{8} + \dfrac{1}{16} + \dfrac{1}{32} = \dfrac{63}{32} = 1.96875$

Since each term of the geometric sequence is smaller than the preceding term, each additional term adds less and less to the sum. Also, the sum seems to be getting closer and closer to 2.00.

Sum of an Infinite Geometric Series

2 Consider the formula for the sum of the first n terms of a geometric series:

$$s_n = \frac{a_1(1 - r^n)}{1 - r}, \qquad r \neq 1$$

What happens to r^n if $|r| < 1$ and n gets larger and larger? Suppose that $r = \frac{1}{2}$; then

$$\left(\frac{1}{2}\right)^1 = \frac{1}{2} = 0.5$$

$$\left(\frac{1}{2}\right)^2 = \frac{1}{4} = 0.25$$

$$\left(\frac{1}{2}\right)^3 = \frac{1}{8} = 0.125$$

$$\left(\frac{1}{2}\right)^{20} \approx 0.000001$$

We can see that when $|r| < 1$ the value of r^n gets exceedingly close to 0 as n gets larger and larger. Thus, when considering the sum of an infinite geometric series, symbolized s_∞, the expression r^n approaches 0 when $|r| < 1$. Therefore, replacing r^n with 0 in the formula $s_n = \dfrac{a_1(1 - r^n)}{1 - r}$ leads to the following formula.

Sum of an Infinite Geometric Series

$$s_\infty = \frac{a_1}{1 - r} \quad \text{where} \quad |r| < 1$$

EXAMPLE 1 Find the sum of the terms of the infinite sequence $1, \frac{1}{2}, \frac{1}{4}, \frac{1}{16}, \cdots$.

Solution: $a_1 = 1$ and $r = \frac{1}{2}$. Note that $\left|\frac{1}{2}\right| < 1$.

$$s_\infty = \frac{a_1}{1 - r} = \frac{1}{1 - \dfrac{1}{2}} = \frac{1}{\dfrac{1}{2}} = 2$$

Thus, $1 + \frac{1}{2} + \frac{1}{4} + \frac{1}{8} + \frac{1}{16} + \cdots + \left(\frac{1}{2}\right)^{n-1} + \cdots = 2$.

EXAMPLE 2   Find the sum of the infinite geometric series

$$5 - 2 + \frac{4}{5} - \frac{8}{25} + \frac{16}{125} - \frac{32}{625} + \cdots$$

Solution: The terms of the corresponding sequence are

$$5, -2, \frac{4}{5}, -\frac{8}{25}, \cdots$$

$$r = -2 \div 5 = -\frac{2}{5} \quad \text{and} \quad a_1 = 5$$

Since $\left|-\frac{2}{5}\right| < 1$,

$$s_\infty = \frac{a_1}{1 - r}$$

$$= \frac{5}{1 - \left(-\dfrac{2}{5}\right)} = \frac{5}{1 + \dfrac{2}{5}} = \frac{5}{\dfrac{7}{5}} = \frac{25}{7}$$

EXAMPLE 3   Write $0.343434\ldots$ as a ratio of integers.

Solution: We can write this decimal as

$$0.34 + 0.0034 + 0.000034 + \cdots + (0.34)(0.01)^{n-1} + \cdots$$

This is an infinite geometric series with $r = 0.01$. Since $|r| < 1$,

$$s_\infty = \frac{a_1}{1 - r} = \frac{0.34}{1 - 0.01} = \frac{0.34}{0.99} = \frac{34}{99}$$

If you divide 34 by 99 on a calculator, you will see .34343434 displayed.

EXAMPLE 4   On each swing, a certain pendulum travels 90% as far as on its previous swing. For example, if the swing to the right is 10 feet, the swing back to the left is $0.9 \times 10 = 9$ feet (see Fig. 12.1). If the first swing is 10 feet long, determine the total distance traveled by the pendulum by the time it comes to rest.

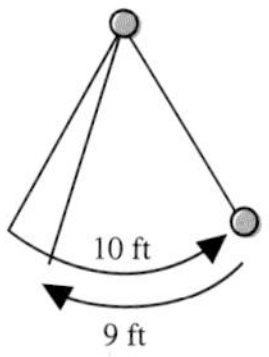

10 ft

9 ft

FIGURE 12.1

Solution: This problem may be considered an infinite geometric series with $a_1 = 10$ and $r = 0.9$. We can therefore use the formula $s_\infty = \dfrac{a_1}{1 - r}$ to find the total distance traveled by the pendulum.

$$s_\infty = \frac{a_1}{1 - r} = \frac{10}{1 - 0.9} = \frac{10}{0.1} = 100 \text{ feet}$$

What is the sum of a geometric series when $|r| > 1$? Consider the geometric sequence in which $a_1 = 1$ and $r = 2$.

$$1, 2, 4, 8, 16, 32, \ldots, 2^{n-1}, \ldots$$

The sum of its terms is

$$1 + 2 + 4 + 8 + 16 + 32 + \cdots + 2^{n-1} + \ldots$$

What is the sum of this series? As n gets larger and larger, the sum gets larger and larger. We therefore say that the sum "does not exist." For $|r| > 1$, the sum of the an infinite geometric series does not exist.

Exercise Set 12.4   **4.** -2 **6.** $\frac{9}{2}$

Find the sum of the terms in each sequence.

1. $6, 3, \dfrac{3}{2}, \dfrac{3}{4}, \dfrac{3}{8}, \ldots$ 12

2. $\dfrac{1}{3}, \dfrac{1}{9}, \dfrac{1}{27}, \dfrac{1}{81}, \ldots$ $\frac{1}{2}$

3. $5, 2, \dfrac{4}{5}, \dfrac{8}{25}, \ldots$ $\frac{25}{3}$

4. $-\dfrac{4}{3}, -\dfrac{4}{9}, -\dfrac{4}{27}, -\dfrac{4}{81}, \ldots$

5. $\dfrac{1}{3}, \dfrac{4}{15}, \dfrac{16}{75}, \dfrac{64}{375}, \ldots$ $\frac{5}{3}$

6. $6, -2, \dfrac{2}{3}, -\dfrac{2}{9}, \dfrac{2}{27}, \ldots$

7. $9, -1, \dfrac{1}{9}, -\dfrac{1}{81}, \ldots$ $\frac{81}{10}$

8. $\dfrac{5}{3}, 1, \dfrac{3}{5}, \dfrac{9}{25}, \ldots$ $\frac{25}{6}$

Find the sum of each infinite series.   **13.** -45 **15.** -15

9. $1 + \dfrac{1}{2} + \dfrac{1}{4} + \dfrac{1}{8} + \cdots$ 2

10. $4 + 2 + 1 + \dfrac{1}{2} + \cdots$ 8

11. $8 + \dfrac{16}{3} + \dfrac{32}{9} + \dfrac{64}{27} + \cdots$ 24

12. $10 - 5 + \dfrac{5}{2} - \dfrac{5}{4} + \cdots$ $\frac{20}{3}$

13. $-60 + 20 - \dfrac{20}{3} + \dfrac{20}{9} - \cdots$

14. $4 + \dfrac{8}{3} + \dfrac{16}{9} + \dfrac{32}{27} + \cdots$ 12

15. $-12 - \dfrac{12}{5} - \dfrac{12}{25} - \dfrac{12}{125} - \cdots$

16. $5 - 1 + \dfrac{1}{5} - \dfrac{1}{25} + \cdots$ $\frac{25}{6}$

Write the repeating decimal as a ratio of integers.

17. $0.2727\ldots$ $\frac{3}{11}$

18. $0.454545\ldots$ $\frac{5}{11}$

19. $0.5555\ldots$ $\frac{5}{9}$

20. $0.375375\ldots$ $\frac{125}{333}$

21. $0.515151\ldots$ $\frac{17}{33}$

22. $0.742742\ldots$ $\frac{742}{999}$

23. On each swing, a pendulum travels 80% as far as on its previous swing. If the first swing is 8 feet long, determine the total distance traveled by the pendulum by the time it comes to rest.   40 ft

24. What is an infinite geometric series?

25. What is the sum of an infinite geometric series when $|r| > 1$?   sum does not exist

24. a geometric series that is not finite

CUMULATIVE REVIEW EXERCISES

[5.3] **26.** Multiply $(4x^2 - 3x - 6)(2x - 3)$.

[5.4] **27.** Divide $(16x^2 + 10x - 18) \div (2x + 5)$.

26. $8x^3 - 18x^2 - 3x + 18$ **27.** $8x - 15 + \dfrac{57}{2x + 5}$

[8.2] **28.** Evaluate $\left(\dfrac{9}{100}\right)^{-1/2}$. $\frac{10}{3}$

[8.6] **29.** Solve $\sqrt{a^2 + 9a + 3} = -a$. $-\frac{1}{3}$

Group Activity/ Challenge Problems

1. A ball is dropped from a height of 10 feet. The ball bounces to a height of 9 feet. On each successive bounce the ball rises to 90% of its previous height. Find the total vertical distance traveled by the ball when it comes to rest (see the figure).   190 ft

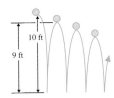

2. A particle follows the path indicated by the wave shown. Find the total vertical distance traveled by the particle.   4 ft

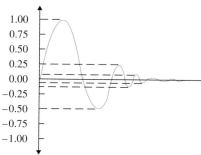

12.5   The Binomial Theorem

Tape 20

1 Evaluate factorials.
2 Use Pascal's triangle to find the numerical coefficients of an expanded binomial.
3 Use the binomial theorem to expand binomials.

Factorials   **1**   To understand the binomial theorem, you must have an understanding of what **factorials** are. The notation $n!$ is read "n factorial." Its definition follows.

> **n Factorial**
>
> $$n! = n(n-1)(n-2)(n-3)\cdots(1)$$
>
> for any positive integer n.

For example,

$$6! = 6\cdot5\cdot4\cdot3\cdot2\cdot1 = 720$$
$$8! = 8\cdot7\cdot6\cdot5\cdot4\cdot3\cdot2\cdot1 = 40{,}320$$

Note that by definition **0! is 1.**

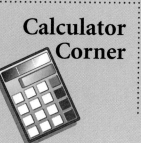

Calculator Corner

Factorials can be found on calculators that contain an $n!$ or $x!$ key. Often, the factorial key is a second function key. In the following, the answers appear after $n!$.

Evaluate 6!   6 2nd $n!$ 720
Evaluate 9!   9 2nd $n!$ 362880

Graphing calculators do not have a factorial key. On most graphing calculators factorials are found under the MATH, Probability function menu.

Pascal's Triangle **2** Using polynomial multiplication, we can obtain the following expansions of the binomial $a + b$:

$$(a + b)^0 = 1$$
$$(a + b)^1 = a + b$$
$$(a + b)^2 = a^2 + 2ab + b^2$$
$$(a + b)^3 = a^3 + 3a^2b + 3ab^2 + b^3$$
$$(a + b)^4 = a^4 + 4a^3b + 6a^2b^2 + 4ab^3 + b^4$$
$$(a + b)^5 = a^5 + 5a^4b + 10a^3b^2 + 10a^2b^3 + 5ab^4 + b^5$$
$$(a + b)^6 = a^6 + 6a^5b + 15a^4b^2 + 20a^3b^3 + 15a^2b^4 + 6ab^5 + b^6$$

Note that when expanding a binomial of the form $(a + b)^n$:

1. There are $n + 1$ terms in the expansion.
2. The first term is a^n and the last term is b^n.
3. Reading from left to right, the exponents on a decrease by 1 from term to term, while the exponents on b increase by 1 from term to term.
4. The sum of the exponents on the variables in each term is n.
5. The coefficients of the terms equidistant from the ends are the same.

If we examine just the variables in $(a + b)^5$, we have a^5, a^4b, a^3b^2, a^2b^3, ab^4, and b^5.

The numerical coefficients of each term in the expansion of $(a + b)^n$ can be found by using **Pascal's triangle**. For example, if $n = 5$, we can determine the numerical coefficients of $(a + b)^5$ as follows.

Exponent on Binomial					*Pascal's Triangle*						
$n = 0$							1				
$n = 1$						1		1			
$n = 2$					1		2		1		
$n = 3$				1		3		3		1	
$n = 4$			1		4		6		4		1
$n = 5$		1		5		10		10		5		1
$n = 6$	1		6		15		20		15		6	1

Examine row 5 ($n = 4$) and row 6 ($n = 5$).

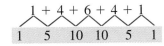

Notice that the first and last numbers in each row are 1, and the inner numbers are obtained by adding the two numbers in the row above (to the right and left). The numerical coefficients of $(a + b)^5$ are 1, 5, 10, 10, 5, and 1. Thus we can write the expansion of $(a + b)^5$ by using the information in 1–5 above for the variables and their exponents and by using Pascal's triangle for the coefficients.

$$(a + b)^5 = a^5 + 5a^4b + 10a^3b^2 + 10a^2b^3 + 5ab^4 + b^5$$

This method of expanding a binomial is not practical when n is large.

The Binomial Theorem   ③ We will shortly introduce a more practical method, called the binomial theorem, to expand expressions of the form $(a + b)^n$. However, before we introduce this formula, we need to explain how to find binomial coefficients of the form $\binom{n}{r}$.

Binomial Coefficients

For n and r nonnegative integers, $n > r$,

$$\binom{n}{r} = \frac{n!}{r! \cdot (n - r)!}$$

The binomial coefficient $\binom{n}{r}$ is read "the number of *combinations* of n items taken r at a time." Combinations are used in the many areas of mathematics, including the study of probability.

EXAMPLE 1   Evaluate $\binom{5}{2}$.

Solution: By the definition, if we substitute 5 for n and 2 for r, we obtain

$$\binom{5}{2} = \frac{5!}{2! \cdot (5 - 2)!} = \frac{5!}{2! \cdot 3!} = \frac{5 \cdot 4 \cdot 3 \cdot 2 \cdot 1}{(2 \cdot 1) \cdot (3 \cdot 2 \cdot 1)} = 10$$

Thus, $\binom{5}{2}$ equals 10.

EXAMPLE 2   Evaluate.   **(a)** $\binom{7}{4}$ **(b)** $\binom{4}{4}$ **(c)** $\binom{5}{0}$

Solution:

(a) $\binom{7}{4} = \dfrac{7!}{4! \cdot (7 - 4)!} = \dfrac{7!}{4! \cdot 3!} = \dfrac{7 \cdot 6 \cdot 5 \cdot 4 \cdot 3 \cdot 2 \cdot 1}{(4 \cdot 3 \cdot 2 \cdot 1)(3 \cdot 2 \cdot 1)} = 35$

(b) $\binom{4}{4} = \dfrac{4!}{4! \cdot (4 - 4)!} = \dfrac{4!}{4! \cdot 0!} = \dfrac{1}{1} = 1$ Remember that $0! = 1$.

(c) $\binom{5}{0} = \dfrac{5!}{0! \cdot (5 - 0)!} = \dfrac{5!}{0! \cdot 5!} = \dfrac{1}{1} = 1$

By studying Examples 2(b) and (c), you can reason that, for any positive integer n,

$$\binom{n}{n} = 1 \quad \text{and} \quad \binom{n}{0} = 1$$

Now we introduce the binomial theorem.

> **Binomial Theorem**
>
> For any positive integer n,
>
> $$(a + b)^n = \binom{n}{0}a^n b^0 + \binom{n}{1}a^{n-1}b^1 + \binom{n}{2}a^{n-2}b^2$$
> $$+ \binom{n}{3}a^{n-3}b^3 + \cdots + \binom{n}{n}a^0 b^n$$

Notice in the binomial theorem that the sum of the exponents on the variables in each term is n, and the bottom number in the combination is always the same as the exponent on the second variable in the term. For example, if we consider the term $\binom{n}{3}a^{n-3}b^3$, the sum of the exponents on the variables is $(n-3) + 3 = n$. Also, the exponent on the variable b is 3, and the bottom number in the combination is also 3.

Now we will expand $(a + b)^5$ using the binomial theorem and see if we get the same expression as we did when we used polynomial multiplication and Pascal's triangle to obtain the expansion.

$$(a + b)^5 = \binom{5}{0}a^5 b^0 + \binom{5}{1}a^{5-1}b^1 + \binom{5}{2}a^{5-2}b^2 + \binom{5}{3}a^{5-3}b^3 + \binom{5}{4}a^{5-4}b^4 + \binom{5}{5}a^{5-5}b^5$$

$$= \binom{5}{0}a^5 b^0 + \binom{5}{1}a^4 b^1 + \binom{5}{2}a^3 b^2 + \binom{5}{3}a^2 b^3 + \binom{5}{4}a^1 b^4 + \binom{5}{5}a^0 b^5$$

$$= \frac{5!}{0! \cdot 5!}a^5 + \frac{5!}{1! \cdot 4!}a^4 b + \frac{5!}{2! \cdot 3!}a^3 b^2 + \frac{5!}{3! \cdot 2!}a^2 b^3 + \frac{5!}{4! \cdot 1!}ab^4 + \frac{5!}{5! \cdot 0!}b^5$$

$$= a^5 + 5a^4 b + 10a^3 b^2 + 10a^2 b^3 + 5ab^4 + b^5$$

This is the same expression as we obtained earlier.

In the binomial theorem, the first and last terms of the expansion contain a factor raised to the zero power. Since any non zero number raised to the 0th power equals one, we could have omitted those factors. These factors were included so that you could see the pattern better.

EXAMPLE 3   Use the binomial theorem to expand $(2x + 3)^6$.

Solution: If we use $2x$ for a and 3 for b, we obtain

$$(2x + 3)^6 = \binom{6}{0}(2x)^6(3)^0 + \binom{6}{1}(2x)^5(3)^1 + \binom{6}{2}(2x)^4(3)^2 + \binom{6}{3}(2x)^3(3)^3 + \binom{6}{4}(2x)^2(3)^4 + \binom{6}{5}(2x)^1(3)^5 + \binom{6}{6}(2x)^0(3)^6$$

$$= 1(2x)^6 + 6(2x)^5(3) + 15(2x)^4(9) + 20(2x)^3(27) + 15(2x)^2(81) + 6(2x)(243) + 1(3)^6$$

$$= 64x^6 + 576x^5 + 2160x^4 + 4320x^3 + 4860x^2 + 2916x + 729$$

EXAMPLE 4   Use the binomial theorem to expand $(5x - 2y)^4$.

Solution: Write $(5x - 2y)^4$ as $[5x + (-2y)]^4$. Use $5x$ in place of a and $-2y$ in place of b in the binomial theorem.

$$[5x + (-2y)]^4 = \binom{4}{0}(5x)^4(-2y)^0 + \binom{4}{1}(5x)^3(-2y)^1 + \binom{4}{2}(5x)^2(-2y)^2 + \binom{4}{3}(5x)^1(-2y)^3 + \binom{4}{4}(5x)^0(-2y)^4$$

$$= 1(5x)^4 + 4(5x)^3(-2y) + 6(5x)^2(-2y)^2 + 4(5x)(-2y)^3 + 1(-2y)^4$$

$$= 625x^4 - 1000x^3 y + 600x^2 y^2 - 160xy^3 + 16y^4$$

11. $x^3 + 12x^2 + 48x + 64$ **12.** $8x^3 + 36x^2 + 54x + 27$ **13.** $a^4 - 4a^3b + 6a^2b^2 - 4ab^3 + b^4$
14. $16r^4 + 32r^3s^2 + 24r^2s^4 + 8rs^6 + s^8$ **15.** $243a^5 - 405a^4b + 270a^3b^2 - 90a^2b^3 + 15ab^4 - b^5$
16. $x^5 + 10x^4y + 40x^3y^2 + 80x^2y^3 + 80xy^4 + 32y^5$

Exercise Set 12.5

Evaluate each of the following combinations.

1. $\binom{4}{2}$ 6

2. $\binom{6}{3}$ 20

3. $\binom{5}{5}$ 1

4. $\binom{9}{3}$ 84

5. $\binom{7}{0}$ 1

6. $\binom{10}{8}$ 45

7. $\binom{8}{4}$ 70

8. $\binom{12}{8}$ 495

9. $\binom{8}{2}$ 28

10. $\binom{7}{5}$ 21

Use the binomial theorem to expand each expression.

17. $16x^4 + 16x^3 + 6x^2 + x + \frac{1}{16}$

11. $(x + 4)^3$

12. $(2x + 3)^3$

13. $(a - b)^4$

14. $(2r + s^2)^4$

15. $(3a - b)^5$

16. $(x + 2y)^5$

17. $\left(2x + \frac{1}{2}\right)^4$

18. $\left(\frac{2}{3}x + \frac{3}{2}\right)^4$

19. $\left(\frac{x}{2} - 3\right)^4$

20. $(3x^2 + y)^5$ $243x^{10} + 405x^8y + 270x^6y^2 + 90x^4y^3 + 15x^2y^4 + y^5$

Write the first four terms of each expansion.   **18.** $\frac{16x^4}{81} + \frac{16x^3}{9} + 6x^2 + 9x + \frac{81}{16}$ **19.** $\frac{x^4}{16} - \frac{3x^3}{2} + \frac{27x^2}{2} - 54x + 81$

21. $(x + y)^{10}$ $x^{10} + 10x^9y + 45x^8y^2 + 120x^7y^3$

22. $(2x + 3)^8$ $256x^8 + 3072x^7 + 16{,}128x^6 + 48{,}384x^5$

23. $(3x - y)^7$ $2187x^7 - 5103x^6y + 5103x^5y^2 - 2835x^4y^3$

24. $(3p + 2q)^{11}$ $3^{11}p^{11} + (22)3^{10}p^{10}q + (220)3^9p^9q^2 + (1320)3^8p^8q^3$

25. $(x^2 - 3y)^8$ $x^{16} - 24x^{14}y + 252x^{12}y^2 - 1512x^{10}y^3$

26. $\left(2x + \frac{y}{5}\right)^9$ $512x^9 + \frac{2304}{5}x^8y + \frac{4608}{25}x^7y^2 + \frac{5376}{125}x^6y^3$

27. Explain how to construct Pascal's triangle. Construct the first five rows of Pascal's triangle.

28. Explain in your own words how to find $n!$ for any positive integer to n. $n! = n(n - 1)(n - 2) \cdots (1)$

29. Is $n!$ equal to $n \cdot (n - 1)!$? Explain and give an example to support your answer.   yes, $4! = 4 \cdot 3!$

30. Is $(n + 1)!$ equal to $(n + 1) \cdot n!$? Explain and give an example to support your answer.

30. yes, $(4 + 1)! = (4 + 1) \cdot 4!$

31. Is $(n - 3)!$ equal to $(n - 3)(n - 4)(n - 5)!$ for $n \geq 5$? Explain and give an example to support your answer.   yes, $(7 - 3)! = (7 - 3)(7 - 4)(7 - 5)!4! = 4 \cdot 3 \cdot 2!$

32. Under what conditions will $\binom{n}{m}$ where n and m are non-negative integers, have a value of 1? $n = m$ or $m = 0$

33. What are the first, second, next to last and last terms of the expansion $(x + 3)^8$? x^8, $24x^7$, $17{,}496x$, 6561

34. What are the first, second, next to last, and last terms of the expansion $(2x + 5)^6$? $64x^6$, $960x^5$, $37{,}500x$, $15{,}625$

27.
```
          1
        1   1
      1   2   1
    1   3   3   1
  1   4   6   4   1
```

CUMULATIVE REVIEW EXERCISES

[2.2] *Solve $s_n - s_nr = a_1 - a_1r^n$ for the indicated variable.*

35. For s_n. $s_n = \frac{a_1 - a_1r^n}{1 - r}$ or $s_n = \frac{a_1(1 - r^n)}{1 - r}$

36. For a_1. $a_1 = \frac{s_n - s_nr}{1 - r^n}$

Factor.

[6.2] **37.** $16x^2 - 8x - 3$. $(4x - 3)(4x + 1)$

[6.4] **38.** $3a^2b^2 - 24ab^2 + 48b^2$. $3b^2(a - 4)^2$

Group Activity/ Challenge Problems

1. Write the binomial theorem using summation notation.

2. (a) Show how to find the rth term in a binomial expansion of the form $(x + y)^n$, where r is any positive integer less than or equal to n, without expanding the bionomial.

(b) Find the third term in the expansion of $(3x + 2)^4$ without expanding the expression. $216x^2$

1. $(a + b)n = \sum_{i=0}^{n} \binom{n}{i} a^{n-i}b^i$ **2. (a)** rth term: $\frac{n!}{[n - (r - 1)]! (r - 1)!}x^{n - (r-1)}y^{r-1}$

Summary

GLOSSARY

Alternating sequence *(672):* A sequence in which the signs of the terms alternate.

Arithmetic sequence (or arithmetic progression) *(678):* A sequence in which each term after the first differs from the preceding term by a constant amount.

Arithmetic series *(679):* The sum of the terms of an arithmetic sequence.

Common difference *(677):* The amount by which each pair of terms differs in an arithmetic sequence.

Common ratio *(684):* The common multiple in a geometric series.

Decreasing sequence *(672):* A sequence in which each term is less than the term that precedes it.

Finite sequence *(671):* A function whose domain includes only the first n natural numbers.

General term of a sequence *(671):* An expression that defines the nth term of the sequence.

Geometric sequence (or geometric progression) *(684):* A sequence in which each term after the first is the same multiple of the preceding term.

Geometric series *(686):* The sum of the terms of a geometric sequence.

Increasing sequence *(671):* A sequence in which each term is greater than the term that precedes it.

Infinite geometric series *(690):* The sum of the terms in an infinite geometric sequence.

Infinite sequence *(670):* A function whose domain is the set of natural numbers.

Partial sum of a series *(673):* The sum of a finite number of consecutive terms of a series, beginning with the first term.

Sequence (or progression) *(670):* A list of numbers arranged in a specific order.

Series *(672):* The sum of a sequence.

IMPORTANT FACTS

$$\sum_{i=1}^{n} x_i = x_1 + x_2 + x_3 + \cdots + x_n$$

Pascal's Triangle

$$
\begin{array}{ccccccccc}
 & & & & 1 & & & & \\
 & & & 1 & & 1 & & & \\
 & & 1 & & 2 & & 1 & & \\
 & 1 & & 3 & & 3 & & 1 & \\
1 & & 4 & & 6 & & 4 & & 1 \\
\end{array}
$$

n Factorial

$$n! = n(n-1)(n-2) \cdots (2)(1)$$

nth Term of an arithmetic sequence

$$a_n = a_1 + (n-1)d$$

nth Partial sum of an arithmetic series

$$s_n = \frac{n(a_1 + a_n)}{2}$$

nth Term of a geometric sequence

$$a_n = a_1 r^{n-1}$$

nth Partial sum of a geometric series

$$s_n = \frac{a_1(1 - r^n)}{1 - r}, \quad r \neq 1$$

Sum of an infinite geometric series

$$s_\infty = \frac{a_1}{1 - r}, \quad |r| < 1$$

Binomial coefficients

$$\binom{n}{r} = \frac{n!}{r! \cdot (n-r)!}$$

Binomial theorem

$$(a + b)^n = \binom{n}{0}a^n b^0 + \binom{n}{1}a^{n-1}b^1 + \binom{n}{2}a^{n-2}b^2 + \binom{n}{3}a^{n-3}b^3 + \cdots + \binom{n}{n}a^0 b^n$$

13. 16, 32, 64; $a_n = 2^{n-1}$ **14.** $-\frac{1}{2}, \frac{1}{4}, -\frac{1}{8}$; $a_n = -8\left(-\frac{1}{2}\right)^{n-1}$ **15.** $\frac{32}{3}, \frac{64}{3}, \frac{128}{3}$; $a_n = \frac{2}{3}(2)^{n-1}$

16. $-3, -6, -9$; $a_n = 9 - 3(n-1) = 12 - 3n$ **17.** $-1, 1, -1$; $a_n = (-1)^n$

Review Exercises

21. $\frac{1}{3} + \frac{4}{3} + \frac{9}{3} + \frac{16}{3} + \frac{25}{3}; \frac{55}{3}$ **22.** $\frac{1}{2} + \frac{2}{3} + \frac{3}{4} + \frac{4}{5}; \frac{163}{60}$

[12.1] *Write the first five terms of the sequence.*

1. $a_n = n + 2$ 3, 4, 5, 6, 7 **2.** $a_n = \dfrac{1}{n}$ $1, \frac{1}{2}, \frac{1}{3}, \frac{1}{4}, \frac{1}{5}$ **3.** $a_n = n(n+1)$ **4.** $a_n = \dfrac{n^2}{n+4}$ $\frac{1}{5}, \frac{2}{3}, \frac{9}{7}, 2, \frac{25}{9}$

2, 6, 12, 20, 30

Find the indicated term of the sequence.

5. $a_n = 3n + 4$, seventh term 25 **6.** $a_n = (-1)^n + 3$, seventh term 2

7. $a_n = \dfrac{n + 7}{n^2}$, ninth term $\frac{16}{81}$ **8.** $a_n = (n)(n-3)$, eleventh term 88

Find the first and third partial sums, s_1 and s_3. **9.** $s_1 = 5$, $s_3 = 24$ **10.** $s_1 = 2$, $s_3 = 28$ **11.** $s_1 = \frac{4}{3}$, $s_3 = \frac{227}{60}$

9. $a_n = 3n + 2$ **10.** $a_n = 2n^2$ **11.** $a_n = \dfrac{n + 3}{n + 2}$ **12.** $a_n = (-1)^n(n+2)$

$s_1 = -3$, $s_3 = -4$

Write the next three terms of each sequence. Then write an expression for the general term, a_n.

13. 1, 2, 4, 8, . . . **14.** $-8, 4, -2, 1, . . .$ **15.** $\dfrac{2}{3}, \dfrac{4}{3}, \dfrac{8}{3}, \dfrac{16}{3}, . . .$ **16.** 9, 6, 3, 0, . . .

17. $-1, 1, -1, 1, . . .$ **18.** 6, 12, 18, 24, . . . 30, 36, 42; $a_n = 6 + 6(n-1) = 6n$

Write out the series. Then find the sum of the series. **19.** $3 + 6 + 11; 20$ **20.** $3 + 8 + 15 + 24; 50$

19. $\displaystyle\sum_{n=1}^{3} (n^2 + 2)$ **20.** $\displaystyle\sum_{k=1}^{4} k(k+2)$ **21.** $\displaystyle\sum_{k=1}^{5} \frac{k^2}{3}$ **22.** $\displaystyle\sum_{n=1}^{4} \frac{n}{n+1}$

For the set of values $x_1 = 3$, $x_2 = 9$, $x_3 = 5$, $x_4 = 10$, evaluate the sum.

23. $\displaystyle\sum_{i=1}^{4} x_i$ 27 **24.** $\displaystyle\sum_{i=1}^{4} x_i^2$ 215 **25.** $\displaystyle\sum_{i=2}^{3} (x_i^2 + 1)$ 108 **26.** $\left(\displaystyle\sum_{i=1}^{4} x_i\right)^2$ 729

[12.2] *Write the first five terms of the arithmetic sequence with first term and common difference as given.*

27. $a_1 = 5$, $d = 2$ 5, 7, 9, 11, 13 **28.** $a_1 = \dfrac{1}{2}$, $d = -2$ $\frac{1}{2}, -\frac{3}{2}, -\frac{7}{2}, -\frac{11}{2}, -\frac{15}{2}$

29. $a_1 = -12$, $d = -\dfrac{1}{2}$ $-12, -\frac{25}{2}, -13, -\frac{27}{2}, -14$ **30.** $a_1 = -100$, $d = \dfrac{1}{5}$ $-100, -\frac{499}{5}, -\frac{498}{5}, -\frac{497}{5}, -\frac{496}{5}$

For each arithmetic sequence, find the value indicated.

31. $a_1 = 2$, $d = 3$; find a_9 26 **32.** $a_1 = -12$, $d = -\dfrac{1}{2}$; find a_7 -15

33. $a_1 = 50$, $a_5 = 34$; find d -4 **34.** $a_1 = -3$, $a_7 = 0$; find d $\frac{1}{2}$

35. $a_1 = 12$, $a_n = -13$, $d = -5$; find n 6 **36.** $a_1 = 80$, $a_n = 152$, $d = 12$; find n 7

Find s_n and d for each arithmetic sequence.

37. $a_1 = 7$, $a_n = 21$, $n = 8$ $s_8 = 112$, $d = 2$ **38.** $a_1 = -12$, $a_n = -48$, $n = 7$ $s_7 = -210$, $d = -6$

39. $a_1 = \dfrac{3}{5}$, $a_n = 3$, $n = 7$ $s_7 = \frac{63}{5}$, $d = \frac{2}{5}$ **40.** $a_1 = -\dfrac{10}{3}$, $a_n = -6$, $n = 9$ $s_9 = -42$, $d = -\frac{1}{3}$

Write the first four terms of each arithmetic sequence. Then find a_{10} and s_{10}. **41.** 2, 6, 10, 14; $a_{10} = 38$, $s_{10} = 200$

41. $a_1 = 2$, $d = 4$ **42.** $a_1 = -8$, $d = -3$ **43.** $a_1 = \dfrac{5}{6}$, $d = \dfrac{2}{3}$ **44.** $a_1 = -80$, $d = 4$

Find the number of terms in each arithmetic sequence and find s_n.

45. 3, 8, 13, . . . , 53 $n = 11$, $s_n = 308$ **46.** $-16, -11, -6, -1, . . . , 24$ $n = 9$, $s_n = 36$

47. $\dfrac{6}{10}, \dfrac{9}{10}, \dfrac{12}{10}, \dfrac{15}{10}, . . . , \dfrac{36}{10}$ $n = 11$, $s_n = \frac{231}{10}$ **48.** $-5, 0, 5, 10, . . . , 85$ $n = 19$, $s_n = 760$

42. $-8, -11, -14, -17$; $a_{10} = -35$, $s_{10} = -215$ **43.** $\frac{5}{6}, \frac{9}{6}, \frac{13}{6}, \frac{17}{6}$; $a_{10} = \frac{41}{6}$, $s_{10} = \frac{115}{3}$ **44.** $-80, -76, -72, -68$; $a_{10} = -44$, $s_{10} = -620$

51. $20, -\frac{40}{3}, \frac{80}{9}, -\frac{160}{27}, \frac{320}{81}$ **52.** $-100, -20, -4, -\frac{4}{5}, -\frac{4}{25}$

2.3] *Determine the first five terms of each geometric sequence.*   **49.** 5, 10, 20, 40, 80   **50.** $-12, -6, -3, -\frac{3}{2}, -\frac{3}{4}$

49. $a_1 = 5, r = 2$ **50.** $a_1 = -12, r = \frac{1}{2}$ **51.** $a_1 = 20, r = -\frac{2}{3}$ **52.** $a_1 = -100, r = \frac{1}{5}$

Find the indicated term of the geometric sequence.   **53.** $\frac{4}{243}$ **54.** 6400   **55.** -2048 **56.** $\frac{160}{6561}$

53. $a_1 = 12, r = \frac{1}{3}$; find a_7 **54.** $a_1 = 25, r = 2$; find a_9 **55.** $a_1 = -8, r = -2$; find a_9 **56.** $a_1 = \frac{5}{12}, r = \frac{2}{3}$; find a_8

Find the sum.   **57.** 3060   **58.** $\frac{37,969}{1215}$ **59.** $-\frac{4305}{64}$ **60.** $\frac{172,539}{256}$

57. $a_1 = 12, r = 2$; find s_8 **58.** $a_1 = \frac{3}{5}, r = \frac{5}{3}$; find s_7 **59.** $a_1 = -84, r = -\frac{1}{4}$; find s_5 **60.** $a_1 = 9, r = \frac{3}{2}$; find s_9

Find the common ratio, r. Then write an expression for the general term, a_n, for each geometric sequence.   **64.** $r = \frac{2}{3}, a_n = \frac{9}{5}\left(\frac{2}{3}\right)^{n-1}$

61. 6, 12, 24, . . .   **62.** $8, \frac{8}{3}, \frac{8}{9}, \ldots$ **63.** $-4, -20, -100, \ldots$ **64.** $\frac{9}{5}, \frac{18}{15}, \frac{36}{45}, \ldots$

61. $r = 2, a_n = 6(2)^{n-1}$ **62.** $r = \frac{1}{3}, a_n = 8\left(\frac{1}{3}\right)^{n-1}$ **63.** $r = 5, a_n = -4(5)^{n-1}$

2.4] *Find the sum of the terms in each infinite geometric sequence.*

65. $7, \frac{7}{2}, \frac{7}{4}, \frac{7}{8}, \ldots$ 14   **66.** $-8, \frac{8}{3}, -\frac{8}{9}, \frac{8}{27}, \ldots$ -6 **67.** $-5, -\frac{10}{3}, -\frac{20}{9}, -\frac{40}{27}, \ldots$ **68.** $\frac{7}{2}, 1, \frac{2}{7}, \frac{4}{49}, \ldots$ $\frac{49}{10}$
-15

Find the sum of each infinite series.

69. $2 + 1 + \frac{1}{2} + \frac{1}{4} + \cdots$ 4   **70.** $7 + \frac{7}{3} + \frac{7}{9} + \frac{7}{27} + \ldots$ $\frac{21}{2}$

71. $-12 - \frac{24}{3} - \frac{48}{9} - \frac{96}{27} - \ldots$ -36 **72.** $5 - 1 + \frac{1}{5} - \frac{1}{25} + \ldots$ $\frac{25}{6}$

Write the repeating decimal as a ratio of integers.
73. $0.5252 \ldots$ $\frac{52}{99}$ **74.** $0.375375 \ldots$ $\frac{125}{333}$

2.5] *Use the binomial theorem to expand the expression.*
75. $(3x + y)^4$ $81x^4 + 108x^3y + 54x^2y^2 + 12xy^3 + y^4$ **76.** $(2x - 3y^2)^3$ $8x^3 - 36x^2y^2 + 54xy^4 - 27y^6$

Write the first four terms of the expansion.   **78.** $256a^{16} + 3072a^{14}b + 16,128a^{12}b^2 + 48,384a^{10}b^3$
77. $(x - 2y)^9$ $x^9 - 18x^8y + 144x^7y^2 - 672x^6y^3$ **78.** $(2a^2 + 3b)^8$

2.2]-   **79.** Find the sum of the integers between 100 and 200
2.4]   inclusive.   15,150

80. Professor Gayvert is offered a job with a starting
        salary of $30,000 with the agreement that his salary
        will increase by $1000 at the end of each of the next
        seven years.
        (a) Write a sequence showing his salary for the first
        five years.   30,000, 31,000, 32,000, 33,000, 34,000
        (b) What is the general term of this sequence?
        (c) If this process were continued, what would his
        salary be after nine years?   39,000

80. (b) $a_n = 30,000 + (n - 1)1000$

81. You begin with $100, double that to get $200, double
        that again to get $400, and so on. How much will you
        have after you perform this process 10 times?

82. If the inflation rate remains constant at 8% per year
        (each year the cost of living is 8% greater than the
        previous year), how much would a product that costs
        $200 now cost after 12 years? $503.63

83. On each swing, a pendulum travels 92% as far as on
        its previous swing. If the first swing is 8 feet long, find
        the distance traveled by the pendulum by the time it
        comes to rest.   100 ft

81. $102,400

Practice Test

1. Write the first five terms of the sequence with
    $a_n = \frac{n + 2}{n^2}$. $3, 1, \frac{5}{9}, \frac{3}{8}, \frac{7}{25}$

2. Find the first and third partial sums of $a_n = \frac{2n + 3}{n}$.

$s_1 = 5, s_3 = \frac{23}{2}$

3. Write out the series and find the sum of the series

$$\sum_{n=1}^{5} (2n^2 + 3)$$ $5 + 11 + 21 + 35 + 53; 125$

4. Write the general term for the arithmetic sequence

$$\frac{1}{3}, \frac{2}{3}, \frac{3}{3}, \frac{4}{3}, \ldots$$ $a_n = \frac{1}{3} + \frac{1}{3}(n-1) = \frac{1}{3}n$

5. Write the general term for the geometric sequence
5, 10, 20, 40, . . . $a_n = 5(2)^{n-1}$

Write the first four terms of the sequence.

6. $a_1 = 12, d = -3$ 12, 9, 6, 3

7. $a_1 = \dfrac{5}{8}, r = \dfrac{2}{3}$ $\frac{5}{8}, \frac{5}{12}, \frac{5}{18}, \frac{5}{27}$

13. $r = \frac{1}{2}, a_n = 12\left(\frac{1}{2}\right)^{n-1}$

8. Find a_8 when $a_1 = 100$ and $d = -12$.   16

9. Find s_8 for an arithmetic series when $a_1 = 3$, $a_8 = -11$. -32

10. Find the number of terms in the arithmetic sequence $-4, -16, -28, \ldots, -148$.   13

11. Find a_7 when $a_1 = 8$ and $r = \frac{2}{3}$. $\frac{512}{729}$

12. Find s_7 when $a_1 = \frac{3}{5}$ and $r = -5$. $\frac{39{,}063}{5}$

13. Find the common ratio and write an expression for the general term of the sequence $12, 6, 3, \dfrac{3}{2}, \ldots$

14. Find the sum of the infinite geometric series

$$3 + \frac{6}{3} + \frac{12}{9} + \frac{24}{27} + \cdots.$$ 9

15. Use the binomial theorem to expand $(x + 2y)^4$.
$x^4 + 8x^3y + 24x^2y^2 + 32xy^3 + 16y^4$

Cumulative Review Test

2. $x < -\frac{11}{2}$ or $x > \frac{17}{2}$ **6.** $\dfrac{4x^2 + 3x - 6}{(x+4)(x-3)(3x+4)}$ **12.** $x < -1$ or $\frac{3}{2} < x < 4$

1. Solve the equation $\dfrac{1}{2}x + \dfrac{1}{3}(x-2) = \dfrac{3}{4}(x-5)$. -37

2. Solve the inequality $|2x - 3| - 4 > 10$.

3. Graph $3x - 5y > 10$.

4. Solve the system of equations

$$5x - 2y = 8$$
$$x - y = 4$$ $(0, -4)$

5. Solve the formula $A = \dfrac{pt}{p + t}$ for p. $p = \dfrac{At}{t - A}$

6. Add $\dfrac{x}{x^2 + x - 12} + \dfrac{x + 2}{3x^2 + 16x + 16}$.

7. Simplify $\dfrac{\sqrt[3]{24x^6y^3}}{\sqrt[3]{2x^2y^5}}$. $\dfrac{x\sqrt[3]{12xy}}{y}$

8. Simplify $\sqrt{28} - 3\sqrt{7} + \sqrt{63}$. $2\sqrt{7}$

9. Solve $\sqrt{5x + 1} - \sqrt{2x - 2} = 2$. $3, \frac{11}{9}$

10. Simplify $\dfrac{5 - 2i}{3 + 4i}$. $\dfrac{7 - 26i}{25}$

11. Solve $-4x^2 - 2x + 8 = 0$. $\dfrac{-1 \pm \sqrt{33}}{4}$

12. Solve the inequality $\dfrac{(2x - 3)(x - 4)}{x + 1} < 0$.

13. Graph $4x^2 + 4y^2 = 36$.

14. Write the following equation in the form $y = a(x - h)^2 + k$ and then sketch the graph of the equation.

$$y = x^2 + 2x - 3$$ $y = (x + 1)^2 - 4$

15. Graph $y = 2^x$ and $y = \log_2 x$ on the same axes.

16. Solve the equation
$\log(4x - 1) + \log 3 = \log(8x + 13)$.   4

17. In an arithmetic series, if $a_1 = 8$ and $a_6 = 28$, find d.   4

18. In a geometric series, if $a_1 = 5$ and $r = 3$, find s_3.   65

19. A train and car leave from South Point heading for the same destination. The car averages 40 miles per hour and the train averages 60 miles per hour. If the train arrives 2 hours ahead of the car, find the distance between the two towns.   240 mi.

20. The Donovan's garden is in the shape of a rectangle. If the area of their garden is 300 square feet and the length of the garden is 20 feet greater than the width, find the dimensions of their garden. $w = 10$ ft, $l = 30$ ft

---

3.

13.
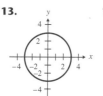

14.

15.
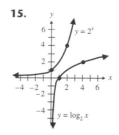

Appendices

Appendix A   Review of Fractions

To be successful in algebra, you must have a thorough understanding of fractions. This appendix gives a brief review of addition, subtraction, multiplication, and division of fractions. For a more complete explanation of fractions, see any arithmetic text.

The top number of a fraction is called the **numerator.** The bottom number is called the **denominator.**

$$\frac{3}{5} \quad \longleftarrow \quad \text{numerator}$$
$$\phantom{\frac{3}{5}} \quad \longleftarrow \quad \text{denominator}$$

Multiplication of Fractions

The product of two or more fractions is obtained by multiplying their numerators and then multiplying their denominators:

$$\frac{a}{b} \cdot \frac{c}{d} = \frac{a \cdot c}{b \cdot d}, \quad b \neq 0, \quad d \neq 0$$

EXAMPLE 1

(a) $\dfrac{3}{5} \cdot \dfrac{4}{7} = \dfrac{3 \cdot 4}{5 \cdot 7} = \dfrac{12}{35}$ **(b)** $\dfrac{5}{12} \cdot \dfrac{3}{10} = \dfrac{5 \cdot 3}{12 \cdot 10} = \dfrac{15}{120} = \dfrac{1}{8}$

In Example 1(b) the fraction $\frac{15}{120}$ was reduced to $\frac{1}{8}$. To avoid having to reduce fractions after multiplication, we often divide out common factors. When any numerator and any denominator in a **multiplication problem** have a common factor, divide both the numerator and the denominator by the common factor before multiplying (see Example 2).

EXAMPLE 2
$$\frac{25}{36} \cdot \frac{8}{15} = \frac{\overset{5}{\cancel{25}}}{36} \cdot \frac{8}{\underset{3}{\cancel{15}}}$$
Divide both 25 and 15 by 5.

$$= \frac{\overset{5}{\cancel{25}}}{\underset{9}{\cancel{36}}} \cdot \frac{\overset{2}{\cancel{8}}}{\underset{3}{\cancel{15}}}$$
Divide both 8 and 36 by 4.

$$= \frac{5 \cdot 2}{9 \cdot 3} = \frac{10}{27}$$

Division of Fractions

To divide fractions, invert the divisor and proceed as in multiplication.

$$\frac{a}{b} \div \frac{c}{d} = \frac{a}{b} \cdot \frac{d}{c} = \frac{a \cdot d}{b \cdot c}, \quad b \neq 0, \quad c \neq 0, \quad d \neq 0$$

EXAMPLE 3   **(a)** $\dfrac{3}{8} \div \dfrac{5}{9} = \dfrac{3}{8} \cdot \dfrac{9}{5} = \dfrac{27}{40}$ **(b)** $\dfrac{7}{15} \div \dfrac{3}{5} = \dfrac{7}{\underset{3}{\cancel{15}}} \cdot \dfrac{\overset{1}{\cancel{5}}}{3} = \dfrac{7}{9}$

EXAMPLE 4

(a) $4 \div \dfrac{3}{5} = \dfrac{4}{1} \cdot \dfrac{5}{3} = \dfrac{20}{3}$ **(b)** $\dfrac{9}{16} \div 6 = \dfrac{9}{16} \div \dfrac{6}{1} = \dfrac{\overset{3}{\cancel{9}}}{16} \cdot \dfrac{1}{\underset{2}{\cancel{6}}} = \dfrac{3}{32}$

Addition and Subtraction of Fractions

Only fractions with the same denominators can be added or subtracted. To add (or subtract) fractions with the same denominators, add (or subtract) the numerators while maintaining the common denominator.

$$\frac{a}{c} + \frac{b}{c} = \frac{a + b}{c}, \quad \frac{a}{c} - \frac{b}{c} = \frac{a - b}{c}, \quad c \neq 0$$

EXAMPLE 5

(a) $\dfrac{3}{7} + \dfrac{2}{7} = \dfrac{3 + 2}{7} = \dfrac{5}{7}$ **(b)** $\dfrac{8}{12} - \dfrac{5}{12} = \dfrac{8 - 5}{12} = \dfrac{3}{12} = \dfrac{1}{4}$

To add or subtract fractions with different denominators, rewrite the fractions so that they have the same, or a common, denominator; then proceed as above.

EXAMPLE 6   **(a)** $\dfrac{3}{4} + \dfrac{5}{6} = \left(\dfrac{3}{4} \cdot \dfrac{3}{3}\right) + \left(\dfrac{5}{6} \cdot \dfrac{2}{2}\right) = \dfrac{9}{12} + \dfrac{10}{12} = \dfrac{19}{12}$

(b) $\dfrac{5}{12} - \dfrac{7}{18} = \left(\dfrac{5}{12} \cdot \dfrac{3}{3}\right) - \left(\dfrac{7}{18} \cdot \dfrac{2}{2}\right) = \dfrac{15}{36} - \dfrac{14}{36} = \dfrac{1}{36}$

EXAMPLE 7 $4 + \dfrac{5}{7} = \dfrac{4}{1} + \dfrac{5}{7} = \dfrac{4}{1} \cdot \dfrac{7}{7} + \dfrac{5}{7} = \dfrac{28}{7} + \dfrac{5}{7} = \dfrac{33}{7}$

COMMON STUDENT ERROR   Dividing out common factors can be performed *only* *when multiplying* fractions; it cannot be performed when adding or subtracting fractions.

Correct

$\dfrac{\overset{1}{\cancel{3}}}{5} \cdot \dfrac{4}{\underset{3}{\cancel{9}}} = \dfrac{1 \cdot 4}{5 \cdot 3} = \dfrac{4}{15}$

$\dfrac{\overset{1}{\cancel{4}} \cdot 7}{\underset{2}{\cancel{8}}} = \dfrac{1 \cdot 7}{2} = \dfrac{7}{2}$

Incorrect

$\dfrac{\overset{1}{\cancel{3}}}{5} + \dfrac{1}{\underset{1}{\cancel{3}}}$ cannot divide out common factors when adding or subtracting

$\dfrac{\overset{1}{\cancel{4}} + 7}{\underset{2}{\cancel{8}}}$

Appendix B Geometric Formulas

AREAS AND PERIMETERS			
Figure	**Sketch**	**Area**	**Perimeter**
Square	s	$A = s^2$	$P = 4s$
Rectangle	w, l	$A = lw$	$P = 2l + 2w$
Parallelogram	h, w, l	$A = lh$	$P = 2l + 2w$
Trapezoid	b_1, s_1, h, s_2, b_2	$A = \dfrac{1}{2}h(b_1 + b_2)$	$P = s_1 + s_2 + b_1 + b_2$
Triangle	s_1, h, s_2, b	$A = \dfrac{1}{2}bh$	$P = s_1 + s_2 + b$

AREA AND CIRCUMFERENCE OF CIRCLE

Circle $A = \pi r^2$ $C = 2\pi r$

VOLUME AND SURFACE AREA OF THREE-DIMENSIONAL FIGURES

Figure	Sketch	Volume	Surface area
Rectangular solid		$V = lwh$	$s = 2lh + 2wh + 2wl$
Right circular cylinder		$V = \pi r^2 h$	$s = 2\pi rh + 2\pi r^2$
Sphere		$V = \dfrac{4}{3}\pi r^3$	$s = 4\pi r^2$
Right circular cone		$V = \dfrac{1}{3}\pi r^2 h$	$s = \pi r\sqrt{r^2 + h^2}$
Square or rectangular pyramid		$V = \dfrac{1}{3}lwh$	

Answers

Chapter 1

EXERCISE SET 1.2

1. $\in$ **3.** $\notin$ **5.** $\notin$ **7.** $\notin$ **9.** $\subseteq$ **11.** $\not\subseteq$ **13.** $\not\subseteq$ **15.** $\not\subseteq$ **17.** $\not\subseteq$ **19.** $\subseteq$ **21.** $A = \{5\}$
23. $C = \{6, 8\}$ **25.** $E = \{0, 1, 2, 3, 4\}$ **27.** $H = \{0, 5, 10, 15, 20, \ldots\}$ **29.** $J = \{-5\}$ **31.** $\subseteq$ **33.** $\subseteq$
35. $\not\subseteq$ **37.** $\subseteq$ **39.** $\not\subseteq$ **41.** $\not\subseteq$ **43.** True   **45.** False   **47.** False   **49.** True   **51.** True   **53. (a)** 4   **(b)** 4, 0
(c) $-6, 4, 0$ **(d)** $-6, 4, \dfrac{1}{2}, \dfrac{5}{9}, 0, -1.23, \dfrac{99}{100}$ **(e)** $\sqrt{7}, \sqrt{5}$ **(f)** $-6, 4, \dfrac{1}{2}, \dfrac{5}{9}, 0, \sqrt{7}, \sqrt{5}, -1.23, \dfrac{99}{100}$
55. $A \cup B = \{1, 2, 3, 4\}$, $A \cap B = \{2, 3\}$ **57.** $A \cup B = \{-1, -2, -4, -5, -6\}$, $A \cap B = \{-2, -4\}$
59. $A \cup B = \{0, 1, 2, 3\}$, $A \cap B = \{\ \}$ **61.** $A \cup B = \{0, 1, 2, 3, 4, 5, 6, 7, 8\}$, $A \cap B = \{\ \}$
63. $A \cup B = \{0.1, 0.2, 0.3, 0.4, \ldots\}$, $A \cap B = \{0.2, 0.3\}$ **65.** The set of natural numbers.   **67.** The set of even natural numbers greater than or equal to 8 and less than or equal to 30.   **69.** The set of odd integers.   **71. (a)** Set A is the set of all x such that x is a natural number less than 8.   **(b)** $A = \{1, 2, 3, 4, 5, 6, 7\}$ **73.** $>$ **75.** $<$ **77.** $<$ **79.** $>$

81. $<$ **83.** $>$ **85.** $>$ **87.** $<$ **89.** $>$ **91.** $>$ **93.** [number line with open circle at 3]   **95.** [number line from 4]   **97.** [number line from -4 to $\frac{3}{7}$]

99. [number line with dots at 3, 4, 5]   **101.** [number line at 0 1 2 3 4 5]   **103.** $\{x \mid x > -2\}$ **105.** $\{x \mid x \geq -3 \text{ and } x \in I\}$

107. $\{x \mid -2 < x \leq 4.6\}$ **109.** $\left\{x \mid x \leq -\dfrac{20}{3}\right\}$ **111.** $\{x \mid -1 \leq x \leq 3 \text{ and } x \in I\}$

113. (a) The digit or group of digits repeats.   **(b)** $0.\overline{6}$ **(c)** $0.\overline{5}$ **(d)** $1.1\overline{6}$ **115.** A letter used to represent a number.
117. A set that contains no elements

GROUP ACTIVITY/CHALLENGE PROBLEMS

1. (a) $0.\overline{1}, 0.\overline{2}, 0.\overline{3}$ **(b)** $\dfrac{4}{9}, \dfrac{5}{9}, \dfrac{6}{9}$ or $\dfrac{2}{3}$ **(c)** 1

EXERCISE SET 1.3

1. 3   **3.** 6   **5.** $\dfrac{1}{2}$ **7.** 0   **9.** -7 **11.** $-\dfrac{5}{9}$ **13.** $=$ **15.** $<$ **17.** $=$ **19.** $>$ **21.** $<$ **23.** $<$

25. $-1, 2, |3|, |-5|, 6$ **27.** $-3, |0|, |-5|, |7|, |-12|$ **29.** $|-9|, 12, 24, |36|, |-45|$

31. $-|2.9|, -2.4, -2.1, -2, |-2.8|$ **33.** $-2, \dfrac{1}{3}, \left|-\dfrac{1}{2}\right|, \left|\dfrac{3}{5}\right|, \left|-\dfrac{3}{4}\right|$ **35.** 1   **37.** 8   **39.** -17 **41.** -18

43. -9.42 **45.** $\dfrac{1}{30}$ **47.** $-\dfrac{41}{12}$ **49.** 11.4 **51.** $-\dfrac{21}{2}$ **53.** 1 **55.** 7 **57.** 0 **59.** -8 **61.** $\dfrac{37}{60}$ **63.** -48

65. $\dfrac{5}{4}$ **67.** 6 **69.** 235.9192 **71.** 8 **73.** 1 **75.** 8 **77.** $\dfrac{4}{3}$ **79.** -2 **81.** 10 **83.** -16 **85.** 5.9

87. 153.216 **89.** 11 **91.** $-\dfrac{9}{5}$ **93.** $\dfrac{81}{16}$ **95.** 1 **97.** $-\dfrac{17}{45}$ **99.** 121 **101.** Commutative property of addition

103. Distributive property   **105.** Associative property of addition   **107.** Identity property of multiplication
109. Commutative property of addition   **111.** Commutative property of addition   **113.** Double negative property
115. Identity property of addition   **117.** Inverse property of addition   **119.** Inverse property of multiplication
121. Identity property of multiplication   **123.** Multiplicative property of zero   **125.** $3 + x$ **127.** $3x + 3y + 12$

129. $x \cdot (3 \cdot 4)$ **131.** $(5x + 2y) + 3x$ **133.** 0 **135.** $-4, \dfrac{1}{4}$ **137.** $\dfrac{2}{3}, -\dfrac{3}{2}$ **139.** $58°\text{F}$ **141.** $\$64.74$

143. (a) owes \$83,000 (or $-\$83,000$)   **(b)** receives \$1,050,000   **145.** all real numbers, $\mathbb{R}$ **147.** $a \le 0$ **149.** $5, -5$
153. (a) $a + b = b + a$ **155. (a)** $a(b + c) = ab + ac$

CUMULATIVE REVIEW EXERCISES

159. True   **160.** $\subseteq$ **161. (a)** $3, 4, -2, 0$ **(b)** $3, 4, -2, \dfrac{5}{6}, 0$ **(c)** $\sqrt{3}$ **(d)** $3, 4, -2, \dfrac{5}{6}, \sqrt{3}, 0$

162. $A \cup B = \{1, 4, 7, 9, 12, 15\}$; $A \cap B = \{4, 7\}$ **163.**

GROUP ACTIVITY/CHALLENGE PROBLEMS

1. -50 **3.** -1

EXERCISE SET 1.4

1. 16 **3.** 16 **5.** -16 **7.** $-\dfrac{81}{625}$ **9.** $-\dfrac{16}{81}$ **11.** -6 **13.** -5 **15.** 0.1 **17.** 0.013 **19.** 0.053

21. -557.060 **23.** 1.710 **25.** 4.160 **27.** -0.723 **29. (a)** 9 **(b)** -9 **31. (a)** 1 **(b)** -1

33. (a) 1 **(b)** -1 **35. (a)** $\dfrac{1}{9}$ **(b)** $-\dfrac{1}{9}$ **37. (a)** 27 **(b)** -27 **39. (a)** -27 **(b)** 27 **41. (a)** -8 **(b)** 8

43. (a) $\dfrac{8}{27}$ **(b)** $-\dfrac{8}{27}$ **45.** 21 **47.** -19 **49.** -30 **51.** 26.874 **53.** $\dfrac{19}{64}$ **55.** 26 **57.** 14 **59.** $\dfrac{27}{4}$

61. $\dfrac{81}{40}$ **63.** 23 **65.** 294 **67.** 64 **69.** 36 **71.** $\dfrac{9}{5}$ **73.** $\dfrac{27}{5}$ **75.** undefined   **77.** -4 **79.** 0 **81.** $-\dfrac{13}{2}$

83. $\dfrac{242}{5}$ **85.** $-\dfrac{3}{4}$ **87.** undefined   **89.** -7 **91.** 44 **93.** $\dfrac{147}{16}$ **95.** 36 **97.** 21 **99.** 33 **101.** -1

103. 2 **105.** $\dfrac{3}{2}$ **107.** $4.2 \, y^x \, 3 \, - \, 3.6 \, x^2 \, = 61.128$ **109.** $4.4 \, - \, 3.5 \, = \div \, (3.7 + 6.2) \, = 0.090909091$

111. $3.6 \times (9.3 - 1.3 \, y^x \, 5) = 20.113452$ **113.** $(3 - 5 \div 8) \, x^2 \, - 2 \div 3 = 4.97395833$ **115.** $(3x + 6)^2, 225$

117. $6(3x + 6) - 9, 81$ **119.** $\left(\dfrac{x + 3}{2y}\right)^2 - 3, 1$ **121.** A positive number raised to an odd power is a positive number.
123. parentheses, exponents or roots, multiplication or division left to right, addition or subtraction left to right.

CUMULATIVE REVIEW EXERCISES

127. (a) $A \cap B = \{b, c, f\}$ **(b)** $A \cup B = \{a, b, c, d, f, g, h\}$ **128.** All real numbers, $\mathbb{R}$ **129.** $a \ge 0$ **130.** $4, -4$
131. $-|6|, -4, -|-2|, 0, |-5|$ **132.** Associative property of addition

GROUP ACTIVITY/CHALLENGE PROBLEMS

1. $\frac{883}{48}$, or 18.396   **2.** $\frac{131{,}072}{5}$ **3.** $-\frac{15}{19}$

GRAPHING CALCULATOR CORNER, PAGE 39

1. (a) -8 **(b)** -1.16 **(c)** -3.24 **3. (a)** 22 **(b)** -20.6 **(c)** -0.33

REVIEW EXERCISES

1. $\{3, 4, 5, 6\}$ **2.** $\{0, 3, 6, 9, \ldots\}$ **3.** $\in$ **4.** $\notin$ **5.** $\notin$ **6.** $\in$ **7.** $\subseteq$ **8.** $\not\subseteq$ **9.** $\not\subseteq$ **10.** $\subseteq$ **11.** $\subseteq$
12. $\subseteq$ **13.** $\subseteq$ **14.** $\not\subseteq$ **15.** $4, 6$ **16.** $4, 6, 0$ **17.** $-3, 4, 6, 0$ **18.** $-3, 4, 6, \frac{1}{2}, 0, \frac{15}{27}, -\frac{1}{5}, 1.47$ **19.** $\sqrt{5}, \sqrt{3}$
20. $-3, 4, 6, \frac{1}{2}, \sqrt{5}, \sqrt{3}, 0, \frac{15}{27}, -\frac{1}{5}, 1.47$ **21.** False   **22.** True   **23.** True   **24.** True   **25.** $A \cup B = \{1, 2, 3, 4, 5\}$,
$A \cap B = \{2, 3, 4, 5\}$ **26.** $A \cup B = \{2, 3, 4, 5, 6, 7, 8, 9\}$, $A \cap B = \varnothing$ **27.** $A \cup B = \{1, 2, 3, 4, \ldots\}$, $A \cap B = \{2, 4, 6, \ldots\}$
28. $A \cup B = \{3, 4, 5, 6, 9, 10, 11, 12\}$, $A \cap B = \{9, 10\}$ **29.**   **30.**
31.   **32.**   **33.** $<$ **34.** $<$ **35.** $<$ **36.** $>$ **37.** $=$ **38.** $=$ **39.** $<$
40. $<$ **41.** $>$ **42.** $>$ **43.** $>$ **44.** $>$ **45.** $-5, -2, 4, |7|$ **46.** $0, \frac{3}{5}, 2.3, |-3|$ **47.** $-2, 3, |-5|, |-7|$
48. $-4, -|3|, -2.1, -2$ **49.** $-4, -|-3|, 5, 6$ **50.** $-3, 0, |1.6|, |-2.3|$ **51.** Distributive property
52. Commutative property of multiplication   **53.** Associative property of addition   **54.** Identity property of addition
55. Associative property of multiplication   **56.** Identity property of multiplication   **57.** Double negative property
58. Multiplication property of zero   **59.** Identity property of multiplication   **60.** Inverse property of addition
61. Inverse property of multiplication   **62.** Identity property of multiplication   **63.** $3 + x$ **64.** $3x + 15$
65. $x + [6 + (-4)]$ **66.** $x \cdot 3$ **67.** $9 \cdot (x \cdot y)$ **68.** $4x - 4y + 20$ **69.** a **70.** 0 **71.** $\frac{22}{5}$ **72.** -1 **73.** 8
74. -1 **75.** -1 **76.** 21 **77.** 19 **78.** -47 **79.** 15 **80.** 31 **81.** 6 **82.** 512 **83.** $-\frac{8}{5}$ **84.** $\frac{8}{3}$
85. 5 **86.** undefined   **87.** $\frac{34}{115}$ **88.** $-\frac{134}{45}$ **89.** 15 **90.** $\frac{26}{3}$ **91.** 7 **92.** 204
93. $1.4 \boxed{-} 6 \div 2.4 \boxed{=} \div 3.6 \boxed{=} -0.305555556$ **94.** $5.3 \, y^x 3 \boxed{-} 4.6 \, x^2 \boxed{=} \div 1.7 \, y^x 5 \boxed{=} 8.99506077$
95. $(3 \div 5) \, x^2 \boxed{-} 4 \, x^2 \boxed{=} -15.64$ **96.** $4.2 \times (1.3 \boxed{-} 4.6 \, y^x 3) \div (5.2 \boxed{-} 3.6) \boxed{=} -252.0945$

PRACTICE TEST

1. $A = \{6, 7, 8, 9, \ldots\}$ **2.** $\not\subseteq$ **3.** $\subseteq$ **4.** True   **5.** False   **6.** True   **7.** $-\frac{3}{5}, 2, -4, 0, \frac{19}{12}, 2.57, -1.92$
8. $-\frac{3}{5}, 2, -4, 0, \frac{19}{12}, 2.57, \sqrt{8}, \sqrt{2}, -1.92$ **9.** $A \cup B = \{5, 7, 8, 9\,10, 11, 14\}$, $A \cap B = \{8, 10\}$
10. $A \cup B = \{1, 3, 5, 7, \ldots\}$, $A \cap B = \{3, 5, 7, 9, 11\}$ **11.**   **12.**   **13.** $<$ **14.** $>$
15. $-|4|, -2, |3|, 6$ **16.** Associative property of addition   **17.** Commutative property of addition
18. Identity property of addition   **19.** -5 **20.** 23 **21.** undefined   **22.** $-\frac{37}{22}$ **23.** $\frac{64}{5}$ **24.** 17 **25.** 39

Chapter 2

EXERCISE SET 2.1

1. Symmetric property   **3.** Transitive property   **5.** Reflexive property   **7.** Addition property
9. Multiplication property   **11.** Multiplication property   **13.** First   **15.** Second   **17.** Fifth   **19.** Zero
21. Thirteenth   **23.** Twelfth   **25.** $15x - 5$ **27.** $5x^2 - x - 5$ **29.** $12.7x^2 + 3.6x$ **31.** Cannot be simplified
33. $4xy + y^2 - 2$ **35.** $-3.56x - 42.76$ **37.** $\frac{8}{3}x + \frac{13}{2}$ **39.** $-17x - 4$ **41.** $6x - 3y$ **43.** $-9b + 93$ **45.** 1

47. $-\dfrac{15}{4}$ **49.** -32 **51.** -5.2 **53.** 3.67 **55.** 3 **57.** $\dfrac{11}{7}$ **59.** -3 **61.** 20 **63.** No solution   **65.** $\dfrac{17}{3}$

67. -12.5 **69.** All real numbers   **71.** No solution   **73.** $\dfrac{27}{11}$ **75.** 0 **77.** No solution   **79.** $-\dfrac{119}{25}$

81. All real numbers   **83.** 4 **85.** -27 **87.** 0 **89.** -5 **91.** 0 **93.** 1 **95.** 0 **97. (b)** $\dfrac{14}{5}$

99. An equation that is true for all real numbers   **101.** An equation that has no solution
103. $2x = 8$, $x + 3 = 7$, $x - 2 = 2$; answers will vary

CUMULATIVE REVIEW EXERCISES

105. $|a| = \begin{cases} a, & a \geq 0 \\ -a, & a < 0 \end{cases}$ **106.** -9 **107.** $-\dfrac{27}{64}$ **108.** -4

GROUP ACTIVITY/CHALLENGE PROBLEMS

1. $\dfrac{306}{157}$ **2.** $-\dfrac{115}{31}$ **3.** $\dfrac{1524}{131}$

EXERCISE SET 2.2

1. 42 **3.** 250 **5.** 625 **7.** 3.33 **9.** 66.67 **11.** 4 **13.** 78.44 **15.** $\dfrac{0.1}{\sqrt{0.025}} \approx 0.63$ **17.** 119.10

19. $y = -3x + 5$ **21.** $y = 2x + 5$ **23.** $y = \dfrac{5x + 4}{3}$ **25.** $y = \dfrac{-x + 12}{4}$ **27.** $y = x + 2$ **29.** $y = \dfrac{3x - 17}{6}$

31. $t = \dfrac{d}{r}$ **33.** $b = \dfrac{2A}{h}$ **35.** $w = \dfrac{P - 2l}{2}$ **37.** $h = \dfrac{V}{lw}$ **39.** $l = \dfrac{3V}{wh}$ **41.** $\mu = x - z\sigma$ **43.** $m = \dfrac{y - b}{x}$

45. $z = \dfrac{kx}{y}$ **47.** $h = \dfrac{2A}{b_1 + b_2}$ **49.** $m = \dfrac{y - y_1}{x - x_1}$ **51.** $n = \dfrac{2S}{f + l}$ **53.** $F = \dfrac{9}{5}C + 32$ **55.** $m_1 = \dfrac{Fd^2}{km_2}$

57. $y = \dfrac{3x - 4}{x - 2}$ **59.** $x = \dfrac{y}{y - 1}$ **61.** $x = \dfrac{2y + 4}{y + 1}$ **63.** $z = \dfrac{-6x}{2x + 3}$ **65.** $r = \dfrac{5s}{6s - 2}$ **67.** $P = \dfrac{A}{1 + rt}$

69. $a = \dfrac{v_2 - v_1}{t_2 - t_1}$ **71.** $r = \dfrac{O + R}{V + D}$ **73.** $S_n = \dfrac{a_1 - a_1 r^n}{1 - r}$ **75.** $L = \dfrac{A - 2HW}{2W + 2H}$ **77.** $\$6341.21$

79. (b) $A = \dfrac{0.5cV^2}{\mu d}$ **81. (a)** $k = 1.15m$ **(b)** The quotient of 6076 to 5280 is about 1.15.

CUMULATIVE REVIEW EXERCISES

82. -22 **83.** $\dfrac{3}{2}$ **84.** 15 **85.** 78

GROUP ACTIVITY/CHALLENGE PROBLEMS

2. (a) $s = \dfrac{rt^2}{u}$ **(b)** $u = \dfrac{rt^2}{s}$

EXERCISE SET 2.3

1. $x + (x + 15) = 41$; 13, 28   **3.** $x + (x + 2) = 78$; 38, 40   **5.** $2x = \dfrac{1}{2}(5x) - 3$; 6, 30

7. $x + (x + 20) + 2x = 180$; $40°, 60°, 80°$ **9.** $42x = 360$; 8.57 yr   **11.** $300 + 32x = 580$; 8.75 yr
13. $1.50x = 40$, more than 26   **15.** $35 + 0.20x = 80$; 225 mi   **17.** $240 + 0.12x = 540$; $\$2500$
19. $17x + 16,000 - x = 1,179,200$; 1970: 72,700 metric tons; 1990: 1,251,900 metric tons

21. $3x + 6000 - x = 18,000$; steel, $6000 \dfrac{\text{lb}}{\text{in}^2}$; bone, $24,000 \dfrac{\text{lb}}{\text{in}^2}$ **23.** $0.10(74)x = 25$; 3.38 mo.

25. $x + 0.08x = 12{,}000$; $11,111.11 **27.** $25 + 10x = 18.50x$; 3 or more hours
29. $x + 0.07x + 0.15x = 10.25$; $8.40 **31. (a)** $563.50x = 538.30x + 0.02(70{,}000) + 200$; 63.49 mo or 5.29 yr
(b) First National **33. (a)** $510x = 420.50x + 2500$; about 28 mo or 2.33 yr **(b)** yes

35. $2\left(\dfrac{x}{2} + 1\right) + 2x = 20$; $l = 6$m, $w = 4$m **37.** $4x + 2(x + 3) = 30$; $w = 4$ ft, $h = 7$ ft

39. $x - 0.10x - 5 = 49$; $60 **41.** $x - \dfrac{1}{4}x - 10 = 290$; $400 **43.** $x + 2x + (3x - 4) = 512$; 86 acres, 172 acres,

254 acres **45.** $x + \dfrac{13}{6}x + \dfrac{4}{3}x + \dfrac{31}{2}x = 240$; Exported, 12 million; Incinerated, 26 million; Recycled, 16 million; Landfill, 186 million **47.** $0.07(375 - x) = 17.50$ or $x + (375 - x) + 0.07(375 - x) = 392.50$; clothing, $125

49. $\dfrac{5}{8}x + \left(\dfrac{3}{8}x + 0.15x\right) = 184.60$; Bulinas, $100.33; Williams, $84.27

51. $x + (x + 4) + (x + 5) + (x + 6) + (2x - 2) + 2(x + 4) = 133$; Stanford, 14; Purdue, 18; Naval Post. Grad., 19;

MIT, 20; Air Force, 26; Naval Academy, 36 **53. (a)** $\dfrac{70 + 83 + 97 + 84 + 74 + x + x}{7} = 80$; 76

(b) No, he would need a score of 111

CUMULATIVE REVIEW EXERCISES

56. $\dfrac{2}{19}$ **57.** $\dfrac{12}{5}$ **58.** $\dfrac{40}{63}$ **59.** $y = \dfrac{4x - 9}{6}$ or $y = \dfrac{2}{3}x - \dfrac{3}{2}$

GROUP ACTIVITY/CHALLENGE PROBLEMS

1. white, 10; brown, 15; yellow, 20; red, 27; blue, 70; purple, 75; green, 100
3. $3(28) + 0.15x + 0.04[(3)(28) + 0.15x] = 121.68$; 220 mi **5. (b)** 0.095

EXERCISE SET 2.4

1. 16.5 bottles per min **3. (a)** 1.91 mph **(b)** 1.78 mph **5.** 16.39 min **7.** 2.17 hr **9. (a)** 26.07 days
(b) 3,345,991.5 mi **11. (a)** 2.29 min **(b)** 1.83 min **(c)** 1.02 min **(d)** 0.81 min **13.** $550t + 650t = 3000$; 2.5 hr
15. $330 = 60(3) + 3r$; 50 mph **17.** $0.1t + 0.05t = 200$; 1333.33 days or 3.65 yr
19. $3(x + 20) = 4.2x$; Freight train, 50 mph; Passenger train, 70 mph **21.** $18 = 2(4x) - 2(x)$; **(a)** 12 mph
(b) 24 mi **23.** $2.6t = 1.2(16 - t)$ **(a)** 5.05 hr **(b)** 26.26 mi **25.** $600(x + 2) + 400x = 15{,}000$; 13.8 hr
27. $0.09x + 0.10(11{,}000 - x) = 1050$; $5000 at 9%, $6000 at 10%
29. (a) $80.75x + 17(4x) = 10{,}000$; Mobil: 67 shares; Limited: 268 shares **(b)** $33.75
31. $0.10x + 0.25(33 - x) = 4.50$; 25 dimes, 8 quarters **33.** $6.20x + 5.80(18) = 6.10(x + 18)$; 54 lb
35. $0.10(40) + 1.00x = 0.25(x + 40)$; 8 oz
37. $0.20x + 0.50(12 - x) = 0.30(12)$; 8 L of 20% solution, 4 L of 50% solution **39.** $12x + 0.76(16) = 0.82(28)$; 90%
41. $28{,}200 - x = 32{,}450 - (6400 - x)$; $1075 for Mr. Hall, $5325 for Mrs. Hall
43. $8.2 = 2x + 2(x + 0.4)$; 1.85 mph, 2.25 mph **45. (a)** $34(3x) + 23(x) = 6000$; 48 shares Walmart, 144 shares Apple
(b) no money left over **47.** $10x + 20x = 15{,}000$; 500 min or $8\frac{1}{3}$ hr **49.** $0.09(2500) + x(1500) = 315$; 6%
51. $300t = 220(11.2 - t)$; **(a)** 4.74 hr **(b)** 1422 mi **53.** $6x + 6.50(18 - x) = 114$; 6 hr at $6, 12 hr at $6.50
55. $4.2t + 7.8(2 - t) = 13.8$; 0.5 hr in 2nd, 1.5 hr in 3rd **57.** $0.03(16) + 0.07(64) = x(80)$; 6.2%
59. $0.80x + 0.00(128 - x) = 0.06(128)$; 9.6 oz 80% solution, 118.4 oz water
61. $0.05(400) + 0.015x = 0.02(x + 400)$; 2400 qt **63.** $70x = 50x + 200$; 10 min

CUMULATIVE REVIEW EXERCISES

69. -5.7 **70.** $\frac{126}{35} = \frac{18}{5}$ **71.** $y = \dfrac{x - 42}{30}$ **72.** 140 mi

GROUP ACTIVITY/CHALLENGE PROBLEMS

1. (a) 440 ft **(b)** 1.45 sec **3.** ≈ 149 mi **5.** 300,000,000 ft **7.** 6 qt

EXERCISE SET 2.5

1. (a) -3 **(b)** $(-\infty, -3)$ **(c)** $\{x|x < -3\}$ **3. (a)** 5.2 **(b)** $[5.2, \infty)$ **(c)** $\{x|x \geq 5.2\}$

5. (a) $2 \quad \frac{12}{5}$ **(b)** $[2, \frac{12}{5})$ **(c)** $\{x|2 \leq x < \frac{12}{5}\}$ **7. (a)** $-6 \quad -4$ **(b)** $(-6, -4]$ **(c)** $\{x|-6 < x \leq -4\}$

9. 5 **11.** 7 **13.** 3.2 **15.** 0 **17.** 0 **19.** $\frac{54}{5}$

21. $(-\infty, 9)$ **23.** $\left(-\infty, \frac{4}{3}\right)$ **25.** $\varnothing$ **27.** $(-\infty, \infty)$ **29.** $(1, 6)$ **31.** $\left(-\frac{3}{5}, \frac{8}{5}\right]$ **33.** $\left[\frac{7}{2}, 5\right)$ **35.** $(2, 5.8]$

37. $\left(-\frac{14}{3}, 6\right)$ **39.** $\left\{x\left|\frac{11}{4} < x \leq \frac{27}{4}\right.\right\}$ **41.** $\{x|0 < x \leq 1\}$ **43.** $\{x|3 < x \leq 33\}$ **45.** $\left\{x\left|-\frac{56}{3} \leq x < \frac{14}{3}\right.\right\}$

47. $\{x|2 < x < 4\}$ **49.** $\varnothing$ **51.** $\{x|-3 < x < 1\}$ **53.** $\{x|x \leq 2 \text{ or } x > 8\}$ **55.** $(-\infty, 4)$ **57.** $[0, 2]$

59. $\left(\frac{13}{5}, \infty\right)$ **61.** $(-\infty, 0) \cup (6, \infty)$ **63.** $80x \leq 900$; 11 boxes **65.** $4.25 + 0.45x \leq 9.50$; 14 minutes

67. $10{,}025 + 1.09x < 6.42x$; 1881 books **69.** $150 + 0.228x < 0.32x$; more than 1630 pieces

71. $8(120) + 6.50x \leq 2000$; 160 hr over 8 weeks or 20 hr/wk **73.** $\dfrac{65 + 72 + 90 + 47 + 62 + x}{6} \geq 60$; 24

75. $\dfrac{2.7 + 3.42 + x}{3} < 3.2$; any value less than 3.48 **77. (a)** \$5050.50 **(b)** \$18,771.87

79. Two inequalities joined by the word *and* or *or* **81.** No real number is both greater than 4 and less than 2

CUMULATIVE REVIEW EXERCISES

83. (a) $\{1, 3, 4, 5, 6, 7, 9\}$ **(b)** $\{1, 4, 6\}$ **84. (a)** 4 **(b)** $0, 4$ **(c)** $-3, 4, \frac{5}{2}, 0, -\frac{29}{80}$ **(d)** $-3, 4, \frac{5}{2}, \sqrt{7}, 0, -\frac{29}{80}$

85. Associative property of addition **86.** Commutative property of addition **87.** $V = \dfrac{R - L + Dr}{r}$

GROUP ACTIVITY/CHALLENGE PROBLEMS

1. $84 \leq x \leq 100$ **3. (b)** $(5, 10)$ **5. (b)** $(-3, \infty)$

EXERCISE SET 2.6

1. $\{-5, 5\}$ **3.** $\{-12, 12\}$ **5.** $\varnothing$ **7.** $\{-12, 2\}$ **9.** $\{-16, 4\}$ **11.** $\left\{\frac{3}{2}, \frac{11}{6}\right\}$ **13.** $\left\{-\frac{5}{2}, \frac{13}{2}\right\}$ **15.** $\left\{-\frac{59}{3}, \frac{49}{3}\right\}$

17. $\left\{-1, \frac{11}{5}\right\}$ **19.** $\{y|-5 \leq y \leq 5\}$ **21.** $\{x|-2 \leq x \leq 16\}$ **23.** $\left\{z\left|0 \leq z \leq \frac{10}{3}\right.\right\}$ **25.** $\{x|-9 \leq x \leq 6\}$

27. $\{x|-1.9 \leq x \leq 2.7\}$ **29.** $\varnothing$ **31.** $\left\{x\left|-4 < x < \frac{52}{3}\right.\right\}$ **33.** $\frac{1}{4}$ **35.** $\{x|x < -3 \text{ or } x > 3\}$

37. $\{x|x < -9 \text{ or } x > 1\}$ **39.** $\left\{x\left|x < -\frac{5}{3} \text{ or } x > 1\right.\right\}$ **41.** $\{z|z < -6 \text{ or } z > 0\}$ **43.** $\{x|x < 2 \text{ or } x > 6\}$ **45.** $\mathbb{R}$

47. $\{x|x \leq -18 \text{ or } x \geq 2\}$ **49.** $\mathbb{R}$ **51.** $\{x|x < 2 \text{ or } x > 2\}$ **53.** $\left(5, \frac{4}{3}\right)$ **55.** $\{-3, 1\}$ **57.** $\left(28, -\frac{12}{5}\right)$

59. $\left\{\frac{2}{5}\right\}$ **61.** $\{7, -7\}$ **63.** $\{x|-2 < x < 8\}$ **65.** $\{x|x < -14 \text{ or } x > 4\}$ **67.** $\left\{y\left|-\frac{5}{2} < y < -\frac{3}{2}\right.\right\}$

69. $\left\{-\frac{11}{4}, \frac{7}{4}\right\}$ **71.** $\left\{x\left|x < -\frac{5}{2} \text{ or } x > -\frac{5}{2}\right.\right\}$ **73.** $\left\{x\left|-\frac{13}{3} \leq x \leq \frac{5}{3}\right.\right\}$ **75.** $\varnothing$ **77.** $\left\{-\frac{22}{3}, \frac{26}{3}\right\}$

79. $\{w|-16 < w < 8\}$ **81.** $\mathbb{R}$ **83.** $\left\{\frac{22}{3}, 2\right\}$ **85.** $\left\{-\frac{3}{2}, \frac{9}{7}\right\}$ **87.** $\left\{x\left|-\frac{3}{2} < x < \frac{15}{2}\right.\right\}$ **89.** $x = -\dfrac{b}{a}$ **91.** $\mathbb{R}$

93. $\{x|x \geq 0\}$ **95. (a)** Write $ax + b = c$ or $ax + b = -c$, then solve each equation for x.

(b) $\left\{x\left|x = \dfrac{c - d}{a} \text{ or } x = \dfrac{-c - b}{a}\right.\right\}$

97. (a) Write $ax + b < -c$ or $ax + b > c$, then solve both inequalities for x. **(b)** $\left\{x\left|x < \dfrac{-c - b}{a} \text{ or } x > \dfrac{c - b}{a}\right.\right\}$

Cumulative Review Exercises

99. $\frac{29}{72}$ **100.** 25   **101.** 1.33 mi   **102.** $\{x \mid x < 4\}$

Group Activity/Challenge Problems

3. $\varnothing$ **5.** $\{3\}$ **7.** $\{-3\}$

Review Exercises

1. Tenth   **2.** First   **3.** Seventh   **4.** Cannot be simplified   **5.** $7x^2 + 2xy - 4$ **6.** 8   **7.** $4x - 3y + 6$

8. $\frac{49}{6}$ **9.** 20   **10.** $-\frac{13}{3}$ **11.** $-\frac{7}{2}$ **12.** $-\frac{9}{2}$ **13.** No solution   **14.** 2   **15.** 200   **16.** $\frac{1}{4}$ **17.** 176   **18.** -4

19. $l = \dfrac{A}{w}$ **20.** $h = \dfrac{A}{\pi r^2}$ **21.** $w = \dfrac{P - 2l}{2}$ **22.** $r = \dfrac{d}{t}$ **23.** $m = \dfrac{y - b}{x}$ **24.** $y = \dfrac{2x - 5}{3}$ **25.** $V_2 = \dfrac{P_1 V_1}{P_2}$

26. $a = \dfrac{2S - b}{3}$ **27.** $l = \dfrac{K - 2d}{2}$ **28.** $n = \dfrac{2s}{f + l}$ **29.** $y = \dfrac{-2x + 2}{3x + 2}$ **30.** $y = \dfrac{6}{x - 1}$ **31.** 5   **32.** 20

33. 13, 15   **34.** \$50   **35.** 3 yr   **36.** \$5833.33   **37.** 9 or more   **38.** \$233.33   **39.** 30.6 rolls per hour

40. \$6000 at 8%, \$4000 at 5%   **41.** $2\frac{2}{3}$ hr   **42. (a)** 3000 mph **(b)** 16,500 mi   **43.** 15 lb at \$6, 25 lb at \$6.80

44. \$25   **45. (a)** 1 hr **(b)** 14.4 mi   **46.** $40°, 65°, 75°$ **47.** 300 gal/hr, 450 gal/hr   **48.** 24, 25   **49.** 7.5 oz

50. \$4500 at 10%, \$7500 at 6%   **51.** more than 5   **52.** 90 mph   **53.** 7   **54.** -3

55. $\frac{5}{2}$ **56.** $\frac{21}{4}$ **57.** $-\frac{9}{2}$ **58.** -10 **59.** $\frac{2}{5}$ **60.** $\frac{20}{9}$

61. 13 boxes   **62.** 7 min   **63.** 17 weeks   **64.** (5, 11)   **65.** $[-3, 3)$ **66.** $\left(\frac{7}{2}, 6\right)$ **67.** $\left(\frac{8}{3}, 6\right)$ **68.** $\left[-\frac{1}{2}, \frac{23}{2}\right)$

69. (2, 14)   **70.** $\{x \mid 81 \le x \le 100\}$ **71.** $\{x \mid -3 < x < 3\}$ **72.** $\mathbb{R}$ **73.** $\{x \mid x > -1\}$ **74.** $\left\{x \mid 2 \le x \le \frac{5}{2}\right\}$

75. $\{x \mid x \le -4\}$ **76.** $\left\{x \mid x \le -4 \text{ or } x > \frac{17}{5}\right\}$ **77.** $\{-4, 4\}$ **78.** $\{x \mid -3 < x < 3\}$ **79.** $\{x \mid x \le -4 \text{ or } x \ge 4\}$

80. $\{-5, 13\}$ **81.** $\{x \mid x \le -3 \text{ or } x \ge 7\}$ **82.** $\left\{-\frac{1}{2}, \frac{9}{2}\right\}$ **83.** $\{x \mid -2 < x < 5\}$ **84.** $\{-1, 4\}$

85. $\{x \mid -14 < x < 22\}$ **86.** $\left\{x \mid x \le \frac{5}{2} \text{ or } x \ge \frac{11}{2}\right\}$ **87.** $\left\{\frac{7}{2}, \frac{1}{4}\right\}$ **88.** $\mathbb{R}$ **89.** $(3, \infty)$ **90.** $[3, \infty)$ **91.** $\varnothing$

92. $[-17, 23]$ **93.** $\left(-\frac{17}{2}, \frac{27}{2}\right]$ **94.** $[-2, 4]$ **95.** $\left[-\frac{7}{2}, -\frac{1}{2}\right]$ **96.** $(-\infty, -11] \cup (14, \infty)$ **97.** $(-\infty, 90)$

98. $\left(-\infty, -\frac{1}{3}\right] \cup [3, \infty)$ **99.** $(-\infty, -12) \cup (20, \infty)$ **100.** $\left(\frac{29}{10}, \frac{43}{8}\right]$

Practice Test

1. Sixth   **2.** $\frac{27}{7}$ **3.** $-\frac{34}{5}$ **4.** 68   **5.** $\frac{13}{3}$ **6.** $b = \dfrac{a - 2c}{3}$ **7.** $b_2 = \dfrac{2A - hb_1}{h}$ **8.** 23, 24   **9.** 200 mi

10. 11.56 mi   **11.** 6.25 L   **12.** \$7000 at 8%, \$5000 at 7%   **13.** 33   **14.** $(-12, 12)$ **15.** $\{-1, 9\}$

16. $\left\{-\dfrac{14}{3}, \dfrac{26}{5}\right\}$ **17.** $\{x \mid x < -1 \text{ or } x > 4\}$ **18.** $\left\{x \mid \frac{1}{2} \le x \le \frac{5}{2}\right\}$

Cumulative Review Test

1. (a) $\{1, 2, 3, 4, 5, 6, 7, 9, 10, 12\}$ **(b)** $\{4, 6, 9, 12\}$ **2. (a)** Commutative property of addition
(b) Associative property of multiplication **(c)** Distributive property   **3.** $<$ **4.** -80 **5.** -15 **6.** -29 **7.** 7
8. $\frac{1}{5}$ **9.** 1.15   **10.** $-\frac{56}{33}$ **11.** 10

12. Conditional linear equation is true only under specific conditions; Identity is true for all values of the variable; Inconsistent equation is never true. **13.** 3 **14.** $t = \dfrac{A - p}{pr}$ **15.** (a) (b) $\left\{x \mid -2 < x < \frac{8}{5}\right\}$

(c) $\left(-2, \frac{8}{5}\right)$ **16.** $\{1, -5\}$ **17.** $\{x \mid x \le -10 \text{ or } x \ge 14\}$ **18.** \$2250 **19.** 40 mph, 50 mph

20. $\frac{2}{3}$ L of 50%, $1\frac{1}{3}$ L of 20%

Chapter 3

Exercise Set 3.1

1. $A(3, 1), B(-6, 0), C(2, -4), D(-2, -4), E(0, 3), F(-8, 1), G\left(\frac{3}{2}, -1\right)$ **3.** **5.** 3 **7.** 9

9. 5 **11.** $\sqrt{90} \approx 9.49$ **13.** $\sqrt{74} \approx 8.60$ **15.** $\sqrt{34.33} \approx 5.86$ **17.** $\sqrt{\dfrac{281}{16}} \approx 4.19$ **19.** $(2, 3)$ **21.** $(0, 0)$

23. $(-4, -5)$ **25.** $\left(-\frac{7}{2}, -5\right)$ **27.** $(-3.05, 9.575)$ **29.** $\left(\frac{9}{4}, \frac{15}{4}\right)$ **31.** $\sqrt{36} + \sqrt{64} + \sqrt{100} = 24$

33. $\sqrt{53} + \sqrt{13} + \sqrt{116} \approx 21.66$

35. (a) $\approx 165 - 10$ or 155 million Btu's per ton (b) ≈ 4650 million Btu or 4,650,000,000 Btu

37. (a) 2400 mi (b) 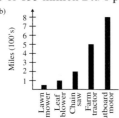 **39.** (a) $\approx 600{,}000$ **41.** (a) ≈ 1.4 billion in 1970 and

(b) 1991, $\approx 1{,}800{,}000$ ≈ 4 billion in 2010. (c)

(c) $\approx 9{,}300{,}000$ (b) ≈ 20 years

Cumulative Review Exercises

43. $\frac{3}{2}$ **44.** 140 mi **45.** $\{x \mid -2 < x \le 4\}$ **46.** $\left\{x \mid x < -\frac{7}{3} \text{ or } x > 1\right\}$

Group Activity/Challenge Problems

1. $P = \sqrt{37} + \sqrt{45} + 4 \approx 16.79, A = 12$ **3.** $P = 12 + \sqrt{20} \approx 16.47, A = 16$

Graphing Calculator Corner, Page 136

The following are from a TI 82 calculator. If you are using a different grapher your answers may differ.

1. **3.**

Exercise Set 3.2

1. **3.** **5.** **7.** **9.** **11.**

13.   **15.** 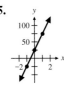 **17.**  **19.** 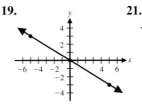 **21.** 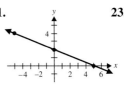 **23.**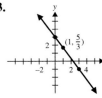

25.   **27.**  **29.** 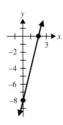 **31.**   **33.**   **35.**

37.   **39.**   **41.**   **43.** 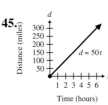 **45.** 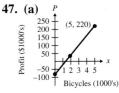 **47. (a)**

47. (b) ≈ 1300 **(c)** ≈ 3800 **49. (a)** $s = 200 + 0.15x$ **49. (b)**  **(c)** $\approx \$800$ **(d)** $\approx \$1300$

51. (a) Maximum ≈ 3773, Minimum ≈ 3752 **(b)** Loss ≈ 13 points **(c)** 78 **53.** The set of points whose coordinates satisfy the equation **57.** Set one side of the equation equal to 0. Then replace the 0 with y and graph the equation. The solution is the x coordinate of the x intercept.

CUMULATIVE REVIEW EXERCISES

59. $\frac{1}{2}$ **60.** $p_2 = \dfrac{E - a_1 p_1 - a_3 p_3}{a_2}$ **61. (a)** ⊙————▶ **(b)** $(3, \infty)$ **(c)** $\{x \mid x > 3\}$ **62.** $-2, 10$

GROUP ACTIVITY/CHALLENGE PROBLEMS

1. (a) 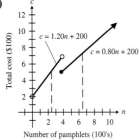 **(b)** $\approx \$500, \approx \$720$ **3. (a)** $f = \begin{cases} 20 + 5n, \, 0 < n < 15 \\ 20 + 10n, \, n \geq 15 \end{cases}$ **(b)**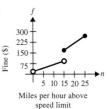

5. (a) $y = 2x - 1$ **(b) & (c)**

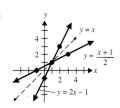

(c) The graphs that are inverses are symmetric to the line $y = x$.

(d) The domain of $y = \dfrac{x + 1}{2}$ is the range of $y = 2x - 1$ and the range of $y = \dfrac{x + 1}{2}$ is the domain of $y = 2x - 1$.

GRAPHING CALCULATOR CORNER, PAGE 143

These screens are from a TI 82 calculator

1.   **3.**

GRAPHING CALCULATOR CORNER, PAGE 147

1. (a) Raises or lowers the graph **(b)** No **3.** The 3 lines fall going from left to right. Decreasing the coefficients makes a line steeper.

GRAPHING CALCULATOR CORNER, PAGE 148

1. 8.0 **3.** 14.7

EXERCISE SET 3.3

1. -8 **3.** $-\dfrac{1}{2}$ **5.** -1 **7.** Undefined **9.** -5 **11.** $-\dfrac{2}{3}$ **13.** $a = 7$ **15.** $b = 2$ **17.** $x = 6$ **19.** $x = 0$

21. $m = -3, y = -3x$ **23.** m is undefined, $x = -2$ **25.** $m = -\dfrac{1}{3}, y = -\dfrac{1}{3}x + 2$ **27.** $m = -\dfrac{3}{2}, y = -\dfrac{3}{2}x + 15$

29. $y = -x + 2$ **31.** $y = \dfrac{2}{3}x - 2$ **33.** $y = \dfrac{5}{2}x + 2$

35. Parallel **37.** Perpendicular **39.** Neither **41.** Parallel **43.** Parallel **45.** Neither **47.** Perpendicular

49. Perpendicular **51.** Neither **53.** $y = 4x - 5$ **55.** $y = -2x - 3$ **57.** $y = \dfrac{1}{2}x - \dfrac{9}{2}$ **59.** $y = -\dfrac{2}{3}x - \dfrac{8}{3}$

61. $y = 2x + 2$ **63.** $y = -\dfrac{3}{2}x - \dfrac{11}{30}$ **65.** $y = -2x - \dfrac{16}{3}$ **67.** $y = -3x + 11$ **69.** $y = -\dfrac{2}{3}x + 6$

71. (a) $m = 0.97, n = 0.97t + 47.2$ **(b)** 66.6 million **73. (a)** $N = 1.33t + 100$ **(b)** $T = -0.22t + 100$

75. The line rises going from left to right **77.** $m = 0$. The line neither rises nor falls; hence no change in y.

79. It does not change. **81. (a)** $4x + 2y = 4$ **(b)** $y = -2x + 2$ **(c)** $y - 2 = -2(x - 0)$

83. (b)

84. 7, 9, 11   **85.** $x = a + b$ or $x = a - b$ **86.** $a - b < x < a + b$ **87.** $x < a - b$ or $x > a + b$

1. (a) 1292.2 in   **(b)** 582.4 in   **(c)** 2.21875   **3. (a)** $n = 8t - 15{,}808$; The slope is the same but the equation is different.
(b) 224; This agrees with the answer in Example 13.   **(c)** There are smaller numbers to work with.

Exercise Set 3.4

1. Function, domain $\{1, 2, 3, 4, 5\}$, range $\{1, 2, 3, 4, 5\}$ **3.** Function, domain $\{1, 2, 3, 4, 5, 7\}$, range $\{-1, 0, 2, 4, 5\}$
5. Relation, domain $\{1, 2, 3, 5\}$, range $\{-4, -1, 0, 1, 2\}$ **7.** Function, domain $\{-2, \frac{1}{2}, 0, 2, 3, 5\}$,
range $\{-3, -1, 0, \frac{2}{3}, 2, 5\}$ **9.** Relation, domain $\{1, 2, 6\}$, range $\{-3, 0, 2, 5\}$ **11.** Relation, domain $\{0, 1, 2\}$, range
$\{-7, -1, 2, 3\}$ **13.** Relation, domain $\{x \mid -2 \le x \le 2\}$, range $\{y \mid -2 \le y \le 2\}$ **15.** Function, domain $\mathbb{R}$, range
$\{y \mid y \ge 0\}$ **17.** Function, domain $\{-1, 0, 1, 2, 3\}$, range $\{-1, 0, 1, 2, 3\}$ **19.** Function, domain $\mathbb{R}$, range $\mathbb{R}$
21. Relation, domain $\{-2\}$, range $\mathbb{R}$ **23.** Function, domain $\mathbb{R}$, range $\{y \mid -5 \le y \le 5\}$ **25. (a)** 13   **(b)** 3
27. (a) -1 **(b)** 2   **29. (a)** $\dfrac{1}{2}$ **(b)** -1 **31. (a)** 4   **(b)** $\dfrac{7}{4}$ or 1.75   **33. (a)** 2.364   **(b)** 35.116
35. (a) 4   **(b)** $\dfrac{5}{3}$ **37. (a)** 7   **(b)** 0.1   **39. (a)** $\dfrac{23}{6}$ or $3.8\overline{3}$ **(b)** $-\dfrac{53}{4}$ or -13.25
41. (a) -3 **(b)** $-\sqrt{108} \approx -10.39$ **43. (a)** 7   **(b)** 9   **45. (a)** $\dfrac{1}{4}$ or 0.25   **(b)** $-\dfrac{1}{4}$ or -0.25
47. (a) 110   **(b)** 240   **49. (a)** $v = 19.65$ m/s, $h = 60.3$ m   **(b)** $v = 8.45$ m/s, $h = 11.125$ m   **51. (a)** 34 m
(b) 10.75 m   **53. (a)** $0.18 million   **(b)** $0.7 million   **55. (a)** $28 thousand   **(b)** $28 thousand
57. (a) each time has a unique amount **(b)** time, number of AIDS cases   **(c)** D: $\{t \mid 1982 \le t \le 1993\}$ R: $\{n \mid 843 \le n \le$
$90{,}000\}$ **(d)** 843, 47,000   **59. (a)** Each age has a unique head circumference.   **(b)** age, head circumference
(c) D: $\{a \mid 2 \le a \le 18\}$ R: $\{c \mid 48 \le c \le 55\}$ **(d)** 52 to 58 cm   **(e)** Circumference is a function of age
(f) 52 cm, 54 cm   **(g)** $y = 0.4375x + 47.125$ **61.** Any set of ordered pairs.   **63.** No, a function has no two distinct
ordered pairs with the same first coordinate.   **65.** If a vertical line can be drawn at any value of x that intersects the graph
more than once, then each x does not have a unique y, and the graph is not a function.   **67.** The set of second coordinates
of the ordered pairs.   **69.** Domain: $\mathbb{R}$, Range: $\mathbb{R}$

70. 19   **71.** $-\frac{92}{5}$ **72.** 2.4   **73.** 60 mph, 75 mph

1. (a) All real numbers except 2   **(b)** 0   **(c)** Undefined   **3.** D: $\{x \mid x \ge 0\}$ R: $\{y \mid y \ge 0\}$ **5.** D: $\{x \mid x \le 4\}$
R: $\{y \mid y \ge 0\}$ **7.** D: $\{x \mid x \le -2$ or $x \ge 2\}$ R: $\{y \mid y \ge 0\}$

Exercises Set 3.5

1. (a) $\mathbb{R}$ **(c)** $\mathbb{R}$
(b)

3. (a) $\mathbb{R}$ **(c)** $\mathbb{R}$
(b)
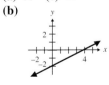

5. (a) $\mathbb{R}$ **(c)** $\mathbb{R}$
(b)
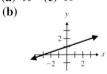

7. (a) $\mathbb{R}$ **(c)** $\{y \mid y \ge 0\}$
(b)
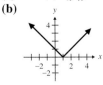

9. (a) $\mathbb{R}$ **(c)** $\{y \mid y \le 0\}$
(b)

11. (a) $\mathbb{R}$ **(c)** $\{y \mid y \ge -2\}$
(b)
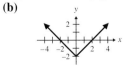

13. (a) $\{x \mid x \geq -1\}$ **(c)** $\{y \mid y \geq 0\}$
(b)

15. (a) $\{x \mid x \geq 2\}$ **(c)** $\{y \mid y \geq 0\}$
(b)
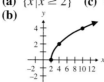

17. (a) $\{x \mid x \geq 0\}$ **(c)** $\{y \mid y \geq 2\}$
(b)
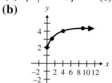

19. (a) $\mathbb{R}$ **(c)** $\{y \mid y \geq 2\}$
(b)

21. (a) $\{x \mid x \geq 0\}$ **(c)** $\{y \mid y \leq 2\}$
(b)
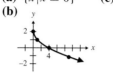

23. (a) $\mathbb{R}$ **(c)** $\{y \mid y \geq 0\}$
(b)
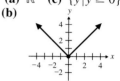

25. (a) $\mathbb{R}$ **(c)** $\{y \mid y \leq 4\}$
(b)
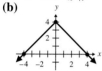

27. (a) $\mathbb{R}$ **(c)** $\{y \mid y \geq 3\}$
(b)

29. (a) $\mathbb{R}$ **(c)** $\mathbb{R}$
(b)

31. (a) $\{x \mid x \geq 3\}$ **(c)** $\{y \mid y \leq 4\}$
(b)

33. (a) $\{x \mid x \geq -6\}$ **(c)** $\{y \mid y \leq -3\}$
(b)

35. (a)

(b) approximately \$2.65 per bushel

37. Functions of the form $f(x) = ax + b$
39. The absolute value of any real number will be greater than or equal to 0.
41. The square root of a negative number is not a real number.

Cumulative Review Exercises

43. $\left(-\infty, \dfrac{2}{5}\right)$ **44.** Reverse the direction of the inequality symbol.   **45.** distance $= \sqrt{162} \approx 12.73$,

midpoint $= \left(-\dfrac{1}{2}, -\dfrac{3}{2}\right)$ **46. (a)** Any set of ordered pairs

(b) A relation in which no two ordered pairs have the same first coordinate
and a different second coordinate.

Group Activity/Challenge Problems

1. (a) $\mathbb{R}$
(b)
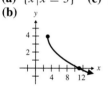

(c) $\{y \mid y \geq -1\}$

3. (a) $\mathbb{R}$
(b)

(c) $\mathbb{R}$

5.
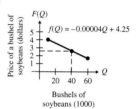

1. 2   **3.** 1   **5.** 0   **7.** 2

EXERCISE SET 3.6

1.   **3.**   **5.** 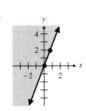   **7.**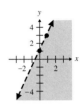

9. 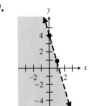   **11.** 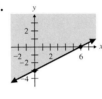   **13.** 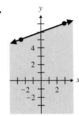   **15.**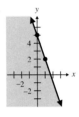

17.   **19.** 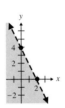   **21.**   **23.**

25. Because ≥ means greater than or *equal to* and ≤ means less than or *equal to*.

CUMULATIVE REVIEW EXERCISES
27. −56   **28.** 81.176   **29.** $15.72   **30.** $x + 2y = 2$ (other forms of the answer are possible)

GROUP ACTIVITY/CHALLENGE PROBLEMS

1. 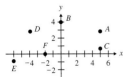   **3.**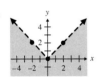

GRAPHING CALCULATOR CORNER, PAGE 192
1. $x > 6.7$ **3.** $x > 7.7$

REVIEW EXERCISES

1.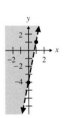

2. 5, $(\frac{3}{2}, -2)$ **3.** 5, $(4, \frac{1}{2})$ **4.** 13, $(\frac{1}{2}, 3)$ **5.** $\sqrt{8} \approx 2.83, (-3, 4)$ **6.** 2, $(4, 4)$

7. 13, $(-3, -\frac{3}{2})$ **8.** (a) 13   (b) 4th quarter of 1992, $\approx 6.0\%$

(c) 1st quarter of 1994, $\approx 3.5\%$

9. **10.** **11.**  **12.**

13. 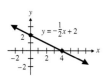 **14.** 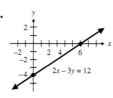 **15.** **16.**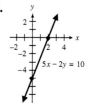

17. **18.** 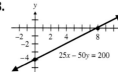 **19.** 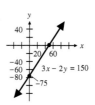 **20.**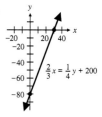

21. $m = -1, b = 5$ **22.** $m = -4, b = \frac{1}{2}$ **23.** $m = -\frac{3}{5}, b = \frac{12}{5}$ **24.** $m = -\frac{9}{7}, b = \frac{15}{7}$ **25.** m is undefined, no y intercept **26.** $m = 0, b = 6$ **27.** -7 **28.** $-\frac{1}{3}$ **29.** Neither **30.** Parallel **31.** Perpendicular **32.** Neither

33. $a = 3$ **34.** $y = 6$ **35.** $y = -37$ **36.** $x = 7$ **37.** $m = 0, y = 3$ **38.** m is undefined, $x = 2$

39. $m = -\frac{1}{2}, y = -\frac{1}{2}x + 2$ **40.** Parallel **41.** Perpendicular **42.** Neither **43.** $y = -\frac{2}{3}x + 4$ **44.** $y = x - 1$

45. $y = 3x + 20$ **46.** $y = \frac{2}{5}x - \frac{18}{5}$ **47.** $y = -\frac{5}{3}x - 4$ **48.** $y = -\frac{1}{2}x + 4$

49. (a) 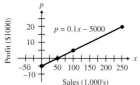 **(b)** Approx. 50,000 bagels **(c)** Approx. 250,000 bagels **50.**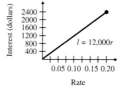

51. (a) 1989, ≈ 205 **(b)** Yes **(c)** Yes **(d)** No **(e)** $y = 21.44x + 12$ **52.** Domain $\{-2, 0, 3, 6\}$, range $\{-1, 4, 5, 9\}$ **53.** Domain $\{\frac{1}{2}, 2, 4, 5\}$, range $\{-6, -1, .2, 3\}$ **54.** Domain $\{x \mid -1 \le x \le 1\}$, range $\{y \mid -1 \le y \le 1\}$

55. Domain $\{x \mid -2 \le x \le 2\}$, range $\{y \mid -1 \le y \le 1\}$ **56.** Domain $\mathbb{R}$, range $\{y \mid y \le 0\}$ **57.** Domain $\mathbb{R}$, range $\mathbb{R}$

58. Function **59.** Function **60.** Function **61.** Not a function **62.** Function **63.** Function **64.** Function

65. Not a function **66. (a)** 7 **(b)** 4 **67. (a)** 2 **(b)** $\frac{5}{2}$ **68. (a)** 3 **(b)** 94 **69. (a)** $\frac{17}{8}$ or 2.125 **(b)** 7

70. (a) 5 **(b)** 30 **71. (a)** 0 **(b)** $\frac{2}{3}$ or $0.\overline{6}$ **72. (a)** $\sqrt{28} \approx 5.29$ **(b)** 7 **73. (a)** 0 **(b)** 7 **74. (a)** 720 **(b)** 1500 **75. (a)** 84 ft **(b)** 36 ft **76. (a)** Each year there is a unique amount of CFC produced. **(b)** Time, Amount of CFC produced **(c)** Domain: $\{t \mid 1940 \le t \le 1990\}$, Range: $\{a \mid 0 \le a \le 815\}$ **(d)** $T(t) = f(t) + g(t)$ **(e)** 400, 200, 600 thousand tonnes

77. (a) $\mathbb{R}$ **(c)** $\mathbb{R}$
(b)

78. (a) $\mathbb{R}$ **(c)** $\{y \,|\, y \geq 0\}$
(b)

79. (a) $\{x \,|\, x \geq 3\}$ **(c)** $\{y \,|\, y \geq 0\}$
(b)

80. (a) $\mathbb{R}$, **(c)** $\{y \,|\, y \leq 0\}$
(b)

81. (a) $\mathbb{R}$ **(c)** $\{y \,|\, y \geq 2\}$
(b)
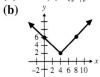

82. (a) $\{x \,|\, x \leq 4\}$ **(c)** $\{y \,|\, y \leq -3\}$
(b)
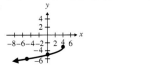

83. (a) 8300 **(c)** 12,500
(b)
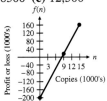

84.

85.

86.

87.
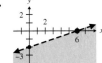

PRACTICE TEST

1. $5, \left(-\frac{1}{2}, 1\right)$ **2.** $m = \frac{4}{9}, b = -\frac{5}{3}$ **3.** $y = 3x - 3$ **4.** $y = 4x + 7$ **5.** $y = -\frac{3}{7}x + \frac{2}{7}$ **6.** $y = \frac{3}{2}x + \frac{11}{2}$

7.

8.
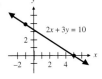

9. (a)
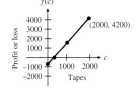

9. (b) ≈ 1900 **(c)** ≈ 320 **10.** Domain $\left\{ \frac{1}{2}, 2, 4, 6 \right\}$, range $\{-3, 0, 2, 9\}$ **11.** a, c **12. (a)** -7 **(b)** -4

13.
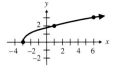

14.

15.

Chapter 4

EXERCISE SET 4.1

1. None **3.** b **5.** b **7.** None **9.** $y = -\frac{1}{2}x + \frac{5}{2}$; **11.** both $y = \frac{1}{2}x + 4$, **13.** $y = x - 3$; **15.** $y = \frac{3}{2}x + \frac{1}{2}$;
$y = \frac{1}{2}x - \frac{1}{2}$ dependent, $y = x + 1$ $y = \frac{3}{2}x + \frac{1}{4}$
consistent—one   infinite number   inconsistent,   inconsistent,
                                  no solution   no solution

17.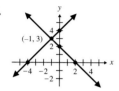

19.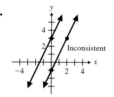

21.

23.

25.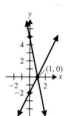

27.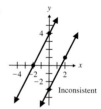

29. $(5, 2)$ **31.** $(3, 3)$ **33.** No solution **35.** $\left(\frac{1}{2}, -\frac{39}{2}\right)$ **37.** Infinite number of solutions **39.** $(-1, 2)$

41. $(5, -3)$ **43.** $(-3, -4)$ **45.** $\left(-\frac{13}{2}, -\frac{23}{6}\right)$ **47.** $(8, 6)$ **49.** $(1, -3)$ **51.** $\left(2, \frac{9}{2}\right)$ **53.** No solution

55. $(4, -2)$ **57.** $(4, -1)$ **59.** Infinite number of solutions **61.** $\left(\frac{59}{7}, \frac{60}{7}\right)$ **63.** $(2, -1)$ **65.** $\left(\frac{29}{22}, -\frac{5}{11}\right)$

67. $(3, 2)$ **69.** $(4, 0)$ **71.** $(14, 66)$ **73.** $\left(\frac{192}{25}, \frac{144}{25}\right)$ **75.** $(10, 4)$ **77.** $(1, 2)$ **79. (e)** ≈ 2003 **(f)** ≈ 1993

81. (a) You will get a false statement like $6 = 0$, **(b)** You will get a true statement like $0 = 0$.

CUMULATIVE REVIEW EXERCISES

83. Rational numbers can be expressed as the quotient of two integers, denominator not 0; irrational numbers cannot.
84. (a) Yes **(b)** Yes **85.** $\mathbb{R}$ **86.** 520 **87.** No, no two ordered pairs can have the same first coordinate and a different second coordinate (each x must have a unique y).

GROUP ACTIVITY/CHALLENGE PROBLEMS

1. $A = 2, B = 5$ **3. (a)** An infinite number. The system is dependent. **(b)** $m = -\frac{1}{2}, y = -\frac{1}{2}x + 5, b = 5,$

4. $(8, -1)$ **5.** $\left(-\frac{105}{41}, \frac{447}{82}\right)$ **7.** $(-3, 1)$ **9.** $\left(\frac{1}{a}, \frac{1}{b}\right)$ **10.** $\left(-\frac{1}{a}, -7\right)$

GRAPHING CALCULATOR CORNER, PAGE 215

1. $(1.1, 2.6)$ **3.** $(-37.3, -75.5)$ **5.** $(13.1, 14.7)$ **7.** $(-101.4, 76.2)$ **9.** $(-0.7, -9.3)$ **11.** $(2.2, 1.3)$

EXERCISE SET 4.2

1. $(1, 2, 5)$ **3.** $\left(-7, -\frac{35}{4}, -3\right)$ **5.** $(0, 3, 6)$ **7.** $(-1, 1, 3)$ **9.** $(-1, 3, 2)$ **11.** $(2, -1, 3)$ **13.** $\left(\frac{2}{3}, -\frac{1}{3}, 1\right)$

15. $(5, -3, -2)$ **17.** $\left(-\frac{11}{17}, \frac{7}{34}, -\frac{49}{17}\right)$ **19.** $(0, 0, 0)$ **21.** $(3, -1, 2)$ **23.** Inconsistent **25.** Dependent

27. Inconsistent   **29. (a)** A line   **(b)** A plane   **31.** No point is common to all 3 planes. Therefore the system is inconsistent.   **33.** A straight line is common to all 3 planes. Therefore there are an infinite number of points common to all 3 planes and the system is dependent.   **35. (b)** $(1, 1, 1)$

CUMULATIVE REVIEW EXERCISES

36. (a) $\frac{5}{12}$ hr (or 25 minutes) after Cameron starts   **(b)** 1.25 miles   **37.** $\left\{x \mid x < -\frac{3}{2} \text{ or } x > \frac{27}{2}\right\}$ **38.** $\left\{x \mid -\frac{8}{3} < x < \frac{16}{3}\right\}$
39. $\varnothing$

GROUP ACTIVITY/CHALLENGE PROBLEMS

1. (a) $a = 1, b = 2, c = -4$ **(b)** $y = x^2 + 2x - 4$ **3.** $(-1, 2, 1, 5)$

EXERCISE SET 4.3

1. $x + y = 73$
$y = 3x - 15$
22, 51

3. $x - y = 25$
$x = 3y - 1$
13, 38

5. $x + y = 180$
$y = 3x - 28$
52°, 128°

7. $x + y = 50$
$y = 3x + 2$
12 ft, 38 ft

9. $2x + 100y = 85$
$3x + 400y = 165$
$35 per day and 15¢ per mi

11. $x + y = 1000$
$0.05x + 0.25y = 0.10(1000)$
750 mL of 5%, 250 mL of 25%

13. $x + y = 100$
$0.05x + 0.00y = 0.035(100)$
70 gal of 5%, 30 gal skim

15. $x + y = 16$
$0.36x + 0.105y = 0.20(16)$
5.96 oz cream, 10.04 oz half and half

17. $4x + 4y = 420$
$y = x + 5$
50 mph, 55 mph

19. $x = 3y$
$35x + 20y = 6250$
150 shares Blockbuster; 50 shares BankAmerica

21. $x + y = 30$
$6.50x + 5.90y = 30(6.30)$
20 lb almonds, 10 lb walnuts

23. $x + y = 25$
$0.10x + 0.25y = 3.55$
18 dimes, 7 quarters

25. $x + y = 12,400$
$26,200 - x = 22,450 - y$
$8075 for Mr., $4325 for Mrs.

27. $7 = -a + b$
$4 = \frac{1}{2}a + b$
$a = -2, b = 5$

29. $2x + \frac{1}{2}y = 35$
$\frac{1}{2}x + \frac{1}{3}y = 15$
10, 30

31. $0.10A + 0.20B = 20$
$0.06A + 0.02B = 6$
80 g A, 60 g B

33. $x + y = 300$
$\left.0.70x + 0.40y = 0.6(300)\right\}$ or $\begin{cases} 0.70x + 0.40y = 0.60(300) \\ 0.30x + 0.60y = 0.40(300) \end{cases}$
200 g 1st alloy, 100 g 2nd alloy

35. $z = 2x$
$x = y - 10$
$x = z - 30$
Balcony $30, floor seats further back $40,
up front seats $60

37. $x + y + z = 180$
$x = \frac{2}{3}y$
$z = 3y - 30$
30°, 45°, 105°

39. $6 = 4a + 2b + c$
$17 = 9a + 3b + c$
$-3 = a - b + c$
$a = 2, b = 1, c = -4$

41. $x + y + z = 8$
$z = x - 2$
$0.10x + 0.12y + 0.20z = 0.13(8)$
4L of 10%, 2L of 12%, 2L of 20%

43. $5x + 4y + 7z = 154$
$3x + 2y + 5z = 94$
$2x + 2y + 4z = 76$
10 children's, 12 standard, 8 executive

45. $I_A = \frac{27}{38}, I_B = -\frac{15}{38}, I_c = -\frac{6}{19}$

CUMULATIVE REVIEW EXERCISES

47. $-\frac{35}{8}$ **48.** $\sqrt{32} \approx 5.66, (4, -6)$ **49.** $y = x - 10$ **50.** Use the vertical line test. If a vertical line cannot be drawn to intersect the graph at more than one point, the graph is a function.

GROUP ACTIVITY/CHALLENGE PROBLEMS

1. (a) $\frac{2400}{17} \approx 141.2$ cm   **(b)** Pull away

EXERCISE SET 4.4

1. 22   **3.** -8 **5.** -12 **7.** -38 **9.** $(3, 1)$ **11.** $(3, 2)$ **13.** $\left(\frac{60}{17}, -\frac{11}{17}\right)$ **15.** $\left(\frac{1}{2}, -4\right)$ **17.** $(2, -3)$

19. $(2, 5)$ **21.** $(-1, 1, 3)$ **23.** $\left(-\frac{1}{2}, \frac{1}{2}, -\frac{3}{2}\right)$ **25.** $\left(\frac{165}{14}, -\frac{153}{14}, -\frac{6}{7}\right)$ **27.** $(-1, 0, 2)$

29. Dependent, infinite number of solutions **31.** $(1, -1, 2)$ **33.** Dependent, infinite number of solutions
35. Inconsistent, no solution **37.** $(3, -2, 1)$ **39.** It will have the opposite sign.
41. If $D = 0$ and D_x, D_y and D_z also equal 0, the system is dependent.

CUMULATIVE REVIEW EXERCISES

43. $\left(-\infty, \frac{14}{11}\right)$ **44.**

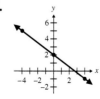

45.

46.

GROUP ACTIVITY/CHALLENGE PROBLEMS

1. 5 **3.** 2 **5.** $x = \dfrac{c_1 b_2 - c_2 b_1}{a_1 b_2 - a_2 b_1}, y = \dfrac{a_1 c_2 - a_2 c_1}{a_1 b_2 - a_2 b_1}$

EXERCISE SET 4.5

1. $\begin{bmatrix} 1 & -\frac{1}{2} & -\frac{5}{4} \\ 3 & 5 & -1 \end{bmatrix}$ **3.** $\begin{bmatrix} 4 & 0 & 3 & 8 \\ -\frac{5}{7} & 1 & -\frac{2}{7} & -2 \\ -1 & 3 & 5 & 12 \end{bmatrix}$ **5.** $\begin{bmatrix} 1 & 3 & 12 \\ 0 & 20 & 42 \end{bmatrix}$ **7.** $\begin{bmatrix} 1 & 0 & 8 & \frac{1}{4} \\ 0 & 2 & -38 & -\frac{13}{4} \\ 6 & -3 & 1 & 0 \end{bmatrix}$ **9.** $(3, 0)$ **11.** $(2, 3)$

13. $\left(0, \frac{1}{2}\right)$ **15.** Dependent system **17.** $(-2, 1)$ **19.** Inconsistent system **21.** $\left(\frac{2}{3}, \frac{1}{4}\right)$ **23.** $\left(\frac{4}{5}, -\frac{7}{8}\right)$

25. $(2, 0, 1)$ **27.** $(3, 1, 2)$ **29.** $\left(-2, \frac{1}{3}, 0\right)$ **31.** Dependent system **33.** $\left(\frac{1}{2}, 2, 4\right)$ **35.** Inconsistent system

37. $\left(5, \frac{1}{3}, -\frac{1}{2}\right)$ **39.** Change the -1 to a 1 by multiplying the second row by -1.

41. (a) All the numbers in a row are zeros. **(b)** All the numbers in a row on the left side of the augmented matrix are zeros, but the number in the same row on the right side is not 0.

CUMULATIVE REVIEW EXERCISES

42. (a) $\{1, 2, 3, 4, 5, 6, 9, 10,\}$ **(b)** $\{4, 6\}$ **43. (a)**

(b) $\{x \,|\, -2 < x \le 4\}$ **(c)** $(-2, 4]$
44. The set of points whose coordinates satisfy an equation. **45.** -76

GROUP ACTIVITY/CHALLENGE PROBLEMS

1. $(2, 0)$ **2.** $\left(1, \frac{1}{2}, -3\right)$

EXERCISE SET 4.6

1.

3.

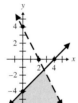

5.

7.

9.

11.

13.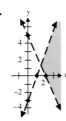

15.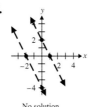
No solution

17.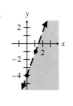

19.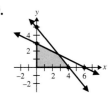

21.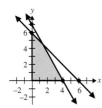

23.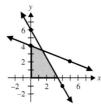

25.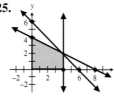

27.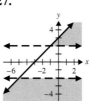

29.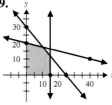

31.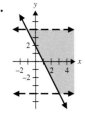

33.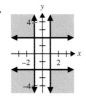

35.

37.

39.

Cumulative Review Exercises

41. $f_2 = \dfrac{f_3 d_3 - f_1 d_1}{d_2}$ **42.** D: $\{-1, 0, 4, 5\}$, R: $\{-5, -2, 2, 3\}$ **43.** D: $\mathbb{R}$, R: $\mathbb{R}$ **44.** D: $\mathbb{R}$, R: $\{y \mid y \geq -1\}$

Group Activity/Challenge Problems

1.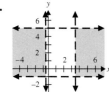

2.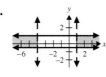

3.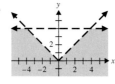

4.
Solution (2, 0)

Review Exercises

1. $y = -\frac{1}{2}x + 4$
$y = -\frac{1}{2}x + 2$
inconsistent, none

2. $y = -3x - 6$
$y = -\frac{2}{3}x + \frac{8}{3}$
consistent, one

3. $y = \frac{1}{2}x + 4$
$y = -\frac{1}{2}x + 4$
consistent, one

4. $y = \frac{3}{2}x + 2$
$y = \frac{2}{3}x - \frac{4}{3}$
consistent, one

5.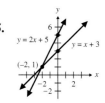
$y = 2x + 5$
$y = x + 3$
$(-2, 1)$

6.

7.

8.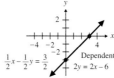
Dependent

9. $(3, 7)$ **10.** $(2, 3)$ **11.** $(5, 2)$ **12.** $(5, 2)$ **13.** $(8, -2)$ **14.** $(1, -2)$ **15.** $(26, -16)$ **16.** $(5, -2)$

17. $\left(\frac{32}{13}, \frac{8}{13}\right)$ **18.** $\left(-1, \frac{13}{3}\right)$ **19.** $(1, 2)$ **20.** $\left(\frac{7}{5}, \frac{13}{5}\right)$ **21.** $(6, -2)$ **22.** $\left(-\frac{78}{7}, -\frac{48}{7}\right)$

23. Dependent, infinite number of solutions   **24.** Inconsistent, no solution   **25.** $\left(2, 5, \frac{34}{5}\right)$ **26.** $\left(5, -\frac{15}{4}, -2\right)$

27. $(1, 2, -1)$ **28.** $(3, 1, 2)$ **29.** $\left(\frac{8}{3}, \frac{2}{3}, 3\right)$ **30.** $(0, 2, -3)$ **31.** Inconsistent, no solution

32. Dependent, infinite number of solutions

33. $x - y = 18$ **34.** $x + y = 600$ **35.** $x + y = 6$ **36.** $x + y = 650$
       $x = 4y$ $x - y = 530$ $0.3x + 0.5y = 0.4(6)$ $15x + 11y = 8790$
       $6, 24$ 565 mph plane, 35 mph wind   3L of each   410 adults, 240 children

37. $x + y + z = 17$ **38.** $x + y + z = 40,000$ **39.** $(4, -1)$ **40.** $(1, -1)$ **41.** $(-1, 2)$ **42.** $(4, 1, 3)$
       $x = y + z + 1$ $y = x - 5,000$
       $y = 3z$ $0.10x + 0.08y + 0.06z = 3500,$
       $(9, 6, 2)$ \$20,000 at 10%, \$15,000 at 8%,
                   \$5,000 at 6%

43. $(-1, 5, -2)$ **44.** No solution   **45.** $\left(-1, \frac{1}{3}\right)$ **46.** $\left(-\frac{5}{2}, -3\right)$ **47.** Infinite number of solutions   **48.** $(2, 1, -2)$
49. No solution   **50.** $(1, -1, 3)$

51. 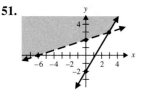   **52.** 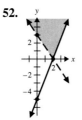   **53.** 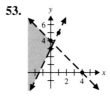   **54.**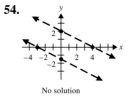

No solution

55. 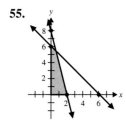   **56.** 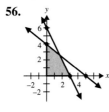   **57.** 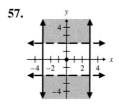   **58.**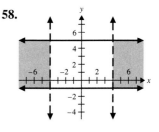

Practice Test

1. Consistent, one   **3.** 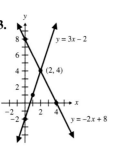   **4.** $\left(\frac{7}{2}, -\frac{5}{2}\right)$ **5.** Dependent,   **6.** $\left(\frac{44}{19}, \frac{48}{19}\right)$ **7.** $(1, 9)$ **8.** $(3, -1)$
2. Dependent,   infinite number   **9.** $(1, -1, 2)$ **10.** $(4, 3, -1$
    infinite number   of solutions   **11.** $x + y = 20$
                                                                                $7.00x + 5.50y = 6.00(20)$

                                                                                $13\frac{1}{3}$ lbs peanuts, $6\frac{2}{3}$ lbs cas

12. 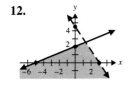   **13.**

CUMULATIVE REVIEW TEST

1. 0   **2. (a)** 9, 1   **(b)** $\frac{1}{2}$, -4, 9, 0, -4.63, 1   **(c)** $\frac{1}{2}$, -4, 9, 0, $\sqrt{3}$, -4.63, 1   **3.** $-\left|-8\right|$, -1, $\frac{5}{8}$, $\frac{3}{4}$, $\left|-4\right|$, $\left|-10\right|$

4. 7   **5.** -15 **6.** $\frac{11}{4}$, $-\frac{5}{4}$ **7.** $b = \dfrac{R - 3a}{3}$ **8.** $\left\{ x \,\middle|\, \frac{2}{3} < x \le \frac{34}{3} \right\}$ **9.** $5, \left(-1, \frac{7}{2}\right)$

10. 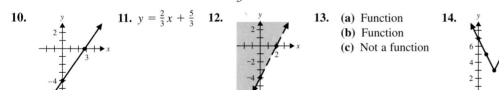   **11.** $y = \frac{2}{3}x + \frac{5}{3}$ **12.**   **13. (a)** Function **(b)** Function **(c)** Not a function   **14.**

15. $(1, 3)$ **16.** $(6, -5)$ **17.** $(2, 1, 3)$ **18.** $10°, 80°, 90°$ **19.** 1 hr   **20.** 600 at \$20, 400 at \$16

Chapter 5

EXERCISE SET 5.1

1. $\dfrac{5}{y^3}$ **3.** x^4 **5.** $2xy^3$ **7.** $\dfrac{5z}{2x^2y^3}$ **9.** $\dfrac{5z^4}{x^2y^3}$ **11.** $\dfrac{1}{4xy}$ **13.** 1   **15.** -2 **17.** -1 **19.** 7   **21.** $\dfrac{1}{16}$ **23.** $-\dfrac{1}{16}$

25. $\dfrac{1}{125}$ **27.** $-\dfrac{1}{125}$ **29.** $\dfrac{1}{6}$ **31.** $\dfrac{5}{4}$ **33.** $\dfrac{11}{48}$ **35.** $\dfrac{1}{6}$ **37.** x^4 **39.** 9   **41.** $\dfrac{1}{49}$ **43.** $\dfrac{1}{x^3}$ **45.** $3y^5$

47. $\dfrac{12}{x^7}$ **49.** $3y$ **51.** $-10x^7z^5$ **53.** $\dfrac{8x^7y^2}{z^3}$ **55.** $\dfrac{3x^2}{y^6}$ **57.** $\dfrac{3x^3}{y^5}$ **59.** $3x^5$ **61.** $\dfrac{-x^2}{y^9}$ **63.** $\dfrac{1}{2x^6y^3}$

65. $6x^6yz^3$ **67.** $\dfrac{2y^5}{3x}$ **69.** $x^6y^6z^3$ **71.** x^{7a+4} **73.** w^{7b+1} **75.** x^{w+7} **77.** x^{p+3} **79.** 3.7×10^3

81. 4.7×10^{-2} **83.** 1.9×10^4 **85.** 1.86×10^{-6} **87.** 5200   **89.** 0.0000213   **91.** 0.312   **93.** 9,000,000

95. 150,000,000   **97.** 0.021   **99.** 2,000,000   **101.** 4.5×10^{-7} **103.** 2.0×10^3 **105.** 2.13×10^{-7}

107. 30,000 hr   **109. (a)** 1,602,739.7 mi.   **(b)** Earth's speed: 1,602,739.7 mi/day or $\dfrac{1,602,739.7}{24} = 66,781$ mph; speed

of bullet: $\dfrac{66,781}{8} \approx 8347.6$ mph   **111. (a)** 5.321×10^{10} or \$53,210,000,000   **(b)** 1.806×10^{12} or \$1,806,000,000,000

(c) 53,210,000,000   **(d)** 18,060,000,000   **(e)** \$1 bills; 35,150,000,000   **113.** $x = \dfrac{1}{5}$ **115. (a)** opposite is $-x$,

reciprocal is $\dfrac{1}{x}$ **(b)** x^{-1}, $\dfrac{1}{x}$ **(c)** $-x$ **117. (a)** add 1 to the exponent   **(b)** add 2 to the exponent   **(c)** add 6 to the

exponent   **(d)** 7.59×10^{13} **119. (a)** 1.00×10^4 or 10,000   **(b)** 472,500   **(c)** the error in part **(b)**

CUMULATIVE REVIEW EXERCISES

120. -5 **121.** -0.75 **122.** 20   **123.** $y = \dfrac{1}{2}x - 2$

GROUP ACTIVITY/CHALLENGE PROBLEMS

2. (a) 0 gives $10^0 = 1$, 1 gives $10^1 = 10$, 2 gives $10^2 = 100$, 3 gives $10^3 = 1000$, 4 gives $10^4 = 10,000$,
5 gives $10^5 = 100,000$, 6 gives $10^6 = 1,000,000$, 7 gives $10^7 = 10,000,000$, 8 gives $10^8 = 100,000,000$,
9 gives $10^9 = 1,000,000,000$, 10 gives $10^{10} = 10,000,000,000$ **(b)** 10,000   **(c)** $10^{1.4} \approx 25.1$
3. (a) about 5.87×10^{12} miles   **(b)** 500 sec or $8\frac{1}{3}$ min.   **(c)** 6.72×10^{11} sec or 21,309 years

EXERCISE SET 5.2

1. 64 **3.** $\dfrac{1}{64}$ **5.** x^6 **7.** $-x^3$ **9.** $-\dfrac{8}{x^6}$ **11.** $\dfrac{9}{25}$ **13.** $\dfrac{25}{4}$ **15.** $\dfrac{9}{4x^2}$ **17.** $\dfrac{16x^4}{y^4}$ **19.** $\dfrac{1}{8x^9y^3}$

21. $-\dfrac{x^{12}}{64y^{15}}$ **23.** $\dfrac{36x^2}{y^4}$ **25.** $8x^6y^{15}$ **27.** $\dfrac{y^6}{64x^3}$ **29.** $64x^9y^3$ **31.** $\dfrac{z^3}{8x^3y^3}$ **33.** $72x^{11}y^7$ **35.** $96x^8y^{13}$

37. $\dfrac{81x^2}{y^{14}}$ **39.** $\dfrac{8z^{13}}{27x^2y^{11}}$ **41.** $\dfrac{2}{27x^{11}y^{36}}$ **43.** $\dfrac{x^9y^{32}}{z^7}$ **45.** $\dfrac{-2}{9x^{10}y^3z^8}$ **47.** $\dfrac{x^7y^6}{576}$ **49.** $\dfrac{3y^3}{10}$ **51.** x^{5m+4}

53. b^{5y^2+3y} **55.** m^{3-7y} **57.** $x^{-3},\,y^1$ **59.** $x^{-4},\,y^{33},\,z^{-22}$ **61.** yes

CUMULATIVE REVIEW EXERCISES

63. $(1, 3)$ **64.** $(2, 4)$ **65.**

66. 15 L of 40%, 5 L of 60%

GROUP ACTIVITY/CHALLENGE PROBLEMS

1. $x^{9/4}$ **3.** $\dfrac{1}{x^{9/2}}$ **5.** x^7y^3 **7.** $x^2,\,y^1$ **9.** $x^4,\,\left(\dfrac{x^4y^5}{xy^7}\right)^2$

EXERCISE SET 5.3

1. Monomial **3.** Monomial **5.** Trinomial **7.** Not polynomial **9.** $-x^2 - 4x - 8$, second
11. $10x^2 + 3xy + 6y^2$, second **13.** In descending order, fourth **15.** $7x - 2$ **17.** $x^2 - 8x - 2$ **19.** $2y^2 + 6y - 9$
21. $-2x^2 - 6x + 3$ **23.** $-7x^2 + x - 12$ **25.** $-x^3 + 3x^2y + 4xy^2$ **27.** $10x^2 - 5x - 6$ **29.** $-3w^2 + 6w$

31. $-x + 11$ **33.** $-3x^2 + 2x - 12$ **35.** $15y^2 - 6y + 4$ **37.** $-7x^2y + 6xy^2$ **39.** $24x^2y^5$ **41.** $\dfrac{1}{9}x^7y^8z^2$

43. $6x^6y^3 - 9x^3y^4 - 12x^2y$ **45.** $2xyz + \dfrac{8}{3}y^2z - 6y^3z$ **47.** $12x^2 - 38x + 30$

49. $-2x^3 + 8x^2 - 3x + 12$ **51.** $\dfrac{2}{15}x^2 + \dfrac{1}{3}xz - \dfrac{1}{5}z^2$ **53.** $x^3 - 7x^2 + 14x - 8$ **55.** $4x^3 + x^2 - 20x + 4$

57. $2a^3 - 7a^2b + 5ab^2 - 6b^3$ **59.** $27x^3 - 27x^2 + 9x - 1$ **61.** $4x^2 - 1$ **63.** $4x^2 - 12xy + 9y^2$
65. $4x^2 + 20xy + 25y^2$ **67.** $25m^4 - 4n^2$ **69.** $y^2 + 8y - 4xy + 16 - 16x + 4x^2$
71. $16 - 8x + 24y + x^2 - 6xy + 9y^2$ **73.** $x^2 - 4xy + 4y^2 - 6x + 12y + 9$ **75.** $-24r^6s^{13}$ **77.** $3x^3 + 9x^2 - 3x$

79. $6y^2 - y - 12$ **81.** $4x^2 - \dfrac{9}{16}$ **83.** $6x^4 - 5x^2y - 6y^2$ **85.** $8x^3 + 2x^2 - 11x + 6$

87. $6x^3 - x^2y - 16xy^2 + 6y^3$ **89.** $\dfrac{2}{5}x^3y^7 - \dfrac{1}{6}x^6y^5 + \dfrac{4}{3}x^3y^7z^5$ **91.** $w^2 - 9x^2 - 24x - 16$ **93.** $a^3 + b^3$

95. $a^3 + 8b^3$ **97.** $9m^2 + 12m + 4 - n^2$ **99.** (a) and (b) $x^2 + 8x + 15$ **101.** $36 - x^2$ **103.** (a) $11x + 12$
(b) 117 sq in, 50 sq in **105.** A finite sum of terms in which all variables have whole-number exponents and no variable
appears in a denominator. **107.** (a) Find the sum of the exponents on the variables in the term.
(b) It is the same as that of the highest degree term. **111.** One example is $x^5 + x + 1$; other answers are possible.

CUMULATIVE REVIEW EXERCISES

113. $-\dfrac{15}{72} = -\dfrac{5}{24}$ **114.** $\left\{x \,\middle|\, -\dfrac{4}{3} \le x < \dfrac{14}{3}\right\}$ **115.** $0.05x + 0.06(10{,}000 - x) = 560$; $6000 at 6%; $4000 at 5%
116. $x + y = 10{,}000$; $0.05x + 0.06y = 560$; $6000 at 6%, $4000 at 5% **117.** $3x - 4y = 8$ **118.** $(2, -1, 6)$

GROUP ACTIVITY/CHALLENGE PROBLEMS

1. $-4xy - 6x^2 + 4y^2 + 2x - 12$　**3.** $6r^{3x} - 7r^{2x} - 5r^x - 9$　**5.** $15x^{3t-1} + 12x^{4t}$　**7.** $12x^{3m} - 18x^m - 10x^{2m} + 15$
9. (b) $a^2 + 2ab + b^2$　**10. (a)**　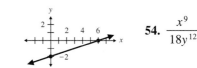　**(b)** $a^3 + 3a^2b + 3ab^2 + b^3$　**11.** $a^2 + 2ab + b^2 + 3a + 3b + 4$
12. $y^2 - 2y - 2xy + 2x + x^2 + 1$
13. $x^4 - 12x^3y + 54x^2y^2 - 108xy^3 + 81y^4$

EXERCISE SET 5.4

1. $3x + 4$　**3.** $2x + 1$　**5.** $3x^2 - x - 2$　**7.** $x^3 - \frac{3}{2}x^2 + 3x - 2$　**9.** $2x^2 - 4xy + \frac{3}{2}y^2$　**11.** $\frac{3x}{y} - 6x^2 + \frac{9y}{2x}$

13. $x + 3$　**15.** $2x + 3$　**17.** $3x + 2$　**19.** $2x - 3 + \frac{2}{4x + 9}$　**21.** $2y^2 + 3y - 1$　**23.** $2x^2 - 8x + 38 - \frac{156}{x + 4}$

25. $3t^2 - 3t + 4$　**27.** $2x^3 - 6x + 4$　**29.** $3x + \frac{3}{2} + \frac{6}{x}$　**31.** $2x + 5$　**33.** $2x + 1 + \frac{1}{2x} + \frac{3}{2x^2}$

35. $\frac{-x^2y^2}{2} + y - \frac{3}{5x}$　**37.** $3x^2 + 2x + 1 + \frac{5}{3x - 2}$　**39.** $\frac{z}{2} + z^2 - \frac{3}{2}x^2y^4z^7$　**41.** $2x^2 - 6x + 3$
43. $x^3 + x^2 - 6$　**45. (b)** $2x - 3$

CUMULATIVE REVIEW EXERCISES

47. (a) $ax + by = c$　**(b)** $y = mx + b$　**(c)** $y - y_1 = m(x - x_1)$　**48.** Every function is a relation, but not every relation is a function. A relation is any set of ordered pairs. A function is a set of ordered pairs no two of which have the same first co-ordinate.　**49.** $\frac{2}{21}$　**50.** 400 bulk, 150 first class

GROUP ACTIVITY/CHALLENGE PROBLEMS

1. $2x^2 + 3xy - y^2$　**3.** $x + \frac{5}{2} + \frac{11}{2(2x - 3)}$　**5.** $3x + 2$　**7.** $x^3 - 6x^2 + 13x - 10$　**9.** $\frac{3}{2}x + 3 - \frac{1}{x}$
11. 24 times greater

EXERCISE SET 5.5

1. $x + 3$　**3.** $x - 1$　**5.** $x + 8 + \frac{12}{x - 3}$　**7.** $3x + 5 + \frac{10}{x - 4}$　**9.** $4x^2 + x + 3 + \frac{3}{x - 1}$

11. $3x^2 - 2x + 2 + \frac{6}{x + 3}$　**13.** $5x^2 - 11x + 14 - \frac{20}{x + 1}$　**15.** $x^3 - 4x^2 + 16x - 64 + \frac{272}{x + 4}$

17. $y^4 - \frac{10}{y + 1}$　**19.** $3x^2 + 3x - 3$　**21.** $2x^3 + 2x - 2$　**23.** 10　**25.** -142　**27.** 0, factor

29. $-\frac{19}{4}$ or -4.75　**31.** 0, factor　**33.** $(x - 1)(6x + 5)$　**35.** $(x - 2)(x^2 + 3x + 5)$　**37.** $(x + 5)(2x^2 - 4x + 2)$

39. $(x + 3)(x^2 - 6x - 4)$　**41.** $\left(x - \frac{1}{2}\right)(x^2 + 2x - 4)$　**43. (b)** $x + 8 + \frac{36}{x - 5}$　**45.** If the remainder is 0 then the
divisor $x - a$ is a factor.　**47.** When $x = d$, $ax^2 + bx + c$ is equal to 0. Therefore, d must be a solution.
49. $x^3 - 2x^2 - 27x - 36$; multiply $(x + 3)(x^2 - 5x - 12)$

CUMULATIVE REVIEW EXERCISES

51. 　**52.** $\sqrt{34} + \sqrt{50} + \sqrt{104}$ (or ≈ 23.1)　**53.**　**54.** $\frac{x^9}{18y^{12}}$

1. No   **3.** Yes   **4.** $0.2x^2 - 3.92x - 1.248 - \dfrac{1.1392}{x - 0.4}$ **5. (a)** $3x^2 - 2x + 5 - \dfrac{13}{3x + 5}$ **(b)** Because we are

expressing the remainder in terms of $3x + 5$ rather than $x + \frac{5}{3}$, the denominator of the remainder is altered rather than the numerator.

EXERCISE SET 5.6

1. Yes, 7th   **3.** No, fractional exponent   **5.** No, negative exponent   **7.** $x = -1$, $(-1, -8)$, up
9. $x = 1$, $(1, 11)$, down   **11.** $x = -1$, $(-1, -8)$, down   **13.** $x = 0.5$, $(0.5, 8.25)$, down
15. $x = -1.5$, $(-1.5, -14)$, up   **17.** $x = 50$, $(50, -200)$, up

19. vertex: $(0, -1)$
Domain: $\mathbb{R}$
Range: $\{y \mid y \geq -1\}$

21. vertex: $(0, 3)$
Domain: $\mathbb{R}$
Range: $\{y \mid y \leq 3\}$

23. vertex: $(-1, -16)$
Domain: $\mathbb{R}$
Range: $\{y \mid y \geq -16\}$

25. vertex: $(2, -1)$
Domain: $\mathbb{R}$
Range: $\{y \mid y \leq -1\}$

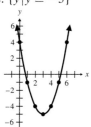
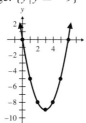
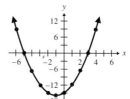
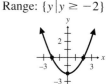

27. vertex: $(3, -5)$
Domain: $\mathbb{R}$
Range: $\{y \mid y \geq -5\}$

29. vertex: $(3, -9)$
Domain: $\mathbb{R}$
Range: $\{y \mid y \geq -9\}$

31. vertex: $(1, -2)$
Domain: $\mathbb{R}$
Range: $\{y \mid y \geq -2\}$

33. vertex: $(0, -2)$
Domain: $\mathbb{R}$
Range: $\{y \mid y \geq -2\}$

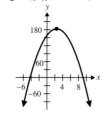
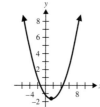
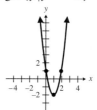
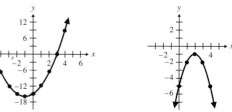
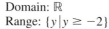

35. vertex: $(2.5, 185)$
Domain: $\mathbb{R}$
Range: $\{y \mid y \leq 185\}$

37. vertex: $(1.5, -1.65)$
Domain: $\mathbb{R}$
Range: $\{y \mid y \geq -1.65\}$

39.
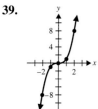

41.
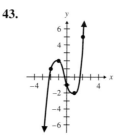

43.

45.

47. 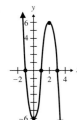   **49.** 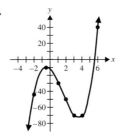   **51.**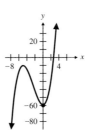

53. A square 12 ft by 12 ft   **55. (a)** 12 in   **(b)** 27 sq in   **57. (a)** 2 sec   **(b)** 64 ft   **59. (a)** and (c), graph must increase both to the right and left   **61.** When the squared term is positive the graph opens up; when negative, it opens down.   **63. (a)** $x = \dfrac{-b}{2a}$ **(b)** Substitute the x value of $\dfrac{-b}{2a}$ into the function to find y value of vertex.

65. y decreases; y increases   **67.** y increases as x goes from -3 to 0; y decreases as x goes from 0 to 3 (y is a maximum when $x = 0$).   **69.** It will be inverted.

CUMULATIVE REVIEW EXERCISES

72. $(-8, 2)$ **73.** $(-8, 2)$ **74.** $(1, 5, 6)$ **75.** $15x^3 + 21x^2 - 38x + 12$

GROUP ACTIVITY/CHALLENGE PROBLEMS

1. $(1, 8)$ **3.** $\left(\dfrac{2}{3}, \dfrac{43}{9}\right)$ or $(0.\overline{6}, 4.\overline{7})$ **5. (a)**   **5. (c)**

7. (b) and (c)   **9.** 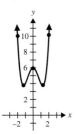   **11.** $x = 35, y = 35$

GRAPHING CALCULATOR CORNER, PAGE 320

1. 2   **3.** 3

REVIEW EXERCISES

1. 64   **2.** x^8 **3.** x^4 **4.** y^9 **5.** x^7 **6.** $\dfrac{1}{x^3}$ **7.** $\dfrac{1}{27}$ **8.** 3   **9.** $\dfrac{23}{12}$ **10.** $\dfrac{63}{8}$ **11.** $9x^4$ **12.** $\dfrac{3}{2}$ **13.** $\dfrac{16}{9}$

14. $\dfrac{y^2}{x}$ **15.** $-21x^3y^9$ **16.** $\dfrac{8}{x^2y}$ **17.** $\dfrac{30x^5}{y^6}$ **18.** $\dfrac{3y^7}{x^5}$ **19.** $\dfrac{3}{xy^9}$ **20.** $\dfrac{y^8}{2}$ **21.** $\dfrac{x^7}{y^8}$ **22.** $125x^3y^3$ **23.** $\dfrac{x^{10}}{9y^2}$

24. x^6y^8 **25.** $\dfrac{-125y^3}{x^6z^9}$ **26.** $\dfrac{z^4}{36x^2y^6}$ **27.** $\dfrac{x^9}{27}$ **28.** $\dfrac{x^{12}}{16y^8}$ **29.** $\dfrac{8x^{10}y^{16}}{z^6}$ **30.** $\dfrac{64y^3}{x^3z^{15}}$ **31.** $\dfrac{4x^5}{z^6}$ **32.** $\dfrac{27x^{21}}{32y^{19}}$

33. $\dfrac{x^9y^{19}}{8z}$ **34.** $\dfrac{x^{26}}{64y^6}$ **35.** 7.42×10^{-5} **36.** 2.6×10^5 **37.** 1.83×10^5 **38.** 1×10^{-6} **39.** 30,000

40. 0.02   **41.** 200,000,000   **42.** 2000   **43.** 1.992×10^{-5} or 0.00001992 sq mi   **44. (a)** 78,190,000   **(b)** ≈ 5.47

45. Trinomial, $x^2 + 5x - 3$, second   **46.** Trinomial, $x^2 + xy - y^2$, second   **47.** Polynomial, $2x^4 - 9x^2y + 6xy^3 - 3$, fourth   **48.** Not polynomial   **49.** $3x^2 + 10x + 3$ **50.** $4x^3 + 8x^2 + 12x$ **51.** $x^2 + 10x + 25$ **52.** $5y^2 + 2$

53. $4x^2 - 9$ **54.** $6x^2y + 9xy^2 + 2x + 3y$ **55.** $2x - 3$ **56.** $2x^3 - 8x^2 - 9$ **57.** $-2x^4y^2 - 2x^3y^7 + 12xy^3$

58. $9x^2 - 12xy + 4y^2$ **59.** $3x^2y + 3xy - 9y^2$ **60.** $25x^2y^2 - 36$ **61.** $3x - 2y + 1$ **62.** $4x^2 - 25y^4$

63. $x + 4 - \dfrac{5}{x - 3}$ **64.** $\dfrac{x^2}{2y} + x + \dfrac{3y}{2}$ **65.** $x^2 + 6xy + 9y^2 + 4x + 12y + 4$ **66.** $x^2 + 6xy + 9y^2 - 4$

67. $-9xy - 9x + 5y^2$ **68.** $6x^3 - x^2 - 24x + 18$ **69.** $2x^3 + x^2 - 3x - 4$ **70.** $4x^4 + 12x^3 + 6x^2 + 13x - 15$

71. $2x^2 + 3x - 4 + \dfrac{2}{2x + 3}$ **72.** $x^3y + 6x^2y + 7xy^2 + x^2y^2 + y^3$ **73.** $3x^2 + 7x + 21 + \dfrac{73}{x - 3}$

74. $2y^4 - 2y^3 - 8y^2 + 8y - 7 + \dfrac{6}{y + 1}$ **75.** $x^4 + 2x^3 + 4x^2 + 8x + 16 + \dfrac{12}{x - 2}$ **76.** $2x^2 + 2x + 6$ **77.** 3

78. -236 **79.** $-\dfrac{53}{9}$ or $-5.\overline{8}$ **80.** 0, factor   **81.** $(x + 2)(x - 5)$ **82.** $(x - 6)(x - 5)$ **83.** $(x - 4)(x^2 - 6x + 3)$

84. $(x + 3)(2x^2 - 3x + 4)$

85. vertex $(1, -1)$ **86.** vertex $(2, -18)$ **87.** vertex $(-1, 1)$
    D: $\mathbb{R}$ D: $\mathbb{R}$ D: $\mathbb{R}$
    R: $\{y \mid y \le -1\}$ R: $\{y \mid y \ge -18\}$ R: $\{y \mid y \ge 1\}$

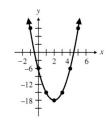
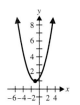

88.   **89.** 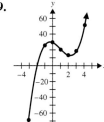   **90.**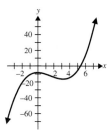

91. (a) 300 **(b)** $\$4,000$

PRACTICE TEST

1. $\dfrac{y^{10}}{9x^6}$ **2.** $\dfrac{9}{x^9y^{11}}$ **3.** $\dfrac{53}{36}$ or $1.47\overline{2}$ **4. (a)** Trinominal   **(b)** $-6x^4 - 4x^2y^3 + 2x$ **(c)** fifth   **5.** $4x^3 - 2x^2 + 2x + 8$

6. $4x^5 - 2y^2 + \dfrac{5}{x}$ **7.** $-6x^2 + xy + y^2$ **8.** $4x^3 + 8x^2y - 9xy^2 - 6y^3$ **9.** $x - 5 + \dfrac{25}{2x + 3}$ **10.** $2x^2y + 3x + 7y^2$

11. $2x^2 + 5x + 20 + \dfrac{53}{x - 3}$ **12.** $-6x^7y^6 + 18x^4y^7 - 9x^3y^4$ **13.** $4x^2 + 12xy + 9y^2$

14. $3x^3 + 3x^2 + 15x + 15 + \dfrac{79}{x - 5}$ **15.** -89 **16.** vertex $(2, 2)$
    D: $\mathbb{R}$
    R: $\{y \mid y \ge 2\}$

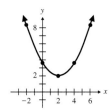

Chapter 6

EXERCISE SET 6.1

1. $8(n + 1)$ **3.** $3(2x^2 + x - 3)$ **5.** $x^3(7x^2 - 9x + 3)$ **7.** $-3y(8y^{14} - 3y^2 + 1)$ **9.** Cannot be factored
11. $3xy(x + 2xy + 1)$ **13.** $-4x^2y(10y^3z - 2x^4yz^2 - x)$ **15.** $x^4y^3z^9(19y^9z^4 - 8x)$ **17.** $-2(26x^2y^2 + 8xy^3 - 13z)$
19. $(4x - 5)^2(12x^2 - 15x + 1)$ **21.** $3(2x + 5)(-3x - 10)$ **23.** $3(3p - q)(p - q)$ **25.** $(x - 2)(-2x + 9)$
27. $(2x + 5)(6x^3 - 2x^2 - 1)$ **29.** $(2r - 3)^6(8pr - 12p - 3)$ **31.** $(x - 5)(x + 3)$ **33.** $(3x + 1)(x + 3)$
35. $(2x - 1)^2$ **37.** $2(2x - 5)(2x - 1)$ **39.** $(7a + 3b)(6c + 5d)$ **41.** $(x^2 + 4)(x - 3)$ **43.** $(2m - 5n)(5m - 6n)$
45. $6(x^2 - 2)(x + 3)$ **47.** $(a^3b - c^2)(2a - 3c)$ **49.** $(3p + 2q)(p^2 + q^2)$ **51.** $2p(10p^2 - 9p + 6)$
53. $4(4xy^2z + x^3y - 2)$ **55.** $(x - 2)(5x + 3)$ **57.** $y^2z^2(14yz^3 - 28yz^4 - 9x)$ **59.** $7y^7(x^4y^2 - 3x^3z^5 - 5yz^9)$
61. $(3a - 4b)(5a - 6b)$ **63.** $(3x + y)(2x - 3y)$ **65.** $(x + 3)(5x^2 + 15x - 3)$ **67.** $(3c - d)(2d)$
69. $x(3x^2 + 2)(x^2 - 5)$ **71. (a)** $(1 - 0.06)(x + 0.06x) = 0.94(1.06x)$
(b) $0.9964x$, slightly less (99.64% of the original cost)   **73. (a)** $(x + 0.15x) - 0.20(x + 0.15x)$
(b) $0.80(1.15x) = 0.92x$ **75.** Determine if all the terms contain a GCF and, if so, factor it out.
77. (b) $2x^2y$ **(c)** $2x^2y(3y^4 - x + 6x^7y^2)$ **79.** Both are correct

CUMULATIVE REVIEW EXERCISES

81. $25   **82.** 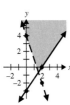   **83.** $\dfrac{3}{4y^2} - \dfrac{3}{2x} + \dfrac{3y}{x}$ **84.** $3x^2 + 2x - 6$

GROUP ACTIVITY/CHALLENGE PROBLEMS

1. $2(x - 3)(2x^4 - 12x^3 + 15x^2 + 9x + 2)$ **3. (a)** $(x + 1)[(x + 1) + 1] = (x + 1)(x + 2)$
(b) $(x + 1)^2[(x + 1) + 1] = (x + 1)^2(x + 2)$ **(c)** $(x + 1)^{n-1}[(x + 1) + 1] = (x + 1)^{n-1}(x + 2)$
(d) $(x + 1)^n[(x + 1) + 1] = (x + 1)^n(x + 2)$ **4.** $x^{1/2}(x + 1)$ **5.** $3x^{1/3}(1 + 2x + 4x^2)$ **6.** $5x^{-3}(2x + 5)$
7. $3n^{-3}(3n^2 + 2n - 4)$ **8.** $2x(x + 5)^{-2}[2 + (x + 5)^1] = 2x(x + 5)^{-2}(x + 7)$
9. $(2x - 3)^{-1/2}[x + 6(2x - 3)^1] = (2x - 3)^{-1/2}(13x - 18)$ **11.** $x^{2mn}(1 + x^{2mn})$ **13.** $r^{y+2}(r^2 + r + 1)$
14. $(b^r - d^r)(a^r + c^r)$ **15.** $(2a^k - 3)(3b^k - c^k)$

EXERCISE SET 6.2

1. $(x + 6)(x + 1)$ **3.** $(y - 1)(y - 11)$ **5.** $(x - 8)^2$ **7.** Cannot be factored   **9.** Cannot be factored
11. $(x - 4y)(x - 2y)$ **13.** $-5(x + 1)(x + 3)$ **15.** $(x - 15y)(x + 3y)$ **17.** $x(x + 14)(x - 3)$
19. $(4w + 1)(w + 3)$ **21.** $-(x + 5)(3x - 1)$ **23.** $(3y - 5)(y + 1)$ **25.** $(2x + 3y)(2x - y)$
27. $2(4x^2 + x - 10)$ **29.** $2(2x - 3y)(2x + y)$ **31.** $(7x - 3)(5x + 4)$ **33.** $2(4x - 5)(x - 3)$
35. $x^3y(x - 6)(x + 3)$ **37.** $ab(a + 7)(a - 5)$ **39.** $6pq^2(p - 5q)(p + q)$ **41.** $(6x - 5)(3x + 4)$
43. $(3x - 20)(2x - 1)$ **45.** $8x^2y^4(x + 4)(x - 1)$ **47.** $(x^2 + 3)(x^2 - 2)$ **49.** $(x^2 + 2)(x^2 + 3)$
51. $(2a^2 + 5)(3a^2 - 5)$ **53.** $(2x + 5)(2x + 3)$ **55.** $(3a + 1)(2a + 5)$ **57.** $(ab - 3)(ab - 5)$
59. $(3xy - 5)(xy + 1)$ **61.** $(2a - 5)(a - 1)(5 - a)$ **63.** $(2x + 3)(x + 2)(x - 3)$ **65.** $(y^2 - 10)(y^2 + 3)$
67. $(x + 2)(x + 1)(x + 3)$ **69.** $a^3b^2(5a - 3b)(a - b)$ **71. (a)** $2x + 3$ **(b)** $(x - 5)(2x + 3)$ **73. (a)** $5x - 8$
(b) $(6x - 1)(5x - 8)$ **75. (a)** $x^2 - 4x + 3$ **(b)** $(x - 2)(x - 1)(x - 3)$ **77. (a)** $2x^2 + 5x + 2$
(b) $(x - 4)(x + 2)(2x + 1)$ **79.** Factor out the GCF, if there is one.   **81.** $24x^2 - 66x + 45$
83. Divide $(x^2 - xy - 6y^2)$ by $(x - 3y)$; $x + 2y$ **85. (b)** $(2x - 3)(3x + 4)$

CUMULATIVE REVIEW EXERCISES

87. 0; change in y is 0.   **88.** Slope is undefined; change in x is 0 and you cannot divide by 0.   **89.** 9.0×10^{10}
90. $-x^2y - 8xy^2 + 6$ **91. (a)** $x = 2$ **(b)**

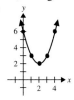

GROUP ACTIVITY/CHALLENGE PROBLEMS

2. (b) $(6x - 5)(5x + 8), (7x - 1)(7x - 13)$ **3. (a)** 6 or -6 **(b)** $c + 1$ or $-(c + 1)$ **5. (a)** 169
(b) $(3y + 2)(2y - 3)$ **7. (a)** 12   **(b)** Not factorable   **8.** $(2a^n + 3)(2a^n - 5)$ **9.** $(a + b)^2(a - 3b)$
10. $(x + y)^2(x - 4y)(x - 3y)$ **11.** $(m - 2n)^2(3m + 2n)$ **13.** $2(3x^ny^n - 1)(2x^ny^n + 1)$
15. $a^m(8a^{2m} + 3)(a^{2m} - 2)$

EXERCISE SET 6.3

1. $(x + 9)(x - 9)$ **3.** Cannot be factored   **5.** $(1 + 3a)(1 - 3a)$ **7.** $(5 + 4y^2)(5 - 4y^2)$
9. $(9a^2 + 4b)(9a^2 - 4b)$ **11.** $(ab + 7c)(ab - 7c)$ **13.** $x^2(3y + 2)(3y - 2)$ **15.** $x(12 - x)$
17. $(a + 3b + 2)(a - 3b - 2)$ **19.** $(x + 5)^2$ **21.** $(2 + x)^2$ **23.** $(2x - 5y)^2$ **25.** $(3a + 2)^2$ **27.** $(w^2 + 8)^2$
29. $(x + y + 1)^2$ **31.** $(a + b)^2(a - b)^2$ **33.** $(x + 3 + y)(x + 3 - y)$ **35.** $(x + 7)(-x + 3)$
37. $(3a - 2b + 3)(3a - 2b - 3)$ **39.** $4xy$ **41.** $(x - 3)(x^2 + 3x + 9)$ **43.** $(x + y)(x^2 - xy + y^2)$
45. $(4 - a)(16 + 4a + a^2)$ **47.** $(3y - 2x)(9y^2 + 6xy + 4x^2)$ **49.** $3(2x - 3y)(4x^2 + 6xy + 9y^2)$
51. $5(x - 5y)(x^2 + 5xy + 25y^2)$ **53.** $(x + 2)(x^2 + x + 1)$ **55.** $(x - y - 3)(x^2 - 2xy + y^2 + 3x - 3y + 9)$
57. $-9(b^2 + 3b + 3)$ **59.** $(11y^2 + 7x)(11y^2 - 7x)$ **61.** $(4y + 9x)(4y - 9x)$
63. $(5x^2 + 9y^3)(5x^2 - 9y^3)$ **65.** $(x - 4)(x^2 + 4x + 16)$ **67.** $(3xy + 4)^2$ **69.** $(a^2 + b^2)^2$
71. $(x - 1 + y)(x - 1 - y)$ **73.** $(x + y + 1)(x^2 + 2xy + y^2 - x - y + 1)$ **75.** $3m(-m + 2n)$ or $-3m(m - 2n)$
77. (a) $a^2 - b^2$ **(b)** $(a + b)(a - b)$ **79. (a)** $6a^3 - 6ab^2$ **(b)** $6a(a + b)(a - b)$ **81. (a)** $\frac{4}{3}\pi R^3 - \frac{4}{3}\pi r^3$
(b) $\frac{4}{3}\pi(R - r)(R^2 + Rr + r^2)$ **83.** $-12, +12$ **85.** $c = 4$ **87.** For the second term to be positive, the factors must
be of the form $(a + b)(a + b)$ or $(a - b)(a - b)$. Neither of these have a product of $a^2 + b^2$.

CUMULATIVE REVIEW EXERCISES

88. (a) $3, 6$ **(b)** $-2, \frac{5}{9}, -1.67, 0, 3, 6$ **(c)** $\sqrt{3}, -\sqrt{6}$ **(d)** $-2, \frac{5}{9}, -1.67, 0, \sqrt{3}, -\sqrt{6}, 3, 6$ **89.** $\in$ **90.** $\subseteq$
91. ≈ 0.44 **92.** $w = 3$ ft, $l = 8$ ft

GROUP ACTIVITY/CHALLENGE PROBLEMS

1. $(x + \sqrt{7})(x - \sqrt{7})$ **3.** $b = 6, (x + 3)^2$ **5.** $c = 4, (x + 2)^2$ **7.** $\left(\frac{b}{2}\right)^2 = c$
9. $(m - 9)(m^2 - 3mn + 3n^2 + 9m - 27n + 81)$ **10.** $(8x^{2a} + 3y^{3a})(8x^{2a} - 3y^{3a})$ **11.** $p^{6w}(4p^w + 7)(4p^w - 7)$
13. $4(6r^{4k} + 1)^2$ **15.** $x^{3m}(3 + 4x^m)(9 - 12x^m + 16x^{2m})$ **16. (a)** $(x^3 + 1)(x^3 - 1)$ **(b)** $(x^2 - 1)(x^4 + x^2 + 1)$

EXERCISE SET 6.4

1. $3(x + 4)(x - 3)$ **3.** $(5s - 3)(2s + 5)$ **5.** $-(4r - 3)(2r - 5)$ **7.** $2(x + 6)(x - 6)$ **9.** $5x(x^2 + 3)(x^2 - 3)$
11. $3x^4(x - 4)(x - 1)$ **13.** $5x^2y^2(x - 3)(x + 4)$ **15.** $x^2(x + y)(x - y)$ **17.** $x^4y^2(x - 1)(x^2 + x + 1)$
19. $x(x^2 + 4)(x + 2)(x - 2)$ **21.** $4(x^2 + 2y)(x^4 - 2x^2y + 4y^2)$ **23.** $2(a + b + 3)(a + b - 3)$ **25.** $6(x + 3y)^2$
27. $x(x + 4)$ **29.** $(2a + b)(a - 2b)$ **31.** $(y + 5)^2$ **33.** $5a^2(3a - 1)^2$ **35.** $(x + \frac{1}{3})(x^2 - \frac{1}{3}x + \frac{1}{9})$
37. $(x + 3)(x - 3)(3x + 2)$ **39.** $ab(a + 4b)(a - 4b)$ **41.** $(9 + x + y)(9 - x - y)$ **43.** $2(3x - 2)(4x - 3)$

45. $(8x + 3)(2x - 5)$ **47.** $(x^2 + 9)(x + 3)(x - 3)$ **49.** $(5c - 6y)(b - 2x)$ **51.** $(3x^2 - 4)(x^2 + 1)$
53. $(y + x - 4)(y - x + 4)$ **55.** $3(2x + 3y)(4a + 3)$ **57.** $(x^3 - 6)(x^3 - 5)$ **59.** $y(1 + y)(1 - y)$
61. $(2xy + 3)^2$ **63.** $(3rs - 1)(2rs + 1)$ **65. (a)** $a(a + b) - b(a + b)$ or $a^2 - b^2$ **(b)** $(a - b)(a + b)$
67. (a) $a^2 + 2ab + b^2$ **(b)** $(a + b)^2$ **69. (a)** $a(a - b) + a(a - b) + b(a - b) + b(a - b)$ **(b)** $2(a + b)(a - b)$

CUMULATIVE REVIEW EXERCISES

71. -8 **72.** No solution   **73.** $\{x \mid x > 6\}$ **74.** $\mathbb{R}$

GROUP ACTIVITY/CHALLENGE PROBLEMS

1. (a) $x^{-4}(x^2 - 5x + 6)$ **(b)** $x^{-4}(x - 3)(x - 2)$ **3. (a)** $x^{1/2}(4x + x^{1/2} - 2)$ **(b)** Not possible

5. (a) $x^{-3/2}(5x^2 + 2x - 3)$ **(b)** $x^{-3/2}(5x - 3)(x + 1)$

EXERCISE SET 6.5

1. $0, -5$ **3.** $0, -9$ **5.** $-\frac{5}{2}, 3, -2$ **7.** 3 **9.** $0, 12$ **11.** $0, -2$ **13.** $-4, 3$ **15.** $2, 10$

17. $3, -6$ **19.** $-4, 6$ **21.** $0, 2, -21$ **23.** $-5, -6$ **25.** $-\frac{1}{6}, \frac{3}{2}$ **27.** $\frac{1}{4}, \frac{2}{7}$ **29.** $0, -\frac{1}{3}, 3$ **31.** $\frac{1}{3}, 7$ **33.** $\frac{2}{3}, -1$

35. $0, -4, 3$ **37.** $0, \frac{8}{3}$ **39.** $0, \frac{4}{5}, -\frac{4}{5}$ **41.** $0, -8$ **43.** $-3, 5$ **45.** $\frac{5}{2}, \frac{4}{3}$ **47.** $-\frac{1}{3}, 2$ **49.** $0, 1$ **51.** $0, -3, -5$

53. $4, -6$ **55.** -7 **57.** $-\frac{2}{5}, \frac{4}{3}$ **59.** $0, \frac{4}{3}, \frac{5}{2}$ **61.** $8, 9$ **63.** $9, 12$ **65.** $w = 3$ ft, $l = 12$ ft

67. $h = 10$ cm, $b = 16$ cm   **69.** 5 sec   **71.** 7 m   **73.** 5 sec   **75. (a)** Zero-factor property only holds when one side of the equation is 0.   **(b)** $-2, -5$ **77. (a)** One possible answer is $f(x) = x^2 + 7x + 10$
(b) One possible answer is $x^2 + 7x + 10 = 0$ **(c)** An infinite number; the function $f(x) = a(x^2 + 7x + 10)$ where a is any non-zero real number, will have x intercepts of -2 and -5 **(d)** An infinite number; the equation $a(x^2 + 7x + 10) = 0$ where a is any non-zero real number, will have solutions -2 and -5 **79. (a)** one; the x intercept is at $(3, 0)$
(b) one; it will have its x intercept at $(r, 0)$ **(c)** $x^2 - 4x + 4 = 0$ **(d)** $y = x^2 - 4x + 4$
81. (a)   **(b)** none, one or two

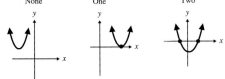

CUMULATIVE REVIEW EXECISES

83. (a) Carmen 4.8 mph, Bob 6 mph   **(b)** 24 miles   **84.**   **85.** $(3, 0)$ **86.** $\varnothing$ **87.** $3x + 4$

GROUP ACTIVITY/CHALLENGE PROBLEMS

1. $x^2 + 3x - 10 = 0$ (other answers are possible for 1 through 4).   **3.** $x^3 - 2x^2 - 3x = 0$

4. $x^4 - 4x^3 - 7x^2 + 22x + 24 = 0$ **5.** $1, -9$ **6.** $2, -\frac{3}{2}$ **7.** $\frac{4}{3}, -\frac{1}{2}$ **9.** $\pm 1, \pm 2$ **10.** $\pm 2, \pm 3$ **11.** $2, 1$

12. $y' = \dfrac{1}{x + y}$ **13.** $y' = \dfrac{xy - 2}{x + 3}$ **14.** $y' = \dfrac{x + xy}{2xy + 3}$

GRAPHING CALCULATOR CORNER, PAGE 370

1. $(x - 4)(x - 12)$ **3.** $(x + 9)(x - 60)$ **5.** $-9, 6$ **7.** $0.2, -12$

REVIEW EXERCISES

1. $4x^2$ **2.** $6xy$ **3.** $3xy^2$ **4.** $2x - 5$ **5.** $x + 5$ **6.** 1 **7.** $4(3x^2 + x + 2)$ **8.** $6x^4(10 + x^5 - 3xy^2)$
9. $(2x + 1)(4x - 3)$ **10.** $(x - 1)(3x^2 - 3x - 2)$ **11.** $(2x - 1)(10x - 3)$ **12.** $3xy^2z^2(4y^2z + 2xy - 5x^2z)$
13. $(x + 2)(x + 3)$ **14.** $(5x - 1)(x + 4)$ **15.** $(x + 4y)(5x - y)$ **16.** $(4x + 5y)(3x - 2y)$
17. $(3x - y)(-4x + 9y)$ **18.** $3(a^2 + 3b)(a^2 - 4b)$ **19.** $(x + 5)(x + 3)$ **20.** $(x - 5)(x - 3)$
21. $-(x - 15)(x + 3)$ **22.** $(x - 10y)(x + 5y)$ **23.** $(x - 18y)(x + 3y)$ **24.** $2(x + 4)^2$ **25.** $(6x + 5)(x - 4)$
26. $x(4x - 5)(2x + 5)$ **27.** $(4x - 5y)(4x + 7y)$ **28.** $x(4x - 5)(x - 1)$ **29.** $x(12x + 1)(x + 5)$
30. $(x^2 - 5)(x^2 + 2)$ **31.** $(x^2 + 4)(x^2 - 5)$ **32.** $(x + 9)(x + 11)$ **33.** $(3x + 8)(x - 4)$ **34. (a)** $8x + 3$
(b) $(3x + 7)(8x + 3)$ **35. (a)** $9x^2 + 15x - 50$ **(b)** $(x - 3)(3x - 5)(3x + 10)$ **36. (a)** $3x^2 - 17x + 20$
(b) $(2x - 5)(x - 4)(3x - 5)$ **37.** $(x + 6)(x - 6)$ **38.** $4(x + 2y^2)(x - 2y^2)$ **39.** $(x^2 + 9)(x - 3)(x + 3)$
40. $(x - 1)(x + 5)$ **41.** $(x - 1)(x - 5)$ **42.** $(2x - 3)^2$ **43.** $(3y + 4)^2$ **44.** $(w^2 - 8)^2$
45. $(a + 3b + 2c)(a + 3b - 2c)$ **46.** $(x - 2)(x^2 + 2x + 4)$ **47.** $(2x + 3)(4x^2 - 6x + 9)$
48. $(3x - 2y)(9x^2 + 6xy + 4y^2)$ **49.** $(2y^2 - 5x)(4y^4 + 10xy^2 + 25x^2)$ **50.** $(x - 1)(x^2 + 4x + 7)$
51. $y^2(x + 3)(x - 5)$ **52.** $3x(x - 4)(x - 2)$ **53.** $3xy^4(x - 2)(x + 6)$ **54.** $3y(y^2 + 3)(y^2 - 3)$
55. $2y(x + 2)(x^2 - 2x + 4)$ **56.** $5x^2y(x + 2)^2$ **57.** $3x(2x + 1)(x - 4)$ **58.** $(x + 5 + y)(x + 5 - y)$
59. $3(x + 2y)(x^2 - 2xy + 4y^2)$ **60.** $(x + 4)^2(x - 1)$ **61.** $(4x + 1)(4x + 5)$ **62.** $(2x^2 - 1)(2x^2 + 3)$
63. $(x + 1)(x - 2)(x - 1)$ **64.** $(3x + 4y)(3a - b)$ **65.** $(2pq - 3)(3pq + 2)$ **66.** $(3x^2 - 2)^2$
67. $(2y + x + 2)(2y - x - 2)$ **68.** $2(3a + 5)(4a + 3)$ **69.** $3x^2y^4(x + 3)(2x - 3)$
70. $(x - \frac{2}{3}y^2)(x^2 + \frac{2}{3}xy^2 + \frac{4}{9}y^4)$ **71. (a)** $a^2 - 4b^2$ **(b)** $(a + 2b)(a - 2b)$ **72. (a)** $2ab + 2b^2$ **(b)** $2b(a + b)$
73. (a) $2a(a + 3b) + b(a + 3b)$ **(b)** $(2a + b)(a + 3b)$ **74. (a)** $a^2 + 2ab + b^2$ **(b)** $(a + b)^2$ **75.** $5, -\frac{2}{3}$
76. $0, \frac{3}{2}$ **77.** $0, -\frac{4}{3}$ **78.** $6, -4$ **79.** $-3, -5$ **80.** $-4, 2$ **81.** $-2, -5$ **82.** $0, 2, 4$ **83.** $0, \frac{4}{3}, -\frac{1}{4}$
84. $\frac{1}{4}, -\frac{3}{2}$ **85.** $2, -2$ **86.** $2, -3$ **87.** $-5, 6$ **88.** $\frac{6}{5}, \frac{5}{4}$ **89.** $\frac{7}{2}, -\frac{4}{7}$ **90.** $5, 6$ **91.** $w = 7$ ft, $l = 9$ ft
92. $h = 4$ m, $b = 11$ m **93.** 5 in, 9 in **94.** 9 sec

PRACTICE TEST

1. $4x(xy - 1)$ **2.** $(3x + 2)(x + 4)$ **3.** $3xy^2(3x^2 + 4xy^3 - 9y^2)$ **4.** $5(x - 2)(x + 1)$ **5.** $(2x + 3y)(x + 2y)$
6. $(x - 4y)(x - 3y)$ **7.** $3x(x - 3)(x + 1)$ **8.** $(3x - 2)(2x - 1)$ **9.** $(5x + 2)(x + 3)$ **10.** $(9x + 4y^2)(9x - 4y^2)$
11. $y^6(3x - 2)(9x^2 + 6x + 4)$ **12.** $(x + 2)(x + 6)$ **13.** $(2x^2 + 9)(x^2 - 2)$ **14. (a)** $4x^2 - 13x - 12$
(b) $(x + 2)(x - 4)(4x + 3)$ **15.** $5, -\frac{2}{3}$ **16.** $-\frac{3}{4}, 6$ **17.** $0, -5, 1$ **18.** $\frac{1}{4}, -\frac{3}{2}$ **19.** $h = 4$ m, $b = 14$ m
20. 7 sec

CUMULATIVE PRACTICE TEST

1. $-\frac{9}{8}$ **2.** $-\frac{15}{14}$ **3.** $L = \dfrac{12P + W}{2}$ **4.** 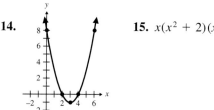 **5. (a)** Yes, each x has a unique y **6.** $\left(\frac{20}{11}, -\frac{14}{11}\right)$
5. (b) no, $(1, 2)$ and $(1, 0)$ have the same x coordinate

7. 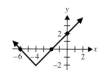 **8.** $4xy^9$ **9.** $\dfrac{27p^{11}q^9}{8}$ **10.** $7x^2 - 9x$ **11.** $2x^3 - 11x^2 + 3x + 30$

12. $3xy^4 - \dfrac{8y^3}{3} - \dfrac{4}{x}$ **13.** -5 **14.** 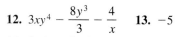 **15.** $x(x^2 + 2)(x - 3)$
16. $3y(x - 2)(4x - 1)$ **17.** $(y + 2)(y - 2)(y^2 + 6)$
18. $(2x - 3y^2)(4x^2 + 6xy^2 + 9y^4)$ **19.** 620 pages
20. $34 \le x < 84$

Chapter 7

EXERCISES SET 7.1

1. 0   **3.** 3, −3   **5.** 5, $\dfrac{5}{2}$ **7.** none   **9.** $\{x \mid x \text{ is a real number and } x \neq -3 \text{ and } x \neq 2\}$ **11.** $\{x \mid x \text{ is a real number}$

and $x \neq 3$ and $x \neq -7\}$ **13.** $\left\{z \mid z \text{ is a real number and } z \neq \dfrac{15}{8}\right\}$ **15.** $\{a \mid a \text{ is a real number and } a \neq \dfrac{1}{2} \text{ and } a \neq -2\}$

17. $1 - y$ **19.** x **21.** $\dfrac{2(x^2 + 2xy - 3y^2)}{3}$ **23.** -1 **25.** $-(x + 4)$ or $-x - 4$ **27.** $\dfrac{x + 3}{x - 5}$

29. $4x^2 + 10xy + 25y^2$ **31.** $\dfrac{2x - 5}{2}$ **33.** $\dfrac{x - 4}{x + 3}$ **35.** $\dfrac{a + 3}{a + 5}$ **37.** $\dfrac{x - 1}{x^2 - 3x + 9}$ **39.** $\dfrac{5x}{x^2}$ **41.** $\dfrac{5y(y - 6)}{y^2 - 3y - 18}$

43. $\dfrac{4p(p - 2)}{p^2(p - 2)}$ **45.** $\dfrac{p(3p + 2)}{6p^2 + p - 2}$ **47.** $\dfrac{xy^2}{4}$ **49.** $12x^3y^2$ **51.** 1   **53.** $\dfrac{12x^{16}yz}{5}$ **55.** $\dfrac{x^6y^{18}}{8}$ **57.** 1

59. 1   **61.** $\dfrac{1}{x - 8}$ **63.** $\dfrac{1}{x - 3}$ **65.** $\dfrac{(a + 1)^2}{9(a + b)^2}$ **67.** $\dfrac{x - 4}{4x + 1}$ **69.** $\dfrac{(x + 2)(x - 2)}{(x^2 + 2x + 4)(x^2 + 4)}$ **71.** $3x^2$

73. $\dfrac{x^3}{x + 2}$ **75.** $\dfrac{x^2(2x - 1)}{(x - 1)(x - 3)}$ **77.** $\dfrac{r + 5s}{2r + 5s}$ **79.** $\dfrac{2p - q^2}{p^2 + q^2}$ **81.** $\dfrac{2x(x + 2)}{x - 5}$ **83.** $x(x^2 - 4x + 16)$

85. $\dfrac{(a + b)^2}{ab}$ **87.** $\dfrac{x - 1}{x + 3}$ **89.** $x + 5$ **91.** $y^2 - 4y - 5$ **93.** (a) $x^2 + x - 2$ (b) Factors must be

$(x - 1)(x + 2)$ **95.** (a) $2x^2 + x - 6$ (b) Factors must be $(x + 2)(2x - 3)$ **97.** One possible answer is

$\dfrac{1}{(x - 2)(x + 3)}$ **99.** (a) 4   (b) 2 and -2 **101.** $\sqrt{x}$ is not a polynomial   **103.** (b) $\dfrac{3x - 4}{2x - 5}$ **105.** (b) 1

CUMULATIVE REVIEW EXERCISES

107. $h = \dfrac{3V}{4\pi r^2}$ **108.** $(0, 4)$ **109.** $m = -\dfrac{1}{3}$, y intercept $= \dfrac{14}{3}$ **110.** $\left(-\dfrac{2}{7}, \dfrac{15}{7}\right)$ **111.** $\dfrac{2x}{3}$

112. $3x^2y - 7xy - 4y^2 - 2x$

GROUP ACTIVITY/CHALLENGE PROBLEMS

1. (a) $\{x \mid x \text{ is a real number and } x \neq 0\}$ (b)
(c)

x	-10	-1	-0.5	-0.1	0.1	0.5	1	10
y	-0.1	-1	-2	-10	10	2	1	0.1

2. (a) $\{x \mid x \text{ is a real number and } x \neq 2\}$ (b)
(c)

x	-2	-1	0	1	1.9	2.1	3	4	5	6
y	0	1	2	3	3.9	4.1	5	6	7	8

3. x^y **4.** $\dfrac{m^x + 1}{m^x + 2}$ **5.** $2(a - b)$ **7.** $\dfrac{x - 2}{x^2}$ **9.** $\dfrac{a^2 + ab + b^2}{a + b}$

GRAPHING CALCULATOR CORNER, PAGE 381

1. (a) $\{x \mid x$ is a real number and $x \neq 0\}$ **3. (a)** $\{x \mid x$ is a real number and $x \neq 0\}$
5. (a) $\{x \mid x$ is a real number and $x \neq -2$ and $x \neq 2\}$

EXERCISE SET 7.2

1. $\dfrac{2x - 11}{3}$ **3.** $\dfrac{1}{x + 2}$ **5.** $\dfrac{x - 3}{x - 1}$ **7.** $\dfrac{5}{r - 2}$ **9.** $\dfrac{x + 5}{x + 3}$ **11.** $(x + 1)(x + 2)$ **13.** $48x^3 y$ **15.** $z - 3$

17. $(a - 8)(a + 3)(a + 8)$ **19.** $(x + 3)(x - 3)(x - 1)$ **21.** $(x - 1)(x + 4)(4x + 9)$ **23.** $\dfrac{10}{3x}$ **25.** $\dfrac{10y + 9}{12y^2}$

27. $\dfrac{25y^2 - 12x^2}{60x^4 y^3}$ **29.** $\dfrac{4 + 6y}{3y}$ **31.** $\dfrac{2x}{x - y}$ **33.** $\dfrac{20z}{(z + 5)(z - 5)}$ **35.** $\dfrac{-4x^2 + 25x - 36}{(x + 3)(x - 3)}$

37. $\dfrac{2m^2 + 5m - 2}{(m - 5)(m + 2)}$ **39.** $\dfrac{1}{x - 5}$ **41.** $\dfrac{2x^2 + 5x + 8}{(x - 1)(x + 4)(x - 2)}$ **43.** $\dfrac{5x^2 + 14x - 49}{(x + 5)(x - 2)}$ **45.** $\dfrac{3a - 1}{4a + 1}$

47. $\dfrac{x - 2}{(x - 1)(2x - 3)}$ **49.** $\dfrac{2x^2 - 4xy + 4y^2}{(x - 2y)^2 (x + 2y)}$ **51.** $\dfrac{12}{x - 3}$ **53.** 0 **55.** 1 **57.** $\dfrac{x - 6}{x - 4}$ **59.** $\dfrac{x^2 - 18x - 30}{(5x + 6)(x - 2)}$

61. $\dfrac{8m^2 + 5mn}{(2m + 3n)(3m + 2n)(2m + n)}$ **63.** 0 **65.** $\dfrac{11m + 6n}{(2m - n)(m - 2n)(3m + 2n)}$ **67.** $\dfrac{1}{2x + 3y}$ **69.** $\dfrac{3x + 10}{x - 2}$

71. $\dfrac{1}{(a - 5)(a + 3)}$ **73.** $\dfrac{x^2 + 5x + 8}{(x + 1)(x + 2)}$ **75.** $\dfrac{4x^2 - 13x + 5}{(2x - 1)(x - 3)}$ **77. (a)** $-5x^2 + x + 6$ **79. (a)** $r^2 - r - 3$

81. (b) $\dfrac{13x - 40}{(x + 2)(3x - 10)}$ **85. (a)** 8 **(b)** $\dfrac{4b^2 - a^2}{b^2}$

CUMULATIVE REVIEW EXERCISES

86. (a) 6 min **(b)** 960 bottles **87.** $\$180$ **88.** $3x^3 y^2 - \dfrac{x^2}{y^2} + \dfrac{5y}{3}$ **89.** $2x - 1$

GROUP ACTIVITY/CHALLENGE PROBLEMS

1. (a) $\dfrac{ax + bn - bx}{n}$ **(b)** 79.2 **3. (a)** $\dfrac{x + 1}{x}$ **(b)** $\dfrac{x^2 + x + 1}{x^2}$ **(c)** $\dfrac{x^5 + x^4 + x^3 + x^2 + x + 1}{x^5}$

(d) $\dfrac{x^n + x^{n - 1} + x^{n - 2} + \cdots + 1}{x^n}$ **5.** $\dfrac{4x^2 - 9x + 27}{(x + 3)(x - 3)(x^2 - 3x + 9)}$ **7.** $\dfrac{4x - 1}{(2x - 4)(2x + 3)}$ **8.** $\dfrac{2ab}{(a - b)(a + b)}$

EXERCISE SET 7.3

1. $\dfrac{8}{11}$ **3.** $\dfrac{57}{32}$ **5.** $\dfrac{25}{1224}$ **7.** $\dfrac{x^3 y}{8}$ **9.** $6xz^2$ **11.** $\dfrac{xy + 1}{x}$ **13.** $\dfrac{3}{x}$ **15.** $\dfrac{3y - 1}{2y - 1}$ **17.** $\dfrac{-a}{b}$ **19.** -1

21. $\dfrac{2(x + 2)}{x^3}$ **23.** $\dfrac{x - 2}{8}$ **25.** $\dfrac{x + 1}{1 - x}$ **27.** $\dfrac{-4a}{a^2 + 4}$ **29.** $\dfrac{5m^2 - m - 1}{2m^2}$ **31.** $\dfrac{2 + a^2 b}{a^2}$ **33.** $\dfrac{ab}{b + a}$

35. $\dfrac{b(1 + a)}{a(1 - b)}$ **37.** $\dfrac{y - x}{x + y}$ **39.** $\dfrac{(a + b)^2}{ab}$ **41.** $\dfrac{6y - x}{3xy}$ **43.** $\dfrac{2y - 3xy + 5y^2}{3y^2 - 2x}$ **45. (a)** $\dfrac{2}{7}$ **(b)** $\dfrac{4}{13}$

47. $R_T = \dfrac{R_1 R_2 R_3}{R_2 R_3 + R_1 R_3 + R_1 R_2}$ **49.** A complex fraction is one that has a fractional expression in its numerator or denominator or both its numerator and denominator.

CUMULATIVE REVIEW EXERCISES

51. $\dfrac{13}{48}$ **52.** $\{x \mid x \leq -\dfrac{5}{2}$ or $x \geq \dfrac{13}{2}\}$ **53.**

54. $x^2 - 5x - 23 - \dfrac{37}{x - 2}$

GROUP ACTIVITY/CHALLENGE PROBLEMS

1. $\dfrac{x+1}{2}$ **2.** $\dfrac{x+15}{x}$ **3.** $\dfrac{7x}{5(x+2)}$ **4.** $\dfrac{4a^2+1}{4a(2a^2+1)}$ **5.** $\dfrac{x^2+x+1}{x^3+x^2+2x+1}$ **6.** $\dfrac{5}{12}$

EXERCISE SET 7.4

1. 4 **3.** -30 **5.** -1 **7.** 4 **9.** 3 **11.** All real numbers **13.** $\frac{1}{4}$ **15.** $\frac{14}{3}$ **17.** No solution **19.** $\frac{6}{5}$

21. ≈ -1.63 **23.** 8 **25.** $-\frac{4}{3}, 1$ **27.** 3.76 **29.** $-2, -3$ **31.** 4 **33.** $-\frac{5}{2}$ **35.** 5 **37.** No solution

39. No solution **41.** $\frac{17}{4}$ **43.** 12, 2 **45.** 12, 4 **47.** $w = \dfrac{fl - df}{d}$ **49.** $p = \dfrac{qf}{q - f}$ **51.** $R_1 = \dfrac{R_T R_2}{R_2 - R_T}$

53. $\bar{x} = \mu + \dfrac{z\sigma}{\sqrt{n}}$ **55.** $q = q' + E_{\in_0} A$ **57.** $C_T = \dfrac{C_1 C_2 C_3}{C_2 C_3 + C_1 C_3 + C_1 C_2}$ **59.** (a) $\dfrac{2x + 7}{(x - 2)(x + 2)}$

(b) $-\dfrac{7}{2}$ **61.** (a) $\dfrac{4}{b + 5}$ **(b)** No solution **63.** 12 in **65.** 17.14 cm **67.** object: 16 cm, image: 48 cm

69. 155.6 ohms **71.** 176.47 ohms **73.** A number obtained when solving an equation which is not a true solution to the original equation. **75.** (a) Multiply both sides of the equation by the LCD, 12. This removes fractions from the equation.
(b) -24 **(c)** Write each term with the common denominator, 12. This step will allow the fractions to be added and

subtracted. **(d)** $\dfrac{-x + 24}{12}$ **77.** You cannot just change an expression to an equation by adding 0.

CUMULATIVE REVIEW EXERCISES

79. $\frac{3}{2}$ **80.** Domain: $\{x \,|\, x \geq 4\}$
Range: $\{y \,|\, y \geq 2\}$

81. Domain: $\mathbb{R}$
Range: $\{y \,|\, y \geq -10\}$

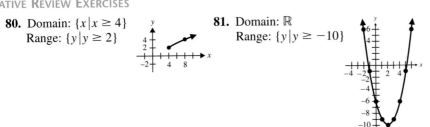

82. $(2x - 4y^2)(4x^2 + 8xy^2 + 16y^4)$

GROUP ACTIVITY/CHALLENGE PROBLEMS

1. There are many answers. One answer is $\dfrac{1}{x - 4} + \dfrac{1}{x + 2} = 0$. **2.** There are many answers. One answer is $\dfrac{x}{2} + \dfrac{x}{2} = x$.

3. There are many answers. One answer is $\dfrac{1}{x} + \dfrac{1}{x} = \dfrac{2}{x}$. **5.** No solution **6.** (a) 9.71% **(b)** Tax free money market
since 9.71% > 7.68% **7.** 115.96 days

GRAPHING CALCULATOR CORNER, PAGE 414

1. -6.0 **3.** 6.0

EXERCISE SET 7.5

1. $\frac{20}{9}$ or 2.22 hr **3.** $\frac{40}{13}$ or 3.08 hr **5.** 100 hr **7.** 7.8 months **9.** 7.3 hr **11.** $\frac{60}{7}$ or 8.57 hr **13.** 18 hr

15. $\frac{160}{3}$ or 53.33 hr **17.** -4 **19.** 5, 6 **21.** 3 **23.** All real numbers except 3 **25.** 1 or 2 **27.** 4 mph

29. 12 mi **31.** (a) $\frac{300}{8}$ or 37.5 mi **(b)** 9:30 AM **33.** ≈ 2.39 mph **35.** train, 60 mph; plane, 300 mph
37. 108,000 mi

38. $-\frac{5}{2}x^2 + \frac{7}{5}xy - 6y^2$ **39.** $12x^3 - 34x^2 + 21x + 4$ **40.** $4x + 5 + \dfrac{22}{3x - 2}$ **41.** $(4x + 3)(2x + 5)$

Group Activity/Challenge Problems

1. (a) 2.5 mi   **(b)** $0.02\overline{7}$ hr. $= 1.\overline{6}$ min   **(c)** 690 mph   **(d)** No   **2.** $2.73\overline{148}$ km/hr

Exercise Set 7.6

1. Direct   **3.** Direct   **5.** Direct   **7.** Inverse   **9.** Direct   **11.** Direct   **13.** Inverse   **15.** Direct
17. (a) $C = kZ^2$ **(b)** 60.75   **19. (a)** $x = k/y$ **(b)** 0.2   **21. (a)** $L = k/P^2$ **(b)** 6.25   **23. (a)** $A = kR_1R_2/L^2$
(b) 57.6   **25. (a)** $x = ky$ **(b)** 18   **27. (a)** $y = kR^2$ **(b)** 20   **29. (a)** $C = k/J$ **(b)** 0.41   **31. (a)** $F = kM_1M_2/d$
(b) 40   **33. (a)** $S = kIT^2$ **(b)** 0.2   **35. (a)** $S = kF$ **(b)** 0.7 in   **37.** $R = kA/P$; 4600   **39.** $W = k/d^2$; 133.25 lb
41. $R = kL/A$; 25 ohms   **45. (a)** Inveresly   **(b)** stays 0.3

46. $d = -5$ **47.** weekly salary $400, commission rate 3%   **48.**   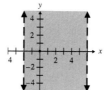   **49.**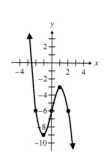

Group Activity/Challenge Problems

1. $\frac{1}{9}$ **3. (a)** $F = km_1m_2/d^2$ **(b)** The force is 24 times greater than the original force.

Review Exercises

1. 4   **2.** -1 **3.** None   **4.** $\{x \mid x \text{ is a real number and } x \neq -3\}$ **5.** $\{x \mid x \text{ is a real number and } x \neq 0\}$
6. $\{x \mid x \text{ is a real number and } x \neq 5 \text{ and } x \neq -2\}$ **7.** x **8.** $x - 3$ **9.** -1 **10.** $\dfrac{x - 1}{x - 2}$ **11.** $\dfrac{x - 3}{x + 1}$
12. $\dfrac{a^2 + 2a + 4}{a + 2}$ **13.** $x(x + 1)$ **14.** $(x + y)(x - y)$ **15.** $(x + 7)(x - 5)(x + 2)$ **16.** $(x + 2)^2(x - 2)(x + 1)$
17. $6xz^2$ **18.** $-\dfrac{1}{2}$ **19.** $4x^3z^5$ **20.** $\dfrac{3}{x}$ **21.** $\dfrac{(x + y)y^2}{2x^3}$ **22.** $3x + 2$ **23.** 1   **24.** $\dfrac{-3.2x^2 - 8.2x + 7.3}{1.7x^2 - 2.4x + 1.5}$
25. $\dfrac{5x^2 - 12y}{3x^2y}$ **26.** $\dfrac{7x + 12}{x + 2}$ **27.** $\dfrac{5x + 12}{x + 3}$ **28.** $\dfrac{x + y}{x}$ **29.** $\dfrac{1}{3(a + 3)}$ **30.** $\dfrac{a^2 + c^2}{ac}$ **31.** $\dfrac{x + 1}{2x - 1}$
32. 1   **33.** $2x(x - 5y)$ **34.** $\dfrac{-3x^2 + 2x - 4}{3x(x - 2)}$ **35.** $\dfrac{x^2 - 2x - 5}{(x + 5)(x - 5)}$ **36.** $\dfrac{3(x - 1)}{(x + 3)(x - 3)}$ **37.** $\dfrac{1}{a + 2}$
38. $\dfrac{16(x - 2y)}{3(x + 2y)}$ **39.** $\dfrac{4}{(x + 2)(x - 3)(x - 2)}$ **40.** $\dfrac{2}{x(x + 2)}$ **41.** $\dfrac{2(x - 4)}{(x - 3)(x - 5)}$ **42.** $\dfrac{22x + 5}{(x - 5)(x - 10)(x + 5)}$
43. $\dfrac{x^3 + y^2}{x^3 - y^2}$ **44.** $\dfrac{-1}{x - 3}$ **45.** $\dfrac{(x + 3)(x - 1)}{4(x + 1)}$ **46.** $\dfrac{x + 3}{x + 5}$ **47.** $\dfrac{x - 4}{x - 6}$ **48.** $\dfrac{x^2 + 6x - 24}{(x - 1)(x + 9)}$ **49.** $\dfrac{55}{26}$
50. $\dfrac{5yz}{6}$ **51.** $\dfrac{xy + 1}{y^3}$ **52.** $\dfrac{4x + 2}{6x^2 - x}$ **53.** $\dfrac{2a + 1}{2}$ **54.** $\dfrac{x + 1}{-x + 1}$ **55.** 9   **56.** 2.8   **57.** 6   **58.** $-7.\overline{5}$
59. 52   **60.** 2.4   **61.** 18   **62.** $\frac{1}{2}$ **63.** -6 **64.** -18 **65.** 120 ohms   **66.** 900 ohms, 1800 ohms   **67.** 3 cm

68. 15 cm   **69.** $\frac{12}{7}$ or 1.71 hr   **70.** $\frac{3000}{35}$ or 85.71 hr   **71.** 2   **72.** $\frac{5}{6}$ **73.** 5 mph   **74.** 50 mph, 150 mph
75. 75   **76.** 20   **77.** 1   **78.** 20   **79.** 426.7   **80.** 5 in   **81.** \$119.88   **82.** 400 ft   **83.** 200.96   **84.** 2.38 min

Practice Test

1. (a) $\{x \mid x$ is a real number and $x \neq 7$ and $x \neq -4\}$ **(b)** $\dfrac{1}{x-7}$ **2.** $\dfrac{12x^4y^2}{z^2}$ **3.** $a + 3$ **4.** $\dfrac{x - 3y}{3}$ **5.** $x^2 + y^2$
6. $\dfrac{10x + 3}{2x^2}$ **7.** $\dfrac{-1}{(x+4)(x-4)}$ **8.** $\dfrac{x(x+10)}{(2x-1)^2(x+3)}$ **9.** $\dfrac{y + x}{y - x}$ **10.** $\dfrac{x^2(y+1)}{y+x}$ **11.** 60   **12.** 12   **13.** 10
14. 1.125   **15.** $\dfrac{40}{13}$ or 3.08 hr

Chapter 8

Exercise Set 8.1

1. 5   **3.** −3   **5.** 5   **7.** Not a real number   **9.** −2   **11.** 23   **13.** −1   **15.** 7   **17.** Not a real number
19. Not a real number   **21.** −4.68   **23.** $\frac{5}{3}$ **25.** Not a real number   **27.** 1.75   **29.** 6   **31.** 1   **33.** 43
35. 147.23   **37.** 0.03   **39.** $\frac{156}{5}$ or 31.2   **41.** $|y - 8|$ **43.** $|x - 3|$ **45.** $|3x + 5|$ **47.** $|6 - 3x|$
49. $|y^2 - 4y + 3|$ **51.** $|8a - b|$ **53.** $|a^4|$ **55.** $|a + b|$ **57. (a)** Two, positive or principal square root and negative square root.   **(b)** 6, −6   **(c)** The positive square root   **(d)** 6   **59.** No real number when squared gives −49.
61. No; example: $\sqrt{-2}$ is not a real number.   **63.** $x = a^n$ **67.** $x \geq -4$ **69.** $x \geq 1$
71. When n is a even integer and $x < 0$ **73.** When n is an even integer, m is an odd integer, and $x < 0$.

Cumulative Review Exercises

74. $3(y - 3 + z)(y - 3 - z)$ **75.** $\left(x + \frac{1}{3}\right)\left(x^2 - \frac{1}{3}x + \frac{1}{9}\right)$ **76.** $(x - 2)(x + 5)$ **77.** $x(x^2 - 3)(2x - 3)$

Group Activity/Challenge Problems

7. $|3m - 5n|$ **9.** $\sqrt{5} \approx 2.24$ **10.** $\sqrt[8]{6} \approx 1.25$ **11.** $\sqrt{x}$

Exercise Set 8.2

1. $x^{3/2}$ **3.** $4^{5/2}$ **5.** $x^{4/5}$ **7.** $x^{3/2}$ **9.** $5^{13/4}$ **11.** $y^{23/8}$ **13.** $\sqrt{x}$ **15.** $\sqrt{z^3}$ **17.** $\sqrt{24y^2}$
19. $\dfrac{1}{\sqrt{19x^2y^4}}$ **21.** $\left(\sqrt[5]{2m^2n^3}\right)^2$ **23.** $\dfrac{1}{(\sqrt[3]{a^2 - 4b^2})^2}$ **25.** y^3 **27.** z^4 **29.** x^3 **31.** $\sqrt{z}$ **33.** 5.1
35. y **37.** $\sqrt[4]{x}$ **39.** x^5 **41.** 4   **43.** $\sqrt[3]{y}$ **45.** 2   **47.** 9   **49.** Not a real number   **51.** $\dfrac{3}{5}$ **53.** −4
55. −3   **57.** $\dfrac{1}{3}$ **59.** $\dfrac{1}{8}$ **61.** $-\dfrac{7}{2}$ **63.** $\dfrac{81}{256}$ **65.** 18   **67.** $\dfrac{10}{21}$ **69.** $x^{11/2}$ **71.** $x^{1/6}$ **73.** $x^{2/15}$ **75.** $\dfrac{1}{x}$
77. 1   **79.** $\dfrac{y^{5/3}}{12}$ **81.** $\dfrac{8}{x^{11/6}}$ **83.** $\dfrac{1}{2x^{1/3}}$ **85.** $\dfrac{121}{x^{1/7}}$ **87.** $\dfrac{64}{y^{66/5}}$ **89.** $\dfrac{x}{y^{16}}$ **91.** $\dfrac{25a^{5/6}}{b^2}$ **93.** $\dfrac{1}{r^{5/4}}$
95. $6z^{7/2} - 3$ **97.** $\dfrac{5}{x^5} + \dfrac{10}{x^{3/2}}$ **99.** $8x^{13/6} - 4x^2$ **101.** $x^{1/2}(x + 1)$ **103.** $y^{1/3}(1 - y)$ **105.** $\dfrac{1+y}{y^{3/5}}$
107. $\dfrac{1 - y^2}{y}$ **109.** $\dfrac{1 + x^2}{x^7}$ **111.** $\dfrac{x^2 + 1}{x^{5/2}}$ **113.** $\dfrac{2(x - 3)}{x^5}$ **115.** $(x^{1/3} + 3)(x^{1/3} - 1)$
117. $(x^{1/2} + 3)(x^{1/2} + 3)$ or $(x^{1/2} + 3)^2$ **119.** $(2x^{1/7} + 3)(x^{1/7} - 1)$ **121.** $(2x^{2/5} + 3)(2x^{2/5} + 1)$
123. $(5x^{1/6} - 3)(3x^{1/6} - 1)$ **125.** 10.95   **127.** 3.32   **129.** 12.65   **131.** 0.03   **133.** 82.06   **135.** −151.48
137. n is odd, or n is even and $a \geq 0$ **139.** $(4^{1/2} + 9^{1/2})^2 \neq 4 + 9$; $25 \neq 13$
141. They are equal; both equal $x^{1/6}$ when converted to exponential form.

CUMULATIVE REVIEW EXERCISES

142. c is a function; a, b, and c are relations.   **143.** $\dfrac{b^2 + a^3 b}{a^3 - b}$ **144.** 0, 3   **145.** 441.67 mph

GROUP ACTIVITY/CHALLENGE PROBLEMS

1. 9   **3.** $x^{3/2} y^4$ **5.** $x^{1/24}$ **7.** 2   **8.** $\dfrac{2(3x - 2)}{(6x - 5)^3}$ **9.** $\dfrac{2x + 4}{(2x + 3)^{1/3}}$

11. $(x^{1/2} + y^{1/2})(x^{1/2} + y^{1/2})$ or $(x^{1/2} + y^{1/2})^2$

EXERCISE SET 8.3

1. $5\sqrt{2}$ **3.** $4\sqrt{2}$ **5.** $2\sqrt[3]{2}$ **7.** $3\sqrt[3]{2}$ **9.** $-x\sqrt{x}$ **11.** $7x^5\sqrt{x}$ **13.** $b^{13}\sqrt{b}$ **15.** $y^2\sqrt[4]{y}$ **17.** $2x\sqrt{6x}$
19. $6y^2\sqrt[3]{3y}$ **21.** $xy^3\sqrt{xy}$ **23.** $3x^2y^2\sqrt[3]{3y^2}$ **25.** $12x^4y^4\sqrt[3]{2y}$ **27.** $-2x^2y\sqrt[5]{2x^2y^2}$ **29.** $2cw^3\sqrt[3]{4cz}$
31. $3x^2y^7z^{16}\sqrt[3]{3xz^2}$ **33.** 5   **35.** 2   **37.** $3\sqrt[3]{6}$ **39.** $3xy^3\sqrt{10y}$ **41.** $4x^3y^2$ **43.** $5xy^4\sqrt[3]{x^2y^2}$ **45.** $x^2y^2\sqrt[3]{4y^2}$
47. $xy^3z\sqrt[4]{24yz^3}$ **49.** $x^7y^7z^3\sqrt[5]{x^2y^3z}$ **51.** $2\sqrt{3} + 2$ **53.** $6 - 3\sqrt{2}$ **55.** 10   **57.** $9y\sqrt{y} - y\sqrt{3}$
59. $2xy^2 + 2x^4\sqrt[3]{3y^2}$ **61.** $4x^5y^3\sqrt[3]{x} + 4xy^4\sqrt[3]{2x^2y^2}$ **63.** $6x^2y^6\sqrt{10x} - 12xy^6\sqrt{3y}$ **65.** $2\sqrt{6}$ **67.** $2\sqrt[3]{4}$
69. $x\sqrt[3]{x^2}$ **71.** $6x^2\sqrt{x}$ **73.** $x^2y^6\sqrt{x}$ **75.** $2b^4c^2\sqrt[4]{abc}$ **77.** $15\sqrt{2}$ **79.** $3x^3\sqrt{10x}$ **81.** $2x^3y^5\sqrt{30y}$
83. $x + 3\sqrt{x}$ **85.** $2y^2\sqrt[3]{2x^2}$ **87.** $2\sqrt[3]{y^2} - y^3$ **89.** $xy\sqrt[3]{12x^2y^2} - 2x^2y^2\sqrt[3]{3}$
91. (a) Square the natural numbers   **(b)** 1, 4, 9, 16, 25, 36   **93. (a)** Raise the natural numbers to the fifth power
(b) 1, 32, 243, 1024, 3125   **95.** If n is even and a or b is negative the numbers are not real numbers, and the rule does not apply.

CUMULATIVE REVIEW EXERCISES

97. A number that can be expressed as a quotient of two integers, denominator not 0.
98. A number that can be represented on the real number line.

99. A real number that cannot be expressed as a quotient of two integers, denominator not 0.   **100.** $|a| = \begin{cases} a, & a \geq 0 \\ -a, & a < 0 \end{cases}$

101. $m = \dfrac{2E}{v^2}$ **102. (a)** ⊶————● $-\frac{1}{2}$ 4   **(b)** $\left(-\frac{1}{2}, 4\right]$ **(c)** $\left\{x \,\middle|\, -\frac{1}{2} < x \leq 4\right\}$

GROUP ACTIVITY/CHALLENGE PROBLEMS

1. (a) ≈ 2.72 sec   **(b)** The new period is $\sqrt{2}$ times the original period, or $\sqrt{2} \cdot T$.
(c) The period is $2\sqrt{6}$ or ≈ 4.90 sec   **5.** $4x - 9y$

EXERCISE SET 8.4

1. 3   **3.** $\dfrac{1}{3}$ **5.** $\dfrac{1}{2}$ **7.** $-2\sqrt{2}$ **9.** $\dfrac{r^2}{5}$ **11.** $-3x^2$ **13.** $\dfrac{\sqrt[3]{7y}}{2x^4}$ **15.** $\dfrac{x\sqrt[4]{20}}{3}$ **17.** $\dfrac{\sqrt{3}}{3}$ **19.** $\dfrac{\sqrt{2m}}{2}$

21. $\dfrac{x\sqrt{6}}{3}$ **23.** $\dfrac{2\sqrt{3w}}{w}$ **25.** $\dfrac{\sqrt{10m}}{4}$ **27.** $\dfrac{2\sqrt{3y}}{y^2}$ **29.** $-\dfrac{\sqrt{10a}}{2a}$ **31.** $\dfrac{2x^2\sqrt{xyz}}{z}$ **33.** $\dfrac{q^2\sqrt{10pr}}{2r}$

35. $\dfrac{y^3\sqrt{30xz}}{6z}$ **37.** $-\dfrac{3x^2y\sqrt{yz}}{z^2}$ **39.** $\dfrac{\sqrt[3]{4}}{2}$ **41.** $\dfrac{\sqrt[3]{9}}{3}$ **43.** $\dfrac{\sqrt[3]{5xy^2}}{y}$ **45.** $-\dfrac{\sqrt[3]{15cy}}{3y}$ **47.** $\dfrac{5m\sqrt[4]{8}}{2}$ **49.** $\dfrac{\sqrt[4]{8x^3y^2}}{2y}$

51. $\dfrac{x^2y^2\sqrt[3]{60yz}}{2z}$ **53.** $\dfrac{2s^3\sqrt[3]{r^2}}{r}$ **55.** 6   **57.** 31   **59.** $x - 25$ **61.** $x - y^2$ **63.** $x^2 - y$ **65.** $25 - y$

67. 10   **69.** $3\sqrt{2} - 3$ **71.** $\dfrac{-3\sqrt{6} - 15}{19}$ **73.** $\dfrac{-4\sqrt{2} - 28}{47}$ **75.** $\dfrac{10 + \sqrt{30}}{14}$ **77.** $\dfrac{\sqrt{17} + 2\sqrt{2}}{9}$

79. $\dfrac{3\sqrt{5a} + 9\sqrt{5}}{a - 9}$ **81.** $\dfrac{4x + 4y\sqrt{x}}{x - y^2}$ **83.** $\dfrac{-13 + 3\sqrt{6}}{23}$ **85.** $\dfrac{c + \sqrt{cd} - \sqrt{2cd} - d\sqrt{2}}{c - d}$ **87.** $\dfrac{x\sqrt{y} - y\sqrt{x}}{x - y}$

89. $\dfrac{\sqrt{x}}{3}$ **91.** $\dfrac{\sqrt{10}}{5}$ **93.** -1 **95.** $\dfrac{2xy^3\sqrt{30xz}}{5z}$ **97.** $\dfrac{\sqrt{6}}{x}$ **99.** $x-9$ **101.** $-\dfrac{\sqrt{2x}}{2}$ **103.** $\dfrac{\sqrt[4]{24x^3y^2}}{2x}$

105. $\dfrac{2y^4z^3\sqrt[3]{2x^2z}}{x}$ **107.** $\dfrac{a\sqrt{r}+2r\sqrt{a}}{a-4r}$ **109.** $\dfrac{\sqrt[3]{150y^2}}{5y}$ **111.** $\dfrac{y^3z\sqrt[3]{54x^2}}{3x}$

115. To remove the radicals from a denominator.   **117. (b)** $-\dfrac{7+2\sqrt{10}}{3}$ **119. (a), (b),** and **(c)** $\dfrac{\sqrt{2}}{6}$ **121.** $\dfrac{3}{\sqrt{3}}$

123. Yes, they are equal.   **125.** (1) No perfect powers in any radicand.   (2) No radicand contains fractions.
(3) No radicals in any denominator.

CUMULATIVE REVIEW EXERCISES

127. 40 mph, 50 mph
128. (a) Solve equation for y. Select values   **(b)** Set $x=0$ and solve for y.   **(c)** Mark y intercept, then use the slope
for x and find corresponding values   Set $y=0$ and solve for x.   to determine a second point.
for y. Plot points.

129. $8x^3-18x^2+5x+6$ **130.** $4, \dfrac{3}{2}$

GROUP ACTIVITY/CHALLENGE PROBLEMS

1. $\dfrac{\sqrt{a+b}}{a+b}$ **3.** $\dfrac{2}{\sqrt{6}}$ **5.** $\dfrac{16x-3}{4x\sqrt{x}+x\sqrt{3}}$ **6.** $\dfrac{1}{\sqrt{x+h}+\sqrt{x}}$

8. (b) $\left(\sqrt[3]{3}-\sqrt[3]{2}\right)\left(\sqrt[3]{3^2}+\sqrt[3]{3}\sqrt[3]{2}+\sqrt[3]{2^2}\right)$ This equals the answer in part (a).   **(c)** $\left(\sqrt[3]{x}+\sqrt[3]{y}\right)\left(\sqrt[3]{x^2}-\sqrt[3]{xy}+\sqrt[3]{y^2}\right)$

EXERCISE SET 8.5

1. $2\sqrt{3}$ **3.** $9\sqrt{10}+2$ **5.** $9\sqrt[3]{15}$ **7.** $-3\sqrt{y}$ **9.** $7\sqrt{5}+2\sqrt[3]{x}$ **11.** $5-4\sqrt[3]{x}$ **13.** $2\left(\sqrt{2}-\sqrt{3}\right)$
15. $-30\sqrt{3}+20\sqrt{5}$ **17.** $-4\sqrt{10}$ **19.** $20x\sqrt{10y}$ **21.** $29\sqrt{2}$ **23.** $-16\sqrt{5x}$ **25.** $-27x\sqrt{2}$ **27.** $-6\sqrt[3]{5}$
29. $7\sqrt[3]{2}$ **31.** $(9x+1)\sqrt{5x}$ **33.** $10a^2b\sqrt{5a}$ **35.** $(6xy-x^2y)\sqrt[3]{3x}$ **37.** $4x^2\sqrt[3]{x^2y^2}$ **39.** 0 **41.** $\sqrt{2}$
43. $\dfrac{2\sqrt{3}}{3}$ **45.** $\dfrac{13\sqrt{6}}{6}$ **47.** $\dfrac{13\sqrt{2}}{2}$ **49.** $2\sqrt{x}\left(2+\dfrac{1}{x}\right)$ **51.** $-15\sqrt{2}$ **53.** $\dfrac{3\sqrt{6}}{4}$ **55.** $23+9\sqrt{3}$
57. $11+7\sqrt{5}$ **59.** $18-\sqrt{2}$ **61.** $8+2\sqrt{15}$ **63.** $1-\sqrt{6}$ **65.** $7-4\sqrt{3}$ **67.** $x-\sqrt{3x}-6$
69. $18x-\sqrt{3xy}-y$ **71.** $8-2\sqrt[3]{18}-\sqrt[3]{12}$ **73.** $3\sqrt{5}$ **75.** $7\sqrt{5}$ **77.** $\dfrac{2\sqrt{6}}{3}$ **79.** $3\sqrt[4]{x}$ **81.** $7-3\sqrt{y}$
83. $-14+11\sqrt{2}$ **85.** $(4x-2)\sqrt{3x}$ **87.** $\sqrt{y}$ **89.** $16ay\sqrt[3]{3y}$ **91.** $(5xy^2+2x)\sqrt[3]{xy}$ **93.** $\dfrac{-301\sqrt{2}}{20}$
95. $\dfrac{-16\sqrt{6}}{3}$ **97.** $P=14\sqrt{5}, A=60$ **99.** Radicals with the same radicands and index.   **101.** ≈ 5.97
103. $2+\sqrt{3}$ **105.** No   **107.** Yes

CUMULATIVE REVIEW EXERCISES

108. $\dfrac{4y^7}{x^3}$ **109.** $\dfrac{3}{5}, -\dfrac{3}{4}$ **110.** $\dfrac{x^{1/2}}{y^{2/3}}$ **111.** $3x^2y\sqrt[3]{y}-x^4y^2\sqrt[3]{3y^2}$

GROUP ACTIVITY/CHALLENGE PROBLEMS

1. $\dfrac{4\sqrt{3}+9}{3}$ **3.** $\sqrt{5}(\sqrt{2}+1)$ **5.** $x^2 - (\sqrt{3})^2 = (x+\sqrt{3})(x-\sqrt{3})$ **7.** $\dfrac{\sqrt[5]{81x^3y}}{x^2y^2}$

EXERCISE SET 8.6

1. 25   **3.** 8   **5.** 81   **7.** 16   **9.** 8   **11.** 9   **13.** No solution   **15.** No solution   **17.** 1   **19.** $-\frac{1}{3}$ **21.** 5

23. $-\frac{9}{4}$ **25.** $\frac{1}{4}$ **27.** 2   **29.** -7 **31.** 5   **33.** 10   **35.** -3 **37.** 4   **39.** $-\frac{5}{2}$ **41.** 2   **43.** $\frac{9}{16}$ **45.** 9

47. 4   **49.** No solution   **51.** 6   **53.** $v = \dfrac{p^2}{2}$ **55.** $g = \dfrac{v^2}{2h}$ **57.** $F = \dfrac{v^2m}{R}$ **59.** $m = \dfrac{x^2k}{v_0^2}$ **61.** 13 in

63. $\sqrt{18.25} \approx 4.27$ m   **65.** $\sqrt{16,200} \approx 127.28$ ft   **67.** $\sqrt{60} \approx 7.75$ m   **69.** $\sqrt{5120} \approx 71.55$ ft/sec   **71.** 3.14 sec

73. $\sqrt{576} = 24$ sq in   **75.** $0.2(\sqrt{149.4})^3 \approx 365.2$ days   **77.** $\sqrt{1,000,000} = 1000$ lb

79. $\sqrt{320} \approx 17.89$ ft/sec   **81.** (a) $2, -2$ (b) $3, -4$ (c) $-4, \frac{3}{2}$ (d) $5, -1$

83. 0, for all other values the left side of the equation is negative and the right side is positive.

85. $\sqrt{x-3}$ cannot equal a negative number, and it must equal -3.

87. (a) 2 (b)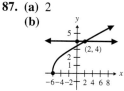
(c) yes

89. (a) 3, 7 (b) yes (c) 3, 7

91. (a) One real, 9; One extraneous (b) one, at $x = 9$ (c) $\{x\,|\,x \geq 0\}$

91. (d)
Yes

CUMULATIVE REVIEW EXERCISES

92. $P_2 = \dfrac{P_1 P_3}{P_1 - P_3}$ **93.** $-2x - 3$ **94.** $\left\{x\,\middle|\,x \text{ is a real number and } x \neq 4 \text{ and } x \neq -\dfrac{2}{3}\right\}$ **95.** $\dfrac{3a}{2b(2a+3b)}$

96. $t(t-4)$ **97.** $\dfrac{3}{x+3}$ **98.** 2

GROUP ACTIVITY/CHALLENGE PROBLEMS

1. $\mathbb{R}$ **3.** 6   **4.** 6   **5.** No solution   **6.** 16   **7.** $\dfrac{1}{4}, 1$ **8.** (a) No solution (b)

9. $n = \dfrac{z^2\sigma^2}{(\bar{x}-\mu)^2}$ **10.** $n = \dfrac{z^2pq}{(p'-p)^2}$ **11.** 1.59 cps

(graph showing $y = \sqrt{2x-3}$ and $y = \sqrt{x-4}$)

GRAPHING CALCULATOR CORNER, PAGE 491

1. 1.5   **3.** $-3.7, 3.7$

EXERCISE SET 8.7

1. $3 + 0i$ **3.** $3 + 2i$ **5.** $(6+\sqrt{3}) + 0i$ **7.** $0 + 5i$ **9.** $4 + 2i\sqrt{3}$ **11.** $0 + 3i$ **13.** $9 - 3i$

15. $0 + (2 - 4\sqrt{5})i$ **17.** $15 - 4i$ **19.** $8 + \frac{47}{36}i$ **21.** $18 - 5i$ **23.** $(4\sqrt{2}+\sqrt{3}) - 2i\sqrt{2}$ **25.** $19 + 12i\sqrt{3}$

27. $(2\sqrt{3}-7) + (7+2\sqrt{3})i$ **29.** $-6 - 4i$ **31.** $-1 + 6i$ **33.** $-6.3 - 22.4i$ **35.** $-4 + 2i\sqrt{3}$

37. $-6 + 3i\sqrt{2}$ **39.** $1 + 5i$ **41.** $6 - 22i$ **43.** $\left(\dfrac{1}{2}+\dfrac{\sqrt{3}}{4}\right) + (2\sqrt{3}+3)i$ **45.** $\dfrac{21}{4} + \dfrac{9}{2}i\sqrt{2}$ **47.** $-\dfrac{5i}{3}$

49. $\dfrac{3-2i}{2}$ **51.** $\dfrac{5-2i}{5}$ **53.** $\dfrac{49+14i}{53}$ **55.** $\dfrac{9-12i}{10}$ **57.** $\dfrac{3+i}{5}$ **59.** $\dfrac{\sqrt{2}+i\sqrt{6}}{4}$

61. $\dfrac{(5\sqrt{10} - 2\sqrt{15}) + (10\sqrt{2} + 5\sqrt{3})i}{45}$ **63.** $\sqrt{30}$ **65.** $10i\sqrt{2}$ **67.** $7 - 7i$ **69.** $6 + i\sqrt{6}$

71. $20.8 - 16.64i$ **73.** $2\sqrt{15} + 3i\sqrt{2}$ **75.** $7\sqrt{2}$ **77.** $\dfrac{-4 - 5i}{2}$ **79.** $\dfrac{4\sqrt{3} + 8i}{7}$ **81.** $3 + \dfrac{2}{45}i$

83. $\dfrac{1}{4} - \dfrac{31}{50}i$ **85.** 2 **87.** $-3.33 - 9.01i$ **89.** 0 **91.** 2 **93.** Yes   **95.** No   **97.** -1 **99.** 1 **101.** i

103. $-i$ **105.** $0.83 - 3i$ **107.** $1.5 - 0.33i$ **109.** $i^{-2} = \dfrac{1}{i^2} = \dfrac{1}{-1} = -1$ **111.** False   **113.** False   **115.** True

Cumulative Review Exercises

118. 15 lb at \$5.00, 25 lb at \$5.80   **119.** $x = 1$, $(1, 0)$, up, D: $\mathbb{R}$, R: $\{y \mid y \geq 0\}$ **120.** $(2x^2 + 3)^2$
121. $(3rs + 2)(5rs - 3)$

Group Activity/Challenge Problems

1. (a) $2 + i\sqrt{2}, 2 - i\sqrt{2}$ **3.** $4 + 3i\sqrt{3}$ **5.** $-3 + 5i\sqrt{3}$ **9.** $\dfrac{-1 + 7i}{5}$

Review Exercises

1. 3 **2.** 5 **3.** -2 **4.** 4 **5.** 3 **6.** -3 **7.** -12 **8.** -16 **9.** 7 **10.** 93.4 **11.** $|x|$ **12.** $|x - 2|$
13. $|x - y|$ **14.** $|x^2 - 4x + 12|$ **15.** $x^{5/2}$ **16.** $x^{5/3}$ **17.** $y^{15/4}$ **18.** $5^{2/7}$ **19.** $\sqrt{a}$ **20.** $\sqrt[5]{y^3}$
21. $\left(\sqrt[5]{2m^2n}\right)^9$ **22.** $\dfrac{1}{\left(\sqrt[4]{2a + 3b}\right)^3}$ **23.** 125 **24.** 9 **25.** $\sqrt[3]{y}$ **26.** x^5 **27.** 16 **28.** 81 **29.** $\sqrt[5]{x}$ **30.** x
31. -5 **32.** not a real number   **33.** $\dfrac{3}{2}$ **34.** $-\dfrac{1}{12}$ **35.** $x^{4/15}$ **36.** $\dfrac{4}{y^2}$ **37.** $\dfrac{1}{y^{8/15}}$ **38.** $\dfrac{9x^{10}}{4y^9z^5}$
39. $2z^2 - 4z^{4/3}$ **40.** $2x^{-4} - 8$ **41.** $-6a^{-3} - 2a^{-3/2}$ **42.** $3r^{-13/6} + 1$ **43.** $x^{1/5}(1 + x)$ **44.** $\dfrac{x - 1}{x^4}$
45. $\dfrac{x^{1/6} + 1}{x^{2/3}}$ **46.** $\dfrac{2(2 + 3x^2)}{x^6}$ **47.** $(x^{1/4} - 1)(x^{1/4} - 4)$ **48.** $(x^{1/5} + 3)(x^{1/5} - 5)$ **49.** $(2x^{1/2} - 3)(3x^{1/2} + 2)$
50. $(4x^{1/3} - 1)(2x^{1/3} + 3)$ **51.** 16.12 **52.** 17.17 **53.** 42.22 **54.** -317.06 **55.** $\dfrac{27r^{9/2}}{p^3}$ **56.** $\dfrac{16x^{56/15}}{81y^2}$
57. $\dfrac{y^{1/5}}{4xz^{1/3}}$ **58.** $\dfrac{15x^{1/2}y^{4/3}z^{1/3}}{2}$ **59.** $2\sqrt{6}$ **60.** $4\sqrt{5}$ **61.** $2\sqrt[3]{2}$ **62.** $3\sqrt[3]{2}$ **63.** $5xy^3\sqrt{2xy}$ **64.** $x^2y\sqrt[3]{9y^2}$
65. $2x^2y^3\sqrt[4]{x}$ **66.** $5x^2y^3\sqrt[3]{xy}$ **67.** $2x^2yz^3\sqrt[5]{x^2y^2z^2}$ **68.** 10 **69.** $2x^3\sqrt{10}$ **70.** $2x^2y^3\sqrt[3]{4x^2}$ **71.** $xy^2\sqrt[3]{25x}$
72. $2x^2y^4\sqrt[4]{x}$ **73.** $6x - 2\sqrt{15x}$ **74.** $4y - y^3\sqrt[3]{y^2}$ **75.** $2x^2y^3\sqrt[3]{y} + xy^5\sqrt[3]{18}$ **76.** $x^2y^2\sqrt[4]{6y^3} + 3x^3y\sqrt[4]{y}$
77. $\dfrac{1}{2}$ **78.** $\dfrac{6}{5}$ **79.** $\dfrac{x}{2}$ **80.** $\dfrac{x}{2}$ **81.** $\dfrac{4y^2}{x}$ **82.** $3xy\sqrt[3]{2}$ **83.** $\dfrac{5}{xy}$ **84.** $\dfrac{\sqrt{2}}{2}$ **85.** $\dfrac{x\sqrt{7}}{7}$ **86.** $\dfrac{\sqrt{10}}{5}$
87. $\dfrac{2\sqrt{15xy}}{5y}$ **88.** $\dfrac{2\sqrt[3]{x^2}}{x}$ **89.** $\dfrac{\sqrt[3]{75xy^2}}{5y}$ **90.** $\dfrac{x\sqrt{3y}}{y}$ **91.** $\dfrac{x^2y^2\sqrt{6yz}}{z}$ **92.** $\dfrac{5xy^2\sqrt{15yz}}{3z}$ **93.** $\dfrac{y\sqrt[4]{4x^3y^2}}{2x}$
94. $\dfrac{y^3z^4\sqrt{30xz}}{3x^2}$ **95.** $\dfrac{y\sqrt[3]{4x^2}}{x}$ **96.** $\dfrac{y^2\sqrt[3]{4x}}{2x}$ **97.** 7 **98.** -2 **99.** $x - y^2$ **100.** $x^2 - y$ **101.** $28 + 10\sqrt{3}$
102. 5 **103.** $x + \sqrt{5xy} - \sqrt{3xy} - y\sqrt{15}$ **104.** $2x + \sqrt{xy} - 3y$ **105.** $\sqrt[3]{6x^2} - \sqrt[3]{4xy} - \sqrt[3]{9xy} + \sqrt[3]{6y^2}$

106. $-10 + 5\sqrt{5}$ **107.** $\dfrac{3x - x\sqrt{x}}{9 - x}$ **108.** $\dfrac{x - \sqrt{xy}}{x - y}$ **109.** $\dfrac{\sqrt{30} + \sqrt{15} + 2\sqrt{3} + \sqrt{6}}{3}$ **110.** $\dfrac{x - \sqrt{xy} - 2y}{x - y}$

111. $\dfrac{4\sqrt{y + 2} + 12}{y - 7}$ **112.** $\dfrac{11\sqrt{6}}{6}$ **113.** $2\sqrt[3]{x}$ **114.** $-\sqrt[3]{2}$ **115.** $-4\sqrt{3}$ **116.** $8 - 13\sqrt[3]{2}$ **117.** $\dfrac{69\sqrt{2}}{8}$

118. $(3x^2y^3 - 4x^3y^4)\sqrt{x}$ **119.** $(2x^2y^2 - x + 3x^3)\sqrt[3]{xy^2}$ **120.** 64   **121.** No solution   **122.** 8   **123.** 2

124. -3 **125.** 2   **126.** 0, 9   **127.** 5   **128.** $L = \dfrac{V^2w}{2}$ **129.** $A = \pi r^2$ **130.** $\sqrt{29} \approx 5.39$ m

131. $\sqrt{1280} \approx 35.78$ ft/sec   **132.** $2\pi\sqrt{2} \approx 2.83\pi \approx 8.89$ sec   **133. (a)** 5 **(b)**   **(c)** Yes

134. (a) 0 **(b)**   **(c)** Yes

135. $5 + 0i$ **136.** $-6 + 0i$ **137.** $2 - 16i$ **138.** $3 + 4i$ **139.** $7 + i$ **140.** $1 - 2i$ **141.** $2 + 5i$
142. $3\sqrt{3} + (\sqrt{5} - \sqrt{7})i$ **143.** $12 + 8i$ **144.** $-2\sqrt{3} + 2i$ **145.** $6\sqrt{2} + 4i$ **146.** $-6 + 6i$ **147.** $17 - 6i$
148. $(24 + 3\sqrt{5}) + (4\sqrt{3} - 6\sqrt{15})i$ **149.** $-2i/3$ **150.** $(-2 - \sqrt{3})i/2$ **151.** $(15 - 10i)/13$
152. $(5\sqrt{3} + 3i\sqrt{2})/31$ **153.** $[(\sqrt{10} - 6\sqrt{2}) + (3\sqrt{2} + 2\sqrt{10})i]/10$ **154.** 0   **155.** 7   **156.** i **157.** $-i$
158. 1   **159.** -1

PRACTICE TEST

1. 26   **2.** $|3x - 4|$ **3.** $\dfrac{1}{y^{7/6}}$ **4.** $(2x^{1/3} + 5)(x^{1/3} - 2)$ **5.** $5x^2y^4\sqrt{2x}$ **6.** $2x^3y^3\sqrt[3]{5x^2y}$ **7.** $\dfrac{x^2y^2\sqrt{yz}}{2z}$

8. $\dfrac{\sqrt[3]{x^2}}{x}$ **9.** $\dfrac{2 - \sqrt{2}}{2}$ **10.** $-20\sqrt{3}$ **11.** $(2xy + 2x^2y^2)\sqrt[3]{y^2}$ **12.** $2\sqrt{5} - 2\sqrt{10} - 6 + 6\sqrt{2}$ **13.** 13

14. -3 **15.** 16   **16.** $h = \dfrac{8w^2}{g}$ **17.** 13 ft   **18.** $(18 + 2\sqrt{2}) + (6\sqrt{2} - 6)i$ **19.** $\dfrac{\sqrt{5} + i\sqrt{10}}{6}$ **20.** 2

CUMULATIVE REVIEW TEST

1. $\dfrac{57}{9}$ **2. (a)** A set of ordered pairs **(b)** A set of ordered pairs no two of which have the same first coordinate.

3. (a) Domain: $\{x \,|\, x \geq -2\}$ **4.** $\left(\dfrac{2}{11}, -\dfrac{16}{11}\right)$ **8.**   **9.** $2(x - 3 + y)(x - 3 - y)$

(b)   **5.** $\dfrac{3}{2xy^3}$ **10.** $\left\{x \,\middle|\, x \text{ is a real number and } x \neq \dfrac{3}{5}\right\}$

(c) Range: $\{y \,|\, y \geq 0\}$ **6.** $6x^3 - 23x^2 + 8x + 30$ **11.** $\dfrac{2x + 1}{3}$

7. $3x + 4 + \dfrac{2}{x + 2}$ **12.** $\dfrac{x + 1}{2x + 3}$

13. $\dfrac{x^2 + 8x + 5}{(x + 3)(x - 1)(2x + 5)}$ **14.** -5 **15.** $x^{7/2}y$ **16.** $2x^4y^9\sqrt[3]{2xy^2}$ **17.** $3, -3$ **18.** $\dfrac{6 - 2i\sqrt{6}}{15}$ **19.** $1\frac{1}{5}$ hr

20. $\sqrt{1300} \approx 36.1$ ft

Chapter 9

EXERCISE SET 9.1

1. $5, -5$ **3.** $5\sqrt{3}, -5\sqrt{3}$ **5.** $2\sqrt{7}, -2\sqrt{7}$ **7.** $8, 0$ **9.** $\frac{1}{3}, -1$ **11.** $-0.9, -2.7$ **13.** $\dfrac{5 \pm 2\sqrt{3}}{2}$

15. $-\frac{1}{20}, -\frac{9}{20}$ **17.** $s = \sqrt{A}$ **19.** $r = \sqrt{\dfrac{A}{\pi}}$ **21.** $F = \sqrt{F_x^2 + F_y^2}$ **23.** $V_x = \sqrt{V^2 - V_y^2}$ **25.** $b = \sqrt{a^2 - L}$

27. $t = \sqrt{\dfrac{v - m}{n}}$ **29.** $b = \sqrt{d + 4ac}$ **31.** $d = \sqrt{\dfrac{w - 3l}{2}}$ **33.** 4   **35.** $\sqrt{175} \approx 13.23$ **37.** $\sqrt{41} \approx 6.40$

39. $\sqrt{128} \approx 11.31$ **41.** 10   **43.** 1, -3 **45.** 5, -1 **47.** $-2, -1$ **49.** 5, 3   **51.** $-1 \pm i\sqrt{14}$ **53.** $-2, -3$

55. $-3, -6$ **57.** 7, 8   **59.** 6, -2 **61.** $-1 \pm \sqrt{7}$ **63.** $-3 \pm \sqrt{3}$ **65.** $\dfrac{5 \pm \sqrt{57}}{2}$ **67.** 0, -2 **69.** $0, \frac{1}{3}$

71. 1, -3 **73.** $\dfrac{-9 \pm \sqrt{73}}{2}$ **75.** $-8, -3$ **77.** $\dfrac{-2 \pm i\sqrt{2}}{2}$ **79.** $1 \pm i$ **81.** $\dfrac{-3 \pm \sqrt{59}}{10}$ **83.** $\dfrac{-1 \pm i\sqrt{191}}{12}$

85. $-a \pm \sqrt{k}$ **87.** 7, 9   **89.** 5 ft, 12 ft   **91.** $12 + 12\sqrt{2} \approx 28.97$ ft

93. Make the coefficient of the squared term equal to 1.

Cumulative Review Exercises

96. $z = \dfrac{3xy}{3y + 1}$ **97.** $|x^2 - 4x|$ **98.** $\frac{1}{5}$ **99.** $\dfrac{x^{1/2}}{y^{3/2}}$

Group Activity/Challenge Problems

1. (a) $S = 32 + 80\sqrt{\pi} \approx 173.80$ sq in   **(b)** $r = \dfrac{4\sqrt{\pi}}{\pi} \approx 2.26$ in   **(c)** $r \approx 2.1$ in   **3.** $(x + 2)^2 + (y - 3)^2 = 16$

4. $4(x - 6)^2 + 9(y + 4)^2 = 144$ **5.** $(x + 2)^2 - 4(y + 2)^2 = 16$ **6.** $2, -2, 2i, -2i$

Exercise Set 9.2

1. Two real solutions   **3.** No real solution   **5.** Two real solutions   **7.** No real solution   **9.** One real solution

11. One real solution   **13.** 1, 2   **15.** 4, 5   **17.** 1, 5   **19.** 6, -6 **21.** 0, 6   **23.** $-8, -9$ **25.** $\dfrac{7 \pm \sqrt{17}}{4}$

27. $\dfrac{1}{3}, -\dfrac{1}{2}$ **29.** $\dfrac{2 \pm i\sqrt{11}}{3}$ **31.** $\dfrac{-7 \pm \sqrt{37}}{2}$ **33.** $\dfrac{9}{2}, -1$ **35.** $\dfrac{1}{2}, -\dfrac{5}{3}$ **37.** $\dfrac{-1 \pm i\sqrt{23}}{4}$ **39.** 0, -3

41. $\dfrac{3 \pm \sqrt{33}}{2}$ **43.** $\dfrac{-6 \pm 2\sqrt{6}}{3}$ **45.** $\dfrac{11 \pm \sqrt{241}}{6}$ **47.** $\dfrac{-0.6 \pm \sqrt{0.84}}{0.2}$ or $-3 \pm \sqrt{21}$

49. $\dfrac{-0.94 \pm \sqrt{32.3116}}{3.24}$ or $\dfrac{-47 \pm \sqrt{80,779}}{162}$ **51.** $y = x^2 - 10x + 24$ **53.** $y = x^2 + x - 12$

55. $y = x^2 + x - 6$ **57.** $y = x^2 - 4x + 4$ **59.** $y = 3x^2 + 4x - 4$ **61.** $y = 6x^2 - x - 2$

63. (a) Upward   **(e)**
(b) 15
(c) $(-4, -1)$
(d) $-3, -5$

65. (a) Upward   **(e)**
(b) 4
(c) $(3, -5)$
(d) $3 + \sqrt{5}, 3 - \sqrt{5}$

67. (a) Upward   **(e)**
(b) 9
(c) $(-3, 0)$
(d) -3

69. (a) Upward   **(e)**
(b) −6
(c) $\left(\frac{1}{4}, -\frac{49}{8}\right)$
(d) $-\frac{3}{2}, 2$

71. (a) Upward   **(e)**
(b) 3
(c) $\left(-\frac{2}{3}, \frac{5}{3}\right)$
(d) No x intercepts

73. (a) Downward   **(e)**
(b) 4
(c) $\left(-\frac{3}{2}, \frac{17}{2}\right)$
(d) $\dfrac{-3 \pm \sqrt{17}}{2}$

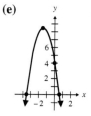

75. (a) Upward   **(e)**
(b) 4
(c) $(0, 4)$
(d) No x intercepts

77. (a) Downward   **(e)**
(b) 0
(c) $(3, 9)$
(d) 0, 6

79. (a) Downward   **(e)**
(b) 5
(c) $(0, 5)$
(d) −1, 1

81. (a) Upward   **(e)**
(b) −6
(c) $\left(-\frac{2}{3}, -\frac{22}{3}\right)$
(d) $\dfrac{-2 \pm \sqrt{22}}{3}$

83. (a) Downward   **(e)**
(b) 6
(c) $\left(\frac{3}{2}, \frac{33}{4}\right)$
(d) $\dfrac{3 \pm \sqrt{33}}{2}$

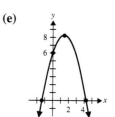

85. (a) Downward   **(e)**
(b) −9
(c) $\left(\frac{3}{4}, -\frac{27}{4}\right)$
(d) No x intercepts

87. 2   **89.** $w = 3$ ft, $l = 7$ ft   **91.** 2 in   **93. (a)** 0.5 thousand   **(b)** 11.3 thousand   **(c)** $1.55
95. (a) 1.2   **(b)** 2.16   **(c)** 7.3 hr   **97. (a)**

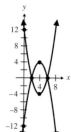

(b) $2
(c) $22
(d) $12
(e) $10,000

99. (a) ≈ 4.57 sec   **(b)** ≈ 4.79 sec
101. ≈ 16.37 sec
105. (a) Graphs are symmetric to x axis
(b) yes   **(c)** no
(d) $y = x^2 - 8x + 12$

CUMULATIVE REVIEW EXERCISES

107. $2xy^2z^4\sqrt[5]{2x^4y^2}$ **108.** $2x^2y^4\sqrt[3]{x} + 2xy^3\sqrt[3]{3}$ **109.** $\dfrac{x^2 + 2x\sqrt{y} + y}{x^2 - y}$

110. 6

$y = -x^2 + 8x - 12$

GROUP ACTIVITY/CHALLENGE PROBLEMS

1. $2\sqrt{5}, -\sqrt{5}$ **3.** 1, −1, 2, −2   **4.** $x = \dfrac{-1 \pm \sqrt{1 + 4y}}{2}$ **5.** $x = \dfrac{-3y \pm \sqrt{9y^2 - 16}}{2}$ **7.** The x coordinate is

$\dfrac{-b}{2a}$. The y coordinate is found by evaluating $f\left(\dfrac{-b}{2a}\right)$.   **8.** $(-0.12 + \sqrt{14.3952})/1.2 \approx 3.0618$ mm   **9. (a)** $\dfrac{1 + \sqrt{5}}{2}$

Exercise Set 9.3

1. $6, -1$ **3.** $1, 2$ **5.** $1, 6$ **7.** $\dfrac{-4 \pm \sqrt{10}}{2}$ **9.** $\dfrac{-1 \pm i\sqrt{3}}{2}$ **11.** $\dfrac{3 \pm i\sqrt{15}}{2}$ **13.** $\sqrt{2}, -\sqrt{2}, 2i, -2i$

15. $2, -2, \sqrt{3}, -\sqrt{3}$ **17.** $\dfrac{\sqrt{3}}{3}, -\dfrac{\sqrt{3}}{3}, i\sqrt{5}, -i\sqrt{5}$ **19.** $-\dfrac{1}{3}$ **21.** $-2, -\dfrac{1}{3}$ **23.** $8, 27$ **25.** $25, 64$

27. 1 **29.** $243, 7776$ **31.** $\dfrac{1}{4}$ **33.** going, 6 mph; returning, 8 mph   **35.** 130 mph   **37.** ≈ 14.64 hr

39. $\approx 4.88\%$

Cumulative Review Exercises

41. $h = \dfrac{S - 2\pi r^2}{2\pi r}$ **42.** No, in a function each x value must have a unique y value.   **43.** If a vertical line can be drawn to intersect a graph at more than one point, the graph is not a function.   **44.** 3   **45.** The set of values that can be used for the independent variable.   **46.** The set of values obtained for the dependent variable.

47. No, $f(x) = \dfrac{1}{x}$ means $f(x) = x^{-1}$. A polynomial function must have whole number exponents on the variable.

Group Activity/Challenge Problems

1. (a) and **(b)** $1 + i, 1 - i$ **2.** $3, \dfrac{5}{4}$ **3.** $2, -3, 3, -4$

Exercise Set 9.4

1. $-2 \quad 5$ **3.** $-4 \quad 0$ **5.** $-4 \quad 4$ **7.** $-3 \quad \frac{1}{2}$ **9.** $-\frac{2}{3} \quad 3$ **11.** $-5 \quad 7$

13. $-\sqrt{6} \quad \sqrt{6}$ **15.** $\frac{1-\sqrt{21}}{2} \quad \frac{1+\sqrt{21}}{2}$ **17.** $\frac{6-3\sqrt{2}}{2} \quad \frac{6+3\sqrt{2}}{2}$ **19.** $-\frac{5}{4} \quad 4$

21. $(-4, -1) \cup (1, \infty)$ **23.** $(-\infty, -3] \cup [1, 4]$ **25.** $[-3, 0] \cup [3, \infty)$ **27.** $\left(-6, -\dfrac{5}{2}\right) \cup (2, \infty)$ **29.** $\left[\dfrac{8}{3}, \infty\right)$

31. $(-\infty, -\frac{3}{5}) \cup (-\frac{3}{5}, 4)$ **33.** $\{x \mid x < -3 \text{ or } x > 1\}$ **35.** $\{y \mid 1 < y \leq 4\}$ **37.** $\{x \mid 1 < x < 2\}$

39. $\{a \mid a \leq -2 \text{ or } a > \frac{1}{2}\}$ **41.** $\{x \mid -4 \leq x < 4\}$ **43.** $\{x \mid x < \frac{1}{2} \text{ or } x > 2\}$ **45.** $(-\infty, -6) \cup (-2, 4)$

47. $[1, 3) \cup [6, \infty)$ **49.** $(-\infty, -4) \cup (1, 6]$ **51.** $\left(-\dfrac{5}{2}, 3\right) \cup (6, \infty)$ **53.** $[-2, 0) \cup \left[\dfrac{3}{2}, \infty\right)$ **55.** $1 \quad 3$

57. $2 \quad 4$ **59.** $1 \quad 4$ **61.** $\frac{4}{7} \quad \frac{2}{3}$ **63.** $\frac{1}{4} \quad \frac{1}{3}$ **65.** -2 **67. (a)** $(4, \infty)$

(b) $(-\infty, 2) \cup (2, 4)$ **69. (a)** $(-\infty, -4] \cup (-2, \infty)$ **(b)** $[-4, -2)$ **71.** $0, 3, -4$ **73.** All real numbers except -1.

Cumulative Review Exercises

75. $\{x \mid x \leq 4\}$ **76.** $\{x \mid x \text{ is a real number and } x \neq 2 \text{ and } x \neq -2\}$ **77.** $\dfrac{a^3 - ab^3}{b^2 + a^3 b}$ **78.** $-8 - 6i$

Group Activity/Challenge Problems

1. $-9 \; -5 \; -1 \quad 3$ **3.** 0 **4.** Many answers are possible. Here are some possibilities.

(a) $x^2 - 8x + 15 \leq 0$ **(b)** $x^2 - 3x > 0$ **(c)** $x^2 - 8x + 16 \leq 0$ **(d)** $x^2 < 0$ **(e)** $x^2 \geq 0$ **5. (a)** Regions 2 and 4
(b) Regions 1 and 3

Graphing Calculator Corner, Page 553

1. $x < 1.42 \text{ or } x > 3.35$ **3.** $-3 < x \leq 1 \text{ or } x > 2$

EXERCISE SET 9.5

1. (a) $x^2 + 3x - 10$ **(b)** 0   **3. (a)** $x^3 - 12x + 16$ **(b)** 27   **5. (a)** $x^2 - 2x - 8$ **(b)** -5 **7. (a)** $x^2 + x - 2$
(b) 0   **9. (a)** $x^3 + 2x^2 - 4x - 8$ **(b)** -24 **11. (a)** $x^2 - 2$ **(b)** 14   **13. (a)** $x - 4 + \sqrt{x + 6}$ **(b)** 2
15. (a) $(x - 4)\sqrt{x + 6}$ **(b)** 24   **17. (a)** $\sqrt{x + 6} - 4$ **(b)** $\sqrt{13} - 4$ **19.** 3   **21.** -4 **23.** -3
25. Undefined   **27.**

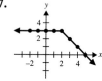

29. (a) For each graph, each year has a unique amount of export.
(c) Share of U.S. export to Asia and the European community.

(b)

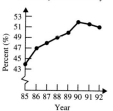

Share of U.S. exports to Asia and European Community

31. No, subtraction is not commutative.   **33.** Yes, multiplication is commutative.   **35.** No, composition of functions is not commutative   **37. (a)** $(f \circ g)(x) = (g \circ f)(x) = x$ **(b)** $\mathbb{R}$ for each of them

CUMULATIVE REVIEW EXERCISES

38. All functions are relations, but not all relations are functions.   **39.** Yes, both linear and quadratic functions are polynomial functions.   **40.** $f(x) = x^3$ **41.** No, if a function is a polynomial function, it cannot be a square root function, and vice versa.   **42.** All polynomial functions are rational functions, but not all rational functions are polynomial functions.

GROUP ACTIVITY/CHALLENGE PROBLEMS

1. (a) $36\pi \approx 113.10$ sq ft   **(b)** $A = 4\pi t^2$ **(c)** $64\pi \approx 201.06$ sq ft   **3. (a)** $\{(1, 7), (2, -1), (3, 10)\}$
(b) $\{(1, -3), (2, 7), (3, -2)\}$ **(c)** $\{(1, 10), (2, -12), (3, 24)\}$ **(d)** $\left\{\left(1, \frac{2}{5}\right), \left(2, -\frac{3}{4}\right), \left(3, \frac{2}{3}\right)\right\}$

GRAPHING CALCULATOR CORNER, PAGE 558

1. It is a straight line with x intercept at $-1.\overline{3}$ and y intercept at 4.
3. It is a straight line with x intercept at 6 and y intercept at -6.

EXERCISE SET 9.6

1. No   **3.** Yes   **5.** No   **7.** Yes   **9.** No   **11.** Yes   **13.** No   **15.** Yes   **17.** Yes
19. $f(x)$: Domain $\{-2, -1, 2, 4, 9\}$, Range $\{0, 3, 4, 6, 7\}$; $f^{-1}(x)$: Domain $\{0, 3, 4, 6, 7\}$, Range $\{-2, -1, 2, 4, 9\}$
21. $f(x)$: Domain $\{-2.9, 0, 1.7, 5.7\}$, Range $\{-3.4, 3, 4, 9.76\}$; $f^{-1}(x)$: Domain $\{-3.4, 3, 4, 9.76\}$,
Range $\{-2.9, 0, 1.7, 5.7\}$ **23.** $f(x)$: Domain $\{-1, 1, 2, 4\}$, Range $\{-3, -1, 0, 2\}$; $f^{-1}(x)$: Domain $\{-3, -1, 0, 2\}$,
Range $\{-1, 1, 2, 4\}$ **25.** $f(x)$: Domain $\{x | x \geq 2\}$, Range $\{y | y \geq 0\}$; $f^{-1}(x)$: Domain $\{x | x \geq 0\}$; Range $\{y | y \geq 2\}$
27. $f^{-1}(x) = (x - 8)/2$ **29.** $f^{-1}(x) = -(x + 10)/3$ **31.** $f^{-1}(x) = (5x + 3)/10$ **33.** $f^{-1}(x) = x^2, x \geq 0$

35. $f^{-1}(x) = x^3$ **37.** $f^{-1}(x) = \dfrac{1}{x}, x > 0$

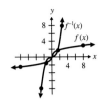

47. $y = f^{-1}(x) = \dfrac{x}{12}$, x is inches, y is feet   **49.** $y = f^{-1}(x) = \dfrac{9}{5}x + 32$, x is degrees Celsius, y is degrees Farenheit

51. Functions for which each value in the range has a unique value in the domain.

53. The domain of the function is the range of the inverse function, and the range of the function is the domain of the inverse function.   **55.** The range of $f^{-1}(x)$ is the domain of $f(x)$.   **57.** 6, $(f \circ f^{-1})(x) = x$

CUMULATIVE REVIEW EXERCISES

59. $(4, 2, -1)$ **60.** $x^2 + 4x - 2 - \dfrac{4}{x+2}$ **61.** $\dfrac{2x\sqrt{2y}}{y}$ **62.** $x = -1 \pm \sqrt{7}$

GROUP ACTIVITY/CHALLENGE PROBLEMS

1. (a) No, not a one-to-one function   **2. (a)** and **(c)**   **3.** $f^{-1}(x) = \sqrt{x^2 + 9}, x \geq 0$

(b) Yes   **(b)** Yes

(c) $f^{-1}(x) = x, x \geq 0$

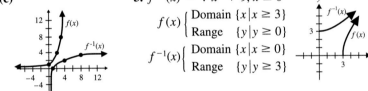

$f(x) \begin{cases} \text{Domain } \{x \mid x \geq 3\} \\ \text{Range } \quad \{y \mid y \geq 0\} \end{cases}$

$f^{-1}(x) \begin{cases} \text{Domain } \{x \mid x \geq 0\} \\ \text{Range } \quad \{y \mid y \geq 3\} \end{cases}$

GRAPHING CALCULATOR CORNER, PAGE 567

1. Yes   **3.** Yes

REVIEW EXERCISES

1. $4 \pm 2\sqrt{5}$ **2.** $\dfrac{4 \pm 2\sqrt{15}}{3}$ **3.** $1, \dfrac{1}{3}$ **4.** $\dfrac{5}{4}, -\dfrac{3}{4}$ **5.** $F_b = \sqrt{F_T^2 - F_a^2}$ **6.** $s = \sqrt{\dfrac{L - 3r}{2}}$

7. $\sqrt{128} \approx 11.31$ **8.** $\sqrt{128} \approx 11.31$ **9.** 3, 5   **10.** 9, -6 **11.** 1, -6 **12.** $-1 + \sqrt{6}, -1 - \sqrt{6}$

13. $(3 + i\sqrt{23})/2, (3 - i\sqrt{23})/2$ **14.** $2 + 2i\sqrt{7}, 2 - 2i\sqrt{7}$ **15.** $(1 + i\sqrt{47})/4, (1 - i\sqrt{47})/4$

16. $-3 + \sqrt{19}, -3 - \sqrt{19}$ **17.** Two real solutions   **18.** No real solution   **19.** No real solution   **20.** No real solution   **21.** One real solution   **22.** No real solution   **23.** Two real solutions   **24.** Two real solutions   **25.** 2, 7

26. 3, -10 **27.** 2, 5   **28.** 2, $-\dfrac{3}{5}$ **29.** $-2, 9$ **30.** $\dfrac{1 \pm i\sqrt{119}}{2}$ **31.** $\dfrac{3}{2}, -\dfrac{5}{3}$ **32.** $\dfrac{-2 \pm \sqrt{10}}{2}$ **33.** $\dfrac{3 \pm \sqrt{57}}{4}$

34. $3 \pm \sqrt{2}$ **35.** $\dfrac{3 \pm \sqrt{33}}{3}$ **36.** $0, \dfrac{5}{2}$ **37.** 3, 8   **38.** 7, 9   **39.** 5, -8 **40.** 3, -9 **41.** 10, -6

42. $\dfrac{1 \pm i\sqrt{167}}{2}$ **43.** $\dfrac{-11 \pm \sqrt{73}}{2}$ **44.** $5, -5$ **45.** $0, -6$ **46.** $\frac{1}{2}, -3$ **47.** $\dfrac{7 \pm \sqrt{89}}{10}$ **48.** $\frac{2}{3}, -\frac{3}{2}$

49. $\dfrac{-3 \pm \sqrt{33}}{2}$ **50.** $2, \frac{5}{3}$ **51.** $\dfrac{-5.7 \pm \sqrt{43.53}}{2.4}$ **52.** $\dfrac{3 \pm 3\sqrt{3}}{2}$ **53.** $0, \frac{5}{2}$ **54.** $3, -\frac{1}{3}$ **55.** $\frac{1}{4}, -\frac{3}{2}$ **56.** $\frac{5}{2}, -\frac{5}{3}$

57. (a) Upward **(e)**
(b) 0
(c) $(-3, -9)$
(d) $0, -6$

58. (a) Upward **(e)**
(b) -8
(c) $(-1, -9)$
(d) $-4, 2$

59. (a) Upward **(e)**
(b) -16
(c) $(-1, -18)$
(d) $-4, 2$

60. (a) Downward **(e)**
(b) -9
(c) $(0, -9)$
(d) No x intercepts

61. (a) Downward **(e)**
(b) 15
(c) $\left(-\frac{1}{4}, \frac{121}{8}\right)$
(d) $\frac{5}{2}, -3$

62. (a) Upward **(e)**
(b) 8
(c) $\left(-\frac{3}{2}, \frac{23}{4}\right)$
(d) No x intercepts

63. $y = x^2 - x - 6$ **64.** $y = 3x^2 + 7x - 6$ **65.** $y = x^2 + 6x + 9$ **66.** $y = 6x^2 - 7x + 2$ **67.** $5, 7$

68. $2, \dfrac{1}{4}$ **69.** $\dfrac{-1 \pm 3\sqrt{5}}{2}$ **70.** $1, \dfrac{1}{4}$ **71.** $2\sqrt{2}, -2\sqrt{2}, i\sqrt{3}, -i\sqrt{3}$ **72.** $\dfrac{3}{2}, -\dfrac{2}{5}$ **73.** $\dfrac{1}{16}$ **74.** $\dfrac{27}{8}, 8$

75. $9, 10$ **76.** $5, 9$ **77.** $w = 6$ in, $l = 11$ in **78.** \$475 **79. (a)** 1656 ft **(b)** $\sqrt{112.5} \approx 10.61$ sec

80. (a) 68 ft **(b)** $\dfrac{4 + \sqrt{416}}{8} \approx 3.05$ sec **81. (a)** 40 mL **(b)** 150°C

82. larger machine 23.51 hr, smaller machine 24.51 hr **83.** 50 mph **84.** 1.6 mph

85.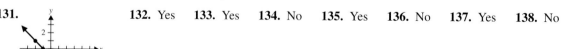

86.

87. $\dfrac{11 - \sqrt{41}}{2} \quad \dfrac{11 + \sqrt{41}}{2}$

88.

89.

90.

91.

92.

93. $\{x \mid x < -2 \text{ or } x > 3\}$ **94.** $\{x \mid -2 < x \le 5\}$

95. $\{x \mid x < -1 \text{ or } x \ge 2\}$ **96.** $\{x \mid -\frac{5}{3} < x < 6\}$ **97.** $\{x \mid -3 < x < -1 \text{ or } x > 2\}$ **98.** $\{x \mid x \le 0 \text{ or } 3 \le x \le 5\}$

99. $\left[-\frac{4}{3}, 1\right] \cup [3, \infty)$ **100.** $(-\infty, -5) \cup (-2, 0)$ **101.** $(-2, 0) \cup (4, \infty)$ **102.** $(-\infty, -3) \cup (2, 5)$

103. $(-2, 3] \cup (5, \infty)$ **104.** $(-\infty, -3) \cup [0, 5]$ **105.** **106.** **107.**

108. **109.** $x^2 - x - 1$ **110.** 5 **111.** $-x^2 + 5x - 9$ **112.** -15 **113.** $2x^3 - 11x^2 + 23x - 20$

114. 70 **115.** $\dfrac{x^2 - 3x + 4}{2x - 5}, x \ne \frac{5}{2}$ **116.** -2 **117.** $4x^2 - 26x + 44$ **118.** 8 **119.** $2x^2 - 6x + 3$ **120.** 39

121. $3x + 2 + \sqrt{x - 4}$ **122.** $3x + 2 - \sqrt{x - 4}$ **123.** $(3x + 2)\sqrt{x - 4}$ **124.** $\dfrac{\sqrt{x - 4}}{3x + 2}, x \ne -\frac{2}{3}$

125. $3\sqrt{x - 4} + 2$ **126.** $\sqrt{3x - 2}, x \ge \frac{2}{3}$ **127.** -1 **128.** 2 **129.** -9 **130.** -3

131. **132.** Yes **133.** Yes **134.** No **135.** Yes **136.** No **137.** Yes **138.** No

139. $f(x)$: domain $\{-4, 0, 5, 6\}$, range $\{-3, 2, 3, 7\}$ **140.** $f(x)$: domain $\{-3, -1, \frac{1}{2}, \sqrt{5}\}$, range $\{2, \sqrt{7}, 3, 8\}$
$f^{-1}(x)$: domain $\{-3, 2, 3, 7\}$, range $\{-4, 0, 5, 6\}$ $f^{-1}(x)$: domain $\{2, \sqrt{7}, 3, 8\}$, range $\{-3, -1, \frac{1}{2}, \sqrt{5}\}$

141. $f(x)$: domain $\{x \,|\, x \le 1\}$, range $\{y \,|\, y \ge 0\}$ **142.** $f(x)$: domain $\{x \,|\, x \ge 0\}$, range $\{y \,|\, y \ge 2\}$
$f^{-1}(x)$: domain $\{x \,|\, x \ge 0\}$, range $\{y \,|\, y \le 1\}$ $f^{-1}(x)$: domain $\{x \,|\, x \ge 2\}$, range $\{y \,|\, \ge 0\}$

143. $f^{-1}(x) = (x + 2)/4$ **144.** $f^{-1}(x) = -\frac{1}{3}(x + 5)$ **145.** $f^{-1}(x) = x^3 + 1$ **146.** $f^{-1}(x) = x^2 + 1, x \ge 0$

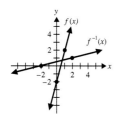 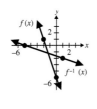 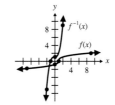

PRACTICE TEST

1. $3, -4$ **2.** $-1 \pm i\sqrt{2}$ **3.** $-1, 6$ **4.** $-4 \pm \sqrt{11}$ **5.** $0, \frac{5}{3}$ **6.** $\frac{1}{2}, -5$

7. $b = \sqrt{3a - P}$ **8.** Two real solutions **9.** $\pm\dfrac{\sqrt{10}}{5}, \pm\dfrac{i\sqrt{10}}{2}$ **10.** $\dfrac{343}{27}, -216$ **11.**

12.   **13.** (a) $\left[-\frac{5}{2}, -2\right)$ (b) $\{x \,|\, -\frac{5}{2} \le x < -2\}$

14. (a) Upward (e)
   (b) -8
   (c) $(1, -9)$
   (d) $4, -2$

15. $y = 2x^2 + 11x - 6$ **16.** $w = 4$ ft, $l = 12$ ft **17.** 5 sec **18.** $-x^2 + 3x - 12$ **19.** $4x^2 - 18x + 28$
20. $f^{-1}(x) = \dfrac{x - 4}{2}$

Chapter 10

EXERCISE SET 10.1

1. $x^2 + y^2 = 9$ **3.** $(x - 3)^2 + y^2 = 1$ **5.** $(x + 6)^2 + (y - 5)^2 = 25$

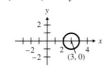

7. $(x - 4)^2 + (y - 7)^2 = 8$

9.

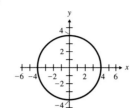

11.

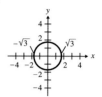

13.

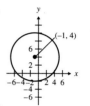

15.

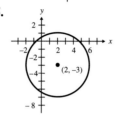

17.

19.

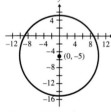

21.

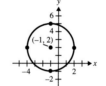

23. $x^2 + y^2 = 16$
25. $(x - 3)^2 + (y - 3)^2 = 9$
27. $(x - 3)^2 + (y + 2)^2 = 9$

29. $x^2 + (y + 5)^2 = 10^2$

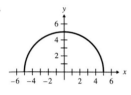

31. $(x + 4)^2 + y^2 = 5^2$

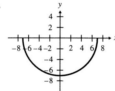

33. $(x + 1)^2 + (y - 2)^2 = 3^2$

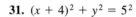

35. $(x + 3)^2 + (y - 1)^2 = 2^2$

37. $(x - 4)^2 + (y + 1)^2 = 2^2$

39. A set of points in a plane that are the same distance from a fixed point

CUMULATIVE REVIEW EXERCISES

40. $8(x - 2)(x^2 + 2x + 4)$ **41.** $T_2 = \dfrac{T_1}{1 - E}$ **42.** $P = \sqrt{2E - V}$ **43.** $\dfrac{5 \pm i\sqrt{35}}{6}$

GROUP ACTIVITY/CHALLENGE PROBLEMS

1. $1 + \sqrt{116} \approx 11.77$ ft   **3.** 4.5 sq units   **5.** 48π

GRAPHING CALCULATOR CORNER, PAGE 578

The following graphs are from a TI 82 calculator

1. 　　**3.**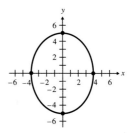

EXERCISE SET 10.2

1. 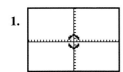　　**3.** 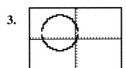　　**5.**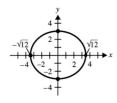

7. 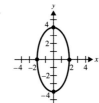　　**9.** 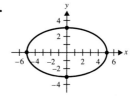　　**11.**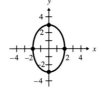

13. The set of points in a plane, the sum of whose distances from two fixed points is a constant.　　**15.** one, at (0,0)
17. The graph becomes more circular as the value of b approaches the value of a. When $a = b$ the graph is a circle.

CUMULATIVE REVIEW EXERCISES

18. 　　**19.** 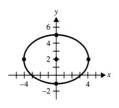　　**20.** 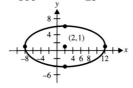　　**21.**

GROUP ACTIVITY/CHALLENGE PROBLEMS

1.　　　　**3.** $\dfrac{(x-2)^2}{100} + \dfrac{(y-1)^2}{25} = 1$

5. **(a)** $(4, 0), (-4, 0)$　**(b)** $(0, 12), (0, -12)$

EXERCISE SET 10.3

1.
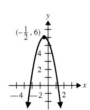

3.
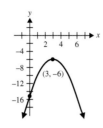

5.
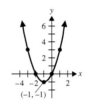

7.
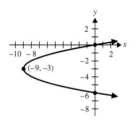

9.
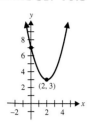

11.
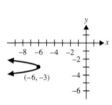

13. $y = (x + 1)^2 - 1$
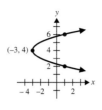

15. $x = (y + 3)^2 - 9$
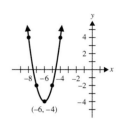

17. $y = (x + 1)^2 - 16$
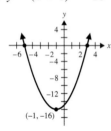

19. $x = -(y - 3)^2$

21. $y = \left(x + \frac{7}{2}\right)^2 - \frac{9}{4}$
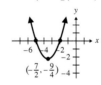

23. $y = 2(x - 1)^2 - 6$
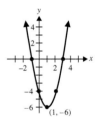

25. $y = a(x - h)^2 + k$ opens up if $a > 0$ and opens down if $a < 0$; $x = a(y - k)^2 + h$ opens to the right if $a > 0$ and opens to the left if $a < 0$.   **27.** Yes, they will all pass the vertical line test. D: $\mathbb{R}$, R: $\{y \mid y \geq k\}$ **29.** Both graphs will have their vertex at (3, 4). The graph of $y = 2(x - 3)^2 + 4$ will open upward and has no x intercepts. The graph of $y = -2(x - 3)^2 + 4$ will open downward and has two x intercepts.

CUMULATIVE REVIEW EXERCISES

30. $y = -\frac{1}{2}x + 1$ **31.** 128   **32.** **(a)** $\frac{13}{9}$ **(b)** $-2a^2 - 4ab - 2b^2 - a - b + 2$ **33.** 15

GROUP ACTIVITY/CHALLENGE PROBLEMS

2. **(a)** 454.5 ft   **(b)** 0, ≈ 8.8, ≈ 35.2, ≈ 182 **(c)** x is the horizontal distance from the vertex of the bridge, y is the vertical distance from the vertex of the bridge.   **5.** **(a)** $y = x^2 + 2x - 8$ **(b)** $y = (x + 1)^2 - 9$ **(c)**

1. $y = \pm \frac{1}{2}x$

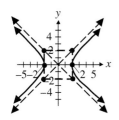

3. $y = \pm \frac{3}{4}x$

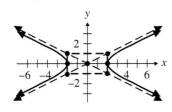

5. $y = \pm \frac{5}{6}x$

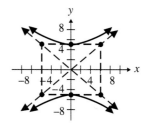

7. $y = \pm x$

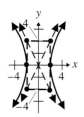

9. $y = \pm \frac{4}{9}x$

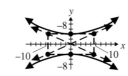

11. $y = \pm \frac{5}{4}x$

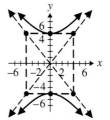

13. $\frac{x^2}{4} - \frac{y^2}{16} = 1, y = \pm 2x$

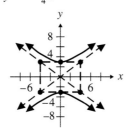

15. $\frac{y^2}{1} - \frac{x^2}{9} = 1, y = \pm \frac{1}{3}x$

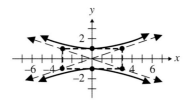

17. $\frac{y^2}{36} - \frac{x^2}{4} = 1, y = \pm 3x$

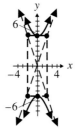

19. $\frac{x^2}{4} - \frac{y^2}{25} = 1, y = \pm \frac{5}{2}x$

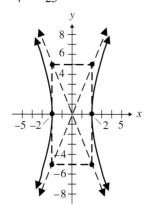

21. $\frac{x^2}{9} - \frac{y^2}{81} = 1, y = \pm 3x$

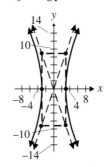

23.

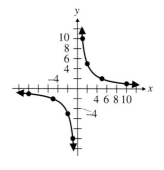

25.

27.

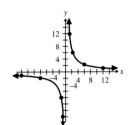

29. Parabola   **31.** Hyperbola   **33.** Parabola   **35.** Hyperbola   **37.** Ellipse   **39.** Parabola   **41.** Circle

43. Circle   **45.** Hyperbola   **47.** Asymptotes are lines that the hyperbola approaches, $y = \pm \frac{b}{a}x$ **49.** Hyperbola with

vertices at $(a, 0)$ and $(-a, 0)$. Transverse axis along x axis. Conjugate axis along y axis. Asymptotes, $y = \pm \frac{b}{a}x$.

CUMULATIVE REVIEW EXERCISES

52. 48   **53.** Domain: the set of values that can be used for the independent variable. Range: the set of values that are obtained for the dependent variable.

54.

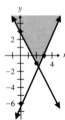

55. $\dfrac{3 \pm i}{2}$

GROUP ACTIVITY/CHALLENGE PROBLEMS

1. The transverse axis of both graphs would be along the x axis. The vertices of the second graph will be closer to the origin, and the hyperbola of the second graph will open wider than that of the first graph.

3. $y + 2 = \pm \frac{2}{3}(x - 3)$

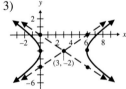

5. $\dfrac{(y + 1)^2}{4} - \dfrac{(x - 1)^2}{1} = 1,\ y + 1 = \pm 2(x - 1)$

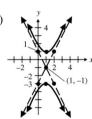

EXERCISE SET 10.5

1. $(0, -2), \left(\frac{8}{5}, -\frac{6}{5}\right)$ **3.** $(-4, 11), \left(\frac{5}{2}, \frac{5}{4}\right)$ **5.** $(2, -4), (-14, -20)$ **7.** No real solution   **9.** $\left(\frac{4}{3}, 3\right), \left(-\frac{1}{2}, -8\right)$

11. $(0, -3), (\sqrt{5}, 2), (-\sqrt{5}, 2),$ **13.** $(2, 0), (-2, 0)$ **15.** $(3, 2), (3, -2), (-3, 2), (-3, -2)$ **17.** $(3, 0), (-3, 0)$

19. $(\sqrt{6}, \sqrt{3}), (\sqrt{6}, -\sqrt{3}), (-\sqrt{6}, \sqrt{3}), (-\sqrt{6}, -\sqrt{3})$ **21.** No real solution

23. $(\sqrt{5}, 2), (\sqrt{5}, -2), (-\sqrt{5}, 2), (-\sqrt{5}, -2)$ **25.** 5, 7   **27.** 5, 3 or 5, -3, 5 or -5, 3, or -5, -3

29. 5, 6 or 5, -6, or -5, 6 or -5, -6 **31.** 7 ft by 12 ft   **33.** 5 ft, 12 ft   **35.** w = 15 m, l = 20 m

37. 6 ft by 8 ft or 4 ft by 12 ft   **39.** 2.35 sec   **41.** $r = 12\%, p = \$600$ **43.** ≈ 26 and ≈ 174 **45.** ≈ 1 and ≈ 14
49. 8 points

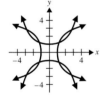

CUMULATIVE REVIEW EXERCISES

50. Parentheses, exponents, multiplication or division from left to right, addition or subtraction from left to right.

51. $\$5849.29$ **52.** $y = \dfrac{2x + 40}{11}$ **53.** $y \le 2$

GROUP ACTIVITY/CHALLENGE PROBLEMS

1. 5 m, 12 m   **3.** $r = 3.5, R = 7$ **4.** $(2, 1), (-2, -1), (-i, 2i), (i, -2i)$

5. $(\sqrt{2}, \sqrt{2}), (-\sqrt{2}, -\sqrt{2}), \left(\sqrt{3}, \dfrac{2\sqrt{3}}{3}\right), \left(-\sqrt{3}, \dfrac{-2\sqrt{3}}{3}\right)$

EXERCISE SET 10.6

1.

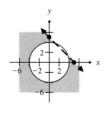

3.

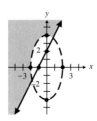

5.

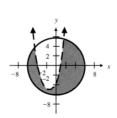

7.

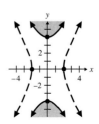

9.

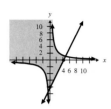

11.

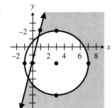

CUMULATIVE REVIEW EXERCISES

13. $16y^{10}$ **14.** $1 \pm \sqrt{5}$ **15.** $1 \pm \sqrt{5}$ **16.**

GROUP ACTIVITY/CHALLENGE PROBLEMS

1.

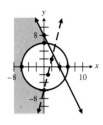

2.

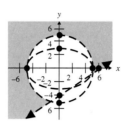

REVIEW EXERCISES

1. $x^2 + y^2 = 5^2$

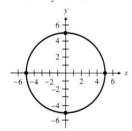

2. $(x + 3)^2 + (y - 4)^2 = 3^2$

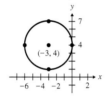

3. $(x - 4)^2 + (y - 2)^2 = (\sqrt{8})^2$

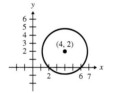

4.

5.

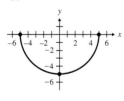

6.

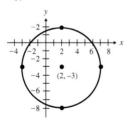

7. $(x + 1)^2 + (y - 1)^2 = 4$

8. $(x - 5)^2 + (y + 3)^2 = 9$

9. $x^2 + (y - 2)^2 = 2^2$

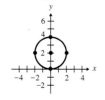

10. $(x - 1)^2 + (y + 3)^2 = 3$

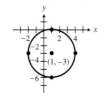

11. $(x - 4)^2 + (y - 5)^2 = 1^2$

12. $(x - 2)^2 + (y + 5)^2 = (\sqrt{1}$

13.
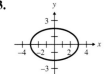

14.
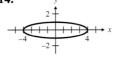

15.
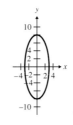

16.

17.
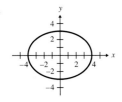

18.
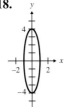

19.
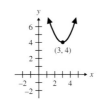

20.
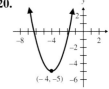

21.
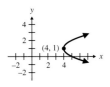

22.
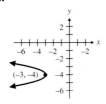

23.
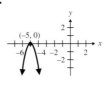

24. $y = (x - 3)^2 - 9$

25. $y = (x - 1)^2 - 4$
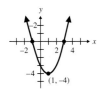

26. $x = -(y + 1)^2 + 9$
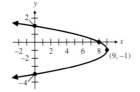

27. $x = (y + \frac{5}{2})^2 - \frac{9}{4}$
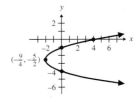

28. $y = 2(x - 2)^2 - 32$
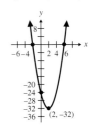

29. $y = \pm \frac{3}{2}x$

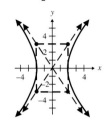

30. $y = \pm 2x$

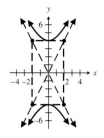

31. $y = \pm \frac{3}{5}x$

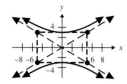

32. $y = \pm 3x$

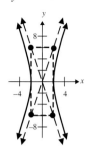

33. $\frac{y^2}{4} - \frac{x^2}{9} = 1,\ y = \pm \frac{2}{3}x$

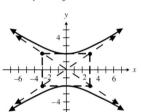

34. $\frac{x^2}{16} - \frac{y^2}{1} = 1,\ y = \pm \frac{1}{4}x$

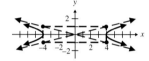

35. $\frac{x^2}{16} - \frac{y^2}{25} = 1,\ y = \pm \frac{5}{4}x$

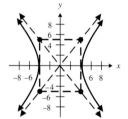

36. $\frac{y^2}{9} - \frac{x^2}{49} = 1,\ y = \pm \frac{3}{7}x$

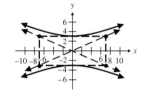

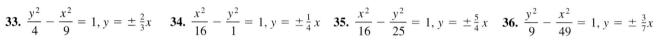

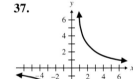

37.

38.

39.

40.

41. Hyperbola

42. Ellipse

43. Circle

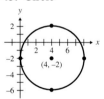

44. Hyperbola

45. Ellipse

46. Parabola

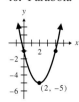

47. Ellipse

48. Parabola

49. Parabola

$(-1, -3)$

50. Hyperbola

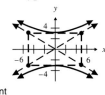

int

51. Circle

$(2, -3)$

52. Parabola

$(2, -3)$

53. $(-3, 0), \left(-\frac{12}{5}, \frac{9}{5}\right)$ **54.** $\left(\frac{1}{3}, 15\right), (5, 1)$ **55.** $(2, 0), (-2, 0)$ **56.** No real solution   **57.** $(4, 0), (-4, 0)$

58. $(4, 3)(4, -3), (-4, 3), (-4, -3)$ **59.** No real solution   **60.** $(\sqrt{3}, 0), (-\sqrt{3}, 0)$ **61.** 4, 5 or 4, -5 or -4, 5 or -4, -5

62. 5 m, 12 m

63.

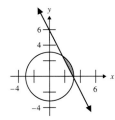

64.

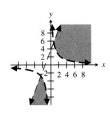

65. No real solution

66.

PRACTICE TEST

1. $(x + 3)^2 + (y + 1)^2 = 9^2$ **2.** $(x - 1)^2 + (y - 3)^2 = 3^2$ **3.**

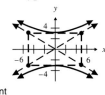

$(-3, -1)$

$(1, 3)$

4.

5.

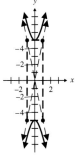

6.

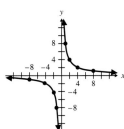

7. $(\sqrt{6}, \sqrt{10}), (\sqrt{6}, -\sqrt{10}), (-\sqrt{6}, \sqrt{10}), (-\sqrt{6}, -\sqrt{10})$

8.

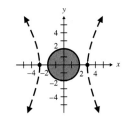

9. 2, 27

CUMULATIVE REVIEW TEST

1. $\left(\frac{16}{7}, -\frac{10}{7}\right)$ **2.** $\dfrac{12y^{10/3}}{x^{11/3}}$ **3.** $a^2 + 8a + 20$ **4.** $\dfrac{(3x + 4)(2x - 1)}{(x - 1)(4x^2 - 2x + 1)}$ **5.** $\dfrac{2x^2 - 9x - 1}{2(x + 3)(x - 4)}$ **6.** 4

7. $\dfrac{x^2y\sqrt{6xyz}}{2z}$ **8.** $-3(\sqrt{3} + \sqrt{5})$ **9.** 3   **10.** 41   **11.** $3 \pm 2\sqrt{7}$ **12.** $\dfrac{2 \pm 2\sqrt{7}}{3}$

13. $\dfrac{64}{27}, -216$ **14.**   **15. (a)** Upward   **(b)** 4   **(c)** $(2, 0)$ **(d)** $(2, 0)$

15. (e)   **16. (a)** $4x^2 - 9$ **(b)** $2x^2 + 12x - 3$ **17.**   **18.**

19. \$225   **20.** $2\frac{1}{2}$ lb cashews, $1\frac{1}{2}$ lb peanuts

Chapter 11

Exercise Set 11.1

1.   **3.**   **5.**   **7.**

9.   **11.**   **13.**   **15.**

17.

19. $\log_2 8 = 3$ **21.** $\log_3 243 = 5$ **23.** $\log_8 2 = \frac{1}{3}$ **25.** $\log_{1/4}\frac{1}{16} = 2$ **27.** $\log_5 \frac{1}{25} = -2$ **29.** $\log_4 \frac{1}{64} = -3$

31. $\log_{16}\frac{1}{4} = -\frac{1}{2}$ **33.** $\log_8\frac{1}{2} = -\frac{1}{3}$ **35.** $2^3 = 8$ **37.** $4^3 = 64$ **39.** $\left(\frac{1}{2}\right)^4 = \frac{1}{16}$ **41.** $5^{-3} = \frac{1}{125}$

43. $125^{1/3} = 5$ **45.** $27^{-1/3} = \frac{1}{3}$ **47.** $10^3 = 1000$ **49.** $6^y = 36, y = 2$ **51.** $2^3 = x, x = 8$

53. $64^y = 8, y = \frac{1}{2}$ **55.** $a^3 = 64, a = 4$ **57.** $\left(\frac{1}{2}\right)^4 = x, x = \frac{1}{16}$ **59.** $a^5 = 32, a = 2$ **61.** $8^{1/3} = x, x = 2$

63. (a) 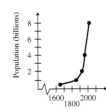 **(b)** 200 yr   **(c)** 40 yr   **(d)** ≈ 2.7 billion

65. 256   **67.** $11,887.58   **69.** 17.4 g   **71. (a)** 5 g   **(b)** 7.28×10^{-11} g   **73.** 10,000,000
75. (b) 85,115.9 gal   **77. (a)** $a > 0, a \neq 1$ **(b)** $\{x \mid x > 0\}$ **(c)** All real numbers   **79. (a)** A horizontal line through
$y = 1$.   **(b)** Yes   **(c)** No, $f(x)$ is not a one-to-one function   **81.** The graphs are symmetric about the line $y = x$. For each
ordered pair (x, y) on the graph of $y = a^x$, the ordered pair (y, x) is on the graph of $y = \log_a x$ **83.** $y = a^x + k$ will have
the same shape as the graph of $y = a^x$. However $y = a^x + k$ will be k units higher if $k > 0$ and k units lower if $k < 0$. The y
intercept of $y = a^x + k$ will be $(0, 1 + k)$.

CUMULATIVE REVIEW EXERCISES
85. $2(3x + 2y)(4x - y)$ **86.** $(2a - 9)(a + 2)$ **87.** $4x^2(x + 3)(x - 3)$ **88.** $\left(2x + \frac{1}{3}\right)\left(4x^2 - \frac{2}{3}x + \frac{1}{9}\right)$

GROUP ACTIVITY/CHALLENGE PROBLEMS
4. (a) $201.36   **(b)** $131.36
5.   **6.**

7. $\frac{2}{3}$ **9. (a)** 2^{n-1} **(b)** $2^{29} = \$536,870,912$ **(c)** $2^0 + 2^1 + 2^2 + \cdots + 2^{29}$

GRAPHING CALCULATOR CORNER, PAGE 624
1.   **3.**   **5.**

EXERCISE SET 11.2

1. $\log_3 7 + \log_3 12$ **3.** $\log_8 7 + \log_8 (x + 3)$ **5.** $\log_4 15 - \log_4 7$ **7.** $\frac{1}{2} \log_{10} x - \log_{10} (x - 3)$ **9.** $4 \log_8 x$

11. $\log_{10} 3 + 2 \log_{10} 8$ **13.** $\frac{1}{2} [5 \log_4 x - \log_4 (x + 4)]$ **15.** $4 \log_{10} x - 3 \log_{10} (a + 2)$

17. $\log_8 x + \log_8 (x - 6) - 3 \log_8 x = -2 \log_8 x + \log_8 (x - 6)$ **19.** $\log_{10} 2 + \log_{10} m - \log_{10} 3 - \log_{10} n$

21. $\log_{10} \dfrac{x^2}{x - 2}$ **23.** $\log_5 \left(\dfrac{x}{4}\right)^2$ **25.** $\log_{10} \dfrac{x(x - 4)}{x + 1}$ **27.** $\log_7 \sqrt{\dfrac{x - 2}{x}}$ **29.** $\log_9 \dfrac{25\sqrt[3]{x + 7}}{\sqrt{x}}$

31. $\log_6 \dfrac{3^4}{(x + 3)^2 x^4}$ **33.** 1   **35.** 0.3980   **37.** 1.3980   **39.** 12   **41.** 5   **43.** 3   **45.** 25   **47.** False   **49.** True
51. False   **53.** False   **55.** False   **57.** True   **59.** $\log_a a^1 = 1$ by property 5

CUMULATIVE REVIEW EXERCISES

CUMULATIVE REVIEW EXERCISES

63. $\dfrac{(2x + 5)^2}{(x - 4)^2}$ **64.** $\dfrac{3x^2 + 28x + 9}{(x - 4)(x - 3)(2x + 5)}$ **65.** $2\frac{2}{9}$ days   **66.** $2x^3y^5\sqrt[3]{6x^2y^2}$

GROUP ACTIVITY/CHALLENGE PROBLEMS

1. $\dfrac{1}{4}\log_2 x + \dfrac{1}{4}\log_2 y + \dfrac{1}{3}\log_2 a - \dfrac{1}{5}\log_2 (a - b)$

2. $2[\log_3(a^2 + b^2) + 2\log_3 c - \log_3 (a - b) - \log_3 (b + c) - \log_3 (c + d)]$

3. (a) 0.864 **(b)** 0.216 **(c)** 0.108 **(d)** 4.32

EXERCISE SET 11.3

1. 2.9395   **3.** 0.9031   **5.** 3.0000   **7.** -4.0671 **9.** 2.0000   **11.** 0.2405   **13.** -0.4260 **15.** -2.0595
17. 3.48   **19.** 209   **21.** 0.0874   **23.** 1.00   **25.** 317   **27.** 0.701   **29.** 100   **31.** 0.000787   **33.** 6610
35. 0.0871   **37.** 0.428   **39.** 0.404   **41.** 3.3747   **43.** -1.3872 **45.** 2.0086   **47.** -2.8928 **49.** 386.0
51. 0.695   **53.** 0.0246   **55.** 15.5   **57.** 0   **59.** -1 **61.** -2 **63.** -3 **65.** 5   **67.** 7   **69.** 31.2   **71.** 47
73. (a) 4.08   **(b)** 19,500   **75. (a)** 6.31×10^{20} **(b)** 2.19   **77.** 2.55   **79.** No, Since 10^3 is 1000, log 6250 must be greater than 3.   **81.** No, since $10^{-2} = 0.01$ and 10^{-3} is 0.001, log 0.0024 must be between -2 and -3.

CUMULATIVE REVIEW EXERCISES

83. $\dfrac{-2 \pm 2i\sqrt{5}}{3}$ **84.** 10 mph   **85.**   **86.**

GROUP ACTIVITY CHALLENGE PROBLEMS

1. 3.46   **3.**

EXERCISE SET 11.4

1. 5   **3.** 3   **5.** 2   **7.** 3   **9.** 1.97   **11.** 3.16   **13.** 5.59   **15.** 6.34   **17.** 3   **19.** 0, -8 **21.** $\frac{13}{2}$ **23.** $\frac{3}{2}$
25. 2   **27.** 2   **29.** 0.91   **31.** 20   **33.** 5   **35.** 3   **37.** 2   **39.** $1932.61   **41.** 139   **43.** $7112.10   **45.** 2.94
47. (a) 1,000,000,000,000   **(b)** 100,000

CUMULATIVE REVIEW EXERCISES

49. (a) $\{x \mid x > 40\}$ **(b)** $(40, \infty)$ **50. (b)** and **(c)** are functions; only **(b)** is a one-to-one function

51. $2x^3 - 11x^2 + 18x - 9$ **52.** $2x + 3 + \dfrac{3}{x + 4}$

GROUP ACTIVITY/CHALLENGE PROBLEMS

2. $(3, 1)$ **3.** $(5, 4)$ **4.** $(54, 46)$ **5.** $(335, 665)$ **6.** $x = 1$ or $x = 2$

1. 2.8   **3.** No solution

EXERCISE SET 11.5

1. 3.9120   **3.** 5.7104   **5.** 4.95   **7.** 0.0721   **9.** 3.6889   **11.** -3.0791 **13.** 2.9300   **15.** 4.3219   **17.** 7.9010
19. -4.8411 **21.** 4   **23.** 4   **25.** 3   **27.** $P = 3410.48$ **29.** $P_0 = 58.09$ **31.** $t = 0.7847$ **33.** $k = 0.2310$
35. $k = -0.2888$ **37.** $A = 4719.77$ **39.** $V_0 = \dfrac{V}{e^{kt}}$ **41.** $t = \dfrac{\ln P - \ln 150}{4}$ **43.** $k = \dfrac{\ln A - \ln A_0}{t}$
45. $y = xe^{2.3}$ **47.** $y = (x + 3)e^6$ **49.** $I_0 = Ie^{x/k}$ **51.** $Q = \dfrac{M}{1 + M}$ **53.** (a) \$5867.55   (b) 8.66 yr
55. (a) 86.47%   (b) 34.66 days   **57.** (a) 5.15 ft/sec   (b) 5.96 ft/sec   (c) 646,000   **59.** (a) 6.38 billion
(b) 34.66 yr   **61.** (a) 6626.62 yr   (b) 5752.26 yr   **63.** $\approx 11.55\%$ **65.** (a) 40.94 watts   (b) 804.72 days
67. 2.5000; find ln 12.183 to obtain the answer.   **69.** 75.1886; press 4.32 inv 1n

CUMULATIVE REVIEW EXERCISES

71. $\dfrac{16y^{14}}{x^4}$ **72.** $\dfrac{12 - 8x}{x^3}$ **73.** $q = \dfrac{fp}{p - f}$ **74.** $4x^2y^3z^4\sqrt[3]{2xz}$

GROUP ACTIVITY/CHALLENGE PROBLEMS

2. $3(e^{\ln 2})^{0.5t} \approx 3e^{0.347t}$ **3.** (a) $y = -\ln x$ (c) yes

(b)   (d) Domain: $\mathbb{R}$, Range: $\{y \mid y > 0\}$
(e) Domain: $\{x \mid x > 0\}$, Range: $\mathbb{R}$

REVIEW EXERCISES

1. 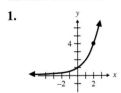   **2.** 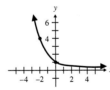   **3.** 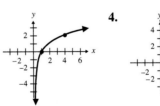   **4.**   **5.**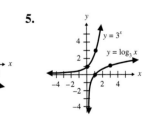

6. $\log_4 16 = 2$ **7.** $\log_8 2 = \frac{1}{3}$ **8.** $\log_6 \frac{1}{36} = -2$ **9.** $\log_{25} 5 = \frac{1}{2}$ **10.** $5^2 = 25$ **11.** $\left(\frac{1}{3}\right)^2 = \frac{1}{9}$ **12.** $3^{-2} = \frac{1}{9}$

13. $2^5 = 32$ **14.** $4^3 = x, 64$ **15.** $4^2 = x, 16$ **16.** $a^3 = 8, 2$ **17.** $\left(\frac{1}{4}\right)^{-3} = x, 64$ **18.** (a) linear   (b) exponential

(c) 8   (d) 1 year   (e) less than 1 year, ≈ 9 months   (f) mid 1994   **19.** $\frac{1}{2}\log_8 12$ **20.** $5 \log (x - 8)$

21. $\log 2 + \log (x - 3) - \log x$ **22.** $4 \log x - \log 39 - \log (2x + 8)$ **23.** $\log_5 \dfrac{x - 2}{x^2}$ **24.** $\log \dfrac{x^2}{(x + 1)^3}$

25. $\log \dfrac{(2x)^3}{y}$ **26.** $\log_8 \dfrac{(x + 3)^2(x - 1)^4}{\sqrt{x}}$ **27.** $\ln \dfrac{\sqrt{\dfrac{x}{x + 2}}}{2}$ **28.** $\ln \dfrac{x^3\sqrt{x + 1}}{(x + 4)^3}$ **29.** 9   **30.** 5   **31.** 6   **32.** 2
33. 3.9138   **34.** -3.1451 **35.** -2.7235 **36.** 4.2455   **37.** 56.6   **38.** 829   **39.** 0.0796   **40.** 0.0422

41. 214   **42.** 0.0594   **43.** 4   **44.** 3   **45.** 31.5   **46.** 5.1   **47.** 2   **48.** $-\dfrac{1}{2}$ **49.** 1   **50.** 5   **51.** 2.605

52. 4.497   **53.** 1.595   **54.** 2.240   **55.** 123   **56.** $\dfrac{298}{3}$ **57.** 2   **58.** 1   **59.** 4   **60.** 3   **61.** $t = 1.1552$

62. $P_0 = 65.50$ **63.** $A_0 = 352.54$ **64.** $w_0 = \dfrac{w}{e^{kt}}$ **65.** $t = \dfrac{\ln A - \ln A_o}{k}$ **66.** $k = \dfrac{\ln 0.25}{t}$ **67.** $y = xe^2$

68. $y = 5x + 2$ **69.** $y = \dfrac{x-3}{e^{1.4}}$ **70.** 6.1092   **71.** 3.5609   **72.** -1.5229 **73.** \$25,723.08   **74.** 566.4 mg

75. (a) 92.88 min   **(b)** 118.14 min   **76. (a)** 72   **(b)** ≈ 63.4 **(c)** ≈ 4 months   **77.** 13.57 lb/in^2 **78.** 9.90 yr

PRACTICE TEST

1.

2.

3. $\log_4 \dfrac{1}{64} = -3$ **4.** $3^5 = 243$ **5.** $2^4 = x,\ 16$ **6.** $27^y = 3,\ \dfrac{1}{3}$ **7.** $\log_3 x + \log_3 (x - 4) - 2\log_3 x$

8. $\log_8 \dfrac{(x-4)^3(x+1)^2}{\sqrt{x}}$ **9.** 3.6646   **10.** -3.1952 **11.** 0.00501   **12.** 4.38   **13.** 3   **14.** \$6349.06

15. 3364.86 yr

Chapter 12

EXERCISE SET 12.1

1. 2, 4, 6, 8, 10   **3.** $6, \dfrac{7}{2}, \dfrac{8}{3}, \dfrac{9}{4}, 2$ **5.** $1, \dfrac{1}{2}, \dfrac{1}{3}, \dfrac{1}{4}, \dfrac{1}{5}$ **7.** $\dfrac{3}{2}, \dfrac{4}{3}, \dfrac{5}{4}, \dfrac{6}{5}, \dfrac{7}{6}$ **9.** $-1, 1, -1, 1, -1$ **11.** 4, -8, 16, -32, 64

13. 27   **15.** 6   **17.** 16   **19.** $\dfrac{81}{19}$ **21.** $s_1 = 7, s_3 = 27$ **23.** $s_1 = 3, s_3 = 17$ **25.** $s_1 = 1, s_3 = 3$

27. $s_1 = \dfrac{1}{2}, s_3 = 7$ **29.** 64, 128, 256   **31.** 15, 17, 19   **33.** $\dfrac{1}{6}, \dfrac{1}{7}, \dfrac{1}{8}$ **35.** 1, -1, 1   **37.** $\dfrac{1}{81}, \dfrac{1}{243}, \dfrac{1}{729}$ **39.** $\dfrac{1}{16}, -\dfrac{1}{32}, \dfrac{1}{64}$

41. $-25, -33, -41$ **43.** $2 + 5 + 8 + 11 + 14;\ 40$ **45.** $-1 + 5 + 15 + 29 + 47 + 69;\ 164$ **47.** $2 + 3 + 4;\ 9$

49. $\displaystyle\sum_{n=1}^{5}(n+3)$ **51.** $\displaystyle\sum_{n=1}^{3}\dfrac{n^2}{4}$ **53.** 13   **55.** 169   **57.** $\dfrac{16}{3}$ **59.** 25   **61.** 42.83   **63.** $x_1 + x_2 + x_3 + \cdots + x_n$

65. A list of numbers arranged in a specific order   **67.** The sum of the first consecutive n terms of a series

CUMULATIVE REVIEW EXERCISES

69. $5, \dfrac{3}{2}$ **70.** Two, $b^2 - 4ac > 0$ **71.**

72. $(2, 1), (-2, -1)$

GROUP ACTIVITY/CHALLENGE PROBLEMS

1. 20   **2.** 162   **3.** 1334   **4.** 26,244   **5.** 8.1

EXERCISE SET 12.2

1. 3, 7, 11, 15, 19; $a_n = 3 + 4(n - 1)$ **3.** $-5, -3, -1, 1, 3; a_n = -5 + 2(n - 1)$

5. $\frac{1}{2}, 2, \frac{7}{2}, 5, \frac{13}{2}; a_n = \frac{1}{2} + \frac{3}{2}(n - 1)$ **7.** 100, 95, 90, 85, 80; $a_n = 100 - 5(n - 1)$ **9.** 22   **11.** -23 **13.** 13

15. 2   **17.** 9   **19.** 6   **21.** $s_{10} = 100, d = 2$ **23.** $s_8 = \frac{52}{5}, d = \frac{1}{5}$ **25.** $s_5 = 20, d = \frac{4}{5}$ **27.** $s_{11} = 407, d = 6$

29. 5, 8, 11, 14; $a_{10} = 32; s_{10} = 185$ **31.** $-8, -13, -18, -23; a_{10} = -53, s_{10} = -305$

33. 100, 93, 86, 79; $a_{10} = 37, s_{10} = 685$ **35.** $\frac{9}{5}, \frac{12}{5}, \frac{15}{5}, \frac{18}{5}; a_{10} = \frac{36}{5}, s_{10} = 45$

37. $n = 15, s_n = 330$ **39.** $n = 10, s_n = 90$ **41.** $n = 9, s_n = -\frac{39}{2}$ **43.** $n = 11, s_n = -352$

45. (a) 20,000; 21,000; 22,000; 23,000; 24,000   **(b)** $a_n = 20,000 + 1000(n - 1)$ **(c)** $a_{13} = 32,000$ **47.** 18,875
49. (a) 304 ft   **(b)** 1600 ft   **51.** $496
53. A sequence in which each term differs from the preceding term by a constant amount.
55. By subtracting any term from the term that follows it.

CUMULATIVE REVIEW EXERCISES

56. One, the slopes of the lines are different. Therefore the lines must intersect once.   **57.** $(7, -6)$
58. 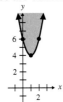   **59.**

GROUP ACTIVITY/CHALLENGE PROBLEMS

3. 0   **5.** 0   **7.** $\frac{4}{3}$

EXERCISE SET 12.3

1. 5, 15, 45, 135, 405   **3.** 90, 30, 10, $\frac{10}{3}, \frac{10}{9}$ **5.** $-15, 30, -60, 120, -240$ **7.** $3, \frac{9}{2}, \frac{27}{4}, \frac{81}{8}, \frac{243}{16}$ **9.** 160   **11.** 13,122

13. $\frac{1}{64}$ **15.** 6144   **17.** 1023   **19.** 10,160   **21.** $-\frac{2565}{128} \approx -20.04$ **23.** $-\frac{9279}{625} \approx -14.85$

25. $r = \frac{1}{2}, a_n = 5\left(\frac{1}{2}\right)^{n-1}$ **27.** $r = -3, a_n = 2(-3)^{n-1}$ **29.** $r = 3, a_n = -1(3)^{n-1}$

31. $r = 2, a_1 = 7$ or $r = -2, a_1 = 7$ **33.** $r = 3, a_1 = 5$ **35.** $572,503.52   **37. (a)** 3 days   **(b)** 1.172 grams

39. (a) 494.31 million   **(b)** 13.9 years   **41. (a)** $\frac{1}{2}, \frac{1}{4}, \frac{1}{8}, \frac{1}{16}, \frac{1}{32}$ **(b)** $a_n = a_1 r^{n-1} = \frac{1}{2}\left(\frac{1}{2}\right)^{n-1} = \left(\frac{1}{2}\right)^n$

(c) amount remaining $= \frac{1}{128}$ or 0.78%   **43.** $\left(\frac{2}{3}\right)^{10}$ or 1.7%   **45. (a)** $7840, $6272, $5017.60   **(b)** $a_n = 7840\left(\frac{4}{5}\right)^{n-1}$
(c) $3211.26   **47.** The sum of the terms of a geometric sequence.

CUMULATIVE REVIEW EXERCISES

49. 12 hr   **50.** $\dfrac{x}{2y^{2/15}}$ **51.** $3x^2y^2\sqrt[3]{y} - 2xy\sqrt[3]{9y^2}$ **52.** $x\sqrt{y}$

GROUP ACTIVITY/CHALLENGE PROBLEMS

1. (a) $3211.26   **(b)** $3211.26   **2.** $n = 21, s = 2,097,151$

EXERCISE SET 12.4

1. 12   **3.** $\frac{25}{3}$ **5.** $\frac{5}{3}$ **7.** $\frac{81}{10}$ **9.** 2   **11.** 24   **13.** -45 **15.** -15 **17.** $\frac{3}{11}$ **19.** $\frac{5}{9}$ **21.** $\frac{17}{33}$ **23.** 40 ft
25. The sum does not exist.

CUMULATIVE REVIEW EXERCISES

26. $8x^3 - 18x^2 - 3x + 18$ **27.** $8x - 15 + \dfrac{57}{2x + 5}$ **28.** $\dfrac{10}{3}$ **29.** $-\dfrac{1}{3}$

GROUP ACTIVITY/CHALLENGE PROBLEMS

1. 190 ft   **2.** 4 ft

EXERCISE SET 12.5

1. 6   **3.** 1   **5.** 1   **7.** 70   **9.** 28   **11.** $x^3 + 12x^2 + 48x + 64$ **13.** $a^4 - 4a^3b + 6a^2b^2 - 4ab^3 + b^4$
15. $243a^5 - 405a^4b + 270a^3b^2 - 90a^2b^3 + 15ab^4 - b^5$ **17.** $16x^4 + 16x^3 + 6x^2 + x + \frac{1}{16}$
19. $(x^4/16) - (3x^3/2) + (27x^2/2) - 54x + 81$ **21.** $x^{10} + 10x^9y + 45x^8y^2 + 120x^7y^3$
23. $2187x^7 - 5103x^6y + 5103x^5y^2 - 2835x^4y^3$ **25.** $x^{16} - 24x^{14}y + 252x^{12}y^2 - 1512x^{10}y^3$
29. yes, $4! = 4 \cdot 3!$ **31.** yes, $(7 - 3)! = (7 - 3)(7 - 4)(7 - 5)!$ **33.** $x^8, 24x^7, 17{,}496x, 6561$
$4! = 4 \cdot 3 \cdot 2!$

CUMULATIVE REVIEW EXERCISE

35. $s_n = \dfrac{a_1 - a_1r^n}{1 - r}$ or $s_n = \dfrac{a_1(1 - r^n)}{1 - r}$ **36.** $a_1 = \dfrac{s_n - s_nr}{1 - r^n}$ **37.** $(4x - 3)(4x + 1)$ **38.** $3b^2(a - 4)^2$

GROUP ACTIVITY/CHALLENGE PROBLEMS

1. $(a + b)^n = \displaystyle\sum_{i=0}^{n} \binom{n}{i} a^{n-i} b^i$ **2. (a)** r^{th} term $= \dfrac{n!}{[n - (r - 1)]!\,(r - 1)!} x^{n-(r-1)} y^{r-1}$ **(b)** $216x^2$

REVIEW EXERCISES

1. 3, 4, 5, 6, 7   **2.** $1, \frac{1}{2}, \frac{1}{3}, \frac{1}{4}, \frac{1}{5}$ **3.** 2, 6, 12, 20, 30   **4.** $\frac{1}{5}, \frac{2}{3}, \frac{9}{7}, 2, \frac{25}{9}$ **5.** 25   **6.** 2   **7.** $\frac{16}{81}$ **8.** 88

9. $s_1 = 5, s_3 = 24$ **10.** $s_1 = 2, s_3 = 28$ **11.** $s_1 = \frac{4}{3}, s_3 = \frac{227}{60}$ **12.** $s_1 = -3, s_3 = -4$

13. 16, 32, 64; $a_n = 2^{n-1}$ **14.** $-\frac{1}{2}, \frac{1}{4}, -\frac{1}{8}; a_n = -8\left(-\frac{1}{2}\right)^{n-1}$ **15.** $\frac{32}{3}, \frac{64}{3}, \frac{128}{3}; a_n = \frac{2}{3}(2)^{n-1}$

16. $-3, -6, -9; a_n = 9 - 3(n - 1) = 12 - 3n$ **17.** $-1, 1, -1; a_n = (-1)^n$
18. 30, 36, 42; $a_n = 6 + 6(n - 1) = 6n$ **19.** $3 + 6 + 11; 20$ **20.** $3 + 8 + 15 + 24; 50$
21. $\frac{1}{3} + \frac{4}{3} + \frac{9}{3} + \frac{16}{3} + \frac{25}{3}, \frac{55}{3}$ **22.** $\frac{1}{2} + \frac{2}{3} + \frac{3}{4} + \frac{4}{5}, \frac{163}{60}$ **23.** 27   **24.** 215   **25.** 108   **26.** 729

27. 5, 7, 9, 11, 13   **28.** $\frac{1}{2}, -\frac{3}{2}, -\frac{7}{2}, -\frac{11}{2}, -\frac{15}{2}$ **29.** $-12, -\frac{25}{2}, -13, -\frac{27}{2}, -14$ **30.** $-100, -\frac{499}{5}, -\frac{498}{5}, -\frac{497}{5}, -\frac{496}{5}$

31. 26   **32.** -15 **33.** -4 **34.** $\frac{1}{2}$ **35.** 6   **36.** 7   **37.** $s_8 = 112, d = 2$ **38.** $s_7 = -210, d = -6$

39. $s_7 = \frac{63}{5}, d = \frac{2}{5}$ **40.** $s_9 = -42, d = -\frac{1}{3}$ **41.** 2, 6, 10, 14; $a_{10} = 38, s_{10} = 200$

42. $-8, -11, -14, -17; a_{10} = -35, s_{10} = -215$ **43.** $\frac{5}{6}, \frac{9}{6}, \frac{13}{6}, \frac{17}{6}; a_{10} = \frac{41}{6}, s_{10} = \frac{115}{3}$
44. $-80, -76, -72, -68; a_{10} = -44, s_{10} = -620$ **45.** $n = 11, s_n = 308$ **46.** $n = 9, s_n = 36$
47. $n = 11, s_n = \frac{231}{10}$ **48.** $n = 19, s_n = 760$ **49.** 5, 10, 20, 40, 80   **50.** $-12, -6, -3, -\frac{3}{2}, -\frac{3}{4}$

51. $20, -\frac{40}{3}, \frac{80}{9}, -\frac{160}{27}, \frac{320}{81}$ **52.** $-100, -20, -4, -\frac{4}{5}, -\frac{4}{25}$ **53.** $\frac{4}{243} \approx 0.02$ **54.** 6400 **55.** -2048 **56.** $\frac{160}{6561}$

57. 3060 **58.** $\frac{37,969}{1215} \approx 31.25$ **59.** $-\frac{4305}{64} \approx -67.27$ **60.** $\frac{172,539}{256} \approx 673.98$ **61.** $r = 2, a_n = 6(2)^{n-1}$

62. $r = \frac{1}{3}, a_n = 8\left(\frac{1}{3}\right)^{n-1}$ **63.** $r = 5, a_n = -4(5)^{n-1}$ **64.** $r = \frac{2}{3}, a_n = \frac{9}{5}\left(\frac{2}{3}\right)^{n-1}$ **65.** 14 **66.** -6 **67.** -15

68. $\frac{49}{10}$ **69.** 4 **70.** $\frac{21}{2}$ **71.** -36 **72.** $\frac{25}{6}$ **73.** $\frac{52}{99}$ **74.** $\frac{125}{333}$ **75.** $81x^4 + 108x^3y + 54x^2y^2 + 12xy^3 + y^4$

76. $8x^3 - 36x^2y^2 + 54xy^4 - 27y^6$ **77.** $x^9 - 18x^8y + 144x^7y^2 - 672x^6y^3$

78. $256a^{16} + 3072a^{14}b + 16,128a^{12}b^2 + 48,384a^{10}b^3$ **79.** $15,150$

80. **(a)** $30,000; $31,000; $32,000; $33,000; $34,000   **(b)** $a_n = 30,000 + (n - 1)1000$ **(c)** $39,000   **81.** $102,400

82. $503.63   **83.** 100 ft

Practice Test

1. $3, 1, \frac{5}{9}, \frac{3}{8}, \frac{7}{25}$ **2.** $s_1 = 5, s_3 = \frac{23}{2}$ **3.** $5 + 11 + 21 + 35 + 53; 125$ **4.** $a_n = \frac{1}{3} + \frac{1}{3}(n - 1) = \frac{1}{3}n$

5. $a_n = 5(2)^{n-1}$ **6.** $12, 9, 6, 3$ **7.** $\frac{5}{8}, \frac{5}{12}, \frac{5}{18}, \frac{5}{27}$ **8.** 16 **9.** -32 **10.** 13 **11.** $\frac{512}{729} \approx 0.0723$

12. $\frac{39,063}{5} \approx 7812.6$ **13.** $r = \frac{1}{2}, a_n = 12\left(\frac{1}{2}\right)^{n-1}$ **14.** 9 **15.** $x^4 + 8x^3y + 24x^2y^2 + 32xy^3 + 16y^4$

Cumulative Review Test

1. -37 **2.** $x < -\frac{11}{2}$ or $x > \frac{17}{2}$ **3.** 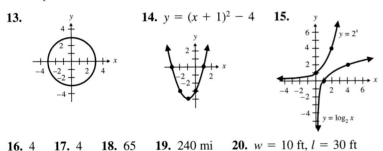   **4.** $(0, -4)$ **5.** $p = \dfrac{At}{t - A}$ **6.** $\dfrac{4x^2 + 3x - 6}{(x + 4)(x - 3)(3x + 4)}$

7. $\dfrac{x\sqrt[3]{12xy}}{y}$ **8.** $2\sqrt{7}$ **9.** $3, \frac{11}{9}$ **10.** $\dfrac{7 - 26i}{25}$ **11.** $\dfrac{-1 \pm \sqrt{33}}{4}$ **12.** $x < -1$ or $\frac{3}{2} < x < 4$

13.   **14.** $y = (x + 1)^2 - 4$ **15.**

16. 4 **17.** 4 **18.** 65 **19.** 240 mi   **20.** $w = 10$ ft, $l = 30$ ft

Index

Photo Credits

	Title Page	Kaplan © Peter B. Kaplan MCMLXX- Peter B. Kaplan Images, Inc., NYC.
Chapter 1	Chapter Opener	James M. McCann/Photo Researchers
	Exercise Set 1.3, Problem 140	Photofest
Chapter 2	Chapter Opener	Sylvain Grandadam/Photo Researchers
	Section 2.2, Example 1	The Photo Works/Monkmeyer Press
	Exercise Set 2.2, Problem 81	Bryan Hemphill/Photo Researchers
	Exercise Set 2.3, Problem 11	Allen R. Angel
	Exercise Set 2.3, Problem 22	Leonard Lee Rue/Monkmeyer Press
	Exercise Set 2.3, Problem 27	Michael Newman/Photo Edit
	Exercise Set 2.3 GA Problem 1	Bettmann Archives
	Section 2.4, Example 1	Cameramann/The Image Works
	Section 2.4, Example 4	Daemmrich/The Image Works
	Section 2.4, Example 8	John Eestoot/ Stock Boston
	Exercise Set 2.4, Problem 3	Allen R. Angel
	Exercise Set 2.4, Problem 4	Lawrence Livermore/National Labratory of Science/Photo Researchers
	Exercise Set 2.4, Problem 23	Allen R. Angel
	Exercise Set 2.4, Problem 43	Allen R. Angel
	Exercise Set 2.4, Problem 62	David Young Wolf/Photo Edit
	Exercise Set 2.4, Problem 68	NASA Headquarters
	Exercise Set 2.4, GA, Problem 4	Dean Ambramson/Stock Boston
	Exercise Set 2.5, Problem 76	Forsyth/Monkmeyer Press
	Review Exercises, Problem 42	NASA Headquarters
Chapter 3	Chapter Opener	Charles V. Angelo/Photo Researchers
	Section 3.1, Example 1	Grantpix/Monkmeyer Press
	Exercise Set 3.1, Problem 41	Allen R. Angel
	Exercise Set 3.2, Problem 51	Reuters/Bettmann Archives
	Exercise Set 3.3, Problem 71	Forsyth/Monkmeyer Press
	Exercise Set 3.3 GA, Problem 1	Allen R. Angel
	Section 3.4, Example 9	NASA Headquarters
	Exercise Set 3.4, Problem 54	Susan Van Etten/Photo Edit
	Exercise Set 3.4, Problem 60	B. Yarvin/The Image Works
Chapter 4	Chapter Opener	Richard Harrington/FPG International
	Exercise Set 4.2, Problem 36	Paul Conklin/Monkmeyer Press
	Exercise Set 4.3, Problem 13	David Sams/Stock Boston
	Exercise Set 4.3 GA, Problem 1	Stock Boston

Chapter 5	Chapter Opener	J. Baum & N. Henbest/Science Photo Library/Photo Researchers
	Exercise Set 5.1, Problem 109	Earl Scott/Photo Researchers
	Exercise Set 5.1 GA, Problem 2	Peter Menzel/Stock Boston
	Exercise 5.4, Problem 50	Lawrence Migdale/Stock Boston
Chapter 6	Chapter Opener	Allen R. Angel
	Exercise Set 6.1, Problem 72	Tony Freeman/Photo Edit
	Exercise Set 6.5, Problem 49	Allen R. Angel
Chapter 7	Chapter Opener	Virginia Division of Tourism
	Exercise Set 7.2, Problem 86	Edward Miller/Stock Boston
	Section 7.4, Example 9	Alfred Pasieka/Science Photo Library/Photo Researchers
	Section 7.5, Example 2	Steve Weber/Stock Boston
	Section 7.5, Example 8	Rafael Macia/Photo Researchers
	Exercise Set 7.5, Problem 7	Allen R. Angel
	Exercise Set 7.5, Problem 27	Bachmann/Photo Edit
	Exercise Set 7.5, Problem 33	Allen R. Angel
	Exercise Set 7.5 GA, Problem 1	Greg Smith
	Exercise Set 7.5 GA, Problem 2	Allen R. Angel
	Exercise 7.6, Problem 34	Grantpix/Monkmeyer Press
	Exercise Set 7.6 GA, Problem 4	Richard Pasley/Stock Boston
Chapter 8	Chapter Opener	Allen R. Angel
	Exercise Set 8.3 GA, Problem 1	Jonathan Nourok/Photo Edit
	Exercise Set 8.6, Problem 65	David Madison/Duomo Photography
	Exercise Set 8.6, Problem 78	NASA Headquarters
Chapter 9	Chapter Opener	Janecek/Monkmeyer Press
	Exercise Set 9.2, Problem 101	NASA Headquarters
	Exercise Set 9.3, Problem 37	Vanessa Vick/Photo Researchers
	Exercise Set 9.5 GA, Problem 1	Martin Dohrn/Science Photo Library/Photo Researchers
Chapter 10	Chapter Opener	Barbara Alper/Stock Boston
	Exercise Set 10.2 GA, Problem 8	Paul Conklin/Photo Edit
	Exercise Set 10.3 GA, Problem 1	Bopp/Monkmeyer Press
	Exercise Set 10.3 GA, Problem 2	Allen R. Angel
	Exercise Set 10.5, Problem 34	Jack Spratt/The Image Works

Chapter 11   Chapter Opener   NASA Headquarters
             Section 11.1, Example 3   Fujiphotos/The Image Works
             Exercise Set 11.1, Problem 71   U.S. Department of Energy/
                                                  Photo Researchers
             Exercise Set 11.1, Problem 76   Daemmrich/The Image Works
             Exercise Set 11.4, Problem 45   Michael Tamborrino/FPG International
             Section 11.5, Example 11   David Weintraub/Photo Researchers
             Exercise Set 11.5, Problem 65   NASA Headquarters

Chapter 12   Chapter Opener   J. Pickerell/The Image Works
             Section 12.1, Example 7   Antman/The Image Works
             Exercise Set 12.3, Problem 41   Hall/The Image Works